SELECTED SOLUTIONS MANUAL

Alicia Paterno Parsi
Duquesne University

Arash Parsi
Duquesne University

Tomislav Pintauer
Duquesne University

Lucio Gelmini
Grant MacEwan College

Robert W. Hilts
Grant MacEwan College

General Chemistry
Principles and Modern Applications
Eleventh Edition

Ralph H. Petrucci
California State University, San Bernardino

F. Geoffrey Herring
University of British Columbia

Jeffry D. Madura
Duquesne University

Carey Bissonnette
University of Waterloo

PEARSON

Toronto

ISBN-13: 978-0-13-338790-2

Executive Acquisitions Editor: Cathleen Sullivan
Developmental Editor: Joanne Sutherland
Project Manager: Sarah Gallagher

The author and publisher of this book have used their best efforts in preparing this book. These efforts include the development, research, and testing of the theories and programs to determine their effectiveness. The author and publisher make no warranty of any kind expressed or implied, with regard to these programs or the documentation contained in this book. The author and publisher shall not be liable in any event for incidental or consequential damages in connection with, or arising out of the furnishing, performance, or use of these programs.

10 9 8 7 6 5 4 3 2 1 [V0RY]

Printed and bound in the United States of America.

ISBN-13: 978-0-13-338790-2

Contents

Preface

The *Selected Solutions Manual* to accompany the 11th edition of *General Chemistry* contains full solutions to all of the in-chapter exercises and red end-of-chapter exercises and problems, namely,

- Practice Examples (A and B)
- Integrative Examples
- Exercises
- Integrative and Advanced Exercises
- Feature Problems
- Self-Assessment Exercises

Each solution is preceded by its level of difficulty, set in parentheses (E = easy, M = moderate, and D = difficult). Solutions for all textbook questions are found in the *Complete Solutions Manual* (978-0-13-292504-4).

Stepwise Approach and Conversion Pathway Approach

Solutions to selected questions are provided in both the Stepwise Approach and Conversion Pathway Approach. These two methods of problem solving are modeled after the text, and their purpose is to help you set up problems as you begin your problem-solving journey in the solutions.

Molecular Formulas of Organic Compounds

For brevity, the molecular formulas of organic compounds in the *Selected Solutions Manual* are often written as CxHyOz rather than as the condensed structural formulas found in the textbook. For example, the formula of glycerol, $HOCH_2CH(OH)CH_2OH$, is written as $C_3H_8O_3$.

CHAPTER 1
MATTER—ITS PROPERTIES AND MEASUREMENT
PRACTICE EXAMPLES

1A (E) Convert the Fahrenheit temperature to Celsius and compare.

$$°C = \left(°F - 32\ °F\right)\frac{5\,°C}{9\,°F} = \left(350\ °F - 32\ °F\right)\frac{5\,°C}{9\,°F} = 177\ °C.$$

1B (E) We convert the Fahrenheit temperature to Celsius. $°C = \left(°F - 32\ °F\right)\frac{5\,°C}{9\,°F} = \left(-15\ °F - 32\ °F\right)\frac{5\,°C}{9\,°F} = -26\ °C.$ The antifreeze only protects to $-22\ °C$ and thus it will not offer protection to temperatures as low as $-15\ °F = -26.1\ °C.$

2A (E) The mass is the difference between the mass of the full and empty flask.

$$\text{density} = \frac{291.4\ \text{g} - 108.6\ \text{g}}{125\ \text{mL}} = 1.46\ \text{g/mL}$$

2B (E) First determine the volume required. $V = (1.000 \times 10^3\ \text{g}) \div (8.96\ \text{g cm}^{-3}) = 111.\underline{6}\ \text{cm}^3.$ Next determine the radius using the relationship between volume of a sphere and radius.

$$V = \frac{4}{3}\pi r^3 = 111.\underline{6}\ \text{cm}^3 = \frac{4}{3}(3.1416)r^3 \qquad r = \sqrt[3]{\frac{111.6 \times 3}{4(3.1416)}} = 2.987\ \text{cm}$$

3A (E) The volume of the stone is the difference between the level in the graduated cylinder with the stone present and with it absent.

$$\text{density} = \frac{\text{mass}}{\text{volume}} = \frac{28.4\ \text{g rock}}{44.1\ \text{mL rock \& water} - 33.8\ \text{mL water}} = 2.76\ \text{g/mL} = 2.76\ \text{g/cm}^3$$

3B (E) The water level will remain unchanged. The mass of the ice cube displaces the same mass of liquid water. A 10.0 g ice cube will displace 10.0 g of water. When the ice cube melts, it simply replaces the displaced water, leaving the liquid level unchanged.

4A (E) The mass of ethanol can be found using dimensional analysis.

$$\text{ethanol mass} = 25\ \text{L gasohol} \times \frac{1000\ \text{mL}}{1\ \text{L}} \times \frac{0.71\ \text{g gasohol}}{1\ \text{mL gasohol}} \times \frac{10\ \text{g ethanol}}{100\ \text{g gasohol}} \times \frac{1\ \text{kg ethanol}}{1000\ \text{g ethanol}}$$

$$= 1.8\ \text{kg ethanol}$$

4B (E) We use the mass percent to determine the mass of the 25.0 mL sample.

$$\text{rubbing alcohol mass} = 15.0\ \text{g (isopropyl alcohol)} \times \frac{100.0\ \text{g rubbing alcohol}}{70.0\ \text{g (isopropyl alcohol)}}$$

$$= 21.4\underline{3}\ \text{g rubbing alcohol}$$

$$\text{rubbing alcohol density} = \frac{21.4\ \text{g}}{25.0\ \text{mL}} = 0.857\ \text{g/mL}$$

5A   (M) For this calculation, the value 0.000456 has the least precision (three significant figures), thus the final answer must also be quoted to three significant figures.

$$\frac{62.356}{0.000456 \times 6.422 \times 10^3} = 21.3$$

5B   (M) For this calculation, the value 1.3×10^{-3} has the least precision (two significant figures), thus the final answer must also be quoted to two significant figures.

$$\frac{8.21 \times 10^4 \times 1.3 \times 10^{-3}}{0.00236 \times 4.071 \times 10^{-2}} = 1.1 \times 10^6$$

6A   (M) The number in the calculation that has the least precision is 102.1 ($\pm$0.1), thus the final answer must be quoted to just one decimal place. $0.236 + 128.55 - 102.1 = 26.7$

6B   (M) This is easier to visualize if the numbers are not in exponential notation.

$$\frac{\left(1.302 \times 10^3\right) + 952.7}{\left(1.57 \times 10^2\right) - 12.22} = \frac{1302 + 952.7}{157 - 12.22} = \frac{2255}{145} = 15.6$$

INTEGRATIVE EXAMPLE

A (D) *Stepwise Approach:* First, determine the density of the alloy by the oil displacement.
Mass of oil displaced = mass of alloy in air – mass of alloy in oil
$$= 211.5 \text{ g} - 135.3 \text{ g} = 76.2 \text{ g}$$

$V_{Oil} = m / d = 76.2 \text{ g} / 0.926 \text{ g/mL} = 82.3 \text{ mL} = V_{Mg-Al}$

$d_{Mg-Al} = 211.5 \text{ g} / 82.3 \text{ mL} = 2.57 \text{ g/cm}^3$

Now, since the density is a linear function of the composition,

$d_{Mg-Al} = mx + b$, where x is the mass fraction of Mg, and b is the y-intercept.

Substituting 0 for x (no Al in the alloy), everything is Mg and the equation becomes:

$1.74 = m\,0 + b$. Therefore, $b = 1.74$

Assuming 1 for x (100% by weight Al):

$2.70 = (m \times 1) + 1.74$, therefore, $m = 0.96$

Therefore, for an alloy:

$2.57 = 0.96x + 1.74$

$x = 0.86 =$ mass % of Al

Mass % of Mg $= 1 - 0.86 = 0.14$, 14%

B (M) *Stepwise approach:*

Mass of seawater $= d \cdot V = 1.027$ g/mL $\times 1500$ mL $= 1540.5$ g

$$1540.5 \text{ g seawater} \times \frac{2.67 \text{ g NaCl}}{100 \text{ g seawater}} \times \frac{39.34 \text{ g Na}}{100 \text{ g NaCl}} = 16.18 \text{ g Na}$$

Then, convert mass of Na to atoms of Na

$$16.18 \text{ g Na} \times \frac{1 \text{ kg Na}}{1000 \text{ g Na}} \times \frac{1 \text{ Na atom}}{3.817 \times 10^{-26} \text{ kg Na}} = 4.239 \times 10^{23} \text{ Na atoms}$$

Conversion Pathway:

$$1540.5 \text{ g seawater} \times \frac{2.67 \text{ g NaCl}}{100 \text{ g seawater}} \times \frac{39.34 \text{ g Na}}{100 \text{ g NaCl}} \times \frac{1 \text{ kg Na}}{1000 \text{ g Na}} \times \frac{1 \text{ Na atom}}{3.8175 \times 10^{-26} \text{ kg Na}}$$

EXERCISES

The Scientific Method

1. **(E)** One theory is preferred over another if it can correctly predict a wider range of phenomena and if it has fewer assumptions.

3. **(E)** For a given set of conditions, a cause, is expected to produce a certain result or effect. Although these cause-and-effect relationships may be difficult to unravel at times ("God is subtle"), they nevertheless do exist ("He is not malicious").

5. **(E)** The experiment should be carefully set up so as to create a controlled situation in which one can make careful observations after altering the experimental parameters, preferably one at a time. The results must be reproducible (to within experimental error) and, as more and more experiments are conducted, a pattern should begin to emerge, from which a comparison to the current theory can be made.

Properties and Classification of Matter

7. **(E)** When an object displays a physical property it retains its basic chemical identity. By contrast, the display of a chemical property is accompanied by a change in composition.

    (a) Physical: The iron nail is not changed in any significant way when it is attracted to a magnet. Its basic chemical identity is unchanged.

    (b) Chemical: The paper is converted to ash, $CO_2(g)$, and $H_2O(g)$ along with the evolution of considerable energy.

    (c)   Chemical: The green patina is the result of the combination of water, oxygen, and carbon dioxide with the copper in the bronze to produce basic copper carbonate.

    (d)   Physical: Neither the block of wood nor the water has changed its identity.

9.   **(E) (a)** Homogeneous mixture: Air is a mixture of nitrogen, oxygen, argon, and traces of other gases. By "fresh," we mean no particles of smoke, pollen, etc., are present. Such species would produce a heterogeneous mixture.

    (b)   Heterogeneous mixture: A silver-plated spoon has a surface coating of the element silver and an underlying baser metal (typically iron). This would make the coated spoon a heterogeneous mixture.

    (c)   Heterogeneous mixture: Garlic salt is simply garlic powder mixed with table salt. Pieces of garlic can be distinguished from those of salt by careful examination.

    (d)   Substance: Ice is simply solid water (assuming no air bubbles).

11.   **(E) (a)** If a magnet is drawn through the mixture, the iron filings will be attracted to the magnet and the wood will be left behind.

    (b)   When the glass-sucrose mixture is mixed with water, the sucrose will dissolve, whereas the glass will not. The water can then be boiled off to produce pure sucrose.

    (c)   Olive oil will float to the top of a container and can be separated from water, which is more dense. It would be best to use something with a narrow opening that has the ability to drain off the water layer at the bottom (i.e., buret).

    (d)   The gold flakes will settle to the bottom if the mixture is left undisturbed. The water then can be decanted (i.e., carefully poured off).

Exponential Arithmetic

13.   **(E) (a)** $8950. = 8.950 \times 10^3$ (4 sig. fig.)

    (b) $10,700. = 1.0700 \times 10^4$ (5 sig. fig.)   **(c)** $0.0240 = 2.40 \times 10^{-2}$

    (d) $0.0047 = 4.7 \times 10^{-3}$ **(e)** $938.3 = 9.383 \times 10^2$ **(f)** $275,482 = 2.75482 \times 10^5$

15.   **(E) (a)** $34,000$ centimeters / second $= 3.4 \times 10^4$ cm/s

    (b) 6378 km $= 6.378 \times 10^3$ km

    (c)   (trillionth $= 1 \times 10^{-12}$) hence, 74×10^{-12} m or 7.4×10^{-11} m

    (d) $\dfrac{(2.2 \times 10^3) + (4.7 \times 10^2)}{5.8 \times 10^{-3}} = \dfrac{2.7 \times 10^3}{5.8 \times 10^{-3}} = 4.6 \times 10^5$

Significant Figures

17. **(E) (a)** An exact number—500 sheets in a ream of paper.

(b) Pouring the milk into the bottle is a process that is subject to error; there can be slightly more or slightly less than one liter of milk in the bottle. This is a measured quantity.

(c) Measured quantity: The distance between any pair of planetary bodies can only be determined through certain astronomical measurements, which are subject to error.

(d) Measured quantity: The internuclear separation quoted for O_2 is an estimated value derived from experimental data, which contains some inherent error.

19. **(E)** Each of the following is expressed to four significant figures.

(a) $3984.6 \approx 3985$ **(b)** $422.04 \approx 422.0$ **(c)** $186{,}000 = 1.860 \times 10^5$

(d) $33{,}900 \approx 3.390 \times 10^4$ **(e)** 6.321×10^4 is correct **(f)** $5.0472 \times 10^{-4} \approx 5.047 \times 10^{-4}$

21. **(E) (a)** $0.406 \times 0.0023 = 9.3 \times 10^{-4}$ **(b)** $0.1357 \times 16.80 \times 0.096 = 2.2 \times 10^{-1}$

(c) $0.458 + 0.12 - 0.037 = 5.4 \times 10^{-1}$ **(d)** $32.18 + 0.055 - 1.652 = 3.058 \times 10^1$

23. **(M) (a)** 2.44×10^4 **(b)** 1.5×10^3 **(c)** 40.0

(d) 2.131×10^3 **(e)** 4.8×10^{-3}

25. **(M) (a)** The average speed is obtained by dividing the distance traveled (in miles) by the elapsed time (in hours). First, we need to obtain the elapsed time, in hours.

$$9 \text{ days} \times \frac{24 \text{ h}}{1 \text{ d}} = 216.000 \text{ h} \quad 3 \text{ min} \times \frac{1 \text{ h}}{60 \text{ min}} = 0.050 \text{ h} \quad 44 \text{ s} \times \frac{1 \text{ h}}{3600 \text{ s}} = 0.012 \text{ h}$$

total time $= 216.000 \text{ h} + 0.050 \text{ h} + 0.012 \text{ h} = 216.062 \text{ h}$

$$\text{average speed} = \frac{25{,}012 \text{ mi}}{216.062 \text{ h}} \times \frac{1.609344 \text{ km}}{1 \text{ mi}} = 186.30 \text{ km/h}$$

(b) First compute the mass of fuel remaining

$$\text{mass} = 14 \text{ gal} \times \frac{4 \text{ qt}}{1 \text{ gal}} \times \frac{0.9464 \text{ L}}{1 \text{ qt}} \times \frac{1000 \text{ mL}}{1 \text{ L}} \times \frac{0.70 \text{ g}}{1 \text{ mL}} \times \frac{1 \text{ lb}}{453.6 \text{ g}} = 82 \text{ lb}$$

Next determine the mass of fuel used, and then finally, the fuel consumption. Notice that the initial quantity of fuel is not known precisely, perhaps at best to the nearest 10 lb, certainly ("nearly 9000 lb") is not to the nearest pound.

$$\text{mass of fuel used} = (9000 \text{ lb} - 82 \text{ lb}) \times \frac{0.4536 \text{ kg}}{1 \text{ lb}} \cong 4045 \text{ kg}$$

$$\text{fuel consumption} = \frac{25{,}012 \text{ mi}}{4045 \text{ kg}} \times \frac{1.609344 \text{ km}}{1 \text{ mi}} = 9.95 \text{ km/kg or } \sim 10 \text{ km/kg}$$

Units of Measurement

27. **(E) (a)** $0.127 \text{ L} \times \dfrac{1000 \text{ mL}}{1 \text{ L}} = 127 \text{ mL}$ **(b)** $15.8 \text{ mL} \times \dfrac{1 \text{ L}}{1000 \text{ mL}} = 0.0158 \text{ L}$

   (c) $981 \text{ cm}^3 \times \dfrac{1 \text{ L}}{1000 \text{ cm}^3} = 0.981 \text{ L}$ **(d)** $2.65 \text{ m}^3 \times \left(\dfrac{100 \text{ cm}}{1 \text{ m}}\right)^3 = 2.65 \times 10^6 \text{ cm}^3$

29. **(E) (a)** $68.4 \text{ in.} \times \dfrac{2.54 \text{ cm}}{1 \text{ in.}} = 174 \text{ cm}$ **(b)** $94 \text{ ft} \times \dfrac{12 \text{ in.}}{1 \text{ ft}} \times \dfrac{2.54 \text{ cm}}{1 \text{ in.}} \times \dfrac{1 \text{ m}}{100 \text{ cm}} = 29 \text{ m}$

   (c) $1.42 \text{ lb} \times \dfrac{453.6 \text{ g}}{1 \text{ lb}} = 644 \text{ g}$ **(d)** $248 \text{ lb} \times \dfrac{0.4536 \text{ kg}}{1 \text{ lb}} = 112 \text{ kg}$

   (e) $1.85 \text{ gal} \times \dfrac{4 \text{ qt}}{1 \text{ gal}} \times \dfrac{0.9464 \text{ dm}^3}{1 \text{ qt}} = 7.00 \text{ dm}^3$

   (f) $3.72 \text{ qt} \times \dfrac{0.9464 \text{ L}}{1 \text{ qt}} \times \dfrac{1000 \text{ mL}}{1 \text{ L}} = 3.52 \times 10^3 \text{ mL}$

31. **(E)** Express both masses in the same units for comparison.

$$3245 \mu g \times \left(\dfrac{1 \text{ g}}{10^6 \mu g}\right) \times \left(\dfrac{10^3 \text{ mg}}{1 \text{ g}}\right) = 3.245 \text{ mg, which is larger than } 0.00515 \text{ mg.}$$

33. **(E)** *Conversion pathway approach:*

$$\text{height} = 15 \text{ hands} \times \dfrac{4 \text{ in.}}{1 \text{ hand}} \times \dfrac{2.54 \text{ cm}}{1 \text{ in.}} \times \dfrac{1 \text{ m}}{100 \text{ cm}} = 1.5 \text{ m}$$

Stepwise approach:

$$15 \text{ hands} \times \dfrac{4 \text{ in.}}{1 \text{ hand}} = 60 \text{ in.}$$

$$60 \text{ in.} \times \dfrac{2.54 \text{ cm}}{1 \text{ in.}} = 152.4 \text{ cm}$$

$$152.4 \text{ cm} \times \dfrac{1 \text{ m}}{100 \text{ cm}} = 1.524 \text{ m} = 1.5 \text{ m}$$

35. **(M) (a)** We use the speed as a conversion factor, but need to convert yards into meters.

$$\text{time} = 100.0 \text{ m} \times \dfrac{9.3 \text{ s}}{100 \text{ yd}} \times \dfrac{1 \text{ yd}}{36 \text{ in.}} \times \dfrac{39.37 \text{ in.}}{1 \text{ m}} = 10. \text{ s}$$

The final answer can only be quoted to a maximum of two significant figures.

(b)   We need to convert yards to meters.

$$\text{speed} = \frac{100 \text{ yd}}{9.3 \text{ s}} \times \frac{36 \text{ in.}}{1 \text{ yd}} \times \frac{2.54 \text{ cm}}{1 \text{ in.}} \times \frac{1 \text{ m}}{100 \text{ cm}} = 9.8\underline{3} \text{ m/s}$$

(c)   The speed is used as a conversion factor.

$$\text{time} = 1.45 \text{ km} \times \frac{1000 \text{ m}}{1 \text{ km}} \times \frac{1 \text{ s}}{9.8\underline{3} \text{ m}} \times \frac{1 \text{ min}}{60 \text{ s}} = 2.5 \text{ min}$$

37.   **(D)** $1 \text{ hectare} = 1 \text{ hm}^2 \times \left(\frac{100 \text{ m}}{1 \text{ hm}} \times \frac{100 \text{ cm}}{1 \text{ m}} \times \frac{1 \text{ in.}}{2.54 \text{ cm}} \times \frac{1 \text{ ft}}{12 \text{ in.}} \times \frac{1 \text{ mi}}{5280 \text{ ft}} \right)^2 \times \frac{640 \text{ acres}}{1 \text{ mi}^2}$

$1 \text{ hectare} = 2.47 \text{ acres}$

39.   **(D)** $\text{pressure} = \frac{32 \text{ lb}}{1 \text{ in.}^2} \times \frac{453.6 \text{ g}}{1 \text{ lb}} \times \left(\frac{1 \text{ in.}}{2.54 \text{ cm}} \right)^2 = 2.2 \times 10^3 \text{ g/cm}^2$

$$\text{pressure} = \frac{2.2 \times 10^3 \text{ g}}{1 \text{ cm}^2} \times \frac{1 \text{ kg}}{1000 \text{ g}} \times \left(\frac{100 \text{ cm}}{1 \text{ m}} \right)^2 = 2.2 \times 10^4 \text{ kg/m}^2$$

Temperature Scales

41.   **(E)**   low: $t(^\circ\text{F}) = \frac{9\,^\circ\text{F}}{5\,^\circ\text{C}} t(^\circ\text{C}) + 32\,^\circ\text{F} = \frac{9\,^\circ\text{F}}{5\,^\circ\text{C}} \left(-10\,^\circ\text{C} \right) + 32\,^\circ\text{F} = 14\,^\circ\text{F}$

high: $t(^\circ\text{F}) = \frac{9\,^\circ\text{F}}{5\,^\circ\text{C}} t(^\circ\text{C}) + 32\,^\circ\text{F} = \frac{9\,^\circ\text{F}}{5\,^\circ\text{C}} \left(50\,^\circ\text{C} \right) + 32\,^\circ\text{F} = 122\,^\circ\text{F}$

43.   **(M)**   Let us determine the Fahrenheit equivalent of absolute zero.

$t(^\circ\text{F}) = \frac{9\,^\circ\text{F}}{5\,^\circ\text{C}} t(^\circ\text{C}) + 32\,^\circ\text{F} = \frac{9\,^\circ\text{F}}{5\,^\circ\text{C}} \left(-273.15\,^\circ\text{C} \right) + 32\,^\circ\text{F} = -459.7\,^\circ\text{F}$

A temperature of $-465\,^\circ\text{F}$ cannot be achieved because it is below absolute zero.

45.   **(D) (a)**   From the data provided we can write down the following relationship:
$-38.9\,^\circ\text{C} = 0\,^\circ\text{M}$ and $356.9\,^\circ\text{C} = 100\,^\circ\text{M}$. To find the mathematical relationship between these two scales, we can treat each relationship as a point on a two-dimensional Cartesian graph:

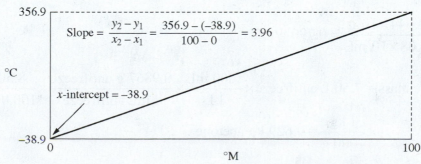

Therefore, the equation for the line is $y = 3.96x - 38.9$. The algebraic relationship between the two temperature scales is

$$t(°C) = 3.96(°M) - 38.9 \text{ or rearranging, } t(°M) = \frac{t(°C) + 38.9}{3.96}$$

Alternatively, note that the change in temperature in °C corresponding to a change of 100 °M is $[356.9 - (-38.9)] = 395.8 \ °C$, hence, $(100 \ °M/395.8 \ °C) = 1 \ °M/3.96 \ °C$. This factor must be multiplied by the number of degrees Celsius above zero on the M scale. This number of degrees is $t(°C) + 38.9$, which leads to the general equation $t(°M) = [t(°C) + 38.9]/3.96$.

The boiling point of water is 100 °C, corresponding to $t(°M) = \dfrac{100 + 38.9}{3.96} = 35.1 \ °M$

(b) $t(°M) = \dfrac{-273.15 + 38.9}{3.96} = -59.2 \ °M$ would be the absolute zero on this scale.

Density

47. (E) butyric acid density $= \dfrac{mass}{volume} = \dfrac{2088 \text{ g}}{2.18 \text{ L}} \times \dfrac{1 \text{ L}}{1000 \text{ mL}} = 0.958$ g/mL

49. (M) The mass of acetone is the difference in masses between empty and filled masses.

Conversion pathway approach:

$$\text{density} = \frac{437.5 \text{ lb} - 75.0 \text{ lb}}{55.0 \text{ gal}} \times \frac{453.6 \text{ g}}{1 \text{ lb}} \times \frac{1 \text{ gal}}{3.785 \text{ L}} \times \frac{1 \text{ L}}{1000 \text{ mL}} = 0.790 \text{ g/mL}$$

Stepwise approach:

$$437.5 \text{ lb} - 75.0 \text{ lb} = 362.5 \text{ lb}$$

$$362.5 \text{ lb} \times \frac{453.6 \text{ g}}{1 \text{ lb}} = 1.644 \times 10^5 \text{g}$$

$$55.0 \text{ gal} \times \frac{3.785 \text{ L}}{1 \text{ gal}} = 208 \text{ L}$$

$$208 \text{ L} \times \frac{1000 \text{ mL}}{1 \text{ L}} = 2.08 \times 10^5 \text{mL}$$

$$\frac{1.644 \times 10^5 \text{g}}{2.08 \times 10^5 \text{mL}} = 0.790 \text{ g/mL}$$

51. (M) acetone mass $= 7.50$ L antifreeze $\times \dfrac{1000 \text{ mL}}{1 \text{ L}} \times \dfrac{0.9867 \text{ g antifreeze}}{1 \text{ mL antifreeze}} \times \dfrac{8.50 \text{ g acetone}}{100.0 \text{ g antifreeze}}$

$$\times \frac{1 \text{ kg}}{1000 \text{ g}} = 0.629 \text{ kg acetone}$$

53. (M) fertilizer mass $= 225$ g nitrogen $\times \dfrac{1 \text{ kg N}}{1000 \text{ g N}} \times \dfrac{100 \text{ kg fertilizer}}{21 \text{ kg N}} = 1.0\underline{7}$ kg fertilizer

55. (M) The calculated volume of the iron block is converted to its mass by using the provided density.

$$\text{mass} = 52.8 \text{ cm} \times 6.74 \text{ cm} \times 3.73 \text{ cm} \times 7.86 \dfrac{\text{g}}{\text{cm}^3} = 1.04 \times 10^4 \text{ g iron}$$

57. (M) We start by determining the mass of each item.

(1) mass of iron bar $= (81.5 \text{ cm} \times 2.1 \text{ cm} \times 1.6 \text{ cm}) \times 7.86 \text{ g/cm}^3 = 2.2 \times 10^3$ g iron

(2) mass of Al foil $= (12.12 \text{ m} \times 3.62 \text{ m} \times 0.003 \times 10^{-2} \text{m}) \times \left(\dfrac{100 \text{ cm}}{1 \text{ m}}\right)^3 \times 2.70 \text{ g Al/cm}^3$

$$= 4 \times 10^3 \text{ g Al}$$

(3) mass of water $= 4.051 \text{ L} \times \dfrac{1000 \text{ cm}^3}{1 \text{ L}} \times 0.998 \text{ g / cm}^3 = 4.04 \times 10^3$ g water

In order of increasing mass, the items are: iron bar < aluminum foil < water. Strictly speaking, the rules for significant figures do not allow us to distinguish between the masses of aluminum and water.

59. (D) First determine the volume of the aluminum foil, then its area, and finally its thickness.

$$\text{volume} = 2.568 \text{ g} \times \dfrac{1 \text{ cm}^3}{2.70 \text{ g}} = 0.951 \text{ cm}^3 ; \qquad \text{area} = (22.86 \text{ cm})^2 = 522.6 \text{ cm}^2$$

$$\text{thickness} = \dfrac{\text{volume}}{\text{area}} = \dfrac{0.951 \text{ cm}^3}{522.6 \text{ cm}^2} \times \dfrac{10 \text{ mm}}{1 \text{ cm}} = 1.82 \times 10^{-2} \text{ mm}$$

61. (D) Here we are asked to calculate the number of liters of whole blood that must be collected in order to end up with 0.5 kg of red blood cells. Each red blood cell has a mass of $90.0 \times 10^{-12} \text{ cm}^3 \times 1.096 \text{ g cm}^{-3} = 9.864 \times 10^{-11}$ g

$$\text{red blood cells (mass per mL)} = \dfrac{9.864 \times 10^{-11} \text{ g}}{1 \text{ cell}} \times \dfrac{5.4 \times 10^9 \text{ cells}}{1 \text{ mL}} = \dfrac{0.533 \text{ g red blood cells}}{1 \text{ mL of blood}}$$

For 0.5 kg or 5×10^2 g of red blood cells, we require

$$= 5 \times 10^2 \text{ g red blood cells} \times \dfrac{1 \text{ mL of blood}}{0.533 \text{ g red blood cells}} = 9 \times 10^2 \text{ mL of blood or 0.9 L blood}$$

Percent Composition

63. **(E)** The percent of students with each grade is obtained by dividing the number of students with that grade by the total number of students. $\%A = \dfrac{7 \text{ A's}}{76 \text{ students}} \times 100\% = 9.2\% \text{ A}$

$\%B = \dfrac{22 \text{ B's}}{76 \text{ students}} \times 100\% = 28.9\% \text{ B}$ $\qquad$ $\%C = \dfrac{37 \text{ C's}}{76 \text{ students}} \times 100\% = 48.7\% \text{ C}$

$\%D = \dfrac{8 \text{ D's}}{76 \text{ students}} \times 100\% = 11\% \text{ D}$ $\qquad$ $\%F = \dfrac{2 \text{ F's}}{76 \text{ students}} \times 100\% = 3\% \text{ F}$

Note that the percentages add to 101% due to rounding effects.

65. **(M)** Use the percent composition as a conversion factor.

Conversion pathway approach:

$$\text{mass of sucrose} = 3.50 \text{ L} \times \frac{1000 \text{ mL}}{1 \text{ L}} \times \frac{1.118 \text{ g soln}}{1 \text{ mL}} \times \frac{28.0 \text{ g sucrose}}{100 \text{ g soln}} = 1.10 \times 10^3 \text{ g sucrose}$$

Stepwise approach:

$$3.50 \text{ L} \times \frac{1000 \text{ mL}}{1 \text{ L}} = 3.50 \times 10^3 \text{ mL}$$

$$3.50 \times 10^3 \text{ mL} \times \frac{1.118 \text{ g soln}}{1 \text{ mL}} = 3.91 \times 10^3 \text{ g soln}$$

$$3.91 \times 10^3 \text{ g soln} \times \frac{28.0 \text{ g sucrose}}{100 \text{ g soln}} = 1.10 \times 10^3 \text{ g sucrose}$$

INTEGRATIVE AND ADVANCED EXERCISES

71. **(D)** *Conversion pathway approach:*

$$\text{NaCl mass} = 330{,}000{,}000 \text{ mi}^3 \times \left(\frac{5280 \text{ ft}}{1 \text{ mi}} \times \frac{12 \text{ in.}}{1 \text{ ft}} \times \frac{2.54 \text{ cm}}{1 \text{ in.}} \right)^3 \times \frac{1 \text{ mL}}{1 \text{ cm}^3} \times \frac{1.03 \text{ g}}{1 \text{ mL}}$$

$$\times \frac{3.5 \text{ g sodium chloride}}{100.0 \text{ g seawater}} \times \frac{1 \text{ lb}}{453.6 \text{ g}} \times \frac{1 \text{ ton}}{2000 \text{ lb}} = 5.5 \times 10^{16} \text{ tons}$$

Stepwise approach:

$$330{,}000{,}000 \text{ mi}^3 \times \left(\frac{5280 \text{ ft}}{1 \text{ mi}} \right)^3 = 4.9 \times 10^{19} \text{ ft}^3$$

$$4.9 \times 10^{19} \text{ ft}^3 \times \left(\frac{12 \text{ in.}}{1 \text{ ft}} \right)^3 = 8.4 \times 10^{22} \text{ in.}^3$$

$$8.4 \times 10^{22} \text{ in.}^3 \times \left(\frac{2.54 \text{ cm}}{1 \text{ in.}} \right)^3 = 1.4 \times 10^{24} \text{ cm}^3$$

$$1.4 \times 10^{24} \text{ cm}^3 \times \frac{1 \text{ mL}}{1 \text{ cm}^3} \times \frac{1.03 \text{ g}}{1 \text{ mL}} = 1.4 \times 10^{24} \text{ g}$$

$$1.4 \times 10^{24} \text{ g} \times \frac{3.5 \text{ g sodium chloride}}{100.0 \text{ g seawater}} = 4.9 \times 10^{22} \text{ g NaCl}$$

$$4.9 \times 10^{22} \text{ g NaCl} \times \frac{1 \text{ lb}}{453.6 \text{ g}} = 1.1 \times 10^{20} \text{ lb}$$

$$1.1 \times 10^{20} \text{ lb} \times \frac{1 \text{ ton}}{2000 \text{ lb}} = 5.4 \times 10^{16} \text{ tons}$$

The answers for the stepwise and conversion pathway approaches differ slightly due to a cumulative rounding error that is present in the stepwise approach.

72. **(D)** First, we find the volume of the wire, then its cross-sectional area, and finally its length. We carry an additional significant figure through the early stages of the calculation to help avoid rounding errors.

$$V = 1 \text{ lb} \times \frac{453.6 \text{ g}}{1 \text{ lb}} \times \frac{1 \text{ cm}^3}{8.92 \text{ g}} = 50.8\underline{5} \text{ cm}^3 \qquad \text{Note: area} = \pi r^2$$

$$\text{area} = 3.1416 \times \left(\frac{0.05082 \text{ in.}}{2} \times \frac{2.54 \text{ cm}}{1 \text{ in.}} \right)^2 = 0.01309 \text{ cm}^2$$

$$\text{length} = \frac{\text{volume}}{\text{area}} = \frac{50.8\underline{5} \text{ cm}^3}{0.01309 \text{ cm}^2} \times \frac{1 \text{ m}}{100 \text{ cm}} = 38.8 \text{ m}$$

74. **(D) (a)**

$$\text{dustfall} = \frac{10 \text{ ton}}{1 \text{ mi}^2 \cdot 1 \text{ mo}} \times \left(\frac{1 \text{ mi}}{5280 \text{ ft}} \times \frac{1 \text{ ft}}{12 \text{ in.}} \times \frac{39.37 \text{ in.}}{1 \text{ m}} \right)^2 \times \frac{2000 \text{ lb}}{1 \text{ ton}} \times \frac{454 \text{ g}}{1 \text{ lb}} \times \frac{1000 \text{ mg}}{1 \text{ g}}$$

$$= \frac{3.5 \times 10^3 \text{ mg}}{1 \text{ m}^2 \cdot 1 \text{ mo}} \times \frac{1 \text{ month}}{30 \text{ d}} \times \frac{1 \text{ d}}{24 \text{ h}} = \frac{5 \text{ mg}}{1 \text{ m}^2 \cdot 1 \text{ h}}$$

(b) This problem is solved by the conversion factor method, starting with the volume that deposits on each square meter, 1 mm deep.

$$\frac{(1.0 \text{ mm} \times 1 \text{ m}^2)}{1 \text{ m}^2} \times \frac{1 \text{ cm}}{10 \text{ mm}} \times \left(\frac{100 \text{ cm}}{1 \text{ m}} \right)^2 \times \frac{2 \text{ g}}{1 \text{ cm}^3} \times \frac{1000 \text{ mg}}{1 \text{ g}} \times \frac{1 \text{ m}^2 \cdot \text{h}}{4.9 \text{ mg}} = 4.1 \times 10^5 \text{ h} = 5 \times 10^1 \text{ y}$$

It would take about half a century to accumulate a depth of 1 mm.

76. **(M)** Let F be the Fahrenheit temperature and C be the Celsius temperature. $C = (F - 32)\frac{5}{9}$

(a) $F = C - 49$ $\quad C = (C - 49 - 32)\frac{5}{9} = \frac{5}{9}(C - 81)$ $\quad C = \frac{5}{9}C - \frac{5}{9}(81)$ $\quad C = \frac{5}{9}C - 45$

$\frac{4}{9}C = -45$ $\quad$ Hence: $C = -101.\underline{25}$

When it is $\sim -101 \,°C$, the temperature in Fahrenheit is $-150 \,°F$ ($49°$ lower).

(b) $F = 2C$ $\qquad C = (2C - 32)\frac{5}{9} = \frac{10}{9}C - 17.8$ $\qquad 17.8 = \frac{10}{9}C - C = \frac{1}{9}C$

$C = 9 \times 17.8 = 160. \,°C$ $\qquad F = \frac{9}{5}C + 32 = \frac{9}{5}(160.) + 32 = 320. \,°F$

(c) $F = \frac{1}{8}C$ $C = (\frac{1}{8}C - 32)\frac{5}{9} = \frac{5}{72}C - 17.8$ $17.8 = \frac{5}{72}C - C = -\frac{67}{72}C$

$C = -\dfrac{72 \times 17.8}{67} = -19.1\,°C$ $F = \frac{9}{5}C + 32 = \frac{9}{5}(-19.1) + 32 = -2.4\,°F$

(d) $F = C + 300$ $C = (C + 300 - 32)\frac{5}{9} = \frac{5}{9}C + 148.9$ $148.9 = \frac{4}{9}C$

$C = \dfrac{9 \times 148.9}{4} = 335\,°C$ $F = \frac{9}{5}C + 32 = \frac{9}{5}(335) + 32 = 635\,°F$

80.   **(D)** We first determine the pycnometer's volume.

pycnometer volume $= (35.55\text{ g} - 25.60\text{ g}) \times \dfrac{1\text{ mL}}{0.9982\text{ g}} = 9.97\text{ mL}$

Then we determine the volume of water present with the lead.

volume of water $= (44.83\text{ g} - 10.20\text{ g} - 25.60\text{ g}) \times \dfrac{1\text{ mL}}{0.9982\text{ g}} = 9.05\text{ mL}$

Difference between the two volumes is the volume of lead, which leads to the density of lead.

density $= \dfrac{10.20\text{ g}}{(9.97\text{ mL} - 9.05\text{ mL})} = 11\text{ g/mL}$

Note that the difference in the denominator has just two significant digits.

81.   **(M)** Water used (in kg/week) $= 1.8 \times 10^6\text{ people} \times \left(\dfrac{750\text{ L}}{1\text{ day}}\right) \times \left(\dfrac{7\text{ day}}{1\text{ week}}\right) \times \dfrac{1\text{ kg}}{1\text{ L}}$

$$= 9.45 \times 10^9\text{ kg water/week}$$

Given: Sodium hypochlorite is NaClO

mass of NaClO $= 9.45 \times 10^9\text{ kg water}\left(\dfrac{1\text{ kg chlorine}}{1 \times 10^6\text{ kg water}}\right) \times \left(\dfrac{100\text{ kg NaClO}}{47.62\text{ kg chlorine}}\right)$

$$= 1.98 \times 10^4\text{ kg sodium hypochlorite}$$

84.   **(D)** First, calculate the volume of the piece of Styrofoam:

$V = 36.0\text{ cm} \times 24.0\text{ cm} \times 5.0\text{ cm} = 4.32 \times 10^3\text{ cm}^3$

Calculate the volume of water displaced (using dimensions in the figure):

$V = 36.0\text{ cm} \times 24.0\text{ cm} \times 3.0\text{ cm} = 2.592 \times 10^3\text{ cm}^3$

The mass of displaced water is given as: $m = d \times V = 1.00\text{ g/cm}^3 \times 2.592 \times 10^3\text{ cm}^3$

$= 2.592 \times 10^3\text{ g}$

Since the object floats, it means that the water is exerting a force equivalent to the mass of Styrofoam/book times the acceleration due to gravity (g). We can factor out g, and are left with masses of Styrofoam and water:

mass of book + mass of Styrofoam = mass of water

$$1.5 \times 10^3 \text{ g} + d \times 4.32 \times 10^3 \text{ cm}^3 = 2.592 \times 10^3 \text{ g}$$

Solving for d, we obtain:

$$d = 0.25 \text{ g/cm}^3$$

85. **(M)** (a) When the mixture is pure benzene, %N = 0, $d = 1/1.153 = 0.867$ g/cm^3
(b) When mixture is pure naphthalene, %N = 100, $d = 1.02$ g/cm^3
(c) %N = 1.15, d = 0.869 g/cm^3
(d) Using $d = 0.952$ g/cm^3 and the quadratic formula to solve for %N. %N = 58.4

86. **(M)** First, calculate the total mass of ice in the Antarctic, which yields the total mass of water which is obtained if all the ice melts:

$$3.01 \times 10^7 \text{ km}^3 \text{ ice} \times \frac{(1 \times 10^5 \text{ cm})^3}{1 \text{ km}^3} \times \frac{0.92 \text{ g ice}}{1 \text{ cm}^3 \text{ ice}} = 2.769 \times 10^{22} \text{ g ice}$$

all of which converts to water. The volume of this extra water is then calculated.

$$2.769 \times 10^{22} \text{ g H}_2\text{O} \times \frac{1 \text{ cm}^3 \text{ H}_2\text{O}}{1 \text{ g H}_2\text{O}} \times \frac{1 \text{ km}^3 \text{ H}_2\text{O}}{(1 \times 10^5 \text{ cm})^3 \text{ H}_2\text{O}} = 2.769 \times 10^7 \text{ km}^3 \text{ H}_2\text{O}$$

Assuming that V (H$_2$O on Earth) $= A \times h = 3.62 \times 10^8$ km$^2 \times h$, the total increase in the height of sea levels with the addition of the melted continental ice will be:

$$h = 2.769 \times 10^7 \text{ km}^3 / 3.62 \times 10^8 \text{ km}^2 = 0.0765 \text{ km} = 76.4 \text{ m.}$$

87. **(M)** First, calculate the mass of wine: 4.72 kg – 1.70 kg = 3.02 kg
Then, calculate the mass of ethyl alcohol (ethanol) in the bottle:

$$3.02 \text{ kg wine} \times \frac{1000 \text{ g wine}}{1 \text{ kg wine}} \times \frac{11.5 \text{ g ethanol}}{100 \text{ g wine}} = 347.3 \text{ g ethanol}$$

Then, use the above amount to determine how much ethanol is in 250 mL of wine:

$$250.0 \text{ mL ethanol} \times \frac{1 \text{ L ethanol}}{1000 \text{ mL ethanol}} \times \frac{347.3 \text{ g ethanol}}{3.00 \text{ L bottle}} = 28.9 \text{ g ethanol}$$

88. **(M)** First, determine the total volume of tungsten:

$$\text{vol W} = m/d = \frac{0.0429 \text{ g W}}{19.3 \text{ g/cm}^3} \times \frac{(10 \text{ mm})^3}{1 \text{ cm}^3} = 2.22 \text{ mm}^3 \text{ W}$$

The wire can be viewed as a cylinder. Therefore:
vol cylinder $= A \times h = \pi(d/2)^2 \times h = \pi(d/2)^2 \times (0.200 \text{ m} \times 1000 \text{ mm}/1 \text{ m}) = 2.22$ mm^3
Solving for d, we obtain: $d = 0.119$ mm

89. (M) First, determine the amount of ethyl alcohol (ethanol) that will cause a BAC of 0.10%:

$$\text{mass of ethanol} = \frac{0.100\ \text{g ethanol}}{100\ \text{mL of blood}} \times 5400\ \text{mL blood} = 5.4\ \text{g ethanol}$$

This person's body metabolizes alcohol at a rate of 10.0 g/h. Therefore, in 3 hours, this person metabolizes 30.0 g of alcohol. For this individual to have a BAC of 0.10% after 3 hours, he must consume 30.0 + 5.4 = 35.4 g of ethanol.

Now, calculate how many glasses of wine are needed for a total intake of 35.4 g of ethanol:

$$35.4\ \text{g ethanol} \times \frac{100\ \text{g wine}}{11.5\ \text{g eth.}} \times \frac{1\ \text{mL wine}}{1.01\ \text{g wine}} \times \frac{1\ \text{glass wine}}{145\ \text{mL wine}} = 2.1\ \text{glasses of wine}$$

FEATURE PROBLEMS

94. (D) In sketch (a), the mass of the plastic block appears to be 50.0 g.
In sketch (b), the plastic block is clearly visible on the bottom of a beaker filled with ethanol, showing that it is both insoluble in and more dense than ethanol (i.e., $> 0.789\ \text{g/cm}^3$).
In sketch (c), because the plastic block floats on bromoform, the density of the plastic must be less than that for bromoform (i.e., $< 2.890\ \text{g/cm}^3$). Moreover, because the block is ~ 40% submerged, the volume of bromoform having the same 50.0 g mass as the block is only about 40% of the volume of the block. Thus, using the expression $V = m/d$, we can write

$$\text{volume of displaced bromoform} \sim 0.40 \times V_{\text{block}}$$

$$\frac{\text{mass of bromoform}}{\text{density of bromoform}} = \frac{0.40 \times \text{mass of block}}{\text{density of plastic}} = \frac{50.0\ \text{g of bromoform}}{2.890\ \dfrac{\text{g bromoform}}{\text{cm}^3}} = 0.40 \times \frac{50.0\ \text{g of plastic}}{\text{density of plastic}}$$

$$\text{density of plastic} \approx \frac{2.890\ \dfrac{\text{g bromoform}}{\text{cm}^3}}{50.0\ \text{g of bromoform}} \times 0.40 \times 50.0\ \text{g of plastic} \approx 1.16\ \frac{\text{g}}{\text{cm}^3}$$

The information provided in sketch (d) provides us with an alternative method for estimating the density of the plastic (use the fact that the density of water is 0.99821 g/cm³ at 20 °C).

$$\text{mass of water displaced} = 50.0\ \text{g} - 5.6\ \text{g} = 44.4\ \text{g}$$

$$\text{volume of water displaced} = 44.4\ \text{g} \times \frac{1\ \text{cm}^3}{0.99821\ \text{g}} = 44.5\ \text{cm}^3$$

Therefore the density of the plastic $= \dfrac{mass}{volume} = \dfrac{50.0\ \text{g}}{44.5\ \text{cm}^3} = 1.12\ \dfrac{\text{g}}{\text{cm}^3}$

This is reasonably close to the estimate based on the information in sketch (c).

SELF-ASSESSMENT EXERCISES

99. **(E)** (c) an experiment

100. **(E)** The answer is (e), a natural law.

101. **(E)** (c) an element. Mixtures (both homogeneous and heterogeneous) can be separated by physical means. A compound can be decomposed into its elements by chemical means. A pure substance can be an element or a compound, and therefore some pure substances can be decomposed by chemical means.

102. **(E)** The answer is (a), because the gas is fully dissolved in the liquid and remains there until the cap is removed. (b) and (c) are pure substances and therefore not mixtures, and material in a kitchen blender is heterogeneous in appearance.

103. **(E)** The answer is (c), the same. Mass is an intrinsic property of matter, and does not change with varying gravitational fields. Weight, which is acceleration due to gravity, does change.

104. **(E)** (d) $(1.172 - 0.4963)(4.193) = 2.83$ has the correct number of significant figures in the answer. (a) should have 3 significant figures and the answer shows 4. (b) should have 3 significant figures and the answer shows 2. (c) should have 2 significant figures and the answer shows 3.

105. **(E)** (d) and (f).

106. **(E)** The answer is (d). To compare, all values are converted to Kelvins.
Converting (c) 217 °F to K: $T(K) = ((217 - 32) \times 5/9) + 273 = 376$ K.
Converting (d) 105 °C to K: $105 + 273 = 378$ K.

107. **(M)** The answer is (b). The results are listed as follows: (a) 752 mL $H_2O \times 1$ g/mL = 752 g.
(b) 1050 mL ethanol $\times$ 0.789 g/mL = 828 g. (c) 750 g as stated.
(d) (19.20 cm $\times$ 19.20 cm $\times$ 19.20 cm) balsa wood $\times$ 0.11 g/mL = 779 g.

109. **(E)** The problem can be solved using dimensional analysis:

(a) g/L: $0.9982 \ \dfrac{g}{cm^3} \times \dfrac{1000 \ cm^3}{1 \ L} = 998.2 \ \dfrac{g}{L}$

(b) kg/m³: $0.9982 \ \dfrac{g}{cm^3} \times \dfrac{1 \ kg}{1000 \ g} \times \dfrac{(100 \ cm)^3}{1 \ m^3} = 998.2 \ \dfrac{kg}{m^3}$

(c) kg/m³: $0.9982 \ \dfrac{g}{cm^3} \times \dfrac{1 \ kg}{1000 \ g} \times \dfrac{(100 \ cm)^3}{1 \ m^3} \times \dfrac{(1000 \ m)^3}{1 \ km^3} = 9.982 \times 10^{11} \ \dfrac{kg}{km^3}$

110. **(E)** Student A is more accurate, Student B more precise.

111. **(E)** The answer is (b). Simply determining the volume from the dimensions (36 cm × 20.2 cm × 0.9 cm, noting that 9 mm = 0.9 cm) gives a volume of 654.48 cm^3. Since one of the dimensions only has one significant figure, the volume is 7×10^2 cm^3.

112. **(E)** (e), (a), (c), (b), (d), listed in order of increasing significant figures, which indicates an increasing precision in the measurement.

113. **(E)** The answer is (d). A 10.0 L solution with a density of 1.295 g/mL has a mass of 12,950 g, 30 mass% of which is an iron compound. Since the iron compound is 34.4% by mass iron, the total Fe content is 12950 × 0.300 × 0.344 = Having an iron content of 34.4 % Fe means that the mass is 1336 g or ~1340 g.

114. **(M)** First, you must determine the volume of copper. To do this, the mass of water displaced by the copper is determined, and the density used to calculate the volume of copper as shown below:

Δm = 25.305 – 22.486 = 2.819 g, mass of displaced water
Vol. of displaced H_2O = m/d = 2.819 g / 0.9982 g·mL^{-1} = 2.824 mL or cm^3 = Vol. of Cu

Vol of Cu = 2.824 cm^3 = surf. Area × thickness = 248 cm^2 × x
Solving for x, the value of thickness is therefore 0.0114 cm or 0.114 mm.

115. **(E)** In short, no, because a pure substance by definition is homogeneous. However, if there are other phases of the same pure substance present (such as pure ice in pure water), we have a heterogeneous mixture from a physical standpoint.

116. **(E)** (a) Physical change: no change of identity, only of physical state. *Air* is a mixture that later is separated physically. *Pressure* is not a form of matter.
(b) Physical change: separation of a mixture. *Temperature* is not a form of matter. *Liquid air* is a mixture that later is separated physically. *Liquid nitrogen* is an element.
(c) Chemical change: two new substances are produced. *Natural gas* is possibly a mixture, more likely a compound, since it is later used in a reaction. *Steam* and *carbon dioxide* are compounds. *Hydrogen* is an element.
(d) Chemical change: a compound is produced from elements. *Ammonia gas* is a compound.
(e) Physical change: separation of a mixture

CHAPTER 2
ATOMS AND THE ATOMIC THEORY
PRACTICE EXAMPLES

1A **(E)** The total mass must be the same before and after reaction.
mass before reaction = 0.382 g magnesium + 2.652 g nitrogen = 3.034 g
mass after reaction = magnesium nitride mass + 2.505 g nitrogen = 3.034 g
magnesium nitride mass = 3.034 g − 2.505 g = 0.529 g magnesium nitride

1B **(E)** Again, the total mass is the same before and after the reaction.
mass before reaction = 7.12 g magnesium + 1.80 g bromine = 8.92 g
mass after reaction = 2.07 g magnesium bromide + magnesium mass = 8.92 g
magnesium mass = 8.92 g − 2.07 g = 6.85 g magnesium

2A **(M)** In Example 2-2 we are told that 0.500 g MgO contains 0.301 g of Mg. With this information, we can determine the mass of magnesium needed to form 2.000 g of magnesium oxide.

$$\text{mass of Mg} = 2.000 \text{ g MgO} \times \frac{0.301 \text{ g Mg}}{0.500 \text{ g MgO}} = 1.20 \text{ g Mg}$$

The remainder of the 2.00 g of magnesium oxide is the mass of oxygen.
mass of oxygen = 2.00 g magnesium oxide − 1.20 g magnesium = 0.80 g oxygen

2B **(M)** In Example 2-2, we see that a 0.500 g sample of MgO has 0.301 g Mg, hence, it must have 0.199 g O_2. From this we see that if we have equal masses of Mg and O_2, the oxygen is in excess. First we find out how many grams of oxygen reacts with 10.00 g of Mg.

$$\text{mass}_{\text{oxygen}} = 10.00 \text{ g Mg} \times \frac{0.199 \text{ g } O_2}{0.301 \text{ g Mg}} = 6.61 \text{ g } O_2 \text{ (used up)}$$

Hence, 10.00 g − 6.61 g = 3.39 g O_2 unreacted. Mg is the limiting reactant.
MgO(s) mass = mass Mg + Mass O_2 = 10.00 g + 6.61 g = 16.61 g MgO.
There are only two substances present, 16.61 g of MgO (product) and 3.39 g of unreacted O_2

3A **(E)** Silver has 47 protons. If the isotope in question has 62 neutrons, then it has a mass number of 109. This can be represented as $^{109}_{47}\text{Ag}$.

3B **(E)** Tin has 50 electrons and 50 protons when neutral, while a neutral cadmium atom has 48 electrons. This means that we are dealing with Sn^{2+}. We do not know how many neutrons tin has. so there can be more than one answer. For instance, $^{116}_{50}\text{Sn}^{2+}$, $^{117}_{50}\text{Sn}^{2+}$, $^{118}_{50}\text{Sn}^{2+}$, $^{119}_{50}\text{Sn}^{2+}$, and $^{120}_{50}\text{Sn}^{2+}$ are all possible answers.

4A **(E)** The ratio of the masses of ^{202}Hg and ^{12}C is: $\dfrac{^{202}\text{Hg}}{^{12}\text{C}} = \dfrac{201.97062 \text{ u}}{12 \text{ u}} = 16.830885$

4B **(E)** Atomic mass is 12 u × 13.16034 = 157.9241 u. The isotope is $_{64}^{158}$Gd. Using an atomic mass of 15.9949 u for ^{16}O, the mass of $_{64}^{158}$Gd relative to ^{16}O is

$$\text{relative mass to oxygen-16} = \frac{157.9241 \text{ u}}{15.9949 \text{ u}} = 9.87340$$

5A **(E)** The atomic mass of boron is 10.81, which is closer to 11.0093054 than to 10.0129370. Thus, boron-11 is the isotope that is present in greater abundance.

5B **(E)** The average atomic mass of indium is 114.82, and one isotope is known to be ^{113}In. Since the weighted-average atomic mass is almost 115, the second isotope must be larger than both In-113 and In-114. Clearly, then, the second isotope must be In-115 (^{115}In). Since the average atomic mass of indium is closest to the mass of the second isotope, In-115, then ^{115}In is the more abundant isotope.

6A **(M)** Weighted-average atomic mass of Si =

$$(27.9769265325 \text{ u} \times 0.9223) \rightarrow 25.8\underline{0} \text{ u}$$
$$(28.976494700 \text{ u} \times 0.04685) \rightarrow 1.35\underline{8} \text{ u}$$
$$\underline{(29.973377017 \text{ u} \times 0.03092) \rightarrow 0.9268 \text{ u}}$$
$$28.08\underline{48} \text{ u}$$

We should report the weighted-average atomic mass of Si as 28.08 u.

6B **(M)** We let x be the fractional abundance of lithium-6.

$$6.941 \text{ u} = [x \times 6.01512 \text{ u}] + [(1-x) \times 7.01600 \text{ u}] = x \times 6.01512 \text{ u} + 7.01600 \text{ u} - x \times 7.01600 \text{ u}$$

$$6.941 \text{ u} - 7.01600 \text{ u} = x \times 6.01512 \text{ u} - x \times 7.01600 \text{ u} = -x \times 1.00088 \text{ u}$$

$$x = \frac{6.941 \text{ u} - 7.01600 \text{ u}}{-1.00088 \text{ u}} = 0.075 \text{ Percent isotopic abundances}: 7.5\% \text{ lithium-6}, \ 92.5\% \text{ lithium-7}$$

7A **(M)** We assume that atoms lose or gain relatively few electrons to become ions. Thus, elements that will form cations will be on the left-hand side of the periodic table, while elements that will form anions will be on the right-hand side. The number of electrons "lost" when a cation forms is usually equal to the last digit of the periodic group number; the number of electrons added when an anion forms is typically eight minus the last digit of the group number.

Li is in group 1(1A); it should form a cation by losing one electron: Li^+.
S is in group 6(6A); it should form an anion by adding two electrons: S^{2-}.
Ra is in group 2(2A); it should form a cation by losing two electrons: Ra^{2+}.
F and I are both group 17(7A); they should form anions by gaining an electron: F^- and I^-.
Al is in group 13(3A); it should form a cation by losing three electrons: Al^{3+}.

7B **(M)** Main-group elements are in the "A" families, while transition elements are in the "B" families. Metals, nonmetals, metalloids, and noble gases are color coded in the periodic table inside the front cover of the textbook.

Na is a main-group metal in group 1(1A).

Re is a transition metal in group 7(7B).

S is a main-group nonmetal in group 16(6A).

I is a main-group nonmetal in group 17(7A).

Kr is a nonmetal in group 18(8A).

Mg is a main-group metal in group 2(2A).

U is an inner transition metal, an actinide.

Si is a main-group metalloid in group 14(4A).

B is a metalloid in group 13(3A).

Al is a main-group metal in group 13(3A).

As is a main-group metalloid in group 15(5A).

H is a main-group nonmetal in group 1(1A). (H is believed to be a metal at extremely high pressures.)

8A **(E)** This is similar to Practice Examples 2-8A and 2-8B.

$$\text{Cu mass} = 2.35\times10^{24}\,\text{Cu atoms}\times\frac{1\,\text{mol Cu}}{6.022\times10^{23}\,\text{atoms}}\times\frac{63.546\,\text{g Cu}}{1\,\text{mol Cu}} = 248\,\text{g Cu}$$

8B **(M)** Of all lead atoms, 24.1% are lead-206, or 241 ^{206}Pb atoms in every 1000 lead atoms. First we need to convert a 22.6 gram sample of lead into moles of lead (below) and then, by using Avogadro's constant, and the percent isotopic abundance, we can determine the number of ^{206}Pb atoms.

$$n_{\text{Pb}} = 22.6\,\text{g Pb}\times\frac{1\,\text{mol Pb}}{207.2\,\text{g Pb}} = 0.109\,\text{mol Pb}$$

$$^{206}\text{Pb atoms} = 0.109\,\text{mol Pb}\times\frac{6.022\times10^{23}\,\text{Pb atoms}}{1\,\text{mol Pb}}\times\frac{241\,^{206}\text{Pb atoms}}{1000\,\text{Pb atoms}} = 1.58\times10^{22}\,^{206}\text{Pb atoms}$$

9A **(M)** Both the density and the molar mass of Pb serve as conversion factors.

$$\text{atoms of Pb} = 0.105\,\text{cm}^3\,\text{Pb}\times\frac{11.34\,\text{g}}{1\,\text{cm}^3}\times\frac{1\,\text{mol Pb}}{207.2\,\text{g}}\times\frac{6.022\times10^{23}\,\text{Pb atoms}}{1\,\text{mol Pb}} = 3.46\times10^{21}\,\text{Pb atoms}$$

9B **(M)** First we find the number of rhenium atoms in 0.100 mg of the element.

$$0.100\,\text{mg}\times\frac{1\,\text{g}}{1000\,\text{mg}}\times\frac{1\,\text{mol Re}}{186.21\,\text{g Re}}\times\frac{6.022\times10^{23}\,\text{Re atoms}}{1\,\text{mol Re}} = 3.23\times10^{17}\,\text{Re atoms}$$

$$\%\text{ abundance }^{187}\text{Re} = \frac{2.02\times10^{17}\,\text{atoms }^{187}\text{Re}}{3.23\times10^{17}\,\text{Re atoms}}\times100\% = 62.5\%$$

INTEGRATIVE EXAMPLE

A. **(M)**

Stepwise approach:

First, determine the total number of Cu atoms in the crystal by determining the volume of the crystal and calculating the mass of Cu from density. Then we can determine the amount of ^{63}Cu by noting its relative abundance

$$\text{Volume of crystal} = (25 \text{ nm})^3 \times \frac{1 \text{ cm}^3}{(1 \times 10^7 \text{ nm})^3} = 1.5625 \times 10^{-17} \text{ cm}^3$$

$$\text{Mass of Cu in crystal} = d \cdot V = 8.92 \text{ g/cm}^3 \times 1.5625 \times 10^{-17} = 1.3938 \times 10^{-16} \text{ g}$$

\# of Cu atoms =

$$1.3938 \times 10^{-16} \text{ g Cu} \times \frac{1 \text{ mol Cu}}{63.546 \text{ g Cu}} \times \frac{6.022 \times 10^{23} \text{ Cu atoms}}{1 \text{ mol Cu}} = 1.3208 \times 10^6 \text{ Cu atoms}$$

Therefore, the number of ^{63}Cu atoms, assuming 69.17% abundance, is 9.14×10^5 atoms.

Conversion pathway approach:

$$\frac{8.92 \text{ g Cu}}{1 \text{ cm}^3} \times \frac{1 \text{ cm}^3}{(1 \times 10^7 \text{ nm})^3} \times \frac{(25 \text{ nm})^3}{\text{crystal}} \times \frac{1 \text{ mol Cu}}{63.546 \text{ g Cu}} \times \frac{6.022 \times 10^{23} \text{ Cu atoms}}{1 \text{ mol Cu}}$$

$$\times \frac{69.17 \text{ atoms of } ^{63}\text{Cu}}{100 \text{ Cu atoms}} = 9.14 \times 10^5 \text{ atoms of } ^{63}\text{Cu}$$

B. **(M)**

Stepwise approach:

Calculate the mass of Fe in a serving of cereal, determine mass of ^{58}Fe in that amount of cereal, and determine how many servings of cereal are needed to reach 58 g of ^{58}Fe.

Amount of Fe in a serving of cereal = 18 mg × 0.45 = 8.1 mg Fe per serving

First calculate the amount of Fe

$$0.0081 \text{ g Fe} \times \frac{1 \text{ mol Fe}}{55.845 \text{ g Fe}} = 1.45 \times 10^{-4} \text{ mol Fe}$$

Then calculate ^{58}Fe amount:

$$1.45 \times 10^{-4} \text{ mol Fe} \times \frac{0.282 \text{ mol } ^{58}\text{Fe}}{100 \text{ mol Fe}} = 4.090 \times 10^{-7} \text{ mol } ^{58}\text{Fe}$$

Converting mol of ^{58}F to # of servings:

$$\frac{4.090\times10^{-7}\ mol\ ^{58}Fe}{1\ serving}\times\frac{57.9333\ g\ ^{58}Fe}{1\ mol\ ^{58}Fe}=2.37\times10^{-5}\ g\ ^{58}Fe\ per\ serving$$

Total # of servings = 58 g total / 2.37×10^{-5} per serving = 2.4477×10^{6} servings

Conversion Pathway Approach:

The number of servings of dry cereal to ingest 58 g of ^{58}Fe =

$$58.0\ g\ ^{58}Fe\times\frac{1\ mol\ ^{58}Fe}{57.9333\ g\ ^{58}Fe}\times\frac{100\ mol\ Fe}{0.282\ mol\ ^{58}Fe}\times\frac{58.845\ g\ Fe}{1\ mol\ Fe}\times\frac{1\ cereal\ serving}{0.018\ g\ Fe\times0.45}$$

$$=2.4477\times10^{6}\ servings$$

$$2.44477\times10^{6}\ servings\times\frac{1\ year}{365\ servings}=6706\ years$$

EXERCISES

Law of Conservation of Mass

1. **(E)** The observations cited do not necessarily violate the law of conservation of mass. The oxide formed when iron rusts is a solid and remains with the solid iron, increasing the mass of the solid by an amount equal to the mass of the oxygen that has combined. The oxide formed when a match burns is a gas and will not remain with the solid product (the ash); the mass of the ash thus is less than that of the match. We would have to collect all reactants and all products and weigh them to determine if the law of conservation of mass is obeyed or violated.

3. **(E)** By the law of conservation of mass, all of the magnesium initially present and all of the oxygen that reacted are present in the product. Thus, the mass of oxygen that has reacted is obtained by difference. mass of oxygen= 0.674 g MgO − 0.406 g Mg = 0.268 g oxygen

5. **(M)** We need to compare the mass before reaction (initial) with that after reaction (final) to answer this question.
initial mass = 10.500 g calcium hydroxide + 11.125 g ammonium chloride = 21.625 g
final mass = 14.336 g solid residue + (69.605 – 62.316) g of gases = 21.625 g

These data support the law of conservation of mass. Note that the gain in the mass of water is equal to the mass of gas absorbed by the water.

Law of Constant Composition

7. **(E)**

(a)   Ratio of O: MgO by mass $= \dfrac{(0.755 - 0.455)\text{ g}}{0.755\text{ g}} = 0.397$

(b)   Ratio of O: Mg in MgO by mass $= \dfrac{0.300\text{ g}}{0.455\text{ g}} = 0.659$

(c)   Percent magnesium by mass $= \dfrac{0.455\text{ g Mg}}{0.755\text{ g MgO}} \times 100\% = 60.3\%$

9.   **(M)** In the first experiment, 2.18 g of sodium produces 5.54 g of sodium chloride. In the second experiment, 2.10 g of chlorine produces 3.46 g of sodium chloride. The amount of sodium contained in this second sample of sodium chloride is given by

mass of sodium $= 3.46$ g sodium chloride $- 2.10$ g chlorine $= 1.36$ g sodium.

We now have sufficient information to determine the % Na in each of the samples of sodium chloride.

$\%\text{Na} = \dfrac{2.18\text{ g Na}}{5.54\text{ g cmpd}} \times 100\% = 39.4\%\text{ Na}$ $\%\text{Na} = \dfrac{1.36\text{ g Na}}{3.46\text{ g cmpd}} \times 100\% = 39.3\%\text{ Na}$

Thus, the two samples of sodium chloride have the same composition. Recognize that, based on significant figures, each percent has an uncertainty of $\pm 0.1\%$.

11.   **(E)** The mass of sulfur (0.312 g) needed to produce 0.623 g sulfur dioxide provides the information required for the conversion factor.

sulfur mass $= 0.842$ g sulfur dioxide $\times \dfrac{0.312\text{ g sulfur}}{0.623\text{ g sulfur dioxide}} = 0.422$ g sulfur

Law of Multiple Proportions

13.   **(M)** By dividing the mass of the oxygen per gram of sulfur in the second sulfur-oxygen compound (compound 2) by the mass of oxygen per gram of sulfur in the first sulfur-oxygen compound (compound 1), we obtain the ratio (shown to the right):

$$\dfrac{\dfrac{1.497\text{ g of O}}{1.000\text{ g of S}}\,(\text{cmpd 2})}{\dfrac{0.998\text{ g of O}}{1.000\text{ g of S}}\,(\text{cmpd 1})} = \dfrac{1.500}{1}$$

To get the simplest whole number ratio we need to multiply both the numerator and the denominator by 2. This gives the simple whole number ratio 3/2. In other words, for a given mass of sulfur, the mass of oxygen in the second compound (SO_3) relative to the mass of oxygen in the first compound (SO_2) is in a ratio of 3:2. These results are entirely consistent with the Law of Multiple Proportions because the same two elements, sulfur and oxygen in this case, have reacted together to give <u>two</u> different compounds that have masses of oxygen that are in the ratio of small positive integers for a fixed amount of sulfur.

15. **(M)**

    (a) First of all we need to fix the mass of nitrogen in all three compounds to some common value, for example, 1.000 g. This can be accomplished by multiplying the masses of hydrogen and nitrogen in compound A by 2 and the amount of hydrogen and nitrogen in compound C by 4/3 (1.333):

    Cmpd A "normalized" mass of nitrogen   = 0.500 g N × 2 = 1.000 g N
              "normalized" mass of hydrogen = 0.108 g H × 2 = 0.216 g H

    Cmpd C "normalized" mass of nitrogen   = 0.750 g N × 1.333 = 1.000 g N
              "normalized" mass of hydrogen = 0.108 g H × 1.333 = 0.144 g H

    Next, we divide the mass of hydrogen in each compound by the smallest mass of hydrogen, namely, 0.0720 g. This gives 3.000 for compound A, 1.000 for compound B, and 2.000 for compound C. The ratio of the amounts of hydrogen in the three compounds is 3 (cmpd A) : 1 (cmpd B) : 2 (cmpd C)

    These results are consistent with the Law of Multiple Proportions because the masses of hydrogen in the three compounds end up in a ratio of small whole numbers when the mass of nitrogen in all three compounds is normalized to a simple value (1.000 g here).

    (b) The text states that compound B is N_2H_2. This means that, based on the relative amounts of hydrogen calculated in part (a), compound A might be N_2H_6 and compound C, N_2H_4. Actually, compound A is NH_3, but we have no way of knowing this from the data. Note that the H:N ratios in NH_3 and N_2H_6 are the same, 3H:1N.

17. **(M)** One oxide of copper has about 20% oxygen by mass. If we assume a 100 gram sample, then ~ 20 grams of the sample is oxygen (~1.25 moles) and 80 grams is copper (~1.26 moles). This would give an empirical formula of CuO (copper(II) oxide). The second oxide has less oxygen by mass, hence the empirical formula must have less oxygen or more copper (Cu:O ratio greater than 1). If we keep whole number ratios of atoms, a plausible formula would be Cu_2O (copper(I) oxide), where the mass percent oxygen is ≈11%.

Fundamental Charges and Mass-to-Charge Ratios

19. **(M)** We can calculate the charge on each drop, express each in terms of 10^{-19} C, and finally express each in terms of $e = 1.6 \times 10^{-19}$ C.

    drop 1: 1.28×10^{-18} $= 12.8 \times 10^{-19}$ C $= 8e$

    drops 2 & 3: $1.28 \times 10^{-18} \div 2 = 0.640 \times 10^{-18}$ C $= 6.40 \times 10^{-19}$ C $= 4e$

    drop 4: $1.28 \times 10^{-18} \div 8 = 0.160 \times 10^{-18}$ C $= 1.60 \times 10^{-19}$ C $= 1e$

    drop 5: $1.28 \times 10^{-18} \times 4 = 5.12 \times 10^{-18}$ C $= 51.2 \times 10^{-19}$ C $= 32e$

    We see that these values are consistent with the charge that Millikan found for that of the electron, and he could have inferred the correct charge from these data, since they are all multiples of e.

21. **(M)**

(a) Determine the ratio of the mass of a hydrogen atom to that of an electron. We use the mass of a proton plus that of an electron for the mass of a hydrogen atom.

$$\frac{\text{mass of proton} + \text{mass of electron}}{\text{mass of electron}} = \frac{1.0073\,u + 0.00055\,u}{0.00055\,u} = 1.8 \times 10^3$$

$$or \quad \frac{\text{mass of electron}}{\text{mass of proton} + \text{mass of electron}} = \frac{1}{1.8 \times 10^3} = 5.6 \times 10^{-4}$$

(b) The only two mass-to-charge ratios that we can determine from the data in Table 2-1 are those for the proton (a hydrogen ion, H^+) and the electron.

$$\text{For the proton}: \quad \frac{\text{mass}}{\text{charge}} = \frac{1.673 \times 10^{-24}\,g}{1.602 \times 10^{-19}\,C} = 1.044 \times 10^{-5}\,g/C$$

$$\text{For the electron}: \quad \frac{\text{mass}}{\text{charge}} = \frac{9.109 \times 10^{-28}\,g}{1.602 \times 10^{-19}\,C} = 5.686 \times 10^{-9}\,g/C$$

The hydrogen ion is the lightest positive ion available. We see that the mass-to-charge ratio for a positive particle is considerably larger than that for an electron.

Atomic Number, Mass Number, and Isotopes

23. **(E) (a)** cobalt-60 $^{60}_{27}Co$ **(b)** phosphorus-32 $^{32}_{15}P$ **(c)** iron-59 $^{59}_{26}Fe$ **(d)** radium-226 $^{226}_{88}Ra$

25.

	Name	Symbol	Number of Protons	Number of Electrons	Number of Neutrons	Mass Number
(E)	Sodium	$^{23}_{11}Na$	11	11	12	23
	Silicon	$^{28}_{14}Si$	14	14[a]	14	28
	Rubidium	$^{85}_{37}Rb$	37	37[a]	48	85
	Potassium	$^{40}_{19}K$	19	19	21	40
	Arsenic[a]	$^{75}_{33}As$	33[a]	33	42	75
	Neon	$^{20}_{10}Ne^{2+}$	10	8	10	20
	Bromine[b]	$^{80}_{35}Br$	35	35	45	80
	Lead[b]	$^{208}_{82}Pb$	82	82	126	208

[a] This result assumes that a neutral atom is involved.
[b] Insufficient data. Does not characterize a specific nuclide; several possibilities exist. The minimum information needed is the atomic number (or some way to obtain it, such as from the name or the symbol of the element involved), the number of electrons (or some way to obtain it, such as the charge on the species), and the mass number (or the number of neutrons).

27. **(E)**

(a) A ^{108}Pd atom has 46 protons and 46 electrons. The atom described is neutral, hence, the number of electrons must equal the number of protons.

Since there are 108 nucleons in the nucleus, the number of neutrons is 62 ($= 108$ nucleons $- 46$ protons).

(b) The ratio of the two masses is determined as follows: $\dfrac{^{108}\text{Pd}}{^{12}\text{C}} = \dfrac{107.90389\,\text{u}}{12\,\text{u}} = 8.9919908$

29. **(E)** The mass of ^{16}O is 15.9949 u.   isotopic mass $= 15.9949\,\text{u} \times 6.68374 = 106.936\,\text{u}$

31. **(E)** Each isotopic mass must be divided by the isotopic mass of ^{12}C, 12 u, an exact number.

(a) $^{35}\text{Cl} \div {}^{12}\text{C} = 34.96885\,\text{u} \div 12\,\text{u} = 2.914071$

(b) $^{26}\text{Mg} \div {}^{12}\text{C} = 25.98259\,\text{u} \div 12\,\text{u} = 2.165216$

(c) $^{222}\text{Rn} \div {}^{12}\text{C} = 222.0175\,\text{u} \div 12\,\text{u} = 18.50146$

33. **(E)** First, we determine the number of protons, neutrons, and electrons in each species.

species:	$^{24}_{12}\text{Mg}^{2+}$	$^{47}_{24}\text{Cr}$	$^{60}_{27}\text{Co}^{3+}$	$^{35}_{17}\text{Cl}^{-}$	$^{124}_{50}\text{Sn}^{2+}$	$^{226}_{90}\text{Th}$	$^{90}_{38}\text{Sr}$
protons:	12	24	27	17	50	90	38
neutrons:	12	23	33	18	74	136	52
electrons:	10	24	24	18	48	90	38

(a) The numbers of neutrons and electrons are equal for $^{35}_{17}\text{Cl}^{-}$.

(b) $^{60}_{27}\text{Co}^{3+}$ has protons (27), neutrons (33), and electrons (24) in the ratio 9:11:8.

(c) The species $^{124}_{50}\text{Sn}^{2+}$ has a number of neutrons (74) equal to its number of protons (50) plus one-half its number of electrons $(48 \div 2 = 24)$.

35. **(E)**

If we let n represent the number of neutrons and p represent the number of protons, then $p + 4 = n$. The mass number is the sum of the number of protons and the number of neutrons: $p + n = 44$. Substitution of $n = p + 4$ yields $p + p + 4 = 44$. From this relation, we see $p = 20$. Reference to the periodic table indicates that 20 is the atomic number of the element calcium.

37. **(M)** The number of protons is the same as the atomic number for iodine, which is 53. There is one more electron than the number of protons because there is a negative charge on the ion. Therefore the number of electrons is 54. The number of neutrons is equal to 70, mass number minus atomic number.

39. **(E)** For americium, $Z = 95$. There are 95 protons, 95 electrons, and $241 - 95 = 146$ neutrons in a single atom of americium-241.

Atomic Mass Units, Atomic Masses

41. **(E)** There are no chlorine atoms that have a mass of 35.45 u. The masses of individual chlorine atoms are close to integers and this mass is about midway between two integers. It is an average atomic mass, the result of averaging two (or more) isotopic masses, each weighted by its isotopic abundance.

43. **(E)** To determine the weighted-average atomic mass, we use the following expression:

$$\text{average atomic mass} = \sum(\text{isotopic mass} \times \text{fractional isotopic abundance})$$

Each of the three percents given is converted to a fractional abundance by dividing it by 100.

$$\text{Mg atomic mass} = (23.985042\,u \times 0.7899) + (24.985837\,u \times 0.1000) + (25.982593\,u \times 0.1101)$$

$$= 18.95\,u + 2.499\,u + 2.861\,u = 24.31\,u$$

45. **(E)** We will use the expression to determine the weighted-average atomic mass.

$$107.868\,u = (106.905092\,u \times 0.5184) + (^{109}Ag \times 0.4816) = 55.42\,u + 0.4816\,^{109}Ag$$

$$107.868\,u - 55.42\,u = 0.4816\,^{109}Ag = 52.45\,u \qquad ^{109}Ag = \frac{52.45\,u}{0.4816} = 108.9\,u$$

47. **(M)** Since the three percent abundances total 100%, the percent abundance of ^{40}K is found by difference. $\%\,^{40}K = 100.0000\% - 93.2581\% - 6.7302\% = 0.0117\%$

Then the expression for the weighted-average atomic mass is used, with the percent abundances converted to fractional abundances by dividing by 100. Note that the average atomic mass of potassium is 39.0983 u.

$$39.0983\,u = (0.932581 \times 38.963707\,u) + (0.000117 \times 39.963999\,u) + (0.067302 \times \text{mass of}\,^{41}K)$$

$$= 36.3368\,u + 0.00468\,u + (0.067302 \times \text{mass of}\,^{41}K)$$

$$\text{mass of}\,^{41}K = \frac{39.0983\,u - (36.3368\,u + 0.00468\,u)}{0.067302} = 40.962\,u$$

Mass Spectrometry

49. **(M)**

(a)

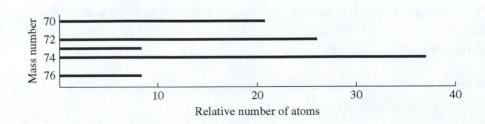

(b) As before, we multiply each isotopic mass by its fractional abundance, after which, we sum these products to obtain the (average) atomic mass for the element.

$$(0.205 \times 70) + (0.274 \times 72) + (0.078 \times 73) + (0.365 \times 74) + (0.078 \times 76)$$

$$= 14 + 20. + 5.7 + 27 + 5.9 = 72.6 = \text{average atomic mass of germanium}$$

The result is only approximately correct because the isotopic masses are given to only two significant figures. Thus, only a two-significant-figure result can be quoted.

The Periodic Table

51. **(E)**

(a) Ge is in group 14 and in the fourth period.

(b) Other elements in group 16(6A) are similar to S: O, Se, and Te. Most of the elements in the periodic table are unlike S, but particularly metals such as Na, K, and Rb.

(c) The alkali metal (group 1), in the fifth period is Rb.

(d) The halogen (group 17) in the sixth period is At.

53. **(E)** If the seventh period of the periodic table is 32 members long, it will be the same length as the sixth period. Elements in the same family (vertical group), will have atomic numbers 32 units higher. The noble gas following radon will have atomic number $= 86 + 32 = 118$. The alkali metal following francium will have atomic number $= 87 + 32 = 119$.

The Avogadro Constant and the Mole

55. **(E)**

(a) atoms of Fe $= 15.8 \text{ mol Fe} \times \dfrac{6.022 \times 10^{23} \text{ atoms Fe}}{1 \text{ mol Fe}} = 9.51 \times 10^{24} \text{ atoms Fe}$

(b) atoms of Ag $= 0.000467 \text{ mol Ag} \times \dfrac{6.022 \times 10^{23} \text{ atoms Ag}}{1 \text{ mol Ag}} = 2.81 \times 10^{20} \text{ atoms Ag}$

(c) atoms of Na $= 8.5 \times 10^{-11} \text{ mol Na} \times \dfrac{6.022 \times 10^{23} \text{ atoms Na}}{1 \text{ mol Na}} = 5.1 \times 10^{13} \text{ atoms Na}$

57. **(E)**

(a) moles of Zn $= 415.0 \text{ g Zn} \times \dfrac{1 \text{ mol Zn}}{65.38 \text{ g Zn}} = 6.347 \text{ mol Zn}$

(b) # of Cr atoms $= 147,400 \text{ g Cr} \times \dfrac{1 \text{ mol Cr}}{51.996 \text{ g Cr}} \times \dfrac{6.022 \times 10^{23} \text{ atoms Cr}}{1 \text{ mol Cr}}$

$$= 1.707 \times 10^{27} \text{ atoms Cr}$$

(c) mass Au $= 1.0 \times 10^{12} \text{atoms Au} \times \dfrac{1 \text{ mol Au}}{6.022 \times 10^{23} \text{ atoms Au}} \times \dfrac{196.97 \text{ g Au}}{1 \text{ mol Au}} = 3.3 \times 10^{-10} \text{g Au}$

(d) mass of F atom = 1 atom F $\times \dfrac{1 \text{ mol F}}{6.022 \times 10^{23} \text{ atoms F}} \times \dfrac{18.998 \text{ g F}}{1 \text{ mol F}} = 3.1547 \times 10^{-23} \text{ g F}$

For exactly 1 F atom, the number of sig figs in the answer is determined by the least precise number in the calculation, namely the mass of fluorine.

59. **(E)** Determine the mass of Cu in the jewelry, then convert to moles and finally to the number of atoms. If sterling silver is 92.5% by mass Ag, it is $100 - 92.5 = 7.5\%$ by mass Cu.

Conversion pathway approach:

number of Cu atoms = 33.24 g sterling $\times \dfrac{7.5 \text{ g Cu}}{100.0 \text{ g sterling}} \times \dfrac{1 \text{ mol Cu}}{63.546 \text{ g Cu}} \times \dfrac{6.022 \times 10^{23} \text{ atoms Cu}}{1 \text{ mol Cu}}$

$= 2.4 \times 10^{22} \text{ Cu atoms}$

Stepwise approach:

33.24 g sterling $\times \dfrac{7.5 \text{ g Cu}}{100.0 \text{ g sterling}} = 2.493 \text{ g Cu}$

2.493 g Cu $\times \dfrac{1 \text{ mol Cu}}{63.546 \text{ g Cu}} = 0.03923 \text{ mol Cu}$

0.03923 mol Cu $\times \dfrac{6.022 \times 10^{23} \text{ atoms Cu}}{1 \text{ mol Cu}} = 2.4 \times 10^{22} \text{ Cu atoms}$

61. **(E)** We first need to determine the number of Pb atoms of all types in 215 mg of Pb, and then use the percent isotopic abundance to determine the number of ^{204}Pb atoms present.

^{204}Pb atoms = 215 mg Pb $\times \dfrac{1 \text{ g}}{1000 \text{ mg}} \times \dfrac{1 \text{ mol Pb}}{207.2 \text{ g Pb}} \times \dfrac{6.022 \times 10^{23} \text{ atoms}}{1 \text{ mol Pb}} \times \dfrac{14 \, ^{204}\text{Pb atoms}}{1000 \text{ Pb atoms}}$

$= 8.7 \times 10^{18} \text{ atoms } ^{204}\text{Pb}$

63. **(E)** We will use the average atomic mass of lead, 207.2 g/mol, to answer this question.

(a) $\dfrac{30 \, \mu\text{g Pb}}{1 \text{ dL}} \times \dfrac{1 \text{ dL}}{0.1 \text{ L}} \times \dfrac{1 \text{ g Pb}}{10^6 \, \mu\text{g Pb}} \times \dfrac{1 \text{ mol Pb}}{207.2 \text{ g}} = 1.4\underline{5} \times 10^{-6} \text{ mol Pb} / \text{L}$

(b) $\dfrac{1.4\underline{5} \times 10^{-6} \text{ mol Pb}}{\text{L}} \times \dfrac{1 \text{ L}}{1000 \text{ mL}} \times \dfrac{6.022 \times 10^{23} \text{ atoms}}{1 \text{ mol}} = 8.7 \times 10^{14} \text{ Pb atoms} / \text{mL}$

65. **(M)** To answer this question, we simply need to calculate the ratio of the mass (in grams) of each sample to its molar mass. Whichever elemental sample gives the largest ratio will be the one that has the greatest number of atoms.

(a) Iron sample: $10 \text{ cm} \times 10 \text{ cm} \times 10 \text{ cm} \times 7.86 \text{ g cm}^{-3} = 7860 \text{ g Fe}$

7860 g Fe $\times \dfrac{1 \text{ mol Fe}}{55.845 \text{ g Fe}} = 141 \text{ mol of Fe atoms}$

(b) Hydrogen sample: $\dfrac{1.00 \times 10^3 \text{g H}_2}{2 \times (1.008 \text{ g H})} \times 1 \text{ mol H} = 496 \text{ mol of H}_2 \text{ molecules} =$

$$992 \text{ mol of H atoms}$$

(c) Sulfur sample: $\dfrac{2.00 \times 10^4 \text{ g S}}{32.06 \text{ g S}} \times 1 \text{ mol S} = 624 \text{ mol of S atoms}$

(d) Mercury sample: $76 \text{ lb Hg} \times \dfrac{454 \text{ g Hg}}{1 \text{ lb Hg}} \times \dfrac{1 \text{ mol Hg}}{200.6 \text{ g Hg}} = 172 \text{ mol of Hg atoms}$

Clearly, then, it is the 1.00 kg sample of hydrogen that contains the greatest number of atoms.

INTEGRATIVE AND ADVANCED EXERCISES

70. (M)

volume of nucleus(single proton) $= \dfrac{4}{3} \pi r^3 = 1.3333 \times 3.14159 \times (0.5 \times 10^{-13} \text{ cm})^3 = 5 \times 10^{-40} \text{ cm}^3$

density $= \dfrac{1.673 \times 10^{-24} \text{ g}}{5 \times 10^{-40} \text{ cm}^3} = 3 \times 10^{15} \text{ g/cm}^3$

72. (M) Let Z = # of protons, N = # of neutrons, E = # of electrons, and A = # of nucleons = $Z + N$.

(a) $Z + N = 234$ The mass number is 234 and the species is an atom.
$N = 1.600 Z$ The atom has 60.0% more neutrons than protons.
Next we will substitute the second expression into the first and solve for Z.
$Z + N = 234 = Z + 1.600 Z = 2.600 Z$

$Z = \dfrac{234}{2.600} = 90 \text{ protons}$

Thus this is an atom of the isotope ^{234}Th.

(b) $Z = E + 2$ The ion has a charge of +2. $Z = 1.100 E$
There are 10.0% more protons than electrons. By equating these two expressions and solving for E, we can find the number of electrons. $E + 2 = 1.100 E$

$2 = 1.100 E - E = 0.100 E$ $E = \dfrac{2}{0.100} = 20 \text{ electrons}$ $Z = 20 + 2 = 22$, (titanium).

The ion is Ti^{2+}. There is not enough information to determine the mass number.

(c) $Z + N = 110$ The mass number is 110.
$Z = E + 2$ The species is a cation with a charge of +2.
$N = 1.25 E$ Thus, there are 25.0% more neutrons than electrons. By substituting the second and third expressions into the first, we can solve for E, the number of electrons.

$(E + 2) + 1.25 E = 110 = 2.25 E + 2$ $108 = 2.25 E$ $E = \dfrac{108}{2.25} = 48$

Then $Z = 48 + 2 = 50$, (the element is Sn) $N = 1.25 \times 48 = 60$ Thus, it is $^{110}\text{Sn}^{2+}$.

74. **(M)** $A = Z + N = 2.50\,Z$ The mass number is 2.50 times the atomic number.
The neutron number of selenium-82 equals $82 - 34 = 48$, since $Z = 34$ for Se. The neutron number of isotope Y also equals 48, which equals 1.33 times the atomic number of isotope Y.

Thus $48 = 1.33 \times Z_Y$ $\quad Z_Y = \dfrac{48}{1.33} = 36$

The mass number of isotope Y = 48 + 36 = 84 = the atomic number of E, and thus, the element is Po. Thus, from the relationship in the first line, the mass number of
$E = 2.50\,Z = 2.50 \times 84 = 210$ $\quad$ The isotope E is ^{210}Po.

76. **(M)** To solve this question, represent the fractional abundance of ^{14}N by x and that of ^{14}N by $(1 - x)$. Then use the expression for determining average atomic mass.
$14.0067 = 14.0031\,x + 15.0001(1 - x)$
$14.0067 - 15.0001 = 14.0031\,x - 15.0001\,x$ $\quad$ OR $\quad$ $-0.9934 = -0.9970\,x$

$x = \dfrac{0.9934}{0.9970} \times 100\% = 99.64\% = $ percent isotopic abundance of ^{14}N.

Thus, $0.36\% = $ percent isotopic abundance of ^{15}N.

77. **(D)** In this case, we will use the expression for determining average atomic mass—the sum of products of nuclidic mass times fractional abundances (from Figure 2-14)—to answer the question.

^{196}Hg: $\quad$ 195.9658 u $\times$ 0.00146 = 0.286 u

^{198}Hg: $\quad$ 197.9668 u $\times$ 0.1002 = 19.84 u $\qquad$ ^{199}Hg: $\quad$ 198.9683 u $\times$ 0.1684 = 33.51 u

^{200}Hg: $\quad$ 199.9683 u $\times$ 0.2313 = 46.25 u $\qquad$ ^{201}Hg: $\quad$ 200.9703 u $\times$ 0.1322 = 26.57 u

^{202}Hg: $\quad$ 201.9706 u $\times$ 0.2980 = 60.19 u $\qquad$ ^{204}Hg: $\quad$ 203.9735 u $\times$ 0.0685 = 14.0 u

Atomic weight = 0.286 u + 19.84 u + 33.51 u + 46.25 u + 26.57 u + 60.19 u + 14.0 u = 200.6 u

82. **(D)** volume $= l \times w \times h - \pi r^2 h$

$$= (15.0\text{ cm} \times 12.5\text{ cm} \times 0.300\text{ cm}) - (3.1416) \times \left(\frac{2.50\text{ cm}}{2}\right)^2 \times (0.300\text{ cm})$$

$$= (56.2\underline{5}\text{ cm} - 1.47\text{ cm}) = 54.8\text{ cm}^3$$

mass of object $= 54.8\text{ cm}^3 \times \dfrac{8.80\text{ g}}{1\text{ cm}^3} = 482$ g Monel metal

Then determine the number of silicon atoms in this quantity of alloy.

482 g Monel metal $\times \dfrac{2.2 \times 10^{-4}\text{ g Si}}{1.000\text{ g metal}} \times \dfrac{1\text{ mol Si}}{28.05\text{ g Si}} \times \dfrac{6.022 \times 10^{23}\text{ Si atoms}}{1\text{ mol Si}} = 2.3 \times 10^{21}$ Si atoms

Finally, determine the number of ^{30}Si atoms in this quantity of silicon.

$$\text{number of } ^{30}Si \text{ atoms} = (2.3 \times 10^{21} \text{ Si atoms}) \times \left(\frac{3.10 \quad ^{30}Si \text{ atoms}}{100 \quad Si \text{ atoms}} \right) = 7.1 \times 10^{19} \ ^{30}Si$$

86. (M) The relative masses of Sn and Pb are 207.2 g Pb (assume one mole of Pb) to $(2.73 \times 118.71 \text{ g/mol Sn} =)$ 324 g Sn. Then the mass of cadmium, on the same scale, is $207.2/1.78 = 116$ g Cd.

$$\% \text{ Sn} = \frac{324\,g\,Sn}{207.2 + 324 + 116\,g\,alloy} \times 100\% = \frac{324\,g\,Sn}{647\,g\,alloy} \times 100\% = 50.1\%\,Sn$$

$$\% \text{ Pb} = \frac{207.2\,g\,Pb}{647\,g\,alloy} \times 100\% = 32.0\%\,Pb \qquad \% \text{ Cd} = \frac{116\,g\,Cd}{647\,g\,alloy} \times 100\% = 17.9\%\,Cd$$

87. (M) We need to apply the law of conservation of mass and convert volumes to masses:

Calculate the mass of zinc:	$125 \text{ cm}^3 \times 7.13 \text{ g/cm}^3 = 891$ g
Calculate the mass of iodine:	$125 \text{ cm}^3 \times 4.93 \text{ g/cm}^3 = 616$ g
Calculate the mass of zinc iodide:	$164 \text{ cm}^3 \times 4.74 \text{ g/cm}^3 = 777$ g
Calculate the mass of zinc unreacted:	$(891 + 616 - 777) \text{ g} = 730$ g
Calculate the volume of zinc unreacted:	$730 \text{ g} \times 1\text{cm}^3 / 7.13 \text{ g} = 102 \text{ cm}^3$

FEATURE PROBLEMS

89. (M) The product mass differs from that of the reactants by $(5.62 - 2.50 =)$ 3.12 grains. In order to determine the percent gain in mass, we need to convert the reactant mass to grains.

$$13 \text{ onces} \times \frac{8 \text{ gros}}{1 \text{ once}} = 104 \text{ gros} \times (104 + 2) \text{ gros} \times \frac{72 \text{ grains}}{1 \text{ gros}} = 7632 \text{ grains}$$

$$\% \text{ mass increase} = \frac{3.12 \text{ grains increase}}{(7632 + 2.50) \text{ grains original}} \times 100\% = 0.0409\% \text{ mass increase}$$

The sensitivity of Lavoisier's balance can be as little as 0.01 grain, which seems to be the limit of the readability of the balance; alternatively, it can be as large as 3.12 grains, which assumes that all of the error in the experiment is due to the (in)sensitivity of the balance. Let us convert 0.01 grains to a mass in grams.

$$\text{minimum error} = 0.01 \text{ gr} \times \frac{1 \text{ gros}}{72 \text{ gr}} \times \frac{1 \text{ once}}{8 \text{ gros}} \times \frac{1 \text{ livre}}{16 \text{ once}} \times \frac{30.59 \text{ g}}{1 \text{ livre}} = 3 \times 10^{-5} g = 0.03 \text{ mg}$$

$$\text{maximum error} = 3.12 \text{ gr} \times \frac{3 \times 10^{-5} \text{ g}}{0.01 \text{ gr}} = 9 \times 10^{-3} \text{ g} = 9 \text{ mg}$$

The maximum error is close to that of a common modern laboratory balance, which has a sensitivity of 1 mg. The minimum error is approximated by a good quality analytical balance. Thus we conclude that Lavoisier's results conform closely to the law of conservation of mass.

92. **(D)** We begin with the amount of reparations and obtain the volume in cubic kilometers with a series of conversion factors.

Conversion pathway approach:

$$V = \$28.8 \times 10^9 \times \frac{1\,\text{troy oz Au}}{\$21.25} \times \frac{31.103\,\text{g Au}}{1\,\text{troy oz Au}} \times \frac{1\,\text{mol Au}}{196.97\,\text{g Au}} \times \frac{6.022 \times 10^{23}\,\text{atoms Au}}{1\,\text{mol Au}}$$

$$\times \frac{1\,\text{ton seawater}}{4.67 \times 10^{17}\,\text{Au atoms}} \times \frac{2000\,\text{lb seawater}}{1\,\text{ton seawater}} \times \frac{453.6\ \text{g seawater}}{1\,\text{lb seawater}} \times \frac{1\,\text{cm}^3\ \text{seawater}}{1.03\,\text{g seawater}}$$

$$\times \left(\frac{1\,\text{m}}{100\,\text{cm}} \times \frac{1\,\text{km}}{1000\,\text{m}} \right)^3 = 2.43 \times 10^5\,\text{km}^3$$

Stepwise approach:

$$\$28.8 \times 10^9 \times \frac{1\,\text{troy oz Au}}{\$21.25} \times \frac{31.103\,\text{g Au}}{1\,\text{troy oz Au}} = 4.22 \times 10^{10}\,\text{g Au}$$

$$4.22 \times 10^{10}\,\text{g Au} \times \frac{1\,\text{mol Au}}{196.97\,\text{g Au}} \times \frac{6.022 \times 10^{23}\,\text{atoms Au}}{1\,\text{mol Au}} = 1.29 \times 10^{32}\,\text{atoms Au}$$

$$1.29 \times 10^{32}\,\text{atoms Au} \times \frac{1\,\text{ton seawater}}{4.67 \times 10^{17}\,\text{Au atoms}} \times \frac{2000\,\text{lb seawater}}{1\,\text{ton seawater}} = 5.52 \times 10^{17}\,\text{lb seawater}$$

$$5.52 \times 10^{17}\,\text{lb seawater} \times \frac{453.6\ \text{g seawater}}{1\,\text{lb seawater}} \times \frac{1\,\text{cm}^3\ \text{seawater}}{1.03\,\text{g seawater}} = 2.43 \times 10^{20}\,\text{cm}^3\ \text{seawater}$$

$$2.43 \times 10^{20}\,\text{cm}^3\ \text{seawater} \times \left(\frac{1\,\text{m}}{100\,\text{cm}} \times \frac{1\,\text{km}}{1000\,\text{m}} \right)^3 = 2.43 \times 10^5\,\text{km}^3$$

93. **(D)** We start by using the percent isotopic abundances for ^{87}Rb and ^{85}Rb along with the data in the "spiked" mass spectrum to find the total mass of Rb in the sample. Then, we calculate the Rb content in the rock sample in ppm by mass by dividing the mass of Rb by the total mass of the rock sample, and then multiplying the result by 10^6 to convert to ppm.

^{87}Rb = 27.83% isotopic abundance ^{85}Rb = 72.17% isotopic abundance

Therefore, $\dfrac{^{87}\text{Rb(isotopic)}}{^{85}\text{Rb(isotopic)}} = \dfrac{27.83\%}{72.17\%} = 0.3856$

For the ^{87}Rb(spiked) sample, the ^{87}Rb peak in the mass spectrum is 1.12 times as tall as the ^{85}Rb peak. Thus, for this sample $\dfrac{^{87}\text{Rb(isotopic)} + {}^{87}\text{Rb(spiked)}}{^{85}\text{Rb(isotopic)}} = 1.12$

Using this relationship, we can now find the masses of both ^{85}Rb and ^{87}Rb in the sample.

So, $\dfrac{^{87}\text{Rb(isotopic)}}{^{85}\text{Rb(isotopic)}} = 0.3856;$ $^{85}\text{Rb(isotopic)} = \dfrac{^{87}\text{Rb(isotopic)}}{0.3856}$

$^{87}\text{Rb(isotopic)} + {}^{87}\text{Rb(spiked)} = \dfrac{1.12 \times {}^{87}\text{Rb (isotopic)}}{0.3856} = 2.905\ {}^{87}\text{Rb(isotopic)}$

$^{87}\text{Rb(spiked)} = 1.905\ {}^{87}\text{Rb(isotopic)}$

and $\dfrac{^{87}Rb(\text{isotopic}) + {}^{87}Rb(\text{spiked})}{^{85}Rb(\text{isotopic})} = \dfrac{^{87}Rb(\text{isotopic}) + {}^{87}Rb(\text{spiked})}{\dfrac{^{87}Rb(\text{isotopic})}{0.3856}} = 1.12$

Since the mass of $^{87}Rb(\text{spiked})$ is equal to 29.45 μg, the mass of $^{87}Rb(\text{isotopic})$ must be $\dfrac{29.45\ \mu g}{1.905} = 15.46\ \mu g$ of $^{87}Rb(\text{isotopic})$

So, the mass of $^{85}Rb(\text{isotopic}) = \dfrac{15.46\ \mu g\ \text{of}\ ^{87}Rb(\text{isotopic})}{0.3856} = 40.09\ \mu g$ of $^{85}Rb(\text{isotopic})$

Therefore, the total mass of Rb in the sample = 15.46 μg of $^{87}Rb(\text{isotopic})$ + 40.09 μg of $^{85}Rb(\text{isotopic})$ = 55.55 μg of Rb. Convert to grams:

$= 55.55\ \mu g\ \text{of}\ Rb \times \dfrac{1\ g\ Rb}{1 \times 10^6\ \mu g\ Rb} = 5.555 \times 10^{-5}\ g\ Rb$

Rb content (ppm) $= \dfrac{5.555 \times 10^{-5}\ g\ Rb}{0.350\ g\ \text{of rock}} \times 10^6 = 159\ ppm\ Rb$

SELF-ASSESSMENT EXERCISES

97. **(E)** $\left(\dfrac{1}{5} \times 10.013\ u\right) + \left(\dfrac{4}{5} \times 11.009\ u\right) = 10.810\ u$. Therefore, the element is boron.

98. **(E)** The answer is (b). If all of the zinc reacts and the total amount of the product (zinc sulfide) is 14.9 g, then 4.9 g of S must have reacted with zinc. Therefore, 3.1 g of S remain.

99. **(E)** The answer is (d). It should be remembered that atoms combine in ratios of whole numbers. Therefore:

(a) 16 g O × (1 mol O/16 g O) = 1 mol O, and 85.5 g Rb × (1 mol Rb/85.5 g Rb) = 1 mol Rb Therefore, the O: Rb ratio is 1:1.

(b) Same calculations as above give an O: Rb ratio of 0.5:0.5, or 1:1.

(c) Same type calculation gives an O: Rb ratio of 2:1.

Because all of the above combine in O and Rb in whole number ratios, they are all at least theoretically possible.

100. (E) If the compound contains 46.7% X, then the percent of oxygen would be 100.0% − 46.7 = 53.3%. Assuming the sample is 100.0 g, there would be 53.3 g of oxygen present.

$$53.3 \text{ g O} \left(\frac{1 \text{ mol O}}{16.00 \text{ g O}} \right) = 3.331 \text{ mol O}$$

The formula of the compound is XO; therefore, X and O are in a 1:1 mole ratio.

$$\frac{46.7 \text{ g X}}{3.331 \text{ mol X}} = 14.0 \text{ g/mol X}$$

101. (E) The answer is (c). Cathode rays are beams of electrons, and as such have identical properties to β particles, although they may not have the same energy.

102. (E) The answer is (a), that the greatest portion of the mass of an atom is concentrated in a small but positively charged nucleus.

103. (E) The answer is (d). A hydrogen atom has one proton and one electron, so its charge is zero. A neutron has the same charge as a proton, but is neutral. Since most of the mass of the atom is at the nucleus, a neutron has nearly the same mass as a hydrogen atom.

104. (E) All the choices in this question are fundamental particles (e).

105. (E) Dalton (d) is correct.

106. (E) There are no particles that have the same mass as the hydrogen atom and a negative charge. The correct answer is (e).

107. (E) $^{35}_{17}Cl^{+}$

108. (E) The answer is (d), calcium, because they are in the same group.

109. (E) (a) Group 18, (b) Group 17, (c) Group 13 and Group 1 (d) Group 18

110. (E) (d) and (f)

111. (E) (c), because it is not close to being a whole number

112. (M) The answer is (d). Even with the mass scale being redefined based on ^{84}Xe, the mass ratio between ^{12}C and ^{84}Xe will remain the same. Using ^{12}C as the original mass scale, the mass ratio of ^{12}C : ^{84}Xe is 12 u/83.9115 u = 0.1430. Therefore, redefining the mass scale by assigning the exact mass of 84 u to ^{84}Xe, the relative mass of ^{12}C becomes 84 × 0.14301 = 12.0127 u.

113. **(M)** The answer is (b)

$$5.585 \text{ kg Fe} \times \frac{1000 \text{ g}}{1 \text{ kg}} \times \frac{1 \text{ mol Fe}}{55.85 \text{ g Fe}} = 1.000 \times 10^{-2} \text{ mol Fe}$$

$$600.6 \text{ g C} \times \frac{1 \text{ mol C}}{12.01 \text{ g C}} = 50.01 \text{ mol C}$$

Therefore, 100 moles of Fe has twice as many atoms as 50 moles of C.

114. **(E)** (d) is correct.

$$91.84 \text{ g} \times \frac{\text{mol}}{47.867 \text{ g}} \times \frac{6.022 \times 10^{23} \text{ atoms}}{\text{mol}} \times \frac{22 \text{ electrons}}{\text{Ti atom}} = 2.542 \times 10^{25} \text{ electrons}$$

115. **(M)**

$$2.327 \text{ g Fe} \times \frac{1 \text{ mol Fe}}{55.8452 \text{ g Fe}} = 0.0417 \text{ mol Fe}$$

$$1.000 \text{ g O} \times \frac{1 \text{ mol O}}{15.999 \text{ g O}} = 0.0625 \text{ mol O}$$

Dividing the two mole values to obtain the mole ratio, we get: 0.0625/0.0417 = 1.50. That is, 1.50 moles (or atoms) of O per 1 mole of Fe, or 3 moles of O per 2 moles of Fe (Fe_2O_3). Performing the above calculations for a compound with 2.618 g of Fe to 1.000 g of O yields 0.0469 mol of Fe and 0.0625 mol of O, or a mole ratio of 1.333, or a 4:3 ratio (Fe_3O_4).

116. **(D)** The weighted-average atomic mass of Sr is expressed as follows:
atomic mass of Sr = 87.62 amu = 83.9134(0.0056) + 85.9093x
$$+ 86.9089[1 - (0.0056 + 0.8258 + x)] + 87.9056(0.8258)$$

Rearrange the above equation and solve for x, which is 0.095 or 9.5%, which is the relative abundance of ^{86}Sr. Therefore, the relative abundance of ^{87}Sr is 0.0735 or 7.3%.

The reason for the imprecision is the low number of significant figures for ^{84}Sr.

117. **(M)** This problem lends itself well to the conversion pathway:

$$\frac{0.15 \text{ mg Au}}{1 \text{ ton seawater}} \times \frac{1 \text{ ton seawater}}{1000 \text{ kg}} \times \frac{1 \text{ kg}}{1000 \text{ mg}} \times \frac{1.03 \text{ g seawater}}{1 \text{ mL seawater}} \times \frac{250 \text{ mL}}{\text{sample}}$$

$$\times \frac{1 \text{ g Au}}{1000 \text{ mg Au}} \times \frac{1 \text{ mol Au}}{196.97 \text{ g Au}} \times \frac{6.02 \times 10^{23} \text{ atoms}}{1 \text{ mol Au}} = 1.2 \times 10^{14} \text{ atoms of Au}$$

CHAPTER 3
CHEMICAL COMPOUNDS
PRACTICE EXAMPLES

1A **(E)** First we convert the number of chloride ions to the mass of $MgCl_2$.

$$mass_{MgCl_2} = 5.0 \times 10^{23} \, Cl^- \times \frac{1 \, MgCl_2}{2 \, Cl^-} \times \frac{1 \, mol \, MgCl_2}{6.022 \times 10^{23} \, MgCl_2} \times \frac{95.205 \, g \, MgCl_2}{1 \, mol \, MgCl_2} = 4.0 \times 10^1 \, g \, MgCl_2$$

1B **(M)** First we convert mass $Mg(NO_3)_2$ to moles $Mg(NO_3)_2$ and formula units $Mg(NO_3)_2$ then finally to NO_3^- ions.

$$1.00 \, \mu g \, Mg(NO_3)_2 \times \frac{1 \, g \, Mg(NO_3)_2}{1,000,000 \, \mu g \, Mg(NO_3)_2} \times \frac{1 \, mol \, Mg(NO_3)_2}{148.313 \, g \, Mg(NO_3)_2} \times \frac{6.022 \times 10^{22} \, formula \, units \, Mg(NO_3)_2}{1 \, mol \, Mg(NO_3)_2}$$

$$= 4.06 \times 10^{15} \, formula \, units \, Mg(NO_3)_2 \times \frac{2 \, NO_3^- \, ions}{1 \, formula \, unit \, Mg(NO_3)_2} = 8.12 \times 10^{15} \, NO_3^- \, ions$$

Next, determine the number of oxygen atoms by multiplying by the appropriate ratio.

$$\# \, O \, atoms = 4.06 \times 10^{15} \, formula \, units \, Mg(NO_3)_2 \times \frac{6 \, atoms \, O}{1 \, formula \, unit \, Mg(NO_3)_2} = 2.44 \times 10^{16} \, O$$

2A **(M)** The volume of gold is converted to its mass and then to the amount in moles.

$$\# \, Au \, atoms = (2.50 \, cm)^2 \times \left(0.100 \, mm \times \frac{1 \, cm}{10 \, mm} \right) \times \frac{19.32 \, g}{1 \, cm^3} \times \frac{1 \, mol \, Au}{196.97 \, g \, Au} \times \frac{6.022 \times 10^{23} \, atoms}{1 \, mol \, Au}$$

$$= 3.69 \times 10^{21} \, Au \, atoms$$

2B **(M)** We need the molar mass of ethyl mercaptan for one conversion factor.
$$M = (2 \times 12.011 \, g \, C) + (6 \times 1.008 \, g \, H) + (1 \times 32.06 \, g \, S) = 62.13 \, g/mol \, C_2H_6S$$

Volume of room: $62 \, ft \times 35 \, ft \times 14 \, ft = 3.0\underline{4} \times 10^4 \, ft^3$. We also need to convert ft^3 to m^3.

$$3.0\underline{4} \times 10^4 \, ft^3 \times \left(\frac{12 \, in}{1 \, ft} \right)^3 \times \left(\frac{2.54 \, cm}{1 \, in} \right)^3 \times \left(\frac{1 \, m}{100 \, cm} \right)^3 = 8.6 \times 10^2 \, m^3$$

$$[C_2H_6S] = \frac{1.0 \, \mu L \, C_2H_6S}{8.6 \times 10^2 \, m^3} \times \frac{1 \, L}{1 \times 10^6 \, \mu L} \times \frac{1000 \, mL}{1 \, L} \times \frac{0.84 \, g}{1 \, mL} \times \frac{1 \, mol \, C_2H_6S}{62.13 \, g} \times \frac{10^6 \, \mu mol}{1 \, mol}$$

$$= 0.016 \, \mu mol/m^3 > 9.0 \times 10^{-4} \, \mu mol/m^3 = the \, detectable \, limit$$

Thus, the vapor will be detectable.

3A **(M)** The molar mass of halothane is given in Example 3-3 *in the textbook* as 197.4 g/mol. The rest of the solution uses conversion factors to change units.

$$\text{mass Br} = 25.0 \text{ mL } C_2HBrClF_3 \times \frac{1.871 \text{ g } C_2HBrClF_3}{1 \text{ mL } C_2HBrClF_3} \times \frac{1 \text{ mol } C_2HBrClF_3}{197.4 \text{ g } C_2HBrClF_3} \times \frac{1 \text{ mol Br}}{1 \text{ mol } C_2HBrClF_3}$$

$$\times \frac{79.904 \text{ g Br}}{1 \text{ mol Br}} = 18.9 \text{ g Br}$$

3B **(M)** Again, the molar mass of halothane is given in Example 3-3 *in the textbook* as 197.4 g/mol.

$$V_{\text{halothane}} = 1.00 \times 10^{24} \text{ Br} \times \frac{1 \text{ mol Br}}{6.022 \times 10^{23} \text{ Br}} \times \frac{1 \text{ mol } C_2HBrClF_3}{1 \text{ mol Br}} \times \frac{197.4 \text{ g } C_2HBrClF_3}{1 \text{ mol } C_2HBrClF_3} \times \frac{1 \text{ mL}}{1.871 \text{ g}}$$

$$= 175 \text{ mL } C_2HBrClF_3$$

4A **(M)** We use the same technique as before: determine the mass of each element in a mole of the compound. Their sum is the molar mass of the compound. The percent composition is determined by comparing the mass of each element with the molar mass of the compound.

$$M = (10 \times 12.011 \text{ g C}) + (16 \times 1.008 \text{ g H}) + (5 \times 14.01 \text{ g N}) + (3 \times 30.97 \text{ g P}) + (13 \times 15.999 \text{ g O})$$

$$= 120.11 \text{ g C} + 16.13 \text{ g H} + 70.05 \text{ g N} + 92.91 \text{ g P} + 207.99 \text{ g O} = 507.19 \text{ g ATP/mol}$$

$$\%C = \frac{120.11 \text{ g C}}{507.19 \text{ g ATP}} \times 100\% = 23.681\% \text{ C} \quad \%H = \frac{16.13 \text{ g H}}{507.19 \text{ g ATP}} \times 100\% = 3.180\% \text{ H}$$

$$\%N = \frac{70.05 \text{ g N}}{507.19 \text{ g ATP}} \times 100\% = 13.81\% \text{ N} \quad \%P = \frac{92.91 \text{ g P}}{507.19 \text{ g ATP}} \times 100\% = 18.32\% \text{ P}$$

$$\%O = \frac{207.99 \text{ g O}}{507.19 \text{ g ATP}} \times 100\% = 41.008\% \text{ O} \quad \text{(NOTE: the mass percents sum to 99.999\%)}$$

4B **(E)** Both (b) and (e) have the same empirical formula, that is, CH_2O. These two molecules have the same percent oxygen by mass.

5A **(M)** Once again, we begin with a 100.00 g sample of the compound. In this way, each elemental mass in grams is numerically equal to its percent. We convert each mass to an amount in moles, and then determine the simplest integer set of molar amounts. This determination begins by dividing all three molar amounts by the smallest.

$$39.56 \text{ g C} \times \frac{1 \text{ mol C}}{12.011 \text{ g C}} = 3.294 \text{ mol C} \div 3.294 \rightarrow 1.000 \text{ mol C} \times 3.000 = 3.000 \text{ mol C}$$

$$7.74 \text{ g H} \times \frac{1 \text{ mol H}}{1.008 \text{ g H}} = 7.68 \text{ mol H} \div 3.294 \rightarrow 2.33 \text{ mol H} \times 3.000 = 6.99 \text{ mol H}$$

$$52.70 \text{ g O} \times \frac{1 \text{ mol O}}{15.999 \text{ g O}} = 3.294 \text{ mol O} \div 3.294 \rightarrow 1.000 \text{ mol O} \times 3.000 = 3.000 \text{ mol O}$$

Thus, the empirical formula of the compound is $C_3H_7O_3$. The empirical molar mass of this compound is:

$$(3 \times 12.011\,g\,C) + (7 \times 1.008\,g\,H) + (3 \times 15.999\,g\,O) = 36.033\,g + 7.056\,g + 47.997\,g = 91.09\,g/mol$$

The empirical mass is almost precisely one half the reported molar mass, leading to the conclusion that the molecular formula must be twice the empirical formula in order to double the molar mass. Thus, the molecular formula is $C_6H_{14}O_6$.

5B **(M)** To answer this question, we start with a 100.00 g sample of the compound. In this way, each elemental mass in grams is numerically equal to its percent. We convert each mass to an amount in moles, and then determine the simplest integer set of molar amounts. This determination begins by dividing all molar amounts by the smallest number of moles in the group of four, which is 1.1025 moles. Multiplication of the resulting quotients by eight produces the smallest possible set of whole numbers.

$$21.51g\,C \times \frac{1\,mol\,C}{12.011g\,C} = 1.791\ mol\,C \div 1.102\underline{5} \rightarrow 1.624\ mol\,C \times 8 = 12.99\,mol\,C$$

$$2.22\ g\,H \times \frac{1\,mol\,H}{1.008\,g\,H} = 2.20\ mol\,H \div 1.102\underline{5} \rightarrow 2.00\,mol\,H \times 8 = 16.0\ mol\,H$$

$$17.64\ g\,O \times \frac{1\,mol\,O}{15.999\,g\,O} = 1.102\underline{5}\,mol\,O \div 1.102\underline{5} \rightarrow 1.000\,mol\,O \times 8 = 8.000\,mol\,O$$

$$58.63g\,Cl \times \frac{1\,mol\,Cl}{35.45\,g\,Cl} = 1.654\ mol\,Cl \div 1.102\underline{5} \rightarrow 1.500\ mol\,C \times 8 = 12.00\,mol\,Cl$$

Thus, the empirical formula of the compound is $C_{13}H_{16}O_8Cl_{12}$. The empirical molar mass of this compound is 725.7 g/mol.

The empirical mass is almost precisely the same as the reported molar mass, leading to the conclusion that the molecular formula must be the same as the empirical formula. Thus, the molecular formula is $C_{13}H_{16}O_8Cl_{12}$.

6A **(M)** We calculate the amount in moles of each element in the sample (determining the mass of oxygen by difference) and transform these molar amounts to the simplest integral amounts, by first dividing all three by the smallest.

$$2.726g\,CO_2 \times \frac{1\,mol\,CO_2}{44.009\,g\,CO_2} \times \frac{1\,mol\,C}{1\,mol\,CO_2} = 0.06194\,mol\,C \times \frac{12.011g\,C}{1\,mol\,C} = 0.7440\,g\,C$$

$$1.116g\,H_2O \times \frac{1\,mol\,H_2O}{18.015g\,H_2O} \times \frac{2\,mol\,H}{1\,mol\,H_2O} = 0.1239\,mol\,H \times \frac{1.008\,g\,H}{1\,mol\,H} = 0.1249\,g\,H$$

$$(1.152\,g\,cmpd - 0.7440\,g\,C - 0.1249\,g\,H) = 0.283\,g\,O \times \frac{1\,mol\,O}{15.999\,g\,O} = 0.0177\,mol\,O$$

$$0.06194 \, \text{mol C} \div 0.0177 \;\rightarrow 3.50$$
$$0.1239 \, \text{mol H} \div 0.0177 \;\rightarrow 7.00$$
$$0.0177 \, \text{mol O} \div 0.0177 \;\rightarrow 1.00$$

All of these amounts in moles are multiplied by 2 to make them integral. Thus, the empirical formula of isobutyl propionate is $C_7H_{14}O_2$.

6B **(M)** Notice that we do not have to obtain the mass of any element in this compound by difference; there is no oxygen present in the compound. We calculate the amount in one mole of each element in the sample and transform these molar amounts to the simplest integral amounts, by first dividing all three by the smallest.

$$3.149 \, \text{g CO}_2 \times \frac{1 \, \text{mol CO}_2}{44.009 \, \text{g CO}_2} \times \frac{1 \, \text{mol C}}{1 \, \text{mol CO}_2} = 0.07155 \, \text{mol C} \div 0.01789 = 3.999 \, \text{mol C}$$

$$0.645 \, \text{g H}_2\text{O} \times \frac{1 \, \text{mol H}_2\text{O}}{18.015 \, \text{g H}_2\text{O}} \times \frac{2 \, \text{mol H}}{1 \, \text{mol H}_2\text{O}} = 0.0716 \, \text{mol H} \div 0.01789 = 4.00 \, \text{mol H}$$

$$1.146 \, \text{g SO}_2 \times \frac{1 \, \text{mol SO}_2}{64.058 \, \text{g SO}_2} \times \frac{1 \, \text{mol S}}{1 \, \text{mol SO}_2} = 0.01789 \, \text{mol S} \div 0.01789 = 1.000 \, \text{mol S}$$

Thus, the empirical formula of thiophene is C_4H_4S.

7A **(E)**

$\underline{S}_8$ For an atom of a free element, the oxidation state is 0 (rule 1).

$\underline{Cr}_2O_7^{2-}$ The sum of all the oxidation numbers in the ion is -2 (rule 2). The O.S. of each oxygen is -2 (rule 6). Thus, the total for all seven oxygens is -14. The total for both chromiums must be $+12$. Thus, each Cr has an O.S. $= +6$.

$\underline{Cl}_2O$ The sum of all oxidation numbers in the compound is 0 (rule 2). The O.S. of oxygen is -2 (rule 6). The total for the two chlorines must be $+2$. Thus, each chlorine must have O.S. $= +1$.

$K\underline{O}_2$ The sum for all the oxidation numbers in the compound is 0 (rule 2). The O.S. of potassium is $+1$ (rule 3). The sum of the oxidation numbers of the two oxygens must be -1. Thus, each oxygen must have O.S. $= -1/2$.

7B **(E)**

$\underline{S}_2O_3^{2-}$ The sum of all the oxidation numbers in the ion is -2 (rule 2). The O.S. of oxygen is -2 (rule 6). Thus, the total for three oxygens must be -6. The total for both sulfurs must be $+4$. Thus, each S has an O.S. $= +2$.

$\underline{Hg}_2Cl_2$ The O.S. of each Cl is -1 (rule 7). The sum of all O.S. is 0 (rule 2). Thus, the total for two Hg is $+2$ and each Hg has O.S. $= +1$.

$K\underline{Mn}O_4$ The O.S. of each O is -2 (rule 6). Thus, the total for 4 oxygens must be -8. The K has O.S. $= +1$ (rule 3). The total of all O.S. is 0 (rule 2). Thus, the O.S. of Mn is $+7$.

$H_2\underline{C}O$ The O.S. of each H is $+1$ (rule 5), producing a total for both hydrogens of $+2$. The O.S. of O is -2 (rule 6). Thus, the O.S. of C is 0, because the total of all O.S. values is 0 (rule 2).

8A **(E)** In each case, we determine the formula *with its accompanying charge* of each ion in the compound. We then produce a formula for the compound in which the total positive charge equals the total negative charge.

lithium oxide	Li^+ and O^{2-}	*two* Li^+ and *one* O^{2-}	Li_2O
tin(II) fluoride	Sn^{2+} and F^-	*one* Sn^{2+} and *two* F^-	SnF_2
lithium nitride	Li^+ and N^{3-}	*three* Li^+ and *one* N^{3-}	Li_3N

8B **(E)** Using a similar procedure as that provided in **8A**

aluminum sulfide	Al^{3+} and S^{2-}	*two* Al^{3+} and *three* S^{2-}	Al_2S_3
magnesium nitride	Mg^{2+} and N^{3-}	*three* Mg^{2+} and *two* N^{3-}	Mg_3N_2
vanadium(III) oxide	V^{3+} and O^{2-}	*two* V^{3+} and *three* O^{2-}	V_2O_3

9A **(E)** The name of each of these ionic compounds is the name of the cation followed by that of the anion. Each anion name is a modified (with the ending *-ide*) version of the name of the element. Each cation name is the name of the metal, with the oxidation state appended in Roman numerals in parentheses if there is more than one type of cation for that metal.

CsI cesium iodide

CaF_2 calcium fluoride

FeO The O.S. of $O = -2$ (rule 6). Thus, the O.S. of $Fe = +2$ (rule 2). The cation is iron(II). The name of the compound is iron(II) oxide.

$CrCl_3$ The O.S. of $Cl = -1$ (rule 7). Thus, the O.S. of $Cr = +3$ (rule 2). The cation is chromium(III). The compound is chromium(III) chloride.

9B **(E)** The name of each of these ionic compounds is the name of the cation followed by that of the anion. Each anion name is a modified (with the ending *-ide*) version of the name of the element. Each cation name is the name of the metal, with the oxidation state appended in Roman numerals in parentheses if there is more than one type of cation for that metal.

The oxidation state of Ca is +2 (rule 3). Hydrogen would therefore have an oxidation number of -1 (which is an exception to rule 5), based on rule 2.
CaH_2 calcium hydride

The oxidation number of sulfur is -2 (rule 7), and therefore silver would be +1 for each silver atom based on rule 2.
Ag_2S silver(I) sulfide

In the next two compounds, the oxidation state of chlorine is -1 (rule 7) and thus the oxidation state of the metal in each cation must be +1 (rule 2).
$CuCl$ copper(I) chloride Hg_2Cl_2 mercury(I) chloride

10A (E)

SF_6	Both S and F are nonmetals. This is a binary molecular compound: sulfur hexafluoride.
HNO_2	The NO_2^- ion is the nitrite ion. Its acid is nitrous acid.
$Ca(HCO_3)_2$	HCO_3^- is the bicarbonate ion or the hydrogen carbonate ion. This compound is calcium bicarbonate or calcium hydrogen carbonate.
$FeSO_4$	The SO_4^{2-} ion is the sulfate ion. The cation is Fe^{2+}, iron(II). This compound is iron(II) sulfate.

10B (E)

NH_4NO_3	The cation is NH_4^+, ammonium ion. The anion is NO_3^-, nitrate ion. This compound is ammonium nitrate.
PCl_3	Both P and Cl are nonmetals. This is a binary molecular compound: phosphorus trichloride.
$HBrO$	BrO^- is hypobromite, this is hypobromous acid.
$AgClO_4$	The anion is perchlorate ion, ClO_4^-. The compound is silver(I) perchlorate.
$Fe_2(SO_4)_3$	The SO_4^{2-} ion is the sulfate ion. The cation is Fe^{3+}, iron(III). This compound is iron(III) sulfate.

11A (E)

boron trifluoride	Both elements are nonmetals. This is a binary molecular compound: BF_3.
potassium dichromate	Potassium ion is K^+, and dichromate ion is $Cr_2O_7^{2-}$. This is $K_2Cr_2O_7$.
sulfuric acid	The anion is sulfate, SO_4^{2-}. There must be two H^+s. This is H_2SO_4.
calcium chloride	The ions are Ca^{2+} and Cl^-. There must be one Ca^{2+} and two Cl^-s: $CaCl_2$.

11B (E)

aluminum nitrate	Aluminum is Al^{3+};; the nitrate ion is NO_3^-. This is $Al(NO_3)_3$.
tetraphosphorus decoxide	Both elements are nonmetals. This is a binary molecular compound, P_4O_{10}.
chromium(III) hydroxide	Chromium(III) ion is Cr^{3+}; the hydroxide ion is OH^-. This is $Cr(OH)_3$.
iodic acid	The halogen "ic" acid has the halogen in a +5 oxidation state. This is HIO_3.

12A (E)

    (a)   Not isomers: molecular formulas are different (C_8H_{18} vs C_9H_{20}).

    (b)   Molecules are isomers (same formula C_7H_{16}).

12B (E)

    (a)   Molecules are isomers (same formula C_7H_{14}).

    (b)   Not isomers: molecular formulas are different (C_4H_8 vs C_5H_{10}).

13A (E)

    (a)   The carbon to carbon bonds are all single bonds in this hydrocarbon. This compound is an alkane.

    (b)   In this compound, there are only single bonds, and a Cl atom has replaced one H atom. This compound is a chloroalkane.

    (c)   The presence of the carboxyl group ($—CO_2H$) in this molecule means that the compound is a carboxylic acid.

    (d)   There is a carbon to carbon double bond in this hydrocarbon. This is an alkene.

13B (E)

    (a)   The presence of the hydroxyl group ($—OH$) in this molecule means that this compound is an alcohol.

    (b)   The presence of the carboxyl group ($—CO_2H$) in this molecule means that the compound is a carboxylic acid. This molecule also contains the hydroxyl group ($—OH$).

    (c)   The presence of the carboxyl group ($—CO_2H$) in this molecule means that the compound is a carboxylic acid. As well, a Cl atom has replaced one H atom. This compound is a chloroalkane. The compound is a chloro carboxylic acid.

    (d)   There is a carbon to carbon double bond in this compound; hence, it is an alkene. There is also one H atom that has been replaced by a Br atom. This compound is also a bromoalkene.

14A (E)

    (a)   The structure is that of an alcohol with the hydroxyl group on the second carbon atom of a three carbon chain. The compound is propan-2-ol (commonly isopropyl alcohol).

    (b)   The structure is that of an iodoalkane molecule with the I atom on the first carbon of a three-carbon chain. The compound is called 1-iodopropane.

    (c)   The carbon chain in this structure is four carbon atoms long with the end C atom in a carboxyl group. There is also a methyl group on the third carbon in the chain. The compound is 3-methylbutanoic acid.

    (d)   The structure is that of a three carbon chain that contains a carbon to carbon double bond. This compound is propene.

14B (E)

    (a)   2-chloropropane   **(b)**   1,4-dichlorobutane   **(c)**   2-methylpropanoic acid

15A **(E)**

(a) pentane: $CH_3(CH_2)_3CH_3$ (b) ethanoic acid: CH_3CO_2H

(c) 1-iodooctane: $ICH_2(CH_2)_6CH_3$ (d) pentan-1-ol: $CH_2(OH)(CH_2)_3CH_3$

15B **(E)**

(a) propene

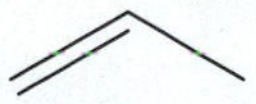

(b) heptan-1-ol

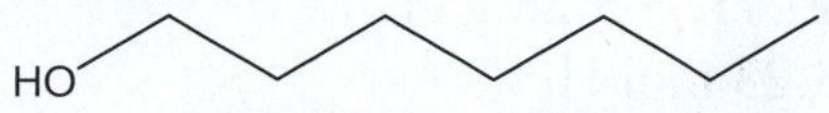

(c) chloroacetic acid

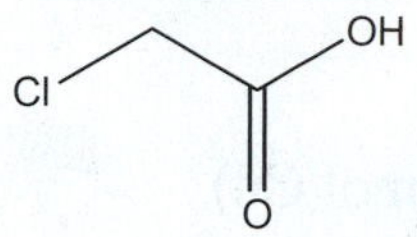

(d) hexanoic acid

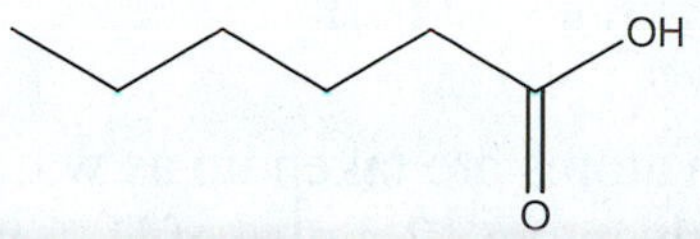

INTEGRATIVE EXAMPLES

A. **(M)**

First, determine the mole ratios of the dehydrated compound:

27.74 g Mg × (1 mol Mg / 24.305 g Mg) = 1.141 mol Mg

23.57 g P × (1 mol P / 30.974 g P) = 0.7610 mol P

48.69 g O × (1 mol O / 15.999 g O) = 3.043 mol O

Mole ratios are determined by dividing by the smallest number:

1.141 mol Mg / 0.7610 mol P ≈ 1.5

0.7610 mol P / 0.7610 mol P ≈ 1.0

3.043 mol O / 0.7610 mol P ≈ 4.0

Multiplying by 2 to get whole numbers, the empirical formula becomes $Mg_3P_2O_8$. The compound is magnesium phosphate, $Mg_3(PO_4)_2$.

To determine the number of waters of hydration, determine the mass of water driven off.

mass of H_2O = 2.4917 g – 1.8558 g = 0.6359 g

mol H_2O = 0.6359 g × (1 mol H_2O/18.015 g H_2O) = 0.0353 mol.

Then, calculate the number of moles of dehydrated $Mg_3(PO_4)_2$ in the same manner above. The number of moles (using 262.86 g/mol for molecular weight) is 0.00706. Dividing the number of moles of H_2O by $Mg_3(PO_4)_2$ gives a ratio of 5. Therefore, the compound is $Mg_3(PO_4)_2 \cdot 5\ H_2O$

B. **(M)**
First, determine the mole ratio of the elements in this compound:

17.15 g Cu $\times$ (1 mol Cu / 63.546 g Cu) = 0.2699 mol Cu
19.14 g Cl $\times$ (1 mol Cl / 35.45 g Cl) = 0.5399 mol Cl
60.45 g O $\times$ (1 mol O / 15.999 g O) = 3.778 mol O

Mass of H: $100 - (17.15 + 19.14 + 60.45) = 3.26$ g H
3.26 g H $\times$ (1 mol H / 1.008 g H) = 3.23 mol H
Mole ratios are determined by dividing by the smallest number:

0.2699 mol Cu / 0.2699 mol Cu = 1.000
0.5399 mol Cl / 0.2699 mol Cu = 2.000 (mol Cl per mol Cu)
3.778 mol O / 0.2699 mol Cu = 14.00 (mol O per mol Cu)
3.23 mol H / 0.2699 mol Cu = 12.0 (mol H per mol Cu)

Now we know that since all the hydrogen atoms are taken up as water, half as many moles of O are also taken up as water. Therefore, if there are 12 moles of H, 6 moles of O are needed, 6 moles of H_2O are generated, and 8 moles of O are left behind.

To determine the oxidation state of Cu and Cl, we note that there are 4 times as many moles of O as there is Cl. If the Cl and O are associated, we have the perchlorate ion (ClO_4^-) and the formula of the compound is $Cu(ClO_4)_2 \cdot 6\,H_2O$. The oxidation state of Cu is +2 and Cl is +7.

EXERCISES

Representing Molecules

1. **(E)**
   (a) H_2O_2 **(b)** CH_3CH_2Cl **(c)** P_4O_{10}

   (d) $CH_3CH(OH)CH_3$ **(e)** HCO_2H

3. **(E)**
   (b) CH_3CH_2Cl **(d)** $CH_3CH(OH)CH_3$ **(e)** HCO_2H

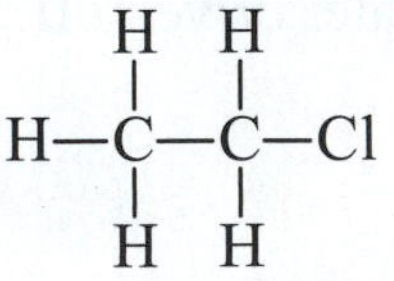

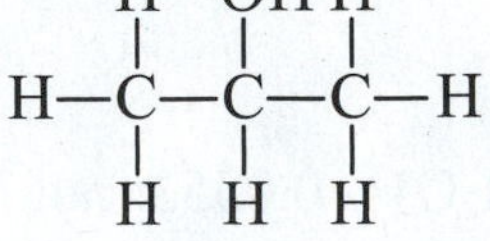

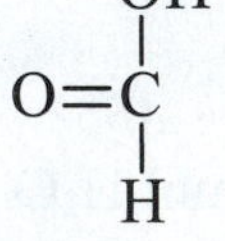

The Avogadro Constant and the Mole

5. **(M)**

(a) A trinitrotoluene molecule, $CH_3C_6H_2(NO_2)_3$, contains 7 C atoms, 5 H atoms, 3 N atoms, and 3×2 O atoms $= 6$ O atoms, for a total of $7 + 5 + 3 + 6 = 21$ atoms.

(b) $CH_3(CH_2)_4CH_2OH$ contains 6 C atoms, 14 H atoms, and 1 O atom, for a total of 21 atoms.

Conversion pathway approach:

$$\text{\# of atoms} = 0.00102 \, \text{mol CH}_3(\text{CH}_2)_4\text{CH}_2\text{OH} \times \frac{6.022 \times 10^{23} \, \text{molecules CH}_3(\text{CH}_2)_4\text{CH}_2\text{OH}}{1 \, \text{mol CH}_3(\text{CH}_2)_4\text{CH}_2\text{OH}}$$

$$\times \frac{21 \, \text{atoms}}{1 \, \text{molecule CH}_3(\text{CH}_2)_4\text{CH}_2\text{OH}} = 1.29 \times 10^{22} \, \text{atoms}$$

Stepwise approach:

$$0.00102 \, \text{mol CH}_3(\text{CH}_2)_4\text{CH}_2\text{OH} \times \frac{6.022 \times 10^{23} \, \text{molecules CH}_3(\text{CH}_2)_4\text{CH}_2\text{OH}}{1 \, \text{mol CH}_3(\text{CH}_2)_4\text{CH}_2\text{OH}} = 6.14 \times 10^{20} \, \text{molecules}$$

$$6.14 \times 10^{20} \, \text{molecules} \times \frac{21 \, \text{atoms}}{1 \, \text{CH}_3(\text{CH}_2)_4\text{CH}_2\text{OH molecule}} = 1.29 \times 10^{22} \, \text{atoms}$$

(c) *Conversion pathway approach:*

$$\text{\# of F atoms} = 12.15 \, \text{mol C}_2\text{HBrClF}_3 \times \frac{3 \, \text{mol F}}{1 \, \text{mol C}_2\text{HBrClF}_3} \times \frac{6.022 \times 10^{23} \, \text{F atoms}}{1 \, \text{mol F atoms}}$$

$$= 2.195 \times 10^{25} \, \text{F atoms}$$

Stepwise approach:

$$12.15 \, \text{mol C}_2\text{HBrClF}_3 \times \frac{3 \, \text{mol F}}{1 \, \text{mol C}_2\text{HBrClF}_3} = 36.45 \, \text{mol F}$$

$$36.45 \, \text{mol F} \times \frac{6.022 \times 10^{23} \, \text{F atoms}}{1 \, \text{mol F atoms}} = 2.195 \times 10^{25} \, \text{F atoms}$$

7. **(M)**

(a) molecular mass (mass of one molecule) of $C_5H_{11}NO_2S$ is:

$$(5 \times 12.011 \, \text{u C}) + (11 \times 1.008 \, \text{u H}) + 14.007 \, \text{u N} + (2 \times 15.999 \, \text{u O}) + 32.06 \, \text{u S}$$

$$= 149.208 \, \text{u/C}_5\text{H}_{11}\text{NO}_2\text{S molecule}$$

(b) Since there are 11 H atoms in each $C_5H_{11}NO_2S$ molecule, there are 11 moles of H atoms in each mole of $C_5H_{11}NO_2S$ molecules.

(c) $\text{mass C} = 1\,\text{mol C}_5\text{H}_{11}\text{NO}_2\text{S} \times \dfrac{5\,\text{mol C}}{1\,\text{mol C}_5\text{H}_{11}\text{NO}_2\text{S}} \times \dfrac{12.011\,\text{g C}}{1\,\text{mol C}} = 60.055\,\text{g C}$

(d) $\text{\# C atoms} = 9.07\,\text{mol C}_5\text{H}_{11}\text{NO}_2\text{S} \times \dfrac{5\,\text{mol C}}{1\,\text{mol C}_5\text{H}_{11}\text{NO}_2\text{S}} \times \dfrac{6.022 \times 10^{23}\,\text{atoms C}}{1\,\text{mol C}}$

$= 2.73 \times 10^{25}\,\text{C atoms}$

9. **(E)** The greatest number of N atoms is found in the compound with the greatest number of moles of N.

The molar mass of $\text{N}_2\text{O} = (2\,\text{mol N} \times 14.0\,\text{g N}) + (1\,\text{mol O} \times 16.0\,\text{g O}) = 44.0\,\text{g/mol N}_2\text{O}$.
Thus, 50.0 g N_2O is slightly more than 1 mole of N_2O, and contains slightly more than 2 moles of N. Each mole of N_2 contains 2 moles of N. The molar mass of NH_3 is 17.0 g. Thus, there is 1 mole of NH_3 present, which contains 1 mole of N.

The molar mass of pyridine is $(5\,\text{mol C} \times 12.0\,\text{g C}) + (5\,\text{mol H} \times 1.01\,\text{g H}) + 14.0\,\text{g N} = 79.1\,\text{g/mol}$. Because each mole of pyridine contains 1 mole of N, we need slightly more than 2 moles of pyridine to have more N than is present in the N_2O. But that would be a mass of about 158 g pyridine, and 150 mL has a mass of less than 150 g. Thus, the greatest number of N atoms is present in 50.0 g N_2O.

11. **(M)**

(a) $\text{moles N}_2\text{O}_4 = 115\,\text{g N}_2\text{O}_4 \times \dfrac{1\,\text{mol N}_2\text{O}_4}{92.01\,\text{g N}_2\text{O}_4} = 1.25\,\text{mol N}_2\text{O}_4$

(b) $\text{moles N} = 43.5\,\text{g Mg(NO}_3)_2 \times \dfrac{1\,\text{mol Mg(NO}_3)_2}{148.31\,\text{g}} \times \dfrac{2\,\text{mol N}}{1\,\text{mol Mg(NO}_3)_2} = 0.587\,\text{mol N atoms}$

(c) $\text{moles N} = 12.4\,\text{g C}_6\text{H}_{12}\text{O}_6 \times \dfrac{1\,\text{mol C}_6\text{H}_{12}\text{O}_6}{180.15\,\text{g}} \times \dfrac{6\,\text{mol O}}{1\,\text{mol C}_6\text{H}_{12}\text{O}_6} \times \dfrac{1\,\text{mol C}_7\text{H}_5(\text{NO}_2)_3}{6\,\text{mol O}}$

$\times \dfrac{3\,\text{mol N}}{1\,\text{mol C}_7\text{H}_5(\text{NO}_2)_3} = 0.206\,\text{mol N}$

13. **(M)** The number of Fe atoms in 6 L of blood can be found using the following conversion pathway.

$= 6\,\text{L blood} \times \dfrac{1000\,\text{mL}}{1\,\text{L}} \times \dfrac{15.5\,\text{g Hb}}{100\,\text{mL blood}} \times \dfrac{1\,\text{mol Hb}}{64{,}500\,\text{g Hb}} \times \dfrac{4\,\text{mol Fe}}{1\,\text{mol Hb}} \times \dfrac{6.022 \times 10^{23}\,\text{atoms Fe}}{1\,\text{mol Fe}}$

$= 3 \times 10^{22}\,\text{Fe atoms}$

Chemical Formulas

15. **(E)** For glucose (blood sugar), $C_6H_{12}O_6$,

    (a) FALSE The percentages by mass of C and O are *different* than in CO. For one thing, CO contains no hydrogen.

    (b) TRUE In dihydroxyacetone, $(CH_2OH)_2\,CO$ or $C_3H_6O_3$, the ratio of C:H:O $= 3:6:3$ or 1:2:1. In glucose, this ratio is C:H:O $= 6:12:6 = 1:2:1$. Thus, the ratios are the same.

    (c) FALSE The proportions, by number of atoms, of C and O are the same in glucose. Since, however, C and O have different molar masses, their proportions by mass must be *different*.

    (d) FALSE Each mole of glucose contains $(12\times1.008=)12.1$ g H. But each mole also contains 72.1 g C and 96.0 g O. Thus, the highest percentage, by mass, is that of O. The highest percentage, by number of atoms, is that of H.

17. **(M)**

    (a) $Cu(UO_2)_2(PO_4)_2 \cdot 8\,H_2O$ has 1 Cu, 2 U, 2 P, 20 O, and 16 H, or a total of 41 atoms.

    (b) By number, $Cu(UO_2)_2(PO_4)_2 \cdot 8\,H_2O$ has a H to O ratio of 16:20 or 4:5 or 0.800 H atoms/O atom.

    (c) By number, $Cu(UO_2)_2(PO_4)_2 \cdot 8\,H_2O$ has a Cu to P ratio of 1:2.

$$\text{The mass ratio of Cu:P is } \frac{1\,\text{mol Cu}\times\dfrac{63.546\text{ g Cu}}{1\,\text{mol Cu}}}{2\text{ mol P}\times\dfrac{30.974\text{ g P}}{1\,\text{mol P}}} = 1.026.$$

    (d) With a mass percent slightly greater than 50%, U has the largest mass percent, with oxygen coming in at ~34%.

$$\text{mass \% U} = \frac{\text{mass U in } Cu(UO_2)_2(PO_4)_2 \cdot 8\,H_2O}{\text{total mass of } Cu(UO_2)_2(PO_4)_2 \cdot 8\,H_2O}\times100\%$$

$$= \frac{2\times238.03\text{ g/mol}}{937.660\text{ g/mol}}\times100\%$$

$$= 50.77\%$$

$$\text{mass \% O} = \frac{\text{mass O in } Cu(UO_2)_2(PO_4)_2 \cdot 8H_2O}{\text{total mass of } Cu(UO_2)_2(PO_4)_2 \cdot 8H_2O}\times100\%$$

$$= \frac{20\times15.999\text{ g/mol}}{937.660\text{ g/mol}}\times100\%$$

$$= 34.13\%$$

    (e) $1.00\text{ g P}\times\dfrac{1\text{ mol P}}{30.974\text{ g P}}\times\dfrac{1\text{ mol } Cu(UO_2)_2(PO_4)_2 \cdot 8\,H_2O}{2\text{ mol P}}\times$

$$\frac{937.666\text{ g } Cu(UO_2)_2(PO_4)_2 \cdot 8\,H_2O}{1\text{ mol } Cu(UO_2)_2(PO_4)_2 \cdot 8\,H_2O}$$

$$= 15.1\text{ g of } Cu(UO_2)_2(PO_4)_2 \cdot 8\,H_2O$$

Percent Composition of Compounds

19. **(E)** The information obtained in the course of calculating the molar mass is used to determine the mass percent of H in decane.

$$\text{molar mass } C_{10}H_{22} = \left(\frac{10\,\text{mol C}}{1\,\text{mol }C_{10}H_{22}}\times\frac{12.011\,\text{g C}}{1\,\text{mol C}}\right)+\left(\frac{22\,\text{mol H}}{1\,\text{mol }C_{10}H_{22}}\times\frac{1.008\,\text{g H}}{1\,\text{mol H}}\right)$$

$$=\frac{120.11\,\text{g C}}{1\,\text{mol }C_{10}H_{22}}+\frac{22.176\,\text{g H}}{1\,\text{mol }C_{10}H_{22}}=\frac{142.29\,\text{g}}{1\,\text{mol }C_{10}H_{22}}$$

$$\%H=\frac{22.176\,\text{g H/mol decane}}{142.29\,\text{g }C_{10}H_{22}\text{/mol decane}}\times100\%=15.369\%\,H$$

21. **(E)** $C(CH_3)_3CH_2CH(CH_3)_2$ has a molar mass of 114.231 g/mol and one mole contains 18.143 g of H.

$$\text{percent hydrogen in sample}=\frac{18.143\,\text{g}}{114.231\,\text{g}}\times100\%=15.88\%\,H$$

23. **(E)** $\text{molar mass}=(20\,\text{mol C}\times12.011\,\text{g C})+(24\,\text{mol H}\times1.008\,\text{g H})+(2\,\text{mol N}\times14.007\,\text{g N})$

$$+(2\,\text{mol O}\times15.999\,\text{g O})=324.42\,\text{g/mol}$$

$$\%C=\frac{240.22}{324.42}\times100\%=74.046\%\,C\qquad \%H=\frac{24.192}{324.42}\times100\%=7.4570\%\,H$$

$$\%N=\frac{28.014}{324.42}\times100\%=8.6351\%\,N\qquad \%O=\frac{31.998}{324.42}\times100\%=9.8631\%\,O$$

25. **(E)** In each case, we first determine the molar mass of the compound, and then the mass of the indicated element in one mole of the compound. Finally, we determine the percent by mass of the indicated element to four significant figures.

(a) $\quad\text{molar mass }Pb(C_2H_5)_4=207.2\,\text{g Pb}+(8\times12.011\,\text{g C})+(20\times1.008\,\text{g H})$

$$=323.448\,\text{g/mol }Pb(C_2H_5)_4$$

$$\text{mass Pb/mol }Pb(C_2H_5)_4=\frac{1\,\text{mol Pb}}{1\,\text{mol }Pb(C_2H_5)_4}\times\frac{207.2\,\text{g Pb}}{1\,\text{mol Pb}}=207.2\,\text{g Pb/mol }Pb(C_2H_5)_4$$

$$\%Pb=\frac{207.2\,\text{g Pb}}{323.448\,\text{g }Pb(C_2H_5)_4}\times100\%=64.06\%\,Pb$$

(b)   molar mass $Fe_4[Fe(CN)_6]_3 = (7 \times 55.845\,g\,Fe) + (18 \times 12.011\,g\,C) + (18 \times 14.007\,g\,N)$

$$= 859.253\,g/mol\,Fe_4[Fe(CN)_6]_3$$

$$\frac{mass\,Fe}{mol\,Fe_4[Fe(CN)_6]_3} = \frac{7\,mol\,Fe}{1\,mol\,Fe_4[Fe(CN)_6]_3} \times \frac{55.845\,g\,Fe}{1\,mol\,Fe} = 390.915\,g\,Fe/mol\,Fe_4[Fe(CN)_6]_3$$

$$\%Fe = \frac{390.915\,g\,Fe}{859.253\,g\,Fe_4[Fe(CN)_6]_3} \times 100\% = 45.495\%\,Fe$$

(c)   molar mass $C_{55}H_{72}MgN_4O_5$

$$= (55 \times 12.011\,g\,C) + (72 \times 1.008\,g\,H) + (1 \times 24.305\,g\,Mg) + (4 \times 14.007\,g\,N) + (5 \times 15.999\,g\,O)$$

$$= 893.509\,g/mol\,C_{55}H_{72}MgN_4O_5$$

$$\frac{mass\,Mg}{mol\,C_{55}H_{72}MgN_4O_5} = \frac{1\,mol\,Mg}{1\,mol\,C_{55}H_{72}MgN_4O_5} \times \frac{24.305\,g\,Mg}{1\,mol\,Mg} = \frac{24.305\,g\,Mg}{mol\,C_{55}H_{72}MgN_4O_5}$$

$$\%Mg = \frac{24.305\,g\,Mg}{893.509\,g\,C_{55}H_{72}MgN_4O_5} \times 100\% = 2.7202\%\,Mg$$

27. **(M)** Oxide with the largest %Cr will have the largest number of moles of Cr per mole of oxygen.

CrO: $\dfrac{1\,mol\,Cr}{1\,mol\,O} = 1\,mol\,Cr/mol\,O$ $\qquad$ Cr_2O_3: $\dfrac{2\,mol\,Cr}{3\,mol\,O} = 0.667\,mol\,Cr/mol\,O$

CrO_2: $\dfrac{1\,mol\,Cr}{2\,mol\,O} = 0.500\,mol\,Cr/mol\,O$ $\qquad$ CrO_3: $\dfrac{1\,mol\,Cr}{3\,mol\,O} = 0.333\,mol\,Cr/mol\,O$

Arranged in order of increasing %Cr: $CrO_3 < CrO_2 < Cr_2O_3 < CrO$

Chemical Formulas from Percent Composition

29. **(M)** SO_3 (40.05% S) and S_2O (80.0% S) (2 O atoms $\approx$ 1 S atom in terms of atomic masses) Note the molar masses are quite close (within 0.05 g/mol).

31. **(M)** Determine the % oxygen by difference.
$\%O = 100.00\% - 45.27\%\,C - 9.50\%\,H = 45.23\%\,O$
The following calculations are based on a 100.00 g sample.

$$mol\,O = 45.23\,g \times \frac{1\,mol\,O}{15.999\,g\,O} = 2.827\,mol\,O \quad \div 2.827 \rightarrow 1.000\,mol\,O$$

$$mol\,C = 45.27\,g\,C \times \frac{1\,mol\,C}{12.011\,g\,C} = 3.769\,mol\,C \quad \div 2.827 \rightarrow 1.333\,mol\,C$$

$$mol\,H = 9.50\,g\,H \times \frac{1\,mol\,H}{1.008\,g\,H} = 9.42\,mol\,H \quad \div 2.827 \rightarrow 3.33\,mol\,H$$

Multiply all amounts by 3 to obtain integers. Empirical formula is $C_4H_{10}O_3$.

33. **(M)**

(a)
$$74.01g\ C\times\frac{1\ mol\ C}{12.011\ g\ C}=6.162\ mol\ C \qquad \div1.298\rightarrow4.747\ mol\ C$$

$$5.23\ g\ H\times\frac{1\ mol\ H}{1.008\ g\ H}=5.19\ mol\ H \qquad \div1.298\rightarrow4.00\ mol\ H$$

$$20.76\ g\ O\times\frac{1\ mol\ O}{15.999\ g\ O}=1.298\ mol\ O \qquad \div1.298\rightarrow1.000\ mol\ O$$

Multiply each of the mole numbers by 4 to obtain an empirical formula of $C_{19}H_{16}O_4$.

(b)
$$39.98g\ C\times\frac{1\ mol\ C}{12.011g\ C}=3.3286\ mol\ C \quad \div0.7399\rightarrow4.499\ mol\ C$$

$$3.73g\ H\times\frac{1\ mol\ H}{1.008\ g\ H}=3.70\ mol\ H \quad \div0.7399\rightarrow5.00\ mol\ H$$

$$20.73g\ N\times\frac{1\ mol\ N}{14.007\ g\ N}=1.480\ mol\ N \quad \div0.7399\rightarrow2.000\ mol\ N$$

$$11.84g\ O\times\frac{1\ mol\ O}{15.999\ g\ O}=0.7400\ mol\ O \div0.7399\rightarrow1.000\ mol\ O$$

$$23.72g\ S\times\frac{1\ mol\ S}{32.06\ g\ S}=0.7399\ mol\ S \quad \div0.7399\rightarrow1.000\ mol\ S$$

Multiply by 2 to obtain the empirical formula

$C_9H_{10}N_4O_2S_2$

35. Convert each percentage into the mass in 100.00 g, and then to the moles of that element.

$$94.34\ g\ C\times\frac{1\ mol\ C}{12.011g\ C}=7.854\ mol\ C \qquad \div5.62=1.40\ mol\ C\times5=7.00$$

$$5.66\ g\ H\times\frac{1\ mol\ H}{1.008\ g\ H}=5.62\ mol\ H \qquad \div5.62=1.00\ mol\ H\times5=5.00$$

Multiply by 5 to achieve whole number ratios. The empirical formula is C_5H_7, and the formula mass $[(7\times12.011\ g\ C)+(5\times1.008\ g\ H)]=89.117\ u$. Since this empirical molar mass is one-half of the 178 u, the correct molecular mass, the molecular formula must be twice the empirical formula. Molecular formula: $C_{14}H_{10}$

37. **(M)** Determine the mass of oxygen by difference. Then convert all masses to amounts in moles. oxygen mass $=100.00g-73.27\ g\ C-3.84\ g\ H-10.68\ g\ N=12.21g\ O$

$$amount\ C=73.27g\ C\times\frac{1mol\ C}{12.011\ g\ C}=6.100\ mol\ C \qquad \div0.7625 \qquad \rightarrow8.000\ mol\ C$$

$$amount\ H=3.84g\ H\times\frac{1\ mol\ H}{1.008\ g\ H}=3.81\ mol\ H \qquad \div0.7625\rightarrow5.00\ mol\ H$$

$$\text{amount N} = 10.68\,\text{g N} \times \frac{1\,\text{mol N}}{14.007\,\text{g N}} = 0.7625\,\text{mol N} \qquad \div 0.7625 \qquad \rightarrow 1.000\,\text{mol N}$$

$$\text{amount O} = 12.21\,\text{g O} \times \frac{1\,\text{mol O}}{15.999\,\text{g O}} = 0.7632\,\text{mol O} \qquad \div 0.7625 \qquad \rightarrow 1.001\,\text{mol O}$$

The empirical formula is C_8H_5NO, which has an empirical mass of 131 u. This is almost exactly half the molecular mass of 262.3 u. Thus, the molecular formula is twice the empirical formula and is $C_{16}H_{10}N_2O_2$.

39. **(M)** The molar mass of element X has the units of grams per mole. We can determine the amount, in moles of Cl, and convert that to the amount of X, equivalent to 25.0 g of X.

$$\text{molar mass} = \frac{25.0\,\text{g X}}{75.0\,\text{g Cl}} \times \frac{35.45\,\text{g Cl}}{1\,\text{mol Cl}} \times \frac{4\,\text{mol Cl}}{1\,\text{mol X}} = \frac{47.3\,\text{g X}}{1\,\text{mol X}}$$

The atomic mass is 47.3 u. This atomic mass is close to that of the element titanium, which therefore is identified as element X.

41. **(M)** Consider 100 g of chlorophyll, which contains 2.72 g of Mg. To answer this problem, we must take note of the fact that 1 mole of chlorophyll contains 1 mole of Mg.

$$\frac{100\,\text{g chlorophyll}}{2.72\,\text{g Mg}} \times \frac{24.305\,\text{g Mg}}{1\,\text{mol Mg}} \times \frac{1\,\text{mol Mg}}{1\,\text{mol chlorophyll}} = 894\,\text{g mol}^{-1}$$

Therefore, the molecular mass of chlorophyll is 894 u.

Combustion Analysis

43. **(M)**
   (a) First we determine the mass of carbon and of hydrogen present in the sample. Remember that a hydrocarbon contains only hydrogen and carbon.

$$0.6260\,\text{g CO}_2 \times \frac{1\,\text{mol CO}_2}{44.009\,\text{g CO}_2} \times \frac{1\,\text{mol C}}{1\,\text{mol CO}_2} = 0.01422\,\text{mol C} \times \frac{12.011\,\text{g C}}{1\,\text{mol C}} = 0.1708\,\text{g C}$$

$$0.1602\,\text{g H}_2\text{O} \times \frac{1\,\text{mol H}_2\text{O}}{18.015\,\text{g H}_2\text{O}} \times \frac{2\,\text{mol H}}{1\,\text{mol H}_2\text{O}} = 0.01779\,\text{mol H} \times \frac{1.008\,\text{g H}}{1\,\text{mol H}} = 0.01793\,\text{g H}$$

Then the % C and % H are found.

$$\%\,\text{C} = \frac{0.1708}{0.1888\,\text{g cmpd}} \times 100\% = 90.47\%\,\text{C} \qquad \%\,\text{H} = \frac{0.01793\,\text{g H}}{0.1888\,\text{g cmpd}} \times 100\% = 9.497\%\,\text{H}$$

   (b) Use the moles of C and H from part (a), and divide both by the smallest value, namely 0.01422 mol. Thus $0.01422\,\text{mol C} \div 0.01422\,\text{mol} = 1.000\,\text{mol H};$
   $0.01779\,\text{mol H} \div 0.01422\,\text{mol} = 1.250\,\text{mol H}.$
   The empirical formula is obtained by multiplying these mole numbers by 4. It is C_4H_5.

   (c) The molar mass of the empirical formula C_4H_5 is $[4 \times 12.011 + 5 \times 1.008]$ = 53.084 g/mol. This value is 1/2 of the actual molar mass. The molecular formula is twice the empirical formula. ∴ Molecular formula: C_8H_{10}.

45. (M) First, determine the mass of carbon and hydrogen present in the sample.

$$0.458 \text{g CO}_2 \times \frac{1 \text{mol CO}_2}{44.01 \text{g CO}_2} \times \frac{1 \text{mol C}}{1 \text{mol CO}_2} = 0.0104 \text{mol C} \times \frac{12.011 \text{ g C}}{1 \text{mol C}} = 0.125 \text{ g C}$$

$$0.374 \text{g H}_2\text{O} \times \frac{1 \text{mol H}_2\text{O}}{18.015 \text{g H}_2\text{O}} \times \frac{2 \text{mol H}}{1 \text{mol H}_2\text{O}} = 0.0415 \text{mol H} \times \frac{1.008 \text{g H}}{1 \text{mol H}} = 0.0418 \text{ g H}$$

Then, the mass of N that this sample would have produced is determined.
(Note that this is also the mass of N_2 produced in the reaction.)

$$0.226 \text{g N}_2 \times \frac{0.312 \text{g 1st sample}}{0.486 \text{g 2nd sample}} = 0.145 \text{g N}_2$$

From which we can calculate the mass of N in the sample.

$$0.145 \text{ g N}_2 \times \frac{1 \text{mol N}_2}{28.014 \text{g N}_2} \times \frac{2 \text{mol N}}{1 \text{mol N}_2} \times \frac{14.007 \text{g N}}{1 \text{mol N}} = 0.145 \text{ g N}$$

We may alternatively determine the mass of N by difference:
$0.312 \text{ g} - 0.125 \text{ g C} - 0.0418 \text{ g H} = 0.145 \text{ g N}$

Then, we can calculate the relative number of moles of each element.

$$0.145 \text{g N} \times \frac{1 \text{mol N}}{14.007 \text{g N}} =$$

0.0104 mol N $\div 0.0104 \rightarrow 1.00$ mol N

0.0104 mol C $\div 0.0104 \rightarrow 1.00$ mol C

0.0415 mol H $\div 0.0104 \rightarrow 4.01$ mol H

Thus, the empirical formula is CH_4N

47. (M) Each mole of CO_2 is produced from a mole of C. Therefore, the compound with the largest number of moles of C per mole of the compound will produce the largest amount of CO_2 and, thus, also the largest mass of CO_2. Of the compounds listed, namely CH_4, C_2H_5OH, $C_{10}H_8$, and C_6H_5OH, $C_{10}H_8$ has the largest number of moles of C per mole of the compound and will produce the greatest mass of CO_2 per mole on complete combustion.

49. (M) The molecular formula for $CH_3CH(OH)CH_2CH_3$ is $C_4H_{10}O$. Here we will use the fact that $C_4H_{10}O$ has a molar mass of 74.123 g/mol to calculate the masses of CO_2 and H_2O:

Mass of CO_2:
Conversion pathway approach:

$$1.562 \text{ g C}_4\text{H}_{10}\text{O} \times \frac{1 \text{ mol C}_4\text{H}_{10}\text{O}}{74.123 \text{ g C}_4\text{H}_{10}\text{O}} \times \frac{4 \text{ mol C}}{1 \text{ mol C}_4\text{H}_{10}\text{O}} \times \frac{1 \text{ mol CO}_2}{1 \text{ mol C}} \times \frac{44.009 \text{ g CO}_2}{1 \text{ mol CO}_2} = 3.710 \text{ g CO}_2$$

Stepwise approach:

$$1.562 \text{ g C}_4\text{H}_{10}\text{O} \times \frac{1 \text{ mol C}_4\text{H}_{10}\text{O}}{74.123 \text{ g C}_4\text{H}_{10}\text{O}} = 0.02107 \text{ mol C}_4\text{H}_{10}\text{O} \times \frac{4 \text{ mol C}}{1 \text{ mol C}_4\text{H}_{10}\text{O}} = 0.08429 \text{ mol C}$$

$$0.08429 \text{ mol C} \times \frac{1 \text{ mol CO}_2}{1 \text{ mol C}} = 0.08429 \text{ mol CO}_2 \times \frac{44.009 \text{ g CO}_2}{1 \text{ mol CO}_2} = 3.710 \text{ g CO}_2$$

Mass of H_2O:
Conversion pathway approach:

$$1.562 \text{ g C}_4\text{H}_{10}\text{O} \times \frac{1 \text{ mol C}_4\text{H}_{10}\text{O}}{74.123 \text{ g C}_4\text{H}_{10}\text{O}} \times \frac{10 \text{ mol H}}{1 \text{ mol C}_4\text{H}_{10}\text{O}} \times \frac{1 \text{ mol H}_2\text{O}}{2 \text{ mol H}} \times \frac{18.015 \text{ g H}_2\text{O}}{1 \text{ mol H}_2\text{O}} = 1.898 \text{ g H}_2\text{O}$$

Stepwise approach:

$$1.562 \text{ g C}_4\text{H}_{10}\text{O} \times \frac{1 \text{ mol C}_4\text{H}_{10}\text{O}}{74.123 \text{ g C}_4\text{H}_{10}\text{O}} = 0.02107 \text{ mol C}_4\text{H}_{10}\text{O}$$

$$0.02107 \text{ mol C}_4\text{H}_{10}\text{O} \times \frac{10 \text{ mol H}}{1 \text{ mol C}_4\text{H}_{10}\text{O}} = 0.2107 \text{ mol H}$$

$$0.2107 \text{ mol H} \times \frac{1 \text{ mol H}_2\text{O}}{2 \text{ mol H}} = 0.1054 \text{ mol H}_2\text{O} \times \frac{18.015 \text{ g H}_2\text{O}}{1 \text{ mol H}_2\text{O}} = 1.898 \text{ g H}_2\text{O}$$

Oxidation States

51. **(E)** The oxidation state (O.S.) is given first, followed by the explanation for its assignment.

(a) $C = -4$ in CH_4 — H has an oxidation state of $+1$ in its nonmetal compounds. (Remember that the sum of the oxidation states in a neutral compound equals 0.)

(b) $S = +4$ in SF_4 — F has O.S. $= -1$ in its compounds.

(c) $O = -1$ in Na_2O_2 — Na has O.S. $= +1$ in its compounds.

(d) $C = 0$ in $C_2H_3O_2^-$ — H has O.S. $= +1$ in its nonmetal compounds; that of $O = -2$ (usually). (Remember that the sum of the oxidation states in a polyatomic ion equals the charge on that ion.)

(e) $Fe = +6$ in FeO_4^{2-} — O has O.S. $= -2$ in most of its compounds (especially metal containing compounds).

53. **(E)** Remember that the oxidation state of oxygen is usually -2 in its compounds. Cr^{3+} and O^{2-} form Cr_2O_3, chromium(III) oxide. Cr^{4+} and O^{2-} form CrO_2, chromium(IV) oxide. Cr^{6+} and O^{2-} form CrO_3, chromium(VI) oxide.

55. **(E)**

(a) $O = +2$ in OF_2 F has an oxidation state of -1 in its compounds.

(b) $O = +1$ in O_2F_2 F has O.S. $= -1$ in its compounds.

(c) $O = \dfrac{-1}{2}$ in CsO_2 Cs has O.S. $= +1$ in its compounds.

(d) $O = -1$ in BaO_2 Ba has O.S. $= +2$ in its compounds.

Nomenclature

57. **(E)**

(a)	SrO	strontium oxide	(b)	ZnS	zinc sulfide
(c)	K_2CrO_4	potassium chromate	(d)	Cs_2SO_4	cesium sulfate
(e)	Cr_2O_3	chromium(III) oxide	(f)	$Fe_2(SO_4)_3$	iron(III) sulfate
(g)	$Mg(HCO_3)_2$	magnesium hydrogen carbonate or magnesium bicarbonate	(h)	$(NH_4)_2HPO_4$	ammonium hydrogen phosphate
(i)	$Ca(HSO_3)_2$	calcium hydrogen sulfite	(j)	$Cu(OH)_2$	copper(II) hydroxide
(k)	HNO_3	nitric acid	(l)	$KClO_4$	potassium perchlorate
(m)	$HBrO_3$	bromic acid	(n)	H_3PO_3	phosphorous acid

59. **(E)**

(a)	CS_2	carbon disulfide	(b)	SiF_4	silicon tetrafluoride
(c)	ClF_5	chlorine pentafluoride	(d)	N_2O_5	dinitrogen pentoxide
(e)	SF_6	sulfur hexafluoride	(f)	I_2Cl_6	diiodine hexachloride

61. **(E)**

(a)	$Al_2(SO_4)_3$	aluminum sulfate	(b)	$(NH_4)_2Cr_2O_7$	ammonium dichromate
(c)	SiF_4	silicon tetrafluoride	(d)	Fe_2O_3	iron(III) oxide
(e)	C_3S_2	tricarbon disulfide	(f)	$Co(NO_3)_2$	cobalt(II) nitrate
(g)	$Sr(NO_2)_2$	strontium nitrite	(h)	$HBr(aq)$	hydrobromic acid
(i)	HIO_3	iodic acid	(j)	PCl_2F_3	phosphorus dichloride trifluoride

63. (E)
   (a) Ti^{4+} and Cl^- produce $TiCl_4$
   (b) Fe^{3+} and SO_4^{2-} produce $Fe_2(SO_4)_3$
   (c) Cl^{7+} and O^{2-} produce Cl_2O_7
   (d) S^{7+} and O^{2-} produce $S_2O_8^{2-}$

65. (E)
   (a) $HClO_2$ chlorous acid
   (b) H_2SO_3 sulfurous acid
   (c) H_2Se hydroselenic acid
   (d) HNO_2 nitrous acid

67. (E)
   (a) OF_2 oxygen difluoride
   (b) XeF_2 xenon difluoride
   (c) $CuSO_3$ copper(II) sulfite
   (d) $(NH_4)_2HPO_4$ ammonium hydrogen phosphate

Both (c) and (d) are ionic compounds.

Hydrates

69. (E) The hydrate with the greatest mass percent H_2O is the one that gives the largest result for the number of moles of water in the hydrate's empirical formula, divided by the mass of one mole of the anhydrous salt for the hydrate.

$$\frac{5\,H_2O}{CuSO_4} = \frac{5\ mol\ H_2O}{159.6\ g} = 0.03133 \qquad \frac{6\,H_2O}{MgCl_2} = \frac{6\ mol\ H_2O}{95.2\ g} = 0.0630$$

$$\frac{18\,H_2O}{Cr_2(SO_4)_3} = \frac{18\ mol\ H_2O}{392.3\ g} = 0.04588 \qquad \frac{2\,H_2O}{LiC_2H_3O_2} = \frac{2\ mol\ H_2O}{66.0\ g} = 0.0303$$

The hydrate with the greatest % H_2O therefore is $MgCl_2 \cdot 6\,H_2O$

71. (M)
molar mass $CuSO_4 = 63.546\ g\ Cu + 32.066\ g\ S + (4 \times 15.9994\ g\ O) = 159.61\ g\ CuSO_4/mol$.
Note that each $CuSO_4$ will pick up 5 equivalents of H_2O to give $CuSO_4\cdot 5H_2O$.

Conversion pathway approach:

$$\text{mass of required } CuSO_4 = 12.6\ g\ H_2O \times \frac{1\ mol\ H_2O}{18.015\ g\ H_2O} \times \frac{1\ mol\ CuSO_4}{5\ mol\ H_2O} \times \frac{159.61\ g\ CuSO_4}{1\ mol\ CuSO_4}$$

$$= 22.3\ g\ CuSO_4 \text{ is the minimum amount required to remove all the water}$$

Stepwise approach:

$$12.6\ g\ H_2O \times \frac{1\ mol\ H_2O}{18.0153\ g\ H_2O} = 0.699\ mol\ H_2O \times \frac{1\ mol\ CuSO_4}{5\ mol\ H_2O} = 0.140\ mol\ CuSO_4$$

$$0.140\ mol\ CuSO_4 \times \frac{159.60\ g\ CuSO_4}{1\ mol\ CuSO_4} = 22.3\ g\ CuSO_4$$

$$= \text{is the minimum amount required to remove all the water}$$

73. **(M)** We start by converting to molar amounts for each element based on 100.0 g:

$$20.3 \text{ g Cu} \times \frac{1 \text{ mol Cu}}{63.546 \text{ g Cu}} = 0.319 \text{ mol Cu} \quad \div\, 0.319 \;\rightarrow\; 1.00 \text{ mol Cu}$$

$$8.95 \text{ g Si} \times \frac{1 \text{ mol Si}}{28.085 \text{ g Si}} = 0.319 \text{ mol Si} \qquad \div\, 0.319 \;\rightarrow\; 1.00 \text{ mol Si}$$

$$36.3 \text{ g F} \times \frac{1 \text{ mol F}}{18.998 \text{ g F}} = 1.91 \text{ mol F} \qquad \div\, 0.319 \;\rightarrow\; 5.99 \text{ mol F}$$

$$34.5 \text{ g H}_2\text{O} \times \frac{1 \text{ mol H}_2\text{O}}{18.015 \text{ g}} = 1.91\underline{5} \text{ mol H}_2\text{O} \;\div\, 0.319 \;\rightarrow\; 6.00 \text{ mol H}_2\text{O}$$

Thus the empirical formula for the hydrate is $CuSiF_6 \cdot 6\,H_2O$.

Organic Compounds and Organic Nomenclature

75. **(E)** Answer is (b), butan-2-ol is the most appropriate name for this molecule. It has a four carbon atom chain with a hydroxyl group on the carbon second from the end.

77. **(E)** Molecules (a), (b), (c), and (d) are structural isomers. They share a common formula, namely $C_5H_{12}O$, but have different molecular structures. Molecule (e) has a different chemical formula ($C_6H_{14}O$) and hence cannot be classified as an isomer. It should be pointed out that molecules (a) and (c) are identical as well as being isomers of (b).

79. **(E)**
(a) $CH_3(CH_2)_5CH_3$ (b) $CH_3CH_2CO_2H$
(c) $CH_3CH_2CH_2CH(CH_3)CH_2OH$ (d) CH_3CH_2F

81. **(M)**
(a)   methanol; CH_3OH Molecular mass = 32.04 u
(b)   2-chlorohexane; $CH_3(CH_2)_3CHClCH_3$ Molecular mass = 120.6 u
(c)   pentanoic acid; $CH_3(CH_2)_3CO_2H$ Molecular mass = 102.1 u
(d)   2-methylprooan-1-ol $CH_3CH(CH_3)CH_2OH$ Molecular mass = 74.12 u

INTEGRATIVE AND ADVANCED EXERCISES

83. (M)

molar mass $= (1 \times 6.94 \text{ g Li}) + (1 \times 26.982 \text{ g Al}) + (2 \times 28.05 \text{ g Si}) + (6 \times 15.999 \text{ g O}) = 186.02 \text{ g/mol}$

Conversion pathway approch:

$$\text{number of Li} - 6 \text{ atoms} = 518 \text{ g spodumene} \times \frac{1 \text{ mol spodumene}}{186.02 \text{ g spodumene}} \times \frac{1 \text{ mol Li}}{1 \text{ mol spodumene}} \times \frac{7.40 \text{ mol Li-6}}{100.00 \text{ mol total Li}}$$

$$\times \frac{6.022 \times 10^{23} \text{ Li-6 atoms}}{1 \text{ mol Li-6}} = 1.24 \times 10^{23} \text{ Li-6 atoms}$$

Stepwise approch:

$$518 \text{ g spodumene} \times \frac{1 \text{ mol spodumene}}{186.02 \text{ g spodumene}} = 2.78 \text{ mol spodumene}$$

$$2.78 \text{ mol spodumene} \times \frac{1 \text{ mol Li}}{1 \text{ mol spodumene}} = 2.78 \text{ mol Li}$$

$$2.78 \text{ mol Li} \times \frac{7.40 \text{ mol Li-6}}{100.00 \text{ mol total Li}} = 0.206 \text{ mol Li-6}$$

$$0.206 \text{ mol Li-6} \times \frac{6.022 \times 10^{23} \text{ Li-6 atoms}}{1 \text{ mol Li-6}} = 1.24 \times 10^{23} \text{ Li-6 atoms}$$

86. (M) First, we determine the amount of each mineral necessary to obtain 1 kg or 1000 g of boron.

$$1000 \text{ g B} \times \frac{1 \text{ mol B}}{10.81 \text{ g B}} \times \frac{1 \text{ mol Na}_2\text{B}_4\text{O}_7 \cdot 4 \text{ H}_2\text{O}}{4 \text{ mol B}} \times \frac{273.28 \text{ g Na}_2\text{B}_4\text{O}_7 \cdot 4 \text{ H}_2\text{O}}{1 \text{ mol Na}_2\text{B}_4\text{O}_7 \cdot 4 \text{ H}_2\text{O}}$$

$$= 6{,}319.\underline{5} \text{ g Na}_2\text{B}_4\text{O}_7 \cdot 4 \text{ H}_2\text{O}$$

$$1000 \text{ g B} \times \frac{1 \text{ mol B}}{10.81 \text{ g B}} \times \frac{1 \text{ mol Na}_2\text{B}_4\text{O}_7 \cdot 4 \text{ H}_2\text{O}}{4 \text{ mol B}} \times \frac{381.372 \text{ g Na}_2\text{B}_4\text{O}_7 \cdot 10 \text{ H}_2\text{O}}{1 \text{ mol Na}_2\text{B}_4\text{O}_7 \cdot 4 \text{ H}_2\text{O}}$$

$$= 8{,}819.\underline{1} \text{ g Na}_2\text{B}_4\text{O}_7 \cdot 10 \text{ H}_2\text{O}$$

The difference between these two masses is the required additional mass. Hence, $8819.\underline{1} \text{ g} - 6319.\underline{5} \text{ g} = 2499.\underline{6} \text{ g}$. Thus, an additional 2.500 kg mass is required.

89. (M) It is not possible to have less than 1 molecule of S_8. In order to determine whether it is possible to have 1.00×10^{-23} g of S_8, determine how many molecules that number is equivalent to.

$$1.00 \times 10^{-23} \text{ g } S_8 \times \frac{1 \text{ mol } S_8}{256.48 \text{ g } S_8} \times \frac{6.022 \times 10^{23} \text{ molecules } S_8}{1 \text{ mol } S_8} = 1.0235 \text{ molecules } S_8$$

Therefore it is **not** possible to have 1.00×10^{-23} g of S_8.

$$1 \text{ molecule } S_8 \times \frac{1 \text{ mol } S_8}{6.022 \times 10^{23} \text{ molecules } S_8} \times \frac{256.48 \text{ g } S}{1 \text{ mol } S_8} = 4.26 \times 10^{-22} \text{ g } S$$

$$4.26 \times 10^{-22} \text{ g } S \times \frac{1 \text{ yg } S}{10^{-24} \text{ g } S} = 426 \text{ yoctograms } S$$

91. (M) We determine the masses of CO_2 and H_2O produced by burning the C_3H_8.

$$\text{mass}_{CO_2} = 6.00 \text{ g } C_3H_8 \times \frac{1 \text{ mol } C_3H_8}{44.097 \text{ g } C_3H_8} \times \frac{3 \text{ mol } C}{1 \text{ mol } C_3H_8} \times \frac{1 \text{ mol } CO_2}{1 \text{ mol } C} \times \frac{44.01 \text{ g } CO_2}{1 \text{ mol } CO_2}$$

$$= 17.96 \text{ g } CO_2$$

$$\text{mass}_{H_2O} = 6.00 \text{ g } C_3H_8 \times \frac{1 \text{ mol } C_3H_8}{44.097 \text{ g } C_3H_8} \times \frac{8 \text{ mol } H}{1 \text{ mol } C_3H_8} \times \frac{1 \text{ mol } H_2O}{2 \text{ mol } H} \times \frac{18.015 \text{ g } H_2O}{1 \text{ mol } H_2O}$$

$$= 9.805 \text{ g } H_2O$$

Then, from the masses of CO_2 and H_2O in the unknown compound, we determine the amounts of C and H in that compound and finally its empirical formula.

$$\text{amount C} = (29.0 - 17.96) \text{ g } CO_2 \times \frac{1 \text{ mol } CO_2}{44.01 \text{ g } CO_2} \times \frac{1 \text{ mol } C}{1 \text{ mol } CO_2} = 0.251 \text{ mol C}$$

$$\text{amount H} = (18.8 - 9.805) \text{ g } H_2O \times \frac{1 \text{ mol } H_2O}{18.015 \text{ g } H_2O} \times \frac{2 \text{ mol } H}{1 \text{ mol } H_2O} = 0.9986 \text{ mol H}$$

The empirical formula of the unknown compound is CH_4. The C:H ratio is $0.9986/0.251 = 3.98$. The molecular formula can be calculated by knowing that we have 0.251 moles, which accounts for the 4.00 g of hydrocarbon (40% of 10.0 g). This gives a molar mass of $4.00 \div 0.251 = 15.9$ g/mol. This is nearly the same as the molar mass of the empirical formula CH_4 (16.04 g/mol).

93. **(M)** Since the compound is composed of H_2SO_4 and H_2O, we will need to determine the percent composition of both H_2SO_4 and water.

$$\% \ H_2SO_4 = \frac{\# \ grams \ H_2SO_4}{total \ mass \ in \ grams} \times 100\%$$

$$\# \ g \ H_2SO_4 = 65.2 \ g \ (NH_4)_2SO_4 \times \frac{1 \ mol \ (NH_4)_2SO_4}{132.15 \ g \ (NH_4)_2SO_4} \times \frac{1 \ mol \ H_2SO_4}{1 \ mol \ (NH_4)_2SO_4} \times \frac{98.08 \ g \ H_2SO_4}{1 \ mol \ H_2SO_4} = 48.4 \ g$$

$$total \ mass = 32.0 \ mL \ mixture \times \frac{1.78 \ g \ mixture}{1 \ mL \ mixture} = 57.0 \ g \ mixture$$

$$\% \ H_2SO_4 = \frac{48.4 \ g}{57.0 \ g} \times 100\% = 85.0\%$$

$$\% \ H_2O = 100.0 - 85.0 = 15.0\%$$

95. **(D)**

$$9.0 \times 10^{-4} \ \frac{\mu \ mol \ C_2H_6S}{m^3 \ air} \times \frac{1 \times 10^{-6} \ mol \ C_2H_6S}{1 \ \mu \ mol \ C_2H_6S} \times \frac{62.13 \ g \ C_2H_6S}{1 \ mol \ C_2H_6S} \times \frac{1 \ m^3}{(100)^3 \ cm^3} \times$$

$$\frac{1 \ cm^3}{1 \ mL} \times \frac{1000 \ mL}{1 \ L} \times \frac{1 \ L \ air}{1.2 \ g \ air} = 4.7 \times 10^{-11} \ \frac{g \ C_2H_6S}{g \ air}$$

$$4.7 \times 10^{-11} \ \frac{g \ C_2H_6S}{g \ air} \times \frac{1 \times 10^9 \ g}{1 \ billion \ grams} = 0.0466 \ ppb = 0.05 \ ppb$$

96. **(D)**
(a) If we have one mole of entities, then we must have 0.7808 mol N_2, 0.2095 mol O_2, 0.0093 mol Ar, and 0.0004 mol CO_2.

$$0.7808 \ mol \ N_2 \times \frac{28.02 \ g \ N_2}{1 \ mol \ N_2} = 21.88 \ g \ N_2$$

$$0.2095 \ mol \ O_2 \times \frac{32.00 \ g \ O_2}{1 \ mol \ O_2} = 6.704 \ g \ O_2$$

$$0.0004 \ mol \ CO_2 \times \frac{44.01 \ g \ CO_2}{1 \ mol \ CO_2} = 0.0176 \ g \ CO_2$$

$$0.0093 \ mol \ Ar \times \frac{39.948 \ g \ Ar}{1 \ mol \ Ar} = 0.3715 \ g \ Ar$$

mass of air sample $= 21.88 g \ N_2 + 6.704 \ g \ O_2 + 0.0176 \ g \ CO_2 + 0.3715 \ g \ Ar = 28.97 \ g$

(b) $1 \ m^3 \times \frac{(100)^3 \ cm^3}{1 \ m^3} \times \frac{1 \ mL}{1 \ cm^3} \times \frac{1 \ L}{1000 \ mL} \times \frac{1.2 \ g}{1 \ L} = 1200 \ g \ dry \ air$

$$1200 \ g \times \frac{1 \ mol \ entities}{28.97 \ g} \times \frac{1.14 \times 10^{-4}}{100} = 4.72 \times 10^{-5} \ mol \ Kr$$

$$4.72 \times 10^{-5} \ mol \ Kr \times \frac{83.80 \ g}{1 \ mol} = 3.96 \times 10^{-3} \ g \ Kr = 4.0 \ mg \ Kr$$

99. (M) $\text{mass } SO_4^{2-} = 1.511 \text{ g } BaSO_4 \times \dfrac{1 \text{ mol } BaSO_4}{233.39 \text{ g } BaSO_4} \times \dfrac{1 \text{ mol } SO_4^{2-}}{1 \text{ mol } BaSO_4} = 0.006474 \text{ mol } SO_4^{2-}$

$$0.006474 \text{ mol } SO_4^{2-} \times \dfrac{96.064 \text{ g } SO_4^{2-}}{1 \text{ mol } SO_4^{2-}} = 0.6219 \text{ g } SO_4^{2-}$$

$$\text{amount M} = 0.006474 \text{ mol } SO_4^{2-} \times \dfrac{2 \text{ mol } M^{3+}}{3 \text{ mol } SO_4^{2-}} = 0.004316 \text{ mol } M^{3+}$$

$$\text{mass M} = 0.738 \text{ g } M_2(SO_4)_2 - 0.6219 \text{ g } SO_4^{2-} = 0.116 \text{ g M}$$

$$\text{atomic mass of M} = \dfrac{0.116 \text{ g M}}{0.004316 \text{ mol M}} = 26.9 \text{ g M/mol} \qquad \text{M is the element aluminum.}$$

103. (M) If we determine the mass of anhydrous $ZnSO_4$ in the hydrate, we then can determine the mass of water, and the formula of the hydrate.

$$\text{mass } ZnSO_4 = 0.8223 \text{ g } BaSO_4 \times \dfrac{1 \text{ mol } BaSO_4}{233.386 \text{ g}} \times \dfrac{1 \text{ mol } ZnSO_4}{1 \text{ mol } BaSO_4} \times \dfrac{161.454 \text{ g } ZnSO_4}{1 \text{ mol } ZnSO_4}$$

$$= 0.5688 \text{ g } ZnSO_4$$

The water present in the hydrate is obtained by difference.

mass $H_2O = 1.013$ g hydrate $- 0.5688$ g $ZnSO_4 = 0.444$ g H_2O

The hydrate's formula is determined by a method similar to that for obtaining an empirical formula.

$$\text{amt. } ZnSO_4 = 0.5688 \text{ g} \times \dfrac{1 \text{ mol } ZnSO_4}{161.436 \text{ mol } ZnSO_4} = 0.003523 \text{ mol } ZnSO_4 \div 0.003523 \longrightarrow 1.00 \text{ mol } ZnSO_4$$

$$\text{amt. } H_2O = 0.444 \text{ g} \times \dfrac{1 \text{ mol } H_2O}{18.015 \text{ g } H_2O} = 0.02465 \text{ mol } H_2O \div 0.003523 \longrightarrow 7.00 \text{ mol } H_2O$$

Thus, the formula of the hydrate is $ZnSO_4 \cdot 7\,H_2O$.

106. (M) First find the mass of carbon, hydrogen, chlorine, and oxygen. From the molar ratios, we determine the molecular formula.

$$2.094 \text{ g } CO_2 \times \dfrac{1 \text{ mol } CO_2}{44.009 \text{ g } CO_2} \times \dfrac{1 \text{ mol C}}{1 \text{ mol } CO_2} = 0.04759 \text{ mol C} \times \dfrac{12.011 \text{ g C}}{1 \text{ mol C}} = 0.5716 \text{ g C}$$

$$0.286 \text{ g } H_2O \times \dfrac{1 \text{ mol } H_2O}{18.015 \text{ g } H_2O} \times \dfrac{2 \text{ mol H}}{1 \text{ mol } H_2O} = 0.03175 \text{ mol H} \times \dfrac{1.008 \text{ g H}}{1 \text{ mol H}} = 0.0320 \text{ g H}$$

$$\text{moles of chlorine} = \dfrac{\text{mol C}}{2} = \dfrac{0.04759}{2} = 0.02380 \text{ mol Cl}$$

$$\text{mass of Cl} = 0.02380 \text{ mol Cl} \times \dfrac{35.45 \text{ g Cl}}{1 \text{ mol Cl}} = 0.8436 \text{ g Cl}$$

mass of oxygen obtained by difference: $1.510 \text{ g} - 0.8436 \text{ g} - 0.5716 \text{ g} - 0.0320 \text{ g} = 0.063 \text{ g O}$

moles of oxygen $= 0.063 \text{ g O} \times \dfrac{1 \text{ mol O}}{15.999 \text{ g O}} = 0.00394 \text{ mol O}$

Divide the number of moles of each element by 0.00394 to give an empirical formula of $C_{12.1} H_{8.06} Cl_{6.04} O_{1.00}$ owing to the fact that the oxygen mass is obtained by difference, and it has only two significant digits and thus a higher degree of uncertainty.

The empirical formula is $C_{12}H_8Cl_6O$, which with a molecular mass of 381 u has the same molecular mass as the molecular formula. Hence, this empirical formula is also the molecular formula.

110. (M)

Calculate the mass of chlorine: $0.244 \text{ L} \times 2.898 \text{ g/L} = 0.707 \text{ g chlorine}$

Calculate the mass of iodine: $1.553 \text{ g} - 0.707 \text{ g} = 0.846 \text{ g iodine}$

Calculate the moles of chlorine: $0.707 \text{ g}/35.45 \text{ g/mol} = 0.0199 \text{ mol chlorine}$

Calculate the moles of iodine: $0.846 \text{ g}/126.90 \text{ g/mol} = 0.00667 \text{ mol iodine}$

Calculate the mole ratio: $0.0199{:}0.00667 = 1{:}2.98 \approx 1{:}3$

Calculate the empirical molar mass: $(126.90 + 3 \times 35.45) \text{ g/mol} = 233.25 \text{ g/mol}$

Because $467/233.25 \approx 2$, the molecular formula is I_2Cl_6.

111. (M)

The coating is a compound of copper and iodine. The mass of iodine reacting with the copper strip is $0.733 \text{ g} - 0.725 \text{ g} = 0.008 \text{ g}$.

The mass of the copper reacted is $0.725 \text{ g} - 0.721 \text{ g} = 0.004 \text{ g}$.

The masses of copper and iodine that reacted are known with very low precision (only one significant figure). Therefore, in the following calculations, we are justified in rounding the molar masses of I and Cu (126.90 and 63.546 g/mol, respectively) to two significant figures and the final result to one significant figure.

The amount of iodine is $0.008 \text{ g} \times \dfrac{1}{130 \text{ g mol}^{-1}} = 6 \times 10^{-5} \text{ mol}$

The amount of copper is $0.004 \text{ g} \times \dfrac{1}{64 \text{ g mol}^{-1}} = 6 \times 10^{-5} \text{ mol}$

The compound contains equal amounts (moles) of Cu and I. Therefore, the empirical formula is CuI.

FEATURE PROBLEMS

112. (D)

(a) "5-10-5" fertilizer contains 5.00 g N (that is, 5.00% N), 10.00 g P_2O_5, and 5.00 g K_2O in 100.00 g fertilizer. We convert the last two numbers into masses of the two elements.

(1) $\% P = 10.00\% \, P_2O_5 \times \dfrac{1\,mol\,P_2O_5}{141.9\,g\,P_2O_5} \times \dfrac{2\,mol\,P}{1\,mol\,P_2O_5} \times \dfrac{30.97\,g\,P}{1\,mol\,P} = 4.37\% \, P$

(2) $\% K = 5.00\% \, K_2O \times \dfrac{1\,mol\,K_2O}{94.20\,g\,K_2O} \times \dfrac{2\,mol\,K}{1\,mol\,K_2O} \times \dfrac{39.10\,g\,K}{1\,mol\,K} = 4.15\% \, K$

(b) First, we determine %P and then convert it to $\% P_2O_5$, given that 10.0% P_2O_5 is equivalent to 4.37% P.

(1) $\% \, P_2O_5 = \dfrac{2\,mol\,P}{1\,mol\,Ca(H_2PO_4)_2} \times \dfrac{30.97\,g\,P}{1\,mol\,P} \times \dfrac{1\,mol\,Ca(H_2PO_4)_2}{234.05\,g\,Ca(H_2PO_4)_2} \times 100\%$

$\times \dfrac{10.0\% \, P_2O_5}{4.37\% \, P} = 60.6\% \, P_2O_5$

(2) $\% \, P_2O_5 = \dfrac{1\,mol\,P}{1\,mol(NH_4)_2HPO_4} \times \dfrac{30.97\,g\,P}{1\,mol\,P} \times \dfrac{1\,mol(NH_4)_2HPO_4}{132.06\,g(NH_4)_2HPO_4} \times 100\%$

$\times \dfrac{10.0\% \, P_2O_5}{4.37\% \, P} = 53.7\% \, P_2O_5$

(c) If the mass ratio of $(NH_4)_2HPO_4$ to KCl is set at 5.00:1.00, then for every 5.00 g of $(NH_4)_2HPO_4$ in the mixture there must be 1.00 g of KCl. Let's start by finding the %N, %P, and %K for the fertilizer mixture.

$\%N\,(by\,mass) = \dfrac{2\,mol\,N}{1\,mol(NH_4)_2HPO_4} \times \dfrac{1\,mol(NH_4)_2HPO_4}{132.06\,g(NH_4)_2PO_4} \times \dfrac{14.007\,g\,N}{1\,mol\,N} \times \dfrac{5.00\,g(NH_4)_2HPO_4}{6.00\,g\,mixture} \times 100\%$

$= 17.7\% \, N$

$\%P\,(by\,mass) = \dfrac{1\,mol\,P}{1\,mol\,(NH_4)_2HPO_4} \times \dfrac{1\,mol\,(NH_4)_2HPO_4}{132.06\,g\,(NH_4)_2HPO_4} \times \dfrac{30.9738\,g\,P}{1\,mol\,P} \times \dfrac{5.00\,g\,(NH_4)_2HPO_4}{6.00\,g\,of\,mixture} \times 100\%$

$= 19.5\% \, P$

$\%K\,(by\,mass) = \dfrac{1\,mol\,K}{1\,mol\,KCl} \times \dfrac{1\,mol\,KCl}{74.55\,g\,KCl} \times \dfrac{39.0983\,g\,K}{1\,mol\,K} \times \dfrac{1.00\,g\,KCl}{6.00\,g\,mixture} \times 100\%$

$= 8.74\% \, K$

Next, we convert %P to %P_2O_5 and %K to %K_2O.

$\%P_2O_5 = 19.5\% \, P \times \dfrac{10.0\,\% \, P_2O_5}{4.37\,\% \, P} = 44.\underline{6}\% \, P_2O_5$

$\%K_2O = 8.\underline{74}\% \, K \times \dfrac{5.00\% \, K_2O}{4.15\% \, K} = 1\underline{0}.5\% \, K_2O$

Thus, the combination of 5.00 g $(NH_4)_2HPO_4$ with 1.00 g KCl affords a "17.7-44.6-10.5" fertilizer, that is, 17.7% N, a percentage of phosphorus expressed as 44.6% P_2O_5, and a percentage of potassium expressed as 10.5% K_2O.

(d) A "5-10-5" fertilizer must possess the mass ratio 5.00 g N: 4.37 g P: 4.15 g K per 100 g of fertilizer. Thus a "5-10-5" fertilizer requires an N:P relative mass ratio of 5.00 g N:4.37 g P = 1.00 g N:0.874 g P. Note specifically that the fertilizer has a somewhat *greater* mass of N than of P.

If all of the N and P in the fertilizer comes solely from $(NH_4)_2HPO_4$, then the atom ratio of N relative to P will remain fixed at 2 N:1 P. Whether or not an inert non-fertilizing filler is present in the mix is immaterial. The relative N:P mass ratio is (2×14.01) g N:30.97 g P, that is, 0.905 g N:1.00 g P. Note specifically that $(NH_4)_2HPO_4$ has a somewhat *lesser* mass of N than of P. Clearly, it is impossible to make a "5-10-5" fertilizer if the only fertilizing components are $(NH_4)_2HPO_4$ and KCl.

114. (D)

(a) The formula for stearic acid, obtained from the molecular model, is $CH_3(CH_2)_{16}CO_2H$. The number of moles of stearic acid in 10.0 grams is

$$= 10.0 \text{ g stearic acid} \times \frac{1 \text{ mol stearic acid}}{284.48 \text{ g stearic acid}} = 3.51\underline{5} \times 10^{-2} \text{ mol of stearic acid.}$$

The layer of stearic acid is one molecule thick. According to the figure provided with the question, each stearic acid molecule has a cross-sectional area of ~0.22 nm^2. In order to find the stearic acid coverage in square meters, we must multiply the total number of stearic acid molecules by the cross-sectional area for an individual stearic acid molecule. The number of stearic acid molecules is:

$$= 3.51\underline{5} \times 10^{-2} \text{ mol of stearic acid} \times \frac{6.022 \times 10^{23} \text{ molecules}}{1 \text{ mol of stearic acid}} = 2.11\underline{7} \times 10^{22} \text{ molecules}$$

$$\text{area in } m^2 = 2.11\underline{7} \times 10^{22} \text{ molecules of stearic acid} \times \frac{0.22 \text{ nm}^2}{\text{molecule}} \times \frac{(1 \text{ m})^2}{(1 \times 10^9 \text{ nm})^2}$$

The area in m^2 = 4657 m^2 or 4.7×10^3 m^2 (with correct number of sig. fig.)

(b) The density for stearic acid is 0.85 g cm^{-3}. Thus, 0.85 grams of stearic acid occupies 1 cm^3. Find the number of moles of stearic acid in 0.85 g of stearic acid

$$= 0.85 \text{ grams of stearic acid} \times \frac{1 \text{ mol stearic acid}}{284.48 \text{ g stearic acid}} = 3.0 \times 10^{-3} \text{ mol of stearic}$$

acid. This number of moles of acid occupies 1 cm^3 of space. So, the number of stearic acid molecules in 1 cm^3

$$= 3.0 \times 10^{-3} \text{ mol of stearic acid} \times \frac{6.022 \times 10^{23} \text{ molecules}}{1 \text{ mol of stearic acid}}$$

$$= 1.8 \times 10^{21} \text{ stearic acid molecules.}$$

Thus, the volume for a single stearic acid molecule in nm^3

$$= 1 \text{ cm}^3 \times \frac{1}{1.8 \times 10^{21} \text{ molecules stearic acid}} \times \frac{(1.0 \times 10^7 \text{ nm})^3}{(1 \text{ cm})^3} = 0.55\underline{6} \text{ nm}^3$$

The volume of a rectangular column is simply the area of its base multiplied by its height (i.e., V = area of base (in nm^2) × height (in nm)).

So, the average height of a stearic acid molecule = $\dfrac{0.556 \text{ nm}^3}{0.22 \text{ nm}^2}$ = 2.5 nm

(c)   The density for oleic acid = 0.895 g mL^{-1}. So, the concentration for oleic acid is

$$= \frac{0.895 \text{ g acid}}{10.00 \text{ mL}} = 0.0895 \text{ g mL}^{-1} \text{ (solution 1)}$$

This solution is then divided by 10, three more times, to give a final concentration of $8.9\underline{5} \times 10^{-5}$ g mL^{-1}. A 0.10 mL sample of this solution contains:

$$= \frac{8.95 \times 10^{-5} \text{ g acid}}{1.00 \text{ mL}} \times 0.10 \text{ mL} = 8.9\underline{5} \times 10^{-6} \text{ g of acid.}$$

The number of acid molecules = 85 cm^2 × $\dfrac{1}{4.6 \times 10^{-15} \text{cm}^2 \text{ per molecule}}$

$$= 1.8\underline{5} \times 10^{16} \text{ oleic acid molecules.}$$

So, 8.95×10^{-6} g of oleic acid corresponds to 1.85×10^{16} oleic acid molecules.

The molar mass for oleic acid, $C_{18}H_{34}O_2$, is 282.47 g mol^{-1}.

The number of moles of oleic acid is

$$= 8.9\underline{5} \times 10^{-6} \text{ g} \times \frac{1 \text{ mol oleic acid}}{282.47 \text{ g}} = 3.1\underline{7} \times 10^{-8} \text{ mol}$$

So, Avogadro's number here would be equal to:

$$= \frac{1.8\underline{5} \times 10^{16} \text{ oleic acid molecules}}{3.1\underline{7} \times 10^{-8} \text{ oleic acid moles}} = 5.8 \times 10^{23} \text{ molecules per mole of oleic acid.}$$

SELF-ASSESSMENT EXERCISES

119. **(E)** The answer is (c), because 12.01 g of H_2O = 0.667 mol H_2O, which equates to $0.667 \times 3 = 2.00$ moles of atoms. One mole of Br_2 also has 2.00 moles of atoms.

120. **(E)** The answer is (b). N_2H_4 can be reduced further to an empirical formula of NH_2.

121. **(E)** The answer is (d), because total atomic mass is 14 for N and 7 for H.

122. **(E)** Answer is (a).
(a) 50.0 g N_2O × (1 mol N_2O/44.0 g N_2O) × (2 mol N/1 mol N_2O) = 2.27 mol
(b) 17.0 g NH_3 × (1 mol NH_3/17.0 g NH_3) × (1 mol N/1 mol NH_3) = 1.00 mol
(c) 150 mL C_5H_5N × (0.983 g/1 mL) × (1 mol Pyr/79.0 g Pyr) × (1 mol N/1 mol Pyr) = 1.87 mol
(d) 1 mol N_2 × (2 mol N/1 mol N_2) = 2.0

123. **(M)**
$$\frac{2.9 \text{ g Fe}}{\text{total blood}} \times \frac{\text{total blood}}{2.6 \times 10^{13} \text{ red blood cells}} \times \frac{1 \text{ mol Fe}}{55.8 \text{ g Fe}} \times \frac{6.02 \times 10^{23} \text{ Fe atoms}}{\text{mol Fe}}$$
$$= 1.2 \times 10^9 \frac{\text{Fe atoms}}{\text{red blood cell}}$$

124. **(E)** Answer is (c).
Mass % of F = (19×3)/(X + 19×3) = 0.65
Solving for X, we get X = 30.7 or 31 u

125. **(E)** Answer is (c). Total formal charge on H: +4. Total charge on O: –12, and the ion has a negative charge. Therefore, oxidation state of I = –12 + 4 + 1 = 7.

126. **(E)** The correct answer is +6 (choice c). Magnesium has a +2 oxidation state, and oxygen is –2. Manganese would need to be +6 for the charge on the molecule to sum to zero.

127. **(M)** Choice (a) is hydrogen periodate and is therefore the correct answer. Na_2SO_3 is sodium sulfite, $KClO_2$ is potassium chlorite, HFO is hydrogen hypofluorite, and NO_2 is nitrogen dioxide.

128. **(E)** $Sr(HCO_3)_2$ is strontium bicarbonate (choice d).

129. **(E)** The answer is (b). Ca is a +2 ion and ClO_2^- is –1 anion.

130. **(E)** Li_3P has a molar mass of 51.79 g mol^{-1} (choice d).

131. **(E)** The answer is (d). Multiplying O atomic mass by 4 (64 u) is nearly the same as the atomic mass of Cu (63.55).

132. **(E)** The answer is (d). Answer (a) isn't correct. While having the correct number of atoms, it is not an isomer because it is only a molecular formula and gives no information on atom bonding. Answer (b) isn't correct, because it's the exact same molecule as stated in the question. Answer (c) isn't correct because it doesn't have enough atoms. Therefore, the answer is (d), because it has the correct number of atoms in a different configuration.

133. **(M)** First, find out the mass of Na_2SO_3, which is 126.0 g/mol. Then:
Mass H_2O $(x) = 0.5$ $(x +$ Mass $Na_2SO_3)$.
$x = 0.5x + 63$.
Solving for x, we obtain $x = 126$ g (mass of H_2O)
Since we have 126 g of water, the number of moles of H_2O is 126 g/18.0 g $mol^{-1} = 7$
Therefore, the formula is $Na_2SO_3 \cdot 7\ H_2O$.

134. **(M)**
(a) Based on this composition, molar mass of malachite is calculated to be 221.18 g/mol. Since there are two moles of Cu per mole of malachite, the %mass of Cu is:

$$1000\ \text{g malachite} \times \frac{1\ \text{mol mal.}}{221.18\ \text{g mal.}} \times \frac{2\ \text{mol Cu}}{1\ \text{mol mal.}} \times \frac{63.546\ \text{g Cu}}{1\ \text{mol Cu}} = 574.61\ \text{g Cu}$$

$$\%\ \text{Cu} = \frac{574.61\ \text{g}}{1000\ \text{g}} \times 100 = 57.46\%$$

(b) The formula for copper(II) oxide is CuO. Therefore, for one mole of malachite, there are two moles of CuO. Therefore,

$$1000\ \text{g malachite} \times \frac{1\ \text{mol mal.}}{221.18\ \text{g mal.}} \times \frac{2\ \text{mol CuO}}{1\ \text{mol mal.}} \times \frac{79.545\ \text{g CuO}}{1\ \text{mol CuO}}$$

mass CuO = 719.5 g

135. **(D)** Molar mass of acetaminophen is 151.2 u, or 151.2 g/mol. To determine the molecular formula, calculate the moles of various constituting elements, as shown below:

mol C = 63.56 g C × (1 mol C/12.011 g C) = 5.292 mol C
mol H = 6.00 g H × (1 mol H/1.008 g H) = 5.95 mol H
mol N = 9.27 g N × (1 mol N/14.01 g N) = 0.662 mol N
mol O = 21.17 g O × (1 mol O/15.999 g O) = 1.323 mol O

Then, divide all values by the smallest to determine mole ratios:
5.92 mol C / 0.662 mol N → 7.99 mol C
5.95 mol H / 0.662 mol N → 8.99 mol H
0.662 mol N / 0.662 mol N → 1.00 mol N
1.323 mol O / 0.662 mol N → 2.00 mol C

The C:H:N:O ratio is 8:9:1:2. The empirical formula is therefore $C_8H_9NO_2$. The molar mass of this formula unit is 151.1, which is the same as the molar mass of acetaminophen. Therefore, the empirical formula obtained is also the same as the molecular formula.

136. **(D)** The first step is to determine the mass of C, H, and O.

$$\text{mol C} = 6.029 \text{ g CO}_2 \times \frac{1 \text{ mol CO}_2}{44.009 \text{ g CO}_2} \times \frac{1 \text{ mol C}}{1 \text{ mol CO}_2} = 0.1370 \text{ mol}$$

mass of C $= 0.1370$ mol C$\times(12.011$ g C/1 mol C$) = 1.646$ g C

$$\text{mol H} = 1.709 \text{ g H}_2\text{O} \times \frac{1 \text{ mol H}_2\text{O}}{18.015 \text{ g H}_2\text{O}} \times \frac{2 \text{ mol H}}{1 \text{ mol H}_2\text{O}} = 0.1897 \text{ mol}$$

mass of H $= 0.1897$ mol H$\times(1.008$ g H/1 mol H$) = 0.1912$ g H

Mass of oxygen is obtained by difference: mass of O $= 2.174$ g $- (1.646 + 0.1912) = 0.337$ g

$$\text{mol O} = 0.337 \text{ g O} \times \frac{1 \text{ mol O}}{16.00 \text{ g O}} = 0.0211 \text{ mol}$$

(a) % Composition:
   1.646 g C / 2.174 g Ibo = 75.71% C
   0.191 g H / 2.174 g Ibo = 8.79% H
   0.337 g O / 2.174 g Ibo = 15.5% O

(b) To determine the empirical formula, divide all mole values by the lowest one:
   0.1370 mol C / 0.0211 mol O $\rightarrow$ 6.49 mol C
   0.1897 mol H / 0.0211 mol O $\rightarrow$ 8.99 mol H
   0.0211 mol O / 0.0211 mol O $\rightarrow$ 1.00 mol O

The empirical formula is obtained by multiplying the above ratios by 2. The formula is $C_{13}H_{18}O_2$.

CHAPTER 4
CHEMICAL REACTIONS

PRACTICE EXAMPLES

1A **(E)**

Unbalanced reaction:	$HgS(s) + CaO(s)$	$\rightarrow CaS(s) + CaSO_4(s) + Hg(l)$
Balance O atoms:	$HgS(s) + 4\,CaO(s)$	$\rightarrow CaS(s) + CaSO_4(s) + Hg(l)$
Balance Ca atoms:	$HgS(s) + 4\,CaO(s)$	$\rightarrow 3\,CaS(s) + CaSO_4(s) + Hg(l)$
Balance S atoms:	$4\,HgS(s) + 4\,CaO(s)$	$\rightarrow 3\,CaS(s) + CaSO_4(s) + Hg(l)$
Balance Hg atoms:	$4\,HgS(s) + 4\,CaO(s)$	$\rightarrow 3\,CaS(s) + CaSO_4(s) + 4\,Hg(l)$
Self Check:	$4\,Hg + 4\,S + 4\,O + 4\,Ca$	$\rightarrow 4\,Hg + 4\,S + 4\,O + 4\,Ca$

1B **(E)**

Unbalanced reaction:	$C_7H_6O_2S(l) + O_2(g)$	$\rightarrow CO_2(g) + H_2O(l) + SO_2(g)$
Balance C atoms:	$C_7H_6O_2S(l) + O_2(g)$	$\rightarrow 7\,CO_2(g) + H_2O(l) + SO_2(g)$
Balance S atoms:	$C_7H_6O_2S(l) + O_2(g)$	$\rightarrow 7\,CO_2(g) + H_2O(l) + SO_2(g)$
Balance H atoms:	$C_7H_6O_2S(l) + O_2(g)$	$\rightarrow 7\,CO_2(g) + 3\,H_2O(l) + SO_2(g)$
Balance O atoms:	$C_7H_6O_2S(l) + 8.5\,O_2(g)$	$\rightarrow 7\,CO_2(g) + 3\,H_2O(l) + SO_2(g)$
Multiply by 2 (whole #):	$2\,C_7H_6O_2S(l) + 17\,O_2(g)$	$\rightarrow 14\,CO_2(g) + 6\,H_2O(l) + 2\,SO_2(g)$
Self Check:	$14\,C + 12\,H + 2\,S + 38\,O$	$\rightarrow 14\,C + 12\,H + 2\,S + 38\,O$

2A **(E)** The balanced chemical equation provides the factor needed to convert from moles $KClO_3$ to moles O_2. Amount $O_2 = 1.76\,mol\,KClO_3 \times \dfrac{3\,mol\,O_2}{2\,mol\,KClO_3} = 2.64\,mol\,O_2$

2B **(E)** First, find the molar mass of Ag_2O.

$$(2\,mol\,Ag \times 107.87\,g/mol\,Ag) + 16.00\,g\,O = 231.74\,g\,Ag_2O\,/\,mol$$

$$amount\ Ag = 1.00\,kg\,Ag_2O \times \frac{1000\,g}{1.00\,kg} \times \frac{1\,mol\,Ag_2O}{231.74\,g\,Ag_2O} \times \frac{2\,mol\,Ag}{1\,mol\,Ag_2O} = 8.63\,mol\,Ag$$

3A **(E)** The balanced chemical equation provides the factor to convert from amount of Mg to amount of Mg_3N_2. First we must determine the molar mass of Mg_3N_2.

$$molar\ mass = (3\,mol\,Mg \times 24.305\,g/mol\,Mg) + (2\,mol\,N \times 14.007\,g/mol\,N) = 100.93\,g\,Mg_3N_2$$

$$mass\ Mg_3N_2 = 3.82\,g\,Mg \times \frac{1\,mol\,Mg}{24.31\,g\,Mg} \times \frac{1\,mol\,Mg_3N_2}{3\,mol\,Mg} \times \frac{100.93\,g\,Mg_3N_2}{1\,mol\,Mg_3N_2} = 5.29\,g\,Mg_3N_2$$

3B **(E)** The pivotal conversion is from $H_2(g)$ to $CH_3OH(l)$. For this we use the balanced equation, which requires that we use the amounts in moles of both substances. The solution involves converting to and from amounts, using molar masses.

$$mass\ H_2(g) = 1.00\,kg\,CH_3OH(l) \times \frac{1000\,g}{1\,kg} \times \frac{1\,mol\,CH_3OH}{32.04\,g\,CH_3OH} \times \frac{2\,mol\,H_2}{1\,mol\,CH_3OH} \times \frac{2.016\ g\,H_2}{1\,mol\,H_2}$$

$$mass\ H_2(g) = 126\ g\,H_2$$

4A **(M)** The equation for the cited reaction is: $2\,NH_3\,(g) + 1.5\,O_2\,(g) \longrightarrow N_2\,(g) + 3\,H_2O\,(l)$

The pivotal conversion is from one substance to another, in moles, with the balanced chemical equation providing the conversion factor.

$$\text{mass } NH_3\,(g) = 1.00\,g\,O_2\,(g) \times \frac{1\,mol\,O_2}{32.00\,g\,O_2} \times \frac{2\,mol\,NH_3}{1.5\ mol\,O_2} \times \frac{17.031\,g\,NH_3}{1\,mol\,H_2} = 0.710\,g\ NH_3$$

4B **(M)** The equation for the combustion reaction is:

$$C_8H_{18}\,(l) + \frac{25}{2}\,O_2\,(g) \rightarrow 8\,CO_2\,(g) + 9\,H_2O\,(l)$$

$$\text{mass } O_2 = 1.00\,g\,C_8H_{18} \times \frac{1\,mol\,C_8H_{18}}{114.23\,g\,C_8H_{18}} \times \frac{12.5\,mol\,O_2}{1\,mol\,C_8H_{18}} \times \frac{32.00\,g\,O_2}{1\,mol\,O_2} = 3.50\,g\,O_2\,(g)$$

5A **(M)** We must convert mass $H_2 \rightarrow$ amount of $H_2 \rightarrow$ amount of $Al \rightarrow$ mass of $Al \rightarrow$ mass of alloy $\rightarrow$ volume of alloy. The calculation is performed as follows: each arrow in the preceding sentence requires a conversion factor.

$$V_{alloy} = 1.000\,g\,H_2 \times \frac{1\,mol\,H_2}{2.016\,g\,H_2} \times \frac{2\,mol\,Al}{3\,mol\,H_2} \times \frac{26.98\,g\,Al}{1\,mol\,Al} \times \frac{100.0\,g\,alloy}{93.7\,g\,Al} \times \frac{1\,cm^3\,alloy}{2.85\,g\,alloy}$$

$$= 3.34\,cm^3\,alloy$$

5B **(M)** In the example, $0.207\,g\,H_2$ is collected from $1.97\,g$ alloy; the alloy is 6.3% Cu by mass. This information provides the conversion factors we need.

$$\text{mass Cu} = 1.31\,g\,H_2 \times \frac{1.97\,g\,alloy}{0.207\,g\,H_2} \times \frac{6.3\,g\,Cu}{100.0\,g\,alloy} = 0.79\,g\,Cu$$

Notice that we do not have to consider each step separately. We can simply use values produced in the course of the calculation as conversion factors.

6A **(M)** The cited reaction is $2\,Al\,(s) + 6\,HCl\,(aq) \rightarrow 2\,AlCl_3\,(aq) + 3\,H_2\,(g)$. The HCl(aq) solution has a density of 1.14 g/mL and contains 28.0% HCl. We need to convert between the substances HCl and H_2; the important conversion factor comes from the balanced chemical equation. The sequence of conversions is: volume of HCl(aq) $\rightarrow$ mass of HCl(aq) $\rightarrow$ mass of pure HCl $\rightarrow$ amount of HCl $\rightarrow$ amount of $H_2 \rightarrow$ mass of H_2.

In the calculation below, each arrow in the sequence is replaced by a conversion factor.

$$\text{mass } H_2 = 0.05\,mL\,HCl\,(aq) \times \frac{1.14\,g\,soln}{1\,mL\,soln} \times \frac{28.0\,g\,HCl}{100.0\,g\,soln} \times \frac{1\,mol\,HCl}{36.46\,g\,HCl} \times \frac{3\,mol\,H_2}{6\,mol\,HCl} \times \frac{2.016\,g\,H_2}{1\,mol\,H_2}$$

$$= 4 \times 10^{-4}\,g\,H_2\,(g) = 0.4\,mg\,H_2\,(g)$$

6B **(M)** Density is necessary to determine the mass of the vinegar, and then the mass of acetic acid.

$$\text{mass } CO_2\,(g) = 5.00\,mL\,vinegar \times \frac{1.01\,g}{1\,mL} \times \frac{0.040\,g\,acid}{1\,g\,vinegar} \times \frac{1\,mol\,CH_3COOH}{60.05\,g\,CH_3COOH} \times \frac{1\,mol\,CO_2}{1\,mol\,CH_3COOH} \times \frac{44.01\,g\,CO_2}{1\,mol\,CO_2}$$

$$= 0.15\,g\,CO_2$$

7A (M) Determine the amount in moles of acetone and the volume in liters of the solution.

$$\text{molarity of acetone} = \frac{22.3\,g\,(CH_3)_2\,CO \times \dfrac{1\,mol\,(CH_3)_2\,CO}{58.08\,g\,(CH_3)_2\,CO}}{1.25\ L\ soln} = 0.307\,M$$

7B (M) The molar mass of acetic acid, CH_3COOH, is 60.05 g/mol. We begin with the quantity of acetic acid in the numerator and that of the solution in the denominator, and transform to the appropriate units for each.

$$\text{molarity} = \frac{15.0\,mL\ CH_3COOH}{500.0\,mL\ soln} \times \frac{1000\,mL}{1L\ soln} \times \frac{1.048\,g\ CH_3COOH}{1\,mL\ CH_3COOH} \times \frac{1\,mol\ CH_3COOH}{60.05\,g\ CH_3COOH} = 0.524\,M$$

8A (E) The molar mass of $NaNO_3$ is 84.99 g/mol. We recall that "M" stands for "mol/L soln."

$$\text{mass } NaNO_3 = 125\,mL\ soln \times \frac{1L}{1000\,mL} \times \frac{10.8\,mol\ NaNO_3}{1L\ soln} \times \frac{84.99\,g\ NaNO_3}{1\,mol\ NaNO_3} = 115\,g\ NaNO_3$$

8B (E) We begin by determining the molar mass of $Na_2SO_4 \cdot 10\,H_2O$. The amount of solute needed is computed from the concentration and volume of the solution.

$$\text{mass } Na_2SO_4 \cdot 10\,H_2O = 355\ mL\ soln \times \frac{1\ L}{1000\ mL} \times \frac{0.445\ mol\ Na_2SO_4}{1\ L\ soln} \times \frac{1\ mol\ Na_2SO_4 \cdot 10\,H_2O}{1\ mol\ Na_2SO_4}$$

$$\times \frac{322.21\ g\ Na_2SO_4 \cdot 10\,H_2O}{1\ mol\ Na_2SO_4 \cdot 10\,H_2O} = 50.9\ g\ Na_2SO_4 \cdot 10\,H_2O$$

9A (E) The amount of solute in the concentrated solution doesn't change when the solution is diluted. We take advantage of an alternative definition of molarity to answer the question: millimoles of solute/milliliter of solution.

$$\text{amount } K_2CrO_4 = 15.00\,mL \times \frac{0.450\,mmol\ K_2CrO_4}{1\,mL\ soln} = 6.75\,mmol\ K_2CrO_4$$

$$K_2CrO_4\ \text{molarity, dilute solution} = \frac{6.75\,mmol\ K_2CrO_4}{100.00\,mL\ soln} = 0.0675\,M$$

9B (E) We know the initial concentration (0.105 M) and volume (275 mL) of the solution, along with its final volume (237 mL). The final concentration equals the initial concentration times a ratio of the two volumes.

$$c_f = c_i \times \frac{V_i}{V_f} = 0.105\,M \times \frac{275\,mL}{237\,mL} = 0.122\ M$$

10A **(M)** The balanced equation is $K_2CrO_4(aq) + 2\,AgNO_3(aq) \rightarrow Ag_2CrO_4(s) + 2\,KNO_3(aq)$.

The molar mass of Ag_2CrO_4 is 331.73 g/mol. The conversions needed are mass $Ag_2CrO_4 \rightarrow$ amount Ag_2CrO_4 (moles) $\rightarrow$ amount K_2CrO_4 (moles) $\rightarrow$ volume K_2CrO_4 (aq).

$$V_{K_2CrO_4} = 1.50\,g\,Ag_2CrO_4 \times \frac{1\,mol\,Ag_2CrO_4}{331.73\,g\,Ag_2CrO_4} \times \frac{1\,mol\,K_2CrO_4}{1\,mol\,Ag_2CrO_4} \times \frac{1\,L\,soln}{0.250\,mol\,K_2CrO_4}$$

$$\times \frac{1000\,mL\,solution}{1\,L\,solution} = 18.1\,mL$$

10B **(M)** Balanced reaction: $2\,AgNO_3(aq) + K_2CrO_4(aq) \rightarrow Ag_2CrO_4(s) + 2\,KNO_3(aq)$

moles of $K_2CrO_4 = c \times V = 0.0855\,M \times 0.175\,L\,sol = 0.0149\underline{6}\,mol\,K_2CrO_4$

moles of $AgNO_3 = 0.0149\underline{6}\,mol\,K_2CrO_4 \times \dfrac{2\,mol\,AgNO_3}{1\,mol\,K_2CrO_4} = 0.0299\,mol\,AgNO_3$

$$V_{AgNO_3} = \frac{n}{c} = \frac{0.0299\,mol\,AgNO_3}{0.150\,\dfrac{mol\,L^{-1}}{L}\,AgNO_3} = 0.199\underline{5}\,L\ \text{or}\ 2.00 \times 10^2\,mL\ (0.200\,L)\ \text{of}\ AgNO_3$$

Mass of Ag_2CrO_4 formed $= 0.0149\underline{6}\,mol\,K_2CrO_4 \times \dfrac{1\,mol\,Ag_2CrO_4}{1\,mol\,K_2CrO_4} \times \dfrac{331.73\,g\,Ag_2CrO_4}{1\,mol\,Ag_2CrO_4}$

Mass of Ag_2CrO_4 formed $= 4.96\,g\,Ag_2CrO_4$

11A **(M)** Reaction: $P_4(s) + 6\,Cl_2(g) \rightarrow 4\,PCl_3(l)$. We must determine the mass of PCl_3 formed by each reactant.

mass PCl_3 from $P_4 = 215\,g\,P_4 \times \dfrac{1\,mol\,P_4}{123.89\,g\,P_4} \times \dfrac{4\,mol\,PCl_3}{1\,mol\,P_4} \times \dfrac{137.32\,g\,PCl_3}{1\,mol\,PCl_3} = 953\,g\,PCl_3$

mass PCl_3 from $Cl_2 = 725\,g\,Cl_2 \times \dfrac{1\,mol\,Cl_2}{70.91\,g\,Cl_2} \times \dfrac{4\,mol\,PCl_3}{6\,mol\,Cl_2} \times \dfrac{137.32\,g\,PCl_3}{1\,mol\,PCl_3} = 936\,g\,PCl_3$

Thus, a maximum of $936\,g\,PCl_3$ can be produced; there is not enough Cl_2 to produce any more.

11B **(M)** Since data are supplied and the answer is requested in kilograms (thousands of grams), we can use kilomoles (thousands of moles) to solve the problem. We calculate the amount in kilomoles of $POCl_3$ that would be produced if each of the reactants were completely converted to product. The smallest of these amounts is the one that is actually produced (this is a limiting reactant question).

amount $POCl_3$ from $PCl_3 = 1.00\,kg\,PCl_3 \times \dfrac{1\,kmol\,PCl_3}{137.32\,kg\,PCl_3} \times \dfrac{10\,kmol\,POCl_3}{6\,kmol\,PCl_3} = 0.0121\,kmol\,POCl_3$

amount $POCl_3$ from $Cl_2 = 1.00\,kg\,Cl_2 \times \dfrac{1\,kmol\,Cl_2}{70.90\,kg\,Cl_2} \times \dfrac{10\,kmol\,POCl_3}{6\,kmol\,Cl_2} = 0.0235\,kmol\,POCl_3$

amount $POCl_3$ from $P_4H_{10} = 1.00\,kg\,P_4O_{10} \times \dfrac{1\,kmol\,P_4O_{10}}{283.89\,kg\,P_4O_{10}} \times \dfrac{10\,kmol\,POCl_3}{1\,kmol\,P_4O_{10}} = 0.0352\,kmol\,POCl_3$

Thus, a maximum of $0.0121\,kmol\,POCl_3$ can be produced.

We next determine the mass of the product.

$$\text{mass POCl}_3 = 0.0121 \, \text{kmol POCl}_3 \times \frac{153.33 \, \text{kg POCl}_3}{1 \, \text{kmol POCl}_3} = 1.86 \, \text{kg POCl}_3$$

12A **(M)** The $725 \, \text{g Cl}_2$ limits the mass of product formed. The $P_4(s)$ therefore is the reactant in excess. From the quantity of excess reactant we can find the amount of product formed: $953 \, \text{g PCl}_3 - 936 \, \text{g PCl}_3 = 17 \, \text{g PCl}_3$. We calculate how much P_4 this is, both in the traditional way and by using the initial $(215 \, \text{g P}_4)$ and final $(953 \, \text{g PCl}_3)$ values of the previous calculation.

$$\text{mass P}_4 = 17 \, \text{g PCl}_3 \times \frac{1 \, \text{mol PCl}_3}{137.33 \, \text{g PCl}_3} \times \frac{1 \, \text{mol P}_4}{4 \, \text{mol PCl}_3} \times \frac{123.89 \, \text{g P}_4}{1 \, \text{mol P}_4} = 3.8 \, \text{g P}_4$$

12B **(M)** Find the amount of $H_2O(l)$ formed by each reactant, to determine the limiting reactant.

$$\text{amount H}_2\text{O from H}_2 = 12.2 \, \text{g H}_2 \times \frac{1 \, \text{mol H}_2}{2.016 \, \text{g H}_2} \times \frac{2 \, \text{mol H}_2\text{O}}{2 \, \text{mol H}_2} = 6.05 \, \text{mol H}_2\text{O}$$

$$\text{amount H}_2\text{O from O}_2 = 154 \, \text{g O}_2 \times \frac{1 \, \text{mol O}_2}{32.00 \, \text{g O}_2} \times \frac{2 \, \text{mol H}_2\text{O}}{1 \, \text{mol O}_2} = 9.63 \, \text{mol H}_2\text{O}$$

Since H_2 is limiting, we must compute the mass of O_2 needed to react with all of the H_2

$$\text{mass O}_2 \text{ reacting} = 6.05 \, \text{mol H}_2\text{O produced} \times \frac{1 \, \text{mol O}_2}{2 \, \text{mol H}_2\text{O}} \times \frac{32.00 \, \text{g O}_2}{1 \, \text{mol O}_2} = 96.8 \, \text{g O}_2 \text{ reacting}$$

$$\text{mass O}_2 \text{ remaining} = 154 \, \text{g originally present} - 96.8 \, \text{g O}_2 \text{ reacting} = 57 \, \text{g O}_2 \text{ remaining}$$

13A **(M)**

(a) The theoretical yield is the calculated maximum mass of product expected if we were to assume that the reaction has no losses (100% reaction).

$$\text{mass CH}_2\text{O}(g) = 1.00 \, \text{mol CH}_3\text{OH} \times \frac{1 \, \text{mol CH}_2\text{O}}{1 \, \text{mol CH}_3\text{OH}} \times \frac{30.03 \, \text{g CH}_2\text{O}}{1 \, \text{mol CH}_2\text{O}} = 30.0 \, \text{g CH}_2\text{O}$$

(b) The actual yield is what is obtained experimentally: $25.7 \, \text{g CH}_2\text{O} \, (g)$.

(c) The percent yield is the ratio of actual yield to theoretical yield, multiplied by 100%:

$$\% \text{ yield} = \frac{25.7 \, \text{g CH}_2\text{O produced}}{30.0 \, \text{g CH}_2\text{O calculated}} \times 100\% = 85.6\% \text{ yield}$$

13B **(M)** First determine the mass of product formed by each reactant.

$$\text{mass PCl}_3 \text{ from P}_4 = 25.0 \, \text{g P}_4 \times \frac{1 \, \text{mol P}_4}{123.89 \, \text{g P}_4} \times \frac{4 \, \text{mol PCl}_3}{1 \, \text{mol P}_4} \times \frac{137.32 \, \text{g PCl}_3}{1 \, \text{mol PCl}_3} = 111 \, \text{g PCl}_3$$

$$\text{mass PCl}_3 \text{ from Cl}_2 = 91.5 \, \text{g Cl}_2 \times \frac{1 \, \text{mol Cl}_2}{70.91 \, \text{g Cl}_2} \times \frac{4 \, \text{mol PCl}_3}{6 \, \text{mol Cl}_2} \times \frac{137.32 \, \text{g PCl}_3}{1 \, \text{mol PCl}_3} = 118 \, \text{g PCl}_3$$

Thus, the limiting reactant is P_4, and $111 \, \text{g PCl}_3$ should be produced. This is the theoretical maximum yield. The actual yield is $104 \, \text{g PCl}_3$. Thus, the percent yield of the reaction is

$$\frac{104 \, \text{g PCl}_3 \text{ produced}}{111 \, \text{g PCl}_3 \text{ calculated}} \times 100\% = 93.7\% \text{ yield.}$$

14A (M) The reaction is $2 NH_3(g) + CO_2(g) \rightarrow CO(NH_2)_2(s) + H_2O(l)$. We need to distinguish between mass of urea produced (actual yield) and mass of urea predicted (theoretical yield).

$$\text{mass } CO_2 = 50.0 \text{ g } CO(NH_2)_2 \text{ produced} \times \frac{100.0 \text{ g predicted}}{87.5 \text{ g produced}} \times \frac{1 \text{ mol } CO(NH_2)_2}{60.1 \text{ g } CO(NH_2)_2} \times \frac{1 \text{ mol } CO_2}{1 \text{ mol } CO(NH_2)_2}$$

$$\times \frac{44.01 \text{ g } CO_2}{1 \text{ mol } CO_2} = 41.8 \text{ g } CO_2 \text{ needed}$$

14B (M) Care must be taken to use the proper units/labels in each conversion factor. Note, you cannot calculate the molar mass of an impure material or mixture.

$$\text{mass } C_6H_{11}OH = 45.0 \text{ g } C_6H_{10} \text{ produced} \times \frac{100.0 \text{ g } C_6H_{10} \text{ cal'd}}{86.2 \text{ g } C_6H_{10} \text{ produc'd}} \times \frac{1 \text{ mol } C_6H_{10}}{82.1 \text{ g } C_6H_{10}} \times \frac{1 \text{ mol } C_6H_{11}OH}{1 \text{ mol } C_6H_{10}}$$

$$\times \frac{100.2 \text{ g pure } C_6H_{11}OH}{1 \text{ mol } C_6H_{11}OH} \times \frac{100.0 \text{ g impure } C_6H_{11}OH}{92.3 \text{ g pure } C_6H_{11}OH} = 69.0 \text{ g impure } C_6H_{11}OH$$

15A (M) We can trace the nitrogen through the sequence of reactions. We notice that 4 moles of N (as 4 mol NH_3) are consumed in the first reaction, and 4 moles of N (as 4 mol NO) are produced. In the second reaction, 2 moles of N (as 2 mol NO) are consumed and 2 moles of N (as 2 mol NO_2) are produced. In the last reaction, 3 moles of N (as 3 mol NO_2) are consumed and just 2 moles of N (as 2 mol HNO_3) are produced.

$$\text{mass } HNO_3 = 1.00 \text{ kg } NH_3 \times \frac{1000 \text{ g } NH_3}{1 \text{ kg } NH_3} \times \frac{1 \text{ mol } NH_3}{17.03 \text{ g } NH_3} \times \frac{4 \text{ mol NO}}{4 \text{ mol } NH_3} \times \frac{2 \text{ mol } NO_2}{2 \text{ mol NO}}$$

$$\times \frac{2 \text{ mol } HNO_3}{3 \text{ mol } NO_2} \times \frac{63.01 \text{ g } HNO_3}{1 \text{ mol } HNO_3} = 2.47 \times 10^3 \text{ g } HNO_3$$

15B (M)

$$\text{mass } KNO_3 = 95 \text{ g } NaN_3 \times \frac{1 \text{ mol } NaN_3}{65.03 \text{ g } NaN_3} \times \frac{2 \text{ mol Na}}{2 \text{ mol } NaN_3} \times \frac{2 \text{ mol } KNO_3}{10 \text{ mol Na}} \times \frac{102 \text{ g } KNO_3}{1 \text{ mol } KNO_3}$$

$$= 29.80 \approx 30 \text{ g } KNO_3$$

$$\text{mass } SiO_2 \text{ (1)} = 1.461 \text{ mol } NaN_3 \times \frac{2 \text{ mol Na}}{2 \text{ mol } NaN_3} \times \frac{1 \text{ mol } K_2O}{10 \text{ mol Na}} \times \frac{1 \text{ mol } SiO_2}{1 \text{ mol } K_2O} \times \frac{64.06 \text{ g } SiO_2}{1 \text{ mol } SiO_2}$$

$$= 9.36 \text{ g} \approx 9.4 \text{ g } SiO_2$$

$$\text{mass } SiO_2 \text{ (2)} = 1.461 \text{ mol } NaN_3 \times \frac{2 \text{ mol Na}}{2 \text{ mol } NaN_3} \times \frac{5 \text{ mol } Na_2O}{10 \text{ mol Na}} \times \frac{1 \text{ mol } SiO_2}{1 \text{ mol } Na_2O} \times \frac{64.06 \text{ g } SiO_2}{1 \text{ mol } SiO_2}$$

$$= 46.80 \text{ g} \approx 47 \text{ g } SiO_2$$

Therefore, the total mass of SiO_2 is the sum of the above two results. Approximately 56 g of SiO_2 and 30 g of KNO_3 are needed.

16A **(D)** Let the mass of Al be m g. Then the mass of Mg will be $(1.00 - m)$ g. We can now write expressions for the mass of H_2 that can be obtained from m g Al and $(1.00 - m)$ g Mg.

$$\text{mass of } H_2 \text{ from Al} = (m) \text{ g Al} \times \frac{1 \text{ mol Al}}{26.98 \text{ g Al}} \times \frac{3 \text{ mol } H_2}{2 \text{ mol Al}} \times \frac{2.016 \text{ g } H_2}{1 \text{ mol } H_2} = (m) \, 0.1121 \text{ g } H_2$$

$$\text{mass of } H_2 \text{ from Mg} = (1.00 - m) \text{ g Al} \times \frac{1 \text{ mol Mg}}{24.305 \text{ g Mg}} \times \frac{1 \text{ mol } H_2}{1 \text{ mol Mg}} \times \frac{2.016 \text{ g } H_2}{1 \text{ mol } H_2}$$

$$= (1.00 - m) \, 0.0829 \text{ g } H_2$$

Now, we note that the total mass of H_2 generated is 0.107 g. Therefore,

Mass $H_2 = (m)(0.1121) + (1.00 - m)(0.0829) = 0.107$
Solving for m gives a value of 0.82 g.

Therefore, mass of Al = 0.83 g. Since the sample is 1.00 g, Mg is 17 wt%.
mass of Mg = 1.00 − 0.83 = 0.17 g, or 17 wt%.

16B **(D)** Mass of CuO and Cu_2O is done in identical fashion to the above problem:

$$\text{mass of Cu from CuO} = (1.500 - x) \text{ g CuO} \times \frac{1 \text{ mol CuO}}{79.545 \text{ g CuO}} \times \frac{1 \text{ mol Cu}}{1 \text{ mol CuO}} \times \frac{63.546 \text{ g Cu}}{1 \text{ mol Cu}}$$

$$= (1.500 - x) \, 0.7989 \text{ g Cu}$$

$$\text{mass of Cu from } Cu_2O = (x) \text{ g } Cu_2O \times \frac{1 \text{ mol } Cu_2O}{143.091 \text{ g } Cu_2O} \times \frac{2 \text{ mol Cu}}{1 \text{ mol } Cu_2O} \times \frac{63.546 \text{ g Cu}}{1 \text{ mol Cu}}$$

$$= (x) \, 0.8882 \text{ g Cu}$$

Now, we note that the total mass of pure Cu is 1.2244 g. Therefore,

Mass Cu = $(1.500 - x)(0.7989) + (x)(0.8882) = 1.2244$
Solving for x gives a value of 0.292 g.

Therefore, mass of Cu_2O = 0.292 g
mass % of Cu_2O = 0.292 g/1.500 × 100 = 19.47

17A **(E)** In the balanced equation, 3 moles of Fe is required to react with 4 moles of H_2O. With 0.500 moles of each available, H_2O is the limiting reactant. Therefore, the maximum possible value for the extent of this reaction if all of the H_2O reacts is

$$\Delta n_{H_2O} = v_{H_2O} \cdot \xi \quad \text{with } v_{H_2O} = -4$$

$$\Delta n_{H_2O} = n_{H_2O, \text{ final}} - n_{H_2O, \text{ initial}} = 0 \text{ mol} - 0.500 \text{ mol} = -0.500 \text{ mol}$$

$$\xi = \Delta n_{H_2O} / v_{H_2O} = -0.500 \text{ mol} / (-4) = 0.125 \text{ mol}$$

17B **(M)** The amount of NOCl generated can be used to determine the extent of the reaction:

$$\Delta n_{NOCl} = v_{NOCl} \cdot \xi \text{ with } v_{NOCl} = +4$$

The amount of NOCl generated is

$$1.17 \text{ g NOCl} \times \frac{1 \text{ mol NOCl}}{65.459 \text{ g NOCl}} = 0.0179 \text{ mol NOCl}$$

$$\Delta n_{NOCl} = n_{NOCl, \text{ final}} - n_{NOCl, \text{ initial}} = 0.0179 \text{ mol} - 0 \text{ mol} = 0.0179 \text{ mol}$$

$$\xi = \Delta n_{NOCl} / v_{NOCl} = 0.0179 \text{ mol} / 4 = 0.00447 \text{ mol}$$

With the extent, ξ, determined, the amount of N_2 that reacted can now be calculated.

$$\Delta n_{N_2} = v_{N_2} \xi = -2(0.00447 \text{ mol}) = -0.00894 \text{ mol}$$

Since the reaction started with 1.00 g N_2, the initial amount of N_2 is

$$1.00 \text{ g N}_2 \times \frac{1 \text{ mol N}_2}{28.02 \text{ g N}_2} = 0.0357 \text{ mol N}_2$$

The percentage of N_2 that reacted is (0.00894 mol N_2 / 0.0357 mol N_2) × 100% = 25.0%.

INTEGRATIVE EXAMPLE

A. **(D)**

Balancing the equation gives the following:

$$C_6H_{10}O_4(l) + 2 NH_3(g) + 4 H_2(g) \rightarrow C_6H_{16}N_2(l) + 4 H_2O(l)$$

Stepwise approach:

The first step is to calculate the number of moles of each reactant from the masses given.

$$\text{mol } C_6H_{10}O_4 = 4.15 \text{ kg} \times \frac{1000 \text{ g}}{1 \text{ kg}} \times \frac{1 \text{ mol}}{146.14 \text{ g}} = 28.30 \text{ mol}$$

$$\text{mol } NH_3 = 0.547 \text{ kg} \times \frac{1000 \text{ g}}{1 \text{ kg}} \times \frac{1 \text{ mol}}{17.03 \text{ g}} = 32.12 \text{ mol}$$

$$\text{mol } H_2 = 0.172 \text{ kg} \times \frac{1000 \text{ g}}{1 \text{ kg}} \times \frac{1 \text{ mol}}{2.016 \text{ g}} = 85.32 \text{ mol}$$

To determine the limiting reactant, calculate the number of moles of product that can be obtained from each of the reactants. The reactant yielding the least amount of product is the limiting reactant.

$$\text{mol of } C_6H_{16}N_2 \text{ from } C_6H_{10}O_4 = 28.30 \text{ mol} \times \frac{1 \text{ mol } C_6H_{16}N_2}{1 \text{ mol } C_6H_{10}O_4} = 28.30 \text{ mol}$$

$$\text{mol of } C_6H_{16}N_2 \text{ from } NH_3 = 32.12 \text{ mol} \times \frac{1 \text{ mol } C_6H_{16}N_2}{2 \text{ mol } NH_3} = 16.06 \text{ mol}$$

$$\text{mol of } C_6H_{16}N_2 \text{ from } H_2 = 85.32 \text{ mol} \times \frac{1 \text{ mol } C_6H_{16}N_2}{4 \text{ mol } H_2} = 21.33 \text{ mol}$$

NH_3 yields the fewest moles of product, and is the limiting reactant.

To calculate the % yield, the theoretical yield must first be calculated using the limiting reactant:

$$\text{Theoretical yield} = 16.06 \text{ mol } C_6H_{16}N_2 \times \frac{116.21 \text{ g } C_6H_{16}N_2}{1 \text{ mol } C_6H_{16}N_2} \times \frac{1 \text{ kg}}{1000 \text{ g}} = 1.866 \text{ kg}$$

$$\text{\% yield} = \frac{1.46 \text{ kg}}{1.866 \text{ kg}} \times 100 = 78.2 \text{ yield}$$

Conversion pathway approach:

$$\text{mol of } C_6H_{16}N_2 \text{ from } C_6H_{10}O_4 = 4.15 \text{ kg } C_6H_{10}O_4 \times \frac{1000 \text{ g}}{1 \text{ kg}} \times \frac{1 \text{ mol}}{146.16 \text{ g}} \times \frac{1 \text{ mol } C_6H_{16}N_2}{1 \text{ mol } C_6H_{10}O_4} = 28.4 \text{ mol}$$

$$\text{mol of } C_6H_{16}N_2 \text{ from } NH_3 = 0.547 \text{ kg } NH_3 \times \frac{1000 \text{ g}}{1 \text{ kg}} \times \frac{1 \text{ mol}}{17.03 \text{ g}} \times \frac{1 \text{ mol } C_6H_{16}N_2}{2 \text{ mol } NH_3} = 16.05 \text{ mol}$$

$$\text{mol of } C_6H_{16}N_2 \text{ from } H_2 = 0.172 \text{ kg } H_2 \times \frac{1000 \text{ g}}{1 \text{ kg}} \times \frac{1 \text{ mol}}{2.016 \text{ g}} \times \frac{1 \text{ mol } C_6H_{16}N_2}{4 \text{ mol } H_2} = 21.3 \text{ mol}$$

NH_3 yields the fewest moles of product and is therefore the limiting reactant.
The % yield is determined exactly as above.

B. (M)

Balancing the equation gives the following:

$$Zn(s) + 2HCl(aq) \rightarrow ZnCl_2(aq) + H_2(g)$$

Stepwise approach:

To determine the amount of zinc in sample, the amount of HCl reacted has to be calculated first:

$$\text{Before reaction: } 0.0179 \text{ M HCl} \times 750.0 \text{ mL} \times \frac{1 \text{ L}}{1000 \text{ mL}} = 0.0134 \text{ mol HCl}$$

After reaction: $0.0043 \text{ M HCl} \times 750.0 \text{ mL} \times \dfrac{1 \text{ L}}{1000 \text{ mL}} = 0.00323 \text{ mol HCl}$

moles of HCl consumed $= 0.0134 - 0.00323 = 0.0102 \text{ mol}$

Based on the number of moles of HCl consumed, the number of moles of Zn reacted can be determined:

$0.0102 \text{ mol HCl} \times \dfrac{1 \text{ mol Zn}}{2 \text{ mol HCl}} \times \dfrac{65.38 \text{ g Zn}}{1 \text{ mol Zn}} = 0.3334 \text{ g Zn}$

$\text{Purity of Zn} = \dfrac{0.3334 \text{ g Zn reacted}}{0.4000 \text{ g Zn in sample}} \times 100\% = 83.3\% \text{ pure}$

Conversion pathway approach:

$(0.0179 - 0.0043 \text{ M HCl}) \times 750.0 \text{ mL} \times \dfrac{1 \text{ L}}{1000 \text{ mL}}$

$\quad = 0.0102 \text{ mol HCl}$

Note that we can only subtract concentrations in the above example because the volume has not changed. Had there been a volume change, we would have to individually convert each concentration to moles first.

$\left(0.0102 \text{ mol HCl} \times \dfrac{1 \text{ mol Zn}}{2 \text{ mol HCl}} \times \dfrac{65.38 \text{ g Zn}}{1 \text{ mol Zn}}\right) / 0.4000 \text{ g} \times 100\% = 83.4\% \text{ Zn}$

EXERCISES

Writing and Balancing Chemical Equations

1. (E) (a) $2 \text{ SO}_3 \longrightarrow 2 \text{ SO}_2 + \text{O}_2$

      (b) $\text{Cl}_2\text{O}_7 + \text{H}_2\text{O} \longrightarrow 2 \text{ HClO}_4$

      (c) $3 \text{ NO}_2 + \text{H}_2\text{O} \longrightarrow 2 \text{ HNO}_3 + \text{NO}$

      (d) $\text{PCl}_3 + 3 \text{ H}_2\text{O} \longrightarrow \text{H}_3\text{PO}_3 + 3 \text{ HCl}$

3. (E) (a) $3 \text{ PbO} + 2 \text{ NH}_3 \longrightarrow 3 \text{ Pb} + \text{N}_2 + 3 \text{ H}_2\text{O}$

      (b) $2 \text{ FeSO}_4 \longrightarrow \text{Fe}_2\text{O}_3 + 2 \text{ SO}_2 + \tfrac{1}{2}\text{O}_2$ or $4 \text{ FeSO}_4 \longrightarrow 2 \text{ Fe}_2\text{O}_3 + 4 \text{ SO}_2 + \text{O}_2$

      (c) $6 \text{ S}_2\text{Cl}_2 + 16 \text{ NH}_3 \longrightarrow \text{N}_4\text{S}_4 + 12 \text{ NH}_4\text{Cl} + \text{S}_8$

(d) $C_3H_7CH(OH)CH(C_2H_5)CH_2OH + \frac{23}{2}O_2 \longrightarrow 8\,CO_2 + 9\,H_2O$

　　or　　$2\,C_3H_7CH(OH)CH(C_2H_5)CH_2OH + 23\,O_2 \longrightarrow 16\,CO_2 + 18\,H_2O$

5. **(E)** **(a)** $2\,Mg(s) + O_2(g) \rightarrow 2\,MgO(s)$

(b) $2\,NO(g) + O_2(g) \rightarrow 2\,NO_2(g)$

(c) $2\,C_2H_6(g) + 7\,O_2(g) \rightarrow 4\,CO_2(g) + 6\,H_2O(l)$

(d) $Ag_2SO_4(aq) + BaI_2(aq) \rightarrow BaSO_4(s) + 2\,AgI(s)$

7. **(E)** **(a)** $2\,C_4H_{10}(g) + 13\,O_2(g) \rightarrow 8\,CO_2(g) + 10\,H_2O(l)$

(b) $2\,CH_3CH(OH)CH_3(l) + 9\,O_2(g) \rightarrow 6\,CO_2(g) + 8\,H_2O(l)$

(c) $CH_3CH(OH)COOH(s) + 3\,O_2(g) \rightarrow 3\,CO_2(g) + 3\,H_2O(l)$

9. **(E)** **(a)** $NH_4NO_3(s) \xrightarrow{\Delta} N_2O(g) + 2\,H_2O(g)$

(b) $Na_2CO_3(aq) + 2\,HCl(aq) \rightarrow 2\,NaCl(aq) + H_2O(l) + CO_2(g)$

(c) $2\,CH_4(g) + 2\,NH_3(g) + 3\,O_2(g) \rightarrow 2\,HCN(g) + 6\,H_2O(g)$

11. **(E)**

Unbalanced reaction:	$N_2H_4(g) + N_2O_4(g)$	$\rightarrow H_2O(g) + N_2(g)$
Balance H atoms:	$N_2H_4(g) + N_2O_4(g)$	$\rightarrow 2\,H_2O(g) + N_2(g)$
Balance O atoms:	$N_2H_4(g) + 1/2\,N_2O_4(g)$	$\rightarrow 2\,H_2O(g) + N_2(g)$
Balance N atoms:	$N_2H_4(g) + 1/2\,N_2O_4(g)$	$\rightarrow 2\,H_2O(g) + 3/2\,N_2(g)$
Multiply by 2 (whole #)	$2\,N_2H_4(g) + N_2O_4(g)$	$\rightarrow 4\,H_2O(g) + 3\,N_2(g)$
Self Check:	$6\,N + 8\,H + 4\,O$	$\rightarrow 6\,N + 8\,H + 4\,O$

Stoichiometry of Chemical Reactions

13. **(E)** In order to write the balanced chemical equation for the reaction, we will need to determine the formula of the chromium oxide product.
First determine the number of moles of chromium and oxygen, and then calculate the mole ratio.

$$\# \text{ mol Cr} = 0.689 \text{ g Cr} \times \frac{1 \text{ mol Cr}}{52.00 \text{ g Cr}} = 0.01325 \text{ mol Cr}$$

$$\# \text{ mol O} = 0.636 \text{ g O}_2 \times \frac{1 \text{ mol O}_2}{32.00 \text{ g O}_2} \times \frac{2 \text{ mol O}}{1 \text{ mol O}_2} = 0.03975 \text{ mol O}$$

$$\frac{0.03975 \text{ mol O}}{0.01325 \text{ mol Cr}} = \frac{3 \text{ mol O}}{1 \text{ mol Cr}} \quad \text{Therefore, the formula for the product is } CrO_3.$$

Balanced equation = 　　　$2\,Cr(s) + 3\,O_2(g) \rightarrow 2\,CrO_3(s)$

15. **(E)** The conversion factor is obtained from the balanced chemical equation.

$$515 \text{ g Cl}_2 \times \frac{1 \text{ mol Cl}_2}{70.90 \text{ g Cl}_2} = 7.26 \text{ mol Cl}_2$$

$$\text{moles FeCl}_3 = 7.26 \text{ mol Cl}_2 \times \frac{2 \text{ mol FeCl}_3}{3 \text{ mol Cl}_2} = 4.84 \text{ mol FeCl}_3$$

17. **(E)**

(a) *Conversion pathway approach:*

$$32.8 \text{ g KClO}_3 \times \frac{1 \text{ mol KClO}_3}{122.6 \text{ g KClO}_3} \times \frac{3 \text{ mol O}_2}{2 \text{ mol KClO}_3} = 0.401 \text{ mol O}_2$$

Stepwise approach:

$$32.8 \text{ g KClO}_3 \times \frac{1 \text{ mol KClO}_3}{122.6 \text{ g KClO}_3} = 0.268 \text{ mol KClO}_3$$

$$0.268 \text{ mol KClO}_3 \times \frac{3 \text{ mol O}_2}{2 \text{ mol KClO}_3} = 0.402 \text{ mol O}_2$$

(b) *Conversion pathway approach:*

$$\text{mass KClO}_3 = 50.0 \text{ g O}_2 \times \frac{1 \text{ mol O}_2}{32.00 \text{ g O}_2} \times \frac{2 \text{ mol KClO}_3}{3 \text{ mol O}_2} \times \frac{122.6 \text{ g KClO}_3}{1 \text{ mol KClO}_3} = 128 \text{ g KClO}_3$$

Stepwise approach:

$$50.0 \text{ g O}_2 \times \frac{1 \text{ mol O}_2}{32.00 \text{ g O}_2} = 1.56 \text{ mol O}_2$$

$$1.56 \text{ mol O}_2 \times \frac{2 \text{ mol KClO}_3}{3 \text{ mol O}_2} = 1.04 \text{ mol KClO}_3$$

$$1.04 \text{ mol KClO}_3 \times \frac{122.6 \text{ g KClO}_3}{1 \text{ mol KClO}_3} = 128 \text{ g KClO}_3$$

(c) *Conversion pathway approach:*

$$\text{mass KCl} = 28.3 \text{ g O}_2 \times \frac{1 \text{ mol O}_2}{32.00 \text{ g O}_2} \times \frac{2 \text{ mol KCl}}{3 \text{ mol O}_2} \times \frac{74.55 \text{ g KCl}}{1 \text{ mol KCl}} = 43.9 \text{ g KCl}$$

Stepwise approach:

$$28.3 \text{ g O}_2 \times \frac{1 \text{ mol O}_2}{32.00 \text{ g O}_2} = 0.884 \text{ mol O}_2$$

$$0.884 \text{ mol O}_2 \times \frac{2 \text{ mol KCl}}{3 \text{ mol O}_2} = 0.589 \text{ mol KCl}$$

$$0.589 \text{ mol KCl} \times \frac{74.55 \text{ g KCl}}{1 \text{ mol KCl}} = 43.9 \text{ g KCl}$$

19. **(M)** Balance the given equation, and then solve the problem.

$$2\,Ag_2CO_3\,(s)\xrightarrow{\Delta}4\,Ag\,(s)+2\,CO_2\,(g)+O_2\,(g)$$

$$\text{mass }Ag_2CO_3 = 75.1\,g\,Ag\times\frac{1\,mol\,Ag}{107.87\,g\,Ag}\times\frac{2\,mol\,Ag_2CO_3}{4\,mol\,Ag}\times\frac{275.75\,g\,Ag_2CO_3}{1\,mol\,Ag_2CO_3}=96.0\,g\,Ag_2CO_3$$

21. **(M)** The balanced equation is $CaH_2(s) + 2\,H_2O(l) \rightarrow Ca(OH)_2(s) + 2\,H_2(g)$

(a) $\text{mass }H_2 = 127\,g\,CaH_2\times\dfrac{1\,mol\,CaH_2}{42.094\,g\,CaH_2}\times\dfrac{2\,mol\,H_2}{1\,mol\,CaH_2}\times\dfrac{2.016\,g\,H_2}{1\,mol\,H_2}=12.2\,g\,H_2$

(b) $\text{mass }H_2O = 56.2\;g\,CaH_2\times\dfrac{1\,mol\,CaH_2}{42.094\;g\,CaH_2}\times\dfrac{2\;mol\,H_2O}{1\,mol\,CaH_2}\times\dfrac{18.015\,g\,H_2O}{1\;mol\,H_2O}=48.1\,g\,H_2O$

(c) $\text{mass }CaH_2 = 8.12\times10^{24}\text{ molecules }H_2\times\dfrac{1\,mol\,H_2}{6.022\times10^{23}\text{ molecules }H_2}\times\dfrac{1\,mol\,CaH_2}{2\,mol\,H_2}\times\dfrac{42.094\,g\,CaH_2}{1\,mol\,CaH_2}$

$$= 284\,g\,CaH_2$$

23. **(M)** The balanced equation is $Fe_2O_3\,(s)+3\,C\,(s)\xrightarrow{\Delta}2\,Fe\,(l)+3\,CO\,(g)$

$$\text{mass }Fe_2O_3 = 523\text{ kg Fe}\times\frac{1\text{ kmol Fe}}{55.85\text{ kg Fe}}\times\frac{1\text{ kmol }Fe_2O_3}{2\text{ kmol Fe}}\times\frac{159.7\text{ kg }Fe_2O_3}{1\text{ kmol }Fe_2O_3}=748\text{ kg }Fe_2O_3$$

$$\%\,Fe_2O_3\text{ in ore} = \frac{748\text{ kg }Fe_2O_3}{938\text{ kg ore}}\times100\%=79.7\%\;Fe_2O_3$$

25. **(M)** $B_{10}H_{14} + 11\,O_2 \rightarrow 5\,B_2O_3 + 7\,H_2O$

$$\%\text{ by mass }B_{10}H_{14} = \frac{\#\text{ g }B_{10}H_{14}}{\#\text{ g }B_{10}H_{14} + \#\text{ g }O_2}\times100$$

1 mol $B_{10}H_{14}$ reacts with 11 mol O_2 (exactly)

$$\text{mass }B_{10}H_{14} = 1\text{ mol }B_{10}H_{14}\times\frac{122.21\text{ g }B_{10}H_{14}}{1\text{ mol }B_{10}H_{14}}=122.22\text{ g }B_{10}H_{14}$$

$$\text{mass }O_2 = 11\text{ mol }O_2\times\frac{32.00\text{ g }O_2}{1\text{ mol }O_2}=352.00\text{ g }O_2$$

$$\%\text{ by mass }B_{10}H_{14} = \frac{122.21\text{ g }B_{10}H_{14}}{122.22\text{ g }B_{10}H_{14} + 352.00\text{ g }O_2}\times100\%=25.8$$

27. (E) $2\,Al(s) + 6\,HCl(aq) \rightarrow 2\,AlCl_3(aq) + 3\,H_2(g)$. First determine the mass of Al in the foil.

$$\text{mass Al} = (10.25\,cm \times 5.50\,cm \times 0.601\,mm) \times \frac{1\,cm}{10\,mm} \times \frac{2.70\,g\,Al}{1\,cm^3} = 9.15\,g\,Al$$

$$\text{mass H}_2 = 9.15\,g\,Al \times \frac{1\,mol\,Al}{26.98\,g\,Al} \times \frac{3\,mol\,H_2}{2\,mol\,Al} \times \frac{2.016\,g\,H_2}{1\,mol\,H_2} = 1.03\,g\,H_2$$

29. (E) First write the balanced chemical equation for each reaction.

$$2\,Na(s) + 2\,HCl(aq) \rightarrow 2\,NaCl(aq) + H_2(g) \qquad Mg(s) + 2\,HCl(aq) \rightarrow MgCl_2(aq) + H_2(g)$$
$$2\,Al(s) + 6\,HCl(aq) \rightarrow 2\,AlCl_3(aq) + 3\,H_2(g) \qquad Zn(s) + 2\,HCl(aq) \rightarrow ZnCl_2(aq) + H_2(g)$$

Three of the reactions—those of Na, Mg, and Zn—produce 1 mole of $H_2(g)$. The one of these three that produces the most hydrogen per gram of metal is the one for which the metal's atomic mass is the smallest, remembering to compare twice the atomic mass for Na. The atomic masses are: 2×23 u for Na, 24.3 u for Mg, and 65.4 u for Zn. Thus, among these three, Mg produces the most H_2 per gram of metal, specifically 1 mol H_2 per 24.3 g Mg. In the case of Al, 3 moles of H_2 are produced by 2 moles of the metal, or 54 g Al. This reduces as follows: 3 mol H_2 / 54 g Al = 1 mol H_2 / 18 g Al. Thus, Al produces the largest amount of H_2 per gram of metal.

Molarity

31. (M)

(a) CH_3OH molarity $(c) = \dfrac{2.92\ \text{mol } CH_3OH}{7.16\ L} = 0.408\ M$

(b) CH_3CH_2OH molarity $(c) = \dfrac{7.69\ \text{mmol } CH_3CH_2OH}{50.00\ mL} = 0.154\ M$

(c) $CO(NH_2)_2$ molarity $(c) = \dfrac{25.2\ g\ CO(NH_2)_2}{275\ mL} \times \dfrac{1\,mol\ CO(NH_2)_2}{60.06\ g\ CO(NH_2)_2} \times \dfrac{1000\ mL}{1\ L} = 1.53\ M$

33. (E)

(a) *Conversion pathway approach:*

$$[C_{12}H_{22}O_{11}] = \frac{150.0\ g\ C_{12}H_{22}O_{11}}{250.0\ mL\ \text{soln}} \times \frac{1000\ mL}{1\ L} \times \frac{1\ mol\ C_{12}H_{22}O_{11}}{342.3\ g\ C_{12}H_{22}O_{11}} = 1.753\ M$$

Stepwise approach:

$$150.0\ g\ C_{12}H_{22}O_{11} \times \frac{1\ mol\ C_{12}H_{22}O_{11}}{342.3\ g\ C_{12}H_{22}O_{11}} = 0.4382\ mol\ C_{12}H_{22}O_{11}$$

$$250.0\ mL\ \text{soln} \times \frac{1\ L}{1000\ mL} = 0.2500\ L$$

$$[C_{12}H_{22}O_{11}] = \frac{0.4382\ mol\ C_{12}H_{22}O_{11}}{0.2500\ L} = 1.753\ M$$

(b) *Conversion pathway approach:*

$$[CO(NH_2)_2] = \frac{98.3 \text{ mg solid}}{5.00 \text{ mL soln}} \times \frac{97.9 \text{ mg CO(NH}_2)_2}{100 \text{ mg solid}} \times \frac{1 \text{ mmol CO(NH}_2)_2}{60.06 \text{ mg CO(NH}_2)_2}$$

$$= 0.320 \text{ M CO(NH}_2)_2$$

Stepwise approach:

$$98.3 \text{ mg solid} \times \frac{97.9 \text{ mg CO(NH}_2)_2}{100 \text{ mg solid}} = 96.2 \text{ mg CO(NH}_2)_2$$

$$96.2 \text{ mg CO(NH}_2)_2 \times \frac{1 \text{ g CO(NH}_2)_2}{1000 \text{ mg CO(NH}_2)_2} = 0.0962 \text{ g CO(NH}_2)_2$$

$$0.0962 \text{ g CO(NH}_2)_2 \times \frac{1 \text{ mol CO(NH}_2)_2}{60.06 \text{ g CO(NH}_2)_2} = 1.60 \times 10^{-3} \text{ mol CO(NH}_2)_2$$

$$5.00 \text{ mL soln} \times \frac{1 \text{ L}}{1000 \text{ mL}} = 0.00500 \text{ L}$$

$$[CO(NH_2)_2] = \frac{1.60 \times 10^{-3} \text{ mol CO(NH}_2)_2}{0.00500 \text{ L}} = 0.320 \text{ M}$$

(c) *Conversion pathway approach:*

$$[CH_3OH] = \frac{125.0 \text{ mL CH}_3\text{OH}}{15.0 \text{ L soln}} \times \frac{0.792 \text{ g}}{1 \text{ mL}} \times \frac{1 \text{ mol CH}_3\text{OH}}{32.04 \text{ g CH}_3\text{OH}} = 0.206 \text{ M}$$

Stepwise approach:

$$[CH_3OH] = \frac{125.0 \text{ mL CH}_3\text{OH}}{15.0 \text{ L soln}} \times \frac{0.792 \text{ g}}{1 \text{ mL}} \times \frac{1 \text{ mol CH}_3\text{OH}}{32.04 \text{ g CH}_3\text{OH}} = 0.206 \text{ M}$$

$$125.0 \text{ mL CH}_3\text{OH} \times \frac{0.792 \text{ g}}{1 \text{ mL}} = 99.0 \text{ g CH}_3\text{OH}$$

$$99.0 \text{ g CH}_3\text{OH} \times \frac{1 \text{ mol CH}_3\text{OH}}{32.04 \text{ g CH}_3\text{OH}} = 3.09 \text{ mol CH}_3\text{OH}$$

$$[CH_3OH] = \frac{3.09 \text{ mol CH}_3\text{OH}}{15.0 \text{ L soln}} = 0.206 \text{ M}$$

35. (E)

(a) $\text{mass C}_6\text{H}_{12}\text{O}_6 = 75.0 \text{ mL soln} \times \dfrac{1 \text{ L}}{1000 \text{ mL}} \times \dfrac{0.350 \text{ mol C}_6\text{H}_{12}\text{O}_6}{1 \text{ L soln}} \times \dfrac{180.16 \text{ g C}_6\text{H}_{12}\text{O}_6}{1 \text{ mol C}_6\text{H}_{12}\text{O}_6} = 4.73 \text{ g}$

(b) $V_{CH_3OH} = 2.25 \text{ L soln} \times \dfrac{0.485 \text{ mol}}{1 \text{ L}} \times \dfrac{32.04 \text{ g CH}_3\text{OH}}{1 \text{ mol CH}_3\text{OH}} \times \dfrac{1 \text{ mL}}{0.792 \text{ g}} = 44.1 \text{ mL CH}_3\text{OH}$

37. (M)

(a) $$\frac{85 \text{ mg C}_6\text{H}_{12}\text{O}_6}{1 \text{ dL blood}} \times \frac{1 \text{ g}}{1000 \text{ mg}} \times \frac{10 \text{ dL}}{1 \text{ L}} \times \frac{1 \text{ mol C}_6\text{H}_{12}\text{O}_6}{180.16 \text{ g C}_6\text{H}_{12}\text{O}_6} \times \frac{1 \text{ mmol C}_6\text{H}_{12}\text{O}_6}{1 \times 10^{-3} \text{ mol C}_6\text{H}_{12}\text{O}_6}$$

$$= 4.7 \frac{\text{mmol C}_6\text{H}_{12}\text{O}_6}{\text{L}}$$

(b) Molarity $= 4.7 \times 10^{-3} \dfrac{\text{mol C}_6\text{H}_{12}\text{O}_6}{\text{L}}$

39. (E) First we determine each concentration in moles per liter and find the 0.500 M solution.

(a) $[\text{KCl}] = \dfrac{0.500 \text{ g KCl}}{1 \text{ mL}} \times \dfrac{1 \text{ mol KCl}}{74.548 \text{ g KCl}} \times \dfrac{1000 \text{ mL}}{1 \text{ L}} = 6.71 \text{ M KCl}$

(b) $[\text{KCl}] = \dfrac{36.0 \text{ g KCl}}{1 \text{ L}} \times \dfrac{1 \text{ mol KCl}}{74.548 \text{ g KCl}} = 0.483 \text{ M KCl}$

(c) $[\text{KCl}] = \dfrac{7.46 \text{ mg KCl}}{1 \text{ mL}} \times \dfrac{1 \text{ g KCl}}{1000 \text{ mg KCl}} \times \dfrac{1 \text{ mol KCl}}{74.548 \text{ g KCl}} \times \dfrac{1000 \text{ mL}}{1 \text{ L}} = 0.100 \text{ M KCl}$

(d) $[\text{KCl}] = \dfrac{373 \text{ g KCl}}{10.00 \text{ L}} \times \dfrac{1 \text{ mol KCl}}{74.548 \text{ g KCl}} = 0.500 \text{ M KCl}$

Solution (d) is a 0.500 M KCl solution.

41. (E) We determine the molar concentration for the 46% by mass sucrose solution.

$$[\text{C}_{12}\text{H}_{22}\text{O}_{11}] = \frac{46 \text{ g C}_{12}\text{H}_{22}\text{O}_{11} \times \dfrac{1 \text{ mol C}_{12}\text{H}_{22}\text{O}_{11}}{342.3 \text{ g C}_{12}\text{H}_{22}\text{O}_{11}}}{100 \text{ g soln} \times \dfrac{1 \text{ mL}}{1.21 \text{ g soln}} \times \dfrac{1 \text{ L}}{1000 \text{ mL}}} = 1.6 \text{ M}$$

The 46% by mass sucrose solution is the more concentrated.

43. (E) $[\text{KNO}_3] = \dfrac{0.01000 \text{ L conc'd soln} \times \dfrac{2.05 \text{ mol KNO}_3}{1 \text{ L}}}{0.250 \text{ L diluted solution}} = 0.0820 \text{ M}$

45. (E) Both the diluted and concentrated solutions contain the same number of moles of K_2SO_4. This number is given in the numerator of the following expression.

$$\text{K}_2\text{SO}_4 \text{ molarity} = \frac{0.125 \text{ L} \times \dfrac{0.198 \text{ mol K}_2\text{SO}_4}{1 \text{ L}}}{0.105 \text{ L}} = 0.236 \text{ M K}_2\text{SO}_4$$

47. (E) Let us compute how many mL of dilute $_{(\text{dil})}$ solution we obtain from each mL of concentrated $_{(\text{conc})}$ solution. $c_{\text{dil}} V_{\text{dil}} = c_{\text{conc}} V_{\text{conc}}$ becomes $x \text{ mL} \times 0.0125 \text{ M} = 1.00 \text{ mL} \times 0.250 \text{ M}$ and $x = 20$ Thus, the ratio of the volume of the volumetric flask to that of the pipet would be 20:1. We could use a 100.0 mL flask and a 5.00 mL pipet, a 1000.0 mL flask and a 50.00 mL pipet, or a 500.0 mL flask and a 25.00 mL pipet. There are many combinations that could be used.

Chemical Reactions in Solutions

49. (M)

(a) $\text{mass Na}_2\text{S} = 27.8\,\text{mL} \times \dfrac{1\,\text{L}}{1000\,\text{mL}} \times \dfrac{0.163\,\text{mol AgNO}_3}{1\,\text{L soln}} \times \dfrac{1\,\text{mol Na}_2\text{S}}{2\,\text{mol AgNO}_3}$

$\times \dfrac{78.05\,\text{g Na}_2\text{S}}{1\,\text{mol Na}_2\text{S}} = 0.177\,\text{g Na}_2\text{S}$

(b) $\text{mass Ag}_2\text{S} = 0.177\,\text{g Na}_2\text{S} \times \dfrac{1\,\text{mol Na}_2\text{S}}{78.04\,\text{g Na}_2\text{S}} \times \dfrac{1\,\text{mol Ag}_2\text{S}}{1\,\text{mol Na}_2\text{S}} \times \dfrac{247.80\,\text{g Ag}_2\text{S}}{1\,\text{mol Ag}_2\text{S}} = 0.562\,\text{g Ag}_2\text{S}$

51. (M) The molarity can be expressed as millimoles of solute per milliliter of solution.

$V_{\text{K}_2\text{CrO}_4} = 415\,\text{mL} \times \dfrac{0.186\,\text{mmol AgNO}_3}{1\,\text{mL soln}} \times \dfrac{1\,\text{mmol K}_2\text{CrO}_4}{2\,\text{mmol AgNO}_3} \times \dfrac{1\,\text{mL K}_2\text{CrO}_4\,(\text{aq})}{0.650\,\text{mmol K}_2\text{CrO}_4}$

$= 59.4\,\text{mL K}_2\text{CrO}_4$

53. (D) The balanced chemical equation for the reaction is:

$2\,\text{HNO}_3(\text{aq}) + \text{Ca(OH)}_2(\text{aq}) \rightarrow \text{Ca(NO}_3)_2(\text{aq}) + 2\,\text{H}_2\text{O}(\text{l})$

$\# \text{mol HNO}_3 = 0.02978\,\text{L soln} \times \dfrac{0.0142\,\text{mol Ca(OH)}_2}{1\,\text{L soln}} \times \dfrac{2\,\text{mol HNO}_3}{1\,\text{mol Ca(OH)}_2} = 8.46 \times 10^{-4}\,\text{mol HNO}_3$

All of the HNO_3 that reacts was contained in the initial, undiluted 1.00 mL sample. Since the moles of HNO_3 are the same in the diluted and undiluted solutions, one can divide the moles of HNO_3 by the volume of the undiluted solution to obtain the molarity.

$\text{Molarity} = \dfrac{8.46 \times 10^{-4}\,\text{mol HNO}_3}{0.00100\,\text{L}} = 8.46 \times 10^{-3}\,\dfrac{\text{mol HNO}_3}{\text{L}}$

55. (M)

(a) We know that the Al forms the AlCl_3.

$\text{mol AlCl}_3 = 1.87\,\text{g Al} \times \dfrac{1\,\text{mol Al}}{26.98\,\text{g Al}} \times \dfrac{1\,\text{mol AlCl}_3}{1\,\text{mol Al}} = 0.0693\,\text{mol AlCl}_3$

(b) $[\text{AlCl}_3] = \dfrac{0.0693\,\text{mol AlCl}_3}{23.8\,\text{mL}} \times \dfrac{1000\,\text{mL}}{1\,\text{L}} = 2.91\,\text{M AlCl}_3$

57. (M) The volume of solution determines the amount of product.

$\text{mass Ag}_2\text{CrO}_4 = 415\,\text{mL} \times \dfrac{1\,\text{L}}{1000\,\text{mL}} \times \dfrac{0.186\,\text{mol AgNO}_3}{1\,\text{L soln}} \times \dfrac{1\,\text{mol Ag}_2\text{CrO}_4}{2\,\text{mol AgNO}_3} \times \dfrac{331.73\,\text{g Ag}_2\text{CrO}_4}{1\,\text{mol Ag}_2\text{CrO}_4}$

$\text{mass Ag}_2\text{CrO}_4 = 12.8\,\text{g Ag}_2\text{CrO}_4$

59. **(M)** mass Na $= 155 \, \text{mL soln} \times \dfrac{1 \, \text{L}}{1000 \, \text{mL}} \times \dfrac{0.175 \, \text{mol NaOH}}{1 \, \text{L soln}} \times \dfrac{2 \, \text{mol Na}}{2 \, \text{mol NaOH}} \times \dfrac{22.99 \, \text{g Na}}{1 \, \text{mol Na}}$

$= 0.624 \, \text{g Na}$

61. **(M)** The mass of oxalic acid enables us to determine the amount of NaOH in the solution.

$[\text{NaOH}] = \dfrac{0.3126 \, \text{g H}_2\text{C}_2\text{O}_4}{26.21 \, \text{mL soln}} \times \dfrac{1000 \, \text{mL}}{1 \, \text{L soln}} \times \dfrac{1 \, \text{mol H}_2\text{C}_2\text{O}_4}{90.04 \, \text{g H}_2\text{C}_2\text{O}_4} \times \dfrac{2 \, \text{mol NaOH}}{1 \, \text{mol H}_2\text{C}_2\text{O}_4} = 0.2649 \, \text{M}$

Determining the Limiting Reactant

63. **(E)** The limiting reactant is NH_3. For every mole of $\text{NH}_3(g)$ that reacts, a mole of $\text{NO}(g)$ forms. Since 3.00 moles of $\text{NH}_3(g)$ reacts, 3.00 moles of $\text{NO}(g)$ forms (1:1 mole ratio).

65. **(M)** First we must determine the number of moles of NO produced by each reactant. The one producing the smaller amount of NO is the limiting reactant.

$\text{mol NO} = 0.696 \, \text{mol Cu} \times \dfrac{2 \, \text{mol NO}}{3 \, \text{mol Cu}} = 0.464 \, \text{mol NO}$

Conversion pathway approach:

$\text{mol NO} = 136 \, \text{mL HNO}_3 \, (\text{aq}) \times \dfrac{1 \, \text{L}}{1000 \, \text{mL}} \times \dfrac{6.0 \, \text{mol HNO}_3}{1 \, \text{L}} \times \dfrac{2 \, \text{mol NO}}{8 \, \text{mol HNO}_3} = 0.204 \, \text{mol NO}$

Stepwise approach:

$136 \, \text{mL HNO}_3 \, (\text{aq}) \times \dfrac{1 \, \text{L}}{1000 \, \text{mL}} = 0.136 \, \text{L HNO}_3$

$0.136 \, \text{L} \times \dfrac{6.0 \, \text{mol HNO}_3}{1 \, \text{L}} = 0.816 \, \text{mol HNO}_3$

$0.816 \, \text{mol HNO}_3 \times \dfrac{2 \, \text{mol NO}}{8 \, \text{mol HNO}_3} = 0.204 \, \text{mol NO}$

Since $\text{HNO}_3(\text{aq})$ is the limiting reactant, it will be completely consumed, leaving some Cu unreacted.

67. **(M)** First we need to determine the amount of Na_2CS_3 produced from each of the reactants.

$n_{\text{Na}_2\text{CS}_3 \, (\text{from CS}_2)} = 92.5 \, \text{mL CS}_2 \times \dfrac{1.26 \, \text{g}}{1 \, \text{mL}} \times \dfrac{1 \, \text{mol CS}_2}{76.13 \, \text{g CS}_2} \times \dfrac{2 \, \text{mol Na}_2\text{CS}_3}{3 \, \text{mol CS}_2} = 1.02 \, \text{mol Na}_2\text{CS}_3$

$n_{\text{Na}_2\text{CS}_3 \, (\text{from NaOH})} = 2.78 \, \text{mol NaOH} \times \dfrac{2 \, \text{mol Na}_2\text{CS}_3}{6 \, \text{mol NaOH}} = 0.927 \, \text{mol Na}_2\text{CS}_3$

Thus, the mass produced is $0.927 \, \text{mol Na}_2\text{CS}_3 \times \dfrac{154.2 \, \text{g Na}_2\text{CS}_3}{1 \, \text{mol Na}_2\text{CS}_3} = 143 \, \text{g Na}_2\text{CS}_3$

69. **(D)**

$$Ca(OH)_2(s) + 2NH_4Cl(s) \rightarrow CaCl_2(aq) + 2H_2O(l) + 2NH_3(g)$$

First compute the amount of NH_3 formed from each reactant in this limiting reactant problem.

$$n_{NH_3 \text{ (from NH}_4\text{Cl)}} = 33.0 \text{ g NH}_4\text{Cl} \times \frac{1 \text{ mol NH}_4\text{Cl}}{53.49 \text{ g NH}_4\text{Cl}} \times \frac{2 \text{ mol NH}_3}{2 \text{ mol NH}_4\text{Cl}} = 0.617 \text{ mol NH}_3$$

$$n_{NH_3 \text{ (from Ca(OH)}_2)} = 33.0 \text{ g Ca(OH)}_2 \times \frac{1 \text{ mol Ca(OH)}_2}{74.09 \text{ g Ca(OH)}_2} \times \frac{2 \text{ mol NH}_3}{1 \text{ mol Ca(OH)}_2} = 0.891 \text{ mol NH}_3$$

Thus, 0.617 mol NH_3 should be produced as NH_4Cl is the limiting reactant.

$$\text{mass NH}_3 = 0.617 \text{ mol NH}_3 \times \frac{17.03 \text{ g NH}_3}{1 \text{ mol NH}_3} = 10.5 \text{ g NH}_3$$

Now we will determine the mass of reactant in excess, $Ca(OH)_2$.

$$Ca(OH)_2 \text{ used} = 0.617 \text{ mol NH}_3 \times \frac{1 \text{ mol Ca(OH)}_2}{2 \text{ mol NH}_3} \times \frac{74.09 \text{ g Ca(OH)}_2}{1 \text{ mol Ca(OH)}_2} = 22.9 \text{ g Ca(OH)}_2$$

$$\text{excess mass Ca(OH)}_2 = 33.0 \text{ g Ca(OH)}_2 - 22.9 \text{ g Ca(OH)}_2 = 10.1 \text{ g excess Ca(OH)}_2$$

71. **(M)** The number of grams of $CrSO_4$ that can be made from the reaction mixture is determined by finding the limiting reactant, and using the limiting reactant to calculate the mass of product that can be formed. The limiting reactant can determined by calculating the amount of product formed from each of the reactants. Whichever reactant produces the smallest amount of product is the limiting reactant.

$$3.2 \text{ mol Zn} \times \frac{2 \text{ mol CrSO}_4}{4 \text{ mol Zn}} \times \frac{148.06 \text{ g CrSO}_4}{1 \text{ mol CrSO}_4} = 236.90 \text{ g CrSO}_4$$

$$1.7 \text{ mol K}_2\text{Cr}_2\text{O}_7 \times \frac{2 \text{ mol CrSO}_4}{1 \text{ mol K}_2\text{Cr}_2\text{O}_7} \times \frac{148.06 \text{ g CrSO}_4}{1 \text{ mol CrSO}_4} = 503.40 \text{ g CrSO}_4$$

$$5.0 \text{ mol H}_2\text{SO}_4 \times \frac{2 \text{ mol CrSO}_4}{7 \text{ mol H}_2\text{SO}_4} \times \frac{148.06 \text{ g CrSO}_4}{1 \text{ mol CrSO}_4} = 211.51 \text{ g CrSO}_4$$

H_2SO_4 is the limiting reactant since it produces the least amount of $CrSO_4$. Therefore, the maximum number of grams of $CrSO_4$ that can be made is 211.51 g.

Theoretical, Actual, and Percent Yields

73. **(M)**

(a) $$277 \text{ g CCl}_4 \times \frac{1 \text{ mol CCl}_4}{153.81 \text{ g CCl}_4} = 1.80 \text{ mol CCl}_4$$

Since the stoichiometry indicates that 1 mole CCl_2F_2 is produced per mole CCl_4, the use of 1.80 mole CCl_4 should produce 1.80 mole CCl_2F_2. This is the theoretical yield of the reaction.

(b) $187 \text{g CCl}_2\text{F}_2 \times \dfrac{1 \text{mol CCl}_2\text{F}_2}{120.91 \text{g CCl}_2\text{F}_2} = 1.55 \text{ mol CCl}_2\text{F}_2$

The actual yield of the reaction is the amount actually produced, $1.55 \text{ mol CCl}_2\text{F}_2$.

(c) $\% \text{ yield} = \dfrac{1.55 \text{ mol CCl}_2\text{F}_2 \text{ obtained}}{1.80 \text{ mol CCl}_2\text{F}_2 \text{ calculated}} \times 100\% = 86.1\% \text{ yield}$

75. **(D)** $\% \text{ yield} = \dfrac{\text{actual yield}}{\text{theoretical yield}} \times 100\%$

The actual yield is given in the problem and is equal to 28.2 g.
In order to determine the theoretical yield, we must find the limiting reactant and do stoichiometry.

Conversion pathway approach:

$7.81 \text{g Al}_2\text{O}_3 \times \dfrac{1 \text{ mol Al}_2\text{O}_3}{101.96 \text{ g Al}_2\text{O}_3} \times \dfrac{2 \text{ mol Na}_3\text{AlF}_6}{1 \text{ mol Al}_2\text{O}_3} \times \dfrac{209.94 \text{ g Na}_3\text{AlF}_6}{1 \text{ mol Na}_3\text{AlF}_6} = 32.2 \text{ g Na}_3\text{AlF}_6$

$3.50 \text{ L} \times \dfrac{0.141 \text{ mol}}{1 \text{ L}} \times \dfrac{2 \text{ mol Na}_3\text{AlF}_6}{6 \text{ mol NaOH}} \times \dfrac{209.94 \text{ g Na}_3\text{AlF}_6}{1 \text{ mol Na}_3\text{AlF}_6} = 34.5 \text{ g Na}_3\text{AlF}_6$

Stepwise approach:

Amount of Na₃AlF₆ produced from Al₂O₃ if all Al₂O₃ reacts

$7.81 \text{g Al}_2\text{O}_3 \times \dfrac{1 \text{ mol Al}_2\text{O}_3}{101.96 \text{ g Al}_2\text{O}_3} = 0.0766 \text{ mol Al}_2\text{O}_3$

$0.0766 \text{ mol Al}_2\text{O}_3 \times \dfrac{2 \text{ mol Na}_3\text{AlF}_6}{1 \text{ mol Al}_2\text{O}_3} = 0.153 \text{ mol Na}_3\text{AlF}_6$

$0.153 \text{ mol Na}_3\text{AlF}_6 \times \dfrac{209.94 \text{ g Na}_3\text{AlF}_6}{1 \text{ mol Na}_3\text{AlF}_6} = 32.1 \text{ g Na}_3\text{AlF}_6$

Amount of Na₃AlF₆ produced from NaOH if all NaOH reacts

$3.50 \text{ L} \times \dfrac{0.141 \text{ mol NaOH}}{1 \text{ L}} = 0.494 \text{ mol NaOH}$

$0.494 \text{ mol NaOH} \times \dfrac{2 \text{ mol Na}_3\text{AlF}_6}{6 \text{ mol NaOH}} = 0.165 \text{ mol Na}_3\text{AlF}_6$

$0.165 \text{ mol Na}_3\text{AlF}_6 \times \dfrac{209.94 \text{ g Na}_3\text{AlF}_6}{1 \text{ mol Na}_3\text{AlF}_6} = 34.5 \text{ g Na}_3\text{AlF}_6$

Al_2O_3 is the limiting reactant. $\qquad \% \text{ yield} = \dfrac{28.2 \text{ g}}{32.2 \text{ g}} \times 100\% = 87.6\%$

81. (E) A less-than-100% yield of desired product in synthesis reactions is always the case. This is because of side reactions that yield products other than those desired and because of the loss of material on the glassware, on filter paper, etc. during the various steps of the procedure. A main criterion for choosing a synthesis reaction is how economically it can be run. In the analysis of a compound, on the other hand, it is essential that all of the material present be detected. Therefore, a 100% yield is required; none of the material present in the sample can be lost during the analysis. Therefore analysis reactions are carefully chosen to meet this 100 % yield criterion; they need not be economical to run.

Consecutive Reactions, Simultaneous Reactions

84. (D) Here we need to determine the amount of CO_2 produced from each reactant.

$$C_3H_8(g) + 5O_2(g) \longrightarrow 3CO_2(g) + 4H_2O(l)$$

$$2C_4H_{10}(g) + 13O_2(g) \longrightarrow 8CO_2(g) + 10H_2O(l)$$

$$n_{CO_2 \, (from \, C_3H_8)} = 406\,g\,mixt. \times \frac{72.7g\,C_3H_8}{100.0\,g\,mixt.} \times \frac{1mol\,C_3H_8}{44.10\,g\,C_3H_8} \times \frac{3\,mol\,CO_2}{1\,mol\,C_3H_8} = 20.1\,mol\,CO_2$$

$$n_{CO_2 \, (from \, C_4H_{10})} = 406\,g\,mixt. \times \frac{27.3g\,C_4H_{10}}{100.0\,g\,mixt} \times \frac{1mol\,C_4H_{10}}{58.12\,g\,C_4H_{10}} \times \frac{8\,mol\,CO_2}{2\,mol\,C_4H_{10}} = 7.63\,mol\,CO_2$$

$$mass\,CO_2 = (20.1 + 7.63)\,mol\,CO_2 \times \frac{44.01g\,CO_2}{1\,mol\,CO_2} = 1.22 \times 10^3\,g\,CO_2$$

86. (M) Balanced Equations:

$$C_2H_6(g) + \tfrac{7}{2}\,O_2(g) \rightarrow 2\,CO_2(g) + 3\,H_2O(l)$$
$$CO_2(g) + Ba(OH)_2(aq) \rightarrow BaCO_3(s) + H_2O(l)$$

Conversion pathway approach:

$$mass_{C_2H_6} = 0.506\,g\,BaCO_3 \times \frac{1mol\,BaCO_3}{197.3\,g\,BaCO_3} \times \frac{1mol\,CO_2}{1mol\,BaCO_3} \times \frac{2\,mol\,C_2H_6}{4\,mol\,CO_2} \times \frac{30.07\,g\,C_2H_6}{1mol\,C_2H_6} = 0.0386\,g\,C_2H_6$$

Stepwise approach:

$$0.506\,g\,BaCO_3 \times \frac{1mol\,BaCO_3}{197.3\,g\,BaCO_3} = 2.56 \times 10^{-3}\,mol\,BaCO_3$$

$$2.56 \times 10^{-3}\,mol\,BaCO_3 \times \frac{1mol\,CO_2}{1mol\,BaCO_3} = 2.56 \times 10^{-3}\,mol\,CO_2$$

$$2.56 \times 10^{-3}\,mol\,CO_2 \times \frac{2\,mol\,C_2H_6}{4\,mol\,CO_2} = 1.28 \times 10^{-3}\,mol\,C_2H_6$$

$$1.28 \times 10^{-3}\,mol\,C_2H_6 \times \frac{30.07\,g\,C_2H_6}{1\,mol\,C_2H_6} = 0.0386\,g\,C_2H_6$$

87. **(D)** $NaI(aq) + AgNO_3(aq) \qquad \rightarrow AgI(s) + NaNO_3(aq)$ (multiply by 4)

$2\ AgI(s) + Fe(s) \qquad \rightarrow FeI_2(aq) + 2\ Ag(s)$ (multiply by 2)

$\underline{2\ FeI_2(aq) + 3\ Cl_2(g) \ \rightarrow 2\ FeCl_3(aq) + 2\ I_2(s)}$ (unchanged)

$4NaI(aq) + 4AgNO_3(aq) + 2Fe(s) + 3Cl_2(g) \rightarrow 4NaNO_3(aq) + 4Ag(s) + 2FeCl_3(aq) + 2I_2(s)$

For every 4 moles of $AgNO_3$, 2 moles of $I_2(s)$ are produced. The mass of $AgNO_3$ required

$$= 1.00\ kg\ I_2(s) \times \frac{1000\ g\ I_2\ (s)}{1\ kg\ I_2\ (s)} \times \frac{1\ mol\ I_2\ (s)}{253.809\ g\ I_2\ (s)} \times \frac{4\ mol\ AgNO_3\ (s)}{2\ mol\ I_2\ (s)} \times \frac{169.873\ g\ AgNO_3\ (s)}{1\ mol\ AgNO_3\ (s)}$$

$= 1338.\underline{59}\ g\ AgNO_3$ per kg of I_2 produced or 1.34 kg $AgNO_3$ per kg of I_2 produced

89. **(M)**

(a)

$SiO_2(s) + 2\ C(s) \xrightarrow{\ \Delta\ } Si(s) + 2\ CO(g)$

$Si(s) + 2\ Cl_2(g) \rightarrow SiCl_4(l)$

$SiCl_4(l) + 2\ H_2(g) \rightarrow Si(s, ultrapure) + 4\ HCl(g)$

(b) $1\ kg\ Si\ (ultrapure,\ s) \times \dfrac{1000\ g}{1kg} \times \dfrac{1\ mol\ Si}{28.09\ g} \times \dfrac{1\ mol\ SiCl_4}{1\ mol\ Si\ ultrapure} \times \dfrac{1\ mol\ Si}{1\ mol\ SiCl_4}$

$\times \dfrac{2\ mol\ C}{1\ mol\ Si} \times \dfrac{12.01\ g\ C}{1\ mol\ C} = 885\ g\ C$

$1\ kg\ Si\ (ultrapure,\ s) \times \dfrac{1000\ g}{1kg} \times \dfrac{1\ mol\ Si}{28.09\ g} \times \dfrac{1\ mol\ SiCl_4}{1\ mol\ Si\ ultrapure} \times \dfrac{2\ mol\ Cl_2}{1\ mol\ SiCl_4}$

$\times \dfrac{70.91\ g\ Cl_2}{1\ mol\ Cl_2} = 5.05 \times 10^3\ g\ Cl_2$

$1\ kg\ Si\ (ultrapure,\ s) \times \dfrac{1000\ g}{1kg} \times \dfrac{1\ mol\ Si}{28.09\ g} \times \dfrac{2\ mol\ H_2}{1\ mol\ Si\ ultrapure} \times \dfrac{2.016\ g\ H_2}{1\ mol\ H_2} = 144\ g\ H_2$

91. (D) The reactions are as follows.

$MgCO_3(s) \rightarrow MgO(s) + CO_2(g)$

$CaCO_3(s) \rightarrow CaO(s) + CO_2(g)$

$$\%\ \text{by mass of}\ MgCO_3 = \frac{g\ MgCO_3}{g\ MgCO_3 + g\ CaCO_3} \times 100\%$$

Let m = mass, in grams, of $MgCO_3$ in the mixture and
let $24.00 - m$ = mass in grams of $CaCO_3$ in the mixture.

Convert from g $MgCO_3$ to g CO_2 to obtain an expression for the mass of CO_2 produced by the first reaction.

$$g\ CO_2\ \text{from}\ MgCO_3 = m\ g\ MgCO_3 \times \frac{1\ mol\ MgCO_3}{84.32\ g\ MgCO_3} \times \frac{1\ mol\ CO_2}{1\ mol\ MgCO_3} \times \frac{44.01\ g\ CO_2}{1\ mol\ CO_2}$$

Convert from g CaCO₃ to g CO₂ to obtain an expression for the mass of CO_2 produced by the second reaction.

$$\text{g CO}_2 \text{ from CaCO}_3 = (24.00 - m) \text{ g CaCO}_3 \times \frac{1 \text{ mol CaCO}_3}{100.09 \text{ g CaCO}_3} \times \frac{1 \text{ mol CO}_2}{1 \text{ mol CaCO}_3} \times \frac{44.01 \text{ g CO}_2}{1 \text{ mol CO}_2}$$

The sum of these two expressions is equal to 12.00 g CO_2. Thus:

$$\left[m \times \frac{44.01}{84.32} \right] + \left[(24.00 - m) \times \frac{44.01}{100.09} \right] = 12.00$$

Solve for m: $m = 17.60$ g

$$\% \text{ by mass of MgCO}_3 = \frac{17.60 \text{ g}}{24.00 \text{ g}} \times 100\% = 73.33\%$$

INTEGRATIVE AND ADVANCED EXERCISES

93. (E)

(a) $CaCO_3(s) \xrightarrow{\Delta} CaO(s) + CO_2(g)$

(b) $2 ZnS(s) + 3 O_2(g) \xrightarrow{\Delta} 2 ZnO(s) + 2 SO_2(g)$

(c) $C_3H_8(g) + 3 H_2O(g) \longrightarrow 3 CO(g) + 7 H_2(g)$

(d) $4 SO_2(g) + 2 Na_2S(aq) + Na_2CO_3(aq) \longrightarrow CO_2(g) + 3 Na_2S_2O_3(aq)$

95. (M) The balanced equation is as follows:

$$2 LiOH(s) + CO_2(g) \rightarrow Li_2CO_3(s) + H_2O(l)$$

Conversion pathway approach:

$$\text{g LiOH} = \frac{1.00 \times 10^3 \text{ g CO}_2}{\text{astronaut day}} \times \frac{1 \text{ mol CO}_2}{44.01 \text{ g CO}_2} \times 3 \text{ astronauts} \times 6 \text{ days} \times \frac{2 \text{ mol LiOH}}{1 \text{ mol CO}_2} \times \frac{23.95 \text{ g LiOH}}{1 \text{ mol LiOH}}$$

$$= 1.96 \times 10^4 \text{ g LiOH}$$

Stepwise approach:

$$\frac{1.00 \times 10^3 \text{ g CO}_2}{\text{astronaut day}} \times \frac{1 \text{ mol CO}_2}{44.01 \text{ g CO}_2} = 22.7 \frac{\text{mol CO}_2}{\text{astronaut day}}$$

$$22.7 \frac{\text{mol CO}_2}{\text{astronaut day}} \times 3 \text{ astronauts} = 68.2 \frac{\text{mol CO}_2}{\text{day}}$$

$$68.2 \frac{\text{mol CO}_2}{\text{day}} \times 6 \text{ days} = 409 \text{ mol CO}_2$$

$$409 \text{ mol CO}_2 \times \frac{2 \text{ mol LiOH}}{1 \text{ mol CO}_2} = 818 \text{ mol LiOH}$$

$$818 \text{ mol LiOH} \times \frac{23.95 \text{ g LiOH}}{1 \text{ mol LiOH}} = 1.96 \times 10^4 \text{ g LiOH}$$

96. **(M)** mass $CaCO_3 = 0.981\,g\ CO_2 \times \dfrac{1\,mol\ CO_2}{44.01\,g\ CO_2} \times \dfrac{1\,mol\ CaCO_3}{1\,mol\ CO_2} \times \dfrac{100.1\,g\ CaCO_3}{1\,mol\ CaCO_3} = 2.23\,g\ CaCO_3$

$\%\ CaCO_3 = \dfrac{2.23\,g\ CaCO_3}{3.28\,g\ sample} \times 100\% = 68.0\%\ CaCO_3\ \text{(by mass)}$

98. **(D)** Assume 100.0 g of the compound Fe_xS_y, then:
Number of moles of S atoms = 36.5g/32.066 g S/mol = 1.13$\underline{8}$ mol
Number of moles of Fe atoms = 63.5g/ 55.847g Fe/mol = 1.137 mol
So the empirical formula for the iron-containing reactant is FeS
Assume 100 g of the compound Fe_xO_y, then:
Number of moles of O atoms = 27.6g/16.0 g O/mol = 1.725 mol
Number of moles of Fe atoms = 72.4g/ 55.847g Fe/mol = 1.296 mol
So the empirical formula for the iron-containing product is Fe_3O_4
Balanced equation: $3\ FeS + 5\ O_2 \rightarrow Fe_3O_4 + 3\ SO_2$

99. **(M)**

$M\ CH_3CH_2OH = \dfrac{mol\ CH_3CH_2OH}{volume\ of\ solution}$

$mol\ CH_3CH_2OH = 50.0\ mL \times 0.7893\ \dfrac{g\ CH_3CH_2OH}{mL} \times \dfrac{1\ mol\ CH_3CH_2OH}{46.07\ g\ CH_3CH_2OH} = 0.857\ mol$

$Molarity = \dfrac{0.857\ mol\ CH_3CH_2OH}{0.0965\ L\ solution} = 8.88\ M\ CH_3CH_2OH$

101. **(D)** Let V be the volume of 0.149 M HCl(aq) that is required.

moles of HCl in solution C = moles HCl in solution A + moles HCl in solution B
$(V + 0.100) \times 0.205\ M =$ $(V \times 0.149\ M\ ) +$ $(0.100 \times 0.285\ M)$

Solve for V: $V = 0.143\ L = 143\ mL$

105. **(D)** We can compute the volume of Al that reacts with the given quantity of HCl.

$V_{Al} = 0.05\ mL \times \dfrac{1\,L}{1000\,mL} \times \dfrac{12.0\,mol\,HCl}{1\,L} \times \dfrac{2\,mol\,Al}{6\,mol\,HCl} \times \dfrac{27.0\,g\,Al}{1\,mol\,Al} \times \dfrac{1\,cm^3}{2.70\,g\,Al} = 0.002\,cm^3$

$area = \dfrac{volume}{thickness} = \dfrac{0.002\,cm^3}{0.10\,mm} \times \dfrac{10\,mm}{1\,cm} = 0.2\,cm^2$

106. (D) Here we need to determine the amount of HCl before and after reaction; the difference is the amount of HCl that reacted.

$$\text{initial amount HCl} = 0.05000\,L \times \frac{1.035\,\text{mol HCl}}{1\,L} = 0.05175\,\text{mol HCl}$$

$$\text{final amount HCl} = 0.05000\,L \times \frac{0.812\,\text{mol HCl}}{1\,L} = 0.0406\,\text{mol HCl}$$

$$\text{mass Zn} = (0.05175 - 0.0406)\,\text{mol HCl} \times \frac{1\,\text{mol Zn}}{2\,\text{mol HCl}} \times \frac{65.38\,\text{g Zn}}{1\,\text{mol Zn}} = 0.365\,\text{g Zn}$$

113. (M)

(a) To determine the extent of the reaction for Cu, the initial and final amounts of Cu need to be calculated.

$$n_{Cu,initial} = 3.177\,\text{g Cu} \times \frac{1\,\text{mol Cu}}{63.546\,\text{g Cu}} = 0.05000\,\text{mol}$$

$$n_{Cu,final} = 0.0739\,\text{g Cu} \times \frac{1\,\text{mol Cu}}{63.546\,\text{g Cu}} = 0.00116\,\text{mol}$$

$$n_{HNO_3,initial} = (0.500\,\text{mol L}^{-1})(0.8000\,L) = 0.400\,\text{mol}$$

The following summary helps us to better visualize the relationships among the initial and final amounts of reactants and products. All the values are in moles.

	3 Cu	+	8 HNO₃	→	3 Cu(NO₃)₂	+	2 NO	+	4 H₂O
initial:	0.05000		0.400		0		0		0
change:	-3ξ		-8ξ		$+3\xi$		$+2\xi$		$+4\xi$
final:	0.00116		0.400-8ξ		3ξ		2ξ		4ξ

The change in the amount of Cu is

$$\Delta n_{Cu} = n_{Cu,\,final} - n_{Cu,\,initial} = 0.00116\,\text{mol} - 0.05000\,\text{mol} = -0.04884\,\text{mol}$$

Since $\Delta n_{Cu} = v_{Cu}\xi$ with $v_{Cu} = -3$, we calculate

$$\xi = \Delta n_{Cu} / v_{Cu} = (-0.04884\,\text{mol}) / (-3) = 0.01628$$

(b) The tabular summary in (a) shows that the amount of NO produced is 2ξ and the amount of HNO₃ remaining is 0.400 mol $- 8\xi$. Since $\xi = 0.01628$ mol, we have:

$$n_{NO,\,final} = 2\xi = 0.03256\,\text{mol}$$

$$n_{HNO_3,\,final} = 0.400\,\text{mol} - 8\,\xi = (0.400 - 8 \times 0.01628)\,\text{mol} = 0.270\,\text{mol}$$

116. (D) $CH_3CH_2OH(l) + 3\ O_2(g) \longrightarrow 2\ CO_2(g) + 3\ H_2O(l)$

$(CH_3CH_2)_2O(l) + 6\ O_2(g) \longrightarrow 4\ CO_2(g) + 5\ H_2O(l)$

Since this is classic mixture problem, we can use the systems of equations method to find the mass percents. First we let x be the mass of $(C_2H_5)_2O$ and y be the mass of $CH3CH2OH$. Thus,

$x + y = 1.005$ g or $y = 1.005$ g $- x$

We then construct a second equation involving x that relates the mass of carbon dioxide formed to the masses of ethanol and diethyl ether.

$$1.963 \text{ g } CO_2 \times \frac{1 \text{ mol } CO_2}{44.010 \text{ g } CO_2} = x \text{ g } (C_2H_5)_2O \times \frac{1 \text{ mol } (C_2H_5)_2O}{74.123 \text{ g } (C_2H_5)_2O} \times \frac{4 \text{ mol } CO_2}{1 \text{ mol } (C_2H_5)_2O}$$

$$+ (1.005 - x) \text{ g } CH_3CH_2OH \times \frac{1 \text{ mol } CH_3CH_2OH}{46.07 \text{ g } CH_3CH_2OH} \times \frac{2 \text{ mol } CO_2}{1 \text{ mol } CH_3CH_2OH}$$

$$0.04460 = 0.05396x + 0.04363 - 0.04341x \qquad x = \frac{0.04460 - 0.04363}{0.05396 - 0.04341} = 0.092 \text{ g } (C_2H_5)_2O$$

$$\% (CH_3CH_2)_2O \text{ (by mass)} = \frac{0.092 \text{ g } (C_2H_5)_2O}{1.005 \text{ g mixture}} \times 100\% = 9.2\% \ (C_2H_5)_2O$$

$$\% \ CH_3CH_2OH \text{ (by mass)} = 100.0\% - 9.2\% \ (C_2H_5)_2O = 90.8\% \ CH_3CH_2OH$$

117. (D) $\% \ Cu \text{ (by mass)} = \dfrac{\# \text{ g Cu}}{0.7391 \text{ g mixture}} \times 100$

Let x = the mass, in grams, of $CuCl_2$ in the mixture.

Let $0.7391 - x$ = mass in grams of $FeCl_3$.

Total moles $AgNO_3$ = mol $AgNO_3$ react with $CuCl_2$ + mol $AgNO_3$ react with $FeCl_3$

Total moles $AgNO_3$ = $0.8691 \text{ L} \times \dfrac{0.1463 \text{ mol}}{1 \text{ L}} = 0.01271 \text{ mol } AgNO_3$

To obtain an expression for the amount of $AgNO_3$ consumed by the first reaction, convert from grams of $CuCl_2$ to moles of $AgCl$:

$$\text{mol } AgNO_3 \text{ that reacts with } CuCl_2 = x \text{ g } CuCl_2 \times \frac{1 \text{ mol } CuCl_2}{134.45 \text{ g } CuCl_2} \times \frac{2 \text{ mol } AgNO_3}{1 \text{ mol } CuCl_2}$$

$$\text{mol } AgNO_3 \text{ that reacts with } CuCl_2 = \frac{2x}{134.45} = 0.014875x$$

To obtain an expression for the amount of $AgNO_3$ consumed by the second reaction, convert from grams of $FeCl_3$ to moles of $AgNO_3$:

$$\text{mol } AgNO_3 \text{ that reacts with } FeCl_3 = (0.7391 - x) \text{ g } FeCl_3 \times \frac{1 \text{ mol } FeCl_3}{162.21 \text{ g } FeCl_3} \times \frac{3 \text{ mol } AgNO_3}{1 \text{ mol } FeCl_3}$$

mol $AgNO_3$ that reacts with $FeCl_3 = (0.7391 - x) \times 0.018496 = 0.013668 - 0.018496x$

The sum of these two expressions is equal to the total number of moles of $AgNO_3$:

Total moles $AgNO_3 = 0.014875x + 0.013668 - 0.018496x = 0.01271$
$$x = 0.2646 \text{ g } CuCl_2$$

This is the mass of $CuCl_2$ in the mixture. We must now convert this to the mass of Cu in the mixture.

$$\# \text{ g } Cu = 0.2646 \text{ g } CuCl_2 \times \frac{1 \text{ mol } CuCl_2}{134.45 \text{ g } CuCl_2} \times \frac{1 \text{ mol } Cu}{1 \text{ mol } CuCl_2} \times \frac{63.546 \text{ g } Cu}{1 \text{ mol } Cu} = 0.1253 \text{ g } Cu$$

$$\% \text{ Cu} = \frac{0.1253 \text{ g } Cu}{0.7391 \text{ g}} \times 100\% = 16.95\%$$

118. (D)

(a) $\text{mol } Cu^{2+} = 48.7 \text{ g } Cu^{2+} \times \dfrac{1 \text{ mol } Cu^{2+}}{63.55 \text{ g } Cu^{2+}} = 0.766 \text{ mol } Cu^{2+} \div 0.307 \longrightarrow 2.50 \text{ mol } Cu^{2+}$

$\text{mol } CrO_4^{2-} = 35.6 \text{ g } CrO_4^{2-} \times \dfrac{1 \text{ mol } CrO_4^{2-}}{115.99 \text{ g } CrO_4^{2-}} = 0.307 \text{ mol } CrO_4^{2-} \div 0.307 \longrightarrow 1.00 \text{ mol } CrO_4^{2-}$

$\text{mol } OH^- = 15.7 \text{ g } OH^- \times \dfrac{1 \text{ mol } OH^-}{17.01 \text{ g } OH^-} = 0.923 \text{ mol } OH^- \div 0.307 \longrightarrow 3.01 \text{ mol } OH^-$

Empirical formula: $Cu_5(CrO_4)_2(OH)_6$

(b) $5 \text{ CuSO}_4(aq) + 2 \text{ K}_2CrO_4(aq) + 6 \text{ H}_2O(l)$

$$\downarrow$$

$$Cu_5(CrO_4)_2(OH)_6(s) + 2 \text{ K}_2SO_4(aq) + 3 \text{ H}_2SO_4(aq)$$

119. (D) We first need to compute the empirical formula of malonic acid.

$34.62 \text{ g C} \times \dfrac{1 \text{ mol C}}{12.01 \text{ g C}} = 2.883 \text{ mol C} \qquad \div 2.883 \longrightarrow 1.000 \text{ mol C}$

$3.88 \text{ g H} \times \dfrac{1 \text{ mol H}}{1.01 \text{ g H}} = 3.84 \text{ mol H} \qquad \div 2.883 \longrightarrow 1.33 \text{ mol H}$

$61.50 \text{ g O} \times \dfrac{1 \text{ mol O}}{16.00 \text{ g O}} = 3.844 \text{ mol O} \qquad \div 2.883 \longrightarrow 1.333 \text{ mol O}$

Multiply each of these mole numbers by 3 to obtain the empirical formula $C_3H_4O_4$.
Combustion reaction: $C_3H_4O_4(l) + 2 \text{ O}_2(g) \longrightarrow 3 \text{ CO}_2(g) + 2 \text{ H}_2O(l)$

120. (D)

$2 \, Al(s) + Fe_2O_3(s) \rightarrow Al_2O_3(s) + 2 \, Fe(s)$

mass of $Fe_2O_3 = 2.5 \text{ g Al} \times \dfrac{1 \text{ mol Al}}{26.982 \text{ g Al}} \times \dfrac{1 \text{ mol Fe}_2O_3}{2 \text{ mol Al}} \times \dfrac{159.69 \text{ g Fe}_2O_3}{2 \text{ mol Fe}_2O_3} = 7.4 \text{ g Fe}_2O_3$ needed

Using 2.5 g Al_2O_3, only 7.4 g of Fe_2O_3 needed, but there are 9.5 g available. Therefore, Al is the limiting reactant.

(a) Mass of $Fe = 2.5 \text{ g Al} \times \dfrac{1 \text{ mol Al}}{26.982 \text{ g Al}} \times \dfrac{2 \text{ mol Fe}}{2 \text{ mol Al}} \times \dfrac{55.85 \text{ g Fe}}{1 \text{ mol Fe}} = 5.2 \text{ g Fe}$

(b) Mass of excess $Fe_2O_3 = 9.5 \text{ g} - 7.4 \text{ g} = 2.1 \text{ g}$

121. (M) Compute the amount of $AgNO_3$ in the solution on hand and the amount of $AgNO_3$ in the desired solution. the difference is the amount of $AgNO_3$ that must be added; simply convert this amount to a mass.

amount $AgNO_3$ present $= 50.00 \text{ mL} \times \dfrac{0.0500 \text{ mmol AgNO}_3}{1 \text{ mL soln}} = 2.50 \text{ mmol AgNO}_3$

amount $AgNO_3$ desired $= 100.0 \text{ mL} \times \dfrac{0.0750 \text{ mmol AgNO}_3}{1 \text{ mL soln}} = 7.50 \text{ mmol AgNO}_3$

mass $AgNO_3 = (7.50 - 2.50) \text{ mmol AgNO}_3 \times \dfrac{1 \text{ mol AgNO}_3}{1000 \text{ mmol AgNO}_3} \times \dfrac{169.9 \text{ g Ag NO}_3}{1 \text{ mol AgNO}_3}$

$= 0.850 \text{ g AgNO}_3$

122. (E) The balanced equation for the reaction is: $S_8(s) + 4 \, Cl_2(g) \rightarrow 4 \, S_2Cl_2(l)$
Both "a" and "b" are consistent with the stoichiometry of this equation. Neither bottom row box is valid. Box (c) does not account for all the S_8, since we started out with 3 molecules, but end up with 1 S_8 molecule and 4 S_2Cl_2 molecules. Box (d) shows a yield of 2 S_8 molecules and 8 S_2Cl_2 molecules so we ended up with more sulfur atoms than we started with. This, of course, violates the Law of Conservation of Mass.

123. (D) The pertinent equations are as follows:

$$C_3N_3(OH)_3 (s) \xrightarrow{\Delta} 3 \, HNCO(g)$$

$$8 \, HNCO(g) + 6 \, NO_2(g) \xrightarrow{\Delta} 7 \, N_2(g) + 8 \, CO_2(g) + 4 \, H_2O(l)$$

The above mole ratios are used to calculate moles of $C_3N_3(OH)_3$ assuming 1.00 g of NO_2.

mass $C_3N_3(OH)_3 = 1.00 \text{ g NO}_2 \times \dfrac{1 \text{ mol NO}_2}{46.00 \text{ g NO}_2} \times \dfrac{8 \text{ mol HNCO}}{6 \text{ mol NO}_2} \times \dfrac{1 \text{ mol C}_3N_3(OH)_3}{3 \text{ mol HNCO}}$

$\times \dfrac{129.1 \text{ g mol C}_3N_3(OH)_3}{1 \text{ mol C}_3N_3(OH)_3} = 1.25 \text{ g C}_3N_3(OH)_3$

125. (D) There are many ways one can go about answering this question. We must use all of the most concentrated solution and dilute this solution down using the next most concentrated solution. Hence, start with 345 mL of 1.29 M then add x L of the 0.775 M solution. The value of x is obtained by solving the following equation.

$$1.25 \text{ M} = \frac{(1.29 \text{ M} \times 0.345 \text{ L}) + (0.775 \text{ M} \times x)}{(0.345 + x) \text{ L}}$$

$$1.25 \text{ M} \times (0.345 + x) \text{ L} = (1.29 \text{ M} \times 0.345 \text{ L}) + (0.775 \text{ M} \times x)$$

$0.43125 + 1.25x = 0.44505 + 0.775x$ Thus, $0.0138 = 0.475x$

$x = 0.029$ L or 29 mL

A total of (29 mL + 345 mL) = 374 mL may be prepared this way.

129. (M)

(a) $2 \text{ C}_3\text{H}_6(g) + 2 \text{ NH}_3(g) + 3 \text{ O}_2(g) \rightarrow 2 \text{ C}_3\text{H}_3\text{N}(l) + 6 \text{ H}_2\text{O}(l)$

(b) For every kilogram of propylene we get 0.73 kilogram of acrylonitrile; we can also say that for every gram of propylene we get 0.73 gram of acrylonitrile. One gram of propylene is 0.023<u>8</u> mol of propylene. The corresponding quantity of NH_3 is 0.023<u>8</u> mol or 0.40<u>5</u> g; then because NH_3 and C_3H_6 are required in the same molar amount (2:2) for the reaction, 0.40<u>5</u> of a kg of NH_3 will be required for every 0.73 of a kg of acrylonitrile. To get 1000 kg of acrylonitrile we need, by simple proportion, $1000 \times (0.40\underline{5})/0.73 = 555$ kg.

130. (D) It is helpful to first calculate the molar masses of all reactants used and the desired product. The molar masses, in g/mol, are as follows.

Desired product: $C_6H_5NH_2$, 93.129

Reaction #1: C_6H_6, 78.114 $(CH_3)_3SiN_3$, 115.211 F_3CSO_3H, 150.07   NaOH, 39.997

Reaction #2: HNO_3, 63.012 H_2, 2.016

We now focus on making 1 mole of $C_6H_5NH_2$ (or 93.129 g $C_6H_5NH_2$) by each reaction. For reaction #1, we need 1 mole (or 78.114 g) of C_6H_6, 1 mole (or 115.221 g) of $(CH_3)_3SiN_3$, 2 moles (or 2 × 150.07 g) of F_3CSO_3H, and 1 mole (or 39.997 g) of NaOH. Therefore,

$$\% \text{ atom economy} = \frac{93.129 \text{ g}}{78.114 \text{ g} + 115.221 \text{ g} + 2 \times 150.07 \text{ g} + 39.997 \text{ g}} \times 100 = 17$$

For reaction #2, we need 1 mole (or 78.114 g) of C_6H_6, 1 mole (or 63.012 g) of HNO_3, and 3 moles (or 3 × 2.016 g) of H_2.

$$\% \text{ atom economy} = \frac{93.129 \text{ g}}{78.114 \text{ g} + 63.012 \text{ g} + 3 \times 2.016 \text{ g}} \times 100 = 63$$

SELF-ASSESSMENT EXERCISES

138. **(E)** The answer is (d). Start balancing in the following order: N, O, H and Cu

$$3\,Cu(s) + 8\,HNO_3(aq) \rightarrow 3\,Cu(NO_3)_2(aq) + 4\,H_2O(l) + 2\,NO(g)$$

139. **(E)** The answer is (d). To determine the number of moles of NH_3, used the balanced equation:

$$\text{\# moles } NH_3 = 1\,mol\,H_2O \times \frac{2\,mol\,NH_3}{3\,mol\,H_2O} = 0.666\,mol\,NH_3$$

140. **(M)** The answer is (a). To determine the number of moles of NH_3, use the balanced equation:

$$2\,KMnO_4 + 10\,KI + 8\,H_2SO_4 \rightarrow 6\,K_2SO_4 + 2\,MnSO_4 + 5\,I_2 + 8\,H_2O$$

$$5\,mol\,KMnO_4 \times \frac{6\,mol\,K_2SO_4}{2\,mol\,KMnO_4} = 15\,mol\,K_2SO_4$$

$$5\,mol\,KI \times \frac{6\,mol\,K_2SO_4}{10\,mol\,KI} = 3\,mol\,K_2SO_4$$

$$5\,mol\,H_2SO_4 \times \frac{6\,mol\,K_2SO_4}{8\,mol\,H_2SO_4} = 3.75\,mol\,K_2SO_4$$

141. **(E)** The answer is (a). To determine the answer, used the balanced equation:
$$2\,Ag_2(CO_3)\,(s) \rightarrow 4\,Ag(s) + 2\,CO_2(g) + O_2(g)$$
The ratio between O_2 and CO_2 is 1:2.

142. **(E)** The answer is (c). To solve this, calculate the number of moles of $NaNO_3$.
mol $NaNO_3$ = 1.00 M × 1.00 L = 1.00 mol

$$1.00\,mol\,NaNO_3 \times \frac{85.0\,g\,NaNO_3}{1\,mol\,NaNO_3} = 85.0\,g\,NaNO_3$$

Concentration = 85.0 g $NaNO_3$/L. While (b) also technically gives you the correct value at 25 °C, it is not the definition of molarity.

143. **(D)** The amount of K+ present in 200.0 mL of 0.240 M K_2SO_4 is

$$0.2000\,L \times \frac{0.240\,mol\,K_2SO_4}{1\,L} \times \frac{2\,mol\,K^+}{1\,mol\,K_2SO_4} = 0.096\,mol\,K^+$$

Let V be the volume of KNO_3 solution added in L.

$$0.400\,M = \frac{0.096\,mol\,K^+ + (0.160\,M \times V)}{V + 0.200\,L}$$
$$V = 0.0667\,L = 66.7\,mL$$

144. **(E)** The answer is (d). There is no need for calculation, because a starting solution of 0.4 M is needed to make a 0.50 M solution, and the only way to make a more concentrated solution is to evaporate off some of the water.

145. **(M)** The answer is (b). To determine the molarity, number of moles of LiBr need to be determined first. Therefore, weight % concentration needs to be converted to number of moles with the aid of the density:

Conc. = 5.30% by mass = 5.30 g LiBr/100 g solution

$$\text{Volume of solution} = \text{mass / Density} = 100\,\text{g sol'n} \times \frac{1\,\text{mL}}{1.040\,\text{g}} = 96.15\,\text{mL}$$

$$\text{mol LiBr} = 5.30\,\text{g LiBr} \times \frac{1\,\text{mol LiBr}}{86.84\,\text{g LiBr}} = 0.0610\,\text{mol}$$

$$\text{Molarity} = \frac{0.0610\,\text{mol}}{96.15\,\text{mL}} \times \frac{1000\,\text{mL}}{1\,\text{L}} = 0.635\,\text{M}$$

146. **(M)** The answer is (d). To determine % yield, calculate the theoretical mole yield:

$$\text{mol CCl}_2\text{F} = 2.00\,\text{mol CCl}_4 \times \frac{1\,\text{mol CCl}_2\text{F}}{1\,\text{mol CCl}_4} = 2.00\,\text{mol CCl}_2\text{F}$$

$$\%\text{ yield} = \frac{1.70\,\text{mol}}{2.00\,\text{mol}} \times 100 = 85.0\%$$

147. **(M)**

$$18.0\,\text{g Fe}_2\text{O}_3 \times \frac{1\,\text{mol Fe}_2\text{O}_3}{159.7\,\text{g Fe}_2\text{O}_3} \times \frac{4\,\text{mol Fe}}{2\,\text{mol Fe}_2\text{O}_3} \times \frac{55.845\,\text{g Fe}}{1\,\text{mol Fe}} = 12.6\,\text{g Fe}$$

(assuming all the Fe_2O_3 reacts)

$$2.5\,\text{g C} \times \frac{1\,\text{mol C}}{12.011\,\text{g C}} \times \frac{4\,\text{mol Fe}}{3\,\text{mol C}} \times \frac{55.845\,\text{g Fe}}{1\,\text{mol Fe}} = 15.5\,\text{g Fe}\ \ \text{(assuming all the C reacts)}$$

Fe_2O_3 is the limiting reactant and 12.6 g of Fe are produced by this reaction.

148. **(D)** We have 26.4 g mixture $\times$ 0.40 g CaO/g mixture = 10.6 g CaO

$$10.5\,\text{g CaO} \times \frac{1\,\text{mol CaO}}{56.08\,\text{g CaO}} \times \frac{1\,\text{mol CaCl}_2}{1\,\text{mol CaO}} \times \frac{110.98\,\text{g CaCl}_2}{1\,\text{mol CaCl}_2} = 20.78\,\text{g CaCl}_2$$

26.4 g mixture – 10.6 g CaO = 15.8 g NaOH

$$15.8\,\text{g NaOH} \times \frac{1\,\text{mol NaOH}}{40.00\,\text{g NaOH}} \times \frac{1\,\text{mol NaCl}}{1\,\text{mol NaOH}} \times \frac{58.44\,\text{g NaCl}}{1\,\text{mol NaCl}} = 23.08\,\text{g NaCl}$$

Total mass of solid = mass of $CaCl_2$ + mass of NaCl = 20.8 g + 23.1 g = 43.9 g

149. **(D)** To balance the below equations, balance C first, then O and finally H.

(a) $2\,C_8H_{18} + 25\,O_2 \rightarrow 16\,CO_2 + 18\,H_2O$

(b) For this part, we note that 25% of the available carbon atoms in C_8H_{18} form CO and the remainder for CO_2. Therefore,
$2\,C_8H_{18} + 25\,O_2 \rightarrow 12\,CO_2 + 4\,CO + 18\,H_2O$

150. **(D)** To determine the compound, the number of moles of each compound needs to be determined, which then helps determine number of moles of emitted CO_2:

$$\text{mass } CO_2 = 1.000 \text{ g } CaCO_3 \times \frac{1 \text{ mol } CaCO_3}{100.08 \text{ g } CaCO_3} \times \frac{1 \text{ mol } CO_2}{1 \text{ mol } CaCO_3} \times \frac{44.0 \text{ g } CO_2}{1 \text{ mol } CO_2}$$

$$= 0.4396 \text{ g } CO_2$$

$$\text{mass } CO_2 = 1.000 \text{ g } MgCO_3 \times \frac{1 \text{ mol } MgCO_3}{84.30 \text{ g } MgCO_3} \times \frac{1 \text{ mol } CO_2}{1 \text{ mol } CaCO_3} \times \frac{44.0 \text{ g } CO_2}{1 \text{ mol } CO_2}$$

$$= 0.5219 \text{ g } CO_2$$

$$\text{mass } CO_2 = 1.000 \text{ g } CaCO_3 \cdot MgCO_3 \times \frac{1 \text{ mol dolomite}}{184.38 \text{ g dolomite}} \times \frac{2 \text{ mol } CO_2}{1 \text{ mol dolomite}}$$

$$\times \frac{44.0 \text{ g } CO_2}{1 \text{ mol } CO_2} = 0.4773 \text{ g } CO_2$$

Therefore, dolomite is the compound.

151. **(D)** The answer is (b). First, the total amount of carbon in our mixture of CH_4 and C_2H_6 must be determined by using the amount of CO_2

$$\text{mass of C} = 2.776 \text{ g } CO_2 \times \frac{1 \text{ mol } CO_2}{44.01 \text{ g } CO_2} \times \frac{1 \text{ mol C}}{1 \text{ mol } CO_2} \times \frac{12.01 \text{ g C}}{1 \text{ mol C}} = 0.758 \text{ g C}$$

Let the mass of CH_4 in the mixture be x g. Then the mass of C_2H_6 will be $(1.000 - x)$ g. Then, the amounts of CH_4 and C_2H_6 can be determined by making sure that the moles of carbon for both add up to 0.0631:

$$x \text{ g } CH_4 \times \left(\frac{1 \text{ mol } CH_4}{16.05 \text{ g } CH_4} \times \frac{1 \text{ mol C}}{1 \text{ mol } CH_4} \times \frac{12.01 \text{ g C}}{1 \text{ mol C}} \right)$$

$$+ (1.000 - x) \times \left(\frac{1 \text{ mol } C_2H_6}{30.08 \text{ g } C_2H_6} \times \frac{2 \text{ mol C}}{1 \text{ mol } C_2H_6} \times \frac{12.01 \text{ g C}}{1 \text{ mol C}} \right) = 0.757 \text{ g C (from } CO_2)$$

$$0.748 x + (1 - x)(0.798) = 0.757$$

$$x = \text{mass of } CH_4 = 0.82 \text{ g, or } 82\% \text{ of a } 1.00 \text{ g sample}$$

152. **(D)** The answer is (c). To do this, perform a stepwise conversion of moles of reactants to moles of products, as shown below:

$$4.00 \text{ mol } NH_3 \times \frac{4 \text{ mol NO}}{4 \text{ mol } NH_3} \times \frac{2 \text{ mol } NO_2}{2 \text{ mol NO}} \times \frac{2 \text{ mol } HNO_3}{3 \text{ mol } NO_2} = 2.67 \text{ mol } HNO_3$$

153. **(E)** All the following reactions incorporate all the reactant atoms into the desired product. Consequently, they all have a percent AE of 100%.

(a) $Rb + \frac{1}{2} Br_2 + 2 O_2 \longrightarrow RbBrO_4$

(b) $S + H_2 + 2 O_2 \longrightarrow H_2SO_4$

(c) $Mg + Cl_2 + 3 O_2 \longrightarrow Mg(ClO_3)_2$

(d) $Na + \frac{1}{2} N_2 + O_2 \longrightarrow NaNO_2$

CHAPTER 5

INTRODUCTION TO REACTIONS
IN AQUEOUS SOLUTIONS

PRACTICE EXAMPLES

1B **(E)**

(a) $\dfrac{1.5 \text{ mg F}^-}{1 \text{ L}} \times \dfrac{1 \text{ g F}^-}{1000 \text{ mg F}^-} \times \dfrac{1 \text{ mol F}^-}{18.998 \text{ F}^-} = 7.9 \times 10^{-5} \text{ M F}^-$

(b) $1.00 \times 10^6 \text{ L} \times \dfrac{7.9 \times 10^{-5} \text{ mol F}^-}{1 \text{ L}} \times \dfrac{1 \text{ mol CaF}_2}{2 \text{ mol F}^-} \times \dfrac{78.074 \text{ g CaF}_2}{1 \text{ mol CaF}_2} \times \dfrac{1 \text{ kg}}{1000 \text{ g}} = 3.1 \text{ kg CaF}_2$

2A **(E)** In each case, we use the solubility rules to determine whether either product is insoluble. The ions in each product compound are determined by simply "switching the partners" of the reactant compounds. The designation "(aq)" on each reactant indicates that it is soluble.

(a) Possible products are potassium chloride, KCl, which is soluble, and aluminum hydroxide, Al(OH)_3, which is not. Net ionic equation:

$$\text{Al}^{3+}(\text{aq}) + 3 \text{ OH}^-(\text{aq}) \rightarrow \text{Al(OH)}_3(\text{s})$$

(b) Possible products are iron(III) sulfate, $\text{Fe}_2(\text{SO}_4)_3$, and potassium bromide, KBr, both of which are soluble. No reaction occurs.

(c) Possible products are calcium nitrate, $\text{Ca(NO}_3)_2$, which is soluble, and lead(II) iodide, PbI_2, which is insoluble. Net ionic equation: $\text{Pb}^{2+}(\text{aq}) + 2 \text{ I}^-(\text{aq}) \rightarrow \text{PbI}_2(\text{s})$

2B **(E)**

(a) Possible products are sodium chloride, NaCl, which is soluble, and aluminum phosphate, AlPO_4, which is insoluble. Net ionic equation:

$$\text{Al}^{3+}(\text{aq}) + \text{PO}_4^{3-}(\text{aq}) \rightarrow \text{AlPO}_4(\text{s})$$

(b) Possible products are aluminum chloride, AlCl_3, which is soluble, and barium sulfate, BaSO_4, which is insoluble. Net ionic equation: $\text{Ba}^{2+}(\text{aq}) + \text{SO}_4^{2-}(\text{aq}) \rightarrow \text{BaSO}_4(\text{s})$

(c) Possible products are ammonium nitrate, NH_4NO_3, which is soluble, and lead(II) carbonate, PbCO_3, which is insoluble. Net ionic equation:

$$\text{Pb}^{2+}(\text{aq}) + \text{CO}_3^{2-}(\text{aq}) \rightarrow \text{PbCO}_3(\text{s})$$

3A **(E)** Propionic acid is a weak acid, not dissociated completely in aqueous solution. Ammonia similarly is a weak base. The acid and base react to form a salt solution of ammonium propionate.

$$NH_3(aq) + CH_3CH_2COOH(aq) \rightarrow NH_4^+(aq) + CH_3CH_2COO^-(aq)$$

3B **(E)** Since acetic acid is a weak acid, it is not dissociated completely in aqueous solution (except at infinite dilution); it is misleading to write it in ionic form. The products of this reaction are the gas carbon dioxide, the molecular compound water, and the ionic solute calcium acetate. Only the latter exists as ions in aqueous solution.

$$CaCO_3(s) + 2\ CH_3COOH(aq) \rightarrow CO_2(g) + H_2O(l) + Ca^{2+}(aq) + 2\ CH_3COO^-(aq)$$

4A **(M)**

(a) Elements do not change oxidation states during this reaction. It is not an oxidation–reduction reaction.

(b) The presence of $O_2(g)$ as a product indicates that this is an oxidation–reduction reaction. Oxygen is oxidized from O.S. $= -2$ in NO_3^- to O.S. $= 0$ in $O_2(g)$. Nitrogen is reduced from O.S. $= +5$ in NO_3^- to O.S. $= +4$ in NO_2.

4B **(M)** Vanadium is oxidized from O.S. $= +4$ in VO^{2+} to an O.S. $= +5$ in VO_2^+ while manganese is reduced from O.S. $= +7$ in MnO_4^- to O.S. $= +2$ in Mn^{2+}.

5A **(M)** Aluminum is oxidized (from an O.S. of 0 to an O.S. of $+3$), while hydrogen is reduced (from an O.S. of $+1$ to an O.S. of 0).

$$Al + HCl \rightarrow AlCl_3 + H_2$$

$$Al + H^+ \rightarrow Al^{3+} + H_2$$

Oxidation: $\{Al(s) \rightarrow Al^{3+}(aq) + 3\ e^-\} \times 2$

Reduction: $\{2\ H^+(aq) + 2\ e^- \rightarrow H_2(g)\} \times 3$

Overall: $2\ Al(s) + 6\ H^+(aq) \rightarrow 2\ Al^{3+}(aq) + 3\ H_2(g)$

5B **(M)** Bromide is oxidized (from -1 to 0), while chlorine is reduced (from 0 to -1).

$$Cl_2 + NaBr \rightarrow Br_2 + NaCl$$

$$Cl_2 + Br^- \rightarrow Br_2 + Cl^-$$

Oxidation: $2\ Br^-(aq) \rightarrow Br_2(l) + 2\ e^-$

Reduction: $Cl_2(g) + 2\ e^- \rightarrow 2\ Cl^-(aq)$

Overall: $2\ Br^-(aq) + Cl_2(g) \rightarrow Br_2(l) + 2\ Cl^-(aq)$

6A **(D)**

Step 1: Write the two skeleton half-equations.

$$Fe^{2+}(aq) \rightarrow Fe^{3+}(aq) \quad and \quad MnO_4^-(aq) \rightarrow Mn^{2+}(aq)$$

Step 2: Balance each skeleton half-equation for O (with H_2O) and for H atoms (with H^+).

$$Fe^{2+}(aq) \rightarrow Fe^{3+}(aq) \quad and \quad MnO_4^-(aq) + 8\,H^+(aq) \rightarrow Mn^{2+}(aq) + 4\,H_2O(l)$$

Step 3: Balance electric charge by adding electrons.

$$Fe^{2+}(aq) \rightarrow Fe^{3+}(aq) + e^- \quad and \quad MnO_4^-(aq) + 8\,H^+(aq) + 5\,e^- \rightarrow Mn^{2+}(aq) + 4\,H_2O(l)$$

Step 4: Combine the two half-equations

$$Oxidation: \left\{ Fe^{2+}(aq) \rightarrow Fe^{3+}(aq) + e^- \right\} \times 5$$

$$Reduction: MnO_4^-(aq) + 8\,H^+(aq) + 5\,e^- \rightarrow Mn^{2+}(aq) + 4\,H_2O(l)$$

$$Overall: MnO_4^-(aq) + 8\,H^+(aq) + 5\,Fe^{2+}(aq) \rightarrow Mn^{2+}(aq) + 4\,H_2O(l) + 5\,Fe^{3+}(aq)$$

6B **(D)**

Step 1: Uranium is oxidized and chromium is reduced in this reaction. The "skeleton" half-equations are: $\quad UO^{2+}(aq) \rightarrow UO_2^{2+}(aq) \quad and \quad Cr_2O_7^{2-}(aq) \rightarrow Cr^{3+}(aq)$

Step 2: First, balance the chromium skeleton half-equation for chromium atoms:

$$Cr_2O_7^{2-}(aq) \rightarrow 2\,Cr^{3+}(aq)$$

Next, balance oxygen atoms with water molecules in each half-equation:

$$UO^{2+}(aq) + H_2O(l) \rightarrow UO_2^{2+}(aq) \quad and \quad Cr_2O_7^{2-}(aq) \rightarrow 2\,Cr^{3+}(aq) + 7\,H_2O(l)$$

Then, balance hydrogen atoms with hydrogen ions in each half-equation:

$$UO^{2+}(aq) + H_2O(l) \rightarrow UO_2^{2+}(aq) + 2\,H^+(aq)$$

$$Cr_2O_7^{2-}(aq) + 14\,H^+(aq) \rightarrow 2\,Cr^{3+}(aq) + 7\,H_2O(l)$$

Step 3: Balance the charge of each half-equation with electrons.

$$UO^{2+}(aq) + H_2O(l) \rightarrow UO_2^{2+}(aq) + 2\,H^+(aq) + 2\,e^-$$

$$Cr_2O_7^{2-}(aq) + 14\,H^+(aq) + 6\,e^- \rightarrow 2\,Cr^{3+}(aq) + 7\,H_2O(l)$$

Step 4: Multiply the uranium half-equation by 3 and add the chromium half-equation to it.

$$Oxidation: \left\{ UO^{2+}(aq) + H_2O(l) \rightarrow UO_2^{2+}(aq) + 2\,H^+(aq) + 2\,e^- \right\} \times 3$$

$$Reduction: Cr_2O_7^{2-}(aq) + 14\,H^+(aq) + 6\,e^- \rightarrow 2\,Cr^{3+}(aq) + 7\,H_2O(l)$$

$$3\,UO^{2+}(aq) + Cr_2O_7^{2-}(aq) + 14\,H^+(aq) + 3\,H_2O(l) \rightarrow 3\,UO_2^{2+}(aq) + 2\,Cr^{3+}(aq) + 7\,H_2O(l) + 6\,H^+(aq)$$

Step 5: Simplify. Subtract $3\,H_2O(l)$ and $6\,H^+(aq)$ from each side of the equation.

$$Overall: 3\,UO^{2+}(aq) + Cr_2O_7^{2-}(aq) + 8\,H^+(aq) \rightarrow 3\,UO_2^{2+}(aq) + 2\,Cr^{3+}(aq) + 4\,H_2O(l)$$

7A **(D)**

Step 1: Write the two skeleton half-equations.

$$S(s) \rightarrow SO_3^{2-}(aq) \quad and \quad OCl^-(aq) \rightarrow Cl^-(aq)$$

Step 2: Balance each skeleton half-equation for O (with H_2O) and for H atoms (with H^+).

$$3\,H_2O(l) + S(s) \rightarrow SO_3^{2-}(aq) + 6\,H^+$$

$$OCl^-(aq) + 2\,H^+ \rightarrow Cl^-(aq) + H_2O(l)$$

Step 3: Balance electric charge by adding electrons.

$$3\,H_2O(l) + S(s) \rightarrow SO_3^{2-}(aq) + 6\,H^+(aq) + 4\,e^-$$

$$OCl^-(aq) + 2\,H^+(aq) + 2\,e^- \rightarrow Cl^-(aq) + H_2O(l)$$

Step 4: Change from an acidic medium to a basic one by adding OH^- to eliminate H^+.

$$3H_2O(l) + S(s) + 6\,OH^-(aq) \rightarrow SO_3^{2-}(aq) + 6\,H^+(aq) + 6\,OH^-(aq) + 4\,e^-$$

$$OCl^-(aq) + 2\,H^+(aq) + 2\,OH^-(aq) + 2\,e^- \rightarrow Cl^-(aq) + H_2O(l) + 2\,OH^-(aq)$$

Step 5: Simplify by removing the items present on both sides of each half-equation, and combine the half-equations to obtain the net redox equation.

Oxidation: $S(s) + 6\,OH^-(aq) \rightarrow SO_3^{2-}(aq) + 3\,H_2O(l) + 4\,e^-$

Reduction: $\{OCl^-(aq) + H_2O(l) + 2\,e^- \rightarrow Cl^-(aq) + 2\,OH^-(aq)\} \times 2$

Overall: $S(s) + 6\,OH^-(aq) + 2\,OCl^-(aq) + 2H_2O(l) \rightarrow SO_3^{2-}(aq) + 3\,H_2O(l) + 2\,Cl^-(aq) + 4\,OH^-$

Simplify by removing the species present on both sides.

Net ionic equation: $S(s) + 2\,OH^-(aq) + 2\,OCl^-(aq) \rightarrow SO_3^{2-}(aq) + H_2O(l) + 2\,Cl^-(aq)$

7B **(D)**
Step 1: Write the two skeleton half-equations.

$$MnO_4^-(aq) \rightarrow MnO_2(s) \ \ and \ \ SO_3^{2-}(aq) \rightarrow SO_4^{2-}(aq)$$

Step 2: Balance each skeleton half-equation for O (with H_2O) and for H atoms (with H^+).

$$MnO_4^-(aq) + 4\,H^+(aq) \rightarrow MnO_2(s) + 2\,H_2O(l)$$

$$SO_3^{2-}(aq) + H_2O(l) \rightarrow SO_4^{2-}(aq) + 2H^+(aq)$$

Step 3:   Balance electric charge by adding electrons.

$$MnO_4^-(aq) + 4\,H^+(aq) + 3\,e^- \rightarrow MnO_2(s) + 2\,H_2O(l)$$

$$SO_3^{2-}(aq) + H_2O(l) \rightarrow SO_4^{2-}(aq) + 2\,H^+(aq) + 2\,e^-$$

Step 4: Change from an acidic medium to a basic one by adding OH^- to eliminate H^+.

$$MnO_4^-(aq) + 4\,H^+(aq) + 4\,OH^-(aq) + 3\,e^- \rightarrow MnO_2(s) + H_2O(l) + 4\,OH^-(aq)$$

$$SO_3^{2-}(aq) + H_2O(l) + 2\,OH^-(aq) \rightarrow SO_4^{2-}(aq) + 2\,H^+(aq) + 2\,OH^-(aq) + 2\,e^-$$

Step 5: Simplify by removing species present on both sides of each half-equation, and combine the half-equations to obtain the net redox equation.

Oxidation: $\{MnO_4^-(aq) + 2\,H_2O(l) + 3\,e^- \rightarrow MnO_2(s) + 4\,OH^-(aq)\} \times 2$

Reduction: $\{SO_3^{2-}(aq) + 2\,OH^-(aq) \rightarrow SO_4^{2-}(aq) + H_2O(l) + 2\,e^-\} \times 3$

$$2\,MnO_4^-(aq) + 3\,SO_3^{2-}(aq) + 6\,OH^-(aq) + 4\,H_2O(l) \rightarrow$$
$$2\,MnO_2(s) + 3\,SO_4^{2-}(aq) + 3\,H_2O(l) + 8\,OH^-(aq)$$

Simplify by removing species present on both sides.

Overall: $2\,MnO_4^-(aq) + 3\,SO_3^{2-}(aq) + H_2O(l) \rightarrow 2\,MnO_2(s) + 3\,SO_4^{2-}(aq) + 2\,OH^-(aq)$

8A **(M)** Since the oxidation state of H is 0 in H_2 (g) and is +1 in both NH_3(g) and H_2O(g), hydrogen is oxidized. A substance that is oxidized is called a reducing agent. In addition, the oxidation state of N in NO_2 (g) is +4, while it is −3 in NH_3; the oxidation state of the element N decreases during this reaction, meaning that NO_2 (g) is reduced. The substance that is reduced is called the oxidizing agent.

8B **(M)** In $\left[Au(CN)_2\right]^-$ (aq), gold has an oxidation state of +1; Au has been oxidized and, thus, Au(s) (oxidization state = 0), is the reducing agent. In OH^- (aq), oxygen has an oxidation state of −2; O has been reduced and thus, O_2(g) (oxidation state = 0) is the oxidizing agent.

9A **(M)** We first determine the amount of NaOH that reacts with 0.500 g $KHC_8H_4O_4$ (or KHP, for short).

$$n_{\text{NaOH}} = 0.5000 \text{ g KHP} \times \frac{1 \text{ mol KHP}}{204.22 \text{ g KHP}} \times \frac{1 \text{ mol OH}^-}{1 \text{ mol KHP}} \times \frac{1 \text{ mol NaOH}}{1 \text{ mol OH}^-} = 0.002448 \text{ mol NaOH}$$

$$\text{molarity of NaOH} = \frac{0.002448 \text{ mol NaOH}}{24.03 \text{ mL soln}} \times \frac{1000 \text{ mL}}{1 \text{ L}} = 0.1019 \text{ M}$$

9B **(M)** The net ionic equation when solid hydroxides react with a strong acid is $OH^- + H^+ \rightarrow H_2O$. There are two sources of OH^-: NaOH and $Ca(OH)_2$. We compute the amount of OH^- from each source and add the results.

$$\text{moles of OH}^- \text{ from NaOH:} = 0.235 \text{ g sample} \times \frac{92.5 \text{ g NaOH}}{100.0 \text{ g sample}} \times \frac{1 \text{ mol NaOH}}{39.997 \text{ g NaOH}} \times$$

$$\frac{1 \text{ mol OH}^-}{1 \text{ mol NaOH}} = 0.00543 \text{ mol OH}^-$$

$$\text{moles of OH}^- \text{ from Ca(OH)}_2 := 0.235 \text{ g sample} \times \frac{7.5 \text{ g Ca(OH)}_2}{100.0 \text{ g sample}} \times$$

$$\frac{1 \text{ mol Ca(OH)}_2}{74.092 \text{ g Ca(OH)}_2} \times \frac{2 \text{ mol OH}^-}{1 \text{ mol Ca(OH)}_2} = 0.00048 \text{ mol OH}^-$$

$$\text{total amount OH}^- = 0.00543 \text{ mol from NaOH} + 0.00048 \text{ mol from Ca(OH)}_2$$

$$= 0.00591 \text{ mol OH}^-$$

$$\text{molarity of HCl} = \frac{0.00591 \text{ mol OH}^-}{45.6 \text{ mL HCl soln}} \times \frac{1 \text{ mol H}^+}{1 \text{ mol OH}^-} \times \frac{1 \text{ mol HCl}}{1 \text{ mol H}^+} \times \frac{1000 \text{ mL soln}}{1 \text{ L soln}}$$

$$= 0.130 \text{ M}$$

10A **(M)** First, determine the mass of iron that has reacted as Fe^{2+} with the titrant. The balanced chemical equation provides the essential conversion factor to answer this question.
Namely: $5 \text{ Fe}^{2+}(\text{aq}) + \text{MnO}_4^-(\text{aq}) + 8 \text{ H}^+(\text{aq}) \longrightarrow 5 \text{ Fe}^{3+}(\text{aq}) + \text{Mn}^{2+}(\text{aq}) + 4 \text{ H}_2\text{O(l)}$

$$\text{mass Fe} = 0.04125 \text{ L titrant} \times \frac{0.02140 \text{ mol MnO}_4^-}{1 \text{ L titrant}} \times \frac{5 \text{ mol Fe}^{2+}}{1 \text{ mol MnO}_4^-} \times \frac{55.845 \text{ g Fe}}{1 \text{ mol Fe}^{2+}}$$

$$= 0.246 \text{ g Fe}$$

Then determine the % Fe in the ore. $\% \text{ Fe} = \dfrac{0.246 \text{ g Fe}}{0.376 \text{ g ore}} \times 100\% = 65.4\% \text{ Fe}$

10B **(M)** The balanced equation provides us with the essential conversion factor to answer this question. See the solution to Exercise 111(b) for detail concerning the balancing of this equation.

Namely: $5\ C_2O_4^{2-}(aq) + 2\ MnO_4^-(aq) + 16\ H^+(aq) \longrightarrow 10\ CO_2(g) + 2\ Mn^{2+}(aq) + 8\ H_2O(l)$

$$\text{amount MnO}_4^- = 0.2482\ g\ Na_2C_2O_4 \times \frac{1\ mol\ Na_2C_2O_4}{134.00\ g\ Na_2C_2O_4} \times \frac{1\ mol\ C_2O_4^{2-}}{1\ mol\ Na_2C_2O_4} \times \frac{2\ mol\ MnO_4^-}{5\ mol\ C_2O_4^{2-}}$$

$$= 0.0007409\ mol\ MnO_4^-$$

$$\text{molarity of KMnO}_4 = \frac{0.0007409\ mol\ MnO_4^-}{23.68\ mL\ soln} \times \frac{1000\ mL}{1\ L} \times \frac{1\ mol\ KMnO_4}{1\ mol\ MnO_4^-} = 0.03129\ M$$

INTEGRATIVE EXAMPLE

A. **(M)** First, balance the equation. Break down the reaction of chlorate and ferrous ion as follows:

$$ClO_3^- + 6\ H^+ + 6\ e^- \rightarrow Cl^- + 3\ H_2O$$

$$6\left(Fe^{2+} \rightarrow Fe^{3+} + e^-\right)$$

Net reaction: $ClO_3^- + 6\ Fe^{2+} + 6\ H^+ \rightarrow Cl^- + 6\ Fe^{3+} + 3\ H_2O$

The reaction between Fe^{2+} and Ce^{4+} is already balanced. To calculate the moles of Fe^{2+} that remains after the reaction with ClO_3^-, determine the moles of Ce^{4+} that react with Fe^{2+}:

mol Ce^{4+} = 0.01259 L × 0.08362 M = 1.0527×10^{-3} mol = mol of excess Fe^{2+}

total mol of Fe^{2+} = 0.0500 L × 0.09101 = 4.551×10^{-3} mol

Therefore, the moles of Fe^{2+} reacted = $4.551 \times 10^{-3} - 1.0527 \times 10^{-3} = 3.498 \times 10^{-3}$ mol. To determine the mass of $KClO_3$, use the mole ratios in the balanced equation in conjunction with the molar mass of $KClO_3$.

$$3.498 \times 10^{-3} mol\ Fe^{2+} \times \frac{1\ mol\ ClO_3^-}{6\ mol\ Fe^{2+}} \times \frac{1\ mol\ KClO_3}{1\ mol\ ClO_3^-} \times \frac{122.54\ g\ KClO_3}{1\ mol\ KClO_3}$$

$$= 0.07144\ g\ KClO_3$$

$$\%KClO_3 = \frac{0.07144\ g}{0.1432\ g} \times 100\% = 49.89\%$$

B. **(M)** First, balance the equation. Break down the reaction of arsenous acid and permanganate as follows:

Oxidation: $\{H_3AsO_3 + H_2O \rightarrow H_3AsO_4 + 2\ e^- + 2H^+\} \times 5$

Reduction: $\{MnO_4^- + 83\ H^+ 5e^- \rightarrow Mn^{2+} + 4\ H_2O\} \times 2$

Overall: $5\ H_3AsO_3 + 2\ MnO_4^- + 6\ H^+ \rightarrow 5\ H_3AsO_4 + 2\ Mn^{2+} + 3\ H_2O$

moles of MnO_4^- = 0.02377 L × 0.02144 M = 5.0963×10^{-4} mol

To calculate the mass of As, use the mole ratios in the balanced equation in conjunction with the molar mass of As:

$$5.0963 \times 10^{-4} \text{ mol MnO}_4^- \times \frac{5 \text{ mol H}_3\text{AsO}_3}{2 \text{ mol MnO}_4^-} \times \frac{1 \text{ mol As}}{1 \text{ mol H}_3\text{AsO}_3} \times \frac{74.922 \text{ g As}}{1 \text{ mol As}}$$

$$= 0.095456 \text{ g As}$$

$$\text{mass\% As} = \frac{0.095456 \text{ g}}{7.25 \text{ g}} \times 100\% = 1.32\%$$

EXERCISES

Strong Electrolytes, Weak Electrolytes, and Nonelectrolytes

1. **(E)**

(a) Because its formula begins with hydrogen, HC_6H_5O is an acid. It is not listed in Table 5-1, so it is a weak acid. A weak acid is a *weak electrolyte*.

(b) Li_2SO_4 is an ionic compound, that is, a salt. A salt is a *strong electrolyte*.

(c) MgI_2 also is a salt, a *strong electrolyte*.

(d) $(CH_3CH_2)_2O$ is a molecular compound whose formula does not begin with H. Thus, it is neither an acid nor a salt. It also is not built around nitrogen, and thus it does not behave as a weak base. This is a *nonelectrolyte*.

(e) $Sr(OH)_2$ is a *strong electrolyte*, one of the strong bases listed in Table 5-2.

3. **(E)** HCl is practically 100% ionized in water. The apparatus should light up brightly. A solution of both HCl and CH_3COOH will yield similar results. In strongly acidic solutions, the weak acid CH_3COOH does not contribute to the conductivity of the solution. However, the strong acid HCl is always 100% ionized in water and is unaffected by the presence of the weak acid CH_3COOH. The apparatus should light up brightly.

5. **(E)**
(a) Barium bromide: strong electrolyte
(b) Propionic acid: weak electrolyte
(c) Ammonia: weak electrolyte

Ion Concentrations

7. **(E)**

(a) $\left[K^+\right] = \frac{0.238 \text{ mol KNO}_3}{1 \text{ L soln}} \times \frac{1 \text{ mol K}^+}{1 \text{ mol KNO}_3} = 0.238 \text{ M}$

(b) $\left[NO_3^-\right] = \dfrac{0.167 \ mol \ Ca(NO_3)_2}{1 \ L \ soln} \times \dfrac{2 \ mol \ NO_3^-}{1 \ mol \ Ca(NO_3)_2} = 0.334 \ M$

(c) $\left[Al^{3+}\right] = \dfrac{0.083 \ mol \ Al_2(SO_4)_3}{1 \ L \ soln} \times \dfrac{2 \ mol \ Al^{3+}}{1 \ mol \ Al_2(SO_4)_3} = 0.166 \ M$

(d) $\left[Na^+\right] = \dfrac{0.209 \ mol \ Na_3PO_4}{1 \ L \ soln} \times \dfrac{3 \ mol \ Na^+}{1 \ mol \ Na_3PO_4} = 0.627 \ M$

9. **(E)**

Conversion pathway approach:

$$\left[OH^-\right] = \dfrac{0.132 \ g \ Ba(OH)_2 \cdot 8 \ H_2O}{275 \ mL \ soln} \times \dfrac{1000 \ mL}{1 \ L} \times \dfrac{1 \ mol \ Ba(OH)_2 \cdot 8 \ H_2O}{315.5 \ g \ Ba(OH)_2 \cdot 8 \ H_2O} \times \dfrac{2 \ mol \ OH^-}{1 \ mol \ Ba(OH)_2 \cdot 8 \ H_2O}$$

$$= 3.04 \times 10^{-3} \ M$$

Stepwise approach:

$$\dfrac{0.132 \ g \ Ba(OH)_2 \cdot 8 \ H_2O}{275 \ mL \ soln} \times \dfrac{1000 \ mL}{1 \ L} = 0.480 \ g/L$$

$$\dfrac{0.480 \ g}{1 \ L} \times \dfrac{1 \ mol \ Ba(OH)_2 \cdot 8 \ H_2O}{315.5 \ g \ Ba(OH)_2 \cdot 8 \ H_2O} = \dfrac{0.00152 \ mol \ Ba(OH)_2 \times 8 \ H_2O}{L}$$

$$\dfrac{0.00152 \ mol \ Ba(OH)_2 \times 8 \ H_2O}{1 \ L} \times \dfrac{2 \ mol \ OH^-}{1 \ mol \ Ba(OH)_2 \cdot 8 \ H_2O} = \dfrac{3.04 \times 10^{-3} \ mol \ OH^-}{L}$$

11. **(E)**

(a) $[Ca^{2+}] = \dfrac{14.2 \ mg \ Ca^{2+}}{1 \ L \ solution} \times \dfrac{1 \ g \ Ca^{2+}}{1000 \ mg \ Ca^{2+}} \times \dfrac{1 \ mol \ Ca^{2+}}{40.078 \ g \ Ca^{2+}} = 3.54 \times 10^{-4} \ M$

(b) $[K^+] = \dfrac{32.8 \ mg \ K^+}{100 \ mL \ solution} \times \dfrac{1 \ g \ K^+}{1000 \ mg \ K^+} \times \dfrac{1000 \ mL \ solution}{1 \ L \ solution} \times \dfrac{1 \ mol \ K^+}{39.0983 \ g \ K^+} = 8.39 \times 10^{-3} \ M$

(c) $[Zn^{2+}] = \dfrac{225 \ \mu g \ Zn^{2+}}{1 \ mL \ solution} \times \dfrac{1 \ g \ Zn^{2+}}{1 \times 10^6 \ \mu g \ Zn^{2+}} \times \dfrac{1000 \ mL \ solution}{1 \ L \ solution} \times \dfrac{1 \ mol \ Zn^{2+}}{65.39 \ g \ Zn^{2+}} = 3.44 \times 10^{-3} \ M$

13. **(E)** In order to determine the solution with the largest concentration of K^+, we begin by converting each concentration to a common concentration unit, namely, molarity of K^+.

$$\dfrac{0.0850 \ mol \ K_2SO_4}{1 \ L \ solution} \times \dfrac{2 \ mol \ K^+}{1 \ mol \ K_2SO_4} = 0.17 \ M$$

$$\dfrac{1.25 \ g \ KBr}{100 \ mL \ solution} \times \dfrac{1000 \ mL \ solution}{1 \ L \ solution} \times \dfrac{1 \ mol \ KBr}{119.0 \ g \ KBr} \times \dfrac{1 \ mol \ K^+}{1 \ mol \ KBr} = 0.105 \ M$$

$$\frac{8.1 \text{ mg K}^+}{1 \text{ mL solution}} \times \frac{1000 \text{ mL solution}}{1 \text{ L solution}} \times \frac{1 \text{ g K}^+}{1000 \text{ mg K}^+} \times \frac{1 \text{ mol K}^+}{39.10 \text{ g K}^+} = 0.207 \text{ M}$$

Clearly, the solution containing 8.1 mg K^+ per mL gives the largest K^+ of the three solutions.

15. **(M)** Determine the amount of I^- in the solution as it now exists, and the amount of I^- in the solution of the desired concentration. The difference in these two amounts is the amount of I^- that must be added. Convert this amount to a mass of MgI_2 in grams.

$$\text{moles of } I^- \text{ in final solution} = 250.0 \text{ mL} \times \frac{1 \text{ L}}{1000 \text{ mL}} \times \frac{0.1000 \text{ mol } I^-}{1 \text{ L soln}} = 0.02500 \text{ mol } I^-$$

$$\text{moles of } I^- \text{ in KI solution} = 250.0 \text{ mL} \times \frac{1 \text{ L}}{1000 \text{ mL}} \times \frac{0.0876 \text{ mol KI}}{1 \text{ L soln}} \times \frac{1 \text{ mol } I^-}{1 \text{ mol KI}} = 0.0219 \text{ mol } I^-$$

$$\text{mass } MgI_2 \text{ required} = (0.02500 - 0.0219) \text{ mol } I^- \times \frac{1 \text{ mol } MgI_2}{2 \text{ mol } I^-} \times \frac{278.11 \text{ g } MgI_2}{1 \text{ mol } MgI_2} \times \frac{1000 \text{ mg}}{1 \text{ g}}$$

$$= 4.3 \times 10^2 \text{ mg } MgI_2$$

17. **(M)** moles of chloride ion

$$= \left(0.225 \text{ L} \times \frac{0.625 \text{ mol KCl}}{1 \text{ L}} \times \frac{1 \text{ mol Cl}^-}{1 \text{ mol KCl}} \right) + \left(0.615 \text{ L} \times \frac{0.385 \text{ mol MgCl}_2}{1 \text{ L}} \times \frac{2 \text{ mol Cl}^-}{1 \text{ mol MgCl}_2} \right)$$

$$= 0.141 \text{ mol Cl}^- + 0.474 \text{ mol Cl}^- = 0.615 \text{ mol Cl}^- \quad [\text{Cl}^-] = \frac{0.615 \text{ mol Cl}^-}{0.225 \text{ L} + 0.615 \text{ L}} = 0.732 \text{ M}$$

Predicting Precipitation Reactions

19. **(E)** In each case, each available cation is paired with the available anions, one at a time, to determine if a compound is produced that is insoluble, based on the solubility rules of Chapter 5. Then a net ionic equation is written to summarize this information.

(a) $Pb^{2+}(aq) + 2 Br^-(aq) \rightarrow PbBr_2(s)$

(b) No reaction occurs (all are spectator ions).

(c) $Fe^{3+}(aq) + 3 OH^-(aq) \rightarrow Fe(OH)_3(s)$

21. **(E)**

Mixture	Result (Net Ionic Equation)
(a) $HI(aq) + Zn(NO_3)_2(aq)$:	No reaction occurs.
(b) $CuSO_4(aq) + Na_2CO_3(aq)$:	$Cu^{2+}(aq) + CO_3{}^{2-}(aq) \rightarrow CuCO_3(s)$
(c) $Cu(NO_3)_2(aq) + Na_3PO_4(aq)$:	$3 Cu^{2+}(aq) + 2 PO_4^{3-}(aq) \rightarrow Cu_3(PO_4)_2(s)$

23. **(E)**

(a) Add K_2SO_4 (aq); $BaSO_4$ (s) will form and $MgSO_4$ will not precipitate.

$$BaCl_2(s) + K_2SO_4(aq) \rightarrow BaSO_4(s) + 2\ KCl(aq)$$

(b) Add $H_2O(l)$; Na_2CO_3 (s) dissolves, but $MgCO_3$ (s) will not dissolve (appreciably).

$$Na_2CO_3(s) \xrightarrow{\text{water}} 2\ Na^+(aq) + CO_3^{2-}(aq)$$

(c) Add KCl(aq); AgCl(s) will form, while $Cu(NO_3)_2$ (s) will dissolve.

$$AgNO_3(s) + KCl(aq) \rightarrow AgCl(s) + KNO_3(aq)$$

25. **(M)**

Mixture	Net Ionic Equation

(a) $Sr(NO_3)_2(aq) + K_2SO_4(aq)$: $Sr^{2+}(aq) + SO_4^{2-}(aq) \rightarrow SrSO_4(s)$

(b) $Mg(NO_3)_2(aq) + NaOH(aq)$: $Mg^{2+}(aq) + 2\ OH^-(aq) \rightarrow Mg(OH)_2(s)$

(c) $BaCl_2(aq) + K_2SO_4(aq)$: $Ba^{2+}(aq) + SO_4^{2-}(aq) \rightarrow BaSO_4(s)$

  (upon filtering, KCl (aq) is obtained)

Acid–Base Reactions

27. **(E)** The type of reaction is given first, followed by the net ionic equation.
(a) Neutralization: $OH^-(aq) + CH_3COOH(aq) \rightarrow H_2O(l) + C_2H_3O_2^-(aq)$
(b) No reaction occurs. This is the physical mixing of two acids.
(c) Gas evolution: $FeS(s) + 2\ H^+(aq) \rightarrow H_2S(g) + Fe^{2+}(aq)$
(d) Gas evolution: $HCO_3^-(aq) + H^+(aq) \rightarrow H_2O(l) + CO_2(g)$
(e) Redox: $Mg(s) + 2\ H^+(aq) \rightarrow Mg^{2+}(aq) + H_2(g)$

29. **(M)**

As a salt: $NaHSO_4(aq) \rightarrow Na^+(aq) + HSO_4^-(aq)$

As an acid: $HSO_4^-(aq) + OH^-(aq) \rightarrow H_2O(l) + SO_4^{2-}(aq)$

31. **(M)** Use (b) NH_3 (aq): NH_3 affords the OH^- ions necessary to form $Mg(OH)_2(s)$.
Applicable reactions: $\{NH_3(aq) + H_2O(l) \rightarrow NH_4^+(aq) + OH^-(aq)\} \times 2$
$MgCl_2(aq) \rightarrow Mg^{2+}(aq) + 2\ Cl^-(aq)$
$Mg^{2+}(aq) + 2\ OH^-(aq) \rightarrow Mg(OH)_2(s)$

Oxidation–Reduction (Redox) Equations

<u>33.</u> **(E)**

(a) The O.S. of H is $+1$, that of O is -2, that of C is $+4$, and that of Mg is $+2$ on each side of this equation. This is not a redox equation.

(b) The O.S. of Cl is 0 on the left and -1 on the right side of this equation. The O.S. of Br is -1 on the left and 0 on the right side of this equation. This is a redox reaction.

(c) The O.S. of Ag is 0 on the left and $+1$ on the right side of this equation. The O.S. of N is $+5$ on the left and $+4$ on the right side of this equation. This is a redox reaction.

(d) On both sides of the equation the O.S. of O is -2, that of Ag is $+1$, and that of Cr is $+6$. Thus, this is not a redox equation.

<u>35.</u> **(E)**

(a) Reduction: $2SO_3^{2-}(aq) + 6\ H^+(aq) + 4\ e^- \rightarrow S_2O_3^{2-}(aq) + 3\ H_2O(l)$

(b) Reduction: $2\ NO_3^-(aq) + 10\ H^+(aq) + 8\ e^- \rightarrow N_2O(g) + 5\ H_2O(l)$

(c) Oxidation: $Al(s) + 4\ OH^-(aq) \rightarrow Al(OH)_4^-(aq) + 3\ e^-$

<u>37.</u> **(M)**

(a) Oxidation: $\{2\ I^-(aq) \rightarrow I_2(s) + 2\ e^-\} \times 5$

Reduction: $\{MnO_4^-(aq) + 8\ H^+(aq) + 5\ e^- \rightarrow Mn^{2+}(aq) + 4\ H_2O(l)\} \times 2$

Overall: $10\ I^-(aq) + 2\ MnO_4^-(aq) + 16\ H^+(aq) \rightarrow 5\ I_2(s) + 2\ Mn^{2+}(aq) + 8\ H_2O(l)$

(b) Oxidation: $\{N_2H_4(l) \rightarrow N_2(g) + 4\ H^+(aq) + 4\ e^-\} \times 3$

Reduction: $\{BrO_3^-(aq) + 6\ H^+(aq) + 6\ e^- \rightarrow Br^-(aq) + 3\ H_2O(l)\} \times 2$

Overall: $3\ N_2H_4(l) + 2\ BrO_3^-(aq) \rightarrow 3\ N_2(g) + 2\ Br^-(aq) + 6\ H_2O(l)$

(c) Oxidation: $Fe^{2+}(aq) \rightarrow Fe^{3+}(aq) + e^-$

Reduction: $VO_4^{3-}(aq) + 6\ H^+(aq) + e^- \rightarrow VO^{2+}(aq) + 3\ H_2O(l)$

Overall: $Fe^{2+}(aq) + VO_4^{3-}(aq) + 6\ H^+(aq) \rightarrow Fe^{3+}(aq) + VO^{2+}(aq) + 3\ H_2O(l)$

(d) Oxidation: $\{UO^{2+}(aq) + H_2O(l) \rightarrow UO_2^{2+}(aq) + 2\ H^+(aq) + 2\ e^-\} \times 3$

Reduction: $\{NO_3^-(aq) + 4\ H^+(aq) + 3\ e^- \rightarrow NO(g) + 2\ H_2O(l)\} \times 2$

Overall: $3\ UO^{2+}(aq) + 2\ NO_3^-(aq) + 2\ H^+(aq) \rightarrow 3\ UO_2^{2+}(aq) + 2\ NO(g) + H_2O(l)$

39. **(M)**

    (a) Oxidation: $\{MnO_2\,(s) + 4\ OH^-\,(aq) \rightarrow MnO_4^-\,(aq) + 2\ H_2O(l) + 3\ e^-\} \times 2$

        Reduction: $ClO_3^-\,(aq) + 3\ H_2O(l) + 6\ e^- \rightarrow Cl^-\,(aq) + 6\ OH^-\,(aq)$

        Overall: $2\ MnO_2\,(s) + ClO_3^-\,(aq) + 2\ OH^-\,(aq) \rightarrow 2\ MnO_4^-\,(aq) + Cl^-\,(aq) + H_2O(l)$

    (b) Oxidation: $\{Fe(OH)_3\,(s) + 5\ OH^-\,(aq) \rightarrow FeO_4^{2-}\,(aq) + 4\ H_2O(l) + 3\ e^-\} \times 2$

        Reduction: $\{OCl^-\,(aq) + H_2O(l) + 2\ e^- \rightarrow Cl^-\,(aq) + 2OH^-\,(aq)\} \times 3$

        Overall: $2\ Fe(OH)_3\,(s) + 3\ OCl^-\,(aq) + 4\ OH^-\,(aq) \rightarrow 2\ FeO_4^{2-}\,(aq) + 3\ Cl^-\,(aq) + 5\ H_2O(l)$

    (c) Oxidation: $\{ClO_2(aq) + 2\ OH^-\,(aq) \rightarrow ClO_3^-\,(aq) + H_2O(l) + e^-\} \times 5$

        Reduction: $ClO_2(aq) + 2\ H_2O(l) + 5\ e^- \rightarrow Cl^-(aq) + 4\ OH^-\,(aq)$

        Overall: $\quad 6\ ClO_2(aq) + 6\ OH^-\,(aq) \rightarrow 5\ ClO_3^-\,(aq) + Cl^-\,(aq) + 3\ H_2O(l)$

    (d) Oxidation: $\{Ag\,(s) \rightarrow Ag^+\,(aq) + 1\ e^-\} \times 3$

        Reduction: $4\ H_2O(l) + CrO_4^{2-} + 3\ e^- \rightarrow Cr(OH)_3(s) + 5\ OH^-$

        Overall: $3\ Ag(s) + CrO_4^{2-} + 4\ H_2O(l) \rightarrow 3\ Ag^+\,(aq) + Cr(OH)_3(s) + 5\ OH^-$

41. **(M)**

    (a) Oxidation: $Cl_2\,(g) + 12\ OH^-\,(aq) \rightarrow 2\ ClO_3^-\,(aq) + 6\ H_2O(l) + 10\ e^-$

        Reduction: $\{Cl_2\,(g) + 2\ e^- \rightarrow 2\ Cl^-\,(aq)\} \times 5$

        Overall: $6\ Cl_2\,(g) + 12\ OH^-\,(aq) \rightarrow 10\ Cl^-(aq) + 2\ ClO_3^-\,(aq) + 6\ H_2O(l)$

        Or: $3\ Cl_2\,(g) + 6\ OH^-\,(aq) \rightarrow 5\ Cl^-\,(aq) + ClO_3^-\,(aq) + 3\ H_2O(l)$

    (b) Oxidation: $S_2O_4^{2-}\,(aq) + 2\ H_2O(l) \rightarrow 2\ HSO_3^-\,(aq) + 2\ H^+\,(aq) + 2\ e^-$

        Reduction: $S_2O_4^{2-}\,(aq) + 2\ H^+\,(aq) + 2\ e^- \rightarrow S_2O_3^{2-}\,(aq) + H_2O(l)$

        Overall: $\quad 2\ S_2O_4^{2-}\,(aq) + H_2O(l) \rightarrow 2\ HSO_3^-\,(aq) + S_2O_3^{2-}\,(aq)$

43. **(M)**

    (a) Oxidation: $\{NO_2^-\,(aq) + H_2O(l) \rightarrow NO_3^-\,(aq) + 2\ H^+\,(aq) + 2\ e^-\} \times 5$

        Reduction: $\{MnO_4^-\,(aq) + 8\ H^+\,(aq) + 5\ e^- \rightarrow Mn^{2+}\,(aq) + 4\ H_2O(l)\} \times 2$

        Overall: $5\ NO_2^-\,(aq) + 2\ MnO_4^-\,(aq) + 6\ H^+\,(aq) \rightarrow 5\ NO_3^-\,(aq) + 2\ Mn^{2+}\,(aq) + 3\ H_2O(l)$

    (b) Oxidation: $\{Mn^{2+}\,(aq) + 4\ OH^-\,(aq) \rightarrow MnO_2\,(s) + 2\ H_2O\,(l) + 2\ e^-\} \times 3$

        Reduction: $\{MnO_4^-\,(aq) + 2\ H_2O\,(l) + 3\ e^- \rightarrow MnO_2\,(s) + 4\ OH^-\,(aq)\} \times 2$

        Overall: $3\ Mn^{2+}\,(aq) + 2\ MnO_4^-\,(aq) + 4\ OH^-\,(aq) \rightarrow 5\ MnO_2\,(s) + 2\ H_2O\,(l)$

    (c) Oxidation: $\{C_2H_5OH \rightarrow CH_3CHO + 2\ H^+\,(aq) + 2\ e^-\} \times 3$

        Reduction: $Cr_2O_7^{2-}\,(aq) + 14\ H^+\,(aq) + 6\ e^- \rightarrow 2\ Cr^{3+}\,(aq) + 7\ H_2O(l)$

        Overall: $Cr_2O_7^{2-}\,(aq) + 8\ H^+\,(aq) + 3\ C_2H_5OH \rightarrow 2\ Cr^{3+}\,(aq) + 7\ H_2O(l) + 3\ CH_3CHO$

45. **(D)** For the purpose of balancing its redox equation, each of the reactions is treated as if it takes place in acidic aqueous solution.

(a) $2 H_2O(g) + CH_4(g) \rightarrow CO_2(g) + 8 H^+(g) + 8 e^-$
$\{2 e^- + 2 H^+(g) + NO(g) \rightarrow \frac{1}{2} N_2(g) + H_2O(g)\} \times 4$
$\overline{CH_4(g) + 4 NO(g) \rightarrow 2 N_2(g) + CO_2(g) + 2 H_2O(g)}$

(b) $\{H_2S(g) \rightarrow 1/8 S_8(s) + 2 H^+(g) + 2 e^-\} \times 2$
$\underline{4 e^- + 4 H^+(g) + SO_2(g) \rightarrow 1/8 S_8(s) + 2 H_2O(g)}$
$2 H_2S(g) + SO_2(g) \rightarrow 3/8 S_8(s) + 2 H_2O(g)$ or
$16 H_2S(g) + 8 SO_2(g) \rightarrow 3 S_8(s) + 16 H_2O(g)$

(c) $\{Cl_2O(g) + 2 NH_4^+(aq) + 2 H^+(aq) + 4 e^- \rightarrow 2 NH_4Cl(s) + H_2O(l)\} \times 3$
$\{2 NH_3(g) \rightarrow N_2(g) + 6 e^- + 6 H^+(aq)\} \times 2$
$\underline{6 NH_3(g) + 6 H^+(aq) \rightarrow 6 NH_4^+(aq)}$
$10 NH_3(g) + 3 Cl_2O(g) \rightarrow 6 NH_4Cl(s) + 2 N_2(g) + 3 H_2O(l)$

Oxidizing and Reducing Agents

47. **(E)** The oxidizing agents experience a decrease in the oxidation state of one of their elements, while the reducing agents experience an increase in the oxidation state of one of their elements.

(a) $SO_3^{2-}(aq)$ is the reducing agent; the O.S. of $S = +4$ in SO_3^{2-} and $= +6$ in SO_4^{2-}.
$MnO_4^-(aq)$ is the oxidizing agent; the O.S. of $Mn = +7$ in MnO_4^- and $+2$ in Mn^{2+}.

(b) $H_2(g)$ is the reducing agent; the O.S. of $H = 0$ in $H_2(g)$ and $= +1$ in $H_2O(g)$.
$NO_2(g)$ is the oxidizing agent; the O.S. of $N = +4$ in $NO_2(g)$ and -3 in $NH_3(g)$.

(c) $\left[Fe(CN)_6\right]^{4-}(aq)$ is the reducing agent; the O.S. of $Fe = +2$ in $\left[Fe(CN)_6\right]^{4-}$
and $= +3$ in $\left[Fe(CN)_6\right]^{3-}$. $H_2O_2(aq)$ is the oxidizing agent; the O.S. of $O = -1$
in H_2O_2 and $= -2$ in H_2O.

Neutralization and Acid–Base Titrations

49. **(E)** The problem is most easily solved with amounts in millimoles.

$$V_{NaOH} = 10.00 \text{ mL HCl(aq)} \times \frac{0.128 \text{ mmol HCl}}{1 \text{ mL HCl(aq)}} \times \frac{1 \text{ mmol H}^+}{1 \text{ mmol HCl}} \times \frac{1 \text{ mmol OH}^-}{1 \text{ mmol H}^+}$$

$$\times \frac{1 \text{ mmol NaOH}}{1 \text{ mmol OH}^-} \times \frac{1 \text{ mL NaOH(aq)}}{0.0962 \text{ mmol NaOH}} = 13.3 \text{ mL NaOH(aq) soln}$$

51. **(E)** The net reaction is $OH^-(aq) + CH_3CH_2COOH(aq) \rightarrow H_2O(l) + CH_3CH_2COO^-(aq)$.

Conversion pathway approach:

$$V_{base} = 25.00 \text{ mL acid} \times \frac{0.3057 \text{ mmol } CH_3CH_2COOH}{1 \text{ mL acid}} \times$$

$$\frac{1 \text{ mmol KOH}}{1 \text{ mmol } CH_3CH_2COOH} \times \frac{1 \text{ mL base}}{2.155 \text{ mmol KOH}}$$

$$= 3.546 \text{ mL KOH solution}$$

Stepwise approach:

$$25.00 \text{ mL acid} \times \frac{0.3057 \text{ mmol } CH_3CH_2COOH}{1 \text{ mL acid}} = 7.643 \text{ mmol } CH_3CH_2COOH$$

$$7.643 \text{ mmol } CH_3CH_2COOH \times \frac{1 \text{ mmol KOH}}{1 \text{ mmol } CH_3CH_2COOH} = 7.643 \text{ mmol KOH}$$

$$7.643 \text{ mmol KOH} \times \frac{1 \text{ mL base}}{2.155 \text{ mmol KOH}} = 3.546 \text{ mL KOH solution}$$

53. **(E)** $NaOH(aq) + HCl(aq) \rightarrow NaCl(aq) + H_2O(l)$ is the titration reaction.

$$\text{molarity of NaOH} = \frac{0.02834 \text{ L} \times \dfrac{0.1085 \text{ mol HCl}}{1 \text{ L soln}} \times \dfrac{1 \text{ mol NaOH}}{1 \text{ mol HCl}}}{0.02500 \text{ L sample}} = 0.1230 \text{ M}$$

55. **(M)** The mass of acetylsalicylic acid is converted to the amount of NaOH, in millimoles, that will react with it.

$$\text{molarity of NaOH} = \frac{0.32 \text{ g } HC_9H_7O_4}{23 \text{ mL NaOH(aq)}} \times \frac{1 \text{ mol } HC_9H_7O_4}{180.2 \text{ g } HC_9H_7O_4}$$

$$\times \frac{1 \text{ mol NaOH}}{1 \text{ mol } HC_9H_7O_4} \times \frac{1000 \text{ mmol NaOH}}{1 \text{ mol NaOH}}$$

$$= 0.077 \text{ M}$$

57. **(M)** The equation for the reaction is $HNO_3(aq) + KOH(aq) \rightarrow KNO_3(aq) + H_2O(l)$.

This equation shows that equal numbers of moles are needed for a complete reaction. We compute the amount of each reactant.

$$\text{mmol } HNO_3 = 25.00 \text{ mL acid} \times \frac{0.132 \text{ mmol } HNO_3}{1 \text{ mL acid}} = 3.30 \text{ mmol } HNO_3$$

$$\text{mmol KOH} = 10.00 \text{ mL base} \times \frac{0.318 \text{ mmol KOH}}{1 \text{ mL base}} = 3.18 \text{ mmol KOH}$$

There is more acid present than base. Thus, the resulting solution is acidic.

59. **(M)** $V_{base} = 5.00 \text{ mL vinegar} \times \dfrac{1.01 \text{ g vinegar}}{1 \text{ mL}} \times \dfrac{4.0 \text{ g CH}_3\text{COOH}}{100.0 \text{ g vinegar}} \times \dfrac{1 \text{ mol CH}_3\text{COOH}}{60.0 \text{ g CH}_3\text{COOH}}$

$\times \dfrac{1 \text{ mol NaOH}}{1 \text{ mol CH}_3\text{COOH}} \times \dfrac{1 \text{ L base}}{0.1000 \text{ mol NaOH}} \times \dfrac{1000 \text{ mL}}{1 \text{ L}} = 34 \text{ mL base}$

61. **(E)** Answer is (d): 120% of necessary titrant added in titration of NH_3

$\left.\begin{array}{l} 5 \text{ NH}_3 \\ + \\ 5 \text{ HCl} \\ + \\ 1 \text{ HCl} \end{array}\right\}$ $\begin{array}{l} \text{required for} \\ \text{equivalence} \\ \text{point} \\ \\ 20\% \text{ excess} \end{array}$ $\longrightarrow$ $5 \text{ NH}_4^+ + 6 \text{ Cl}^- + \text{H}_3\text{O}^+$
(depicted in question's drawing)

Stoichiometry of Oxidation–Reduction Reactions

63. **(M)**

Conversion pathway approach

$[MnO_4^-] = \dfrac{0.1078 \text{ g As}_2\text{O}_3 \times \dfrac{1 \text{ mol As}_2\text{O}_3}{197.84 \text{ g As}_2\text{O}_3} \times \dfrac{4 \text{ mol MnO}_4^-}{5 \text{ mol As}_2\text{O}_3} \times \dfrac{1 \text{ mol KMnO}_4}{1 \text{ mol MnO}_4^-}}{22.15 \text{ mL} \times \dfrac{1 \text{ L}}{1000 \text{ mL}}} = 0.01968 \text{ M}$

Stepwise approach:

$0.1078 \text{ g As}_2\text{O}_3 \times \dfrac{1 \text{ mol As}_2\text{O}_3}{197.84 \text{ g As}_2\text{O}_3} = 5.449 \times 10^{-4} \text{ mol As}_2\text{O}_3$

$5.449 \times 10^{-4} \text{ mol As}_2\text{O}_3 \times \dfrac{4 \text{ mol MnO}^{-4}}{5 \text{ mol As}_2\text{O}_3} = 4.359 \times 10^{-4} \text{ mol MnO}_4^-$

$4.359 \times 10^{-4} \text{ mol MnO}_4^- \times \dfrac{1 \text{ mol KMnO}_4}{1 \text{ mol MnO}_4^-} = 4.359 \times 10^{-4} \text{ mol KMnO}_4$

$22.15 \text{ mL} \times \dfrac{1 \text{ L}}{1000 \text{ mL}} = 0.02215 \text{ L solution}$

$\text{molarity of KMnO}_4 = \dfrac{4.359 \times 10^{-4} \text{ mol KMnO}_4}{0.02215 \text{ L solution}} = 1.968 \times 10^{-2} \text{ M}$

65. **(M)** First, we will determine the mass of Fe, then the percentage of iron in the ore.

$\text{mass Fe} = 28.72 \text{ mL} \times \dfrac{1 \text{ L}}{1000 \text{ mL}} \times \dfrac{0.05051 \text{ mol Cr}_2\text{O}_7^{2-}}{1 \text{ L soln}} \times \dfrac{6 \text{ mol Fe}^{2+}}{1 \text{ mol Cr}_2\text{O}_7^{2-}} \times \dfrac{55.85 \text{ g Fe}}{1 \text{ mol Fe}^{2+}}$

$\text{mass Fe} = 0.4861 \text{ g Fe}$ $\qquad$ $\% \text{Fe} = \dfrac{0.4861 \text{ g Fe}}{0.9132 \text{ g ore}} \times 100\% = 53.23\% \text{ Fe}$

67. **(M)** First, balance the titration equation:

Oxidation: $\{C_2O_4^{2-}(aq) \rightarrow 2\ CO_2(g) + 2\ e^-\} \times 5$

Reduction: $\{MnO_4^-(aq) + 8\ H^+(aq) + 5\ e^- \rightarrow Mn^{2+}(aq) + 4\ H_2O(l)\} \times 2$

Overall: $5\ C_2O_4^{2-}(aq) + 2\ MnO_4^-(aq) + 16\ H^+(aq) \rightarrow 10\ CO_2(g) + 2\ Mn^{2+}(aq) + 8\ H_2O(l)$

$$mass_{Na_2C_2O_4} = 1.00\ L\ satd\ soln\ Na_2C_2O_4 \times \frac{1000\ mL}{1\ L} \times \frac{25.8\ mL\ satd\ soln\ KMnO_4}{5.00\ mL\ satd\ soln\ Na_2C_2O_4} \times \frac{0.02140\ mol\ KMnO_4}{1000\ mL\ KMnO_4}$$

$$\times \frac{1\ mol\ MnO_4^-}{1\ mol\ KMnO_4} \times \frac{5\ mol\ C_2O_4^{2-}}{2\ mol\ MnO_4^-} \times \frac{1\ mol\ Na_2C_2O_4}{1\ mol\ C_2O_4^{2-}} \times \frac{134.0\ g\ Na_2C_2O_4}{1\ mol\ Na_2C_2O_4}$$

$$mass_{Na_2C_2O_4} = 37.0\ g\ Na_2C_2O_4$$

INTEGRATIVE AND ADVANCED EXERCISES

71. **(M)** A possible product, based on solubility rules, is $Ca_3(PO_4)_2$. We determine the % Ca in this compound.

molar mass $= 3 \times 40.078\ g\ Ca + 2 \times 30.974\ g\ P + 8 \times 15.999\ g\ O$

$\qquad\qquad = 120.23\ g\ Ca + 61.948\ g\ P + 127.99\ g\ O = 310.17\ g$

$$\%\ Ca = \frac{120.23\ g\ Ca}{310.17\ g\ Ca_3(PO_4)_2} \times 100\% = 38.763\%$$

Thus, $Ca_3(PO_4)_2$ is the predicted product. The net ionic equation follows.

$$3\ Ca^{2+}(aq) + 2\ HPO_4^{2-}(aq) \longrightarrow Ca_3(PO_4)_2(s) + 2\ H^+(aq)$$

74. **(D)** Let us first determine the mass of Mg in the sample analyzed.

Conversion pathway approach:

$$mass\ Mg = 0.0549\ g\ Mg_2P_2O_7 \times \frac{1\ mol\ Mg_2P_2O_7}{222.55\ g\ Mg_2P_2O_7} \times \frac{2\ mol\ Mg}{1\ mol\ Mg_2P_2O_7} \times \frac{24.305\ g}{1\ mol\ Mg} = 0.0120\ g\ Mg$$

$$ppm\ Mg = 10^6\ g\ sample \times \frac{0.0120\ g\ Mg}{110.520\ g\ sample} = 108\ ppm\ Mg$$

Stepwise approach:

$$0.0549\ g\ Mg_2P_2O_7 \times \frac{1\ mol\ Mg_2P_2O_7}{222.55\ g\ Mg_2P_2O_7} = 2.47 \times 10^{-4}\ mol\ Mg_2P_2O_7$$

$$2.47 \times 10^{-4}\ mol\ Mg_2P_2O_7 \times \frac{2\ mol\ Mg}{1\ mol\ Mg_2P_2O_7} = 4.93 \times 10^{-4}\ mol\ Mg$$

$$4.93 \times 10^{-4}\ mol\ Mg \times \frac{24.305\ g}{1\ mol\ Mg} = 0.0120\ g\ Mg$$

$$ppm\ Mg = 10^6\ g\ sample \times \frac{0.0120\ g\ Mg}{110.520\ g\ sample} = 108\ ppm\ Mg$$

75. (M) Let V represent the volume of 0.248 M $CaCl_2$ that must be added.

We know that $[Cl^-]$ = 0.250 M, but also,

$$[Cl^-] = \dfrac{0.335\,L\,\dfrac{0.186\;mol\;KCl}{1\,L\;soln}\times\dfrac{1\;mol\;Cl^-}{1\;mol\;KCl}+V\times\dfrac{0.248\;mol\;CaCl_2}{1\,L\;soln}\times\dfrac{2\;mol\;Cl^-}{1\;mol\;CaCl_2}}{0.335\,L+V}$$

$$0.250\,(0.335+V)=0.0838+0.250\;V=0.0623+0.496\,V \qquad V=\dfrac{0.0838-0.0623}{0.496-0.250}=0.0874\,L$$

80. (D)

(a) Oxidation: $\{FeS_2 +8\,H_2O \rightarrow Fe^{3+} +2\,SO_4^{2-} +16\,OH^+ +15\,e^-\}\times 4$

Reduction: $\{O_2 +4\,H^+ +4\,e^- \rightarrow 2\,H_2O\}\times 15$

Overall: $4\,FeS_2(s)+15\,O_2(g)+2\,H_2O(l)\rightarrow 4\,Fe^{3+}(aq)+8\,SO_4^{2-}(aq)+4\,H^+(aq)$

(b) One kilogram of tailings contains 0.03 kg (30 g) of S. We have

$$\text{moles of } FeS_2 = 30\;g\;S\times\dfrac{1\;mol\;S}{32.06\;g\;S}\times\dfrac{1\;mol\;FeS_2}{2\;mol\;S}=0.468\;mol\;FeS_2$$

$$\text{moles of } H^+ = 0.468\;mol\;FeS_2\times\dfrac{4\;mol\;H^+}{4\;mol\;FeS_2}=0.467\;mol\;H^+$$

$$\text{moles of } CaCO_3 = 0.467\;mol\;H^+\times\dfrac{1\;mol\;CaCO_3}{2\;mol\;H^+}=0.234\;mol\;CaCO_3$$

$$\text{mass of } CaCO_3 = 0.234\;mol\;CaCO_3\times\dfrac{100.09\;g\;CaCO_3}{1\;mol\;CaCO_3}=23.4\;g\;CaCO_3$$

83. (M)

Oxidation: $\{2\,Cl^-(aq)\longrightarrow Cl_2(g)+2\,e^-\}\qquad \times 3$

Reduction: $Cr_2O_7^{2-}(aq)+14\,H^+(aq)+6\,e^-\longrightarrow 2\,Cr^{3+}(aq)+7\,H_2O$

Overall: $6\,Cl^-(aq)+Cr_2O_7^{2-}(aq)+14\,H^+(aq)\longrightarrow 2\,Cr^{3+}(aq)+7\,H_2O+3\,Cl_2(g)$

We need to determine the amount of $Cl_2(g)$ produced from each of the reactants. The limiting reactant is the one that produces the lesser amount of Cl_2.

$$\text{amount } Cl_2 = 325\;mL\times\dfrac{1.15\;g}{1\;mL}\times\dfrac{30.1\;g\;HCl}{100.\;g\;soln}\times\dfrac{1\;mol\;HCl}{36.46\;g\;HCl}\times\dfrac{1\;mol\;Cl^-}{1\;mol\;HCl}\times\dfrac{3\;mol\;Cl_2}{6\;mol\;Cl^-}$$

$$=1.54\;mol\;Cl_2$$

$$\text{amount Cl}_2 = 62.6 \text{ g} \times \frac{98.5 \text{ g K}_2\text{Cr}_2\text{O}_7}{100. \text{ g sample}} \times \frac{1 \text{ mol K}_2\text{Cr}_2\text{O}_7}{294.2 \text{ g K}_2\text{Cr}_2\text{O}_7} \times \frac{1 \text{ mol Cr}_2\text{O}_7{}^{2-}}{1 \text{ mol K}_2\text{Cr}_2\text{O}_7} \times \frac{3 \text{ mol Cl}_2}{1 \text{ mol Cr}_2\text{O}_7{}^{2-}}$$

$= 0.629 \text{ mol Cl}_2$, the amount produced from the limiting reactant

Then we determine the mass of $\text{Cl}_2(g)$ produced. $= 0.629 \text{ mol Cl}_2 \times \dfrac{70.90 \text{ g Cl}_2}{1 \text{ mol Cl}_2} = 44.6 \text{ g Cl}_2$

85. (M)

$$\text{Cl}_2(g) + \text{NaClO}_2(aq) \longrightarrow \text{NaCl}(aq) + \text{ClO}_2(g) \quad \text{(not balanced)}$$

$$\text{Cl}_2(g) + 2 \text{NaClO}_2(aq) \longrightarrow 2 \text{NaCl}(aq) + 2 \text{ClO}_2(g)$$

$$\text{amount ClO}_2 = 1 \text{ gal} \times \frac{3.785 \text{ L}}{1 \text{ gal}} \times \frac{2.0 \text{ mol NaClO}_2}{1 \text{ L soln}} \times \frac{2 \text{ mol ClO}_2}{2 \text{ mol NaClO}_2} \times \frac{67.45 \text{ g ClO}_2}{1 \text{ mol ClO}_2}$$

$$\times \frac{97 \text{ g ClO}_2 \text{ produced}}{100 \text{ g ClO}_2 \text{ calculated}} = 5.0 \times 10^2 \text{ g ClO}_2(g)$$

88. (D)

(a) First, balance the redox equations needed for the calculation.

Oxidation: $\{\text{HSO}_3{}^-(aq) + \text{H}_2\text{O}(l) \to \text{SO}_4{}^{2-}(aq) + 3 \text{ H}^+(aq) + 2 \text{ e}^-\} \times 3$

Reduction: $\text{IO}_3{}^-(aq) + 6 \text{ H}^+(aq) + 6 \text{ e}^- \to \text{I}^-(aq) + 3 \text{ H}_2\text{O}(l)$

Overall: $3 \text{ HSO}_3{}^-(aq) + \text{IO}_3{}^-(aq) \to 3 \text{ SO}_4{}^{2-}(aq) + 3 \text{ H}^+(aq) + \text{I}^-(aq)$

The solution volume of 5.00 L contains 29.0 g NaIO_3. This represents 29.0 g/197.9 g/mol NaIO_3 = 0.147 mol NaIO_3.

(b) From the above equation, we need 3 times that molar amount of NaHSO_3, which is 3(0.147 mol) = 0.441 mol NaHSO_3; the molar mass of NaHSO_3 is 104.06 g/mol.

The required mass then is 0.441(104.06) = 45.9 g.

For the second process:

Oxidation: $\{2 \text{ I}^-(aq) \to \text{I}_2(aq) + 2 \text{ e}^-\} \times 5$

Reduction: $2 \text{ IO}_3{}^-(aq) + 12 \text{ H}^+(aq) + 10 \text{ e}^- \to \text{I}_2(aq) + 6 \text{ H}_2\text{O}(l)$

Overall: $5 \text{ I}^-(aq) + \text{IO}_3{}^-(aq) + 6 \text{ H}^+(aq) \to 3 \text{ I}_2(aq) + 3 \text{ H}_2\text{O}(l)$

In Step 1, we produced 1 mol of I^- for every mole of $\text{IO}_3{}^-$ reactant; therefore we had 0.147 mol I^-.

In step 2, we require 1/5 mol $\text{IO}_3{}^-$ for every mol of I^-.

We require only 1.00 L of the solution in the question instead of the 5.00 L in the first step.

89. **(D)**

$$Mg(OH)_2(aq) + 2\ HCl(aq) \rightarrow MgCl_2(aq) + 2\ H_2O(l) \qquad (1)$$
$$Al(OH)_3(aq) + 3\ HCl(aq) \rightarrow AlCl_3(aq) + 3\ H_2O(l) \qquad (2)$$
$$HCl(aq) + NaOH(aq) \rightarrow NaCl(aq) + H_2O(l) \qquad (3)$$

initial moles of HCl $= 0.0500\ L \times \dfrac{0.500\ mol}{1\ L} = 0.0250\ mol$

moles of HCl that reacted with NaOH = moles of HCl left over from reaction with active ingredients

$$= 0.0165\ L \times \frac{0.377\ mol\ NaOH}{1\ L} \times \frac{1\ mol\ HCl}{1\ mol\ NaOH} = 6.22 \times 10^{-3}\ mol$$

moles of HCl that react with active ingredients $= 0.0250\ mol - 6.22 \times 10^{-3}\ mol = 0.0188\ mol$

$$\begin{bmatrix} \#\ moles\ HCl\ that \\ react\ with\ Mg(OH)_2 \end{bmatrix} + \begin{bmatrix} \#\ moles\ HCl\ that \\ react\ with\ Al(OH)_3 \end{bmatrix} = total\ moles\ of\ HCl\ reacted/used$$

Let the mass of $Mg(OH)_2$ in the sample be X grams.

\# moles HCl that react with $Mg(OH)_2$ =

$$\left[X\ g\ Mg(OH)_2 \times \frac{1\ mol\ Mg(OH)_2}{58.32\ g\ Mg(OH)_2} \times \frac{2\ mol\ HCl}{1\ mol\ Mg(OH)_2} \right]$$

\# moles HCl that react with $Al(OH)_3$ =

$$\left[(0.500 - X)\ g\ Al(OH)_3 \times \frac{1\ mol\ Al(OH)_3}{78.00\ g\ Al(OH)_3} \times \frac{3\ mol\ HCl}{1\ mol\ Al(OH)_3} \right]$$

$$\frac{2X}{58.32} + \frac{3(0.500 - X)}{78.00} = 0.0188$$

X = 0.108, therefore the mass of $Mg(OH)_2$ in the sample is 0.108 grams.

% $Mg(OH)_2 = (0.108/0.500) \times 100 = 21.6$ \qquad %$Al = 100 - \%Mg(OH)_2 = 78.4$

91. **(M)**

$$0.1386\ g\ AgI \times \frac{1\ mol\ AgI}{234.77\ g\ AgI} \times \frac{1\ mol\ CHI_3}{3\ mol\ AgI} \times \frac{1\ mol\ C_{19}H_{16}O_4}{1\ mol\ CHI_3} \times \frac{308.33\ g\ C_{19}H_{16}O_4}{1\ mol\ C_{19}H_{16}O_4}$$

$$= 0.06068\ g\ C_{19}H_{16}O_4$$

$$\%\ C_{19}H_{16}O_4 = \frac{0.06068\ g}{13.96\ g} \times 100 = 0.4346$$

93. **(M) (a)**

$$CaO(s) + H_2O(l) \rightarrow Ca^{2+}(aq) + 2\,OH^-(aq)$$

$$H_2PO_4^-(aq) + 2\,OH^-(aq) \rightarrow PO_4^{3-}(aq) + 2\,H_2O(l)$$

$$HPO_4^-(aq) + OH^-(aq) \rightarrow PO_4^{3-}(aq) + H_2O(l)$$

$$5\,Ca^{2+}(aq) + 3\,PO_4^{3-}(aq) + OH^-(aq) \rightarrow Ca_5(PO_4)_3OH(s)$$

(b) $1.00 \times 10^4 \text{ L} \times \dfrac{10.0 \times 10^{-3} \text{ g P}}{\text{L}} \times \dfrac{1 \text{ mol P}}{30.97 \text{ g P}} \times \dfrac{1 \text{ mol PO}_4^{3-}}{1 \text{ mol P}} \times \dfrac{5 \text{ mol Ca}^{2+}}{3 \text{ mol PO}_4^{3-}}$

$\times \dfrac{1 \text{ mol CaO}}{1 \text{ mol Ca}^{2+}} \times \dfrac{56.08 \text{ g CaO}}{1 \text{ mol CaO}} = 301.80 \text{ g CaO} = 302 \text{ g} = 0.302 \text{ kg}$

FEATURE PROBLEMS

94. **(D)** From the volume of titrant, we can calculate both the amount in moles of NaC_5H_5 and (through its molar mass of 88.08 g/mol) the mass of NaC_5H_5 in a sample. The remaining mass in a sample is that of C_4H_8O (72.11 g/mol), whose amount in moles we calculate. The ratio of the molar amount of C_4H_8O in the sample to the molar amount of NaC_5H_5 is the value of x.

Conversion pathway approach:

moles of $NaC_5H_5 = 0.01492 \text{ L} \times \dfrac{0.1001 \text{ mol HCl}}{1 \text{ L soln}} \times \dfrac{1 \text{ mol NaOH}}{1 \text{ mol HCl}} \times \dfrac{1 \text{ mol NaC}_5\text{H}_5}{1 \text{ mol NaOH}}$

$= 0.001493 \text{ mol NaC}_5\text{H}_5$

mass of $C_4H_8O = 0.242 \text{ g sample} - \left(0.001493 \text{ mol NaC}_5\text{H}_5 \times \dfrac{88.09 \text{ g NaC}_5\text{H}_5}{1 \text{ mol NaC}_5\text{H}_5} \right)$

$= 0.111 \text{ g C}_4\text{H}_8\text{O}$

$x = \dfrac{0.110 \text{ g C}_4\text{H}_8\text{O} \times \dfrac{1 \text{ mol C}_4\text{H}_8\text{O}}{72.11 \text{ g C}_4\text{H}_8\text{O}}}{0.001493 \text{ mol NaC}_5\text{H}_5} = 1.03$

Stepwise approach:

$0.01492 \text{ L} \times \dfrac{0.1001 \text{ mol HCl}}{1 \text{ L soln}} = 1.493 \times 10^{-3} \text{ mol HCl}$

$1.493 \times 10^{-3} \text{ mol HCl} \times \dfrac{1 \text{ mol NaOH}}{1 \text{ mol HCl}} = 1.493 \times 10^{-3} \text{ mol NaOH}$

$1.493 \times 10^{-3} \text{ mol NaOH} \times \dfrac{1 \text{ mol NaC}_5\text{H}_5}{1 \text{ mol NaOH}} = 1.493 \times 10^{-3} \text{ mol NaC}_5\text{H}_5$

$1.493 \times 10^{-3} \text{ mol NaC}_5\text{H}_5 \times \dfrac{88.09 \text{ g NaC}_5\text{H}_5}{1 \text{ mol NaC}_5\text{H}_5} = 0.1315 \text{ g NaC}_5\text{H}_5$

mass of $C_4H_8O = 0.242 \text{ g sample} - 0.1315 \text{ g NaC}_5\text{H}_5 = 0.111 \text{ g C}_4\text{H}_8\text{O}$

$$0.111 \text{g C}_4\text{H}_8\text{O} \times \frac{1 \text{mol C}_4\text{H}_8\text{O}}{72.11 \text{g C}_4\text{H}_8\text{O}} = 1.54 \times 10^{-3} \text{ mol C}_4\text{H}_8\text{O}$$

$$\frac{1.54 \times 10^{-3} \text{ mol C}_4\text{H}_8\text{O}}{0.001493 \text{ mol NaC}_5\text{H}_5} = 1.03$$

For the second sample, parallel calculations give 0.001200 mol NaC_5H_5, 0.093 g C_4H_8, $x = 1.1$. There is rounding error in this second calculation because it is limited to two significant figures. The best answer is from the first run $x \sim 1.03$ or 1. The formula is $NaC_5H_5(THF)_1$.

95. **(D)** First, we balance the two equations.

Oxidation: $H_2C_2O_4(aq) \rightarrow 2 \text{ CO}_2(g) + 2 \text{ H}^+(aq) + 2 \text{ e}^-$

Reduction: $MnO_2(s) + 4 \text{ H}^+(aq) + 2 \text{ e}^- \rightarrow \text{ Mn}^{2+}(aq) + 2 \text{ H}_2O(l)$

Overall: $H_2C_2O_4(aq) + \text{ MnO}_2(s) + 2 \text{ H}^+(aq) \rightarrow 2 \text{ CO}_2(g) + \text{ Mn}^{2+}(aq) + 2 \text{ H}_2O(l)$

Oxidation: $\{ H_2C_2O_4(aq) \rightarrow 2 \text{ CO}_2(g) + 2 \text{ H}^+(aq) + 2 \text{ e}^- \} \times 5$

Reduction: $\{ MnO_4^-(aq) + 8 \text{ H}^+(aq) + 5 \text{ e}^- \rightarrow \text{ Mn}^{2+}(aq) + 4 \text{ H}_2O(l) \} \times 2$

Overall: $5 \text{ H}_2C_2O_4(aq) + 2 \text{ MnO}_4^-(aq) + 6 \text{ H}^+(aq) \rightarrow 10 \text{ CO}_2(g) + 2 \text{ Mn}^{2+}(aq) + 8 \text{ H}_2O(l)$

Next, we determine the mass of the excess oxalic acid.

$$\text{mass H}_2\text{C}_2\text{O}_4 \cdot 2 \text{ H}_2\text{O} = 0.03006 \text{ L} \times \frac{0.1000 \text{ mol KMnO}_4}{1 \text{L}} \times \frac{1 \text{mol MnO}_4^-}{1 \text{mol KMnO}_4} \times \frac{5 \text{mol H}_2\text{C}_2\text{O}_4}{2 \text{mol MnO}_4^-}$$

$$\times \frac{1 \text{ mol H}_2\text{C}_2\text{O}_4 \cdot 2 \text{ H}_2\text{O}}{1 \text{ mol H}_2\text{C}_2\text{O}_4} \times \frac{126.06 \text{ g H}_2\text{C}_2\text{O}_4 \cdot 2 \text{ H}_2\text{O}}{1 \text{ mol H}_2\text{C}_2\text{O}_4 \cdot 2 \text{ H}_2\text{O}} = 0.9474 \text{ g H}_2\text{C}_2\text{O}_4 \cdot 2 \text{ H}_2\text{O}$$

The mass of $H_2C_2O_4 \cdot 2H_2O$ that reacted with $MnO_2 = 1.651 \text{ g} - 0.9474 \text{ g} = 0.704 \text{ g H}_2\text{C}_2\text{O}_4 \cdot 2 \text{ H}_2\text{O}$

$$\text{mass MnO}_2 = 0.704 \text{ g H}_2\text{C}_2\text{O}_4 \cdot 2 \text{ H}_2\text{O} \times \frac{1 \text{ mol H}_2\text{C}_2\text{O}_4}{126.06 \text{ g H}_2\text{C}_2\text{O}_4 \cdot 2 \text{ H}_2\text{O}} \times \frac{1 \text{ mol MnO}_2}{1 \text{ mol H}_2\text{C}_2\text{O}_4} \times \frac{86.9 \text{ g MnO}_2}{1 \text{ mol MnO}_2}$$

$$= 0.485 \text{ g MnO}_2$$

$$\% \text{MnO}_2 = \frac{0.485 \text{ g MnO}_2}{0.533 \text{ g sample}} \times 100\% = 91.0\% \text{ MnO}_2$$

97. **(D)**

The molecular formula for CH_3CH_2OH is C_2H_6O and for CH_3COOH is $C_2H_4O_2$.

The first step is to balance the oxidation–reduction reaction.

Oxidation: $\{C_2H_6O + H_2O \rightarrow C_2H_4O_2 + 4 \text{ H}^+ + 4 \text{ e}^-\} \times 3$

Reduction: $\{Cr_2O_7^{2-} + 14 \text{ H}^+ + 6 \text{ e}^- \rightarrow 2 \text{ Cr}^{3+} + 7 \text{ H}_2O\} \times 2$

Overall: $3 \text{ C}_2\text{H}_6\text{O} + 2 \text{ Cr}_2\text{O}_7^{2-} + 16 \text{ H}^+ \rightarrow 3 \text{ C}_2\text{H}_4\text{O}_2 + 4 \text{ Cr}^{3+} + 11 \text{ H}_2\text{O}$

Before the breath test:

$$\frac{0.75\ \text{mg K}_2\text{Cr}_2\text{O}_7}{3\ \text{mL}} \times \frac{1\ \text{g}}{1000\ \text{mg}} \times \frac{1\ \text{mol}}{294.18\ \text{g}} \times \frac{1000\ \text{mL}}{1\ \text{L}} = 8.498 \times 10^{-4}\ \text{M}$$

$$= 8 \times 10^{-4}\ \text{M (to 1 sig fig)}$$

For the breath sample:

$$\text{BrAC} = \frac{0.05\ \text{g C}_2\text{H}_6\text{O}}{100\ \text{mL blood}} \times \frac{1\ \text{mL blood}}{2100\ \text{mL breath}} = \frac{2.38 \times 10^{-7}\ \text{g C}_2\text{H}_6\text{O}}{\text{mL breath}}$$

$$\text{mass C}_2\text{H}_6\text{O} = \frac{2.38 \times 10^{-7}\ \text{g C}_2\text{H}_6\text{O}}{\text{mL breath}} \times 500.\ \text{mL breath} = 1.19 \times 10^{-4}\ \text{g C}_2\text{H}_6\text{O}$$

Calculate the amount of $K_2Cr_2O_7$ that reacts:

$$1.19 \times 10^{-4}\text{g C}_2\text{H}_6\text{O} \times \frac{1\ \text{mol C}_2\text{H}_6\text{O}}{46.069\ \text{g C}_2\text{H}_6\text{O}} \times \frac{2\ \text{mol Cr}_2\text{O}_7^{2-}}{3\ \text{mol C}_2\text{H}_6\text{O}} \times \frac{1\ \text{mol K}_2\text{Cr}_2\text{O}_7}{1\ \text{mol Cr}_2\text{O}_7^{2-}}$$

$$= 1.72 \times 10^{-6}\text{mol K}_2\text{Cr}_2\text{O}_7$$

mol $K_2Cr_2O_7$ remaining = moles $K_2Cr_2O_7$ before – moles $K_2Cr_2O_7$ that reacts

$$\text{moles K}_2\text{Cr}_2\text{O}_7\ \text{before} = 0.75\ \text{mg K}_2\text{Cr}_2\text{O}_7 \times \frac{1\ \text{g}}{1000\ \text{mg}} \times \frac{1\ \text{mol}}{294.18\ \text{g}} = 2.5 \times 10^{-6}\ \text{mol}$$

mol $K_2Cr_2O_7$ remaining = 2.5×10^{-6} mol $- 1.72 \times 10^{-6}$ mol $= 0.78 \times 10^{-6}$ mol

concentration of $K_2Cr_2O_7$ after the breath test = 0.78×10^{-6} mol$/0.003$ L $= 2.6 \times 10^{-4}$ mol/L $= 3 \times 10^{-4}$ mol/L (to 1 sig fig)

SELF-ASSESSMENT EXERCISES

102. (E) The answer is (b).
Conversion pathway approach:

$$0.300\ \text{L} \times \frac{0.0050\ \text{mol Ba(OH)}_2}{1\ \text{L}} \times \frac{2\ \text{mol OH}^-}{1\ \text{mol Ba(OH)}_2} = 0.0030\ \text{mol}$$

Stepwise approach:

$$0.300\ \text{L} \times \frac{0.0050\ \text{mol Ba(OH)}_2}{1\ \text{L}} = 1.5 \times 10^{-3}\ \text{mol Ba(OH)}_2$$

$$1.5 \times 10^{-3}\ \text{mol Ba(OH)}_2 \times \frac{2\ \text{mol OH}^-}{1\ \text{mol Ba(OH)}_2} = 0.0030\ \text{mol}$$

103. (E) The answer is (d). A 0.10 M solution of H_2SO_4 will have $[H^+]$ somewhere between 0.10 M and 0.20 M, which is greater than $[H^+]$ for any of the other solutions. As pointed out in Table 5.2, H_2SO_4 is a strong acid only in its first ionization step. So the first ionization step will provide 0.10 mol/L of H^+ and the second ionization will make an additional contribution that is less than 0.10 mol/L.

104. **(E)** The answer is (c). Based on the solubility guidelines in Table 5-1, carbonates (CO_3^{2-}) are insoluble.

105. **(M)** The answer is (a). Reaction with ZnO gives $ZnCl_2$ (soluble) and H_2O. There is no reaction with NaBr and Na_2SO_4, since all species are aqueous. By the process of elimination, (a) is the answer.

106. **(D)** The solid compound dissolves readily to yield a solution that conducts electricity. Therefore, the unknown compound is either a soluble ionic compound or a strong acid or base. The pungent odor produced by treatment with NaOH(aq) indicates the presence of NH_4^+ ions in the solid compound, which react with OH^- as follows: $NH_4^+ + OH^- \rightarrow NH_3 + H_2O$. The presence of NH_4^+ ions in the compound is consistent with the observation that the compound is soluble. Treatment of the solid with $Cu(NO_3)_2$ does not produce a precipitate, an observation that indicates the anion of the unknown solid combines with Cu^{2+} to produce a soluble copper compound. Focusing on information provided in Table 5.1, the anion could be NO_3^-, CH_3COO^-, ClO_4^-, Cl^-, Br^-, I^-, or SO_4^{2-}. However, treatment of the unknown solid with Ba^{2+}(aq) yields a precipitate, an observation that eliminates all these anions except for sulfate. The unknown might be $(NH_4)_2SO_4$.

107. **(E)**
Balanced equation: $2\,KI + Pb(NO_3)_2 \rightarrow 2\,KNO_3 + PbI_2$

Net ionic equation: $2\,I^- + Pb^{2+} \rightarrow PbI_2$ (s)

108. **(E)**
Balanced equation: $Na_2CO_3 + 2\,HCl \rightarrow 2\,NaCl + H_2O + CO_2$

Net ionic equation: $CO_3^{2-} + 2\,H^+ \rightarrow H_2O$ (l) $+ CO_2$ (g)

109. **(M)**
  (a) Balanced equation: $2\,Na_3PO_4 + 3\,Zn(NO_3)_2 \rightarrow 6\,NaNO_3 + Zn_3(PO_4)_2$

  Net ionic equation: $3\,Zn^{2+} + PO_4^{3-} \rightarrow Zn_3(PO_4)_2$(s)

  (b) Balanced equation: $2\,NaOH + Cu(NO_3)_2 \rightarrow Cu(OH)_2 + 2\,NaNO_3$

  Net ionic equation: $Cu^{2+} + 2\,OH^- \rightarrow Cu(OH)_2$ (s)

  (c) Balanced equation: $NiCl_2 + Na_2CO_3 \rightarrow NiCO_3 + 2\,NaCl$

  Net ionic equation: $Ni^{2+} + CO_3^{2-} \rightarrow NiCO_3$(s)

110. **(M)**
  (a) Species oxidized: N in NO
  (b) Species reduced: O_2
  (c) Oxidizing agent: O_2
  (d) Reducing agent: NO
  (e) Gains electrons: O_2
  (f) Loses electrons: NO

111. (M)

(a)

Oxidation:	$Cl_2(aq) + 4 OH^-(aq)$	$\rightarrow$	$2 ClO^-(aq) + 2 H_2O(l) + 2e^-$
Reduction:	$Cl_2(aq) + 2e^-$	$\rightarrow$	$2 Cl^-(aq)$
Overall:	$2 Cl_2(aq) + 4 OH^-(aq)$	$\rightarrow$	$2 ClO^-(aq) + 2 Cl^-(aq) + 2 H_2O(l)$
or:	$Cl_2(aq) + 2 OH^-(aq)$	$\rightarrow$	$ClO^-(aq) + Cl^-(aq) + H_2O(l)$

(b)

Oxidation: $\{ C_2O_4^{2-}(aq)$	$\rightarrow 2 CO_2(g) + 2e^- \} \times 5$
Reduction: $\{ MnO_4^-(aq) + 8 H^+(aq) + 5e^-$	$\rightarrow Mn^{2+}(aq) + 4 H_2O(l) \} \times 2$
Overall: $\quad 5 C_2O_4^{2-}(aq) + 2 MnO_4^-(aq) + 16 H^+(aq) \rightarrow 10 CO_2(g) + 2 Mn^{2+}(aq) + 8 H_2O(l)$	

112. (M) The answer is (b). The charges need to be balanced on both sides. Using a coefficient of 4, the charges on both sides of the reaction becomes +12.

113. (D) The answer is (d), $5 ClO^-$ to $1 I_2$. The work to balance the half-equations is shown below:

Reduction: $5 ClO^- + 2 H^+ + 2 e^- \rightarrow Cl^- + H_2O$

Oxidation: $I_2 + 6 H_2O \rightarrow 2 IO_3^- + 10 e^- + 12 H^+$

To combine the above reactions, the oxidation reaction should be multiplied by 5. The combined equation is:

Combined: $5 ClO^- + I_2 + H_2O \rightarrow 5 Cl^- + 2 IO_3^- + 2 H^+$

114. (M) The answer is (a). The balanced half-equation is as follows:

$NpO_2^+ + 4 H^+ + e^- \rightarrow Np^{4+} + 2 H_2O$

115. (E) The answer is (c). SO_2 is a gas and a nonelectrolyte, HF is a weak acid and therefore a weak electrolyte, and $FeSO_4$ fully dissociates in water and therefore is a strong electrolyte.

116. (E) The answer is (b). HNO_2 is a weak acid, NH_3 is a weak base, and NH_4NO_2 is the salt formed by their reaction.

117. (E) The answer is (d). $Pb(ClO_4)_2$ and $Ca(NO_3)_2$ are soluble in water. Hg_2SO_4 is insoluble.

118. (M)

(a) False. Based on solubility rules, $BaCl_2$ dissolves well in water. Therefore, it is a strong electrolyte.

(b) True. Since H^- is a base, H_2O is by necessity an acid. It also reduces H^- (–1) to H_2 (0).

(c) False. The product of such a reaction would be NaCl and H_2CO_3, neither of which precipitates out.

(d) False. HF is among the strongest of weak acids. It is not a strong acid, because it doesn't completely ionize in water.

(e) True. For every mole of $Mg(NO_3)_2$, there are 3 moles of ions, in contrast to 2 moles of ions for $NaNO_3$.

119. (M)

(a) No. Oxidation states of C, H or O do not change throughout the reaction.

(b) Yes. Li is oxidized to Li^+ and H in H_2O is reduced from +1 to 0 in H_2.

(c) Yes. Ag is oxidized and Pt is reduced.

(d) No. Oxidation states of Cl, Ca, H, and O remain unchanged.

CHAPTER 6
GASES

PRACTICE EXAMPLES

1A **(E)** The pressure measured by each liquid must be the same. They are related through $P = g\,h\,d$ Thus, we have the following: $g\,h_{DEG}\,d_{DEG} = g\,h_{Hg}\,d_{Hg}$. The g's cancel; we substitute known values:

$9.25\ m_{DEG} \times 1.118\ g/cm^3\,{}_{DEG} = h_{Hg} \times 13.6\ g/cm^3$

$h_{Hg} = 9.25\ m \times \dfrac{1.118\,g/cm^3}{13.6\,g/cm^3} = 0.760\ m\ Hg,\ \ P = 0.760\ mHg = 760\ mmHg$

1B **(E)** The solution is found through the expression relating density and height: $h_{TEG}d_{TEG} = h_{Hg}d_{Hg}$

We substitute known values and solve for triethylene glycol's density:

$9.14\ m_{TEG} \times d_{TEG} = 757\ mmHg \times 13.6\ g/cm^3$. Using unit conversions, we get

$d_{TEG} = \dfrac{0.757\ m}{9.14\ m} \times 13.6\ g/cm^3 = 1.13\ g/cm^3$

2A **(E)** We know that $P_{gas} = P_{bar.} + \Delta P$ with $P_{bar.} = 748.2$ mmHg. We are told that $\Delta P = 7.8$ mmHg. Thus, $P_{gas} = 748.2$ mmHg + 7.8 mmHg = 756.0 mmHg.

2B **(M)** The difference in pressure between the two levels must be the same, just expressed in different units. Hence, this problem is almost a repetition of Practice Example 6-1.

$h_{Hg} = 748.2$ mmHg – 739.6 mmHg = 8.6 mmHg. Again we have $g\,h_g\,d_g = g\,h_{Hg}\,d_{Hg}$. This becomes $h_g \times 1.26\ g/cm^3 = 8.6\ mmHg \times 13.6\ g/cm^3$

$h_g = 8.6\,mmHg \times \dfrac{13.6\,g/cm^3}{1.26\,g/cm^3} = 93\ mm\ glycerol$

3A **(M)** $A = \pi r^2$ (here $r = \frac{1}{2}(2.60\ cm \times \dfrac{1\ m}{100\ cm}) = 0.0130\ m$)

$A = \pi (0.0130\ m)^2 = 5.31 \times 10^{-4}\ m^2$

$F = m \times g = (1.000\ kg)(9.81\ m\ s^{-2}) = 9.81\ kg\ m\ s^{-2} = 9.81\ N$

$P = \dfrac{F}{A} = \dfrac{9.81\ N}{5.31 \times 10^{-4}\ m^2} = 18{,}475\ N\ m^{-2}$ or 1.85×10^4 Pa

$P\ (Torr) = 1.85 \times 10^4\ Pa \times = 139\ Torr$

3B **(M)** Final pressure = 100 mbar. $100 \text{ mbar} \times \dfrac{101{,}325 \text{ Pa}}{1013.25 \text{ mbar}} = 1.000 \times 10^4 \text{ Pa}$

The area of the cylinder is unchanged from that in Example 6-3, $(1.32 \times 10^{-3} \text{ m}^2)$.

$$P = \frac{F}{A} = 1.000 \times 10^4 \text{ Pa} = \frac{F}{1.32 \times 10^{-3} \text{ m}^2}$$

Solving for F, we find $F = 13.2 \text{ (Pa)m}^2 = 13.2 \text{ (N m}^{-2})\text{m}^2 = 13.2 \text{ N}$

$F = m \times g = 13.2 \text{ kg m s}^{-2} = m \times 9.81 \text{ m s}^{-2}$

total mass = mass of cylinder + mass added weight $= m = \dfrac{F}{g} = \dfrac{13.2 \text{ kg m s}^{-2}}{9.81 \text{ m s}^{-2}} = 1.35 \text{ kg}$

An additional 350 grams must be added to the top of the 1.000 kg (1000 g) red cylinder to increase the pressure to 100 mbar. It is not necessary to add a mass with the same cross-sectional area. The pressure will only be exerted over the area that is the base of the cylinder on the surface beneath it.

4A **(M)** The ideal gas equation is solved for volume. Conversions are made within the equation.

$$V = \frac{nRT}{P} = \frac{\left(20.2 \text{ g NH}_3 \times \dfrac{1 \text{ mol NH}_3}{17.03 \text{ g NH}_3} \right) \times \dfrac{0.08206 \, \text{L} \cdot \text{atm}}{\text{mol} \cdot \text{K}} \times (-25 + 273) \text{ K}}{752 \text{ mmHg} \times \dfrac{1 \text{ atm}}{760 \text{ mmHg}}} = 24.4 \text{ L NH}_3$$

4B **(E)** The amount of $Cl_2(g)$ is 0.193 mol Cl_2 and the pressure is 0.993 bar. This information is substituted into the ideal gas equation after it has been solved for temperature.

$$T = \frac{PV}{nR} = \frac{0.993 \text{ bar} \times 7.50 \text{ L}}{0.193 \text{ mol} \times 0.08314 \text{ bar L mol}^{-1} \text{ K}^{-1}} = 464 \text{ K}$$

5A **(E)** The ideal gas equation is solved for amount and the quantities are substituted.

$$n = \frac{PV}{RT} = \frac{10.5 \text{ atm} \times 5.00 \text{ L}}{\dfrac{0.08206 \text{ atm L}}{\text{mol K}} \times (30.0 + 273.15) \text{ K}} = 2.11 \text{ mol}$$

5B **(M)**

$$n = \frac{PV}{RT} = \frac{(6.67 \times 10^{-7} \text{ Pa})(3.45 \text{ m}^3)}{(8.3145 \text{ Pa m}^3 \text{ L mol}^{-1} \text{ K}^{-1})(25 + 273.15)\text{K}} = 9.28 \times 10^{-10} \text{ mol N}_2$$

molecules of $N_2 = 9.28 \times 10^{-10} \text{ mol N}_2 \times \dfrac{6.022 \times 10^{23} \text{ molecules N}_2}{1 \text{ mol N}_2}$

$$= 5.59 \times 10^{14} \text{ molecules N}_2$$

6A (E) The general gas equation is solved for volume, after the constant amount in moles is cancelled. Temperatures are converted to kelvin.

$$V_2 = \frac{V_1 P_1 T_2}{P_2 T_1} = \frac{1.00\,\text{mL} \times 2.14\,\text{atm} \times (37.8 + 273.2)\,\text{K}}{1.02\,\text{atm} \times (36.2 + 273.2)\,\text{K}} = 2.11\,\text{mL}$$

6B (M) The flask has a volume of 1.00 L and initially contains $O_2(g)$ at STP. The mass of $O_2(g)$ that must be released is obtained from the difference in the amount of $O_2(g)$ at the two temperatures, 273 K and 373 K. We also could compute the masses separately and subtract them.

$$\text{mass released} = (n_{\text{STP}} - n_{100\,^\circ\text{C}}) \times M_{O_2} = \left(\frac{PV}{R \times 273\,\text{K}} - \frac{PV}{R \times 373\,\text{K}}\right) \times M_{O_2} = \frac{PV}{R}\left(\frac{1}{273\,\text{K}} - \frac{1}{373\,\text{K}}\right) \times M_{O_2}$$

$$= \frac{1.00\,\text{bar} \times 1.00\,\text{L}}{0.08314\,\text{bar L mol}^{-1}\,\text{K}^{-1}}\left(\frac{1}{273\,\text{K}} - \frac{1}{373\,\text{K}}\right) \times \frac{32.00\,\text{g}}{1\,\text{mol O}_2} = 0.378\,\text{g O}_2$$

7A (M) The volume of the vessel is 0.09841 L. The mass of the O_2 sample is $(40.4868 - 40.1305)\,\text{g} = 0.3563\,\text{g}$. We substitute other values into the expression for molar mass.

$$M = \frac{mRT}{PV} = \frac{(0.3563\,\text{g}) \times \dfrac{0.08206\,\text{atm L}}{\text{mol} \cdot \text{K}} \times (22.4 + 273.2)\,\text{K}}{\left(772\,\text{mmHg} \times \dfrac{1\,\text{atm}}{760\,\text{mmHg}}\right) \times 0.09841\,\text{L}} = 86.5\,\text{g/mol}$$

7B (M) The gas's molar mass is the mass of the sample (1.27 g) divided by the amount of the gas in moles. The amount can be determined from the ideal gas equation.

$$n = \frac{PV}{RT} = \frac{0.982\,\text{bar} \times 1.07\,\text{L}}{\dfrac{0.08314\,\text{bar L}}{\text{mol K}} \times (25 + 273)\,\text{K}} = 0.0424\,\text{mol gas}$$

$$M = \frac{1.27\,\text{g}}{0.0424\,\text{mol}} = 30.0\,\text{g/mol}$$

This answer is in good agreement with the molar mass of NO, 30.006 g/mol.

8A (M) The molar mass of He is 4.003 g/mol. This is substituted into the expression for density.

$$d = \frac{MP}{RT} = \frac{4.003\,\text{g mol}^{-1} \times 0.987\,\text{bar}}{0.08314\,\text{bar L mol}^{-1}\,\text{K}^{-1} \times 298\,\text{K}} = 0.159\,\text{g/L}$$

When compared to the density of air under the same conditions (1.16 g/L, based on the "average molar mass of air" = 28.8 g/mol) the density of He is only about one seventh as much. Thus, helium is less dense ("lighter") than air.

8B (M) The suggested solution is a simple one; we merely need to solve for mass of gas from density and its moles from the ideal gas equation.

$$m(\text{gas}) = d \times V = (1.00 \text{ g/L})(1.00 \text{ L}) = 1.00 \text{ g}$$

$$n = \frac{PV}{RT} = \frac{\left(745 \text{ mmHg} \times \dfrac{1 \text{ atm}}{760 \text{ mmHg}}\right)(1.00 \text{ L})}{0.08206 \text{ L atm mol}^{-1} \text{ K}^{-1} \times 382 \text{ K}} = 0.0312 \text{ mol}$$

Therefore, the molar mass of the gas is as follows:

$$M = \frac{m}{n} = \frac{1.00 \text{ g}}{0.0312 \text{ mol}} = 32.0 \text{ g/mol}$$

The molecular weight suggests that the gas is O_2.

9A (M) The balanced equation is $2 \text{ NaN}_3(s) \xrightarrow{\Delta} 2 \text{ Na}(l) + 3 \text{ N}_2(g)$

$$\text{moles N}_2 = \frac{PV}{RT} = \frac{\left(776 \text{ mmHg} \times \dfrac{1 \text{ atm}}{760 \text{ mmHg}}\right) \times 20.0 \text{ L}}{0.08206 \dfrac{\text{atm L}}{\text{mol K}} \times (30.0 + 273.2) \text{ K}} = 0.821 \text{ mol N}_2$$

Now, solve the stoichiometry problem.

$$\text{mass NaN}_3 = 0.821 \text{ mol N}_2 \times \frac{2 \text{ mol NaN}_3}{3 \text{ mol N}_2} \times \frac{65.01 \text{ g NaN}_3}{1 \text{ mol NaN}_3} = 35.6 \text{ g NaN}_3$$

9B (M) Here we are not dealing with gaseous reactants; the law of combining volumes cannot be used. From the ideal gas equation we determine the amount of $N_2(g)$ per liter under the specified conditions. Then we determine the amount of $Na(l)$ produced simultaneously, and finally the mass of that $Na(l)$.

$$\text{mass of Na}(l) = \frac{1.00 \text{ bar} \times 1.000 \text{ L}}{0.08314 \dfrac{\text{bar L}}{\text{mol K}} \times (25 + 273) \text{ K}} \times \frac{2 \text{ mol Na}}{3 \text{ mol N}_2} \times \frac{22.99 \text{ g Na}}{1 \text{ mol Na}} = 0.619 \text{ g Na}(l)$$

10A (E) The law of combining volumes permits us to use stoichiometric coefficients for volume ratios.

$$O_2 \text{ volume} = 1.00 \text{ L NO}(g) \times \frac{5 \text{ L O}_2}{4 \text{ L NO}} = 1.25 \text{ L O}_2(g)$$

10B (E) The first task is to balance the chemical equation. There must be three moles of hydrogen for every mole of nitrogen in both products (because of the formula of NH_3) and reactants: $N_2(g) + 3 H_2(g) \rightarrow 2 NH_3(g)$. The volumes of gaseous reactants and products are related by their stoichiometric coefficients, as long as all gases are at the same temperature and pressure.

$$\text{volume NH}_3(g) = 225 \text{ L H}_2(g) \times \frac{2 \text{ L NH}_3(g)}{3 \text{ L H}_2(g)} = 150. \text{ L NH}_3$$

11A **(M)** We can work easily with the ideal gas equation, with a the new temperature of $T = (55 + 273)\,K = 328\,K$. The amount of Ne added is readily computed.

$$n_{Ne} = 12.5\,g\,Ne \times \frac{1\,mol\,Ne}{20.18\,g\,Ne} = 0.619\,mol\,Ne$$

$$P = \frac{n_{tot}RT}{V} = \frac{(1.75 + 0.619)\,mol \times \dfrac{0.08314\,bar\,L}{mol\,K} \times 328\,K}{5.0\,L} = 13\,bar$$

11B **(E)** The total volume initially is $2.0\,L + 8.0\,L = 10.0\,L$. These two mixed ideal gases then obey the general gas equation as if they were one gas.

$$P_2 = \frac{P_1 V_1 T_2}{V_2 T_1} = \frac{1.00\,atm \times 10.0\,L \times 298\,K}{2.0\,L \times 273\,K} = 5.5\,atm$$

12A **(M)** The partial pressures are proportional to the mole fractions.

$$P_{H_2O} = \frac{n_{H_2O}}{n_{tot}} \times P_{tot} = \frac{0.00278\,mol}{0.197\,mol + 0.00278\,mol} \times 2.50\,atm = 0.0348\,atm$$

$$P_{CO_2} = P_{tot} - P_{H_2O} = 2.50\,atm - 0.0348\,atm = 2.47\,atm$$

12B **(M)** Expression (6.17) indicates that, in a mixture of gases, the mole percent equals the volume percent, which in turn equals the pressure percent. Thus, we can apply these volume percents—converted to fractions by dividing by 100—directly to the total pressure.

N_2 pressure $= 0.7808 \times 748\,mmHg\ \ = 584\,mmHg,$

O_2 pressure $= 0.2095 \times 748\,mmHg\ \ = 157\,mmHg,$

Ar pressure $= 0.0093 \times 748\,mmHg\ \ = 7.0\,mmHg,$

CO_2 pressure $= 0.00036 \times 748\,mmHg = 0.27\,mmHg$

13A **(M)** First compute the moles of $H_2(g)$, then use stoichiometry to convert to moles of HCl.

$$\text{amount HCl} = \frac{\left((755 - 25.2)\,mmHg \times \dfrac{1\,atm}{760\,mmHg}\right) \times 0.0355\,L}{\dfrac{0.08206\,atm\,L}{mol\,K} \times (26 + 273)\,K} \times \frac{6\,mol\,HCl}{3\,mol\,H_2} = 0.00278\,mol\,HCl$$

13B **(M)** The volume occupied by the $O_2(g)$ at its partial pressure is the same as the volume occupied by the mixed gases: water vapor and $O_2(g)$. The partial pressure of $O_2(g)$ is found by difference.

$$P_{O_2} = P_{bar.} - P_{H_2O} = 749.2\,mmHg - 23.8\,mmHg = 725.4\,mmHg$$

$$mol\,O_2 = \frac{PV}{RT} = \frac{\left(725.4\,mmHg \times \dfrac{1\,atm}{760\,mmHg}\right)(0.395\,L)}{\left(0.08206\,atm\,L\,mol^{-1}K^{-1}\right)(298\,K)} = 0.0154\,mol\,O_2$$

$$\text{mass Ag}_2\text{O} = 0.0154 \text{ mol O}_2 \times \frac{2 \text{ mol Ag}_2\text{O}}{1 \text{ mol O}_2} \times \frac{231.74 \text{ g Ag}_2\text{O}}{1 \text{ mol Ag}_2\text{O}} = 7.14 \text{ g Ag}_2\text{O}$$

Mass% Ag_2O = 7.14 g/8.07 g × 100 = 88.4

14A **(M)** The gas with the smaller molar mass, NH_3 at 17.0 g/mol, has the greater root-mean-square speed.

$$u_{rms} = \sqrt{\frac{3RT}{M}} = \sqrt{\frac{3 \times 8.3145 \text{ kg m}^2 \text{ s}^{-2}\text{mol}^{-1}\text{ K}^{-1} \times 298 \text{ K}}{0.0170 \text{ kg mol}^{-1}}} = 661 \text{ m/s}$$

14B **(D)**

$$\text{bullet speed} = \frac{2180 \text{ mi}}{1 \text{ h}} \times \frac{1 \text{ h}}{3600 \text{ s}} \times \frac{5280 \text{ ft}}{1 \text{ mi}} \times \frac{12 \text{ in.}}{1 \text{ ft}} \times \frac{2.54 \text{ cm}}{1 \text{ in.}} \times \frac{1 \text{ m}}{100 \text{ cm}} = 974.5 \text{ m/s}$$

Solve the rms-speed equation (6.20) for temperature by first squaring both sides.

$$\left(u_{rms}\right)^2 = \frac{3RT}{M} \qquad T = \frac{\left(u_{rms}\right)^2 M}{3R} = \frac{\left(\frac{974.5 \text{ m}}{1 \text{ s}}\right)^2 \times \frac{2.016 \times 10^{-3} \text{ kg}}{1 \text{ mol H}_2}}{3 \times \frac{8.3145 \text{ kg m}^2}{\text{s}^2 \text{ mol K}}} = 76.75 \text{ K}$$

We expected the temperature to be lower than 298 K. Note that the speed of the bullet is about half the speed of a H_2 molecule at 298 K. To halve the speed of a molecule, its temperature must be divided by four.

15A **(M)** The only difference is the gas's molar mass. 2.2×10^{-4} mol N_2 effuses through the orifice in 105 s.

$$\frac{? \text{ mol O}_2}{2.2 \times 10^{-4} \text{ mol N}_2} = \sqrt{\frac{M_{N_2}}{M_{O_2}}} = \sqrt{\frac{28.014 \text{ g/mol}}{31.998 \text{ g/mol}}} = 0.9357$$

moles $O_2 = 0.9357 \times \left(2.2 \times 10^{-4}\right) = 2.1 \times 10^{-4}$ mol O_2

15B **(M)** Rates of effusion are related by the square root of the ratio of the molar masses of the two gases. H_2, effuses faster (by virtue of being lighter), and thus requires a shorter time for the same amount of gas to effuse.

$$\text{time}_{H_2} = \text{time}_{N_2} \times \sqrt{\frac{M_{H_2}}{M_{N_2}}} = 105 \text{ s} \times \sqrt{\frac{2.016 \text{ g/mol}}{28.014 \text{ g/mol}}} = 28.2 \text{ s}$$

16A **(M)** Effusion times are related as the square root of the molar mass. It requires 87.3 s for Kr to effuse.

$$\frac{\text{unknown time}}{\text{Kr time}} = \sqrt{\frac{M_{unk}}{M_{Kr}}} \qquad \text{substitute in values} \qquad \frac{131.3 \text{ s}}{87.3 \text{ s}} = \sqrt{\frac{M_{unk}}{83.80 \text{ g/mol}}} = 1.50$$

$$M_{unk} = \left(1.504\right)^2 \times 83.80 \text{ g/mol} = 1.90 \times 10^2 \text{ g/mol}$$

16B **(M)** This problem is solved in virtually the same manner as Practice Example 18B. The lighter gas is ethane, with a molar mass of 30.07 g/mol.

$$C_2H_6 \text{ time} = Kr \text{ time} \times \sqrt{\frac{M_{C_2H_6}}{M_{Kr}}} = 87.3 \text{ s} \times \sqrt{\frac{30.07 \text{ g/mol}}{83.80 \text{ g/mol}}} = 52.3 \text{ s}$$

17A **(D)** Because one mole of gas is being considered, the value of an^2 is numerically the same as the value of a, and the value of nb is numerically the same as the value of b.

$$P = \frac{nRT}{V - nb} - \frac{an^2}{V^2} = \frac{1.00 \text{ mol} \times \dfrac{0.083145 \text{ bar L}}{\text{mol K}} \times 273 \text{ K}}{(2.00 - 0.0429) \text{ L}} - \frac{3.66 \text{ L}^2 \text{ bar mol}^{-2}}{(2.00 \text{ L})^2} = 11.6 \text{ bar} - 0.915 \text{ bar}$$

$$= 10.7 \text{ bar } CO_2(g) \quad \text{compared with 10.1 bar for } Cl_2(g)$$

$$P_{ideal} = \frac{nRT}{V} = \frac{1.00 \text{ mol} \times \dfrac{0.083145 \text{ bar L}}{\text{mol K}} \times 273 \text{ K}}{2.00 \text{ L}} = 11.35 \sim 11.4 \text{ bar}$$

$Cl_2(g)$ shows a greater deviation from ideal gas behavior than does $CO_2(g)$.

17B **(M)** Because one mole of gas is being considered, the value of an^2 is numerically the same as the value of a, and the value of nb is numerically the same as the value of b.

$$P = \frac{nRT}{V - nb} - \frac{an^2}{V^2} = \frac{1.00 \text{ mol} \times \dfrac{0.083145 \text{ bar L}}{\text{mol K}} \times 273 \text{ K}}{(2.00 - 0.0395) \text{ L}} - \frac{1.47 \text{ L}^2 \text{ bar}}{(2.00 \text{ L})^2} = 11.6 \text{ bar} - 0.368 \text{ bar}$$

$$= 11.2 \text{ bar } CO(g)$$

compared to 10.1 bar for $Cl_2(g)$, 11.2 bar for CO, and 10.7 bar for $CO_2(g)$.

Thus, $Cl_2(g)$ displays the greatest deviation from ideality, 11.4 bar.

INTEGRATIVE EXERCISE

A. **(M)** Let's assume that we have a 1 L sample of this gas, and determine how many moles of gas are present:

$$n = \frac{PV}{RT} = \frac{(101.3 \text{ kPa})(1.00 \text{ L})}{(8.3145 \text{ kPa L} \cdot \text{K}^{-1}\text{mol}^{-1})(298 \text{ K})} = 0.0409 \text{ mol}$$

Knowing the density of the gas (1.637 g/L) and its volume (1 L) gives us the mass of 1 L of gas, or 1.637 g. Therefore, the molar mass of this gas is:

$$(1.637 \text{ g} / 0.0409 \text{ mol}) = 40.03 \text{ g/mol}$$

Now, determine the number of moles of C and H to ascertain the empirical formula:

$$\text{mol C: } 1.687 \text{ g CO}_2 \times \frac{1 \text{ mol CO}_2}{44.01 \text{ g CO}_2} \times \frac{1 \text{ mol C}}{1 \text{ mol CO}_2} = 0.0383 \text{ mol}$$

$$\text{mol H: } 0.4605 \text{ g H}_2\text{O} \times \frac{1 \text{ mol H}_2\text{O}}{18.02 \text{ g H}_2\text{O}} \times \frac{2 \text{ mol H}}{1 \text{ mol H}_2\text{O}} = 0.05111 \text{ mol}$$

Dividing by the smallest value (0.0383 mol C), we get a H:C ratio of 1.33:1, or 4:3. Therefore, the empirical formula is C_3H_4, which has a molar mass of 40.07, which is essentially the same as the molar mass calculated. Therefore, the actual formula is also C_3H_4. Below are three possible Lewis structures:

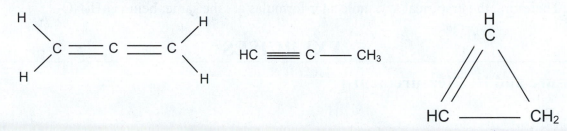

B. **(D)** First, let us determine the amount of each element in the compound:

$$\text{mol C: } 151.2 \times 10^{-3} \text{ g CO}_2 \times \frac{1 \text{ mol CO}_2}{44.01 \text{ g CO}_2} \times \frac{1 \text{ mol C}}{1 \text{ mol CO}_2} = 0.003435 \text{ mol C}$$

$$\text{g C: } 0.003435 \text{ mol C} \times \frac{12.01 \text{ g C}}{1 \text{ mol C}} = 0.04126 \text{ g C}$$

$$\text{mol H: } 69.62 \times 10^{-3} \text{ g H}_2\text{O} \times \frac{1 \text{ mol H}_2\text{O}}{18.02 \text{ g H}_2\text{O}} \times \frac{2 \text{ mol H}}{1 \text{ mol H}_2\text{O}} = 0.007727 \text{ mol H}$$

$$\text{g H: } 0.007727 \text{ mol H} \times \frac{1.008 \text{ g H}}{1 \text{ mol H}} = 0.007789 \text{ g H}$$

$$\text{mol N: } \frac{PV}{RT} = \frac{1 \text{ atm} \cdot 9.62 \times 10^{-3} \text{ L}}{\left(0.08206 \text{ atm L mol}^{-1} \text{ K}^{-1}\right)(273 \text{ K})} \times \frac{2 \text{ mol N}}{1 \text{ mol N}_2} = 0.0008589 \text{ mol N}$$

$$\text{g N: } 0.0008589 \text{ mol N} \times \frac{14.01 \text{ g N}}{1 \text{ mol N}} = 0.01203 \text{ g N}$$

Therefore, the mass of O is determined by subtracting the sum of the above masses from the mass of the compound:

$$\text{g O: } 0.1023 - (0.04126 + 0.007804 + 0.01203) = 0.04121 \text{ g O}$$

$$\text{mol O: } 0.04121 \text{ g O} \times \frac{1 \text{ mol O}}{16.00 \text{ g O}} = 0.002576 \text{ mol O}$$

To determine the empirical formula, all of the calculated moles above should be divided by the smallest value. Doing so will give the following ratios:
C: 0.003435/0.0008589 = 4
H: 0.007727/0.0008589 = 9
N: 1
O: 0.002576/0.0008589 = 3

The empirical formula is $C_4H_9NO_3$, and has a molar mass of 119.14 g/mol.

To determine the actual formula, we have to calculate its molecular mass. We know the density at a given volume and therefore we need to find out the number of moles. Before that, we should convert the experimental conditions to more convenient units. $T = 127\ ^\circ\text{C} + 273 = 400$ K, and $P = 748$ mmHg/760 mmHg $= 0.9842$ atm.

$$n = \frac{PV}{RT} = \frac{(0.9842\ \text{atm})1\ \text{L}}{\left(0.08206\ \text{atm L mol}^{-1}\ \text{K}^{-1}\right)(400\ \text{K})} = 0.02999\ \text{mol}$$

M = g/mol = 3.57 g/0.02999 mol = 119 g/mol

Therefore, the empirical and molecular formulas are the same, being $C_4H_9NO_3$.

EXERCISES

Pressure and Its Measurement

1. **(E)**

    (a) $P = 736\ \text{mmHg} \times \dfrac{1\ \text{atm}}{760\ \text{mmHg}} = 0.968\ \text{atm}$

    (b) $P = 0.776\ \text{bar} \times \dfrac{1\ \text{atm}}{1.01325\ \text{bar}} = 0.766\ \text{atm}$

    (c) $P = 892\ \text{torr} \times \dfrac{1\ \text{atm}}{760\ \text{Torr}} = 1.17\ \text{atm}$

    (d) $P = 225\ \text{kPa} \times \dfrac{1000\ \text{Pa}}{1\ \text{kPa}} \times \dfrac{1\ \text{atm}}{101{,}325\ \text{Pa}} = 2.22\ \text{atm}$

3. **(E)** We use: $h_{\text{bnz}}d_{\text{bnz}} = h_{\text{Hg}}d_{\text{Hg}}$

$$h_{\text{bnz}} = 0.970\ \text{atm} \times \frac{0.760\ \text{mHg}}{1\ \text{atm}} \times \frac{13.6\ \text{g/cm}^3\ \text{Hg}}{0.879\ \text{g/cm}^3\ \text{benzene}} = 11.4\ \text{m benzene}$$

5. **(E)** $P = P_{\text{bar}} - h_1 = 740\ \text{mmHg} - 30\ \text{mm}\ (h_1) = 710\ \text{mmHg}$

7. $F = m \times g$ and 1 atm = 101,325 Pa = 101,325 kg m^{-1} s^{-2} = $P = \dfrac{F}{A} = \dfrac{m \times 9.81\ \text{m s}^{-2}}{1\ \text{m}^2}$

mass (per m^2) = $\dfrac{101{,}325\ \text{kg m}^{-1}\ \text{s}^{-2} \times 1\ \text{m}^2}{9.81\ \text{m s}^{-2}} = 10{,}329\ \text{kg}$

(Note: 1 m^2 = (100 cm)^2 = 10,000 cm^2)

P (kg cm^{-2}) = $\dfrac{m}{A} = \dfrac{10{,}329\ \text{kg}}{10{,}000\ \text{cm}^2} = 1.03\ \text{kg cm}^{-2}$

The Simple Gas Laws

9. **(E)**

(a) $V = 26.7 \text{ L} \times \dfrac{762 \text{ Torr}}{385 \text{ Torr}} = 52.8 \text{ L}$

(b) $V = 26.7 \text{ L} \times \dfrac{762 \text{ Torr}}{3.68 \text{ atm } \times \dfrac{760 \text{ Torr}}{1 \text{ atm}}} = 7.27 \text{ L}$

11. **(E)** Charles's law states that $V_1/T_1 = V_2/T_2$. Since $V_1 = 3.0$ L, $T_1 = (177 + 273)$, K = 450 K, and $V_2 = 1.50$ L, then:

$\dfrac{3.0 \text{ L}}{450 \text{ K}} = \dfrac{1.50 \text{ L}}{T_2}$, and $T_2 = 225$ K

13. **(E)** $P_i = P_f \times \dfrac{V_f}{V_i} = \left(721 \text{ mmHg} \times \dfrac{35.8 \text{ L} + 1875 \text{ L}}{35.8 \text{ L}} \right) \times \dfrac{1 \text{ atm}}{760 \text{ mmHg}} = 50.6 \text{ atm}$

15. **(E)** Combining Boyle's and Charles's law, we get the following expression: $\dfrac{P_1 V_1}{T_1} = \dfrac{P_2 V_2}{T_2}$

Therefore,

$T_2 = \dfrac{P_2 V_2 T_1}{P_1 V_1} = \dfrac{(0.340 \text{ atm})(5.00 \times 10^3 \text{ m}^3)(300 \text{ K})}{(1.000 \text{ atm})(2.00 \times 10^3 \text{ m}^3)} = 255 \text{ K}$

17. **(E)** STP: $P = 1$ bar and $T = 273.15$ K

$\text{mass}_{Ar} = \dfrac{1.00 \text{ bar} \times 0.0750 \text{ L}}{0.08314 \dfrac{\text{bar L}}{\text{K mol}} \times 273.15 \text{ K}} \times \dfrac{39.948 \text{ g Ar}}{1 \text{ mol Ar}} = 0.132 \text{ g Ar}$

19. **(M)** In the following calculations, we use the fact that 1 mol of gas occupies 22.711 L at STP (1 bar and 273.15 K). See equation (6.10).

(a) *Conversion pathway approach:*

$\text{mass} = 27.6 \text{ mL} \times \dfrac{1 \text{ L}}{1000 \text{ mL}} \times \dfrac{1 \text{ mol PH}_3}{22.711 \text{ L (at STP)}} \times \dfrac{34.0 \text{ g PH}_3}{1 \text{ mol PH}_3} \times \dfrac{1000 \text{ mg PH}_3}{1 \text{ g}}$

$= 41.3 \text{ mg PH}_3$

Stepwise approach:

$27.6 \text{ mL} \times \dfrac{1 \text{ L}}{1000 \text{ mL}} = 0.0276 \text{ L}$

$0.0276 \text{ L} \times \dfrac{1 \text{ mol PH}_3}{22.711 \text{ L (at STP)}} = 1.215 \times 10^{-3} \text{ mol PH}_3$

$$1.215 \times 10^{-3} \text{ mol PH}_3 \times \frac{34.00 \text{ g PH}_3}{1 \text{ mol PH}_3} = 0.0413 \text{ g PH}_3$$

$$0.0413 \text{ g PH}_3 \times \frac{1000 \text{ mg PH}_3}{1 \text{ g}} = 41.3 \text{ mg PH}_3$$

(b) number of molecules of $PH_3 = 1.215 \times 10^{-3} \text{ mol PH}_3 \times \frac{6.022 \times 10^{23} \text{ molecules}}{1 \text{ mol PH}_3}$

$$= 7.32 \times 10^{20} \text{ molecules}$$

21. **(M)** At the higher elevation of the mountains, the atmospheric pressure is lower than at the beach. However, the bag is virtually leak proof; no gas escapes. Thus, the gas inside the bag expands in the lower pressure until the bag is filled to near bursting. (It would have been difficult to predict this result. The temperature in the mountains is usually lower than at the beach. The lower temperature would *decrease* the pressure of the gas.)

General Gas Equation

23. **(E)** Because the number of moles of gas does not change, $\dfrac{P_i V_i}{T_i} = nR = \dfrac{P_f V_f}{T_f}$ is

obtained from the ideal gas equation. This expression can be rearranged as follows.

$$V_f = \frac{V_i \, P_i \, T_f}{P_f \, T_i} = \frac{4.25 \text{ L} \times 748 \text{ mmHg} \times (26.8 + 273.2)\text{K}}{742 \text{ mmHg} \times (25.6 + 273.2)\text{K}} = 4.30 \text{ L}$$

25. **(E)** Volume and pressure are constant. Hence $n_i T_i = \dfrac{PV}{R} = n_f T_f$

$$\frac{n_f}{n_i} = \frac{T_i}{T_f} = \frac{(21 + 273) \text{ K}}{(210 + 273) \text{ K}} = 0.609 \ (60.9\% \text{ of the gas remains})$$

Hence, 39.1% of the gas must be released. Mass of gas released $= 12.5 \text{ g} \times \dfrac{39.1}{100} = 4.89 \text{ g}$

Ideal Gas Equation

27. **(M)** Assume that the $CO_2(g)$ behaves ideally and use the ideal gas equation: $PV = nRT$

$$V = \frac{nRT}{P} = \frac{\left(89.2 \text{ g} \times \dfrac{1 \text{ mol CO}_2}{44.01 \text{ g}}\right) 8.314 \dfrac{\text{kPa L}}{\text{mol K}} (37 + 273.2) \text{ K}}{98.3 \text{ kPa}} = 53.2 \text{ L}$$

29. **(E)** mass $= n \times M = \dfrac{PV}{RT}$ $\qquad M = \dfrac{11.2 \text{ atm} \times 18.5 \text{ L} \times 83.80 \text{ g/mol}}{0.08206 \dfrac{\text{atm L}}{\text{mol K}} (28.2 + 273.2) \text{ K}} = 702 \text{ g Kr}$

31. **(M)** $n_{gas} = 5.0 \times 10^9$ molecules gas $\times \dfrac{1 \text{ mol gas}}{6.022 \times 10^{23} \text{ molecules gas}} = 8.3 \times 10^{-15}$ mol gas

We next determine the pressure that the gas exerts at 25 °C in a cubic meter

$$P = \dfrac{8.3 \times 10^{-15} \text{ mol gas} \times 8.314 \dfrac{\text{kPa L}}{\text{mol K}} \times 273 \text{ K}}{1 \text{ m}^3 \times \left(\dfrac{10 \text{ dm}}{1 \text{ m}}\right)^3 \times \left(\dfrac{1 \text{ L}}{1 \text{ dm}^3}\right)} \times \dfrac{1000 \text{ Pa}}{1 \text{ kPa}} = 2.1 \times 10^{-11} \text{ Pa}$$

33. **(E)** The basic ideal law relationship applies here. Molar volume is the amount of volume that one mole of a gas occupies. If $PV = nRT$, then molar volume is V/n, and the relationship rearranges to:

$$\dfrac{V}{n} = V_m = \dfrac{RT}{P}$$

   (a) $V_m = (298 \text{ K} \cdot 0.08206 \text{ atm L mol}^{-1} \text{ K}^{-1})/1.00 \text{ atm} = 24.4 \text{ L mol}^{-1}$

   (b) $P_{atm} = 748 \text{ mmHg} / 760 \text{ mmHg} = 0.978 \text{ atm}$

       $V_m = (373 \text{ K} \cdot 0.08206 \text{ atm L mol}^{-1} \text{ K}^{-1})/0.978 \text{ atm} = 31.3 \text{ L mol}^{-1}$

Determining Molar Mass

35. **(M)** Use the ideal gas equation to determine the amount in moles of the given quantity of gas.

$$M = \dfrac{mRT}{PV} = \dfrac{0.418 \text{ g} \times 8.314 \dfrac{\text{kPa L}}{\text{mol K}} \times 339.5 \text{ K}}{99.0 \text{ kPa} \times 0.115 \text{ L}} = 104 \text{ g mol}^{-1}$$

Alternatively

$$n = \dfrac{PV}{RT} = \dfrac{99.0 \text{ kPa}\left(115 \text{ mL} \times \dfrac{1 \text{ L}}{1000 \text{ mL}}\right)}{8.314 \dfrac{\text{kPa L}}{\text{mol K}}(66.3 + 273.2)\text{ K}} = 0.00403 \text{ mol} \quad M = \dfrac{0.418 \text{ g}}{0.00403 \text{ mol}} = 104 \text{ g/mol}$$

37. **(M)** First we determine the empirical formula for the sulfur fluoride. Assume a 100 g sample of S_xF_y.

moles S $= 29.6 \text{ g S} \times \dfrac{1 \text{ mol S}}{32.06 \text{ g S}} = 0.923 \text{ mol S}$ moles F $= 70.4 \text{ g F} \times \dfrac{1 \text{ mol F}}{18.998 \text{ g F}} = 3.706 \text{ mol F}$

Dividing the number of moles of each element by 0.923 moles gives the empirical formula SF_4.

To find the molar mass we use the relationship:

$$\text{Molar mass} = \dfrac{dRT}{P} = \dfrac{4.5 \text{ g L}^{-1} \times 0.08206 \text{ atm L mol}^{-1} \text{ K}^{-1} \times 293 \text{ K}}{1.0 \text{ atm}} = 10\underline{8} \text{ g mol}^{-1}$$

Thus, molecular formula $= (\text{empirical formula}) \times \left(\dfrac{\text{molecular formula mass}}{\text{empirical formula mass}}\right)$

$$= \dfrac{10\underline{8} \text{ g mol}^{-1}}{108.06 \text{ g mol}^{-1}} \times SF_4 = SF_4$$

39. (M)

$$M = \frac{mRT}{PV} = \frac{0.231 \text{ g} \times 0.08206 \dfrac{\text{atm L}}{\text{mol K}} \times (23+273) \text{ K}}{\left(749 \text{ mmHg} \times \dfrac{1 \text{ atm}}{760 \text{ mmHg}}\right) \times \left(102 \text{ mL} \times \dfrac{1 \text{ L}}{1000 \text{ mL}}\right)} = 55.8 \text{ g/mol}$$

The formula contains 4 atoms of carbon. (5 atoms of carbon gives a molar mass of at least 60—too high—and 3 C atoms gives a molar mass of 36—too low to be made up by adding H's.) To produce a molar mass of 56 with 4 carbons requires the inclusion of 8 atoms of H in the formula of the compound. Thus the formula is C_4H_8.

Gas Densities

41. (E) $d = \dfrac{MP}{RT} \quad \rightarrow \quad P = \dfrac{dRT}{M} = \dfrac{1.80 \text{ g/L} \times 0.08206 \dfrac{\text{L atm}}{\text{mol K}} \times (32+273) \text{ K}}{28.0 \text{ g/mol}} \times \dfrac{760 \text{ mmHg}}{1 \text{ atm}}$

$$P = 1.21 \times 10^3 \text{ mmHg}$$

Molar volume $= \dfrac{1 \text{ L}}{1.8 \text{ g}} \text{ N}_2 \times \dfrac{28.0 \text{ g N}_2}{1 \text{ mol N}_2} = 15.6 \text{ L/mol}$

43. (M)

(a) $d = \dfrac{MP}{RT} = \dfrac{28.96 \text{ g/mol} \times 1.00 \text{ atm}}{0.08206 \dfrac{\text{atm L}}{\text{mol K}} \times (25+273)\text{K}} = 1.18 \text{ g/L air}$

(b) $d = \dfrac{MP}{RT} = \dfrac{44.0 \text{ g/mol CO}_2 \times 1.00 \text{ atm}}{0.08206 \dfrac{\text{atm L}}{\text{mol K}} \times (25+273)\text{K}} = 1.80 \text{ g/L CO}_2$

Since this density is greater than that of air, the balloon will not rise in air when filled with CO_2 at 25 °C; instead, it will sink!

45. (E) $d = \dfrac{MP}{RT}$ becomes $M = \dfrac{dRT}{P} = \dfrac{2.64 \text{ g/L} \times 0.08314 \dfrac{\text{bar L}}{\text{mol K}} \times (310+273)\text{K}}{1.03 \text{ bar}} = 124 \text{ g/mol}$

Since the atomic mass of phosphorus is 31.0, the formula of phosphorus molecules in the vapor must be P_4. (4 atoms / molecule $\times 31.0 = 124$)

Gases in Chemical Reactions

47. (E) Balanced equation: $C_3H_8(g) + 5 O_2(g) \rightarrow 3 CO_2(g) + 4 H_2O(l)$

Use the law of combining volumes. O_2 volume $= 75.6 \text{ L C}_3\text{H}_8 \times \dfrac{5 \text{ L O}_2}{1 \text{ L C}_3\text{H}_8} = 378 \text{ L O}_2$

49. **(M)** Determine the moles of $SO_2(g)$ produced and then use the ideal gas equation.

Conversion pathway approach:

$$\text{mol } SO_2 = \left(1.2\times10^6 \text{ kg coal}\times\frac{3.28 \text{ kg S}}{100.00 \text{ kg coal}}\times\frac{1000 \text{ g S}}{1 \text{ kg S}}\times\frac{1 \text{ mol S}}{32.1 \text{ g S}}\times\frac{1 \text{ mol } SO_2}{1 \text{ mol S}}\right)$$

$$= 1.23\times10^6 \text{ mol } SO_2$$

$$V = \frac{nRT}{P}$$

$$V = \frac{1.23\times10^6 \text{ mol } SO_2 \times0.08206\dfrac{\text{atm L}}{\text{mol K}}\times296 \text{ K}}{738 \text{ mmHg}\times\dfrac{1 \text{ atm}}{760 \text{ mmHg}}} = 3.1\times10^7 \text{ L } SO_2$$

Stepwise approach:

$$1.2\times10^6 \text{ kg coal}\times\frac{3.28 \text{ kg S}}{100.00 \text{ kg coal}} = 3.94\times10^4 \text{ kg S}$$

$$3.94\times10^4 \text{ kg S}\times\frac{1000 \text{ g S}}{1 \text{ kg S}} = 3.94\times10^7 \text{ g S}$$

$$3.94\times10^7 \text{ g S}\times\frac{1 \text{ mol S}}{32.1 \text{ g S}} = 1.23\times10^6 \text{ mol S}$$

$$1.23\times10^6 \text{ mol S}\times\frac{1 \text{ mol } SO_2}{1 \text{ mol S}} = 1.23\times10^6 \text{ mol } SO_2$$

$$V = \frac{nRT}{P} = \frac{1.23\times10^6 \text{ mol } SO_2 \times0.08206\dfrac{\text{atm L}}{\text{mol K}}\times296 \text{ K}}{738 \text{ mmHg}\times\dfrac{1 \text{ atm}}{760 \text{ mmHg}}} = 3.1\times10^7 \text{ L } SO_2$$

51. **(M)** Determine the moles of O_2, and then the mass of $KClO_3$ that produced this amount of O_2.

$$n_{O_2} = \frac{98.3 \text{ kPa}\times\left(119 \text{ mL}\times\dfrac{1 \text{ L}}{1000 \text{ mL}}\right)}{8.314 \text{ kPa L K}^{-1}\text{ mol}^{-1}\times(22.4+273.2) \text{ K}} = 0.00476 \text{ mol}$$

$$\text{mass } KClO_3 = 0.00476 \text{ mol } O_2 \times\frac{2 \text{ mol } KClO_3}{3 \text{ mol } O_2}\times\frac{122.5 \text{ g } KClO_3}{1 \text{ mol } KClO_3} = 0.389 \text{ g } KClO_3$$

$$\% \text{ } KClO_3 = \frac{0.389 \text{ g } KClO_3}{3.57 \text{ g sample}}\times100\% = 10.9\% \text{ } KClO_3$$

53. **(M)** First we need to find the number of moles of CO(g)

Reaction is $3\,CO(g) + 7\,H_2(g) \longrightarrow C_3H_8(g) + 3\,H_2O(l)$

$$n_{CO} = \frac{PV}{RT} = \frac{28.5\,L \times 760\,Torr \times \dfrac{1\,atm}{760\,Torr}}{0.08206\,\dfrac{L\,atm}{K\,mol} \times 273.15\,K} = 1.27\,mol$$

$$V_{H_2\,(required)} = \frac{n_{H_2}RT}{P} = \frac{1.27\,mol\,CO \times \dfrac{7\,mol\,H_2}{3\,mol\,CO} \times 0.08206\,\dfrac{atm\,L}{mol\,K} \times 299\,K}{751\,mmHg \times \dfrac{1\,atm}{760\,mmHg}} = 73.7\,L\,H_2$$

Mixtures of Gases

55. **(M)** Determine the total amount of gas; then use the ideal gas equation, assuming that the gases behave ideally.

$$n_{total} = n_{Ne} + n_{Ar} = 15.2\,g \times \frac{1\,mol}{20.18\,g} + 34.8\,g \times \frac{1\,mol}{39.95\,g} = 0.753\,mol + 0.871\,mol = 1.624\,mol$$

$$V_{total} = \frac{1.624\,mol \times 0.08314\,bar\,L\,mol^{-1}\,K^{-1} \times (273.2 + 26.7)\,K}{7.24\,bar} = 5.59\,L$$

57. **(M)** The two pressures are related, as are the number of moles of $N_2(g)$ to the total number of moles of gas.

$$moles\,N_2 = \frac{PV}{RT} = \frac{28.2\,atm \times 53.7\,L}{0.08206\,\dfrac{atm\,L}{mol\,K} \times (26 + 273)\,K} = 61.7\,mol\,N_2$$

$$total\,moles\,of\,gas = 61.7\,mol\,N_2 \times \frac{75.0\,atm}{28.2\,atm} = 164\,mol\,gas$$

$$mass\,Ne = (164\,mol\,total - 61.7\,mol\,N_2) \times \frac{20.18\,g\,Ne}{1\,mol\,Ne} = 2.06 \times 10^3\,g\,Ne$$

59. **(M)** Initial pressure of the cylinder

$$P = \frac{nRT}{V} = \frac{\left(1.60\,g\,O_2 \times \dfrac{1\,mol\,O_2}{31.998\,g\,O_2}\right)(0.08206\,atm\,L\,mol^{-1}\,K^{-1})(273.15\,K)}{2.24\,L} = 0.500\,atm$$

We need to quadruple the pressure from 0.500 atm to 2.00 atm.

The mass of O_2 needs to quadruple as well from 1.60 g $\rightarrow$ 6.40 g or add 4.80 g O_2

(this answer eliminates answer (a) and (b) as being correct).

One could also increase the pressure by adding the same number of another gas (e.g. He)

mass of He = $n_{He} \times M_{He}$

(Note: moles of O_2 needed = 4.80 g $\times \dfrac{1 \text{ mol } O_2}{31.998 \text{ g } O_2} = 0.150$ mol $O_2 = 0.150$ mol He)

mass of He = 0.150 mol He $\times \dfrac{4.0026 \text{ g He}}{1 \text{ mol He}} = 0.600$ g He ((d) is correct, add 0.600 g He)

61. **(M)**

(a) $P_{benzene} = \dfrac{nRT}{V} = \dfrac{\left(0.728 \text{ g} \times \dfrac{1 \text{ mol } C_6H_6}{78.11 \text{ g } C_6H_6}\right) \times 0.08206 \dfrac{\text{atm L}}{\text{mol K}} \times (35+273) \text{ K}}{2.00 \text{ L}} \times \dfrac{760 \text{ mmHg}}{1 \text{ atm}}$

$\qquad = 89.5$ mmHg

$P_{total} = 89.5$ mmHg $C_6H_6(g) + 752$ mmHg Ar$(g) = 842$ mmHg

(b) $P_{Ar} = 752$ mmHg $\qquad\qquad P_{benzene} = 89.5$ mmHg

63. **(E)** 1.00 g $H_2 \approx 0.50$ mol H_2 $\qquad 1.00$ g He ≈ 0.25 mol He

Adding 1.00 g of He to a vessel that only contains 1.00 g of H_2 results in the number of moles of gas being increased by 50%. Situation (b) best represents the resulting mixture, as the volume has increased by 50%.

65. **(M)** For each gas, use $P_iV_i = P_fV_f$ to calculate the final partial pressure and then add the resulting partial pressures to obtain the total pressure.

For O_2: $\qquad\qquad P_f = \dfrac{V_i}{V_f} \times P_i = \dfrac{4.0 \text{ L}}{2.0 \text{ L}} \times 1.0 \text{ bar} = 2.0 \text{ bar}$

For N_2: $\qquad\qquad P_f = \dfrac{V_i}{V_f} \times P_i = \dfrac{2.0 \text{ L}}{2.0 \text{ L}} \times 2.0 \text{ bar} = 2.0 \text{ bar}$

Total pressure = 2.0 bar + 2.0 bar = 4.0 bar

Collecting Gases over Liquids

67. **(M)** The pressure of the liberated $H_2(g)$ is 744 mmHg $- 23.8$ mmHg $= 720.$ mmHg

$V = \dfrac{nRT}{P} = \dfrac{\left(1.65 \text{ g Al} \times \dfrac{1 \text{ mol Al}}{26.98 \text{ g Al}} \times \dfrac{3 \text{ mol } H_2}{2 \text{ mol Al}}\right) 0.08206 \dfrac{\text{atm L}}{\text{mol K}} (25+273)\text{K}}{720. \text{ mmHg} \times \dfrac{1 \text{ atm}}{760 \text{ mmHg}}} = 2.37 \text{ L } H_2(g)$

This is the total volume of both gases, each with a different partial pressure.

69. (M) We first determine the pressure of the gas collected. This would be its "dry gas" pressure and, when added to 22.4 mmHg, gives the barometric pressure.

$$P = \frac{nRT}{V} = \frac{\left(1.46 \text{ g} \times \dfrac{1 \text{ mol } O_2}{32.0 \text{ g } O_2}\right) 0.08206 \dfrac{\text{atm L}}{\text{mol K}} \times 297 \text{ K}}{1.16 \text{ L}} \times \frac{760 \text{ mmHg}}{1 \text{ atm}} = 729 \text{ mmHg}$$

$P_{\text{bar.}} = 729 \text{ mm Hg} + 22.4 \text{ mmHg} = 751 \text{ mmHg}$

71. (M) The first step is to balance the equation:

$$2 \text{ NaClO}_3 \xrightarrow{\Delta} 2 \text{ NaCl} + 3 \text{ O}_2$$

The pressure of O_2 is determined by subtracting the known vapor pressure of water at the given temperature from the measured total pressure.

$P_{O_2} = P_{\text{tot}} - P_{H_2O} = 734 \text{ Torr} - 21.07 \text{ Torr} = 713 \text{ Torr}$

$P_{\text{atm}} = 713 \text{ mmHg} \times 1 \text{ atm} / 760 \text{ Torr} = 0.938 \text{ atm}$

$$\text{mol } O_2 = \frac{PV}{RT} = \frac{(0.938 \text{ atm})(0.0572 \text{ L})}{(0.08206 \text{ atm L mol}^{-1} \text{ K}^{-1})(296 \text{ K})} = 0.00221 \text{ mol}$$

Mass of $NaClO_3$ is then determined as follows:

$$0.00221 \text{ mol } O_2 \times \frac{2 \text{ mol NaClO}_3}{3 \text{ mol } O_2} \times \frac{106.44 \text{ g}}{1 \text{ mol NaClO}_3} = 0.1568 \text{ g NaClO}_3$$

$$\%\text{NaClO}_3 = \frac{0.157 \text{ g}}{0.8765 \text{ g}} \times 100 = 17.9$$

Kinetic-Molecular Theory

73. (M) Recall that $1 \text{ J} = \text{kg m}^2 \text{ s}^{-2}$

$$u_{\text{rms}} = \sqrt{\frac{3RT}{M}} = \sqrt{\frac{3 \times 8.3145 \dfrac{\text{J}}{\text{mol K}} \times \dfrac{1 \text{ kg m}^2 \text{ s}^{-2}}{1 \text{ J}} \times 303 \text{ K}}{\dfrac{70.91 \times 10^{-3} \text{ kg Cl}_2}{1 \text{ mol Cl}_2}}} = 326 \text{ m/s}$$

75. (M) $M = \dfrac{3RT}{(u_{\text{rms}})^2} = \dfrac{3 \times 8.3145 \dfrac{\text{J}}{\text{mol K}} \times 298 \text{ K}}{\left(2180 \dfrac{\text{mi}}{\text{hr}} \times \dfrac{1 \text{ hr}}{3600 \text{ sec}} \times \dfrac{5280 \text{ ft}}{1 \text{ mi}} \times \dfrac{12 \text{ in.}}{1 \text{ ft}} \times \dfrac{1 \text{ m}}{39.37 \text{ in.}}\right)^2} = 0.00783 \text{ kg/mol}$

$= 7.83 \text{ g/mol or } 7.83 \, u$

77. **(E)** We equate the two expressions for root mean square speed, cancel the common factors, and solve for the temperature of Ne. Note that the units of molar masses do not have to be in kg/mol in this calculation; they simply must be expressed in the same units.

$$\sqrt{\frac{3RT}{M}} = \sqrt{\frac{3R \times 300 \text{ K}}{4.003}} = \sqrt{\frac{3R \times T_{Ne}}{20.18}} \quad \text{Square both sides:} \quad \frac{300 \text{ K}}{4.003} = \frac{T_{Ne}}{20.18}$$

Solve for T_{Ne}: $\quad T_{Ne} = 300 \text{ K} \times \frac{20.18}{4.003} = 1.51 \times 10^3 \text{ K}$

79. **(D)** The greatest pitfall of this type of problem is using improper units. Therefore, convert everything to SI units.

$M_{O_2} = 32.0 \text{ g/mol} = 32.0 \times 10^{-3} \text{ kg/mol}$

Mass of O_2 molecule: $\dfrac{32.0 \times 10^{-3} \text{ kg}}{\text{mol}} \times \dfrac{1 \text{ mol}}{6.022 \times 10^{23} \text{ molecule}} = 5.314 \times 10^{-26} \text{ kg}$

$R = 8.3145 \text{ J mol}^{-1} \text{ K}^{-1}$, or kg m^2/(s^2 mol K)

Now, we must determine the u_{rms} first to determine kinetic energy:

$$\overline{u}_{rms} = \sqrt{\frac{3RT}{M}} = \sqrt{\frac{3 \times 8.3145 \times 298}{32.0 \times 10^{-3}}} = 482 \text{ m/s}$$

Kinetic energy of an O_2 molecule is as follows:

$$\overline{e}_k = \frac{1}{2} m \, \overline{u}_{rms}^2 = \frac{1}{2} (5.314 \times 10^{-26} \text{ kg})(482 \text{ m/s})^2 = 6.17 \times 10^{-21} \text{ J}$$

Diffusion and Effusion of Gases

81. **(M)**

$$\frac{\text{rate (NO}_2)}{\text{rate (N}_2\text{O)}} = \sqrt{\frac{M(N_2O)}{M(NO_2)}} = \sqrt{\frac{44.02 \text{ g/mol}}{46.01 \text{ g/mol}}} = 0.9781 = \frac{x \text{ mol NO}_2/t}{0.00484 \text{ mol N}_2O/t}$$

mol $NO_2 = 0.00484 \text{ mol} \times 0.9781 = 0.00473 \text{ mol NO}_2$

83. **(M)**

(a) $\dfrac{\text{rate (N}_2)}{\text{rate (O}_2)} = \sqrt{\dfrac{M(O_2)}{M(N_2)}} = \sqrt{\dfrac{32.00 \text{ g/mol}}{28.01 \text{ g/mol}}} = 1.07$

(b) $\dfrac{\text{rate (H}_2\text{O)}}{\text{rate (D}_2\text{O)}} = \sqrt{\dfrac{M(D_2O)}{M(H_2O)}} = \sqrt{\dfrac{20.0 \text{ g/mol}}{18.02 \text{ g/mol}}} = 1.05$

(c) $\dfrac{\text{rate }(^{14}CO_2)}{\text{rate }(^{12}CO_2)} = \sqrt{\dfrac{M\,(^{12}CO_2)}{M(^{14}CO_2)}} = \sqrt{\dfrac{44.0 \text{ g/mol}}{46.0 \text{ g/mol}}} = 0.978$

(d) $\dfrac{\text{rate }(^{235}UF_6)}{\text{rate }(^{238}UF_6)} = \sqrt{\dfrac{M\,(^{238}UF_6)}{M(^{235}UF_6)}} = \sqrt{\dfrac{352 \text{ g/mol}}{349 \text{ g/mol}}} = 1.004$

85. **(M)** For ideal gases, the effusion rate is inversely proportional to their molecular mass. As such, the rate of effusion of a known gas can be determined if the rate of effusion for another gas is known:

$$\frac{\text{rate of effusion of Ne}}{\text{rate of effusion of He}} = \sqrt{\frac{M_{He}}{M_{Ne}}}$$

Since effusion is loosely defined as movement of a fixed number of atoms per unit time, and since in this problem we are looking at the time it takes for the same number of moles of both Ne and He to effuse, the above equation can be rearranged as follows:

$$\frac{mol_{Ne}}{time_{Ne}} \bigg/ \frac{mol_{He}}{time_{He}} = \frac{time_{He}}{time_{Ne}} = \sqrt{\frac{M_{He}}{M_{Ne}}}$$

$$\frac{x}{22 \text{ h}} = \sqrt{\frac{4.00 \text{ g/mol}}{20.18 \text{ g/mol}}}$$

Solving for x,

$x = 22 \times \sqrt{4.00/20.18} = 9.80 \text{ h}$

Nonideal Gases

87. **(M)** To facilitate calculations at several temperatures, substitute the given quantities into the ideal gas and van der Waals equations, and simplify the resulting expressions.

$$P_{\text{ideal}} = \frac{1.00 \text{ mol} \times 0.08314 \text{ bar L mol}^{-1}\text{ K}^{-1} \times T}{2.00 \text{ L}} = 0.0415\underline{7} \text{ bar K}^{-1} \times T$$

$$P_{\text{vdw}} = \frac{1.00 \text{ mol} \times 0.08314 \text{ bar L mol}^{-1}\text{ K}^{-1} \times T}{2.00 \text{ L} - 1.00 \text{ mol} \times 0.0542 \text{ L mol}^{-1}} - \frac{(1.00 \text{ mol})^2 \times 6.34 \text{ bar L}^2\text{ mol}^{-2}}{(2.00 \text{ L})^2}$$

$$= 0.0427\underline{3} \text{ bar K}^{-1} \times T - 1.58\underline{5} \text{ bar}$$

(a) For $T = (100 + 273) \text{ K} = 373 \text{ K}$:

$P_{\text{ideal}} = 0.0415\underline{7} \text{ bar K}^{-1} \times 373 \text{ K} = 15.5 \text{ bar}$

$P_{\text{vdw}} = 0.0427\underline{3} \text{ bar K}^{-1} \times 373 \text{ K} - 1.58\underline{5} \text{ bar} = 14.4 \text{ bar}$

(b) For $T = (200 + 273) \text{ K} = 473 \text{ K}$:

$P_{\text{ideal}} = 0.0415\underline{7} \text{ bar K}^{-1} \times 473 \text{ K} = 19.7 \text{ bar}$

$P_{\text{vdw}} = 0.0427\underline{3} \text{ bar K}^{-1} \times 473 \text{ K} - 1.58\underline{5} \text{ bar} = 18.6 \text{ bar}$

(c) For $T = (400 + 273)$ K $= 673$ K:

$P_{ideal} = 0.0415\underline{7}$ bar K^{-1} $\times 673$ K $= 28.0$ bar

$P_{vdw} = 0.0427\underline{3}$ bar K^{-1} $\times 673$ K $- 1.58\underline{5}$ bar $= 27.1$ bar

These results, along with those of Example 6-17, are summarized below. The percent difference is calculated as $100\% \times (P_{ideal} - P_{vdw})/P_{vdw}$. These results support the observation that a gas tends to behave more like an ideal gas as temperature increases.

t, °C	P_{ideal}, bar	P_{vdw}, bar	$P_{ideal} - P_{vdw}$	Percent difference
0	11.3	10.1	1.2 bar	12%
100	15.5	14.4	1.1 bar	7.6%
200	19.7	18.6	1.1 bar	5.9%
400	28.0	27.1	0.9 bar	3.3%

89. **(E)** The van der Waals parameter b is defined as the excluded volume per mole, or the volume that is taken up by 1 mole of gas once converted to a liquid.

From Table 6-5, $b_{He} = 0.0238$ L/mol. Therefore, the volume of a single He atom is:

Conversion pathway approach:

$$\frac{0.0238 \text{ L}}{1 \text{ mol He}} \times \frac{1 \text{ mol He}}{6.022 \times 10^{23} \text{ atoms}} \times \frac{1 \text{ m}^3}{1000 \text{ L}} \times \frac{(1 \times 10^{12} \text{ pm})^3}{1 \text{ m}^3} = 3.95 \times 10^7 \text{ pm}^3/\text{He atom}$$

Step-wise approach:

$$\frac{0.0238 \text{ L}}{\text{mol He}} \times \frac{1 \text{ mol He}}{6.022 \times 10^{23} \text{ atoms}} = 3.95 \times 10^{-26} \text{ L/atom}$$

$$3.95 \times 10^{-26} \text{ L/atom} \times \frac{1 \text{ m}^3}{1000 \text{ L}} = 3.95 \times 10^{-29} \text{ m}^3/\text{atom}$$

$$3.95 \times 10^{-29} \text{ m}^3/\text{atom} \times \frac{(1 \times 10^{12} \text{ pm})^3}{1 \text{ m}^3} = 3.95 \times 10^7 \text{ pm}^3/\text{He atom}$$

$V = (4/3)\pi r^3$. Rearranging to solve for r gives $r = \sqrt[3]{(3V)/(4\pi)}$. Solving for r gives an atomic radius of 211.3 pm.

INTEGRATIVE AND ADVANCED EXERCISES

94. **(M)** We know that the sum of the moles of gas in the two bulbs is 1.00 mol, and that both bulbs have the same volume, and are at the same pressure because they are connected. Therefore,

$$n_1 T_1 = n_2 T_2$$

$$\frac{n_1}{n_2} = \frac{T_2}{T_1} = \frac{350\ \text{K}}{225\ \text{K}} = 1.556$$

$$n_1 = 1.556\ n_2$$

Therefore,

$n_2(1.556) + n_2 = 1.00$. Solving the equation gives (Flask 2) $n_2 = 0.391$, and (Flask 1) $n_1 = 0.609$.

97. **(M)**

Stepwise approach:

Note that three moles of gas are produced for each mole of NH_4NO_3 that decomposes.

$$\text{amount of gas} = 3.05\ \text{g } NH_4NO_3 \times \frac{1\ \text{mol } NH_4NO_3}{80.04\ \text{g } NH_4NO_3} \times \frac{3\ \text{mol gas}}{1\ \text{mol } NH_4NO_3} = 0.114\ \text{mol gas}$$

$$T = 250 + 273 = 523\ \text{K}$$

$$P = \frac{nRT}{V} = \frac{0.114\ \text{mol} \times 0.08206\ \text{L atm mol}^{-1}\ \text{K}^{-1} \times 523\ \text{K}}{2.18\ \text{L}} = 2.25\ \text{atm}$$

Conversion pathway:

$$P = \frac{nRT}{V}$$

$$P = \frac{3.05\ \text{g } NH_4NO_3 \times \dfrac{1\ \text{mol } NH_4NO_3}{80.04\ \text{g } NH_4NO_3} \times \dfrac{3\ \text{mol gas}}{1\ \text{mol } NH_4NO_3} \times 0.08206\ \text{atm L mol}^{-1}\ \text{K}^{-1} \times 523\ \text{K}}{2.18\ \text{L}}$$

$$= 2.25\ \text{atm}$$

98. **(M)** First, let's convert the given units to those easier used:

$P = 101\ \text{kPa} \times (1\ \text{bar}/101\ \text{kPa}) \times (1\ \text{atm}/1.01\ \text{bar}) = 0.9901\ \text{atm}$

$T = (819 + 273)\ \text{K} = 1092\ \text{K}$

$$\text{mol } NH_4NO_2 = 128\ \text{g } NH_4NO_2 \times \frac{1\ \text{mol } NH_4NO_2}{64.052\ \text{g } NH_4NO_2} = 1.998\ \text{mol}$$

$$\text{mol gas} = 1.998\ \text{mol } NH_4NO_2 \times \frac{3\ \text{mol gas}}{1\ \text{mol } NH_4NO_2} = 5.994$$

$$V = \frac{nRT}{P} = \frac{(5.994\ \text{mol})(0.08206\ \text{atm L mol}^{-1}\ \text{K}^{-1})(1092\ \text{K})}{0.9901\ \text{atm}} = 542\ \text{L}$$

99. (M)

(a) $n_{H_2} = 1.00 \text{ g}/2.02 \text{ g/mol} = 0.495 \text{ mol } H_2$; $n_{O_2} = 8.60 \text{ g}/32.0 \text{ g/mol} = 0.269 \text{ mol } O_2$

The limiting reactant in the production of water is H_2 (0.495 mol) with O_2 (0.269 mol) in excess.

	$2 H_2(g)$	$+$	$O_2(g)$	$\rightarrow$	$2 H_2O(l)$
(initial)	0.495 mol		0.269 mol		0
after reaction	~0		0.022 mol		0.495 mol

so $P_{total} = P_{O_2} + P_{H_2O}$

$$P_{O_2} = \frac{n_{O_2} RT}{V} = \frac{0.022 \text{ mol} \times 0.08206 \text{ atm L mol}^{-1} \text{ K}^{-1} \times 298 \text{ K}}{1.50 \text{ L}} = 0.359 \text{ atm}$$

$$P_{total} = 0.359 \text{ atm} + \frac{23.8 \text{ mmHg}}{760 \text{ mmHg/atm}} = 0.39 \text{ atm}$$

101. (E) Determine relative numbers of moles, and the mole fractions of the 3 gases in the 100.0 g gaseous mixture.

amount $N_2 = 46.5 \text{ g } N_2 \times \dfrac{1 \text{ mol } N_2}{28.01 \text{ g } N_2} = 1.66 \text{ mol } N_2 \quad \div 2.86 \text{ mol} \longrightarrow 0.580$

amount $Ne = 12.7 \text{ g } Ne \times \dfrac{1 \text{ mol } Ne}{20.18 \text{ g } Ne} = 0.629 \text{ mol } Ne \quad \div 2.86 \text{ mol} \longrightarrow 0.220$

amount $Cl_2 = 40.8 \text{ g } Cl_2 \times \dfrac{1 \text{ mol } Cl_2}{70.9 \text{ g } Cl_2} = 0.575 \text{ mol } Cl_2 \quad \div 2.86 \text{ mol} \longrightarrow 0.201$

total amount $= 1.66 \text{ mol } N_2 + 0.629 \text{ mol } Ne + 0.575 \text{ mol } Cl_2 = 2.86 \text{ mol}$

Since the total pressure of the mixture is 1 atm, the partial pressure of each gas is numerically very close to its mole fraction. Thus, the partial pressure of Cl_2 is 0.201 atm or 153 mmHg.

102. (M) Let us first determine the molar mass of this mixture.

$$M = \frac{dRT}{P} = \frac{0.518 \text{ g L}^{-1} \times 0.08206 \text{ atm L mol}^{-1} \text{ K}^{-1} \times 298 \text{ K}}{721 \text{ mmHg} \times \dfrac{1 \text{ atm}}{760 \text{ mmHg}}} = 13.4 \text{ g/mol}$$

Then we let x be the mole fraction of He in the mixture.

$$13.4 \text{ g/mol} = x \times 4.003 \text{ g/mol} + (1.000 - x) \times 32.00 \text{ g/mol} = 4.003x + 32.00 - 32.00x$$

$$x = \frac{32.00 - 13.4}{32.00 - 4.003} = 0.664$$

Thus, one mole of the mixture contains 0.664 mol He. We determine the mass of that He and then the % He by mass in the mixture.

mass He $= 0.664 \text{ mol He} \times \dfrac{4.003 \text{ g He}}{1 \text{ mol He}} = 2.66 \text{ g He} \qquad$ % He $= \dfrac{2.66 \text{ g He}}{13.4 \text{ g mixture}} \times 100\% = 19.9\% \text{ He}$

105. (M) The amount of $N_2(g)$ plus the amount of $He(g)$ equals the total amount of gas. We use this equality, and substitute with the ideal gas equation, letting V symbolize the volume of cylinder B.

$n_{N_2} + n_{He} = n_{total}$ becomes $P_{N_2} \times V_{N_2} + P_{He} \times V_{He} = P_{total} \times V_{total}$

since all temperatures are equal.

Substitution gives $(8.35 \text{ atm} \times 48.2 \text{ L}) + (9.50 \text{ atm} \times V) = 8.71 \text{ atm} (48.2 \text{ L} + V)$

$$402 \text{ atm L} + 9.50 \, V = 420 \text{ atm L} + 8.71 \, V \qquad V = \frac{420 - 402}{9.50 - 8.71} = 23 \text{ L}$$

108. (M) First, balance the equation:

$2 C_8H_{18} + 25 O_2 \rightarrow 16 CO_2 + 18 H_2O$

mol O_2 needed: $2000 \text{ g } C_8H_{18} \times \dfrac{1 \text{ mol } C_8H_{18}}{114.23 \text{ g } C_8H_{18}} \times \dfrac{25 \text{ mol } O_2}{2 \text{ mol } C_8H_{18}}$

$= 218.86 \text{ mol } O_2$

Using the ideal gas equation, we can determine the volume of 177.33 mol of O_2:

$$\text{vol } O_2 = \frac{nRT}{P} = \frac{(218.86 \text{ mol})(8.314 \text{ kPa L mol}^{-1} \text{ K}^{-1})(298 \text{ K})}{101 \text{ kPa}} = 5369 \text{ L}$$

To determine the volume of air needed, we note that O_2 represents 20.9% of air by volume: $V_{air} \times 0.209 = 5369 \text{ L}$. Solving for V_{air} gives $2.57 \times 10^4 \text{ L}$.

109. (M) First recognize that 3 moles of gas are produced from every 2 moles of water, and compute the number of moles of gas produced. Then determine the partial pressure these gases would exert.

amount of gas $= 1.32 \text{ g } H_2O \times \dfrac{1 \text{ mol } H_2O}{18.02 \text{ g } H_2O} \times \dfrac{3 \text{ mol gas}}{2 \text{ mol } H_2O} = 0.110 \text{ mol gas}$

$$P = \frac{nRT}{V} = \frac{0.110 \text{ mol} \times 0.08206 \text{ atm L mol}^{-1} \text{ K}^{-1} \times 303 \text{ K}}{2.90 \text{ L}} \times \frac{760 \text{ mmHg}}{1 \text{ atm}} = 717 \text{ mmHg}$$

Then the vapor pressure (partial pressure) of water is determined by difference.

$P_{water} = 748 \text{ mmHg} - 717 \text{ mmHg} = 31 \text{ mmHg}$

111. (M) The total pressure of the mixture of O_2 and H_2O is 737 mmHg, and the partial pressure of H_2O is 25.2 mmHg.

(a) The percent of water vapor by volume equals its percent pressure.

$$\%H_2O = \frac{25.2 \text{ mmHg}}{737 \text{ mmHg}} \times 100\% = 3.42\% \text{ H}_2\text{O by volume}$$

(b) The % water vapor by number of molecules equals its percent pressure, 3.42% by number.

(c) One mole of the combined gases contains 0.0342 mol H_2O and 0.9658 mol O_2.

$$\text{molar mass} = 0.0342 \text{ mol } H_2O \times \frac{18.02 \text{ g } H_2O}{1 \text{ mol } H_2O} + 0.9658 \text{ mol } O_2 \times \frac{31.999 \text{ g } O_2}{1 \text{ mol } O_2}$$

$$= 0.616 \text{ g } H_2O + 30.90 \text{ g } O_2 = 31.52 \text{ g}$$

$$\% \ H_2O = \frac{0.616 \text{ g } H_2O}{31.52 \text{ g}} \times 100\% = 1.95\% \ H_2O \text{ by mass}$$

112. (D)

(a) 1 mol of the mixture at 0 °C and 1 atm occupies a volume of 22.414 L. It contains 0.79 mol He and 0.21 mol O_2.

$$\text{density} = \frac{\text{mass}}{\text{volume}} = \frac{0.79 \text{ mol He} \times \dfrac{4.003 \text{ g He}}{1 \text{ mol He}} + 0.21 \text{ mol } O_2 \times \dfrac{32.00 \text{ g } O_2}{1 \text{ mol } O_2}}{22.414 \text{ L}} = 0.44 \text{ g/L}$$

Let's focus on a 1.00 L sample (which weighs 0.44 g), and calculate the final volume when the temperature is increased from 0 °C to 25 °C.

$$V_{final} = 1.00 \text{ L} \times \frac{(25 + 273.2) \text{ K}}{273.2 \text{ K}} = 1.09 \text{ L} \qquad d_{final} = \frac{0.44 \text{ g}}{1.09 \text{ L}} = 0.40 \text{ g/L}$$

We determine the apparent molar masses of each mixture by multiplying the mole fraction (numerically equal to the volume fraction) of each gas by its molar mass, and then summing these products for all gases in the mixture.

$$M_{air} = \left(0.7808 \times \frac{28.01 \text{ g } N_2}{1 \text{ mol } N_2} \right) + \left(0.2095 \times \frac{32.00 \text{ g } O_2}{1 \text{ mol } O_2} \right) + \left(0.0093 \times \frac{39.95 \text{ g Ar}}{1 \text{ mol Ar}} \right) + \left(0.00036 \times \frac{44.01 \text{ g } CO_2}{1 \text{ mol } CO_2} \right)$$

$$= 21.87 \text{ g } N_2 + 6.704 \text{ g } O_2 + 0.374 \text{ g Ar} + 0.016 \text{ g } CO_2 = 28.96 \text{ g/mol air}$$

$$M_{mix} = \left(0.79 \times \frac{4.003 \text{ g He}}{1 \text{ mol He}} \right) + \left(0.21 \times \frac{32.00 \text{ g } O_2}{1 \text{ mol } O_2} \right) = 3.2 \text{ g He} + 6.7 \text{ g } O_2 = \frac{9.9 \text{ g mixture}}{\text{mol}}$$

(b) In order to prepare two gases with the same density, the volume of the gas of smaller molar mass must be smaller by a factor equal to the ratio of the molar masses. According to Boyle's law, this means that the pressure on the less dense gas must be larger by a factor equal to a ratio of molar masses.

$$P_{mix} = \frac{28.95}{9.9} \times 1.00 \text{ atm} = 2.9 \text{ atm}$$

114. (M) First, determine the moles of $Na_2S_2O_3$, then use the chemical equations given to determine the moles of O_3 in the mixture:

$$\text{mol } Na_2S_2O_3 = 0.0262 \text{ L} \times 0.1359 \text{ M} = 0.003561 \text{ mol}$$

$$\text{mol } O_3\text{: } 0.003561 \text{ mol } Na_2S_2O_3 \times \frac{1 \text{ mol } I_3^-}{2 \text{ mol } Na_2S_2O_3} \times \frac{1 \text{ mol } O_3}{1 \text{ mol } I_3^-} = 0.001780 \text{ mol } O_3$$

Using the ideal gas equation, we can determine the moles of gas:

$$\text{moles of gas} = \frac{(0.993 \text{ atm})(53.2 \text{ L})}{(0.08206 \text{ atm L mol}^{-1} \text{ K}^{-1})(291 \text{ K})} = 2.21 \text{ mol}$$

$$X_{O_3} = \frac{0.001780 \text{ mol}}{2.21 \text{ mol total}} = 8.05 \times 10^{-4}$$

115. (D) First compute the pressure of the water vapor at 30.1°C.

$$P_{H_2O} = \frac{\left(0.1052 \text{ g} \times \dfrac{1 \text{ mol H}_2\text{O}}{18.015 \text{ g H}_2\text{O}}\right) \times \dfrac{0.08206 \text{ atm L}}{\text{mol K}} \times (30.1 + 273.2) \text{ K}}{8.050 \text{ L}} \times \frac{760 \text{ mmHg}}{1 \text{ atm}} = 13.72 \text{ mmHg}$$

The water vapor is kept in the 8.050-L container, which means that its pressure is proportional to the absolute temperature in the container. Thus, for each of the six temperatures, we need to calculate two numbers: (1) the pressure due to this water (because gas pressure varies with temperature), and (2) 80% of the vapor pressure. The temperature we are seeking is where the two numbers agree.

$$P_{\text{water}}(T) = 13.72 \text{ mmHg} \times \frac{(T + 273.2) \text{ K}}{(30.1 + 273.2) \text{ K}}$$

For example, $P(20 \text{ °C}) = 13.72 \text{ mmHg} \times \dfrac{(20. + 273.2) \text{ K}}{(30.1 + 273.2) \text{ K}} = 13.3 \text{ mmHg}$

T	20. °C	19. °C	18. °C	17. °C	16. °C	15. °C
P_{water}, mmHg	13.3	13.2	13.2	13.1	13.1	13.0
80.0% v.p., mmHg	14.0	13.2	12.4	11.6	10.9	10.2

At approximately 19 °C, the relative humidity of the air will be 80.0%.

118. (M)

(a) $\overline{u}_{\text{rms}}$ is determined as follows:

$$\overline{u}_{\text{rms}} = \sqrt{\frac{3RT}{M}} = \sqrt{\frac{3(8.3145 \text{ J} \cdot \text{mol}^{-1} \cdot \text{K}^{-1})(298 \text{ K})}{32.00 \times 10^{-3} \text{kg}}} = 482 \text{ m/s}$$

(b) $F_u = 4\pi \left(\dfrac{M}{2\pi RT}\right)^{3/2} u^2 \exp\left(-Mu^2 / 2RT\right)$

$$F_u = 4\pi \left(\frac{32.00 \times 10^{-3}}{2\pi(8.3145)(298)}\right)^{3/2} (498)^2 \exp\left(-\frac{(32.00 \times 10^{-3})(498)^2}{2(8.3145)(298)}\right) = 1.92 \times 10^{-3}$$

119. (D) Potential energy (E_p) of an object is highest when the kinetic energy (E_k) of the object is zero and the object has attained its maximum height. Therefore, we must determine the kinetic energy. But first, we have to determine the velocity of the N_2 molecule.

$$\overline{u_{rms}} = \sqrt{\frac{3RT}{M}} = \sqrt{\frac{3\left(8.3145 \text{ J}\cdot\text{mol}^{-1}\cdot\text{K}^{-1}\right)\left(300 \text{ K}\right)}{28.00\times10^{-3} kg}} = 517 \text{ m/s}$$

$$\overline{E_k} = \frac{1}{2}\overline{mu_{rms}^2} = \frac{1}{2}\left(28.00\times10^{-3} kg\right)\left(517 \text{ m/s}\right)^2 = 3742 \text{ J}$$

$$\overline{E_k} = \overline{E_p} \Rightarrow m\cdot g\cdot h = \left(28.00\times10^{-3} \text{kg}\right)\left(9.8 \text{ m/s}^2\right)\cdot h = 3742 \text{ J}$$

Solving for h, the altitude reached by an N_2 molecule is 13637 m or 13.6 km.

121. (D) (a) First multiply out the left-hand side of the equation.

$$\left(P+\frac{an^2}{V^2}\right)(V - nb) = nRT = PV - Pnb + \frac{an^2}{V} - \frac{abn^3}{V^2}$$

Now multiply the entire equation through by V^2, and collect all terms on the right-hand side.

$$0 = - nRTV^2 + PV^3 - PnbV^2 + an^2V - abn^3$$

Finally, divide the entire equation by P and collect terms with the same power of V, to obtain:

$$0 = V^3 - n\left(\frac{RT + bP}{P}\right)V^2 + \left(\frac{n^2a}{P}\right)V - \frac{n^3ab}{P} = 0$$

(b)

$$n = 185 \text{ g } CO_2 \times \frac{1 \text{ mol } CO_2}{44.0 \text{ g } CO_2} = 4.20 \text{ mol } CO_2$$

$$0 = V^3 - 4.20 \text{ mol}\left(\frac{(0.08206 \text{ atm L mol}^{-1} \text{ K}^{-1}\times286 \text{ K})+(0.0429 \text{ L mol}^{-1}\times12.5 \text{ atm})}{12.5 \text{ atm}}\right)V^2$$

$$+ \left(\frac{(4.20 \text{ mol})^2 \ 3.61 \text{ L}^2 \text{ atm mol}^{-2}}{12.5 \text{ atm}}\right)V - \frac{(4.20 \text{ mol})^3 \ 3.61 \text{ L}^2 \text{ atm mol}^{-2}\times0.0429 \text{ L mol}^{-1}}{12.5 \text{ atm}}$$

$$= V^3 - 8.07 V^2 + 5.09 V - 0.918$$

We can solve this equation by the method of successive approximations. As a first value, we use the volume obtained from the ideal gas equation:

$$V = \frac{nRT}{P} = \frac{4.20 \text{ mol}\times0.08206 \text{ atm L mol}^{-1} \text{ K}^{-1}\times286 \text{ K}}{12.5 \text{ atm}} = 7.89 \text{ L}$$

$$V = \frac{nRT}{P} = \frac{4.20 \text{ mol}\times0.08206 \text{ L atm mol}^{-1} \text{ K}^{-1}\times286 \text{ K}}{12.5 \text{ atm}} = 7.89 \text{ L}$$

	$V = 7.89$ L	$(7.89)3 - 8.06 (7.89)2 + (5.07 \times 7.89) - 0.909 = 28.5$	> 0
Try	$V = 7.00$ L	$(7.00)3 - 8.06 (7.00)2 + (5.07 \times 7.00) - 0.909 = -17.4$	< 0
Try	$V = 7.40$ L	$(7.40)3 - 8.06 (7.40)2 + (5.07 \times 7.40) - 0.909 = 0.47$	> 0
Try	$V = 7.38$ L	$(7.38)3 - 8.06 (7.38)2 + (5.07 \times 7.38) - 0.909 = -0.53$	< 0
Try	$V = 7.39$ L	$(7.39)3 - 8.06 (7.39)2 + (5.07 \times 7.39) - 0.909 = -0.03$	< 0

The volume of CO_2 is very close to 7.39 L.

A second way is to simply disregard the last term and to solve the equation
$0 = V_3 - 8.06\ V_2 + 5.07\ V$

This equation simplifies to the following quadratic equation:
$0 = V_2 - 8.06\ V + 5.07$, which is solved with the quadratic formula.

$$V = \frac{8.06 \pm \sqrt{(8.06)^2 - 4 \times 5.07}}{2} = \frac{+8.06 \pm 6.68}{2} = \frac{14.74}{2} = 7.37\ \text{L}$$

The other root, 0.69 L does not appear to be reasonable, due to its small size.

FEATURE PROBLEMS

126. (E) Nitryl Fluoride $\quad 65.01\ \text{u} \left(\dfrac{49.4}{100}\right) = 32.1\ \text{u of X}$

Nitrosyl Fluoride $\quad 49.01\ \text{u} \left(\dfrac{32.7}{100}\right) = 16.0\ \text{u of X}$

Thionyl Fluoride $\quad 86.07\ \text{u} \left(\dfrac{18.6}{100}\right) = 16.0\ \text{u of X}$

Sulfuryl Fluoride $\quad 102.07\ \text{u} \left(\dfrac{31.4}{100}\right) = 32.0\ \text{u of X}$

The atomic mass of X is 16 u which corresponds to the element oxygen.
The number of atoms of X (oxygen) in each compound is given below:
Nitryl Fluoride $\quad = 2$ atoms of O $\qquad$ Nitrosyl Fluoride $\quad = 1$ atom of O
Thionyl Fluoride $\quad = 1$ atom of O $\qquad$ Sulfuryl Fluoride $\quad = 2$ atoms of O

127. (M)
    (a) The $N_2(g)$ extracted from liquid air has some Ar(g) mixed in. Only $O_2(g)$ was removed from liquid air in the oxygen-related experiments.

    (b) Because of the presence of Ar(g) [39.95 g/mol], the $N_2(g)$ [28.01 g/mol] from liquid air will have a greater density than $N_2(g)$ from nitrogen compounds.

    (c) Magnesium will react with molecular nitrogen $\left[3\ Mg(s) + N_2(g) \rightarrow Mg_3N_2(s)\right]$ but not with Ar. Thus, magnesium reacts with all the nitrogen in the mixture, but leaves the relatively inert Ar(g) unreacted.

    (d) The "nitrogen" remaining after oxygen is extracted from each mole of air (Rayleigh's mixture) contains $0.78084 + 0.00934 = 0.79018$ mol and has the mass calculated below.

        mass of gaseous mixture $= \left(0.78084 \times 28.013\ \text{g/mol}\ N_2\right) + \left(0.00934 \times 39.948\ \text{g/mol}\ Ar\right)$

        mass of gaseous mixture $= 21.874\ \text{g}\ N_2 + 0.373\ \text{g}\ Ar = 22.247\ \text{g mixture}$

Then, the molar mass of the mixture can be computed: 22.247 g mixture / 0.79018 mol = 28.154 g/mol. Since the STP molar volume of an ideal gas is 22.414 L, we can compute the two densities.

$$d(N_2) = \frac{28.013 \text{ g/mol}}{22.414 \text{ L/mol}} = 1.2498 \text{ g/mol} \qquad d(\text{mixture}) = \frac{28.154 \text{ g/mol}}{22.414 \text{ L/mol}} = 1.2561 \text{ g/mol}$$

These densities differ by 0.50%.

129. **(D)** Total mass = mass of payload + mass of balloon + mass of H_2

Use ideal gas equation to calculate mass of H_2: $PV = nRT = \dfrac{mRT}{M}$

$$m = \frac{PVM}{RT} = \frac{1 \text{ atm}\left(120 \text{ ft}^3 \times \dfrac{(12 \text{ in.})^3}{(1 \text{ ft})^3} \times \dfrac{(2.54 \text{ cm})^3}{(1 \text{ in.})^3} \times \dfrac{1\times10^{-3} \text{ L}}{1 \text{ cm}^3}\right) 2.016 \dfrac{\text{g}}{\text{mol}}}{\left(0.08206 \dfrac{\text{atm L}}{\text{K mol}}\right) 273 \text{ K}} = 306 \text{ g}$$

Total mass = 1200 g + 1700 g + 306 g ≈ 3200 g

We know at the maximum height, the balloon will be 25 ft in diameter. Need to find out what mass of air is displaced. We need to make one assumption – the volume percent of air is unchanged with altitude. Hence we use an apparent molar mass for air of 29 g mol^{-1} (question 99). Using the data provided, we find the altitude at which the balloon displaces 3200 g of air.

Note: balloon radius = $\dfrac{25}{2}$ = 12.5 ft. volume = $\dfrac{4}{3}(\pi)r^3 = \dfrac{4}{3}(3.1416)(12.5)^3 = 8181 \text{ ft}^3$

Convert to liters: $8181 \text{ ft}^3 \times \dfrac{(12 \text{ in.})^3}{(1 \text{ ft})^3} \times \dfrac{(2.54 \text{ cm})^3}{(1 \text{ in.})^3} \times \dfrac{1\times10^{-3} \text{ L}}{1 \text{ cm}^3} = 231,660 \text{ L}$

At 10 km: $m = \dfrac{PVM}{RT} = \dfrac{\left(2.7\times10^2 \text{ mbar} \times \dfrac{1 \text{ atm}}{1013.25 \text{ mbar}}\right)(231,660 \text{ L})\left(29\dfrac{\text{g}}{\text{mol}}\right)}{\left(0.08206 \dfrac{\text{atm L}}{\text{mol K}}\right) 223 \text{ K}} = 97,827 \text{ g}$

At 20 km: $m = \dfrac{PVM}{RT} = \dfrac{\left(5.5\times10^1 \text{ mbar} \times \dfrac{1 \text{ atm}}{1013.25 \text{ mbar}}\right)(231,660 \text{ L})\left(29\dfrac{\text{g}}{\text{mol}}\right)}{\left(0.08206 \dfrac{\text{atm L}}{\text{mol K}}\right) 217 \text{ K}} = 20,478 \text{ g}$

At 30 km: $m = \dfrac{PVM}{RT} = \dfrac{\left(1.2\times10^1 \text{ mbar} \times \dfrac{1 \text{ atm}}{1013.25 \text{ mbar}}\right)(231,660 \text{ L})\left(29\dfrac{\text{g}}{\text{mol}}\right)}{\left(0.08206 \dfrac{\text{atm L}}{\text{mol K}}\right) 230 \text{ K}} = 4,215 \text{ g}$

At 40 km: $m = \dfrac{PVM}{RT} = \dfrac{\left(2.9 \times 10^0 \, \text{mbar} \times \dfrac{1 \, \text{atm}}{1013.25 \, \text{mbar}}\right)(231{,}660 \, \text{L})\left(29 \, \dfrac{\text{g}}{\text{mol}}\right)}{\left(0.08206 \, \dfrac{\text{atm L}}{\text{mol K}}\right)250 \, \text{K}} = 93\underline{7} \, \text{g}$

The lifting power of the balloon will allow it to rise to an altitude of just over 30 km.

SELF-ASSESSMENT EXERCISES

133. **(E)** The answer is (d). The following shows the values for each:
   (a) $P = g \times h \times d = (9.8 \, \text{m/s}^2)(0.75 \, \text{m})(13{,}600 \, \text{kg/m}^3) = 99{,}960 \, \text{Pa}$
   (b) Just for a rough approximation, we assume the density of air to be the same as that of nitrogen (this underestimates it a bit, but is close enough). The density of N_2 is 0.02802 g/22.7 L = 1.234 g/L = 1.234 kg/m^3. $P = (9.8 \, \text{m/s}^2)(16{,}093 \, \text{m})(1.234 \, \text{kg/m}^3) = 194{,}616 \, \text{Pa}$.
   (c) $P = g \times h \times d = (9.8 \, \text{m/s}^2)(5.0 \, \text{m})(1590 \, \text{kg/m}^3) = 77{,}910 \, \text{Pa}$
   (d) $P = nRT/V = (10.00 \, \text{g H}_2 \times 1 \, \text{mol}/2.02 \, \text{g})(0.083145 \, \text{L bar mol}^{-1} \, \text{K}^{-1})(273 \, \text{K})/(22.7 \, \text{L}) = 4.95 \, \text{bar} = 495{,}032 \, \text{Pa}$

134. **(E)** The answer is (c), because the temperature decreases from 100 °C (373 K) to 200 K.

135. **(M)** First, determine the amounts of both gases:

$$n_A = \dfrac{PV}{RT}$$

$$= \dfrac{(1.081 \, \text{bar})(14.20 \, \text{L})}{(0.083145 \, \text{bar L mol}^{-1} \, \text{K}^{-1})(303.1 \, \text{K})}$$

$$= 0.6092 \, \text{mol}$$

$$n_A = \dfrac{PV}{RT}$$

$$= \dfrac{(26.77 \, \text{bar})(1.251 \, \text{L})}{(0.083145 \, \text{bar L mol}^{-1} \, \text{K}^{-1})(327.5 \, \text{K})}$$

$$= 1.230 \, \text{mol}$$

Therefore,
$n_{\text{total}} = 0.6092 \, \text{mol} + 1.230 \, \text{mol} = 1.839 \, \text{mol}$

(a) With n_{total} obtained, we can now calculate total pressure, P_{total}:

$$P_{total} = \frac{n_{total}RT}{V}$$

$$= \frac{(1.839 \text{ mol})(0.083145 \text{ atm L mol}^{-1} \text{ K}^{-1})(327.5 \text{ K})}{5.000 \text{ L}}$$

$$= 8.90 \text{ bar}$$

(b) The partial pressure of gas A can now be obtained from P_{total} and the mole fraction of A:

$$P_A = \left(\frac{n_A}{n_{total}}\right) P_{total} = \frac{0.6092 \text{ mol}}{1.839 \text{ mol}} \times 8.90 \text{ bar}$$

$$= 2.95 \text{ bar}$$

(c) The partial pressure of gas B can likewise be calculated:

$$P_B = \left(\frac{n_B}{n_{total}}\right) P_{total} = \frac{1.230 \text{ mol}}{1.839 \text{ mol}} \times 8.90 \text{ bar}$$

$$= 5.95 \text{ bar}$$

136. (M) $P_1/T_1 = P_2/T_2$. To calculate T_2, rearrange the formula:
$T_2 = 2.0 \text{ bar} / (1.0 \text{ bar} \times 273 \text{ K}) = 546 \text{ K}$

137. (M) The answer is (d).
$$\frac{P_1 V_1}{T_1} = \frac{P_2 V_2}{T_2}$$

$$V_2 = \frac{1 \text{ atm} \times 22.4 \text{ L} \times 298 \text{ K}}{273 \text{ K} \times 1.5 \text{ atm}} = 16.3 \text{ L}$$

138. (M) The answer is (b). Since the same number of moles of all ideal gases occupy the same volume, density is driven by the molar mass of the gas. Therefore, Kr has the highest density because it has a molar mass of 83 g/mol.

139. (E)

(a) Helium has the greater average molecular speed because it is the lighter gas. Recall,

$$\bar{u} = \sqrt{\frac{3RT}{M}}$$

(b) Helium strikes the container walls more frequently because it is the lighter of the two gases and has the greater average molecular speed.

(c) The gases exert equal pressure.

140. **(E)** The correct answer is (d). If the Kelvin temperature of the gas is doubled, it has twice the kinetic energy. Recall that

$$\bar{u}_1 = \sqrt{\frac{3RT}{M}}$$

Doubling T yields

$$\bar{u}_2 = \sqrt{\frac{3R(2T)}{M}} = \sqrt{2} \times \sqrt{\frac{3RT}{M}} = \bar{u}_1\sqrt{2}$$

We know that the average kinetic energy is given by

$$E_{k,1} = \frac{1}{2}m\bar{u}_1^2$$

Substituting $\bar{u}_2$,

$$E_{k,2} = \frac{1}{2}m\left(\bar{u}_1\sqrt{2}\right)^2 = \frac{1}{2}m \times 2 \times \bar{u}_1^2$$

which is twice that of $E_{k,1}$.

141. **(E)** The **correct** answer is (c). The statement that molecules are repelled by the walls of the container is not a postulate of the kinetic-molecular theory.

142. **(M)**
  (a) False. They both have the same kinetic energy.
  (b) True. All else being equal, the heavier molecule is slower.
  (c) False. The formula $PV = nRT$ can be used to confirm this. The answer is 24.4 L.
  (d) True. There is ~1.0 mole of each gas present. All else being equal, the same number of moles of any ideal gas occupies the same volume.
  (e) False. Total pressure is the sum of the partial pressures. So long as there is nothing else but H_2 and O_2, the total pressure is equal to the sum of the individual partial pressures.

143. **(E)** The answer is (c). Partial pressures are additive, so:
$$P_{tot} = P_{H_2O} + P_{O_2}$$

$$P_{O_2}(atm) = \frac{P_{tot} - P_{H_2O}}{760 \text{ mmHg}} = \frac{751 - 21}{760} = 0.96 \text{ atm}$$

144. **(M)** The answer is (a). First, determine the # moles of NH_3 using the ideal gas equation relationship:
$n = PV/RT = (0.500 \text{ atm} \times 4.48 \text{ L}) / (0.08206 \text{ atm L K}^{-1} \times 273 \text{ K}) = 0.100 \text{ mol}$
If 1.0 mol of a substance has 6.022×10^{23} atoms, 0.100 mol has 6.022×10^{22} atoms.

145. **(M)** The answer is (b). Since $PV = nRT$, the number of moles of O_2 needed to satisfy the conditions of the problem is:

$n = (2.00 \text{ atm} \times 2.24 \text{ L}) / (0.08206 \text{ L atm K}^{-1} \times 273 \text{ K}) = 0.200 \text{ mol}$

The amount of O_2 available $= 1.60 \text{ g } O_2 \times \dfrac{1 \text{ mol } O_2}{32.0 \text{ g } O_2} = 0.050 \text{ mol } O_2$

To have an additional 0.150 mol in the system, we would add 0.6 g of He to the container.

146. **(E)** The volumes of both gases were measured at the same temperature and pressure. Therefore, the proportionality constant between volume and moles for both gases is the same (that is, volume can essentially replace moles in the following calculations):

$25.0 \text{ L } H_2 \times \dfrac{3 \text{ L CO}}{7 \text{ L } H_2} = 10.7 \text{ L CO needed}$

So, all of the H_2 and 10.7 L of CO are consumed, and 1.3 L of CO remain.

148. **(M)** The answer is (c). Gases behave more ideally at high temperatures and low pressures.

149. **(M)**
(a) He or Ne: Ne has higher a and b values.
(b) CH_4 or C_3H_8: C_3H_8 has higher a and b values.
(c) H_2 or Cl_2: Cl_2 has higher a and b values.

150. **(D)** We know that pressure is force per unit area, that is:

$P = \dfrac{F}{A} = \dfrac{m \times g}{A}$

Using the fact that area $A = \pi r^2 = \pi (D/2)^2$ and that mass $m = d \times V$, and that volume of a cylindrical tube is $V = A \times h$ (where h is the height of the liquid in the tube), we can express the pressure formula as follows:

$P = \dfrac{F}{A} = \dfrac{d \times V \times g}{A} = \dfrac{d \times (A \times h) \times g}{A} = d \times h \times g$

Therefore,

$\dfrac{d \times V \times g}{A} = d \times h \times g$, and,

$h = \dfrac{d}{A} = \dfrac{d}{\pi (D/2)^2}$

As we can see, the height, h is inversely proportional to D. That is, the larger the diameter of the tube, the shorter the height of the liquid.

151. **(D)** First, convert the given information to more useful units. The pressure (752 Torr) is equivalent to 0.9895 atm (752 mmHg/760 mmHg), and the temperature is 298 K. Then, use the ideal gas relationship to determine how many moles of gas are present, assuming 1 L of gas:

$$PV = nRT$$

$$n = \frac{PV}{RT} = \frac{(0.9895 \text{ atm})(1 \text{ L})}{(0.08206 \text{ atm L mol}^{-1} \text{ K}^{-1})(298 \text{ K})} = 0.04046 \text{ mol}$$

From the problem, we know the mass of the gas per 1 L, as expressed in the density. Mass of 1 L of this gas is 2.35 g. The molar mass of this substance is therefore:

$$M = \text{g/mol} = 2.35 \text{ g}/0.04046 \text{ mol} = 58.082 \text{ g/mol}$$

Now, let's calculate the empirical formula for the hydrocarbon gas and see how it compares to the molar mass:

$$\text{mol C} = 82.7 \text{ g C} \times \frac{1 \text{ mol C}}{12.01 \text{ g C}} = 6.89 \text{ mol}$$

$$\text{mol H} = 17.3 \text{ g H} \times \frac{1 \text{ mol H}}{1.01 \text{ g H}} = 17.1 \text{ mol}$$

H:C ratio = 17.1 mol / 6.89 mol = 2.5

The ratio between H and C is 2.5:1 or 5:2, making the empirical formula C_2H_5. The mass of this formula unit is 29.07 g/mol. Comparing to the calculated molar mass of the gas, it is smaller by a factor of 2. Therefore, the gas in question in C_4H_{10} or butane.

152. **(E)** N_2 comprises 78.084% of atmosphere, oxygen 20.946%, argon 0.934%, and CO_2 0.0379%. To graphically show the scale of this difference, divide all values by the smallest one (CO_2). Therefore, for every single mark representing CO_2, we need 2060 marks for N_2, 553 marks for O_2, and 25 for Ar.

CHAPTER 7
THERMOCHEMISTRY
PRACTICE EXAMPLES

1A **(E)** The heat absorbed is the product of the mass of water, its specific heat capacity $\left(4.18 \text{ J g}^{-1}\,°\text{C}^{-1}\right)$, and the temperature change that occurs.

$$\text{heat energy} = 237 \text{ g} \times \frac{4.18 \text{ J}}{\text{g}\,°\text{C}} \times \left(37.0\,°\text{C} - 4.0\,°\text{C}\right) \times \frac{1 \text{ kJ}}{1000 \text{ J}} = 32.7 \text{ kJ of heat energy}$$

1B **(E)** The heat absorbed is the product of the amount of mercury, its molar heat capacity, and the temperature change that occurs.

$$\text{heat energy} = \left(2.50 \text{ kg} \times \frac{1000 \text{ g}}{1 \text{ kg}} \times \frac{1 \text{ mol Hg}}{200.59 \text{ g Hg}}\right) \times \frac{28.0 \text{ J}}{\text{mol}\,°\text{C}} \times \left[-6.0 - (-20.0)\right]°\text{C} \times \frac{1 \text{ kJ}}{1000 \text{ J}}$$

$$= 4.89 \text{ kJ of heat energy}$$

2A **(E)** First calculate the quantity of heat lost by the lead. This heat energy must be absorbed by the surroundings (water). We assume 100% efficiency in the energy transfer.

$$q_{\text{lead}} = 1.00 \text{ kg} \times \frac{1000 \text{ g}}{1 \text{ kg}} \times \frac{0.13 \text{ J}}{\text{g}\,°\text{C}} \times \left(35.2\,°\text{C} - 100.0\,°\text{C}\right) = -8.4 \times 10^3 \text{ J} = -q_{\text{water}}$$

$$8.4 \times 10^3 \text{ J} = m_{\text{water}} \times \frac{4.18 \text{ J}}{\text{g}\,°\text{C}} \times \left(35.2\,°\text{C} - 28.5\,°\text{C}\right) = 28 m_{\text{water}} \qquad m_{\text{water}} = \frac{8.4 \times 10^3 \text{ J}}{28 \text{ J g}^{-1}} = 3.0 \times 10^2 \text{ g}$$

2B **(M)** We use the same equation, equating the heat lost by the copper to the heat absorbed by the water, except now we solve for final temperature.

$$q_{\text{Cu}} = 100.0 \text{ g} \times \frac{0.385 \text{ J}}{\text{g}\,°\text{C}} \times \left(x\,°\text{C} - 100.0\,°\text{C}\right) = -50.0 \text{ g} \times \frac{4.18 \text{ J}}{\text{g}\,°\text{C}} \times \left(x\,°\text{C} - 26.5\,°\text{C}\right) = -q_{\text{water}}$$

$$38.5x - 3850 = -209x + 5539 \qquad 38.5x + 209x = 5539 + 3850 \;\rightarrow\; 247.5x = 9389$$

$$x = \frac{9389 \text{ J}}{247.5 \text{ J}\,°\text{C}^{-1}} = 37.9\,°\text{C}$$

3A **(E)** The molar mass of $C_8H_8O_3$ is 152.14 g/mol. The calorimeter has a heat capacity of 4.90 kJ/°C.

$$q_{\text{calorim}} = \frac{4.90 \text{ kJ}\,°\text{C}^{-1} \times \left(30.09\,°\text{C} - 24.89\,°\text{C}\right)}{1.013 \text{ g}} \times \frac{152.14 \text{ g}}{1 \text{ mol}} = 3.83 \times 10^3 \text{ kJ/mol}$$

$$\Delta_c H = -q_{\text{calorim}} = -3.83 \times 10^3 \text{ kJ/mol}$$

3B (E) The heat that is liberated by the benzoic acid's combustion serves to raise the temperature of the assembly. We designate the calorimeter's heat capacity by C.

$$q_r = 1.176 \text{ g} \times \frac{-26.42 \text{ kJ}}{1 \text{ g}} = -31.07 \text{ kJ} = -q_{calorim}$$

4A (M) The heat that is liberated by the reaction raises the temperature of the reaction mixture. We assume that this reaction mixture has the same density and specific heat capacity as pure water.

$$q_{calorim} = \left(200.0 \text{ mL} \times \frac{1.00 \text{ g}}{1 \text{ mL}}\right) \times \frac{4.18 \text{ J}}{\text{g} \,^{\circ}\text{C}} \times (30.2 - 22.4) \,^{\circ}\text{C} = 6.5 \times 10^3 \text{ J} = -q_r$$

Owing to the 1:1 stoichiometry of the reaction, the number of moles of AgCl(s) formed is equal to the number of moles of $AgNO_3$(aq) in the original sample.

$$\text{mol AgCl} = 100.0 \text{ mL} \times \frac{1 \text{ L}}{1000 \text{ mL}} \times \frac{1.00 \text{ mol AgNO}_3}{1 \text{ L}} \times \frac{1 \text{ mol AgCl}}{1 \text{ mol AgNO}_3} = 0.100 \text{ mol AgCl}$$

$$q_r = \frac{-6.5 \times 10^3 \text{ J}}{0.100 \text{ mol}} \times \frac{1 \text{ kJ}}{1000 \text{ J}} = -65. \text{ kJ/mol}$$

Because q_r is a negative quantity, the precipitation reaction is exothermic.

4B (M) The assumptions include no heat loss to the surroundings or to the calorimeter, a solution density of 1.00 g/mL, a specific heat capacity of $4.18 \text{ J g}^{-1}\,^{\circ}\text{C}^{-1}$, and that the initial and final solution volumes are the same. The equation for the reaction that occurs is $NaOH(aq) + HCl(aq) \rightarrow NaCl(aq) + H_2O(l)$. Since the two reactants combine in a one to one mole ratio, the limiting reactant is the one present in smaller amount (i.e. the one with a smaller molar quantity).

$$\text{amount HCl} = 100.0 \text{ mL} \times \frac{1.020 \text{ mmol HCl}}{1 \text{ mL soln}} = 102.0 \text{ mmol HCl}$$

$$\text{amount NaOH} = 50.0 \text{ mL} \times \frac{1.988 \text{ mmol NaOH}}{1 \text{ mL soln}} = 99.4 \text{ mmol NaOH}$$

Thus, NaOH is the limiting reactant.

$$q_{neutr} = 99.4 \text{ mmol NaOH} \times \frac{1 \text{ mmol H}_2\text{O}}{1 \text{ mmol NaOH}} \times \frac{1 \text{ mol H}_2\text{O}}{1000 \text{ mmol H}_2\text{O}} \times \frac{-56 \text{ kJ}}{1 \text{ mol H}_2\text{O}} = -5.57 \text{ kJ}$$

Let the final temperature of the calorimeter and its contents be $t \,^{\circ}\text{C}$. The heat absorbed by the calorimeter is:

$$q_{calorim} = (100.0 + 50.0) \text{ mL} \times \frac{1.00 \text{ g}}{1 \text{ mL}} \times \frac{4.18 \text{ J}}{\text{g} \,^{\circ}\text{C}} \times \frac{1 \text{ kJ}}{1000 \text{ J}} \times (t - 24.52) \,^{\circ}\text{C} = (0.627t - 15.37) \text{ kJ}$$

Since $q_{calorim} = -q_{neutr} = 5.57$ kJ, we have $0.627t - 15.37 = 5.57$ and

thus, $t = \dfrac{5.57 + 15.37}{0.627} = 33.4$.

5A (E) $w = -P\Delta V = -0.750 \text{ atm}(+1.50 \text{ L}) = -1.12\underline{5} \text{ atm L} \times \dfrac{101.33 \text{ J}}{1 \text{ atm L}} = -114 \text{ J}$

114 J of work is done by system.

5B **(M)** Determine the initial number of moles:

$$n = 50.0 \text{ g N}_2 \times \frac{1 \text{ mol N}_2}{28.014 \text{ g N}_2} = 1.78\underline{5} \text{ mol N}_2$$

$$V = \frac{nRT}{P} = \frac{(1.785 \text{ mol N}_2)(0.08206 \text{ atm L mol}^{-1} \text{ K}^{-1})(293.15 \text{ K})}{2.50 \text{ atm}} = 17.2 \text{ L}$$

$$\Delta V = 17.2 - 75.0 \text{ L} = -57.8 \text{ L}$$

$$w = -P\Delta V = -2.50 \text{ atm}(-57.8 \text{ L}) \times \frac{101.33 \text{ J}}{1 \text{ atm L}} \times \frac{1 \text{ kJ}}{1000 \text{ J}} = +14.6 \text{ kJ work done on system}$$

6A **(E)** The work is $w = +355$ J. The heat flow is $q = -185$ J. These two are related to the energy change of the system by the first law equation: $\Delta U = q + w$, which becomes

$$\Delta U = +355 \text{ J} - 185 \text{ J} = +1.70 \times 10^2 \text{ J}$$

6B **(E)** The internal energy change is $\Delta U = -125$ J. The heat flow is $q = +54$ J. These two are related to the work done on the system by the first law equation: $\Delta U = q + w$, which becomes $-125 \text{ J} = +54 \text{ J} + w$. The solution to this equation is $w = -125 \text{ J} - 54 \text{ J} = -179$ J, which means that 179 J of work is done by the system to the surroundings.

7A **(E)** Heat that is given off has a negative sign. In addition, we use the molar mass of sucrose, 342.30 g/mol.

$$\text{sucrose mass} = -1.00 \times 10^3 \text{ kJ} \times \frac{1 \text{ mol C}_{12}\text{H}_{22}\text{O}_{11}}{-5.65 \times 10^3 \text{ kJ}} \times \frac{342.30 \text{ g C}_{12}\text{H}_{22}\text{O}_{11}}{1 \text{ mol C}_{12}\text{H}_{22}\text{O}_{11}} = 60.6 \text{ g C}_{12}\text{H}_{22}\text{O}_{11}$$

7B **(E)** Although the equation does not say so explicitly, the reaction of $H^+(aq) + OH^-(aq) \rightarrow H_2O(l)$ gives off 56 kJ of heat per mole of water formed. The equation then is the source of a conversion factor.

$$\text{heat flow} = 25.0 \text{ mL} \times \frac{1 \text{ L}}{1000 \text{ mL}} \times \frac{0.1045 \text{ mol HCl}}{1 \text{ L soln}} \times \frac{1 \text{ mol H}_2\text{O}}{1 \text{ mol HCl}} \times \frac{56 \text{ kJ evolved}}{1 \text{ mol H}_2\text{O}}$$

$$= 0.15 \text{ kJ heat evolved}$$

8A **(M)** $V_{ice} = (2.00 \text{ cm})^3 = 8.00 \text{ cm}^3$

$$m_{ice} = m_{water} = 8.00 \text{ cm}^3 \times 0.917 \text{ g cm}^{-3} = 7.34 \text{ g ice} = 7.34 \text{ g H}_2\text{O}$$

$$\text{moles of ice} = 7.34 \text{ g ice} \times \frac{1 \text{ mol H}_2\text{O}}{18.015 \text{ g H}_2\text{O}} = 0.407 \text{ mol ice}$$

$$q_{overall} = q_{ice}(-10 \text{ to } 0 \text{ °C}) + q_{fus} + q_{water}(0 \text{ to } 23.2 \text{ °C})$$

$$q_{overall} = m_{ice} \, c_p(\text{ice}) \, \Delta T + n_{ice}\Delta H_{fus} + m_{water} \, c_p(\text{water}) \, \Delta T$$

$$q_{overall} = (7.34 \text{ g})(10.0 \text{ °C})(2.01 \times 10^{-3} \frac{\text{kJ}}{\text{g °C}}) + (0.407 \text{ mol ice})(6.01 \frac{\text{kJ}}{\text{mol}}) + (7.34 \text{ g})$$

$$(23.2 \text{ °C}) (4.184 \times 10^{-3} \frac{\text{kJ}}{\text{g °C}})$$

$$q_{overall} = 0.148 \text{ kJ} + 2.45 \text{ kJ} + 0.712 \text{ kJ}$$

$$q_{overall} = +3.31 \text{ kJ (the system absorbs this much heat)}$$

8B   **(M)** $5.00 \times 10^3 \text{ kJ} = q_{\text{ice}}(-15 \text{ to } 0 \text{ °C}) + q_{\text{fus}} + q_{\text{water}}(0 \text{ to } 25 \text{ °C}) + q_{\text{vap}}$

$5.00 \times 10^3 \text{ kJ} = m_{\text{ice}} c_p(\text{ice}) \Delta T + n_{\text{ice}}\Delta H_{\text{fus}} + m_{\text{water}} c_p(\text{water}) \Delta T + n_{\text{water}}\Delta H_{\text{vap}}$

$5.00 \times 10^6 \text{ J} = m(15.0 \text{ °C})(2.01 \dfrac{\text{J}}{\text{g °C}}) + \dfrac{m}{18.015 \text{ g H}_2\text{O/mol H}_2\text{O}} (6.01 \times 10^3 \dfrac{\text{J}}{\text{mol}})$

$+ m(25.0 \text{ °C})(4.184 \dfrac{\text{J}}{\text{g °C}}) + \dfrac{m}{18.015 \text{ g H}_2\text{O/mol}} (44.0 \times 10^3 \dfrac{\text{J}}{\text{mol}})$

$5.00 \times 10^6 \text{ J} = m(30.1\underline{5} \text{ J/g}) + m(333.\underline{6} \text{ J/g}) + m(104.\underline{5} \text{ J/g}) + m(2.4\underline{4} \times 10^3 \text{ J/g})$

$5.00 \times 10^6 \text{ J} = m(2.91 \times 10^3 \text{ J/g}) \quad m = \dfrac{5.00 \times 10^6 \text{ J}}{2.91 \times 10^3 \text{ J/g}} = 1718 \text{ g or } 1.72 \text{ kg H}_2\text{O}$

9A   **(M)** We combine the three combustion reactions to produce the hydrogenation reaction.

$\text{C}_3\text{H}_6(g) + \tfrac{9}{2}\text{O}_2(g) \rightarrow 3\,\text{CO}_2(g) + 3\,\text{H}_2\text{O}(l) \qquad \Delta_c H = \Delta H_1 = -2058 \text{ kJ/mol}$

$\text{H}_2(g) + \tfrac{1}{2}\text{O}_2(g) \rightarrow \text{H}_2\text{O}(l) \qquad \Delta_c H = \Delta H_2 = -285.8 \text{ kJ/mol}$

$3\,\text{CO}_2(g) + 4\,\text{H}_2\text{O}(l) \rightarrow \text{C}_3\text{H}_8(g) + 5\,\text{O}_2(g) \qquad -\Delta_c H = \Delta H_3 = +2219.9 \text{ kJ/mol}$

$\overline{\text{C}_3\text{H}_6(g) + \text{H}_2(g) \rightarrow \text{C}_3\text{H}_8(g) \qquad \Delta_r H = \Delta H_1 + \Delta H_2 + \Delta H_3 = -124 \text{ kJ/mol}}$

9B   **(M)** The combustion reaction has propanol and $\text{O}_{2(g)}$ as reactants; the products are $\text{CO}_2(g)$ and $\text{H}_2\text{O}(l)$. Reverse the reaction given and combine it with the combustion reaction of $\text{C}_3\text{H}_6(g)$.

$\text{C}_3\text{H}_7\text{OH}(l) \rightarrow \text{C}_3\text{H}_6(g) + \text{H}_2\text{O}(l) \qquad \Delta H_1 = +52 \text{ kJ/mol}$

$\text{C}_3\text{H}_6(g) + \tfrac{9}{2}\text{O}_2(g) \rightarrow 3\,\text{CO}_2(g) + 3\,\text{H}_2\text{O}(l) \qquad \Delta H_2 = -2058 \text{ kJ/mol}$

$\overline{\text{C}_3\text{H}_7\text{OH}(l) + \tfrac{9}{2}\text{O}_2(g) \rightarrow 3\,\text{CO}_2(g) + 4\,\text{H}_2\text{O}(l) \qquad \Delta_r H = \Delta H_1 + \Delta H_2 = -2006 \text{ kJ/mol}}$

10A **(M)** The enthalpy of formation is the enthalpy change for the reaction in which one mole of the product, $\text{C}_6\text{H}_{13}\text{O}_2\text{N}(s)$, is produced from appropriate amounts of the reference forms of the elements (in most cases, the most stable form of the elements).

$6\,\text{C}(\text{graphite}) + \tfrac{13}{2}\text{H}_2(g) + \text{O}_2(g) + \tfrac{1}{2}\text{N}_2(g) \rightarrow \text{C}_6\text{H}_{13}\text{O}_2\text{N}(s)$

10B **(M)** The enthalpy of formation is the enthalpy change for the reaction in which one mole of the product, $\text{NH}_3(g)$, is produced from appropriate amounts of the reference forms of the elements, in this case from 0.5 mol $\text{N}_2(g)$ and 1.5 mol $\text{H}_2(g)$, that is, for the reaction:

$\tfrac{1}{2}\text{N}_2(g) + \tfrac{3}{2}\text{H}_2(g) \rightarrow \text{NH}_3(g)$

The specified reaction is twice the reverse of the formation reaction, and its enthalpy change is minus two times the enthalpy of formation of $\text{NH}_3(g)$:

$-2 \times (-46.11 \text{ kJ}) = +92.22 \text{ kJ/mol}$

11A (M)

$$\Delta_r H^\circ = 2 \times \Delta_f H^\circ \left[CO_2(g) \right] + 3 \times \Delta_f H^\circ \left[H_2O(l) \right] - \Delta_f H^\circ \left[CH_3CH_2OH(l) \right] - 3 \times \Delta_f H^\circ \left[O_2(g) \right]$$

$$= \left[2 \times (-393.5 \text{ kJ/mol}) \right] + \left[3 \times (-285.8 \text{ kJ/mol}) \right] - \left[-277.7 \text{ kJ/mol} \right] - \left[3 \times 0.00 \text{ kJ/mol} \right]$$

$$= -1367 \text{ kj/mol}$$

11B (D) We write the combustion reaction for each compound, and use that reaction to determine the compound's heat of combustion.

$$C_3H_8(g) + 5 O_2(g) \rightarrow 3 CO_2(g) + 4 H_2O(l)$$

$$\Delta_c H^\circ = 3 \times \Delta_f H^\circ \left[CO_2(g) \right] + 4 \times \Delta_f H^\circ \left[H_2O(l) \right] - \Delta_f H^\circ \left[C_3H_8(g) \right] - 5 \times \Delta_f H^\circ \left[O_2(g) \right]$$

$$= \left[3 \times (-393.5 \text{ kJ}) \right] + \left[4 \times (-285.8 \text{ kJ}) \right] - \left[-103.8 \text{ kJ} \right] - \left[5 \times 0.00 \text{ kJ mol}^{-1} \right]$$

$$= -1181 \text{ kJ} - 1143 \text{ kJ} + 103.8 - 0.00 \text{ kJ mol}^{-1} = -2220. \text{ kJ/mol } C_3H_8$$

$$C_4H_{10}(g) + \tfrac{13}{2} O_2(g) \rightarrow 4 CO_2(g) + 5 H_2O(l)$$

$$\Delta_c H^\circ = 4 \times \Delta_f H^\circ \left[CO_2(g) \right] + 5 \times \Delta_f H^\circ \left[H_2O(l) \right] - \Delta_f H^\circ \left[C_4H_{10}(g) \right] - 6.5 \times \Delta_f H^\circ \left[O_2(g) \right]$$

$$= \left[4 \times (-393.5 \text{ kJ}) \right] + \left[5 \times (-285.8 \text{ kJ}) \right] - \left[-125.6 \right] - \left[6.5 \times 0.00 \text{ kJ mol}^{-1} \right]$$

$$= -1574 \text{ kJ} - 1429 \text{ kJ} + 125.6 \text{ kJ} - 0.00 \text{ kJ} = -2877 \text{ kJ/mol } C_4H_{10}$$

In 1.00 mole of the mixture there are 0.62 mol $C_3H_8(g)$ and 0.38 mol $C_4H_{10}(g)$.

$$\text{heat of combustion} = \left(0.62 \text{ mol } C_3H_8 \times \frac{-2220. \text{ kJ}}{1 \text{ mol } C_3H_8} \right) + \left(0.38 \text{ mol } C_4H_{10} \times \frac{-2877 \text{ kJ}}{1 \text{ mol } C_4H_{10}} \right)$$

$$= -1.4 \times 10^3 \text{ kJ} - 1.1 \times 10^3 \text{ kJ} = -2.5 \times 10^3 \text{ kJ/mol mixture}$$

12A (M) $6 CO_2(g) + 6 H_2O(l) \rightarrow C_6H_{12}O_6(s) + 6 O_2(g)$

$$\Delta_r H^\circ = 2803 \text{ kJ} = \Sigma v \Delta_f H^\circ_{\text{products}} - \Sigma |v| \Delta_f H^\circ_{\text{reactants}}$$

$$2803 \text{ kJ/mol} = [1 \, (\Delta_f H^\circ [C_6H_{12}O_6(s)]) + 6 \, (0 \tfrac{kJ}{mol})] - [6 \, (-393.5 \tfrac{kJ}{mol}) + 6 \, (-285.8 \tfrac{kJ}{mol})]$$

$$2803 \text{ kJ/mol} = \Delta_f H^\circ [C_6H_{12}O_6(s)] - [-4075.8 \text{ kJ}]. \text{ Thus, } \Delta_f H^\circ [C_6H_{12}O_6(s)]$$

$$= -1273 \text{ kJ/mol } C_6H_{12}O_6(s)$$

12B (M) $\Delta_c H^\circ [CH_3OCH_3(g)] = -31.70 \tfrac{kJ}{g}$ molar mass of $CH_3OCH_3 = 46.069 \text{ g mol}^{-1}$

$$\Delta_c H^\circ [CH_3OCH_3(g)] = -31.70 \tfrac{kJ}{g} \times 46.069 \tfrac{g}{mol} = -1460 \tfrac{kJ}{mol} = \Delta_r H^\circ$$

$$\Delta_r H^\circ = \Sigma v \Delta_f H^\circ_{\text{products}} - \Sigma |v| \Delta_f H^\circ_{\text{reactants}} \quad \text{Reaction: } CH_3OCH_3(g) + 3 O_2(g) \rightarrow 2 CO_2(g) + 3 H_2O(l)$$

$$-1460 \text{ kJ/mol} = [2 \, (-393.5 \tfrac{kJ}{mol}) + 3 \, (-285.8 \tfrac{kJ}{mol})] - [1 \, (\Delta_f H^\circ [CH_3OCH_3(g)]) + 3 \, (0 \tfrac{kJ}{mol})]$$

$$-1460 \text{ kJ/mol} = -1644.4 \text{ kJ/mol} - \Delta_f H^\circ [CH_3OCH_3(g)]$$

Hence, $\Delta_f H^\circ [CH_3OCH_3(g)] = -184 \text{ kJ/mol } CH_3OCH_3(g)$

13A **(M)** The net ionic equation is: $Ag^+(aq) + I^-(aq) \rightarrow AgI(s)$ and we have the following:

$$\Delta_r H° = \Delta_f H°\left[AgI(s)\right] - \left[\Delta_f H°\left[Ag^+(aq)\right] + \Delta_f H°\left[I^-(aq)\right]\right]$$

$$= -61.84 \text{ kJ/mol} - \left[\left(+105.6 \text{ kJ/mol}\right) + \left(-55.19 \text{ kJ/mol}\right)\right] = -112.3 \text{ kJ/mol AgI(s) formed}$$

13B **(M)** $2 Ag^+(aq) + CO_3^{2-}(aq) \rightarrow Ag_2CO_3(s)$

$\Delta_r H° = -39.9 \text{ kJ} = \Sigma \upsilon \Delta_f H°_{\text{products}} - \Sigma |\upsilon| \Delta_f H°_{\text{reactants}}$

$-39.9 \text{ kJ/mol} = \Delta_f H°[Ag_2CO_3(s)] - [2 (105.6 \text{ kJ/mol}) + 1 (-677.1 \text{ kJ/mol})]$

$-39.9 \text{ kJ/mol} = \Delta_f H°[Ag_2CO_3(s)] + 465.9 \text{ kJ/mol}$

Hence, $\Delta_f H°[Ag_2CO_3(s)] = -505.8 \text{ kJ/mol Ag}_2CO_3(s)$ formed.

INTEGRATIVE EXAMPLE

A. **(M)** The combustion reactions of $C_{16}H_{32}$ and $C_{16}H_{34}$ are shown below

(1) $C_{16}H_{32} + 24 O_2 \rightarrow 16 CO_2 + 16 H_2O$ $\qquad$ $\Delta_r H_1 = -10{,}539 \text{ kJ/mol}$
(2) $2 C_{16}H_{34} + 49 O_2 \rightarrow 32 CO_2 + 34 H_2O$ $\qquad$ $\Delta_r H_2 = -10{,}699 \text{ kJ/mol} = -21{,}398 \text{ kJ/mol}$

Since we are studying the hydrogenation of $C_{16}H_{32}$ to give $C_{16}H_{34}$, the final equation has to include the former as the reactant and the latter as the product. This is done by doubling equation 1 and reversing equation 2:

(3) $2 C_{16}H_{32} + 48 O_2 \rightarrow 32 CO_2 + 32 H_2O$ $\qquad$ $\Delta_r H_3 = -21{,}078 \text{ kJ/mol}$
(4) $32 CO_2 + 34 H_2O \rightarrow 2 C_{16}H_{34} + 49 O_2$ $\qquad$ $\Delta_r H_4 = +21{,}398 \text{ kJ/mol}$

(5) $2 C_{16}H_{32} + 2 H_2O \rightarrow 2 C_{16}H_{34} + O_2$ $\qquad$ $\Delta_r H_5 = +320 \text{ kJ/mol}$

Since a hydrogenation reaction involves hydrogen as a reactant, and looking at equation (5), we add the following reaction to (5):

(6) $H_2 + \frac{1}{2} O_2 \rightarrow H_2O$ $\qquad$ $\Delta_r H_6 = -285.5 \text{ kJ/mol}$

Double equation (6) and add it to equation 5:

(5) $2 C_{16}H_{32} + 2 H_2O \rightarrow 2 C_{16}H_{34} + O_2$ $\qquad$ $\Delta_r H_5 = +320 \text{ kJ/mol}$
(7) $2 H_2 + O_2 \rightarrow 2 H_2O$ $\qquad$ $\Delta_r H_7 = -571 \text{ kJ/mol}$

(8) $2 C_{16}H_{32} + 2 H_2 \rightarrow 2 C_{16}H_{34}$ $\qquad$ $\Delta_r H_8 = -251 \text{ kJ/mol}$

Since (8) is for 2 moles, $\Delta_f H$ is -125.5 kJ/mol

EXERCISES

Heat Capacity (Specific Heat Capacity)

1. **(E)**

   (a) $q = 9.25 \text{ L} \times \dfrac{1000 \text{ cm}^3}{1 \text{ L}} \times \dfrac{1.00 \text{ g}}{1 \text{ cm}^3} \times \dfrac{4.18 \text{ J}}{1 \text{ g} \, ^\circ\text{C}} \times \dfrac{1 \text{ kJ}}{1000 \text{ J}} \left(29.4 \, ^\circ\text{C} - 22.0 \, ^\circ\text{C}\right) = +2.9 \times 10^2 \text{ kJ}$

   (b) $q = 5.85 \text{ kg} \times \dfrac{1000 \text{ g}}{1 \text{ kg}} \times \dfrac{0.903 \text{ J}}{\text{g} \, ^\circ\text{C}} \times \left(-33.5 \, ^\circ\text{C}\right) \times \dfrac{1 \text{ kJ}}{1000 \text{ J}} = -177 \text{ kJ}$

3. **(E)** heat gained by the water = heat lost by the metal; heat = mass $\times c_p \times \Delta T$

   (a) $50.0 \text{ g} \times 4.18 \dfrac{\text{J}}{\text{g} \, ^\circ\text{C}} \left(38.9 - 22.0\right) ^\circ\text{C} = 3.53 \times 10^3 \text{ J} = -150.0 \text{ g} \times c_p \times \left(38.9 - 100.0\right) ^\circ\text{C}$

        $c_p = \dfrac{3.53 \times 10^3 \text{ J}}{150.0 \text{ g} \times 61.1 ^\circ\text{C}} = 0.385 \text{ J g}^{-1} \, ^\circ\text{C}^{-1}$ for Zn

   (b) $50.0 \text{ g} \times 4.18 \dfrac{\text{J}}{\text{g} \, ^\circ\text{C}} \left(28.8 - 22.0\right) ^\circ\text{C} = 1.4 \times 10^3 \text{ J} = -150.0 \text{ g} \times c_p \times \left(28.8 - 100.0\right) ^\circ\text{C}$

        $c_p = \dfrac{1.4 \times 10^3 \text{ J}}{150.0 \text{ g} \times 71.2 ^\circ\text{C}} = 0.13 \text{ J g}^{-1} \, ^\circ\text{C}^{-1}$ for Pt

   (c) $50.0 \text{ g} \times 4.18 \dfrac{\text{J}}{\text{g} \, ^\circ\text{C}} \left(52.7 - 22.0\right) ^\circ\text{C} = 6.42 \times 10^3 \text{ J} = -150.0 \text{ g} \times c_p \times \left(52.7 - 100.0\right) ^\circ\text{C}$

        $c_p = \dfrac{6.42 \times 10^3 \text{ J}}{150.0 \text{ g} \times 47.3 ^\circ\text{C}} = 0.905 \text{ J g}^{-1} \, ^\circ\text{C}^{-1}$ for Al

5. **(M)** $q_{\text{water}} = 375 \text{ g} \times 4.18 \dfrac{\text{J}}{\text{g} \, ^\circ\text{C}} \left(87 - 26\right) ^\circ\text{C} = 9.5\underline{6} \times 10^4 \text{ J} = -q_{\text{iron}}$

$q_{\text{iron}} = -9.5\underline{6} \times 10^4 \text{ J} = 465 \text{ g} \times 0.449 \dfrac{\text{J}}{\text{g} \, ^\circ\text{C}} \left(87 - T_1\right) = 1.81\underline{6} \times 10^4 \text{ J} - 2.08\underline{8} \times 10^2 \, T_1$

$T_1 = \dfrac{-9.56 \times 10^4 - 1.81\underline{6} \times 10^4}{-2.08\underline{8} \times 10^2} = \dfrac{-11.3\underline{8} \times 10^4}{-2.08\underline{8} \times 10^2} = 5.4\underline{4}8 \times 10^2 \, ^\circ\text{C}$ or 545 $^\circ$C

The number of significant figures in the final answer is limited by the two significant figures for the given temperatures.

7. (M) heat lost by Mg = heat gained by water

$$-\left(1.00 \text{ kg Mg} \times \frac{1000 \text{ g}}{1 \text{ kg}}\right)1.024\frac{J}{g\,°C}\left(T_f - 40.0\,°C\right) = \left(1.00 \text{ L} \times \frac{1000 \text{ cm}^3}{1 \text{ L}} \times \frac{1.00 \text{ g}}{1 \text{ cm}^3}\right)4.18\frac{J}{g\,°C}\left(T_f - 20.0\,°C\right)$$

$$-1.024 \times 10^3\, T_f + 4.10 \times 10^4 = 4.18 \times 10^3\, T_f - 8.36 \times 10^4$$

$$4.10 \times 10^4 + 8.36 \times 10^4 = \left(4.18 \times 10^3 + 1.024 \times 10^3\right)T_f \quad \rightarrow \quad 12.46 \times 10^4 = 5.20 \times 10^3\, T_f$$

$$T_f = \frac{12.46 \times 10^4}{5.20 \times 10^3} = 24.0\,°C$$

9. (M) heat lost by copper = heat gained by glycerol

$$-74.8 \text{ g} \times \frac{0.385 \text{ J}}{g\,°C} \times (31.1\,°C - 143.2\,°C) = 165 \text{ mL} \times \frac{1.26 \text{ g}}{1 \text{ mL}} \times c_p \times (31.1\,°C - 24.8\,°C)$$

$$3.23 \times 10^3 = 1.3 \times 10^3 \times (c_p) \qquad c_p = \frac{3.23 \times 10^3}{1.3 \times 10^3} = 2.5 \text{ J g}^{-1}\,°C^{-1}$$

molar heat capacity $= 2.5 \text{ J g}^{-1}\,°C^{-1} \times \dfrac{92.1 \text{ g}}{1 \text{ mol C}_3\text{H}_8\text{O}_3} = 2.3 \times 10^2 \text{ J mol}^{-1}\,°C^{-1}$

11. (M) heat capacity $= \dfrac{\text{energy transferred}}{\Delta T} = \dfrac{6.052 \text{ J}}{(25.0 - 20.0\,°C)} = 1.21 \text{ J/K}$

*Note: since 1 K = 1 °C, it is not necessary to convert the temperatures to Kelvin. The change in temperature in both K and °C is the same.

Heats of Reaction

13. (E) heat $= 283 \text{ kg} \times \dfrac{1000 \text{ g}}{1 \text{ kg}} \times \dfrac{1 \text{ mol Ca(OH)}_2}{74.09 \text{ g Ca(OH)}_2} \times \dfrac{65.2 \text{ kJ}}{1 \text{ mol Ca(OH)}_2} = 2.49 \times 10^5 \text{ kJ of heat evolved}$

15. (M)

(a) heat evolved $= 1.325 \text{ g C}_4\text{H}_{10} \times \dfrac{1 \text{ mol C}_4\text{H}_{10}}{58.123 \text{ g C}_4\text{H}_{10}} \times \dfrac{2877 \text{ kJ}}{1 \text{ mol C}_4\text{H}_{10}} = 3.65 \times 10^3 \text{ kJ}$

(b) heat evolved $= 28.4 \text{ L C}_4\text{H}_{10}\,(\text{at STP}) \times \dfrac{1 \text{ mol C}_4\text{H}_{10}}{22.711 \text{ L C}_4\text{H}_{10}\,(\text{at STP})} \times \dfrac{2877 \text{ kJ}}{1 \text{ mol C}_4\text{H}_{10}}$

$$= 3.65 \times 10^3 \text{ kJ}$$

(c) Use the ideal gas equation to determine the amount of propane in moles and multiply this amount by 2877 kJ heat produced per mole.

$$\text{heat evolved} = \frac{\left(738 \text{ mmHg} \times \dfrac{1 \text{ atm}}{760 \text{ mmHg}}\right) \times 12.6 \text{ L}}{\dfrac{0.08206 \text{ atm L}}{\text{mol K}} \times (23.6 + 273.2)} \times \frac{2877 \text{ kJ}}{1 \text{ mol C}_4\text{H}_{10}} = 1.45 \times 10^3 \text{ kJ}$$

17. **(M)**

(a) $\text{mass} = 2.80 \times 10^7 \text{ kJ} \times \dfrac{1 \text{ mol CH}_4}{890.3 \text{ kJ}} \times \dfrac{16.04 \text{ g CH}_4}{1 \text{ mol CH}_4} \times \dfrac{1 \text{ kg}}{1000 \text{ g}} = 504 \text{ kg CH}_4$

(b) First determine the moles of CH_4 present, with the ideal gas equation.

$$\text{mol CH}_4 = \dfrac{\left(768 \text{ mmHg} \times \dfrac{1 \text{ atm}}{760 \text{ mmHg}}\right) 1.65 \times 10^4 \text{ L}}{0.08206 \dfrac{\text{L atm}}{\text{mol K}} \times (18.6 + 273.2) \text{ K}} = 696 \text{ mol CH}_4$$

$$\text{heat energy} = 696 \text{ mol CH}_4 \times \dfrac{-890.3 \text{ kJ}}{1 \text{ mol CH}_4} = -6.20 \times 10^5 \text{ kJ of heat energy}$$

(c) $V_{\text{H}_2\text{O}} = \dfrac{6.21 \times 10^5 \text{ kJ} \times \dfrac{1000 \text{ J}}{1 \text{ kJ}}}{4.18 \dfrac{\text{J}}{\text{g} \, ^\circ\text{C}} (60.0 - 8.8) \, ^\circ\text{C}} \times \dfrac{1 \text{ mL H}_2\text{O}}{1 \text{ g}} = 2.90 \times 10^6 \text{ mL} = 2.90 \times 10^3 \text{ L H}_2\text{O}$

19. **(M)** Since the molar mass of H_2 (2.0 g/mol) is $\frac{1}{16}$ of the molar mass of O_2 (32.0 g/mol) and only twice as many moles of H_2 are needed as O_2, we see that $O_2(g)$ is the limiting reactant in this reaction.

$\dfrac{180.}{2} \text{ g O}_2 \times \dfrac{1 \text{ mol O}_2}{32.00 \text{ g O}_2} \times \dfrac{241.8 \text{ kJ heat}}{0.500 \text{ mol O}_2} = 1.36 \times 10^3 \text{ kJ heat}$

21. **(M)**

(a) We first compute the heat produced by this reaction, then determine the value of ΔH in kJ/mol KOH.

$$q_{\text{calorim}} = (0.205 + 55.9) \text{ g} \times 4.18 \dfrac{\text{J}}{\text{g} \, ^\circ\text{C}} (24.4 \, ^\circ\text{C} - 23.5 \, ^\circ\text{C}) = 2 \times 10^2 \text{ J heat} = -q_{\text{rxn}}$$

$$\Delta H = -\dfrac{2 \times 10^2 \text{ J} \times \dfrac{1 \text{ kJ}}{1000 \text{ J}}}{0.205 \text{ g} \times \dfrac{1 \text{ mol KOH}}{56.1 \text{ g KOH}}} = -5 \times 10^1 \text{ kJ/mol}$$

(b) The ΔT here is known to just one significant figure (0.9 °C). Doubling the amount of KOH should give a temperature change known to two significant figures (1.6 °C) and using twenty times the mass of KOH should give a temperature change known to three significant figures (16.0 °C). This would require 4.10 g KOH rather than the 0.205 g KOH actually used, and would increase the precision from one part in five to one part in 500, or ~0.2 %. Note that as the mass of KOH is increased and the mass of H_2O stays constant, the assumption of a constant specific heat capacity becomes less valid.

23. **(M)** Let x be the mass, (in grams), of NH_4Cl added to the water. heat $= mass \times c_p \times \Delta T$

$$x \times \frac{1 \text{ mol } NH_4Cl}{53.49 \text{ g } NH_4Cl} \times \frac{14.7 \text{ kJ}}{1 \text{ mol } NH_4Cl} \times \frac{1000 \text{ J}}{1 \text{ kJ}} = -\left(\left(1400 \text{ mL} \times \frac{1.00 \text{ g}}{1 \text{ mL}}\right) + x\right) 4.18 \frac{J}{g \, ^\circ C} (10. - 25) \, ^\circ C$$

$$275 \, x = 8.8 \times 10^4 + 63 \, x; \qquad x = \frac{8.8 \times 10^4}{275 - 63} = 4.2 \times 10^2 \text{ g } NH_4Cl$$

Our final value is approximate because of the assumed density (1.00 g/mL). The solution's density probably is a bit larger than 1.00 g/mL. Many aqueous solutions are somewhat more dense than water.

25. **(E)** We assume that the solution volumes are additive; that is, that 200.0 mL of solution is formed. Then we compute the heat needed to warm the solution and the cup, and finally ΔH for the reaction.

$$\text{heat} = \left(200.0 \text{ mL} \times \frac{1.02 \text{ g}}{1 \text{ mL}}\right) 4.02 \frac{J}{g \, ^\circ C} (27.8 \, ^\circ C - 21.1 \, ^\circ C) + 10 \frac{J}{^\circ C} (27.8 \, ^\circ C - 21.1 \, ^\circ C)$$

$$= 5.6 \times 10^3 \text{ J}$$

$$\Delta_{neutr.} H = \frac{-5.6 \times 10^3 \text{ J}}{0.100 \text{ mol}} \times \frac{1 \text{ kJ}}{1000 \text{ J}} = -56 \text{ kJ/mol} \left(-55.6 \text{ kJ/mol to three significant figures}\right)$$

27. **(M)**

$$5.0 \text{ L } C_2H_2 \times \frac{1 \text{ m}^3}{1000 \text{ L}} \times \frac{1.0967 \text{ kg}}{1 \text{ m}^3} \times \frac{1000 \text{ g}}{1 \text{ kg}} \times \frac{1 \text{ mol}}{26.04 \text{ g}} = 0.2106 \text{ mol } C_2H_2$$

$$\frac{1299.5 \text{ kJ heat evolved}}{1 \text{ mol } C_2H_2} \times 0.2106 \text{ mol } C_2H_2 = 272.9 \text{ kJ} = 2.7 \times 10^2 \text{ kJ heat evolved}$$

Enthalpy Changes and States of Matter

29. **(M)** $q_{water} = q_{ice}$ $\qquad mc_p(water)\Delta T_{water} = n_{ice}\Delta_{fus}H(ice)$

$$(3.50 \text{ mol } H_2O \times \frac{18.015 \text{ g } H_2O}{1 \text{ mol } H_2O})(4.184 \frac{J}{g \, ^\circ C})(50.0 \, ^\circ C) = (\frac{m}{\frac{18.015 \text{ g } H_2O}{1 \text{ mol } H_2O}} \times 6.01 \times 10^3 \frac{J}{mol})$$

$13.2 \times 10^3 \text{ J} = m(333.\underline{6} \text{ J g}^{-1})$ Hence, $m = 39.6$ g

31. **(M)** Assume $H_2O(l)$ density $= 1.00 \text{ g mL}^{-1}$ (at 28.5 °C) $\quad -q_{lost by ball} = q_{gained by water} + q_{vap water}$

$$-[(125 \text{ g})(0.50 \frac{J}{g \, ^\circ C})(100 \, ^\circ C - 525 \, ^\circ C)] = [(75.0 \text{ g})(4.184 \frac{J}{g \, ^\circ C})(100.0 \, ^\circ C - 28.5 \, ^\circ C)] + n_{H_2O}\Delta_{vap}H^\circ$$

$$26,\underline{5}62.5 \text{ J} = 22,4\underline{3}6.7 \text{ J} + n_{H_2O}\Delta_{vap}H^\circ \quad \text{(Note: } n_{H_2O} = \frac{mass_{H_2O}}{molar \ mass_{H_2O}})$$

$$4\underline{1}25.8 \text{ J} = (m_{H_2O})(\frac{1 \text{ mol } H_2O}{18.016 \text{ g } H_2O})(40.6 \times 10^3 \frac{J}{mol})$$

$m_{H_2O} = 1.\underline{8}3 \text{ g } H_2O \cong 2 \text{ g } H_2O$ (1 sig. fig.)

33. **(E)**

$$\frac{571 \text{ kJ}}{1 \text{ kg}} \times \frac{1000 \text{ J}}{1 \text{ kJ}} = 5.71 \times 10^5 \text{ J/kg}$$

$$125.0 \text{ J} \times \frac{1 \text{ kg}}{5.71 \times 10^5 \text{ J}} \times \frac{1000 \text{ g}}{1 \text{ kg}} \times \frac{1 \text{ L}}{1.98 \text{ g}} = 0.111 \text{ L}$$

Calorimetry

35. **(E)** Heat capacity $= \dfrac{\text{heat absorbed}}{\Delta T} = \dfrac{5228 \text{ cal}}{4.39 \,^\circ\text{C}} \times \dfrac{4.184 \text{ J}}{1 \text{ cal}} \times \dfrac{1 \text{ kJ}}{1000 \text{ J}} = 4.98 \text{ kJ/}^\circ\text{C}$

37. **(M)**

(a) $\dfrac{\text{heat}}{\text{mass}} = \dfrac{\text{heat cap.} \times \Delta t}{\text{mass}} = \dfrac{4.728 \text{ kJ/}^\circ\text{C} \times (27.19 - 23.29)^\circ\text{C}}{1.183 \text{ g}} = 15.6 \text{ kJ/g xylose}$

$\Delta H = \text{heat given off/g} \times M\,(\text{g/mol}) = \dfrac{-15.6 \text{ kJ}}{1 \text{ g C}_5\text{H}_{10}\text{O}_5} \times \dfrac{150.13 \text{ g C}_5\text{H}_{10}\text{O}_5}{1 \text{ mol}}$

$\Delta H = -2.34 \times 10^3 \text{ kJ/mol C}_5\text{H}_{10}\text{O}_5$

(b) $C_5H_{10}O_5(g) + 5\,O_2(g) \rightarrow 5\,CO_2(g) + 5\,H_2O(l) \qquad \Delta_r H = -2.34 \times 10^3 \text{ kJ mol}^{-1}$

39. **(M)**

(a) Because the temperature of the mixture decreases, the reaction molecules (the system) must have absorbed heat from the reaction mixture (the surroundings). Consequently, the reaction must be endothermic.

(b) We assume that the specific heat capacity of the solution is $4.18 \text{ J g}^{-1}\,^\circ\text{C}^{-1}$. The enthalpy of solution in kJ/mol KCl is obtained by the heat absorbed per gram KCl.

$$\Delta_{\text{soln}} H = -\frac{(0.75 + 35.0) \text{ g} \dfrac{4.18 \text{ J}}{\text{g }^\circ\text{C}} (23.6 - 24.8)\,^\circ\text{C}}{0.75 \text{ g KCl}} \times \frac{1 \text{ kJ}}{1000 \text{ J}} \times \frac{74.55 \text{ g KCl}}{1 \text{ mol KCl}} = +18 \text{ kJ/mol}$$

41. **(M)** To determine the heat capacity of the calorimeter, recognize that the heat evolved by the reaction is the negative of the heat of combustion.

$$\text{heat capacity} = \frac{\text{heat evolved}}{\Delta T} = \frac{1.620 \text{ g C}_{10}\text{H}_8 \times \dfrac{1 \text{ mol C}_{10}\text{H}_8}{128.2 \text{ g C}_{10}\text{H}_8} \times \dfrac{5156.1 \text{ kJ}}{1 \text{ mol C}_{10}\text{H}_8}}{8.44\,^\circ\text{C}} = 7.72 \text{ kJ/}^\circ\text{C}$$

43. **(M)** The temperature should increase as the result of an exothermic combustion reaction.

$$\Delta T = 1.227 \text{ g C}_{12}\text{H}_{22}\text{O}_{11} \times \frac{1 \text{ mol C}_{12}\text{H}_{22}\text{O}_{11}}{342.3 \text{ g C}_{12}\text{H}_{22}\text{O}_{11}} \times \frac{5.65 \times 10^3 \text{ kJ}}{1 \text{ mol C}_{12}\text{H}_{22}\text{O}_{11}} \times \frac{1\,^\circ\text{C}}{3.87 \text{ kJ}} = 5.23\,^\circ\text{C}$$

45. **(M)**

$$5.0 \text{ g NaCl} \times \frac{1 \text{ mol}}{58.44 \text{ g}} = 0.08556 \text{ mol NaCl}$$

$$0.08556 \text{ mol NaCl} \times \frac{3.76 \text{ kJ}}{1 \text{ mol}} = 0.322 \text{ kJ} = 322 \text{ J}$$

$$q = m_{H_2O} \times \text{specific heat capacity} \times \Delta T$$

$$322 \text{ J} = m_{H_2O} \times 4.18 \text{ J/g }°C \times 5.0 °C$$

$$m_{H_2O} = 15 \text{ g}$$

Pressure-Volume Work

47. **(M)**

(a) $-P\Delta V = 3.5 \text{ L} \times (748 \text{ mmHg}) \left(\dfrac{1 \text{ atm}}{760 \text{ mmHg}} \right) = -3.4\underline{4} \text{ atm L or } -3.4 \text{ atm L}$

(b) 1 kPa L = 1 J, hence,

$$-3.4\underline{4} \text{ atm L} \times \left(\frac{101.325 \text{ kPa}}{1 \text{ atm}} \right) \times \left(\frac{1 \text{ J}}{1 \text{ kPa L}} \right) = -3.4\underline{9} \times 10^2 \text{ J or } -3.5 \times 10^2 \text{ J}$$

(c) $-3.4\underline{9} \times 10^2 \text{ J} \times \left(\dfrac{1 \text{ cal}}{4.184 \text{ J}} \right) = -83.\underline{4} \text{ cal or } -83 \text{ cal}$

49. **(E)** When the Ne(g) sample expands into an evacuated vessel it does not push aside any matter, hence no work is done.

51. **(M)**

(a) No pressure–volume work is done (no gases are formed or consumed).

(b) $2 NO_2(g) \rightarrow N_2O_4(g)$ $\Delta n_{gas} = -1$ mol. Work is done on the system by the surroundings (compression).

(c) $CaCO_3(s) \rightarrow CaO(s) + CO_2(g)$. Formation of a gas, $\Delta n_{gas} = +1$ mol, results in an expansion. The system does work on the surroundings.

53. **(E)** We can either convert pressure from atm to Pascals, or convert work from joules to atm L. We opt for the latter. Since the conversion between J and atm L is 101.33 J/(atm L), the amount of work is 325 J × (1 atm L/101.33 J) = 3.207 atm L. Therefore,

$$w = -P\Delta V$$

3.207 atm L = (1.0 atm) ΔV. Solving for ΔV, we get a volume of 3.21 L.

First Law of Thermodynamics

55. **(E)**
  (a) $\Delta U = q + w = +58 \text{ J} + (-58 \text{ J}) = 0$
  (b) $\Delta U = q + w = +125 \text{ J} + (-687 \text{ J}) = -562 \text{ J}$
  (c) $280 \text{ cal} \times (4.184 \dfrac{\text{J}}{\text{cal}}) = 1171.52 \text{ J} = 1.17 \text{ kJ } \Delta U = q + w = -1.17 \text{ kJ} + 1.25 \text{ kJ} = 0.08 \text{ kJ}$

57. **(E)**
  (a) Yes, the gas does work ($w < 0$).
  (b) Yes, the gas exchanges energy with the surroundings, it absorbs energy.
  (c) The temperature of the gas stays the same if the process is isothermal.
  (d) ΔU for the gas must equal zero by definition (temperature is not changing).

59. **(E)** This situation is impossible. An ideal gas expanding isothermally means that $\Delta U = 0 = q + w$, or $w = -q$, not $w = -2q$.

61. **(E)** We note that since the charge of the system is going from 10 to 5, the net flow of the charge is negative. Therefore, w = (5 C – 10 C) = –5 C. Voltage (V) is J/C. The internal energy of the system is:
  $\Delta U = q + w = -45 \text{ J} + (100 \text{ J/C})(-5 \text{ C}) = -45 \text{ J} + (-500 \text{ J}) = -545 \text{ J}$

Relating ΔH and ΔU

63. **(E)** According to the First Law of Thermodynamics, the answer is (c). Both (a) q_v and (b) q_p are heats of chemical reaction carried out under conditions of constant volume and constant pressure, respectively. Both ΔU and ΔH incorporate terms related to work as well as heat.

65. **(M)** $C_3H_8O(l) + 9/2 \, O_2(g) \rightarrow 3 \, CO_2(g) + 4 \, H_2O(l)$ $\quad \Delta n_{gas} = -1.5 \text{ mol}$
  (a) $\Delta U = - 33.41 \dfrac{\text{kJ}}{\text{g}} \times \dfrac{60.096 \text{ g C}_3\text{H}_8\text{O}}{1 \text{ mol C}_3\text{H}_8\text{O}} = -2008 \dfrac{\text{kJ}}{\text{mol}}$
  (b) $\Delta H = \Delta U - w, = \Delta U - (-P\Delta V) = \Delta U - (-\Delta n_{gas}RT) = \Delta U + \Delta n_{gas}RT$
  $\Delta H = -2008 \dfrac{\text{kJ}}{\text{mol}} + (-1.5)(\dfrac{8.3145 \times 10^{-3} \text{ kJ}}{\text{mol K}})(298.15 \text{ K}) = -2012 \dfrac{\text{kJ}}{\text{mol}}$

Hess's Law

67. **(E)** The formation reaction for $NH_3(g)$ is $\frac{1}{2}N_2(g) + \frac{3}{2}H_2(g) \rightarrow NH_3(g)$. The given reaction is two-thirds the reverse of the formation reaction. The sign of the enthalpy is changed and it is multiplied by two-thirds. Thus, the enthalpy of the given reaction is $-(-46.11 \text{ kJ/mol}) \times \frac{2}{3} = +30.74 \text{ kJ/mol}$.

69. **(M)**

$$-(3) \quad 3\,CO_2(g) + 4\,H_2O(l) \rightarrow C_3H_8(g) + 5\,O_2(g) \qquad \Delta_r H° = +2219.1 \text{ kJ/mol}$$

$$+(2) \quad C_3H_4(g) + 4\,O_2(g) \rightarrow 3\,CO_2(g) + 2\,H_2O(l) \qquad \Delta_r H° = -1937 \text{ kJ/mol}$$

$$2(1) \quad 2\,H_2(g) + O_2(g) \rightarrow 2\,H_2O(l) \qquad \Delta_r H° = -571.6 \text{ kJ/mol}$$

$$\overline{\quad\quad\quad C_3H_4(g) + 2\,H_2(g) \rightarrow C_3H_8(g) \qquad\qquad\qquad \Delta_r H° = -290. \text{ kJ/mol}}$$

71. **(M)**

$$2\,HCl(g) + C_2H_4(g) + \tfrac{1}{2}O_2(g) \rightarrow C_2H_4Cl_2(l) + H_2O(l) \qquad \Delta_r H° = -318.7 \text{ kJ mol}^{-1}$$

$$Cl_2(g) + H_2O(l) \rightarrow 2\,HCl(g) + \tfrac{1}{2}O_2(g) \qquad \Delta_r H° = 0.5(+202.4) = +101.2 \text{ kJ mol}^{-1}$$

$$\overline{\quad C_2H_4(g) + Cl_2(g) \rightarrow C_2H_4Cl_2(l) \qquad\qquad\qquad \Delta_r H° = -217.5 \text{ kJ mol}^{-1}}$$

73. **(M)**

$$CO(g) + \tfrac{1}{2}O_2(g) \rightarrow CO_2(g) \qquad \Delta_r H° = -283.0 \text{ kJ mol}^{-1}$$

$$3\,C(\text{graphite}) + 6\,H_2(g) \rightarrow 3\,CH_4(g) \qquad \Delta_r H° = 3(-74.81) = -224.43 \text{ kJ mol}^{-1}$$

$$2\,H_2(g) + O_2(g) \rightarrow 2\,H_2O(l) \qquad \Delta_r H° = 2(-285.8) = -571.6 \text{ kJ mol}^{-1}$$

$$3\,CO(g) \rightarrow \tfrac{3}{2}O_2(g) + 3\,C(\text{graphite}) \qquad \Delta_r H° = 3(+110.5) = +331.5 \text{ kJ mol}^{-1}$$

$$\overline{4\,CO(g) + 8\,H_2(g) \rightarrow CO_2(g) + 3\,CH_4(g) + 2\,H_2O(l) \quad \Delta_r H° = -747.5 \text{ kJ mol}^{-1}}$$

75. **(M)**

$$CH_4(g) + CO_2(g) \rightarrow 2\,CO(g) + 2\,H_2(g) \qquad \Delta_r H° = +247 \text{ kJ mol}^{-1}$$

$$2\,CH_4(g) + 2\,H_2O(g) \rightarrow 2\,CO(g) + 6\,H_2(g) \qquad \Delta_r H° = 2(+206 \text{ kJ}) = +412 \text{ kJ mol}^{-1}$$

$$CH_4(g) + 2\,O_2(g) \rightarrow CO_2(g) + 2\,H_2O(g) \qquad \Delta_r H° = -802 \text{ kJ mol}^{-1}$$

$$\overline{4\,CH_4(g) + 2\,O_2(g) \rightarrow 4\,CO(g) + 8\,H_2(g) \qquad \Delta_r H° = -143 \text{ kJ mol}^{-1}}$$

$$\div 4 \text{ produces } CH_4(g) + \tfrac{1}{2}O_2(g) \rightarrow CO(g) + 2\,H_2(g) \quad \Delta_r H° = -35.8 \text{ kJ mol}^{-1}$$

77. **(M)**

$$C_6H_{12}O_6(s) + 6\,O_2(g) \rightarrow 6\,CO_2(g) + 6\,H_2O(l) \qquad \Delta_r H° = -2808 \text{ kJ mol}^{-1}$$

$$6\,CO_2(g) + 6\,H_2O(l) \rightarrow 2\,CH_3CH(OH)COOH(s) + 6\,O_2(g) \qquad \Delta_r H° = 2(1344) \text{ kJ mol}^{-1}$$

$$\overline{C_6H_{12}O_6(s) \rightarrow 2\,CH_3CH(OH)COOH(s) \qquad\qquad \Delta_r H° = -120. \text{ kJ mol}^{-1}}$$

Standard Enthalpies of Formation

<u>79.</u> **(M)**

(a) $\Delta_r H^\circ = \Delta_f H^\circ[C_2H_6(g)] + \Delta_f H^\circ[CH_4(g)] - \Delta_f H^\circ[C_3H_8(g)] - \Delta_f H^\circ[H_2(g)]$

$\Delta_r H^\circ = (-84.68 - 74.81 - (-103.8) - 0.00) \text{ kJ mol}^{-1} = -55.7 \text{ kJ mol}^{-1}$

(b) $\Delta_r H^\circ = 2\Delta_f H^\circ[SO_2(g)] + 2\Delta_f H^\circ[H_2O(l)] - 2\Delta_f H^\circ[H_2S(g)] - 3\Delta_f H^\circ[O_2(g)]$

$\Delta_r H^\circ = (2(-296.8) + 2(-285.8) - 2(-20.63) - 3(0.00)) \text{ kJ mol}^{-1} = -1124 \text{ kJ mol}^{-1}$

<u>81.</u> **(M)**

$ZnO(s) + SO_2(g) \rightarrow ZnS(s) + \tfrac{3}{2}O_2(g); \qquad \Delta_r H^\circ = -(-878.2 \text{ kJ})/2 = +439.1 \text{ kJ mol}^{-1}$

$439.1 \text{ kJ mol}^{-1} = \Delta_f H^\circ[ZnS(s)] + \tfrac{3}{2}\Delta_f H^\circ[O_2(g)] - \Delta_f H^\circ[ZnO(s)] - \Delta_f H^\circ[SO_2(g)]$

$439.1 \text{ kJ mol}^{-1} = \Delta_f H^\circ[ZnS(s)] + \tfrac{3}{2}(0.00 \text{ kJ mol}^{-1}) - (-348.3 \text{ kJ mol}^{-1}) - (-296.8 \text{ kJ mol}^{-1})$

$\Delta_f H^\circ[ZnS(s)] = (439.1 - 348.3 - 296.8) \text{ kJ mol}^{-1} = -206.0 \text{ kJ/mol}$

<u>83.</u> **(E)** $\Delta_r H^\circ = 4\Delta_f H^\circ[HCl(g)] + \Delta_f H^\circ[O_2(g)] - 2\Delta_f H^\circ[Cl_2(g)] - 2\Delta_f H^\circ[H_2O(l)]$

$= 4(-92.31) + (0.00) - 2(0.00) - 2(-285.8) = +202.4 \text{ kJ mol}^{-1}$

<u>85.</u> **(E)** Balanced equation: $C_2H_5OH(l) + 3 O_2(g) \rightarrow 2 CO_2(g) + 3 H_2O(l)$

$\Delta_r H^\circ = 2\Delta_f H^\circ[CO_2(g)] + 3\Delta_f H^\circ[H_2O(l)] - \Delta_f H^\circ[C_2H_5OH(l)] - 3\Delta_f H^\circ[O_2(g)]$

$= 2(-393.5) + 3(-285.8) - (-277.7) - 3(0.00) = -1366.7 \text{ kJ mol}^{-1}$

<u>87.</u> **(E)**

$\Delta_r H^\circ = -397.3 \text{ kJ mol}^{-1} = \Delta_f H^\circ[CCl_4(g)] + 4\Delta_f H^\circ[HCl(g)] - \Delta_f H^\circ[CH_4(g)] - 4\Delta_f H^\circ[Cl_2(g)]$

$= \Delta_f H^\circ[CCl_4(g)] + (4(-92.31) - (-74.81) - 4(0.00)) \text{ kJ mol}^{-1} = \Delta_f H^\circ[CCl_4(g)] - 294.4 \text{ kJ mol}^{-1}$

$\Delta_f H^\circ[CCl_4(g)] = (-397.3 + 294.4) \text{ kJ mol}^{-1} = -102.9 \text{ kJ/mol}^{-1}$

<u>89.</u> **(E)** $\Delta_r H^\circ = \Delta_f H^\circ[Al(OH)_3(s)] - \Delta_f H^\circ[Al^{3+}(aq)] - 3\Delta_f H^\circ[OH^-(aq)]$

$= ((-1276) - (-531) - 3(-230.0)) \text{ kJ mol}^{-1} = -55 \text{ kJ mol}^{-1}$

<u>91.</u> **(M)** Balanced equation: $CaCO_3(s) \rightarrow CaO(s) + CO_2(g)$

$\Delta_r H^\circ = \Delta_f H^\circ[CaO(s)] + \Delta_f H^\circ[CO_2(g)] - \Delta_f H^\circ[CaCO_3(s)]$

$= (-635.1 - 393.5 - (-1207)) \text{ kJ mol}^{-1} = +178 \text{ kJ mol}^{-1}$

$\text{heat} = 1.35 \times 10^3 \text{ kg CaCO}_3 \times \dfrac{1000 \text{ g}}{1 \text{ kg}} \times \dfrac{1 \text{ mol CaCO}_3}{100.09 \text{ g CaCO}_3} \times \dfrac{178 \text{ kJ}}{1 \text{ mol CaCO}_3} = 2.40 \times 10^6 \text{ kJ}$

93. **(M)** The reaction for the combustion of formic acid is:

$HCOOH(s) + ½ O_2(g) \rightarrow CO_2(g) + H_2O(l)$

$\Delta_r H° = \Delta_f H°\left[CO_2(g)\right] + \Delta_f H°\left[H_2O(l)\right] - \Delta_f H°\left[HCOOH(s)\right] - \tfrac{1}{2}\Delta_f H°\left[O_2(g)\right]$

$-255 \text{ kJ mol}^{-1} = \left(1(-393.5) + 1(-285.8) - \Delta_f H°\left[HCOOH(s)\right] - 0.5(0.00)\right) \text{kJ mol}^{-1}$

$-424 \text{ kJ mol}^{-1} = \Delta_f H°\left[HCOOH(s)\right]$

INTEGRATIVE AND ADVANCED EXERCISES

97. **(M)** Potential energy $= mgh = 7.26 \text{ kg} \times 9.81 \text{ ms}^{-2} \times 168 \text{ m} = 1.20 \times 10^4 \text{ J}$. This potential energy is converted entirely into kinetic energy just before the object hits, and this kinetic energy is converted entirely into heat when the object strikes.

$$\Delta t = \frac{\text{heat}}{\text{mass} \times c_p} = \frac{1.20 \times 10^4 \text{ J}}{7.26 \text{ kg} \times \dfrac{1000 \text{ g}}{1 \text{ kg}} \times \dfrac{0.47 \text{ J}}{\text{g °C}}} = 3.5 \text{ °C}$$

This large a temperature rise is unlikely, as some of the kinetic energy will be converted into forms other than heat, such as sound and the fracturing of the object along with the surface it strikes. In addition, some heat energy would be transferred to the surface.

98. **(M)** $\text{heat} = \Delta t \left[\text{heat cap.} + (\text{mass } H_2O \times 4.184 \dfrac{\text{J}}{\text{g °C}})\right]$

$\text{heat cap.} = \dfrac{\text{heat}}{\Delta t} - (\text{mass } H_2O \times 4.184 \dfrac{\text{J}}{\text{g °C}})$

The heat of combustion of anthracene is -7067 kJ/mol, meaning that burning one mole of anthracene releases $+7067$ kJ of heat to the calorimeter.

$$\text{heat cap.} = \frac{1.354 \text{ g } C_{14}H_{10} \times \dfrac{1 \text{ mol } C_{14}H_{10}}{178.23 \text{ g } C_{14}H_{10}} \times \dfrac{7067 \text{ kJ}}{1 \text{ mol } C_{14}H_{10}}}{(35.63 - 24.87) \text{ °C}} - \left(983.5 \text{ g} \times 4.184 \times 10^{-3} \frac{\text{kJ}}{\text{g °C}}\right)$$

$= (4.990 - 4.115) \text{ kJ/°C} = 0.875 \text{ kJ/°C}$

$\text{heat} = (27.19 - 25.01) \text{ °C} [0.875 \text{ kJ/°C} + (968.6 \text{ g } H_2O \times 4.184 \times 10^{-3} \text{ kJ g}^{-1} \text{ °C}^{-1})] = 10.7 \text{ kJ}$

$q_{rxn} = \dfrac{-10.7 \text{ kJ}}{1.053 \text{ g } C_6H_8O_7} \times \dfrac{192.1 \text{ g } C_6H_8O_7}{1 \text{ mol } C_6H_8O_7} = -1.95 \times 10^3 \text{ kJ/mol } C_6H_8O_7$

104. **(D)** We first compute the heats of combustion of the combustible gases.

$CH_4(g) + 2O_2(g) \longrightarrow CO_2(g) + 2H_2O(l)$

$\Delta_r H° = \Delta_f H°[CO_2(g)] + 2\Delta_f H°[H_2O(l)] - \Delta_f H°[CH_4(g)] - 2\Delta_f H°[O_2(g)]$

$\qquad = -393.5 \text{ kJ mol}^{-1} + 2 \times (-285.8 \text{ kJ mol}^{-1}) - (-74.81 \text{ kJ mol}^{-1}) - 2 \times 0.00 \text{ kJ mol}^{-1}$

$\qquad = -890.3 \text{ kJ mol}^{-1}$

$$C_3H_8(g) + 5O_2(g) \longrightarrow 3CO_2(g) + 4H_2O(l)$$

$$\Delta_r H° = 3\Delta_f H°[CO_2(g)] + 4\Delta_f H°[H_2O(l)] - \Delta_f H°[C_3H_8(g)] - 5\Delta_f H°[O_2(g)]$$

$$= 3 \times (-393.5\,kJ\,mol^{-1}) + 4 \times (-285.8\,kJ\,mol^{-1}) - (-103.8\,kJ\,mol^{-1}) - 5 \times 0.00\,kJ\,mol^{-1}$$

$$= -2220.\,kJ\,mol^{-1}$$

$$H_2(g) + \tfrac{1}{2}O_2(g) \longrightarrow H_2O(l) \quad \Delta_r H° = \Delta_f H°[H_2O(l)] = -285.8\,kJ\,mol^{-1}$$

$$CO(g) + \tfrac{1}{2}O_2(g) \longrightarrow CO_2(g)$$

$$\Delta_r H° = \Delta_f H°[CO_2(g)] - \Delta H°_f[CO(g)] - 0.5\Delta_f H°[O_2(g)] = -393.5 + 110.5 = -283.0\,kJ\,mol^{-1}$$

Then, for each gaseous mixture, we compute the enthalpy of combustion per mole of gas. The enthalpy of combustion per liter at STP (i.e., at 1 bar and 0 °C) is 1/22.711 of this value. Recall that volume percents are equal to mole percents.

(a) H_2 combustion $= 0.497\,mol\,H_2 \times \dfrac{-285.8\,kJ}{1\,mol\,H_2} = -142\,kJ$

CH_4 combustion $= 0.299\,mol\,CH_4 \times \dfrac{-890.3\,kJ}{1\,mol\,CH_4} = -266\,kJ$

CO combustion $= 0.069\,mol\,CO \times \dfrac{-283.0\,kJ}{1\,mol\,CO} = -20\,kJ$

C_3H_8 combustion $= 0.031\,mol\,C_3H_8 \times \dfrac{-2220.\,kJ}{1\,mol\,C_3H_8} = -69\,kJ$

total enthalpy of combustion $= \dfrac{(-142 - 266 - 20 - 69)\,kJ}{1\,mol\,gas} \times \dfrac{1\,mol\,gas}{22.711\,L} = -21.9\,kJ/L$

(b) CH_4 is the only combustible gas present in sewage gas.

total enthalpy of combustion $= 0.660\,mol\,CH_4 \times \dfrac{-890.3\,kJ}{1\,mol\,CH_4} \times \dfrac{1\,mol\,gas}{22.711\,L} = -39.2\,kJ/L$

Thus, sewage gas produces more heat per liter at STP than does coal gas.

108. (M) work $= 4 \times mgh = 4 \times 58.0\,kg \times 9.807\,m\,s^{-2} \times 1450\,m \times \dfrac{1\,kJ}{1000\,J} = 3.30 \times 10^3\,kJ$

We compute the enthalpy change for the metabolism (combustion) of glucose.

$$C_6H_{12}O_6(s) + 6O_2(g) \longrightarrow 6CO_2(g) + 6H_2O(l)$$

$$\Delta_r H° = 6\Delta_f H°[CO_2(g)] + 6\Delta_f H°[H_2O(l)] - \Delta_f H°[C_6H_{12}O_6(s)] - 6\Delta_f H°[O_2(g)]$$

$$= 6 \times (-393.5\,kJ\,mol^{-1}) - 6 \times (-285.8\,kJ\,mol^{-1}) - (-1273.3\,kJ\,mol^{-1}) - 6 \times 0.00\,kJ\,mol^{-1}$$

$$= -2802.5\,kJ\,mol^{-1}$$

Then we compute the mass of glucose needed to perform the necessary work.

$$\text{mass } C_6H_{12}O_6 = 3.30\times10^3 \text{ kJ} \times \frac{1\,\text{kJ heat}}{0.70\,\text{kJ work}} \times \frac{1\,\text{mol}\,C_6H_{12}O_6}{2802.5\,\text{kJ}} \times \frac{180.2\,\text{g}\,C_6H_{12}O_6}{1\,\text{mol}\,C_6H_{12}O_6}$$

$$= 303\,\text{g } C_6H_{12}O_6$$

110. **(M)** First determine the molar heats of combustion for CH_4 and C_2H_6.

$$CH_4(g) + 2O_2(g) \longrightarrow CO_2(g) + 2H_2O(l)$$
$$\Delta_r H° = \Delta_f H°[CO_2(g)] + 2\Delta_f H°[H_2O(l)] - \Delta_f H°[CH_4(g)] - 2\Delta_f H°[O_2(g)]$$
$$= \left((-393.5) + 2(-285.8) - (-74.81) - 2(0.00)\right)\text{kJ mol}^{-1} = -890.3\,\text{kJ/mol}$$

$$C_2H_6(g) + \tfrac{7}{2}O_2(g) \longrightarrow 2CO_2(g) + 3H_2O(l)$$
$$\Delta_r H° = 2\Delta_f H°[CO_2(g)] + 3\Delta_f H°[H_2O(l)] - \Delta_f H°[C_2H_6(g)] - \tfrac{7}{2}\Delta_f H°[O_2(g)]$$
$$= \left(2(-393.5) + 3(-285.8) - (-84.68) - \tfrac{7}{2}(0.00)\right)\text{kJ mol}^{-1} = -1559.7\,\text{kJ mol}^{-1}$$

Since the STP molar volume of an ideal gas is 22.7 L, there is 1/22.7 of a mole of gas present in the sample. We first compute the heat produced by one mole (that is 22.7 L at STP) of the mixed gas.

$$\text{heat} = \frac{43.6\,\text{kJ}}{1.00\,\text{L}} \times \frac{22.7\,\text{L}}{1\,\text{mol}} = 990\,\text{kJ/mol}$$

Then, if we let the number of moles of CH_4 be represented by x, the number of moles of C_2H_6 is represented by $(1.00-x)$. Now we can construct an equation for the heat evolved per mole of mixture and solve this equation for x.

$$990\,\text{kJ} = 890.3\,x + 1559.7(1.00-x) = 1559.7 + (890.3 - 1559.7)x = 1559.7 - 669.4\,x$$

$$x = \frac{1559.7 - 990}{669.4} = 0.851\,\text{mol}\,CH_4/\text{mol mixture}$$

By the ideal gas law, gases at the same temperature and pressure have the same volume ratio as their molar ratio. Hence, this gas mixture contains 85.1% CH_4 and 14.9% C_2H_6, both by volume.

113. **(M)**

$$n = \frac{(765.5/760)\text{atm}}{(0.08206\,\text{atm L mol}^{-1}\,\text{K}^{-1})} \frac{0.582\,\text{L}}{298.15\,\text{K}} = 2.40\times10^{-2}\,\text{mol}$$

$$\text{molar mass} = \frac{1.103\,\text{g}}{2.40\times10^{-2}\,\text{mol}} = 46.0\,\text{g/mol}$$

moles of $CO_2 = 2.108/44.01 = 0.04790(0.0479\,\text{mol C in unknown})$

moles of $H_2O = 1.294/18.02 = 0.0719(0.144\,\text{mol H in unknown})$

$$\text{moles of O in unknown} = \frac{1.103\,\text{g} - 0.04790\,\text{g}\,(12.011\text{g C/mol}) - 0.144(1.00794)}{15.9994\,\text{g mol}^{-1}} = 0.0239\,\text{mol O}$$

So C:H:O ratio is 2:6:1 and the molecular formula is C_2H_6O.

$\Delta_r T = 31.94 - 25.00 = 6.94\ °C \quad q = 6.94\ °C \times 5.015\ kJ/°C = -34.8\ kJ$

$\Delta_r H = \dfrac{-34.8\ kJ}{0.0240\ mol} = -1.45 \times 10^3\ kJ/mol \quad 3\ O_2(g) + C_2H_6O(g) \rightarrow 2\ CO_2(g) + 3\ H_2O(l)$

114. **(M)** Energy needed $= mc\Delta T = (250\ g)((50-4)\ °C)(4.2\ J/g\ °C) = 4.8 \times 10^4\ J$
A 700-watt oven delivers a joule of energy/sec.

$\text{time} = \left(\dfrac{4.8 \times 10^4\ J}{700\ J\ sec^{-1}}\right) = 69\ \text{seconds}$

115. **(M)** $w = -P\Delta V,\ V_i = (3.1416)(6.00\ cm)^2(8.10\ cm) = 916.1\ cm^3$

$P = \dfrac{\text{force}}{\text{area}} = \dfrac{(3.1416)(5.00)^2(25.00)(7.75\ g/cm^3)(1\ kg/1000\ g)(9.807\ m/sec^2)}{(3.1416)(6.00\ cm)^2(1\ m/100cm)^2}$

$= 1.320 \times 10^4\ Pa = \dfrac{1.320 \times 10^4\ Pa}{101325\ Pa\ atm^{-1}} = 0.130\ \text{atm difference} \qquad \text{Use: } P_1V_1 = P_2V_2$

$V_2 = \dfrac{((745/760) + 0.130\ atm) \times (0.9161\ L)}{(745/760)\,atm} = 1.04\ L$

$-P\Delta V = \dfrac{745}{760} \times (1.04 - 0.916) = -0.121\ atm\ L = w = (-0.121\ atm\ L)\dfrac{101\ J}{atm\ L} = -12\ J$

122. **(M)** First, write the balanced equation for reaction of 1 mole of each substance and determine the enthalpy of reaction (or combustion), $\Delta_r H°$, for each reaction. Then use the calculated $\Delta_r H°$ and the reaction stoichiometry to determine moles of CO_2 generated per kJ of energy. Recall that

$$\Delta_r H° = \sum_{\text{products}} \text{coefficient} \times \Delta_f H° - \sum_{\text{reactants}} \text{coefficient} \times \Delta_f H°$$

To determine which fuel produces the least amount of CO_2 per equivalent energy output, divide the moles of CO_2 generated per combustion of 1 mole of each fuel by the energy produced by the combustion of 1 mole of that fuel.

Methanol

$CH_3OH(l) + 3/2\ O_2(g) \rightarrow CO_2(g) + 2\ H_2O(l)$

$\Delta_r H° = \left(-393.5\ kJ\ mol^{-1} + 2 \times -285.8\ kJ\ mol^{-1}\right) - \left(-238.7\ kJ\ mol^{-1}\right) = -726.4\ kJ\ mol^{-1}$

Since the combustion of 1 mol CH_3OH produces 1 mol CO_2 and 726.4 kJ of energy, we have 1 mol CO_2/726.4 kJ $= 1.377 \times 10^{-3}$ mol CO_2/kJ.

Cetane

$$C_{16}H_{34}(l) + 49/2\ O_2(l) \rightarrow 16\ CO_2(g) + 17\ H_2O(l)$$

$$\Delta_r H^\circ = \left[\left(16\times -393.5\ kJ\ mol^{-1}\right)+\left(17\times -285.8\ kJ\ mol^{-1}\right)\right]-\left(-456.3\ kJ\ mol^{-1}\right)=-10{,}698\ kJ\ mol^{-1}$$

The combustion of 1 mol $C_{16}H_{34}$ produces 16 mol CO_2, 10,698 kJ of energy, and, therefore, 16 mol CO_2/10,698 kJ $= 1.4956\times 10^{-3}$ mol CO_2/kJ.

Methyl linoleate

$$C_{19}H_{34}O_2(l) + 53/2\ O_2(g) \rightarrow 19\ CO_2(g) + 17\ H_2O(l)$$

$$\Delta_r H^\circ = \left(19\times -393.5\ kJ\ mol^{-1}+17\times -285.8\ kJ\ mol^{-1}\right)-\left(-604.9\ kJ\ mol^{-1}\right)=-11{,}73\underline{0}\ kJ\ mol^{-1}$$

The combustion of 1 mol $C_{19}H_{34}O_2$ produces 19 mol CO_2, 11,730 kJ of energy, and, therefore, 19 mol CO_2/11,730 kJ $= 1.6198\times 10^{-3}$ mol CO_2/kJ.

Octane

$$C_8H_{18}(l) + 25/2\ O_2(g) \rightarrow 8\ CO_2(g) + 9\ H_2O(l)$$

$$\Delta_r H^\circ = \left(8\times -393.5\ kJ\ mol^{-1}+9\times -285.8\ kJ\ mol^{-1}\right)-\left(-250.1\ kJ\ mol^{-1}\right)=-547\underline{0}\ kJ\ mol^{-1}$$

The combustion of 1 mol C_8H_{18} produces 8 mol CO_2, 5470 kJ of energy, and, therefore, 8 mol CO_2/5470 kJ $= 1.463\times 10^{-3}$ mol CO_2/kJ.

Therefore, methanol produces the least amount of CO_2 per unit of energy generated.

FEATURE PROBLEMS

125. **(D)** The plot's maximum is the equivalence point. (Assume $\Delta T = 0$ at 0 mL of added NaOH, (i.e., only 60 mL of citric acid are present), and that $\Delta T = 0$ at 60 mL of NaOH (i.e. no citric acid added).

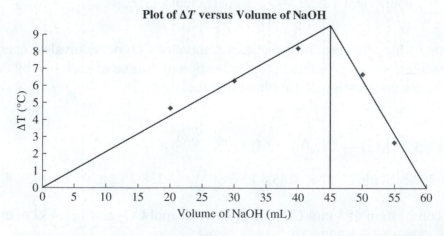

Plot of ΔT versus Volume of NaOH

(a) The equivalence point occurs with 45.0 mL of 1.00 M NaOH(aq) [45.0 mmol NaOH] added and 15.0 mL of 1.00 M citric acid [15.0 mmol citric acid]. Again, we assume that ΔT = zero if no NaOH added (V_{NaOH} = 0 mL) and $\Delta T = 0$ if no citric acid is added (V_{NaOH} = 60 mL).

(b) Heat is a product of the reaction, as are chemical species (products). Products are maximized at the exact stoichiometric proportions. Since each reaction mixture has the same volume, and thus about the same mass to heat, the temperature also is a maximum at this point.

(c) $H_3C_6H_5O_7(s) + 3OH^-(aq) \rightarrow 3H_2O(l) + C_6H_5O_7^{3-}(aq)$

128. (D)

(a) Here we must determine the volume between 2.40 atm and 1.30 atm using $PV = nRT$

$$V = \frac{0.100 \text{ mol} \times 0.08206 \text{ L mol}^{-1} \text{ K}^{-1} \times 298 \text{ K}}{P} = \frac{2.44\underline{5} \text{ atm L}}{P}$$

For $P = 2.40$ atm: $V = 1.02$ L
For $P = 2.30$ atm: $V = 1.06$ L $P\Delta V = 2.30$ atm $\times -0.04$ L $= -0.09\underline{2}$ atm L
For $P = 2.20$ atm: $V = 1.11$ L $P\Delta V = 2.20$ atm $\times -0.05$ L $= -0.1\underline{1}$ atm L
For $P = 2.10$ atm: $V = 1.16$ L $P\Delta V = 2.10$ atm $\times -0.05$ L $= -0.1\underline{1}$ atm L
For $P = 2.00$ atm: $V = 1.22$ L $P\Delta V = 2.00$ atm $\times -0.06$ L $= -0.1\underline{2}$ atm L
For $P = 1.90$ atm: $V = 1.29$ L $P\Delta V = 1.90$ atm $\times -0.06$ L $= -0.1\underline{2}$ atm L
For $P = 1.80$ atm: $V = 1.36$ L $P\Delta V = 1.80$ atm $\times -0.07$ L $= -0.1\underline{3}$ atm L
For $P = 1.70$ atm: $V = 1.44$ L $P\Delta V = 1.70$ atm $\times -0.08$ L $= -0.1\underline{4}$ atm L
For $P = 1.60$ atm: $V = 1.53$ L $P\Delta V = 1.60$ atm $\times -0.09$ L $= -0.1\underline{4}$ atm L
For $P = 1.50$ atm: $V = 1.63$ L $P\Delta V = 1.50$ atm $\times -0.10$ L $= -0.1\underline{5}$ atm L
For $P = 1.40$ atm: $V = 1.75$ L $P\Delta V = 1.40$ atm $\times -0.12$ L $= -0.1\underline{7}$ atm L
For $P = 1.30$ atm: $V = 1.88$ L $P\Delta V = 1.30$ atm $\times -0.13$ L $= -0.1\underline{7}$ atm L

total work $= -\Sigma P\Delta V = -1.4\underline{5}$ atm L

Expressed in joules, the work is $-1.4\underline{5}$ atm L $\times 101.325$ J/atm L $= -14\underline{7}$ J.

(b)

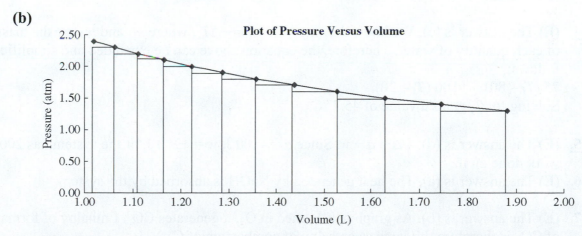

(c) The total work done in the two-step expansion is minus one times the total of the area of the two rectangles under the graph, which turns out to be -1.29 atm L or -131 J. In the 11-step expansion in (b), the total area of the rectangles is 1.45 atm L or $-14\underline{7}$ J. If the expansion were divided into a larger number of stages, the total area of the rectangles would be even greater. The maximum amount of work is for an expansion with an infinite number of stages and is equal to the area under the pressure–volume curve between $V = 1.02$ L and 1.88 L. This area is also obtained as the integral of the expression:

$dw = -PdV = -nRT(dV/V)$. The value obtained is:
$w = -nRT \times \ln V_f/V_i = 0.100$ mol $\times 8.3145$ J mol^{-1} K$^{-1} \times 298$ K $\times \ln (1.88$ L$/1.02$ L$)$
$w = -152$ J

(d) The maximum work of compression is for a one-stage compression using an external pressure of 2.40 atm and producing a compression in volume of 1.02 L $- 1.88$ L $= -0.86$ L: $w = -P\Delta V = (2.40$ atm $\times 0.86$ L$) \times 101.33$ J/atm L $= 209$ J

The minimum work would be that done in an infinite number of steps and would be the same as the work determined in (c) but with a positive sign, namely, $+152$ J.

(e) Because the internal energy of an ideal gas is a function only of temperature, and the temperature remains constant, $\Delta U = 0$. Because $\Delta U = q + w = 0$, $q = -w$. This means that -209 J corresponds to the maximum work of compression, and -152 J corresponds to the minimum work of compression.

(f) For the expansion described in part in (c),
$q = -w = nRT \ln V_f/V_i$ and $q/T = nR \ln V_f/V_i$
Because the terms on the right side are all constants or functions of state, so too is the term on the left, q/T. In Chapter 19, we learn that q/T is equal to ΔS, the change in a state function called *entropy*.

SELF-ASSESSMENT EXERCISES

133. **(E)** The answer is (b), Al, because it has the highest heat capacity.

134. **(E)** The answer is (c). We know that $m_1 c\Delta T_1 = -m_2 c\Delta T_2$, where m_1 and m_2 are the masses of each quantity of water. Therefore, the equation above can be expanded and simplified as follows:
$75 (T_f - 80) = -100 (T_f - 20)$
Solving for T_f gives a value of 45.7 °C.

135. **(E)** The answer is (d). $U\Delta = q + w$. Since $q = -100$ J, $w = +200$ J, or the system has 200 J of work done on it.

136. **(E)** The answer is (a). The heat generated by NaOH is absorbed by the system.

137. **(E)** The answer is (b). As graphite is burned in O_2, it generates CO_2. Enthalpy of formation of CO_2 is therefore the same as enthalpy of combustion of C.

138. **(E)**
(a) $Sn(s) + Cl_2(g) \rightarrow SnCl_2(s)$
(b) $7 C(graphite) + 3 H_2(g) + O_2(g) \rightarrow C_6H_5COOH(s)$
(c) $C(graphite) + \frac{1}{2} O_2(g) + Cl_2(g) \rightarrow COCl_2(g)$

139. (E)

(a)

$$\Delta_r H° = \Delta_f H°[SiF_4(g)] + 2 \times \Delta_f H°[H_2O(g)] - (\Delta_f H°[SiO_2(s)] + 4 \times \Delta_f H°[HF(g)])$$
$$= -1615.0 + 2(-241.8) - [-910.9 + 4(-271.1)]$$
$$= -103.3 \text{ kJ mol}^{-1}$$

(b)

$$\Delta_r H° = 2 \times \Delta_f H°[CuO(s)] + 2 \times \Delta_f H°[SO_2(g)] - (2 \times \Delta_f H°[CuS(s)] + 3 \times \Delta_f H°[O_2(g)])$$
$$= 2(-157.3) + 2(-296.8) - [2(-53.1) + 3(0.0)]$$
$$= -802.0 \text{ kJ mol}^{-1}$$

140. (M) First, determine the number of moles of LiCl:

$$n = 5.00 \text{ g} \times \frac{1 \text{ mol}}{42.391 \text{ g}} = 0.1179 \text{ mol}$$

Then, determine the amount of heat generated:

$$q = \frac{37.2 \text{ kJ}}{1 \text{ mol}} \times 0.1179 \text{ mol} = 4.3859 \text{ kJ}$$

Now use $q = C\Delta T$:

$$4.3859 \text{ kJ} = 115.0 \text{ g} \times (4.00 \times 10^{-3} \text{ kJ g}^{-1} °C^{-1}) \times (T_f - 20.00 °C)$$

Solving for the final temperature, T_f:

$$T_f - 20.00 °C = \frac{4.3859 \text{ kJ}}{115.0 \text{ g} \times (4.00 \times 10^{-3} \text{ kJ g}^{-1} °C^{-1})}$$

$$T_f = 9.53 °C + 20.00 °C$$
$$= 29.53 °C$$

141. (E) The correct answer is (b).

142. (E) The correct answer is (a).

143. (E) The answer is (a), because q_V and q_P are not the same.

144. (E)

(a) We have to solve for the heat capacity (c_p) of Fe:

$$m \cdot c_p \cdot \Delta T (H_2O) = - m \cdot c_p \cdot \Delta T (Fe)$$
$$(981 \text{ g})(4.189)(12.3 °C) = -(1220)(c_{P(Fe)})(-92.1 °C)$$

Solving for c_{pFe}, the heat capacity of Fe is 0.449 J g^{-1} °C^{-1}

(b) Now knowing the c_p of Fe, we can calculate the T_f of the iron–glycerol system:

$$(409.5)(2.378)(T_f - 26.2) = -(1220)(0.4489)(T_f - 99.8)$$

Solving $T_f = 52.7\ °C$

145. **(E)**
(a) $2\ N_2 + O_2 \rightarrow 2\ N_2O$
(b) $S + O_2 + Cl_2 \rightarrow SO_2Cl_2$
(c) $2\ CH_3CH_2COOH + 7\ O_2 \rightarrow 6\ CO_2 + 6\ H_2O$

146. **(D)** First, determine the $\Delta_f H°$ of CO, which is used to make $COCl_2$. This is done by using the equations for the combustion of C and CO gases:

(1) $CO + ½\ O_2 \rightarrow CO_2$	-283.0 kJ/mol
(2) $C + O_2 \rightarrow CO_2$	-393.5 kJ/mol

To determine the $\Delta_f H°$ of CO, reverse and double equation (1) and double equation (2):

(1) $2\ CO_2 \rightarrow 2\ CO + O_2$	566.0 kJ mol^{-1}
(2) $2\ C + 2\ O_2 \rightarrow 2\ CO_2$	-787.0 kJ mol^{-1}
(3) $2\ C + O_2 \rightarrow 2\ CO$	-221.0 kJ mol^{-1}

or -110.5 kJ/mol if only 1 mole of CO is formed

Now, we use the equation for the formation of $COCl_2$ with equation (3) times ½.

(4) $CO + Cl_2 \rightarrow COCl_2$	-108.0 kJ mol^{-1}
(3) $C + ½\ O_2 \rightarrow CO$	-110.5 kJ mol^{-1}
(3) $C + ½\ O_2 + Cl_2 \rightarrow COCl_2$	-218.5 kJ mol^{-1}

Therefore, $\Delta_f H°$ of $COCl_2$ is -218 kJ/mol.

147. **(E)** Enthalpy of formation for elements (even molecular ones, such as O_2 or Cl_2) is by convention set to 0. While it is possible for the enthalpy of formation of a compound to be near zero, it is unlikely.

148. **(M)** We note that $\Delta H = \Delta U + \Delta(PV)$. From a theoretical standpoint, one can have a situation where the $\Delta U < 0$, but there is enough work done on the system that makes $\Delta H > 0$. In reality, because the $\Delta(PV)$ is relatively small, ΔH and ΔU often have the same sign.

149. **(M)** A gas stove works by combustion of a flammable fuel. The amount of heat can be controlled by a valve. Once shut off, the heat source instantly disappears. However, an electric stove works by the principle of heat conduction, where the heat coil on the stove transfers heat to the pot through direct contact. Even after the electricity is shut off to the heating coil, it takes time for the coil to cool because of its heat capacity, and therefore it continues to supply heat to the pot.

150. **(E)** The answer is (a), 0. This is because there is no loss of energy to or gain of energy from the surroundings.

151. **(M)** The answer is (b), the temperature decreases (or at least it increases at a slower rate than it would if there was no moisture on the outside). The moisture put outside of the pot evaporates mainly because of removing heat from the pot. Therefore, the moisture on the outside of the pot removes heat from the pot as it evaporates, therefore slightly cooling the pot.

CHAPTER 8
ELECTRONS IN ATOMS
PRACTICE EXAMPLES

1A (E) Use $c = \lambda v$, solve for frequency. $v = \dfrac{2.9979 \times 10^8 \text{ m/s}}{690 \text{ nm}} \times \dfrac{10^9 \text{ nm}}{1 \text{ m}} = 4.34 \times 10^{14}$ Hz

1B (E) Wavelength and frequency are related through the equation $c = \lambda v$, which can be solved for either one.

$$\lambda = \frac{c}{v} = \frac{2.9979 \times 10^8 \text{ m/s}}{91.5 \times 10^6 \text{ s}^{-1}} = 3.28 \text{ m} \quad \text{Note that Hz} = \text{ s}^{-1}$$

2A (E) The relationship $v = c/\lambda$ can be substituted into the equation $E = hv$ to obtain $E = hc/\lambda$. This energy, in J/photon, can then be converted to kJ/mol.

$$E = \frac{hc}{\lambda} = \frac{6.626 \times 10^{-34} \text{ J s photon}^{-1} \times 2.998 \times 10^8 \text{ m s}^{-1}}{230 \text{ nm} \times \dfrac{1 \text{ m}}{10^9 \text{ nm}}} \times \frac{6.022 \times 10^{23} \text{ photons}}{1 \text{ mol}} \times \frac{1 \text{ kJ}}{1000 \text{ J}} = 520 \text{ kJ/mol}$$

With a similar calculation one finds that 290 nm corresponds to 410 kJ/mol. Thus, the energy range is from 410 to 520 kJ/mol, respectively.

2B (M) The equation $E = hv$ is solved for frequency and the two frequencies are calculated.

$$v = \frac{E}{h} = \frac{3.056 \times 10^{-19} \text{ J/photon}}{6.626 \times 10^{-34} \text{ Js/photon}} \qquad v = \frac{E}{h} = \frac{4.414 \times 10^{-19} \text{ J/photon}}{6.626 \times 10^{-34} \text{ Js/photon}}$$

$$= 4.612 \times 10^{14} \text{ Hz} \qquad\qquad = 6.662 \times 10^{14} \text{ Hz}$$

To determine color, we calculate the wavelength of each frequency and compare it with *text* Figure 8-3.

$$\lambda = \frac{c}{v} = \frac{2.9979 \times 10^8 \text{ m/s}}{4.612 \times 10^{14} \text{ Hz}} \times \frac{10^9 \text{ nm}}{1 \text{ m}} \qquad \lambda = \frac{c}{v} = \frac{2.9979 \times 10^8 \text{ m/s}}{6.662 \times 10^{14} \text{ Hz}} \times \frac{10^9 \text{ nm}}{1 \text{ m}}$$

$$= 650 \text{ nm} \qquad \text{orange} \qquad\qquad = 450 \text{ nm} \qquad \text{indigo}$$

The colors of the spectrum that are not absorbed are what we see when we look at a plant, namely in this case blue, green, and yellow. The plant appears green.

3A (E) We solve the Rydberg equation for n to see if we obtain an integer.

$$n = \sqrt{n^2} = \sqrt{\frac{-R_H}{E_n}} = \sqrt{\frac{-2.179 \times 10^{-18} \text{ J}}{-2.69 \times 10^{-20} \text{ J}}} = \sqrt{81.00} = 9.00 \qquad \text{This is } E_9 \text{ for } n = 9.$$

3B (E)

$$E_n = \frac{-R_H}{n^2}$$

$$-4.45 \times 10^{-20} \text{ J} = \frac{-2.179 \times 10^{-18} \text{ J}}{n^2}$$

$$n^2 = 48.97 \Rightarrow n = 7$$

4A (E) We first determine the energy difference, and then the wavelength of light for that energy.

$$\Delta E = R_{\mathrm{H}}\left(\frac{1}{2^2} - \frac{1}{4^2}\right) = 2.179 \times 10^{-18} \text{ J}\left(\frac{1}{2^2} - \frac{1}{4^2}\right) = 4.086 \times 10^{-19} \text{ J}$$

$$\lambda = \frac{hc}{\Delta E} = \frac{6.626 \times 10^{-34} \text{ J s} \times 2.998 \times 10^8 \text{ m s}^{-1}}{4.086 \times 10^{-19} \text{ J}} = 4.862 \times 10^{-7} \text{ m} \text{ or } 486.2 \text{ nm}$$

4B (M) The longest wavelength light results from the transition that spans the smallest difference in energy. Since all Lyman series emissions end with $n_f = 1$, the smallest energy transition has $n_i = 2$. From this, we obtain the value of ΔE.

$$\Delta E = R_{\mathrm{H}}\left(\frac{1}{n_i^2} - \frac{1}{n_f^2}\right) = 2.179 \times 10^{-18} \text{ J}\left(\frac{1}{2^2} - \frac{1}{1^2}\right) = -1.634 \times 10^{-18} \text{ J}$$

From this energy emitted, we can obtain the wavelength of the emitted light: $\Delta E = hc/\lambda$

$$\lambda = \frac{hc}{\Delta E} = \frac{6.626 \times 10^{-34} \text{ J s} \times 2.998 \times 10^8 \text{ m s}^{-1}}{1.634 \times 10^{-18} \text{ J}} = 1.216 \times 10^{-7} \text{ m} \text{ or } 121.6 \text{ nm (1216 angstroms)}$$

5A (M)

$$E_f = \frac{-Z^2 \times R_{\mathrm{H}}}{n_f^2} = \frac{-4^2 \times 2.179 \times 10^{-18} \text{ J}}{3^2}$$

$$E_f = -3.874 \times 10^{-18} \text{ J}$$

$$E_i = \frac{-Z^2 \times R_{\mathrm{H}}}{n_i^2} = \frac{-4^2 \times 2.179 \times 10^{-18} \text{ J}}{5^2}$$

$$E_i = -1.395 \times 10^{-18} \text{ J}$$

$$\Delta E = E_f - E_i$$

$$\Delta E = (-3.874 \times 10^{-18} \text{ J}) - (-1.395 \times 10^{-18} \text{ J})$$

$$\Delta E = -2.479 \times 10^{-18} \text{ J}$$

To determine the wavelength, use $E = h\nu = \dfrac{hc}{\lambda}$; Rearrange for λ:

$$\lambda = \frac{hc}{E} = \frac{(6.626 \times 10^{-34} \text{ J s})\left(2.998 \times 10^8 \frac{\text{m}}{\text{s}}\right)}{2.479 \times 10^{-18} \text{ J}} = 8.013 \times 10^{-8} \text{ m or } 80.13 \text{ nm}$$

5B (E) Since $E = \dfrac{-Z^2 \times R_{\mathrm{H}}}{n^2}$, the transitions are related to Z^2, hence, if the frequency is 16 times greater, then the value of the ratio $\dfrac{Z^2(\text{?-atom})}{Z^2(\text{H-atom})} = \dfrac{Z_?^2}{1^2} = 16$.

We can see $Z^2 = 16$ or $Z = 4$ This is a Be nucleus. The hydrogen-like ion must be Be^{3+}.

6A (E) Superman's de Broglie wavelength is given by the relationship $\lambda = h/mu$.

$$\lambda = \frac{h}{mu} = \frac{6.626 \times 10^{-34} \text{ Js}}{91 \text{ kg} \times \frac{1}{5} \times 2.998 \times 10^8 \text{ m/s}} = 1.21 \times 10^{-43} \text{ m}$$

6B (M) The de Broglie wavelength is given by $\lambda = h/mu$, which can be solved for u.

$$u = \frac{h}{m\lambda} = \frac{6.626 \times 10^{-34} \text{ J s}}{1.673 \times 10^{-27} \text{ kg} \times 10.0 \times 10^{-12} \text{ m}} = 3.96 \times 10^{4} \text{ m/s}$$

We used the facts that $1 \text{ J} = \text{ kg m}^2\text{s}^{-2}$, $1 \text{pm} = 10^{-12}$ m, and $1 \text{ g} = 10^{-3}$ kg

7A (M) $p = (91 \text{ kg})(5.996 \times 10^{7} \text{ m s}^{-1}) = 5.4\underline{6} \times 10^{9} \text{ kg m s}^{-1}$
$\Delta p = (0.015)(5.4\underline{6} \times 10^{9} \text{ kg m s}^{-1}) = 8.2 \times 10^{7} \text{ kg m s}^{-1}$

$$\Delta x = \frac{h}{4\pi \Delta p} = \frac{6.626 \times 10^{-34} \text{ J s}}{(4\pi)(8.2 \times 10^{7} \frac{\text{kg m}}{\text{s}})} = 6.4 \times 10^{-43} \text{ m}$$

7B (M) $24 \text{ nm} = 2.4 \times 10^{-8} \text{ m} = \Delta x = \dfrac{h}{4\pi \Delta p} = \dfrac{6.626 \times 10^{-34} \text{ J s}}{(4\pi)(\Delta p)}$

Solve for Δp: $\Delta p = 2.2 \times 10^{-27}$ kg m s^{-1}
$(\Delta u)(m) = \Delta p = 2.2 \times 10^{-27}$ kg m s$^{-1} = (\Delta u)(1.67 \times 10^{-27}$ kg) Hence, $\Delta u = 1.3$ m s^{-1}.

8A (M) To calculate the probability percentage of finding an electron between 50 and 75 pm for an electron in level 6 ($n = 6$, # nodes = $6 - 1 = 5$), one must integrate the probability function, which is the square of the wave function between 50 and 75 pm:

Probability function:

$$\psi_6^2 = \frac{2}{L} \sin^2\left(\frac{n\pi}{L} x\right)$$

$$\int_{50}^{75} \frac{2}{L} \sin^2\left(\frac{n\pi}{L} x\right) dx = \left[\frac{2}{L}\left(\frac{x}{2} - \frac{1}{2(n\pi/L)} \cdot \sin\left(\frac{n\pi}{L} x\right) \cdot \cos\left(\frac{n\pi}{L} x\right)\right)\right]_{50}^{75}$$

$$= 0.499\overline{9} - 0.333\overline{3} = 0.166\overline{6}$$

The probability is 0.167 out of 1, or 16.7%. Of course, we could have done this without any use of calculus by following the simple algebra used in Example 8-8. However, it is just more fun to integrate the function. The above example was made simple by giving the limits of integration at two nodes. Had the limits been in locations that were *not* nodes, you would have had no choice but to integrate.

8B (E) We simply note here that at $n = 3$, the number of nodes is $n - 1 = 2$. Therefore, a box that is 300 pm long will have two nodes at 100 and 200 pm.

9A   **(D)** $50. \text{ pm} \times \dfrac{1 \text{ m}}{1 \times 10^{12} \text{ pm}} = 5.0 \times 10^{-11} \text{ m}$

$\Delta E = E_{\text{excited state}} - E_{\text{ground state}}$

$E = \dfrac{n^2 h^2}{8mL^2}$ Where n = energy level, h = Planck's constant, m = mass, L = length of box

$\Delta E = \dfrac{3^2 h^2}{8 \ mL^2} - \dfrac{5^2 h^2}{8 \ mL^2}$

$\Delta E = \dfrac{-16 \ h^2}{8 \ mL^2} = \dfrac{-16(6.626 \times 10^{-34} \ Js)^2}{8(9.109 \times 10^{-31})(5.0 \times 10^{-11})^2}$

$\Delta E = -3.86 \times 10^{-16} \text{ J}$

The negative sign indicates that energy was released / emitted.

$\lambda = \dfrac{hc}{\Delta E} = \dfrac{(6.626 \times 10^{-34} \text{ J s})(3.00 \times 10^8 \text{ m s}^{-1})}{3.86 \times 10^{-16} \ J}$

$\lambda = 5.15 \times 10^{-10} \text{ m} \times \dfrac{10^9 \text{ nm}}{1 \text{ m}} = 0.515 \text{ nm} = 0.52 \text{ nm}$

9B   **(D)**

$24.9 \text{ nm} \times \dfrac{1 \text{ m}}{10^9 \text{ nm}} = 2.49 \times 10^{-8} \text{ m}$

$\Delta E = \dfrac{hc}{\lambda} = \dfrac{(6.626 \times 10^{-34} \text{ J s})(3.00 \times 10^8 \text{ m s}^{-1})}{2.49 \times 10^{-8} \text{ m}}$

$\Delta E = 7.98 \times 10^{-18} \text{ J}$

$\Delta E = \dfrac{2^2 h^2}{8 \ mL^2} - \dfrac{1^2 h^2}{8 \ mL^2}$

$\Delta E = \dfrac{3 \ h^2}{8 \ mL^2}$

$7.98 \times 10^{-18} \text{ J} = \dfrac{3(6.626 \times 10^{-34} \text{ J s})^2}{8(9.109 \times 10^{-31})(L)^2}$

$L^2 = 2.265 \times 10^{-20}$

$L = 1.50 \times 10^{-10} \text{ m} = 150. \text{ pm}$

10A **(E)** Yes, an orbital can have the quantum numbers n = 3, $\ell = 0$ and $m_\ell = 0$. The values of ℓ can be between 0 and $n - 1$. The values of m_ℓ m_ℓ can be between ℓ and $-\ell$ encompassing zero. The three quantum numbers given in this question represent a $3s$ orbital.

10B **(E)** For an orbital with $n = 3$ the possible values of ℓ are 0, 1, and 2. However, when $m_\ell = 0$, this would omit $\ell = 0$ because when $\ell = 0$, m_ℓ must be 0. Therefore in order for both quantum numbers of $n = 3$ and $m_\ell = 1$ to be fulfilled, the only m_ℓ values allowed would be $\ell = 1$ and 2.

11A **(E)** The magnetic quantum number, m_ℓ, is not reflected in the orbital designation. Because $\ell = 1$, this is a p orbital. Because $n = 3$, the designation is $3p$.

11B **(M)** The H-atom orbitals $3s$, $3p$, and $3d$ are degenerate. Therefore, the 9 quantum number combinations are:

	n	ℓ	m_ℓ
$3s$	3	0	0
$3p$	3	1	$-1, 0, +1$
$3d$	3	2	$-2, -1, 0, +1, +2$

Hence, $n = 3$; $l = 0, 1, 2$; $m_1 = -2, -1, 0, 1, 2$

12A **(E)**

$(3, 2, -2, 1)$	$m_s = 1$ is incorrect. The values of m_s can only be $+\frac{1}{2}$ or $-\frac{1}{2}$.
$(3, 1, -2, \frac{1}{2})$	$m_\ell = -2$ is incorrect. The values of m_ℓ can be $+1, 0, +1$ when $\ell = 1$
$(3, 0, 0, \frac{1}{2})$	All quantum numbers are allowed.
$(2, 3, 0, \frac{1}{2})$	$\ell = 3$ is incorrect. The value for ℓ can not be larger than n.
$(1, 0, 0, -\frac{1}{2})$	All quantum numbers are allowed.
$(2, -1, -1, \frac{1}{2})$	$\ell = -1$ is incorrect. The value for ℓ can not be negative.

12B **(E)**

$(2, 1, 1, 0)$	$m_s = 0$ is incorrect. The values of m_s can only be $+\frac{1}{2}$ or $-\frac{1}{2}$.
$(1, 1, 0, \frac{1}{2})$	$\ell = 1$ is incorrect. The value for ℓ is 0 when $n = 1$.
$(3, -1, 1, \frac{1}{2})$	$\ell = -1$ is incorrect. The value for ℓ can not be negative.
$(0, 0, 0, -\frac{1}{2})$	$n = 0$ is incorrect. The value for n can not be zero.
$(2, 1, 2, \frac{1}{2})$	$m_\ell = 2$ is incorrect. The values of m_ℓ can be $+1, 0, +1$ when $\ell = 1$.

13A **(E)** (a) and (c) are equivalent. The valence electrons are in two different degenerate p orbitals and the electrons are spinning in the same direction in both orbital diagrams.

13B **(E)** This orbital diagram represents an excited state of a neutral species. The ground state would follow Hund's rule and there would be one electron in each of the three degenerate p orbitals.

14A **(E)** We can simply sum the exponents to obtain the number of electrons in the neutral atom and thus the atomic number of the element. $Z = 2 + 2 + 6 + 2 + 6 + 2 + 2 = 22$, which is the atomic number for Ti.

14B **(E)** Iodine has an atomic number of 53. The first 36 electrons have the same electron configuration as Kr: $1s^2 2s^2 2p^6 3s^2 3p^6 3d^{10} 4s^2 4p^6$. The next two electrons go into the $5s$ subshell $(5s^2)$, then 10 electrons fill the $4d$ subshell $(4d^{10})$, accounting for a total of 48 electrons. The last five electrons partially fill the $5p$ subshell $(5p^5)$.

The electron configuration of I is therefore $1s^2 2s^2 2p^6 3s^2 3p^6 3d^{10} 4s^2 4p^6 4d^{10} 5s^2 5p^5$. Each iodine atom has ten $3d$ electrons and one unpaired $5p$ electron.

15A **(E)** Iron has 26 electrons, of which 18 are accounted for by the [Ar] core configuration. Beyond [Ar] there are two $4s$ electrons and six $3d$ electrons, as shown in the following orbital diagram.

15B **(E)** Bismuth has 83 electrons, of which 54 are accounted for by the [Xe] configuration. Beyond [Xe] there are two $6s$ electrons, fourteen $4f$ electrons, ten $5d$ electrons, and three $6p$ electrons, as shown in the following orbital diagram.

16A **(E)**
  (a) Tin is in the 5th period, hence, five electronic shells are filled or partially filled.
  (b) The $3p$ subshell was filled with Ar; there are six $3p$ electrons in an atom of Sn.
  (c) The electron configuration of Sn is [Kr] $4d^{10} 5s^2 5p^2$. There are no $5d$ electrons.
  (d) Both of the $5p$ electrons are unpaired, thus there are two unpaired electrons in a Sn atom.

16B **(E)**
  (a) The $3d$ subshell was filled at Zn, thus each Y atom has ten $3d$ electrons.
  (b) Ge is in the $4p$ row; each germanium atom has two $4p$ electrons.
  (c) We would expect each Au atom to have ten $5d$ electrons and one $6s$ electron. Thus each Au atom should have one unpaired electron.

INTEGRATIVE EXAMPLE

A. **(M)**
(a) First, we must find the u_{rms} speed of the He atom

$$u_{rms} = \sqrt{\frac{3\,RT}{M}} = \sqrt{\frac{3\left(8.3145\ \text{J mol}^{-1}\,\text{K}^{-1}\right)\left(298\ \text{K}\right)}{4.003\times10^{-3}\,\text{kg mol}^{-1}}} = 1363\ \text{m/s}$$

Using the rms speed and the mass of the He atom, we can determine the momentum, and therefore the de Broglie's wavelength:

$$\text{mass He atom} = 4.003 \times 10^{-3} \frac{\text{kg}}{\text{mol}} \times \frac{1 \text{ mol}}{6.022 \times 10^{23} \text{ atoms}} = 6.647 \times 10^{-27} \text{ kg}$$

$$p = mu = (6.647 \times 10^{-27} \text{ kg})(1363 \text{ m/s}) = 9.06 \times 10^{-24} \text{ kg m/s}$$

$$\lambda = \frac{h}{p} = \frac{6.6261 \times 10^{-34} \text{ J/s}}{9.06 \times 10^{-24} \text{ kg m/s}} = 7.3135 \times 10^{-11} \text{ m} = 73.14 \text{ pm}$$

(b) Since the de Broglie wavelength is known to be ~300 pm, we have to perform the above solution backwards to determine the temperature:

$$p = \frac{h}{\lambda} = \frac{6.6261 \times 10^{-34} \text{ J/s}}{300 \times 10^{-12} m} = 2.209 \times 10^{-24} \text{ kg m/s}$$

$$u = u_{rms} = \frac{p}{m} = \frac{2.209 \times 10^{-24} \text{ kg m/s}}{6.647 \times 10^{-27} \text{ kg}} = 332.33 \text{ m/s}$$

Since $u_{rms} = \sqrt{3RT/M}$, solving for T yields the following:

$$T = \frac{(332.33 \text{ m/s})^2 (4.003 \times 10^{-3} \text{ kg})}{3(8.3145 \text{ J mol}^{-1}\text{K}^{-1})} = 17.7 \text{ K}$$

B. (M) The possible combinations are $1s \rightarrow np \rightarrow nd$, for example, $1s \rightarrow 3p \rightarrow 5d$. The frequencies of these transitions are calculated as follows:

$$1s \rightarrow 3p : \nu = 3.2881 \times 10^{15} \text{ s}^{-1} \left(\frac{1}{n_i^2} - \frac{1}{n_f^2} \right) = 3.2881 \times 10^{15} \text{ s}^{-1} \left(\frac{1}{1^2} - \frac{1}{3^2} \right) = 2.92 \times 10^{15} \text{ Hz}$$

$$3p \rightarrow 5d : \nu = 3.2881 \times 10^{15} \text{ s}^{-1} \left(\frac{1}{3^2} - \frac{1}{5^2} \right) = 2.34 \times 10^{14} \text{ Hz}$$

The emission spectrum will have lines representing $5d \rightarrow 4p$, $5d \rightarrow 3p$, $5d \rightarrow 2p$, $4p \rightarrow 3s$, $4p \rightarrow 2s$, $4p \rightarrow 1s$, $3p \rightarrow 2s$, $3p \rightarrow 1s$, and $2p \rightarrow 1s$. The difference between the sodium atoms is that the positions of the lines will be shifted to higher frequencies by 11^2.

EXERCISES

Electromagnetic Radiation

1. **(E)** The wavelength is the distance between successive peaks.
    Thus, 4×1.17 nm $= \lambda = 4.68$ nm.

3. **(E)**
    (a) TRUE   Since frequency and wavelength are inversely related to each other,
                radiation of shorter wavelength has higher frequency.
    (b) FALSE   Light of wavelengths between 390 nm and 790 nm is visible to the eye.
    (c) FALSE   All electromagnetic radiation has the same speed in a vacuum.
    (d) TRUE   The wavelength of an X-ray is approximately 0.1 nm.

5. **(E)** The light having the highest frequency also has the shortest wavelength. Therefore, choice (c) 80 nm has the highest frequency.

7. **(E)** The speed of light is used to convert the distance into an elapsed time.
    $$\text{time} = 93 \times 10^6 \text{ mi} \times \frac{5280 \text{ ft}}{1 \text{ mi}} \times \frac{12 \text{ in.}}{1 \text{ ft}} \times \frac{2.54 \text{ cm}}{1 \text{ in.}} \times \frac{1 \text{ s}}{3.00 \times 10^{10} \text{ cm}} \times \frac{1 \text{ min}}{60 \text{ s}} = 8.3 \text{ min}$$

Photons and the Photoelectric Effect

9. **(E)**
    (a) $E = h\nu = 6.626 \times 10^{-34} \text{ J s} \times 7.39 \times 10^{15} \text{ s}^{-1} = 4.90 \times 10^{-18}$ J/photon

    (b) $E_m = 6.626 \times 10^{-34} \text{ J s} \times 1.97 \times 10^{14} \text{ s}^{-1} \times \dfrac{6.022 \times 10^{23} \text{ photons}}{1 \text{ mol}} \times \dfrac{1 \text{ kJ}}{1000 \text{ J}} = 78.6$ kJ/mol

11. **(E)**
    (a) Here we combine $E = h\nu$ and $c = \nu\lambda$ to obtain $E = hc / \lambda$
    $$E = \frac{6.626 \times 10^{-34} \text{J s} \times 2.998 \times 10^8 \text{ m/s}}{574 \text{ nm} \times \dfrac{1 \text{m}}{10^9 \text{ nm}}} = 3.46 \times 10^{-19} \text{ J/photon}$$

    (b) $E_m = 3.46 \times 10^{-19} \dfrac{\text{J}}{\text{photon}} \times 6.022 \times 10^{23} \dfrac{\text{photons}}{\text{mol}} = 2.08 \times 10^5$ J/mol

13. **(E)** The easiest way to answer this question is to convert all of (b) through (d) into nanometers. The radiation with the smallest wavelength will have the greatest energy per photon, while the radiation with the largest wavelength has the smallest amount of energy per photon.

    (a) 6.62×10^2 nm

(b) 2.1×10^{-5} cm $\times \dfrac{1 \times 10^7 \text{ nm}}{1 \text{ cm}} = 2.1 \times 10^2$ nm

(c) 3.58 μm $\times \dfrac{1 \times 10^3 \text{ nm}}{1 \ \mu\text{m}} = 3.58 \times 10^3$ nm

(d) 4.1×10^{-6} m $\times \dfrac{1 \times 10^9 \text{ nm}}{1 \text{ m}} = 4.1 \times 10^3$ nm

So, 2.1×10^{-5} nm radiation, by virtue of possessing the smallest wavelength in the set, has the greatest energy per photon. Conversely, since 4.1×10^3 nm has the largest wavelength, it possesses the least amount of energy per photon.

15. **(E)** Notice that energy and wavelength are inversely related: $E = \dfrac{hc}{\lambda}$. Therefore radiation that is 100 times as energetic as radiation with a wavelength of 988 nm will have a wavelength one-hundredth as long, namely 9.88 nm. The frequency of this radiation is found by employing the wave equation.

$$v = \frac{c}{\lambda} = \frac{2.998 \times 10^8 \text{ m/s}}{9.88 \text{ nm} \times \dfrac{1\text{m}}{10^9\text{nm}}} = 3.03 \times 10^{16} \text{ s}^{-1} \text{ From Figure 8-3, we can see that this is UV radiation.}$$

17. **(M)**

(a) $E = hv = 6.63 \times 10^{-34}$ J s $\times 9.96 \times 10^{14}$ s$^{-1} = 6.60 \times 10^{-19}$ J/photon

(b) Indium will display the photoelectric effect when exposed to ultraviolet light since ultraviolet light has a maximum frequency of 1×10^{16} s^{-1}, which is above the threshold frequency of indium. It will not display the photoelectric effect when exposed to infrared light since the maximum frequency of infrared light is $\sim 3 \times 10^{14}$ s^{-1}, which is below the threshold frequency of indium.

Atomic Spectra

19. **(M)**

(a) $v = 3.2881 \times 10^{15} \text{s}^{-1} \left(\dfrac{1}{2^2} - \dfrac{1}{5^2} \right) = 6.9050 \times 10^{14} \text{s}^{-1}$

(b) $v = 3.2881 \times 10^{15} \text{ s}^{-1} \left(\dfrac{1}{2^2} - \dfrac{1}{7^2} \right) = 7.5492 \times 10^{14} \text{ s}^{-1}$

$\lambda = \dfrac{2.9979 \times 10^8 \text{ m/s}}{7.5492 \times 10^{14} \text{ s}^{-1}} = 3.9711 \times 10^{-7} \text{ m} \times \dfrac{10^9 \text{ nm}}{1 \text{ m}} = 397.11 \text{ nm}$

(c) $v = \dfrac{3.00 \times 10^8 \text{ m/s}}{380 \text{ nm} \times \dfrac{1\text{m}}{10^9\text{nm}}} = 7.89 \times 10^{14} \text{s}^{-1} = 3.2881 \times 10^{15} \text{s}^{-1} \left(\dfrac{1}{2^2} - \dfrac{1}{n^2} \right)$

$0.250 - \dfrac{1}{n^2} = \dfrac{7.89 \times 10^{14} \text{ s}^{-1}}{3.2881 \times 10^{15} \text{ s}^{-1}} = 0.240 \qquad \dfrac{1}{n^2} = 0.250 - 0.240 = 0.010 \qquad n = 10$

21. **(E)** $\Delta E = -2.179 \times 10^{-18} \text{ J} \left(\dfrac{1}{n_f^2} - \dfrac{1}{n_i^2} \right) = -2.179 \times 10^{-18} \text{ J} \left(\dfrac{1}{3^2} - \dfrac{1}{6^2} \right) = -1.816 \times 10^{-19} \text{ J}$

$E_{\text{photon emitted}} = 1.816 \times 10^{-19} \text{ J} = h\nu \qquad \nu = \dfrac{E}{h} = \dfrac{1.550 \times 10^{-19} \text{ J}}{6.626 \times 10^{-34} \text{ J s}} = 2.740 \times 10^{14} \text{ s}^{-1}$

23. **(M)** First we determine the frequency of the radiation, and then match it with the Balmer equation.

$\nu = \dfrac{c}{\lambda} = \dfrac{2.9979 \times 10^8 \text{ m s}^{-1} \times \dfrac{10^9 \text{nm}}{1\text{m}}}{389 \text{ nm}} = 7.71 \times 10^{14} \text{ s}^{-1} = 3.2881 \times 10^{15} \text{ s}^{-1} \left(\dfrac{1}{2^2} - \dfrac{1}{n^2} \right)$

$\left(\dfrac{1}{2^2} - \dfrac{1}{n^2} \right) = \dfrac{7.71 \times 10^{14} \text{ s}^{-1}}{3.2881 \times 10^{15} \text{ s}^{-1}} = 0.234 = 0.2500 - \dfrac{1}{n^2} \qquad \dfrac{1}{n^2} = 0.016 \qquad n = 7.9 \approx 8$

25. **(M)** The longest wavelength component has the lowest frequency (and thus, the smallest energy).

$\nu = 3.2881 \times 10^{15} \text{ s}^{-1} \left(\dfrac{1}{2^2} - \dfrac{1}{3^2} \right) = 4.5668 \times 10^{14} \text{ s}^{-1} \quad \lambda = \dfrac{c}{\nu} = \dfrac{2.9979 \times 10^8 \text{ m/s}}{4.5668 \times 10^{14} \text{ s}^{-1}} = 6.5646 \times 10^{-7} \text{m}$

$= 656.46 \text{ nm}$

$\nu = 3.2881 \times 10^{15} \text{ s}^{-1} \left(\dfrac{1}{2^2} - \dfrac{1}{4^2} \right) = 6.1652 \times 10^{14} \text{ s}^{-1} \quad \lambda = \dfrac{c}{\nu} = \dfrac{2.9979 \times 10^8 \text{ m/s}}{6.1652 \times 10^{14} \text{ s}^{-1}} = 4.8626 \times 10^{-7} \text{m}$

$= 486.26 \text{ nm}$

$\nu = 3.2881 \times 10^{15} \text{ s}^{-1} \left(\dfrac{1}{2^2} - \dfrac{1}{5^2} \right) = 6.9050 \times 10^{14} \text{ s}^{-1} \quad \lambda = \dfrac{c}{\nu} = \dfrac{2.9979 \times 10^8 \text{ m/s}}{6.9050 \times 10^{14} \text{ s}^{-1}} = 4.3416 \times 10^{-7} \text{m}$

$= 434.16 \text{ nm}$

$\nu = 3.2881 \times 10^{15} \text{ s}^{-1} \left(\dfrac{1}{2^2} - \dfrac{1}{6^2} \right) = 7.3069 \times 10^{14} \text{ s}^{-1} \quad \lambda = \dfrac{c}{\nu} = \dfrac{2.9979 \times 10^8 \text{ m/s}}{7.3069 \times 10^{14} \text{ s}^{-1}} = 4.1028 \times 10^{-7} \text{m}$

$= 410.28 \text{ nm}$

Energy Levels and Spectrum of the Hydrogen Atom

27. **(E)**

$E = -R_\text{H}/n^2 = \dfrac{-2.179 \times 10^{-18} \text{ J}}{6^2} = -6.053 \times 10^{-18} \text{ J}$

29. **(M)**

(a) $\quad \nu = \dfrac{2.179 \times 10^{-18} \text{ J}}{6.626 \times 10^{-34} \text{ J s}} \left(\dfrac{1}{4^2} - \dfrac{1}{7^2} \right) = 1.384 \times 10^{14} \text{ s}^{-1}$

(b) $\quad \lambda = \dfrac{c}{\nu} = \dfrac{2.998 \times 10^8 \text{ m/s}}{1.384 \times 10^{14} \text{ s}^{-1}} = 2.166 \times 10^{-6} \text{ m} \times \dfrac{10^9 \text{ nm}}{1 \text{ m}} = 2166 \text{ nm}$

(c) This is infrared radiation.

31. **(E)**

    (a) To determine the energy corresponding to $n = 8$,

$$E = -R_H / n^2 = \frac{-2.179 \times 10^{-18} \text{J}}{8^2} = -3.405 \times 10^{-20} \text{J}$$

    (b) Recall that the quantum number n must be an integer. We have

$$E = -R_H / n^2$$

$$-2.500 \times 10^{-19} \text{J} = \frac{-2.179 \times 10^{-18} \text{J}}{n^2}$$

$$n = \sqrt{0.08716} = 0.2952$$

Since n is not an integer, there is no energy level at -2.500×10^{-19} J.

    (c) In this case, the ionization energy corresponds to the energy change for the transition from $n_i = 6$ to $n_f = \infty$. Therefore,

$$\Delta E = -R_H \left(\frac{1}{n_f^2} - \frac{1}{n_i^2} \right) = -2.179 \times 10^{-18} \text{J} \left(0 - \frac{1}{6^2} \right) = 6.053 \times 10^{-20} \text{J}$$

$$E_{i(\text{from } n = 6)} = 6.053 \times 10^{-20} \text{J}$$

33. **(M)** If infrared light is produced, the quantum number of the final state must have a lower value (i.e., be of lower energy) than the quantum number of the initial state. First we compute the frequency of the transition being considered (from $v = c / \lambda$), and then solve for the final quantum number.

$$v = \frac{c}{\lambda} = \frac{3.00 \times 10^8 \text{ m/s}}{410 \text{ nm} \times \dfrac{1 \text{ m}}{10^9 \text{ nm}}} = 7.32 \times 10^{14} \text{ s}^{-1}$$

$$7.32 \times 10^{14} \text{ s}^{-1} = \frac{2.179 \times 10^{-18} \text{ J}}{6.626 \times 10^{-34} \text{ J s}} \left(\frac{1}{n^2} - \frac{1}{7^2} \right) = 3.289 \times 10^{15} \text{ s}^{-1} \left(\frac{1}{n^2} - \frac{1}{7^2} \right)$$

$$\left(\frac{1}{n^2} - \frac{1}{7^2} \right) = \frac{7.32 \times 10^{14} \text{ s}^{-1}}{3.289 \times 10^{15} \text{ s}^{-1}} = 0.2226 \qquad \frac{1}{n^2} = 0.2226 + \frac{1}{7^2} = 0.2429 \qquad n = 2$$

35. **(M)**

    (a) Line A is for the transition $n = 3 \rightarrow n = 1$, while Line B is for the transition $n = 4 \rightarrow n = 1$

    (b) This transition corresponds to the $n = 3$ to $n = 1$ transition. Hence, $\Delta E = hc / \lambda$

$$\Delta E = (6.626 \times 10^{-34} \text{ J s}) \times (2.998 \times 10^8 \text{ m s}^{-1}) \div (103 \times 10^{-9} \text{ m}) = 1.929 \times 10^{-18} \text{ J}$$

$$\Delta E = -Z^2 R_H / n_1^2 - -Z^2 R_H / n_2^2$$

$$1.929 \times 10^{-18} \text{ J} = -Z^2 (2.179 \times 10^{-18}) / (3)^2 + Z^2 (2.179 \times 10^{-18}) / (1)^2$$

$Z^2 = 0.996$ and $Z = 0.998$ Thus, this is the spectrum for the hydrogen atom.

37. (M)

(a) Line A is for the transition $n = 5 \rightarrow n = 2$, while Line B is for the transition $n = 6 \rightarrow n = 2$

(b) This transition corresponds to the $n = 5$ to $n = 2$ transition. Hence, $\Delta E = hc/\lambda$.

$\Delta E = (6.626 \times 10^{-34}\text{ J s}) \times (2.998 \times 10^8\text{ m s}^{-1}) \div (27.1 \times 10^{-9}\text{ m}) = 7.33 \times 10^{-18}\text{ J}$

$\Delta E = -Z^2 R_H/n_1^2 - -Z^2 R_H/n_2^2$

$4.57\underline{7} \times 10^{-19}\text{ J} = -Z^2(2.179 \times 10^{-18})/(5)^2 + Z^2(2.179 \times 10^{-18})/(2)^2$

$Z^2 = 16.0\underline{2}$ and $Z = 4.00$ Thus, this is the spectrum for the Be^{3+} cation.

Wave–Particle Duality

39. (E) The de Broglie equation is $\lambda = h/mu$. This means that, for a given wavelength to be produced, a lighter particle would have to be moving faster. Thus, electrons would have to move faster than protons to display matter waves of the same wavelength.

41. (M)

$$\lambda = \frac{h}{mu} = \frac{6.626 \times 10^{-34}\text{ J s}}{\left(145\text{ g} \times \dfrac{1\text{ kg}}{1000\text{ g}}\right)\left(168\text{ km/h} \times \dfrac{1\text{ h}}{3600\text{ s}} \times \dfrac{1000\text{ m}}{1\text{ km}}\right)} = 9.79 \times 10^{-35}\text{ m}$$

The diameter of a nucleus approximates 10^{-15} m, which is far larger than the baseball's wavelength.

The Heisenberg Uncertainty Principle

43. (M) The uncertainty relation $\Delta r \Delta p \geq h/(4\pi)$ indicates that it is not possible to specify exact values for both the radial position and the momentum along the radial direction. For an electron moving in a circular orbit, the radius is exactly known. And because the distance between the electron and the nucleus is constant, the electron has no velocity along the radial direction. Therefore, the momentum along the radial direction is exactly zero. The specification of exact values for both r and p violates the Heisenberg uncertainty principle. Alternatively, because r and p are precisely known, the corresponding uncertainties are $\Delta r = 0$ and $\Delta p = 0$. However, the product of Δr and Δp cannot be zero; it must be greater than or equal to $h/(4\pi)$.

45. (M)

$$\Delta u = \left(\frac{1}{100}\right)(0.1)\left(2.998 \times 10^8\,\frac{\text{m}}{\text{s}}\right) = 2.99\underline{8} \times 10^5\text{ m/s} \quad m = 1.673 \times 10^{-27}\text{ kg}$$

$$\Delta p = m\Delta u = (1.673 \times 10^{-27}\text{ kg})(2.998 \times 10^5\text{ m/s}) = 5.\underline{0} \times 10^{-22}\text{ kg m s}^{-1}$$

$$\Delta x = \frac{h}{4\pi \Delta p} = \frac{6.626 \times 10^{-34}\text{ J s}}{(4\pi)(5.0 \times 10^{-22}\,\dfrac{\text{kg m}}{\text{s}})} = {\sim}1 \times 10^{-13}\text{ m} \ ({\sim}100\text{ times the diameter of a nucleus})$$

47. **(M)** Electron mass = 9.109×10^{-31} kg, $\lambda = 0.53$ Å (1 Å = 1×10^{-10} m), hence $\lambda = 0.53 \times 10^{-10}$ m

$$\lambda = \frac{h}{mu} \text{ or } u = \frac{h}{m\lambda} = \frac{6.626 \times 10^{-34} \text{ J s}}{(9.109 \times 10^{-31} \text{ kg})(0.53 \times 10^{-10} \text{ m})} = 1.4 \times 10^{7} \text{ m s}^{-1}$$

Wave Mechanics

49. **(M)** A sketch of this situation is presented at right. We see that 2.50 waves span the space of the 42 cm. Thus, the length of each wave is obtained by equating: $2.50\lambda = 42$ cm, giving $\lambda \approx 17$ cm.

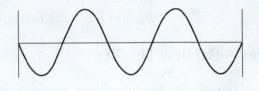

51. **(D)** $50. \text{ pm} \times \dfrac{1 \text{ m}}{1 \times 10^{12} \text{ pm}} = 5.0 \times 10^{-11}$ m

$$E = \frac{n^2 h^2}{8\, mL^2}$$

Where n = energy level, h = Planck's constant, m = mass, L = length of box

$$\Delta E = E_{\text{excited state}} - E_{\text{ground state}}$$

$$\Delta E = \left[\frac{4^2 (6.626 \times 10^{-34} \text{ J s})^2}{8(9.109 \times 10^{-31} \text{ kg})(5.0 \times 10^{-11} \text{ m})^2} \right] - \left[\frac{1^2 (6.626 \times 10^{-34} \text{ J s})^2}{8(9.109 \times 10^{-31} \text{ kg})(5.0 \times 10^{-11} \text{ m})^2} \right]$$

$$\Delta E = 3.856 \times 10^{-16} \text{ J} - 2.410 \times 10^{-17} \text{ J}$$

$$\Delta E = 3.615 \times 10^{-16} \text{ J}$$

$$\lambda = \frac{hc}{\Delta E} = \frac{(6.626 \times 10^{-34} \text{ J s})(3.00 \times 10^{8} \text{ m s}^{-1})}{3.615 \times 10^{-16} \text{ J}}$$

$$\lambda = 5.499 \times 10^{-10} \text{ m} \times \frac{10^9 \text{ nm}}{1 \text{ m}} = 0.5499 \text{ nm} = 0.55 \text{ nm}$$

53. **(D)**

$$20.0 \text{ nm} \times \frac{1 \text{ m}}{10^9 \text{ nm}} = 2.00 \times 10^{-8} \text{ m} = \text{length of the box}$$

$$\Delta E = \frac{hc}{\lambda} = \frac{(6.626 \times 10^{-34} \text{ J s})(3.00 \times 10^{8} \text{ m s}^{-1})}{8.60 \times 10^{-5} \text{ m}}$$

$$\Delta E = 2.311 \times 10^{-21} \text{ J}$$

$$\Delta E = E_{\text{excited state}} - E_{\text{ground state}}$$

$$E = \frac{n^2 h^2}{8\, mL^2}$$

Where n = energy level, h = Planck's constant, m = mass, L = length of box

$$2.311\times10^{-21}\,J = \left[\frac{n^2(6.626\times10^{-34}\,J\,s)^2}{8(9.109\times10^{-31}\,kg)(2.00\times10^{-8}\,m)^2}\right] - \left[\frac{1^2(6.626\times10^{-34}\,J\,s)^2}{8(9.109\times10^{-31}\,kg)(2.00\times10^{-8}\,m)^2}\right]$$

$$2.311\times10^{-21}\,J = 1.506\times10^{-22}n^2 - 1.506\times10^{-22}$$

$$n^2 = 16.35$$

$$n = 4.0$$

55. **(M)** The differences between Bohr orbits and wave mechanical orbitals are given below.

(a) The first difference is that of shape. Bohr orbits, as originally proposed, are circular (later, Sommerfeld proposed elliptical orbits). Orbitals, on the other hand, can be spherical, or shaped like two tear drops or two squashed spheres, or shaped like four tear drops meeting at their points.

(b) Bohr orbits are planar pathways, while orbitals are three-dimensional regions of space in which there is a high probability of finding electrons.

(c) The electron in a Bohr orbit has a definite trajectory. Its position and velocity are known at all times. The electron in an orbital, however, does not have a well-known position or velocity. In fact, there is a small but definite probability that the electron may be found outside the boundaries generally drawn for the orbital. Orbits and orbitals are similar in that the radius of a Bohr orbit is comparable to the average distance of the electron from the nucleus in the corresponding wave mechanical orbital.

Quantum Numbers and Electron Orbitals

57. **(E)** Answer (a) is incorrect because the values of m_s may be either $+\frac{1}{2}$ or $-\frac{1}{2}$. Answers (b) and (d) are incorrect because the value of ℓ may be any integer $\geq |m_\ell|$, and less than n. Thus, answer (c) is the only one that is correct.

59. **(E)**

(a) $n=5,\ \ell=1,\ m_\ell=0$ designates a $5p$ orbital. ($\ell=1$ for all p orbitals.)

(b) $n=4,\ \ell=2,\ m_\ell=-2$ designates a $4d$ orbital. ($\ell=2$ for all d orbitals.)

(c) $n=2,\ \ell=0,\ m_\ell=0$ designates a $2s$ orbital. ($\ell=0$ for all s orbitals.)

61. **(E)**

(a) 1 electron   (All quantum numbers are allowed and each electron has a unique set of four quantum numbers)

(b) 2 electrons   ($m_s = +\frac{1}{2}$ and $-\frac{1}{2}$)

(c) 10 electrons   ($m_\ell = -2, -1, 0, 1, 2$ and $m_s = +\frac{1}{2}$ and $-\frac{1}{2}$ for each m_ℓ orbital)

(d) 32 electrons   ($\ell=0, 1, 2, 3$ so there are one s, three p, five d, and seven f orbitals in $n=4$ energy level. Each orbital has 2 electrons.)

(e) 5 electrons   (There are five electrons in the $4d$ orbital that are spin up.)

The Shapes of Orbitals and Radial Probabilities

63. **(M)** The wave function for the 2s orbital of a hydrogen atom is:

$$\psi_{2s} = \frac{1}{4}\left(\frac{1}{2\pi a_0^3}\right)^{1/2}\left(2 - \frac{r}{a_0}\right)e^{-\frac{r}{2a_0}}$$

Where $r = 2a_0$, the $\left(2 - \dfrac{r}{a_0}\right)$ term becomes zero, thereby making $\psi_{2s} = 0$. At this point, the

wave function has a radial node (i.e., the electron density is zero). The finite value of r is $2\ a_0$ at the node, which is equal to 2×53 pm or 106 pm. Thus at 106 pm, there is a nodal surface with zero electron density.

65. **(M)** The angular part of the $2p_y$ wave function is $Y(\theta, \phi)_{py} = \sqrt{\dfrac{3}{4\pi}}\sin\theta\sin\phi$. The two lobes

of the $2p_y$ orbital lie in the xy plane and perpendicular to this plane is the xz plane. For all points in the xz plane $\phi = 0$, and since the sine of $0°$ is zero, this means that the entire xz plane is a node. Thus, the probability of finding a $2p_y$ electron in the xz plane is zero.

67. **(D)** The $2p_y$ orbital $Y(\theta, \phi) = \sqrt{\dfrac{3}{4\pi}}\sin\theta\sin\phi$, however, in the xy plane $\theta = 90°$ and

$\sin\theta = 1$. Plotting in the xy plane requires that we vary only ϕ.

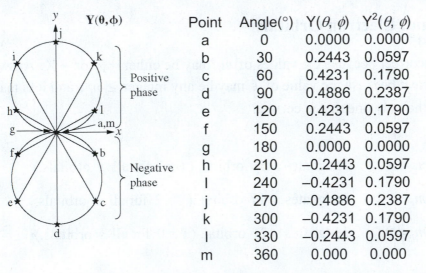

Point	Angle(°)	$Y(\theta, \phi)$	$Y^2(\theta, \phi)$
a	0	0.0000	0.0000
b	30	0.2443	0.0597
c	60	0.4231	0.1790
d	90	0.4886	0.2387
e	120	0.4231	0.1790
f	150	0.2443	0.0597
g	180	0.0000	0.0000
h	210	−0.2443	0.0597
I	240	−0.4231	0.1790
j	270	−0.4886	0.2387
k	300	−0.4231	0.1790
l	330	−0.2443	0.0597
m	360	0.000	0.000

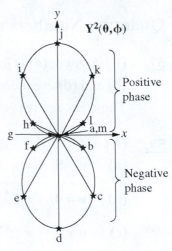

69. **(D)** A plot of radial probability distribution versus r/a_0 for a H$_{1s}$ orbital shows a maximum at 1.0 (that is, $r = a_0$ or $r = 53$ pm). The plot is shown below:

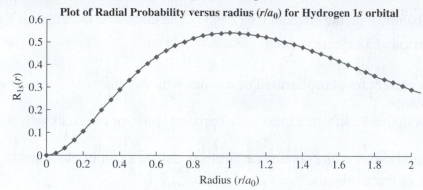

Plot of Radial Probability versus radius (r/a_0) for Hydrogen 1s orbital

(y-axis: $R_{1s}(r)$; x-axis: Radius (r/a_0))

71. **(E)**
    (a)   To answer this question, we must keep two simple rules in mind.
        1. Value of ℓ is the number of angular nodes.
        2. Total number of nodes = $n - 1$.

        From this we see that this is a *p*-orbital (1 angular node → $\ell = 1$) and because there are a total of 2 nodes, $n = 3$. This must be a 3*p* orbital.

    (b)   From this we see that this is a *d*-orbital (2 angular nodes → $\ell = 2$) and because there are a total of 2 nodes, $n = 3$. This must be a 3*d* orbital.

    (c)   From this we see that this is an *f*-orbital (3 angular nodes → $\ell = 2$) and because there are a total of 5 nodes, $n = 6$. This must be a 6*f* orbital.

73. **(E)** The orbital is in the xy plane and has two angular nodes (*d*-orbital) and 2 spherical nodes (total nodes = 4, hence $n = 5$). Since the orbital points between the *x*-axis and *y*-axis, this is a 5d_{xy} orbital. The second view of the same orbital is just a 90° rotation about the *x*-axis.

Electron Configurations

75. **(E)**
    (a)   N is the third element in the *p* block of the second period. It has three 2*p* electrons.

    (b)   Rb is the first element in the *s* block of the *fifth* period. It has two 4*s* electrons.

    (c)   As is in the *p* block of the fourth period. The 3*d* subshell is filled with ten electrons, but no 4*d* electrons have been added.

    (d)   Au is in the *d* block of the sixth period; the 4*f* subshell is filled. Au has fourteen 4*f* electrons.

    (e)   Pb is the second element in the *p* block of the sixth period; it has two 6*p* electrons. Since these two electrons are placed in separate 6*p* orbitals, they are unpaired. There are two unpaired electrons.

(f) Group 14 of the periodic table is the group with the elements C, Si, Ge, Sn, and Pb. This group currently has five named elements.

(g) The sixth period begins with the element Cs $(Z = 55)$ and ends with the element Rn $(Z = 86)$. This period is 32 elements long.

77. **(E)** Configuration (b) is correct for phosphorus. The reasons why the other configurations are incorrect are given below.

(a) The two electrons in the $3s$ subshell must have opposed spins, or different values of m_s.

(c) The three $3p$ orbitals must each contain one electron, before a pair of electrons is placed in any one of these orbitals.

(d) The three unpaired electrons in the $3p$ subshell must all have the same spin, either all spin up or all spin down.

79. **(E)** We write the correct electron configuration first in each case.

(a) P: $[Ne]3s^2 3p^3$ There are 3 unpaired electrons in each P atom.

(b) Br: $[Ar]3d^{10}4s^2 4p^5$ There are ten $3d$ electrons in an atom of Br.

(c) Ge: $[Ar]3d^{10}4s^2 4p^2$ There are two $4p$ electrons in an atom of Ge.

(d) Ba: $[Xe]6s^2$ There are two $6s$ electrons in an atom of Ba.

(e) Au: $[Xe]4f^{14}5d^{10}6s^1$ (exception)   There are fourteen $4f$ electrons in an atom of Au.

81. **(E)** Since the periodic table is based on electron structure, two elements in the same group (Pb and element 114) should have similar electron configurations.

(a) Pb: $[Xe]\ 4f^{14}5d^{10}6s^2 6p^2$ **(b)**   114: $[Rn]\ 5f^{14}6d^{10}7s^2 7p^2$

83. **(E)**

(a) This is an excited state; the $2s$ orbital should fill before any electrons enter the $2p$ orbital.

(b) This is an excited state; the electrons in the $2p$ orbitals should have the same spin (Hund's rule).

(c) This is the ground-state configuration of N.

(d) This is an excited state; there should be one set of electrons paired up in the $2p$ orbital (Hund's rule is violated).

85. **(E)**

(a) Hg: $[Xe]4f^{14}5d^{10}6s^2$ **(d)** Sn: $[Kr]4d^{10}5s^2 5p^2$

(b) Ca: $[Ar]4s^2$ **(e)** Ta: $[Xe]4f^{14}5d^3 6s^2$

(c) Po: $[Xe]4f^{14}5d^{10}6s^2 6p^4$ **(f)** I: $[Kr]4d^{10}5s^2 5p^5$

87. **(E)** **(a)** rutherfordium; **(b)** carbon; **(c)** vanadium; **(d)** tellurium; **(e)** not an element

INTEGRATIVE AND ADVANCED EXERCISES

91. (M)

(a) We first must determine the wavelength of light that has an energy of 435 kJ/mol and compare that wavelength with those known for visible light.

$$E = \frac{435 \dfrac{kJ}{mol} \times \dfrac{1000\ J}{1\ kJ}}{6.022 \times 10^{23}\ photons/mol} = 7.22 \times 10^{-19}\ \text{J/photon} = h\nu$$

$$\nu = \frac{E}{h} = \frac{7.22 \times 10^{-19}\ J}{6.626 \times 10^{-34}\ J\,s} = 1.09 \times 10^{15}\ s^{-1} \qquad \lambda = \frac{c}{\nu} = \frac{2.9979 \times 10^{8}\ m\ s^{-1}}{1.09 \times 10^{15}\ s^{-1}} \times \frac{10^{9}\ nm}{1\ m} = 275\ nm$$

Because the shortest wavelength of visible light is 390 nm, the photoelectric effect for mercury cannot be obtained with visible light.

(b) We first determine the energy per photon for light with 215 nm wavelength.

$$E = h\nu = \frac{hc}{\lambda} = \frac{6.626 \times 10^{-34}\ J\,s \times 2.9979 \times 10^{8}\ m\ s^{-1}}{215\ nm \times \dfrac{1\ m}{10^{9}\ nm}} = 9.24 \times 10^{-19}\ \text{J/photon}$$

Excess energy, over and above the threshold energy, is imparted to the electron as kinetic energy. Electron kinetic energy $= 9.24 \times 10^{-19}$ J $- 7.22 \times 10^{-19}$ J $= 2.02 \times 10^{-19}$ J $= \dfrac{mu^{2}}{2}$

(c) We solve for the velocity $u = \sqrt{\dfrac{2 \times 2.02 \times 10^{-19}\ J}{9.109 \times 10^{-31}\ kg}} = 6.66 \times 10^{5}\ m\ s^{-1}$

92. (M) We first determine the energy of an individual photon.

$$E = \frac{hc}{\lambda} = \frac{6.626 \times 10^{-34}\ J\,s \times 2.998 \times 10^{8}\ m/s}{1525\ nm \times \dfrac{1\ m}{10^{9}\ nm}} = 1.303 \times 10^{-19}\ J$$

$$\frac{\text{no. photons}}{\text{sec}} = \frac{95\ J}{s} \times \frac{1\ photon}{1.303 \times 10^{-19}\ J} \times \frac{14\ \text{photons produced}}{100\ \text{photons theoretically possible}} = 1.0 \times 10^{20}\ \text{photons/s}$$

93. (M) A watt = joule/second, so joules = watts × seconds J = 75 watts × 5.0 seconds = 37$\underline{5}$ Joules
$E =$ (number of photons) $h\nu$ and $\nu = c/\lambda$, so $E =$ (number of photons)hc/λ and

$$\lambda = (\text{number of photons})\frac{hc}{E} = \frac{(9.91 \times 10^{20}\ \text{photons})(6.626 \times 10^{-34}\ J\,s)(3.00 \times 10^{8}\ m/s)}{37\underline{5}\ \text{watts}}$$

$\lambda = 5.3 \times 10^{-7}$ m or 530 nm   The light will be green in color.

96. (M) First we determine the frequency of the radiation. Then rearrange the Rydberg equation (generalized from the Balmer equation) and solve for the parenthesized expression.

$$v = \frac{c}{\lambda} = \frac{2.9979 \times 10^8 \text{ m s}^{-1} \times \dfrac{10^9 \text{ nm}}{1 \text{ m}}}{1876 \text{ nm}} = 1.598 \times 10^{14} \text{ s}^{-1} = 3.2881 \times 10^{15} \text{ s}^{-1} \left(\frac{1}{m^2} - \frac{1}{n^2}\right)$$

$$\left(\frac{1}{m^2} - \frac{1}{n^2}\right) = \frac{1.598 \times 10^{14} \text{ s}^{-1}}{3.2881 \times 10^{15} \text{ s}^{-1}} = 0.0486$$

We know that $m < n$, and both numbers are integers. Furthermore, we know that $m \neq 2$ (the Balmer series) which is in the visible region, and $m \neq 1$ which is in the ultraviolet region, since the wavelength 1876 nm is in the infrared region. Let us try $m = 3$ and $n = 4$.

$$\frac{1}{3^2} - \frac{1}{4^2} = 0.04861 \text{ These are the values we want.}$$

101. (D) First we must determine the energy per photon of the radiation, and then calculate the number of photons needed, (i.e., the number of ozone molecules (with the ideal gas equation)). (Parts per million O_3 are assumed to be by volume.) The product of these two numbers is total energy in joules.

$$E_1 = hv = \frac{hc}{\lambda} = \frac{6.626 \times 10^{-34} \text{ J s} \times 2.9979 \times 10^8 \text{ m s}^{-1}}{254 \text{ nm} \times \dfrac{1 \text{ m}}{10^9 \text{ nm}}} = 7.82 \times 10^{-19} \text{ J/photon}$$

$$\text{no. photons} = \frac{\left(748 \text{ mmHg} \times \dfrac{1 \text{ atm}}{760 \text{ mmHg}}\right) \times \left(1.00 \text{ L} \times \dfrac{0.25 \text{ L O}_3}{10^6 \text{ L air}}\right)}{\dfrac{0.08206 \text{ atm L}}{\text{mol K}} \times (22 + 273) \text{ K}} \times \frac{6.022 \times 10^{23} \text{ molecules}}{1 \text{ mol O}_3}$$

$$\times \frac{1 \text{ photon}}{1 \text{ molecule O}_3} = 6.1 \times 10^{15} \text{ photons}$$

energy needed $= 7.82 \times 10^{-19}$ J/photon $\times 6.1 \times 10^{15}$ photons $= 4.8 \times 10^{-3}$ J or 4.8 mJ

102. (M) First we compute the energy per photon, and then the number of photons received per second.

$$E = hv = 6.626 \times 10^{-34} \text{ J s} \times 8.4 \times 10^9 \text{ s}^{-1} = 5.6 \times 10^{-24} \text{ J/photon}$$

$$\frac{\text{photons}}{\text{second}} = \frac{4 \times 10^{-21} \text{ J/s}}{5.6 \times 10^{-24} \text{ J/photon}} = 7 \times 10^2 \text{ photons/s}$$

105. (D) First, we must calculate the energy of the 300 nm photon. Then, using the amount of energy required to break a Cl–Cl bond energy and the energy of the photon, we can determine how much excess energy there is after bond breakage.

Energy of a single photon at 300 nm is:

$$E = h\nu = hc/\lambda$$

$$E = \frac{\left(6.6261 \times 10^{-34} \text{ J s}\right)\left(2.998 \times 10^{8} \text{ m s}^{-1}\right)}{\left(300 \times 10^{-9} \text{ m}\right)} = 6.6215 \times 10^{-19} \text{ J}$$

The bond energy of a single Cl–Cl bond is determined as follows:

$$\text{Cl} - \text{Cl B.E.} = \frac{242.6 \times 10^{3} \text{ J}}{1 \text{ mol Cl}_2} \times \frac{1 \text{ mol Cl}_2}{6.022 \times 10^{23} \text{ molecules}} \times \frac{1 \text{ molecules}}{1 \text{ Cl} - \text{Cl bond}} = 4.028 \times 10^{-19} \text{ J/Cl} - \text{Cl bond}$$

Therefore, the excess energy after splitting a Cl–Cl bond is $(6.6215 - 4.028) \times 10^{-19}$ J $= 2.59 \times 10^{-19}$ J. Statistically, this energy is split evenly between the two Cl atoms and imparts a kinetic energy of 1.29×10^{-19} J to each.

The velocity of each atom is determined as follows:

$$\text{mass of Cl} = \frac{35.45 \text{ g Cl}}{1 \text{ mol Cl}} \times \frac{1 \text{ mol Cl}}{6.022 \times 10^{23} \text{ molecules}} = 5.887 \times 10^{-23} \text{ kg/Cl atom}$$

$$E_k = \frac{1}{2} m u^2$$

$$u = \sqrt{\frac{2E_k}{m}} = \sqrt{\frac{2\left(1.29 \times 10^{-19} \text{ J}\right)}{5.887 \times 10^{-23} \text{ kg}}} = 66.2 \text{ m/s}$$

FEATURE PROBLEMS

114. (D) The equation of a straight line is $y = mx + b$, where m is the slope of the line and b is its y-intercept. The Balmer equation is $\nu = 3.2881 \times 10^{15} \text{ Hz}\left(\frac{1}{2^2} - \frac{1}{n^2}\right) = \frac{c}{\lambda}$. In this equation, one plots ν on the vertical axis, and $1/n^2$ on the horizontal axis. The slope is $b = -3.2881 \times 10^{15}$ Hz and the intercept is $3.2881 \times 10^{15} \text{ Hz} \div 2^2 = 8.2203 \times 10^{14}$ Hz. The plot of the data for Figure 8-12 follows.

λ	656.3 nm	486.1 nm	434.0 nm	410.1 nm
ν	4.568×10^{14} Hz	6.167×10^{14} Hz	6.908×10^{14} Hz	7.310×10^{14} Hz
n	3	4	5	6

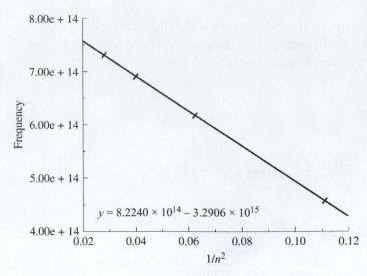

We see that the slope (-3.2906×10^{15}) and the y-intercept (8.2240×10^{14}) are almost exactly what we had predicted from the Balmer equation.

116. (D)

(a) First we calculate the range of energies for the incident photons used in the absorption experiment. Remember: $E_{photon} = h\nu$ and $\nu = c/\lambda$. At one end of the range, $\lambda = 100$ nm.

Therefore, $\nu = 2.998 \times 10^8$ m s$^{-1} \div (1.00 \times 10^{-7}$ m$) = 2.998 \times 10^{15}$ s^{-1}.

So $E_{photon} = 6.626 \times 10^{-34}$ J s$(2.998 \times 10^{15}$ s$^{-1}) = 1.98 \times 10^{-18}$ J.

At the other end of the range, $\lambda = 1000$ nm.

Therefore, $\nu = 2.998 \times 10^8$ m s$^{-1} \div 1.00 \times 10^{-6}$ m $= 2.998 \times 10^{14}$ s^{-1}.

So $E_{photon} = 6.626 \times 10^{-34}$ J s$(2.998 \times 10^{14}$ s$^{-1}) = 1.98 \times 10^{-19}$ J.

Next, we will calculate what excitations are possible using photons with energies between 1.98×10^{-18} J and 1.98×10^{-19} J and the electron initially residing in the $n = 1$ level. The photon energies for these transitions can be found with the equation

$\Delta E = E_f - E_i = -2.179 \times 10^{-18}\left(\dfrac{1}{(1)^2} - \dfrac{1}{(n_f)^2}\right)$. For the lowest energy photon

1.98×10^{-19} J $= -2.179 \times 10^{-18}\left(\dfrac{1}{(1)^2} - \dfrac{1}{(n_f)^2}\right)$ or $0.0904 = 1 - \dfrac{1}{(n_f)^2}$

From this $-0.9096 = -\dfrac{1}{(n_f)^2}$ and $n_f = 1.05$

Thus, the lowest energy photon is not capable of promoting the electron above the $n = 1$ level. For the highest energy level:

1.98×10^{-18} J $= -2.179 \times 10^{-18}\left(\dfrac{1}{(1)^2} - \dfrac{1}{(n_f)^2}\right)$ or $0.9114 = 1 - \dfrac{1}{(n_f)^2}$

From this $-0.0886 = -\dfrac{1}{(n_f)^2}$ and $n_f = 3.35$ Thus, the highest energy photon can promote a ground-state electron to both the $n = 2$ and $n = 3$ levels. This means that we would see <u>two</u> lines in the absorption spectrum, one corresponding to the $n = 1 \rightarrow n = 2$ transition and the other to the $n = 1 \rightarrow n = 3$ transition.

Energy for the $n = 1 \rightarrow n = 2$ transition $= -2.179 \times 10^{-18}\left(\dfrac{1}{(1)^2} - \dfrac{1}{(2)^2}\right) = 1.634 \times 10^{-18}$ J

$$\nu = \frac{1.634 \times 10^{-18} \text{ J}}{6.626 \times 10^{-34} \text{ J s}} = 2.466 \times 10^{15} \text{ s}^{-1} \quad \lambda = \frac{2.998 \times 10^{8} \text{ m s}^{-1}}{2.466 \times 10^{15} \text{ s}^{-1}} = 1.215 \times 10^{-7} \text{ m}$$

$\lambda = 121.5$ nm

Thus, we should see a line at 121.5 nm in the absorption spectrum.

Energy for the $n = 1 \rightarrow n = 3$ transition $= -2.179 \times 10^{-18}\left(\dfrac{1}{(1)^2} - \dfrac{1}{(3)^2}\right) = 1.937 \times 10^{-18}$ J

$$\nu = \frac{1.937 \times 10^{-18} \text{ J}}{6.626 \times 10^{-18} \text{ J}} = 2.923 \times 10^{15} \text{ s}^{-1} \quad \lambda = \frac{2.998 \times 10^{8} \text{ m s}^{-1}}{2.923 \times 10^{15} \text{ s}^{-1}} = 1.025 \times 10^{-7} \text{ m}$$

$\lambda = 102.5$ nm

Consequently, the second line should appear at 102.6 nm in the absorption spectrum.

(b) An excitation energy of 1230 kJ mol^{-1} to 1240 kJ mol^{-1} works out to 2×10^{-18} J per photon. This amount of energy is sufficient to raise the electron to the $n = 4$ level. Consequently, six lines will be observed in the emission spectrum. The calculation for each emission line is summarized below:

$$E_{4 \rightarrow 1} = \frac{2.179 \times 10^{-18} \text{ J}}{6.626 \times 10^{-34} \text{ J s}}\left(\frac{1}{1^2} - \frac{1}{4^2}\right) = 3.083 \times 10^{15} \text{ s}^{-1} \quad \lambda = \frac{2.998 \times 10^{8} \text{ m s}^{-1}}{3.083 \times 10^{15} \text{ s}^{-1}} = 9.724 \times 10^{-8} \text{ m}$$

$\lambda = 97.2$ nm

$$E_{4 \rightarrow 2} = \frac{2.179 \times 10^{-18} \text{ J}}{6.626 \times 10^{-34} \text{ J s}}\left(\frac{1}{2^2} - \frac{1}{4^2}\right) = 6.167 \times 10^{14} \text{ s}^{-1} \quad \lambda = \frac{2.998 \times 10^{8} \text{ m s}^{-1}}{6.167 \times 10^{14} \text{ s}^{-1}} = 4.861 \times 10^{-7} \text{ m}$$

$\lambda = 486.1$ nm

$$E_{4 \rightarrow 3} = \frac{2.179 \times 10^{-18} \text{ J}}{6.626 \times 10^{-34} \text{ J s}}\left(\frac{1}{3^2} - \frac{1}{4^2}\right) = 1.599 \times 10^{14} \text{ s}^{-1} \quad \lambda = \frac{2.998 \times 10^{8} \text{ m s}^{-1}}{1.599 \times 10^{14} \text{ s}^{-1}} = 1.875 \times 10^{-6} \text{ m}$$

$\lambda = 1875$ nm

$$E_{3 \rightarrow 1} = \frac{2.179 \times 10^{-18} \text{ J}}{6.626 \times 10^{-34} \text{ J s}}\left(\frac{1}{1^2} - \frac{1}{3^2}\right) = 2.924 \times 10^{15} \text{ s}^{-1} \quad \lambda = \frac{2.998 \times 10^{8} \text{ m s}^{-1}}{2.924 \times 10^{15} \text{ s}^{-1}} = 1.025 \times 10^{-7} \text{ m}$$

$\lambda = 102.5$ nm

$$E_{3 \rightarrow 2} = \frac{2.179 \times 10^{-18} \text{ J}}{6.626 \times 10^{-34} \text{ J s}}\left(\frac{1}{2^2} - \frac{1}{3^2}\right) = 4.568 \times 10^{14} \text{ s}^{-1} \quad \lambda = \frac{2.998 \times 10^{8} \text{ m s}^{-1}}{4.568 \times 10^{14} \text{ s}^{-1}} = 6.563 \times 10^{-7} \text{ m}$$

$\lambda = 656.3$ nm

$$E_{2 \rightarrow 1} = \frac{2.179 \times 10^{-18} \text{ J}}{6.626 \times 10^{-34} \text{ J s}}\left(\frac{1}{1^2} - \frac{1}{2^2}\right) = 2.467 \times 10^{15} \text{ s}^{-1} \quad \lambda = \frac{2.998 \times 10^{8} \text{ m s}^{-1}}{2.467 \times 10^{15} \text{ s}^{-1}} = 1.215 \times 10^{-7} \text{ m}$$

$\lambda = 121.5$ nm

(c) The number of lines observed in the two spectra is not the same. The absorption spectrum has two lines, while the emission spectrum has six lines. Notice that the 102.5 nm and 1021.5 nm lines are present in both spectra. This is not surprising since the energy difference between each level is the same whether it is probed by emission or absorption spectroscopy.

SELF-ASSESSMENT EXERCISES

124. (M) Atomic orbitals of multielectron atoms resemble those of the H atom in having both angular and radial nodes. They differ in that subshell energy levels are not degenerate and their radial wave functions no longer conform to the expressions in Table 8.2.

125. (E) Effective nuclear charge is the amount of positive charge from the nucleus that the valence shell of the electrons actually experiences. This amount is less than the actual nuclear charge, because electrons in other shells shield the full effect.

126. (E) The p_x, p_y, and p_z orbitals are triply degenerate (they are the same energy), and they have the same shape. Their difference lies in their orientation with respect to the arbitrarily assigned x, y, and z axes of the atom, as shown in Figure 8-28 of the textbook.

127. (E) The difference between the 2p and 3p orbitals is that the 2p orbital has only one node ($n = 2 - 1$) which is angular, whereas the 3p orbital has two nodes, angular and radial. See the figures below, which are extracted from the text.

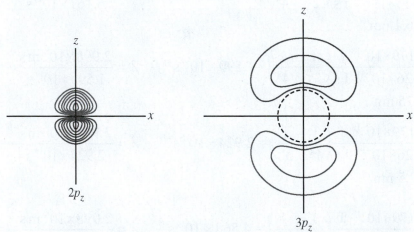

$2p_z$

$3p_z$

128. (E) The answer is (a). If the speed is the same for all particles, the lightest particle will have the longest wavelength.

129. (E)
(a) Velocity of the electromagnetic radiation is fixed at the speed of light in a vacuum.
(b) Wavelength is inversely proportional to frequency, because $v = c/\lambda$.
(c) Energy is directly proportional to frequency, because $E = hv$.

CHAPTER 9
THE PERIODIC TABLE AND SOME ATOMIC PROPERTIES
PRACTICE EXAMPLES

1A   **(E)** Atomic size decreases from left to right across a period, and from bottom to top in a family. We expect the smallest elements to be in the upper right corner of the periodic table. S is the element closest to the upper right corner and thus should have the smallest atom.
S = 104 pm As = 121 pm I = 133 pm

1B   **(E)** From the periodic table inside the front cover, we see that Na is in the same period as Al (period 3), but in a different group from K, Ca, and Br (period 4), which might suggest that Na and Al are about the same size. However, there is a substantial decrease in size as one moves from left to right in a period due to an increase in effective nuclear charge. Enough in fact, that Ca should be about the same size as Na.

2A   **(E)** Ti^{2+} and V^{3+} are isoelectronic; the one with higher positive charge should be smaller: $V^{3+} < Ti^{2+}$. Sr^{2+} and Br^- are isoelectronic; again, the one with higher positive charge should be smaller: $Sr^{2+} < Br^-$. In addition Ca^{2+} and Sr^{2+} both are ions of Group 2A; the one of lower atomic number should be smaller. $Ca^{2+} < Sr^{2+} < Br^-$. Finally, we know that the size of atoms decreases from left to right across a period; we expect sizes of like-charged ions to follow the same trend: $Ti^{2+} < Ca^{2+}$. The species are arranged below in order of increasing size.

$$V^{3+}(64\ pm) < Ti^{2+}(86\ pm) < Ca^{2+}(100\ pm) < Sr^{2+}(113\ pm) < Br^-(196\ pm)$$

2B   **(E)** Br^- clearly is larger than As since Br^- is an anion in the same period as As. In turn, As is larger than N since both are in the same group, with As lower down in the group. As also should be larger than P, which is larger than Mg^{2+}, an ion smaller than N. All that remains is to note that Cs is a truly large atom, one of the largest in the periodic table. The As atom should be in the middle. Data from Figure 9-11 shows:
$$Mg^{2+} < N < As < Br^- < Cs$$

3A   **(E)** Ionization increases from bottom to top of a group and from left to right through a period. The first ionization energy of K is less than that of Mg and the first ionization energy of S is less than that of Cl. We would expect also that the first ionization energy of Mg is smaller than that of S, because Mg is a metal.

3B   **(E)** We would expect an alkali metal (Rb) or an alkaline earth metal (Sr) to have a low first ionization energy and nonmetals (e.g., Br) to have relatively high first ionization energies. Metalloids (such as Sb and As) should have intermediate ionization energies. Since the first ionization energy for As is larger than that for Sb, the first ionization energy of Sb should be in the middle.

4A **(E)** Cl and Al must be paramagnetic, since each has an odd number of electrons. The electron configurations of K^+ ([Ar]) and O^{2-} ([Ne]) are those of the nearest noble gas. Because all of the electrons are paired, they are diamagnetic species. In Zn: [Ar] $3d^{10}4s^2$ all electrons are paired and so the atom is diamagnetic.

4B **(E)** The electron configuration of Cr is [Ar] $3d^5 4s^1$; it has six unpaired electrons. The electron configuration of Cr^{2+} is [Ar] $3d^4$; it has four unpaired electrons. The electron configuration of Cr^{3+} is [Ar] $3d^3$; it has three unpaired electrons. Thus, of the two ions, Cr^{2+} has the greater number of unpaired electrons.

INTEGRATIVE EXAMPLE

A **(M)** The physical properties of elements in the same period follow general trends. Below is a tabulation of the melting points, densities, and atomic radii of the alkali earth metals.

	Z	M.P. (°C)	Density (g/cc)	Metallic Radii (Å)
Li	3	180.54	0.53	2.05
Na	11	97.81	0.97	2.23
K	19	63.25	0.86	2.77
Rb	37	38.89	1.53	2.98
Cs	55	28.4	1.87	3.34

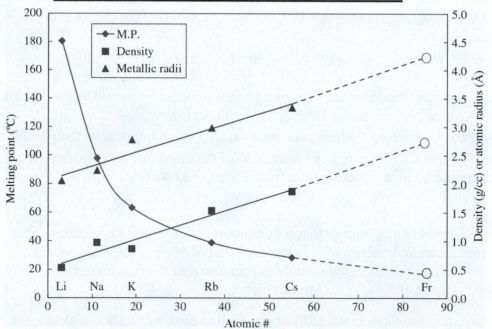

Accompanying this table is the plot of the data. Based on rough approximations of the trends of data, the properties of francium can be approximated as follows:

Melting point: 22 °C, density: 2.75 g/cc, atomic radius: 4.25 Å

B **(M)** Element 168 should be a solid since the trend in boiling point and melting point would put the boiling point temperature above 298 K. The electronic configuration is [Unk] $8s^2$ $5g^{18}\ 6f^{14}\ 7d^{10}\ 8p^6$, where Unk represents element 118.

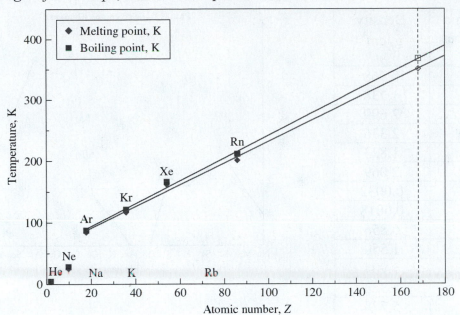

EXERCISES

The Periodic Law

1. **(E)** Element 114 will be a metal in the same group as Pb, element 82 ($18\ cm^3$/mol); Sn, element 50 ($18\ cm^3$/mol); and Ge, element 32 ($14\ cm^3$/mol). We note that the atomic volumes of Pb and Sn are essentially equal, probably due to the lanthanide contraction. If there is also an actinide contraction, element 114 will have an atomic volume of $18\ cm^3$ / mol. If there is no actinide contraction, we would predict a molar volume of ~ $22\ cm^3$ / mol. This need to estimate atomic volume is what makes the value for density questionable.

$$\text{density}\left(\frac{g}{cm^3}\right) = \frac{298\ \dfrac{g}{mol}}{18\ \dfrac{cm^3}{mol}} = 16\ \frac{g}{cm^3} \qquad \text{density}\left(\frac{g}{cm^3}\right) = \frac{298\ \dfrac{g}{mol}}{22\ \dfrac{cm^3}{mol}} = 14\ \frac{g}{cm^3}$$

3. **(M)** The following data are plotted at right. Density clearly is a periodic property for these two periods of main group elements. It rises, falls a bit, rises again, and falls back to the axis, in both cases.

Element	Atomic Number Z	Density g/cm^3
Na	11	0.968
Mg	12	1.738
Al	13	2.699
Si	14	2.336
P	15	1.823
S	16	2.069
Cl	17	0.0032
Ar	18	0.0018
K	19	0.856
Ca	20	1.550
Ga	31	5.904
Ge	32	5.323
As	33	5.778
Se	34	4.285
Br	35	3.100
Kr	36	0.0037

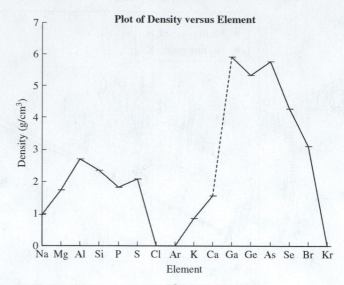

The Periodic Table

5. **(E)** Mendeleev arranged elements in the periodic table in order of increasing atomic weight. Of course, atomic masses with non-integral values are permissible. Hence, there always is room for an added element between two elements that already are present in the table. On the other hand, Moseley arranged elements in order of increasing atomic number. Only integral (whole number) values of atomic number are permitted. Thus, when elements with all possible integral values in a certain range have been discovered, no new elements are possible in that range.

7. **(E)**

The noble gas following radon $(Z = 86)$ will have an atomic number of $(86 + 32 =)\,118$.

The alkali metal following francium $(Z = 87)$ will have an atomic number of $(87 + 32 =)\,119$.

The mass number of radon $(A = 222)$ is $(222 \div 86 =)\,2.58$ times its atomic number. The mass number of Lr $(A = 262)$ is $(262 \div 103 =)\,2.54$ times its atomic number. Thus, we would expect the mass numbers, and hence approximate atomic masses, of elements 118 and 119 to be about 2.5 times their atomic numbers, that is, $A_{119} \approx 298$ u and $A_{118} \approx 295$ u.

Atomic Radii and Ionic Radii

9. **(E)** In general, atomic size in the periodic table increases from top to bottom for a group and increases from right to left through a period, as indicated in Figure 9-4. The larger element is indicated first, followed by the reason for making the choice.
    (a) Te: Te is to the left of Br and also in the period below that of Br in the 4th period.
    (b) K: K is to the left of Ca within the same period, Period 4.
    (c) Cs : Cs is both below and to the left of Ca in the periodic table.
    (d) N: N is to the left of O within the same period, Period 2.
    (e) P: P is both below and to the left of O in the periodic table.
    (f) Au: Au is both below and to the left of Al in the periodic table.

11. **(E)** Sizes of atoms do not simply increase with atomic number is because electrons often are added successively to the same subshell. These electrons do not fully screen each other from the nuclear charge (they do not effectively get between each other and the nucleus). Consequently, as each electron is added to a subshell and the nuclear charge increases by one unit, all of the electrons in this subshell are drawn more closely into the nucleus, because of the ineffective shielding.

13. **(E)**
    (a)   The smallest atom in Group 13 is the first: B
    (b)   Po is in the sixth period, and is larger than the others, which are rewritten in the following list from left to right in the fifth period, that is, from largest to smallest: Sr, In, Sb, Te. Thus, Te is the smallest of the elements given.

17. **(M)** In the literal sense, isoelectronic means having the same number and types of electrons. (In another sense, not used in the text, it means having the same electron configuration.) We determine the total number of electrons and the electron configuration for each species and make our decisions based on this information.

Fe^{2+}	24 electrons	$[Ar]\,3d^6$	Sc^{3+}	18 electrons	$[Ar]$
Ca^{2+}	18 electrons	$[Ar]$	F^-	10 electrons	$[He]\,2s^2 2p^6$
Co^{2+}	25 electrons	$[Ar]\,3d^7$	Co^{3+}	24 electrons	$[Ar]\,3d^6$
Sr^{2+}	36 electrons	$[Ar]\,3d^{10} 4s^2 4p^6$	Cu^+	28 electrons	$[Ar]\,3d^{10}$
Zn^{2+}	28 electrons	$[Ar]\,3d^{10}$	Al^{3+}	10 electrons	$[He]\,2s^2 2p^6$

Thus the species with the same number of electrons and the same electron configuration are the following. Fe^{2+} and Co^{3+} Sc^{3+} and Ca^{2+} F^- and Al^{3+} Zn^{2+} and Cu^+

19. **(E)** Ions can be isoelectronic without having noble-gas electron configurations. Take, for instance, Cu^+ and Zn^{2+}. Both of these ions have the electron configuration $[Ar]3d^{10}$.

Ionization Energies; Electron Affinities

21. **(E)** Ionization energy in the periodic table decreases from top to bottom for a group, and increases from left to right for a period, as summarized in Figure 9-12. Cs has the lowest ionization energy as it is farthest to the left and nearest to the bottom of the periodic table. Next comes Sr, followed by As, then S, and finally F, the most nonmetallic element in the group (and in the periodic table). Thus, the elements listed in order of increasing ionization energy are:

$$Cs < Sr < As < S < F$$

23. **(E)** In the case of a first electron affinity, a negative electron is being added to a neutral atom. This process may be either exothermic or endothermic depending upon the electronic configuration of the atom. Energy tends to be released when filled shells or filled subshells are generated. In the case of an ionization potential, however, a negatively charged electron is being separated from a positively charged cation, a process that must always require energy, because unlike charges attract each other.

25. **(E)** Ionization energies for Si: $E_{i,1} = 786.5$ kJ/mol, $E_{i,2} = 1577$ kJ/mol, $E_{i,3} = 3232$ kJ/mol, $E_{i,4} = 4356$ kJ/mol. To remove all four electrons from the third shell ($3s^2 3p^2$) would require the sum of all four ionization energies or 9951.5 kJ/mol. This would be 9.952×10^6 J per mole of Si atoms.

27. **(M)** The electron affinity of bromine is -324.6 kJ/mol (Figure 9-15). We use Hess's law to determine the heat of reaction for $Br_2(g)$ becoming $2\,Br^-(g)$.

$$
\begin{array}{lll}
Br_2(g) \rightarrow 2\,Br(g) & \Delta H = +193 \text{ kJ} & \\
2\,Br(g) + 2\,e^- \rightarrow 2\,Br^-(g) & 2 \times \Delta_{ea}H = 2(-324.6) \text{ kJ} & \\
\hline
Br_2(g) + 2\,e^- \rightarrow 2\,Br^-(g) & \Delta H = -456 \text{ kJ} & \text{Overall process is } \textit{exothermic.}
\end{array}
$$

29. **(E)** The electron is being removed from a species with a neon electron configuration. But in the case of Na^+, the electron is being removed from a species that is left with a +2 charge, while in the case of Ne, the electron is being removed from a species with a +1 charge. The more highly charged the resulting species, the more difficult it is to produce it by removing an electron.

31. **(M)**

   (a) Ionization energy in the periodic table decreases from top to bottom in a group, and increases from left to right across a period, as summarized in Figure 9-12. Therefore, the elements listed in order of increasing ionization energy are:

$$Al < Si < S < Cl$$

   (b) Electron affinity is the measure of the energy change that occurs when a gaseous atom gains an electron. If energy is given off when this occurs, the process is exothermic and the electron affinity is negative. It is harder to make generalizations about electron affinities. If an atom has a high affinity for an electron, the electon affinity tends to be a large negative value. Chlorine hast the greatest affinity for an electron, becuase it will have a noble gas configuration when this occurs. In this series, aluminum has the smallest affinity for an electron. Therefore, the elements listed in order of increasing electron affinity are:

$$Al < Si < S < Cl$$

(c) Polarizability increases with increasing size and, as a result, decreases going from left to right across a period. Al, Si, S, and Cl are in the same period, which means that Cl has the lowest polarizability and Al has the highest polarizability. Since Si is larger than S, Si will have a larger polarizability. Therefore, polarizability increases as follows: Cl $< S < Si < Al$.

Magnetic Properties

33. **(E)** Three of the ions have noble gas electron configurations and thus have no unpaired electrons: F^- is $1s^2 2s^2 2p^6$ Ca^{2+} and S^{2-} are $[Ne]3s^2 3p^6$

Only Fe^{2+} has unpaired electrons. Its electron configuration is $[Ar]3d^6$.

35. **(E)**
(a) K^+ is isoelectronic with Ar, with no unpaired electrons. It is diamagnetic.
(b) Cr^{3+} has the configuration $[Ar]3d^3$ with three unpaired electrons. It is paramagnetic.
(c) Zn^{2+} has the configuration $[Ar]3d^{10}$, with no unpaired electrons. It is diamagnetic.
(d) Cd has the configuration $[Kr]4d^{10}5s^2$, with no unpaired electrons. It is diamagnetic.
(e) Co^{3+} has the configuration $[Ar]3d^6$ with four unpaired electrons. It is paramagnetic
(f) Sn^{2+} has the configuration $[Kr]4d^{10}5s^2$, with no unpaired electrons. It is diamagnetic.
(g) Br has the configuration $[Ar]3d^{10}4s^25p^5$ with one unpaired electron. It is paramagnetic.
From this, we see that (a), (c), (d), and (f) are diamagnetic and (b), (e), and (g) are paramagnetic.

37. **(E)** All atoms with an odd number of electrons must be paramagnetic. There is no way to pair all of the electrons up if there is an odd number of electrons. Many atoms with an even number of electrons are diamagnetic, but some are paramagnetic. The one of lowest atomic number is carbon $(Z = 6)$, which has two unpaired p-electrons producing the paramagnetic behavior: $[He] 2s^2 2p^2$.

Predictions Based on the Periodic Table

39. **(M)**
(a) Elements that one would expect to exhibit the photoelectric effect with visible light should be ones that have a small value of their first ionization energy. Based on Figure 9-12, the alkali metals have the lowest first ionization energies of these. Cs, Rb, and K are three suitable metals. Metals that would not exhibit the photoelectric effect with visible light are those that have high values of their first ionization energy. Again from Figure 9-9, Zn, Cd, and Hg seem to be three metals that would not exhibit the photoelectric effect with visible light.

(b) From Figure 9-1, we notice that the atomic (molar) volume increases for the solid forms of the noble gases as we travel down the group (the data points just before the alkali metal peaks). But it seems to increase less rapidly than the molar mass. This means that the density should increase with atomic mass, and Rn should be the densest solid in the group. We expect densities of liquids to follow the same trend as densities of solids.

(c) To estimate the first ionization energy of fermium, we note in Figure 9-12 that the ionization energies of the lanthanides (following the Cs valley) are approximately the same. We expect similar behavior of the actinides, and estimate a first ionization energy of about +600 kJ/mol.

(d) We can estimate densities of solids from the information in Figure 9-1. Radium has $Z = 88$ and an approximate atomic volume of 40 cm^3/mol. Then we use the molar mass of radium to determine its density:

$$\text{density} = \frac{1 \text{ mol}}{40 \text{ cm}^3} \times \frac{226 \text{ g Ra}}{1 \text{ mol}} = 5.7 \text{ g} / \text{cm}^3$$

41. **(E)** Polarizability decreases going from left to right across a period and increases from top to bottom within a group. Be, N, and O are in the same period, which means that O has the lowest polarizability and Be has the highest polarizability in this period. S belongs to period 3 and is farther to the right than K, which is in period 4. Therefore, S is less polarizable than K. In order of increasing polarizability, they are O < N < Be < S < K.

43. **(E)** Polarizability increases with increasing size. Given the choice of Cl, Cl$^+$, and Cl$^-$, size depends on the number of electrons in the valence shell. Therefore, polarizability increases as follows: Cl$^+$ < Cl < Cl$^-$.

45. **(E)**

(a) $Z = 32$ 1. This is the element Ge, with an outer electron configuration of $3s^2 3p^2$. Thus, Ge has two unpaired p electrons.

(b) $Z = 8$ 1. This is the element O. Each atom has an outer electron configuration of $2s^2 2p^4$. Thus, O has two unpaired p electrons.

(c) $Z = 53$ 3. This is the element I, with an electron affinity more negative than that of the adjacent atoms: Xe and Te.

(d) $Z = 38$ 4. This is the element Sr, which has two 5s electrons. It is easier to remove one 5s electron than to remove the outermost 4s electron of Ca, but harder than removing the outermost 6s electron of Cs.

(e) $Z = 48$ 2. This is the element Cd. Its outer electron configuration is s^2 and thus it is diamagnetic.

(f) $Z = 20$ 2. This is the element Ca. Since its electron configuration is [Ar]$4s^2$, all electrons are paired and it is diamagnetic.

47. **(M)** Ga^{4+} and Ge^{5+} are unlikely to be found in chemical compounds because these ions are unstable. Atoms tend to gain or lose electrons such that they achieve a noble gas configuration. Removing 4 electrons from Ga and 5 from Ge does not give these atoms a noble gas configuration due to the electrons populating the 3d orbitals.

INTEGRATIVE AND ADVANCED EXERCISES

49. **(M)**
  (a) Not possible. C (77 pm) and Ca^{2+} (100 pm) are very different in size; however the diagram requires these to be nearly the same size
  (b) This is a possibility. Na^+ is approximately the same size as Sr (99 pm). Cl^- (181 pm) and Br^- (196 pm) are comparable, with one being slightly smaller.
  (c) Not possible. There are three large atoms and only one small one. Y (165 pm), K (227 pm), Ca (197 pm), and the small Na^+ (99 pm).
  (d) Not possible. The smaller atoms are of noticeably different sizes: Zr^{2+} (95 pm) and Mg^{2+} (72 pm).

  (e) This is possible. Fe and Co are of comparable sizes (~125 pm), and Cs (265 pm) and Rb (248 pm) are comparable, with one being slightly smaller.

Thus, the answer is that both (b) and (e) are compatible with the sketch.

51. **(E)** We can determine the atomic mass of indium by beginning with the atomic mass of oxygen, and using the chemical formula of InO to determine the amount (in moles) of indium in 100.0 g InO. In 100.0 g InO there is 82.5 g In and also 17.5 g O.

$$17.5 \text{ g O} \times \frac{1 \text{ mol O}}{16.0 \text{ g O}} \times \frac{1 \text{ mol In}}{1 \text{ mol O}} = 1.09 \text{ mol In} \quad \text{Atomic mass of In} = \frac{82.5 \text{ g In}}{1.09 \text{ mol In}} = 75.7 \text{ g In/mol}$$

With this value of the atomic mass of indium, Mendeleev might well have placed the element between As (75 g/mol) and Se (78 g/mol), that is, in his "Gruppe V" or "Gruppe VI."

56. **(M)**
  (a) A; Element "A" (Sr) has an electron configuration consistent with group 1A which are the alkali metals.
  (b) B; Element "B" (Br) is a non-metal. It is easier for Br to gain an electron to form Br^- than for Sr to form Sr^-.
  (c) A; Element "A" (Sr) has the larger atomic radius. In general, size decreases going across a period and increases going down a group. Since Sr has valence electrons in the higher energy 5s orbital, which is farther from the nucleus, Sr will be larger in size.
  (d) B; Element "B" (Br) has the greatest electron affinity. Non-metals have a greater tendency to add electrons compared with metals.

57. **(M)** Since $(0.3734)(382) = 143$ u, if we subtract 143 from 382 we get 239 u, which is very close to the correct value of 238 u.

First we convert to J/g°C: $0.0276 \times (4.184 \text{ J/cal}) = 0.1155$ J/g°C, so:

$0.1155 = 0.011440 + (23.967/\text{atomic mass})$; solving for atomic mass yields a value of 230 u, which is within about 3% of the correct value.

60. **(M)** $E_1 = \dfrac{-2^2 \times 2.179 \times 10^{-18} \text{ J}}{1^2} = -8.716 \times 10^{-18}$ J/atom

This is the energy released when an electron combines with an alpha particle (He^{2+}) to the nucleus to form the ion He^+. The energy absorbed when the electron is removed from He^+, the ionization energy of He^+ or the *second* ionization energy of He, is the negative of this value. Then the energy per mole, E_m, is computed.

$E_1 = E_1 \times N_A = \dfrac{8.716 \times 10^{-18}}{1 \text{ atom}} \times \dfrac{6.022 \times 10^{23}}{1 \text{ mol}} \times \dfrac{1 \text{ kJ}}{1000 \text{J}} = 5249$ kJ/mol

This is in excellent agreement with the tabulated value of 5251 kJ/mol.

61. **(M)** It should be relatively easy to remove electrons from ions of metallic elements, if the metal does not have a noble gas electron configuration. Thus, the third ionization energy for Sc and the second ionization energy for Ba should be small, with the second being smaller, since the electron is being removed from a more highly charged species in the case of the third ionization energy for Sc. The first ionization energy for F might be smaller than either of these because it involves removing only the first electron from a neutral atom, rather than removing an electron from a cation. There is some uncertainty here, because the electron being removed from F is not well shielded from the nuclear charge, and this value could be larger than the other two. The remaining ionization energies are both substantially larger than the other three because they both involve disrupting a noble gas electron configuration. The third ionization energy for Mg is larger than the second ionization energy for Na because it is more difficult to remove an electron from a more highly charged species.

Literature values (in kJ/mol) are given below.

First ionization energy of F:	1681 kJ/mol
Second ionization energy of Ba:	965
Third ionization energy of Sc:	2389
Second ionization energy of Na:	4562
Third ionization energy of Mg:	7733

Note that we have overestimated the difficulty of removing a second electron from Ba (a relatively large metal atom) and underestimated the difficulty of removing the first electron from F (a small nonmetal atom).

FEATURE PROBLEMS

66. **(D)** The Moseley equation, $v = A(Z - b)^2$, where v is the frequency of the emitted X-ray radiation, Z is the atomic number, and A and b are constants, relates the frequency of emitted X-rays to the nuclear charge for the atoms that make up the target of the cathode ray tube. X-rays are emitted by the element after one of its K-level electrons has been knocked out of the atom by collision with a fast moving electron. In this question, we have been asked to determine the values for the constants A and b. The simplest way to find these values is to plot $\sqrt{v}$ vs. Z. This plot provides $\sqrt{A}$ as the slope and $-\sqrt{A}$ (b) as the $y-$ intercept. Starting with $v = A(Z - b)^2$, we first take the square root of both sides. This affords $\sqrt{v} = \sqrt{A}$ $(Z - b)$. Multiplying out this expression gives $\sqrt{v} = \sqrt{A}$ $(Z) - \sqrt{A}$ (b). This expression follows the equation of a straight line $y = mx + b$, where $y = \sqrt{v}$, m $= \sqrt{A}$, $x = Z$ and $b = -\sqrt{A}$ (b). So a plot of $\sqrt{v}$ vs. Z will provide us with A and b, after a small amount of mathematical manipulation. Before we can construct the plot, we need to convert the provided X-ray wavelengths into their corresponding frequencies. For instance, Mg has an X-ray wavelength = 987 pm. The corresponding frequency for this radiation = c/λ, hence,

$$v = \frac{2.998 \times 10^8 \text{ m s}^{-1}}{9.87 \times 10^{-10} \text{ m}} = 3.04 \times 10^{17} \text{ s}^{-1}$$

Performing similar conversions on the rest of the data allows for the construction of the following table and plot (below).

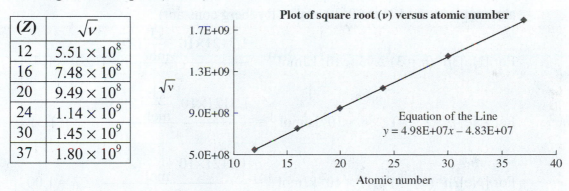

(Z)	$\sqrt{v}$
12	5.51×10^8
16	7.48×10^8
20	9.49×10^8
24	1.14×10^9
30	1.45×10^9
37	1.80×10^9

Plot of square root (v) versus atomic number

Equation of the Line
$y = 4.98\text{E}+07x - 4.83\text{E}+07$

The slope of the line is $4.98 \times 10^7 = \sqrt{A}$ and the y-intercept is $-4.83 \times 10^7 = -\sqrt{A}$ (b).

Thus, $A = 2.30 \times 10^{15}$ Hz and $b = \dfrac{-4.83 \times 10^7}{-4.98 \times 10^7} = 0.969$.

According to Bohr's theory, the frequencies that correspond to the lines in the emission

spectrum are given by the equation: $(3.2881 \times 10^{15} \text{ s}^{-1}) \left(\dfrac{1}{(n_i)^2} - \dfrac{1}{(n_f)^2} \right)$,

where $(3.2881 \times 10^{15} \text{ s}^{-1})$ represents the frequency for the lowest energy photon that is capable of completely removing (ionizing) an electron from a hydrogen atom in its ground state. The value of A (calculated in this question) is close to the Rydberg frequency $(3.2881 \times 10^{15} \text{ s}^{-1})$, so it is probably the equivalent term in the Moseley equation. The constant b, which is close to unity, could represent the number of electrons left in the K shell after one K-shell electron has been ejected by a cathode ray. Thus, one can think of b as representing the screening afforded by the remaining electron in the K-shell. Of course screening of the nucleus is only be possible for those elements with $Z > 1$.

67. **(D)**

(a) The table provided in this question shows the energy changes associated with the promotion of the outermost valence electron of sodium into the first four excited states above the highest occupied ground state atomic orbital. In addition, we have been told that the energy needed to completely remove one mole of $3s$ electrons from one mole of sodium atoms in the ground state is 496 kJ. The ionization energy for each excited state can be found by subtracting the "energy quanta" entry for the excited state from 496 kJ mol^{-1}.

e.g., for $[Ne]3p^1$, the first ionization energy = $496 \dfrac{kJ}{mol} - 203 \dfrac{kJ}{mol} = 293 \dfrac{kJ}{mol}$

Thus, the rest of the ionization energies are:
$[Ne]4s^1, = 496$ kJ $mol^{-1} - 308$ kJ $mol^{-1} = 188$ kJ mol^{-1}
$[Ne]3d^1, = 496$ kJ $mol^{-1} - 349$ kJ $mol^{-1} = 147$ kJ mol^{-1}
$[Ne]4p^1, = 496$ kJ $mol^{-1} - 362$ kJ $mol^{-1} = 134$ kJ mol^{-1}

(b) Z_{eff} (the effective nuclear charge) for each state can be found by using the equation:

ionization energy in kJ mol^{-1} (I.E.) $= \dfrac{A(Z_{eff})^2}{n^2}$

Where n = starting principal quantum level for the electron that is promoted out of the atom and $A = 1.3121 \times 10^3$ kJ mol^{-1} (Rydberg constant).

For $[Ne]3p^1$ ($n = 3$) 2.93×10^2 kJmol^{-1} = $\dfrac{1.3121 \times 10^3 \dfrac{kJ}{mol} (Z_{eff})^2}{3^2}$ $Z_{eff} = 1.42$

For $[Ne]4s^1$ ($n = 4$) 1.88×10^2 kJmol^{-1} = $\dfrac{1.3121 \times 10^3 \dfrac{kJ}{mol} (Z_{eff})^2}{4^2}$ $Z_{eff} = 1.51$

For $[Ne]3d^1$ ($n = 3$) 1.47×10^2 kJmol^{-1} = $\dfrac{1.3121 \times 10^3 \dfrac{kJ}{mol} (Z_{eff})^2}{3^2}$ $Z_{eff} = 1.00$

For $[Ne]4p^1$ ($n = 4$) 1.34×10^2 kJmol^{-1} = $\dfrac{1.3121 \times 10^3 \dfrac{kJ}{mol} (Z_{eff})^2}{4^2}$ $Z_{eff} = 1.28$

(c) $\bar{r}_{nl}$, which is the average distance of the electron from the nucleus for a particular orbital, can be calculated with the equation:

$$\bar{r}_{nl} = \dfrac{n^2 a_o}{Z_{eff}}\left(1 + \dfrac{1}{2}\left(1 - \dfrac{\ell(\ell+1)}{n^2}\right)\right)$$ Where $a_o = 52.9$ pm,
n = principal quantum number
ℓ = angular quantum number for the orbital

For [Ne]$3p^1$ $(n = 3, \ \ell = 1\ )\ \bar{r}_{3p} = \dfrac{3^2(52.9 \ \text{pm})}{1.42}\left(1 + \dfrac{1}{2}\left(1 - \dfrac{1(1+1)}{3^2}\right)\right) = 466 \ \text{pm}$

For [Ne]$4s^1$ $(n = 4, \ \ell = 0\ )\ \bar{r}_{4s} = \dfrac{4^2(52.9 \ \text{pm})}{1.51}\left(1 + \dfrac{1}{2}\left(1 - \dfrac{0(0+1)}{4^2}\right)\right) = 823 \ \text{pm}$

For [Ne]$3d^1$ $(n = 3, \ \ell = 2\ )\ \bar{r}_{3d} = \dfrac{3^2(52.9 \ \text{pm})}{1.00}\left(1 + \dfrac{1}{2}\left(1 - \dfrac{2(2+1)}{3^2}\right)\right) = 555 \ \text{pm}$

For [Ne]$4p^1$ $(n = 4, \ \ell = 1\ )\ \bar{r}_{4p} = \dfrac{4^2(52.9 \ \text{pm})}{1.28}\left(1 + \dfrac{1}{2}\left(1 - \dfrac{1(1+1)}{4^2}\right)\right) = 950 \ \text{pm}$

(d) The results from the Z_{eff} calculations show that the greatest effective nuclear charge is experienced by the $4s$ orbital ($Z_{eff} = 1.51$). Next are the two p-orbitals, $3p$ and $4p$, which come in at 1.42 and 1.28 respectively. Coming in last is the $3d$ orbital, which has a $Z_{eff} = 1.00$. These results are precisely in keeping with what we would expect. First of all, only the s-orbital penetrates all the way to the nucleus. Both the p- and d-orbitals have nodes at the nucleus. Also p-orbitals penetrate more deeply than do d-orbitals. Recall that the more deeply an orbital penetrates (i.e., the closer the orbital is to the nucleus), the greater is the effective nuclear charge felt by the electrons in that orbital. It follows then that the $4s$ orbital will experience the greatest effective nuclear charge and that the Z_{eff} values for the $3p$ and $4p$ orbitals should be larger than the Z_{eff} for the $3d$ orbital.

The results from the $\bar{r}_{nl}$ calculations for the four excited state orbitals show that the largest orbital in the set is the $4p$ orbital. This is exactly as expected because the $4p$ orbital is highest in energy and hence, on average farthest from the nucleus. The $4s$ orbital has an average position closer to the nucleus because it experiences a larger effective nuclear charge. The $3p$ orbital, being lowest in energy and hence on average closest to the nucleus, is the smallest orbital in the set. The $3p$ orbital has an average position closer to the nucleus than the $3d$ orbital (which is in the same principal quantum level), because it penetrates more deeply into the atom.

SELF-ASSESSMENT EXERCISES

73. **(E)** The answer is (b). The element in question is antimony (Sb), which is in the same group as Bi.

74. **(E)** The answer is (a), K. This is because atomic radius decreases going from left to right of the period.

75. **(E)** The answer is (a), Cl⁻. All of the choices have the same electron configuration as Ar, but Cl⁻ has an extra electron in the valence shell, which expands the ionic radius.

76. **(E)** The answer is (b), because it has the smallest radius (and highest electron affinity). Therefore, the valence electrons are held more tightly, hence a higher first ionization energy.

77. **(E)** The answer is (a), Br. This is because electron affinities increase across the periodic table, and are greatest for nonmetallic main group elements. The other choices are for metals.

78. **(E)** The answer is (d), Sr^{2+}. They both have the electron configuration of Kr.

79. **(E)** The electron configurations of the first and second ionization products of Cs are as follows:
Cs^+: [Xe], or [Kr] $4d^{10}5s^25p^6$
Cs^{2+}: [Kr] $4d^{10}5s^25p^5$
The second ionization energy is much greater because one has to overcome the extra energy required to remove an electron from a stable, filled subshell (which resembles that of a stable noble atom Xe).

80. **(M)** The first ionization energy of Mg is higher than Na because, in the case of Mg, an electron from a filled subshell is being removed, whereas in Na, the removing of one electron leads to the highly stable electron configuration of Ne. The second ionization of Mg is lower than Na, because in the case Mg, the electron configuration of Ar is achieved, whereas in Na, the stable [Ar] configuration is being lost by removing an electron from the filled subshell.

81. **(M)**
(a) As, because it is the left and bottom-most element in the choices given
(b) F^-. Xe valence shell has $n = 5$, so it would be the largest and therefore not correct. Among those with shells with $n = 2$, F^- is the smallest because it has the highest nuclear charge and therefore more attraction of the orbitals to the nucleus
(c) Cl^-, because it is the most electronegative, and is being farther removed from the ideal filled subshell electron configuration
(d) Carbon, because it is the smallest, and hence has the least amount of shielding of the nucleus from the valence band, and the greatest attraction between the valence electrons and the nucleus
(e) Carbon, because electron affinity increases going across a period

82. **(M)** The trends would generally follow higher first ionization energy values for a fuller subshell. The exception is the case of S and P, where P has a slightly higher value. This is because there is a slight energy advantage to having a half-filled subshell with an electron in each of p_x, p_y and p_z orbitals as in the case of P, whereas the S has one extra electron.

83. **(E)** The pairs are Ar/Ca, Co/Ni, Te/I and Th/Pa. The periodic table must be arranged by atomic number because only this order is consistent with the regularity in electron configurations that is the ultimate basis of the table.

84. **(E)**
(a) protons = 50
(b) neutrons = 69
(c) $4d$ electrons = 10
(d) $3s$ electrons = 2
(e) $5p$ electrons = 2
(f) valence shell electrons = 4

85. **(E)**
  (a) F; it is the top right-most reactive element and has the highest electron affinity
  (b) Sc
  (c) Si

86. **(E)**
  (a) C
  (b) Rb
  (c) At

87. **(E)**
  (a) Ba
  (b) S
  (c) Bi, because Ba < Ca < Bi < As < S

88. **(E)** Rb > Ca > Sc > Fe > Te > Br > O > F

89. **(M)**
  (a) False. The s orbitals have a higher probability of being near the nucleus (whereas the probability is zero for p and d orbitals), so they are more effective at shielding.
  (b) True. The s orbitals have much better penetration than p or d orbitals and therefore are better at shielding nuclear charge.
  (c) True for all atoms except hydrogen. Z_{eff} has a maximum theoretical value equal to Z. In practice, it is always less in a multi-electron atom, because there is always some shielding of the nuclear charge by the electrons.
  (d) True. Electrons in p orbitals penetrate better than those in d orbitals.
  (e) True. To understand this, remember that ionization energy, $I = R_H \times Z_{eff}^2 / n^2$. Use the data in Table 9-4 to determine Z_{eff} for these elements.

90. **(M)**
  (a) False. The $1s$ orbital has more penetration with the nucleus than the $2s$ orbital, and feels nearly the entire charge of the nucleus (higher Z_{eff}).
  (b) False. The Z_{eff} of a $2s$ orbital is greater than a $2p$ because of higher penetration.
  (c) True, because the electron in an s orbital has greater penetration with the nucleus and is more tightly attracted.
  (d) True. Because of the greater penetration of the $2s$ electrons, the $2p$ electrons are more effectively shielded from the full nuclear charge.

91. **(M)** These ionization energies are the reverse of electron affinities, for example, the ionization energy, E_i, for Li^- is $-(-59.6 \text{ kJ/mol})$. The variations in these anions follow those seen in Figure 9-15.

92. **(M)** Ionization energy generally increases with Z for a given period (and decreases going to higher periods) with the exception of the small deviation observed going from N to O because N has a slightly more stable configuration for N where half of the orbitals are filled.

93. **(E)** Mg would have the more positive electron affinity, because Mg has the electron configuration $[Ne]3s^2$. Therefore, the added electron would need to occupy the higher energy level $3p$ orbitals, compared with Na (electron configuration = $[Ne]3s^1$), which has a partially filled $3s$ orbital that can accommodate another electron.

94. **(E)** The addition of an electron is usually exothermic because the electrons in the atom do not completely shield the incoming electron from the nuclear charge. Therefore, the incoming electron is attracted by the nucleus. The reverse process is easier to think about. Given a valence electron in an atom, energy (such as heat) must be expended to remove that electron.

95. **(E)** The answer is (a) atomic radius. Metallic elements tend to occur earlier in a period than nonmetals and have larger radii. Compared with nonmetal atoms, metal atoms have low ionization energies and low electron affinities.

96. **(E)** The answer is (c) Ge. On the periodic table, Ge falls on the boundary line between metals and nonmetals.

97. **(E)** The answer is (a) Mg^{2+}. When Mg loses two electrons, it has the same electron configuration as Ne. However, because Mg has two more protons in its nucleus, the effective nuclear charge experienced by the electrons is greater, which decreases the radius.

98. **(E)** The answer is (c) the outer-shell electron with the highest orbital quantum number. When an atom is ionized, it loses its electron(s) in the valence shell, which is the outer-most shell and has the highest energy electrons. The electrons in the valence shell with highest orbital quantum number have the highest energy. Hence, they are the first to be removed when the atom ionizes.

99. **(E)** The answer is (c) $4s$. The electron configuration of Fe is $1s^2 2s^2 2p^6 3s^2 3p^6 3d^6 4s^2$, whereas the electron configuration for Fe^{2+} is $1s^2 2s^2 2p^6 3s^2 3p^6 3d^6$.

CHAPTER 10
CHEMICAL BONDING I: BASIC CONCEPTS
PRACTICE EXAMPLES

1A **(E)** Mg is in group 2(2A), and thus has 2 valence electrons and 2 dots in its Lewis symbol. Ge is in group 14(4A), and thus has 4 valence electrons and 4 dots in its Lewis symbol. K is in group 1(1A), and thus has 1 valence electron and 1 dot in its Lewis symbol. Ne is in group 18(8A), and thus has 8 valence electrons and 8 dots in its Lewis symbol.

$$\cdot \text{Mg} \cdot \qquad \cdot \text{Ge} \cdot \qquad \text{K} \cdot \qquad :\!\text{Ne}\!:$$

1B **(E)** Sn is in Family 4A, and thus has 4 electrons and 4 dots in its Lewis symbol. Br is in Family 7A with 7 valence electrons. Adding an electron produces an ion with 8 valence electrons. Tl is in Family 3A with 3 valence electrons. Removing an electron produces a cation with 2 valence electrons.

S is in Family 6A with 6 valence electrons. Adding 2 electrons produces an anion with 8 valence electrons.

$$\cdot \text{Sn} \cdot \qquad [:\!\ddot{\text{Br}}\!:]^- \qquad [\cdot\text{Tl}\cdot]^+ \qquad [:\!\ddot{\text{S}}\!:]^{2-}$$

2A **(E)** The Lewis structures for the cation, the anion, and the compound follows the explanation.

(a) Na loses one electron to form Na^+, while S gains two to form S^{2-}.

$$\text{Na}\cdot -1\,e^- \rightarrow [\,\text{Na}\,]^+ \qquad \cdot\ddot{\text{S}}\cdot + 2e^- \rightarrow [:\!\ddot{\text{S}}\!:]^{2-} \quad \text{Lewis Structure: } [\text{Na}]^+[:\!\ddot{\text{S}}\!:]^{2-}[\text{Na}]^+$$

(b) Mg loses two electrons to form Mg^{2+}, while N gains three to form N^{3-}.

$$\cdot\text{Mg}\cdot -2\,e^- \rightarrow [\,\text{Mg}\,]^{2+} \qquad \cdot\dot{\ddot{\text{N}}}\cdot + 3e^- \rightarrow [:\!\ddot{\text{N}}\!:]^{3-}$$

Lewis Structure: $[\text{Mg}]^{2+}[:\!\ddot{\text{N}}\!:]^{3-}[\text{Mg}]^{2+}[:\!\ddot{\text{N}}\!:]^{3-}[\text{Mg}]^{2+}$

2B **(E)** Below each explanation are the Lewis structures for the cation, the anion, and the compound.

(a) In order to acquire a noble-gas electron configuration, Ca loses two electrons, and I gains one, forming the ions Ca^{2+} and I^-. The formula of the compound is CaI_2.

$$\cdot\text{Ca}\cdot -2\,e^- \rightarrow [\,\text{Ca}\,]^{2+} \qquad :\!\dot{\ddot{\text{I}}}\cdot + e^- \rightarrow [:\!\ddot{\text{I}}\!:]^- \quad \text{Lewis Structure: } [:\!\ddot{\text{I}}\!:]^-[\text{Ca}]^{2+}[:\!\ddot{\text{I}}\!:]^-$$

(b) Ba loses two electrons and S gains two to acquire a noble-gas electron configuration, forming the ions Ba^{2+} and S^{2-}. The formula of the compound is BaS.

$$\cdot\text{Ba}\cdot -2\,e^- \rightarrow [\,\text{Ba}\,]^{2+} \qquad \cdot\ddot{\text{S}}\cdot + 2e^- \rightarrow [:\!\ddot{\text{S}}\!:]^{2-} \quad \text{Lewis Structure: } [\text{Ba}]^{2+}[:\!\ddot{\text{S}}\!:]^{2-}$$

(c) Each Li loses one electron and each O gains two to attain a noble-gas electron configuration, producing the ions Li^+ and O^{2-}. The formula of the compound is Li_2O.

$$\cdot\text{Li} - 1\,e^- \rightarrow [\,\text{Li}\,]^+ \quad \cdot\ddot{\text{O}}\cdot + 2e^- \rightarrow [:\!\ddot{\text{O}}\!:]^{2-} \quad \text{Lewis Structure: } [\text{Li}]^+[:\!\ddot{\text{O}}\!:]^{2-}[\text{Li}]^+$$

3A **(M)** In the Br_2 molecule, the two Br atoms are joined by a single covalent bond. This bonding arrangement gives each Br atom a closed valence shell configuration that is equivalent to that for a Kr atom.

In CH_4, the carbon atom is covalently bonded to four hydrogen atoms. This arrangement gives the carbon atom a valence shell octet and each H atom a valence shell duet.

In HOCl, the hydrogen and chlorine atoms are attached to the central oxygen atom through single covalent bonds. This bonding arrangement provides each atom in the molecule with a closed valence shell.

3B **(M)** The Lewis structure for NI_3 is similar to that of NH_3. The central nitrogen atom is attached to each iodine atom by a single covalent bond. All of the atoms in this structure get a closed valence shell.

The Lewis diagram for N_2H_4 has each nitrogen with one lone pair of electrons, two covalent bonds to hydrogen atoms, and one covalent bond to the other nitrogen atom. With this arrangement, the nitrogen atoms complete their octets while the hydrogen atoms complete their duets.

In the Lewis structure for C_2H_6, each carbon atom shares four pairs of electrons with three hydrogen atoms and the other carbon atom. With this arrangement, the carbon atoms complete their octets while the hydrogen atoms complete their duets.

4A **(E)** The bond with the most ionic character is the one in which the two bonded atoms are the most different in their electronegativities. We find electronegativities in Figure 10-6 and calculate ΔEN for each bond.

Electronegativities: H = 2.1 Br = 2.8 N = 3.0 O= 3.5 P = 2.1 Cl = 3.0
Bonds :   H—Br   N—H   N—O   P—Cl
ΔEN values :   0.7   0.9   0.5   0.9
Therefore, the N—H and P—Cl bonds are the most polar of the four bonds cited.

4B **(E)** The most polar bond is the one with the greatest electronegativity difference.
Electronegativities: C = 2.5 S = 2.5 P = 2.1 O = 3.5 F = 4.0
Bonds: C—S   C—P   P—O   O—F
ΔEN values: 0.0   0.4   1.4   0.5
Therefore, the P—O bond is the most polar of the four bonds cited.

5A **(E)** The electrostatic potential map that corresponds to IF is the one with the most red in it. This suggests polarization in the molecule. Specifically, the red region signifies a build-up of negative charge that one would expect with the very electronegative fluorine. The other electrostatic potential map corresponds to IBr. The electronegativities are similar, resulting in a relatively non-polar molecule (i.e., little in the way of charge build-up in the molecule).

5B **(E)** The electrostatic potential map that corresponds to CH_3OH is the one with the most red in it. This suggests polarization in the molecule. Specifically, the red region signifies a build-up of negative charge that one would expect with the very electronegative oxygen atom. The other electrostatic potential map corresponds to CH_3SH. The carbon and sulfur electronegativities are similar, resulting in a relatively non-polar molecule (i.e., little in the way of charge build-up in the molecule).

6A **(M)**

(a) C has 4 valence electrons and each S has 6 valence electrons: $4 + (2 \times 6) = 16$ valence electrons or 8 pairs of valence electrons. We place C between two S, and use two electron pairs to hold the molecule together, one between C and each S. We complete the octet on each S with three electron pairs for each S. This uses up six more electron pairs, for a total of eight electron pairs used. $:\ddot{S} - C - \ddot{S}:$ But C does not have an octet. We correct this situation by moving one lone pair from each S into a bonding position between C and S. $:\ddot{S} = C = \ddot{S}:$

(b) C has 4 valence electrons, N has 5 valence electrons and hydrogen has 1 valence electron: Total number of valence electrons $= 4 + 5 + 1 = 10$ valence electrons or 5 pairs of valence electrons. We place C between H and N, and use two electron pairs to hold the molecule together, one between C and N, as well as one between C and H. We complete the octet on N using three lone pairs. This uses up all five valence electron pairs ($H - C - \ddot{N}:$). But C does not yet have an octet. We correct this situation by moving two lone pairs from N into bonding position between C and N. $H - C \equiv N:$

(c) C has 4 valence electrons, each Cl has 7 valence electrons, and oxygen has 6 valence electrons: Thus, the total number of valence electrons $= 4 + 2(7) + 6 = 24$ valence electrons or 12 pairs of valence electrons. We choose C as the central atom, and use three electron pairs to hold the molecule together, one between C and O, as well as one between C and each Cl. We complete the octet on Cl and O using three lone pairs. This uses all twelve electron pairs. But C does not have an octet. We correct this situation by moving one lone pair from O into a bonding position between C and O.

$$:\ddot{O} - C(-\ddot{C}l:)_2 \rightarrow :\ddot{O} = C(-\ddot{C}l:)_2$$

6B **(a)**

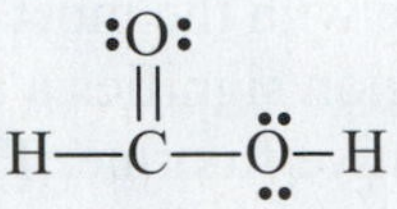

$$:\overset{\displaystyle ..}{\underset{\displaystyle ..}{O}}:$$
$$\| $$
$$H—C—\overset{\displaystyle ..}{\underset{\displaystyle ..}{O}}—H$$

(b)

$$H \quad :\overset{\displaystyle ..}{O}:$$
$$| \qquad \|$$
$$H—C—C—H$$
$$|$$
$$H$$

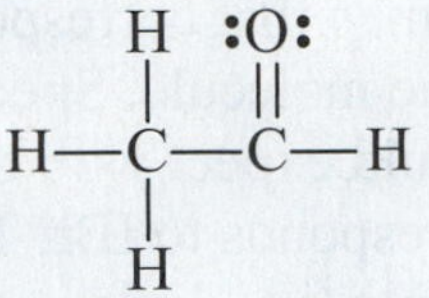

7A **(E)**

(a) A plausible Lewis structure for the nitrosonium cation, NO^+, is drawn below:

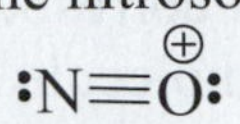

$$:N\!\equiv\!\overset{\oplus}{\underset{\displaystyle ..}{O}}:$$

The nitrogen atom is triply bonded to the oxygen atom and both atoms in the structure possess a lone pair of electrons. This gives each atom an octet and a positive formal charge appears on the oxygen atom.

(b) A plausible Lewis structure for $N_2H_5^+$ is given below:

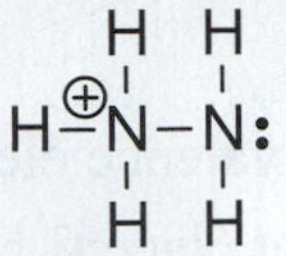

$$H \quad H$$
$$| \quad |$$
$$H—\overset{\oplus}{N}—N:$$
$$| \quad |$$
$$H \quad H$$

The two nitrogen atoms have each achieved an octet. The right hand side N atom is surrounded by three bonding pairs and one lone pair of electrons, while the left hand side N atom is surrounded by four bonding pairs of electrons. Each hydrogen atom has completed its duet by sharing a pair of electrons with a nitrogen atom. A formal 1+ charge has been assigned to the left hand side nitrogen atom because it is bonded to four atoms (one more than its usual number) in this structure.

(c) In order to achieve a noble gas configuration, oxygen gains two electrons, forming the stable dianion. The Lewis structure for O^{2-} is shown below.

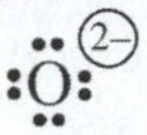

$$:\overset{\displaystyle ..}{\underset{\displaystyle ..}{O}}:^{\textcircled{2-}}$$

7B **(M)**

(a) The most likely Lewis structure for BF_4^- is drawn below:

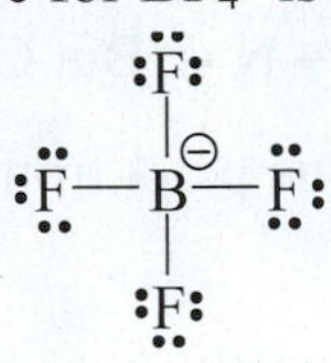

$$:\overset{\displaystyle ..}{\underset{\displaystyle ..}{F}}:$$
$$|$$
$$:\overset{\displaystyle ..}{\underset{\displaystyle ..}{F}}—\overset{\ominus}{B}—\overset{\displaystyle ..}{\underset{\displaystyle ..}{F}}:$$
$$|$$
$$:\overset{\displaystyle ..}{\underset{\displaystyle ..}{F}}:$$

Four bonding pairs of electrons surround the central boron atom in this structure. This arrangement gives the boron atom a complete octet and a formal charge of -1. By virtue of being surrounded by three lone pairs and one bonding electron pair, each fluorine achieves a full octet.

(b) A plausible Lewis structural form for NH_3OH^+, the hydroxylammonium ion, has been provided below:

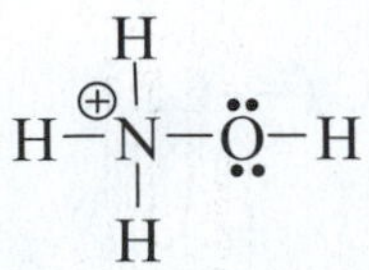

$$H$$
$$|$$
$$H—\overset{\oplus}{N}—\overset{\displaystyle ..}{\underset{\displaystyle ..}{O}}—H$$
$$|$$
$$H$$

By sharing bonding electron pairs with three hydrogen atoms and the oxygen atom, the nitrogen atom acquires a full octet and a formal charge of 1+. The oxygen atom shares one bonding electron pair with the nitrogen and a second bonding pair with a hydrogen atom.

(c) Three plausible resonance structures can be drawn for the isocyanate ion, NCO⁻. The nitrogen contributes five electrons, the carbon four, oxygen six, and one more electron is added to account for the negative charge, giving a total of 16 electrons or eight pairs of electrons. In the first resonance contributor, structure 1 below, the carbon atom is joined to the nitrogen and oxygen atoms by two double bonds, thereby creating an octet for carbon. To complete the octet of nitrogen and oxygen, each atom is given a lone pair of electrons. Since nitrogen is sharing just two bonding pairs of electrons in this structure, it must be assigned a formal charge of 1−. In structure 2, the carbon atom is again surrounded by four bonding pairs of electrons, but this time, the carbon atom forms a triple bond with oxygen and just a single bond with nitrogen. The octet for the nitrogen atom is closed with three lone pairs of electrons, while that for oxygen is closed with one lone pair of electrons. This bonding arrangement necessitates giving nitrogen a formal charge of 2− and the oxygen atom a formal charge of 1+. In structure 3, which is the dominant contributor because it has a negative formal charge on oxygen (the most electronegative element in the anion), the carbon achieves a full octet by forming a triple bond with the nitrogen atom and a single bond with the oxygen atom. The octet for oxygen is closed with three lone pairs of electrons, while that for nitrogen is closed with one lone pair of electrons.

$$:\overset{\ominus}{\ddot{N}}{=}C{=}\ddot{O}: \qquad \overset{(-2)}{:\ddot{N}}{-}C{\equiv}\overset{\oplus}{O}: \qquad :N{\equiv}C{-}\overset{\ominus}{\ddot{O}}:$$

| Structure 1 | Structure 2 | Structure 3 |

8A **(M)** The total number of valence electrons in NOCl is 18 (5 from nitrogen, 6 from oxygen and 7 from chlorine). Four electrons are used to covalently link the central oxygen atom to the terminal chlorine and nitrogen atoms in the skeletal structure: N—O—Cl. Next, we need to distribute the remaining electrons to achieve a noble gas electron configuration for each atom. Since four electrons were used to form the two covalent single bonds, fourteen electrons remain to be distributed. By convention, the valence shells for the terminal atoms are filled first. If we follow this convention, we can close the valence shells for both the nitrogen and the chlorine atoms with twelve electrons.

$$:\ddot{N}{-}O{-}\ddot{\ddot{C}}l:$$

Oxygen is moved closer to a complete octet by placing the remaining pair of electrons on oxygen as a lone pair.

$$:\ddot{N}{-}\ddot{O}{-}\ddot{\ddot{C}}l:$$

The valence shell for the oxygen atom can then be closed by forming a double bond between the nitrogen atom and the oxygen atom.

$$:\overset{\ominus}{\ddot{N}}{=}\overset{\oplus}{\ddot{O}}{-}\ddot{\ddot{C}}l:$$

225

This structure obeys the requirement that all of the atoms end up with a filled valence shell, but is much poorer than the one derived in Example 10-8 because it has a positive formal charge on oxygen, which is the most electronegative atom in the molecule. In other words, this structure can be rejected on the grounds that it does not conform to the third rule for determining plausibility of a Lewis structure based on formal charges, which states that "negative formal charges should appear on the most electronegative atom, while any positive formal charge should appear on the least electronegative atom."

8B **(D)** There are a total of sixteen valence electrons in the cyanamide molecule (five from each nitrogen atom, four from carbon and one electron from each hydrogen atom). The formula has been written as NH_2CN to remind us that carbon, the most electropositive p-block element in the compound, should be selected as the central atom in the skeletal structure.

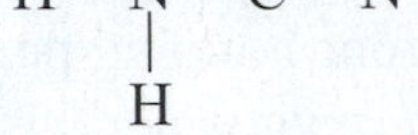

$$H-N-C-N$$
$$\underset{\displaystyle H}{|}$$

To construct this skeletal structure we use 8 electrons. Eight electrons remain to be added to the structure. Note: each hydrogen atom at this stage has achieved a duet by forming a covalent bond with the nitrogen atom in the NH_2 group. The octet for the NH_2 nitrogen is completed by giving it a lone pair of electrons.

$$H-\overset{\displaystyle ..}{N}-C-N$$
$$\underset{\displaystyle H}{|}$$

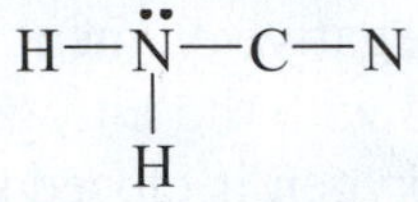

The remaining six electrons can then be given to the terminal nitrogen atom, affording structure 1, shown below. Alternatively, four electrons can be assigned to the terminal nitrogen atom and the last two electrons can be given to the central carbon atom, to produce structure 2 below:

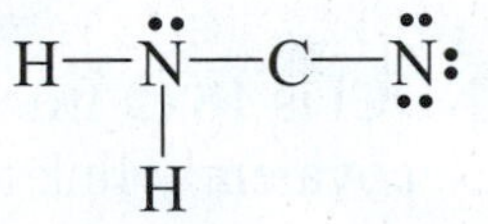

$$H-\overset{..}{N}-C-\overset{..}{N}:\qquad H-\overset{..}{N}-\overset{..}{C}-\overset{..}{N}:$$
$$\;\;|\qquad\qquad\qquad\qquad\;\;|$$
$$\;\;H\qquad\qquad\qquad\qquad\;\;H$$

(Structure 1) 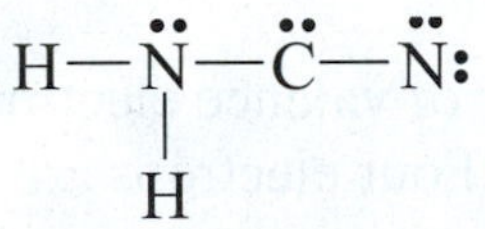   (Structure 2)

The octet for the carbon atom in structure 1 can be completed by converting two lone pairs of electrons on the terminal nitrogen atom into two more covalent bonds to the central carbon atom.

$$H-\overset{..}{N}-C\equiv N:$$
$$\;\;|$$
$$\;\;H$$

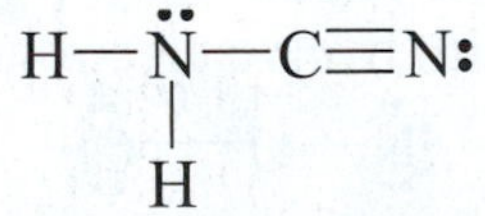

Structure 3

Each atom in structure 3 has a closed-shell electron configuration and a formal charge of zero. We can complete the octet for the carbon and nitrogen atoms in structure 1 by converting a lone pair of electrons on each nitrogen atom into a covalent bond to the central carbon atom.

$$H-\overset{\oplus}{N}=C=\overset{..}{\underset{..}{N}}{}^{\ominus}$$
$$\;\;|$$
$$\;\;H$$

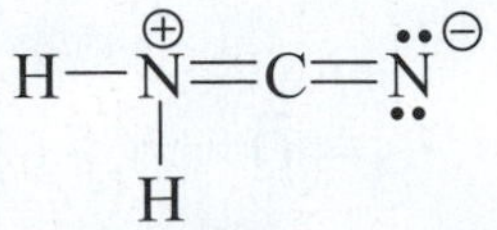

Structure 4

The resulting structure has a formal charge of 1– on the terminal nitrogen atom and a 1+ formal charge on the NH_2 nitrogen atom. Although structures 3 and 4 both satisfy the octet and duet rules, structure 3 is the better of the two structures because it has no formal charges. A third structure which obeys the octet rule (depicted below), can be rejected on the grounds that it has formal charges of the same type (two 1+ formal charges) on adjacent atoms, as well as negative formal charges on carbon, which is not the most electronegative element in the molecule.

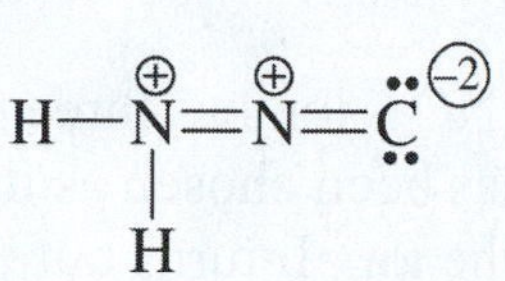

$$H—\overset{\oplus}{N}=\overset{\oplus}{N}=\overset{\ominus}{\underset{..}{C}}$$
$$|$$
$$H$$

9A   **(D)** The skeletal structure for SO_2 has two terminal oxygen atoms bonded to a central sulfur atom. Sulfur has been selected as the central atom by virtue of its being the most electropositive atom in the molecule. Two different Lewis structures of identical energy can be derived from the skeletal structure described above. First we determine that SO_2 has 18 valence electrons (6 from each atom). Four of the valence electrons must be used to covalently bond the three atoms together. The remaining 14 electrons are used to close the valence shell of each atom. Twelve electrons are used to give the terminal oxygen atoms a closed shell. The remaining two electrons ($14 - 12 = 2$) are placed on the sulfur atom, affording the structure depicted below:

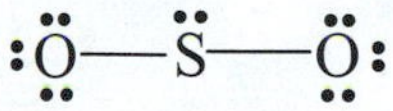

$$:\overset{..}{\underset{..}{O}}—\overset{..}{S}—\overset{..}{\underset{..}{O}}:$$

At this stage, the valence shells for the two oxygen atoms are closed, but the sulfur atom is two electrons short of a complete octet. If we complete the octet for sulfur by converting a lone pair of electrons on the right hand side oxygen atom into a sulfur-to-oxygen π-bond, we end up generating the resonance contributor (A) shown below:

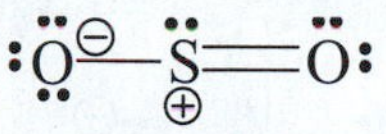

$$:\overset{..}{\underset{..}{O}}{}^{\ominus}—\underset{\oplus}{\overset{..}{S}}=\overset{..}{O}:\qquad (A)$$

Notice that the structure has a positive formal charge on the sulfur atom (most electropositive element) and a negative formal charge on the left-hand oxygen atom. Remember that oxygen is more electronegative than sulfur, so these charges are plausible.
The second completely equivalent contributor, (B), is produced by converting a lone pair on the left-most oxygen atom in the structure into a π-bond, resulting in conversion of a sulfur-oxygen single bond into a sulfur-oxygen double bond:

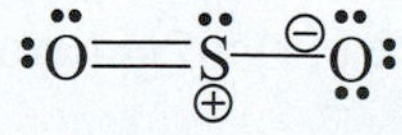

$$:\overset{..}{O}=\underset{\oplus}{\overset{..}{S}}—\overset{..}{\underset{..}{O}}{}^{\ominus}:\qquad (B)$$

We obtain a third resonance structure from structure A or B by converting a lone pair on the singly bonded oxygen to a π bonding pair. In the resulting structure, the formal charges on the S and O atoms are all zero.

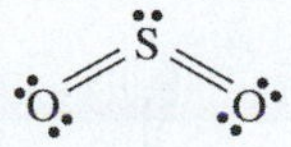

$$\cdot\overset{..}{\underset{..}{O}}\cdot=\overset{..}{S}=\cdot\overset{..}{\underset{..}{O}}\cdot$$

The true structure of SO_2 is a resonance hybrid of three structures.

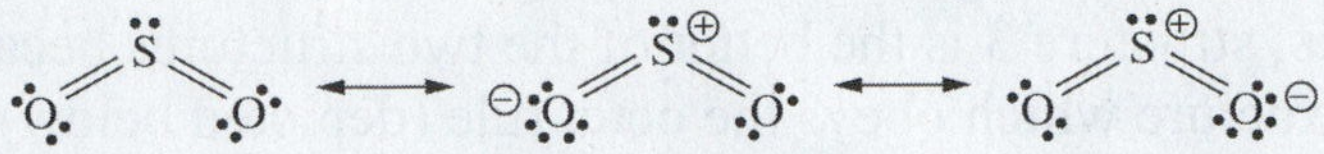

As a result, the sulfur–oxygen bond lengths in SO_2 are equal. (In the first structure, the two sulfur–oxygen bond lengths are the same, and the last two structures contribute equally.)

9B **(D)** The skeletal structure for the NO_3^- ion has three terminal oxygen atoms bonded to a central nitrogen atom. Nitrogen has been chosen as the central atom by virtue of being the most electropositive atom in the ion. It turns out that three contributing resonance structures of identical energy can be derived from the skeletal structure described here. We begin the process of generating these three structures by counting the total number of valence electrons in the NO_3^- anion. The nitrogen atom contributes five electrons, each oxygen contributes six electrons, and an additional electron must be added to account for the 1– charge on the ion. In total, we must account for 24 electrons. Six electrons are used to draw single covalent bonds between the nitrogen atom and three oxygen atoms. The remaining 18 electrons are used to complete the octet for the three terminal oxygen atoms:

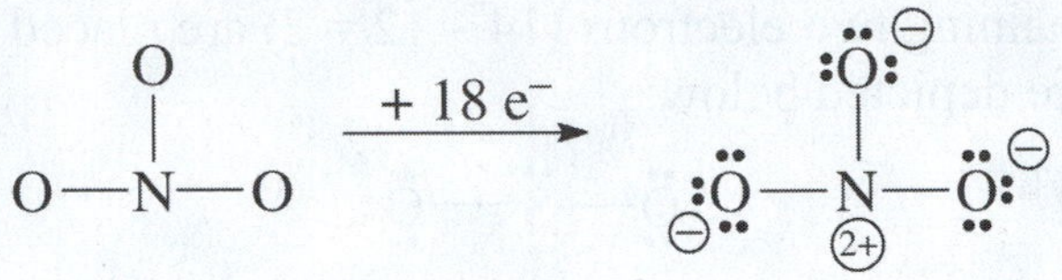

At this stage the valence shells for the oxygen atoms are filled, but the nitrogen atom is two electrons short of a complete octet. If we complete the octet for nitrogen by converting a lone pair on O_1 into a nitrogen-to-oxygen π-bond, we end up generating resonance contributor (A):

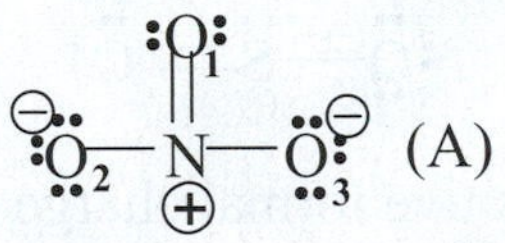

Notice the structure has a 1+ formal charge on the nitrogen atom and a 1– on two of the oxygen atoms (O_2 and O_3). These formal charges are quite reasonable energetically. The second and third equivalent structures are generated similarly; by moving a lone pair from O_2 to form a nitrogen to oxygen (O_2) double bond, we end up generating resonance contributor (B), shown below. Likewise, by converting a lone pair from oxygen (O_3) into a π-bond with the nitrogen atom, we end up generating resonance contributor (C), also shown below.

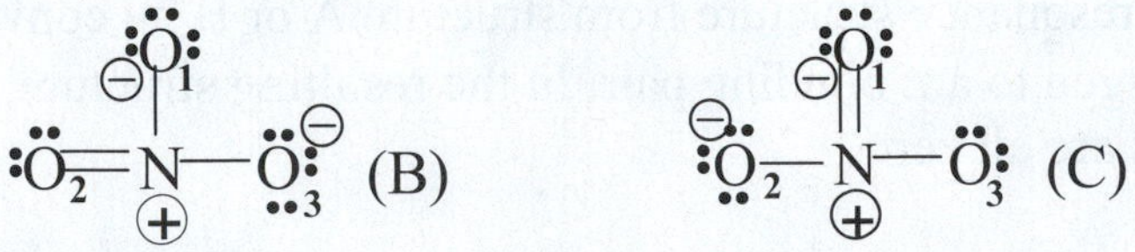

None of these individual structures ((A), (B), or (C)) correctly represents the actual bonding in the nitrate anion. The actual structure, called the resonance hybrid, is the equally weighted average of all three structures (i.e. 1/3(A) + 1/3(B) + 1/3(C)):

These three resonance forms give bond lengths that are comparable to nitrogen- nitrogen double bonds.

10A **(E)** The Lewis structure of NCl_3 has three Cl atoms bonded to N and one lone pair attached to N. These four electron groups around N produce a tetrahedral electron-group geometry. The fact that one of the electron groups is a lone pair means that the molecular geometry is trigonal pyramidal.

10B **(E)** The Lewis structure of $POCl_3$ has three single P—Cl bonds and one P—O bond. These four electron groups around P produce a tetrahedral electron-group geometry. No lone pairs are attached to P and thus the molecular geometry is tetrahedral.

11A **(E)** The Lewis structure of COS has one S doubly-bonded to C and an O doubly-bonded to C. There are no lone pairs attached to C. The electron-group and molecular geometries are the same: linear. $\ddot{S} = C = \ddot{O}$. We can draw other resonance forms, however, the molecular geometry is unaffected.

11B **(E)** N is the central atom. $:N \equiv N - \ddot{O}:$ This gives an octet on each atom, a formal charge of 1+ on the central N, and a 1– on the O atom. There are two bonding pairs of electrons and no lone pairs on the central N atom. The N_2O molecule is linear. We can draw other resonance forms, however, the molecular geometry is unaffected.

12A **(E)** In the Lewis structure of methanol, each H atom contributes 1 valence electron, the C atom contributes 4, and the O atom contributes 6, for a total of $(4 \times 1) + 4 + 6 = 14$ valence electrons, or 7 electron pairs. 4 electron pairs are used to connect the H atoms to the C and the O, 1 electron pair is used to connect C to O, and the remaining 2 electron pairs are lone pairs on O, completing its octet.

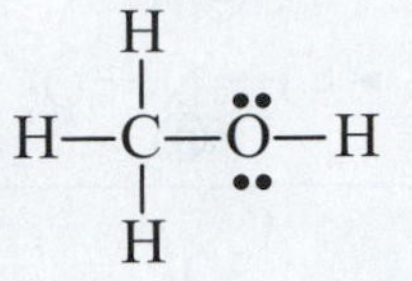

$$
\begin{array}{c}
\text{H} \\
| \\
\text{H--C--\overset{..}{\underset{..}{O}}--H} \\
| \\
\text{H}
\end{array}
$$

The resulting molecule has two central atoms. Around the C there are four bonding pairs, resulting in a tetrahedral electron-group geometry and molecular geometry. The H—C—H bond angles are ~109.5°, as are the H—C—O bond angles. Around the O there are two bonding pairs of electrons and two lone pairs, resulting in a tetrahedral electron-group geometry and a bent molecular shape around the O atom, with a C—O—H bond angle of slightly less than 109.5°.

12B **(M)** The Lewis structure is drawn below. With four electron groups surrounding each, the electron-group geometries of N, the central C, and the right-hand O are all tetrahedral. The H—N—H bond angle and the H—N—C bond angles are almost the tetrahedral angle of 109.5°, made a bit smaller by the lone pair. The H—C—N angles, the H—C—H angle and the H—C—C angles all are very close to 109.5°. The C—O—H bond angle is made somewhat smaller than 109.5° by the presence of two lone pairs on O. Three electron groups surround the right-hand C, making its electron-group and molecular geometries trigonal planar. The O—C—O bond angle and the O—C—C bond angles all are very close to 120°.

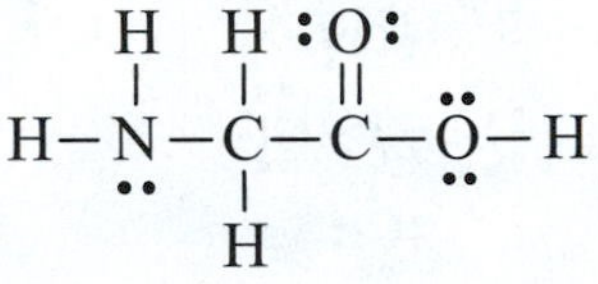

$$
\begin{array}{ccccc}
\text{H} & \text{H} & \overset{..}{\underset{}{:}}\overset{..}{O}: & & \\
| & | & || & & \\
\text{H--}\underset{..}{N}\text{--C--C--}\overset{..}{\underset{..}{O}}\text{--H} \\
& | & & \\
& \text{H} & &
\end{array}
$$

13A **(M)** Lewis structures of the three molecules are drawn below. Around the S in the SF_6 molecule are six bonding pairs of electrons, and no lone pairs. The molecule is octahedral; each of the S-F bond moments is cancelled by one on the other side of the molecule.

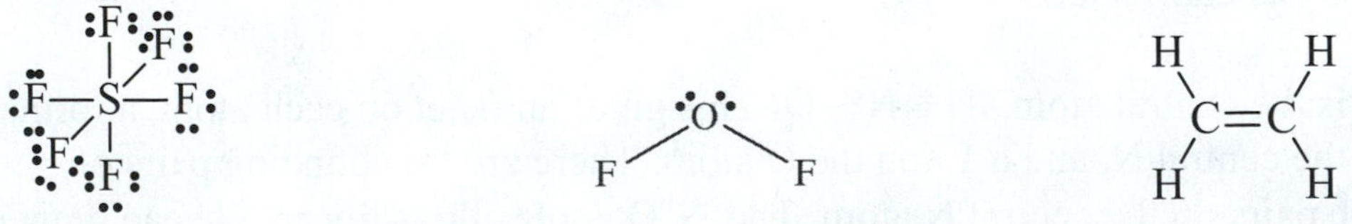

SF_6 is nonpolar. The OF_2 molecule is bent; the bond dipoles do not cancel. OF_2 is polar. H_2O_2 is polar. Around each C in C_2H_4 are three bonding pairs of electrons; the molecule is planar around each C and planar overall. The polarity of each —CH_2 group is cancelled by the polarity of the other H_2C— group. C_2H_2 is nonpolar.

13B **(M)** Lewis structures of the four molecules are drawn below and we can consider the three C—H bonds and the one C—C bond to be nonpolar. The three C—Cl bonds are tetrahedrally oriented.

$$
\begin{array}{cccc}
\ddot{:}\!Cl\!:\ \ H & :\!\ddot{C}l\!: & :\!\ddot{C}l\!: & H \\
\ |\ \ \ \ | & \ \ | & \ \ | & | \\
:\!\ddot{C}l\!-\!C\!-\!C\!-\!H & :\!\ddot{C}l\!\!>\!\!P\!-\!\ddot{C}l\!: & H\!-\!C\!-\!H & :\!N\!-\!H \\
\ |\ \ \ \ | & :\!\ddot{C}l\ \ :\!\ddot{C}l\!: & \ \ | & | \\
:\!\ddot{C}l\!:\ \ H & & :\!\ddot{C}l\!: & H
\end{array}
$$

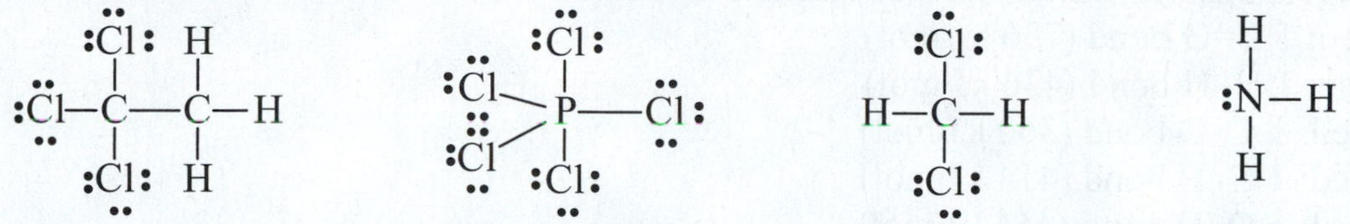

If there were a fourth C—Cl bond on the left-hand C, the bond dipoles would cancel out, producing a nonpolar molecule. Since it is not there, the molecule is polar. A similar argument is made for NH_3, where three tetrahedrally-oriented N—H polar bonds are not balanced by a fourth, and for CH_2Cl_2, where two tetrahedrally oriented C—Cl bonds are not balanced by two others. This leaves PCl_5 as the only nonpolar species; it is a highly symmetrical molecule in which individual bond dipoles cancel out.

14A **(M)** The Lewis structure of CH_3Br has all single bonds. From Table 10.2, the length of a C—H bond is 110 pm. The length of a C—Br bond is not given in the table. A reasonable value is the average of the C—C and Br—Br bond lengths.

$$
C\!-\!Br = \frac{C\!-\!C + Br\!-\!Br}{2} = \frac{154\ pm + 228\ pm}{2} = 191\ pm \qquad
\begin{array}{c}
H \\
| \\
H\!-\!C\!-\!\ddot{B}r\!: \\
| \\
H
\end{array}
$$

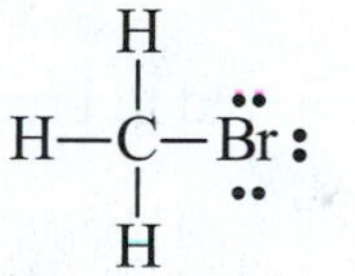

14B **(E)** In Table 10.2, the C=N bond length is 128 pm, while the C≡N bond length is 116 pm. The observed C—N bond length of 115 pm is much closer to a carbon-nitrogen triple bond. This can be explained by using the following Lewis structure: $:\!\ddot{S}\!-\!C\!\equiv\!N\!:$ where there is a formal negative charge on the sulfur atom. This molecule is linear according to VSEPR theory.

15A **(M)** We first draw Lewis structures for all of the molecules involved in the reaction.

$$
2\ H\!-\!H\ +\ \ddot{O}\!=\!\ddot{O}\ \longrightarrow\ 2H\!-\!\ddot{O}\!-\!H
$$

Break 1 O=O + 2H—H = 498 kJ/mol + (2 × 436 kJ/mol) = 1370 kJ/mol absorbed
Form 4H—O = (4 × 464 kJ/mol) = 1856 kJ/mol given off
Enthalpy change = 1370 kJ / mol − 1856 kJ / mol = −486 kJ / mol

15B **(M)** The chemical equation, with Lewis structures, is:

$$
1/2\ :\!N\!\equiv\!N\!:\ +\ 3/2\ H\!-\!H\ \longrightarrow\
\begin{array}{c}
H\!-\!\ddot{N}\!-\!H \\
| \\
H
\end{array}
$$

Energy required to break bonds $= \frac{1}{2}N \equiv N + \frac{3}{2}H\!-\!H$

$= (0.5 \times 946\ kJ / mol) + (1.5 \times 436\ kJ / mol) = 1.13 \times 10^3\ kJ / mol$

Energy realized by forming bonds $= 3\ N\!-\!H = 3 \times 389\ kJ / mol = 1.17 \times 10^3\ kJ / mol$

$\Delta_r H = 1.13 \times 10^3\ kJ/mol - 1.17 \times 10^3\ kJ/mol = -4 \times 10^1\ kJ/mol$ of NH_3.

Thus, $\Delta_f H = -4 \times 10^1\ kJ/mol\ NH_3$ (Appendix D value is $\Delta_f H = -46.11\ kJ/mol\ NH_3$)

16A **(M)** The reaction below,

$$CH_3(C=O)CH_3(g) + H_2(g) \rightarrow (CH_3)_2 CH(OH)(g)$$

Involves the following bond breakages and formations:
Broken: 1 C=O bond (736 kJ/mol)
Broken: 1 H–H bond (436 kJ/mol)
Formed: 1 C–O bond (360 kJ/mol)
Formed: 1 C–H bond (414 kJ/mol)
Formed: 1 O–H bond (464 kJ/mol)

Therefore, the energy of the reaction is:
$$\Delta_r H = \Delta H(\text{bond breakage}) + \Delta_r H(\text{bond formation})$$
$$\Delta_r H = \left[(1 \times 736) + (1 \times 436) \right] + \left[(1 \times -360) + (1 \times -414) + (1 \times -464) \right]$$
$$\Delta_r H = -66 \text{ kJ/mol}$$
Therefore, the reaction is exothermic.

16B **(E)** First we will double the chemical equation, and represent it in terms of Lewis structures:

2 H—Ö—H + 2 :Ċl–Ċl: → Ö=Ö + 4 H—Ċl:

Energy required to break bonds = 2 Cl — Cl + 4H — O
$$= (2 \times 243 \text{ kJ/mol}) + (4 \times 464 \text{ kJ/mol}) = 2342 \text{ kJ/mol}$$

Energy realized by forming bonds = 1 O = O + 4 × H — Cl
$$= 498 \text{ kJ/mol} + (4 \times 431 \text{ kJ/mol}) = 2222 \text{ kJ/mol}$$

$$\Delta H = \tfrac{1}{2}(2342 \text{ kJ/mol} - 2222 \text{ kJ/mol}) = +60 \text{ kJ/mol}; \text{ The reaction is endothermic.}$$

Important Note: In this and subsequent chapters, a lone pair of electrons in a Lewis structure can be shown as a line or a pair of dots. Thus, the Lewis structure of Be is both Be| and Be:.

INTEGRATIVE EXAMPLE

A   **(M)** The reaction is as follows: $PCl_3 + Cl_2 \rightarrow PCl_5$

$$\Delta_r H = \sum \Delta H_{\text{prod}} - \sum \Delta H_{\text{react}}$$
$$\Delta_r H = (-374.9) - (-287.0 + 0) = -87.9 \text{ kJ/mol}$$

To determine the P–Cl bond energy, we must first deduce the Cl–Cl bond energy:
$2Cl \rightarrow Cl_2 \ \Delta_r H = 0 - 2(121.7) = -243$ kJ/mol, which is in reasonable agreement with Table 10-3.

Using the above information, we can determine the P–Cl bond energy:

$$\Delta_r H = \Delta_r H(\text{bond breakage}) + \Delta_r H(\text{bond formation})$$
$$\Delta_r H = 1\ \text{Cl}-\text{Cl (broken)} + 2\ \text{P}-\text{Cl (formed)}$$
$$-87.9 = 243 + 2\ \text{P}-\text{Cl}$$
$$\text{P}-\text{Cl} = -165.5\ \text{kJ/mol}$$

The Lewis structures of PCl_3 and PCl_5 are shown below.

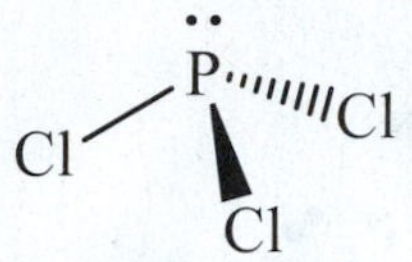

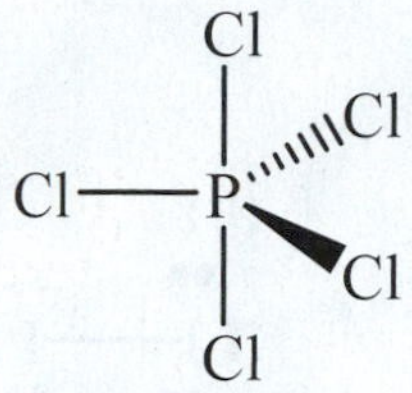

Since the geometries of the two molecules differ, the orbital overlap between P and the surrounding Cl atoms will be different and therefore the P–Cl bonds in these two compounds will also be slightly different.

B **(M)**
(a) The structures are shown below, with appropriate geometries:

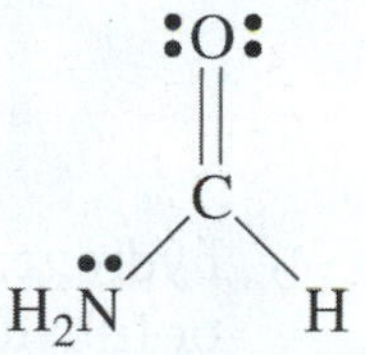

Formamide

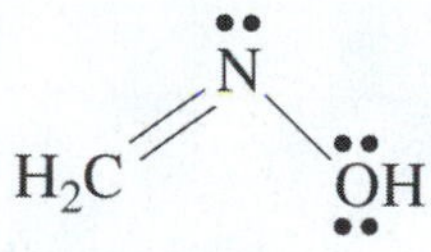

Formaldoxime

$$\sum \text{BE}(\text{H}_2\text{NCOH}) = 2(\text{N}-\text{H}) + 1(\text{N}-\text{C}) + 1(\text{C}=\text{O}) + 1(\text{C}-\text{H})$$
$$= 2(389) + 305 + 736 + 414 = 2233\ \text{kJ/mol}$$

$$\sum \text{BE}(\text{H}_2\text{C}=\text{N}-\text{OH}) = 2(\text{C}-\text{H}) + 1(\text{C}=\text{N}) + 1(\text{N}-\text{O}) + 1(\text{O}-\text{H})$$
$$= 2(414) + 615 + 222 + 464 = 2129\ \text{kJ / mol}$$

Since BE of formamide is greater than that of formaldoxime, it is more stable, and its conversion endothermic.

(b) The experiment shows that the geometry around C is trigonal planar, and around N is trigonal pyramidal.

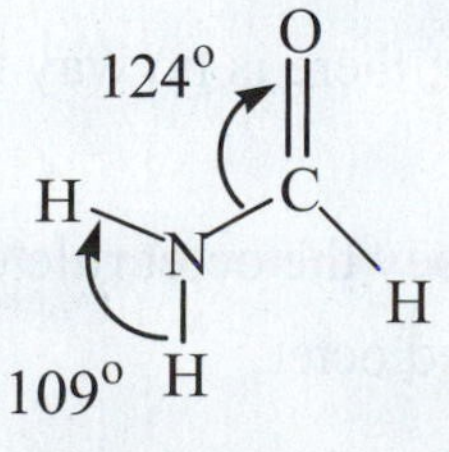

EXERCISES

Lewis Theory

1. **(E)**

(a) $:\ddot{\text{Kr}}:$ (b) $\cdot\dot{\text{Ge}}\cdot$ (c) $\cdot\dot{\text{N}}:$ (d) $\cdot\dot{\text{Ga}}$ (e) $\cdot\dot{\text{As}}:$ (f) $\text{Rb}\cdot$

3. **(E)**

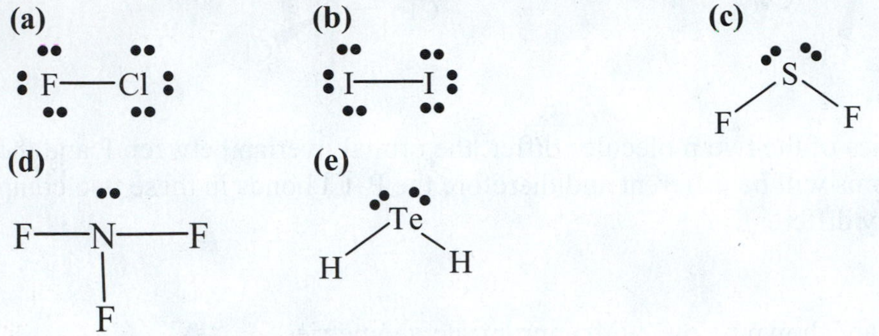

(a)

$:\ddot{\text{F}}\!-\!\ddot{\text{Cl}}:$

(b)

$:\ddot{\text{I}}\!-\!\ddot{\text{I}}:$

(c)

$\underset{\text{F}\qquad\text{F}}{\ddot{\text{S}}}$

(d)

$\text{F}\!-\!\ddot{\text{N}}\!-\!\text{F}$
$\quad\;\;\overset{|}{\text{F}}$

(e)

$\underset{\text{H}\qquad\text{H}}{\ddot{\text{Te}}}$

5. **(E)**

(a) 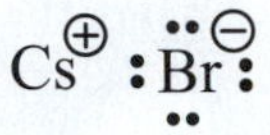 $\text{Cs}^{\oplus} \; :\!\ddot{\text{Br}}\!:^{\ominus}$ CsBr, cesium bromide

(b) 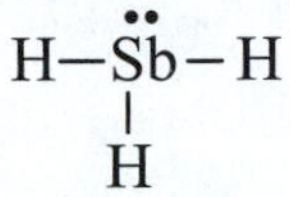 $\begin{array}{c}\text{H}\!-\!\ddot{\text{Sb}}\!-\!\text{H}\\ \overset{|}{\text{H}}\end{array}$ H_3Sb, hydrogen antimonide or trihydrogen antimonide

(c) 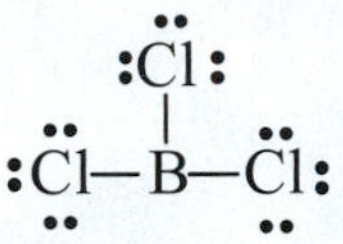 $:\ddot{\text{Cl}}\!-\!\overset{\displaystyle:\ddot{\text{Cl}}:}{\underset{|}{\text{B}}}\!-\!\ddot{\text{Cl}}:$ BCl_3, boron trichloride

(d)  $\text{Cs}^{\oplus} \; :\!\ddot{\text{Cl}}\!:^{\ominus}$ CsCl, cesium chloride

(e) 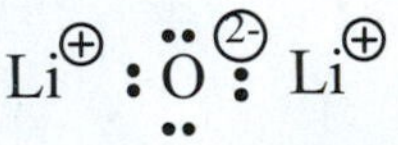 $\text{Li}^{\oplus} \; :\!\ddot{\text{O}}\!:^{\oslash 2} \; \text{Li}^{\oplus}$ Li_2O, lithium oxide

(f) 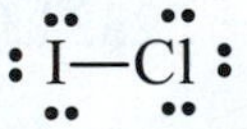 $:\ddot{\text{I}}\!-\!\ddot{\text{Cl}}:$ ICl, iodine monochloride

7. **(M)**

$NH_3 \quad 5+(3\times1)=8 \text{ v.e.}=4 \text{ pairs}$ $BF_3 \quad 3+(3\times7)=24 \text{ v.e.}=12 \text{ pairs}$

$SF_6 \quad 6+(6\times7)=48 \text{ v.e.}=24 \text{ pairs}$ $SO_3 \quad 6+(3\times6)=24 \text{ v.e.}=12 \text{ pairs}$

$NH_4^+ \; 5+(4\times1)-1=8 \text{ v.e.}=4 \text{ pairs}$ $SO_4^{2-} \; 6+(4\times6)+2=32 \text{ v.e.}=16 \text{ pairs}$

$NO_2 \quad 5+(2\times6)=17 \text{ v.e.}=8.5 \text{ pairs}$

NO_2 cannot obey the octet rule; there is no way to pair all electrons when the number of electrons is odd.

All of these Lewis structures obey the octet rule except for BF_3, which is electron deficient, and SF_6, which has an expanded octet.

9. **(E)**

(a) $H-H-N-\ddot{O}-H$ has two bonds to (four electrons around) the
second hydrogen, and only six electrons around the nitrogen.
A better Lewis structure is shown below.

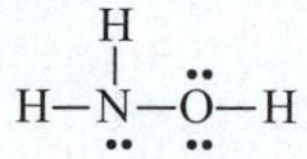

$$H-\underset{|}{\overset{H}{N}}-\ddot{O}-H$$

(b) Here, $Ca-\ddot{O}:$ is improperly written as a covalent Lewis structure. CaO is actually an

ionic compound. $[Ca]^{2+}\left[:\ddot{O}:\right]^{2-}$ is a more plausible Lewis structure for CaO.

11. **(M)** The answer is (c), hypochlorite ion. The flaws with the other answers are as follows:

(a) $\ominus:\ddot{O}-\overset{\oplus}{C}=\ddot{N}\ominus$ – does not have an octet of electrons around C.

(b) $[C=C:]^-$ does not have an octet around either C. Moreover, it has only 6 valence
electrons in total while it should have 10, and finally, the sum of the formal charges
on the two carbons doesn't equal the charge on the ion.

(d) The total number of valence electrons in NO is incorrect. No, being an odd-electron
species should have 11 valence electrons, not 12.

Ionic Bonding

13. **(E)**

(a) $[:\ddot{Cl}:]^-[Ca]^{2+}[:\ddot{Cl}:]^-$ (b) $[Ba]^{2+}[:\ddot{S}:]^{2-}$ (c) $[Li]^+[:\ddot{O}:]^{2-}[Li]^+$ (d) $[Na]^+[:\ddot{F}:]^-$

15. **(E)**

(a) $[Li]^+[:\ddot{S}:]^{2-}[Li]^+ : Li_2S$ (b) $[Na]^+[:\ddot{F}:]^- : NaF$

(c) $[:\ddot{I}:]^-[Ca]^{2+}[:\ddot{I}:]^- : CaI_2$ (d) $[:\ddot{Cl}:]^-[Sc]^{3+}[:\ddot{Cl}:]^- : ScCl_3$

Formal Charge

17. **(M)**

(a)	computations for:	H—	—C≡	≡C
	no. valence e⁻	1	4	4
	− no. lone-pair e⁻	−0	−0	−2
	− ½ no. bond-pair e⁻	−1	−4	−3
	formal charge	0	0	−1

(b)	computations for:	=O	—O(×2)	C
	no. valence e⁻	6	6	4
	− no. lone-pair e⁻	−4	−6	−0
	− ½ no. bond-pair e⁻	−2	−1	−4
	formal charge	−0	−1	0

(c)	computations for:	—H(×7)	side C(×2)	central C
	no. valence e⁻	1	4	4
	− no. lone-pair e⁻	−0	−0	−0
	− ½ no. bond-pair e⁻	−1	−4	−3
	formal charge	0	0	+1

19. **(M)** There are three features common to formal charge and oxidation state. First, both indicate how the bonding electrons are distributed in the molecule. Second, negative formal charge (in the most plausible Lewis structure) and negative oxidation state are generally assigned to the more electronegative atoms. And third, both numbers are determined by a set of rules, rather than being determined experimentally. Bear in mind, however, that there are also significant differences. For instance, there are cases where atoms of the same type with the same oxidation state have different formal charges, such as oxygen in ozone, O_3. Another is that formal charges are used to decide between alternative Lewis structures, while oxidation state is used in balancing equations and naming compounds. Also, the oxidation state in a compound is invariant, while the formal charge can change. The most significant difference, though, is that whereas the oxidation state of an element in its compounds is usually not zero, its formal charge usually is.

21. **(E)** FC = # valence e⁻ in free atom − number lone-pair e⁻ − ½ # bond pair e⁻
(a) Central O in O_3: $6 - 2 - 3 = +1$
(b) Al in AlH_4^- : $3 - 0 - 4 = -1$
(c) Cl in ClO_3^- : $7 - 2 - 5 = 0$
(d) Si in SiF_6^{2-} : $4 - 0 - 6 = -2$
(e) Cl in ClF_3 : $7 - 4 - 3 = 0$

23. **(M)** We begin by counting the total number of valence electrons that must appear in the Lewis structure of the ion CO_2H^+: one from hydrogen, four from carbon, and six from each of the two oxygen (12 in all from the oxygen atoms). One electron is lost to establish the 1+ charge on the ion. In all, sixteen electrons are in the valence shell of the cation. If the usual rules for constructing valid Lewis structures are applied to HCO_2^+, we come up with the following structures:

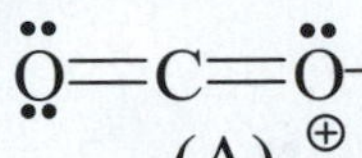

$$\overset{..}{\underset{..}{O}}=C=\overset{..}{\underset{..}{O}}-H \qquad :O\equiv C-\overset{..}{\underset{..}{O}}-H \qquad :\overset{..}{\underset{..}{O}}-C\equiv\overset{\oplus}{O}-H$$

$$\text{(A)}^{\oplus} \qquad\qquad \underset{\oplus}{}\text{(B)} \qquad\qquad \underset{\ominus}{}\text{(C)}^{\oplus}$$

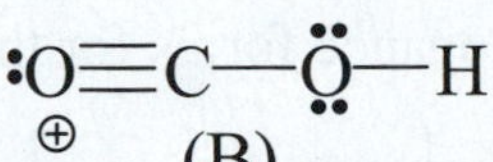

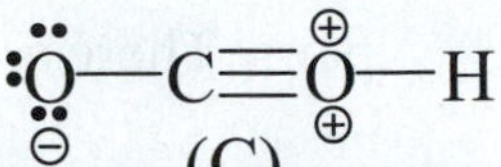

In structure (A), the internal oxygen atom caries the positive charge, while in structure (B), the positive charge is located on the terminal oxygen atom. A third structure can also be drawn, however, due to an unacceptably large charge build-up, this form can be neglected. Thus, in this case of A and B, we cannot use the concept of formal charge to pick one structure over the other because the positive formal charge in both structures is located on the same type of atom, namely, an oxygen atom. In other words, based on formal charge rules alone, we must conclude that structures (A) and (B) are equally plausible.

Lewis Structures

25. **(E)**

(a)

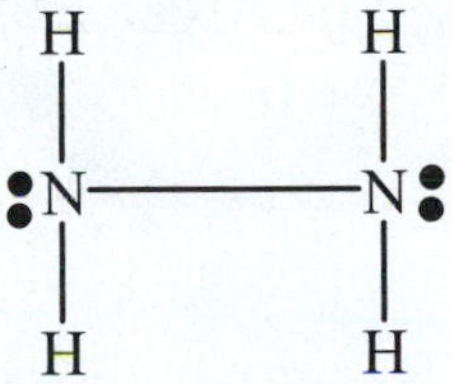

$$H \qquad\qquad H$$
$$| \qquad\qquad\qquad |$$
$$:N\text{————}N:$$
$$| \qquad\qquad\qquad |$$
$$H \qquad\qquad H$$

(b)

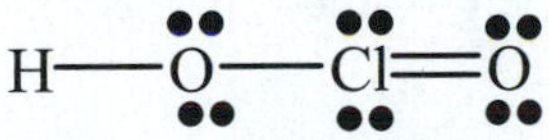

$$H\text{—}\overset{..}{\underset{..}{O}}\text{—}\overset{..}{Cl}=\overset{..}{\underset{..}{O}}$$

(c)

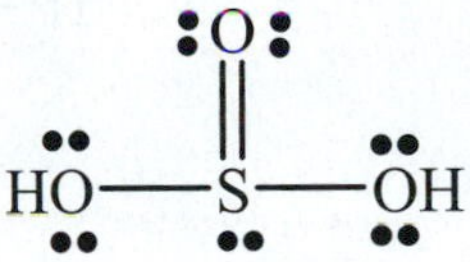

$$:\overset{..}{\underset{..}{O}}:$$
$$\|$$
$$HO\text{—}S\text{—}OH$$
$$\overset{..}{} \qquad \overset{..}{}$$

(d)

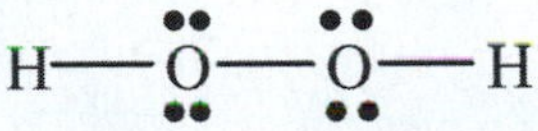

$$H\text{—}\overset{..}{\underset{..}{O}}\text{—}\overset{..}{\underset{..}{O}}\text{—}H$$

(e)

$$:\overset{..}{\underset{..}{O}}:$$
$$\|$$
$$^{-}:\overset{..}{\underset{..}{O}}\text{—}S\text{—}\overset{..}{\underset{..}{O}}:^{-}$$
$$\|$$
$$:\overset{..}{\underset{..}{O}}:$$

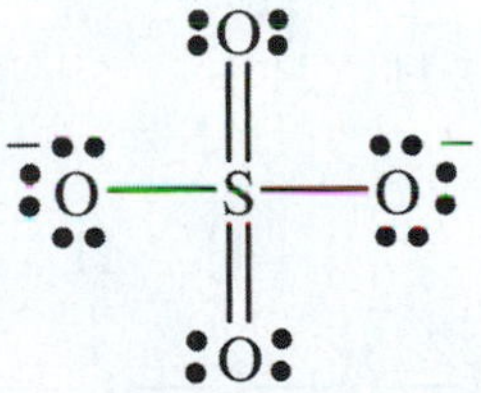

27. **(M)**

(a) The total number of valence electrons in SO_3^{2-} is $6+(3\times6)+2=26$, or 13 electron pairs. A plausible Lewis structure is:

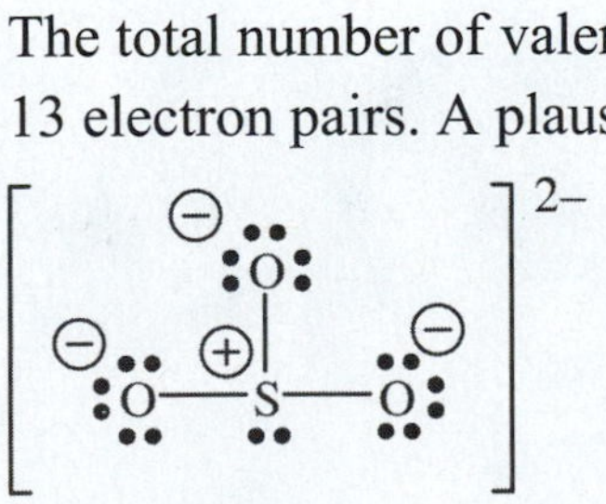

$$\left[\begin{array}{c} \overset{\ominus}{}\overset{..}{\underset{..}{O}}: \\ | \\ \overset{\ominus}{}\overset{..}{\underset{..}{O}}\text{—}\overset{\oplus}{S}\text{—}\overset{..}{\underset{..}{O}}:\overset{\ominus}{} \end{array} \right]^{2-}$$

(b) The total number of valence electrons in NO_2^- is $5 + (2 \times 6) + 1 = 18$, or 9 electron pairs. There are two resonance forms for the nitrite ion:

$$\overset{\ominus}{\underset{..}{\overset{..}{O}}} - \overset{..}{N} = \overset{..}{\underset{..}{O}} \quad \longleftrightarrow \quad \overset{..}{\underset{..}{O}} = \overset{..}{N} - \overset{..}{\underset{..}{O}} \overset{\ominus}{:}$$

(c) The total number of valence electrons in CO_3^{2-} is $4 + (3 \times 6) + 2 = 24$, or 12 electron pairs. There are three resonance forms for the carbonate ion:

$$\left[\begin{array}{c} \ominus :O: \\ | \\ \overset{..}{O} = C - \overset{..}{\underset{..}{O}}: \ominus \end{array} \right]^{2-} \longleftrightarrow \left[\begin{array}{c} :O: \\ || \\ \ominus :\overset{..}{O} - C - \overset{..}{\underset{..}{O}}: \ominus \end{array} \right]^{2-} \longleftrightarrow \left[\begin{array}{c} :\overset{..}{O}: \ominus \\ | \\ \ominus :\overset{..}{O} - C = \overset{..}{O} \end{array} \right]^{2-}$$

(d) The total number of valence electrons in HO_2^- is $1 + (2 \times 6) + 1 = 14$, or 7 electron pairs. A plausible Lewis structure is $H - \overset{..}{\underset{..}{O}} - \overset{..}{\underset{..}{O}} : \overset{\ominus}{}$

29. **(M)** In $CH_3CHCHCHO$ there are $(4 \times 4) + (6 \times 1) + 6 = 28$ valence electrons, or 14 electron pairs. We expect that the carbon atoms bond to each other. A plausible Lewis structure is:

$$\begin{array}{ccccc} & H & H & H & :O: \\ & | & | & | & || \\ H - & C - & C = & C - & C - H \\ & | & & & \\ & H & & & \end{array}$$

31. **(a)**
$$\begin{array}{c} :O: \\ || \\ H - C - H \end{array}$$

(E)

(b)
$$\begin{array}{ccc} H & :\overset{..}{Cl}: & H \\ | & | & | \\ H - C - & C - & C - \overset{..}{\underset{..}{O}} - H \\ | & | & | \\ H & H & H \end{array}$$

33. **(a)**
$$\begin{array}{ccc} H & :O: \\ | & || \\ :\overset{..}{\underset{..}{Cl}} - C - & C - \overset{..}{\underset{..}{O}} - H \\ | \\ H \end{array}$$

(E)

(b)
$$\begin{array}{cccc} & H & H & :O: \\ & | & | & || \\ H - \overset{..}{\underset{..}{O}} - & C - & C - & C - \overset{..}{\underset{..}{O}} - H \\ & | & | \\ & H & H \end{array}$$

35. (M)
(a) Group 16 **(b)** Group 16 except oxygen

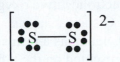

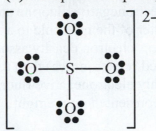

(c) Group 17 except fluorine **(d)** Group 13

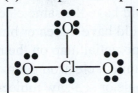

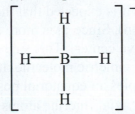

Polar Covalent Bonds and Electrostatic Potential Maps

37. (M) Na—Cl and K—F both possess bonds between a metal and a nonmetal. Thus, they have the largest ionic character, with the ionic character of K—F being greater than that of Na—Cl, both because K is more metallic (closer to the lower left of the periodic table) than Na and because F is more nonmetallic (closer to the upper right) than Cl. The remaining three bonds are covalent bonds to H. Since H and C have about the same electronegativity (a fact you need to memorize), the H—C bond is the most covalent (or the least ionic). Br is somewhat more electronegative than is C, while F is considerably more electronegative than C, making the F—H bond the most polar of the three covalent bonds. Thus, ranked in order of increasing ionic character, these five bonds are:

$$C—H < Br—H < F—H < Na—Cl < K—F$$

The actual electronegativity differences follow:

$$C(2.5)—H(2.1) < Br(2.8)—H(2.1) < F(4.0)—H(2.1) < Na(0.9)—Cl(3.0) < K(0.8)—F(4.0)$$
$$\Delta EN = 0.4 = 0.7 = 1.9 = 2.1 = 3.2$$

39. (M) The percent ionic character of a bond is based on the difference in electronegativity of its constituent atoms and Figure 10.7.

(a) $S(2.5)—H(2.1)$	**(b)** $O(3.5)—Cl(3.0)$	**(c)** $Al(1.5)—O(3.5)$	**(d)** $As(2.0)—O(3.5)$
$\Delta EN\ 0.4$	0.5	2.0	1.5
%ionic = 4%	= 5%	=60%	=33%

43. (M) When looking at the electrostatic potential map, we expect similar structures. However, in the case of $F_2C = O$, the carbon should be more electropositive than in $H_2C=O$ due to the presence of very electronegative fluorine atoms (as opposed to H atoms). Thus, for $F_2C = O$, one expects the center of the molecule to appear blue. As well, the electronegative oxygen atom should have less electron density associated with it, thus $H_2C = O$ should have a greater amount of red (electron rich) than the corresponding $F_2C = O$, again, as a result of the presence of highly electronegative fluorine atoms. $F_2C = O$ is represented on the left, while $H_2C = O$ is represented on the right.

45. (M) The molecular formulas for the compounds are SF_4 and SiF_4. SiF_4 is a symmetric molecule (tetrahedral). It is expected that the fluorine atoms should have the same electron density (same coloration). Since Si is more electropositive, it should have a greater blue coloration (more positive center). This suggests the electrostatic potential map on the right is for SiF_4. SF_4 is not a symmetric molecule. It has a trigonal bipyramidal electron geometry, where a lone pair occupies an equatorial position. It has a saw-horse or see-saw molecular shape in which the two axial fluorine atoms are nearly 180° to one another, while the two equatorial fluorine atoms are ~120° to one another. Owing to the lone pair in the equatorial position, the equatorial fluorine atoms will not be as electronegative as the axial fluorine atoms. This is certainly the case for the representation on the left.

Resonance

47. (E) In NO_2^-, the total number of valence electrons is $1+5+(2\times6)=18$ valence electrons, or 9 electron pairs. N is the central atom. The two resonance forms are shown below:

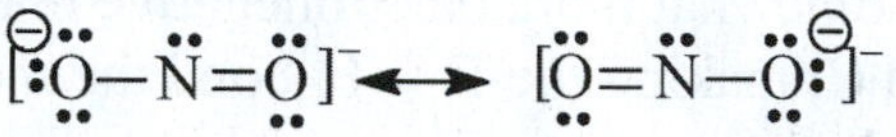

$$[\overset{\ominus}{\underset{\cdot\cdot}{\overset{\cdot\cdot}{O}}}-\overset{\cdot\cdot}{N}=\underset{\cdot\cdot}{\overset{\cdot\cdot}{O}}]^- \longleftrightarrow [\underset{\cdot\cdot}{\overset{\cdot\cdot}{O}}=\overset{\cdot\cdot}{N}-\overset{\ominus}{\underset{\cdot\cdot}{\overset{\cdot\cdot}{O}}}]^-$$

49. (M) Bond length data from Table 10.2 follow:

$N \equiv N$ 109.8 pm $N = N$ 123 pm $N — N$ 145 pm $N = O$ 120 pm $N — O$ 136 pm

The experimental $N — N$ bond length of 113 pm approximates that of the $N \equiv N$ triple bond, which appears in structure (a). The experimental $N — O$ bond length of 119 pm approximates that of the $N=O$ double bond, which appears in structure (b). Structure (d) is highly unlikely because it contains no nitrogen-to-nitrogen bonds, and a $N—N$ bond was found experimentally. Structure (c) also is unlikely, because it contains a very long (145 pm) $N — N$ single bond, which does not agree at all well with the experimental N-to-N bond length. The molecule seems best represented as a resonance hybrid of (a) and (b).

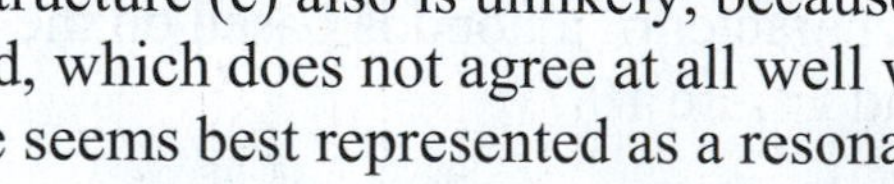

$$\overset{\ominus}{\underset{\cdot\cdot}{N}}=\overset{\oplus}{\underset{\cdot\cdot}{N}}=\underset{\cdot\cdot}{\overset{\cdot\cdot}{O}} \longleftrightarrow :N\equiv\overset{\oplus}{N}-\underset{\cdot\cdot}{\overset{\cdot\cdot}{\overset{\ominus}{O}}}:$$

51. (M)

(a) Both structures are equivalent.

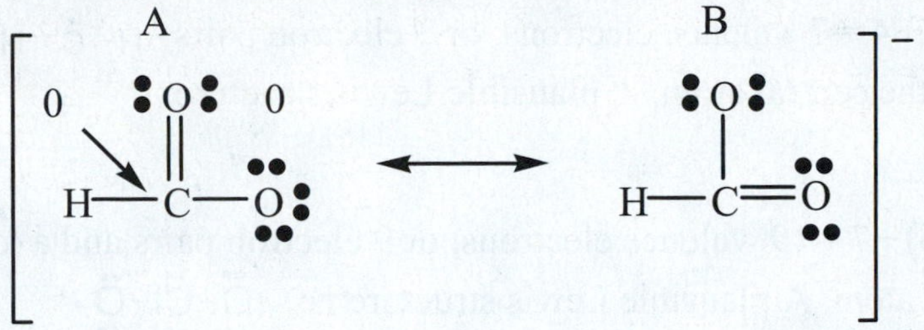

A ⟷ B (resonance structures of the formate-type ion, shown in brackets with a − charge)

(b) Structures A and C are the most important, and they are equivalent.

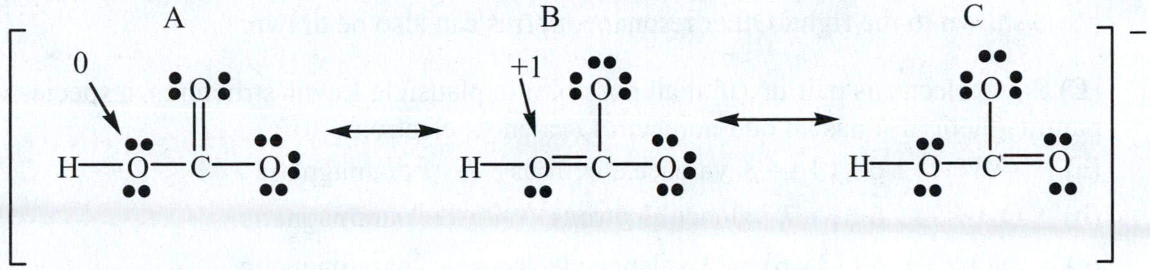

A ⟷ B ⟷ C (resonance structures shown in brackets with a − charge)

(c) Structure B is more important.

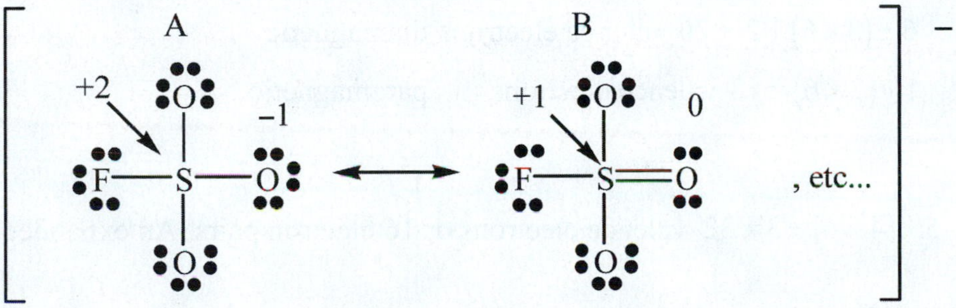

A ⟷ B , etc... (resonance structures shown in brackets with a − charge)

(d) Structure A is more important.

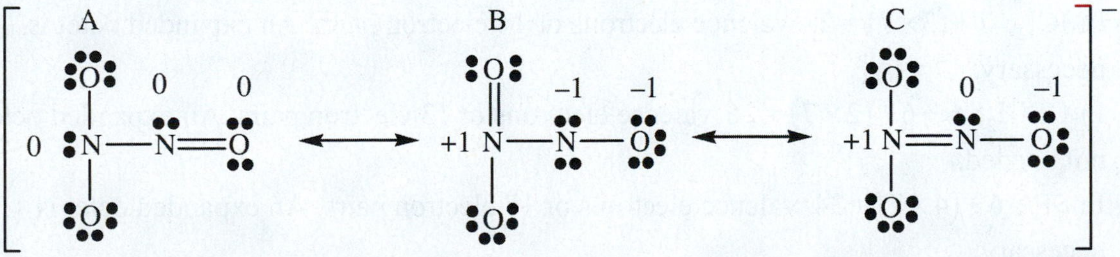

A ⟷ B ⟷ C (resonance structures shown in brackets with a − charge)

Odd-electron species

<u>53.</u> **(M)**

(a) CH_3 has a total of $(3 \times 1) + 4 = 7$ valence electrons, or 3 electron pairs $\quad H-\overset{.}{\underset{|}{C}}-H$ and a lone electron. C is the central atom. A plausible Lewis structure $\qquad\quad H$ is shown on the right.

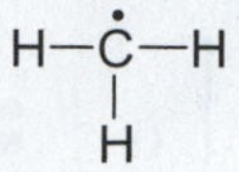

(b) ClO_2 has a total of $(2 \times 6) + 7 = 19$ valence electrons, or 9 electron pairs and a lone electron. Cl is the central atom. A plausible Lewis structure is: $\;:\overset{..}{\underset{..}{O}}-\overset{..}{\underset{..}{Cl}}-\overset{..}{\underset{..}{O}}\cdot$

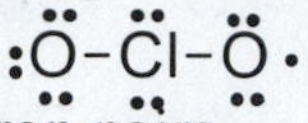

(c) NO_3 has a total of $(3 \times 6) + 5 = 23$ valence electrons, or 11 electron pairs, plus a lone electron. N is the central atom. A plausible Lewis structure is shown to the right. Other resonance forms can also be drawn.

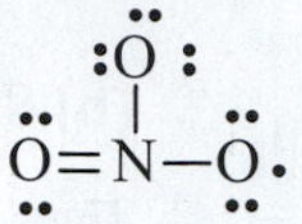

$$\begin{array}{c} :\overset{..}{\underset{..}{O}}: \\ | \\ \overset{..}{\underset{..}{O}}=N-\overset{..}{\underset{..}{O}}\cdot \end{array}$$

<u>55.</u> **(E)** Since electrons pair up (if at all possible) in plausible Lewis structures, a species will be paramagnetic if it has an odd number of (valence) electrons.

(a) OH^- $\qquad 6 + 1 + 1 = 8$ valence electrons $\qquad$ diamagnetic

(b) OH $\qquad 6 + 1 = 7$ valence electrons $\qquad$ paramagnetic

(c) NO_3 $\qquad 5 + (3 \times 6) = 23$ valence electrons $\qquad$ paramagnetic

(d) SO_3 $\qquad 6 + (3 \times 6) = 24$ valence electrons $\qquad$ diamagnetic

(e) SO_3^{2-} $\qquad 6 + (3 \times 6) + 2 = 26$ valence electrons diamagnetic

(f) HO_2 $\qquad 1 + (2 \times 6) = 13$ valence electrons $\qquad$ paramagnetic

Expanded Octets

<u>57.</u> **(M)** In PO_4^{3-}: $5 + (4 \times 6) + 3 = 32$ valence electrons or 16 electron pairs. An expanded octet is not needed.

In PI_3: $5 + (3 \times 7) = 26$ valence electrons or 13 electron pairs. An expanded octet is not needed.

In ICl_3: $7 + (3 \times 7) = 28$ valence electrons or 14 electron pairs. An expanded octet is necessary.

In $OSCl_2$: $6 + 6 + (2 \times 7) = 26$ valence electrons or 13 electron pairs. An expanded octet is not needed.

In SF_4: $6 + (4 \times 7) = 34$ valence electrons or 17 electron pairs. An expanded octet is necessary.

In ClO_4^-: $7 + (4 \times 6) + 1 = 32$ valence electrons or 16 electron pairs. An expanded octet is not needed.

Molecular Shapes

59. **(M)** The AX_nE_m designations that are cited below are to be found in Table 10.1 of the text, along with a sketch and a picture of a model of each type of structure.

(a) Dinitogen is linear; two points define a line.

$:N\equiv N:$

(b) Hydrogen cyanide is linear. The molecule belongs to the AX_2 category, and these species are linear.

$H\!-\!C\equiv N:$

(c) NH_4^+ is tetrahedral. The ion is of the AX_4 type, which has a tetrahedral electron-group geometry and a tetrahedral shape.

$$\left[\begin{array}{c} H \\ | \\ H\!-\!\overset{\oplus}{N}\!-\!H \\ | \\ H \end{array}\right]^+$$

(d) NO_3^- is trigonal planar. The ion is of the AX_3 type, which has a trigonal planar electron-group geometry and a trigonal planar shape. The other resonance forms are of the same type.

$$\left[\,{}^{\ominus}\!:\!\ddot{O}\!-\!\overset{\oplus}{N}\overset{\overset{\textstyle :\ddot{O}:}{||}}{}\!-\!\ddot{O}\!:^{\ominus}\,\right]^-$$

(e) NSF is bent. The molecule is of the AX_2E type, which has a trigonal planar electron-group geometry and a bent shape.

$${}^{\ominus}\ddot{N}\!=\!\underset{\oplus}{\ddot{S}}\!-\!\ddot{F}\!:$$

61. **(M)** We first draw all the Lewis structures. From each, we can deduce the electron-group geometry and the molecular shape.

(a) H_2S tetrahedral electron-group geometry, bent (angular) molecular geometry

$H\!-\!\ddot{\underset{..}{S}}\!-\!H$

(b) N_2O_4 trigonal planar electron-group geometry around each N, (planar molecule)

$${}^{\ominus}\!:\!\ddot{O}\!-\!\underset{\oplus}{N}\overset{\overset{\textstyle :\ddot{O}:}{||}}{}\!-\!\underset{\oplus}{N}\overset{\overset{\textstyle :\ddot{O}:}{||}}{}\!-\!\ddot{O}\!:^{\ominus}$$

(c) HCN   linear electron-group geometry, linear molecular geometry

$H\!-\!C\equiv N:$

(d) $SbCl_6^-$ octahedral electron-group geometry, octahedral geometry

$$\left[\begin{array}{c} :\ddot{C}l: \quad \ddot{C}l: \\ | \quad / \\ :\ddot{C}l\!-\!\underset{\ominus}{Sb}\!-\!\ddot{C}l: \\ / \quad | \\ :\ddot{C}l \quad :\ddot{C}l: \end{array}\right]^-$$

(e) BF_4^- tetrahedral electron-group geometry, tetrahedral molecular geometry

$$\left[\begin{array}{c} :\ddot{F}: \\ | \\ :\ddot{F}\!-\!\underset{\ominus}{B}\!-\!\ddot{F}: \\ | \\ :\ddot{F}: \end{array}\right]^-$$

63. **(M)** A trigonal planar shape requires that three groups and no lone pairs be bonded to the central atom. Thus PF_6^- cannot have a trigonal planar shape, since six atoms are attached to the central atom. In addition, PO_4^{3-} cannot have a trigonal planar shape, since four O atoms are attached to the central P atom. We now draw the Lewis structure of each of the remaining ions, as a first step in predicting their shapes. The SO_3^{2-} ion is of the AX_3E type. It has a tetrahedral electron-group geometry and a trigonal pyramidal shape. The CO_3^{2-} ion is of the AX_3 type, and has a trigonal planar electron-group geometry and a trigonal planar shape.

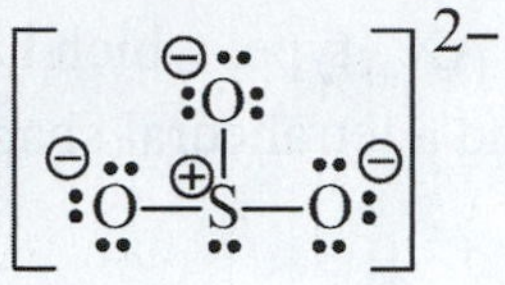

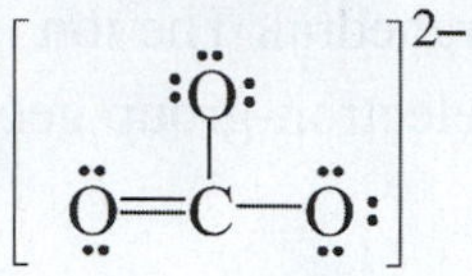

65. **(M)**
   (a) In CO_2 there are a total of $4 + (2 \times 6) = 16$ valence electrons, or 8 electron pairs. The following Lewis structure is plausible. $\ddot{O} = C = \ddot{O}$ This is a molecule of type AX_2. CO_2 has a linear electron-shape geometry and a linear shape.

   (b) In Cl_2CO there are a total of $(2 \times 7) + 4 + 6 = 24$ valence electrons, or 12 electron pairs. The molecule can be represented by a Lewis structure with C as the central atom. This molecule is of the AX_3 type. It has a trigonal planar electron-group geometry and molecular shape.

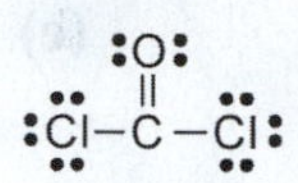

   (c) In $ClNO_2$ there are a total of $7 + 5 + (2 \times 6) = 24$ valence electrons, or 12 electron pairs. N is the central atom. A plausible Lewis structure is shown to the right: This molecule is of the AX_3 type. It has a trigonal planar electron-group geometry and a trigonal planar shape.

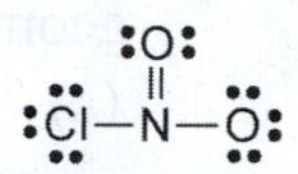

67. **(M)** First we draw the Lewis structure of each species, then use it to predict the molecular shape. The structures are provided below.

   (a) In ClO_4^- there are $7 + (4 \times 6) + 1 = 32$ valence electrons or 16 electron pairs. A plausible Lewis structure follows. Since there are four atoms and no lone pairs bonded to the central atom, the molecular shape and the electron-group geometry are the same: tetrahedral.

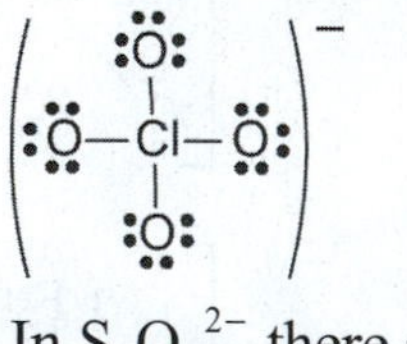

   (b) In $S_2O_3^{2-}$ there are $(2 \times 6) + (3 \times 6) + 2 = 32$ valence electrons or 16 electron pairs. A plausible Lewis structure follows. Since there are four atoms and no lone pairs bonded to the central atom, both the electron-group geometry and molecular shape are tetrahedral.

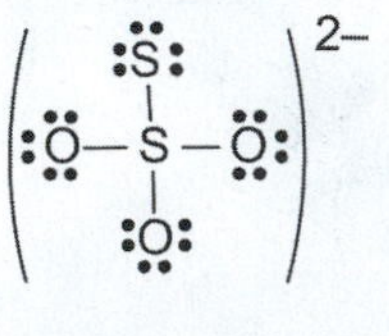

(c) In PF_6^- there are $5+(6\times7)+1=48$ valence electrons or 24 electron pairs. Since there are six atoms and no lone pairs bonded to the central atom, the electron-group geometry and molecular shape are octahedral.

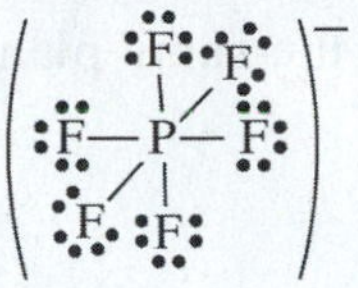

$$\left(\begin{array}{c}:\ddot{F}\cdot\cdot\ddot{F}:\\ \ |\ /\\ :\ddot{F}-P-\ddot{F}:\\ /\ \ |\\ :\ddot{F}\cdot\ :\ddot{F}:\end{array}\right)^{-}$$

(d) In I_3^- there are $(3\times7)+1=22$ valence electrons or 11 electron pairs. There are three lone pairs and two atoms bound to the central atom. The electron-group geometry is trigonal bipyramidal, thus, the molecular shape is linear.

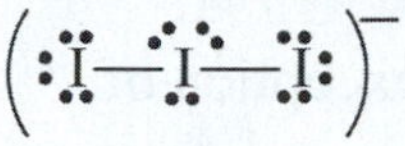

$$\left(:\ddot{I}-\ddot{I}-\ddot{I}:\right)^{-}$$

69. **(M)** In BF_4^-, there are a total of $1+3+(4\times7)=32$ valence electrons, or 16 electron pairs. A plausible Lewis structure has B as the central atom. This ion is of the type AX_4. It has a tetrahedral electron-group geometry and a tetrahedral shape.

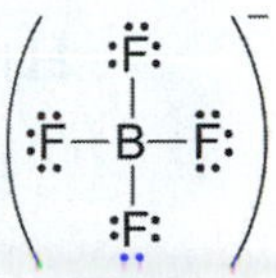

$$\left(\begin{array}{c}:\ddot{F}:\\ |\\ :\ddot{F}-B-\ddot{F}:\\ |\\ :\ddot{F}:\end{array}\right)^{-}$$

71. **(M)** Looking at the structures, the molecular angle/shape depends on the number of valence electron pairs on the central atom. The more pairs there are, the more acute the angle becomes.

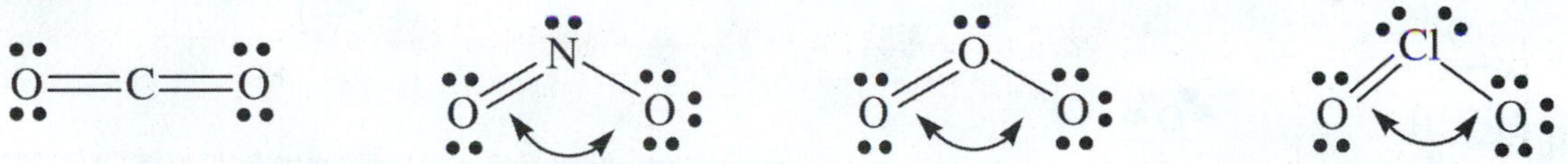

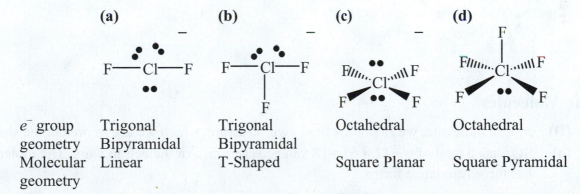

$180°$ $120°$ $120°$ $109°$

73. **(M)**

	(a)	**(b)**	**(c)**	**(d)**
e^- group geometry	Trigonal Bipyramidal	Trigonal Bipyramidal	Octahedral	Octahedral
Molecular geometry	Linear	T-Shaped	Square Planar	Square Pyramidal

245

Shapes of Molecules with More Than One Central Atom

75. **(E)**

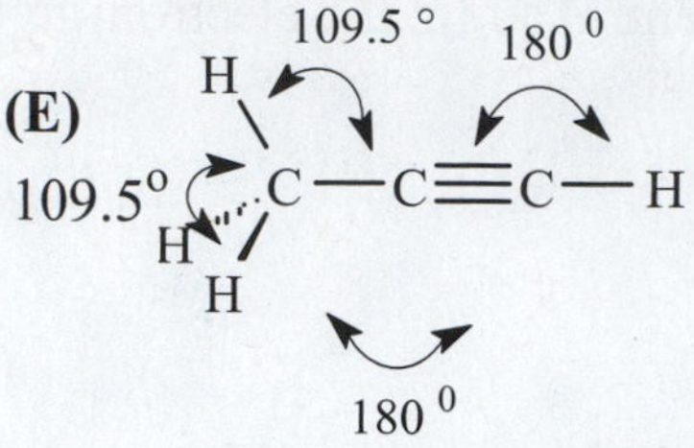

A maximum of 5 atoms can be in the same plane

77. **(E)**

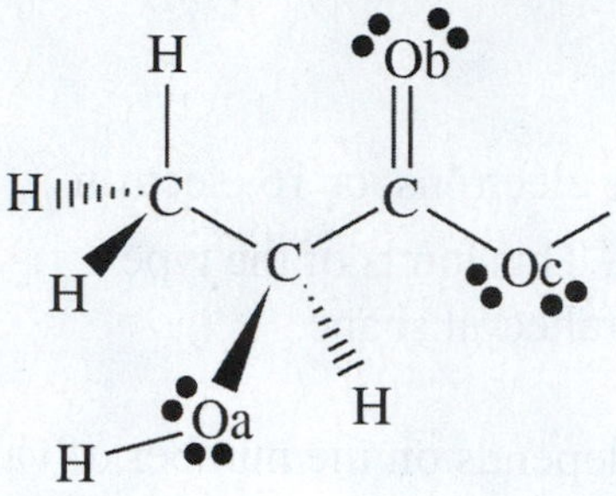

All angles ~109.5° with the exception of

$$\left. \begin{array}{l} O_c - C - C \\ O_b = C - C \\ O_b = C - O_c \end{array} \right\} 120°$$

79. **(M)**

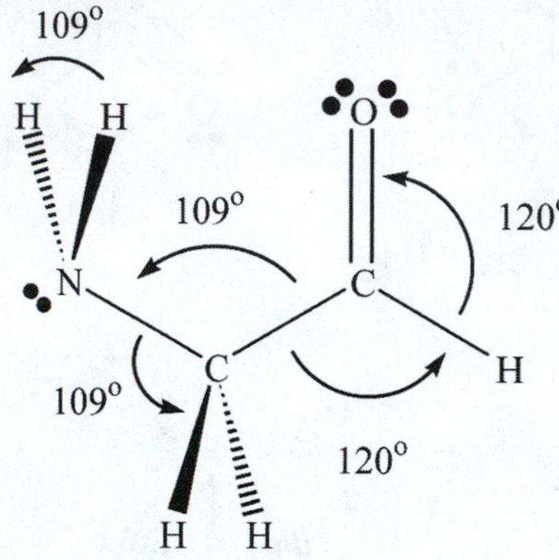

Polar Molecules

81. **(D)** For each molecule, we first draw the Lewis structure, which we use to predict the shape.

(a) SO_2 has a total of $6 + (2 \times 6) = 18$ valence electrons, or 9 electron pairs. The molecule has three resonance forms.

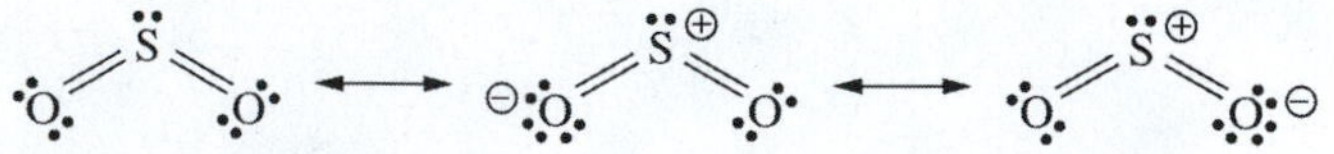

Each of these resonance forms is of the type AX_2E. Thus it has a trigonal planar electron-group geometry and a bent shape. Since each S—O bond is polar toward O, and since the bond dipoles do not cancel, the molecule has a resultant dipole moment, pointing from S through a point midway between the two O atoms. Consequently, SO_2 is polar.

(b) NH_3 has a total of $5 + (3 \times 1) = 8$ valence electrons, or 4 electron pairs. N is the central atom. A plausible Lewis structure is shown to the right. The molecule is of the AX_3E type; it has a tetrahedral electron-group geometry and a trigonal pyramidal shape. Each N-H bond is polar toward N. Since the bonds do not symmetrically oppose each other, there is a resultant molecular dipole moment, pointing from the triangular base (formed by the three H atoms) through N. Consequently, the molecule is polar.

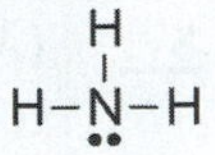

$$H-\overset{\displaystyle H}{\underset{\displaystyle ..}{N}}-H$$

(c) H_2S has a total of $6 + (2 \times 1) = 8$ valence electrons, or 4 electron pairs. S is the central atom and a plausible Lewis structure is H-$\ddot{\underset{..}{S}}$-H. This molecule is of the AX_2E_2 type; it has a tetrahedral electron-group geometry and a bent shape. Each H — S bond is polar toward S. Since the bonds do not symmetrically oppose each other, the molecule has a net dipole moment, pointing through S from a point midway between the two H atoms. H_2S is polar.

(d) C_2H_4 consists of atoms that all have about the same electronegativities. Of course, the C — C bond is not polar and essentially neither are the C — H bonds. The molecule is planar. Thus, the entire molecule is nonpolar.

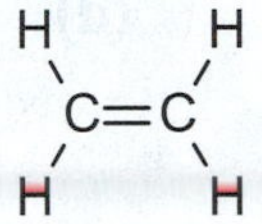

$$\overset{H}{\underset{H}{>}}C=C\overset{H}{\underset{H}{<}}$$

(e) SF_6 has a total of $6 + (6 \times 7) = 48$ valence electrons, or 24 electron pairs. S is the central atom. All atoms have zero formal charge in the Lewis structure. This molecule is of the AX_6 type. It has an octahedral electron-group geometry and an octahedral shape. Even though each S — F bond is polar toward F, the bonds symmetrically oppose each other, resulting in a molecule that is nonpolar.

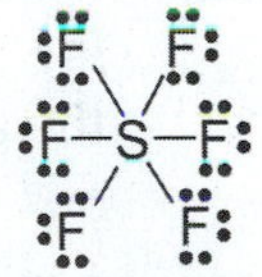

(f) CH_2Cl_2 has a total of $4 + (2 \times 1) + (2 \times 7) = 20$ valence electrons, or 10 electron pairs. A plausible Lewis structure is shown to the right. The molecule is tetrahedral and polar, since the two polar bonds $(C — Cl)$ do not cancel the effect of each other.

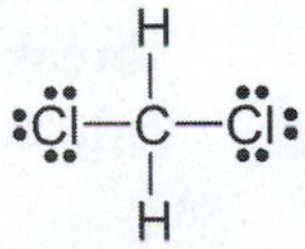

$$:\ddot{\underset{..}{C}l}-\overset{\displaystyle H}{\underset{\displaystyle H}{C}}-\ddot{\underset{..}{C}l}:$$

83. **(M)** In H_2O_2, there are a total of $(2 \times 1) + (2 \times 6) = 14$ valence electrons, 7 electron pairs. The two O atoms are central atoms. A plausible Lewis structure has zero formal charge on each atom: H-$\ddot{\underset{..}{O}}$-$\ddot{\underset{..}{O}}$-H. In the hydrogen peroxide molecule, the O — O bond is non-polar, while the H — O bonds are polar, with the dipole moment pointing toward O. Since the molecule has a resultant dipole moment, it cannot be linear, for, if it were linear the two polar bonds would oppose each other and their polarities would cancel.

Bond Lengths

85. **(E)** The answer is c. Br_2 possess the longest bond. Single bonds are generally longer than multiple bonds. Of the two molecules with single bonds, Br_2 is expected to have longer bonds than BrCl, since Br is larger than Cl.

(a) $\ddot{\underset{..}{O}}=\ddot{\underset{..}{O}}$ (b) :N≡N: (c) :$\ddot{\underset{..}{B}r}$—$\ddot{\underset{..}{B}r}$:   (d) :$\ddot{\underset{..}{B}r}$—$\ddot{\underset{..}{C}l}$:

87. **(M)** A heteronuclear bond length (one between two different atoms) is approximately equal to the average of two homonuclear bond lengths (one between two like atoms) of the same order (both single, both double, or both triple).

(a) I—Cl bond length $= [(\text{I—I bond length}) + (\text{Cl—Cl bond length})] \div 2$
$= [266 \text{ pm} + 199 \text{ pm}] \div 2 = 233 \text{ pm}$

(b) O—Cl bond length $= [(\text{O—O bond length}) + (\text{Cl—Cl bond length})] \div 2$
$= [145 \text{ pm} + 199 \text{ pm}] \div 2 = 172 \text{ pm}$

(c) C—F bond length $= [(\text{C—C bond length}) + (\text{F—F bond length})] \div 2$
$= [154 \text{ pm} + 143 \text{ pm}] \div 2 = 149 \text{ pm}$

(d) C—Br bond length $= [(\text{C—C bond length}) + (\text{Br—Br bond length})] \div 2$
$= [154 \text{ pm} + 228 \text{ pm}] \div 2 = 191 \text{ pm}$

89. **(E)** The N—F bond is a single bond. Its bond length should be the average of the N—N single bond (145 pm) and the F—F single bond (143 pm). Thus, the average N—F bond length $= (145 + 143) \div 2 = 144$ pm.

Bond Energies

91. **(E)** The reaction $O_2(g) \rightarrow 2\,O(g)$ is an endothermic reaction since it requires the breaking of the bond between two oxygen atoms without the formation of any new bonds. Since bond breakage is endothermic and the process involves only bond breakage, the entire process must be endothermic.

93. **(M)**

Analysis of the Lewis structures of products and reactants indicates that a C—H bond and a Cl—Cl bond are broken, and a C—Cl and a H—Cl bond are formed.

Energy required to break bonds $= \text{C—H} + \text{Cl—Cl} = 414\ \dfrac{\text{kJ}}{\text{mol}} + 243\ \dfrac{\text{kJ}}{\text{mol}} = 657\ \dfrac{\text{kJ}}{\text{mol}}$

Energy realized by forming bonds $= \text{C—Cl} + \text{H—Cl} = 339\ \dfrac{\text{kJ}}{\text{mol}} + 431\ \dfrac{\text{kJ}}{\text{mol}} = 770\ \dfrac{\text{kJ}}{\text{mol}}$

$\Delta_r H = 657 \text{ kJ/mol} - 770 \text{ kJ/mol} = -113 \text{ kJ/mol}$

95. **(M)** In each case we write the formation reaction, but specify reactants and products with their Lewis structures. All species are assumed to be gases.

(a) $\frac{1}{2}(\overset{\cdot\cdot}{\underset{\cdot\cdot}{O}}=\overset{\cdot\cdot}{\underset{\cdot\cdot}{O}})+\frac{1}{2}(H-H)\longrightarrow \cdot\overset{\cdot\cdot}{\underset{\cdot\cdot}{O}}-H$

Bonds broken: $\frac{1}{2}(O=O)+\frac{1}{2}(H-H)=0.5(498\ kJ+436\ kJ)=467\ kJ$

Bonds formed: $O-H=464\ kJ$ $\qquad \Delta_r H°=467\ kJ-464\ kJ=3\ kJ/mol$

If the $O-H$ bond dissociation energy of 428.0 kJ/mol from Figure 10-16 is used, $\Delta_f H°=39\ kJ/mol$.

(b) $:N\equiv N: + \ 2\ H-H \longrightarrow \overset{\displaystyle H\ \ \ \ H}{\underset{\displaystyle H\ \ \ H}{H-\overset{\cdot\cdot}{N}-\overset{\cdot\cdot}{N}-H}}$

Bonds broken $= N\equiv N+2\,H-H=946\ kJ+2\times436\ kJ=1818\ kJ$

Bonds formed $= N-N+4\,N-H=163\ kJ+4\times389\ kJ=1719\ kJ$

$\Delta_f H°=1818\ kJ-1719\ kJ=99\ kJ$

97.
(M)

$H-C\equiv C-H \ + \ H-H \longrightarrow \ \overset{\displaystyle H\ \ \ \ \ \ \ H}{\underset{\displaystyle H\ \ \ \ \ \ \ H}{\diagdown C=C\diagup}}$

Bonds broken	Energy change	Bonds formed	Energy change
1 mol C≡C	1×837 kJ	1 mol C=C	1×-611 kJ
1 mol H—H	1×436 kJ	2 mol C-H	2×-414 kJ

Energy required to break bonds	+1273 kJ	Energy obtained upon bond formation	−1439 kJ

Overall energy change $=1273\ kJ - 1439\ kJ = -166\ kJ/mol = \Delta_r H°$

99. **(M)**

$\left[\ \overset{\cdot\cdot}{\underset{\cdot\cdot}{O}}=N-\overset{\cdot\cdot}{\underset{\cdot\cdot}{O}}: \longleftrightarrow :\overset{\cdot\cdot}{\underset{\cdot\cdot}{O}}-N=\overset{\cdot\cdot}{\underset{\cdot\cdot}{O}}\ \right] + \ O \longrightarrow \ \cdot N=\overset{\cdot\cdot}{\underset{\cdot\cdot}{O}} \ + \ \overset{\cdot\cdot}{\underset{\cdot\cdot}{O}}=\overset{\cdot\cdot}{\underset{\cdot\cdot}{O}}$

$\Delta_f H°=(0+90.25)-(33.18+249.2)=-192.1\ kJ/mol$

Using the calculated $\Delta_f H°$, we can calculate the $O\text{----}N$ bond energy:

$\Delta_f H°=-(2\times(O:::N)+0)-(590+498)=-192.1$

$O\text{----}N = 448\ kJ/mol$, which is between the values for N=O and N–O.

INTEGRATIVE AND ADVANCED EXERCISES

Important Note: In this and subsequent chapters, a lone pair of electrons in a Lewis structure often is shown as a line rather than a pair of dots. Thus, the Lewis structure of Be is Be| or Be:

101. **(M)** Recall that bond breaking is endothermic, while bond making is exothermic.

Break 2 N-O bonds requires 2(631 kJ/mol) =   +1262 kJ

Break 5 H-H bonds requires 4(436 kJ/mol) =   +2180 kJ

Make 6 N-H bonds yields 5(–389 kJ/mol) =   –2334 kJ

Make 4 O-H bonds yields 4(–463 kJ/mol) =   –1852 kJ

$$\Sigma(\text{bond energies}) = \Delta_r H = -744 \text{ kJ/mol reaction}$$

104. **(M)**

$$\text{amount of gas} = \frac{PV}{RT} = \frac{749 \text{ mmHg} \times \dfrac{1 \text{ atm}}{760 \text{ mmHg}} \times 0.193 \text{ L}}{0.08206 \text{ L atm mol}^{-1} \text{ K}^{-1} \times 299.3 \text{ K}} = 0.00774 \text{ mol}$$

$$M = \frac{\text{mass}}{\text{amount}} = \frac{0.325 \text{ g}}{0.00774 \text{ mol}} = 42.0 \text{ g/mol}$$

Because three moles of C weigh 36.0 g and four moles weigh 48.0 g, this hydrocarbon is

C_3H_6. A possible Lewis structure is

$$\begin{array}{ccc} & \text{H} & \\ & | & \\ \text{H}-\text{C}-\text{C} & = & \text{C}-\text{H} \\ & | \quad | \quad | & \\ & \text{H} \quad \text{H} \quad \text{H} & \end{array}$$

There is another possible Lewis structure: the three C atoms are arranged in a ring, with two H atoms bonded to each C atom (see below).

106. **(M)** The two isomers are

$$\begin{array}{cc} \text{H} & \text{H} \\ \diagdown & \diagup \\ \text{C}=\text{C}=\text{C} \\ \diagup & \diagdown \\ \text{H} & \text{H} \end{array} \quad \text{and} \quad \begin{array}{c} \text{H} \\ | \\ \text{H}-\text{C}-\text{C}\equiv\text{C}-\text{H} \\ | \\ \text{H} \end{array}$$

The left-hand isomer is planar around the first and third C atoms, but we cannot predict with VSEPR theory whether the molecule is planar overall; in other words, the two H—C—H planes may be at 90° to each other. (They are, in fact.) In the right-hand isomer, the C—C≡C—H chain is linear, but the H_3C— molecular geometry is tetrahedral.

109. **(M)** The HN_3 molecule has $1+(3\times5)=16$ valence electrons, or 8 pairs. Average bond lengths are 136 pm for N—N, 123 pm for N=N, and 110 pm for N≡N. Thus it seems that one nitrogen-to-nitrogen bond is a double bond, while the other is a triple bond. A plausible Lewis structure is H—N̈=N=N̈ The three N's lie on a line, with a 120° H—N—N

bond angle: $\overset{H}{\underset{}{\backslash}}$N=N=N̈ Another valid resonance form is H—N̈—N≡N: which would have one N—N separation consistent with a nitrogen-nitrogen triple bond. It would also predict a tetrahedral H—N—N bond angle of 109.5°. Thus, the resulting resonance hybrid should have a bond angle between 120° and 109.5°, which is in good agreement with the observed 112° H—N—N bond angle.

110. **(D)** For N_5^+ the number of valence electrons is $(5\times5)-1=24$. There are four possible Lewis structures with formal charges that are not excessive $(<\pm2)$.

$$:\overset{}{\underset{\ominus}{\ddot{N}}}=\overset{\oplus}{N}=\overset{\oplus}{N}=\overset{\oplus}{N}=\overset{}{\underset{}{\ddot{N}}}: \qquad :\overset{}{\underset{\ominus}{\ddot{N}}}=\overset{\oplus}{N}=\overset{\oplus}{\underset{\ominus}{\ddot{N}}}-\overset{\oplus}{N}\equiv N: \qquad :N\equiv\overset{\oplus}{N}-\overset{}{\underset{\ominus}{\ddot{N}}}=\overset{\oplus}{N}=\overset{\oplus}{\ddot{N}}: \qquad :N\equiv\overset{\oplus}{N}-\overset{\ominus}{\underset{}{\ddot{N}}}-\overset{\oplus}{N}\equiv N:$$

$$\qquad\qquad 1 \qquad\qquad\qquad 2 \qquad\qquad\qquad\qquad 3 \qquad\qquad\qquad\qquad 4$$

Structure **1** has three adjacent atoms possessing formal charges of the same sign. Energetically, this is highly unfavorable. Structure **2** & **3** are similar and highly unsymmetrical but these are energetically more favorable than **1** Structure **4** is probably best of all, as all of the charges are close together with no two adjacent charges of the same sign. If structures **2**, **3**, or **4** are chosen, the central nitrogen has one or two lone pairs. Thus, the structure of N_5^+ will be angular. (Note: an angle of 107.9° has been experimentally observed for the angle about the central nitrogen, as well, the bond length of the terminal N—N bonds is very close to that seen in N_2, suggesting a triple bond. This suggests that resonance form **4** best describes the structure of the ion.)

113. **(D)** The Lewis structures for the species in reaction (2) follow.

$$:N\equiv N: \quad H—H \quad H—C\equiv N:$$

Bonds broken $=\frac{1}{2}$ N≡N$+\frac{1}{2}$ H—H $=\frac{1}{2}$ (946 kJ + 436 kJ) = 691 kJ

Bonds formed = H—C + C≡N = 414 kJ + 891 kJ = 1305 kJ

$\Delta_r H = 691$ kJ -1305 kJ $= -614$ kJ

Then we determine $\Delta_f H°$

(a) $C(s)\longrightarrow C(g)$ $\qquad\qquad\qquad\qquad\qquad \Delta_r H = +717$ kJ

(b) $C(g)+\frac{1}{2} N_2(g)+\frac{1}{2} H_2(g)\longrightarrow HCN(g)$ $\Delta_r H = -614$ kJ

Net: $C(s)+\frac{1}{2} N_2(g)+\frac{1}{2} H_2(g)\longrightarrow HCN(g)$ $\Delta_f H° = +103$ kJ

This compares favorably to the value of 135.1 kJ/mol given in Appendix D-2.

116. **(M)** We first compute the heats of reaction.

$$CH_3OH(g) + H_2S(g) \longrightarrow CH_3SH(g) + H_2O(g)$$

$$\Delta_r H° = \Delta_f H°[H_2O(g)] + \Delta_f H°[CH_3SH(g)] - [\Delta_f H°[CH_3OH(g)] + \Delta_f H°[H_2S(g)]]$$
$$= -241.8 \text{ kJ} + (-22.9 \text{ kJ}) - [(-200.7 \text{ kJ} - 20.63 \text{ kJ})] = -43.4 \text{ kJ}$$

Breaking of one mole of C—O bond requires (360 kJ) = +360 kJ

Breaking of one mole of H—S bond requires (368 kJ) = + 368 kJ

Breaking of one mole of O—H bond requires (464 kJ) = + 464 kJ

Making of one mole of C—S bond yields = $-x$ kJ

Making 2 moles of O—H bonds yields 2(−464 kJ) = −928 kJ

$$\Sigma(\text{bond energies}) = \Delta_r H° = -43.4 \text{ kJ}$$

Then... $264 - x = -43.4$ and $x = 307$ kJ, which is the C-S bond energy for the C—S bond in methanethiol (estimate only).

120. **(M)** For the halogens we have the following data:

Atom	Electronegativity	Ionization Energy (kJ/mol)	Electron Affinity (kJ/mol)
F	4.0	1680	−328
Cl	3.0	1256	−349
Br	2.8	1143	−324.6
I	2.5	1009	−295.2

From the data above and using $EN = k \times (E_i - E_{ea})$ for the halogens, we find the following values for k: F = 0.00199   Cl = 0.00187   Br = 0.00191   I = 0.00192
We shall assume that the value of k for astatine is 0.0019. As well, from the data, we can estimate the ionization energy for astatine to be ~900 kJ/mol. The text gives 2.2 as the electronegativity for astatine. We can now estimate a value for the electron affinity of At.

$$EN = k \times (E_i - E_{ea}) = 2.2 = 0.0019(900 - E_{ea}) \quad E_{ea} \sim -260 \text{ kJ/mol}$$

127. (M)

(a)

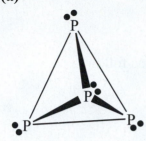

(b)

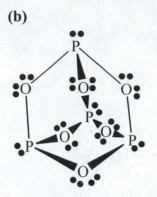

The O atoms in P_4O_6 are surrounded by 4 groups and so the geometry around each O is tetrahedral. Therefore, the P–O–P bond is not linear.

FEATURE PROBLEMS

128. (M) The table containing the constants used for the calculations is reproduced below. In all cases, a sample calculation is worked out for F.

			Atom, X		
	H	F	Cl	Br	I
E_i/eV	13.5985	17.423	12.9677	11.8139	10.4513
E_{ea}/eV	0.7542	3.399	3.617	3.365	3.059
$D(HX)$/eV	4.44	6.39	4.38	3.74	3.07
$D(X_2)$/eV	4.44	2.80	2.468	1.962	1.535
Z_{eff}	1	4.86	5.75	7.25	7.25
r_{cov}/pm	–	71.5	99.4	114.2	133.4

Electronegativity using the Pauling method, calculated relative to EN_H, is shown below. The general formula is

$$EN_A - EN_B = \sqrt{\frac{D_{A-B} - \frac{1}{2}(D_{A-A} + D_{B-B})}{1\ eV}}$$

For F, the relative χ is calculated as follows:

$$EN_F - EN_H = \sqrt{\frac{6.39\ eV - \frac{1}{2}(2.8 + 4.44)\ eV}{1\ eV}} = 1.7$$

Assuming a EN_H value of 2.1, the absolute value of EN_F is 3.8. Electronegativity values for Cl, Br, and I are provided in the table at the end of this problem.

Electronegativity using the Mullikan method for F is calculated as follows:

$$EN = 0.336 \times \left(\frac{E_i + E_{ea}}{2 \text{ eV}} \right) - 0.165$$

$$EN = 0.336 \times \left(\frac{17.423 + 3.399}{2 \text{ eV}} \right) - 0.165 = 3.33$$

Electronegativity using the Allred-Rochow method is calculated as follows:

$$EN = \frac{3590 \, Z_{eff}}{(r_{cov}/1 \text{ pm})^2} + 0.744$$

$$EN = \frac{3590 \times 4.86}{(71.5/1 \text{ pm})^2} + 0.744 = 4.16$$

The calculation results for all the nuclides are shown in the table below.

	F	Cl	Br	I
Pauling (relative to H)	1.7	1.0	0.7	0.3
Pauling (absolute)	3.8	3.1	2.8	2.4
Mulliken	3.3	2.6	2.3	2.1
Allred–Rochow	4.2	2.8	2.7	2.2

129. (D)

(a) The two bond dipole moments can be added geometrically, by placing the head of one at the tail of the other, as long as we do not change the direction or the length of the moved dipole. The resultant molecular dipole moment is represented by the arrow drawn from the tail of one bond dipole to the head of the other. This is shown in the figure to the right. The 52.0° angle in the figure is one-half of the 104° bond angle in water. The length is given as 1.84 D. We can construct a right angled triangle by bisecting the 76.0° angle. The right angled triangle has a hypotenuse = O — H bond dipole and the two other angles are 52° and 38°. The side opposite the bisected 76.0 ° angles is ½ (1.84 D) = 0.92 D. We can calculate the bond dipole using: sin

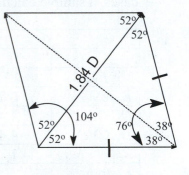

$$38.0° = \frac{0.92 \text{ D}}{\text{O-H bond dipole}} = 0.61566, \text{ hence O—H bond dipole} = 1.49 \text{ D.}$$

(b) For H_2S, we do not know the bond angle. We shall represent this bond angle as 2α. Using a similar procedure to that described in part (a), above, a diagram can be constructed and the angle 2α calculated as follows:

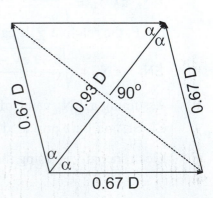

$$\cos \alpha = \frac{\frac{1}{2}(0.93 \text{ D})}{0.67 \text{ D}} = 0.694 \quad \alpha = 46.05°$$

or $2\alpha = 92.1$ °

The H—S—H angle is approximately 92°.

(c)

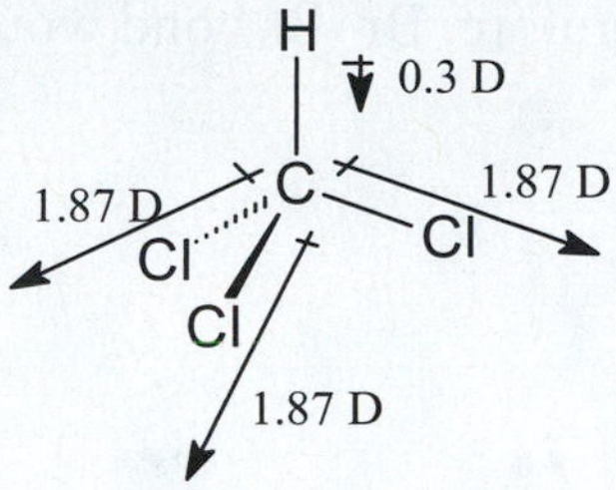

$$\mu_{C\text{-}H} + 3(\mu_{C\text{-}Cl}) = 1.04\ D$$
$$0.30\ D + 3(x) = 1.04\ D$$
$$x = 0.25\ D = \mu_{C\text{-}Cl}$$

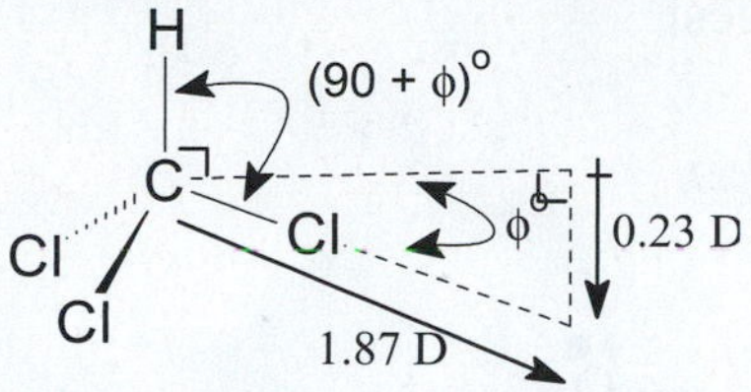

Molecule and associated individual bond dipoles

Relationship between dipole moment(molecular) and bond dipoles(Vector addition)

Geometric Relationship

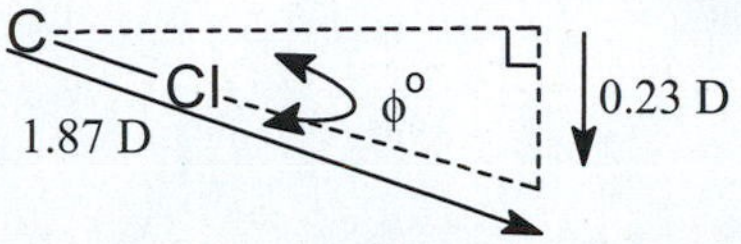

Mathematical solution:
$$\sin(\phi) = \frac{0.25\ D}{1.87\ D}$$
$$\phi = 7.6^{\circ}$$

The H-C-Cl bond angle is $(90 + \phi)^{\circ} = 90^{\circ} + 7.6^{\circ} = 97.6^{\circ}$

SELF-ASSESSMENT EXERCISES

134. (E) The answer is (b). The structure is:

$$:C\equiv N:^{\ominus}$$

135. (E) The answer is (c). The structure is:

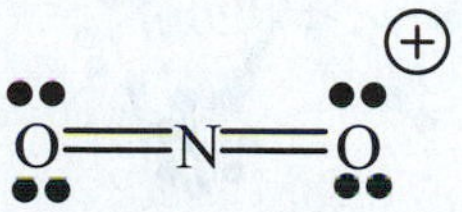

$$O=\overset{\oplus}{N}=O$$

Formal charge on O is: $6 - 4 - 2 = 0$

136. (E) The answer is (a), SO_2. NO is linear by definition since there are only two atoms. The other molecules have the following structures:

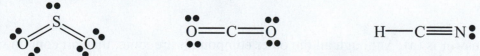

137. (E) The answer is (a), SO_3. The other choices are polar, because: CH_2Cl_2 has a non-uniform field of atoms around the central atom C; NH_3 is trigonal pyramidal and has a lone pair of electrons on N; in FNO, there are three different atoms with different electron affinities.

138. (E) The answer is (b), N_2, because one has to break three covalent bonds to dissociate the two N atoms from each other.

139. (E) The answer is (c), Br₂, because the greater the covalent radii of the two atoms involved, the longer the bond length. Br is the largest atom and therefore, Br–Br bond would be the longest.

140. (M)

(a)

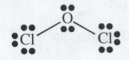

(b)

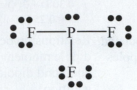

(c)

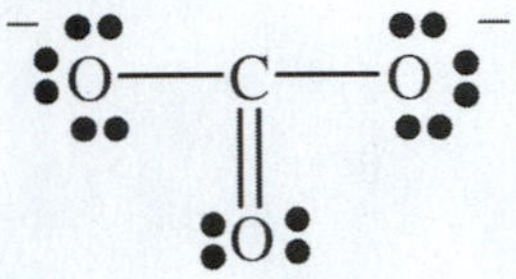

(d)

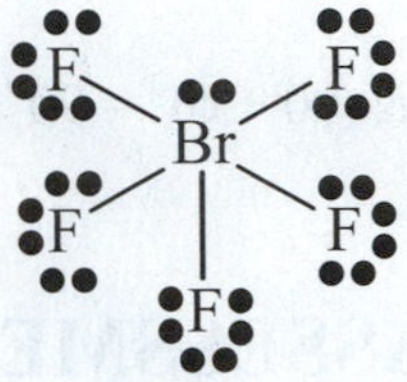

141. (M)

	(a)	(b)	(c)
e⁻ group geometry	Trigonal planar	Tetrahedral	Tetrahedral
Molecular geometry	Bent	Trigonal pyramidal	Tetrahedral

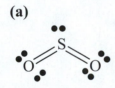 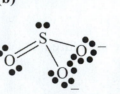

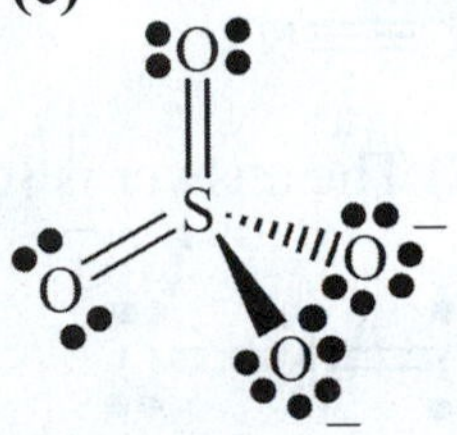

142. (E) The answer is (a). Although all the other compounds are ionic, they all contain metals.

143. (E) The answer is (d). It does not obey the octet rule since a full set of eight electrons cannot be placed around both atoms because of the odd number of valence electrons. The correct Lewis structure is shown below.

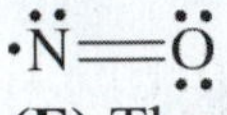

·N̈＝Ö

144. (E) The answer is (e). All the choices have a polar bond, which arises from the difference in electronegativity of the atoms making the bond.

145. **(E)** The answer is (a). The oxygen atom in H_2O has two bonds and two lone electron pairs, which gives it a tetrahedral electron-group geometry and a bent (or V-shaped) molecular structure.

146. **(M)**

Electron-Group Geometry	Molecular Shape	
Tetrahedral	Tetrahedral	
Tetrahedral	Trigonal pyramidal	
Trigonal pyramidal	Seesaw	
Octahedral	Square pyramidal	

147. **(M)** The total enthalpy change for each reaction is the difference between the sum of the bond energies for the bonds broken (which consumes energy) and the sum of the bond energies for the bonds formed (which releases energy).

(a) CH_4: The balanced reaction is

$$CH_4(g) + 2\,O_2 \rightarrow CO_2(g) + 2\,H_2O(g)$$

The oxidation state of C in CH_4 is −4.

Bonds broken: $4\,C–H + 2\,O=O$

$$4(414) + 2(498) = 2642 \text{ kJ mol}^{-1}$$

Bonds formed: $2\,C=O + 4\,H–O$

$$2(799) + 4(464) = 3454 \text{ kJ mol}^{-1}$$

Therefore, the total energy change for combustion of CH_4 is $2642 - 3454 = -812 \text{ kJ mol}^{-1}$
The energy change per gram is

$$\frac{-812 \text{ kJ mol}^{-1}}{16.05 \text{ g mol}^{-1}} = -50.6 \text{ kJ g}^{-1}$$

(b) CH_3OH: The balanced reaction is

$$2\ CH_3OH(g) + 3\ O_2 \rightarrow 2\ CO_2(g) + 4\ H_2O(g)$$

The oxidation state of C in CH_3OH is -2.

Bonds broken: $6\ C–H + 2\ C–O + 2\ O–H + 3\ O{=}O$

$$6(414) + 2(360) + 2(464) + 3(498) = 5626\ kJ\ mol^{-1}$$

Bonds formed: $4\ C{=}O + 8\ H–O$

$$4(799) + 8(464) = 6908\ mol^{-1}$$

Therefore, the total energy change for combustion of CH_4 is $5626 - 6908 = -1282\ kJ\ mol^{-1}$
The energy change per gram is

$$\frac{-1282\ kJ\ mol^{-1}}{2(32.05\ g\ mol^{-1})} = -20.0\ kJ\ g^{-1}$$

(c) H_2CO: The balanced reaction is

$$H_2CO(g) + O_2 \rightarrow CO_2(g) + H_2O(g)$$

The oxidation state of C in H_2CO is 0.

Bonds broken: $2\ C–H + C{=}O + O{=}O$

$$2(414) + (799) + (498) = 2125\ kJ\ mol^{-1}$$

Bonds formed: $2\ C{=}O + 2\ H–O$

$$2(707) + 2(464) = 2342\ kJ\ mol^{-1}$$

Therefore, the total energy change for combustion of CH_4 is $2125 - 2342 = -217\ kJ\ mol^{-1}$
The energy change per gram is

$$\frac{-217\ kJ\ mol^{-1}}{30.03\ g\ mol^{-1}} = -7.23\ kJ\ g^{-1}$$

(d) H_2CO: The balanced reaction is

$$2\ HCOOH(g) + O_2 \rightarrow 2\ CO_2(g) + 2\ H_2O(g)$$

The oxidation state of C in HCOOH is $+2$.

Bonds broken: $2\ O–H + C–H + 2\ C{=}O + 2\ C–O + O{=}O$

$$2(464) + 2(414) + 2(799) + 2(360) + (498) = 4572\ kJ\ mol^{-1}$$

Bonds formed: $4\ C{=}O + 4\ H–O$

$$4(707) + 4(464) = 4684\ kJ\ mol^{-1}$$

Therefore, the total energy change for combustion of HCOOH is
$$4572 - 4684 = -112\ kJ\ mol^{-1}$$

The energy change per gram is

$$\frac{-112\ kJ\ mol^{-1}}{2(46.03\ g\ mol^{-1})} = -1.22\ kJ\ g^{-1}$$

There is a trend with respect to the oxidation state and energy generated per gram of combusted material. The more negative the oxidation state of the carbon, the more energy that is generated per gram of that material.

148. **(E)** Ba and Mg are active metals with a low EN, S is the most electronegative (and non-metallic) as indicated by its location on the periodic table. In comparing Bi and As, Bi has lower EN value, as indicated by its location. Therefore, Bi has the middle position: Ba < Mg < Bi < As < S

149. **(E)**

Bond	Bond Energy (kJ/mol)	Bond Length (pm)
C–H	414	110
C=O	736	120
C–C	347	154
C–Cl	339	178

150. **(E)** VSEPR theory is valence shell electron pair repulsion theory. It is based on the premise that electron pairs assume orientations about an atom to minimize electron pair repulsions.

151. **(E)** The structure of the NH_3 molecule and the arrangement of electrons is shown below:

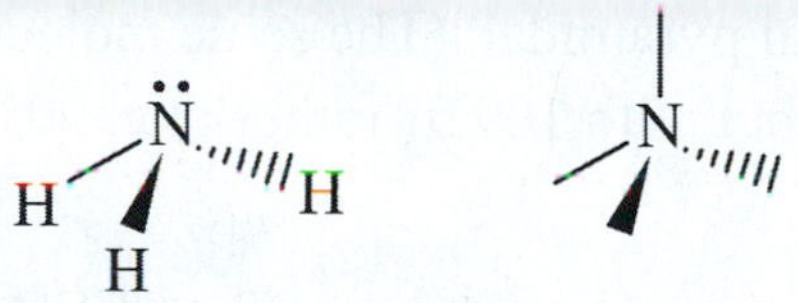

As can be seen, there are 4 pairs of electrons around the nitrogen atoms. Three pairs are in the form of covalent bonds with hydrogen atoms, and one is a lone pair. Since there are 4 electron pairs around the central atom, the way to maximize the distance between them is to set up a tetrahedral electron group geometry. However, since there are only three atoms bonding to the central atom, the molecular geometry is trigonal pyramidal.

152. **(E)** A pyramidal geometry is observed when an atom has one lone pair and is bonded to three other atoms (AX_3E). A bent geometry is observed when an atom has two lone pairs and is bonded to two other atoms (AX_2E_2). For both, the bond angles will be approximately (usually smaller than) $109°$.

153. **(M)**

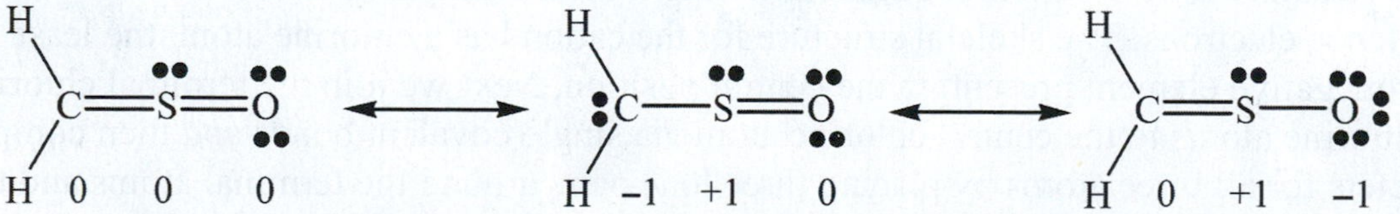

CHAPTER 11
CHEMICAL BONDING II: VALENCE BOND AND MOLECULAR ORBITAL THEORIES
PRACTICE EXAMPLES

1A **(M)** The valence-shell orbital diagrams of N and I are as follows:

N [He]$_{2s}$ [↑↓] $_{2p}$ [↑][↑][↑]   I [Kr] $4d^{10}$ $_{5s}$[↑↓] $_{5p}$[↑↓][↑↓][↑]

There are three half-filled 2p orbitals on N, and one half-filled 5p orbital on I. Each half-filled 2p orbital from N will overlap with one half-filled 5p orbital of an I. Thus, there will be three N—I bonds. The I atoms will be oriented in the same direction as the three 2p orbitals of N: toward the $x-$, $y-$, and z-directions of a Cartesian coordinate system. Thus, the I—N—I angles will be approximately 90° (probably larger because the I atoms will repel each other). The three I atoms will lie in the same plane at the points of a triangle, with the N atom centered above them. The molecule is trigonal pyramidal. (The same molecular shape is predicted if N is assumed to be sp^3 hybridized, but with 109.5° rather than 90° bond angles.)

1B **(M)** The valence-shell orbital diagrams of N and H are as follows. N: [He]$_{2s}$[↑↓] $_{2p}$[↑][↑][↑] H: $_{1s}$[↑] There are three half-filled orbitals on N and one half-filled orbital on each H. There will be three N—H bonds, with bond angles of approximately 90°. The molecule is trigonal pyramid. (We obtain the same molecular shape if N is assumed to be sp^3 hybridized, but bond angles are closer to 109.5°, the tetrahedral bond angle.) VSEPR theory begins with the Lewis structure and notes that there are three bond pairs and one lone pair attached to N. This produces a tetrahedral electron pair geometry and a trigonal pyramidal molecular shape with bond angles a bit less than the tetrahedral angle of 109.5° because of the lone pair downward repulsion of the bonded pairs of electrons. Since VSEPR theory makes a prediction closer to the experimental bond angle of 107°, it seems more appropriate in this case.

2A **(D)** Following the strategy outlined in the textbook, we begin by drawing a plausible Lewis structure for the cation in question. In this case, the Lewis structure must contain 20 valence electrons. The skeletal structure for the cation has a chlorine atom, the least electronegative element present, in the central position. Next we join the terminal chlorine and fluorine atoms to the central chlorine atom via single covalent bonds and then complete the octets for all three atoms by placing three lone pairs around the terminal atoms and two lone pairs around the central atom.

$$:\!\ddot{F}\!\!-\!\!\overset{\oplus}{\underset{..}{\ddot{C}l}}\!\!-\!\!\ddot{C}l\!:$$

With this bonding arrangement, the central chlorine atom ends up with a 1+ formal charge.

Once the Lewis diagram is complete, we can then use the VSEPR method to establish the geometry for the electron pairs on the central atom. The Lewis structure has two bonding electron pairs and two lone pairs of electrons around the central chlorine atom. These four pairs of electrons assume a tetrahedral geometry to minimize electron-electron repulsions. The VSEPR notation for the Cl_2F^+ ion is AX_2E_2. According to Table 10.1, molecules of this type exhibit a tetrahedral geometric orientation. Our next task is to select a hybridization scheme that is consistent with the predicted shape. It turns out that the only way we can end up with a tetrahedral array of electron groups is if the central chlorine atom is sp^3 hybridized. In this scheme, two of the sp^3 hybrid orbitals are filled, while the remaining two are half occupied.

$\boxed{\uparrow\downarrow\,|\,\uparrow\downarrow\,|\,\uparrow\,|\,\uparrow}$ sp^3 hybridized central chlorine atom (Cl^+)

The Cl—F and Cl—Cl bonds in the cation are then formed by the overlap of the half-filled sp^3 hybrid orbitals of the central chlorine atom with the half-filled p-orbitals of the terminal Cl and F atoms. Thus, by using sp^3 hybridization, we end up with the same <u>bent</u> molecular geometry for the ion as that predicted by VSEPR theory (when the lone pairs on the central atom are ignored)

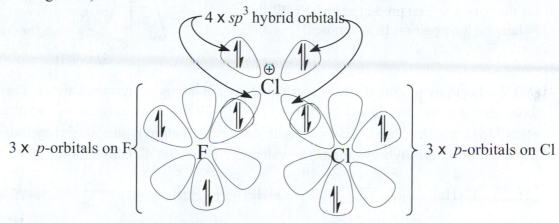

2B **(D)** As was the case in 2A, we begin by drawing a plausible Lewis structure for the cation in question. This time, the Lewis structure must contain 34 valence electrons. The skeletal structure has bromine, the least electronegative element present, as the central atom. Next, we join the four terminal fluorine atoms to the central bromine atom via single covalent bonds and complete the octets for all of the fluorine atoms by assigning three lone pairs to each fluorine atom. Placing the last two electrons on the central bromine atom completes the diagram.

In order to accommodate ten electrons, the bromine atom is forced to expand its valence shell. Notice that the Br ends up with a 1+ formal charge in this structure. With the completed Lewis structure in hand, we can then use VSEPR theory to establish the geometry for the electron pairs around the central atom. The Lewis structure has four bonding pairs and one lone pair of electrons around the central bromine atom. These five pairs of electrons assume a trigonal bipyramidal geometry to minimize electron-electron repulsions. The VSEPR notation for the BrF_4^+ cation is AX_4E. According to Table 10.1, molecules of this type exhibit a see-saw molecular geometry.

Next we must select a hybridization scheme for the Br atom that is compatible with the predicted shape. It turns out that only sp^3d hybridization will provide the necessary trigonal bipyramidal distribution of electron pairs around the bromine atom. In this scheme, one of the sp^3d hybrid orbitals is filled, while the remaining four are half-occupied.

| ↑↓ | ↑ | ↑ | ↑ | ↑ | sp^3d hybridized central bromine atom (Br⁺)

The four Br—F bonds in the cation are then formed by the overlap of the four half-filled sp^3d hybrid orbitals of the bromine atom with the half-filled p-orbitals of the four separate terminal fluorine atoms. Thus, by using sp^3d hybridization, we end up with the same see-saw molecular geometry for the cation as that predicted by VSEPR theory (when the lone pair on Br is ignored).

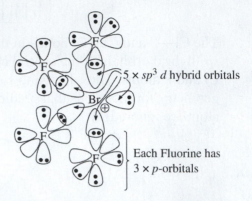

5 × sp^3d hybrid orbitals

Each Fluorine has 3 × p-orbitals

3A **(M)** We begin by writing the Lewis structure. The H atoms are terminal atoms. There are three central atoms and $(3 \times 1) + 4 + 6 + 4 + (3 \times 1) = 20$ valence electrons, or 10 pairs. A plausible Lewis structure is drawn at right. Each central atom is surrounded by four electron pairs, requiring sp^3 hybridization. The valence-shell orbital diagrams for the atoms follow.

H_{1s} ↑   C [He] $_{2s}$ ↑↓ $_{2p}$ ↑ ↑ __   O [He] $_{2s}$ ↑↓ $_{2p}$ ↑↓ ↑ ↑

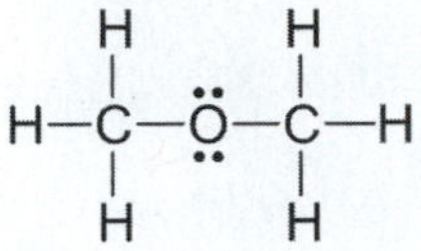

The valence-shell orbital diagrams for the hybridized central atoms then are:

C_{sp^3} ↑ ↑ ↑ ↑ O_{sp^3} ↑↓ ↑↓ ↑ ↑

All bonds in the molecule are σ bonds. The H—C—H bond angles are 109.5°, as are the H—C—O bond angles. The C—O—C bond angle is possibly a bit smaller than 109.5° because of the repulsion of the two lone pairs of electrons on O. A wedge-and-dash sketch of the molecule is at right.

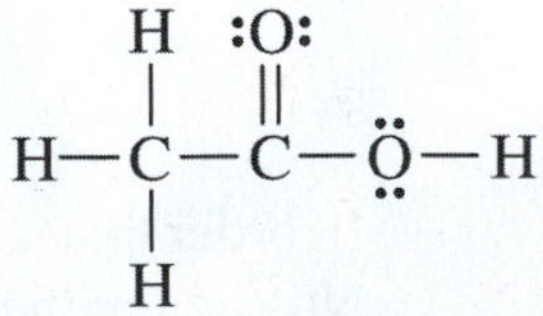

3B **(M)** The H atoms and one O are terminal atoms in the Lewis structure, which has $3 \times 1 + 4 + 4 + 2 \times 6 + 1 = 24$ valence electrons, or 12 pairs. The left-most C and the right-most O are surrounded by four electron pairs, and thus require sp^3 hybridization. The central carbon is surrounded by three electron groups and is sp^2 hybridized. The orbital diagrams for the un-hybridized atoms are:

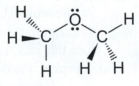

$H_{:1s}$ ↑   C: [He] $_{2s}$ ↑↓ $_{2p}$ ↑ ↑ __   O: [He] $_{2s}$ ↑↓ $_{2p}$ ↑↓ ↑ ↑

Hybridized orbital diagrams:

C: [He]$sp3$ ⊞⊞⊞⊞   C: [He] $sp2$ ⊞⊞⊞ $2p$ ⊞   O:[He] $sp3$ ⊞⊞⊞⊞   terminal O: [He] $sp2$ ⊞⊞⊞ $2p$ ⊞

There is one π bond in the molecule: between the $2p$ on the central C and the $2p$ on the terminal O. The remaining bonds are σ bonds. The H—C—H and H—C—C bond angles are 109.5°. The H—O—C angle is somewhat less, perhaps 105° because of lone pair repulsion. The C—C—O bond angles and O—C—O bond angles are all 120°.

4A   (M) There are four bond pairs around the left-hand C, requiring sp^3 hybridization. Three of the bonds that form are C—H sigma bonds resulting from the overlap of a half-filled sp^3 hybrid orbital on C with a half-filled $1s$ orbital on H. The other C has two attached electron groups, utilizing sp hybridization.

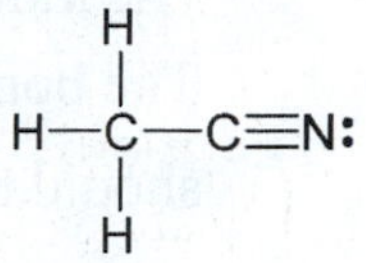

$$\begin{array}{c} H \\ | \\ H{-}C{-}C{\equiv}N{:} \\ | \\ H \end{array}$$

C: [He]sp ⊞⊞ $2p$ ⊞⊞ The N atom is sp hybridized. N: [He] $2s$ ⊞ $2p$ ⊞⊞⊞

The two C atoms join with a sigma bond: overlap of sp^2 on the left-hand C with sp on the right-hand C. The three bonds between C and N consist of a sigma bond (sp on C with sp on N), and two pi bonds ($2p$ on C with $2p$ on N).

4B   (M) The bond lengths that are given indicate that N is the central atom. The molecule has $(2 \times 5) + 6 = 16$ valence electrons, or 8 pairs. Average bond lengths are: N—N = 145 pm, N=N = 123 pm, N≡N = 110 pm, N—O = 136 pm, N=O = 120 pm. Plausible resonance structures are (with subscripts for identification):

$$\text{structure(1)} \quad :N_a{\equiv}N_b^{\oplus}{-}\overset{..}{\underset{..}{O}}{:}^{\ominus} \longleftrightarrow {}^{\ominus}\overset{..}{\underset{..}{N}}_a{=}N_b^{\oplus}{=}\overset{..}{\underset{..}{O}} \quad \text{structure(2)}$$

In both structures, the central N is attached to two other atoms, and possesses no lone pairs. The geometry of the molecule thus is linear and the hybridization on this central N is sp. N_b [He] sp ⊞⊞ $2p$ ⊞⊞ We will assume that the terminal atoms are not hybridized in either structure. Their valence-shell orbital diagrams are:
terminal N: N_a [He] $2s$ ⊞ $2p$ ⊞⊞⊞ O [He] $2s$ ⊞ $2p$ ⊞⊞⊞

In structure (1) the N≡N bond results from the overlap of three pairs of half-filled orbitals: (1) sp_x on N_b with $2p_x$ on N_a forming a σ bond, (2) $2p_y$ on N_b with $2p_y$ on N_a forming a π bond, and (3) $2p_z$ on N_b with $2p_z$ on N_a also forming a π bond. The N—O bond is a coordinate covalent bond, and requires that the electron configuration of O be written as O [He] $2s$ ⊞ $2p$ ⊞⊞⊞. The N—O bond then forms by the overlap of the full sp_x orbital on N_b with the empty $2p$ orbital on O.

In structure (2) the N=O bond results from the overlap of two pairs of half-filled orbitals: (1) sp_y on N_b with $2p_y$ on O forming a σ bond and (2) $2p_z$ on N_b with $2p_z$ on O forming a π bond. The N=N σ bond is a coordinate covalent bond, and requires that the configuration of N_a be written N_a [He] $2s$ ⊞ $2p$ ⊞⊞⊞. The N=N bond is formed by two overlaps: (1) the overlap of the full sp_y orbital on N_b with the empty $2p_y$ orbital on N_a to form a σ bond, and (2) the overlap of the half-filled $2p_x$ orbital on N_b with the half-filled $2p_x$ orbital on N_a to form a π bond. Based on formal charge arguments, structure (1) is preferred, because the negative formal charge is on the more electronegative atom, O.

5A **(M)** An electron from Li_2 removes a bonding electron because the valence molecular orbital diagram for Li_2 is the same as that for H_2, only it is just moved up a principal quantum level: $\sigma_{1s}b$ ⥮ $\sigma_{1s}*$ ⥮ $\sigma_{2s}b$ ⥮ $\sigma_{2s}*$ ☐ .

The molecular orbital diagram for Li2+ is: $\sigma_{1s}b$ ⥮ $\sigma_{1s}*$ ⥮ $\sigma_{2s}b$ ↿ $\sigma_{2s}*$ ☐

The bond order in Li_2^+ is: 1/2, while that in Li_2 is one. Thus, the Li-Li bond in Li_2^+ should be one half as strong as the Li-Li bond in Li_2: 106 kJ/mol $\div 2 = 53$ kJ/mol Li_2^+

5B **(M)** The H_2^- ion contains 1 electron from each H plus 1 electron for the negative charge for a total of three electrons. Its molecular orbital diagram is σ_{1s} ⥮ $\sigma_{1s}*$ ↿ . There are two bonding and one antibonding electrons. The bond order in H_2^- is obtained as follows:

$$\text{bond order} = \frac{(2 \text{ bonding } e^- - 1 \text{ antibonding } e^-)}{2} = \frac{1}{2}$$

Thus, we would expect the ion H_2^- to be stable, with a bond strength about half that of a hydrogen molecule.

6A **(M)** For each case, the empty molecular-orbital diagram has the following appearance. (KK indicates that the molecular orbitals formed from $1s$ atomic orbitals are full: KK = σ_{1s} ⥮ $\sigma_{1s}*$ ⥮

KK = σ_{2s} ☐ $\sigma_{2s}*$ ☐ σ_{2p} ☐ π_{2p} ☐☐ $\pi_{2p}*$ ☐☐ $\sigma_{2p}*$ ☐ . We need to simply count up the total number of electrons in each species, and place them in the valence-shell molecular-orbital diagram.

(a) N_2^+ has $(2 \times 5) - 1 = 9$ valence electrons. Its molecular orbital diagram is

N_2^+ KK σ_{2s} ⥮ $\sigma_{2s}*$ ⥮ π_{2p} ⥮⥮ σ_{2p} ↿ $\pi_{2p}*$ ☐☐ $\sigma_{2p}*$ ☐

bond order = (7 bonding electrons -2 antibonding electrons) $\div 2 = 2.5$

(b) Ne_2^+ has $(2 \times 8) - 1 = 15$ valence electrons. Its molecular orbital diagram is

Ne_2^+ KK σ_{2s} ⥮ $\sigma_{2s}*$ ⥮ σ_{2p} ⥮ π_{2p} ⥮⥮ $\pi_{2p}*$ ⥮⥮ $\sigma_{2p}*$ ↿

bond order = (8 bonding electrons -7 antibonding electrons) $\div 2 = 0.5$

(c) C_2^{2-} has $(2 \times 4) + 2 = 10$ valence electrons. Its molecular orbital diagram is

C_2^{2-} KK σ_{2s} ⥮ $\sigma_{2s}*$ ⥮ π_{2p} ⥮⥮ σ_{2p} ⥮ $\pi_{2p}*$ ☐☐ $\sigma_{2p}*$ ☐

bond order = (8 bonding electrons -2 antibonding electrons) $\div 2 = 3.0$

6B   **(M)** For each case, the empty molecular-orbital diagram has the following appearance. (KK indicates that the molecular orbitals formed from $1s$ atomic orbitals are full:

KK = σ_{1s} ⥮ σ_{1s}* ⥮

KK σ_{2s} ☐ σ_{2s}* ☐ σ_{2p} ☐ π_{2p} ☐☐ π_{2p}* ☐☐ σ_{2p}* ☐ . All of these species are based on O_2, which has $2\times6 = 12$ valence electrons. We simply put the appropriate number of valence electrons in each diagram and determine the bond order.

O_2^{+} 11 v.e.   KK σ_{2s}⥮ σ_{2s}*⥮ σ_{2p}⥮ π_{2p}⥮⥮ π_{2p}*⥯☐ σ_{2p}*☐

bond order = (8 bonding e^- –5 antibonding e^-) $\div 2 = 2.5$; bond length = 112 pm

O_2 12 v.e.   KK

bond order = (8 bonding e^- – 4 antibonding e^-) $\div 2 = 2.0$; bond length = 121 pm

O_2^{-} 13 v.e.   KK

bond order = (8 bonding e^- –5 antibonding e^-) $\div 2 = 1.5$; bond length = 128 pm

O_2^{2-} 14 v.e.   KK σ_{2s}⥮ σ_{2s}*⥮ σ_{2p}⥮ π_{2p}⥮⥮ π_{2p}*⥮⥮ σ_{2p}*☐

bond order = (8 bonding electrons –6 antibonding electrons) $\div 2 = 1.0$; bond length = 149 pm. We see that the bond length does indeed increase as the bond order decreases. Longer bonds are weaker bonds.

7A   **(M)** There are 8 valence electrons that must be placed into the valence molecular orbital diagram for CN^+ (5 electrons from nitrogen, four electrons from carbon and one electron is removed to produce the positive charge). Since both C and N precede oxygen and they are not far apart in atomic number, we must use the modified molecular-orbital energy-level diagram to get the correct configuration. We will assume that the order of the MOs for CN^+ is the same as for CO, shown in Figure 11-28. By following the Aufbau orbital filling method, one obtains the ground state diagram asked for in the question.

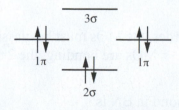

Before calculating the bond order, the MOs must be classified as bonding, antibonding, or nonbonding. Both the 3σ and 1π MOs are bonding, the 2π and 6σ are antibonding, and the 4σ and 5σ are nonbonding.

The bond order for the C–N bond in CN^+ is

$$\frac{6 \text{ bonding electrons} - 0 \text{ antibonding electrons}}{2} = 3$$

Thus, the C and N atoms in CN^+ are joined by a triple bond.

7B   **(M)** There are 8 valence electrons that must be placed in the molecular orbital diagram for BN (3 electrons from boron and five electrons from nitrogen). Since both B and N precede oxygen and they are not far apart in atomic numbers, we must use the modified molecular-orbital energy-level diagram to get the correct configuration. We will assume that the order of the MOs for CN^+ is the as as for CO, shown in Figure 11-28. By following the Aufbau orbital filling method, one obtains the ground-state diagram asked for in the question.

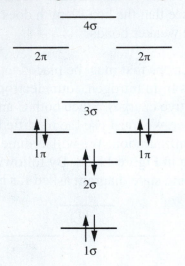

Before calculating the bond order, the MOs must be classified as bonding, antibonding, or nonbonding. Both the 3σ and 1π MOs are bonding, the 2π and 6σ are antibonding, and the 4σ and 5σ are nonbonding.

The bond order for the B–N bond in BN is

$$\frac{6 \text{ bonding electrons} - 0 \text{ antibonding electrons}}{2} = 3$$

Thus, the B and N atoms in BN are joined by a triple bond.

8A **(D)** In this exercise we will combine valence-bond and molecular orbital methods to describe the bonding in the SO_3 molecule. By invoking a π-bonding scheme, we can replace the three resonance structures for SO_3 (shown below) with just one structure that exhibits both σ-bonding and delocalized π-bonding.

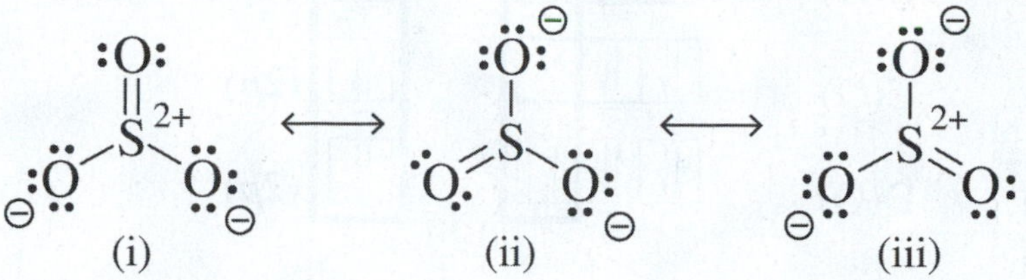

We will use structure (i) to develop a combined localized/delocalized bonding description for the molecule. (Note: Any one of the three resonance contributors can be used as the starting structure.) We begin by assuming that every atom in the molecule is sp^2 hybridized to produce the σ-framework for the molecule. The half-filled sp^2 hybrid orbitals of the oxygen atoms will each be overlapped with a half-filled sp^2 hybrid orbital on sulfur. By contrast, to generate a set of π molecular orbitals, we will combine one unhybridized $2p$ orbital from each of three oxygen atoms with an unhybridized $3p$ orbital on the sulfur atom. This will generate four π molecular orbitals: a bonding molecular orbital, two nonbonding molecular orbitals, and an antibonding orbital. Remember that the number of valence electrons assigned to each atom must reflect the formal charge for that atom. Accordingly, sulfur, with a 2+ formal charge, can have only four valence electrons, the two oxygens with a 1– formal charge must each end up with 7 electrons and the oxygen with a zero formal charge must have its customary six electrons

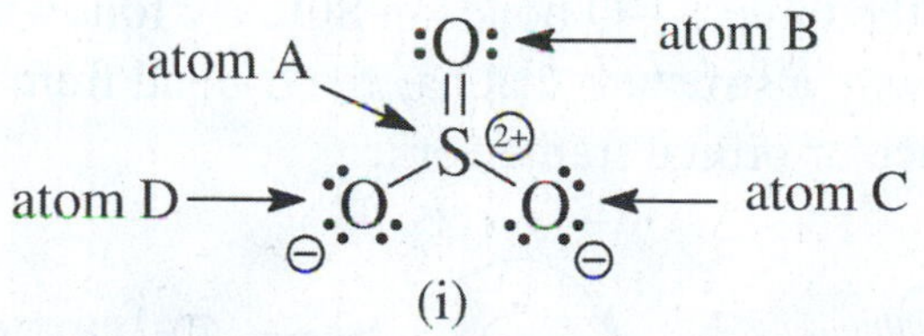

Let's begin assigning valence electrons by half-filling three sp^2 hybrid orbitals on the sulfur atom (atom A) and one sp^2 hybrid orbital on each of the oxygen atoms (atoms B, C and D) Next, we half-fill the lone unhybridized $3p$ orbital on sulfur and the lone $2p$ orbital on the oxygen atom with a formal charge of zero (atom B). Following this, the $2p$ orbital of the other two oxygen atoms (atoms C and D), are filled and then lone pairs are placed in the sp^2 hybrid orbitals that are still empty. At this stage, then, all 24 valence electrons have been put into atomic and hybrid orbitals on the four atoms. Now we overlap the six half-filled sp^2 hybrid orbitals to generate the σ-bond framework and combine the three $2p$ orbitals (2 filled, one half-filled) and the $3p$ orbital (half-filled) to form the four π-molecular orbitals, as shown below

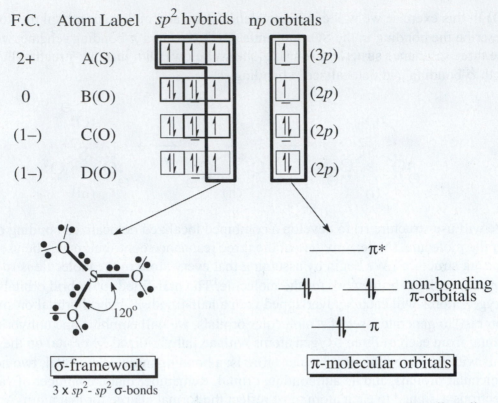

Overall bond order for this set of π-molecular orbitals $\dfrac{2\,\text{bonding e}^- - 0\,\text{antibonding e}^-}{2} = 1$

The π-bond is spread out evenly over the three S—O linkages. This leads to an average bond order of 1.33 for the three S—O bonds in SO_3. By following this "combined approach", we end up with a structure that has the σ-bond framework sandwiched within the delocalized π-molecular orbital framework:

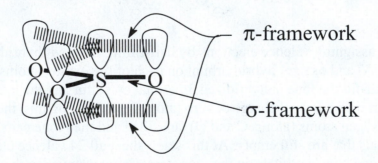

This is a much more accurate description of the bonding in SO_3 than that provided by any one of the three Lewis diagrams shown above.

8B **(D)** We will use the same basic approach to answer this question as was used to solve Practice Example 11-8A. This time, the bonding in NO_2^- will be described by combining valence-bond and molecular orbital theory. With this approach we will be able to generate a structure that more accurately describes the bonding in NO_2^- than either of the two equivalent Lewis diagrams that can be drawn for the nitrate ion (below).

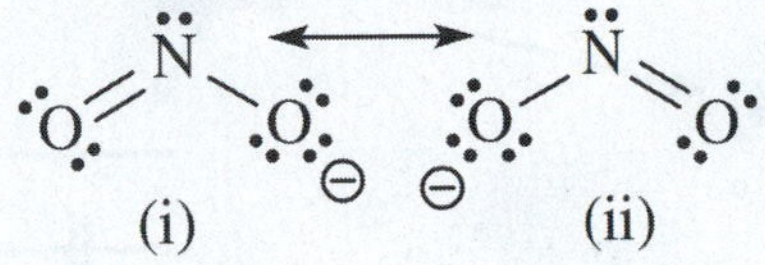

(i)   (ii)

Structure (i) will be used to develop the combined localized/delocalized bonding description for the anion (either structure could have been used as the starting structure). We begin by assuming that every atom in the molecule is sp^2 hybridized. To produce the σ-framework for the molecule, the half-filled sp^2 orbitals of the oxygen atoms will be overlapped with a half-filled sp^2 orbital of nitrogen. By contrast, to generate a set of π-molecular orbitals, we will combine one unhybridized $2p$ orbital from each of the two oxygen atoms with an unhybridized $2p$ orbital on nitrogen. This will generate three π molecular orbitals: a bonding molecular orbital, a non-bonding molecular orbital and an antibonding molecular orbital.

(i)

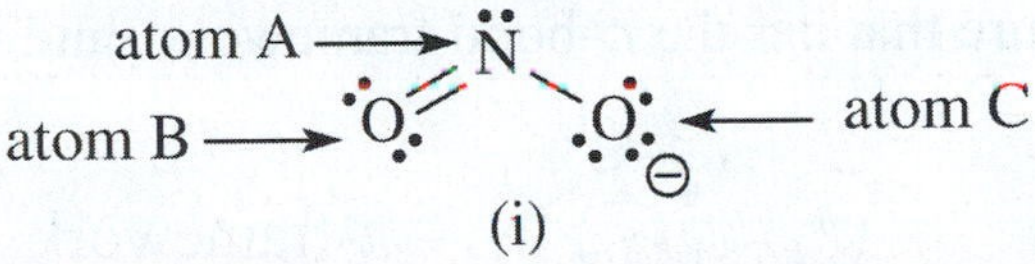

Remember that the number of valence electrons assigned to each atom must reflect the formal charge for that atom. Accordingly, the oxygen with a 1– formal charge (atom C) must end up with 7 electrons, whereas the nitrogen atom (atom A) and the other oxygen atom (atom B) must have their customary 5 and 6 valence electrons, respectively, because they both have a formal charge of zero. Let's begin assigning valence electrons by half-filling two sp^2 hybrid orbitals on the nitrogen atom (atom A) and one sp^2 hybrid orbital on each oxygen atom (atoms B and C). Next, we half-fill the lone $2p$ orbital on nitrogen (atom A) and the lone $2p$ orbital on the oxygen atom with a formal charge of zero (atom B). Following this, the $2p$ orbital for the remaining oxygen (atom C) is filled and then lone pairs are placed in the sp^2 hybrid orbitals that are still empty. At this stage, then, all 18 valence electrons have been put into atomic and hybrid orbitals on the three atoms. Now, we overlap the four half-filled sp^2 hybrid orbitals to generate the σ-bond framework and combine the three $2p$ orbitals (two half-filled, one filled) to form three π-molecular orbitals as shown below

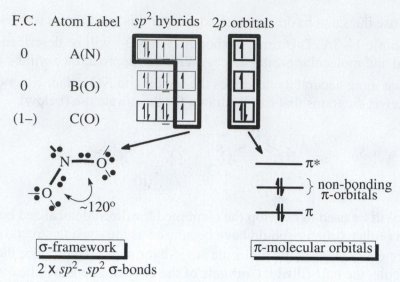

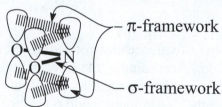

σ-framework
2 x sp^2- sp^2 σ-bonds

Overall bond order for this set of π-molecular orbitals $\dfrac{2 \text{ bonding e}^- - 0 \text{ antibonding e}^-}{2} = 1$

The π bond is spread out evenly over the two N—O linkages. This leads to an average bond order of 1.5 for each of the two N—O bonds in NO_2^-. By following this combined approach, we end up with a structure that has the σ-bond framework sandwiched within the delocalized π-molecular orbitals:

This is a much more accurate depiction of the bonding in NO_2^- than that provided by one of the two Lewis diagrams given above.

INTEGRATIVE EXAMPLE

A   **(M)** We will assume 100 g of the compound and find the empirical formula in the usual way.
moles of carbon = 28.57 g(1 mole/12.011g) = 2.379 mol
moles of hydrogen = 4.80 g(1 mole/1.008g) = 4.76 mol
moles of nitrogen = 66.64 g(mole/14.01g) = 4.76 mol
Dividing by 2.38 we get: C(2.379/2.379) H(4.76/2.379) N(4.758/2.379)
This yields an empirical formula of: CH_2N_2 From the description given in the question (6-membered ring), the molecular formula is $C_3N_6H_6$

(a) Lewis structure for $C_3N_6H_6$ (based on info provided)

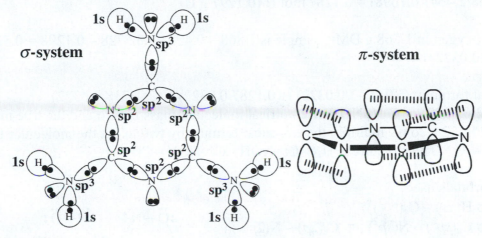

(b) Valence-bond description σ and π-bonding systems

σ-system   π-system

(c) The bonds in the ring system are similar to those seen in benzene (C_6H_6) and pyridine (C_5H_5N), namely

Antibonding π moleclar orbitals {

Bonding π moleclar orbitals {

One 2p electron from each of the atoms is placed in the ring

B   **(D)** We have 2.464 g of dimethylglyoxime (DMG)

moles C : 3.735(1 mol C/44.01g CO_2) = 0.08487 mol C
moles H : 1.530(2 mol H/18.01g H_2O) = 0.1698 mol C

First we find the excess H_2SO_4 left over from the nitrogen determination:

$$18.6 \text{ mL} \times \frac{1 \text{ L}}{1000 \text{ mL}} \times \frac{0.2050 \text{ mol NaOH}}{1 \text{ L}} \times \frac{1 \text{ mol } H_2SO_4}{2 \text{ mol NaOH}} = 1.910 \times 10^{-3} \text{ mol } H_2SO_4 \text{ in excess}$$

Next we find the moles of H_2SO_4 initially added:

$$50.00 \text{ mL} \times \frac{1 \text{ L}}{1000 \text{ mL}} \times \frac{0.3600 \text{ mol } H_2SO_4}{1 \text{ L}} = 1.800 \times 10^{-2} \text{ mol } H_2SO_4 \text{ used}$$

Now we find the moles of H_2SO_4 that reacted:

mole H_2SO_4 that reacted $= 1.800 \times 10^{-2}$ mol H_2SO_4

initially used -1.910×10^{-3} mol H_2SO_4 in excess

mole H_2SO_4 that reacted $= 1.609 \times 10^{-2}$ mol H_2SO_4

Therefore, the moles of nitrogen in 1.868 g DMG sample is:

$$0.01609 \text{ mol } H_2SO_4 \times \frac{2 \text{ mol } NH_3}{1 \text{ mol } H_2SO_4} \times \frac{1 \text{ mol } N}{1 \text{ mol } NH_3} = 3.218 \times 10^{-2} \text{ mol nitrogen } (0.4507 \text{ g nitrogen})$$

For a 1.868 g sample, moles of C and H present are fond as follows:

C: (1.868/2.464)(0.08487) = 0.06434 mol C (0.7728 g C)

H: (1.868/2.464)(0.1698) = 0.1287 mol H (0.1297 g H)

Mass of oxygen in 1.868 g DMG sample is 1.868 − 0.4507 − 0.7728 − 0.1298 = 0.515 g oxygen (0.0322 mol O)

Empirical formula: C(0.0643/0.0322) H(0.1287/0.0322) N(0.03218/0.0322) O(0.0322/0.0322) or C_2H_4NO The empirical molecular mass is 58 u and the true molecular mass is 116.12 u so we multiply the empirical formula by two to get the molecular formula of $C_4H_8N_2O_2$. The structure is $H_3C-C(=NOH)-C(=NOH)-CH_3$

Hybrid orbitals used:

$C_b - H$: σ $H(1s) - C_b(sp^3)$

$C_a = N$: σ $C_a(sp^2) - N(sp^2)$, π: $C_a(2p) - N(2p)$

$N - O$: σ $N(sp^2) - O(sp^3)$

$C_a - C_a$: σ $C_a(sp^2) - C_a(sp^2)$

$C_a - C_b$: σ $C_b(sp^3) - C_a(sp^2)$

$O - H$: σ $H(1s) - O(sp^3)$

EXERCISES

Valence-Bond Method

Note: In VSEPR theory, the term *"bond pair"* is used for a single bond, a double bond, or a triple bond, even though a single bond consists of one pair of electrons, a double bond two pairs of electrons, and a triple bond three pairs of electrons. To avoid any confusion between the number of electron pairs actually involved in the bonding to a central atom, and the number of atoms bonded to that central atom, we shall occasionally use the term *"ligand"* to indicate an atom or a group of atoms attached to the central atom.

1. **(E)** There are several ways in which valence-bond theory is superior to Lewis structures in describing covalent bonds. *First*, valence-bond theory clearly distinguishes between sigma and pi bonds. In Lewis theory, a double bond appears to be just two bonds and it is not clear why a double bond is not simply twice as strong as a single bond. In valence-bond theory, it is clear

that a sigma bond must be stronger than a pi bond, for the orbitals overlap more effectively in a sigma bond (end-to-end) than they do in a pi bond (side-to-side). *Second*, molecular geometries are more directly obtained in valence-bond theory than in Lewis theory. Although valence-bond theory requires the introduction of hybridization to explain these geometries, Lewis theory does not predict geometries at all; it simply provides the basis from which VSEPR theory predicts geometries. *Third*, Lewis theory does not explain hindered rotation about double bonds. With valence-bond theory, any rotation about a double bond involves cleavage of the π-bond, which would require the input of considerable energy.

3. **(E)**
   (a) Lewis theory does not describe the shape of the water molecule. It does indicate that there is a single bond between each H atom and the O atom, and that there are two lone pairs attached to the O atom, but it says nothing about molecular shape.

   (b) In valence-bond theory using simple atomic orbitals, each H—O bond results from the overlap of a $1s$ orbital on H with a $2p$ orbital on O. The angle between $2p$ orbitals is 90° so this method initially predicts a 90° bond angle. The observed 104° bond angle is explained as arising from repulsion between the two slightly positively charged H atoms.

   (c) In VSEPR theory the H_2O molecule is categorized as being of the AX_2E_2 type, with two atoms and two lone pairs attached to the central oxygen atom. The lone pairs repel each other more than do the bond pairs, explaining the smaller than 109.5° tetrahedral bond angle.

   (d) In valence-bond theory using hybrid orbitals, each H—O bond results from the overlap of a $1s$ orbital on H with an sp^3 orbital on O. The angle between sp^3 orbitals is 109.5°. The observed bond angle of 104° is rationalized based on the greater repulsion of lone pair electrons when compared to bonding pair electrons.

5. **(M)** Determining hybridization is made easier if we begin with Lewis structures. Only one resonance form is drawn for CO_3^{2-}, SO_2, and NO_2^-.

The C atom is attached to three ligands and no lone pairs and thus is sp^2 hybridized in CO_3^{2-}. The S atom is attached to two ligands and one lone pair and thus is sp^2 hybridized in SO_2. The C atom is attached to four ligands and no lone pairs and thus is sp^3 hybridized in CCl_4. Both the oxygen and the carbon in CO are sp hybridized. The N atom is attached to two ligands and one lone pair in NO_2^- and thus is sp^2 hybridized. Thus, the central atom is sp^2 hybridized in SO_2, CO_3^{2-}, and NO_2^-.

273

7. (M) For each species, we first draw the Lewis structure, to help explain the bonding.

(a) In CO_2, there are a total of $4+(2\times6)=16$ valence electrons, or 8 pairs. C is the central atom. The Lewis structure is $\ddot{\text{O}}{=}\text{C}{=}\ddot{\text{O}}$ The molecule is linear and C is sp hybridized.

(b) In $HONO_2$, there are a total of $1+5+(3\times6)=24$ valence electrons, or 12 pairs. N is the central atom, and a plausible Lewis structure is shown on the right The molecule is trigonal planar around N which is sp^2 hybridized. The O in the H—O—N portion of the molecule is sp^3 hybridized.

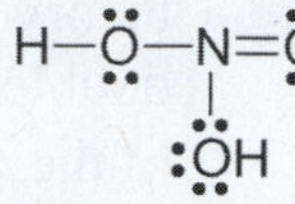

(c) In ClO_3^-, there are a total of $7+(3\times6)+1=26$ valence electrons, or 13 pairs. Cl is the central atom, and a plausible Lewis structure is shown on the right. The electron-group geometry around Cl is tetrahedral, indicating that Cl is sp^3 hybridized.

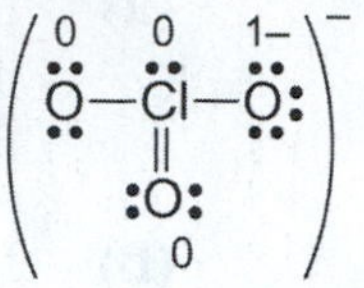

(d) In BF_4^-, there are a total of $3+(4\times7)+1=32$ valence electrons, or 16 pairs. B is the central atom, and a plausible Lewis structure is shown on the right. The electron-group geometry is tetrahedral, indicating that B is sp^3 hybridized.

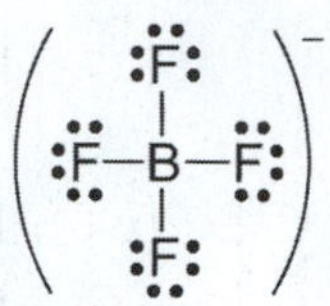

9. (E) The Lewis structure of ClF_3 is shown on the right. There are three atoms and two lone pairs attached to the central atom, its hybridization is sp^3d, which is achieved as follows. $Cl_{unhyb}[\text{Ne}]\,3s\,\boxed{\uparrow\downarrow}\;3p\,\boxed{\uparrow\downarrow\,|\uparrow\downarrow\,|\uparrow}\;3d\,\boxed{\;|\;|\;|\;|\;} \longrightarrow Cl_{hyb}\,[\text{Ne}]\;_{dsp^3}\boxed{\uparrow\downarrow\,|\uparrow\downarrow\,|\uparrow\,|\uparrow\,|\uparrow}\;_{3d}\boxed{\;|\;|\;|\;}$ The orbital diagram of F is $[\text{He}]\,2s\,\boxed{\uparrow\downarrow}\;2p\,\boxed{\uparrow\downarrow\,|\uparrow\downarrow\,|\uparrow}$ Each of the three sigma bonds are formed by the overlap of a $2p$ orbital on F with one of the half-filled sp^3d orbitals on Cl. Since the sp^3d hybridization has a trigonal bipyramidal shape, and the two lone pairs occupy the equatorial positions in the molecule, ClF_3 is T-shaped.

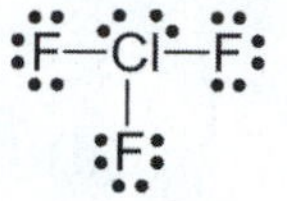

11. (M)

(a) In PF_6^- there are a total of $5+(6\times7)+1=48$ valence electrons, or 24 pairs. A plausible Lewis structure is shown below. In order to form the six P—F bonds, the hybridization on P must be sp^3d^2.

(b) In COS there are a total of $4+6+6=16$ valence electrons, or 8 pairs. A plausible Lewis structure is shown below. In order to bond two atoms to the central C, the hybridization on that C atom must be sp.

(c) In $SiCl_4$, there are a total of $4+(4\times7)=32$ valence electrons, or 16 pairs. A plausible Lewis structure is given below. In order to form four Si—Cl bonds, the hybridization on Si must be sp^3.

(d) In NO_3^-, there are a total of $5+(3\times6)+1=24$ valence electrons, or 12 pairs. A plausible Lewis structure is given below. In order to bond three O atoms to the central N atom, with no lone pairs on that N atom, its hybridization must be sp^2.

(e) In AsF_5, there are a total of $5+(5\times7)=40$ valence electrons, or 20 pairs. A plausible Lewis structure is shown below. This is a molecule of the type AX_5. To form five As—F bonds, the hybridization on As must be sp^3d.

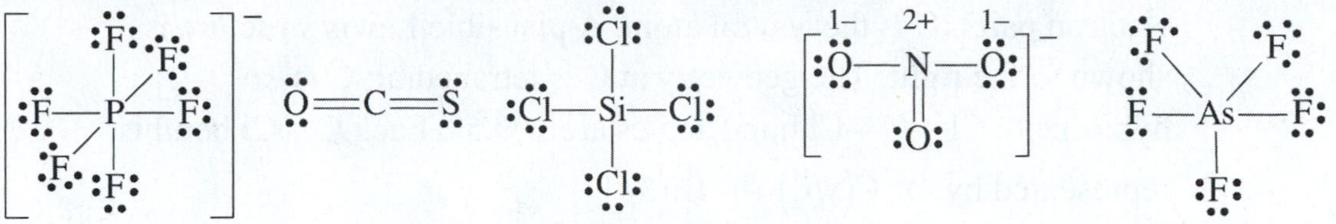

13. **(M)**

(a) This is a planar molecule. The hybridization on C is sp^2 (one bond to each of the three attached atoms).

(b) $:N\equiv C-C\equiv N:$ is a linear molecule. The hybridization for each C is sp (one bond to each of the N atoms).

(c) Trifluoroacetonitrile is neither linear nor planar. The shape around the left-hand C is tetrahedral and that C has sp^3 hybridization. The shape around the right-hand carbon is linear and that C has sp hybridization.(N atom is sp hybridized).

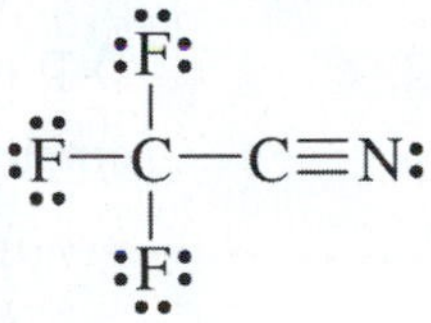

(d) $[:\ddot{S}-C\equiv N:]^-$ is a linear molecule. The hybridization for C is sp and N is sp. The hybridization of the S atom is sp^3.

15. **(M)**

(a) In HCN, there are a total of $1+4+5=10$ valence electrons, or 5 electron pairs. A plausible Lewis structure follows. $H-C\equiv N|$. The H—C bond is a σ bond, and the $C\equiv N$ bond is composed of 1 σ and 2 π bonds.

(b) In C_2N_2, there are a total of $(2\times4)+(2\times5)=18$ valence electrons, or 9 electron pairs. A plausible Lewis structure follows: $|N\equiv C-C\equiv N|$ The C—C bond is a σ bond, and each $C\equiv N$ bond is composed of 1 σ and 2 π bonds.

(c) In $CH_3CHCHCCl_3$, there are a total of 42 valence electrons, or 21 electron pairs. A plausible Lewis structure is shown to the right. All bonds are σ bonds except one of the bonds that comprise the C=C bond. The C=C bond is composed of one σ and one π bond.

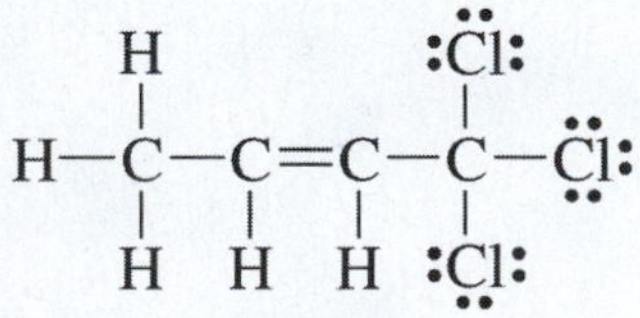

(d) In HONO, there is a total of $1+5+(2\times6)=18$ valence electrons, or 9 electron pairs. A plausible Lewis structure is shown below:

$$H—\ddot{\underset{\cdot\cdot}{O}}—\ddot{N}=\ddot{\underset{\cdot\cdot}{O}}$$

All single bonds in this structure are σ bonds. The double bond is composed of one σ and one π bond.

17. (M)

(a) In CCl_4, there are a total of $4+(4\times7)=32$ valence electrons, or 8 electron pairs. C is the central atom. A plausible Lewis structure is shown to the right. The geometry at C is tetrahedral; C is sp^3 hybridized. Cl—C—Cl bond angles are 109.5°. Each C—Cl bond is represented by σ: $C(sp^3)^1 — Cl(3p)^1$

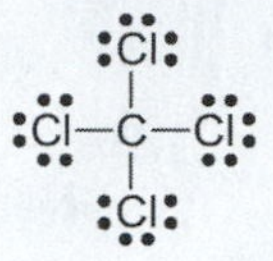

(b) In ONCl, there are a total of $6+5+7=18$ valence electrons, or 9 electron pairs. N is the central atom. A plausible Lewis structure is $\ddot{\underset{\cdot\cdot}{O}}=\ddot{N}—\ddot{\underset{\cdot\cdot}{Cl}}\mathbf{:}$. The e^- group geometry around N is triangular planar, and N is sp^2 hybridized. The O—N—C bond angle is about 120°. The bonds are: σ: $O(2p_y)^1—N(sp^2)^1$ σ: $N(sp^2)^1—Cl(3p_z)^1$

π: $O(2p_z)^1—N(2p_z)^1$

(c) In HONO, there are a total of $1+(2\times6)+5=18$ valence electrons, or 9 electron pairs. A plausible Lewis structure is $H-\ddot{\underset{\cdot\cdot}{O}}_a-\ddot{N}=\ddot{\underset{\cdot\cdot}{O}}_b$. The geometry of O_a is tetrahedral, O_a is sp^3 hybridized and the $H—O_a—N$ bond angle is (at least close to)109.5°. The e^- group geometry at N is trigonal planar, N is sp^2 hybridized, and the $O_a—N—O_b$ bond angle is 120°. The four bonds are represented as follows. σ: $H(1s)^1—O_a(sp^3)^1$

σ: $O_a(sp^3)^1—N(sp^2)^1$ σ: $N(sp^2)^1—O_b(2p_y)^1$ π: $N(2p_z)^1—O_b(2p_z)^1$.

(d) In $COCl_2$, there are a total of $4+6+(2\times7)=24$ valence electrons, or 12 electron pairs. A plausible Lewis structure is shown to the right. The e^- group geometry around C is trigonal planar; all bond angles around C are 120°, and the hybridization of C is sp^2. The four bonds in the molecules are:

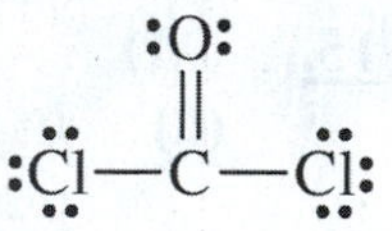

$2\times\sigma$: $Cl(3p_x)^1—C(sp^2)^1$ σ: $O(2p_y)^1—C(sp^2)^1$ π: $O(2p_z)^1—C(2p_z)^1$.

19. **(D)** Citric acid has the molecular structure shown below. Using Figure 11-18 as a guide, the flowing hybridization and bonding scheme is obtained for citric acid:

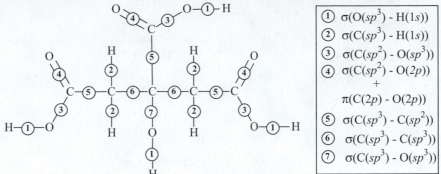

21. **(M)**
(a)

Central S is sp^2 hybridized and the terrminal O and S atoms are unhybridized

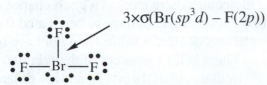

$\sigma(S(sp^2)\text{-}S(3p_x))$

$\pi(S(3p_z)\text{-}S(3p_z))$

$\sigma(S(sp^2)\text{-}O(2p_z))$

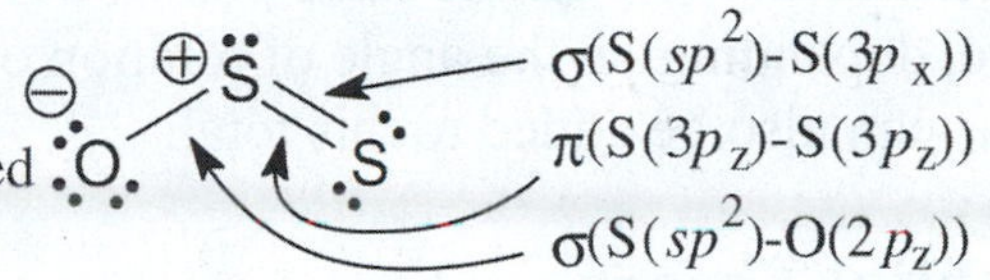

(b)

Br atom is sp^3d hybridized. F atoms are unhybridized.

$$3 \times \sigma(Br(sp^3d) - F(2p))$$

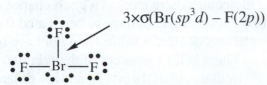

23. **(M)** The bond lengths are consistent with the left-hand C—C bond being a triple bond (120 pm), the other C—C bond being a single bond (154 pm) rather than a double bond (134 pm), and the C—O bond being a double bond (123 pm). Of course, the two C—H bonds are single bonds (110 pm). All of this is depicted in the Lewis structure on the right. The overlap that produces each bond follows:

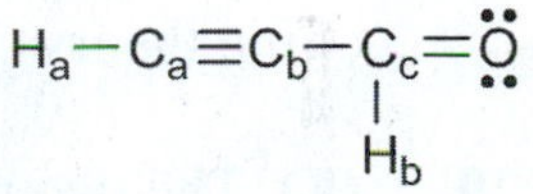

$H_a - C_a \equiv C_b - C_c = \overset{..}{\underset{..}{O}}$
$\quad\quad\quad\quad\quad\quad\mid$
$\quad\quad\quad\quad\quad\quad H_b$

$$\sigma : C_a(sp)^1 - H_a(1s)^1 \quad\quad \sigma : C_a(sp)^1 - C_b(sp)^1 \quad\quad \sigma : C_b(sp)^1 - C_c(sp^2)^1$$

$$\sigma : C_c(sp^2)^1 - H_b(1s)^1 \quad\quad \sigma : C_c(sp^2)^1 - O(2p_y)^1 \quad\quad \pi : C_a(2p_y)^1 - C_b(2p_y)^1$$

$$\pi : C_a(2p_z)^1 - C_b(2p_z)^1 \quad\quad \pi : C_c(2p_z)^1 - O(2p_z)^1$$

25. **(M)** The structure is shown below:

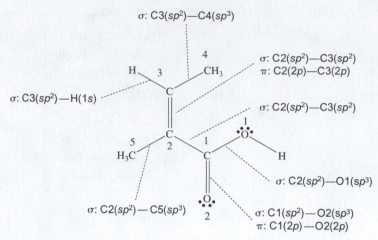

There are 8 atoms that are on the same plane (O1, O2, C1-5, H attached to C3). Furthermore, depending on the angle of rotation of the –CH$_3$ groups (C4 and C5), two H atoms can also be added to this total.

Molecular–Orbital Theory

27. **(E)** The valence-bond method describes a covalent bond as the result of the overlap of atomic orbitals. The more complete the overlap, the stronger the bond. Molecular orbital theory describes a bond as a region of space in a molecule where there is a good chance of finding the bonding electrons. The molecular orbital bond does not have to be created from atomic orbitals (although it often is) and the orientations of atomic orbitals do not have to be manipulated to obtain the correct geometric shape. There is little concept of the relative energies of bonding in valence-bond theory. In molecular orbital theory, bonds are ordered energetically. These energy orderings, in fact, provide a means of checking the predictions of the theory through the spectroscopic analysis of the molecules.

29. **(E)** The two molecular orbital diagrams follow:

N_2^- no. valence e$^-$ = $(2 \times 5) + 1 = 11$

N_2^{2-} no. valence e$^-$ = $(2 \times 5) + 2 = 12$

N_2^- bond order = $(8-3) \div 2 = 2.5$ (stable) N_2^{2-} bond order = $(8-4) \div 2 = 2$ (stable)

31. **(E)** In order to have a bond order higher than three, there would have to be a region in a molecular orbital diagram where four bonding orbitals occur together in order of increasing energy, with no intervening antibonding orbitals. No such region exists in any of the molecular orbital diagrams in Figure 11-25. Alternatively, three bonding orbitals would have to occur together energetically, following an electron configuration which, when full, results in a bond order greater than zero. This arrangement does not occur in any of the molecular orbital diagrams in Figure 11-25.

33. **(M)**

(a) A σ_{1s} orbital must be lower in energy than a σ_{1s}^{*} orbital. The bonding orbital always is lower than the antibonding orbital if they are derived from the same atomic orbitals.

(b) The σ_{2s} orbital is derived from the $2s$ atomic orbitals while the σ_{2p} molecular orbitals are constructed from $2p$ atomic orbitals. Since the 2s orbitals are lower in energy than the 2p orbitals, we would expect that the σ_{2p} molecular orbital would be higher in energy than the σ_{2s} molecular orbital.

(c) A σ_{1s}^{*} orbital should be lower than a σ_{2s} orbital, since the $1s$ atomic orbital is considerably lower in energy than is the $2s$ orbital.

(d) A σ_{2p} orbital should be lower in energy than a σ_{2p}^{*} orbital. Both orbitals are from atomic orbitals in the same subshell but we expect a bonding orbital to be more stable than an antibonding orbital.

35. **(M)** For each of the following heteronuclear species, the valence number of electrons and valence molecular orbital diagram is listed.

(a) NO: $5+6=11$

(b) NO^{+}: $5+6-1=10$

(c) CO: $4+6=10$

(d) CN: $4+5=9$

(e) CN^{-}: $4+5+1=10$

(f) CN^{+}: $4+5-1=8$

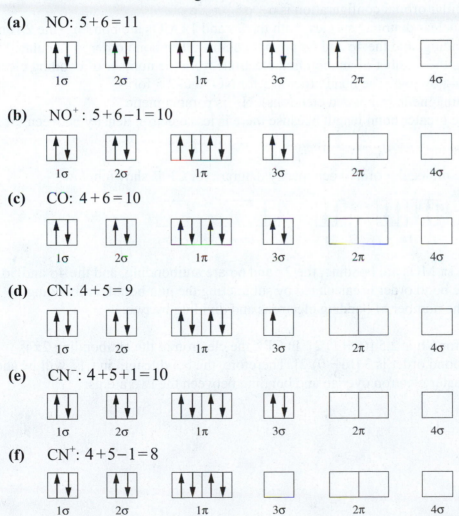

(g) BN: $3 + 5 = 8$

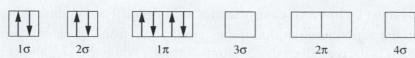

1σ   2σ   1π   3σ   2π   4σ

37. **(M)**

(a) NO^+ has ten valence electrons ($5 + 6 - 1 = 10$). The valence MO diagram is

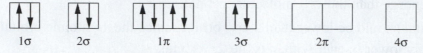

1σ   2σ   1π   3σ   2π   4σ

The molecular orbital configuration is $1\sigma^2 2\sigma^2 1\pi^4 3\sigma^2$.

N_2^+ has nine valence electrons ($5 + 5 - 1 = 9$). The valence MO occupancy diagram is

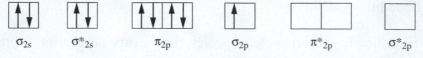

σ$_{2s}$ σ*$_{2s}$ π$_{2p}$ σ$_{2p}$ π*$_{2p}$ σ*$_{2p}$

The molecular orbital configuration is $\sigma_{2s}^2 \sigma*_{2s}^2 \pi_{2p}^4 \sigma_{2p}^1$.

(b) In heteronuclear diatomic species, both the 3σ and 1π MOs are bonding, the 2π and 6σ are antibonding, and the 4σ and 5σ are nonbonding. The bond order is calculated by subtracting the number of antibonding electrons from the number of bonding electrons and dividing by two. The bond order is 3 for NO^+ and 2.5 for N_2^+.

(c) NO^+ is diamagnetic (all paired electrons). N_2^+ is paramagnetic.

(d) N_2^+ has the greater bond length because there is less electron density between the two nuclei.

39. **(M)** The valence molecular orbital occupancy diagram for CF is shown below:

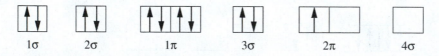

1σ   2σ   1π   3σ   2π   4σ

Both the 3σ and 1π MOs are bonding, the 2π and 6σ are antibonding, and the 4σ and 5σ are nonbonding. The bond order is calculated by subtracting the number of antibonding electrons from the number of bonding electrons and dividing by two.

The bond order for CF is 2.5 [$(6 - 1)/2$]. In CF^+, the electron in the antibonding 2π is removed, so its bond order is 3 [$(6 - 0)/2$]. Therefore, the bond length in CF^+ will be shorter, which means greater electron overlap and bonding between the two atoms.

Delocalized Molecular Orbitals

41. **(E)** With either Lewis structures or the valence-bond method, two structures must be drawn (and "averaged") to explain the π bonding in C_6H_6. The σ bonding is well explained by assuming sp^2 hybridization on each C atom. But the π bonding requires that all six C—C π bonds must be equivalent. This can be achieved by creating six π molecular orbitals—three bonding and three antibonding—into which the 6π electrons are placed. This creates a single delocalized structure for the C_6H_6 molecule.

43. **(M)** We expect to find delocalized orbitals in those species for which bonding cannot be represented thoroughly by one Lewis structure, that is, for compounds that require several resonance forms.

(a) In C_2H_4, there are a total of $(2 \times 4) + (4 \times 1) = 12$ valence electrons, or 6 pairs. C atoms are the central atoms. Thus, the bonding description of C_2H_4 does not involve the use of delocalized orbitals.

(b) In SO_2, there are a total of $6 + (2 \times 6) = 18$ valence electrons, or 9 pairs. N is the central atom. A plausible Lewis structure has two resonance forms. The bonding description of SO_2 will require the use of delocalized molecular orbitals.

$$\ddot{O}=\ddot{S}-\ddot{O}: \longleftrightarrow :\ddot{O}-\ddot{S}=\ddot{O}$$

(c) In H_2CO, there are a total of $(2 \times 1) + 4 + 6 = 12$ valence electrons, or 6 pairs. C is the central atom. A plausible Lewis structure is shown on the right. Because one Lewis structure adequately represents the bonding in the molecule, the bonding description of H_2CO does not involve the use of delocalized molecular orbitals.

INTEGRATIVE AND ADVANCED EXERCISES

46. **(M)** We begin with the orbital diagram for Na to describe the valence-bond picture for Na_2. [Ne]$_{3s}$ The half-filled 3s orbital on each Na overlaps with another to form a σ covalent bond. There are 22 electrons in Na_2. These electrons are distributed in the molecular orbitals as follows.

σ_{1s} σ_{1s}^* σ_{2s} σ_{2s}^* $\sigma_{2p}\ \pi_{2p}$ $\pi_{2p}^*\ \sigma_{2p}^*$ σ_{3s}

Again, we predict a single bond for Na_2. A Lewis-theory picture of the bonding would have the two Lewis symbols for two Na atoms uniting to form a bond: $Na\cdot\ \cdot Na \longrightarrow Na - Na$ Thus, the bonding in Na_2 is very much like that in H_2.

48. **(M)** The superoxide ion, O_2^-, has 17 electrons, while the peroxide ion, O_2^{2-}, has a total of 18 electrons,

(a) The molecular orbital diagrams for these two ions are given below.

$$O_2^- \quad \sigma_{1s} \quad \sigma_{1s}^* \quad \sigma_{2s} \quad \sigma_{2s}^* \quad \sigma_{2p} \ \pi_{2p} \quad \pi_{2p}^* \quad \sigma_{2p}^*$$

bond order = (no. bonding electrons − no. antibonding electrons) $\div 2 = (10 - 7) \div 2 = 1.5$

$$O_2^{2-} \quad \sigma_{1s} \quad \sigma_{1s}^* \quad \sigma_{2s} \quad \sigma_{2s}^* \quad \sigma_{2p} \ \pi_{2p} \quad \pi_{2p}^* \quad \sigma_{2p}^*$$

bond order = (no. bonding electrons − no. antibonding electrons) $\div 2 = (10 - 8) \div 2 = 1.0$

(b) O_2^- has $(2 \times 6) + 1 = 13$ electrons, 6 electron pairs plus one electron. O_2^{2-} has $(2 \times 6) + 2 = 14$ electrons, 7 electron pairs. Plausible Lewis structures are

$O_2^- \quad [:\ddot{O}-\ddot{O}\cdot]^-$ $O_2^{2-} \quad [:\ddot{O}-\ddot{O}:]^{2-}$

52. **(M)** The O—N—O bond angle of 125° indicates the N atom is sp^2 hybridized, while the F—O_a—N bond angle of 105° indicates the O_a atom is sp^3 hybridized. The orbital diagrams of the atoms follow.

O [He] $2s$ ☐ $2p$ ☐ ☐ ☐ O_a [He] $2s$ ☐ $2p$ ☐ ☐ ☐ $\longrightarrow$ O [He] sp^3 ☐ ☐ ☐ ☐

F [He] $2s$ ☐ $2p$ ☐ ☐ ☐   N [He] $2s$ ☐ $2p$ ☐ ☐ ☐ $\longrightarrow$ N [He] sp^2 ☐ ☐ ☐ $2p$ ☐

Electron transfer from N to the terminal oxygen atom results in a +1 formal charge for nitrogen, a −1 formal charge for O and a single bond being developed between these two atoms.

O [He] $2s$ ☐ $2p$ ☐ ☐ ☐ $\xrightarrow{+ e^-}$ O^- [He] $2s$ ☐ $2p$ ☐ ☐ ☐

Bonds are formed by the following overlaps.

$$\sigma: F(2p) - O_a(sp^3) \quad \sigma: O_a(sp^3) - N(sp^2) \quad \sigma: N(sp^2) - O(2p_y) \quad \pi: N(2p_z) - O(2p_z)$$

Two resonance structures are required $:\overset{\displaystyle :\ddot{O}:}{\underset{||}{}}$ $:\ddot{F}-\ddot{O}_a-N-\ddot{O}:$ $\longleftrightarrow$ $:\ddot{F}-\ddot{O}_a-N=\ddot{O}$

With the wedge-and-dash representation, a sketch of the molecule is
$$\begin{array}{c} O \\ \diagdown \\ \quad N-O \\ \diagup \quad \diagdown \\ O \qquad F \end{array}$$

54. **(M)** Let us begin by drawing Lewis structures for the species concerned.

$$:\ddot{F}-\overset{:\ddot{F}:}{Cl}-\ddot{F}: \ + \ :\ddot{F}-\overset{:\ddot{F}\quad:\ddot{F}:}{\underset{:\ddot{F}:}{As}}-\ddot{F}: \ \longrightarrow \ [:\ddot{F}-Cl-\ddot{F}:]^+ \ + \ \left[:\ddot{F}-\overset{:\ddot{F}\quad:\ddot{F}:}{\underset{:\ddot{F}\quad:\ddot{F}:}{As}}-\ddot{F}:\right]^-$$

(a) For ClF_3 there are three atoms and two lone pairs attached to the central atom. The electron group geometry is trigonal bipyramidal and the molecule is T-shaped. For As F_5 there are five atoms attached to the central atom; its electron group geometry and molecular shape are trigonal bipyramidal. For ClF_2^+ there are two lone pairs and two atoms attached to the central atom. Its electron group geometry is tetrahedral and the ion is bent in shape. For AsF_6^- there are six atoms and no lone pairs attached to the central atom. Thus, the electron group geometry and the shape of this ion are octahedral.

(b) Trigonal bipyramidal electron group geometry is associated with sp^3d hybridization. The Cl in ClF_3 and As in AsF_5 both have sp^3d hybridization. Tetrahedral electron group geometry is associated with sp^3 hybridization. Thus Cl in ClF_2^+ has sp^3 hybridization. Octahedral electron group geometry is associated with sp^3d^2 hybridization, which is the hybridization adopted by As in AsF_6^-.

56. **(E)** Suppose two He atoms in the excited state $1s^1 2s^1$ unite to form an He_2 molecule. One possible configuration is $\sigma_{1s}^2 \sigma_{1s}^{*0} \sigma_{2s}^2 \sigma_{2s}^{0*}$. The bond order would be (4-0)/2 = 2.

58. **(E)** The orbital diagrams for C and N are as follows. C [He] sp^2 ⊡⊡⊡ $2p$ ⊡
N [He] sp^2 ⊡⊡⊡ $2p$ ⊡
The sp^2 electrons are involved in σ bonding. For N, the lone pair is in one sp^2 orbital; the remaining two half-filled sp^2 orbitals bond to adjacent C atoms. For C, one sp^2 orbital forms a σ bond by overlap with a half-filled $1s$ H atom. The remaining two half-filled sp^2 orbitals bond to either adjacent C or N atoms. The $2p$ electrons are involved in π bonding. The six $2p$ orbitals form six delocalized π molecular orbitals, three bonding and three antibonding. These six π orbitals are filled as shown in the π molecular orbital diagram sketched below.

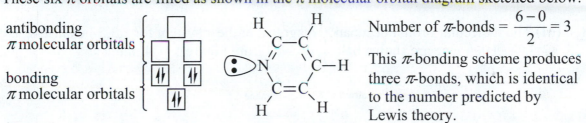

Number of π-bonds = $\dfrac{6-0}{2} = 3$

This π-bonding scheme produces three π-bonds, which is identical to the number predicted by Lewis theory.

61. **(M)** We will assume 100 g of the compound and find the empirical formula in the usual way.

$$\text{moles of carbon} = 53.09 \text{ g C} \times \frac{1 \text{ mole C}}{12.011 \text{g C}} = 4.424 \text{ mol}$$

$$\text{moles of hydrogen} = 6.24 \text{ g} \times \frac{1 \text{ mole H}}{1.008 \text{ g H}} = 6.19 \text{ mol}$$

$$\text{moles of nitrogen} = 12.39 \text{ g} \times \frac{1 \text{ mole N}}{14.007 \text{ g N}} = 0.885 \text{ mol}$$

$$\text{moles of oxygen} = 28.29 \text{ g} \times \frac{1 \text{ mole O}}{15.999 \text{ g O}} = 1.768 \text{ mol}$$

Dividing all result by 0.885 we get: 5.00 moles C, 6.99$\underline{5}$ moles H, 1.00 mole N and 2.00 moles O. This yields an empirical formula of $C_5H_7NO_2$.
Structure: $N{\equiv}C{-}CH_2(C{=}O)OC_2H_5$

Hybrid orbitals used:

C_b—H , C_d—H, C_e—H : σ H($1s$) – C(sp^3) (all tetrahedral carbon uses sp^3 hybrid orbitals)

C_c=O_b: σ C_c (sp^2) –O_b ($2p$ or sp^2), π: C_c ($2p$) – O_b ($2p$)

C_c-O_a: σ C_c (sp^2) –O_a ($2p$ or sp^3)

C_d-O_a: σ C_d (sp^3) –O_a ($2p$ or sp^3)

C_a≡N: σ C_a (sp) –N(sp)

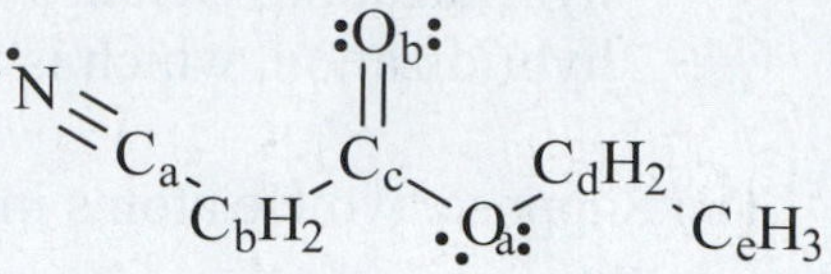

Two mutually perpendicular π-bonds: C($2p$) – N($2p$)

C_a-C_b: σ C_b(sp^3) –C_a(sp)

C_d-C_e: σ C_d(sp^3) –C_e(sp^3)

C_c-C_b: σ C_b(sp^3) –C_c(sp^2)

63. **(E)**

(a) power output = $(1.00 \text{ kW/m}^2) \times (1000 \text{ W/kW}) \times (40.0 \text{ cm}^2) \times (1\text{m}^2/10^4 \text{ cm}^2) = 4.00$ watts

amps = $i = w/v$

(b) $i = 4.00$ watts $= 4.00\dfrac{\text{J}}{\text{s}} \times \dfrac{1 \text{ C}}{0.45 \text{ J}} = 8.9\dfrac{\text{C}}{\text{s}} = 8.9$ amps

68. **(M)** The molecular orbital occupancy diagrams of the π system for 1,3,5-hexatriene and 1,3,5,7-octatetraene are shown below, derived from Figure 11-34 in the text.

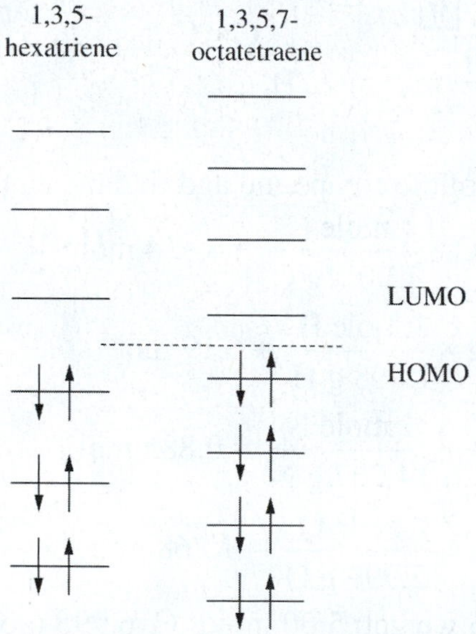

As is seen, the HOMO-LUMO gap in 1,3,5,7-ocatetraene is smaller than the gap for 1,3,5-hexatriene. Therefore, the wavelength corresponding to the excitation of the former would be longer. Energy and wavelength are related according to the equation below:

$$E = hc/\lambda$$

Since the energy gap needed to overcome for 1,3,5,7-octatetraene is smaller, the wavelength is larger.

For 1,3,4-hexatriene, there are three bonding orbitals and three nonbonding molecular orbitals. The combinations of atomic orbitals contributing to each molecular orbital are illustrated below.

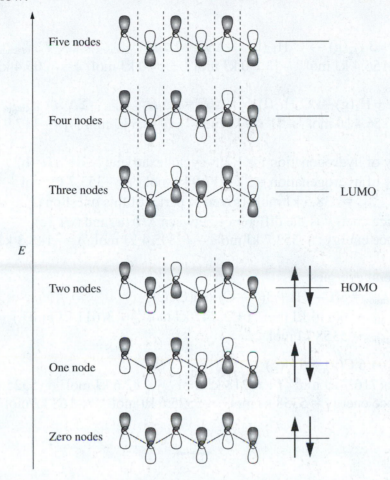

FEATURE PROBLEMS

70. (D)

(a) $C_6H_6(l) + 3 H_2(g) \rightarrow C_6H_{12}(l)$ $\Delta H° = \Sigma\Delta_f H°$, products $- \Sigma\Delta_f H°$, reactants
$\Delta H° = -156.4 \text{ kJ mol}^{-1} - [3 \times 0 \text{ kJ mol}^{-1} + 49.0 \text{ kJ mol}^{-1}] = -205.4 \text{ kJ mol}^{-1} = \Delta H°(a)$

(b) $C_6H_{10}(l) + H_2(g) \rightarrow C_6H_{12}(l)$ $\Delta H° = \Sigma\Delta_f H°$ products $- \Sigma\Delta_f H°$ reactants
$\Delta H° = -156.4 \text{ kJ mol}^{-1} - [1 \times 0 \text{ kJ mol}^{-1} + (-38.5 \text{ kJ mol}^{-1})] = -117.9 \text{ kJ mol}^{-1} = \Delta H°(b)$

(c) Enthalpy of hydrogenation for 1,3,5-cyclohexatriene $= 3 \times \Delta H°(b)$
Enthalpy of hydrogenation $= 3 (-117.9 \text{ kJ mol}^{-1}) = -353.7 \text{ kJ mol}^{-1} = \Delta H°(c)$
$\Delta_f H°_{cyclohexene} = -38.5 \text{ kJ/mole}$ (given in part b of this question).
Resonance energy is the difference between $\Delta H°(a)$ and $\Delta H°(c)$.
Resonance energy $= -353.7 \text{ kJ mol}^{-1} - (-205.4 \text{ kJ mol}^{-1}) = -148.3 \text{ kJ mol}^{-1}$

(d) Using bond energies: 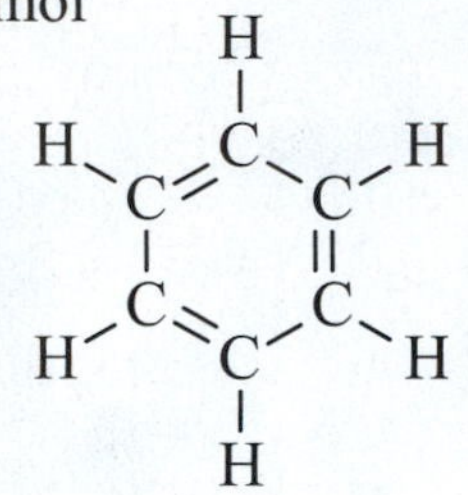
$\Delta H°_{atomization} = 6(C\text{—}H) + 3(C\text{—}C) + 3(C\text{=}C)$
$\Delta H°_{atomization} = 6(414 \text{ kJ mol}^{-1}) + 3(347 \text{ kJ mol}^{-1}) + 3(611 \text{ kJ mol}^{-1})$
$\Delta H°_{atomization} = 5358 \text{ kJ mol}^{-1}$

$C_6H_6(g) \rightarrow 6 C(g) + 6 H(g)$
$\Delta H° = [6(716.7 \text{ kJ mol}^{-1}) + 6(218 \text{ kJ mol}^{-1})] - 82.6 \text{ kJ mol}^{-1} = 5525.6 \text{ kJ mol}^{-1}$
Resonance energy $= 5358 \text{ kJ mol}^{-1} - 5525.6 \text{ kJ mol}^{-1} = -168 \text{ kJ mol}^{-1}$

76. (M)

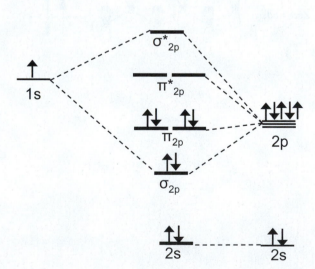

SELF-ASSESSMENT EXERCISES

80. (E) The answer is (c). It is the only option that has three species attached to the central atom (2 oxygen atoms, one electron pair):

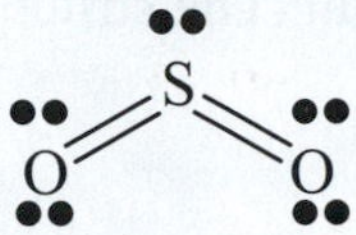

81. (E) The answer is (c). H_2Se has the same geometry as H_2S and H_2O. It is sp^3 hybridized, and the bond angle is less than H_2S, but it has to be more than $90°$.

82. (E) The answer is (a). The Lewis structure of I_3^- is shown below:

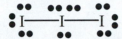

There are $1s$ and $3p$ orbitals for the central atom. However, there are 5 groups around it, which means that the d orbital needs to be used.

83. (E) The answer is (b). Each Li has the electron configuration $1s^2 2s^1$. The valence electrons occupy the 2σ orbitals completely, and not the $2\sigma^*$, so the bond order is $(4 - 2)/2 = 1$

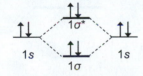

84. (E) The answer is (c). The Lewis structure of XeF_2 is shown below:

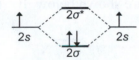

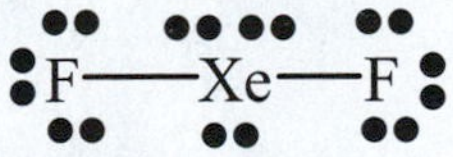

Since there are 5 groups around Xe, the hybridization is sp^3d.

85. (E) The answer is (d). The Lewis structure of CO_3^{2-} and the result of its resonance structures is shown below:

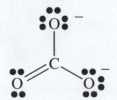   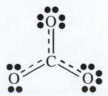

86. **(M)** From VSEPR theory we conclude that BF_3 is a trigonal planar molecule (as seen in Table 11.1). The valence-bond method using pure s and p orbitals incorrectly predicts a trigonal pyramidal shape with 90° F—B—F bond angles.

87. **(E)** BrF_5 has six constituents around it; five are fluorine atoms, and the sixth is a lone pair. Therefore, the hybridization is sp^3d^2, but the geometry is square pyramidal. The structure is shown below:

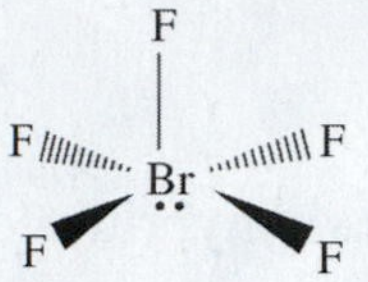

88. **(E)** The structure of CH_3NCO is shown below:

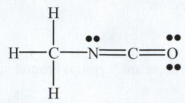

There are two double bonds, which means that there are **(a)** 6 σ and **(b)** 2 π bonds.

89. **(M)** All three are paramagnetic, because all three have unpaired electrons. The one with the strongest bond is **(b)**. The molecular orbital diagrams for all three are shown below. B_2^- has 3 bonding electrons, so the B–B bond is the strongest.

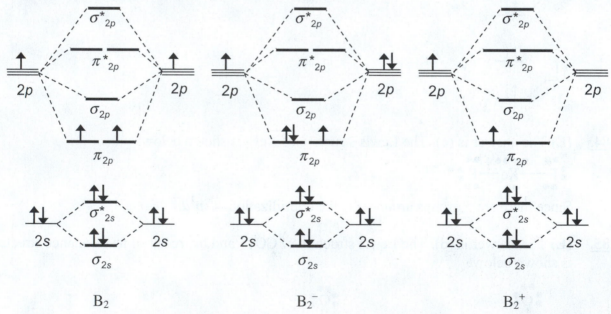

90. **(E)** The answer is (c), because C_2^- has an unpaired electron in the $\sigma(2s)$ bonding orbital, which is easier to remove than one already paired up because there is no pairing energy to overcome. The other two have electrons in the bonding orbitals, which require more energy to remove.

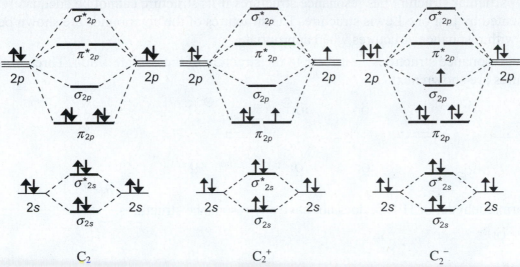

91. **(M)**

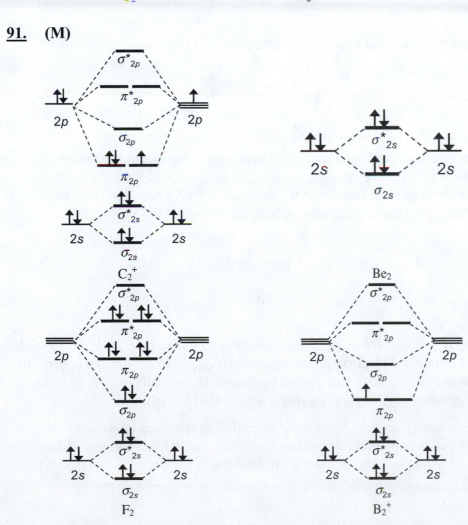

92. **(E)** The bond order in C_2 is two, whereas in Li_2 is one. Therefore, C_2 has the greater bond energy.

93. **(E)** A particular structure has resonance structures if its structure cannot be adequately represented by just one Lewis structure. The structures of the compounds are shown below, along with resonance structures when appropriate.

Several resonance structures contribute to the structure of the oxalate anion. Three of the structures are shown next.

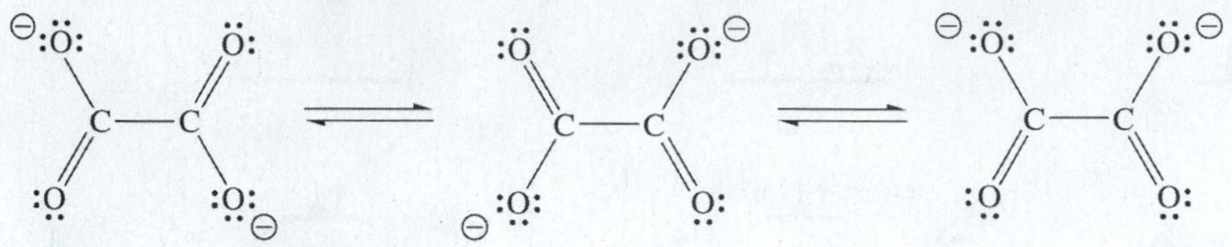

(a) Formaldehyde, or H_2CO, does not have any resonance structures.

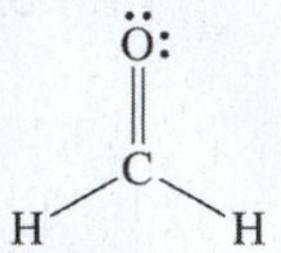

(b) The structure of the NO_3^- ion is a hybrid of three equivalent resonance structures.

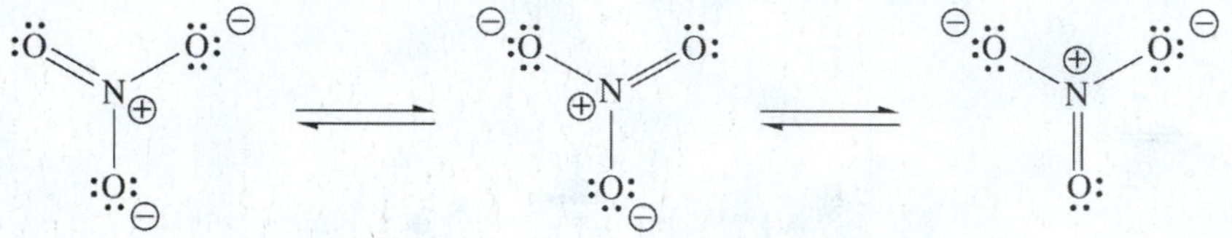

94. **(M)** We draw the Lewis structure of each species to account for the electron pairs around each central atom, which is N in both species. In NO_2^-, there are $5 + (2 \times 6) + 1 = 18$ valence electrons, or 9 pairs of electrons. In NO_2^+, there are $5 + (2 \times 6) - 1 = 16$ valence electron, or 8 pairs. Plausible Lewis structures for both are shown below:

In case of NO_2^-, N has two bonds (one single, one double) and a lone pair. Its hybridization is sp^2, it has a trigonal pyramidal electron group geometry (AX_2E), and its shape is bent. In case of NO_2^+, N has two double bonds and a positive charge. Its hybridization is sp, it has a linear electron group geometry (AX_2), and its shape is therefore straight.

95. **(E)** The answer is (a). In $BeCl_2$, sp hybridization involves the combination of an s and a p orbital from Be, which yields two hybrid sp orbitals for bonding with two Cl atoms. Boron is sp^2 hybridized in BCl_3. Carbon and nitrogen are sp^3 hybridized in CCl_4 and NCl_3, respectively.

96. **(E)** Bond hybridization is used to explain why all C—H bonds have equal length in methane. The C atom in CH_4 is sp^3 hybridized. The sp^3 hybrid orbitals have identical shapes and point toward the corners of a tetrahedron. Therefore, the C—H bonds are identical, and the H—C—H bond angles are all 109.5°. Since the C—H bonds are single bonds, there is no resonance or delocalization of electrons. The electronegativities of C and H are similar, and therefore the C—H bond is considered nonpolar.

97. **(M)** The peroxide ion has $(6 + 6 + 2) = 14$ valence electrons. The valence molecular orbital occupancy diagram is shown below:

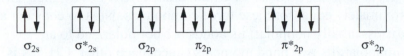

$$\sigma_{2s} \qquad \sigma^*_{2s} \qquad \sigma_{2p} \qquad \pi_{2p} \qquad \pi^*_{2p} \qquad \sigma^*_{2p}$$

All the electrons are paired. The bond order of O_2^{2-} is one $[(8 - 6)/2]$. The highest energy electron is in the π^*_{2p} MO. The correct answer is (d) because the peroxide ion has electrons in sigma and pi molecular orbitals and not in an atomic $2s$ orbital.

98. **(E)** The correct answer is (b). Three hybrid orbitals (sp^2) are formed by the combination of a $2s$ and two $2p$ orbitals; therefore, the angle between the hybrid orbitals is 120°.

99. **(M)**

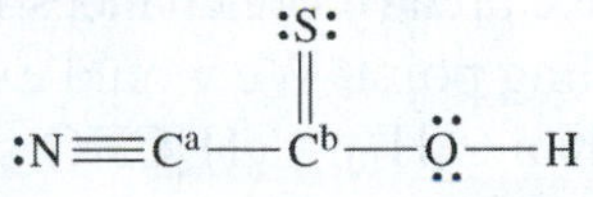

$$:N{\equiv}C^a{-}C^b{-}\overset{..}{\underset{..}{O}}{-}H$$
with $:\!S\!:$ double bonded above C^b

(a) There are five σ bonds: C^a—C^b, C^b—O, O—H are easily assigned. The double and triple bonds have one sigma bond each. There are three π bonds: one in the C=S bond and two in the C≡N bond.

(b) C^a is sp hybridized, C^b is sp^2 hybridized and O is sp^3 hybridized.

(c) The lone pairs on the nitrogen are in one of the sp hybrid orbitals on N. Both lone pairs on the sulfur are in two of the sp^2 hybrid orbitals on S. The lone pairs on the oxygen are in two of the sp^3 hybrid orbitals on O.

(d) N—C^a—C^b; C^a is sp hybridized, so the ideal bond angle is 180°
C^a—C^b=S; C^b is sp^2 hybridized, so the ideal bond angle is 120°.
C^a—C^b—O; C^b is sp^2 hybridized, so the ideal bond angle is 120°.
C^b—O—H; O is sp^3 hybridized, so the ideal bond angle is 109.5°.

CHAPTER 12
INTERMOLECULAR FORCES: LIQUIDS AND SOLIDS
PRACTICE EXAMPLES

1A **(E)** The substance with the highest boiling point will have the strongest intermolecular forces. The weakest of van der Waals forces are London forces, which depend on molar mass (and surface area): C_3H_8 is 44 g/mol, CO_2 is 44 g/mol, and CH_3CN is 41 g/mol. Thus, the London forces are approximately equal for these three compounds. Next to consider are dipole–dipole forces. C_3H_8 is essentially nonpolar; its bonds are not polarized to an appreciable extent. CO_2 is nonpolar; its two bond moments cancel each other. CH_3CN is polar and thus has the strongest intermolecular forces and should have the highest boiling point. The actual boiling points are $-78.44°$ C for CO_2, $-42.1°$ C for C_3H_8, and $81.6°$ C for CH_3CN.

1B **(M)** Dispersion forces, which depend on the number of electrons (molar mass) and structure, are one of the determinants of boiling point. The molar masses are: C_8H_{18} (114.2 g/mol), $CH_3CH_2CH_2CH_3$ (58.1 g/mol), $(CH_3)_3CH$ (58.1 g/mol), C_6H_5CHO (106.1 g/mol), and SO_3 (80.1 g/mol). We would expect $(CH_3)_3CH$ to have the lowest boiling point because it has the lowest molar mass and the most compact (ball-like) shape, whereas $CH_3CH_2CH_2CH_3$, which has the same mass but is longer and hence has more surface area (more chances for intermolecular interactions), should have the second highest boiling point. We would expect SO_3 to be next in line as it is also non-polar, but more massive than C_4H_{10}. C_6H_5CHO should have a boiling point higher than the more massive C_8H_{18} because benzaldehyde is polar while octane is not. Actual boiling points are given in parentheses in the following ranking. $(CH_3)_3CH$ $(-11.6\ °C) < CH_3CH_2CH_2CH_3$ $(-0.5\ °C) < SO_3$ (44.8 °C) C_8H_{18} (125.7 °C) $< C_6H_5CHO$ (178°C)

2A **(E)** Values of $\Delta_{vap}H$ are in kJ/mol so we first determine the amount in moles of diethyl ether.

$$\text{Heat} = 2.35\ \text{g}\ (C_2H_5)_2O \times \frac{1\ \text{mol}\ (C_2H_5)_2O}{74.12\ \text{g}\ (C_2H_5)_2O} \times \frac{29.1\ \text{kJ}}{1\ \text{mol}\ (C_2H_5)_2O} = 0.923\ \text{kJ}$$

2B **(M)**

$$\Delta_{overall}H = \Delta_{cond}H + \Delta_{cooling}H$$

$$\Delta_{cond}H = 0.0245\ \text{mol} \times (-40.7\ \text{kJ mol}^{-1}) = -0.997\ \text{kJ} = -997\ \text{J}$$

$$\Delta_{cooling}H = 0.0245\ \text{mol} \times (4.21\ \text{J g}^{-1}\ °C^{-1})(85.0\ °C - 100.0\ °C)(18.0153\ \text{g mol}^{-1}) = -27.9\ \text{J}$$

$$\Delta_{overall}H = -997\ \text{J} + -27.9\ \text{J} = -1025\ \text{J or } -1.025\ \text{kJ}$$

3A **(E)** $d = 0.701$ g/L at 25 °C for C_6H_{14} (molar mass = 86.177 g mol^{-1})
Consider a 1.00 L sample. This contains 0.701 g C_6H_{14}.

$$\text{moles } C_6H_{14} \text{ in 1.00 L sample} = 0.701 \text{ g } C_6H_{14} \times \frac{1 \text{ mol } C_6H_{14}}{86.177 \text{ g } C_6H_{14}} = 8.13 \times 10^{-3} \text{ mol } C_6H_{14}$$

Find pressure using the ideal gas law: $P = \dfrac{nRT}{V} = \dfrac{(8.31 \times 10^{-3} \text{ mol}) \left(\dfrac{0.08206 \text{ L atm}}{\text{K mol}} \right) (298 \text{ K})}{1.00 \text{ L}}$

$$P = 0.199 \text{ atm or } 151 \text{ Torr}$$

3B **(M)** From Figure 12-9, the vapor pressure is ≈ 420 mmHg or

$$420 \text{ mmHg} \times \frac{1 \text{ atm}}{760 \text{ Torr}} = 0.553 \text{ atm}$$

molar mass = 74.123 g mol^{-1}. $P = \dfrac{nRT}{V} = \dfrac{\left(\dfrac{\text{mass}}{\text{molar mass}} \right) RT}{V} = \dfrac{(\text{density}) RT}{\text{molar mass}}$

or $d = \dfrac{(\text{molar mass}) P}{RT} = \dfrac{\left(74.123 \dfrac{\text{g}}{\text{mol}} \right)(0.553 \text{ atm})}{\left(0.08206 \dfrac{\text{L atm}}{\text{K mol}} \right)(293 \text{ K})} = 1.70 \text{ g } L^{-1} \approx 1.7 \text{ g/L}$

4A **(E)** We first calculate pressure created by the water at 80.0 °C, assuming all 0.132 g H_2O vaporizes.

$$P_2 = \frac{nRT}{V} = \frac{\left(0.132 \text{ g } H_2O \times \dfrac{1 \text{ mol } H_2O}{18.02 \text{ g } H_2O} \right) \times 0.08206 \dfrac{\text{L atm}}{\text{mol K}} \times 353.2 \text{ K}}{0.525 \text{ L}} \times \frac{760 \text{ mmHg}}{1 \text{ atm}}$$

$$= 307 \text{ mmHg}$$

At 80.0 °C, the vapor pressure of water is 355.1 mmHg, thus, all the water exists as vapor.

4B **(E)** The result of Example 12-3 is that 0.132 g H_2O would exert a pressure of 281 mmHg if it all existed as a vapor. Since that 281 mmHg is greater than the vapor pressure of water at this temperature, some of the water must exist as liquid. The calculation of the example is based on the equation $P = nRT/V$, which means that the pressure of water is proportional to its mass. Thus, the mass of water needed to produce a pressure of 92.5 mmHg under this situation is

$$\text{mass of water vapor} = 92.5 \text{ mmHg} \times \frac{0.132 \text{ g } H_2O}{281 \text{ mmHg}} = 0.0435 \text{ g } H_2O$$

$$\text{mass of liquid water} = 0.132 \text{ g } H_2O \text{ total} - 0.0435 \text{ g } H_2O \text{ vapor} = 0.089 \text{ g liquid water}$$

5A **(M)** From Table 12.4 we know that $\Delta_{vap}H = 37.4$ kJ/mol for methanol. We now can use the Clausius-Clapeyron equation to determine the vapor pressure at $25.0\,°C = 298.2$ K.

$$\ln\frac{P}{100 \text{ mmHg}} = -\frac{37.4\times10^3 \text{ J mol}^{-1}}{8.3145 \text{ J mol}^{-1}\text{ K}^{-1}}\left(\frac{1}{298.2\text{K}} - \frac{1}{(273.2+21.2)\text{ K}}\right) = +0.195$$

$$\frac{P}{100 \text{ mmHg}} = e^{+0.195} = 1.22 \qquad P = 1.22\times100 \text{ mmHg} = 122 \text{ mmHg}$$

5B **(M)** The vapor pressure at the normal boiling point $(99.2\,°C = 372.4$ K$)$ is 760 mmHg precisely. We can use the Clausius-Clapeyron equation to determine the vapor pressure at $25\,°C = 298$ K.

$$\ln\frac{P}{760 \text{ mmHg}} = -\frac{35.76\times10^3 \text{ J mol}^{-1}}{8.3145 \text{ J mol}^{-1}\text{ K}^{-1}}\left(\frac{1}{298.2\text{ K}} - \frac{1}{372.4\text{ K}}\right) = -2.874$$

$$\frac{P}{760 \text{ mmHg}} = e^{-2.874} = 0.0565 \quad P = 0.0565\times760 \text{ mmHg} = 42.9 \text{ mmHg}$$

6A **(M)** We first look to molar masses: Ne (20.2 g/mol), He (4.0 g/mol), Cl_2 (70.9 g/mol), $(CH_3)_2 CO$ (58.1 g/mol), O_2 (32.0 g/mol), and O_3 (48.0 g/mol). Both $(CH_3)_2 CO$ and O_3 are polar, O_3 weakly so (because of its uneven distribution of electrons). We expect $(CH_3)_2 CO$ to have the highest boiling point, followed by Cl_2, O_3, O_2, Ne, and He. In the following ranking, actual boiling points are given in parentheses. He (–268.9 °C), Ne (–245.9 °C), O_2 (–183.0 °C), O_3 (–111.9 °C), Cl_2 (–34.6 °C), and $(CH_3)_2CO$ (56.2 °C).

6B **(M)** The magnitude of the enthalpy of vaporization is strongly related to the strength of intermolecular forces: the stronger these forces, the more endothermic the vaporization process. The first three substances all are nonpolar and, therefore, their only intermolecular forces are London forces, whose strength primarily depends on molar mass. The substances are arranged in order of increasing molar mass: $H_2 = 2.0$ g / mol, $CH_4 = 16.0$ g / mol, $C_6H_6 = 78.1$ g/mol, and also in order of increasing heat of vaporization. The last substance has a molar mass of 61.0 g/mol, which would produce intermolecular forces smaller than those of C_6H_6 if CH_3NO_2 were nonpolar. But the molecule is definitely polar. Thus, the strong dipole–dipole forces developed between CH_3NO_2 molecules make the enthalpy of vaporization for CH_3NO_2 larger than that for C_6H_6, which is, of course, essentially non-polar.

7A **(M)** Moving from point R to P we begin with $H_2O(g)$ at high temperature (>100 °C). When the temperature reaches the point on the vaporization curve, OC, water condenses at constant temperature (100 °C). Once all of the water is in the liquid state, the temperature drops. When the temperature reaches the point on the fusion curve, OD, ice begins to form at constant temperature (0 °C). Once all of the water has been converted to $H_2O(s)$, the temperature of the sample decreases slightly until point P is reached.

Since solids are not very compressible, very little change occurs until the pressure reaches the point on the fusion curve OD. Here, melting begins. A significant decrease in the volume occurs ($\approx 10\%$) as ice is converted to liquid water. After melting, additional pressure produces very little change in volume because liquids are not very compressible.

7B **(M)**

1.00 mol H_2O. At point R, $T = 374.1\ ^\circ C$ or 647.3 K

$$V_{\text{point R}} = \frac{nRT}{P} = \frac{(1.00\,\text{mol})\left(0.08206\,\dfrac{\text{L atm}}{\text{K mol}}\right)(647.3\,\text{K})}{1.00\,\text{atm}} = 53.1\ \text{L}$$

51. 3 L
At
Point
R

1.00 mol H_2O on P-R line, if 1/2 of water is vaporized, $T = 100\ ^\circ C$ (273.015 K)

$$V_{1/2\,\text{vap(PR)}} = \frac{nRT}{P} = \frac{(0.500\,\text{mol})\left(0.08206\,\dfrac{\text{L atm}}{\text{K mol}}\right)(373.15\,\text{K})}{1.00\,\text{atm}} = 15.3\ \text{L}$$

15.3 L at 100C 1/2 vap

A much smaller volume results when just 1/2 of the sample is vaporized (moles of gas smaller as well, temperature is smaller). 53.1 L vs 15.3 L (about 28.8 % of the volume as that seen at point R).

6A **(M)** We first look to molar masses: Ne (20.2 g/mol), He (4.0 g/mol), Cl_2 (70.9 g/mol), $(CH_3)_2 CO$ (58.1 g/mol), O_2 (32.0 g/mol), and O_3 (48.0 g/mol). Both $(CH_3)_2 CO$ and O_3 are polar, O_3 weakly so (because of its uneven distribution of electrons). We expect $(CH_3)_2 CO$ to have the highest boiling, followed by Cl_2, O_3, O_2, Ne, and He. In the following ranking, actual boiling points are given in parentheses. He ($-268.9\ ^\circ C$), Ne ($-245.9\ ^\circ C$), O_2 ($-183.0\ ^\circ C$), O_3 ($-111.9\ ^\circ C$), Cl_2 ($-34.6\ ^\circ C$), and $(CH_3)_2CO$ ($56.2 ^\circ C$)

6B **(M)** The magnitude of the enthalpy of vaporization is strongly related to the strength of intermolecular forces: the stronger these forces are, the more endothermic the vaporization process. The first three substances all are nonpolar and, therefore, their only intermolecular forces are London forces, whose strength primarily depends on molar mass. The substances are arranged in order of increasing molar mass: $H_2 = 2.0\ \text{g/mol}$, $CH_4 = 16.0\ \text{g/mol}$, $C_6H_6 = 78.1\ \text{g/mol}$, and also in order of increasing heat of vaporization. The last substance has a molar mass of 61.0 g/mol, which would produce intermolecular forces smaller than those of C_6H_6 if CH_3NO_2 were nonpolar. But the molecule is definitely polar. Thus, the strong dipole-dipole forces developed between CH_3NO_2 molecules make the enthalpy of vaporization for CH_3NO_2 larger than that for C_6H_6, which is, of course, essentially non-polar.

8A **(E)** Strong interionic forces lead to high melting points. Strong interionic forces are created by ions with high charge and of small size. Thus, for a compound to have a lower melting point than KI it must be composed of ions of larger size, such as RbI or CsI. A compound with a melting point higher than CaO would have either smaller ions, such as MgO, or more highly charged ions, such as Ga_2O_3 or Ca_3N_2, or both, such as AlN or Mg_3N_2.

8B **(E)** Mg^{2+} has a higher charge and a smaller size than does Na^+. In addition, Cl^- has a smaller size than I^-. Thus, interionic forces should be stronger in $MgCl_2$ than in NaI. We expect $MgCl_2$ to have lower solubility and, in fact, 12.3 mol (1840 g) of NaI dissolves in a liter of water, compared to just 5.7 mol (543 g) of $MgCl_2$, confirming our prediction.

9A **(E)** The length (l) of a bcc unit cell and the radius (r) of the atom involved are related by $4r = l\sqrt{3}$. For potassium, $r = 227$ pm. Then $l = 4 \times 227$ pm$/\sqrt{3} = 524$ pm

9B **(M)** Consider just the face of Figure 12-46. Note that it is composed of one atom at each of the four corners and one in the center. The four corner atoms touch the atom in the center, but not each other. Thus, the atoms are in contact across the diagonal of the face. If each atomic radius is designated r, then the length of the diagonal is $4r$ ($= r$ for one corner atom $+2r$ for the center atom $+r$ for the other corner atom). The diagonal also is related to the length of a side, l, by the Pythagorean theorem: $d^2 = l^2 + l^2 = 2l^2$ or $d = \sqrt{2}l$. We have two quantities equal to the diagonal, and thus to each other.

$$\sqrt{2}l = \text{diagonal} = 4r = 4 \times 143.1 \text{ pm} = 572.4 \text{ pm}$$

$$l = \frac{572.4}{\sqrt{2}} = 404.7 \text{ pm}$$

The cubic unit cell volume, V, is equal to the cube of one side.

$$V = l^3 = (404.7 \text{ pm})^3 = 6.628 \times 10^7 \text{ pm}^3$$

10A **(M)** In a bcc unit cell, there are eight corner atoms, of which $\frac{1}{8}$ of each is apportioned to the unit cell. There is also one atom in the center. The total number of atoms per unit cell is:

$= 1$ center $+ 8$ corners $\times \frac{1}{8} = 2$ atoms. The density, in g/cm^3, for this cubic cell:

$$\text{density} = \frac{2 \text{ atoms}}{(524 \text{ pm})^3} \times \left(\frac{10^{12} \text{ pm}}{10^2 \text{ cm}}\right)^3 \times \frac{1 \text{ mol}}{6.022 \times 10^{23} \text{ atoms}} \times \frac{39.10 \text{ g K}}{1 \text{ mol K}} = 0.903 \text{ g}/\text{cm}^3$$

The tabulated density of potassium at 20 °C is 0.86 g/cm^3.

10B **(M)** In a fcc unit cell the number of atoms is computed as 1/8 atom for each of the eight corner atoms (since each is shared among eight unit cells) plus 1/2 atom for each of the six face atoms (since each is shared between two unit cells). This gives the total number of atoms per unit cell as: atoms/unit cell = (1/8 corner atom × 8 corner atoms/unit cell) + (1/2 face atom × 6 face atoms/unit cell) = 4 atoms/unit cell

Now we can determine the mass per Al atom, and a value for the Avogadro constant.

$$\frac{\text{mass}}{\text{Al atom}} = \frac{2.6984 \text{ g Al}}{1 \text{ cm}^3} \times \left(\frac{100 \text{ cm}}{1 \text{ m}} \times \frac{1 \text{ m}}{10^{12} \text{ pm}}\right)^3 \times \frac{6.628 \times 10^7 \text{ pm}^3}{1 \text{ unit cell}} \times \frac{1 \text{ unit cell}}{4 \text{ Al atoms}}$$

$$= 4.471 \times 10^{-23} \text{ g/Al atom}$$

Therefore,

$$N_A = \frac{26.9815 \text{ g Al}}{1 \text{ mol Al}} \times \frac{1 \text{ Al atom}}{4.471 \times 10^{-23} \text{ g Al}} = 6.035 \times 10^{23} \frac{\text{atoms Al}}{\text{mol Al}}$$

11A **(E)** Across the diagonal of a CsCl unit cell are Cs^+ and Cl^- ions, so that the body diagonal equals $2r(Cs^+) + 2r(Cl^-)$. This body diagonal equals $\sqrt{3}l$, where l is the length of the unit cell.

$$l = \frac{2r(Cs^+) + 2r(Cl^-)}{\sqrt{3}} = \frac{2(167 + 181) \text{ pm}}{\sqrt{3}} = 402 \text{ pm}$$

11B **(M)** Since NaCl is fcc, the Na^+ ions are in the same locations as were the Al atoms in Practice Example 12-10B, and there are 4 Na^+ ions per unit cell. For stoichiometric reasons, there must also be 4 Cl^- ions per unit cell. These are accounted for as follows: there is one Cl^- along each edge, and each of these edge Cl^- ions are shared among four unit cells, and there is one Cl^- precisely in the body center of the unit cell, not shared with any other unit cells. Thus, the number of Cl^- ions is given by: Cl^- ions/unit cell = $\left(1/4 \ Cl^- \text{ on edge} \times 12 \text{ edges per unit cell}\right) + 1 \ Cl^- \text{ in body center} = 4 \ Cl^-/\text{unit cell}$.

The volume of this cubic unit cell is the cube of its length. The density is:

$$\text{NaCl density} = \frac{4 \text{ formula units}}{1 \text{ unit cell}} \times \frac{1 \text{ unit cell}}{(560 \text{ pm})^3} \times \left(\frac{10^{12} \text{ pm}}{1 \text{ m}} \times \frac{1 \text{ m}}{100 \text{ cm}}\right)^3 \times \frac{1 \text{ mol NaCl}}{6.022 \times 10^{23} \text{ f.u.}}$$

$$\times \frac{58.44 \text{ g NaCl}}{1 \text{ mol NaCl}} = 2.21 \text{ g/cm}^3$$

12A **(M)** Sublimation of Cs: $\quad Cs(s) \rightarrow Cs(g)$ $\qquad\qquad \Delta_{sub}H = +78.2 \text{ kJ/mol}$

Ionization of Cs(g): $\quad Cs(g) \rightarrow Cs^+(g) + e^-$ $\qquad E_1 = +375.7 \text{ kJ/mol (Table 9.3)}$

$\frac{1}{2}$ Dissociation of $Cl_2(g)$: $\quad \frac{1}{2} Cl_2(g) \rightarrow Cl(g)$ $\qquad DE = \frac{1}{2} \times 243 \text{ kJ} = 121.5 \text{ kJ/mol}$

$\qquad\qquad\qquad\qquad\qquad\qquad\qquad\qquad\qquad\qquad\qquad$ (Table 10.3)

Cl(g) electron affinity: $\quad Cl(g) + e^- \rightarrow Cl^-(g)$ $\qquad E_{ea\ 1} = -349.0 \text{ kJ/mol}$

$\qquad\qquad\qquad\qquad\qquad\qquad\qquad\qquad\qquad\qquad\qquad$ (Figure 9–12)

Lattice energy: $\quad Cs^+(g) + Cl^-(g) \rightarrow CsCl(s) \quad \text{L.E.}$

Enthalpy of formation: $\quad Cs(s) + \frac{1}{2} Cl_2(s) \rightarrow CsCl(s) \quad \Delta_f H° = -442.8 \text{ kJ/mol}$

$-442.8 \text{ kJ/mol} = +78.2 \text{ kJ/mol} + 375.7 \text{ kJ/mol} + 121.5 \text{ kJ/mol} - 349.0 \text{ kJ/mol} + \text{L.E.}$

$$= +226.4 \text{ kJ/mol} + \text{L.E.}$$

$\text{L.E.} = -442.8 \text{ kJ} - 226.4 \text{ kJ} = -669.2 \text{ kJ/mol}$

12B **(M)** Sublimation: $Ca(s) \rightarrow Ca(g)$ $\Delta_{sub}H = +178.2$ kJ/mol

First ionization energy: $Ca(g) \rightarrow Ca^+(g) + e^-$ $E_{i,1} = +590$ kJ/mol

Second ionization energy: $Ca^+(g) \rightarrow Ca^{2+}(g) + e^-$ $E_{i,2} = +1145$ kJ/mol

Dissociation energy: $Cl_2(g) \rightarrow 2Cl(g)$ $D.E. = (2 \times 122)$ kJ/mol

Electron Affinity: $2\,Cl(g) + 2\,e^- \rightarrow 2Cl^-(g)$ $2 \times E_{ea} = 2(-349)$ kJ/mol

Lattice energy: $Ca^{2+}(g) + 2\,Cl^-(g) \rightarrow CaCl_2(s)$ $L.E. = -2223$ kJ/mol

Enthalpy of formation: $Ca(s) + Cl_2(s) \rightarrow CaCl_2(s)$ $\Delta_f H^\circ = ?$

$\Delta_f H^\circ = \Delta_{sub}H + E_{i,1} + E_{i,2} + D.E. + (2 \times E_{ea}) + L.E.$

$= 178.2$ kJ/mol $+ 590$ kJ/mol $+ 1145$ kJ/mol $+ 244$ kJ/mol $- 698$ kJ/mol $- 2223$ kJ/mol

$= -764$ kJ/mol

INTEGRATIVE EXAMPLE

A **(M)** At 25.0 °C, the vapor pressure of water is 23.8 mmHg. We calculate the vapor pressure for isooctane with the Clausius-Clapeyron equation.

$$\ln \frac{P}{760 \text{ mmHg}} = \frac{35.76 \times 10^3 \text{J/mol}}{8.3145 \text{ J mol}^{-1}\text{K}^{-1}} \left(\frac{1}{(99.2 + 273.2)\text{ K}} - \frac{1}{298.2 \text{ K}} \right) = -2.87$$

$P = e^{-2.87} \times 760$ mmHg $= 43.1$ mmHg which is higher than H_2O's vapor pressure.

B **(D) (a)** and **(b)** We will work both parts simultaneously.

Sublimation of Mg(s): $Mg(s) \longrightarrow Mg(g)$ $\Delta_{sub}H = +146$ kJ

First ionization of Mg(g): $Mg(g) \longrightarrow Mg^+(g) + e^-$ $E_{i,1} = +737.7$ kJ

Second ionization of Mg(g): $Mg^+(g) \longrightarrow Mg^{2+}(g) + e^-$ $E_{i,2} = +1451$ kJ

$\frac{1}{2}$ Dissociation of O_2(g): $\frac{1}{2} O_2(g) \longrightarrow O(g)$ $\Delta_{dis}H = +249$ kJ

First electron affinity: $O(g) + e^- \longrightarrow O^-(g)$ $E_{ea,1} = -141.0$ kJ

Second electron affinity: $O^-(g) + e^- \rightarrow O^{2-}(g)$ $E_{ea,2}$

Lattice energy: $Mg^{2+}(g) + O^{2-}(g) \longrightarrow MgO(s)$ $L.E. = -3925$ kJ

Enthalpy of formation: $Mg(s) + \frac{1}{2} O_2(g) \longrightarrow MgO(s)$ $\Delta_f H^\circ = -601.7$ kJ

-601.7 kJ $= +146$ kJ $+ 737.7$ kJ $+ 1451$ kJ $+ 249$ kJ $- 141.0$ kJ $+ E_{ea,2} - 3925$ kJ

$E_{ea,2} = +881$ kJ

EXERCISES

Intermolecular Forces

1. **(M)**
    (a) HCl is not a very heavy diatomic molecule. Thus, the London forces between HCl molecules are expected to be relatively weak. Hydrogen bonding is weak in the case of H–Cl bonds; Cl is not one of the three atoms (F, O, N) that form strong hydrogen bonds. Finally, because Cl is an electronegative atom, and H is only moderately electronegative, dipole–dipole interactions should be relatively strong.

    (b) In Br_2 neither hydrogen bonds nor dipole–dipole attractions can occur (there are no H atoms in the molecule, and homonuclear molecules are nonpolar). London forces are more important in Br_2 than in HCl since Br_2 has more electrons (i.e., is more polarizable).

    (c) In ICl there are no hydrogen bonds since there are no H atoms in the molecule. The London forces are as strong as in Br_2 since the two molecules have the same number of electrons. However, dipole–dipole interactions are important in ICl, due to the polarity of the I–Cl bond.

    (d) In HF London forces are not very important; the molecule has only 10 electrons and thus is quite small. Hydrogen bonding is obviously the most important interaction developed between HF molecules.

    (e) In CH_4, H bonds are not important (the H atoms are not bonded to F, O, or N). In addition the molecule is not polar, so there are no dipole-dipole interactions. Finally, London forces are quite weak since the molecule contains only 10 electrons. For these reasons CH_4 has a very low critical temperature.

3. **(E)**

(c)	<	(b)	<	(d)	<	(a)
(ethane thiol)		(ethanol)		(butanol)		(acetic acid)

    Viscosity will depend on the intermolecular forces. The stronger the intermolecular bonding, the more viscous the substance.

5. **(E)** We expect CH_3OH to be a liquid from among the four substances listed. Of these four molecules, C_3H_8 has the most electrons and should have the strongest London forces. However, only CH_3OH satisfies the conditions for hydrogen bonding (H bonded to and attracted to N, O, or F) and thus its intermolecular attractions should be much stronger than those of the other substances.

7. **(M)** Three water molecules: the two lone pairs on the oxygen will interact with two hydrogens on two different water molecules, and one will interact with the hydrogen attached to O itself.

9. **(M)** There are three H-bonds:

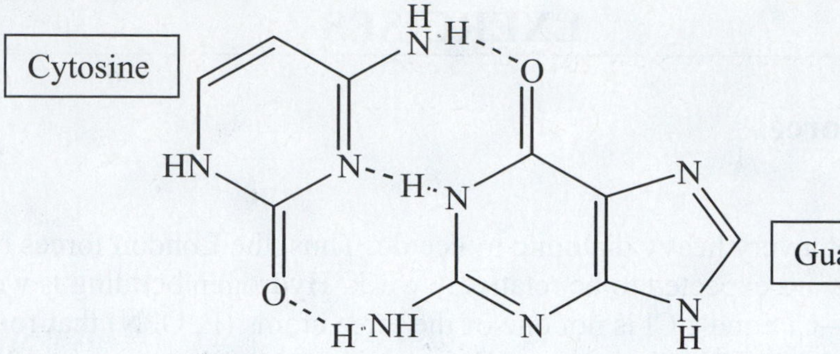

Surface Tension and Viscosity

11. **(E)** Since both the silicone oil and the cloth or leather are composed of relatively nonpolar molecules, they attract each other. The oil thus adheres well to the material. Water, on the other hand is polar and adheres very poorly to the silicone oil (actually, the water is repelled by the oil), much more poorly, in fact, than it adheres to the cloth or leather. This is because the oil is more nonpolar than is the cloth or the leather. Thus, water is repelled from the silicone-treated cloth or leather.

13. **(E)**
Molasses, like honey, is a very viscous liquid (high resistance to flow). The coldest temperatures are generally in January (in the northern hemisphere). Viscosity generally increases as the temperature decreases. Hence, molasses at low temperature is a very slow flowing liquid. Thus there is indeed a scientific basis for the expression "slower than molasses in January."

15. **(E)** $CH_3CH_2OCH_2CH_3 < CH_3OH < HOCH_2CH_2OH$. Although all three substances are polar, the CH_3OH and $HOCH_2CH_2OH$ molecules form hydrogen bonds. The surface tension of $HOCH_2CH_2OH$ is higher than that of CH_3OH because the $HOCH_2CH_2OH$ molecule is not only larger (more polarizable) but also able to form a greater number of hydrogen bonds because of the OH groups at either end.

17. **(E)** The intermolecular interactions in butanol are dominated by H-bonding, which is much stronger than the London dispersion forces dominant in pentane.

Vaporization

19. **(E)** The process of evaporation is endothermic, meaning it requires energy. If evaporation occurs from an uninsulated container, this energy is obtained from the surroundings, through the walls of the container. However, if the evaporation occurs from an insulated container, the only source of the needed energy is the liquid that is evaporating. Therefore, the temperature of the liquid will decrease as the liquid evaporates.

21. **(E)** We use the quantity of heat to determine the number of moles of benzene that vaporize.

$$V = \frac{nRT}{P} = \frac{\left(1.54\,kJ \times \dfrac{1\,mol}{33.9\,kJ}\right) \times 0.08206\dfrac{L\,atm}{mol\,K} \times 298\,K}{95.1\,mmHg \times \dfrac{1\,atm}{760\,mmHg}} = 8.88\,L\,C_6H_6(l)$$

23. **(M)**

25.00 mL of N_2H_4 (25 °C) density (25°C) = 1.0036 g mL^{-1} (molar mass = 32.0452 g mol^{-1})

mass of N_2H_4 = (volume) × (density) = (25.00 mL) × (1.0036 g mL^{-1}) = 25.09 g N_2H_4

$$n_{N2H4} = 25.09\,g\,N_2H_4 \times \frac{1\,mol\,N_2H_4}{32.0452\,g\,N_2H_4} = 0.7830\,mol$$

Energy required to increase temperature from 25.0 °C to 113.5 °C (Δt = 88.5 °C)

$$q_{heating} = (n)(C)(\Delta t) = (0.78295\,mol\,N_2H_4)\left(\frac{98.84\,J}{1\,mol\,N_2H_4\,°C}\right)(88.5\,°C) = 6848.7\,J\ or\ 6.85\,kJ$$

$$q_{vap} = (n_{N_2H_4})(\Delta_{vap}H) = (0.78295\,mol\,N_2H_4)\left(\frac{43.0\,kJ}{1\,mol\,N_2H_4}\right) = 33.7\,kJ$$

$$q_{overall} = q_{heating} + q_{vap} = 6.85\,kJ + 33.7\,kJ = 40.5\,kJ$$

25. **(M)**

$$heat\ needed = 3.78\,L\,H_2O \times \frac{1000\,cm^3}{1\,L} \times \frac{0.958\,g\,H_2O}{1\,cm^3} \times \frac{1\,mol\,H_2O}{18.02\,g\,H_2O} \times \frac{40.7\,kJ}{1\,mol\,H_2O} = 8.18 \times 10^3\,kJ$$

$$amount\ CH_4\ needed = 8.18 \times 10^3\,kJ \times \frac{1\,mol\,CH_4}{890\,kJ} = 9.19\,mol\,CH_4$$

$$V = \frac{nRT}{P} = \frac{9.19\,mol \times 0.08206\,L\,atm\,mol^{-1}K^{-1} \times 296.6\,K}{768\,mmHg \times \dfrac{1\,atm}{760\,mmHg}} = 221\,L\ methane$$

Vapor Pressure and Boiling Point

27. **(E)**

(a) We read up the 100 °C line until we arrive at C_6H_7N curve (e). This occurs at about 45 mmHg.

(b) We read across the 760 mmHg line until we arrive at the C_7H_8 curve (d). This occurs at about 110 °C.

29. **(E)** Use the ideal gas equation, $n = $ moles $Br_2 = 0.486$ g $Br_2 \times \dfrac{1 \text{ mol } Br_2}{159.8 \text{ g } Br_2} = 3.04 \times 10^{-3}$ mol Br_2.

$$P = \frac{nRT}{V} = \frac{3.04 \times 10^{-3} \text{ mol } Br_2 \times 0.08206 \text{ L atm mol}^{-1} \text{ K}^{-1} \times 298.2 \text{ K}}{0.2500 \text{ L}} \times \frac{760 \text{ mmHg}}{1 \text{ atm}}$$

$$= 226 \text{ mmHg}$$

31. **(E)**

(a) In order to vaporize water in the outer container, heat must be applied (i.e., vaporization is an endothermic process). When this vapor (steam) condenses on the outside walls of the inner container, that same heat is liberated. Thus condensation is an exothermic process.

(b) Liquid water, condensed on the outside wall, is in equilibrium with the water vapor that fills the space between the two containers. This equilibrium exists at the boiling point of water. We assume that the pressure is 1.000 atm, and thus, the temperature of the equilibrium must be 373.15 K or 100.00 °C. This is the maximum temperature that can be realized without pressurizing the apparatus.

33. **(M)** Use the Clausius-Clapeyron equation, and the vapor pressure of water at 100.0 °C (373.2 K) and 120.0 °C (393.2 K) to determine $\Delta_{vap}H$ of water near its boiling point. We then use the equation again, to determine the temperature at which water's vapor pressure is 2.00 atm.

$$\ln \frac{1489.1 \text{ mmHg}}{760.0 \text{ mmHg}} = \frac{\Delta_{vap}H}{8.3145 \text{ J mol}^{-1} \text{ K}^{-1}} \left(\frac{1}{373.2 \text{ K}} - \frac{1}{393.2 \text{ K}} \right) = 0.6726 = 1.639 \times 10^{-5} \Delta_{vap}H$$

$$\Delta_{vap}H = 4.104 \times 10^4 \text{ J/mol} = 41.04 \text{ kJ/mol}$$

$$\ln \frac{2.00 \text{ atm}}{1.00 \text{ atm}} = 0.6931 = \frac{41.04 \times 10^3 \text{ J mol}}{8.3145 \text{ J mol}^{-1} \text{ K}^{-1}} \left(\frac{1}{373.2 \text{ K}} - \frac{1}{T} \right)$$

$$\left(\frac{1}{373.2 \text{ K}} - \frac{1}{T_{bp}} \right) = 0.6931 \times \frac{8.1345 \text{ K}^{-1}}{41.03 \times 10^3} = 1.404 \times 10^{-4} \text{ K}^{-1}$$

$$\frac{1}{T_{bp}} = \frac{1}{373.2 \text{ K}} - 1.404 \times 10^{-4} \text{ K}^{-1} = 2.539 \times 10^{-3} \text{ K}^{-1} \quad T_{bp} = 393.9 \text{ K} = 120.7 \text{ °C}$$

35. **(M)** The 25.0 L of He becomes saturated with aniline vapor, at a pressure equal to the vapor pressure of aniline.

$$n_{aniline} = (6.220 \text{ g} - 6.108 \text{ g}) \times \frac{1 \text{ mole aniline}}{93.13 \text{ g aniline}} = 0.00120 \text{ mol aniline}$$

$$P = \frac{nRT}{V} = \frac{0.00120 \text{ mol} \times 0.08206 \text{ L atm mol}^{-1}\text{K}^{-1} \times 303.2 \text{ K}}{25.0 \text{ L}} = 0.00119 \text{ atm} = 0.907 \text{ mmHg}$$

37.
(M)

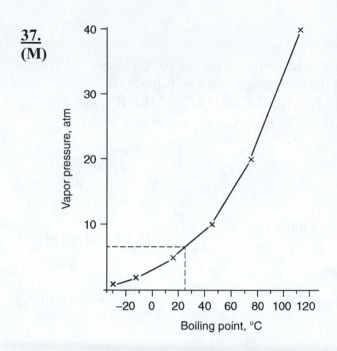

The graph of pressure vs. boiling point for Freon-12 is shown.

At a temperature of 25 °C the vapor pressure is approximately 6.5 atm for Freon-12. Thus the compressor must be capable of producing a pressure greater than 6.5 atm.

The Clausius-Clapeyron Equation

39. **(M)** We use the Clausius-Clapeyron equation (12.2) to answer this question.

$T_1 = (56.0 + 273.2) \text{ K} = 329.2 \text{ K}$ $\qquad$ $T_2 = (103.7 + 273.2) \text{ K} = 376.9 \text{ K}$

$$\ln \frac{10.0 \text{ mmHg}}{100.0 \text{ mmHg}} = \frac{\Delta_{vap}H}{8.3145 \text{ J mol}^{-1} \text{ K}^{-1}} \left(\frac{1}{376.9 \text{ K}} - \frac{1}{329.2 \text{ K}} \right) = -2.30 = -4.624 \times 10^{-5} \Delta_{vap}H$$

$\Delta_{vap}H = 4.97 \times 10^4 \text{ J/mol} = 49.7 \text{ kJ/mol}$

41. **(M)** Once again, we will employ the Clausius-Clapeyron equation.

$T = 56.2 \text{ °C is } T = 329.4 \text{ K}$

$$\ln \frac{760 \text{ mmHg}}{375 \text{ mmHg}} = \frac{25.5 \times 10^3 \text{ J/mol}}{8.3145 \text{ J mol}^{-1} \text{ K}^{-1}} \left(\frac{1}{T} - \frac{1}{329.4 \text{ K}} \right) = 0.706$$

$$\left(\frac{1}{T} - \frac{1}{329.4 \text{ K}} \right) = \frac{0.706 \times 8.3145}{25.5 \times 10^3} \text{ K}^{-1} = 2.30 \times 10^{-4} \text{ K}^{-1} = 1/T - 3.03_6 \times 10^{-3} \text{ K}^{-1}$$

$1/T = (3.03_6 + 0.230) \times 10^{-3} \text{ K}^{-1} = 3.266 \times 10^{-3} \text{ K}^{-1}$ $\qquad$ $T = 306 \text{ K} = 33 \text{ °C}$

43. **(M)** Normal boiling point = 179 °C and critical point = 422 °C and 45.9 atm

$$\ln \left(\frac{P_2}{P_1} \right) = \frac{\Delta_{vap}H}{R} \left(\frac{1}{T_1} - \frac{1}{T_2} \right) \qquad \ln \left(\frac{45.9}{1} \right) = \frac{\Delta_{vap}H}{8.3145 \text{ J K}^{-1}\text{mol}^{-1}} \left(\frac{1}{452.2 \text{ K}} - \frac{1}{695.2 \text{ K}} \right)$$

$\Delta_{vap}H = 41.2 \text{ kJ mol}^{-1}$

$$\ln \left(\frac{1}{P} \right) = \frac{41,200 \text{ J mol}^{-1}}{8.3145 \text{ J K}^{-1}\text{mol}^{-1}} \left(\frac{1}{373.2} - \frac{1}{452.2 \text{ K}} \right) \qquad P = 0.0981 \text{ atm or } 74.6 \text{ Torr}$$

Critical Point

45. **(E)** Substances that can exist as a liquid at room temperature (about 20 °C) have critical temperature above 20 °C, 293 K. Of the substances listed in Table 12.6, $CO_2(T_c = 304.2$ K), $HCl(T_c = 324.6$ K), $NH_3(T_c = 405.7$ K), $SO_2(T_c = 431.0$ K), and $H_2O(T_c = 647.3$ K) can exist in liquid form at 20 °C. In fact, CO_2 exists as a liquid in CO_2 fire extinguishers.

Melting and Freezing

47. **(M)**

(a) heat evolved $= 3.78$ kg Cu $\times \dfrac{1000 \text{ g}}{1 \text{ kg}} \times \dfrac{1 \text{ mol Cu}}{63.55 \text{ g Cu}} \times \dfrac{13.05 \text{ kJ}}{1 \text{ mol Cu}} = 776$ kJ evolved or

$\Delta H = -776$ kJ

(b) heat absorbed $= (75 \text{ cm} \times 15 \text{ cm} \times 12 \text{ cm}) \times \dfrac{8.92 \text{ g}}{1 \text{ cm}^3} \times \dfrac{1 \text{ mol Cu}}{63.55 \text{ g Cu}} \times \dfrac{13.05 \text{ kJ}}{1 \text{ mol Cu}} = 2.5 \times 10^4$ kJ

States of Matter and Phase Diagrams

49. **(M)** Let us use the ideal gas law to determine the final pressure in the container, assuming that all of the dry ice vaporizes. We then locate this pressure, at a temperature of 25 °C, on the phase diagram of Figure 12-28.

$$P = \frac{nRT}{V} = \frac{\left(80.0 \text{ g CO}_2 \times \dfrac{1 \text{ mol CO}_2}{44.0 \text{ g CO}_2}\right) \times 0.08206 \dfrac{\text{L atm}}{\text{mol K}} \times 298 \text{ K}}{0.500 \text{ L}} = 88.9 \text{ atm}$$

Although this point (25 °C and 88.9 atm) is most likely in the region labeled "liquid" in Figure 12–28, we computed its pressure assuming the CO_2 is a gas. Some of this gas should condense to a liquid. Thus, both liquid and gas are present in the container. Solid would not be present unless the temperature is below –50 °C at ~ 88.9 atm.

52. **(D)**

(a) As heat is added initially, the temperature of the ice rises from –20 °C to 0 °C. At (or just slightly below) 0 °C, ice begins to melt to liquid water. The temperature remains at 0 °C until all of the ice has melted. Adding heat then warms the liquid until a temperature of about 93.5 °C is reached, at which point the liquid begins to vaporize to steam. The temperature remains fixed until all the water is converted to steam. Adding heat then warms the steam to 200 °C. (Data for this part are taken from Figure 12-30 and Table 12.5.)

(b) As the pressure is raised, initially gaseous iodine is just compressed. At ~ 91 mmHg, liquid iodine appears. As the pressure is pushed above 99 mmHg, more liquid condenses until eventually all the vapor is converted to a liquid. Increasing the pressure further simply compresses the liquid until a high pressure is reached, perhaps 50 atm, where solid iodine appears. Again the pressure remains fixed with further compression until all of the iodine is converted to its solid form. After this has occurred further compression raises the pressure on the solid until 100 atm is reached. (Data for this part are from Figure 12-27 and the surrounding text.)

(c)   Cooling of gaseous CO_2 simply lowers the temperature until a temperature of perhaps 20 °C is reached. At this point, liquid CO_2 appears. The temperature remains constant as more heat is removed until all the gas is converted to liquid. Further cooling then lowers the temperature of the liquid until a temperature of slightly higher than –56.7 °C is reached, where solid CO_2 appears. At this point, further cooling simply converts liquid to solid at constant temperature, until all liquid has been converted to solid. From this point, further cooling lowers the temperature of the solid. (Data for this part are taken from Figure 12-28 and Table 12.6.)

53.   (D) 0.240 g of H_2O corresponds to 0.0133 mol H_2O. If the water does not vaporize completely, the pressure of the vapor in the flask equals the vapor pressure of water at the indicated temperature. However, if the water vaporizes completely, the pressure of the vapor is determined by the ideal gas law.

(a)   30.0°C, vapor pressure of H_2O = 31.8 mmHg = 0.0418 atm

$$n = \frac{PV}{RT} = \frac{0.0418 \text{ atm} \times 3.20 \text{ L}}{0.08206 \text{ L atm mol}^{-1} \text{ K}^{-1} \times 303.2 \text{ K}}$$

n = 0.00538 mol H_2O vapor, which is less than 0.0133 mol H_2O;

This represents a non-eqilibrium condition, since not all the H_2O vaporizes.

The pressure in the flask is 0.0418 atm. (from tables)

(b)   50.0 °C, vapor pressure of H_2O = 92.5 mmHg = 0.122 atm

$$n = \frac{PV}{RT} = \frac{0.122 \text{ atm} \times 3.20 \text{ L}}{0.08206 \text{ L atm mol}^{-1} \text{ K}^{-1} \times 323.2 \text{ K}}$$

$= 0.0147$ mol H_2O vapor > 0.0133 mol H_2O; all the H_2O vaporizes. Thus,

$$P = \frac{nRT}{V} = \frac{0.0133 \text{ mol} \times 0.08206 \text{ L atm mol}^{-1} \text{ K}^{-1} \times 323.2 \text{ K}}{3.20 \text{ L}} = 0.110 \text{ atm} = 83.8 \text{ mmHg}$$

(c)   70.0 °C All the H_2O must vaporize, as this temperature is higher than that of part (b). Thus,

$$P = \frac{nRT}{V} = \frac{0.0133 \text{ mol} \times 0.08206 \text{ L atm mol}^{-1} \text{ K}^{-1} \times 343.2 \text{ K}}{3.20 \text{ L}} = 0.117 \text{ atm} = 89.0 \text{ mmHg}$$

55.   (M)
(a)   According to Figure 12-28, $CO_2(s)$ exists at temperatures below –78.5 °C when the pressure is 1 atm or less. We do not expect to find temperatures this low and partial pressures of $CO_2(g)$ of 1 atm on the surface of Earth.

(b)   According to Table 12.6, the critical temperature of CH_4, the maximum temperature at which $CH_4(l)$ can exist, is 191.1 K = –82.1 °C. We do not expect to find temperatures this low on the surface of Earth.

(c)   Since, according to Table 12.6, the critical temperature of SO_2 is 431.0 K = 157.8 °C, $SO_2(g)$ can be found on the surface of Earth.

(d) According to Figure 12–27, $I_2(l)$ can exist at pressures less than 1.00 atm between the temperatures of 114 °C and 184 °C. There are very few places on Earth that reach temperatures this far above the boiling point of water at pressures below 1 atm. One example of such a place would be the mouth of a volcano high above sea level. Essentially, $I_2(l)$ is not found on the surface of Earth.

(e) According to Table 12.6, the critical temperature – the maximum temperature at which $O_2(l)$ exists – is 154.8 K = −118.4 °C. Temperatures this low do not exist on the surface of Earth.

<u>57.</u> (D)

(a) heat lost by water = $q_{water} = (m)(C)(\Delta t)$

$$q_{water} = (100.0 \text{ g})\left(4.18\frac{J}{g\,°C}\right)(0.00 \text{ °C} - 20.00 \text{ °C})\left(\frac{1\,kJ}{1000\,J}\right) = -8.36 \text{ kJ}$$

Using $\Delta_{cond}H = -\Delta_{vap}H$ and heat lost by system = heat loss of condensation + cooling

$$q_{steam} = (175 \text{ g } H_2O)\left(4.18\frac{J}{g\,°C}\right)(0.0 \text{ °C} - 100.0 \text{ °C})\left(\frac{1\,kJ}{1000\,J}\right)$$

$$+ (175 \text{ g } H_2O)\left(\frac{1\,mol\,H_2O}{18.015\,g\,H_2O}\right)\left(\frac{-40.7\,kJ}{1\,mol\,H_2O}\right)$$

$q_{steam} = -395.4 \text{ kJ} + (-73.2 \text{ kJ}) = -468.6 \text{ kJ or} \sim -469 \text{ kJ}$

total energy to melt the ice = $q_{water} + q_{steam} = -8.37 \text{ kJ} + (-469 \text{ kJ}) = -477 \text{ kJ}$

$$\text{moles of ice melted} = (477 \text{ kJ})\left(\frac{1\,mol\,ice}{6.01\,kJ}\right) = 79.4 \text{ mol ice melted}$$

$$\text{mass of ice melted} = (79.4 \text{ mol } H_2O)\left(\frac{18.015\,g\,H_2O}{1\,mol\,H_2O}\right)\left(\frac{1\,kg\,H_2O}{1000\,g\,H_2O}\right) = 1.43 \text{ kg}$$

mass of unmelted ice = 1.65 kg − 1.43 kg = 0.22 kg

(b) mass of unmelted ice = 0.22 kg (from above)

heat required to melt ice = $n\,\Delta_{fus}H$

$$\text{heat required} = (0.22 \text{ kg ice})\left(\frac{1000\,g\,H_2O}{1\,kg\,H_2O}\right)\left(\frac{1\,mol\,H_2O}{18.015\,g\,H_2O}\right)\left(\frac{6.01\,kJ}{1\,mol\,H_2O}\right) = 73.\underline{4} \text{ kJ}$$

Next we need to determine heat produced when 1 mole of steam (18.015 g) condenses and cools from 100 °C to 0.0 °C.
Heat evolved can be calculated as shown below:

$$= (1 \text{ mol } H_2O)\left(\frac{-40.7\,kJ}{1\,mol\,H_2O}\right) + (18.015 \text{ g})\left(4.18\frac{J}{g\,°C}\right)(0.0 \text{ °C} - 100 \text{ °C})\left(\frac{1\,kJ}{1000\,J}\right)$$

$= -40.7 \text{ kJ} + -7.53 \text{ kJ} = -48.2 \text{ kJ per mole of } H_2O(g) \text{ or per 18.015 g } H_2O(g)$

$$\text{mass of steam required} = (73.\underline{4} \text{ kJ})\left(\frac{1\,mol\,H_2O(g)}{48.2\,kJ}\right)\left(\frac{18.015\,g\,H_2O}{1\,mol\,H_2O}\right) = 27 \text{ g steam}$$

59. **(E)** The liquid in the can is supercooled. When the can is opened, gas bubbles released from the carbonated beverage serve as sites for the formation of ice crystals. The condition of supercooling is destroyed and the liquid reverts to the solid phase. An alternative explanation follows. The process of the gas coming out of solution is endothermic (heat is required). (We know this to be true because the reaction solution of gas in water → gas + liquid water proceeds to the right as the temperature is raised, a characteristic direction of an endothermic reaction.) The required heat is taken from the cooled liquid, causing it to freeze.

Network Covalent Solids

61. **(E)** One would expect diamond to have a greater density than graphite. Although the bond distance in graphite, "one-and-a-half" bonds, would be expected to be shorter than the single bonds in diamond, the large spacing between the layers of C atoms in graphite makes its crystals much less dense than those of diamond.

63. **(a)** We expect Si and C atoms to alternate in the
(M) structure, as shown at the right. The C atoms are on the corners ($8 \times 1/8 = 1$ C atom) and on the faces ($6 \times 1/2 = 3$ C atoms), a total of four C atoms/unit cell. The Si atoms are each totally within the cell, a total of four Si atoms/unit cell.

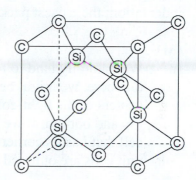

(b) To have a graphite structure, we expect sp^2 hybridization for each atom. The hybridization schemes for B and N atoms are shown to the right. The half-filled sp^2 hybrid orbitals of the boron and nitrogen atoms overlap to form the σ bonding structure, and a hexagonal array of atoms. The $2p_z$ orbitals then overlap to form the π bonding orbitals. Thus, there will be as many π electrons in a sample of BN as there are in a sample of graphite, assuming both samples have the same number of atoms.

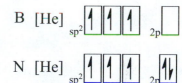

Ionic Bonding and Properties

65. **(E)** We expect forces in ionic compounds to increase as the sizes of ions become smaller and as ionic charges become greater. As the forces between ions become stronger, a higher temperature is required to melt the crystal. In the series of compounds NaF, NaCl, NaBr, and NaI, the anions are progressively larger, and thus the ionic forces become weaker. We expect the melting points to decrease in this series from NaF to NaI. This is precisely what is observed.

67. (E) NaF will give the highest Mohs value, because hardness is a function of stronger bonds, which are also shorter, and are affected by charge and size of the anions and cations. NaF has the smallest anion, and is tied with NaCl for the smallest cation.

Crystal Structures

69. (M) In each layer of a closest packing arrangement of spheres, there are six spheres surrounding and touching any given sphere. A second similar layer then is placed on top of this first layer so that its spheres fit into the indentations in the layer below. The two different closest packing arrangements arise from two different ways of placing the third layer on top of these two, with its spheres fitting into the indentations of the layer below. In one case, one can look down into these indentations and see a sphere of the bottom (first) layer. If these indentations are used, the closest packing arrangement *abab* results (hexagonal closest packing). In the other case, no first layer sphere is visible through the indentation; the closest packing arrangement *abcabc* results (cubic closest packing).

71. (M)

(a) We naturally tend to look at crystal structures in right-left, up-down terms. If we do that here, we might be tempted to assign a unit cell as a square, with its corners at the centers of the light-colored squares. But the crystal does not "know" right and left or top and bottom. If we look at this crystal from the lower right corner, we see a unit cell that has its corners at the centers of dark-colored diamonds. These two types of unit cells are outlined at the top of the diagram below.

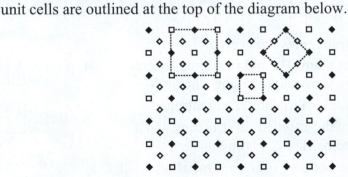

(b) The unit cell has one light-colored square fully inside it. It has four light-colored "circles" (which the computer doesn't draw as very round) on the edges, each shared with one other unit cell. So there are a total of $4 \times 1/2 = 2$ circles per unit cell. Also, the unit cell has four dark-colored diamonds, one at each corner, and each diamond is shared with four other unit cells, for a total of $4 \times 1/4 = 1$ diamond per unit cell.

(c) One example of an erroneous unit cell is the small square outlined near the center of the figure drawn in part (a). Notice that simply repeatedly translating this unit cell toward the right, so that its left edge sits where its right edge is now, will not generate the lattice.

73. **(M)** In Figure 12–45 we see that the body diagonal of a cube has a length of $\sqrt{3}l$, where l is the length of one edge of the cube. The length of this body diagonal also equals $4r$, where r is the radius of the atom in the structure. Hence $4r = \sqrt{3}l$ or $l = 4r \div \sqrt{3}$. Recall that the volume of a cube is l^3, and $\sqrt{3} = 1.732$.

$$\text{density} = \frac{\text{mass}}{\text{volume}} = \frac{\dfrac{2\ \text{W atoms}}{1\ \text{unit cell}} \times \dfrac{1\ \text{mol W}}{6.022 \times 10^{23}\ \text{W atoms}} \times \dfrac{183.85\ \text{g W}}{1\ \text{mol W}}}{\left(\dfrac{4 \times 139\ \text{pm}}{1.732} \times \dfrac{1\ \text{m}}{10^{12}\ \text{pm}} \times \dfrac{100\ \text{cm}}{1\ \text{m}} \right)^3} = 18.5\ \text{g/cm}^3$$

This compares well with a tabulated density of 19.25 g/cm^3.

75. **(M)**

(a) 335 pm = 2 radii or 1 diameter. Hence Po diameter = 335 pm

(b) 1 Po unit cell = $(335\ \text{pm})^3 = 3.76 \times 10^7\ \text{pm}^3$ ($3.76 \times 10^{-23}\ \text{cm}^3$) per unit cell.

$$\text{density} = \frac{m}{V} = \frac{3.47 \times 10^{-22}\ \text{g}}{3.76 \times 10^{-23}\ \text{cm}^3} = 9.23\ \text{g cm}^{-3}$$

(c) $n = 1$, $d = 335$ pm and $\lambda = 1.785 \times 10^{-10}$ or 178.5 pm Solve for sin θ, then determine θ.

$$\sin \theta = \frac{n\lambda}{2d} = \frac{(1)(1.785 \times 10^{-10})}{2(335 \times 10^{-12})} = 0.2664 \text{ or } \theta = 15.45°$$

77. **(E)** There are 8 SiF$_4$ molecules with the Si atoms at each corner of the cube, and one molecule in the center. Therefore, there are 8 × 1/8 + 1 = 2 Si atoms per unit cell.

Ionic Crystal Structures

79. **(M)** CaF$_2$: There are eight Ca^{2+} ions on the corners, each shared among eight unit cells, for a total of one $(8 \times 1/8)$ corner ion per unit cell. There are six Ca^{2+} ions on the faces, each shared between two unit cells, for a total of three $(6 \times 1/2)$ face ions per unit cell. This gives a total of four Ca^{2+} ions per unit cell. There are eight F$^-$ ions, each wholly contained within the unit cell. The ratio of Ca^{2+} ions to F$^-$ ions is 4 Ca^{2+} ions per 8 F$^-$ ions: Ca$_4$F$_8$ or CaF$_2$. TiO$_2$. There are eight Ti^{4+} ions on the corners, each shared among eight unit cells, for a total of one Ti^{4+} corner ion $(8 \times 1/8)$ per unit cell. There is one Ti^{4+} ion in the center, wholly contained within the unit cell. Thus, there are a total of two Ti^{4+} ions per unit cell. There are four O^{2-} ions on the faces of the unit cell, each shared between two unit cells, for a total of two $(4 \times 1/2)$ face atoms per unit cells. There are two O^{2-} ions totally contained within the unit cell. This gives a total of four O^{2-} ions per unit cell. The ratio of Ti^{4+} ions to O^{2-} ions is 2 Ti^{4+} ions per 4 O^{2-} ion: Ti$_2$O$_4$ or TiO$_2$.

81. (D)

(a) In a sodium chloride type of lattice, there are six cations around each anion and six anions around each cation. These oppositely charged ions are arranged as follows: one above, one below, one in front, one in back, one to the right, and one to the left. Thus the coordination number of Mg^{2+} is 6 and that of O^{2-} is 6 also.

(b) In the unit cell, there is an oxide ion at each of the eight corners; each of these is shared between eight unit cells. There also is an oxide ion at the center of each of the six faces; each of these oxide ions is shared between two unit cells. Thus, the total number of oxide ions is computed as follows.

$$\text{total \# of oxide ions} = 8 \text{ corners} \times \frac{1 \text{ oxide ion}}{8 \text{ unit cells}} + 6 \text{ faces} \times \frac{1 \text{ oxide ion}}{2 \text{ unit cells}} = 4 \text{ } O^{2-} \text{ ions}$$

There is a magnesium ion on each of the twelve edges; each of these is shared between four unit cells. There also is a magnesium ion in the center which is not shared with another unit cell.

$$\text{total \# of } Mg^{2+} \text{ ions} = 12 \text{ adjoining cells} \times \frac{1 \text{ magnesium ion}}{4 \text{ unit cells}} + 1 \text{ central } Mg^{2+} \text{ ion}$$

$$= 4 \text{ } Mg^{2+} \text{ ions (Thus, there are four formula units per unit cell of MgO.)}$$

(c) Along the edge of the unit cell, Mg^{2+} and O^{2-} ions are in contact. The length of the edge is equal to the radius of one O^{2-}, plus the diameter of Mg^{2+}, plus the radius of another O^{2-}.

$$\text{edge length} = 2 \times O^{2-} \text{ radius} + 2 \times Mg^{2+} \text{ radius} = 2 \times 140 \text{ pm} + 2 \times 72 \text{ pm} = 424 \text{ pm}$$

The unit cell is a cube; its volume is the cube of its length.

$$\text{volume} = (424 \text{ pm})^3 = 7.62 \times 10^7 \text{ pm} \left(\frac{1 \text{ m}}{10^{12} \text{ pm}} \times \frac{100 \text{ cm}}{1 \text{ m}} \right)^3 = 7.62 \times 10^{-23} \text{ cm}^3$$

(d) $\text{density} = \dfrac{\text{mass}}{\text{volume}} = \dfrac{4 \text{ MgO f.u.}}{7.62 \times 10^{-23} \text{ cm}^3} \times \dfrac{1 \text{ mol MgO}}{6.022 \times 10^{23} \text{ f.u.}} \times \dfrac{40.30 \text{ g MgO}}{1 \text{ mol MgO}} = 3.51 \text{ g/cm}^3$

Thus, there are four formula units (f.u.) per unit cell of MgO.

83. (M)

(a) $CaO \rightarrow$ radius ratio $= \dfrac{r_{Ca^{2+}}}{r_{O^{2-}}} = \dfrac{100 \text{ pm}}{140 \text{ pm}} = 0.714$

Cations occupy octahedral holes of a face centered cubic array of anions.

(b) $CuCl \rightarrow$ radius ratio $= \dfrac{r_{Cu^+}}{r_{Cl^-}} = \dfrac{96 \text{ pm}}{181 \text{ pm}} = 0.530$

Cations occupy octahedral holes of a face centered cubic array of anions.

(c) $LiO_2 \rightarrow$ radius ratio $= \dfrac{r_{Li^+}}{r_{O_2^-}} = \dfrac{59 \text{ pm}}{128 \text{ pm}} = 0.461$

Cations occupy octahedral holes of a face centered cubic array of anions.

Lattice Energy

85. **(E)** Lattice energies of a series such as LiCl(s), NaCl(s), KCl(s), RbCl(s), and CsCl(s) will vary approximately with the size of the cation. A smaller cation will produce a more exothermic lattice energy. Thus, the lattice energy for LiCl(s) should be the most exothermic and CsCl(s) the least in this series, with NaCl(s) falling in the middle of the series.

87. **(D)**

Second ionization energy:	$Mg^+(g) \rightarrow Mg^{2+}(g) + e^-$	$E_{i,2} = 1451 \, kJ/mol$
Lattice energy:	$Mg^{2+}(g) + 2 \, Cl^-(g) \rightarrow MgCl_2(s)$	$L.E. = -2526 \, kJ/mol$
Sublimation:	$Mg(s) \rightarrow Mg(g)$	$\Delta_{sub}H = 146 \, kJ/mol$
First ionization energy	$Mg(g) \rightarrow Mg^+(g) + e^-$	$E_{i,1} = 738 \, kJ/mol$
Dissociation energy:	$Cl_2(g) \rightarrow 2 \, Cl(g)$	$D.E. = (2 \times 122) kJ/mol$
Electron affinity:	$2 \, Cl(g) + 2 \, e^- \rightarrow 2 \, Cl^-(g)$	$2 \times E_{ea} = 2(-349) \, kJ/mol$

$$\Delta_f H^\circ = \Delta_{sub}H + E_{i,1} + E_{i,2} + D.E. + (2 \times E_{ea}) + L.E.$$

$$= 146 \, \frac{kJ}{mol} + 738 \, \frac{kJ}{mol} + 1451 \, \frac{kJ}{mol} - 1 + 244 \, \frac{kJ}{mol} - 698 \, \frac{kJ}{mol} - 2526 \, \frac{kJ}{mol} = -645 \, \frac{kJ}{mol}$$

In Example 12-12, the value of $\Delta_f H^\circ$ for MgCl is calculated as -19 kJ/mol. Therefore, $MgCl_2$ is much more stable than MgCl, since considerably more energy is released when it forms. $MgCl_2(s)$ is more stable than MgCl(s)

INTEGRATIVE AND ADVANCED EXERCISES

91. **(M)** In many instances, with CO_2 being one, the substance in the tank is not present as a gas only, but as a liquid in equilibrium with its vapor. As gas is released from the tank, the liquid will vaporize to replace it, maintaining a pressure in the tank equal to the vapor pressure of the substance at the temperature at which the cylinder is stored. This will continue until all of the liquid vaporizes, after which only gas will be present. The remaining gas will be quickly consumed, and hence the reason for the warning. However, the situation of gas in equilibrium with liquid only applies to substances that have a critical temperature above room temperature (20 °C or 293 K). Thus, substances in Table 12.6 for which gas pressure does serve as a measure of the quantity of gas in the tank are H_2 ($T_c = 33.3$ K), N_2 ($T_c = 126.2$ K), O_2 ($T_c = 154.8$ K), and CH_4 ($T_c = 191.1$ K).

93. **(M)** For this question we need to determine the quantity of heat required to vaporize 1.000 g H_2O at each temperature. At 20 °C, 2447 J of heat is needed to vaporize each 1.000 g H_2O. At 100 °C., the quantity of heat needed is

$$\frac{10.00 \, kJ}{4.430 \, g \, H_2O} \times \frac{1000 \, J}{1 \, kJ} = 2257 \, J/g \, H_2O$$

Thus, less heat is needed to vaporize 1.000 g of H_2O at the higher temperature of 100 °C. This makes sense, for at the higher temperature the molecules of the liquid already are in rapid motion. Some of this energy of motion or vibration will contribute to the energy needed to break the cohesive forces and vaporize the molecules.

96. **(E)** If only gas were present, the final pressure would be 10 atm. This is far in excess of the vapor pressure of water at 30.0 °C (~ 0.042 atm). Most of the water vapor condenses to liquid water. (It cannot all be liquid, because the liquid volume is only about 20 mL and the system volume is 2.61 L.) The final condition is a point on the vapor pressure curve at 30.0 °C.

99. **(M)**

(a) $\text{pressure} = \dfrac{\text{force}}{\text{area}} = \dfrac{80.\ \text{kg} \times 9.8067\ \text{m s}^{-2}}{2.5\ \text{cm}^2 \times \dfrac{1\ \text{m}^2}{10^4\ \text{cm}^2}} \times \dfrac{1\ \text{N}}{1\ \text{kg m s}^{-2}} \times \dfrac{1\ \text{atm}}{101325\ \text{N m}^{-2}} = 31\ \text{atm}$

(b) $\text{decrease in melting point} = 31\ \text{atm} \times \dfrac{1.0\ °\text{C}}{125\ \text{atm}} = 0.25\ °\text{C}$

The ice under the skates will melt at −0.25 °C.

100. **(M)** $P = P_0 \times 10^{-Mgh/2.303\,RT}$ Assume ambient temperature is $10.0\,°\text{C} = 283.2\,\text{K}$

$\dfrac{Mgh}{2.303\ RT} = \dfrac{0.02896\ \text{kg/mol air} \times 9.8067\ \text{m/s}^2 \times 3170\ \text{m}}{2.303 \times 8.3145\ \text{Jmol}^{-1}\text{K}^{-1} \times 283.2\ \text{K}} = 0.166$

$P = P_0 \times 10^{-0.166} = 1\ \text{atm} \times 0.682 = 0.682\ \text{atm} = \text{atmospheric pressure in Leadville, CO.}$

$\ln \dfrac{0.682\ \text{atm}}{1.000\ \text{atm}} = \dfrac{-41 \times 10^3\ \text{J/mol}}{8.3145\ \text{J mol}^{-1}\ \text{K}^{-1}} \left(\dfrac{1}{T} - \dfrac{1}{373.2\ \text{K}} \right) = -0.383$

$\left(\dfrac{1}{T} - \dfrac{1}{373.2\ \text{K}} \right) = -0.383 \times \dfrac{8.3145}{-41 \times 10^3} = +7.8 \times 10^{-5}\ \text{K}^{-1}$

$\dfrac{1}{T} = +7.77 \times 10^{-5}\ \text{K}^{-1} + 2.68 \times 10^{-3}\ \text{K}^{-1} = 2.76 \times 10^{-3}\ \text{K}^{-1}$ $T = 360\ \text{K} = 87\ °\text{C}$

103. **(M)** The molar mass of acetic acid monomer is 60.05 g/mol. We first determine the volume occupied by 1 mole of molecules of the vapor, assuming that the vapor consists only of monomer molecules: CH_3COOH.

$\text{volume of vapor} = 60.05\ \text{g} \times \dfrac{1\ \text{L}}{3.23\ \text{g}} = 18.6\ \text{L}$

Next, we can determine the actual number of moles of vapor in 18.6 L at 350 K.

$\text{moles of vapor} = \dfrac{PV}{RT} = \dfrac{1\ \text{atm} \times 18.6\ \text{L}}{0.08206\ \text{L atm mol}^{-1}\ \text{K}^{-1} \times 350\ \text{K}} = 0.648\ \text{mol vapor}$

Then, we can determine the number of moles of dimer and monomer, starting with 1.00 mole of monomer, and producing a final mixture of 0.648 moles total (monomer and dimer together).

Reaction: $2\ CH_3COOH(g) \longrightarrow (CH_3COOH)_2(g)$

Initial: 1.00 mol   0

Changes: $-2x$ mol $+x$ mol

Final: $(1.00-2x)$ mol x mol

The line labeled "Changes" indicates that 2 moles of monomer are needed to form each mole of dimer. The line labeled "Final" results from adding the "Initial" and "Changes" lines.

total number of moles $=1.00-2x+x=1.00-x=0.648$ mol

$x=0.352$ mol dimer $(1.00-2x)=0.296$ mol monomer

% dimer $=\dfrac{0.352\text{ mol dimer}}{0.648\text{ mol total}}\times100\% = 54.3\%$ dimer

We would expect the % dimer to decrease with temperature. Higher temperatures will provide the energy (as translational energy (heat)) needed to break the relatively weak hydrogen bonds that hold the dimers together.

104. **(M)** First we compute the mass and the amount of mercury.

mass Hg $= 685$ mL$\times\dfrac{13.6\text{ g}}{1\text{ mL}}=9.32\times10^3$ g $n_{Hg}=9.32\times10^3\text{ g}\times\dfrac{1\text{ mol Hg}}{200.59\text{ g}}=46.5$ mol Hg

Then we calculate the heat given up by the mercury in lowering its temperature, as the sum of the following three terms.

cool liquid $=9.32\times10^3\text{ g}\times0.138\text{ J g}^{-1}\,^{\circ}\text{C}^{-1}\times(-39\,^{\circ}\text{C}-20\,^{\circ}\text{C})=-7.6\times10^4$ J $=-76$ kJ

freeze liquid $=46.5$ mol Hg$\times(-2.30\text{ kJ/mol})=-107$ kJ

cool solid $=9.32\times10^3\text{ g}\times0.126\text{ J g}^{-1}\,^{\circ}\text{C}^{-1}\times(-196+39)\,^{\circ}\text{C}=-1.84\times10^5$ J $=-184$ kJ

total heat lost by Hg $=-76$ kJ-107 kJ-184 kJ$=-367$kJ $=-$heat gained by N_2

mass of N_2(l) vaporized$=367$ kJ$\times\dfrac{1\text{mol }N_2}{5.58\text{ kJ}}\times\dfrac{28.0\text{g }N_2}{1\text{ mol }N_2}=1.84\times10^3$ g N_2(l)$=1.84$ kg N_2

107. **(D)** We need to calculate five different times. They are shown below.

1. Heat the solid to its melting point $T_1 =$

$$1\text{ mol Bi}\times\dfrac{0.028\text{ kJ}}{\text{K mol}}\times(554.5\text{ K}-300\text{ K})\times\dfrac{1\text{ min}}{1.00\text{ kJ}}=7.1\text{ min}$$

2. Melt the solid $T_2 =1$ mol Bi$\times\dfrac{10.9\text{ kJ}}{\text{K mol}}\times\dfrac{1\text{ min}}{1.00\text{ kJ}}=10.9$ min

3. Heat the liquid to its boiling point $T_3 =$

$$1\text{ mol Bi}\times\dfrac{0.031\text{ kJ}}{\text{K mol}}\times(1832\text{ K}-554.5\text{ K})\times\dfrac{1\text{ min}}{1.00\text{ kJ}}=39.6\text{ min}$$

4. Vaporize the liquid $\quad T_4 = 1 \text{ mol Bi} \times \dfrac{151.5 \text{ kJ}}{\text{K mol}} \times \dfrac{1 \text{ min}}{1.00 \text{ kJ}} = 151.5 \text{ min}$

5. Heat the gas to 2000 K $\quad T_5 = 1 \text{ mol Bi} \times \dfrac{0.021 \text{ kJ}}{\text{K mol}} \times (2000 \text{ K } - 1832 \text{ K}) \times \dfrac{1 \text{ min}}{1.00 \text{ kJ}} = 3.5 \text{ min}$

A plot is shown below:

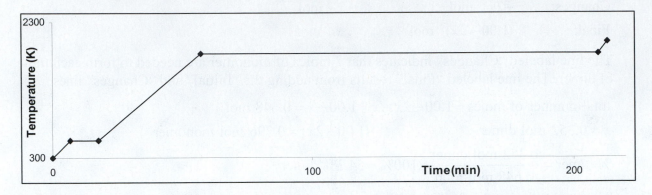

109. **(E)** The edge length of the NaCl unit cell is 560 pm (from Example 12-11), and thus the distance between the top and the middle layers in the NaCl unit cell is 560 pm $\div$ 2 = 280 pm. This is equal to the value of d in the Bragg equation (12.5). We first solve for $\sin\theta$ and then for θ.

$$\sin\theta = \frac{n\lambda}{2d} = \frac{1 \times 154.1 \text{ pm}}{2 \times 280 \text{ pm}} = 0.275 \qquad \theta = \sin^{-1}(0.275) = 16.0°$$

121. Sample calculations are shown below for the H_2O—H_2O interaction.

$$E(\text{dipole-dipole}) = -\frac{1}{r^6} \cdot \frac{2\mu^4}{3k_sT} \cdot \frac{1}{(4\pi\varepsilon_0)^2}$$

$$= -\frac{1}{(4 \times 10^{-10} \text{m})^6} \times \frac{2(1.8546 \text{ D} \times 3.33564 \times 10^{-30} \text{C m D}^{-1})^4}{3(1.3807 \times 10^{-23} \text{ JK}^{-1})(298 \text{ K})}$$

$$\times \frac{1}{(4\pi(8.854 \times 10^{-12} \text{J}^{-1}\text{C}^2\text{m}^{-1}))^2} - 4.86 \times 10^{-21} \text{ J}$$

(a) To express this result in kJ mol^{-1}, we multiply by the Avogadro constant and divide by 1000:

$$E(\text{dipole-dipole}) = -4.68 \times 10^{-21} \text{J} \times 6.022 \times 10^{23} \text{ mol}^{-1} \times \frac{\text{kJ}}{1000 \text{ J}} = -2.82 \text{ kJ mol}^{-1}$$

$$E(\text{dipole-induced dipole}) = -\frac{1}{r^6} \cdot 2\mu^2 a' \cdot \frac{1}{4\pi\varepsilon_0}$$

$$= -\frac{1}{(4.00 \times 10^{-10} \text{ m})^6} \times 2(1.8546 \text{ D} \times 3.33564$$

$$\times 10^{-30} \text{ C m D}^{-1})^2 (14.5 \times 10^{-25} \text{ cm}^3 \times 10^{-6} \text{m}^3\text{cm}^3)$$

$$\times \frac{1}{4\pi(8.854 \times 10^{-12} \text{ J}^{-1}\text{C}^2\text{m}^{-1})}$$

$$= -2.435 \times 10^{-22} \text{ J}$$

$$E(\text{dipole-induced dipole}) = -2.435 \times 10^{-22} \, \text{J} \times 6.022 \times 10^{23} \, \text{mol}^{-1} \times \frac{\text{kJ}}{1000 \, \text{J}} = -0.15 \, \text{kJ mol}^{-1}$$

$$E(\text{dispersion}) = -\frac{1}{r^6} \cdot \frac{3}{4} \alpha^2 E_i$$

$$= \frac{1}{(4.00 \times 10^{-10} \, \text{m})^6} \times \frac{3}{4} (14.5 \times 10^{-25} \, \text{cm}^3 \times 10^{-6} \, \text{m}^3 \text{cm}^{-3})^2 \times (1218 \, \text{kJ mol}^{-1})$$

$$= -4.689 \times 10^{-2} \, \text{kJ mol}^{-1}$$

The sum of these contributions is
$$E = E(\text{dipole-dipole}) + E(\text{dipole-induced dipole}) + E(\text{dispersion})$$

$$= -2.82 \, \text{kJ mol}^{-1} + (-0.15 \, \text{kJ mol}^{-1}) + (-0.47 \, \text{kJ mol}^{-1})$$

$$= -3.44 \, \text{kJ mol}^{-1}$$

(b) The percent contributions are

$$\% \text{ dipole} = \frac{E(\text{dipole-dipole})}{E} \times 100\%$$

$$= \frac{-2.82 \, \text{kJ mol}^{-1}}{-3.44 \, \text{kJ mol}^{-1}} \times 100\%$$

$$= 82.0\%$$

$$\% \text{ induced} = \frac{E(\text{dipole-induced dipole})}{E} \times 100\%$$

$$= \frac{2(-0.15 \, \text{kJ mol}^{-1})}{-3.44 \, \text{kJ mol}^{-1}} \times 100\%$$

$$= 4.3\%$$

$$\% \text{ dispersion} = \frac{E(\text{dispersion})}{E} \times 100\%$$

$$= \frac{-0.47 \, \text{kJ mol}^{-1})}{-3.44 \, \text{kJ mol}^{-1}} \times 100\%$$

$$= 13.6\%$$

Results for the other substances are obtained in the same manner. The following table provides a summary:

	E(dipole–dipole) (kJ mol^{-1})	E(dipole–induced dipole) (kJ mol^{-1})	E(dispersion) (kJ mol^{-1})	E (kJ mol^{-1})	Dipole–Dipole	Percent Dipole–Induced Dipole	Dispersion
HF	−2.65	−0.08	−0.18	−2.91	91.1	2.7	6.2
HCl	−0.36	−0.10	−1.56	−2.01	17.9	4.7	77.4
HBr	−0.11	−0.07	−2.68	−2.87	3.9	2.5	93.6
HI	−0.01	−0.03	−5.43	−5.47	0.2	0.6	99.2
H$_2$O	−2.82	−0.15	−0.47	−3.44	82.1	4.3	13.6
CH$_3$OH	−1.99	−0.28	−2.08	−4.35	45.8	6.4	47.8
CH$_3$CH$_2$OH	−1.94	−0.45	−5.39	−7.79	25.0	5.8	69.2
CH$_3$(CH$_2$)$_2$OH	−1.38	−0.48	−8.17	−10.02	3.7	4.8	81.5
CH$_3$(CH$_2$)$_3$OH	−1.81	−0.72	−13.92	−16.45	11.0	4.4	84.6
CH$_4$	0.00	0.00	−1.50	−1.50	0.0	0.0	100.0
CH$_3$CH$_3$	0.00	0.00	−4.08	−4.08	0.0	0.0	100.0
CH$_3$CH$_2$CH$_3$	0.00	0.00	−7.66	−7.66	0.0	0.0	100.0
(CH$_3$)$_3$CH	0.00	0.00	−12.38	−12.38	0.0	0.0	100.0
CH$_3$(CH$_2$)$_2$CH$_3$	0.00	0.00	−12.51	−12.51	0.0	0.0	100.0
CH$_3$(CH$_2$)$_3$CH$_3$	0.00	0.00	−18.13	−18.13	0.0	0.0	100.0

(c) The values range from tenths of a kJ mol^{-1} to approximately 12 kJ mol^{-1}. Intermolecular interaction energies are typically ten to a thousand times smaller than the energy of a covalent bond.

(d) The plot for the halogen acids   The plot for the alcohols

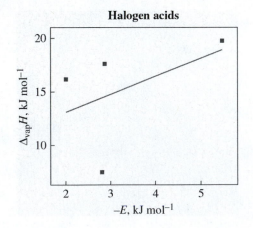

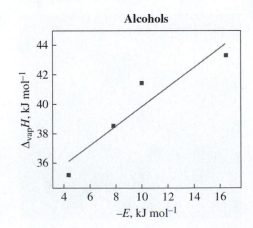

The plot for the alkanes

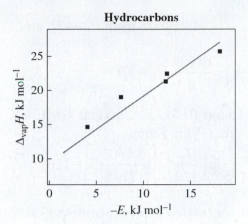

Hydrocarbons

These plots show that as the magnitude of the intermolecular interaction energies increases, the enthalpy of vaporization increases.

(e) For low temperatures, when thermal energies are small, the majority of molecules will have their dipoles aligned head-to-tail. However, at very high temperatures, the molecules will be able to "jiggle" out of alignment, with their dipoles more randomly oriented. Therefore, the attraction arising from a favorable alignment of dipole moments is less significant at higher temperatures.

FEATURE PROBLEMS

122. (E) We obtain the surface tension by substituting the experimental values into the equation for surface tension.

$$h = \frac{2\gamma}{dgr} \qquad \gamma = \frac{hdgr}{2} = \frac{1.1 \text{ cm} \times 0.789 \text{ g cm}^{-3} \times 981 \text{ cm s}^{-2} \times 0.050 \text{ cm}}{2} = 21 \text{ g/s}^2 = 0.021 \text{ J/m}^2$$

123. (D)

(a) $$\frac{dP}{dT} = \frac{\Delta_{vap}H}{T(V_g - V_l)} = \frac{\Delta_{vap}H}{T(V_g)} \quad \text{Note: } V_l \approx 0 \text{ Rearrange expression, Use } V_g = \frac{nRT}{P}$$

$$\frac{dP}{dT} = \frac{\Delta_{vap}H}{T(\frac{nRT}{P})} = \frac{\Delta_{vap}H}{\frac{nRT^2}{P}} = \frac{P\Delta_{vap}H}{nRT^2} \quad or \quad \frac{dP}{P} = \frac{\Delta_{vap}H \times dT}{nRT^2}$$

Consider 1 mole (n = 1) and substitute in $\Delta_{vap}H = 15{,}971 + 14.55 \, T - 0.160 \, T^2$

$$\frac{dP}{P} = \frac{(15{,}971 + 14.55 \, T - 0.160 \, T^2)dT}{RT^2} = \frac{(15{,}971)dT}{RT^2} + \frac{(14.55 \, T)dT}{RT^2} - \frac{(0.160 \, T^2)dT}{RT^2}$$

Simplify and collect constants

$$\frac{dP}{P} = \frac{(15{,}971)}{R}\frac{dT}{T^2} + \frac{(14.55)}{R}\frac{dT}{T} - \frac{(0.160)}{R}dT$$ Integrate from $P_1 \rightarrow P_2$ and $T_1 \rightarrow T_2$

$$\ln\left(\frac{P_2}{P_1}\right) = \frac{(15{,}971)}{R}\left(\frac{1}{T_1} - \frac{1}{T_2}\right) + \frac{(14.55)}{R}\ln\left(\frac{T_2}{T_1}\right) - \frac{(0.160)}{R}(T_2 - T_1)$$

(b) First we consider 1 mole ($n = 1$) $P_1 = 10.16$ torr (0.01337 atm) and $T_1 = 120$ K
Find the boiling point (T_2) when the pressure (P_2) is 1 atm.

$$\ln\left(\frac{1}{0.01337}\right) = \frac{15971}{8.3145}\left(\frac{1}{120} - \frac{1}{T_2}\right) + \frac{14.55}{8.3145}\ln\left(\frac{T_2}{120}\right) - \frac{0.160}{8.3145}(T_2 - 120)$$

Then we solve for T_2 using the method of successive approximations: $T_2 = 169$ K

SELF-ASSESSMENT EXERCISES

131. **(E)** The correct answer is (e) because all these properties depend on the strength of the intermolecular forces. Surface tension is the tendency of matter to resist an external force. The strength of the intermolecular attractions directly affects the ability of the surface of the liquid to resist this external force. Boiling point and vapor pressure are likewise affected. The stronger the intermolecular attractions, the lower the vapor pressure at a given temperature, and the higher the boiling temperature. Similarly, heat of vaporization of the substance also increases with increasing intermolecular attractions; more energy is required to take 1 mole of the substance from the liquid phase to the gas phase.

132. **(E)** The correct answer is (b). As the vapor escapes, there is less vapor over the liquid available for condensation, so the immediate effect is a lower condensation rate. More vapor will be produced because the vaporization rate will be greater than the condensation rate, until finally the two rates are again equal.

133. **(E)** The answer is (c). As temperature increases, more molecules from a liquid get sufficient energy to escape, and thus vapor pressure increases.

134. **(E)** The answer is (c). HF, CH_3OH, and N_2H_4 all participate in hydrogen bonding.

135. **(E)** The correct answer is (d). The normal boiling point is defined as the temperature where the vapor pressure of the liquid is the same as normal (sea level) atmospheric pressure of 760 Torr (1 atm).

136. **(M)** The van't Hoff equation can be used to calculate the vapor pressure at a given temperature:

$$\ln\frac{p_2}{p_1} = -\frac{\Delta_{vap}H°}{R}\left[\frac{1}{T_2} - \frac{1}{T_1}\right]$$

Plugging in the values, we have

$$\ln\frac{50.0 \text{ mmHg}}{760 \text{ mmHg}} = -\frac{\Delta_{vap}H°}{8.3145 \text{ J}\cdot\text{K}^{-1}\cdot\text{mol}^{-1}}\left[\frac{1}{T_2} - \frac{1}{329.4 \text{ K}}\right]$$

T_2 can be solved as follows:

$$T_2 = -\left[\ln\frac{50.0 \text{ mmHg}}{760 \text{ mmHg}}\times\left(\frac{8.3145 \text{ J}\cdot\text{K}^{-1}\cdot\text{mol}^{-1}}{32.0\times10^3 \text{ J mol}^{-1}}\right) - \frac{1}{329.4 \text{ K}}\right]^{-1} = 267.17 \text{ K or } -6.0 \text{ °C}$$

137. **(E)** The answer is (b). Refer to Table 12.9.

138. **(E)** The anwer is (a).

139. **(M)** The answers are (d) and (f).

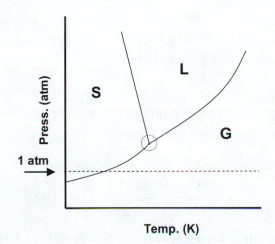

140. **(E)** The species with higher boiling points are underlined.
    (a) C_7H_{16} v. $\underline{C_{10}H_{22}}$: The only interaction is London dispersion. $C_{10}H_{22}$ because it has the higher mass
    (b) C_3H_8 v. $\underline{(CH_3)_2O}$: because dipole–dipole interactions are predominant, versus just London dispersion for C_3H_8.
    (c) CH_3CH_2SH v. $\underline{CH_3CH_2OH}$: because H-bonding dominates the inter-molecular interactions. This interaction is much weaker for CH_3CH_2SH.

141. **(E)** O_3 is the one that is out of place. The correct order of boiling points based on molar masses is: $N_2 < F_2 < Ar < O_3 < Cl_2$. O_3 is the only polar molecule in the group, but this is not important enough to put it after the more massive Cl_2 (bp: 162 K for O_3 and 239 K for Cl_2).

142. **(E)** The following listing reflects that organic compounds are lower melting than inorganic compounds, hydrogen bonding is an important intermolecular force for two of the compounds, and melting points of inorganic compounds are affected by ionic sizes and charges:
$Ne < C_3H_8 < CH_3CH_2OH < CH_2OHCHOHCH_2OH < KI < K_2SO_4 < MgO$

143. **(M)** Refer to the photograph on page 538 of water boiling under a reduced pressure. If the vapor is evacuated fast enough, to supply the required $\Delta_{vap}H$, the water may cool to 0 °C and ice may begin to form.

144. **(M)** If there is too little benzene(l) in the sealed tube in Figure 12-22 initially, the liquid will all be converted to benzene(g) before T_c is reached. If too much is present initially, the liquid will expand and cause the benzene(l) to condense, and therefore only benzene(l) will be present at the time T_c is reached.

145. **(D)**

(a) To determine whether any CCl_4 remains in the flask, we have to determine how many moles of CCl_4 are placed in the vessel, determine that how much CCl_4 is in the vapor phase in a vessel of 8.21 L if the vapor pressure is 110 Torr at 25 °C, and then figure out if there will be more or less CCl_4 in the vapor phase compared to the amount of liquid given.

$$mol\ CCl_4 = 3.50\ g\ CCl_4 \times \frac{1\ mol\ CCl_4}{153.80\ g\ CCl_4} = 0.02276\ mol\ \text{of liquid } CCl_4 \text{ placed in vessel}$$

Assuming a pressure of 110 Torr at 25 °C:

vol CCl_4 = 3.50 g CCl_4 / 1.59 g mol^{-1} = 2.20 mL
vol inside the vessel = 8.210 L – 0.00220 L = 8.208 L
pressure = 110 Torr/760 Torr = 0.145 atm.

Moles of CCl_4 in the gas phase in this closed container:

$$mol\ CCl_4 = \frac{PV}{RT} = \frac{(0.145\ atm)(8.208\ L)}{(0.08206\ L\ atm\ K^{-1})(298\ K)} = 0.0486\ mol\ \text{in the vapor phase}$$

0.0486 mol > 0.0228 mol, therefore at equilibrium, all the CCl_4 will be in the vapor phase.

(b) To determine the amount of energy required to vaporize a certain amount of CCl_4, we have to first determine the enthalpy of vaporization, or $\Delta_{vap}H$:

$$\ln\left(\frac{P_2}{P_1}\right) = \frac{-\Delta_{vap}H}{R}\left(\frac{1}{T_2} - \frac{1}{T_1}\right)$$

$$\ln\left(\frac{760}{110}\right) = \frac{-\Delta_{vap}H}{8.3145\ J\ mol^{-1}\ K}\left(\frac{1}{350\ K} - \frac{1}{298\ K}\right)$$

$\Delta_{vap}H = 32.2$ kJ/mol

The energy required to vaporize 2.00 L of CCl_4 is therefore determined as follows:

$$energy = 2000\ mL \times \frac{1.59\ g}{1\ mL} \times \frac{1\ mol\ CCl_4}{153.80\ g\ CCl_4} \times \frac{32.2\ kJ}{mol} = 666\ kJ$$

146. **(M)**

(a) Unit cell length: we note from the picture that the hypotenuse of the right triangle equals $4 \times r$.

$$L^2 + L^2 = (4r)^2 = 16 \cdot (128 \text{ pm})^2 = 2.621 \times 10^5$$

$$L = \sqrt{2.621 \times 10^5 / 2} = 362 \text{ pm}$$

(b) volume $= (362 \text{ pm})^3 = 4.74 \times 10^7 \text{ pm}^3$

(c) 8 corners $\times$ 1/8 + 6 faces $\times$ ½ = 4 atoms/unit cell.

(d) Volume % is the ratio between the volume taken up by the atoms and the volume of the unit cell.

$$\frac{\text{vol of atoms}}{\text{vol of cells}} = \frac{4 \times (4/3) \pi (128 \text{ pm})^3}{4.74 \times 10^7 \text{ pm}^3} \times 100 = 74\%$$

(e)

$$\frac{\text{mass of Cu}}{\text{unit cell}} = \frac{4 \text{ atoms}}{\text{unit cell}} \times \frac{1 \text{ mol Cu}}{6.022 \times 10^{23} \text{ atoms}} \times \frac{63.546 \text{ g Cu}}{1 \text{ mol Cu}} = 4.221 \times 10^{-22} \text{ g}$$

(f) $D = m/V$

$$D = \frac{4.221 \times 10^{-22} \text{ g Cu}}{4.74 \times 10^7 \text{ pm}^3} \times \frac{(1 \times 10^{-10} \text{ pm})^3}{(1 \text{ cm})^3} = 8.91 \text{ g/cm}^3$$

147. **(E)** The answer is (a). All of these liquids participate in hydrogen-bonding. Therefore, higher van der Waals interactions translate to higher surface tensions. Methanol, CH_3OH, is the smallest molecule, therefore has the least amount of van der Waals forces, and the lowest surface tension.

148. **(E)** The answer is (d). All of these compounds are straight-chain hydrocarbons. Their only major intermolecular interaction is London dispersion. The lower this interaction, the lower the viscosity. *N*-pentane is the lightest, and therefore has the lowest viscosity.

149. **(E)** A network covalent solid will have a higher melting point, because it takes much more energy to overcome the covalent bonds in the solid (such as, for example, diamond) than to overcome ionic interactions.

150. **(E)** MgO would have the highest melting point. Lattice energy increases as the magnitude of the charges on the ions increases and as the sizes (radii) of the ions decreases. Large lattice energies result from highly charged ions with the smallest radii.

151. **(M)** The Li$^+$ and I$^-$ have an fcc structure. Because I$^-$ is much larger, the iodide ions touch. The structure is shown below.

Since the Li–I distance is 3.02 Å, the length of the cube is 2 × 3.02 Å = 6.04 Å. Therefore, the face diagonal of the cube is:

$$D = \sqrt{6.04^2 + 6.04^2} = 8.54 \ \text{Å}$$

Since there are 4 I$^-$ radii in the face diagonal of the cube, radius of I$^-$ is 8.54/4 = 2.13 Å

152. **(E)** The answer is (c), because of increased pressure, the transformation that causes the greatest packing efficiency change is likely to prevail.

CHAPTER 13

SPONTANEOUS CHANGE:
ENTROPY AND GIBBS ENERGY

PRACTICE EXAMPLES

1A **(M)** The energy levels for the 905 pm box are given by

$$E_n = \frac{(6.626 \times 10^{-34} \text{ J})^2}{(8)(3.351 \times 10^{-26} \text{ kg})(905 \times 10^{-12} \text{ m})^2} \times n^2 = (2.00 \times 10^{-24} \text{ J}) \times n^2$$

The energies of the first four levels are

n	$E_n / 10^{-24}$ J
1	2.00
2	8.00
3	18.0
4	32.0

As shown in the diagrams below, there are 6 microstates for the system in the 905 pm box, each having a total energy of 20.0×10^{-24} J.

1B **(E)** A volume change causes a change in the spacing between the (translational) energy levels. (In an expansion, the energy levels become more closely spaced. In a compression, the energy levels are more widely spaced.) Part (b) illustrates a volume change, more specifically a compression. There is a larger change in energy from one level to the next in the final state than in the initial state. Part (a) represents an increase in total energy, with no change in volume.

2A **(E)** In general, $\Delta_r S > 0$ if $\Delta n_{gas} > 0$. This is because gases are very dispersed compared to liquids or solids; (gases possess large entropies). Recall that Δn_{gas} is the difference between the sum of the stoichiometric coefficients of the gaseous products and a similar sum for the reactants.

(a) $\Delta n_{gas} = 2 + 0 - (2+1) = -1.$ One mole of gas is consumed here. We predict $\Delta_r S < 0.$

(b) $\Delta n_{gas} = 1 + 0 - 0 = +1.$ Since one mole of gas is produced, we predict $\Delta_r S > 0.$

2B (E) (a) The outcome is uncertain in the reaction between ZnS(s) and $Ag_2O(s)$. We have used Δn_{gas} to estimate the sign of entropy change. There is no gas involved in this reaction and thus our prediction is uncertain.

(b) In the chlor-alkali process the entropy increases because two moles of gas have formed where none were originally present ($\Delta n_{gas} = (1+1+0) - (0+0) = 2$

3A **(E)** For a vaporization which occurs at the usual or normal vaporization temperature, $\Delta_{vap} G° = 0 = \Delta_{vap} H° - T\Delta_{vap} S°.$ Thus, $\Delta_{vap} S° = \Delta_{vap} H°/T_{vap}.$

We substitute the given values. $\Delta_{vap} S° = \dfrac{\Delta_{vap} H°}{T_{vap}} = \dfrac{20.2 \text{ kJ mol}^{-1}}{(-29.79 + 273.15) \text{ K}} = 83.0 \text{ J mol}^{-1} \text{ K}^{-1}$

3B **(E)** For a phase change which occurs at the normal transition temperature, $\Delta_{tr} G° = 0 = \Delta_{tr} H° - T\Delta_{tr} S°.$ Thus, $\Delta_{tr} H° = T\Delta_{tr} S°.$ We substitute in the given values.
$\Delta_{tr} H° = T\Delta_{tr} S° = (95.5 + 273.15) \text{ K} \times 1.09 \text{ J mol}^{-1} \text{ K}^{-1} = 402 \text{ J/mol}$

4A **(E)** Here, we use $\Delta S = C_p \ln(T_f/T_i) - nR \ln(P_f/P_i),$ with $T_f = 240$ K, $T_i = 300$ K, $P_f = 0.50$ bar, $P_i = 1.00$ bar, $C_p = 20.8$ J K^{-1}, and $n = 1$ mol. We get $\Delta S = 20.8$ J K^{-1} ln(240/300) $-$ 8.3145 J K^{-1} ln(0.5) $= -4.64$ J K^{-1} $+ 5.76$ J K^{-1} $= 1.12$ J K^{-1}

4B **(E)** We use $\Delta S = C_p \ln(T_f/T_i) - nR \ln(P_f/P_i),$ with $T_f = T_i = 300$ K, $P_f = 0.5$ bar, $P_i = 1.00$ bar, $C_p = 20.8$ J K^{-1}, and $n = 1$ mol. We get
$\Delta S = 20.8$ J K^{-1} ln(300/300) $-$ 8.3145 J K^{-1} ln(0.5) $= (0 + 5.76)$ J K^{-1} $= 5.76$ J K^{-1}

5A **(M)** The standard entropy change of reaction is expressed in terms of the standard entropies of the reagents.
$\Delta_r S° = 2S°[NH_3(g)] - S°[N_2(g)] - 3S°[H_2(g)]$
$= 2 \times 192.5 \text{ J mol}^{-1}\text{K}^{-1} - 191.6 \text{ J mol}^{-1} \text{ K}^{-1} - 3 \times 130.7 \text{ J mol}^{-1} \text{ K}^{-1} = -198.7 \text{ J mol}^{-1} \text{ K}^{-1}$
Thus to form *one* mole of $NH_3(g)$, the standard entropy of reaction is -99.4 J mol^{-1} K^{-1}

5B **(M)** The standard entropy of reaction is expressed in terms of the standard entropies of the reagents.

$$\Delta_r S^\circ = S^\circ\left[NO(g)\right] + S^\circ\left[NO_2(g)\right] - S^\circ\left[N_2O_3(g)\right]$$

$$138.5 \text{ J mol}^{-1}\text{ K}^{-1} = 210.8 \text{ J mol}^{-1}\text{ K}^{-1} + 240.1 \text{ J mol}^{-1}\text{ K}^{-1} - S^\circ\left[N_2O_3(g)\right]$$

$$= 450.9 \text{ J mol}^{-1}\text{ K}^{-1} - S^\circ\left[N_2O_3(g)\right]$$

$$S^\circ\left[N_2O_3(g)\right] = 450.9 \text{ J mol}^{-1}\text{ K}^{-1} - 138.5 \text{ J mol}^{-1}\text{ K}^{-1} = 312.4 \text{ J mol}^{-1}\text{ K}$$

6A **(E)**

(a) Because $\Delta n_{gas} = 2 - (1 + 3) = -2$ for the synthesis of ammonia, we would predict $\Delta_r S < 0$ for the reaction. We already know that $\Delta_r H < 0$. Thus, the reaction falls into case 2, namely, a reaction that is spontaneous at low temperatures and non-spontaneous at high temperatures.

(b) For the formation of ethylene $\Delta n_{gas} = 1 - (2 + 0) = -1$ and thus $\Delta_r S < 0$. We are given that $\Delta_r H > 0$ and, thus, this reaction corresponds to case 4, namely, a reaction that is non-spontaneous at all temperatures.

6B **(E)**

(a) Because $\Delta n_{gas} = +1$ for the decomposition of calcium carbonate, $CaCO_3(s) \rightarrow CaO(s) + CO_2(g)$, we would predict $\Delta_r S > 0$ for the reaction, favoring the reaction at high temperatures. High temperatures also favor this endothermic $(\Delta H^\circ > 0)$ reaction.

(b) The "roasting" of ZnS(s), $2 ZnS(s) + 3 O_2(g) \rightarrow 2 ZnO(s) + 2 SO_2(g)$, has $\Delta n_{gas} = 2 - 3 = -1$ and, thus, $\Delta_r S < 0$. We are given that $\Delta_r H < 0$; thus, this reaction corresponds to case 2, namely, a reaction that is spontaneous at low temperatures, and non-spontaneous at high ones.

7A **(E)** The expression $\Delta_r G^\circ = \Delta_r H^\circ - T\Delta_r S^\circ$ is used with $T = 298.15$ K.

$$\Delta_r G^\circ = \Delta_r H^\circ - T\Delta_r S^\circ = -1648 \text{ kJ} - 298.15 \text{ K} \times \left(-549.3 \text{ J K}^{-1}\right) \times \left(1 \text{ kJ}/1000 \text{ J}\right)$$

$$= -1648 \text{ kJ} + 163.8 \text{ kJ} = -1484 \text{ kJ}$$

7B **(M)** We just need to substitute values from Appendix D into the supplied expression.

$$\Delta_r G^\circ = 2\Delta_f G^\circ\left[NO_2(g)\right] - 2\Delta_f G^\circ\left[NO(g)\right] - \Delta_f G^\circ\left[O_2(g)\right]$$

$$= 2 \times 51.31 \text{ kJ mol}^{-1} - 2 \times 86.55 \text{ kJ mol}^{-1} - 0.00 \text{ kJ mol}^{-1} = -70.48 \text{ kJ mol}^{-1}$$

8A **(M)** Using equation (13.14), the balanced chemical equation, and the value for $\Delta_r G^\circ = -32.90$ kJ mol^{-1} from Example 13-8, it follows that

$$\Delta_r G = \Delta_r G^\circ + RT \ln Q$$

$$-82.00 \text{ kJ mol}^{-1} = -32.90 \text{ kJ mol}^{-1} + 8.314 \times 10^{-3} \text{ kJ K}^{-1} \text{ mol}^{-1} \times 298.15 \text{ K} \ln \frac{\left(P_{NH_3}/P^\circ\right)^2}{\left(P_{N_2}/P^\circ\right)\left(P_{H_2}/P^\circ\right)^3}$$

$$-82.00 \text{ kJ mol}^{-1} = -32.90 \text{ kJ mol}^{-1} + 2.479 \text{ kJ mol}^{-1} \ln \frac{\left(P_{NH_3}/P^\circ\right)^2}{(0.5)(0.5)^3}$$

Solving for P_{NH_3}, we obtain

$$2.478 \text{ kJ mol}^{-1} \ln \frac{\left(P_{NH_3}/P^\circ\right)^2}{(0.5)(0.5)^3} = -82.00 \text{ kJ mol}^{-1} + 32.90 \text{ kJ mol}^{-1} = -49.10 \text{ kJ mol}^{-1}$$

$$\ln \frac{\left(P_{NH_3}/P^\circ\right)^2}{(0.5)(0.5)^3} = \frac{-49.10 \text{ kJ mol}^{-1}}{2.478 \text{ kJ mol}^{-1}} = 19.81 \Rightarrow \frac{\left(P_{NH_3}/P^\circ\right)^2}{(0.5)(0.5)^3} = e^{19.81} = 4.0 \times 10^8$$

$$P_{NH_3} = P^\circ \times \sqrt{4.0 \times 10^8} = 5.0 \times 10^3 \text{ bar}$$

8B **(M)** The minimum value of Q required to make the reverse reaction, conversion of ammonia to nitrogen and hydrogen, spontaneous implies that $\Delta_r G > 0$. Therefore,

$$\Delta_r G^\circ + RT \ln Q > 0 \Rightarrow RT \ln Q > -\Delta_r G^\circ \Rightarrow \ln Q > \frac{-\Delta_r G^\circ}{RT}$$

$$Q > e^{\frac{-\Delta_r G^\circ}{RT}} = e^{\frac{+32.90 \text{ kJ mol}^{-1}}{8.314 \times 10^{-3} \text{ kJ mol}^{-1} \text{ K}^{-1} \times 298.15 \text{ K}}} = 5.81 \times 10^5$$

9A **(M)** Pressures of gases and molarities of solutes in aqueous solution appear in thermodynamic equilibrium constant expressions. Pure solids and liquids (including solvents) do not appear.

(a) $\quad K = \dfrac{P_{SiCl_4}}{P_{Cl_2}^2} = K_p$ **(b)** $\quad K = \dfrac{\left[HOCl\right]\left[H^+\right]\left[Cl^-\right]}{P_{Cl_2}}$

$K = K_p$ for (a) because all terms in the K expression are gas pressures.

9B **(M)** We need the balanced chemical equation in order to write the equilibrium constant expression. We start by translating names into formulas.

$$PbS(s) + HNO_3(aq) \rightarrow Pb(NO_3)_2(aq) + S(s) + NO(g)$$

The equation then is balanced with the ion-electron method.

oxidation : $\{PbS(s) \rightarrow Pb^{2+}(aq) + S(s) + 2e^- \qquad\qquad \} \times 3$

reduction : $\{NO_3^-(aq) + 4H^+(aq) + 3e^- \rightarrow NO(g) + 2H_2O(l) \quad \} \times 2$

net ionic : $3\,PbS(s) + 2NO_3^-(aq) + 8H^+(aq) \rightarrow 3Pb^{2+}(aq) + 3S(s) + 2NO(g) + 4H_2O(l)$

In writing the thermodynamic equilibrium constant, recall that neither pure solids (PbS(s) and S(s)) nor pure liquids $(H_2O(l))$ appear in the thermodynamic equilibrium constant expression. Note also that we have written $H^+(aq)$ here for brevity even though we understand that $H_3O^+(aq)$ is the acidic species in aqueous solution.

$$K = \frac{[Pb^{2+}]^3 \, p_{NO}^2}{[NO_3^-]^2 [H^+]^8}$$

10A (D) We first determine the value of $\Delta_r G°$ and then set $\Delta_r G° = -RT \ln K$ to determine K.

$$\Delta_r G = \Delta_f G°\left[Ag^+(aq)\right] + \Delta_f G°[I^-(aq)] - \Delta_f G°\left[AgI(s)\right]$$

$$= [(77.11 - 51.57) - (-66.19)] \text{ kJ/mol} = +91.73 \text{ kJ mol}^{-1}$$

$$\ln K = \frac{-\Delta_r G°}{RT} = -\frac{-91.73 \text{ kJ/mol}}{8.3145 \text{ J mol}^{-1} \text{ K}^{-1} \times 298.15 \text{ K}} \times \frac{1000 \text{ J}}{1 \text{ kJ}} = -37.00$$

$$K = e^{-37.00} = 8.5 \times 10^{-17}$$

This is precisely equal to the value for the K_{sp} of AgI listed in Appendix D.

10B (D) We begin by translating names into formulas: $MnO_2(s) + HCl(aq) \rightarrow Mn^{2+}(aq) + Cl_2(aq)$. Then we produce a balanced net ionic equation with the ion–electron method:

oxidation: $2Cl^-(aq) \qquad\qquad\qquad \rightarrow Cl_2(aq) + 2e^-$

reduction: $MnO_2(s) + 4YH^+(aq) + 2e^- \rightarrow Mn^{2+}(aq) + 2H_2O(l)$

net ionic: $MnO_2(s) + 4H^+(aq) + 2Cl^-(aq) \rightarrow Mn^{2+}(aq) + Cl_2(g) + 2H_2O(l)$

Next we determine the value of $\Delta_r G°$ for the reaction and then the value of K.

$$\Delta_r G° = \Delta_f G°\left[Mn^{2+}(aq)\right] + \Delta_f G°\left[Cl_2(g)\right] + 2\Delta_f G°\left[H_2O(l)\right]$$

$$- \Delta_f G°\left[MnO^{2+}(s)\right] - 4\Delta_f G°\left[H^+(aq)\right] - 2\Delta_f G°\left[Cl^-(aq)\right]$$

$$= -228.1 \text{ kJ mol}^{-1} + 0.0 \text{ kJ mol}^{-1} + 2 \times (-237.1 \text{ kJ mol}^{-1})$$

$$- (-465.1 \text{ kJ mol}^{-1}) - 4 \times 0.0 \text{ kJ mol}^{-1} - 2 \times (-131.2 \text{ kJ mol}^{-1})$$

$$= +25.2 \text{ kJ mol}^{-1}$$

$$\ln K = \frac{-\Delta_r G°}{RT} = \frac{-(+25.2 \times 10^3 \text{ J mol}^{-1})}{8.3145 \text{ J mol}^{-1} \text{ K}^{-1} \times 298.15 \text{ K}} = -10.1\underline{7} \quad K = e^{-10.2} = 4 \times 10^{-5}$$

Because the value of K is so much smaller than 1, we expect the reaction will proceed in the forward direction only to a very limited extent.

11A **(M)** We set equal the two expressions for $\Delta_r G^\circ$ and solve for the absolute temperature.

$$\Delta_r G^\circ = \Delta_r H^\circ - T\Delta_r S^\circ = -RT \ln K \qquad \Delta_r H^\circ = T\Delta_r S^\circ - RT \ln K = T\left(\Delta_r S^\circ - R \ln K\right)$$

$$T = \frac{\Delta_r H^\circ}{\Delta_r S^\circ - R \ln K} = \frac{-114.1\times10^3 \text{ J/mol}}{\left[-146.4 - 8.3145 \ \ln(150)\right] \text{J mol}^{-1}\text{ K}^{-1}} = 607 \text{ K}$$

11B **(D)** We expect the value of the equilibrium constant to increase as the temperature decreases since this is an exothermic reaction and exothermic reactions will have a larger equilibrium constant (shift right to form more products), as the temperature decreases. Thus, we expect K to be larger than 1000, which is its value at 4.3×10^2 K.

(a) The value of the equilibrium constant at 25 °C is obtained directly from the value of $\Delta_r G^\circ$, since that value is also for 25 °C. Note:

$$\Delta_r G^\circ = \Delta_r H^\circ - T\Delta_r S^\circ = -77.1 \text{ kJ/mol} - 298.15 \text{ K}\left(-0.1213 \text{ kJ/mol}\cdot\text{K}\right) = -40.9 \text{ kJ/mol}$$

$$\ln K = \frac{-\Delta_r G^\circ}{RT} = \frac{-\left(-40.9\times10^3 \text{ J mol}^{-1}\right)}{8.3145 \text{ J mol}^{-1}\text{ K}^{-1}\times298.15 \text{ K}} = 16.5 \qquad K = e^{+16.5} = 1.5\times10^7$$

(b) First, we solve for $\Delta_r G^\circ$ at 75 °C = 348 K

$$\Delta_r G^\circ = \Delta_r H^\circ - T\Delta_r S^\circ = -77.1 \ \frac{\text{kJ}}{\text{mol}}\times\frac{1000 \text{ J}}{1 \text{ kJ}} - \left(348.15 \text{ K}\times\left(-121.3 \ \frac{\text{J}}{\text{mol K}}\right)\right)$$

$$= -34.87\times10^3 \text{ J/mol}$$

Then we use this value to obtain the value of the equilibrium constant, as in part (a).

$$\ln K = \frac{-\Delta_r G^\circ}{RT} = \frac{-\left(-34.87\times10^3 \text{ J mol}^{-1}\right)}{8.3145 \text{ J mol}^{-1}\text{ K}^{-1}\times348.15 \text{ K}} = 12.05 \qquad K = e^{+12.05} = 1.7\times10^5$$

As expected, K for this exothermic reaction decreases with increasing temperature.

12A **(M)** We use the value of $K = 9.1\times10^2$ at 800 K and $\Delta_r H^\circ = -1.8\times10^5$ J/mol, for the appropriate terms, in the van't Hoff equation.

$$\ln\frac{5.8\times10^{-2}}{9.1\times10^2} = \frac{-1.8\times10^5 \text{ J/mol}}{8.3145 \text{ J mol}^{-1}\text{ K}^{-1}}\left(\frac{1}{800 \text{ K}} - \frac{1}{T \text{ K}}\right) = -9.66; \ \frac{1}{T} = \frac{1}{800} - \frac{9.66\times8.3145}{1.8\times10^5}$$

$$1/T = 1.25\times10^{-3} - 4.5\times10^{-4} = 8.0\times10^{-4} \qquad T = 1240 \text{ K} \approx 970 \text{ °C}$$

This temperature is an estimate because it is an extrapolated point beyond the range of the data supplied.

12B **(M)** The temperature we are considering is 235 °C = 508 K. We substitute the value of $K = 9.1\times10^2$ at 800 K and $\Delta_r H^\circ = -1.8\times10^5$ J/mol, for the appropriate terms, in the van't Hoff equation.

$$\ln\frac{K}{9.1\times10^2} = \frac{-1.8\times10^5 \text{ J/mol}}{8.3145 \text{ J mol}^{-1}\text{ K}^{-1}}\left(\frac{1}{800 \text{ K}} - \frac{1}{508 \text{ K}}\right) = +15._6; \ \frac{K}{9.1\times10^2} = e^{+15._6} = 6\times10^6$$

$$K = 6\times10^6\times9.1\times10^2 = 5\times10^9$$

INTEGRATIVE EXAMPLE

A **(D)** The value of $\Delta_r G°$ can be calculated by finding the value of the equilibrium constant K_p at 25 °C. The equilibrium constant for the reaction is simply given by $K_p = p\{N_2O_5(g)\}$. The vapor pressure of $N_2O_5(g)$ can be determined from the Clausius-Clapeyron eqution, which is a specialized version of the van't Hoff equation.

$$\ln\frac{P_2}{P_1} = \frac{\Delta_{sub}H}{R}\left(\frac{1}{T_1}-\frac{1}{T_2}\right) \Rightarrow \Delta_{sub}H = \frac{R\ln\frac{P_2}{P_1}}{\left(\frac{1}{T_1}-\frac{1}{T_2}\right)}$$

$$\Delta_{sub}H = \frac{8.314\ \text{Jmol}^{-1}\text{K}^{-1} \times \ln\dfrac{760\ \text{mmHg}}{100\ \text{mmHg}}}{\left(\dfrac{1}{7.5+273.15}-\dfrac{1}{32.4+273.15}\right)} = \frac{2.028}{3.49\times10^{-5}}\ \text{Jmol}^{-1} = 5.81\times10^4\ \text{Jmol}^{-1}$$

$$\ln\frac{P_3}{P_1} = \frac{\Delta_{sub}H}{R}\left(\frac{1}{T_1}-\frac{1}{T_2}\right) \Rightarrow P_3 = P_1 e^{\frac{\Delta_{sub}H}{R}\left(\frac{1}{T_1}-\frac{1}{T_2}\right)}$$

$$P_3 = 100\ \text{mmHg}\times e^{\frac{5.81\times10^4\ \text{mol}^{-1}}{8.314\ \text{J K}^{-1}\ \text{mol}^{-1}}\left(\frac{1}{280.7}-\frac{1}{298.2}\right)\text{K}^{-1}} = 431\ \text{mmHg}\times\frac{1\ \text{atm}}{760\ \text{mmHg}} = 0.567\ \text{atm} = K_p$$

$$\Delta_r G° = -RT\ln K_p = -(8.314\times10^{-3}\ \text{kJ mol}^{-1}\ \text{K}^{-1}\times298.15\ \text{K})\ln(0.567) = 1.42\ \text{kJ/mol}$$

B **(D)** The standard entropy change for the reaction $(\Delta_r S°)$ can be calculated from the known values of $\Delta_r H°$ and $\Delta_r G°$.

$$\Delta_r G° = \Delta_r H° - T\Delta_r S° \Rightarrow \Delta_r S° = \frac{\Delta_r H° - \Delta_r G°}{T} = \frac{-454.8\ \text{kJ mol}^{-1} - (-323.1\ \text{kJ mol}^{-1})}{298.15\ \text{K}}$$

$$= -441.7\ \text{J K}^{-1}\ \text{mol}^{-1}$$

Plausible chemical reaction for the production of ethylene glycol can also be written as:

$$2\ C(s) + 3\ H_2(g) + O_2(g) \longrightarrow CH_2OHCH_2OH(l)$$

Since $\Delta_r S° = \sum\{S°_{products}\} - \sum\{S°_{reactants}\}$ it follows that:

$$\Delta_r S° = S°(CH_2OHCH_2OH(l)) - [2\times S°(C(s)) + 3\times S°(H_2(g)) + S°(O_2(g))]$$

$$-441.7\ \text{J K}^{-1}\ \text{mol}^{-1} = S°(CH_2OHCH_2OH(l)) - [2\times5.74\ \text{J K}^{-1}\ \text{mol}^{-1} + 3\times130.7\ \text{J K}^{-1}\ \text{mol}^{-1}$$

$$+ 205.1\ \text{J K}^{-1}\ \text{mol}^{-1}]$$

$$S°(CH_2OHCH_2OH(l)) = -441.7\ \text{J K}^{-1}\ \text{mol}^{-1} + 608.68\ \text{J K}^{-1}\ \text{mol}^{-1} = 167\ \text{J K}^{-1}\ \text{mol}^{-1}$$

EXERCISES

Spontaneous Change and Entropy

1. **(M)**
   (a) If the length of the box is increased to $2L$ for a fixed total energy, the number of possible microstates will increase. Hence, the entropy of the system will increase.
   (b) If the total energy of the box is increased at constant length L, the number of energy levels and number of microstates that are accessible to the particles will increase. Hence, the entropy of the system will increase.

3. **(D)** The entropy of one mole of H_2 molecules is $S = 1 \text{ mol} \times 130.7 \text{ J K}^{-1} \text{ mol}^{-1} = 130.7 \text{ J K}^{-1}$. Using $S = k \ln W$, we can calculate

   $$\ln W = \frac{S}{k} = \frac{130.7 \text{ J K}^{-1}}{1.3807 \times 10^{-23} \text{ J K}^{-1}} = 9.466 \times 10^{24}$$

   Since $\ln W = 2.030 \log W$, we can write $\log W = \dfrac{9.466 \times 10^{24}}{2.303} = 4.111 \times 10^{24}$. Therefore,

   $W = 10^{4.111 \times 10^{24}}$. To write this number in decimal form, we would write 1 followed by 4.111×10^{24} zeros.

5. **(E)**
   (a) The freezing of ethanol involves a *decrease* in the entropy of the system. There is a reduction in mobility and in the number of forms in which their energy can be stored when they leave the solution and arrange themselves into a crystalline state.
   (b) The sublimation of dry ice involves converting a solid that has little mobility into a highly dispersed vapor which has a number of ways in which energy can be stored (rotational, translational). Thus, the entropy of the system *increases* substantially.
   (c) The burning of rocket fuel involves converting a liquid fuel into the highly dispersed mixture of the gaseous combustion products. The entropy of the system *increases* substantially.

7. **(E)** The first law of thermodynamics states that energy is neither created nor destroyed (thus, "The energy of the universe is constant"). A consequence of the second law of thermodynamics is that entropy of the universe increases for all spontaneous, that is, naturally occurring, processes (and therefore, "the entropy of the universe increases toward a maximum").

9. **(E)**
   (a) Increase in entropy because a gas has been created from a liquid, a condensed phase.
   (b) Decrease in entropy as a condensed phase, a solid, is created from a solid and a gas.
   (c) For this reaction we cannot be certain of the entropy change. Even though the number of moles of gas produced is the same as the number that reacted, we cannot conclude that the entropy change is zero because not all gases have the same molar entropy.

(d) $2H_2S(g) + 3O_2(g) \rightarrow 2H_2O(g) + 2SO_2(g)$ Decrease in entropy since five moles of gas with high entropy become only four moles of gas, with about the same quantity of entropy per mole.

11. (E)

(a) Negative; A liquid (moderate entropy) combines with a solid to form another solid.

(b) Positive; One mole of high entropy gas forms where no gas was present before.

(c) Positive; One mole of high entropy gas forms where no gas was present before.

(d) Uncertain; The number of moles of gaseous products is the same as the number of moles of gaseous reactants.

(e) Negative; Two moles of gas (and a solid) combine to form just one mole of gas.

13. (M)

(a) For an isothermal gas expansion, $\Delta S = nR \ln\left(\dfrac{V_f}{V_i}\right)$. Therefore,

$$\Delta S = 1 \text{ mol} \times 8.3145 \text{ J K}^{-1} \text{ mol}^{-1} \ln\left(\frac{2V}{V}\right) = 5.76 \text{ J K}^{-1}$$

(b) This is a constant pressure heating process for which $\Delta S \approx C_p \ln\left(\dfrac{T_f}{T_i}\right)$. Using 35.69 J K^{-1} mol^{-1} for the molar heat capacity of methane, we obtain

$$\Delta S \approx 1 \text{ mol} \times 35.69 \text{ J K}^{-1} \text{ mol}^{-1} \times \ln\left(\frac{325 \text{ K}}{298 \text{ K}}\right) = 3.10 \text{ J K}^{-1}$$

15. (M)

(a)
$$\Delta_{vap}H = \Delta_f H°[H_2O(g)] - \Delta_f H°[H_2O(l)] = -241.8 \text{ kJ/mol} - (-285.8 \text{ kJ/mol})$$
$$= +44.0 \text{ kJ/mol}$$
$$\Delta_{vap}S° = S°\left[H_2O(g)\right] - S°\left[H_2O(l)\right] = 188.8 \text{ J mol}^{-1} \text{ K}^{-1} - 69.91 \text{ J mol}^{-1}\text{K}^{-1}$$
$$= 118.9 \text{ J mol}^{-1} \text{ K}^{-1}$$

There is an alternate, but incorrect, method of obtaining $\Delta_{vap}S°$.

$$\Delta_{vap}S° = \frac{\Delta_{vap}H°}{T} = \frac{44.0 \times 10^3 \text{ J/mol}}{298.15 \text{ K}} = 148 \text{ J mol}^{-1} \text{ K}^{-1}$$

This method is invalid because the temperature in the denominator of the equation must be the temperature at which the liquid-vapor transition is at equilibrium. Liquid water and water vapor at 1 atm pressure (standard state, indicated by °) are in equilibrium only at $100\,°C = 373$ K.

(b) The reason why $\Delta_{vap}H°$ is different at 25 °C from its value at 100 °C has to do with the heat required to bring the reactants and products down to 298 K from 373 K. The specific heat of liquid water is higher than the heat capacity of steam. Thus, more heat is given off by lowering the temperature of the liquid water from 100 °C to 25 °C than is given off by lowering the temperature of the same amount of steam. Another way to think of this is that hydrogen bonding is more disrupted in water at 100 °C than at 25 °C (because the molecules are in rapid—thermal—motion), and hence, there is not as much energy needed to convert liquid to vapor (thus $\Delta_{vap}H°$ has a smaller value at 100 °C. The reason why $\Delta_{vap}S°$ has a larger value at 25 °C than at 100 °C has to do with dispersion. A vapor at 1 atm pressure (the case at both temperatures) has about the same entropy. On the other hand, liquid water is more disordered (better able to disperse energy) at higher temperatures since more of the hydrogen bonds are disrupted by thermal motion. (The hydrogen bonds are totally disrupted in the two vapors).

17. **(M)** Trouton's rule is obeyed most closely by liquids that do not have a high degree of order within the liquid. In both HF and CH_3OH, hydrogen bonds create considerable order within the liquid. In $C_6H_5CH_3$, the only attractive forces are non-directional London forces, which have no preferred orientation as hydrogen bonds do. Thus, of the three choices, liquid $C_6H_5CH_3$ would most closely follow Trouton's rule.

19. **(M)** The liquid water-gaseous water equilibrium H_2O (l, 0.50 atm) $\rightleftharpoons$ H_2O (g, 0.50 atm) can only be established at <u>one temperature</u>, namely the boiling point for water under 0.50 atm external pressure. We can estimate the boiling point for water under 0.50 atm external pressure by using the Clausius-Clapeyron equation:

$$\ln\frac{P_2}{P_1} = \frac{\Delta_{vap}H°}{R}\left(\frac{1}{T_1} - \frac{1}{T_2}\right)$$

We know that at 373 K, the pressure of water vapor is 1.00 atm. Let's make P_1 = 1.00 atm, P_2 = 0.50 atm and T_1 = 373 K. Thus, the boiling point under 0.50 atm pressure is T_2. To find T_2 we simply insert the appropriate information into the Clausius-Clapeyron equation and solve for T_2:

$$\ln\frac{0.50 \text{ atm}}{1.00 \text{ atm}} = \frac{40.7 \text{ kJ mol}^{-1}}{8.3145\times10^{-3} \text{ kJ K}^{-1} \text{ mol}^{-1}}\left(\frac{1}{373 \text{ K}} - \frac{1}{T_2}\right)$$

$$-1.4\underline{16} \times 10^{-4} \text{ K} = \left(\frac{1}{373 \text{ K}} - \frac{1}{T_2}\right)$$

Solving for T_2 we find a temperature of 354 K or 81°C. Consequently, to achieve an equilibrium between gaseous and liquid water under 0.50 atm pressure, the temperature must be set at 354 K.

Gibbs Energy and Spontaneous Change

21. **(E)** Answer (b) is correct. Br—Br bonds are broken in this reaction, meaning that it is endothermic, with $\Delta_r H > 0$. Since the number of moles of gas increases during the reaction, $\Delta_r S > 0$. And, because $\Delta_r G = \Delta_r H - T \Delta_r S$, this reaction is non-spontaneous ($\Delta_r G > 0$) at low temperatures where the $\Delta_r H$ term predominates and spontaneous ($\Delta_r G < 0$) at high temperatures where the $T \Delta_r S$ term predominates.

23. **(E)**

    (a) $\Delta_r H^\circ < 0$ and $\Delta_r S^\circ < 0$ (since $\Delta n_{gas} < 0$) for this reaction. Thus, this reaction is case 2 of Table 13-3. It is spontaneous at low temperatures and non-spontaneous at high temperatures.

    (b) We are unable to predict the sign of $\Delta_r S^\circ$ for this reaction, since $\Delta n_{gas} = 0$. Thus, no strong prediction as to the temperature behavior of this reaction can be made. Since $\Delta_r H^\circ > 0$, we can, however, conclude that the reaction will be non-spontaneous at low temperatures.

    (c) $\Delta_r H^\circ > 0$ and $\Delta_r S^\circ > 0$ (since $\Delta n_{gas} > 0$) for this reaction. This is case 3 of Table 13-3. It is non-spontaneous at low temperatures, but spontaneous at high temperatures.

25. **(E)** First of all, the process is clearly spontaneous, and therefore $\Delta G < 0$. In addition, the gases are more dispersed when they are at a lower pressure and therefore $\Delta S > 0$. We also conclude that $\Delta H = 0$ because the gases are ideal and thus there are no forces of attraction or repulsion between them.

27. **(M)**

    (a) An exothermic reaction (one that gives off heat) may not occur spontaneously if, at the same time, the system becomes more ordered (concentrated) that is, $\Delta_r S^\circ < 0$. This is particularly true at a high temperature, where the $T \Delta_r S$ term dominates the $\Delta_r G$ expression. An example of such a process is freezing water (clearly exothermic because the reverse process, melting ice, is endothermic), which is not spontaneous at temperatures above $0\,^\circ C$.

    (b) A reaction in which $\Delta_r S > 0$ need not be spontaneous if that process also is endothermic. This is particularly true at low temperatures, where the $\Delta_r H$ term dominates the $\Delta_r G$ expression. An example is the vaporization of water (clearly an endothermic process, one that requires heat, and one that produces a gas, so $\Delta_r S > 0$), which is not spontaneous at low temperatures, that is, below $100\,^\circ C$ (assuming $P_{ext} = 1.00$ atm).

Standard Gibbs Energy of Reaction, $\Delta_r G°$

29. **(M)** $\Delta_r H° = \Delta_f H° \left[NH_4Cl(s) \right] - \Delta_f H° \left[NH_3(g) \right] - \Delta_f H° \left[HCl(g) \right]$

$$= -314.4 \text{ kJ/mol} - (-46.11 \text{ kJ/mol} - 92.31 \text{ kJ/mol}) = -176.0 \text{ kJ/mol}$$

$\Delta_r G° = \Delta_f G° \left[NH_4Cl(s) \right] - \Delta_f G° \left[NH_3(g) \right] - \Delta_f G° \left[HCl(g) \right]$

$$= -202.9 \text{ kJ/mol} - (-16.48 \text{ kJ/mol} - 95.30 \text{ kJ/mol}) = -91.1 \text{ kJ/mol}$$

$\Delta_r G° = \Delta_r H° - T\Delta_r S°$

$$\Delta_r S° = \frac{\Delta_r H° - \Delta_r G°}{T} = \frac{-176.0 \text{ kJ/mol} + 91.1 \text{ kJ/mol}}{298 \text{ K}} \times \frac{1000 \text{ J}}{1 \text{ kJ}} = -285 \text{ J mol}^{-1}$$

31. **(M)**

(a) $\Delta_r S° = 2 S° \left[POCl_3(l) \right] - 2 S° \left[PCl_3(g) \right] - S° \left[O_2(g) \right]$

$$= 2 \left(222.4 \text{ J K}^{-1} \text{ mol}^{-1} \right) - 2 \left(311.7 \text{ J K}^{-1} \text{ mol}^{-1} \right) - 205.1 \text{ J K}^{-1} \text{ mol}^{-1}$$

$$= -383.7 \text{ J K}^{-1} \text{ mol}^{-1}$$

$\Delta_r G° = \Delta_r H° - T\Delta_r S° = -620.2 \times 10^3 \text{ J mol}^{-1} - (298 \text{ K})(-383.7 \text{ J K}^{-1} \text{ mol}^{-1})$

$$= -506 \times 10^3 \text{ J mol}^{-1} = -506 \text{ kJ mol}^{-1}$$

(b) The reaction proceeds spontaneously in the forward direction when reactants and products are in their standard states, because the value of $\Delta_r G°$ is less than zero.

33. **(M)** We combine the reactions in the same way as for any Hess's law calculations.

(a) $N_2O(g) \rightarrow N_2(g) + \frac{1}{2}O_2(g)$ $\qquad \Delta_r G° = -\frac{1}{2}(+208.4 \text{ kJ mol}^{-1}) = -104.2 \text{ kJ mol}^{-1}$

$N_2(g) + 2O_2(g) \rightarrow 2NO_2(g)$ $\qquad \Delta_r G° = +102.6 \text{ kJ mol}^{-1}$

Net: $N_2O(g) + \frac{3}{2}O_2(g) \rightarrow 2NO_2(g)$ $\quad \Delta_r G° = -104.2 + 102.6 = -1.6 \text{ kJ mol}^{-1}$

This reaction reaches an equilibrium condition, with significant amounts of all species being present. This conclusion is based on the relatively small absolute value of $\Delta_r G°$.

(b) $2N_2(g) + 6H_2(g) \rightarrow 4NH_3(g)$ $\qquad \Delta_r G° = 2(-33.0 \text{ kJ mol}^{-1}) = -66.0 \text{ kJ mol}^{-1}$

$4NH_3(g) + 5O_2(g) \rightarrow 4NO(g) + 6H_2O(l)$ $\quad \Delta_r G° = -1010.5 \text{ kJ mol}^{-1}$

$4NO(g) \rightarrow 2N_2(g) + 2O_2(g)$ $\qquad \Delta_r G° = -2(+173.1 \text{ kJ mol}^{-1}) = -346.2 \text{ kJ mol}^{-1}$

Net: $6H_2(g) + 3O_2(g) \rightarrow 6H_2O(l)$ $\quad \Delta_r G° = -66.0 \text{ kJ mol}^{-1} - 1010.5 \text{ kJ mol}^{-1}$

$$-346.2 \text{ kJ mol}^{-1} = -1422.7 \text{ kJ mol}^{-1}$$

This reaction is three times the desired reaction, which therefore has
$\Delta_r G° = -1422.7 \text{ kJ mol}^{-1} \div 3 = -474.3 \text{ kJ mol}^{-1}$.

The large negative $\Delta_r G°$ value indicates that this reaction will go to completion at 25 °C.

(c) $4NH_3(g) + 5O_2(g) \rightarrow 4NO(g) + 6H_2O(l)$ $\Delta_rG° = -1010.5\,kJ\,mol^{-1}$

$4NO(g) \rightarrow 2N_2(g) + 2O_2(g)$ $\Delta_rG° = -2(+173.1\,kJ\,mol^{-1}) = -346.2\,kJ\,mol^{-1}$

$2N_2(g) + O_2(g) \rightarrow 2N_2O(g)$ $\Delta_rG° = +208.4\,kJ\,mol^{-1}$

$\overline{4NH_3(g) + 4O_2(g) \rightarrow 2N_2O(g) + 6H_2O(l)}$ $\Delta_rG° = -1010.5\,kJ\,mol^{-1} - 346.2\,kJ\,mol^{-1}$

$$+208.4\,kJ\,mol^{-1}$$

$$= -1148.3\,kJ\,mol^{-1}$$

This reaction is twice the desired reaction, which, therefore, has $\Delta_rG° = -574.2\,kJ\,mol^{-1}$. The very large negative value of the $\Delta_rG°$ for this reaction indicates that it will go to completion.

35. **(D)** The combustion reaction is : $C_6H_6(l) + \frac{15}{2}O_2(g) \rightarrow 6CO_2(g) + 3H_2O(g\ or\ l)$

(a) $\Delta_rG° = 6\Delta_fG°[CO_2(g)] + 3\Delta_fG°[H_2O(l)] - \Delta_fG°[C_6H_6(l)] - \frac{15}{2}\Delta_fG°[O_2(g)]$

$= 6(-394.4\,kJ\,mol^{-1}) + 3(-237.1\,kJ\,mol^{-1}) - (+124.5\,kJ\,mol^{-1}) - \frac{15}{2}(0.00\,kJ\,mol^{-1})$

$= -3202\,kJ\,mol^{-1}$

(b) $\Delta_rG° = 6\Delta_fG°[CO_2(g)] + 3\Delta_fG°[H_2O(g)] - \Delta_fG°[C_6H_6(l)] - \frac{15}{2}\Delta_fG°[O_2(g)]$

$= 6(-394.4\,kJ\,mol^{-1}) + 3(-228.6\,kJ\,mol^{-1}) - (+124.5\,kJ\,mol^{-1}) - \frac{15}{2}(0.00\,kJ\,mol^{-1})$

$= -3177\,kJ\,mol^{-1}$

We could determine the difference between the two values of $\Delta_rG°$ by noting the difference between the two products: $3H_2O(l) \rightarrow 3H_2O(g)$ and determining the value of $\Delta_rG°$ for this difference:

$\Delta_rG° = 3\Delta_fG°[H_2O(g)] - 3\Delta_fG°[H_2O(l)] = 3[-228.6 - (-237.1)]\,kJ\,mol^{-1}$

$= 25.5\,kJ\,mol^{-1}$

37. **(M)**

(a) $\Delta_rS° = \Sigma S°_{products} - \Sigma S°_{reactants}$

$= [1 \times 301.2\,J\,K^{-1}mol^{-1} + 2 \times 188.8\,J\,K^{-1}mol^{-1}] - [2 \times 247.4\,J\,K^{-1}mol^{-1}$
$+ 1 \times 238.5\,J\,K^{-1}mol^{-1}] = -54.5\,J\,K^{-1}\,mol^{-1}$

$\Delta_rS° = -0.0545\,kJ\,K^{-1}\,mol^{-1}$

(b) $\Delta_rH° = \Sigma$ (bonds broken in reactants (kJ/mol)) $-\Sigma$(bonds broken in products(kJ/ mol))

$= [4 \times (389\,kJ\,mol^{-1})_{N-H} + 4 \times (222\,kJ\,mol^{-1})_{O-F}] -$
$[4 \times (301\,kJ\,mol^{-1})_{N-F} + 4 \times (464\,kJ\,mol^{-1})_{O-H}]$

$\Delta_rH° = -616\,kJ\,mol^{-1}$

(c) $\Delta_r G° = \Delta_r H° - T\Delta_r S° = -616 \text{ kJ mol}^{-1} - 298 \text{ K}(-0.0545 \text{ kJ K}^{-1}\text{mol}^{-1})$

$= -600 \text{ kJ mol}^{-1}$

Since the $\Delta_r G°$ is <u>negative</u>, the reaction is <u>spontaneous</u>, and hence feasible (at 25 °C). Because both the entropy and enthalpy changes are *negative*, this reaction will be more highly favored at low temperatures (i.e., the reaction is enthalpy driven)

The Thermodynamic Equilibrium Constant

39. **(E)** In all three cases, $K_{eq} = K_p$ because only gases, pure solids, and pure liquids are present in the chemical equations. There are no factors for solids and liquids in K_{eq} expressions, and gases appear as partial pressures in atmospheres. That makes K_{eq} the same as K_p for these three reactions.

We now recall that $K_p = K_c (RT)^{\Delta n}$. Hence, in these three cases we have:

(a) $2SO_2(g) + O_2(g) \rightleftharpoons 2SO_3(g); \quad \Delta n_{gas} = 2 - (2+1) = -1; \quad K = K_p = K_c (RT)^{-1}$

(b) $HI(g) \rightleftharpoons \frac{1}{2}H_2(g) + \frac{1}{2}I_2(g); \qquad \Delta n_{gas} = 1 - (\frac{1}{2} + \frac{1}{2}) = 0; \quad K = K_p = K_c$

(c) $NH_4HCO_3(s) \rightleftharpoons NH_3(g) + CO_2(g) + H_2O(l);$

$\Delta n_{gas} = 2 - (0) = +2 \qquad K = K_p = K_c (RT)^2$

41. **(M)** In this problem, we are asked to determine the equilibrium constant and $\Delta_r G°$ for the reaction between carbon monoxide and hydrogen to yield methanol. To calculate $\Delta_r G°$, we can use the equation $\Delta_r G° = -RT \ln K$, where K is the thermodynamic equilibrium constant. Our first task is to use the given data to obtain the value of the thermodynamic equilibrium constant. For the given reaction, $K = \dfrac{(P_{CH_3OH}/P°)}{(P_{CO}/P°)(P_{H_2}/P°)^2}$ with $P° = 1$ bar

The equilibrium concentrations at 483 K were provided. We proceed by first converting the molar concentrations to pressures by using the ideal gas law, and then calculating the thermodynamic equilibrium constant, K. The standard Gibbs energy of reaction can then be determined by using $\Delta_r G° = -RT \ln K$.

$CO(g) + 2H_2(g) \rightleftharpoons CH_3OH(g)$

$pV = nRT \Rightarrow p = \dfrac{n}{V}RT = cRT$

$p_{CO} = 0.0911 \text{ mol L}^{-1} \times 0.08314 \text{ L bar K}^{-1} \text{ mol}^{-1} \times 483 \text{ K} = 3.66 \text{ bar}$

$p_{H_2} = 0.0822 \text{ mol L}^{-1} \times 0.08314 \text{ L bar K}^{-1} \text{ mol}^{-1} \times 483 \text{ K} = 3.30 \text{ bar}$

$p_{CH_3OH} = 0.00892 \text{ mol L}^{-1} \times 0.08314 \text{ L bar K}^{-1} \text{ mol}^{-1} \times 483 \text{ K} = 0.358 \text{ bar}$

$$K = \frac{p_{CH_3OH}}{p_{CO} \times p_{H_2}{}^2} = \frac{0.358}{3.66 \times 3.30^2} = 8.98 \times 10^{-3}$$

$$\Delta_r G° = -RT \ln K = -8.314 \text{ J K}^{-1} \text{ mol}^{-1} \times 483 \text{ K} \times \ln(8.98 \times 10^{-3}) = 1.89 \times 10^4 \text{ J mol}^{-1}$$

$$\Delta_r G° = 18.9 \text{ kJ mol}^{-1}$$

Relationships Involving $\Delta_r G$, $\Delta_r G°$, Q and K

43. **(M)** $\Delta_r G° = 2\Delta_f G°\left[NO(g)\right] - \Delta_f G°\left[N_2O(g)\right] - 0.5\,\Delta_f G°\left[O_2(g)\right]$

$$= 2(86.55 \text{ kJ/mol}) - (104.2 \text{ kJ/mol}) - 0.5(0.00 \text{ kJ/mol}) = 68.9 \text{ kJ/mol}$$

$$= -RT \ln K = -\left(8.3145 \times 10^{-3} \text{ kJ mol}^{-1} \text{ K}^{-1}\right)(298 \text{ K}) \ln K$$

$$\ln K = -\frac{68.9 \text{ kJ/mol}}{8.3145 \times 10^{-3} \text{ kJ mol}^{-1} \text{ K}^{-1} \times 298 \text{ K}} = -27.8 \qquad K = e^{-27.8} = 8 \times 10^{-13}$$

45. **(M)** We first balance each chemical equation, then calculate the value of $\Delta_r G°$ with data from Appendix D, and finally calculate the value of K_{eq} with the use of $\Delta_r G° = -RT \ln K$.

(a) $4HCl(g) + O_2(g) \rightleftharpoons 2H_2O(g) + 2Cl_2(s)$

$$\Delta_r G° = 2\Delta_f G°\left[H_2O(g)\right] + 2\Delta_f G°\left[Cl_2(g)\right] - 4\Delta_f G°\left[HCl(g)\right] - \Delta_f G°\left[O_2(g)\right]$$

$$= 2 \times \left(-228.6 \frac{\text{kJ}}{\text{mol}}\right) + 2 \times 0 \frac{\text{kJ}}{\text{mol}} - 4 \times \left(-95.30 \frac{\text{kJ}}{\text{mol}}\right) - 0 \frac{\text{kJ}}{\text{mol}} = -76.0 \frac{\text{kJ}}{\text{mol}}$$

$$\ln K = \frac{-\Delta_r G°}{RT} = \frac{+76.0 \times 10^3 \text{ J/mol}}{8.3145 \text{ J mol}^{-1} \text{ K}^{-1} \times 298 \text{ K}} = +30.7 \qquad K = e^{+30.7} = 2 \times 10^{13}$$

(b) $3Fe_2O_3(s) + H_2(g) \rightleftharpoons 2Fe_3O_4(s) + H_2O(g)$

$$\Delta_r G° = 2\Delta_f G°\left[Fe_3O_4(s)\right] + \Delta_f G°\left[H_2O(g)\right] - 3\Delta_f G°\left[Fe_2O_3(s)\right] - \Delta_f G°\left[H_2(g)\right]$$

$$= 2 \times (-1015 \text{ kJ/mol}) - 228.6 \text{ kJ/mol} - 3 \times (-742.2 \text{ kJ/mol}) - 0.00 \text{ kJ/mol}$$

$$= -32 \text{ kJ/mol}$$

$$\ln K = \frac{-\Delta_r G°}{RT} = \frac{32 \times 10^3 \text{ J/mol}}{8.3145 \text{ J mol}^{-1} \text{ K}^{-1} \times 298 \text{ K}} = 13; \quad K = e^{+13} = 4 \times 10^5$$

(c) $2Ag^+(aq) + SO_4{}^{2-}(aq) \rightleftharpoons Ag_2SO_4(s)$

$$\Delta_r G° = \Delta_f G°\left[Ag_2SO_4(s)\right] - 2\Delta_f G°\left[Ag^+(aq)\right] - \Delta_f G°\left[SO_4{}^{2-}(aq)\right]$$

$$= -618.4 \text{ kJ/mol} - 2 \times 77.11 \text{ kJ/mol} - (-744.5 \text{ kJ/mol}) = -28.1 \text{ kJ/mol}$$

$$\ln K = \frac{-\Delta_r G°}{RT} = \frac{28.1 \times 10^3 \text{ J/mol}}{8.3145 \text{ J mol}^{-1} \text{ K}^{-1} \times 298 \text{ K}} = 11.3; \quad K = e^{+11.3} = 8 \times 10^4$$

47. **(M)** In this problem we need to determine in which direction the reaction
$2SO_2(g) + O_2(g) \rightleftharpoons 2SO_3(g)$ is spontaneous when the partial pressure of SO_2, O_2, and SO_3
are 1.0×10^{-4}, 0.20 and 0.10 atm, respectively. We proceed by first determining the standard
Gibbs energy of reaction ($\Delta_r G°$) using tabulated data in Appendix D. The Gibbs energy of
reaction for the given conditions ($\Delta_r G$) is then calculated by employing the equation
$\Delta_r G = \Delta_r G° + RT \ln Q$, where Q is the reaction quotient. Based on the sign of $\Delta_r G$, we can
determine in which direction is the reaction spontaneous.

$\Delta_r G° = 2\Delta_f G°[SO_3(g)] - 2\Delta_f G°[SO_2(g)] - \Delta_f G°[O_2(g)]$

$\Delta_r G° = 2 \times (-371.1 \text{ kJ/mol}) - 2 \times (-300.2 \text{ kJ/mol}) - 0.0 \text{ kJ/mol} = -141.8 \text{ kJ mol}^{-1}$

$\Delta_r G = \Delta_r G° + RT \ln Q$

$Q_p = \dfrac{P\{SO_3(g)\}^2}{P\{O_2(g)\}P\{SO_2(g)\}^2} = \dfrac{(0.10)^2}{(0.20)(1.0 \times 10^{-4})^2} = 5.0 \times 10^6$

$\Delta_r G = -141.8 \text{ kJ mol}^{-1} + (8.3145 \times 10^{-3} \text{ kJ/K·mol})(298 \text{ K}) \ln(5.0 \times 10^6) = -104 \text{ kJ mol}^{-1}$.
Since $\Delta_r G$ is negative, the reaction is spontaneous in the forward direction.

49. **(M)** In order to determine the direction in which the reaction is spontaneous, we need to
calculate the non-standard Gibbs energy of reaction. To accomplish this, we will employ
the equation $\Delta_r G = \Delta_r G° + RT \ln Q$, where

$Q = \dfrac{[H_3O^+(aq)][CH_3CO_2^-(aq)]}{[CH_3CO_2H(aq)]}; \quad Q = \dfrac{(1.0 \times 10^{-3})^2}{(0.10)} = 1.0 \times 10^{-5}$

$\Delta_r G = 27.07 \text{ kJ mol}^{-1} + (8.3145 \times 10^{-3} \text{ kJ/K·mol})(298 \text{ K}) \ln(1.0 \times 10^{-5})$

$\Delta_r G = 27.07 \text{ kJ mol}^{-1} + (-28.53 \text{ kJ mol}^{-1}) = -1.46 \text{ kJ mol}^{-1}$.

Since $\Delta_r G$ is negative, the reaction is spontaneous in the forward direction.

51. **(E)** The relationship $\Delta_r S° = (\Delta_r G° - \Delta_r H°)/T$ (Equation (b)) is incorrect. Rearranging this
equation to put $\Delta_r G°$ on one side by itself gives $\Delta_r G° = \Delta_r H° + T\Delta_r S°$. This equation is not
valid. The $T\Delta_r S°$ term should be subtracted from the $\Delta_r H°$ term, not added to it.

53. **(M)**

(a) To determine K_p we need the equilibrium partial pressures. In the ideal gas law,
each partial pressure is defined by $P = nRT/V$. Because R, T, and V are the same
for each gas, and because there are the same number of partial pressure factors in the
numerator as in the denominator of the K_p expression, we can use the ratio of amounts
to determine K_p.

$K_p = \dfrac{P\{CO(g)\}P\{H_2O(g)\}}{P\{CO_2(g)\}P\{H_2(g)\}} = \dfrac{n\{CO(g)\}n\{H_2O(g)\}}{n\{CO_2(g)\}n\{H_2(g)\}} = \dfrac{0.224 \text{ mol CO} \times 0.224 \text{ mol H}_2O}{0.276 \text{ mol CO}_2 \times 0.276 \text{ mol H}_2} = 0.659$

(b) $\Delta_r G°_{1000K} = -RT \ln K_p = -8.3145 \text{ J mol}^{-1}\text{K}^{-1} \times 1000. \text{ K} \times \ln(0.659)$

$\quad = 3.467 \times 10^3 \text{ J/mol} = 3.467 \text{ kJ/mol}$

(c) $Q_p = \dfrac{0.0340 \text{ mol CO} \times 0.0650 \text{ mol H}_2\text{O}}{0.0750 \text{ mol CO}_2 \times 0.095 \text{ mol H}_2} = 0.31 < 0.659 = K_p$

Since Q_p is smaller than K_p, the reaction will proceed to the right, forming products, to attain equilibrium, i.e., $\Delta_r G = 0$.

55. (E)

(a) $\Delta_r G^\circ = -RT \ln K = -8.314 \times 10^{-3} \text{ kJ mol}^{-1}\text{K}^{-1} \times 718.15 \text{ K} \times \ln(50.2) = -23.4 \text{ kJ mol}^{-1}$

(b) $\Delta_r G^\circ = -RT \ln K = -8.314 \times 10^{-3} \text{ kJ mol}^{-1}\text{K}^{-1} \times 298.15 \text{ K} \times \ln(8.5 \times 10^{-13})$
$$= +69 \text{ kJ mol}^{-1}$$

(c) $\Delta_r G^\circ = -RT \ln K = -8.314 \times 10^{-3} \text{ kJ mol}^{-1}\text{K}^{-1} \times 298.15 \text{ K} \times \ln(0.114) = +5.38 \text{ kJ mol}^{-1}$

(d) $\Delta_r G^\circ = -RT \ln K = -8.314 \times 10^{-3} \text{ kJ mol}^{-1}\text{K}^{-1} \times 298.15 \text{ K} \times \ln(9.14 \times 10^{-6})$
$$= +28.8 \text{ kJ mol}^{-1}$$

57. (E) $\Delta_r G^\circ = -RT \ln K = -\left(8.3145 \times 10^{-3} \text{ kJ mol}^{-1}\text{K}^{-1}\right)(298 \text{ K}) \ln\left(6.5 \times 10^{11}\right) = -67.4 \text{ kJ/mol}$

$CO(g) + Cl_2(g) \rightarrow COCl_2(g)$	$\Delta_r G^\circ = -67.4 \text{ kJ/mol}$
$C(\text{graphite}) + \frac{1}{2}O_2(g) \rightarrow CO(g)$	$\Delta_f G^\circ = -137.2 \text{ kJ/mol}$

$C(\text{graphite}) + \frac{1}{2}O_2(g) + Cl_2(g) \rightarrow COCl_2(g)$ $\Delta_f G^\circ = -204.6 \text{ kJ/mol}$

$\Delta_f G^\circ$ of $COCl_2(g)$ given in Appendix D is -204.6 kJ/mol, thus the agreement is excellent.

59. (M)
(a) We can determine the equilibrium partial pressure from the value of the equilibrium constant.

$$\Delta_r G^\circ = -RT \ln K \qquad \ln K = -\frac{\Delta_r G^\circ}{RT} = -\frac{58.54 \times 10^3 \text{ J/mol}}{8.3145 \text{ J mol}^{-1}\text{K}^{-1} \times 298.15 \text{ K}} = -23.63$$

$$K = P\{O_2(g)\}^{1/2} = e^{-23.63} = 5.5 \times 10^{-11} \qquad P\{O_2(g)\} = \left(5.5 \times 10^{-11}\right)^2 = 3.0 \times 10^{-21} \text{ atm}$$

(b) Lavoisier did two things to increase the quantity of oxygen that he obtained. First, he ran the reaction at a high temperature, which favors the products (i.e., the side with molecular oxygen.) Second, the molecular oxygen was removed immediately after it was formed, which causes the equilibrium to shift to the right continuously (the shift towards products as result of the removal of the O_2 is an example of Le Châtelier's principle).

$\Delta_r G°$ and K as Function of Temperature

61. **(M)**

(a) $\Delta_r S° = S°\left[Na_2CO_3(s)\right] + S°\left[H_2O(l)\right] + S°\left[CO_2(g)\right] - 2S°\left[NaHCO_3(s)\right]$

$$= 135.0\frac{J}{K\ mol} + 69.91\frac{J}{K\ mol} + 213.7\frac{J}{K\ mol} - 2\left(101.7\frac{J}{K\ mol}\right) = +215.2\frac{J}{K\ mol}$$

(b) $\Delta_r H° = \Delta_f H°\left[Na_2CO_3(s)\right] + \Delta_f H°\left[H_2O(l)\right] + \Delta_f H°\left[CO_2(g)\right] - 2\Delta_f H°\left[NaHCO_3(s)\right]$

$$= -1131\frac{kJ}{mol} - 285.8\frac{kJ}{mol} - 393.5\frac{kJ}{mol} - 2\left(-950.8\frac{kJ}{mol}\right) = +91\frac{kJ}{mol}$$

(c) $\Delta_r G° = \Delta_r H° - T\Delta_r S° = 91\,kJ/mol - (298\,K)\left(215.2\times10^{-3}\,kJ\ mol^{-1}K^{-1}\right)$

$$= 91\,kJ/mol - 64.13\,kJ/mol = 27\,kJ/mol$$

(d) $\Delta_r G° = -RT\ln K$ $\qquad \ln K = -\dfrac{\Delta_r G°}{RT} = -\dfrac{27\times10^3\,J/mol}{8.3145\,J\ mol^{-1}\,K^{-1}\times298\,K} = -10.9$

$$K = e^{-10.9} = 2\times10^{-5}$$

63. **(E)** In this problem we are asked to determine the temperature for the reaction between iron(III) oxide and carbon monoxide to yield iron and carbon dioxide given $\Delta_r G°$, $\Delta_r H°$, and $\Delta_r S°$. We proceed by rearranging $\Delta_r G° = \Delta_r H° - T\Delta_r S°$ in order to express the temperature as a function of $\Delta_r G°$, $\Delta_r H°$, and $\Delta_r S°$.

$$\Delta_r G° = \Delta_r H° - T\Delta_r S° \Rightarrow T = \frac{\Delta_r H° - \Delta_r G°}{\Delta_r S°} = \frac{-24.8\times10^3\,J - \left(-45.5\times10^3\,J\right)}{15.2\,J/K} = 1.36\times10^3\,K$$

65. **(M)** We first determine the value of $\Delta_r G°$ at $400\,°C$, from the values of $\Delta_r H°$ and $\Delta_r S°$, which are calculated from information listed in Appendix D.

$\Delta_r H° = 2\Delta_f H°\left[NH_3(g)\right] - \Delta_f H°\left[N_2(g)\right] - 3\Delta_f H°\left[H_2(g)\right]$

$$= 2\left(-46.11\,kJ/mol\right) - \left(0.00\,kJ/mol\right) - 3\left(0.00\ kJ/mol\right) = -92.22\,kJ/mol\ N_2$$

$\Delta_r S° = 2S°\left[NH_3(g)\right] - S°\left[N_2(g)\right] - 3S°\left[H_2(g)\right]$

$$= 2\left(192.5\,J\ mol^{-1}\,K^{-1}\right) - \left(191.6\,J\ mol^{-1}\,K^{-1}\right) - 3\left(130.7\,J\ mol^{-1}\,K^{-1}\right) = -198.7\,J\ mol^{-1}\,K^{-1}$$

$\Delta_r G° = \Delta_r H° - T\Delta_r S° = -92.22\ kJ/mol - 673\ K\times\left(-0.1987\ kJ\ mol^{-1}\,K^{-1}\right)$

$$= +41.51\,kJ/mol = -RT\ln K_p$$

$$\ln K_p = \frac{-\Delta_r G°}{RT} = \frac{-41.51\times10^3\,J/mol}{8.3145\,J\ mol^{-1}K^{-1}\times673K} = -7.42; \quad K_p = e^{-7.42} = 6.0\times10^{-4}$$

67. **(M)** We assume that both $\Delta_r H°$ and $\Delta_r S°$ are constant with temperature.

$$\Delta_r H° = 2\Delta_f H°\left[SO_3(g)\right] - 2\Delta_f H°\left[SO_2(g)\right] - \Delta_f H°\left[O_2(g)\right]$$

$$= 2(-395.7 \text{ kJ/mol}) - 2(-296.8 \text{ kJ/mol}) - (0.00 \text{ kJ/mol}) = -197.8 \text{ kJ/mol}$$

$$\Delta_r S° = 2S°\left[SO_3(g)\right] - 2S°\left[SO_2(g)\right] - S°\left[O_2(g)\right]$$

$$= 2(256.8 \text{ J mol}^{-1}\text{ K}^{-1}) - 2(248.2 \text{ J mol}^{-1}\text{ K}^{-1}) - (205.1 \text{ J mol}^{-1}\text{ K}^{-1})$$

$$= -187.9 \text{ J mol}^{-1}\text{ K}^{-1}$$

$$\Delta_r G° = \Delta_r H° - T\Delta_r S° = -RT\ln K \qquad \Delta_r H° = T\Delta_r S° - RT\ln K \qquad T = \frac{\Delta_r H°}{\Delta_r S° - R\ln K}$$

$$T = \frac{-197.8 \times 10^3 \text{ J/mol}}{-187.9 \text{ J mol}^{-1}\text{ K}^{-1} - 8.3145 \text{ J mol}^{-1}\text{ K}^{-1}\ln(1.0 \times 10^6)} \approx 650 \text{ K}$$

This value compares very favorably with the value of $T = 6.37 \times 10^2$ K that was obtained in Example 13-10.

69. **(M)**

(a) $\ln\dfrac{K_2}{K_1} = \dfrac{\Delta_r H°}{R}\left(\dfrac{1}{T_1} - \dfrac{1}{T_2}\right) = \dfrac{57.2 \times 10^3 \text{ J/mol}}{8.3145 \text{ J mol}^{-1}\text{ K}^{-1}}\left(\dfrac{1}{298 \text{ K}} - \dfrac{1}{273 \text{ K}}\right) = -2.11$

$\dfrac{K_2}{K_1} = e^{-2.11} = 0.121 \qquad K_2 = 0.121 \times 0.113 = 0.014$ at 273 K

(b) $\ln\dfrac{K_2}{K_1} = \dfrac{\Delta_r H°}{R}\left(\dfrac{1}{T_1} - \dfrac{1}{T_2}\right) = \dfrac{57.2 \times 10^3 \text{ J/mol}}{8.3145 \text{ J mol}^{-1}\text{ K}^{-1}}\left(\dfrac{1}{T_1} - \dfrac{1}{298 \text{ K}}\right) = \ln\dfrac{0.113}{1.00} = -2.180$

$\left(\dfrac{1}{T_1} - \dfrac{1}{298\text{K}}\right) = \dfrac{-2.180 \times 8.3145}{57.2 \times 10^3}\text{ K}^{-1} = -3.17 \times 10^{-4}\text{K}^{-1}$

$\dfrac{1}{T_1} = \dfrac{1}{298} - 3.17 \times 10^{-4} = 3.36 \times 10^{-3} - 3.17 \times 10^{-4} = 3.04 \times 10^{-3} \text{ K}^{-1}; \qquad T_1 = 329 \text{ K}$

71. **(M)** First, the van't Hoff equation is used to obtain a value of $\Delta_r H°$. $200°\text{C} = 473\text{ K}$ and $260°\text{C} = 533\text{ K}$.

$$\ln\dfrac{K_2}{K_1} = \dfrac{\Delta_r H°}{R}\left(\dfrac{1}{T_1} - \dfrac{1}{T_2}\right) = \ln\dfrac{2.15 \times 10^{11}}{4.56 \times 10^8} = 6.156 = \dfrac{\Delta_r H°}{8.3145 \text{ J mol}^{-1}\text{ K}^{-1}}\left(\dfrac{1}{533 \text{ K}} - \dfrac{1}{473 \text{ K}}\right)$$

$$6.156 = -2.9 \times 10^{-5}\Delta_r H° \qquad \Delta_r H° = \dfrac{6.156}{-2.9 \times 10^{-5}} = -2.1 \times 10^5 \text{ J/mol} = -2.1 \times 10^2 \text{ kJ/mol}$$

Another route to $\Delta_r H^\circ$ is the combination of standard enthalpies of formation.

$$CO(g) + 3 H_2(g) \rightleftharpoons CH_4(g) + H_2O(g)$$

$$\Delta_r H^\circ = \Delta_f H^\circ \left[CH_4(g) \right] + \Delta_f H^\circ \left[H_2O(g) \right] - \Delta_f H^\circ \left[CO(g) \right] - 3\Delta_f H^\circ \left[H_2(g) \right]$$

$$= -74.81 \text{ kJ/mol} - 241.8 \text{ kJ/mol} - (-110.5) - 3 \times 0.00 \text{ kJ/mol} = -206.1 \text{ kJ/mol}$$

Within the precision of the data supplied, the results are in good agreement.

Coupled Reactions

73. **(E)**

(a) We compute $\Delta_r G^\circ$ for the given reaction in the following manner

$$\Delta_r H^\circ = \Delta_f H^\circ \left[TiCl_4(1) \right] + \Delta_f H^\circ \left[O_2(g) \right] - \Delta_f H^\circ \left[TiO_2(s) \right] - 2\Delta_f H^\circ \left[Cl_2(g) \right]$$

$$= -804.2 \text{ kJ/mol} + 0.00 \text{ kJ/mol} - (-944.7 \text{ kJ/mol}) - 2(0.00 \text{ kJ/mol})$$

$$= +140.5 \text{ kJ/mol}$$

$$\Delta_r S^\circ = S^\circ \left[TiCl_4(1) \right] + S^\circ \left[O_2(g) \right] - S^\circ \left[TiO_2(s) \right] - 2S^\circ \left[Cl_2(g) \right]$$

$$= 252.3 \text{ J mol}^{-1} \text{ K}^{-1} + 205.1 \text{ J mol}^{-1} \text{ K}^{-1} - (50.33 \text{ J mol}^{-1} \text{ K}^{-1}) - 2(223.1 \text{ J mol}^{-1} \text{ K}^{-1})$$

$$= -39.1 \text{ J mol}^{-1} \text{ K}^{-1}$$

$$\Delta_r G^\circ = \Delta_r H^\circ - T \Delta_r S^\circ = +140.5 \text{ kJ/mol} - (298 \text{ K})(-39.1 \times 10^{-3} \text{ kJ mol}^{-1} \text{ K}^{-1})$$

$$= +140.5 \text{ kJ/mol} + 11.6 \text{ kJ/mol} = +152.1 \text{ kJ/mol}$$

Thus the reaction is non-spontaneous at $25 \,^\circ\text{C}$. (we also could have used values of $\Delta_f G^\circ$ to calculate $\Delta_r G^\circ$).

(b) For the cited reaction, $\Delta_r G^\circ = 2\Delta_f G^\circ \left[CO_2(g) \right] - 2\Delta_f G^\circ \left[CO(g) \right] - \Delta_f G^\circ \left[O_2(g) \right]$

$$\Delta_r G^\circ = 2(-394.4 \text{ kJ/mol}) - 2(-137.2 \text{ kJ/mol}) - 0.00 \text{ kJ/mol} = -514.4 \text{ kJ/mol}$$

Then we couple the two reactions.

$$TiO_2(s) + 2Cl_2(g) \longrightarrow TiCl_4(1) + O_2(g) \qquad\qquad \Delta_r G^\circ = +152.1 \text{ kJ/mol}$$

$$2CO(g) + O_2(g) \longrightarrow 2CO_2(g) \qquad\qquad \Delta_r G^\circ = -514.4 \text{ kJ/mol}$$

$$TiO_2(s) + 2Cl_2(g) + 2CO(g) \longrightarrow TiCl_4(1) + 2CO_2(g); \Delta_r G^\circ = -362.3 \text{ kJ/mol}$$

The coupled reaction has $\Delta_r G^\circ < 0$, and, therefore, is spontaneous.

75. **(E)** In this problem we need to determine if the phosphorylation of arginine with ATP is a spontaneous reaction. We proceed by coupling the two given reactions in order to calculate $\Delta_r G^\circ$ for the overall reaction. The sign of $\Delta_r G^\circ$ can then be used to determine whether the reaction is spontaneous or not.

Stepwise approach:

First determine $\Delta_r G°$ for the coupled reaction:

$$ATP + H_2O \longrightarrow ADP + P \qquad \Delta_r G° = -31.5 \text{ kJ mol}^{-1}$$

$$arginine + P \longrightarrow phosphorarginine + H_2O \qquad \Delta_r G° = -33.2 \text{ kJ mol}^{-1}$$

$$ATP + arginine \longrightarrow phosphorarginine + ADP$$

$$\Delta_r G° = (-31.5 + 33.2) \text{ kJ mol}^{-1} = 1.7 \text{ kJ mol}^{-1}$$

Examine the sign of $\Delta_r G°$:

$\Delta_r G° > 0$. Therefore, the reaction is not spontaneous.

Conversion pathway approach:

$\Delta_r G°$ for the coupled reaction is:

$$ATP + arginine \longrightarrow phosphorarginine + ADP$$

$$\Delta_r G° = (-31.5 + 33.2) \text{ kJ mol}^{-1} = 1.7 \text{ kJ mol}^{-1}$$

Since $\Delta_r G° > 0$, the reaction is not spontaneous.

INTEGRATIVE AND ADVANCED EXERCISES

<u>**77.**</u>   **(D)** For an adiabatic process, no heat is exchanged with the surroundings. Therefore, $\Delta S_{surr} = 0$. For a spontaneous process, we must have $\Delta S_{univ} = \Delta S + \Delta S_{surr} > 0$. If $\Delta S_{surr} = 0$, then we must have $\Delta S > 0$ for the system. In other words, for a spontaneous adiabatic process only, ΔS_{surr} is equal to zero. Only for a reversible, adiabatic process will we have $\Delta S = \Delta S_{surr} = \Delta S_{univ} = 0$.

<u>**82.**</u>   **(M)**

(a) **TRUE;**   It is the change in Gibbs energy for a process in which reactants and products are all in their standard states (regardless of whatever states might be mentioned in the statement of the problem). When liquid and gaseous water are each at 1 atm at 100 °C (the normal boiling point), they are in equilibrium, so that $\Delta_r G = \Delta_r G° = 0$ is only true when the difference of the standard free energies of products and reactants is zero. A reaction with $\Delta_r G° = 0$ would be at equilibrium when products and reactants were all present under standard state conditions and the pressure of $H_2O(g) = 2.0$ atm is not the standard pressure for $H_2O(g)$.

(b) **FALSE;** $\Delta_r G \ne 0$. The system is not at equilibrium.

(c) **FALSE;** $\Delta_r G°$ can have only one value at any given temperature, and that is the value corresponding to all reactants and products in their standard states, so at the normal boiling point $\Delta_r G° = 0$ [as was also the case in answering part (a)]. Water will not vaporize spontaneously under standard conditions to produce water vapor with a pressure of 2 atm.

(d) TRUE; $\Delta_r G > 0$. The process of transforming water to vapor at 2.0 atm pressure at 100°C is not a spontaneous process; the condensation (reverse) process is spontaneous. (i.e. for the system to reach equilibrium, some $H_2O(l)$ must form)

83. **(D)** $\Delta_r G° = +\frac{1}{2}\Delta_f G°[Br_2(g)] + \frac{1}{2}\Delta_f G°[Cl_2(g)] - \Delta_f G°[BrCl(g)]$

$= +\frac{1}{2}(3.11\,kJ/mol) + \frac{1}{2}(0.00\,kJ/mol) - (-0.98\,kJ/mol) = +2.54\,kJ/mol = -RT \ln K_p$

$\ln K_p = -\dfrac{\Delta G°}{RT} = -\dfrac{2.54\times10^3\,J/mol}{8.3145\,J\,mol^{-1}\,K^{-1}\times298.15\,K} = -1.02 \quad K_p = e^{-1.02} = 0.361$

For ease of solving the problem, we double the reaction, which squares the value of the equilibrium constant. $K_{eq} = (0.357)^2 = 0.130$

Reaction: $2\,BrCl(g) \rightleftharpoons Br_2(g) + Cl_2(g)$

Initial: 1.00 mol   0 mol   0 mol

Changes: $-2x$ mol $+x$ mol $+x$ mol

Equil: $(1.00-2x)$mol x mol x mol

$K_p = \dfrac{P\{Br_2(g)\}\,P\{Cl_2(g)\}}{P\{BrCl(g)\}^2} = \dfrac{[n\{Br_2(g)\}RT/V][n\{Cl_2(g)\}RT/V]}{[n\{BrCl(g)\}RT/V]^2}$

$= \dfrac{n\{Br_2(g)\}\,n\{Cl_2(g)\}}{n\{BrCl(g)\}^2}$

$= \dfrac{x^2}{(1.00-2x)^2} = (0.361)^2 \qquad \dfrac{x}{1.00-2x} = 0.361 \qquad x = 0.361 - 0.722\,x$

$x = \dfrac{0.361}{1.722} = 0.210\,mol\,Br_2 = 0.210\,mol\,Cl_2 \quad 1.00-2x = 0.580\,mol\,BrCl$

87. **(M)** First we need a value for the equilibrium constant. 1% conversion means that 0.99 mol $N_2(g)$ are present at equilibrium for every 1.00 mole present initially.

$K = K_p = \dfrac{P_{NO(g)}^2}{P_{N_2(g)}\,P_{O_2(g)}} = \dfrac{[n\{NO(g)\}RT/V]^2}{[n\{N_2(g)\}RT/V][n\{O_2(g)\}RT/V]} = \dfrac{n\{NO(g)\}^2}{n\{N_2(g)\}\,n\{O_2(g)\}}$

Reaction: $N_2(g) + O_2(g) \rightleftharpoons 2\,NO(g)$

Initial: 1.00 mol   1.00 mol   0 mol

Changes(1% rxn): -0.010 mol -0.010 mol $+0.020$ mol $K = \dfrac{(0.020)^2}{(0.99)(0.99)} = 4.1\times10^{-4}$

Equil: 0.99 mol   0.99 mol   0.020 mol

The cited reaction is twice the formation reaction of NO(g), and thus

$\Delta_r H° = 2\Delta_f H°[NO(g)] = 2\times90.25\,kJ/mol = 180.50\,kJ/mol$

$\Delta_r S° = 2\,S°[NO(g)] - S°[N_2(g)] - S°[O_2(g)]$

$= 2(210.7\,J\,mol^{-1}\,K) - 191.5\,J\,mol^{-1}\,K^{-1} - 205.0\,J\,mol^{-1}\,K^{-1} = 24.9\,J\,mol^{-1}\,K^{-1}$

$$\Delta_r G° = -RT \ln K = -8.31447 \, JK^{-1}mol^{-1}(T)\ln(4.1\times10^{-4}) = 64.85(T)$$

$$\Delta_r G° = \Delta_r H° - T\Delta_r S° = 64.85(T) = 180.5 \, kJ/mol - (T)24.9 \, J \, mol^{-1} \, K^{-1}$$

$$180.5\times10 \, J/mol = 64.85(T) + (T)24.9 \, J \, mol^{-1} \, K^{-1} = 89.75(T) \qquad T = 2.01\times10^3 \, K$$

90. **(M)** In this problem we are asked to estimate the temperature at which the vapor pressure of cyclohexane is 100 mmHg. We begin by using Trouton's rule to determine the value of $\Delta_{vap}H$ for cyclohexane. The temperature at which the vapor pressure is 100.00 mmHg can then be determined using Clausius–Clapeyron equation.

Use Trouton's rule to find the value of $\Delta_{vap}H$:

$$\Delta_{vap}H = T_{nbp} \Delta_{vap}S = 353.9 \, K \times 87 \, J \, mol^{-1} \, K^{-1} = 31\times10^3 \, J/mol$$

Next, use Clausius–Clapeyron equation to find the required temperature:

$$\ln\frac{P_2}{P_1} = \frac{\Delta_{vap}H}{R}\left(\frac{1}{T_1} - \frac{1}{T_2}\right) = \ln\frac{100 \, mmHg}{760 \, mmHg}$$

$$= \frac{31\times10^3 \, J/mol}{8.3145 \, J \, mol^{-1} \, K^{-1}}\left(\frac{1}{353.9 \, K} - \frac{1}{T}\right) = -2.028$$

$$\frac{1}{353.9} - \frac{1}{T} = \frac{-2.028\times8.3145}{31\times10^3} = -5.4\times10^{-4} =$$

$$2.826\times10^{-3} - \frac{1}{T} \qquad \frac{1}{T} = 3.37\times10^{-3} \, K^{-1}$$

$$T = 297 \, K = 24 \, °C$$

93. **(D)** First we determine the value of K_p that corresponds to 15% dissociation. We represent the initial pressure of phosgene as x atm.

Reaction: $\quad COCl_2(g) \quad \rightleftharpoons \quad CO(g) \quad + \quad Cl_2(g)$

Initial: $\qquad x$ atm $\qquad\qquad$ 0 atm $\qquad\qquad$ 0 atm

Changes: $\quad -0.15\,x$ atm $\qquad +0.15\,x$ atm $\qquad +0.15\,x$ atm

Equil: $\qquad 0.85\,x$ atm $\qquad 0.15\,x$ atm $\qquad 0.15\,x$ atm

$$P_{total} = 0.85\,x \, atm + 0.15\,x \, atm + 0.15\,x \, atm = 1.15\,x \, atm = 1.00 \, atm$$

$$x = \frac{1.00}{1.15} = 0.870 \, atm$$

$$K_p = \frac{P_{CO}\,P_{Cl_2}}{P_{COCl_2}} = \frac{(0.15\times0.870)^2}{0.85\times0.870} = 0.0230$$

Next we find the value of $\Delta H°$ for the decomposition reaction.

$$\ln\frac{K_1}{K_2} = \frac{\Delta_r H°}{R}\left(\frac{1}{T_2} - \frac{1}{T_1}\right) = \ln\frac{6.7\times10^{-9}}{4.44\times10^{-2}} = -15.71 = \frac{\Delta_r H°}{R}\left(\frac{1}{668} - \frac{1}{373.0}\right)$$

$$= \frac{\Delta_r H°}{R}(-1.18\times10^{-3})$$

$$\frac{\Delta_r H^\circ}{R} = \frac{-15.71}{-1.18 \times 10^{-3}} = 1.33 \times 10^4,$$

$$\Delta_r H^\circ = 1.33 \times 10^4 \times 8.3145 = 111 \times 10^3 \text{ J/mol} = 111 \text{ kJ/mol}$$

And finally we find the temperature at which $K = 0.0230$.

$$\ln \frac{K_1}{K_2} = \frac{\Delta_r H^\circ}{R}\left(\frac{1}{T_2} - \frac{1}{T_1}\right) = \ln \frac{0.0230}{0.0444} = \frac{111 \times 10^3 \text{ J/mol}}{8.3145 \text{ J mol}^{-1}\text{K}^{-1}}\left(\frac{1}{668 \text{ K}} - \frac{1}{T}\right) = -0.658$$

$$\frac{1}{668} - \frac{1}{T} = \frac{-0.658 \times 8.3145}{111 \times 10^3} = -4.93 \times 10^{-5} = 1.497 \times 10^{-3} - \frac{1}{T} \qquad \frac{1}{T} = 1.546 \times 10^{-3}$$

$$T = 647 \text{ K} = 374\,^\circ\text{C}$$

94. From the given expressions for S° and $\Delta_r H^\circ$, we can calculate the ratio $S^\circ/\Delta_r H^\circ$:

$$\frac{S^\circ}{\Delta_r H^\circ} = \frac{\displaystyle\int_{0K}^{298.15K} \frac{aT}{T}dT}{\displaystyle\int_{0K}^{298.15K} aT\,dT} = \frac{a\displaystyle\int_{0K}^{298.15K} dT}{a\displaystyle\int_{0K}^{298.15K} T\,dt} = \frac{aT\,|_{0K}^{298.15K}}{\frac{1}{2}aT^2\,|_{0K}^{298.15K}} =$$

$$= \frac{2}{T}\Big|_{0K}^{298.15K} = \frac{2}{(298.15 \text{ K} - 0 \text{ K})} = \frac{2}{298.15 \text{ K}} = 0.0067 \text{ K}^{-1}$$

98. **(M) (a)** In the solid as the temperature increases, so do the translational, rotational, and vibrational degrees of freedom. In the liquid, most of the vibrational degrees of freedom are saturated and only translational and rotational degrees of freedom can increase. In the gas phase, all degrees of freedom are saturated. **(b)** The increase in translation and rotation on going from solid to liquid is much less than on going from liquid to gas. This is where most of the change in entropy is derived.

FEATURE PROBLEMS

100. **(M)**

(a) The first method involves combining the values of $\Delta_f G^\circ$. The second uses

$$\Delta_r G^\circ = \Delta_r H^\circ - T\Delta_r S^\circ$$

$$\Delta_r G^\circ = \Delta_f G^\circ\big[H_2O(g)\big] - \Delta_f G^\circ\big[H_2O(l)\big]$$

$$= -228.572 \text{ kJ/mol} - (-237.129 \text{ kJ/mol}) = +8.557 \text{ kJ/mol}$$

$$\Delta_r H^\circ = \Delta_f H^\circ\big[H_2O(g)\big] - \Delta_f H^\circ\big[H_2O(l)\big]$$

$$= -241.818 \text{ kJ/mol} - (-285.830 \text{ kJ/mol}) = +44.012 \text{ kJ/mol}$$

$$\Delta_r S° = S°\big[H_2O(g)\big] - S°\big[H_2O(l)\big]$$

$$= 188.825 \text{ J mol}^{-1}\text{ K}^{-1} - 69.91 \text{ J mol}^{-1}\text{ K}^{-1} = +118.92 \text{ J mol}^{-1}\text{ K}^{-1}$$

$$\Delta_r G° = \Delta_r H° - T\Delta_r S°$$

$$= 44.012 \text{ kJ/mol} - 298.15 \text{ K} \times 118.92 \times 10^{-3} \text{ kJ mol}^{-1}\text{ K}^{-1} = +8.556 \text{ kJ/mol}$$

(b) We use the average value: $\Delta_r G° = +8.558 \times 10^3 \text{ J/mol} = -RT \ln K$

$$\ln K = -\frac{8558 \text{ J/mol}}{8.3145 \text{ J mol}^{-1}\text{ K}^{-1} \times 298.15 \text{ K}} = -3.452; \quad K = e^{-3.452} = 0.0317 \text{ bar}$$

(c) $$P\{H_2O\} = 0.0317 \text{ bar} \times \frac{1 \text{ atm}}{1.01325 \text{ bar}} \times \frac{760 \text{ mmHg}}{1 \text{ atm}} = 23.8 \text{ mmHg}$$

(d) $$\ln K = -\frac{8590 \text{ J/mol}}{8.3145 \text{ J mol}^{-1}\text{ K}^{-1} \times 298.15 \text{ K}} = -3.465;$$

$$K = e^{-3.465} = 0.0312_7 \text{ atm};$$

$$P\{H_2O\} = 0.0313 \text{ atm} \times \frac{760 \text{ mmHg}}{1 \text{ atm}} = 23.8 \text{ mmHg}$$

101. (D)

(a) When we combine two reactions and obtain the overall value of $\Delta_r G°$, we subtract the value on the plot of the reaction that becomes a reduction from the value on the plot of the reaction that is an oxidation. Thus, to reduce ZnO with elemental Mg, we subtract the values on the line labeled "$2Zn + O_2 \rightarrow 2ZnO$" from those on the line labeled "$2Mg + O_2 \rightarrow 2MgO$". The result for the overall $\Delta G°$ will always be negative because every point on the "zinc" line is above the corresponding point on the "magnesium" line

(b) In contrast, the "carbon" line is only below the "zinc" line at temperatures above about 1000 °C. Thus, only at these elevated temperatures can ZnO be reduced by carbon.

(c) The decomposition of zinc oxide to its elements is the reverse of the plotted reaction, the value of $\Delta_r G°$ for the decomposition becomes negative, and the reaction becomes spontaneous, where the value of $\Delta_r G°$ for the plotted reaction becomes positive. This occurs above about 1850 °C.

(d) The "carbon" line has a negative slope, indicating that carbon monoxide becomes more stable as temperature rises. The point where CO(g) would become less stable than 2C(s) and O_2(g) looks to be below −1000 °C (by extrapolating the line to lower temperatures). Based on this plot, it is not possible to decompose CO(g) to C(s) and $O_2(g)$ in a spontaneous reaction.

(e)

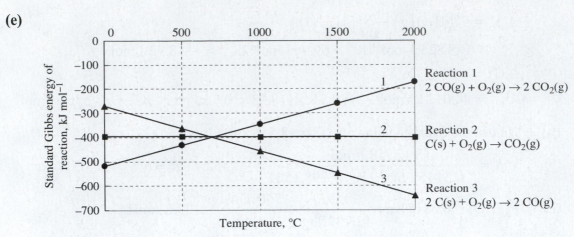

All three lines are straight-line plots of $\Delta_r G^\circ$ vs. T following the equation $\Delta_r G^\circ = \Delta_r H^\circ - T\Delta_r S^\circ$.

The general equation for a straight line is given below with the slightly modified Gibbs energy equation as a reference: $\Delta_r G^\circ = -\Delta_r S^\circ\, T + \Delta_r H^\circ$ (here $\Delta_r H^\circ$ assumed constant)

$y = mx + b$ (m $= -\Delta_r S^\circ$ = slope of the line)

Thus, the slope of each line multiplied by minus one is equal to the $\Delta_r S^\circ$ for the oxide formation reaction. It is hardly surprising, therefore, that the slopes for these lines differ so markedly because these three reactions have quite different $\Delta_r S^\circ$ values ($\Delta_r S^\circ$ for Reaction 1 = –173 J K^{-1}, $\Delta_r S^\circ$ for Reaction 2 = 2.86 J K^{-1}, $\Delta_r S^\circ$ for Reaction 3 = 178.8 J K^{-1})

(f)   Since other metal oxides apparently have positive slopes similar to Mg and Zn, we can conclude that in general, the stability of metal oxides <u>decreases</u> as the temperature increases. Put another way, the decomposition of metal oxides to their elements becomes more spontaneous as the temperature is increased. By contrast, the two reactions involving elemental carbon, namely Reaction 2 and Reaction 3, have negative slopes, indicating that the formation of $CO_2(g)$ and $CO(g)$ from graphite becomes more favorable as the temperature rises. This means that the $\Delta_r G^\circ$ for the reduction of metal oxides by carbon becomes more and more negative with increasing temperature. Moreover, there must exist a threshold temperature for each metal oxide above which the reaction with carbon will occur spontaneously. Carbon would appear to be an excellent reducing agent, therefore, because it will reduce virtually <u>any</u> metal oxide to its corresponding metal as long as the temperature chosen for the reaction is higher than the threshold temperature (the threshold temperature is commonly referred to as the transition temperature).

Consider for instance the reaction of MgO(s) with graphite to give $CO_2(g)$ and Mg metal:
$2\ MgO(s) + C(s) \rightarrow 2\ Mg(s) + CO_2(g)$ $\Delta_r S^\circ$ = 219.4 J/K and $\Delta_r H^\circ$ = 809.9 kJ

$$T_{transition} = \frac{\Delta_r H^\circ}{\Delta_r S^\circ} = \frac{809.9\ kJ}{0.2194\ kJ\ K^{-1}} = 3691\ K = T_{threshold}$$

Consequently, above 3691 K, carbon will spontaneously reduce MgO to Mg metal.

104. (E)

(a) In this case CO can exist in two states, therefore, W = 2. There are N of these states in the crystal, and so we have

$$S = k \ln 2^N = 1.381 \times 10^{-23} JK^{-1} \times 6.022 \times 10^{23} mol^{-1} \ln 2 = 5.8 JK^{-1} mol^{-1}$$

(b) Let's calculate W. A mole of water has $2N$ hydrogen nuclei. If each nucleus had the choice of two positions along its O—O axis, one close to one oxygen atom and the other closer to the second oxygen atom, there would be 2^{2N} configurations. Many of these configurations are ruled out by the condition that each oxygen atom has two attached hydrogen atoms. Let's consider a particular oxygen atom and the four surrounding hydrogen atoms. There are a total of 16 arrangements of this OH_4 group; one with all four hydrogen atoms close to the oxygen atom, corresponding to $(H_4O)^{2+}$, four corresponding to $(H_3O)^+$, six to H_2O, four to OH^-, and one to O^{2-}. The acceptable arrangements assigning two strongly bonded hydrogen atoms to this oxygen atom accordingly compose six-sixteenths or three-eighths of the total possible arrangements. Of these arrangements, only three-eighths are suitable with respect to the second oxygen atom, and so forth; the number of configurations, W, is hence $2^{2N}(3/8)^N = (3/2)^N$. For water, W = 3/2, which leads to

$$S = k \ln(\frac{3}{2})^N = 1.381 \times 10^{-23} JK^{-1} \times 6.022 \times 10^{23} mol^{-1} \ln 1.5 = 3.4 JK^{-1} mol^{-1}$$

SELF-ASSESSMENT EXERCISES

108. (E) Second law of thermodynamics states that all spontaneous processes produce an increase in the entropy of the universe. In other words, $\Delta_r S_{univ} = \Delta_r S_{sys} + \Delta_r S_{surr} > 0$. Therefore, the correct answer is (d).

109. (E) The Gibbs energy of reaction cannot be used to determine how much heat is absorbed from the surroundings or how much work the system does on the surroundings. Furthermore, it also cannot be used to determine the proportion of the heat evolved in an exothermic reaction that can be converted to various forms of work. The sign of the Gibbs energy of reaction can be used to determine whether forward or reverse reaction is spontaneous for the given conditions: If $\Delta_r G < 0$, the forward reaction is spontaneous; if $\Delta_r G > 0$, the reverse reaction is spontaneous. Therefore, the correct answer is (c).

110. (M) To answer this question, we must first determine whether the entropy change for the given reaction is positive or negative. The reaction produces three moles of gas from two moles; therefore, the entropy change for the reaction is positive. The Gibbs energy of a reaction is a function of enthalpy, entropy, and temperature ($\Delta_r G = \Delta_r H - T\Delta_r S$). Since $\Delta_r H < 0$ and $\Delta_r S > 0$, this reaction will be spontaneous at any temperature. The correct answer is (a).

111. (M) Recall that $\Delta_r G° = -RT \ln K$. If $\Delta_r G° = 0$, then it follows that $\Delta_r G° = -RT \ln K = 0$. Solving for K yields: $\ln K = 0 \Rightarrow K = e^0 = 1$. Therefore, the correct answer is (b).

112. **(E)** In this reaction, the number of moles of reactants equals the number of moles of products. Therefore, K is equal to K_p and K_c. The correct answers are (a) and (d).

113. **(M)**

(a) No reaction is expected because of the decrease in entropy and the expectation that the reaction is endothermic. As a check with data from Appendix D, $\Delta_r G° = 326.4$ kJ mol^{-1} for the reaction as written—a very large value.

(b) Based on the increase in entropy, the forward reaction should occur, at least to some extent. For this reaction $\Delta_r G° = 75.21$ kJ mol^{-1}.

(c) $\Delta_r S$ is probably small, and $\Delta_r H$ is probably also small (one Cl-Cl bond and one Br-Br bonds are broken and two Br-Cl bonds are formed). $\Delta_r G°$ should be small and the forward reaction should occur to a significant extent. For this reaction $\Delta_r G° = -5.07$ kJmol^{-1}.

114. **(M)**

(a) Entropy change must be accessed for the system and its surroundings ($\Delta_r S_{univ}$), not just for the system alone.

(b) Equilibrium constant can be calculated from $\Delta_r G°$ ($\Delta_r G° = -RT \ln K$), and K permits equilibrium calculations for nonstandard conditions.

115. **(D)**

(a) First we need to determine $\Delta_{vap} H°$ which is simply equal to:

$$\Delta_{vap} H° = \Delta_f H°[(C_5 H_{10}(g)] - \Delta_f H°[(C_5 H_{10}(l)] = -77.2\text{kJ/mol} - (-105.9\text{kJ/mol})$$

$= 28.7$kJ/mol. Now we use Trouton's rule to calculate the boiling point of cyclopentane:

$$\Delta_{vap}S° = \frac{\Delta_{vap} H°}{T_{bp}} = 87 \text{ J mol}^{-1}\text{ K}^{-1} \Rightarrow T_{bp} = \frac{\Delta_{vap} H°}{87 \text{ J mol}^{-1}\text{ K}^{-1}} = \frac{28.7 \times 1000 \text{ J mol}^{-1}}{87 \text{ J mol}^{-1}\text{ K}^{-1}} = 330 \text{ K}$$

$T_{bp} = 330 \text{ K} - 273.15 \text{ K} = 57 °C$

(b) If we assume that $\Delta_{vap} H°$ and $\Delta_{vap} S°$ are independent of T we can calculate $\Delta_{vap} G°$:

$$\Delta_{vap, 298 \text{ K}}G° = \Delta_{vap}H° - T\Delta_{vap}S° = 28.7 \text{ kJ mol}^{-1} - 298.15K \times \frac{87}{1000} \text{kJK}^{-1}\text{mol}^{-1} = 2.8 \text{ kJ mol}^{-1}$$

(c) Because $\Delta_{vap, 298 \text{ K}}G° > 0$, the vapor pressure is less than 1 atm at 298 K, consistent with $T_{bp} = 57 °C$.

116. (M)

(a) We can use the data from Appendix D to determine the change in enthalpy and entropy for the reaction:

$\Delta_r H° = \Delta_f H°(N_2O(g)) + 2\Delta_f H°(H_2O(l)) - \Delta_f H°(NH_4NO_3(s))$

$\Delta_r H° = 82.05 \text{ kJ mol}^{-1} + 2 \times (-285.8 \text{ kJ mol}^{-1}) - (-365.6 \text{ kJ mol}^{-1}) = -124 \text{ kJ mol}^{-1}$

$\Delta_r S° = S°(N_2O(g)) + 2S°(H_2O(l)) - S°(NH_4NO_3(s))$

$\Delta_r S° = 219.9 \text{ J K}^{-1} \text{ mol}^{-1} + 2 \times 69.91 \text{ J K}^{-1} \text{ mol}^{-1} - 151.1 \text{ J K}^{-1} \text{ mol}^{-1} = 208.6 \text{ J K}^{-1} \text{ mol}^{-1}$

The forward reaction is exothermic.

(b) From the values of $\Delta_r H°$ and $\Delta_r S°$ determined in part (a) we can calculate $\Delta_r G°$ at 298 K:

$\Delta_r G° = \Delta_r H° - T\Delta_r S°$

$\Delta_r G° = -124 \text{ kJ mol}^{-1} - 298 \text{ } K \times \dfrac{208.6 \text{ kJ mol}^{-1} \text{ K}^{-1}}{1000} = -186.1 \text{ kJ mol}^{-1}$

Alternatively, $\Delta_r G°$ can also be calculated directly using $\Delta_f G°$ values tabulated in Appendix D.

(c) The equilibrium constant for the reaction is calculated using $\Delta_r G° = -RT \ln K$:

$\Delta_r G° = -RT \ln K \Rightarrow -186.1 \times 1000 \text{ J mol}^{-1} = -8.314 \text{ J K}^{-1} \text{ mol}^{-1} \times 298 \text{ K} \times \ln K$

$-186100 \text{ J mol}^{-1} = -2477.6 \ln K \Rightarrow \ln K = 75.1 \Rightarrow K = e^{75.1} = 4.1 \times 10^{32}$

(d) The reaction has $\Delta_r H° < 0$ and $\Delta_r S° > 0$. Because $\Delta_r G° = \Delta_r H° - T\Delta_r S°$, the reaction will be spontaneous at all temperatures.

117. (M) Recall from Exercise 110 that $\Delta_r G° = 0$ when K = 1. Therefore, we are looking for the graph with smallest change in Gibbs energy between the products and the reactants. The correct answer is graph (a). Notice that graphs (b) and (c) represent chemical reactions with small and large values of equilibrium constants, respectively.

118. (M) Carbon dioxide is a gas at room temperature. The melting point of carbon dioxide is expected to be very low. At room temperature and normal atmospheric pressure this process is spontaneous. The entropy of the universe if positive.

CHAPTER 14
SOLUTIONS AND THEIR PHYSICAL PROPERTIES
PRACTICE EXAMPLES

1A **(E)** To determine mass percent, we need both the mass of ethanol and the mass of solution. From volume percent, we know that 100.0 mL of solution contains 20.0 mL pure ethanol. The density of pure ethanol is 0.789 g/mL. We now can determine the mass of solute (ethanol) and solution. We perform the calculation in one step.

$$\text{mass percent ethanol} = \frac{20.0 \text{ mL ethanol} \times \dfrac{0.789 \text{ g}}{1 \text{ mL ethanol}}}{100.0 \text{ mL soln} \times \dfrac{0.977 \text{ g}}{1 \text{ mL soln}}} \times 100\% = 16.2\% \text{ ethanol by mass}$$

2A **(M)** First we need to find the amount of each component in solution. Let us consider a 100.00-g sample of solution, in which there are 16.00 g glycerol and 84.00 g water. The shorthand notation for glycerol, $HOCH_2CH(OH)CH_2OH$, is $C_3H_5(OH)_3$

$$\text{amount of glycerol} = 16.00 \text{ g } C_3H_5(OH)_3 \times \frac{1 \text{ mol } C_3H_5(OH)_3}{92.10 \text{ g } C_3H_5(OH)_3} = 0.1737 \text{ mol } C_3H_5(OH)_3$$

$$\text{amount of water} = 84.00 \text{ g } H_2O \times \frac{1 \text{ mol } H_2O}{18.02 \text{ g } H_2O} = 4.661 \text{ mol } H_2O$$

$$\text{mole fraction of } C_3H_5(OH)_3 = \frac{n_{C_3H_5(OH)_3}}{n_{C_3H_5(OH)_3} + n_{H_2O}} = \frac{0.1737 \text{ mol } C_3H_5(OH)_3}{0.1737 \text{ mol } C_3H_5(OH)_3 + 4.661 \text{ mol } H_2O}$$

$$\text{mole fraction of } C_3H_5(OH)_3 = 0.03593$$

2B **(M)** First we need the amount of sucrose in solution. We use a 100.00-g sample of solution, in which there are 10.00 g sucrose and 90.00 g water.

$$\text{amount } C_{12}H_{22}O_{11} = 10.00 \text{ g } C_{12}H_{22}O_{11} \times \frac{1 \text{ mol } C_{12}H_{22}O_{11}}{342.30 \text{ g } C_{12}H_{22}O_{11}} = 0.02921 \text{ mol } C_{12}H_{22}O_{11}$$

(a) Molarity is amount of solute in moles per liter of solution. Convert the 100.00 g of solution to L with density as a conversion factor.

$$C_{12}H_{22}O_{11} \text{ molarity} = \frac{0.02921 \text{ mol } C_{12}H_{22}O_{11}}{1000. \text{ g soln}} \times \frac{1.040 \text{ g soln}}{1 \text{ mL}} \times \frac{1000 \text{ mL}}{1 \text{ L}} = 0.3038 \text{ M}$$

(b) Molality is amount of solute in moles per kilogram of solvent. Convert 90.00 g of solvent to kg.

$$C_{12}H_{22}O_{11} \text{ molality} = \frac{0.02921 \text{ mol } C_{12}H_{22}O_{11}}{90.00 \text{ g } H_2O} \times \frac{1000 \text{ g}}{1 \text{ kg}} = 0.3246 \text{ mol kg}^{-1}$$

(c) Mole fraction is the moles of solute per moles of solution. First compute the moles in 90.00 g H_2O.

$$n_{H_2O} = 90.00 \text{ g} \times \frac{1 \text{ mol } H_2O}{18.02 \text{ g } H_2O} = 4.994 \text{ mol } H_2O$$

$$\text{mole fraction } C_{12}H_{22}O_{11} = \frac{0.02921 \text{ mol } C_{12}H_{22}O_{11}}{0.02921 \text{ mol } C_{12}H_{22}O_{11} + 4.994 \text{ mol } H_2O} = 0.005815$$

3A **(E)** Water is a highly polar compound. In fact, water molecules bond to each other through hydrogen bonds, which are unusually strong dipole–dipole interactions. Thus, water should mix well with other polar, hydrogen bonding compounds. (a) Toluene is nonpolar and should not be very soluble in water. (c) Benzaldehyde can form hydrogen bonds to water through its O atom. However, most of the molecule is nonpolar and, as a result, it has limited solubility in water. (b) Oxalic acid is polar and can form hydrogen bonds. Of these three compounds, oxalic acid should be the most readily soluble in water. Actual solubilities (w/w%) are: toluene (0.067%) < benzaldehyde (0.28%) < oxalic acid (14%).

3B **(E)** Both I_2 and CCl_4 are nonpolar molecules. It does not take much energy to break the attractions among I_2 molecules, or among CCl_4 molecules. Also, there is not a strong I_2–CCl_4 attraction created when a solution forms. Thus, I_2 should dissolve well in CCl_4 by simple mixing. H_2O is extensively hydrogen bonded with strong intermolecular forces that are difficult to break, but there is not a strong I_2–H_2O attraction created when a solution forms. Thus, we expect I_2 to dissolve poorly in water. Actual solubilities are: 2.603 g I_2/100 g CCl_4 and 0.033 g I_2/100 g H_2O.

4A **(M)** The two suggestions are quoted first, followed by the means for achieving each one.
(1) Dissolve the 95 g NH_4Cl in just enough water to produce a saturated solution (55 g NH_4Cl/100 g H_2O) at 60 °C.

$$\text{mass of water needed} = 95 \text{ g } NH_4Cl \times \frac{100 \text{ g } H_2O}{55 \text{ g } NH_4Cl} = 173 \text{ g } H_2O$$

The mass of NH_4Cl in the saturated solution at 20 °C will be smaller.

$$\text{mass } NH_4Cl \text{ dissolved} = 173 \text{ g } H_2O \times \frac{37 \text{ g } NH_4Cl}{100 \text{ g } H_2O} = 64 \text{ g } NH_4Cl \text{ dissolved}$$

$$\text{crystallized mass } NH_4Cl = 95 \text{ g } NH_4Cl \text{ total} - 64 \text{ g } NH_4Cl \text{ dissolved at } 20\,°C$$
$$= 31 \text{ g } NH_4Cl \text{ crystallized}$$

(2) Lower the final temperature to 0 °C, rather than 20 °C. From Figure 14-10, at 0 °C, the solubility of NH_4Cl is 28.5 g NH_4Cl/100 g H_2O. From this (and knowing that there are 173 g H_2O present in the solution) we calculate the mass of NH_4Cl dissolved at this lower temperature.

$$\text{mass dissolved } NH_4Cl = 173 \text{ g } H_2O \times \frac{28.5 \text{ g } NH_4Cl}{100 \text{ g } H_2O} = 49.3 \text{ g } NH_4Cl \text{ dissolved}$$

The mass of NH_4Cl recrystallized is $95\ g - 49.3\ g = 46\ g$

$$\text{yield} = \left(\frac{46\ g}{95\ g}\right) \times 100\% = 48\%$$

4B (M) Percent yield for the recrystallization can be defined as:

$$\% \text{ yield} = \frac{\text{mass crystallized}}{\text{mass dissolved}(40\ °C)} \times 100\%$$

$$\% \text{ yield} = \frac{\text{mass dissolved}(\text{at } 40\ °C) - \text{mass dissolved}(\text{at } 20\ °C)}{\text{mass dissolved}(\text{at } 40\ °C)} \times 100\%$$

Figure 14-10 solubilities per 100 g H_2O are followed by percent yield calculations. Solubility of $KClO_4$: 4.84 g at 40 °C and 3.0 g at 20 °C

$$\text{Percent yield of } KClO_4 = \frac{4.84\ g - 3.0\ g}{4.84\ g} \times 100\% = 38\%\ KClO_4$$

Solubility of KNO_3: 60.7 g at 40 °C and 32.3 g at 20 °C

$$\text{Percent Yield of } KNO_3 = \frac{60.7\ g - 32.3\ g}{60.7\ g} \times 100\% = 47\%\ KNO_3$$

Solubility of K_2SO_4: 15.1 g at 40 °C and 11.9 g at 20 °C

$$\text{Percent yield of } K_2SO_4 = \frac{15.1\ g - 11.9\ g}{15.1\ g} \times 100\% = 21\%\ K_2SO_4$$

Ranked in order of decreasing percent yield we have:
$$KNO_3\ (47\%) > KClO_4\ (38\%) > K_2SO_4\ (21\%)$$

5A (M) From Example 14-5, we know that the Henry's law constant for O_2 dissolved in water is $k = 2.18 \times 10^{-3}$ M atm^{-1}. Consequently,

$$P_{gas} = \frac{C}{k} = \frac{\dfrac{5.00 \times 10^{-3}\ g \times \dfrac{1\ mol\ O_2}{32.00\ g\ O_2}}{0.100\ L}}{2.18 \times 10^{-3}\ M\ atm^{-1}} = 0.717\ atm\ O_2\ \text{pressure}$$

5B (M) We note the relationship between gas partial pressure over a liquid and its dissolved concentration is given by Henry's Law: $C = k\,P$, where C is the concentration (given in dimensionless "volume parts" unit, akin to ppm), k is the Henry's Law constant, and P is the partial pressure of CO, which is given by: $\chi_{P_{CO}} = P_{CO}/P_{total}$. Assuming a CO partial pressure of 1.0 (100% at 1 atm), the k value is 0.0354 mL mL^{-1} atm^{-1}.

First, since the concentration of CO is 0.0100 mol CO/1.000 L of H_2O, we have to find the volume of CO at 273 K and 1.0 atm. That is,

$$V = \frac{nRT}{P} = \frac{(0.0100 \text{ mol})(0.08206 \text{ L} \cdot \text{atm} \cdot \text{mol}^{-1} \cdot \text{K}^{-1})(273 \text{ K})}{1.0 \text{ atm}} = 0.224 \text{ L}$$

Therefore, the concentration of CO may be expressed as follows:

$$\text{v/v concentration of CO} = \frac{0.224 \text{ L CO}}{1.000 \text{ L H}_2\text{O}} = 0.224 \text{ L/L or } 0.224 \text{ mL/mL}$$

$$P_{\text{CO}} = \frac{C}{k} = \frac{0.224 \text{ mL} \cdot \text{mL}^{-1}}{0.0354 \text{ mL} \cdot \text{mL}^{-1} \cdot \text{atm}^{-1}} = 6.328 \text{ atm}$$

6A **(E)** Raoult's law enables us to determine the vapor pressure of each component.

$$P_{\text{hex}} = \chi_{\text{hex}} P^*_{\text{hex}} = 0.750 \times 149.1 \text{ mmHg} = 112 \text{ mmHg}$$

$$P_{\text{pen}} = \chi_{\text{pen}} P^*_{\text{pen}} = 0.250 \times 508.5 \text{ mmHg} = 127 \text{ mmHg}.$$

We use Dalton's law to determine the total vapor pressure:

$$P_{\text{total}} = P_{\text{hex}} + P_{\text{pen}} = 112 \text{ mmHg} + 127 \text{ mmHg} = 239 \text{ mmHg}$$

6B **(M)** The masses of solution components need to be converted to amounts in moles through the use of molar masses. Let us choose as our amount precisely 1.0000 mole of $C_6H_6 = 78.11$ g C_6H_6 and an equal mass of toluene.

$$\text{amount of toluene} = 78.11 \text{ g C}_7\text{H}_8 \times \frac{1 \text{ mol C}_7\text{H}_8}{92.14 \text{ g C}_7\text{H}_8} = 0.8477 \text{ mol C}_7\text{H}_8$$

$$\text{mole fraction toluene} = \chi_{\text{tol}} = \frac{0.8477 \text{ mol C}_7\text{H}_8}{0.8477 \text{ mol C}_7\text{H}_8 + 1.0000 \text{ mol C}_6\text{H}_6} = 0.4588$$

$$\text{toluene vapor pressure} = \chi_{\text{tol}} P^*_{\text{tol}} = 0.4588 \times 28.4 \text{ mmHg} = 13.0 \text{ mmHg}$$

$$\text{benzene vapor pressure} = \chi_{\text{benz}} P^*_{\text{benz}} = (1.0000 - 0.4588) \times 95.1 \text{ mmHg} = 51.5 \text{ mmHg}.$$

$$\text{total vapor pressure} = 13.0 \text{ mmHg} + 51.5 \text{ mmHg} = 64.5 \text{ mmHg}$$

7A **(E)** The vapor pressure composition of each component is that component's partial pressure divided by the total pressure. Again, we note that the vapor is richer in the more volatile component.

$$y_{\text{hexane}} = \frac{P_{\text{hexane}}}{P_{\text{total}}} = \frac{112 \text{ mmHg hexane}}{239 \text{ mmHg total}} = 0.469 \quad y_{\text{pentane}} = \frac{P_{\text{pentane}}}{P_{\text{total}}} = \frac{127 \text{ mmHg pentane}}{239 \text{ mmHg total}} = 0.531$$

or simply $1.000 - 0.469 = 0.531$

7B **(E)** The vapor pressure composition of each component is that component's partial pressure divided by the total pressure. Again we note that the vapor is richer in the more volatile component.

$$y_t = \frac{P_{\text{toluene}}}{P_{\text{total}}} = \frac{13.0 \text{ mmHg toluene}}{64.5 \text{ mmHg total}} = 0.202 \quad y_b = \frac{P_{\text{benzene}}}{P_{\text{total}}} = \frac{51.5 \text{ mmHg benzene}}{64.5 \text{ mmHg total}} = 0.798$$

or simply $1.000 - 0.202 = 0.798$

8A **(M)** We use the osmotic pressure equation, converting the mass of solute to amount in moles, the temperature to Kelvin, and the solution volume to liters.

$$\pi = \frac{nRT}{V} = \frac{\left(1.50 \text{ g C}_{12}\text{H}_{22}\text{O}_{11} \times \dfrac{1 \text{ mol C}_{12}\text{H}_{22}\text{O}_{11}}{342.3 \text{ g C}_{12}\text{H}_{22}\text{O}_{11}}\right) \times 0.08206 \dfrac{\text{L atm}}{\text{mol K}} \times 298 \text{ K}}{125 \text{ mL} \times \dfrac{1 \text{ L}}{1000 \text{ mL}}} = 0.857 \text{ atm}$$

8B **(E)** We use the osmotic pressure equation to determine the molarity of the solution.

$$\frac{n}{V} = \frac{\pi}{RT} = \frac{0.015 \text{ atm}}{0.08206 \text{ L atm mol}^{-1} \text{ K}^{-1} \times 298 \text{ K}} = 6.1 \times 10^{-4} \text{ M}$$

Now, we can calculate the mass of urea.

$$\text{urea mass} = 0.225 \text{ L} \times \frac{6.1 \times 10^{-4} \text{ mol urea}}{1 \text{ L soln}} \times \frac{60.06 \text{ g CO}(\text{NH}_2)_2}{1 \text{ mol CO}(\text{NH}_2)_2} = 8.2\underline{4} \times 10^{-3} \text{ g}$$

9A **(M)** We could substitute directly into the equation for molar mass derived in Example 14-9, but let us rather think our way through each step of the process. First, we find the concentration of the solution, by rearranging $\pi = \dfrac{n}{V} RT$. We need to convert the osmotic pressure to atmospheres.

$$\frac{n}{V} = \frac{\pi}{RT} = \frac{8.73 \text{ mmHg} \times \dfrac{1 \text{ atm}}{760 \text{ mmHg}}}{0.08206 \dfrac{\text{L atm}}{\text{mol K}} \times 298 \text{ K}} = 4.70 \times 10^{-4} \text{ M}$$

Next we determine the amount in moles of dissolved solute.

$$\text{amount of solute} = 100.0 \text{ mL} \times \frac{1 \text{ L}}{1000 \text{ mL}} \times \frac{4.70 \times 10^{-4} \text{ mol solute}}{1 \text{ L solution}} = 4.70 \times 10^{-5} \text{ mol solute}$$

We use the mass of solute, 4.04 g, to determine the molar mass. $\rightarrow$ $M = \dfrac{4.04 \text{ g}}{4.70 \times 10^{-5} \text{ mol}} = 8.60 \times 10^{4} \text{ g/mol}$

9B **(M)** We use the osmotic pressure equation along with the molarity of the solution.

$$\pi = \frac{n}{V} RT = \frac{2.12 \text{ g} \times \dfrac{1 \text{ mol}}{6.86 \times 10^{4} \text{ g}}}{75.00 \text{ mL} \times \dfrac{1 \text{ L}}{1000 \text{ mL}}} \times \frac{0.08206 \text{ L atm}}{\text{mol K}} \times (310.2) \text{K} = 0.0105 \text{ atm} = 7.97 \text{ mmHg}$$

10A (M)

(a) The freezing point depression constant for water is $K_f = 1.86\,^\circ\text{C mol}^{-1}\text{ kg}$.

$$\text{molality} = \frac{\Delta T_f}{-K_f} = \frac{-0.227\,^\circ\text{C}}{-1.86\,^\circ\text{C mol}^{-1}\text{ kg}} = 0.122\text{ mol kg}^{-1}$$

(b) We will use the definition of molality to determine the number of moles of riboflavin in 0.833 g of dissolved riboflavin.

$$\text{amount of riboflavin} = 18.1\text{ g solvent } H_2O \times \frac{1\text{ kg solvent}}{1000\text{ g}} \times \frac{0.122\text{ mol solute}}{1\text{ kg solvent}}$$

$$\text{amount of riboflavin} = 2.21\times10^{-3}\text{ mol riboflavin}$$

$$\text{molar mass} = \frac{0.833\text{ g riboflavin}}{2.21\times10^{-3}\text{ mol}} = 377\text{ g/mol}$$

(c) We use the method of Chapter 3 to find riboflavin's empirical formula, starting with a 100.00-g sample.

$$54.25\text{ g C} \times \frac{1\text{ mol C}}{12.01\text{ g C}} = 4.517\text{ mol C } \div1.063 \quad \rightarrow \quad = 4.249\text{ mol C}$$

$$5.36\text{ g H} \times \frac{1\text{ mol H}}{1.008\text{ g H}} = 5.32\text{ mol H} \quad \div1.063 \quad \rightarrow \quad = 5.00\text{ mol H}$$

$$25.51\text{ g O} \times \frac{1\text{ mol O}}{16.00\text{ g O}} = 1.594\text{ mol O} \div1.063 \quad \rightarrow \quad = 1.500\text{ mol O}$$

$$14.89\text{ g N} \times \frac{1\text{ mol N}}{14.01\text{ g N}} = 1.063\text{ mol N} \div1.063 \quad \rightarrow \quad = 1.000\text{ mol N}$$

If we multiply each of these amounts by 4 (because 4.249 is almost equal to $4\frac{1}{4}$), the empirical formula is found to be $C_{17}H_{20}O_6N_4$ with a molar mass of 376 g/mol. The molecular formula is $C_{17}H_{20}O_6N_4$.

10B (M) The boiling point of pure water at 760.0 mmHg is 100.000 °C. For higher pressures, the boiling point occurs at a higher temperature; for lower pressures, a lower boiling point is observed. The boiling point elevation for the urea solution is calculated as follows.

$\Delta T_b = K_b \times m = 0.512\,^\circ\text{C mol}^{-1}\text{ kg} \times 0.205\text{ m} = 0.105\,^\circ\text{C}$

We would expect this urea solution to boil at $(100.00 + 0.105 =)100.105\,^\circ\text{C}$ under 760.0 mmHg atmospheric pressure. Since it boils at a lower temperature, the atmospheric pressure must be lower than 760.0 mmHg.

11A **(M)** We assume a van't Hoff factor of $i = 3.00$ and convert the temperature to Kelvin, 298 K.

$$\pi = iMRT = \frac{3.00 \text{ mol ions}}{1 \text{ mol MgCl}_2} \times \frac{0.0530 \text{ mol MgCl}_2}{1 \text{ L soln}} \times 0.08206 \frac{\text{L atm}}{\text{mol K}} \times 298 \text{ K} = 3.89 \text{ atm}$$

11B **(M)** We first determine the molality of the solution, and assume a van't Hoff factor of $i = 2.00$.

$$m = \frac{\Delta T_f}{-K_f \times i} = \frac{-0.100 \,°\text{C}}{-1.86 \,°\text{C mol}^{-1} \text{ kg} \times 2.00} = 0.0269 \text{ mol kg}^{-1} \approx 0.0269 \text{ M}$$

$$\text{volume of HCl(aq)} = 250.0 \text{ mL final soln} \times \frac{0.0269 \text{ mmol HCl}}{1 \text{ mL soln}} \times \frac{1 \text{ mL conc soln}}{12.0 \text{ mmol HCl}}$$

$$\text{volume of HCl(aq)} = 0.560 \text{ mL conc soln}$$

INTEGRATIVE EXAMPLE

A **(M)** We determine the mass of each component in the water-rich phase.

$$\text{mass H}_2\text{O} = 32.8 \text{ g phase} \times \frac{92.50 \text{ g H}_2\text{O}}{100.00 \text{ g phase}} = 30.3 \text{ g H}_2\text{O}$$

mass phenol = 32.8 g phase − 30.3 g H_2O = 2.5 g phenol
Then we determine the mass of each component in the other phase.
mass phenol = 50.0 g − 2.5 g = 47.5 g phenol   mass H_2O = 50.0 g − 30.3 g = 19.7 g H_2O

$$\% \,\text{H}_2\text{O} = \frac{19.7 \text{ g H}_2\text{O}}{19.7 \text{ g H}_2\text{O} + 47.5 \text{ g phenol}} \times 100\% = 29.3\% \text{ H}_2\text{O}$$

Above 66.8 °C phenol and water are completely miscible. Consequently, for temperatures above 66.8 °C, the mixture will be a homogeneous solution consisting of 50.0 g of H_2O and 50.0 g of phenol. To calculate the mole fraction of phenol in the mixture, we must first determine the number of moles of each component.

$$\text{number of moles of H}_2\text{O} = 50.0 \text{ g H}_2\text{O} \times \frac{1 \text{ mol H}_2\text{O}}{18.016 \text{ g H}_2\text{O}} = 2.77\underline{5} \text{ mol H}_2\text{O}$$

$$\text{number of moles of H}_2\text{O} = 50.0 \text{ g phenol} \times \frac{1 \text{ mol phenol}}{94.11 \text{ g phenol}} = 0.531 \text{ mol phenol}$$

$$\text{Thus } \chi_{\text{phenol}}(\text{mol fraction}) = \frac{0.531 \text{ mol phenol}}{2.775 \text{ mol phenol} + 0.531 \text{ mol H}_2\text{O}} = 0.161$$

B **(D)**

(a) Rearrange Raoult's law (Equation 14.3) to the form $\dfrac{P_A^* - P_A}{P_A^*} = x_B$. The mass of H_2O absorbed by D_1 is proportional to P_A^*; the mass of H_2O absorbed by D_2 is proportional to P_A. Hence,

$$\frac{11.7458 - 11.5057}{11.7458} = x_B = 0.0204. \ x_A = 0.9796,$$

$P_A = 0.9796 \times 23.76 \text{ mmHg} = 23.28 \text{ mmHg}.$

The observed vapor pressure lowering = 23.76 mmHg − 23.28 mmHg = 0.48 mmHg.

(b) Calculate x_B for a 1.00 mol kg^{-1} solution. $x_B = 0.0177$; $x_A = 0.9823$. Calculate P_A using Raoult's law. $P_A = 23.34$ mmHg. The expected vapor pressure lowering = (23.76 − 23.34) mmHg = 0.42 mmHg.

EXERCISES

Homogeneous and Heterogeneous Mixtures

1. **(E)** $NH_2OH(s)$ should be the most soluble in water. Both $C_6H_6(l)$ and $C_{10}H_8(s)$ are composed of essentially nonpolar molecules, which are barely (if at all) soluble in water. Both $NH_2OH(s)$ and $CaCO_3(s)$ should be able to interact with water molecules. But $CaCO_3(s)$ contains ions of high charge, and thus it dissolves with great difficulty because of the high lattice energy. (Recall the solubility rules of Chapter 5: most carbonates are insoluble in water.)

3. **(E)** (b) Salicyl alcohol probably is moderately soluble in both benzene and water. The reason for this assertion is that salicyl alcohol contains a benzene ring, which would make it soluble in benzene, and also can use its −OH groups to hydrogen bond to water molecules. On the other hand, (c) diphenyl contains only nonpolar benzene rings; it should be soluble in benzene but not in water. (a) *para*-dichlorobenzene contains a benzene ring, making it soluble in benzene, and two polar C−Cl bonds, which oppose each other, producing a nonpolar—and thus water-insoluble—molecule. (d) Hydroxyacetic acid is a very polar molecule with many opportunities for hydrogen bonding. Its polar nature would make it insoluble in benzene, while the prospective hydrogen bonding will enhance aqueous solubility.

5. **(E)** (c) Formic acid and (f) propylene glycol are soluble in water. They both can form hydrogen bonds with water, and they both have small nonpolar portions. (b) Benzoic acid and (d) 1-Butanol are only slightly soluble in water. Although they both can form hydrogen bonds with water, both molecules contain reasonably large nonpolar portions, which will not interact strongly with water. (a) Iodoform and (e) chlorobenzene are insoluble in water. Although both molecules have polar groups, their influence is too small to enable the molecules to disrupt the hydrogen bonds in water and form a homogeneous liquid mixture.

7. **(M)** We expect small, highly charged ions to form crystals with large lattice energies, which tends to decrease their solubility in water. Based on this information, we would expect MgF_2 to be insoluble and KF to be soluble. It is also probable that CaF_2 is insoluble due to its high lattice energy, but that NaF, with smaller lattice energy, is soluble. Of all of the fluorides listed, KF is probably the most water soluble. The actual solubilities at 25 °C are:
0.00020 M CaF_2 < 0.0021 M MgF_2 < 0.95 M NaF < 16 M KF.

Percent Concentration

9. **(E)** $\% \text{ NaBr} = \dfrac{116 \text{ g NaBr}}{116 \text{ g NaBr} + 100 \text{ g H}_2\text{O}} \times 100\% = 53.7\% = 53.7 \text{ g NaBr/100 g solution}$

11. **(E)**

$$\text{soln. volume} = 725 \text{ kg NaCl} \times \frac{1000 \text{ g NaCl}}{1 \text{ kg NaCl}} \times \frac{100.00 \text{ g soln}}{3.87 \text{ g NaCl}} \times \frac{75.0 \text{ mL soln}}{76.9 \text{ g soln}} \times \frac{1 \text{ L soln}}{1000 \text{ mL soln}}$$

$$= 1.83 \times 10^4 \text{ L sol'n}$$

13. **(E)** For water, the mass in grams and the volume in mL are about equal; the density of water is close to 1.0 g/mL. For ethanol, on the other hand, the density is about 0.8 g/mL. As long as the final solution volume after mixing is close to the sum of the volumes for the two pure liquids, the percent by volume of ethanol will have to be larger than its percent by mass. This would not necessarily be true of other ethanol solutions. It would only be true in those cases where the density of the other component is greater than the density of ethanol.

15. **(E)**

$$\text{mass HC}_2\text{H}_3\text{O}_2 = 355 \text{ mL vinegar} \times \frac{1.01 \text{ g vinegar}}{1 \text{ mL}} \times \frac{6.02 \text{ g HC}_2\text{H}_3\text{O}_2}{100.00 \text{ g vinegar}} = 21.6 \text{ g HC}_2\text{H}_3\text{O}_2$$

17. **(M)** $46.1 \text{ ppm} = \dfrac{46.1 \text{ mg SO}_4^{2-}}{1 \text{ L solution}}$ (Assumes density of water ~1.00 g mL^{-1})

$$[SO_4^{2-}] = \frac{46.1 \text{ mg SO}_4^{2-}}{1 \text{ L solution}} \times \frac{1 \text{ g SO}_4^{2-}}{1000 \text{ mg SO}_4^{2-}} \times \frac{1 \text{ mol SO}_4^{2-}}{96.06 \text{ g SO}_4^{2-}} = 4.80 \times 10^{-4} \text{ M}$$

Molarity

19. **(E)**

$$\text{molarity} = \frac{6.00 \text{ g CH}_3\text{OH} \times \dfrac{1 \text{ mol CH}_3\text{OH}}{32.04 \text{ g CH}_3\text{OH}}}{100.00 \text{ g soln} \times \dfrac{1 \text{ mL}}{0.988 \text{ g}} \times \dfrac{1 \text{ L}}{1000 \text{ mL}}} = 1.85 \text{ M} = [\text{CH}_3\text{OH}]$$

21. **(E)** The solution of Example 14-1 is 1.71 M C_2H_5OH, or 1.71 mmol C_2H_5OH in each mL of solution.

$$\text{volume conc. soln} = 825 \text{ mL} \times \frac{0.235 \text{ mmol C}_2\text{H}_5\text{OH}}{1 \text{ mL soln}} \times \frac{1 \text{ mL conc. soln}}{1.71 \text{ mmol C}_2\text{H}_5\text{OH}}$$

$$= 113 \text{ mL conc. soln}$$

23. **(E)** The easiest way to work with ppm is to think of it in terms of mg of a substance in a kg of solvent. The molarity of CO_2 is calculated as follows:

$$\text{molarity of CO}_2 = \frac{280 \text{ mg CO}_2}{1 \text{ kg H}_2\text{O}} \times \frac{1 \text{ g CO}_2}{1000 \text{ mg CO}_2} \times \frac{1 \text{ mol CO}_2}{44.0 \text{ g CO}_2} \times \frac{1027 \text{ kg H}_2\text{O}}{1000 \text{ L H}_2\text{O}}$$

$$= 0.00654 \text{ M}$$

Molality

25. **(M)**

$$\text{molality} = \frac{2.65 \text{ g C}_6\text{H}_4\text{Cl}_2 \times \dfrac{1 \text{ mol C}_6\text{H}_4\text{Cl}_2}{147.0 \text{ g C}_6\text{H}_4\text{Cl}_2}}{50.0 \text{ mL} \times \dfrac{0.879 \text{ g}}{1 \text{ mL}} \times \dfrac{1 \text{ kg}}{1000 \text{ g}}} = 0.410 \text{ mol kg}^{-1}$$

27. **(E)** The mass of solvent in kg multiplied by the molality gives the amount in moles of the solute.

$$\text{mass I}_2 = \left(725.0 \text{ mL CS}_2 \times \frac{1.261 \text{ g}}{1 \text{ mL}} \times \frac{1 \text{ kg}}{1000 \text{ g}}\right) \times \frac{0.236 \text{ mol I}_2}{1 \text{ kg CS}_2} \times \frac{253.8 \text{ g I}_2}{1 \text{ mol I}_2} = 54.8 \text{ g I}_2$$

29. (M)

$$\text{H}_3\text{PO}_4 \text{ molarity} = \frac{34.0 \text{ g H}_3\text{PO}_4 \times \dfrac{1 \text{ mol H}_3\text{PO}_4}{98.00 \text{ g H}_3\text{PO}_4}}{100.0 \text{ g soln} \times \dfrac{1 \text{ mL}}{1.209 \text{ g}} \times \dfrac{1 \text{ L}}{1000 \text{ mL}}} = 4.19 \text{ M}$$

$$\text{H}_3\text{PO}_4 \text{ molality} = \frac{34.0 \text{ g H}_3\text{PO}_4 \times \dfrac{1 \text{ mol H}_3\text{PO}_4}{98.00 \text{ g H}_3\text{PO}_4}}{66.0 \text{ g solvent} \times \dfrac{1 \text{ kg}}{1000 \text{ g}}} = 5.26 \text{ mol kg}^{-1}$$

Mole Fraction, Mole Percent

31. (M) The total number of moles
$$= 1.28 \text{ mol C}_7\text{H}_{16} + 2.92 \text{ mol C}_8\text{H}_{18} + 2.64 \text{ mol C}_9\text{H}_{20} = 6.84 \text{ moles}$$

(a) $\chi_{\text{C}_7\text{H}_{16}} = \dfrac{1.28 \text{ mol C}_7\text{H}_{16}}{6.84 \text{ moles total}} = 0.187$ **(b)** $\times 100\% = 18.7 \text{ mol\% C}_7\text{H}_{16}$

$\chi_{\text{C}_8\text{H}_{18}} = \dfrac{2.92 \text{ mol C}_8\text{H}_{18}}{6.84 \text{ moles total}} = 0.427$ $\times 100\% = 42.7 \text{ mol\% C}_8\text{H}_{18}$

$\chi_{\text{C}_9\text{H}_{20}} = \dfrac{2.64 \text{ mol C}_9\text{H}_{20}}{6.84 \text{ moles total}} = 0.386$ $\times 100\% = 38.6 \text{ mol\% C}_9\text{H}_{20}$

or $1.00 - 0.187 - 0.427 = 0.386$ or $100 - 18.7 - 42.7 = 38.6 \%$

33. (M)

(a)   The amount of solvent is found after the solute's mass is subtracted from the total mass of the solution.

$$\begin{aligned} \text{solvent} \atop \text{amount} &= \left(\left(1 \text{ L soln} \times \frac{1000 \text{ mL}}{1 \text{ L}} \times \frac{1.006 \text{ g}}{1 \text{ mL}} \right) - \left(0.112 \text{ mol C}_6\text{H}_{12}\text{O}_6 \times \frac{180.2 \text{ g C}_6\text{H}_{12}\text{O}_6}{1 \text{ mol C}_6\text{H}_{12}\text{O}_6} \right) \right) \end{aligned}$$

$$= \left[1006 \text{ g solution} - 20.2 \text{ g C}_6\text{H}_{12}\text{O}_6 \right] \times \frac{1 \text{ mol H}_2\text{O}}{18.02 \text{ g H}_2\text{O}} = 54.7 \text{ mol H}_2\text{O}$$

$$\chi_{\text{solute}} = \frac{0.112 \text{ mol C}_6\text{H}_{12}\text{O}_6}{0.112 \text{ mol C}_6\text{H}_{12}\text{O}_6 + 54.7 \text{ mol H}_2\text{O}} = 0.00204$$

(b) First we must determine the mass and the number of moles of ethanol. The number of moles of solvent is calculated after the ethanol's mass is subtracted from the solution's mass. Use a 100.00-mL sample of solution for computation. This 100 mL sample would contain 3.20 mL of C_2H_5OH. We calculate the mass of ethanol first, followed by the number of moles:

$$\text{mass}_{C_2H_5OH} = 3.20 \text{ mL } C_2H_5OH \times \frac{0.789 \text{ g}}{1 \text{ mL } C_2H_5OH} = 2.52 \text{ g } C_2H_5OH$$

$$\text{moles } C_2H_5OH = 2.52 \text{ g } C_2H_5OH \times \frac{1 \text{ mol } C_2H_5OH}{46.07 \text{ g } C_2H_5OH} = 0.0547 \text{ mol } C_2H_5OH$$

$$\text{mass of } H_2O = \left(\left(100.0 \text{ mL soln} \times \frac{0.993 \text{ g}}{1 \text{ mL}}\right) - 2.52 \text{ g } C_2H_5OH\right) = 96.8 \text{ g } H_2O$$

$$\text{amount of } H_2O = 96.8 \text{ g } H_2O \times \frac{1 \text{ mol } H_2O}{18.02 \text{ g } H_2O} = 5.37 \text{ mol } H_2O$$

$$\chi_{\text{solute}} = \frac{0.0547 \text{ mol } C_2H_5OH}{0.0547 \text{ mol } C_2H_5OH + 5.37 \text{ mol } H_2O} = 0.0101$$

35. **(M)** The amount of water present in 1 kg is

$$n_{\text{water}} = 1000 \text{ g } H_2O \times \frac{1 \text{ mol } H_2O}{18.02 \text{ g } H_2O} = 55.49 \text{ mol } H_2O.$$ Now, solve the following expression

for n_{gly}, the amount of glycerol. $4.85\% = 0.0485$ mole fraction.

$$\chi_{\text{gly}} = 0.0485 = \frac{n_{\text{gly}}}{n_{\text{gly}} + 55.49} \qquad\qquad n_{\text{gly}} = 0.0485 \ n_{\text{gly}} + 2.69$$

$$n_{\text{gly}} = \frac{2.69}{(1.0000 - 0.0485)} = 2.83 \text{ mol glycerol}$$

$$\text{volume glycerol} = 2.83 \text{ mol } C_3H_8O_3 \times \frac{92.09 \text{ g } C_3H_8O_3}{1 \text{ mol } C_3H_8O_3} \times \frac{1 \text{ mL}}{1.26 \text{ g}} = 207 \text{ mL glycerol}$$

37. **(E)** We assume that density of water is 1.000 g/mL, so 1.000 mL of water has a mass of 1.000 g. First, determine the number of moles of Pb, and then the number of moles of water:

$$\text{mol Pb} = \frac{15 \ \mu g \text{ Pb}}{1.000 \text{ g } H_2O} \times \frac{1 \text{ g}}{10^6 \ \mu g} \times \frac{1 \text{ mol Pb}}{207.19 \text{ g Pb}} = 7.24 \times 10^{-8} \text{ mol}$$

$$\text{\# Pb atoms} = 7.24 \times 10^{-8} \text{ mol Pb} \times \frac{6.022 \times 10^{23} \text{ atoms}}{1 \text{ mol Pb}} = 4.36 \times 10^{16} \text{ atoms}$$

$$\text{mol } H_2O = 1.000 \text{ g } H_2O \times \frac{1 \text{ mol } H_2O}{18.0 \text{ g } H_2O} = 0.05556 \text{ mol}$$

$$\chi_{\text{Pb}} = \frac{7.24 \times 10^{-8}}{0.05556 + 7.24 \times 10^{-8}} = 1.303 \times 10^{-6}$$

Solubility Equilibrium

39. **(E)** At 40 °C the solubility of NH_4Cl is 46.3 g per 100 g of H_2O. To determine molality, we calculate amount in moles of the solute and the solvent mass in kg.

$$\text{molality} = \frac{46.3 \text{ g} \times \dfrac{1 \text{ mol } NH_4Cl}{53.49 \text{ g } NH_4Cl}}{100 \text{ g } H_2O \times \dfrac{1 \text{ kg}}{1000 \text{ g}}} = 8.66 \text{ mol kg}^{-1}$$

41. **(M)**

(a) The concentration for $KClO_4$ in this mixture is calculated first.

$$\frac{\text{mass solute}}{100 \text{ g } H_2O} = 100 \text{ g } H_2O \times \frac{20.0 \text{ g } KClO_4}{500.0 \text{ g water}} = 4.00 \text{ g } KClO_4$$

At 40 °C a saturated $KClO_4$ solution has a concentration of about 4.6 g $KClO_4$ dissolved in 100 g water. Thus, the solution is unsaturated.

(b) The mass of $KClO_4$ that must be added is the difference between the mass now present in the mixture and the mass that is dissolved in 500 g H_2O to produce a saturated solution.

$$\text{mass to be added} = \left(500.0 \text{ g } H_2O \times \frac{4.6 \text{ g } KClO_4}{100 \text{ g } H_2O} \right) - 20.0 \text{ g } KClO_4 = 3.0 \text{ g } KClO_4$$

Solubility of Gases

43. **(M)** We first determine the number of moles of O_2 that have dissolved.

$$\text{moles of } O_2 = \frac{PV}{RT} = \frac{1.00 \text{ atm} \times 0.02831 \text{ L}}{0.08206 \text{ L atm mol}^{-1} \text{ K}^{-1} \times 298 \text{ K}} = 1.16 \times 10^{-3} \text{ mol } O_2$$

$$\left[O_2 \right] = \frac{1.16 \times 10^{-3} \text{ mol } O_2}{1.00 \text{ L soln}} = 1.16 \times 10^{-3} \text{ M}$$

The oxygen concentration now is computed at the higher pressure.

$$\left[O_2 \right] = \frac{1.16 \times 10^{-3} \text{ M}}{1 \text{ atm } O_2} \times 3.86 \text{ atm } O_2 = 4.48 \times 10^{-3} \text{ M}$$

45. **(E)** mass of $CH_4 = 1.00 \times 10^3$ kg $H_2O \times \dfrac{0.02 \text{ g } CH_4}{1 \text{ kg } H_2O \cdot \text{ atm}} \times 20 \text{ atm} = 4 \times 10^2 \text{g } CH_4$ (natural gas)

47. **(M)** We use the STP molar volume $(22.414 \text{ L} = 22,414 \text{ mL})$ to determine the molarity of Ar under 1 atmosphere of pressure and then use Henry's law.

$$k_{Ar} = \frac{C}{P_{Ar}} = \frac{\dfrac{33.7 \text{ mL Ar}}{1 \text{ L soln}} \times \dfrac{1 \text{ mol Ar}}{22,414 \text{ mL at STP}}}{1 \text{ atm pressure}} = \frac{0.00150 \text{ M}}{\text{atm}}$$

In the atmosphere, the partial pressure of argon is $P_{Ar} = 0.00934$ atm. (Recall that pressure fractions equal volume fractions for ideal gases.) We now compute the concentration of argon in aqueous solution.

$$C = k_{Ar}P_{Ar} = \frac{0.00150 \text{ M}}{\text{atm}} \times 0.00934 \text{ atm} = 1.40 \times 10^{-5} \text{ M Ar}$$

49. **(M)** Because of the low density of molecules in the gaseous state, the solution volume remains essentially constant as a gas dissolves in a liquid. Changes in concentrations in the solution result from changes in the number of dissolved gas molecules (recall Figure 14-13). This number is directly proportional to the mass of dissolved gas.

Raoult's Law and Liquid–Vapor Equilibrium

51. **(M)** First we determine the number of moles of each component, its mole fraction in the solution, the partial pressure due to that component above the solution, and finally the total pressure.

$$\text{amount benzene} = n_b = 35.8 \text{ g C}_6\text{H}_6 \times \frac{1 \text{ mol C}_6\text{H}_6}{78.11 \text{ g C}_6\text{H}_6} = 0.458 \text{ mol C}_6\text{H}_6$$

$$\text{amount toluene} = n_t = 56.7 \text{ g C}_7\text{H}_8 \times \frac{1 \text{ mol C}_7\text{H}_8}{92.14 \text{ g C}_7\text{H}_8} = 0.615 \text{ mol C}_7\text{H}_8$$

$$\chi_b = \frac{0.458 \text{ mol C}_6\text{H}_6}{(0.458+0.615) \text{ total moles}} = 0.427 \qquad \chi_t = \frac{0.615 \text{ mol C}_7\text{H}_8}{(0.458+0.615) \text{ total moles}} = 0.573$$

$$P_b = 0.427 \times 95.1 \text{ mmHg} = 40.6 \text{ mmHg} \qquad P_t = 0.573 \times 28.4 \text{ mmHg} = 16.3 \text{ mmHg}$$

$$\text{total pressure} = 40.6 \text{ mmHg} + 16.3 \text{ mmHg} = 56.9 \text{ mmHg}$$

53. **(M)** We need to determine the mole fraction of water in this solution.

$$n_{glucose} = 165 \text{ g C}_6\text{H}_{12}\text{O}_6 \times \frac{1 \text{ mol C}_6\text{H}_{12}\text{O}_6}{180.2 \text{ g C}_6\text{H}_{12}\text{O}_6} = 0.916 \text{ mol C}_6\text{H}_{12}\text{O}_6$$

$$n_{water} = 685 \text{ g H}_2\text{O} \times \frac{1 \text{ mol H}_2\text{O}}{18.02 \text{ g H}_2\text{O}} = 38.0 \text{ mol H}_2\text{O} \qquad \chi_{water} = \frac{38.0 \text{ mol H}_2\text{O}}{(38.0+0.916) \text{ total moles}} = 0.976$$

$$P_{sol'n} = \chi_{water} P_{water}^* = 0.976 \times 23.8 \text{ mmHg} = 23.2 \text{ mmHg}$$

55. **(M)** We consider a sample of 100.0 g of the solution and determine the number of moles of each component in this sample. From this information and the given vapor pressures, we determine the vapor pressure of each component. Note that styrene, $C_6H_5C=CH_2$, is represented by the simplified molecular formula C_8H_8 in this problem.

$$\text{amount of styrene} = n_s = 38 \text{ g styrene} \times \frac{1 \text{ mol } C_8H_8}{104 \text{ g } C_8H_8} = 0.37 \text{ mol } C_8H_8$$

$$\text{amount of ethylbenzene} = n_e = 62 \text{ g ethylbenzene} \times \frac{1 \text{ mol } C_8H_{10}}{106 \text{ g } C_8H_{10}} = 0.58 \text{ mol } C_8H_{10}$$

$$X_s = \frac{n_s}{n_s + n_e} = \frac{0.37 \text{ mol styrene}}{(0.37 + 0.58) \text{ total moles}} = 0.39; \quad P_s = 0.39 \times 134 \text{ mmHg} = 52 \text{ mmHg for } C_8H_8$$

$$X_e = \frac{n_e}{n_s + n_e} = \frac{0.58 \text{ mol ethylbenzene}}{(0.37 + 0.58) \text{ total moles}} = 0.61; \quad P_e = 0.61 \times 182 \text{ mmHg} = 111 \text{ mmHg for } C_8H_{10}$$

Then the mole fraction in the vapor can be determined.

$$y_e = \frac{P_e}{P_e + P_s} = \frac{111 \text{ mmHg}}{(111 + 52) \text{ mmHg}} = 0.68 \qquad y_s = 1.00 - 0.68 = 0.32$$

57. **(M)** The total vapor pressure above the solution at its normal boiling point is 760 mm Hg. The vapor pressure due to toluene is given by the following equation.

$P_{toluene} = \chi_{toluene} \cdot P^*_{toluene} = 0.700 \times 533 \text{ mm Hg} = 373 \text{ mm Hg}$. Next, the vapor pressure due to benzene is determined, followed by the vapor pressure for pure benzene.

$P_{benzene} = P_{total} - P_{toluene} = 760 \text{ mm Hg} - 373 \text{ mm Hg} = 387 \text{ mm Hg} = \chi_{benzene} \cdot P^*_{benzene}$

$387 \text{ mm Hg} = 0.300 \times P^*_{benzene}$ and hence, $P^*_{benzene} = 1.29 \times 10^3$ mm Hg

Osmotic Pressure

59. **(M)** We first compute the concentration of the solution. Then, assuming that the solution volume is the same as that of the solvent (0.2500 L), we determine the amount of solute dissolved, and finally the molar mass.

$$\frac{n}{V} = \frac{\pi}{RT} = \frac{1.67 \text{ mmHg} \times \dfrac{1 \text{ atm}}{760 \text{ mmHg}}}{0.08206 \text{ L atm mol}^{-1} \text{ K}^{-1} \times 298.2 \text{ K}} = 8.98 \times 10^{-5} \text{ M}$$

$$\text{solute amount} = 0.2500 \text{ L} \times \frac{8.98 \times 10^{-5} \text{ mol}}{1 \text{ L}} = 2.25 \times 10^{-5} \text{ mol}$$

$$M = \frac{0.72 \text{ g}}{2.25 \times 10^{-5} \text{ mol}} = 3.2 \times 10^4 \text{ g/mol}$$

61. **(E)** Both the flowers and the cucumber contain ionic solutions (plant sap), but both of these solutions are less concentrated than the salt solution. Thus, the solution in the plant material moves across the semipermeable membrane in an attempt to dilute the salt solution, leaving behind wilted flowers and shriveled pickles (wilted/shriveled plants have less water in their tissues).

63. **(E)** We first determine the molarity of the solution. Let's work the problem out with three significant figures.

$$\frac{n}{V} = \frac{\pi}{RT} = \frac{1.00 \text{ atm}}{0.08206 \text{ L atm mol}^{-1} \text{ K}^{-1} \times 273 \text{ K}} = 0.0446 \text{ M}$$

$$\text{volume} = 1 \text{ mol} \times \frac{1 \text{ L}}{0.0446 \text{ mol solute}} = 22.4 \text{ L solution} \approx 22.4 \text{ L solvent}$$

We have assumed that the solution is so dilute that its volume closely approximates the volume of the solvent constituting it. Note that this volume corresponds to the STP molar volume of an ideal gas. The osmotic pressure equation also resembles the ideal gas equation.

65. **(M)** First we determine the concentration of the solution from the osmotic pressure, then the amount of solute dissolved, and finally the molar mass of that solute.

$$\pi = 5.1 \text{ mm soln} \times \frac{0.88 \text{ mmHg}}{13.6 \text{ mm soln}} \times \frac{1 \text{ atm}}{760 \text{ mmHg}} = 4.3 \times 10^{-4} \text{ atm}$$

$$\frac{n}{V} = \frac{\pi}{RT} = \frac{4.3 \times 10^{-4} \text{ atm}}{0.08206 \text{ L atm mol}^{-1} \text{ K}^{-1} \times 298 \text{ K}} = 1.8 \times 10^{-5} \text{ M}$$

$$\text{amount solute} = 100.0 \text{ mL} \times \frac{1 \text{ L}}{1000 \text{ mL}} \times 1.8 \times 10^{-5} \text{ M} = 1.8 \times 10^{-6} \text{ mol solute}$$

$$\text{molar mass} = \frac{0.50 \text{ g}}{1.8 \times 10^{-6} \text{ mol}} = 2.8 \times 10^{5} \text{ g/mol}$$

67. **(M)** The reverse osmosis process requires a pressure equal to or slightly greater than the osmotic pressure of the solution. We assume that this solution has a density of 1.00 g/mL. First, we determine the molar concentration of ions in the solution.

$$[\text{ions}] = \frac{2.5 \text{ g NaCl} \times \dfrac{1 \text{ mol NaCl}}{58.4 \text{ g NaCl}} \times \dfrac{2 \text{ mol ions}}{1 \text{ mol NaCl}}}{100.0 \text{ mL soln} \times \dfrac{1 \text{ mL soln}}{1.00 \text{ g}} \times \dfrac{1 \text{ L soln}}{1000 \text{ mL}}} = 0.86 \text{ M}$$

$$\pi = \frac{n}{V} RT = 0.86 \frac{\text{mol}}{\text{L}} \times 0.08206 \text{ L atm mol}^{-1} \text{ K}^{-1} \times (25 + 273.2) \text{ K} = 21 \text{ atm}$$

Freezing-Point Depression and Boiling-Point Elevation

69. **(M)** First we compute the molality of the benzene solution, then the number of moles of solute dissolved, and finally the molar mass of the unknown compound.

$$m = \frac{\Delta T_f}{-K_f} = \frac{4.92 \,^{\circ}\text{C} - 5.53 \,^{\circ}\text{C}}{-5.12 \,^{\circ}\text{C/mol kg}^{-1}} = 0.12 \text{ mol kg}^{-1}$$

$$\text{amount solute} = 0.07522 \text{ kg benzene} \times \frac{0.12 \text{ mol solute}}{1 \text{ kg benzene}} = 9.0 \times 10^{-3} \text{ mol solute}$$

$$\text{molecular weight} = \frac{1.10 \text{ g unknown compound}}{9.0 \times 10^{-3} \text{ mol}} = 1.2 \times 10^{2} \text{ g/mol}$$

71. **(M)**

   (a) First we determine the molality of the solution, then the value of the freezing-point depression constant.

$$m = \frac{1.00 \text{ g C}_6\text{H}_6 \times \dfrac{1 \text{ mol C}_6\text{H}_6}{78.11 \text{ g C}_6\text{H}_6}}{80.00 \text{ g solvent} \times \dfrac{1 \text{ kg}}{1000 \text{ g}}} = 0.160 \text{ mol kg}^{-1}$$

$$K_f = \frac{\Delta T_f}{-m} = \frac{3.3 \,°\text{C} - 6.5 \,°\text{C}}{-0.160 \text{ mol kg}^{-1}} = 20. \,°\text{C/(mol kg}^{-1})$$

   (b) For benzene, $K_f = 5.12 \,°\text{C mol}^{-1}$ kg. Cyclohexane is the better solvent for freezing-point depression determinations of molar mass, because a less concentrated solution will still give a substantial freezing-point depression. For the same concentration, cyclohexane solutions will show a freezing-point depression approximately four times that of benzene. Also, one should steer clear of benzene because it is a known carcinogen.

73. **(M)** Here we determine the molality of the solution, then the number of moles of solute present, and the molar mass of the solute to start things off. Then we determine the compound's empirical formula, and combine this with the molar mass to determine the molecular formula.

$$m = \frac{\Delta T_f}{-K_f} = \frac{1.37 \,°\text{C} - 5.53 \,°\text{C}}{-5.12 \,°\text{C/(mol kg}^{-1})} = 0.813 \text{ mol kg}^{-1}$$

$$\text{amount} = \left(50.0 \text{ mL C}_6\text{H}_6 \times \frac{0.879 \text{ g}}{1 \text{ mL}} \times \frac{1 \text{ kg}}{1000 \text{ g}}\right) \times \frac{0.813 \text{ mol solute}}{1 \text{ kg C}_6\text{H}_6} = 3.57 \times 10^{-2} \text{ mol}$$

$$\text{molecular weight} = \frac{6.45 \text{ g}}{3.57 \times 10^{-2} \text{ mol}} = 181 \text{ g/mol}$$

Now, calculate the empirical formula from the provided mass percents for C, H, N, and O.

$$42.9 \text{ g C} \times \frac{1 \text{ mol C}}{12.01 \text{ g C}} = 3.57 \text{ mol C} \qquad \div 1.19 \rightarrow 3.00 \text{ mol C}$$

$$2.4 \text{ g H} \times \frac{1 \text{ mol H}}{1.01 \text{ g H}} = 2.4 \text{ mol H} \qquad \div 1.19 \rightarrow 2.0 \text{ mol H}$$

$$16.7 \text{ g N} \times \frac{1 \text{ mol N}}{14.01 \text{ g N}} = 1.19 \text{ mol N} \qquad \div 1.19 \rightarrow 1.00 \text{ mol N}$$

$$38.1 \text{ g O} \times \frac{1 \text{ mol O}}{16.00 \text{ g O}} = 2.38 \text{ mol O} \qquad \div 1.19 \rightarrow 2.00 \text{ mol O}$$

The empirical formula is $C_3H_2NO_2$, with a formula mass of 84.0 g/mol. This is one-half the experimentally determined molar mass. Thus, the molecular formula is $C_6H_4N_2O_4$.

75. (M) We determine the molality of the benzene solution first, then the molar mass of the solute.

$$m = \frac{\Delta T_f}{-K_f} = \frac{-1.183\,°C}{-5.12\,°C/(mol\,kg^{-1})} = 0.231\,mol\,kg^{-1}$$

$$amount\ solute = 0.04456\ kg\ benzene \times \frac{0.231\ mol\ solute}{1\ kg\ benzene} = 0.0103\ mol\ solute$$

$$molar\ mass = \frac{0.867\ g\ thiophene}{0.0103\ mol\ thiophene} = 84.2\ g/mol$$

Next, we determine the empirical formula from the masses of the combustion products.

$$amount\ C = 4.913\ g\ CO_2 \times \frac{1\ mol\ CO_2}{44.010\ g\ CO_2} \times \frac{1\ mol\ C}{1\ mol\ CO_2} = 0.1116\ mol\ C \div 0.02791 \rightarrow 4.000\ mol\ C$$

$$amount\ H = 1.005\ g\ H_2O \times \frac{1\ mol\ H_2O}{18.015\ g\ H_2O} \times \frac{2\ mol\ H}{1\ mol\ H_2O} = 0.1116\ mol\ H \div 0.02791 \rightarrow 4.000\ mol\ H$$

$$amount\ S = 1.788\ g\ SO_2 \times \frac{1\ mol\ SO_2}{64.065\ g\ SO_2} \times \frac{1\ mol\ S}{1\ mol\ SO_2} = 0.02791\ mol\ S \div 0.02791 \rightarrow 1.000\ mol\ S$$

A reasonable empirical formula is C_4H_4S, which has an empirical mass of 84.1 g/mol. Since this is the same as the experimentally determined molar mass, the molecular formula of thiophene is C_4H_4S.

77. (M) The boiling point must go up by 2 °C, so $\Delta T_b = 2\,°C$. We know that $K_b = 0.512\,°C/(mol\,kg^{-1})$ for water. We assume that the mass of a liter of water is 1.000 kg and the van't Hoff factor for NaCl is $i = 2$. We first determine the molality of the saltwater solution and then the mass of solute needed.

$$m = \frac{\Delta T_b}{i\,K_b} = \frac{2\,°C}{2.00 \times 0.512\,°C/(mol\,kg^{-1})} = 2\,mol\,kg^{-1}$$

$$solute\ mass = 1.00\ L\ H_2O \times \frac{1\ kg\ H_2O}{1\ L\ H_2O} \times \frac{2\ mol\ NaCl}{1\ kg\ H_2O} \times \frac{58.4\ g\ NaCl}{1\ mol\ NaCl} = 12\underline{0}\ g\ NaCl$$

This is at least ten times the amount of salt one would typically add to a liter of water for cooking purposes!

79. (E) Assume that 1 L of ocean water has a mass of 1 kg.
$$\Delta T_f = -K_f\,m$$
$$-1.94\,°C = (-1.86\,°C \cdot mol^{-1}\,kg) \times mol\,kg^{-1}$$
$$m = 1.04$$

Strong Electrolytes, Weak Electrolytes, and Nonelectrolytes

81. (M) The freezing-point depression is given by $\Delta T_f = -iK_f m$. Since $K_f = 1.86\,°C/(mol\,kg^{-1})$ for water, $\Delta T_f = -i0.186\,°C$ for this group of 0.10 mol kg^{-1} solutions.

(a) $T_f = -0.186\,°C$. Urea is a nonelectrolyte, and $i = 1$.

(b) $T_f = -0.372\ °C$. NH_4NO_3 is a strong electrolyte, composed of two ions per formula unit; $i = 2$.

(c) $T_f = -0.372\ °C$. HCl is a strong electrolyte, composed of two ions per formula unit; $i = 2$.

(d) $T_f = -0.558\ °C$. $CaCl_2$ is a strong electrolyte, composed of three ions per formula unit; $i = 3$.

(e) $T_f = -0.372\ °C$. $MgSO_4$ is a strong electrolyte, composed of two ions per formula unit; $i = 2$.

(f) $T_f = -0.186\ °C$. Ethanol is a nonelectrolyte; $i = 1$.

(g) $T_f < -0.186\ °C$. $HC_2H_3O_2$ is a weak electrolyte; i is somewhat larger than 1.

83. **(E)** The combination of $NH_3(aq)$ with $CH_3COOH(aq)$, results in the formation of $NH_4CH_3COO(aq)$, which is a solution of the ions NH_4^+ and CH_3COO^-.
$$NH_3(aq) + CH_3COOH(aq) \rightarrow NH_4CH_3COO(aq) \rightarrow NH_4^+(aq) + CH_3COO^-(aq)$$
This solution of ions or strong electrolytes conducts a current very well.

85. **(E)** The answer is (d). The other options describe (a) a gas dissolved in water, (b) a gas which when dissolved in water fully dissociates and is a strong electrolyte, and (c) a non-electrolytic molecule in water.

INTEGRATIVE AND ADVANCED EXERCISES

88. **(M)** The molarity of the original solution is computed first.
$$\text{KOH molarity} = \frac{109.2\ \text{g KOH} \times \dfrac{1\ \text{mol KOH}}{56.010\ \text{g KOH}}}{1\ \text{L soln}} = 1.950\ M$$
A 1.950 M solution is more concentrated than a 0.250 m solution. Thus, we must dilute the original solution. First we determine the mass of water produced in the final solution, and the mass of water present in the original solution, and finally the mass of water we must add.

$$\text{mass } H_2O \text{ in final soln} = 0.1000\ \text{L orig. soln} \times \frac{1.950\ \text{mol KOH}}{1\ \text{L soln}} \times \frac{1\ \text{kg } H_2O}{0.250\ \text{mol KOH}} = 0.780\ \text{kg } H_2O$$

$$\text{mass original solution} = 100.0\ \text{mL} \times \frac{1.09\ \text{g}}{1\ \text{mL}} = 109\ \text{g original solution}$$

$$\text{mass KOH} = 100.0\ \text{mL} \times \frac{1\ \text{L}}{1000\ \text{mL}} \times \frac{109.2\ \text{g KOH}}{1\ \text{L soln}} = 10.92\ \text{g KOH}$$

original mass of water = 109 g soln − 10.92 g KOH = 98 g H_2O

mass added H_2O = 780.g H_2O − 98 g H_2O = 682 g H_2O

92. **(M)** First determine the molality of the solution with the desired freezing point and compare it to the molality of the supplied solution to determine if ethanol or water needs to be added.

$$m = \frac{\Delta T_f}{-K_f} = \frac{-2.0\ ^\circ C}{-1.86\ ^\circ C/(mol\ kg^{-1})} = 1.1\ mol\ kg^{-1}\quad (desired)$$

$$mass\ solution = 2.50\ L \times \frac{1000\ mL}{1\ L} \times \frac{0.9767\ g}{1\ mL} = 2.44 \times 10^3\ g\ solution$$

$$mass\ solute = 2.44 \times 10^3\ g\ soln \times \frac{13.8\ g\ C_2H_5OH}{100.0\ g\ soln} = 337\ g\ C_2H_5OH$$

$$mass\ H_2O = (2.44 \times 10^3\ g - 337\ g\ C_2H_5OH) \times \frac{1\ kg}{1000\ g} = 2.10\ kg\ H_2O$$

$$m = \frac{337\ g\ C_2H_5OH \times \dfrac{1\ mol\ C_2H_5OH}{46.07\ g\ C_2H_5OH}}{2.10\ kg\ H_2O} = 3.48\ mol\ kg^{-1}\quad (available)$$

Thus we need to dilute the solution with water. We must find the mass of water needed in the final solution, then the mass of water that must be added.

$$final\ mass\ of\ water = 337\ g\ C_2H_5OH \times \frac{1\ mol\ C_2H_5OH}{46.07\ g\ C_2H_5OH} \times \frac{1\ kg\ H_2O}{1.1\ mol\ C_2H_5OH} = 6.6\ kg\ H_2O$$

mass H_2O needed = 6.6 kg total − 2.10 kg already present = 4.5 kg H_2O

95. **(M)** First we determine the molality of the benzene solution, and then the number of moles of solute in the sample.

$$molality = \frac{\Delta T_f}{-K_f} = \frac{5.072\ ^\circ C - 5.533\ ^\circ C}{-5.12\ ^\circ C/(mol\ kg^{-1})} = 0.0900\ mol\ kg^{-1}$$

$$amount\ of\ solute = 50.00\ mL \times \frac{0.879\ g}{1\ mL} \times \frac{1\ kg}{1000\ g} \times \frac{0.0900\ mol\ solute}{1\ kg\ solvent} = 0.00396\ mol\ solute$$

We use this amount to determine the amount of each acid in the solute, with the added data of the molar masses of stearic acid, $C_{18}H_{36}O_2$, 284.5 g/mol, and palmitic acid, $C_{18}H_{36}O_2$, 256.4 g/mol. We let x represent the amount in moles of palmitic acid.

$$1.115\ g\ sample = (0.00396 - x)\ mol \times 284.5\ g/mol + x\ mol \times 256.4\ g/mol$$
$$= 1.12\underline{7} - 284.5\ x + 256.4\ x = 1.12\underline{7} - 28.1\ x$$

$$x = \frac{1.12\underline{7} - 1.115}{28.1} = 0.0004\underline{3}\ mol\ palmitic\ acid$$

$$0.00396 - 0.0004\underline{3} = 0.0035\underline{3}\ mol\ stearic\ acid$$

mass palmitic acid = 0.00043 mol × 256.4 g/mol = 0.11 g palmitic acid

$$\%\ palmitic\ acid = \frac{0.11\ g\ palmitic\ acid}{1.115\ g\ sample} \times 100\% = 1 \times 10^1\ \%\ palmitic\ acid\ (about\ 10\%)$$

97. **(D)** We first determine the amount in moles of each substance.

$$\text{amount } CO(NH_2)_2 = 0.515 \text{ g } CO(NH_2)_2 \times \frac{1 \text{ mol } CO(NH_2)_2}{60.06 \text{ g } CO(NH_2)_2} = 0.00858 \text{ mol } CO(NH_2)_2$$

$$\text{amount } H_2O \text{ with urea} = 92.5 \text{ g } H_2O \times \frac{1 \text{ mol } H_2O}{18.015 \text{ g } H_2O} = 5.13 \text{ mol } H_2O$$

$$\text{amount } C_{12}H_{22}O_{11} = 2.50 \text{ g } C_{12}H_{22}O_{11} \times \frac{1 \text{ mol } C_{12}H_{22}O_{11}}{342.3 \text{ g } C_{12}H_{22}O_{11}} = 0.00730 \text{ mol } C_{12}H_{22}O_{11}$$

$$\text{amount } H_2O \text{ with sucrose} = 85.0 \text{ g } H_2O \times \frac{1 \text{ mol } H_2O}{18.015 \text{ g } H_2O} = 4.72 \text{ mol } H_2O$$

The vapor pressure of water will be the same above both solutions when their mole fractions are equal. We assume that the amount of water present as water vapor is negligible. The total amount of water in the two solutions is $(4.72 + 5.13 =) 9.85$ mol H_2O. We let n_{water} designate the amount of water in the urea solution. The amount of water in the sucrose solution is then 9.85 mole $- n_{water}$. Note that we can compute the mole fraction of solute for comparison, since when the two solute mole fractions are equal, the mole fractions of solvent will also be equal $(\chi_{solvent} = 1.0000 - \chi_{solute})$.

$$\frac{0.00857 \text{ mol urea}}{0.00858 \text{ mol urea} + n_{water}} = \chi_{urea} = \chi_{sucrose} = \frac{0.00730 \text{ mol sucrose}}{0.00730 \text{ mol sucrose} + (9.85 - n_{water})}$$

We "cross multiply" to begin the solution for n_{water}.

$$0.00857 (0.00730 + 9.85 - n_{water}) = 0.00730 (0.00858 + n_{water})$$
$$0.0845 - 0.00857 \, n_{water} = 0.0000626 + 0.00730 \, n_{water} \quad n_{water}$$
$$= \frac{0.0845 - 0.0000626}{0.00730 + 0.00857} = 5.32 \text{ mol}$$

We check the answer by substitution into the mole fraction equation.

$$\frac{0.00857 \text{ mol urea}}{0.00857 \text{ mol} + 5.32\underline{4}} = 0.00161 = \chi_{urea} = \chi_{sucrose} = 0.00161$$

$$= \frac{0.00730 \text{ mol sucrose}}{0.00730 \text{ mol} + (9.85 - 5.32\underline{4}) \text{mol}}$$

The mole fraction of water in each solution is $(1.00000 - 0.00161 =) 0.99839$, or 99.839 mol%

100. (M)

(a) Surface area of a particle $= 4\pi r^2 = 4(3.1416)(1 \times 10^{-7} \text{m})^2 = 1.26 \times 10^{-13} \text{ m}^2/\text{particle}$. We need to find the number of particles present (not atoms!)

$$\text{particle volume} = \frac{4\pi r^3}{3} = \frac{4\pi (1 \times 10^{-7}\text{m})^3}{3} = 4.19 \times 10^{-21} \text{m}^3$$

$$\text{particle mass} = \text{density} \times \text{volume} = \frac{19.3 \text{ g}}{\text{cm}^3} \frac{(100 \text{ cm})^3}{(1 \text{ m})^3} \times 4.19 \times 10^{-21} \text{m}^3 = 8.09 \times 10^{-14} \text{g/particle}$$

$$\text{number of Au particles} = \frac{\text{mass of Au}}{\text{particle mass}} = \frac{1.00 \times 10^{-3}\text{g Au}}{8.09 \times 10^{-14}\text{g/particle}} = 1.24 \times 10^{10} \text{particles}$$

$$\text{total surface area} = (1.26 \times 10^{-13}\text{m}^2/\text{particle})(1.24 \times 10^{10}\text{particles}) = 1.56 \times 10^{-3} \text{m}^2$$

(b) For 1.00 mg Au the volume of Au is: $(1.00 \text{ mg})(1 \text{ g}/1000 \text{ mg})/(19.3 \text{ g/cm}^3) = 5.18 \times 10^{-5}$ cm^3

Volume of a cube is L^3, where L is the edge length, so
$$L = \sqrt[3]{5.18 \times 10^{-5} \text{ cm}^3} = 3.73 \times 10^{-2} \text{ cm}$$

$$\text{area} = 6 \times L^2 = 6 \times (3.73 \times 10^{-2} \text{ cm})^2 \times (1 \text{ m}/100 \text{ cm})^2 = 8.34 \times 10 - 7 \text{ m}^2$$

105. (M) We determine the volumes of O_2 and N_2 that dissolve under the appropriate partial pressures of 0.2095 atm O_2 and 0.7808 atm N_2, and at a temperature of 25 °C, with both gas volumes measured at 1.00 atm and 25 °C. Note that the percent by volume and the percent by partial pressures of the atmosphere are numerically the same.

$$O_2 \text{ volume} = 0.2095 \text{ atm} \times \frac{28.31 \text{ mL O}_2}{1 \text{ atm} \cdot \text{L soln}} = 5.931 \frac{\text{mL O}_2}{\text{L solution}}$$

$$N_2 \text{ volume} = 0.7808 \text{ atm} \times \frac{14.34 \text{ mL N}_2}{1 \text{ atm} \cdot \text{L soln}} = 11.20 \frac{\text{mL N}_2}{\text{L solution}}$$

$$\text{volume \% N}_2 = \frac{11.20 \frac{\text{mL N}_2}{\text{L solution}}}{11.20 \frac{\text{mL N}_2}{\text{L solution}} + 5.931 \frac{\text{mL O}_2}{\text{L solution}}} \times 100\% = 65.38\% \text{ N}_2 \text{ by volume}$$

$$\text{volume \% O}_2 = 100.00\% - 65.38\% = 34.62\% \text{ O}_2 \text{ by volume}$$

107. **(D)** First we determine the mass of $CuSO_4$ in the original solution at 70 °C.

$$335 \text{ g sample} \times \frac{32.0 \text{ g } CuSO_4}{100.0 \text{ g soln}} = 107.\underline{2} \text{ g } CuSO_4$$

We let x represent the mass of $CuSO_4$ removed from the solution when the $CuSO_4 \cdot 5H_2O$ recrystallizes. Then the mass of H_2O present in the recrystallized product is determined.

$$\text{mass } H_2O = x \text{ g } CuSO_4 \times \frac{1 \text{ mol } CuSO_4}{159.6 \text{ g } CuSO_4} \times \frac{5 \text{ mol } H_2O}{1 \text{ mol } CuSO_4} \times \frac{18.02 \text{ g } H_2O}{1 \text{ mol } H_2O} = 0.565x \text{ g } H_2O$$

Now we determine the value of x by using the concentration of the saturated solution at 0 °C.

$$\frac{12.5 \text{ g } CuSO_4}{100.0 \text{ g soln}} = \frac{(107.\underline{2} - x) \text{ g } CuSO_4}{(335 - x - 0.565\,x) \text{ g soln}}$$

$$1.07\underline{2} \times 10^4 - 100.0x = 4.19 \times 10^3 - 19.6x \qquad (100.0 - 19.6)x = 1.07\underline{2} \times 10^4 - 4.19 \times 10^3$$

$$x = \frac{6.5\underline{3} \times 10^3}{80.4} = 81 \text{ g } CuSO_4$$

$$\text{mass } CuSO_4 \cdot 5H_2O = 81 \text{ g } CuSO_4 \times \frac{1.000 \text{ g } CuSO_4 + 0.565 \text{ g } H_2O}{1.000 \text{ g } CuSO_4} = 1.3 \times 10^2 \text{ g } CuSO_4 \cdot 5H_2O$$

109. **(E)**

$$\text{mass of } N_2 = \frac{445 \times 10^{-6} \text{mol } N_2}{L \text{ of seawater}} \times \frac{28.0 \text{ g } N_2}{1 \text{ mol}} = 0.0125 \frac{g}{L}$$

Partial pressure can be determined from Henry's Law:

$C = k_H P$

$$445 \times 10^{-6} \text{ M} = \left(0.61 \times 10^{-3} \text{M} \cdot \text{atm}^{-1}\right) P$$

$$P = \frac{445 \times 10^{-6} \text{ M}}{0.61 \times 10^{-3} \text{M} \cdot \text{atm}^{-1}} = 0.73 \text{ atm}$$

111. **(M)** First, determine what molality of ethylene glycol is needed to cause a 10.0 °C freezing-point depression:

$\Delta T_f = -K_f \cdot m$

$-10 \,°C = \left(-1.86 \,°C \cdot \text{mol}^{-1} \text{kg}\right) \cdot m$

$m = 5.376 \text{ mol kg}^{-1}$

Then, determine the number of moles of ethylene glycol, which can be used to determine the volume, using the density of ethylene glycol given.

$$\text{mol C}_2\text{H}_6\text{O}_2 = \frac{5.376 \text{ mol C}_2\text{H}_6\text{O}_2}{1 \text{ kg H}_2\text{O}} \times \frac{1 \text{ kg H}_2\text{O}}{1 \text{ L H}_2\text{O}} \times \frac{20.0 \text{ L H}_2\text{O}}{\text{sample}} = 107.53 \text{ mol C}_2\text{H}_6\text{O}_2/\text{sample}$$

$$\text{Vol C}_2\text{H}_6\text{O}_2 = 107.53 \text{ mol C}_2\text{H}_6\text{O}_2 \times \frac{62.09 \text{ g C}_2\text{H}_6\text{O}_2}{1 \text{ mol C}_2\text{H}_6\text{O}_2} \times \frac{1 \text{ mL}}{1.12 \text{ g C}_2\text{H}_6\text{O}_2} = 5961 \text{ mL} = 5.96 \text{ L}$$

FEATURE PROBLEMS

114. (D)

(a) A solution with $\chi_{HCl} = 0.50$ begins to boil at about 18 °C. At that temperature, the composition of the vapor is about $\chi_{HCl} = 0.63$, reading directly across the tie line at 18 °C. The vapor has $\chi_{HCl} > 0.50$.

(b) The composition of HCl(aq) changes as the solution boils in an open container because the vapor has a different composition than does the liquid. Thus, the component with the lower boiling point is depleted as the solution boils. The boiling point of the remaining solution must change as the vapor escapes due to changing composition.

(c) The azeotrope occurs at the maximum of the curve: at $\chi_{HCl} = 0.12$ and a boiling temperature of 110 °C.

(d) We first determine the amount of HCl in the sample.

$$\text{amount HCl} = 30.32 \text{ mL NaOH} \times \frac{1 \text{ L}}{1000 \text{ mL}} \times \frac{1.006 \text{ mol NaOH}}{1 \text{ L}} \times \frac{1 \text{ mol HCl}}{1 \text{ mol NaOH}}$$

amount HCl = 0.03050 mol HCl

The mass of water is the difference between the mass of solution and that of HCl.

$$\text{mass H}_2\text{O} = \left(5.00 \text{ mL soln} \times \frac{1.099 \text{ g}}{1 \text{ mL}}\right) - \left(0.03050 \text{ mol HCl} \times \frac{36.46 \text{ g HCl}}{1 \text{ mol HCl}}\right) = 4.38 \text{ g}$$

Now we determine the amount of H$_2$O and then the mole fraction of HCl.

$$\text{amount H}_2\text{O} = 4.38 \text{ g H}_2\text{O} \times \frac{1 \text{ mol H}_2\text{O}}{18.02 \text{ g H}_2\text{O}} = 0.243 \text{ mol H}_2\text{O}$$

$$\chi_{HCl} = \frac{0.03050 \text{ mol HCl}}{0.03050 \text{ mol HCl} + 0.243 \text{ mol H}_2\text{O}} = 0.112$$

115. (M)

    (a)   At 20 °C, the solubility of NaCl is 35.9 g NaCl / 100 g H_2O. We determine the mole fraction of H_2O in this solution

$$\text{amount } H_2O = 100 \text{ g } H_2O \times \frac{1 \text{ mol } H_2O}{18.02 \text{ g } H_2O} = 5.549 \text{ mol } H_2O$$

$$\text{amount NaCl} = 35.9 \text{ g NaCl} \times \frac{1 \text{ mol NaCl}}{58.44 \text{ g NaCl}} = 0.614 \text{ mol NaCl}$$

$$\chi_{water} = \frac{5.549 \text{ mol } H_2O}{0.614 \text{ mol NaCl} + 5.549 \text{ mol } H_2O} = 0.9004$$

The approximate relative humidity then will be 90% (90.04%), because the water vapor pressure above the NaCl saturated solution will be 90.04% of the vapor pressure of pure water at 20 °C.

    (b) $CaCl_2 \cdot 6 \, H_2O$ deliquesces if the relative humidity is over 32%. Thus, $CaCl_2 \cdot 6 \, H_2O$ will deliquesce (i.e., it will absorb water from the atmosphere).

    (c)   If the substance in the bottom of the desiccator has high water solubility, its saturated solution will have a low χ_{water}, which in turn will produce a low relative humidity. Thus, a relative humidity lower than 32% is needed to keep $CaCl_2 \cdot 6 \, H_2O$ dry.

116. (D)

    (a)   We first compute the molality of a 0.92% mass/volume solution, assuming the solution's density is about 1.00 g/mL, meaning that 100.0 mL solution has a mass of 100.0 g.

$$\text{molality} = \frac{0.92 \text{ g NaCl} \times \dfrac{1 \text{ mol NaCl}}{58.44 \text{ g NaCl}}}{(100.0 \text{ g soln} - 0.92 \text{ g NaCl}) \times \dfrac{1 \text{ kg solvent}}{1000 \text{ g}}} = 0.16 \text{ mol kg}^{-1}$$

Then we compute the freezing-point depression of this solution.

$$\Delta T_f = -i K_f m = \frac{-2.0 \text{ mol ions}}{\text{mol NaCl}} \times \frac{1.86 \, °C}{\text{mol kg}^{-1}} \times 0.16 \text{ mol kg}^{-1} = -0.60 \, °C$$

The van't Hoff factor of NaCl most likely is not equal to 2.0, but a bit less and thus the two definitions are in fair agreement.

    (b)   We calculate the amount of each solute, assume 1.00 L of solution has a mass of 1000 g, and subtract the mass of all solutes to determine the mass of solvent.

$$\text{amount NaCl ions} = 3.5 \text{ g NaCl} \times \frac{1 \text{ mol NaCl}}{58.44 \text{ g NaCl}} \times \frac{2 \text{ mol ions}}{1 \text{ mol NaCl}} = 0.12 \text{ mol ions}$$

$$\text{amount KCl ions} = 1.5 \text{ g} \times \frac{1 \text{ mol KCl}}{74.55 \text{ g KCl}} \times \frac{2 \text{ mol ions}}{1 \text{ mol KCl}} = 0.040 \text{ mol ions}$$

$$\text{amount Na}_3\text{C}_6\text{H}_5\text{O}_7 \text{ ions} = 2.9 \text{ g} \times \frac{1 \text{ mol Na}_3\text{C}_6\text{H}_5\text{O}_7}{258.07 \text{ g Na}_3\text{C}_6\text{H}_5\text{O}_7} \times \frac{4 \text{ mol ions}}{1 \text{ mol Na}_3\text{C}_6\text{H}_5\text{O}_7}$$

$$\text{amount Na}_3\text{C}_6\text{H}_5\text{O}_7 \text{ ions} = 0.045 \text{ mol ions}$$

$$\text{amount C}_6\text{H}_{12}\text{O}_6 = 20.0 \text{ g C}_6\text{H}_{12}\text{O}_6 \times \frac{1 \text{ mol C}_6\text{H}_{12}\text{O}_6}{180.2 \text{ g C}_6\text{H}_{12}\text{O}_6} = 0.111 \text{ mol C}_6\text{H}_{12}\text{O}_6$$

$$\text{solvent mass} = 1000.0 \text{ g} - (3.5 \text{ g} + 1.5 \text{ g} + 2.9 \text{ g} + 20.0 \text{ g}) = 972.1 \text{ g H}_2\text{O}$$

$$\text{solvent mass} = 0.9721 \text{ kg H}_2\text{O}$$

$$\text{solution molality} = \frac{(0.120 + 0.040 + 0.045 + 0.111) \text{ mol}}{0.9721 \text{ kg H}_2\text{O}} = 0.325 \text{ mol kg}^{-1}$$

$$\Delta T_f = -K_f m = -1.86 \text{ °C/(mol kg}^{-1}) \times 0.325 \text{ mol kg}^{-1} = -0.60 \text{ °C}$$

This again is close to the defined freezing point of –0.52 °C, with the error most likely arising from the van't Hoff factors not being integral. We can conclude, therefore, that the solution is isotonic.

SELF-ASSESSMENT EXERCISES

120. (E) The answer is (b), because molality is moles solute/kg solvent. Assuming a volume of 1 L, at a concentration of 0.010 mol kg^{-1}, the mass of the water (~1 kg) is nearly the same as the volume of the solution (1 L).

121. (E) The answer is (d), because the other choices either don't form a solution because they are immiscible or the solute-solvent interactions are stronger than intramolecular interactions for either of the pure solute or solvent, making the solutions non-ideal.

122. (E) The answer is (a). As partial pressure of the gas increases, Henry's law shows that the concentration of the gas in the liquid increases.

123. (E) The answer is (d), because the total ionic molality of the $MgCl_2$ is 3×0.008 = 0.024.

124. (E) The answer is (c), because nothing in the information provided suggests that the concentrations of the two volatile species are the same.

125. (E) The mass of NH_4Cl is: 1.12 mol NH_4Cl × (53.45 g/mol) = 59.86 g.
Solubility is 59.86 g NH_4Cl / 150.0 g H_2O = 0.40, or 40 g/100 g.

Based on Figure 14-10, the concentration limit of NH_4Cl is 42 g per 100 g of H_2O. Since the amount of NH_4Cl is below the solubility limit, no crystals will form.

126. (M)

(a) $\left[Na^+\right] = \dfrac{0.92 \text{ g NaCl}}{100 \text{ mL H}_2\text{O}} \times \dfrac{1 \text{ mol NaCl}}{58.45 \text{ g NaCl}} \times \dfrac{1 \text{ mol Na}^+}{1 \text{ mol NaCl}} \times \dfrac{1000 \text{ mL}}{1 \text{ L}} = 0.16 \text{ M Na}^+$

(b) total molarity of ions $= 0.16 \text{ M Na}^+ \times \dfrac{1 \text{ mol NaCl}}{1 \text{ mol Na}^+} \times \dfrac{2 \text{ mol ions}}{1 \text{ mol NaCl}} = 0.32 \text{ M}$

(c) $\Pi = MRT = (0.32 \text{ M})(0.08206 \text{ L}\cdot\text{atm}\cdot\text{K}^{-1})(310 \text{ K}) = 8.1 \text{ atm}$

(d) Determine the solution molality first, then determine freezing-point depression.

molality $H_2O = \dfrac{0.32 \text{ mol ions}}{1 \text{ L solution}} \times \dfrac{1 \text{ L solution}}{1.005 \text{ kg solution}} = 0.3184 \text{ mol kg}^{-1}$

$\Delta_f T = \left(-1.86 \text{ °C/(mol kg}^{-1})\right)(0.3184 \text{ mol kg}^{-1}) = -0.60 \text{ °C}$

127. (M)

(a)

$[C_3H_8O_3] = \dfrac{62.0 \text{ g C}_3\text{H}_8\text{O}_3}{100 \text{ g sol'n}} \times \dfrac{1.159 \text{ g sol'n}}{1 \text{ mL sol'n}} \times \dfrac{1000 \text{ mL sol'n}}{1 \text{ L sol'n}} \times \dfrac{1 \text{ mol C}_3\text{H}_8\text{O}_3}{92.11 \text{ g C}_3\text{H}_8\text{O}_3}$

$= 7.80 \text{ M C}_3\text{H}_8\text{O}_3$

(b)

$[H_2O] = \dfrac{38.0 \text{ g H}_2\text{O}}{100 \text{ g sol'n}} \times \dfrac{1.159 \text{ g sol'n}}{1 \text{ mL sol'n}} \times \dfrac{1000 \text{ mL sol'n}}{1 \text{ L sol'n}} \times \dfrac{1 \text{ mol H}_2\text{O}}{18.02 \text{ g H}_2\text{O}}$

$= 24.4 \text{ M H}_2\text{O}$

(c)

$[H_2O] = \dfrac{38.0 \text{ g H}_2\text{O}}{62.0 \text{ g C}_3\text{H}_8\text{O}_3} \times \dfrac{1000 \text{ g solvent}}{1 \text{ kg solvent}} \times \dfrac{1 \text{ mol H}_2\text{O}}{18.02 \text{ g H}_2\text{O}}$

$= 34.0 \text{ mol kg}^{-1} \text{ H}_2\text{O}$

(d) We can determine the moles of $C_3H_8O_3$ and H_2O from the above calculations.
moles of $H_2O = 2.11$ mol.
moles of $C_3H_8O_3 = 0.673$ mol.

$x_{C_3H_8O_3} = \dfrac{0.673 \text{ mol C}_3\text{H}_8\text{O}_3}{0.673 \text{ mol C}_3\text{H}_8\text{O}_3 + 2.11 \text{ mol H}_2\text{O}} = 0.242$

(e) mol% $H_2O = \dfrac{2.11 \text{ mol H}_2\text{O}}{0.673 \text{ mol C}_3\text{H}_8\text{O}_3 + 2.11 \text{ mol H}_2\text{O}} \times 100 = 75.8\%$

128. (E)

    (1)   Solution with lowest conductivity: (b) 0.15 mol kg^{-1} $C_{12}H_{22}O_{11}$, because it is a non-electrolyte.

    (2)   Lowest boiling point: (d) 0.05 m NaCl, because the molality of the solute is the smallest, making boiling-point elevation the smallest value.

    (3)   Highest vapor pressure of water: (d), for the same reason as above.

    (4)   Lowest freezing point: (a), because the molality of the solute is the highest (2×0.10 m).

130. (E) The answer is (b), because the charge in all species is the same (+1 or –1), but F$^-$ is the smallest.

131. (E) The magnitude of $\Delta_{lattice}H$ is larger than the sum of the $\Delta_{hydration}H$ of the individual ions. Since the test tube turns cold, dissolution of NH_4Cl is endothermic, which means that the final energy of the system, the hydration of the individual ions, is less than the initial energy of the system, which is the lattice energy of the solid.

132. (E) The answer is (a). Lowering temperature increases solubility.

133. (E) The answer is (b), because the HCl dissociates completely in water to give cations and anions, which are then solvated with water (and are none-volatile).

134. (E) The answer is (a). Use Raoult's law to determine the mole fraction:

$$P_{C_6H_{14}} = x_{C_6H_{14}} \cdot P^*_{C_6H_{14}}$$

$$600 \text{ mmHg} = x_{solute} \cdot 760 \text{ mmHg}$$

$$x_{C_6H_{14}} = 0.789$$

$$x_{solute} = 1 - 0.789 = 0.211$$

135. (M) The answer is (e).

$$\Pi = MRT$$

$$\Pi = (0.312 \text{ M})(0.08206 \text{ L atm K}^{-1})(348 \text{ K}) = 8.91 \text{ atm}$$

$$P(\text{bar}) = 8.91 \text{ atm} \times \frac{1.0135 \text{ bar}}{1 \text{ atm}} = 9.03 \text{ bar}$$

136. (E) The answer is (c).

$$10.5 \text{ mol H}_2\text{O} \times \frac{18.0 \text{ g H}_2\text{O}}{1 \text{ mol H}_2\text{O}} = 189 \text{ g H}_2\text{O}$$

$$\text{mass}\% = \frac{23.4}{189 + 23.4} \times 100 = 11.0\%$$

CHAPTER 15
PRINCIPLES OF CHEMICAL EQUILIBRIUM
PRACTICE EXAMPLES

1A **(E)** The reaction is as follows:

$$2\,Cu^{2+}(aq) + Sn^{2+}(aq) \rightleftharpoons 2\,Cu^{+}(aq) + Sn^{4+}(aq)$$

Therefore, the equilibrium expression is as follows:

$$K = \frac{\left[Cu^{+}\right]^{2}\left[Sn^{4+}\right]}{\left[Cu^{2+}\right]^{2}\left[Sn^{2+}\right]}$$

Rearranging and solving for Cu^{2+}, the following expression is obtained:

$$\left[Cu^{2+}\right] = \left(\frac{\left[Cu^{+}\right]^{2}\left[Sn^{4+}\right]}{K\left[Sn^{2+}\right]}\right)^{1/2} = \left(\frac{x^{2}\cdot x}{(1.48)\,x}\right)^{1/2} = \frac{x}{1.22}$$

1B **(E)** The reaction is as follows:

$$2\,Fe^{3+}(aq) + Hg_{2}^{2+}(aq) \rightleftharpoons 2\,Fe^{2+}(aq) + 2\,Hg^{2+}(aq)$$

Therefore, the equilibrium expression is as follows:

$$K = \frac{\left[Fe^{2+}\right]^{2}\left[Hg^{2+}\right]^{2}}{\left[Fe^{3+}\right]^{2}\left[Hg_{2}^{2+}\right]} = \frac{(0.0025)^{2}(0.0018)^{2}}{(0.015)^{2}(x)} = 9.14\times10^{-6}$$

Rearranging and solving for Hg_{2}^{2+}, the following expression is obtained:

$$\left[Hg_{2}^{2+}\right] = \frac{\left[Fe^{2+}\right]^{2}\left[Hg^{2+}\right]^{2}}{\left[Fe^{3+}\right]^{2}\cdot K} = \frac{(0.0025)^{2}(0.0018)^{2}}{(0.015)^{2}(9.14\times10^{-6})} = 0.009847 \approx 0.0098\ M$$

2A **(E)** The example gives $K_c = 5.8\times10^{5}$ for the reaction $N_2(g) + 3\,H_2(g) \rightleftharpoons 2\,NH_3(g)$. The reaction we are considering is one-third of this reaction. If we divide the reaction by 3, we should take the cube root of the equilibrium constant to obtain the value of the equilibrium constant for the "divided" reaction: $K_{c3} = \sqrt[3]{K_c} = \sqrt[3]{5.8\times10^{5}} = 8.3\times10^{2}$

2B **(E)** First we reverse the given reaction to put $NO_2(g)$ on the reactant side. The new equilibrium constant is the inverse of the given one.

$$NO_2(g) \rightleftharpoons NO(g) + \tfrac{1}{2}O_2(g) \qquad K_c' = 1/(1.2\times10^{2}) = 0.0083$$

Then we double the reaction to obtain 2 moles of $NO_2(g)$ as reactant. The equilibrium constant is then raised to the second power.

$$2\,NO_2(g) \rightleftharpoons 2\,NO(g) + O_2(g) \qquad K_c = (0.00833)^{2} = 6.9\times10^{-5}$$

3A **(E)** We remember that neither solids, such as $Ca_5(PO_4)_3OH(s)$, nor liquids, such as $H_2O(l)$, appear in the equilibrium constant expression. Concentrations of products appear in the numerator, those of reactants in the denominator. $K_c = \dfrac{\left[Ca^{2+}\right]^5\left[HPO_4^{2-}\right]^3}{\left[H^+\right]^4}$

3B **(E)** First we write the balanced chemical equation for the reaction. Then we write the equilibrium constant expression, remembering that gases appear in the K_p expression, but pure liquids and pure solids do not.

$$3\,Fe(s) + 4\,H_2O(g) \rightleftharpoons Fe_3O_4(s) + 4\,H_2(g)$$

$$K_p = \frac{\{P(H_2)\}^4}{\{P(H_2O)\}^4}$$

4A **(E)** We use the expression $K_p = K_c(RT)^{\Delta v_{gas}}$. In this case, $\Delta v_{gas} = 3 + 1 - 2 = 2$ and thus we have

$$K_p = K_c(RT)^2 = 2.8 \times 10^{-9} \times (0.08314 \times 298)^2 = 1.7 \times 10^{-6}$$

4B **(M)** We begin by writing the K_p expression. We then substitute $P = (n/V)RT = [concentration]RT$ for each pressure. We collect terms to obtain an expression relating K_c and K_p, into which we substitute to find the value of K_c.

$$K_p = \frac{\{P(H_2)\}^2\{P(S_2)\}}{\{P(H_2S)\}^2} = \frac{([H_2]RT)^2([S_2]RT)}{([H_2S]RT)^2} = \frac{[H_2]^2[S_2]}{[H_2S]^2}RT = K_c RT$$

The same result can be obtained by using $K_p = K_c(RT)^{\Delta v_{gas}}$, since $\Delta v_{gas} = 2 + 1 - 2 = +1$.

$$K_c = \frac{K_p}{RT} = \frac{1.2 \times 10^{-2}}{0.08314 \times (1065 + 273)} = 1.1 \times 10^{-4}$$

But the reaction has been reversed and halved. Thus $K_{final} = \sqrt{\dfrac{1}{K_c}} = \sqrt{\dfrac{1}{1.1 \times 10^{-4}}} = \sqrt{9091} = 95$

5A **(M)** We compute the value of Q_c. Each concentration equals the mass (m) of the substance divided by its molar mass (this quotient is the amount of the substance in moles) and further divided by the volume of the container.

$$Q_c = \frac{[CO_2][H_2]}{[CO][H_2O]} = \frac{\dfrac{m \times \dfrac{1\ mol\ CO_2}{44.0\ g\ CO_2}}{V} \times \dfrac{m \times \dfrac{1\ mol\ H_2}{2.0\ g\ H_2}}{V}}{\dfrac{m \times \dfrac{1\ mol\ CO}{28.0\ g\ CO}}{V} \times \dfrac{m \times \dfrac{1\ mol\ H_2O}{18.0\ g\ H_2O}}{V}} = \frac{\dfrac{1}{44.0 \times 2.0}}{\dfrac{1}{28.0 \times 18.0}} = \frac{28.0 \times 18.0}{44.0 \times 2.0} = 5.7 > 1.00 = K_c$$

(In evaluating the expression above, we cancelled the equal values of V, and we also cancelled the equal values of m.) Because the value of Q_c is larger than the value of K_c, the reaction will proceed to the left to reach a state of equilibrium. Thus, at equilibrium there will be greater quantities of reactants, and smaller quantities of products than there were initially.

5B **(M)** We compare the value of the reaction quotient, Q_p, to that of K_p.

$$Q_p = \frac{\{P(PCl_3)\}\{P(Cl_2)\}}{\{P(PCl_5)\}} = \frac{2.19 \times 0.88}{19.7} = 0.098$$

$$K_p = K_c (RT)^{2-1} = K_c (RT)^1 = 0.0454 \times (0.08314 \times (261 + 273))^1 = 2.02$$

Because $Q_c < K_c$, the net reaction will proceed to the right, forming products and consuming reactants.

6A **(E)** $O_2(g)$ is a reactant. The equilibrium system will shift right, forming product in an attempt to consume some of the added $O_2(g)$ reactant. Looked at in another way, $[O_2]$ is increased above its equilibrium value by the addition of oxygen. This makes Q_c smaller than K_c. (The $[O_2]$ is in the denominator of the expression.) And the system shifts right to drive Q_c back up to K_c, at which point equilibrium will have been achieved.

6B **(M)**

(a) The position of an equilibrium mixture is affected only by changing the concentration of substances that appear in the equilibrium constant expression, $K_c = [CO_2]$. Since CaO(s) is a pure solid, its concentration does not appear in the equilibrium constant expression and thus adding extra CaO(s) will have no direct effect on the position of equilibrium.

(b) The addition of $CO_2(g)$ will increase $[CO_2]$ above its equilibrium value. The reaction will shift left to alleviate this increase, causing some $CaCO_3(s)$ to form.

(c) Since $CaCO_3(s)$ is a pure solid like CaO(s), its concentration does not appear in the equilibrium constant expression and thus the addition of any solid $CaCO_3$ to an equilibrium mixture will not have an effect upon the position of equilibrium.

7A **(E)** We know that a decrease in volume or an increase in pressure of an equilibrium mixture of gases causes a net reaction in the direction producing the smaller number of moles of gas. In the reaction in question, that direction is to the left: one mole of $N_2O_4(g)$ is formed when two moles of $NO_2(g)$ combine. Thus, decreasing the cylinder volume would have the initial effect of doubling both $[N_2O_4]$ and $[NO_2]$. In order to reestablish equilibrium, some NO_2 will then be converted into N_2O_4. Note, however, that the NO_2 concentration will still ultimately end up being higher than it was prior to pressurization.

7B **(E)** In the balanced chemical equation for the chemical reaction, $\Delta v_{\text{gas}} = (1+1) - (1+1) = 0$.

As a consequence, a change in overall volume or total gas pressure will have no effect on the position of equilibrium. In the equilibrium constant expression, the two partial pressures in the numerator will be affected to exactly the same degree, as will the two partial pressures in the denominator, and, as a result, Q_p will continue to equal K_p.

8A **(E)** The cited reaction is endothermic. Raising the temperature on an equilibrium mixture favors the endothermic reaction. Thus, $N_2O_4(g)$ should decompose more completely at higher temperatures and the amount of $NO_2(g)$ formed from a given amount of $N_2O_4(g)$ will be greater at high temperatures than at low ones.

8B **(E)** The $NH_3(g)$ formation reaction is $\frac{1}{2}N_2(g) + \frac{3}{2}H_2(g) \rightarrow NH_3(g)$, $\Delta_r H° = -46.11$ kJ/mol. This reaction is an exothermic reaction. Lowering temperature causes a shift in the direction of this exothermic reaction to the right toward products. Thus, the equilibrium $[NH_3(g)]$ will be greater at 100°C.

9A **(E)** We write the expression for K_c and then substitute expressions for molar concentrations.

$$K_c = \frac{[H_2]^2[S_2]}{[H_2S]^2} = \frac{\left(\dfrac{0.22}{3.00}\right)^2 \dfrac{0.11}{3.00}}{\left(\dfrac{2.78}{3.00}\right)^2} = 2.3 \times 10^{-4}$$

9B **(M)** We write the equilibrium constant expression and solve for $[N_2O_4]$.

$$K_c = 4.61 \times 10^{-3} = \frac{[NO_2]^2}{[N_2O_4]} \qquad [N_2O_4] = \frac{[NO_2]^2}{4.61 \times 10^{-3}} = \frac{(0.0236)^2}{4.61 \times 10^{-3}} = 0.121 \text{ M}$$

Then we determine the mass of N_2O_4 present in 2.26 L.

$$N_2O_4 \text{ mass} = 2.26 \text{ L} \times \frac{0.121 \text{ mol } N_2O_4}{1 \text{ L}} \times \frac{92.01 \text{ g } N_2O_4}{1 \text{ mol } N_2O_4} = 25.2 \text{ g } N_2O_4$$

10A **(M)** We use the initial-change-equilibrium setup to establish the amount of each substance at equilibrium. We then label each entry in the table in the order of its determination (1st, 2nd, 3rd, 4th, 5th), to better illustrate the technique. We know the initial amounts of all substances (1st). There are no products at the start.

Because ''initial'' + ''change'' = ''equilibrium'', the equilibrium amount (2nd) of $Br_2(g)$ enables us to determine "change" (3rd) for $Br_2(g)$. We then use stoichiometry to write other entries (4th) on the "change" line. And finally, we determine the remaining equilibrium amounts (5th).

Reaction:	2 NOBr(g)	$\rightleftharpoons$	2 NO(g)	+	$Br_2(g)$
Initial:	1.86 mol (1st)		0.00 mol (1st)		0.00 mol (1st)
Change:	−0.164 mol (4th)		+0.164 mol (4th)		+0.082 mol (3rd)
Equil.:	1.70 mol (5th)		0.164 mol (5th)		0.082 mol (2nd)

$$K_c = \frac{[NO]^2[Br_2]}{[NOBr]^2} = \frac{\left(\dfrac{0.164}{5.00}\right)^2\left(\dfrac{0.082\ mol}{5.00}\right)}{\left(\dfrac{1.70}{5.00}\right)^2} = 1.5\times10^{-4}$$

Here, $\Delta v_{gas} = 2+1-2 = +1$. $K_p = K_c(RT)^{+1} = 1.5\times10^{-4}\times(0.08314\times298) = 3.7\times10^{-3}$

10B **(M)** Use the amounts stated in the problem to determine the equilibrium concentration for each substance.

Reaction:	$2\ SO_3(g)$	$\rightleftharpoons$	$2\ SO_2(g)$	$+$	$O_2(g)$
Initial:	0 mol		0.100 mol		0.100 mol
Changes:	+0.0916 mol		−0.0916 mol		−0.0916/2 mol
Equil.:	0.0916 mol		0.0084 mol		0.0542 mol
Concentrations:	$\dfrac{0.0916\ mol}{1.52\ L}$		$\dfrac{0.0084\ mol}{1.52\ L}$		$\dfrac{0.0542\ mol}{1.52\ L}$
Concentrations:	0.0603 M		0.0055 M		0.0357 M

We use these values to compute K_c for the reaction and then the relationship $K_p = K_c(RT)^{\Delta v_{gas}}$ (with $\Delta v_{gas} = 2+1-2 = +1$) to determine the value of K_p.

$$K_c = \frac{[SO_2]^2[O_2]}{[SO_3]^2} = \frac{(0.0055)^2(0.0357)}{(0.0603)^2} = 3.0\times10^{-4}$$

$$K_p = 3.0\times10^{-4}\times(0.08314\times900) \simeq 0.022$$

11A **(M)** The equilibrium constant expression is $K_p = P\{H_2O\}P\{CO_2\} = 0.231$ at 100 °C. From the balanced chemical equation, we see that one mole of $H_2O(g)$ is formed for each mole of $CO_2(g)$ produced. Consequently, $P\{H_2O\} = P\{CO_2\}$ and $K_p = (P\{CO_2\})^2$. We solve this expression for $P\{CO_2\}$: $P\{CO_2\} = \sqrt{(P\{CO_2\})^2} = \sqrt{K_p} = \sqrt{0.231} = 0.481$ bar.

11B **(M)** The equation for the reaction is $NH_4HS(s) \rightleftharpoons NH_3(g) + H_2S(g)$, $K_p = 0.108$ at 25 °C. The two partial pressures do not have to be equal at equilibrium. The only instance in which they must be equal is when the two gases come solely from the decomposition of $NH_4HS(s)$. In this case, some of the $NH_3(g)$ has come from another source. We can obtain the pressure of $H_2S(g)$ by substitution into the equilibrium constant expression, since we are given the equilibrium pressure of $NH_3(g)$.

$$K_p = P\{H_2S\}\ P\{NH_3\} = 0.108 = P\{H_2S\}\times0.500\ bar\ NH_3 \qquad P\{H_2S\} = \frac{0.108}{0.500} = 0.216\ bar$$

So, $P_{total} = P_{H_2S} + P_{NH_3} = 0.216\ bar + 0.500\ bar = 0.716\ bar$

12A **(M)** We set up this problem in the same manner that we have previously employed, namely designating the equilibrium amount of HI as $2x$. (Note that we have used the same multipliers for x as the stoichiometric coefficients.)

Equation: $H_2(g)$ $+$ $I_2(g)$ $\rightleftharpoons$ $2\,HI(g)$

Initial:	0.150 mol	0.200 mol	0 mol
Changes:	$-x$ mol	$-x$ mol	$+2x$ mol
Equil:	$(0.150-x)$ mol	$(0.200-x)$ mol	$2x$ mol

$$K_c = \frac{\left(\dfrac{2x}{15.0}\right)^2}{\dfrac{0.150-x}{15.0}\times\dfrac{0.200-x}{15.0}} = \frac{(2x)^2}{(0.150-x)(0.200-x)} = 50.2$$

We substitute these terms into the equilibrium constant expression and solve for x.

$$4x^2 = (0.150-x)(0.200-x)50.2 = 50.2\left(0.0300-0.350x+x^2\right) = 1.51-17.6x+50.2x^2$$

$0 = 46.2x^2 - 17.6x + 1.51$ Now we use the quadratic equation to determine the value of x.

$$x = \frac{-b\pm\sqrt{b^2-4ac}}{2a} = \frac{17.6\pm\sqrt{(17.6)^2-4\times46.2\times1.51}}{2\times46.2} = \frac{17.6\pm5.54}{92.4} = 0.250 \text{ or } 0.131$$

The first root cannot be used because it would afford a negative amount of H_2 (namely, $0.150 - 0.250 = -0.100$). Thus, we have $2\times0.131 = 0.262$ mol HI at equilibrium. We check by substituting the amounts into the K_c expression. (Notice that the volumes cancel.) The slight disagreement in the two values (52 compared to 50.2) is the result of rounding error.

$$K_c = \frac{(0.262)^2}{(0.150-0.131)(0.200-0.131)} = \frac{0.0686}{0.019\times0.069} = 52$$

12B **(D)**

(a) The equation for the reaction is $N_2O_4(g)\rightleftharpoons 2\,NO_2(g)$ and $K_c = 4.61\times10^{-3}$ at 25 °C.

In the example, this reaction is conducted in a 0.372 L flask. The effect of moving the mixture to the larger, 10.0 L container is that the reaction will be shifted to produce a greater number of moles of gas. Thus, $NO_2(g)$ will be produced and $N_2O_4(g)$ will dissociate. Consequently, the amount of N_2O_4 will decrease.

(b) The equilibrium constant expression, substituting 10.0 L for 0.372 L, follows.

$$K_c = \frac{[NO_2]^2}{[N_2O_4]} = \frac{\left(\dfrac{2x}{10.0}\right)^2}{\dfrac{0.0240-x}{10.0}} = \frac{4x^2}{10.0(0.0240-x)} = 4.61\times10^{-3}$$

This can be solved with the quadratic equation, and the sensible result is $x = 0.0118$ moles. We can attempt the method of successive approximations. *First*, assume that $x \ll 0.0240$. We obtain:

$$x = \frac{\sqrt{4.61 \times 10^{-3} \times 10.0 \ (0.0240 - 0)}}{4} = \sqrt{4.61 \times 10^{-3} \times 2.50 \ (0.0240 - 0)} = 0.0166$$

Clearly x is not much smaller than 0.0240. So, *second*, assume $x \approx 0.0166$. We obtain:

$$x = \sqrt{4.61 \times 10^{-3} \times 2.50 (0.0240 - 0.0166)} = 0.00925$$

This assumption is not valid either. So, *third*, assume $x \approx 0.00925$. We obtain:

$$x = \sqrt{4.61 \times 10^{-3} \times 2.50 (0.0240 - 0.00925)} = 0.0130$$

Notice that after each cycle the value we obtain for x gets closer to the value obtained from the roots of the equation. The values from the next several cycles follow.

Cycle	4th	5th	6th	7th	8th	9th	10th	11th
x value	0.0112	0.0121	0.0117	0.0119	0.01181	0.01186	0.01183	0.01184

The amount of N_2O_4 at equilibrium is 0.0118 mol, less than the 0.0210 mol N_2O_4 at equilibrium in the 0.372 L flask, as predicted.

13A **(M)** Again we base our solution on the balanced chemical equation.

Equation: $Ag^+(aq) + Fe^{2+}(aq) \rightleftharpoons Fe^{3+}(aq) + Ag(s)$ $K_c = 2.98$

Initial: 0 M   0 M   1.20 M

Changes: +x M   +x M   −x M

Equil: x M   x M $(1.20 - x)$ M

$$K_c = \frac{\left[Fe^{3+}\right]}{\left[Ag^+\right]\left[Fe^{2+}\right]} = 2.98 = \frac{1.20 - x}{x^2} \quad 2.98 \ x^2 = 1.20 - x \quad 0 = 2.98x^2 + x - 1.20$$

We use the quadratic formula to obtain a solution.

$$x = \frac{-b \pm \sqrt{b^2 - 4ac}}{2a} = \frac{-1.00 \pm \sqrt{(1.00)^2 + 4 \times 2.98 \times 1.20}}{2 \times 2.98} = \frac{-1.00 \pm 3.91}{5.96} = 0.488 \text{ M or} - 0.824 \text{ M}$$

A negative root makes no physical sense. We obtain the equilibrium concentrations from x.

$$\left[Ag^+\right] = \left[Fe^{2+}\right] = 0.488 \text{ M} \qquad \left[Fe^{3+}\right] = 1.20 - 0.488 = 0.71 \text{ M}$$

13B **(M)** We first calculate the value of Q_c to determine the direction of the reaction.

$$Q_c = \frac{\left[V^{2+}\right]\left[Cr^{3+}\right]}{\left[V^{3+}\right]\left[Cr^{2+}\right]} = \frac{0.150 \times 0.150}{0.0100 \times 0.0100} = 225 < 7.2 \times 10^2 = K_c$$

Because the reaction quotient has a smaller value than the equilibrium constant, a net reaction to the right will occur. We now set up this solution as we have others, heretofore, based on the balanced chemical equation.

$$V^{3+}(aq) \quad + \quad Cr^{2+}(aq) \quad \rightleftharpoons \quad V^{2+}(aq) \quad + \quad Cr^{3+}(aq)$$

initial	0.0100 M	0.0100 M	0.150 M	0.150 M
changes	$-x$ M	$-x$ M	$+x$ M	$+x$ M
equil	$(0.0100-x)$M	$(0.0100-x)$M	$(0.150+x)$M	$(0.150+x)$M

$$K_c = \frac{\left[V^{2+}\right]\left[Cr^{3+}\right]}{\left[V^{3+}\right]\left[Cr^{2+}\right]} = \frac{(0.150+x) \times (0.150+x)}{(0.0100-x) \times (0.0100-x)} = 7.2 \times 10^2 = \left(\frac{0.150+x}{0.0100-x}\right)^2$$

If we take the square root of both sides of this expression, we obtain

$$\sqrt{7.2 \times 10^2} = \frac{0.150+x}{0.0100-x} = 27$$

$0.150 + x = 0.27 - 27x$ which becomes $28x = 0.12$ and yields 0.0043 M. Then the equilibrium concentrations are: $\left[V^{3+}\right] = \left[Cr^{2+}\right] = 0.0100\ M - 0.0043\ M = 0.0057\ M$

$$\left[V^{2+}\right] = \left[Cr^{3+}\right] = 0.150\ M + 0.0043\ M = 0.154\ M$$

INTEGRATIVE EXAMPLE

<u>**A**</u> **(E)** We will determine the concentration of F6P and the final enthalpy by adding the two reactions:

$$C_6H_{12}O_6 + ATP \rightleftharpoons G6P + ADP$$
$$\underline{\qquad\qquad\qquad G6P \rightleftharpoons F6P \qquad\qquad\qquad}$$
$$C_6H_{12}O_6 + ATP \rightleftharpoons ADP + F6P$$

$\Delta H_{TOT} = -19.74\ kJ\ mol^{-1} + 2.84\ kJ\ mol^{-1} = -16.9\ kJ\ mol^{-1}$

Since the overall reaction is obtained by adding the two individual reactions, then the overall reaction equilibrium constant is the product of the two individual K values. That is, $K = K_1 \cdot K_2 = 1278$

The equilibrium concentrations of the reactants and products is determined as follows:

	$C_6H_{12}O_6$	+	ATP	$\rightleftharpoons$	ADP	+	F6P
Initial	1.20×10^{-6}		1×10^{-4}		1×10^{-2}		0
Change	$-x$		$-x$		$+x$		$+x$
Equil	$1.20 \times 10^{-6} - x$		$1 \times 10^{-4} - x$		$1 \times 10^{-2} + x$		x

$$K = \frac{[ADP][F6P]}{[C_6H_{12}O_6][ATP]}$$

$$1278 = \frac{(1\times10^{-2}+x)(x)}{(1.20\times10^{-6}-x)(1\times10^{-4}-x)} = \frac{1.0\times10^{-2}x+x^2}{1.2\times10^{-10}-1.012\times10^{-4}x+x^2}$$

Expanding and rearranging the above equation yields the following second-order polynomial:
$1277\,x^2 - 0.1393\,x + 1.534\times10^{-7} = 0$
Using the quadratic equation to solve for x, we obtain two roots: $x = 1.113\times10^{-6}$ and 1.080×10^{-4}. Only the first one makes physical sense, because it is less than the initial value of $C_6H_{12}O_6$. Therefore, $[F6P]_{eq} = 1.113\times10^{-6}$.

During a fever, the body generates heat. Since the net reaction above is exothermic, Le Châtelier's principle would force the equilibrium to the left, reducing the amount of F6P generated.

B **(E)**
(a) The ideal gas law can be used for this reaction, since we are relating vapor pressure and concentration. Since $K = 3.3\times10^{-29}$ for decomposition of Br_2 to Br (very small), then it can be ignored.

$$V = \frac{nRT}{P} = \frac{(0.100\text{ mol})(0.08206\text{ L atm K}^{-1})(298.15\text{ K})}{0.289\text{ atm}} = 8.47\text{ L}$$

(b) At 1000 K, there is much more Br being generated from the decomposition of Br_2. However, K is still rather small, and this decomposition does not notably affect the volume needed.

EXERCISES

Writing Equilibrium Constants Expressions

1. **(E)**

(a) $2\,COF_2(g) \rightleftharpoons CO_2(g) + CF_4(g)$ $\qquad K_c = \dfrac{[CO_2][CF_4]}{[COF_2]^2}$

(b) $Cu(s) + 2\,Ag^+(aq) \rightleftharpoons Cu^{2+}(aq) + 2\,Ag(s)$ $\qquad K_c = \dfrac{[Cu^{2+}]}{[Ag^+]^2}$

(c) $S_2O_8^{2-}(aq) + 2\,Fe^{2+}(aq) \rightleftharpoons 2\,SO_4^{2-}(aq) + 2\,Fe^{3+}(aq)$ $\qquad K_c = \dfrac{[SO_4^{2-}]^2[Fe^{3+}]^2}{[S_2O_8^{2-}][Fe^{2+}]^2}$

3. **(E)**

(a) $K_c = \dfrac{[NO_2]^2}{[NO]^2[O_2]}$ (b) $K_c = \dfrac{[Zn^{2+}]}{[Ag^+]^2}$ (c) $K_c = \dfrac{[OH^-]^2}{[CO_3^{2-}]}$

5. **(E)** In each case we write the equation for the formation reaction and then the equilibrium constant expression, K_c, for that reaction.

(a) $\frac{1}{2}H_2(g) + \frac{1}{2}F_2(g) \rightleftharpoons HF(g)$ $K_c = \dfrac{[HF]}{[H_2]^{1/2}[F_2]^{1/2}}$

(b) $N_2(g) + 3H_2(g) \rightleftharpoons 2NH_3(g)$ $K_c = \dfrac{[NH_3]^2}{[N_2][H_2]^3}$

(c) $2N_2(g) + O_2(g) \rightleftharpoons 2N_2O(g)$ $K_c = \dfrac{[N_2O]^2}{[N_2]^2[O_2]}$

(d) $\frac{1}{2}Cl_2(g) + \frac{3}{2}F_2(g) \rightleftharpoons ClF_3(l)$ $K_c = \dfrac{1}{[Cl_2]^{1/2}[F_2]^{3/2}}$

7. **(E)** Since $K_p = K_c(RT)^{\Delta v_{gas}}$, it is also true that $K_c = K_p(RT)^{-\Delta v_{gas}}$.

(a) $K_c = \dfrac{[SO_2][Cl_2]}{[SO_2Cl_2]} = K_p(RT)^{-(+1)} = 2.9\times10^{-2}(0.08314\times303)^{-1} = 0.0012$

(b) $K_c = \dfrac{[NO_2]^2}{[NO]^2[O_2]} = K_p(RT)^{-(-1)} = 1.48\times10^4 \times(0.08314\times457.15) = 5.63\times10^5$

(c) $K_c = \dfrac{[H_2S]^3}{[H_2]^3} = K_p(RT)^0 = K_p = 0.429$

9. **(E)** The equilibrium reaction is $H_2O(l) \rightleftharpoons H_2O(g)$ with $\Delta v_{gas} = +1$. $K_p = K_c(RT)^{\Delta v_{gas}}$ gives

$K_c = K_p(RT)^{-\Delta v_{gas}}$. $K_p = P\{H_2O\} = 23.8\text{ mmHg}\times\dfrac{1\text{ atm}}{760\text{ mmHg}} = 0.0313$

$K_c = K_p(RT)^{-1} = \dfrac{K_p}{RT} = \dfrac{0.0313}{0.08206\times298} = 1.28\times10^{-3}$

11. **(E)** Add one-half of the reversed 1st reaction with the 2nd reaction to obtain the desired reaction.

$\frac{1}{2}N_2(g) + \frac{1}{2}O_2(g) \rightleftharpoons NO(g)$ $K_c = \dfrac{1}{\sqrt{2.1\times10^{30}}}$

$NO(g) + \frac{1}{2}Br_2(g) \rightleftharpoons NOBr(g)$ $K_c = 1.4$

net : $\frac{1}{2}N_2(g) + \frac{1}{2}O_2(g) + \frac{1}{2}Br_2(g) \rightleftharpoons NOBr(g)$ $K_c = \dfrac{1.4}{\sqrt{2.1\times10^{30}}} = 9.7\times10^{-16}$

13. **(M)** We combine the K_c values to obtain the value of K_c for the overall reaction, and then convert this to a value for K_p.

$$2\,CO_2(g) + 2H_2(g) \rightleftharpoons 2\,CO(g) + 2\,H_2O(g) \qquad K_c = (1.4)^2$$

$$2\,C(graphite) + O_2(g) \rightleftharpoons 2\,CO(g) \qquad K_c = (1\times10^8)^2$$

$$4\,CO(g) \rightleftharpoons 2\,C(graphite) + 2\,CO_2(g) \qquad K_c = \dfrac{1}{(0.64)^2}$$

Overall: $2H_2(g) + O_2(g) \rightleftharpoons 2H_2O(g) \qquad K_{c(overall)} = \dfrac{(1.4)^2(1\times10^8)^2}{(0.64)^2} = 5\times10^6$

$$K_p = K_c(RT)^{\Delta v} = \dfrac{K_c}{RT} = \dfrac{5\times10^{16}}{0.08206\times1200} = 5\times10^{14}$$

15. **(E)** $CO_2(g) + H_2O(l) \rightleftharpoons H_2CO_3(aq)$

In terms of activities, $K = a(H_2CO_3)/a(CO_2)$

In terms of concentration and partial pressure, $K = \dfrac{[H_2CO_3]/c^\circ}{P_{CO_2}/P^\circ}$

Experimental Determination of Equilibrium Constants

17. **(M)** First, we determine the concentration of PCl_5 and of Cl_2 present initially and at equilibrium, respectively. Then we use the balanced equation to help us determine the concentration of each species present at equilibrium.

$$[PCl_5]_{initial} = \dfrac{1.00\times10^{-3}\text{ mol }PCl_5}{0.250\text{ L}} = 0.00400\text{ M} \qquad [Cl_2]_{equil} = \dfrac{9.65\times10^{-4}\text{ mol }Cl_2}{0.250\text{ L}} = 0.00386\text{ M}$$

Equation: $PCl_5(g) \rightleftharpoons PCl_3(g) + Cl_2(g)$

Initial: 0.00400M   0 M   0 M

Changes: $-x$M $+x$M $+x$M

Equil: 0.00400M-xM xM xM $\leftarrow$ 0.00386 M (from above)

At equilibrium, $[Cl_2] = [PCl_3] = 0.00386$ M and $[PCl_5] = 0.00400M - x$M $= 0.00014$ M

$$K_c = \dfrac{[PCl_3][Cl_2]}{[PCl_5]} = \dfrac{(0.00386\,M)(0.00386\,M)}{0.00014\,M} = 0.10\underline{6}$$

19. **(M)**

(a) $K_c = \dfrac{[PCl_5]}{[PCl_3][Cl_2]} = \dfrac{\dfrac{0.105\text{ g }PCl_5}{2.50\text{ L}}\times\dfrac{1\text{ mol }PCl_5}{208.2\text{ g}}}{\left(\dfrac{0.220\text{ g }PCl_3}{2.50\text{ L}}\times\dfrac{1\text{ mol }PCl_3}{137.3\text{ g}}\right)\times\left(\dfrac{2.12\text{ g }Cl_2}{2.50\text{ L}}\times\dfrac{1\text{ mol }Cl_2}{70.9\text{ g}}\right)} = 26.3$

(b) $K_p = K_c(RT)^{\Delta v} = 26.3\,(0.08206\times523)^{-1} = 0.613$

21. (E)

$$K = \frac{\left[Fe^{3+}\right]}{\left[H^+\right]^3} \Rightarrow 9.1 \times 10^3 = \frac{\left[Fe^{3+}\right]}{\left(1.0 \times 10^{-7}\right)^3}$$

$$\left[Fe^{3+}\right] = 9.1 \times 10^{-18} \text{ M}$$

Equilibrium Relationships

23. (M) $K_c = 281 = \dfrac{\left[SO_3\right]^2}{\left[SO_2\right]^2\left[O_2\right]} = \dfrac{\left[SO_3\right]^2}{\left[SO_2\right]^2} \times \dfrac{0.185 \text{ L}}{0.00247 \text{ mol}}$ $\dfrac{\left[SO_2\right]}{\left[SO_3\right]} = \sqrt{\dfrac{0.185}{0.00247 \times 281}} = 0.516$

25. (M)

(a) A possible equation for the oxidation of $NH_3(g)$ to $NO_2(g)$ follows.

$$NH_3(g) + \tfrac{7}{4}O_2(g) \rightleftharpoons NO_2(g) + \tfrac{3}{2}H_2O(g)$$

(b) We obtain K_p for the reaction in part (a) by appropriately combining the values of K_p given in the problem.

$NH_3(g) + \tfrac{5}{4}O_2(g) \rightleftharpoons NO(g) + \tfrac{3}{2}H_2O(g)$ $\qquad K_p = 2.11 \times 10^{19}$

$NO(g) + \tfrac{1}{2}O_2(g) \rightleftharpoons NO_2(g)$ $\qquad\qquad K_p = \dfrac{1}{0.524}$

net: $NH_3(g) + \tfrac{7}{4}O_2(g) \rightleftharpoons NO_2(g) + \tfrac{3}{2}H_2O(g)$ $\qquad K_p = \dfrac{2.11 \times 10^{19}}{0.524} = 4.03 \times 10^{19}$

27. (M)

(a) $K_c = \dfrac{[CO][H_2O]}{[CO_2][H_2]} = \dfrac{\dfrac{n\{CO\}}{V} \times \dfrac{n\{H_2O\}}{V}}{\dfrac{n\{CO_2\}}{V} \times \dfrac{n\{H_2\}}{V}}$

Since V is present in both the denominator and the numerator, it can be stricken from the expression. This happens here because $\Delta v_g = 0$. Therefore, K_c is independent of V.

(b) Note that $K_p = K_c$ for this reaction, since $\Delta v_{gas} = 0$.

$$K_c = K_p = \frac{0.224 \text{ mol CO} \times 0.224 \text{ mol H}_2O}{0.276 \text{ mol CO}_2 \times 0.276 \text{ mol H}_2} = 0.659$$

Direction and Extent of Chemical Change

29. **(M)** We compute the value of Q_c for the given amounts of product and reactants.

$$Q_c = \frac{[SO_3]^2}{[SO_2]^2[O_2]} = \frac{\left(\dfrac{1.8\,\text{mol SO}_3}{7.2\,\text{L}}\right)^2}{\left(\dfrac{3.6\,\text{mol SO}_2}{7.2\,\text{L}}\right)^2 \dfrac{2.2\,\text{mol O}_2}{7.2\,\text{L}}} = 0.82 < K_c = 100$$

The mixture described cannot be maintained indefinitely. In fact, because $Q_c < K_c$, the reaction will proceed to the right, that is, toward products, until equilibrium is established. We do not know how long it will take to reach equilibrium.

31. **(M)**

(a) We determine the concentration of each species in the gaseous mixture, use these concentrations to determine the value of the reaction quotient, and compare this value of Q_c with the value of K_c.

$$[SO_2] = \frac{0.455\,\text{mol SO}_2}{1.90\,\text{L}} = 0.239\,\text{M} \qquad [O_2] = \frac{0.183\,\text{mol O}_2}{1.90\,\text{L}} = 0.0963\,\text{M}$$

$$[SO_3] = \frac{0.568\,\text{mol SO}_3}{1.90\,\text{L}} = 0.299\,\text{M} \qquad Q_c = \frac{[SO_3]^2}{[SO_2]^2[O_2]} = \frac{(0.299)^2}{(0.239)^2\,0.0963} = 16.3$$

Since $Q_c = 16.3 \neq 2.8 \times 10^2 = K_c$, this mixture is not at equilibrium.

(b) Since the value of Q_c is smaller than that of K_c, the reaction will proceed to the right, forming product and consuming reactants to reach equilibrium.

33. **(M)** The information for the calculation is organized around the chemical equation. Let $x = \text{mol H}_2$ (or I_2) that reacts. Then use stoichiometry to determine the amount of HI formed, in terms of x, and finally solve for x.

Equation:	$H_2(g)$	+	$I_2(g) \rightleftharpoons$	$2\,HI(g)$
Initial:	0.150 mol		0.150 mol	0.000 mol
Changes:	$-x$ mol		$-x$ mol	$+2x$ mol
Equil:	$0.150-x$		$0.150-x$	$2x$

$$K_c = \frac{[HI]^2}{[H_2][I_2]} = \frac{\left(\dfrac{2x}{3.25\,\text{L}}\right)^2}{\dfrac{0.150-x}{3.25\,\text{L}} \times \dfrac{0.150-x}{3.25\,\text{L}}}$$

Then take the square root of both sides: $\sqrt{K_c} = \sqrt{50.2} = \dfrac{2x}{0.150-x} = 7.09$

$$2x = 1.06 - 7.09x \qquad x = \frac{1.06}{9.09} = 0.117\,\text{mol}, \quad \text{amount HI} = 2x = 2 \times 0.117\,\text{mol} = 0.234\,\text{mol HI}$$

amount $H_2 =$ amount $I_2 = (0.150 - x)\,\text{mol} = (0.150 - 0.117)\,\text{mol} = 0.033\,\text{mol H}_2$ (or I_2)

35. **(M)** We use the chemical equation as a basis to organize the information provided about the reaction, and then determine the final number of moles of $Cl_2(g)$ present.

Equation: $CO(g)$ $+$ $Cl_2(g)$ $\rightleftharpoons$ $COCl_2(g)$

Initial: 0.3500 mol   0.0000 mol   0.05500 mol

Changes: $+x$ mol $+x$ mol $-x$ mol

Equil.: $(0.3500+x)$ mol x mol $(0.05500-x)$ mol

$$K_c = 1.2\times10^3 = \frac{[COCl_2]}{[CO][Cl_2]} = \frac{\dfrac{(0.0550-x)\,mol}{3.050\,L}}{\dfrac{(0.3500+x)mol}{3.050\,L}\times\dfrac{x\,mol}{3.050\,L}}$$

$$\frac{1.2\times10^3}{3.050} = \frac{0.05500-x}{(0.3500+x)x} \quad \text{Assume } x \ll 0.0550 \text{ This produces the following expression.}$$

$$\frac{1.2\times10^3}{3.050} = \frac{0.05500}{0.3500\,x} \qquad x = \frac{3.050\times0.05500}{0.3500\times1.2\times10^3} = 4.0\times10^{-4} \text{ mol } Cl_2$$

We use the first value we obtained, 4.0×10^{-4} $(= 0.00040)$, to arrive at a second value.

$$x = \frac{3.050\times(0.0550-0.00040)}{(0.3500+0.00040)\times1.2\times10^3} = 4.0\times10^{-4} \text{ mol } Cl_2$$

Because the value did not change on the second iteration, we have arrived at a solution.

37. **(D)** We base each of our solutions on the balanced chemical equation.

(a)

Equation: $PCl_5(g)$ $\rightleftharpoons$ $PCl_3(g)$ $+$ $Cl_2(g)$

Initial: $\dfrac{0.550 \text{ mol}}{2.50 \text{ L}}$ $\dfrac{0.550 \text{ mol}}{2.50 \text{ L}}$ $\dfrac{0 \text{ mol}}{2.50 \text{ L}}$

Changes: $\dfrac{-x \text{ mol}}{2.50 \text{ L}}$ $\dfrac{+x \text{ mol}}{2.50 \text{ L}}$ $\dfrac{+x \text{ mol}}{2.50 \text{ L}}$

Equil: $\dfrac{(0.550-x) \text{ mol}}{2.50 \text{ L}}$ $\dfrac{(0.550+x) \text{ mol}}{2.50 \text{ L}}$ $\dfrac{x \text{ mol}}{2.50 \text{ L}}$

$$K_c = \frac{[PCl_3][Cl_2]}{[PCl_5]} = 3.8\times10^{-2} = \frac{\dfrac{(0.550+x)\,mol}{2.50L}\times\dfrac{x\,mol}{2.50L}}{\dfrac{(0.550-x)mol}{2.50L}} \qquad \frac{x(0.550+x)}{2.50(0.550-x)} = 3.8\times10^{-2}$$

$$x^2 + 0.550x = 0.052 - 0.095x \qquad x^2 + 0.645x - 0.052 = 0$$

$$x = \frac{-b\pm\sqrt{b^2-4ac}}{2a} = \frac{-0.645\pm\sqrt{0.416+0.208}}{2} = 0.0725 \text{ mol}, \ -0.717 \text{ mol}$$

The second answer gives a negative quantity. of Cl_2, which makes no physical sense.

$$n_{PCl_5} = (0.550 - 0.0725) = 0.478 \text{ mol } PCl_5 \quad n_{PCl_3} = (0.550 + 0.0725) = 0.623 \text{ mol } PCl_3$$

$$n_{Cl_2} = x = 0.0725 \text{ mol } Cl_2$$

(b)

Equation:	$PCl_5 (g)$	$\rightleftharpoons$	$PCl_3 (g)$	+	$Cl_2 (g)$
Initial:	$\dfrac{0.610 \text{ mol}}{2.50 \text{ L}}$		0 M		0 M
Changes:	$\dfrac{-x \text{ mol}}{2.50 \text{ L}}$		$\dfrac{+x \text{ mol}}{2.50 \text{ L}}$		$\dfrac{+x \text{ mol}}{2.50 \text{ L}}$
Equil:	$\dfrac{0.610 - x \text{ mol}}{2.50 \text{ L}}$		$\dfrac{(x \text{ mol})}{2.50 \text{ L}}$		$\dfrac{(x \text{ mol})}{2.50 \text{ L}}$

$$K_c = \frac{[PCl_3][Cl_2]}{[PCl_5]} = 3.8 \times 10^{-2} = \frac{\dfrac{(x \text{ mol})}{2.50 \text{ L}} \times \dfrac{(x \text{ mol})}{2.50 \text{ L}}}{\dfrac{0.610 - x \text{ mol}}{2.50 \text{ L}}}$$

$$2.50 \times 3.8 \times 10^{-2} = \frac{x^2}{0.610 - x} = 0.095 \quad 0.058 - 0.095x = x^2 \quad x^2 + 0.095x - 0.058 = 0$$

$$x = \frac{-b \pm \sqrt{b^2 - 4ac}}{2a} = \frac{-0.095 \pm \sqrt{0.0090 + 0.23}}{2} = 0.20 \text{ mol}, \ -0.29 \text{ mol}$$

amount $PCl_3 = 0.20 \text{ mol} =$ amount Cl_2; amount $PCl_5 = 0.610 - 0.20 = 0.41 \text{ mol}$

39. **(D)**

(a) We calculate the initial amount of each substance.

$$n \{C_2H_5OH\} = 17.2 \text{ g } C_2H_5OH \times \frac{1 \text{ mol } C_2H_5OH}{46.07 \text{ g } C_2H_5OH} = 0.373 \text{ mol } C_2H_5OH$$

$$n\{CH_3CO_2H\} = 23.8 \text{ g } CH_3CO_2H \times \frac{1 \text{ mol } CH_3CO_2H}{60.05 \text{ g } CH_3CO_2H} = 0.396 \text{ mol } CH_3CO_2H$$

$$n\{CH_3CO_2C_2H_5\} = 48.6 \text{ g } CH_3CO_2C_2H_5 \times \frac{1 \text{ mol } CH_3CO_2C_2H_5}{88.11 \text{ g } CH_3CO_2C_2H_5}$$

$$n\{CH_3CO_2C_2H_5\} = 0.552 \text{ mol } CH_3CO_2C_2H_5$$

$$n\{H_2O\} = 71.2 \text{ g } H_2O \times \frac{1 \text{ mol } H_2O}{18.02 \text{ g } H_2O} = 3.95 \text{ mol } H_2O$$

Since we would divide each amount by the total volume, and since there are the same numbers of product and reactant stoichiometric coefficients, we can use moles rather than concentrations in the Q_c expression.

$$Q_c = \frac{n\{CH_3CO_2C_2H_5\} n\{H_2O\}}{n\{C_2H_5OH\} n\{CH_3CO_2H\}} = \frac{0.552 \text{ mol} \times 3.95 \text{ mol}}{0.373 \text{ mol} \times 0.396 \text{ mol}} = 14.8 > K_c = 4.0$$

Since $Q_c > K_c$ the reaction will shift to the left, forming reactants, as it attains equilibrium.

(b)

Equation:	C_2H_5OH	+	CH_3CO_2H	$\rightleftharpoons$	$CH_3CO_2C_2H_5$	+	H_2O
Initial	0.373 mol		0.396 mol		0.552 mol		3.95 mol
Changes	$+x$ mol		$+x$ mol		$-x$ mol		$-x$ mol
Equil	$(0.373 + x)$ mol		$(0.396 + x)$ mol		$(0.552 - x)$ mol		$(3.95 - x)$ mol

$$K_c = \frac{(0.552 - x)(3.95 - x)}{(0.373 + x)(0.396 + x)} = \frac{2.18 - 4.50x + x^2}{0.148 + 0.769x + x^2} = 4.0$$

$$x^2 - 4.50x + 2.18 = 4x^2 + 3.08x + 0.59 \qquad 3x^2 + 7.58x - 1.59 = 0$$

$$x = \frac{-b \pm \sqrt{b^2 - 4ac}}{2a} = \frac{-7.58 \pm \sqrt{57 + 19}}{6} = 0.19 \text{ moles}, -2.72 \text{ moles}$$

Negative amounts do not make physical sense. We compute the equilibrium amount of each substance with $x = 0.19$ moles.

$$n\{C_2H_5OH\} = 0.373 \text{ mol} + 0.19 \text{ mol} = 0.56 \text{ mol } C_2H_5OH$$

$$\text{mass } C_2H_5OH = 0.56 \text{ mol } C_2H_5OH \times \frac{46.07 \text{ g } C_2H_5OH}{1 \text{ mol } C_2H_5OH} = 26 \text{ g } C_2H_5OH$$

$$n\{CH_3CO_2H\} = 0.396 \text{ mol} + 0.19 \text{ mol} = 0.59 \text{ mol } CH_3CO_2H$$

$$\text{mass } CH_3CO_2H = 0.59 \text{ mol } CH_3CO_2H \times \frac{60.05 \text{ g } CH_3CO_2H}{1 \text{ mol } CH_3CO_2H} = 35 \text{ g } CH_3CO_2H$$

$$n\{CH_3CO_2C_2H_5\} = 0.552 \text{ mol} - 0.19 \text{ mol} = 0.36 \text{ } CH_3CO_2C_2H_5$$

$$\text{mass } CH_3CO_2C_2H_5 = 0.36 \text{ mol } CH_3CO_2C_2H_5 \times \frac{88.10 \text{ g } CH_3CO_2C_2H_5}{1 \text{ mol } CH_3CO_2C_2H_5} = 32 \text{ g } CH_3CO_2C_2H_5$$

$$n\{H_2O\} = 3.95 \text{ mol} - 0.19 \text{ mol} = 3.76 \text{ mol } H_2O$$

$$\text{mass } H_2O = 3.76 \text{ mol } H_2O \times \frac{18.02 \text{ g } H_2O}{1 \text{ mol } H_2O} = 68 \text{ g } H_2O$$

$$\text{To check } K_c = \frac{n\{CH_3CO_2C_2H_5\}n\{H_2O\}}{n\{C_2H_5OH\}n\{CH_3CO_2H\}} = \frac{0.36 \text{ mol} \times 3.76 \text{ mol}}{0.56 \text{ mol} \times 0.59 \text{ mol}} = 4.1$$

41. **(M)** $\left[HCONH_2\right]_{init} = \dfrac{0.186 \text{ mol}}{2.16 \text{ L}} = 0.0861 \text{ M}$

Equation:	$HCONH_2(g)$	$\rightleftharpoons$	$NH_3(g)$	+	$CO(g)$
Initial:	0.0861 M		0 M		0 M
Changes:	$-x$ M		$+x$ M		$+x$ M
Equil:	$(0.0861 - x)$ M		x M		x M

$$K_c = \frac{[NH_3][CO]}{[HCONH_2]} = \frac{x \cdot x}{0.0861 - x} = 4.84 \qquad x^2 = 0.417 - 4.84x \qquad 0 = x^2 + 4.84x - 0.417$$

$$x = \frac{-b \pm \sqrt{b^2 - 4ac}}{2a} = \frac{-4.84 \pm \sqrt{23.4 + 1.67}}{2} = 0.084 \text{ M}, \ -4.92 \text{ M}$$

The negative concentration obviously is physically meaningless. We determine the total concentration of all species, and then the total pressure with $x = 0.084$.

$$[\text{total}] = [NH_3] + [CO] + [HCONH_2] = x + x + 0.0861 - x = 0.0861 + 0.084 = 0.170 \text{ M}$$

$$P_{\text{tot}} = 0.170 \text{ mol L}^{-1} \times 0.08206 \text{ L atm mol}^{-1} \text{ K}^{-1} \times 400. \text{ K} = 5.58 \text{ atm}$$

45. **(M)** We are told in this question that the reaction $SO_2(g) + Cl_2(g) \rightleftharpoons SO_2Cl_2(g)$ has $K_c = 4.0$ at a certain temperature T. This means that at the temperature T, $[SO_2Cl_2] = 4.0 \times [Cl_2] \times [SO_2]$. Careful scrutiny of the three diagrams reveals that sketch (b) is the best representation because it contains numbers of SO_2Cl_2, SO_2, and Cl_2 molecules that are consistent with the K_c for the reaction. In other words, sketch (b) is the best choice because it contains 12 SO_2Cl_2 molecules (per unit volume), 1 Cl_2 molecule (per unit volume) and 3 SO_2 molecules (per unit volume), which is the requisite number of each type of molecule needed to generate the expected K_c value for the reaction at temperature T.

47. **(E)**

$$K = \frac{[\text{aconitate}]}{[\text{citrate}]}$$

$$Q = \frac{4.0 \times 10^{-5}}{(0.00128)} = 0.031$$

Since $Q = K$, the reaction is at equilibrium,

Partial Pressure Equilibrium Constant, K_p

49. **(M)** The $I_2(s)$ maintains the presence of I_2 in the flask until it has all vaporized. Thus, if enough HI(g) is produced to completely consume the $I_2(s)$, equilibrium will not be achieved.

$$P\{H_2S\} = 747.6 \text{ mmHg} \times \frac{1 \text{ atm}}{760 \text{ mmHg}} = 0.9837 \text{ atm}$$

Equation:	$H_2S(g) +$	$I_2(s)$	$\rightleftharpoons$	$2 HI(g) + S(s)$
Initial:	0.9837 atm			0 atm
Changes:	$-x$ atm			$+2x$ atm
Equil:	$(0.9837 - x)$ atm			$2x$ atm

$$K_p = \frac{P\{HI\}^2}{P\{H_2S\}} = \frac{(2x)^2}{(0.9837-x)} = 1.34 \times 10^{-5} = \frac{4x^2}{0.9837}$$

$$x = \sqrt{\frac{1.34 \times 10^{-5} \times 0.9837}{4}} = 1.82 \times 10^{-3} \text{ atm}$$

The assumption that $0.9837 \gg x$ is valid. Now we verify that sufficient $I_2(s)$ is present by computing the mass of I_2 needed to produce the predicted pressure of HI(g). Initially, 1.85 g I_2 is present (given).

$$\text{mass } I_2 = \frac{1.82 \times 10^{-3} \text{ atm} \times 0.725 \text{ L}}{0.08206 \text{ L atm mol}^{-1} \text{ K}^{-1} \times 333 \text{ K}} \times \frac{1 \text{ mol } I_2}{2 \text{ mol HI}} \times \frac{253.8 \text{ g } I_2}{1 \text{ mol } I_2} = 0.00613 \text{ g } I_2$$

$$P_{tot} = P\{H_2S\} + P\{HI\} = (0.9837 - x) + 2x = 0.9837 + x = 0.9837 + 0.00182 = 0.9855 \text{ atm}$$

$$P_{tot} = 749.0 \text{ mmHg}$$

51. **(M)** We substitute the given equilibrium pressure into the equilibrium constant expression and solve for the other equilibrium pressure. $K_p = \dfrac{P\{O_2\}^3}{P\{CO_2\}^2} = 28.5 = \dfrac{P\{O_2\}^3}{(0.0721 \text{ atm } CO_2)^2}$

$$P\{O_2\} = \sqrt[3]{P\{O_2\}^3} = \sqrt[3]{28.5(0.0712 \text{ atm})^2} = 0.529 \text{ atm } O_2$$

$$P_{total} = P\{CO_2\} + P\{O_2\} = 0.0721 \text{ atm } CO_2 + 0.529 \text{ atm } O_2 = 0.601 \text{ atm total}$$

53. **(M)**

(a) We first determine the initial pressure of each gas.

$$P\{CO\} = P\{Cl_2\} = \frac{nRT}{V} = \frac{1.00 \text{ mol} \times 0.08206 \text{ L atm mol}^{-1} \text{ K}^{-1} \times 668 \text{ K}}{1.75 \text{ L}} = 31.3 \text{ atm}$$

Then we calculate equilibrium partial pressures, organizing our calculation around the balanced chemical equation. We see that the equilibrium constant is not very large, meaning that we must solve the polynomial exactly (or by successive approximations).

Equation	$CO(g)$	$+ \ Cl_2(g)$	$\rightleftharpoons$	$COCl_2(g)$	$K_p = 22.5$
Initial:	31.3 atm	31.3 atm		0 atm	
Changes:	$-x$ atm	$-x$ atm		$+x$ atm	
Equil:	$31.3 - x$ atm	$31.3 - x$ atm		x atm	

$$K_p = \frac{P\{COCl_2\}}{P\{CO\}P\{Cl_2\}} = 22.5 = \frac{x}{(31.3-x)^2} = \frac{x}{(979.7 - 62.6x + x^2)}$$

$$22.5(979.7 - 62.6x + x^2) = x = 22043 - 1408.5x + 22.5x^2 = x$$

$$22043 - 1409.5x + 22.5x^2 = 0 \quad \text{(Solve by using the quadratic equation)}$$

$$x = \frac{-b \pm \sqrt{b^2 - 4ac}}{2a} = \frac{-(-1409.5) \pm \sqrt{(-1409.\underline{5})^2 - 4(22.5)(220\underline{43})}}{2(22.5)}$$

$$x = \frac{1409.5 \pm \sqrt{2818}}{45} = 30.1\underline{4}, \ 32.5 \text{ (too large)}$$

$$P\{CO\} = P\{Cl_2\} = 31.3 \text{ atm} - 30.1\underline{4} \text{ atm} = 1.16 \text{ atm} \qquad P\{COCl_2\} = 30.1\underline{4} \text{ atm}$$

(b) $P_{\text{total}} = P\{CO\} + P\{Cl_2\} + P\{COCl_2\} = 1.1\underline{6} \text{ atm} + 1.1\underline{6} \text{ atm} + 30.1\underline{4} \text{ atm} = 32.46 \text{ atm}$

Le Châtelier's Principle

55. **(E)** Continuous removal of the product, of course, has the effect of decreasing the concentration of the products below their equilibrium values. Thus, the equilibrium system is disturbed by removing the products and the system will attempt (in vain, as it turns out) to re-establish the equilibrium by shifting toward the right, that is, to generate more products.

57. **(M)**

(a) This reaction is exothermic with $\Delta_r H^\circ = -150.$ kJ mol^{-1}. Thus, high temperatures favor the reverse reaction (endothermic reaction). The amount of $H_2(g)$ present at high temperatures will be less than that present at low temperatures.

(b) $H_2O(g)$ is one of the reactants involved. Introducing more will cause the equilibrium position to shift to the right, favoring products. The amount of $H_2(g)$ will increase.

(c) Doubling the volume of the container will favor the side of the reaction with the largest sum of gaseous stoichiometric coefficients. The sum of the stoichiometric coefficients of gaseous species is the same (4) on both sides of this reaction. Therefore, increasing the volume of the container will have no effect on the amount of $H_2(g)$ present at equilibrium.

(d) A catalyst merely speeds up the rate at which a reaction reaches the equilibrium position. The addition of a catalyst has no effect on the amount of $H_2(g)$ present at equilibrium.

59. **(M)**

(a) The formation of $NO(g)$ from its elements is an endothermic reaction ($\Delta_r H^\circ = +$ 181 kJ/mol). Since the equilibrium position of endothermic reactions is shifted toward products at higher temperatures, we expect more $NO(g)$ to be formed from the elements at higher temperatures.

(b) Reaction rates always are enhanced by higher temperatures, since a larger fraction of the collisions will have an energy that surmounts the activation energy. This enhancement of rates affects both the forward and the reverse reactions. Thus, the position of equilibrium is reached more rapidly at higher temperatures than at lower temperatures.

61. **(E)** If the total pressure of a mixture of gases at equilibrium is doubled by compression, the equilibrium will shift to the side with fewer moles of gas to counteract the increase in pressure. Thus, if the pressure of an equilibrium mixture of $N_2(g)$, $H_2(g)$, and $NH_3(g)$ is doubled, the reaction involving these three gases, i.e., $N_2(g) + 3\ H_2(g) \rightleftharpoons 2\ NH_3(g)$, will proceed in the forward direction to produce a new equilibrium mixture that contains additional ammonia and less molecular nitrogen and molecular hydrogen. In other words, $P\{N_2(g)\}$ will have decreased when equilibrium is re-established. It is important to note, however, that the final equilibrium partial pressure for the N_2 will, nevertheless, be higher than its original partial pressure prior to the doubling of the total pressure.

63. **(M)** Increasing the volume of an equilibrium mixture causes that mixture to shift toward the side (reactants or products) where the sum of the stoichiometric coefficients of the gaseous species is the larger. That is: shifts to the right if $\Delta v_{gas} > 0$, shifts to the left if $\Delta v_{gas} > 0$, and does not shift if $\Delta v_{gas} > 0$.

(a) $C(s) + H_2O(g) \rightleftharpoons CO(g) + H_2(g)$, $\Delta v_{gas} > 0$, shift right, toward products

(b) $Ca(OH)_2(s) + CO_2(g) \rightleftharpoons CaCO_3(s) + H_2O(g)$, $\Delta v_{gas} = 0$, no shift, no change in equilibrium position.

(c) $4\ NH_3(g) + 5\ O_2(g) \rightleftharpoons 4\ NO(g) + 6\ H_2O(g)$, $\Delta v_{gas} > 0$, shifts right, towards products

65. **(E)**

(a) $Hb{:}O_2$ is reduced, because the reaction is exothermic and heat is like a product.

(b) No effect, because the equilibrium involves O_2 (aq). Eventually it will reduce the $Hb{:}O_2$ level because removing $O_2(g)$ from the atmosphere also reduces O_2 (aq) in the blood.

(c) $Hb{:}O_2$ level increases to use up the extra Hb.

67. **(E)** The pressure on N_2O_4 will initially increase as the crystal melts and then vaporizes, but over time the new concentration decreases as the equilibrium is shifted toward NO_2.

69. **(E)** Since $\Delta_r H$ is >0, the reaction is endothermic. If we increase the temperature of the reaction, we are adding heat to the reaction, which shifts the reaction toward the decomposition of calcium carbonate. While the amount of calcium carbonate will decrease, its concentration will remain the same because it is a solid.

Reactions with Very Small or Very Large *K* Values

71. **(M)** The reaction of excess CuS in acid is shown below:

$$CuS(s) + 2 H_3O^+(aq) \rightleftharpoons Cu^{2+}(aq) + H_2S(aq) + 2 H_2O(l)$$

We can set up the equilibrium table as follows:

	CuS(s) +	2 H$_3$O$^+$(aq)	$\rightleftharpoons$	Cu^{2+}(aq)	+ H$_2$S(aq)	+ H$_2$O(l)
I		0.30 M		0 M	0.10 M	
C		$-2x$ M		$+x$ M	$+x$ M	
E		$(0.30 - 2x)$ M		x M	$(0.10 + x)$ M	

At equilibrium, we must have

$$K = \frac{(x)(0.10 + x)}{(0.30 - 2x)^2}$$

Although the expression can be solved exactly by rearranging it and using the quadratic formula, an approximate solution can be obtained rather easily. Because $K \ll 1$, we expect only a small amount of CuS to dissolve. Thus, we anticipate that $2x \ll 0.30$ and $x \ll 0.10$. Therefore,

$$K \approx \frac{(x)(0.10)}{(0.30)^2} \implies x \approx \frac{(0.30)^2 K}{(0.10)} = 5.4 \times 10^{-16} \text{ M}$$

The calculated value of x is much less than 0.10, so the assumption is justified. The equilibrium concentration of Cu^{2+} is about 5×10^{-16} M.

73. **(M)** The reaction of CH$_3$COOH and NaOH is shown below:

$$CH_3COOH(aq) + OH^-(aq) \rightleftharpoons CH_3COO^-(aq) + H_2O(l)$$

Because $K_c \gg 1$, for simplicity we can view the reaction reaching true equilibrium by going to completion by completely forming CH$_3$COO$^-$ and H$_2$O, and then backing up by a small extent as the CH$_3$COO$^-$ and H$_2$O react to form CH$_3$COOH and OH$^-$.

	CH$_3$COOH(aq) +	OH$^-$(aq)	$\rightleftharpoons$	CH$_3$COO$^-$(aq)	+ H$_2$O (l)
I	0.100 M	0.100 M		0 M	
100%	0 M	0 M		0.100 M	
C	$+x$ M	$+x$ M		$-x$ M	
E	x M	x M		$(0.100 - x)$ M	

Because K_c is so large for the forward reaction, we expect the reverse reaction to happen to a very slight extent. Therefore, the concentration of CH$_3$COO$^-$ remains essentially unchanged from its final value; that is, $0.100 - x \approx 0.100$.

$$K_c = \frac{[CH_3COO^-]}{[CH_3COOH][OH^-]} \approx \frac{0.100}{(x)(x)} \implies x \approx \sqrt{\frac{0.100}{K_c}} = 7.45 \times 10^{-6}$$

$[CH_3COOH]_{eq} = 7.5 \times 10^{-6}$ M

Of the initial amount of CH$_3$COOH, only 0.0075% remains at equilibrium. Therefore, 99.9925% reacted. The reaction goes essentially to completion.

INTEGRATIVE AND ADVANCED EXERCISES

77. **(M)** Dilution makes Q_c larger than K_c. Thus, the reaction mixture will shift left in order to regain equilibrium. We organize our calculation around the balanced chemical equation.

Equation: $\quad$ $Ag^+(aq)$ $\quad + \quad$ $Fe^{2+}(aq)$ $\quad \rightleftharpoons \quad$ $Fe^{3+}(aq) + Ag(s)$ $\quad K_c = 2.98$

Equil: $\qquad$ 0.31 M $\qquad\qquad$ 0.21 M $\qquad\qquad$ 0.19 M $\quad -$

Dilution: $\quad$ 0.12 M $\qquad\qquad$ 0.084 M $\qquad\qquad$ 0.076 M $\quad -$

Changes: $\quad$ $+x$ M $\qquad\qquad$ $+x$ M $\qquad\qquad$ $-x$ M $\quad -$

New equil: $(0.12 + x)$ M $\quad (0.084 + x)$ M $\quad (0.076 - x)$ M $\quad -$

$$K_c = \frac{[Fe^{3+}]}{[Ag^+][Fe^{2+}]} = 2.98 = \frac{0.076 - x}{(0.12 + x)(0.084 + x)} \quad 2.98(0.12 + x)(0.084 + x) = 0.076 - x$$

$$0.076 - x = 0.030 + 0.61x + 2.98\,x^2 \qquad 2.98\,x^2 + 1.61 \quad x - 0.046 = 0$$

$$x = \frac{-1.61 \pm \sqrt{2.59 + 0.55}}{5.96} = 0.027, -0.57 \quad \text{Note that the negative root makes no physical}$$

sense; it gives $[Fe^{2+}] = 0.084 - 0.57 = -0.49$ M.

Thus, the new equilibrium concentrations are

$[Fe^{2+}] = 0.084 + 0.027 = 0.111$ M $\qquad [Ag^+] = 0.12 + 0.027 = 0.15$ M

$[Fe^{3+}] = 0.076 - 0.027 = 0.049$ M $\qquad$ We can check our answer by substitution.

$$K_c = \frac{0.049 \text{ M}}{0.111 \text{ M} \times 0.15 \text{ M}} = 2.94 \approx 2.98 \text{ (within precision limits)}$$

78. **(M)** The percent dissociation should increase as the pressure is lowered, according to Le Châtelier's principle. Thus the total pressure in this instance should be more than in Example 15-12, where the percent dissociation is 12.5%. The total pressure in Example 15-12 was computed starting from the total number of moles at equilibrium.

The total amount = $(0.0240 - 0.00300)$ moles $N_2O_4 + 2 \times 0.00300$ mol $NO_2 = 0.027$ mol gas.

$$P_{total} = \frac{nRT}{V} = \frac{0.0270 \text{ mol} \times 0.08206 \text{ L atm K}^{-1} \text{ mol}^{-1} \times 298 \text{ K}}{0.372 \text{ L}} = 1.77 \text{ atm (Example 15-12)}$$

We base our solution on the balanced chemical equation. We designate the initial pressure of N_2O_4 as P. The change in $P\{N_2O_4\}$ is given as $-0.10\,P$ atm. to represent the 10.0 % dissociation.

Equation: $\qquad$ $N_2O_4(g)$ $\quad \rightleftharpoons \quad$ $2\,NO_2(g)$

Initial: $\qquad\qquad$ P atm $\qquad\qquad$ 0 atm

Changes: $\qquad\quad$ $-0.10\,P$ atm $\qquad$ $+2(0.10\,P$ atm$)$

Equil: $\qquad\qquad$ $0.90\,P$ atm $\qquad$ $0.20\,P$ atm

$$K_p = \frac{P\{NO_2\}^2}{P\{N_2O_4\}} = \frac{(0.20\,P)^2}{0.90\,P} = \frac{0.040\,P}{0.90} = 0.113 \qquad\qquad P = \frac{0.113 \times 0.90}{0.040} = 2.54 \text{ atm.}$$

Thus, the total pressure at equilibrium is $0.90\,P + 0.20\,P$ and $1.10\,P$ (where $P = 2.54$ atm) Therefore, total pressure at equilibrium = 2.79 atm.

80. **(M)** Let us start with one mole of air, and let $2x$ be the amount in moles of NO formed.

Equation:	$N_2(g)$	+	$O_2(g)$	$\rightleftharpoons$	$2\,NO(g)$
Initial:	0.79 mol		0.21 mol		0 mol
Changes:	$-x$ mol		$-x$ mol		$+2x$ mol
Equil:	$(0.79-x)$mol		$(0.21-x)$mol		$2x$ mol

$$\chi_{NO} = \frac{n\{NO\}}{n\{N_2\} + n\{O_2\} + n\{NO\}} = \frac{2x}{(0.79-x) + (0.21-x) + 2x} = 0.018 = \frac{2x}{1.00}$$

$x = 0.0090$ mol $\qquad 0.79 - x = 0.78$ mol $N_2 \qquad 0.21 - x = 0.20$ mol O_2

$2x = 0.018$ mol NO

$$K_p = \frac{P\{NO\}^2}{P(N_2)\,P\{O_2\}} = \frac{\left(\dfrac{n\{NO\}\,RT}{V_{total}}\right)^2}{\dfrac{n\{N_2\}\,RT}{V_{total}}\,\dfrac{n\{O_2\}\,RT}{V_{total}}} = \frac{n\{NO\}^2}{n\{N_2\}\,n\{O_2\}} = \frac{(0.018)^2}{0.78 \times 0.20} = 2.1 \times 10^{-3}$$

82. **(M)**

Equation: $\quad HOC_6H_4COOH(g) \rightleftharpoons C_6H_5OH(g) + CO_2(g)$

$$n_{co_2} = \frac{PV}{RT} = \left(\frac{\dfrac{730 \text{ mmHg}}{760 \text{ mmHg/atm}}}{0.0821 \text{ L atm mol K}}\right)\left(\frac{\left(\dfrac{48.2 + 48.5}{2}\right) \times \dfrac{1 \text{ L}}{1000 \text{ mL}}}{(293 \text{ K})}\right) = 1.93 \times 10^{-3} \text{ mol } CO_2$$

Note that moles of CO_2 = moles phenol

$$n_{salicylic\ acid} = \frac{0.300 \text{ g}}{138 \text{ g/mol}} = 2.17 \times 10^{-3} \text{ mol salicylic acid}$$

$$K_c = \frac{[C_6H_5OH]\,[CO_2(g)]}{[HOC_6H_4COOH]} = \frac{\left(\dfrac{1.93 \text{ mmol}}{50.0 \text{ mL}}\right)^2}{\dfrac{(2.17 - 1.93) \text{ mmol}}{50.0 \text{ mL}}} = 0.310$$

$$K_p = K_c(RT)^{(2-1)} = (0.310) \times (0.08206) \times (473) = 12.0$$

85. **(M)** We assume that the entire 5.00 g is N_2O_4 and reach equilibrium from this starting point.

$$[N_2O_4]_i = \frac{5.00\ g}{0.500\ L} \times \frac{1\ mol\ N_2O_4}{92.01\ g\ N_2O_4} = 0.109\ M$$

Equation: $N_2O_4\ (g)$ $\rightleftharpoons$ $2\ NO_2\ (g)$
Initial: 0.109   0 M
Changes: $-x$ M $+2x$ M
Equil: $(0.0109 - x)$ M $2x$ M

$$K_c = \frac{[NO_2]^2}{[N_2O_4]} = 4.61\times10^{-3} = \frac{(2x)^2}{0.109 - x} \qquad 4x^2 = 5.02\times10^{-4} - 4.61\times10^{-3}\ x$$

$$4x^2 + 0.00461\ x - 0.000502 = 0$$

$$x = \frac{-b \pm \sqrt{b^2 - 4ac}}{2a} = \frac{-0.00461 \pm \sqrt{2.13\times10^{-5} + 8.03\times10^{-3}}}{8} = 0.0106\ M,\ -0.0118\ M$$

(The method of successive approximations yields 0.0106 after two iterations)

amount $N_2O_4 = 0.500\ L\ (0.109 - 0.0106)\ M = 0.0492\ mol\ N_2O_4$

amount $NO_2 = 0.500\ L \times 2 \times 0.0106\ M = 0.0106\ mol\ NO_2$

$$mol\ fraction\ NO_2 = \frac{0.0106\ mol\ NO_2}{0.0106\ mol\ NO_2 + 0.0492\ mol\ N_2O_4} = 0.177$$

86. **(M)** We let P be the initial pressure in atmospheres of $COCl_2(g)$.

Equation: $COCl_2(g)$ $\rightleftharpoons$ $CO(g)$ + $Cl_2(g)$
Initial: P 0 M   0 M
Changes: $-x$ $+x$ $+x$
Equil: $P-x$ x x

Total pressure $= 3.00\ atm = P - x + x + x = P + x \quad P = 3.00 - x$

$P\{COCl_2\} = P - x = 3.00 - x - x = 3.00 - 2x$

$$K_p = \frac{P\{CO\}P\{Cl_2\}}{P(COCl_2)} = 0.0444 = \frac{x \cdot x}{3.00 - 2x}$$

$$x^2 = 0.133 - 0.0888\ x \quad x^2 + 0.0888\ x - 0.133 = 0$$

$$x = \frac{-b \pm \sqrt{b^2 - 4ac}}{2a} = \frac{-0.0888 \pm \sqrt{0.00789 + 0.532}}{2} = 0.323,\ -0.421$$

Since a negative pressure is physically meaningless, $x = 0.323$ atm.
(The method of successive approximations yields $x = 0.323$ after four iterations.)
$P\{CO\} = P\{Cl_2\} = 0.323$ atm

$P\{COCl_2\} = 3.00 - 2 \times 0.323 = 2.35$ atm

The mole fraction of each gas is its partial pressure divided by the total pressure. And the contribution of each gas to the apparent molar mass of the mixture is the mole fraction of that gas multiplied by the molar mass of that gas.

$$M_{avg} = \frac{P\{CO\}}{P_{tot}} M\{CO\} + \frac{P\{Cl_2\}}{P_{tot}} M\{Cl_2\} + \frac{P\{COCl_2\}}{P_{tot}} M\{COCl_2\}$$

$$= \left(\frac{0.323 \text{ atm}}{3.00 \text{ atm}} \times 28.01 \text{ g/mol}\right) + \left(\frac{0.323 \text{ atm}}{3.00 \text{ atm}} \times 70.91 \text{ g/mol}\right) + \left(\frac{2.32 \text{ atm}}{3.00 \text{ atm}} \times 98.92 \text{ g/mol}\right)$$

$$= 87.1 \text{ g/mol}$$

89. **(M)** Since the initial mole ratio is 2 $H_2S(g)$ to 1 $CH_4(g)$, the reactants remain in their stoichiometric ratio when equilibrium is reached. Also, the products are formed in their stoichiometric ratio.

$$\text{amount } CH_4 = 9.54 \times 10^{-3} \text{ mol } H_2S \times \frac{1 \text{ mol } CH_4}{2 \text{ mol } H_2S} = 4.77 \times 10^{-3} \text{ mol } CH_4$$

$$\text{amount } CS_2 = 1.42 \times 10^{-3} \text{ mol } BaSO_4 \times \frac{1 \text{ mol } S}{1 \text{ mol } BaSO_4} \times \frac{1 \text{ mol } CS_2}{2 \text{ mol } S} = 7.10 \times 10^{-4} \text{ mol } CS_2$$

$$\text{amount } H_2 = 7.10 \times 10^{-4} \text{ mol } CS_2 \times \frac{4 \text{ mol } H_2}{1 \text{ mol } CS_2} = 2.84 \times 10^{-3} \text{ mol } H_2$$

total amount $= 9.54 \times 10^{-3} \text{ mol } H_2S + 4.77 \times 10^{-3} \text{ mol } CH_4 + 7.10 \times 10^{-4} \text{ mol } CS_2 + 2.84 \times 10^{-3} \text{ mol } H_2$

$$= 17.86 \times 10^{-3} \text{ mol}$$

The partial pressure of each gas equals its mole fraction times the total pressure.

$$P\{H_2S\} = 1.00 \text{ atm} \times \frac{9.54 \times 10^{-3} \text{ mol } H_2S}{17.86 \times 10^{-3} \text{ mol total}} = 0.534 \text{ atm}$$

$$P\{CH_4\} = 1.00 \text{ atm} \times \frac{4.77 \times 10^{-3} \text{ mol } CH_4}{17.86 \times 10^{-3} \text{ mol total}} = 0.267 \text{ atm}$$

$$P\{CS_2\} = 1.00 \text{ atm} \times \frac{7.10 \times 10^{-4} \text{ mol } CS_2}{17.86 \times 10^{-3} \text{ mol total}} = 0.0398 \text{ atm}$$

$$P\{H_2\} = 1.00 \text{ atm} \times \frac{2.84 \times 10^{-3} \text{ mol } H_2}{17.86 \times 10^{-3} \text{ mol total}} = 0.159 \text{ atm}$$

$$K_p = \frac{P\{H_2\}^4 \, P\{CS_2\}}{P\{H_2S\}^2 \, P\{CH_4\}} = \frac{0.159^4 \times 0.0398}{0.534^2 \times 0.267} = 3.34 \times 10^{-4}$$

91. **(D)** Again we base our calculation on an I.C.E. table. In the course of solving the Integrative Example, we found that we could obtain the desired equation by reversing equation (2) and adding the result to equation (1)

$-(2) \quad H_2O(g) + CH_4(g) \rightleftharpoons CO(g) + 3 H_2(g) \quad K = 1/190$

$+(1) \quad \underline{CO(g) + H_2O(g) \rightleftharpoons CO_2(g) + H_2(g) \quad K = 1.4}$

Equation: $CH_4(g) + 2 H_2O(g) \rightleftharpoons CO_2(g) + 4 H_2(g) \quad K = 1.4/190 = 0.0074$

$-(2)$	$H_2O(g)$	$+$	$CH_4(g)$	$\rightleftharpoons$	$CO(g)$	$+ \ 3\,H_2(g)$	$K = 1/190$
$+(1)$	$CO(g)$	$+$	$H_2O(g)$	$\rightleftharpoons$	$CO_2(g)$	$+ \ H_2(g)$	$K = 1.4$

Equation:	$CH_4(g) \ + \ 2\,H_2O(g) \rightleftharpoons CO_2(g) \ + \ 4\,H_2(g)$			$K = 1.4/190 = 0.0074$
Initial:	0.100 mol	0.100 mol	0.100 mol	0.100 mol
$\longleftarrow$	$+0.025$ mol	$+0.050$ mol	-0.025 mol	-0.100 mol
To left:	0.125 mol	0.150 mol	0.075 mol	0.000 mol
Concns:	0.0250 M	0.0300 M	0.015 M	0.000 M
Changes:	$-x$ M	$-2x$ mol	$+x$ mol	$+4x$ mol
Equil:	$(0.0250 - x)$ M	$(0.0300 - 2x)$	$(0.015 + x)$ M	$4x$ mol

Notice that we have a fifth order polynomial to solve. Hence, we need to try to approximate its final solution as closely as possible. The reaction favors the reactants because of the small size of the equilibrium constant. Thus, we approach equilibrium from as far to the left as possible.

$$K_c = 0.0074 = \frac{[CO_2][H_2]^4}{[CH_4][H_2O]^2} = \frac{(0.0150 + x)(4x)^4}{(0.0250 - x)\,(0.0300 - 2x)^2} \approx \frac{0.0150\,(4x)^4}{0.0250\,(0.0300)^2}$$

$$x \approx \sqrt[4]{\frac{0.0250\,(0.0300)^2\,0.0074}{0.0150 \times 256}} = 0.014 \text{ M}$$

Our assumption is terrible. We substitute to continue successive approximations.

$$0.0074 = \frac{(0.0150 \ + \ 0.014)\,(4x)^4}{(0.0250 - 0.014)\,(0.0300 - 2 \times 0.014)^2} = \frac{(0.029)(4x)^4}{(0.011)(0.002)^2}$$

Next, try $x_2 = 0.0026$

$$0.074 = \frac{(0.0150 + 0.0026)(4x)^4}{(0.0250 - 0.0026)\,(0.0300 - 2 \times 0.0026)^2}$$

then, try $x_3 = 0.0123$.

After 18 iterations, the x value converges to 0.0080.

Considering that the equilibrium constant is known to only two significant figures, this is a pretty good result. Recall that the total volume is 5.00 L. We calculate amounts in moles.

$CH_4(g)$	$(0.0250 - 0.0080) \times 5.00 \text{ L} = 0.017 \text{ M} \times 5.00 \text{ L} = 0.085 \text{ moles } CH_4(g)$
$H_2O(g)$	$(0.0300 - 2 \times 0.0080) \text{ M} \times 5.00 \text{ L} = 0.014 \text{ M} \times 5.00 \text{ L} = 0.070 \text{ moles } H_2O(g)$
$CO_2(g)$	$(0.015 \ + \ 0.0080) \text{ M} \times 5.00 \text{ L} = 0.023 \text{ M} \times 5.00 \text{ L} = 0.12 \text{ mol } CO_2$
$H_2(g)$	$(4 \times 0.0080) \text{ M} \times 5.00 \text{ L} = 0.032 \text{ M} \times 5.00 \text{ L} = 0.16 \text{ mol } H_2$

95. **(D)**

Equation: $2\,H_2(g) \quad + \quad CO(g) \quad \rightleftharpoons \quad CH_3OH(g) \quad K_c = 14.5$ at 483 K

$$K_p = K_c(RT)^{\Delta v} = 14.5 \left(0.08206\, \frac{L\,atm}{mol\,K} \times 483K\right)^{-2} = 9.23 \times 10^{-3}$$

We know that mole percents equal pressure percents for ideal gases.

$P_{CO} = 0.350 \times 100\,atm = 35.0\,atm$

$P_{H_2} = 0.650 \times 100\,atm = 65.0\,atm$

Equation: $2\,H_2(g) \quad + \quad CO(g) \quad \rightleftharpoons \quad CH_3OH(g)$

Initial: $\qquad$ 65 atm $\qquad$ 35 atm

Changes: $\quad -2\,P$ atm $\qquad -P$ atm $\qquad\qquad +P$ atm

Equil: $\qquad 65 - 2P \qquad 35 - P \qquad\qquad\quad P$

$$K_p = \frac{P_{CH_3OH}}{P_{CO} \times P_{H_2}{}^2} = \frac{P}{(35.0 - P)(65.0 - 2P)^2} = 9.23 \times 10^{-3}$$

By successive approximations, $P = 24.6\,atm = P_{CH_3OH}$ at equilibrium.

Mathematica (version 4.0, Wolfram Research, Champaign, IL) gives a root of 24.5.

98. **(M)** Let c_0 represent the initial concentrations of A and B. The equilibrium summary (ICE table) is as follows.

	A	+	B	$\rightleftharpoons$	C	+	D
I	c_0		c_0		0		0
C	$-x$		$-x$		$+x$		$+x$
E	$c_0 - x$		$c_0 - x$		x		x

The equilibrium condition is $K = \dfrac{[C]_{eq}[D]_{eq}}{[A]_{eq}[B]_{eq}} = \dfrac{x^2}{(c_0 - x)^2}$, which may be solved for x as follows.

$$\sqrt{K} = \frac{x}{c_0 - x} \Rightarrow c_0\sqrt{K} - x\sqrt{K} = x \Rightarrow c_0\sqrt{K} = x(\sqrt{K} + 1) \Rightarrow x = \frac{c_0\sqrt{K}}{\sqrt{K} + 1}$$

The fraction of A and B that reacts is calculated as follows.

$$\text{fraction that reacts} = \frac{x}{c_0} = \frac{\sqrt{K}}{\sqrt{K} + 1}$$

Therefore,

$$\text{percentage that reacts} = c \times 100 = \frac{100\sqrt{K}}{\sqrt{K} + 1}$$

For $K = 10^{10}$: $\quad$ percentage that reacts $= \dfrac{100\sqrt{10^{10}}}{\sqrt{10^{10}} + 1} = \dfrac{100 \times 10^5}{10^5 + 1} = 99.999$

For $K = 10^{-10}$:   percentage that reacts $= \dfrac{100\sqrt{10^{-10}}}{\sqrt{10^{-10}}+1} = \dfrac{100\times10^{-5}}{10^{-5}+1} = 0.001$, and therefore

99.999% of the reactants remain.

FEATURE PROBLEMS

99. **(M)** We first determine the amount in moles of acetic acid in the equilibrium mixture.

amount $CH_3CO_2H = 28.85 \text{ mL} \times \dfrac{1 \text{ L}}{1000 \text{ mL}} \times \dfrac{0.1000 \text{ mol Ba(OH)}_2}{1 \text{ L}} \times \dfrac{2 \text{ mol } CH_3CO_2H}{1 \text{ mol Ba(OH)}_2}$

$\times \dfrac{\text{complete equilibrium mixture}}{0.01 \text{ of equilibrium mixture}} = 0.5770 \text{ mol } CH_3CO_2H$

$K_c = \dfrac{[CH_3CO_2C_2H_5][H_2O]}{[C_2H_5OH][CH_3CO_2H]} = \dfrac{\dfrac{0.423 \text{ mol}}{V} \times \dfrac{0.423 \text{ mol}}{V}}{\dfrac{0.077 \text{ mol}}{V} \times \dfrac{0.577 \text{ mol}}{V}} = \dfrac{0.423 \times 0.423}{0.077 \times 0.577} = 4.0$

100. **(D)** In order to determine whether or not equilibrium has been established in each bulb, we need to calculate the concentrations for all three species at the time of opening. The results from these calculations are tabulated below and a typical calculation is given beneath this table.

Bulb No.	Time Bulb Opened (hours)	Initial Amount HI(g) (in mmol)	Amount of I_2(g) and H_2(g) at Time of Opening (in mmol)	Amount HI(g) at Time of Opening (in mmol)	[HI] (mM)	$[I_2]$ & $[H_2]$ (mM)	$\dfrac{[H_2][I_2]}{[HI]^2}$
1	2	2.34$\underline{5}$	0.1572	2.03	5.08	0.393	0.00599
2	4	2.51$\underline{8}$	0.2093	2.10	5.25	0.523	0.00992
3	12	2.46$\underline{3}$	0.2423	1.98	4.95	0.606	0.0150
4	20	3.17$\underline{4}$	0.3113	2.55	6.38	0.778	0.0149
5	40	2.18$\underline{9}$	0.2151	1.76	4.40	0.538	0.0150

Consider, for instance, bulb #4 (opened after 20 hours).

Initial moles of HI(g) $= 0.406 \text{ g HI(g)} \times \dfrac{1 \text{ mole HI}}{127.9 \text{ g HI}} = 0.00317\underline{4} \text{ mol HI(g)}$ or $3.17\underline{4}$ mmol

moles of I_2(g) present in bulb when opened.

$= 0.04150 \text{ L Na}_2S_2O_3 \times \dfrac{0.0150 \text{ mol Na}_2S_2O_3}{1 \text{ L Na}_2S_2O_3} \times \dfrac{1 \text{ mol } I_2}{2 \text{ mol Na}_2S_2O_3} = 3.113 \times 10^{-4} \text{mol } I_2$

millimoles of I_2(g) present in bulb when opened $= 3.113 \times 10^{-4} \text{mol } I_2$

moles of H_2 present in bulb when opened = moles of I_2(g) present in bulb when opened.

HI reacted $= 3.113 \times 10^{-4}$ mol $I_2 \times \dfrac{2 \text{ mole HI}}{1 \text{ mol } I_2} = 6.226 \times 10^{-4}$ mol HI (0.6226 mmol HI)

moles of HI(g) in bulb when opened $= 3.17\underline{4}$ mmol HI $- 0.6226$ mmol HI $= 2.55$ mmol HI

Concentrations of HI, I_2, and H_2
[HI] $= 2.55$ mmol HI $\div 0.400$ L $= 6.38$ mM

$[I_2] = [H_2] = 0.3113$ mmol $\div 0.400$ L $= 0.778$ mM

Ratio: $\dfrac{[H_2][I_2]}{[HI]^2} = \dfrac{(0.778 \text{ mM})(0.778 \text{ mM})}{(6.38 \text{ mM})^2} = 0.0149$

As the time increases, the ratio $\dfrac{[H_2][I_2]}{[HI]^2}$ initially climbs sharply, but then plateaus at

0.0150 somewhere between 4 and 12 hours. Consequently, it seems quite reasonable to conclude that the reaction 2 HI(g) $\rightleftharpoons$ H_2(g) + I_2(g) has a $K_c \sim 0.015$ at 623 K.

101. **(D)** We first need to determine the number of moles of ammonia that were present in the sample of gas that left the reactor. This will be accomplished by using the data from the titrations involving HCl(aq).
Original number of moles of HCl(aq) in the 20.00 mL sample

$= 0.01872$ L of KOH $\times \dfrac{0.0523 \text{ mol KOH}}{1 \text{ L KOH}} \times \dfrac{1 \text{ mol HCl}}{1 \text{ mol KOH}}$

$= 9.79\underline{06} \times 10^{-4}$ moles of HCl$_{(initially)}$

Moles of unreacted HCl(aq)

$= 0.01542$ L of KOH $\times \dfrac{0.0523 \text{ mol KOH}}{1 \text{ L KOH}} \times \dfrac{1 \text{ mol HCl}}{1 \text{ mol KOH}} =$

$8.06\underline{47} \times 10^{-4}$ moles of HCl$_{(unreacted)}$

Moles of HCl that reacted and /or moles of NH_3 present in the sample of reactor gas
$= 9.79\underline{06} \times 10^{-4}$ moles $- 8.06\underline{47} \times 10^{-4}$ moles $= 1.73 \times 10^{-4}$ mole of NH_3 (or HCl).

The remaining gas, which is a mixture of N_2(g) and H_2(g) gases, was found to occupy 1.82 L at 273.2 K and 1.00 atm. Thus, the total number of moles of N_2 and H_2 can be found via the ideal gas law:

$n_{H_2 + N_2} = \dfrac{PV}{RT} = \dfrac{(1.00 \text{ atm})(1.82 \text{ L})}{(0.08206 \dfrac{\text{L atm}}{\text{K mol}})(273.2 \text{ K})} = 0.0811\underline{8}$ moles of ($N_2 + H_2$)

According to the stoichiometry for the reaction, 2 parts NH_3 decompose to give 3 parts H_2 and 1 part N_2. Thus the non-reacting mixture must be 75% H_2 and 25% N_2.

So, the number of moles of $N_2 = 0.25 \times 0.0811\underline{8}$ moles $= 0.0203$ moles N_2 and the number of moles of $H_2 = 0.75 \times 0.0811\underline{8}$ moles $= 0.0609$ moles H_2.

Before we can calculate K_c, we need to determine the volume that the NH_3, N_2, and H_2 molecules occupied in the reactor. Once again, the ideal gas law ($PV = nRT$) will be employed. $v_{gas} = 0.08118$ moles ($N_2 + H_2$) + 1.73×10^{-4} moles $NH_3 = 0.08135$ moles

$$V_{gases} = \frac{nRT}{P} = \frac{(0.08135 \text{ mol})(0.08206 \frac{\text{L atm}}{\text{K mol}})(1174.2 \text{ K})}{30.0 \text{ atm}} = 0.2613 \text{ L}$$

So, $K_c = \dfrac{\left[\dfrac{1.73 \times 10^{-4} \text{ moles}}{0.2613 \text{ L}}\right]^2}{\left[\dfrac{0.0609 \text{ moles}}{0.2613 \text{ L}}\right]^3 \left[\dfrac{0.0203 \text{ moles}}{0.2613 \text{ L}}\right]^1} = 4.46 \times 10^{-4}$

To calculate K_p at 901 °C, we need to employ the equation $K_p = K_c (RT)^{\Delta v_{gas}}$; $\Delta v_{gas} = -2$

$K_p = 4.46 \times 10^{-4}$ [(0.08206 L atm K^{-1}mol^{-1})] $\times (1174.2$ K)]$^{-2} = 4.80 \times 10^{-8}$ at 901°C for the reaction $N_2(g) + 3 H_2(g) \rightleftharpoons 2 NH_3(g)$

103. **(D)** First, it is most important to get a general feel for the direction of the reaction by determining the reaction quotient:

$$Q = \frac{[C(aq)]}{[A(aq)]\cdot[B(aq)]} = \frac{0.1}{0.1 \times 0.1} = 10$$

Since $Q >> K$, the reaction proceeds toward the reactants. Looking at the reaction in the aqueous phase only, the equilibrium can be expressed as follows:

	A(aq)	+	B(aq)	⇌	C(aq)
Initial	0.1		0.1		0.1
Change	$-x$		$-x$		$+x$
Equil.	$0.1 - x$		$0.1 - x$		$0.1 + x$

We will do part (b) first, which assumes the absence of an organic layer for extraction:

$$K = \frac{(0.1+x)}{(0.1-x)(0.1-x)} = 0.01$$

Expanding the above equation and using the quadratic formula, $x = -0.0996$. Therefore, the concentration of C(aq) and equilibrium is $0.1 + (-0.0996) = 4 \times 10^{-4}$ M.

If the organic layer is present for extraction, we can add the two equations together, as shown below:

$$
\begin{array}{rcl}
A(aq) \;+\; B(aq) & \rightleftharpoons & C(aq) \\
C(aq) & \rightleftharpoons & C(or) \\
\hline
A(aq) \;+\; B(aq) & \rightleftharpoons & C(or)
\end{array}
$$

$K = K_1 \times K_2 = 0.1 \times 15 = 0.15$.

Since the organic layer is present with the aqueous layer, and K_2 is large, we can expect that the vast portion of C initially placed in the aqueous phase will go into the organic phase. Therefore, the initial $[C] = 0.1$ can be assumed to be for C(or). The equilibrium can be expressed as follows

	A(aq)	+ B(aq)	$\rightleftharpoons$ C(or)
Initial	0.1	0.1	0.1
Change	$-x$	$-x$	$+x$
Equil.	$0.1 - x$	$0.1 - x$	$0.1 + x$

We will do part (b) first, which assumes the absence of an organic layer for extraction:

$$
K = \frac{(0.1+x)}{(0.1-x)(0.1-x)} = 0.15
$$

Expanding the above equation and using the quadratic formula, $x = -0.0943$. Therefore, the concentration of C(or) and equilibrium is $0.1 + (-0.0943) = 6 \times 10^{-4}$ M. This makes sense because the K for the overall reaction is < 1, which means that the reaction favors the reactants.

SELF-ASSESSMENT EXERCISES

107. **(E)** The answer is (c). Because the limiting reagent is I_2 at one mole, the theoretical yield of HI is 2 moles. However, because there is an established equilibrium, there is a small amount of HI which will decompose to yield H_2 and I_2. Therefore the total moles of HI created is close, but less than 2.

108. **(E)** The answer is (d). The equilibrium expression is:

$$
K = \frac{P(SO_3)^2}{P(SO_2)^2\, P(O_2)} = 100
$$

If equilibrium is established, moles of SO_3 and SO_2 cancel out of the equilibrium expression. Therefore, if $K = 100$, the moles of O_2 have to be 0.01 to make $K = 100$.

109. **(E)** The answer is (a). As the volume of the vessel is expanded (i.e., pressure is reduced), the equilibrium shifts $_{\text{toward}}$ the side with more moles of gas.

110. **(E)** The answer is (b). At half the stoichiometric values, the equilibrium constant is $K^{1/2}$. If the equation is reversed, it is K^{-1}. Therefore, the $K' = K^{-1/2} = (1.8 \times 10^{-6})^{-1/2} = 7.5 \times 10^{-2}$.

111. **(E)** The answer is (a). We know that $K_p = K_c (RT)^{\Delta v}$. Since $\Delta v = (3-2) = 1$, $K_p = K_c (RT)$. Therefore, $K_p > K_c$.

112. **(E)** The answer is (c). Since the number of moles of gas of products is more than the reactants, increasing the vessel volume will drive the equilibrium more toward the product side. The other options: (a) has no effect, and (b) drives the equilibrium to the reactant side.

113. **(E)** The equilibrium expression is:

$$K = \frac{[C]^2}{[B]^2[A]} = \frac{(0.43)^2}{(0.55)^2(0.33)} = 1.9$$

114. **(E)**
   (a) As more O_2 (a reactant) is added, more Cl_2 is produced.
   (b) As HCl (a reactant) is removed, equilibrium shifts to the left and less Cl_2 is made.
   (c) Since there are more moles of reactants, equilibrium shifts to the left and less Cl_2 is made.
   (d) No change. However, the equilibrium is reached faster.
   (e) Since the reaction is exothermic, increasing the temperature causes less Cl_2 to be made.

115. **(E)** SO_2 (g) will be less than SO_2 (aq), because $K > 1$, so the equilibrium lies to the product side, SO_2 (aq).

116. **(E)** Since $K \gg 1$, there will be much more product than reactant

117. **(M)** The equilibrium expression for this reaction is:

$$K = \frac{[SO_3]^2}{[SO_2]^2[O_2]} = 35.5$$

   (a) If $[SO_3]_{eq} = [SO_2]_{eq}$, then $[O_2] = 1/35.5 = 0.0282$ M.
   moles of $O_2 = 0.0282 \times 2.05$ L $= 0.0578$ mol
   (b) Plugging in the new concentration values into the equilibrium expression:

$$K = \frac{[SO_3]^2}{[SO_2]^2[O_2]} = \frac{[2\times SO_2]^2}{[SO_2]^2[O_2]} = \frac{4}{[O_2]} = 35.5$$

   $[O_2] = 0.113$ M
   moles of $O_2 = 0.113 \times 2.05$ L $= 0.232$ mol

CHAPTER 16
ACIDS AND BASES

PRACTICE EXAMPLES

1A **(E)**

(a) In the forward direction, HF is the acid (proton donor; forms F^-), and H_2O is the base (proton acceptor; forms H_3O^+). In the reverse direction, F^- is the base (forms HF), accepting a proton from H_3O^+, which is the acid (forms H_2O).

(b) In the forward direction, HSO_4^- is the acid (proton donor; forms SO_4^{2-}), and NH_3 is the base (proton acceptor; forms NH_4^+). In the reverse direction, SO_4^{2-} is the base (forms HSO_4^-), accepting a proton from NH_4^+, which is the acid (forms NH_3).

(c) In the forward direction, HCl is the acid (proton donor; forms Cl^-), and CH_3COO^- is the base (proton acceptor; forms CH_3COOH). In the reverse direction, Cl^- is the base (forms HCl), accepting a proton from CH_3COOH, which is the acid (forms CH_3COO^-).

1B **(E)** We know that the formulas of most acids begin with H. Thus, we identify HNO_2 and HCO_3^- as acids.

$$HNO_2(aq) + H_2O(l) \rightleftharpoons NO_2^-(aq) + H_3O^+(aq);$$
$$HCO_3^-(aq) + H_2O(l) \rightleftharpoons CO_3^{2-}(aq) + H_3O^+(aq)$$

A negatively charged species will attract a positively charged proton and act as a base. Thus PO_4^{3-} and HCO_3^- can act as bases. We also know that PO_4^{3-} must be a base because it cannot act as an acid—it has no protons to donate—and we know that all three species have acid-base properties.

$$PO_4^{3-}(aq) + H_2O(l) \rightleftharpoons HPO_4^{2-}(aq) + OH^-(aq);$$
$$HCO_3^-(aq) + H_2O(l) \rightleftharpoons H_2CO_3(aq) \rightarrow CO_2 \cdot H_2O(aq) + OH^-(aq)$$

Notice that HCO_3^- is the amphiprotic species, acting as both an acid and a base.

2A **(M)** $[H_3O^+]$ is readily computed from pH: $[H_3O^+] = 10^{-pH}$ $[H_3O^+] = 10^{-2.85} = 1.4 \times 10^{-3}$ M.

$[OH^-]$ can be found in two ways: (1) from $K_w = [H_3O^+][OH^-]$, giving

$$[OH^-] = \frac{K_w}{[H_3O^+]} = \frac{1.0 \times 10^{-14}}{1.4 \times 10^{-3}} = 7.1 \times 10^{-12} \text{ M, or (2) from pH + pOH = 14.00, giving}$$

$pOH = 14.00 - pH = 14.00 - 2.85 = 11.15$, and then $[OH^-] = 10^{-pOH} = 10^{-11.15} = 7.1 \times 10^{-12}$ M.

2B **(M)** The concentration of hydronium ion can be calculated from the pH.
$[H_3O^+] = 10^{-pH} = 10^{-2.50} = 3.16 \times 10^{-6}$ M

The concentration of hydroxide ion can be calculated by using the formula
$K_w = [H_3O^+][OH^-]$
$[OH^-] = K_w/[H_3O^+] = 1.0 \times 10^{-14}/3.16 \times 10^{-6} = 3.16 \times 10^{-9}$ M

Because the self-ionization of water produces equal amounts of H_3O^+ and OH^-, we can say $[H_3O^+]_{water} = [OH^-] = 3.16 \times 10^{-9}$ M. Therefore,

$$\% \text{ H}_3\text{O}^+ \text{ from water} \quad = \quad ([H_3O^+]_{water}/[H_3O^+]) \times 100$$
$$= \quad (3.16 \times 10^{-9} \text{ M})/(3.16 \times 10^{-6} \text{ M}) \times 100$$
$$= \quad 0.1$$

Only 0.1% of the H_3O^+ in solution comes from the self-ionization of water. Also, because $[H_3O^+]_{water} = [OH^-] = 3.16 \times 10^{-9}$ M $< 1.0 \times 10^{-7}$ M, we have also verified that the self-ionization of water is partially suppressed by the addition of acid to the water.

3A **(M)** Since the K_a of HNO_3 is larger than 1, we expect most of the nitric acid to react and x is a very small number.

	HNO₃ +	**H₂O(l)** ⇌	**NO₃⁻ (aq) +**	**H₃O⁺(aq)**
initial concns:	0.010 M			
to completion:	0 M		0.010 M	0.010 M
to the left:	+ x M		− x M	− x M
equil concns:	x M		(0.010 − x) M	(0.010 − x) M

By assuming $x \ll 1$, we can write the following:

$$K_a = \frac{[NO_3^-][H_3O^+]}{[HNO_3]} = \frac{(0.010)(0.010)}{x}$$

$$x = \frac{(0.010)^2}{K_a} = \frac{0.0001}{20} = 5.0 \times 10^{-6}$$

Therefore, $[HNO_3] = x$ M $= 5.0 \times 10^{-6}$ M

$$\text{fraction of HNO}_3 \text{ remaining} = \frac{5.0 \times 10^{-6} \text{M}}{0.010 \text{ M}} \times 100\% = 0.05 \%$$

percent ionization $= 100 - 0.05 = 99.95\%$

3B **(M)** Since the K_a of HOCl is $\ll 1$, HOCl is considered a weak acid and we expect a very small amount of the HOCl to react.

	HOCl (aq) +	**H$_2$O(l)** $\rightleftharpoons$	**OCl$^-$ (aq) +**	**H$_3$O$^+$ (aq)**
initial concns:	0.010 M			
change:	$-x$ M		$+x$ M	$+x$ M
equil concns:	$(0.010 - x)$ M		x M	x M

$$K_a = \frac{[H_3O^+][OCl^-]}{[HOCl]} = \frac{(x)(x)}{0.010 - x}$$

By assuming that $x \ll 0.010$, we can simplify the denominator of the equation:

$$K_a = \frac{(x)(x)}{0.010}$$

$$x = \sqrt{0.010 \times K_a}$$

$$x = \sqrt{0.010 \times 2.9 \times 10^{-8}}$$

$$x = 1.70 \times 10^{-5}$$

Therefore, $[HOCl] = (0.010 - 1.70 \times 10^{-5}) = 9.98 \times 10^{-3}$ M

$$\text{percent ionization} = \frac{1.70 \times 10^{-5} \text{M}}{0.010 \text{ M}} \times 100\% = 0.17\%$$

$$pH = -\log(1.70 \times 10^{-5} \text{ M}) = 4.77$$

4A **(M)** We organize the solution around the balanced chemical equation; a M is $[HF]_{initial}$.

Equation:	HF(aq)	+ H$_2$O(l)	$\rightleftharpoons$	H$_3$O$^+$(aq)	+ F$^-$(aq)
	a M	$-$		≈ 0 M	0 M
Initial:	$-x$ M	$-$		$+x$ M	$+x$ M
Changes:	$(a-x)$ M	$-$		x M	x M
Equil:					

$$K_a = \frac{[H_3O^+][F^-]}{[HF]} = \frac{(x)(x)}{a-x} \approx \frac{x^2}{a} = 6.6 \times 10^{-4} \quad x = \sqrt{a \times 6.6 \times 10^{-4}}$$

For 0.20 M HF, $a = 0.20$ M $\qquad x = \sqrt{0.20 \times 6.6 \times 10^{-4}} = 0.011$ M

$$\% \text{ dissoc} = \frac{0.011 \text{ M}}{0.20 \text{ M}} \times 100\% = 5.5\%$$

For 0.020 M HF, $a = 0.020$ M $\qquad x = \sqrt{0.020 \times 6.6 \times 10^{-4}} = 0.0036$ M

We need another cycle of approximation: $x = \sqrt{(0.020 - 0.0036) \times 6.6 \times 10^{-4}} = 0.0033$ M

Yet another cycle with $x \approx 0.0033$ M : $x = \sqrt{(0.020 - 0.0033) \times 6.6 \times 10^{-4}} = 0.0033$ M

$$\% \text{ dissoc} = \frac{0.0033 \text{M}}{0.020 \text{M}} \times 100\% = 17\%$$

As expected, the weak acid is more dissociated.

4B **(E)** Since both H_3O^+ and $CH_3CH(OH)COO^-$ come from the same source in equimolar amounts, their concentrations are equal. $\left[H_3O^+\right] = \left[CH_3CH(OH)COO^-\right] = 0.067 \times 0.0284 \text{M} = 0.0019 \text{M}$

$$K_a = \frac{\left[H_3O^+\right]\left[CH_3CH(OH)COO^-\right]}{\left[HC_3H_5CH_3CH(OH)COOH\right]} = \frac{(0.0019)(0.0019)}{0.0284 - 0.0019} = 1.4 \times 10^{-4}$$

5A **(E)** pH is computed directly from the equation below.

$\left[H_3O^+\right]$, $\text{pH} = -\log\left[H_3O^+\right] = -\log(0.0025) = 2.60$.

We know that HI is a strong acid and, thus, is completely dissociated into H_3O^+ and I^-. The consequence is that $\left[I^-\right] = \left[H_3O^+\right] = 0.0025$ M. $\left[OH^-\right]$ is most readily computed from pH: $\text{pOH} = 14.00 - \text{pH} = 14.00 - 2.60 = 11.40$; $\left[OH^-\right] = 10^{-\text{pOH}} = 10^{-11.40} = 4.0 \times 10^{-12}$ M

5B **(M)** The number of moles of HCl(g) is calculated from the ideal gas law. Then $\left[H_3O^+\right]$ is calculated, based on the fact that HCl(aq) is a strong acid (1 mol H_3O^+ is produced from each mole of HCl).

$$\text{moles HCl(g)} = \frac{\left(747 \text{ mmHg} \times \dfrac{1 \text{ atm}}{760 \text{ mmHg}}\right) \times 0.535 \text{ L}}{\dfrac{0.08206 \text{ L atm}}{\text{mol K}} \times (26.5 + 273.2) \text{ K}}$$

moles HCl(g) = 0.0214 mol HCl(g) = 0.0214 mol H_3O^+ when dissolved in water

$[H_3O^+] = 0.0214$ mol $\qquad$ pH $= -\log(0.0214) = 1.670$

6A **(E)** pH is most readily determined from $\text{pOH} = -\log\left[OH^-\right]$. Assume $Mg(OH)_2$ is a strong base.

$$\left[OH^-\right] = \frac{9.63 \text{ mg Mg}(OH)_2}{100.0 \text{ mL soln}} \times \frac{1000 \text{ mL}}{1 \text{ L}} \times \frac{1 \text{ g}}{1000 \text{ mg}} \times \frac{1 \text{ mol Mg}(OH)_2}{58.32 \text{ g Mg}(OH)_2} \times \frac{2 \text{ mol OH}^-}{1 \text{ mol Mg}(OH)_2}$$

$\left[OH^-\right] = 0.00330 \text{M}$; $\text{pOH} = -\log(0.00330) = 2.481$

$\qquad$ pH $= 14.000 - \text{pOH} = 14.000 - 2.481 = 11.519$

6B **(E)** KOH is a strong base, which means that each mole of KOH that dissolves produces one mole of dissolved $OH^-(aq)$. First we calculate $\left[OH^-\right]$ and the pOH. We then use $pH + pOH = 14.00$ to determine pH.

$$\left[OH^-\right] = \frac{3.00\,g\,KOH}{100.00\,g\,soln} \times \frac{1\,mol\,KOH}{56.11\,g\,KOH} \times \frac{1\,mol\,OH^-}{1\,mol\,KOH} \times \frac{1.0242\,g\,soln}{1\,mL\,soln} \times \frac{1000\,mL}{1\,L} = 0.548\,M$$

$$pOH = -\log(0.548) = 0.261 \quad pH = 14.000 - pOH = 14.000 - 0.261 = 13.739$$

7A **(M)** $\left[H_3O^+\right] = 10^{-pH} = 10^{-4.18} = 6.6 \times 10^{-5}$ M.

Organize the solution using the balanced chemical equation.

Equation:	$HOCl(aq)$	+	$H_2O(l)$	$\rightleftharpoons$	$H_3O^+(aq)$ +	$OCl^-(aq)$
Initial:	0.150 M		$-$		≈ 0 M	0 M
Changes:	-6.6×10^{-5} M		$-$		$+6.6 \times 10^{-5}$ M	$+6.6 \times 10^{-5}$ M
Equil:	≈ 0.150 M		$-$		6.6×10^{-5} M	6.6×10^{-5} M

$$K_a = \frac{\left[H_3O^+\right]\left[OCl^-\right]}{\left[HOCl\right]} = \frac{(6.6 \times 10^{-5})(6.6 \times 10^{-5})}{0.150} = 2.9 \times 10^{-8}$$

7B **(M)** First, we use pH to determine $\left[OH^-\right]$. $pOH = 14.00 - pH = 14.00 - 10.08 = 3.92$.

$\left[OH^-\right] = 10^{-pOH} = 10^{-3.92} = 1.2 \times 10^{-4}$ M. We determine the initial concentration of cocaine and then organize the solution around the balanced equation in the manner we have used before.

$$[C_{17}H_{21}O_4N] = \frac{0.17\,g\,C_{17}H_{21}O_4N}{100\,mL\,soln} \times \frac{1000\,mL}{1\,L} \times \frac{1\,mol\,C_{17}H_{21}O_4N}{303.36\,g\,C_{17}H_{21}O_4N} = 0.0056\,M$$

Equation:	$C_{17}H_{21}O_4N(aq)$ +	$H_2O(l)$	$\rightleftharpoons$	$C_{17}H_{21}O_4NH^+(aq)$ +	$OH^-(aq)$
Initial:	0.0056 M	$-$		0 M	≈ 0 M
Changes:	-1.2×10^{-4} M	$-$		$+1.2 \times 10^{-4}$ M	$+1.2 \times 10^{-4}$ M
Equil:	≈ 0.0055 M	$-$		1.2×10^{-4} M	1.2×10^{-4} M

$$K_b = \frac{\left[C_{17}H_{21}O_4NH^+\right]\left[OH^-\right]}{\left[C_{17}H_{21}O_4N\right]} = \frac{(1.2 \times 10^{-4})(1.2 \times 10^{-4})}{0.0055} = 2.6 \times 10^{-6}$$

8A (M) Again we organize our solution around the balanced chemical equation.

Equation: $CH_2FCOOH(aq) + H_2O(l) \rightleftharpoons H_3O^+(aq) + CH_2FCOO^-(aq)$

Initial:	0.100 M	—	≈ 0 M	0 M
Changes:	$-x$ M	—	$+x$ M	$+x$ M
Equil:	$(0.100-x)$ M	—	x M	x M

$$K_a = \frac{[H_3O^+][C_2H_2FO_2^-]}{[HC_2H_2FO_2^-]} \; ; \text{ therefore, } 2.6 \times 10^{-3} = \frac{x \cdot x}{(0.100-x)}$$

We can use the 5% rule to ignore x in the denominator. Therefore, $x = [H_3O^+] = 0.016$ M, and pH $= -\log(0.016) = 1.8$. Thus, the calculated pH is considerably lower than 2.89 (Example 16-6).

8B (M) We first determine the concentration of undissociated acid. We then use this value in a set-up that is based on the balanced chemical equation.

$$2 \text{ aspirin tablets} \times \frac{0.500 \text{ g } C_6H_4(OOCCH_3)COOH}{\text{tablet}} \times \frac{1 \text{ mol } C_6H_4(OOCCH_3)COOH}{180.155 \text{ g } C_6H_4(OOCCH_3)COOH}$$

$$\times \frac{1}{0.325 \text{ L}} = 0.0171 \text{ M}$$

Equation: $C_6H_4(OOCCH_3)COOH(aq) + H_2O(l) \rightleftharpoons H_3O^+(aq) + C_6H_4(OOCCH_3)COO^-(aq)$

Initial:	0.0171 M	—	0 M	0 M
Changes:	$-x$ M	—	$+x$ M	$+x$ M
Equil:	$(0.0171-x)$ M	—	x M	x M

$$K_a = \frac{[H_3O^+][C_6H_4(OOCCH_3)COO^-]}{[C_6H_4(OOCCH_3)COOH]} = 3.3 \times 10^{-4} = \frac{x \cdot x}{0.0171-x}$$

$$x^2 + 3.3 \times 10^{-4} - 5.64 \times 10^{-6} = 0 \text{ (find the physically reasonable roots of the quadratic equation)}$$

$$x = \frac{-3.3 \times 10^{-4} \pm \sqrt{1.1 \times 10^{-7} + 2.3 \times 10^{-5}}}{2} = 0.0022 \text{ M}; \quad \text{pH} = -\log(0.0022) = 2.66$$

9A (M) Again we organize our solution around the balanced chemical equation.

Equation: $CH_2FCOOH(aq) + H_2O(l) \rightleftharpoons H_3O^+(aq) + CH_2FCOOH^-(aq)$

Initial:	0.015 M	—	≈ 0 M	0 M
Changes	$-x$ M	—	$+x$ M	$+x$ M
Equil:	$(0.015-x)$ M	—	x M	x M

$$K_a = \frac{[H_3O^+][CH_2FCOOH^-]}{[CH_2FCOOH]} = 2.6 \times 10^{-3} = \frac{(x)(x)}{0.015-x} \approx \frac{x^2}{0.015}$$

$$x = \sqrt{x^2} = \sqrt{0.015 \times 2.6 \times 10^{-3}} = 0.0062 \text{ M} = [H_3O^+] \qquad \text{Our assumption is invalid:}$$

0.0062 is not quite small enough compared to 0.015 for the 5% rule to hold. Thus we use another cycle of successive approximations.

$$K_a = \frac{(x)(x)}{0.015 - 0.0062} = 2.6 \times 10^{-3} \quad x = \sqrt{(0.015 - 0.0062) \times 2.6 \times 10^{-3}} = 0.0048 \text{ M} = [\text{H}_3\text{O}^+]$$

$$K_a = \frac{(x)(x)}{0.015 - 0.0048} = 2.6 \times 10^{-3} \quad x = \sqrt{(0.015 - 0.0048) \times 2.6 \times 10^{-3}} = 0.0051 \text{ M} = [\text{H}_3\text{O}^+]$$

$$K_a = \frac{(x)(x)}{0.015 - 0.0051} = 2.6 \times 10^{-3} \quad x = \sqrt{(0.015 - 0.0051) \times 2.6 \times 10^{-3}} = 0.0051 \text{ M} = [\text{H}_3\text{O}^+]$$

Two successive identical results indicate that we have the solution.

$\text{pH} = -\log[\text{H}_3\text{O}^+] = -\log(0.0051) = 2.29$. The quadratic equation gives the same result (0.0051 M) as this method of successive approximations.

9B **(M)** First we find $[\text{C}_5\text{H}_{10}\text{NH}]$. We then use this value as the starting base concentration in a set-up based on the balanced chemical equation.

$$[\text{C}_5\text{H}_{10}\text{NH}] = \frac{114 \text{ mg C}_5\text{H}_{10}\text{NH}}{315 \text{ mL soln}} \times \frac{1 \text{ mmol C}_5\text{H}_{10}\text{NH}}{85.15 \text{ mg C}_5\text{H}_{10}\text{NH}} = 0.00425 \text{ M}$$

Equation: $\text{C}_5\text{H}_{10}\text{NH(aq)} + \text{H}_2\text{O(l)} \rightleftharpoons \text{C}_5\text{H}_{10}\text{NH}_2^+\text{(aq)} + \text{OH}^-\text{(aq)}$

	C₅H₁₀NH(aq)	+ H₂O(l)	⇌ C₅H₁₀NH₂⁺(aq)	+ OH⁻(aq)
Initial:	0.00425 M	—	0 M	≈ 0 M
Changes:	−x M	—	+x M	+x M
Equil:	(0.00425 − x) M	—	x M	x M

$$K_b = \frac{[\text{C}_5\text{H}_{10}\text{NH}_2^+][\text{OH}^-]}{[\text{C}_5\text{H}_{10}\text{NH}]} = 1.6 \times 10^{-3} = \frac{x \cdot x}{0.00425 - x} \approx \frac{x \cdot x}{0.00425} \quad \text{We assumed that } x \ll 0.00425$$

$x = \sqrt{0.0016 \times 0.00425} = 0.0026 \text{ M}$ The assumption is not valid. Let's assume $x \approx 0.0026$

$x = \sqrt{0.0016(0.00425 - 0.0026)} = 0.0016$ Let's try again, with $x \approx 0.0016$

$x = \sqrt{0.0016(0.00425 - 0.0016)} = 0.0021$ Yet another try, with $x \approx 0.0021$

$x = \sqrt{0.0016(0.00425 - 0.0021)} = 0.0019$ The last time, with $x \approx 0.0019$

$x = \sqrt{0.0016(0.00425 - 0.0019)} = 0.0019 \text{ M} = [\text{OH}^-]$

$\text{pOH} = -\log[\text{H}_3\text{O}^+] = -\log(0.0019) = 2.72 \quad \text{pH} = 14.00 - \text{pOH} = 14.00 - 2.72 = 11.28$

We could have solved the problem with the quadratic formula roots equation rather than by successive approximations. The same answer is obtained. In fact, if we substitute $x = 0.0019$ into the K_b expression, we obtain $(0.0019)^2 / (0.00425 - 0.0019) = 1.5 \times 10^{-3}$ compared to $K_b = 1.6 \times 10^{-3}$. The error is due to rounding, not to an incorrect value. Using $x = 0.0020$ gives a value of 1.8×10^{-3}, while using $x = 0.0018$ gives 1.3×10^{-3}.

10A (M) For an aqueous solution of a diprotic acid, the concentration of the divalent anion is very close to the second ionization constant: $\left[^-OOCCH_2COO^- \right] \approx K_{a_2} = 2.0 \times 10^{-6} M$. We organize around the chemical equation.

Equation:	$CH_2(COOH)_2$ (aq)	+	$H_2O(l)$	$\rightleftharpoons$	H_3O^+ (aq)	+	$HOOCCH_2COO^-$ (aq)
Initial:	1.0 M		—		≈ 0 M		0 M
Change:	$-x$ M		—		$+x$ M		$+x$ M
Equil:	$(1.0 - x)$ M		—		x M		x M

$$K_a = \frac{[H_3O^+][HOOCCH_2COO^-]}{[CH_2(COOH)_2]} = \frac{(x)(x)}{1.0 - x} \approx \frac{x^2}{1.0} = 1.4 \times 10^{-3}$$

$x = \sqrt{1.4 \times 10^{-3}} = 3.7 \times 10^{-2} \ M = [H_3O^+] = [HOOCCH_2COO^-] \quad x \ll 1.0 \ M$ is a valid assumption.

10B (M) We know $K_{a_2} \approx \left[\text{doubly charged anion} \right]$ for a polyprotic acid. Thus, $K_{a_2} = 5.3 \times 10^{-5} \approx \left[C_2O_4^{2-} \right]$. From the pH, $\left[H_3O^+ \right] = 10^{-pH} = 10^{-0.67} = 0.21$ M. We also recognize that $\left[HC_2O_4^- \right] = \left[H_3O^+ \right]$, since the second ionization occurs to only a very small extent. We note as well that $HC_2O_4^-$ is produced by the ionization of $H_2C_2O_4$. Each mole of $HC_2O_4^-$ present results from the ionization of 1 mole of $H_2C_2O_4$. Now we have sufficient information to determine the K_{a_1}.

$$K_{a_1} = \frac{\left[H_3O^+ \right]\left[HC_2O_4^- \right]}{\left[H_2C_2O_4 \right]} = \frac{0.21 \times 0.21}{1.05 - 0.21} = 5.3 \times 10^{-2}$$

11A (M) H_2SO_4 is a strong acid in its first ionization, and somewhat weak in its second, with $K_{a_2} = 1.1 \times 10^{-2} = 0.011$. Because of the strong first ionization step, this problem involves determining concentrations in a solution that initially is 0.20 M H_3O^+ and 0.20 M HSO_4^-. We base the set-up on the balanced chemical equation.

Equation:	HSO_4^- (aq)	+	H_2O (l)	$\rightleftharpoons$	H_3O^+ (aq)	+	SO_4^{2-} (aq)
Initial:	0.20 M		—		0.20 M		0 M
Changes:	$-x$ M		—		$+x$ M		$+x$ M
Equil:	$(0.20 - x)$ M		—		$(0.20 + x)$ M		x M

$$K_{a_2} = \frac{[H_3O^+][SO_4^{2-}]}{[HSO_4^-]} = \frac{(0.20 + x)x}{0.20 - x} = 0.011 \approx \frac{0.20 \times x}{0.20}, \text{ assuming that } x \ll 0.20 M.$$

$x = 0.011M$ Try one cycle of approximation:

$$0.011 \approx \frac{(0.20 + 0.011)x}{(0.20 - 0.011)} = \frac{0.21x}{0.19} \qquad x = \frac{0.19 \times 0.011}{0.21} = 0.010 \text{ M}$$

The next cycle of approximation produces the same answer $0.010M = \left[SO_4^{2-} \right]$,

$\left[H_3O^+ \right] = 0.010 + 0.20 \text{ M} = 0.21 \text{ M}, \qquad \left[HSO_4^- \right] = 0.20 - 0.010 \text{ M} = 0.19 \text{ M}$

11B **(M)** We know that H_2SO_4 is a strong acid in its first ionization, and a somewhat weak acid in its second, with $K_{a_2} = 1.1 \times 10^{-2} = 0.011$. Because of the strong first ionization step, the problem essentially reduces to determining concentrations in a solution that initially is 0.020 M H_3O^+ and 0.020 M HSO_4^-. We base the set-up on the balanced chemical equation. The result is solved using the quadratic equation.

Equation: $HSO_4^-(aq) + H_2O(l) \rightleftharpoons H_3O^+(aq) \quad + \quad SO_4^{2-}(aq)$

Initial: 0.020 M   —   0.020 M   0 M

Changes: $-x$ M   — $+x$ M $+x$ M

Equil: $(0.020 - x)$ M   — $(0.020 + x)$ M x M

$$K_{a_2} = \frac{\left[H_3O^+ \right]\left[SO_4^{2-} \right]}{\left[HSO_4^- \right]} = \frac{(0.020 + x)x}{0.020 - x} = 0.011 \qquad\qquad 0.020x + x^2 = 2.2 \times 10^{-4} - 0.011x$$

$$x^2 + 0.031x - 0.00022 = 0 \qquad x = \frac{-0.031 \pm \sqrt{0.00096 + 0.00088}}{2} = 0.0060 \text{ M} = \left[SO_4^{2-} \right]$$

$\left[HSO_4^- \right] = 0.020 - 0.0060 = 0.014 \text{ M} \qquad\qquad \left[H_3O^+ \right] = 0.020 + 0.0060 = 0.026 \text{ M}$

(The method of successive approximations converges to $x = 0.006$ M in 8 cycles.)

12A **(E)**

(a) $CH_3NH_3^+NO_3^-$ is the salt of the cation of a weak base. The cation, $CH_3NH_3^+$, will hydrolyze to form an acidic solution $\left(CH_3NH_3^+ + H_2O \rightleftharpoons CH_3NH_2 + H_3O^+ \right)$, while NO_3^-, by virtue of being the conjugate base of a strong acid will not hydrolyze to a detectable extent. The aqueous solutions of this compound will thus be acidic.

(b)   NaI is the salt composed of the cation of a strong base and the anion of a strong acid, neither of which hydrolyzes in water. Solutions of this compound will be pH neutral.

(c) $NaNO_2$ is the salt composed of the cation of a strong base that will not hydrolyze in water and the anion of a weak acid that will hydrolyze to form an alkaline solution $\left(NO_2^- + H_2O \rightleftharpoons HNO_2 + OH^- \right)$. Thus aqueous solutions of this compound will be basic (alkaline).

12B **(E)** Even without referring to the K values for acids and bases, we can predict that the reaction that produces H_3O^+ occurs to the greater extent. This, of course, is because the pH is less than 7, thus acid hydrolysis must predominate.

We write the two reactions of $H_2PO_4^-$ with water, along with the values of their equilibrium constants.

$$H_2PO_4^-(aq) + H_2O(l) \rightleftharpoons H_3O^+(aq) + HPO_4^{2-}(aq) \qquad K_{a_2} = 6.3 \times 10^{-8}$$

$$H_2PO_4^-(aq) + H_2O(l) \rightleftharpoons OH^-(aq) + H_3PO_4(aq) \qquad K_b = \frac{K_w}{K_{a_1}} = \frac{1.0 \times 10^{-14}}{7.1 \times 10^{-3}} = 1.4 \times 10^{-12}$$

As predicted, the acid ionization occurs to the greater extent.

13A **(M)** From the value of pK_b we determine the value of K_b and then K_a for the cation.

cocaine: $\qquad K_b = 10^{-pK} = 10^{-8.41} = 3.9 \times 10^{-9} \qquad K_a = \frac{K_w}{K_b} = \frac{1.0 \times 10^{-14}}{3.9 \times 10^{-9}} = 2.6 \times 10^{-6}$

codeine: $\qquad K_b = 10^{-pK} = 10^{-7.95} = 1.1 \times 10^{-8} \qquad K_a = \frac{K_w}{K_b} = \frac{1.0 \times 10^{-14}}{1.1 \times 10^{-8}} = 9.1 \times 10^{-7}$

(This method may be a bit easier:

$pK_a = 14.00 - pK_b = 14.00 - 8.41 = 5.59$, $K_a = 10^{-5.59} = 2.6 \times 10^{-6}$) The acid with the larger K_a will produce the higher $[H^+]$, and that solution will have the lower pH. Thus, the solution of codeine hydrochloride will have the higher pH (i.e., codeine hydrochloride is the weaker acid).

13B **(E)** Both of the ions of $NH_4CN(aq)$ react with water in hydrolysis reactions.

$$NH_4^+(aq) + H_2O(l) \rightleftharpoons NH_3(aq) + H_3O^+(aq) \qquad K_a = \frac{K_w}{K_b} = \frac{1.0 \times 10^{-14}}{1.8 \times 10^{-5}} = 5.6 \times 10^{-10}$$

$$CN^-(aq) + H_2O(l) \rightleftharpoons HCN(aq) + OH^-(aq) \qquad K_b = \frac{K_w}{K_a} = \frac{1.0 \times 10^{-14}}{6.2 \times 10^{-10}} = 1.6 \times 10^{-5}$$

Since the value of the equilibrium constant for the hydrolysis reaction of cyanide ion is larger than that for the hydrolysis of ammonium ion, the cyanide ion hydrolysis reaction will proceed to a greater extent and thus the solution of $NH_4CN(aq)$ will be basic (alkaline).

14A **(M)** NaF dissociates completely into sodium ions and fluoride ions. The released fluoride ion hydrolyzes in aqueous solution to form hydroxide ion. The determination of the equilibrium pH is organized around the balanced equation.

Equation:	$F^-(aq)$	+	$H_2O(l)$	$\rightleftharpoons$	$HF(aq)$	+	$OH^-(aq)$
Initial:	0.10 M		—		0 M		0 M
Changes:	$-x$ M		—		$+x$ M		$+x$ M
Equil:	$(0.10 - x)$ M		—		x M		x M

$$K_b = \frac{K_w}{K_a} = \frac{1.0 \times 10^{-14}}{6.6 \times 10^{-4}} = 1.5 \times 10^{-11} = \frac{[HF][OH^-]}{[F^-]} = \frac{(x)(x)}{(0.10-x)} \approx \frac{x^2}{0.10}$$

$$x = \sqrt{0.10 \times 1.5 \times 10^{-11}} = 1.2 \times 10^{-6} \ M = [OH^-]; \ pOH = -\log(1.2 \times 10^{-6}) = 5.92$$

$$pH = 14.00 - pOH = 14.00 - 5.92 = 8.08 \quad (\text{As expected, pH} > 7)$$

14B **(M)** The cyanide ion hydrolyzes in solution, as indicated in Practice Example 16-12B. As a consequence of the hydrolysis, $[OH^-] = [HCN]$ $[OH^-]$ can be found from the pH of the solution, and then values are substituted into the K_b expression for CN^-, which is then solved for $[CN^-]$.

$$pOH = 14.00 - pH = 14.00 - 10.38 = 3.62$$

$$[OH^-] = 10^{-pOH} = 10^{-3.62} = 2.4 \times 10^{-4} M = [HCN]$$

$$K_b = \frac{[HCN][OH^-]}{[CN^-]} = 1.6 \times 10^{-5} = \frac{(2.4 \times 10^{-4})^2}{[CN^-]} \qquad [CN^-] = \frac{(2.4 \times 10^{-4})^2}{1.6 \times 10^{-5}} = 3.6 \times 10^{-3} M$$

15A **(M)** First we draw the Lewis structures of the four acids. Lone pairs have been omitted since we are interested only in the arrangements of atoms.

$HClO_4$ should be stronger than HNO_3. Although Cl and N have similar electronegativities, there are more terminal oxygen atoms attached to the chlorine in perchloric acid than to the nitrogen in nitric acid. By virtue of having more terminal oxygens, perchloric acid, when ionized, affords a more stable conjugate base. The more stable the anion, the more easily it is formed and hence the stronger is the conjugate acid from which it is derived. CH_2FCOOH will be a stronger acid than $CH_2BrCOOH$ because F is a more electronegative atom than Br. The F atom withdraws additional electron density from the O—H bond, making the bond easier to break, which leads to increased acidity.

15B **(M)** First we draw the Lewis structures of the first two acids. Lone pairs are not depicted since we are interested in the arrangements of atoms.

422

H_3PO_4 and H_2SO_3 both have one terminal oxygen atom, but S is more electronegative than P. This suggests that $H_2SO_3 (K_{a_1} = 1.3 \times 10^{-2})$ should be a stronger acid than $H_3PO_4 (K_{a_1} = 7.1 \times 10^{-3})$, and it is. The only difference between CCl_3CH_2COOH and CCl_2FCH_2COOH is the replacement of Cl by F. Since F is more electronegative than Cl, CCl_3CH_2COOH should be a weaker acid than CCl_2FCH_2COOH.

16A **(M)** We draw Lewis structures to help us decide.

(a) Clearly, BF_3 is an electron pair acceptor, a Lewis acid, and NH_3 is an electron pair donor, a Lewis base.

(b) H_2O certainly has electron pairs (lone pairs) to donate and thus it can be a Lewis base. It is unlikely that the cation Cr^{3+} has any accessible valence electrons that can be donated to another atom, thus, it is the Lewis acid. The product of the reaction, $[Cr(H_2O)_6]^{3+}$, is described as a water adduct of Cr^{3+}.

16B **(M)** The Lewis structures of the six species follow.

Both the hydroxide ion and the chloride ion have lone pairs of electrons that can be donated to electron-poor centers. These two are the electron pair donors, or the Lewis bases. $Al(OH)_3$ and $SnCl_4$ have additional spaces in their structures to accept pairs of electrons, which is what occurs when they form the complex anions $[Al(OH)_4]^-$ and $[SnCl_6]^{2-}$. Thus, $Al(OH)_3$ and $SnCl_4$ are the Lewis acids in these reactions.

INTEGRATIVE EXAMPLE

A   **(D)** To confirm the pH of rainwater, we have to calculate the concentration of $CO_2(aq)$ in water and then use simple acid-base equilibrium to calculate pH.

Concentration of CO_2 in water at 1 atm pressure and 298 K is 1.45 g/L, or

$$\frac{1.45 \text{ g } CO_2}{L} \times \frac{1 \text{ mol } CO_2}{44.0 \text{ g } CO_2} = 0.0329 \text{ M } CO_2$$

Furthermore, if the atmosphere is 0.037% by volume CO_2, then the mole fraction of CO_2 is 0.00037, and the partial pressure of CO_2 also becomes 0.00037 atm (because $P_{CO_2} = x_{CO_2} P_{atm}$)

From Henry's law, we know that concentration of a gas in a liquid is proportional to its partial pressure.
$C(\text{mol/L}) = k P_{CO_2}$, which can rearrange to solve for k:

$$k = \frac{0.0329 \text{ M}}{1 \text{ atm}} = 0.0329 \text{ mol L}^{-1} \text{ atm}^{-1}$$

Therefore, concentration of $CO_2(aq)$ in water under atmospheric pressures at 298 K is:

$$C_{CO_2} (\text{mol / L}) = 0.0329 \text{ mol L}^{-1} \text{ atm}^{-1}\; 0.00037 \text{ atm} = 1.217 \times 10^{-5} \text{ M}$$

Using the equation for reaction of CO_2 with water:

CO_2 (aq)	+	2 H_2O (l)	$\rightleftharpoons$	H_3O^+ (aq)	+	HCO_3^- (aq)
1.217×10^{-5}				0		0
$-x$				$+x$		$+x$
$1.217 \times 10^{-5} - x$				x		x

$$K_{a1} = \frac{x^2}{\left(1.217 \times 10^{-5} - x\right)} = 4.4 \times 10^{-7}$$

Using the quadratic formula, $x = 2.104 \times 10^{-6}$ M

$$\text{pH} = -\log\left[H_3O^+\right] = -\log\left(2.104 \times 10^{-6}\right) = 5.68 \approx 5.7$$

We can, of course, continue to refine the value of $[H_3O^+]$ further by considering the dissociation of HCO_3^-, but the change is too small to matter.

B **(D)**

(a) For the acids given, we determine values of m and n in the formula $EO_m(OH)_n$. HOCl or Cl(OH) has $m = 0$ and $n = 1$. We expect $K_a \approx 10^{-7}$ or $pK_a \approx 7$, which is in good agreement with the accepted $pK_a = 7.52$. HOClO or ClO(OH) has $m = 1$ and $n = 1$. We expect $K_a \approx 10^{-2}$ or $pK_a \approx 2$, in good agreement with the $pK_a = 1.92$. HOClO$_2$ or ClO$_2$(OH) has $m = 2$ and $n = 1$. We expect K_a to be large and in good agreement with the accepted value of $pK_a = -3$, $K_a = 10^{-pKa} = 10^3$. HOClO$_3$ or ClO$_3$(OH) has $m = 3$ and $n = 1$. We expect K_a to be very large and in good agreement with the accepted $K_a = -8$, $pK_a = -8$, $K_a = 10^{-pKa} = 10^8$ which turns out to be the case.

(b) The formula H_3AsO_4 can be rewritten as $AsO(OH)_3$, which has $m = 1$ and $n = 3$. The expected value is $K_a = 10^{-2}$.

(c) The value of $pK_a = 1.1$ corresponds to $K_a = 10^{-pKa} = 10^{-1.1} = 0.08$, which indicates that $m = 1$. The following Lewis structure is consistent with this value of m.

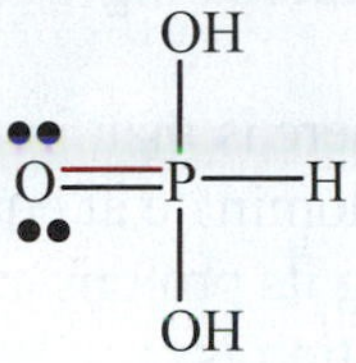

EXERCISES

Brønsted-Lowry Theory of Acids and Bases

1. **(E)**

(a) HNO_2 is an acid, a proton donor. Its conjugate base is NO_2^-.

(b) OCl^- is a base, a proton acceptor. Its conjugate acid is HOCl.

(c) NH_2^- is a base, a proton acceptor. Its conjugate acid is NH_3.

(d) NH_4^+ is an acid, a proton donor. It's conjugate base is NH_3.

(e) $CH_3NH_3^+$ is an acid, a proton donor. It's conjugate base is CH_3NH_2.

3. **(E)** The acids (proton donors) and bases (proton acceptors) are labeled below their formulas. Remember that a proton, in Brønsted-Lowry acid-base theory, is H^+.

(a) $HOBr(aq) + H_2O(l) \rightleftharpoons H_3O^+(aq) + OBr^-(aq)$

      acid   base   acid   base

(b) $HSO_4^-(aq) + H_2O(l) \rightleftharpoons H_3O^+(aq) + SO_4^{2-}(aq)$

$\qquad$ acid $\qquad\qquad$ base $\qquad\qquad$ acid $\qquad\qquad$ base

(c) $HS^-(aq) + H_2O(l) \rightleftharpoons H_2S(aq) + OH^-(aq)$

$\qquad$ base $\qquad\qquad$ acid $\qquad\qquad$ acid $\qquad\qquad$ base

(d) $C_6H_5NH_3^+(aq) + OH^-(aq) \rightleftharpoons C_6H_5NH_2(aq) + H_2O(l)$

$\qquad$ acid $\qquad\qquad\qquad$ base $\qquad\qquad\qquad$ base $\qquad\qquad\qquad$ acid

5. **(E)** Answer (b), NH_3, is correct. CH_3COOH will react most completely with the strongest base. NO_3^- and Cl^- are very weak bases. H_2O is a weak base, but it is amphiprotic, acting as an acid (donating protons), as in the presence of NH_3. Thus, NH_3 must be the strongest base and the most effective in deprotonating CH_3COOH.

7. **(E)** The principle we will follow here is that, in terms of their concentrations, the weaker acid and the weaker base will predominate at equilibrium. The reason for this is that a strong acid will do a good job of donating its protons and, having done so, its conjugate base will be left behind. The preferred direction is:

strong acid + strong base → weak (conjugate) base + weak (conjugate) acid

(a) The reaction will favor the forward direction because OH^- (a strong base) $> NH_3$ (a weak base) and NH_4^+ (relatively strong weak acid) $> H_2O$ (very weak acid).

(b) The reaction will favor the reverse direction because $HNO_3 > HSO_4^-$ (a weak acid in the second ionization) (acting as acids), and $SO_4^{2-} > NO_3^-$ (acting as bases).

(c) The reaction will favor the reverse direction because $CH_3COOH > CH_3OH$ (not usually thought of as an acid) (acting as acids), and $CH_3O^- > CH_3COO^-$ (acting as bases).

Strong Acids, Strong Bases, and pH

9. **(M)** All of the solutes are strong acids or strong bases.

(a) $[H_3O^+] = 0.00165 \text{ M } HNO_3 \times \dfrac{1 \text{ mol } H_3O^+}{1 \text{ mol } HNO_3} = 0.00165 \text{ M}$

$[OH^-] = \dfrac{K_w}{[H_3O^+]} = \dfrac{1.0 \times 10^{-14}}{0.00165 \text{M}} = 6.1 \times 10^{-12} \text{ M}$

(b) $\left[OH^-\right] = 0.0087 \text{ M KOH} \times \dfrac{1 \text{ mol OH}^-}{1 \text{ mol KOH}} = 0.0087 \text{ M}$

$\left[H_3O^+\right] = \dfrac{K_w}{\left[OH^-\right]} = \dfrac{1.0 \times 10^{-14}}{0.0087 \text{M}} = 1.1 \times 10^{-12} \text{ M}$

(c) $\left[OH^-\right] = 0.00213 \text{ M Sr}(OH)_2 \times \dfrac{2 \text{ mol OH}^-}{1 \text{ mol Sr}(OH)_2} = 0.00426 \text{ M}$

$\left[H_3O^+\right] = \dfrac{K_w}{\left[OH^-\right]} = \dfrac{1.0 \times 10^{-14}}{0.00426 \text{ M}} = 2.3 \times 10^{-12} \text{ M}$

(d) $\left[H_3O^+\right] = 5.8 \times 10^{-4} \text{M HI} \times \dfrac{1 \text{mol H}_3O^+}{1 \text{mol HI}} = 5.8 \times 10^{-4} \text{ M}$

$\left[OH^-\right] = \dfrac{K_w}{\left[H_3O^+\right]} = \dfrac{1.0 \times 10^{-14}}{5.8 \times 10^{-4} \text{ M}} = 1.7 \times 10^{-11} \text{ M}$

11. **(E)**

$\left[OH^-\right] = \dfrac{3.9 \text{ g Ba}(OH)_2 \cdot 8H_2O}{100 \text{ mL soln}} \times \dfrac{1000 \text{ mL}}{1 \text{ L}} \times \dfrac{1 \text{ mol Ba}(OH)_2 \cdot 8H_2O}{315.5 \text{ g Ba}(OH)_2 \cdot 8H_2O} \times \dfrac{2 \text{ mol OH}^-}{1 \text{ mol Ba}(OH)_2 \cdot 8H_2O}$

$= 0.25 \text{ M}$

$\left[H_3O^+\right] = \dfrac{K_w}{\left[OH^-\right]} = \dfrac{1.0 \times 10^{-14}}{0.25 \text{ M OH}^-} = 4.0 \times 10^{-14} \text{ M} \qquad \text{pH} = -\log(4.0 \times 10^{-14}) = 13.40$

13. **(M)** First we determine the moles of HCl, and then its concentration.

$\text{moles HCl} = \dfrac{PV}{RT} = \dfrac{\left(751 \text{ mmHg} \times \dfrac{1 \text{ atm}}{760 \text{ mmHg}}\right) \times 0.205 \text{ L}}{0.08206 \text{ L atm mol}^{-1} \text{ K}^{-1} \times 296 \text{ K}} = 8.34 \times 10^{-3} \text{ mol HCl}$

$\left[H_3O^+\right] = \dfrac{8.34 \times 10^{-3} \text{ mol HCl}}{4.25 \text{ L soln}} \times \dfrac{1 \text{ mol H}_3O^+}{1 \text{ mol HCl}} = 1.96 \times 10^{-3} \text{ M}$

15. **(E)** First determine the amount of HCl, and then the volume of the concentrated solution required.

$\text{amount HCl} = 12.5 \text{ L} \times \dfrac{10^{-2.10} \text{ mol H}_3O^+}{1 \text{ L soln}} \times \dfrac{1 \text{ mol HCl}}{1 \text{ mol H}_3O^+} = 0.099 \text{ mol HCl}$

$V_{\text{solution}} = 0.099 \text{ mol HCl} \times \dfrac{36.46 \text{ g HCl}}{1 \text{ mol HCl}} \times \dfrac{100.0 \text{ g soln}}{36.0 \text{ g HCl}} \times \dfrac{1 \text{ mL soln}}{1.18 \text{ g soln}} = 8.5 \text{ mL soln}$

17. **(E)** The volume of HCl(aq) needed is determined by first finding the amount of $NH_3(aq)$ present, and then realizing that acid and base react in a 1:1 molar ratio.

$$V_{HCl} = 1.25 \text{ L base} \times \frac{0.265 \text{ mol NH}_3}{1 \text{ L base}} \times \frac{1 \text{ mol H}_3O^+}{1 \text{ mol NH}_3} \times \frac{1 \text{ mol HCl}}{1 \text{ mol H}_3O^+} \times \frac{1 \text{ L acid}}{6.15 \text{ mol HCl}}$$

$$= 0.0539 \text{ L acid or } 53.9 \text{ mL acid.}$$

19. **(M)** Here we determine the amounts of H_3O^+ and OH^- and then the amount of the one that is in excess. We express molar concentration in millimoles/milliliter, equivalent to mol/L.

$$50.00 \text{ mL} \times \frac{0.0155 \text{ mmol HI}}{1 \text{ mL soln}} \times \frac{1 \text{ mmol H}_3O^+}{1 \text{ mmol HI}} = 0.775 \text{ mmol H}_3O^+$$

$$75.00 \text{ mL} \times \frac{0.0106 \text{ mmol KOH}}{1 \text{ mL soln}} \times \frac{1 \text{ mmol OH}^-}{1 \text{ mmol KOH}} = 0.795 \text{ mmol OH}^-$$

The net reaction is $H_3O^+(aq) + OH^-(aq) \rightarrow 2 H_2O(l)$.

There is an excess of OH^- of $(0.795 - 0.775 =) 0.020 \text{ mmol OH}^-$.

Thus, this is a basic solution. The total solution volume is $(50.00 + 75.00 =) 125.00 \text{ mL}$.

$$\left[OH^-\right] = \frac{0.020 \text{ mmol OH}^-}{125.00 \text{ mL}} = 1.6 \times 10^{-4} \text{ M}, \quad pOH = -\log(1.6 \times 10^{-4}) = 3.80, \quad pH = 10.20$$

Weak Acids, Weak Bases, and pH

21. **(M)** We organize the solution around the balanced chemical equation.

Equation:	$HNO_2(aq)$	+	$H_2O(l)$	$\rightleftharpoons$	$H_3O^+(aq)$	+	$NO_2^-(aq)$
Initial:	0.143 M		—		≈ 0 M		0 M
Changes:	$-x$ M		—		$+x$ M		$+x$ M
Equil:	$(0.143 - x)$ M		—		x M		x M

$$K_a = \frac{\left[H_3O^+\right]\left[NO_2^-\right]}{\left[HNO_2\right]} = \frac{x^2}{0.143 - x} = 7.2 \times 10^{-4} \gg \frac{x^2}{0.143} \quad (\text{if } x \ll 0.143)$$

$$x = \sqrt{0.143 \times 7.2 \times 10^{-4}} = 0.010 \text{ M}$$

We have assumed that $x \ll 0.143$ M, an almost acceptable assumption. Another cycle of approximations yields:

$$x = \sqrt{(0.143 - 0.010) \times 7.2 \times 10^{-4}} = 0.0098 \text{ M} = [H_3O^+] \qquad pH = -\log(0.0098) = 2.01$$

This is the same result as is determined with the quadratic formula roots equation.

23. **(M)**

(a) The set-up is based on the balanced chemical equation.

Equation: $C_6H_5CH_2CO_2H(aq)$ + $H_2O(l)$ $\rightleftharpoons$ $H_3O^+(aq)$ + $C_6H_5CH_2CO_2^-(aq)$

Initial: $\quad$ 0.186 M $\quad\quad\quad\quad\quad\quad\quad\quad\quad\quad$ ≈ 0 M $\quad$ 0 M

Changes: $\quad$ $-x$ M $\quad\quad\quad\quad\quad\quad\quad\quad\quad\quad$ $+x$ M $\quad$ $+x$ M

Equil: $\quad$ $(0.186-x)$ M $\quad\quad\quad\quad\quad\quad\quad\quad$ x M $\quad$ x M

$$K_a = 4.9\times10^{-5} = \frac{[H_3O^+][C_6H_5CH_2CO_2^-]}{[C_6H_5CH_2CO_2H]} = \frac{x\cdot x}{0.186-x} \approx \frac{x^2}{0.186}$$

$$x = \sqrt{0.186\times 4.9\times10^{-5}} = 0.0030\ M = [H_3O^+] = [C_6H_5CH_2CO_2^-]$$

0.0030 M is less than 5 % OF 0.186 M, thus, the approximation is valid.

(b) The set-up is based on the balanced chemical equation.

Equation: $\quad C_6H_5CH_2CO_2H(aq)$ + $H_2O(l)$ $\rightleftharpoons$ $H_3O^+(aq)$ + $C_6H_5CH_2CO_2^-(aq)$

Initial: $\quad\quad$ 0.121 M $\quad\quad\quad\quad\quad\quad\quad\quad\quad\quad$ ≈ 0 M $\quad\quad$ 0 M

Changes: $\quad\quad$ $-x$ M $\quad\quad\quad\quad\quad\quad\quad\quad\quad\quad$ $+x$ M $\quad\quad$ $+x$ M

Equil: $\quad\quad$ $(0.121-x)$ M $\quad\quad\quad\quad\quad\quad\quad\quad$ x M $\quad\quad$ x M

$$K_a = 4.9\times10^{-5} = \frac{\left[H_3O^+\right]\left[C_6H_5CH_2CO_2^-\right]}{\left[C_6H_5CH_2CO_2H\right]} = \frac{x^2}{0.121-x} \approx \frac{x^2}{0.121} \quad x = 0.0024\ M = \left[H_3O^+\right]$$

Assumption $x \ll 0.121$, is correct. $pH = -\log(0.0024) = 2.62$

25. **(M)** We base our solution on the balanced chemical equation.

$$\left[H_3O^+\right] = 10^{-1.56} = 2.8\times10^{-2}\ M$$

Equation: $\quad CH_2FCOOH(aq)$ + $H_2O(l)$ $\rightleftharpoons$ $CH_2FCOO^-(aq)$ + $H_3O^+(aq)$

Initial: $\quad\quad\quad$ 0.318 M $\quad\quad\quad$ – $\quad\quad\quad\quad$ 0 M $\quad\quad\quad\quad$ ≈ 0 M

Changes: $\quad\quad\quad$ -0.028 M $\quad\quad$ – $\quad\quad\quad\quad$ $+0.028$ M $\quad\quad$ $+0.028$ M

Equil: $\quad\quad\quad$ 0.290 M $\quad\quad\quad$ – $\quad\quad\quad\quad$ 0.028 M $\quad\quad\quad$ ≈ 0.028 M

$$K_a = \frac{\left[H_3O^+\right]\left[CH_2FCOO^-\right]}{\left[CH_2FCOOH\right]} = \frac{(0.028)(0.028)}{0.290} = 2.7\times10^{-3}$$

27. **(M)** Here we need to find the molarity S of the acid needed that yields

$$\left[H_3O^+\right] = 10^{-2.85} = 1.4\times10^{-3}\ M$$

Equation: $\quad C_6H_5COOH(aq)$ + $H_2O(l)$ $\rightleftharpoons$ $H_3O^+(aq)$ + $C_6H_5COO^-(aq)$

Initial: $\quad\quad\quad$ S $\quad\quad\quad\quad\quad$ – $\quad\quad\quad\quad$ 0 M $\quad\quad\quad\quad$ 0 M

Changes: $\quad$ -0.0014 M $\quad\quad\quad$ – $\quad\quad\quad\quad$ $+0.0014$M $\quad\quad$ $+0.0014$M

Equil: $\quad\quad$ $S-0.0014$M $\quad\quad$ – $\quad\quad\quad\quad$ 0.0014 M $\quad\quad\quad$ 0.0014 M

$$K_a = \frac{\left[H_3O^+\right]\left[C_6H_5COO^-\right]}{\left[C_6H_5COOH\right]} = \frac{(0.0014)^2}{S - 0.0014} = 6.3 \times 10^{-5} \quad S - 0.0014 = \frac{(0.0014)^2}{6.3 \times 10^{-5}} = 0.031$$

$$S = 0.031 + 0.0014 = 0.032 \text{ M} = \left[C_6H_5COOH\right]$$

$$350.0 \text{ mL} \times \frac{1 \text{ L}}{1000 \text{ mL}} \times \frac{0.032 \text{ mol } C_6H_5COOH}{1 \text{ L soln}} \times \frac{122.1 \text{ g } C_6H_5COOH}{1 \text{ mol } C_6H_5COOH} = 1.4 \text{ g } C_6H_5COOH$$

29. **(M)** We use the balanced chemical equation, then solve using the quadratic formula.

Equation :	$HClO_2(aq)$	$+$	$H_2O(l)$	$\rightleftharpoons$	$H_3O^+(aq)$	$+$	$ClO_2^-(aq)$
Initial :	0.55 M		—		≈ 0 M		0 M
Changes :	$-x$ M		—		$+x$ M		$+x$ M
Equil :	$(0.55 - x)$ M		—		x M		x M

$$K_a = \frac{\left[H_3O^+\right]\left[ClO_2^-\right]}{\left[HClO_2\right]} = \frac{x^2}{0.55 - x} = 1.1 \times 10^{-2} = 0.011 \quad x^2 = 0.0061 - 0.011x$$

$$x^2 + 0.011x - 0.0061 = 0$$

$$x = \frac{-b \pm \sqrt{b^2 - 4ac}}{2a} = \frac{-0.011 \pm \sqrt{0.000121 + 0.0244}}{2} = 0.073 \text{ M} = \left[H_3O^+\right]$$

The method of successive approximations converges to the same answer in four cycles.

$$pH = -\log\left[H_3O^+\right] = -\log(0.073) = 1.14 \qquad pOH = 14.00 - pH = 14.00 - 1.14 = 12.86$$

$$\left[OH^-\right] = 10^{-pOH} = 10^{-12.86} = 1.4 \times 10^{-13} \text{ M}$$

31. **(M)**

$$\left[C_{10}H_7NH_2\right] = \frac{1 \text{ g } C_{10}H_7NH_2}{590 \text{ g } H_2O} \times \frac{1.00 \text{ g } H_2O}{1 \text{ mL}} \times \frac{1000 \text{ mL}}{1 \text{ L}} \times \frac{1 \text{ mol } C_{10}H_7NH_2}{143.2 \text{ g } C_{10}H_7NH_2}$$

$$= 0.012 \text{ M } C_{10}H_7NH_2$$

$$K_b = 10^{-pK_b} = 10^{-3.92} = 1.2 \times 10^{-4}$$

Equation:	$C_{10}H_7NH_2(aq)$	$+$	$H_2O(l)$	$\rightleftharpoons$	$OH^-(aq)$	$+$	$C_{10}H_7NH_3^+(aq)$
Initial:	0.012 M		—		≈ 0 M		0 M
Changes:	$-x$ M		—		$+x$ M		$+x$ M
Equil:	$(0.012 - x)$ M		—		x M		x M

$$K_b = \frac{\left[OH^-\right]\left[C_{10}H_7NH_3^+\right]}{\left[C_{10}H_7NH_2\right]} = \frac{x^2}{0.012 - x} = 1.2 \times 10^{-4} \approx \frac{x^2}{0.012} \qquad \text{assuming } x \ll 0.012$$

$$x = \sqrt{0.012 \times 1.2 \times 10^{-4}} = 0.0012 \text{ This is an almost acceptable assumption. Another}$$
approximation cycle gives:

$x = \sqrt{(0.012 - 0.0012) \times 1.2 \times 10^{-4}} = 0.0011$ Yet another cycle seems necessary.

$x = \sqrt{(0.012 - 0.0011) \times 1.2 \times 10^{-4}} = 0.0011 \, M = [OH^-]$

The quadratic equation roots formula provides the same answer.

$pOH = -\log[OH^-] = -\log(0.0011) = 2.96 \quad pH = 14.00 - pOH = 11.04$

$H_3O^+ = 10^{-pH} = 10^{-11.04} = 9.1 \times 10^{-12} \, M$

33. **(M)** Here we determine $[H_3O^+]$ which, because of the stoichiometry of the reaction, equals $[CH_3COO^-]$. $[H_3O^+] = 10^{-pH} = 10^{-4.52} = 3.0 \times 10^{-5} \, M = [CH_3COO^-]$

We solve for S, the concentration of $HC_2H_3O_2$ in the 0.750 L solution before it dissociates.

Equation: $\quad CH_3COOH \, (aq) \; + \; H_2O(l) \; \rightleftharpoons \; CH_3COO^- (aq) \; + \; H_3O^+ (aq)$

Initial:	$S \, M$	—	$0 \, M$	$\approx 0 M$
Changes:	$-3.0 \times 10^{-5} M$	—	$+3.0 \times 10^{-5} M$	$+3.0 \times 10^{-5} M$
Equil:	$(S - 3.0 \times 10^{-5}) \, M$	—	$3.0 \times 10^{-5} M$	$3.0 \times 10^{-5} M$

$K_a = \dfrac{[H_3O^+][CH_3COO^-]}{[CH_3COOH]} = 1.8 \times 10^{-5} = \dfrac{(3.0 \times 10^{-5})^2}{(S - 3.0 \times 10^{-5})}$

$(3.0 \times 10^{-5})^2 = 1.8 \times 10^{-5}(S - 3.0 \times 10^{-5}) = 9.0 \times 10^{-10} = 1.8 \times 10^{-5} S - 5.4 \times 10^{-10}$

$S = \dfrac{9.0 \times 10^{-10} + 5.4 \times 10^{-10}}{1.8 \times 10^{-5}} = 8.0 \times 10^{-5} M$ Now we determine the mass of vinegar needed.

mass vinegar $= 0.750 \, L \times \dfrac{8.0 \times 10^{-5} \, mol \, CH_3COOH}{1 \, L \, soln} \times \dfrac{60.05 \, g \, CH_3COOH}{1 \, mol \, CH_3COOH} \times \dfrac{100.0 \, g \, vinegar}{5.7 \, g \, CH_3COO^-}$

$CH_3COO^- = 0.063$ g vinegar

35. **(D)**

(a) $n_{proplyamine} = \dfrac{PV}{RT} = \dfrac{\left(316 \, Torr \times \dfrac{1 \, atm}{760 \, Torr}\right)(0.275 \, L)}{(0.08206 \, atm \, L \, K^{-1} \, mol^{-1})(298.15 \, K)} = 4.67 \times 10^{-3}$ mol propylamine

$[propylamine] = \dfrac{n}{V} = \dfrac{4.67 \times 10^{-3} \, moles}{0.500 \, L} = 9.35 \times 10^{-3} \, M$

$K_b = 10^{-pK_b} = 10^{-3.43} = 3.7 \times 10^{-4}$

$\quad\quad CH_3CH_2CH_2NH_2(aq) + H_2O(l) \; \rightleftharpoons \; CH_3CH_2CH_2NH_3^+(aq) \; + \; OH^-(aq)$

Initial	$9.35 \times 10^{-3} \, M$	—	$0 \, M$	$\approx 0 \, M$
Change	$-x \, M$	—	$+x \, M$	$+x \, M$
Equil.	$(9.35 \times 10^{-3} - x) \, M$	—	$x \, M$	$x \, M$

$$K_b = 3.7 \times 10^{-4} = \frac{x^2}{9.35 \times 10^{-3} - x} \text{ or } 3.5 \times 10^{-6} - 3.7 \times 10^{-4}(x) = x^2$$

$$x^2 + 3.7 \times 10^{-4}x - 3.5 \times 10^{-6} = 0$$

Find x with the roots formula:

$$x = \frac{-3.7 \times 10^{-4} \pm \sqrt{(3.7 \times 10^{-4})^2 - 4(1)(-3.5 \times 10^{-6})}}{2(1)}$$

Therefore $x = 1.6\underline{95} \times 10^{-3}$ M $= [OH^-]$ pOH $= 2.77$ and pH $= 11.23$

(b) $[OH^-] = 1.7 \times 10^{-3}$ M $= [NaOH]$ $\qquad$ (MM$_{NaOH} = 39.997$ g mol^{-1})

$n_{NaOH} = (C)(V) = 1.7 \times 10^{-3}$ M $\times 0.500$ L $= 8.5 \times 10^{-4}$ moles NaOH

mass of NaOH $= (n)(MM_{NaOH}) = 8.5 \times 10^{-4}$ mol NaOH $\times 39.997$ g NaOH/mol NaOH

mass of NaOH $= 0.034$ g NaOH (34 mg of NaOH)

37. **(M)** If the molarity of acetic acid is doubled, we expect a lower initial pH (more H_3O^+(aq) in solution) and a lower percent ionization as a result of the increase in concentration. The ratio between $[H_3O^+]$ of concentration, c, and concentration $2c$ is $\sqrt{2} \approx 1.4$. Therefore, (b), containing 14 H_3O^+ symbols best represents the conditions (~1.4 times greater).

Percent Ionization

39. **(M)** Let us first compute the $\left[H_3O^+\right]$ in this solution.

Equation: $\quad$ CH$_3$CH$_2$COOH(aq) $\quad + \quad$ H$_2$O(l) $\quad \rightleftharpoons \quad$ H$_3$O$^+$(aq) $\quad + \quad$ CH$_3$CH$_2$COO$^-$(aq)

Initial: $\qquad$ 0.45 M $\qquad\qquad\qquad$ — $\qquad\qquad\qquad \approx 0$ M $\qquad\qquad \approx 0$ M

Changes: $\qquad$ $-x$ M $\qquad\qquad\qquad$ — $\qquad\qquad\qquad +x$ M $\qquad\qquad +x$ M

Equil: $\qquad$ $(0.45 - x)$ M $\qquad\qquad$ — $\qquad\qquad\qquad\qquad x$ M $\qquad\qquad\qquad x$ M

$$K_a = \frac{\left[H_3O^+\right]\left[CH_3CH_2COO^-\right]}{\left[CH_3CH_2COOH\right]} = \frac{x^2}{0.45 - x} = 10^{-4.89} = 1.3 \times 10^{-5} \approx \frac{x^2}{0.45}$$

$x = 2.4 \times 10^{-3}$ M; We have assumed that $x \ll 0.45$ M, an assumption that clearly is correct.

(a) $\quad \alpha = \dfrac{\left[H_3O^+\right]_{equil}}{\left[CH_3CH_2COOH\right]_{initial}} = \dfrac{2.4 \times 10^{-3} \text{ M}}{0.45 \text{ M}} = 0.0053 = $ degree of ionization

(b) $\quad$ % ionization $= \alpha \times 100\% = 0.0053 \times 100\% = 0.53\%$

41. **(M)** Let x be the initial concentration of NH_3, hence, the amount dissociated is $0.042\,x$

	$NH_3(aq)$	$+ H_2O(l)$	$\xrightarrow{K_a=1.8\times10^{-5}}$	$NH_4^+(aq)$	$+$	$OH^-(aq)$
Initial	x M	—		0 M		≈ 0 M
Change	$-0.042x$ M	—		$+0.042x$ M		$+0.042x$ M
Equil.	$(1x - 0.042x)$ M $= 0.958\,x$	—		$0.042x$ M		$0.042x$ M

$$K_b = 1.8\times10^{-5} = \frac{[NH_4^+][OH^-]}{[NH_3]_{equil}} = \frac{[0.042x]^2}{[0.958x]} = 0.00184\underline{x}$$

$$[NH_3]_{initial} = x = \frac{1.8\times10^{-5}}{0.00184} = 0.00978\,M = 0.0098\,M$$

43. **(E)** We would not expect these ionizations to be correct because the calculated degree of ionization is based on the assumption that the $[CH_3COOH]_{initial} \approx [CH_3COOH]_{initial} - [CH_3COOH]_{equil.}$, which is invalid at the 13 and 42 percent levels of ionization seen here.

Polyprotic Acids

45. **(E)** Because H_3PO_4 is a weak acid, there is little HPO_4^{2-} (produced in the 2nd ionization) compared to the H_3O^+ (produced in the 1st ionization). In turn, there is little PO_4^{3-} (produced in the 3rd ionization) compared to the HPO_4^{2-}, and very little compared to the H_3O^+.

47. **(D)**

(a)

Equation:	$H_2S(aq)$	$+$	$H_2O(l)$	$\rightleftharpoons$	$HS^-(aq)$	$+$	$H_3O^+(aq)$
Initial:	0.075 M		—		0 M		≈ 0 M
Changes:	$-x$ M		—		$+x$ M		$+x$ M
Equil:	$(0.075 - x)$ M		—		x M		$+x$ M

$$K_{a_1} = \frac{[HS^-][H_3O^+]}{[H_2S]} = 1.0\times10^{-7} = \frac{x^2}{0.075-x} \approx \frac{x^2}{0.075} \quad x = 8.7\times10^{-5}\,M = [H_3O^+]$$

$$[HS^-] = 8.7\times10^{-5}\,M \text{ and } [S^{2-}] = K_{a_2} = 1\times10^{-19}\,M$$

(b) The set-up for this problem is the same as for part (a), with 0.0050 M replacing 0.075 M as the initial value of $[H_2S]$.

$$K_{a_1} = \frac{[HS^-][H_3O^+]}{[H_2S]} = 1.0\times10^{-7} = \frac{x^2}{0.0050-x} \approx \frac{x^2}{0.0050} \quad x = 2.2\times10^{-5}\,M = [H_3O^+]$$

$$[HS^-] = 2.2\times10^{-5}\,M \text{ and } [S^{2-}] = K_{a_2} = 1\times10^{-19}\,M$$

(c) The set-up for this part is the same as for part (a), with 1.0×10^{-5} M replacing 0.075 M as the initial value of $[H_2S]$. The solution differs in that we cannot assume $x \ll 1.0 \times 10^{-5}$. Solve the quadratic equation to find the desired equilibrium concentrations.

$$K_{a_1} = \frac{[HS^-][H_3O^+]}{[H_2S]} = 1.0 \times 10^{-7} = \frac{x^2}{1.0 \times 10^{-5} - x}$$

$$x^2 + 1.0 \times 10^{-7}x - 1.0 \times 10^{-12} = 0$$

$$x = \frac{-1.0 \times 10^{-7} \pm \sqrt{1.0 \times 10^{-14} + (4 \times 1.0 \times 10^{-12})}}{2 \times (1)}$$

$$x = 9.5 \times 10^{-7} M = [H_3O^+] \qquad [HS^-] = 9.5 \times 10^{-7} M \qquad [S^{2-}] = K_{a_2} = 1 \times 10^{-19} M$$

49. **(D)** In all cases, of course, the first ionization of H_2SO_4 is complete, and establishes the initial values of $[H_3O^+]$ and $[HSO_4^-]$. Thus, we need only deal with the second ionization in each case.

(a) Equation: $\quad HSO_4^-(aq) \; + \; H_2O(l) \rightleftharpoons \; SO_4^{2-}(aq) \; + \; H_3O^+(aq)$

Initial:	0.75 M	–	0 M	0.75 M
Changes:	$-x$ M	–	$+x$ M	$+x$ M
Equil:	$(0.75-x)$ M	–	x M	$(0.75+x)$ M

$$K_{a_2} = \frac{[SO_4^{2-}][H_3O^+]}{[HSO_4^-]} = 0.011 = \frac{x(0.75+x)}{0.75-x} \approx \frac{0.75x}{0.75} \qquad x = 0.011 \text{ M} = [SO_4^{2-}]$$

We have assumed that $x \ll 0.75$ M, an assumption that clearly is correct.

$$[HSO_4^-] = 0.75 - 0.011 = 0.74 \text{ M} \qquad [H_3O^+] = 0.75 + 0.011 = 0.76 \text{ M}$$

(b) The set-up for this part is similar to part (a), with the exception that 0.75 M is replaced by 0.075 M.

$$K_{a_2} = \frac{[SO_4^{2-}][H_3O^+]}{[HSO_4^-]} = 0.011 = \frac{x(0.075+x)}{0.075-x} \qquad 0.011(0.075-x) = 0.075x + x^2$$

$$x^2 + 0.086x - 8.3 \times 10^{-4} = 0$$

$$x = \frac{-b \pm \sqrt{b^2 - 4ac}}{2a} = \frac{-0.086 \pm \sqrt{0.0074 + 0.0033}}{2} = 0.0087 \text{ M}$$

$$x = 0.0087 \text{ M} = [SO_4^{2-}]$$

$$[HSO_4^-] = 0.075 - 0.0087 = 0.066 M \qquad [H_3O^+] = 0.075 + 0.0088M = 0.084 \text{ M}$$

(c) Again, the set-up is the same as for part (a), with the exception that 0.75 M is replaced by 0.00075 M

$$K_{a_2} = \frac{[SO_4^{2-}][H_3O^+]}{[HSO_4^-]} = 0.011 = \frac{x(0.00075 + x)}{0.00075 - x} \quad 0.011(0.00075 - x) = 0.00075\,x + x^2$$

$$x^2 + 0.0118x - 8.3 \times 10^{-6} = 0$$

$$x = \frac{-b \pm \sqrt{b^2 - 4ac}}{2a} = \frac{-0.0118 \pm \sqrt{1.39 \times 10^{-4} + 3.3 \times 10^{-5}}}{2} = 6.6 \times 10^{-4}$$

$$x = 6.6 \times 10^{-4}\,M = \left[SO_4^{2-}\right] \quad \left[HSO_4^-\right] = 0.00075 - 0.00066 = 9 \times 10^{-5}\,M$$

$$\left[H_3O^+\right] = 0.00075 + 0.00066\,M = 1.41 \times 10^{-3}\,M \quad \left[H_3O^+\right]$$ is almost twice the initial value of $\left[H_2SO_4\right]$. Thus, the second ionization of H_2SO_4 is nearly complete in this dilute solution.

51. (M)

(a) Recall that a base is a proton acceptor, in this case, accepting H^+ from H_2O.

First ionization : $C_{20}H_{24}O_2N_2 + H_2O \rightleftharpoons C_{20}H_{24}O_2N_2H^+ + OH^- \qquad pK_{b_1} = 6.0$

Second ionization : $C_{20}H_{24}O_2N_2H^+ + H_2O \rightleftharpoons C_{20}H_{24}O_2N_2H_2^{2+} + OH^- \quad pK_{b_2} = 9.8$

(b) $$[C_{20}H_{24}O_2N_2] = \frac{1.00\,g\,quinine \times \dfrac{1\,mol\,quinine}{324.4\,g\,quinine}}{1900\ mL \times \dfrac{1L}{1000\,mL}} = 1.62 \times 10^{-3}\,M$$

$$K_{b_1} = 10^{-6.0} = 1 \times 10^{-6}$$

Because the OH^- produced in (a) suppresses the reaction in (b) and since $K_{b1} \gg K_{b2}$, the solution's pH is determined almost entirely by the first base hydrolysis of the first reaction. Once again, we set-up the I.C.E. table and solve for the $[OH^-]$ in this case:

Equation:	$C_{20}H_{24}O_2N_2\,(aq)$	$+$	$H_2O(l)$	$\rightleftharpoons$	$C_{20}H_{24}O_2N_2H^+\,(aq)$	$+$	$OH^-\,(aq)$
Initial:	0.00162 M		—		0 M		≈ 0 M
Changes:	$-x$ M		—		$+x$ M		$+x$ M
Equil:	$(0.00162 - x)$ M		—		x M		x M

$$K_{b_1} = \frac{\left[C_{20}H_{24}O_2N_2H^+\right]\left[OH^-\right]}{\left[C_{20}H_{24}O_2N_2\right]} = 1 \times 10^{-6} = \frac{x^2}{0.00162 - x} \approx \frac{x^2}{0.00162} \quad x \approx 4 \times 10^{-5}\,M$$

The assumption $x \ll 0.00162$, is valid. $pOH = -\log\left(4 \times 10^{-5}\right) = 4.4 \quad pH = 9.6$

53. **(E)** Protonated codeine hydrolyzes water according to the following reaction:

$$C_{18}H_{21}O_3NH^+ + H_2O \rightleftharpoons C_{18}H_{21}O_3N + H_3O^+$$

$$pK_a = 6.05$$

$$pK_b = 14 - pK_a = 7.95$$

Molecular Structure and Acid-Base Behavior

65. **(M)**

(a) $HClO_3$ should be a stronger acid than is $HClO_2$. In each acid there is an $H - O - Cl$ grouping. The remaining oxygen atoms are bonded directly to Cl as terminal O atoms. Thus, there are two terminal O atoms in $HClO_3$ and only one in $HClO_2$. For oxoacids of the same element, the one with the higher number of terminal oxygen atoms is the stronger. With more oxygen atoms, the negative charge on the conjugate base is more effectively spread out, which affords greater stability. $HClO_3 : K_{a_1} = 5 \times 10^2$ and $HClO_2 : K_a = 1.1 \times 10^{-2}$.

(b) HNO_2 and H_2CO_3 each have one terminal oxygen atom. The difference? N is more electronegative than C, which makes $HNO_2 \left(K_a = 7.2 \times 10^{-4} \right)$, a stronger acid than $H_2CO_3 \left(K_{a_1} = 4.4 \times 10^{-7} \right)$.

(c) H_3PO_4 and H_2SiO_3 have the same number (one) of terminal oxygen atoms. They differ in P being more electronegative than Si, which makes H_3PO_4 $\left(K_{a_1} = 7.1 \times 10^{-3} \right)$ a stronger acid than $H_2SiO_3 \left(K_{a_1} = 1.7 \times 10^{-10} \right)$.

67. **(E)**

(a) HI is the stronger acid because the $H - I$ bond length is longer than the $H - Br$ bond length and, as a result, $H - I$ is easier to cleave.

(b) HOClO is a stronger acid than HOBr because
(i) there is a terminal O in HOClO but not in HOBr
(ii) Cl is more electronegative than Br.

(c) $H_3CCH_2CCl_2COOH$ is a stronger acid than $I_3CCH_2CH_2COOH$ both because Cl is more electronegative than is I and because the Cl atoms are closer to the acidic hydrogen in the COOH group and thus can exert a stronger e^- withdrawing effect on the O–H bond than can the more distant I atoms.

69. **(E)** The largest K_b (most basic) belongs to (c) $CH_3CH_2CH_2NH_2$ (hydrocarbon chains have the lowest electronegativity). The smallest K_b (least basic) is that of (a) o-chloroaniline (the nitrogen lone pair is delocalized (spread out over the ring), hence, less available to accept a proton (i.e., it is a poorer Brønsted base)).

Lewis Theory of Acids and Bases

71. **(E)**
(a) acid is CO_2 and base is H_2O

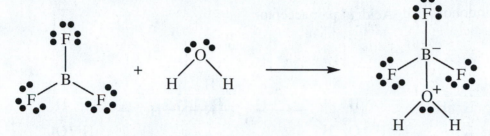

(b) acid is BF_3 and base is H_2O

(c) acid is H_2O and base is O^{2-}

(d) acid is SO_3 and base is S^{2-}

73. **(E)** A Lewis base is an electron pair donor, while a Lewis acid is an electron pair acceptor. We draw Lewis structures to assist our interpretation.

(a) $\left[\ddot{\ddot{O}} - H \right]^{-}$ The lone pairs on oxygen can readily be donated; this is a Lewis base.

(b)   The incomplete octet of B provides a site for acceptance of an electron pair, this is a Lewis acid.

(c)   The incomplete octet of B provides a site for acceptance of an electron pair, this is a Lewis acid.

75. **(M) (a)**

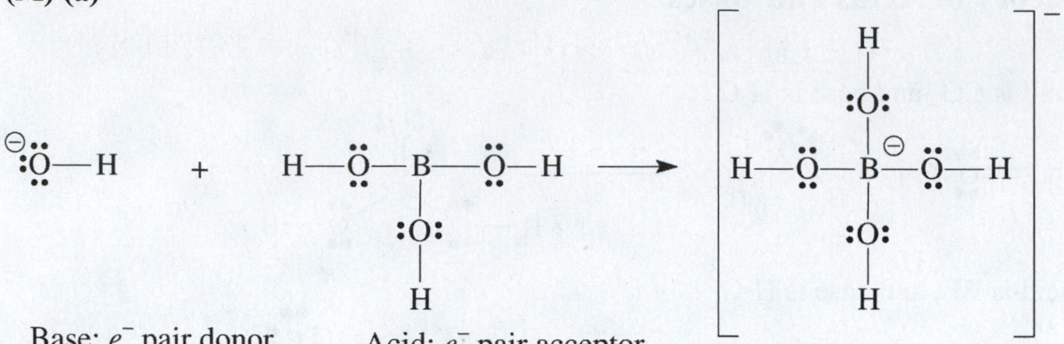

Base: e^- pair donor   Acid: e^- pair acceptor

(b)

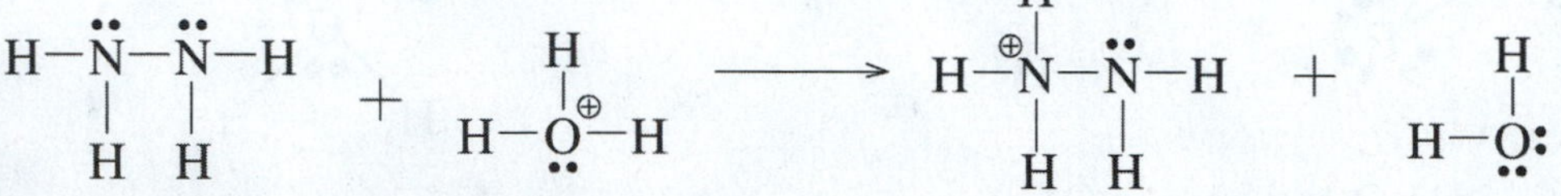

The actual Lewis acid is H^+, which is supplied by H_3O^+

(c)

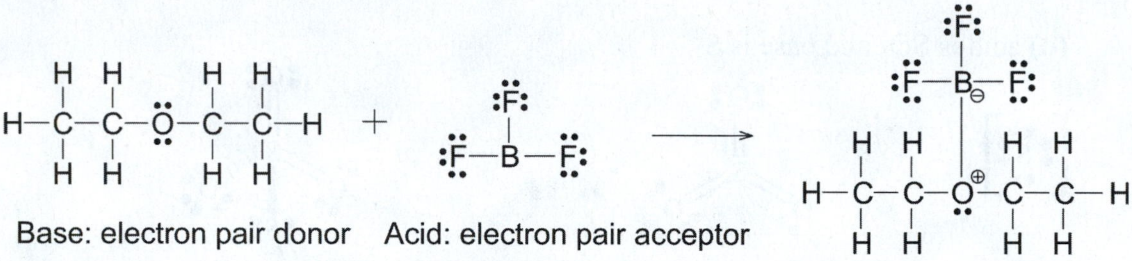

Base: electron pair donor   Acid: electron pair acceptor

77. **(E)** $I_2(aq) + I^-(aq) \rightleftharpoons I_3^-(aq)$

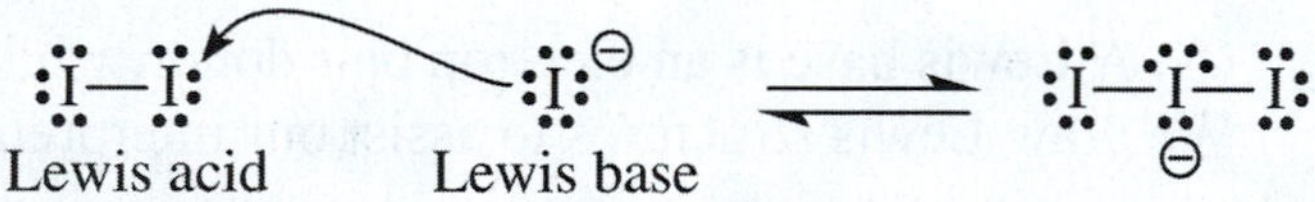

Lewis acid   Lewis base
(e^- pair acceptor)   (e^- pair donor)

79. **(E)**

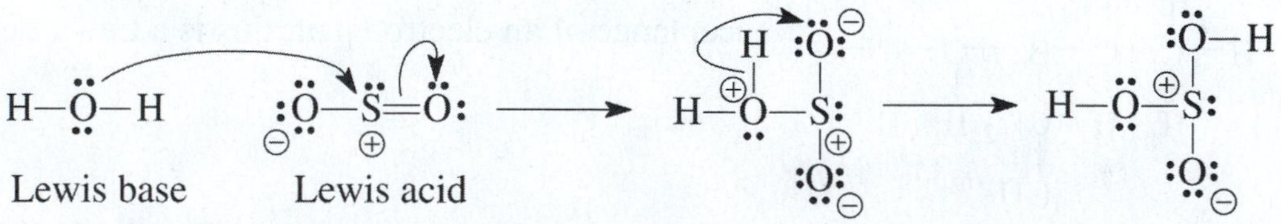

Lewis base   Lewis acid
(e^- pair donor)   (e^- pair acceptor)

INTEGRATIVE AND ADVANCED EXERCISES

81. **(E)** We use (ac) to represent the fact that a species is dissolved in acetic acid.

(a) $CH_3COO^-(ac) + CH_3COOH(l) \rightleftharpoons CH_3COOH(l) + CH_3COO^-(ac)$

CH_3COO^- is a base in acetic acid.

(b) $H_2O(ac) + CH_3COOH(l) \rightleftharpoons H_3O^+(ac) + CH_3COO^-(ac)$

Since H_2O is a weaker acid than CH_3COOH in aqueous solution, it will also be weaker in acetic acid. Thus H_2O will accept a proton from the solvent acetic acid, making H_2O a base in acetic acid.

(c) $CH_3COOH(l) + CH_3COOH(l) \rightleftharpoons CH_3COOH_2^+(ac) + CH_3COO^-(aq)$

Acetic acid can act as an acid or a base in acetic acid.

(d) $HClO_4(ac) + CH_3COOH(l) \rightleftharpoons ClO_4^-(ac) + CH_3COOH_2^+(ac)$

Since $HClO_4$ is a stronger acid than CH_3COOH in aqueous solution, it will also be stronger in acetic acid. Thus, $HClO_4$ will donate a proton to the solvent acetic acid, making $HClO_4$ an acid in acetic acid.

82. **(E)** $Sr(OH)_2 \text{ (sat'd)} \rightleftharpoons Sr^{2+}(aq) + 2\,OH^-(aq)$

$$pOH = 14 - pH = 14 - 13.12 = 0.88$$

$$[OH^-] = 10^{-0.88} = 0.132 \text{ M}$$

$$[OH^-]_{dilute} = 0.132M \times (\frac{10.0 \text{ ml concentrate}}{250.0 \text{ mL total}}) = 5.28 \times 10^{-3} M$$

$$[HCl] = \frac{(10.0 \text{ ml}) \times (5.28 \times 10^{-3} M)}{25.1 \text{ mL}} = 2.10 \times 10^{-3} M$$

83. **(M)**

(a) H_2SO_4 is a diprotic acid. A pH of 2 assumes no second ionization step and the second ionization step alone would produce a pH of less than 2, so it is not matched.

(b) pH 4.6 matched; ammonium salts hydrolyze to form solutions with pH < 7.

(c) KI should be nearly neutral (pH 7.0) in solution since it is the salt of a strong acid (HI) and a strong base (KOH). Thus it is not matched.

(d) pH of this solution is not matched as it's a weak base. A 0.002 M solution of KOH has a pH of about 11.3. A 0.0020 M methylamine solution would be lower (~10.9).

(e) pH of this solution is matched, since a 1.0 M hypochlorite salt hydrolyzes to form a solution of ~ pH 10.8.

(f) pH of this solution is not matched. Phenol is a very weak acid (pH >5).

(g) pH of this solution is 4.3. It is close, but not matched.

(h) pH of this solution is 2.1, matched as it's a strong organic acid.

(i) pH of this solution is 2.5, it is close, but not matched.

87. **(M)** First we assume that molarity and molality are numerically equal in this dilute aqueous solution. Then compute the molality that creates the observed freezing point, with $K_f = 1.86$ °C/m for water.

$$m = \frac{\Delta T_f}{K_f} = \frac{0.096\,°C}{1.86\,°C/m} = 0.052\,m \qquad \text{concentration} = 0.052\,M$$

This is the total concentration of all species in solution, ions as well as molecules. We base our remaining calculation on the balanced chemical equation.

Equation: $CH_2 = CHCH_2CO_2H(aq) + H_2O \rightleftharpoons CH_2 = CHCH_2CO_2^-(aq) + H_3O^+(aq)$

Initial:	0.0500 M	0 M	0 M
Changes:	$-x$ M	$+x$ M	$+x$ M
Equil:	$(0.0500 - x)$ M	x M	x M

Total conc. $= 0.051_6$ M $= (0.0500 - x)$ M $+ x$ M $+ x$ M $= 0.0500$ M $+ x$M $x = 0.001_6$ M

$$K_a = \frac{[CH_2\text{=}CHCH_2CO_2^-][H_3O^+]}{[CH_2\text{=}CHCH_2CO_2H]} = \frac{x \cdot x}{0.0500 - x} = \frac{(0.001_6)^2}{0.0500 - 0.001_6} = 5 \times 10^{-5}$$

88. **(D)** $[H_3O^+] = 10^{-pH} = 10^{-5.50} = 3.2 \times 10^{-6}$ M

(a) To produce this acidic solution we could not use 15 M NH_3(aq), because aqueous solutions of NH_3 have pH values greater than 7 owing to the fact that NH_3 is a weak base.

(b) $V_{HCl} = 100.0\,\text{mL} \times \dfrac{3.2 \times 10^{-6} \text{mol } H_3O^+}{1\,L} \times \dfrac{1 \text{ mol HCl}}{1 \text{ mol } H_3O^+} \times \dfrac{1\,L \text{ soln}}{12 \text{ mol HCl}}$

$= 2.7 \times 10^{-5}$ mL 12 M HCl soln

(very unlikely, since this small a volume is hard to measure)

(c) $K_a = 5.6 \times 10^{-10}$ for NH_4^+(aq)

Equation: $NH_4^+(aq) \quad + \quad H_2O(l) \rightleftharpoons NH_3(aq) + H_3O^+(aq)$

Initial:	c M	–	0 M	≈ 0 M
Changes:	-3.2×10^{-6} M	–	$+3.2 \times 10^{-6}$ M	$+3.2 \times 10^{-6}$ M
Equil:	$(c - 3.2 \times 10^{-6})$ M	–	3.2×10^{-6} M	3.2×10^{-6} M

$$K_a = 5.6 \times 10^{-10} = \frac{[NH_3][H_3O^+]}{[NH_4^+]} = \frac{(3.2 \times 10^{-6})(3.2 \times 10^{-6})}{(c - 3.2 \times 10^{-6})}$$

$$c - 3.2 \times 10^{-6} = \frac{(3.2 \times 10^{-6})^2}{5.6 \times 10^{-10}} = 1.8 \times 10^{-2} \text{ M} \qquad [NH_4^+] = 0.018 \text{ M}$$

$$NH_4Cl \text{ mass} = 100.0 \text{ mL} \times \frac{0.018 \text{ mol } NH_4^+}{1000 \text{ mL}} \times \frac{1 \text{ mol } NH_4Cl}{1 \text{ mol } NH_4^+} \times \frac{53.49 \text{ g } NH_4Cl}{1 \text{ mol } NH_4Cl} = 0.096 \text{ g } NH_4Cl$$

This would be an easy mass to measure on a laboratory three decimal place balance.

(d) The set-up is similar to that for NH_4Cl.

Equation:	$CH_3COOH \text{ (aq)}$	$+$	$H_2O \text{ (l)}$	$\rightleftharpoons$	$CH_3COO^- \text{(aq)}$	$+$	$H_3O^+ \text{(aq)}$
Initial:	c M		$-$		0 M		≈ 0 M
Changes:	-3.2×10^{-6} M		$-$		$+3.2 \times 10^{-6}$ M		$+3.2 \times 10^{-6}$ M
Equil:	$(c - 3.2 \times 10^{-6})$ M		$-$		3.2×10^{-6} M		3.2×10^{-6} M

$$K_a = 1.8 \times 10^{-5} = \frac{[CH_3COO^-][H_3O^+]}{[CH_3COOH]} = \frac{(3.2 \times 10^{-6})(3.2 \times 10^{-6})}{c - 3.2 \times 10^{-6}}$$

$$c - 3.2 \times 10^{-6} = \frac{(3.2 \times 10^{-6})^2}{1.8 \times 10^{-5}} = 5.7 \times 10^{-7} \qquad c = [CH_3COOH] = 3.8 \times 10^{-6}$$

$$CH_3COOH \text{ mass} = 100.0 \text{ mL} \times \frac{3.8 \times 10^{-6} \text{ mol } CH_3COOH}{1000 \text{ mL}} \times \frac{60.05 \text{ g } CH_3COOH}{1 \text{ mol } CH_3COOH}$$

$$= 2.3 \times 10^{-5} \text{ g } CH_3COOH \text{ (an almost impossibly small mass to measure}$$
$$\text{using conventional laboratory scales)}$$

94. (M) To have the same freezing point, the two solutions must have the same total concentration of all particles of solute—ions and molecules. We first determine the concentrations of all solute species in 0.150 M $ClCH_2COOH$, $K_a = 1.4 \times 10^{-3}$

Equation:	$ClCH_2COOH_2 \text{ (aq)} + H_2O \text{(l)}$	$\rightleftharpoons$	$H_3O^+ \text{(aq)}$	$+$	$ClCH_2COO^- \text{(aq)}$
Initial:	0.150 M		≈ 0 M		0 M
Changes:	$-x$ M		$+x$ M		$+x$ M
Equil:	$(0.150 - x)$ M		$+x$ M		x M

$$K_a = \frac{[H_3O^+][ClCH_2COO^-]}{[ClCH_2COOH]} = \frac{x \cdot x}{0.150 - x} = 1.4 \times 10^{-3}$$

$$x^2 = 2.1 \times 10^{-4} - 1.4 \times 10^{-3} x \qquad x^2 + 1.4 \times 10^{-3} x - 2.1 \times 10^{-4} = 0$$

$$x = \frac{-b \pm \sqrt{b^2 - 4ac}}{2a} = \frac{-1.4 \times 10^{-3} \pm \sqrt{2.0 \times 10^{-6} + 8.4 \times 10^{-4}}}{2} = 0.014 \text{ M}$$

total concentration $= (0.150 - x) + x + x = 0.150 + x = 0.150 + 0.014 = 0.164$ M

Now we determine the $[CH_3COOH]$ that has this total concentration.

Equation: $\quad CH_3COOH(aq) \; + \; H_2O(l) \; \rightleftharpoons \; H_3O^+(aq) \; + \; CH_3COO^-(aq)$

Initial: $\qquad z$ M $\qquad\qquad\qquad - \qquad\qquad\quad \approx 0$ M $\qquad\quad 0$ M

Changes: $\quad -y$ M $\qquad\qquad\qquad - \qquad\qquad\quad +y$ M $\qquad\quad +y$ M

Equil: $\qquad (z-y)$ M $\qquad\qquad\quad - \qquad\qquad\qquad y$ M $\qquad\qquad y$ M

$$K_a = \frac{[H_3O^+][CH_3COO^-]}{[CH_3COOH]} = 1.8 \times 10^{-5} = \frac{y \cdot y}{z - y}$$

We also know that the total concentration of all solute species is 0.164 M.

$$z - y + y + y = z + y = 0.164 \qquad z = 0.164 - y \qquad \frac{y^2}{0.164 - 2y} = 1.8 \times 10^{-5}$$

We assume that $2y \ll 0.164 \qquad y^2 = 0.164 \times 1.8 \times 10^{-5} = 3.0 \times 10^{-6} \qquad y = 1.7 \times 10^{-3}$

The assumption is valid: $2y = 0.0034 \ll 0.164$

Thus, $[CH_3COOH] = z = 0.164 - y = 0.164 - 0.0017 = 0.162$ M

$$\text{mass } CH_3COOH = 1.000 \text{ L} \times \frac{0.162 \text{ mol } CH_3COOH}{1 \text{ L soln}} \times \frac{60.05 \text{ g } CH_3COOH}{1 \text{ mol } CH_3COOH} = 9.73 \text{ g } CH_3COOH$$

95. **(M)** The first ionization of H_2SO_4 makes the major contribution to the acidity of the solution. Then the following two ionizations, both of which are repressed because of the presence of H_3O^+ in the solution, must be solved simultaneously.

Equation: $\quad HSO_4^-(aq) \quad + \quad H_2O(l) \; \rightleftharpoons \; SO_4^{2-}(aq) \; + \; H_3O^+(aq)$

Initial: $\qquad 0.68$ M $\qquad\qquad\qquad\quad - \qquad\qquad\quad 0$ M $\qquad\quad 0.68$ M

Changes: $\quad -x$ M $\qquad\qquad\qquad\qquad - \qquad\qquad\quad +x$ M $\qquad\quad +x$ M

Equil: $\qquad (0.68 - x)$ M $\qquad\qquad - \qquad\qquad\qquad x$ M $\qquad (0.68 + x)$M

$$K_2 = \frac{[H_3O^+][SO_4^{2-}]}{[HSO_4^-]} = 0.011 = \frac{x(0.68 + x)}{0.68 - x}$$

Let us solve this expression for x. $0.011(0.68 - x) = 0.68x + x^2 = 0.0075 - 0.011x$

$$x^2 + 0.69x - 0.0075 = 0 \qquad x = \frac{-b \pm \sqrt{b^2 - 4ac}}{2a} = \frac{-0.69 \pm \sqrt{0.48 + 0.030}}{2} = 0.01 \text{ M}$$

This gives $[H_3O^+] = 0.68 + 0.01 = 0.69$ M. Now we solve the second equilibrium.

Equation: $\quad HCOOH(aq) \; + \; H_2O(l) \; \rightleftharpoons \; HCOO^-(aq) \; + \; H_3O^+(aq)$

Initial: $\qquad 1.5$ M $\qquad\qquad\qquad\quad - \qquad\qquad\qquad 0$ M $\qquad\qquad 0.69$ M

Changes: $\quad -x$ M $\qquad\qquad\qquad\quad - \qquad\qquad\qquad +x$ M $\qquad\qquad +x$ M

Equil: $\qquad (1.5 - x)$ M $\qquad\qquad\quad - \qquad\qquad\qquad x$ M $\qquad (0.69 + x)$ M

$$K_a = \frac{[H_3O^+][HCOO^-]}{[HCOOH]} = 1.8 \times 10^{-4} = \frac{x(0.69 + x)}{1.5 - x} \approx \frac{0.69x}{1.5} \qquad x = 3.9 \times 10^{-4} \text{ M}$$

We see that the second acid does not significantly affect the $[H_3O^+]$, for which a final value is now obtained. $[H_3O^+] = 0.69$ M, pH $= -\log(0.69) = 0.16$

98. **(M)** The structure of H_3PO_3 is shown on the right.

The two ionizable protons are bound to oxygen.

99. **(M)**
(a) $H_2SO_3 > HF > N_2H_5^+ > CH_3NH_3^+ > H_2O$
(b) $OH^- > CH_3NH_2 > N_2H_4 > F^- > HSO_3^-$
(c) **(i)** to the right and **(ii)** to the left.

100. **(M)**
(a) Because $K_a \gg 1$, the following equilibrium summary is appropriate.

$$HI(aq) + H_2O(l) \rightleftharpoons I^-(aq) + H_3O^+(aq)$$

initial concns:	1.00 M		
to completion:	0 M	1.00 M	1.00 M
to the left:	$+x$ M	$-x$ M	$-x$ M
equil concns:	x M	$(1.00 - x)$ M ≈ 1.00 M	$(1.00 - x)$ M ≈ 1.00 M

The equilibrium condition is $K_a = (1.00)(1.00)/x$. When this expression is solved for x, we get $x = 1.00/K_a = 10^{-9}$ M.

Therefore, $[HI] \approx 10^{-9}$ M and $[I^-] \approx 1.0$ M

(b) Because K_b (I^-) is so small, we expect very little I^- to react. Therefore, $[I^-] \approx 1.0$ M and the $[H_3O^+]$ and $[OH^-]$ in 1.0 M NaI(aq) will be no different than they are in pure water. Use these values in the K_a expression and solve for $[HI]$.

$$K_a = [H_3O^+] [I^-]/[HI] \approx (10^{-7})(1.0)/[HI]$$

$$[HI] = 10^{-7}/K_a = 10^{-7}/10^9 = 10^{-16} \text{ M}$$

103. **(M)** In general, the stronger the acid, the more stable its conjugate base. Therefore, we can rank a series of compounds in order of increasing acid strength by focusing on the relative stabilities of the anions formed on ionization. When comparing the stabilities of various anions, we consider the sizes and electronegativities of the atoms bearing the negative charge. The greater the electronegativity and the larger the atom, the more stable it is with a negative charge.

(a) The anions formed on ionization are CH_3COO^-, CH_3O^-, and $C_6H_5O^-$. For CH_3O^-, the negative charge is localized on the O atom (no resonance structures can be drawn). For CH_3COO^-, the negative charge is shared equally by the two O atoms.

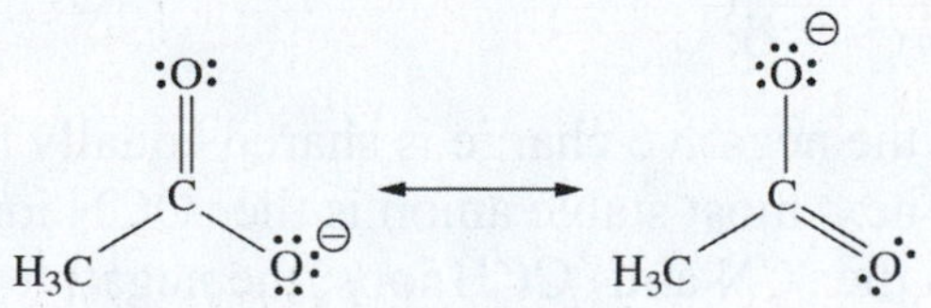

For $C_6H_5O^-$, we can draw several resonance structures.

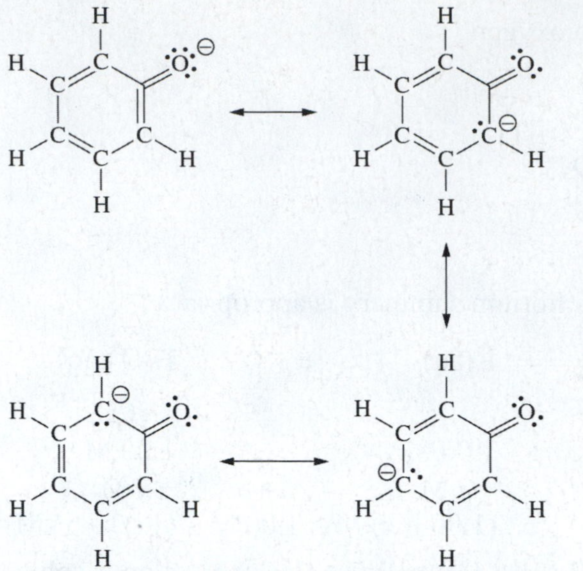

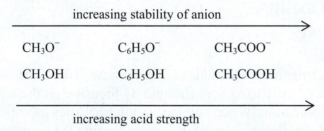

Although we can draw a greater number of resonance structures for $C_6H_5O^-$ than for CH_3COO^-, the CH_3COO^- ion is more stable than $C_6H_5O^-$ because in CH_3COO^-, the negative charge is shared equally by two electronegative atoms. Most of the resonance structures we can draw for $C_6H_5O^-$ place the negative charge on a C atom, which is much less electronegative (and therefore less stable with a negative charge) than an O atom.

Therefore,

increasing stability of anion →

CH_3O^- $\qquad\qquad$ $C_6H_5O^-$ $\qquad\qquad$ CH_3COO^-

CH_3OH $\qquad\qquad$ C_6H_5OH $\qquad\qquad$ CH_3COOH

increasing acid strength →

(b) The anions formed upon ionization are ^-CN, ^-OCN, ^-CCH, and ^-OClO. The Lewis structures are as follows.

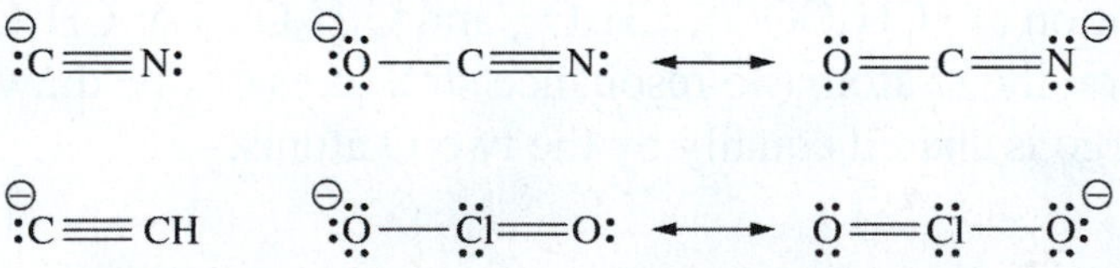

The ^-OClO ion is the most stable because the negative charge is shared equally by two O atoms, and O is very electronegative. The next most stable anion is the ^-OCN ion, with the negative charge shared by O and N. In the ^-CN and ^-CCH ions, the negative charge is localized on C. However, the ^-CN ion is more stable than the ^-CCH ion because the N atom is more electronegative than C and draws some electron density away from the C atom toward itself, which helps to stabilize the anion.

Therefore,

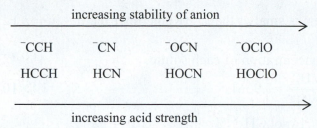

increasing stability of anion →

| ⁻CCH | ⁻CN | ⁻OCN | ⁻OClO |
| HCCH | HCN | HOCN | HOClO |

increasing acid strength →

FEATURE PROBLEMS

107. (D)

(a) Here, two equilibria must be satisfied simultaneously. The common variable, $\left[H_3O^+\right] = z$.

Equation:	$CH_3COOH(aq)$	+ $H_2O(l)$	$\rightleftharpoons$	$CH_3COO^-(aq)$	+	$H_3O^+(aq)$
Initial:	0.315 M	–		0 M		≈ 0 M
Changes:	$+x$ M	–		$+x$ M		$+z$ M
Equil:	$(0.315-x)$ M	–		x M		z M

$$K_a = \frac{\left[H_3O^+\right]\left[CH_3COO^-\right]}{\left[CH_3COOH\right]} = \frac{xz}{0.315-x} = 1.8\times10^{-5}$$

Equation:	$HCOOH(aq)$	+ $H_2O(l)$	$\rightleftharpoons$	$HCOO^-(aq)$	+	$H_3O^+(aq)$
Initial:	0.250 M	–		0 M		0 M
Changes:	$-y$ M	–		$+y$ M		$+z$ M
Equil:	$(0.250-y)$ M	–		y M		z M

$$K_a = \frac{\left[H_3O^+\right]\left[HCOO^-\right]}{\left[HCOOH\right]} = \frac{yz}{0.250-y} = 1.8\times10^{-4}$$

In this system, there are three variables, $x = \left[CH_3COO^-\right]$, $y = \left[HCOO^-\right]$, and $z = \left[H_3O^+\right]$. These three variables also represent the concentrations of the only charged species in the reaction, and thus $x + y = z$. We solve the two K_a expressions for x and y. (Before we do so, however, we take advantage of the fact that x and y are quite small: $x \ll 0.315$ and $y \ll 0.250$.) Then we substitute these results into the expression for z, and solve to obtain a value of that variable.

$$xz = 1.8\times10^{-5}\left(0.315\right) = 5.7\times10^{-6} \qquad yz = 1.8\times10^{-4}\left(0.250\right) = 4.5\times10^{-5}$$

$$x = \frac{5.7\times10^{-6}}{z} \qquad\qquad y = \frac{4.5\times10^{-5}}{z}$$

$$z = x + y = \frac{5.7\times10^{-6}}{z} + \frac{4.5\times10^{-5}}{z} = \frac{5.07\times10^{-5}}{z} \qquad z = \sqrt{5.07\times10^{-5}} = 7.1\times10^{-3}\,M = [H_3O^+]$$

$pH = -\log(7.1 \times 10^{-3}) = 2.15$ We see that our assumptions about the sizes of x and y must be valid, since each of them is smaller than z. (Remember that $x + y = z$.)

(b) We first determine the initial concentration of each solute.

$$[NH_3] = \frac{12.5g\ NH_2}{0.375L\ soln} \times \frac{1\ mol\ NH_3}{17.03g\ NH_3} = 1.96M \qquad\qquad K_b = 1.8 \times 10^{-5}$$

$$[CH_3NH_2] = \frac{1.55\ g\ CH_3NH_2}{0.375L\ soln} \times \frac{1\ mol\ CH_3NH_2}{31.06\ g\ CH_3NH_2} = 0.133M \qquad K_b = 4.2 \times 10^{-4}$$

Now we solve simultaneously two equilibria, which have a common variable, $[OH^-] = z$.

Equation:	$NH_3\ (aq)$	$+$	H_2O	$\rightleftharpoons$	$NH_4^+\ (aq)$	$+$	$OH^-\ (aq)$
Initial:	1.96 M		$-$		0 M		$\approx 0\,M$
Changes:	$-x\,M$		$-$		$+x\,M$		$+z\,M$
Equil:	$(1.96 - x)\,M$		$-$		$x\,M$		$z\,M$

$$K_b = \frac{[NH_4^+][OH^-]}{[NH_3]} = 1.8 \times 10^{-5} = \frac{xz}{1.96 - x} \approx \frac{xz}{1.96}$$

Eqn:	$CH_3NH_2\ (aq)$	$+$	$H_2O(l)$	$\rightleftharpoons$	$CH_3NH_3^+\ (aq)$	$+$	$OH^-\ (aq)$
Initial:	0.133 M		$-$		0 M		$\approx 0\,M$
Changes:	$-y\,M$		$-$		$+y\,M$		$+z\,M$
Equil:	$(0.133 - y)\,M$		$-$		$y\,M$		$z\,M$

$$K_b = \frac{[CH_3NH_3^+][OH^-]}{[CH_3NH_2]} = 4.2 \times 10^{-4} = \frac{yz}{0.133 - y} = \frac{yz}{0.133}$$

In this system, there are three variables, $x = [NH_4^+]$, $y = [CH_3NH_3^+]$, and $z = [OH^-]$. These three variables also represent the concentrations of the only charged species in solution in substantial concentration, and thus $x + y = z$. We solve the two K_b expressions for x and y. Then we substitute these results into the expression for z, and solve to obtain the value of that variable.

$$xz = 1.96 \times 1.8 \times 10^{-5} = 3.53 \times 10^{-5} \qquad\qquad yz = 0.133 \times 4.2 \times 10^{-5} = 5.59 \times 10^{-5}$$

$$x = \frac{3.53 \times 10^{-5}}{z} \qquad\qquad\qquad y = \frac{5.59 \times 10^{-5}}{z}$$

$$z = x + y = \frac{3.53 \times 10^{-5}}{z} + \frac{5.59 \times 10^{-5}}{z} = \frac{9.12 \times 10^{-5}}{z} \qquad z^2 = 9.12 \times 10^{-5}$$

$$z = 9.5 \times 10^{-3}\,M = [OH^-] \quad pOH = -\log(9.5 \times 10^{-3}) = 2.02 \quad pH = 14.00 - 2.02 = 11.98$$

We see that our assumptions about x and y (that $x \ll 1.96$ M and $y \ll 0.133$ M) must be valid, since each of them is smaller than z. (Remember that $x + y = z$.)

(c) In 1.0 M NH_4CN there are the following species: $NH_4^+(aq), NH_3(aq), CN^-(aq),$ $HCN(aq), H_3O^+(aq),$ and $OH^-(aq)$. These species are related by the following six equations.

(1) $K_w = 1.0\times10^{-14} = [H_3O^+][OH^-]$ $\qquad [OH^-] = \dfrac{1.0\times10^{-14}}{[H_3O^+]}$

(2) $K_a = 6.2\times10^{-10} = \dfrac{[H_3O^+][CN^-]}{[HCN]}$ $\qquad$ (3) $K_b = 1.8\times10^{-5} = \dfrac{[NH_4^+][OH^-]}{[NH_3]}$

(4) $[NH_3]+[NH_4^+] = 1.0$ M $\qquad [NH_3] = 1.0-[NH_4^+]$

(5) $[HCN]+[CN^-] = 1.0$ M $\qquad [HCN] = 1.0-[CN^-]$

(6) $[NH_4^+]+[H_3O^+] = [CN^-]+[OH^-]$ or $[NH_4^+] \approx [CN^-]$

Equation (6) is the result of charge balance, that there must be the same quantity of positive and negative charge in the solution. The approximation is the result of remembering that not much H_3O^+ or OH^- will be formed as the result of hydrolysis of ions. Substitute equation (4) into equation (3), and equation (5) into equation (2).

(3') $K_b = 1.8\times10^{-5} = \dfrac{[NH_4^+][OH^-]}{1.0-[NH_4^+]}$ $\qquad$ (2') $K_a = 4.0\times10^{-10} = \dfrac{[H_3O^+][CN^-]}{1.0-[CN^-]}$

Now substitute equation (6) into equation (2'), and equation (1) into equation (3').

(2'') $K_a = 6.2\times10^{-10} = \dfrac{[H_3O^+][NH_4^+]}{1.0-[NH_4^+]}$ $\qquad$ (3'') $\dfrac{1.8\times10^{-5}}{1.0\times10^{-14}} = \dfrac{[NH_4^+]}{[H_3O^+](1.0-[NH_4^+])}$

Now we solve both of these equations for $[NH_4^+]$.

(2'): $6.2\times10^{-10} - 6.2\times10^{-10}[NH_4^+] = [H_3O^+][NH_4^+]$ $\qquad [NH_4^+] = \dfrac{6.2\times10^{-10}}{6.2\times10^{-10}+[H_3O^+]}$

(3'): $1.8\times10^9[H_3O^+] - 1.8\times10^9[H_3O^+][NH_4^+] = [NH_4^+]$ $\qquad [NH_4^+] = \dfrac{1.8\times10^9[H_3O^+]}{1.00+1.8\times10^9[H_3O^+]}$

We equate the two results and solve for $[H_3O^+]$.

$$\dfrac{6.2\times10^{-10}}{6.2\times10^{-10}+[H_3O^+]} = \dfrac{1.8\times10^9[H_3O^+]}{1.00+1.8\times10^9[H_3O^+]}$$

$6.2\times10^{-10}+1.1[H_3O^+] = 1.1[H_3O^+]+1.8\times10^9[H_3O^+]^2$ $\qquad 6.2\times10^{-10} = 1.8\times10^9[H_3O^+]^2$

$[H_3O^+] = \sqrt{\dfrac{6.2\times10^{-10}}{1.8\times10^9}} = 5.9\times10^{-10}$ M $\quad$ pH $= -\log(5.9\times10^{-10}) = 9.23$

Note that $[H_3O^+] = \sqrt{\dfrac{K_a \times K_w}{K_b}}$ $\qquad$ or $\qquad$ pH $= 0.500(pK_a + pK_w - pK_b)$

SELF-ASSESSMENT EXERCISES

__111.__ **(E)** The answer is (a), HCO_3^-, because it can donate a proton to water to become CO_3^{2-}, or it can be protonated to give H_2CO_3 (i.e., $CO_2(aq)$).

__112.__ **(E)** The answer is (c). CH_3CH_2COOH is a weak acid, so the concentration of H_3O^+ ions it will generate upon dissociation in water will be significantly less than 0.1 M. Therefore, its pH will be higher the 0.10 M HBr, which dissociates completely in water to give 0.10 M H_3O^+.

__113.__ **(E)** The answer is (d). CH_3NH_2 is a weak base.

__114.__ **(M)** The answer is (e) because CO_3^{2-} is the strongest base and it drives the dissociation of acetic acid furthest.

__115.__ **(M)** The answer is (c). H_2SO_4 dissociation is nearly complete for the first proton, giving a $[H_3O^+] = 0.10$ M (pH = 1). The second dissociation is not complete. Therefore, (c) is the only choice that makes sense.

__116.__ **(M)** The answer is (b). You can write down the stepwise equations for dissociation of H_2SO_3 and then the dissociation of HSO_3^- and calculate the equilibrium concentration of each species. However, you can determine the answer without detailed calculations. The two dissociation equations are given below:

Eq.1 $\quad H_2SO_3 + H_2O \rightleftharpoons HSO_3^- + H_3O^+ \qquad\qquad K_{a1} = 1.3\times10^{-2}$

Eq.2 $\quad HSO_3^- + H_2O \rightleftharpoons SO_3^{2-} + H_3O^+ \qquad\qquad K_{a2} = 6.3\times10^{-8}$

For the first equation, the equilibrium concentrations, of HSO_3^- and H_3O^+ are the same. They become the initial concentration values for the second dissociation reaction. Since K_{a2} is 6.3×10^{-8} (which is small), the equilibrium concentrations of HSO_3^- and H_3O^+ won't change significantly. So, the equation simplifies as follows:

$$K_{a2} = 6.8\times10^{-8} = \frac{\left[SO_3^{2-}\right]\left(\left[H_3O^+\right] + \cancel{x}\right)}{\left[HSO_3^-\right] - \cancel{x}} = \frac{\left[SO_3^{2-}\right]\cancel{\left[H_3O^+\right]}}{\cancel{\left[HSO_3^-\right]}}$$

$$K_{a2} = 6.8\times10^{-8} = \left[SO_3^{2-}\right]$$

__117.__ **(E)** Since both the acid and the base in this reaction are strong, they dissociate completely. The pH is determined as follows:

mol H_3O^+: 0.248 M $HNO_3 \times 0.02480$ L = 0.00615 mol

mol OH^-: 0.394 M KOH $\times 0.01540$ L = 0.00607 mol

Final: $0.00615 - 0.00607 = 8.00\times10^{-5}$ mol H_3O^+

$$\left[H_3O^+\right] = \frac{8.00\times10^{-5}\,\text{mol}}{(0.02480\,\text{L} + 0.01540\,\text{L})} = 0.00200\ \text{M}$$

$$\text{pH} = -\log(0.00200) = 2.70$$

118. **(M)** Since pH = 3.25, the $[H_3O^+] = 10^{-3.25} = 5.62 \times 10^{-4}$ M. Using the reaction below, we can determine the equilibrium concentrations of other species

$$CH_3COOH \; + \; H_2O \; \rightleftharpoons \; CH_3COO^- \; + \; H_3O^+$$

Initial	c	0	0
Change	$-x$	$+x$	$+x$
Equil.	$c - x$	x	x

Since $x = 5.62 \times 10^{-4}$ M, the equilibrium expression becomes

$$K_a = \frac{[H_3O^+][CH_3COO^-]}{[CH_3COOH]} = \frac{(5.62 \times 10^{-4}\,M)(5.62 \times 10^{-4}\,M)}{(c - 5.62 \times 10^{-4}\,M)} = 1.8 \times 10^{-5}$$

Solving c gives a concentration of 0.0181 M.
Since concentrated acetic acid is 35% by weight and the density is 1.044 g/mL, the concentration of acetic acid is:

$$\frac{35\ g\ CH_3COOH}{100\ g\ Conc.\ CH_3COOH\ sol'n} \times \frac{1.044\ g\ Conc.\ CH_3COOH}{0.001\ L\ sol'n} \times \frac{1\ mol\ CH_3COOH}{60.06\ g\ CH_3COOH} = 6.084\ M\ CH_3COOH$$

Therefore, volume of concentrated solution required to make 12.5 L of 0.0181 M HAc solution is:

$$(c \cdot V)_{dilute} = (c \cdot V)_{conc.}$$

$$V_{conc} = \frac{(0.0181\ M)(12500\ mL)}{6.084\ M} = 37.2\ mL$$

119. **(M)** Table 16.3 has the K_a value for chloroacetic acid, $CH_2ClCOOH$. $K_a = 1.4 \times 10^{-3}$. The solution is made of the chloroacetate salt, which hydrolyzes water according to the following reaction:

$$CH_2ClCOO^- \; + \; H_2O \; \rightleftharpoons \; CH_2ClCOOH \; + \; OH^-$$

Initial	2.05	0	0
Change	$-x$	$+x$	$+x$
Equil.	$2.05 - x$	x	x

$$K_b = \frac{K_W}{K_a} = \frac{1.0 \times 10^{-14}}{1.4 \times 10^{-3}} = 7.14 \times 10^{-12}$$

$$K_b = \frac{[CH_2ClCOOH][OH^-]}{[CH_2ClCOO^-]}$$

$$7.14 \times 10^{-12} = \frac{x \cdot x}{2.05 - x}$$

Solving the above simplified expression for x, we get $x = [OH^-] = 3.826 \times 10^{-6}$

$$pH = 14 - pOH = 14 - \left[-\log(3.826 \times 10^{-6}) \right] = 8.58$$

120. (M) 0.10 M HI < 0.05 M H_2SO_4 < 0.05 M $CH_2ClCOOH$ < 0.50 M CH_3COOH < 0.50 M NH_4Cl < 1.0 M NaBr < 0.05 M KCH_3COO < 0.05 M NH_3 < 0.06 M NaOH < 0.05 M $Ba(OH)_2$

121. (M) Since pH = 5 pOH, pH must be significantly larger than pOH, which means that the solution is basic. The pH can be determined by solving two simultaneous equations:

pH − 5 pOH = 0

pH + pOH = 14

pOH = 2.333

pH = 14 − 2.33 = 11.67.

Therefore, $[H_3O^+]$ = 2.14×10^{-12} M (and $[OH^-]$ = 4.67×10^{-3} M)

The solute must be NH_3, because it is the only basic species (the other two are acidic and neutral, respectively). Since the K_b of NH_3 is 1.8×10^{-5}, the concentration of NH_3 can be determined as follows:

$$K_b = \frac{[OH^-][NH_4^+]}{[NH_3]} = \frac{(4.67\times10^{-3})(4.67\times10^{-3})}{x} = 1.8\times10^{-5}$$

x = 1.2 M

122. (M) The answer is (a). If CH_3CH_2COOH is 0.42% ionized for a 0.80 M solution, then the concentration of the acid at equilibrium is (0.80×0.0042) = 0.00336. The equilibrium expression for the dissociation of CH_3CH_2COOH is:

$$K_a = \frac{[H_3O^+][CH_3CH_2COO^-]}{[CH_3CH_2COOH]} = \frac{(0.00336)(0.00336)}{(0.800-0.00336)} = 1.42\times10^{-5}$$

123. (E) The answer is (b), because $H_2PO_4^-$ is a result of addition of one proton to the base HPO_4^{2-}.

124. (M) The answer is (d). This is because, in the second equation, HNO_2 will give its proton to ClO^- (rather than HClO giving its proton to NO_2^-), which means that HNO_2 is a stronger acid than HClO. Also, in the first equation, ClO^- is not a strong enough base to abstract a proton from water, which means that HClO is in turn a stronger acid than H_2O.

125. (M) The dominant species in the solution is ClO_2^-. Therefore, determine its concentration and ionization constant first:

$$[ClO_2^-] = \frac{3.00 \text{ mol } CaClO_2}{2.50 \text{ L}} \times \frac{2 \text{ mol } ClO_2^-}{1 \text{ mol } CaClO_2} = 2.40 \text{ M}$$

$$K_b = \frac{K_w}{K_a} = \frac{1.0\times10^{-14}}{1.1\times10^{-2}} = 9.1\times10^{-13}$$

The reaction and the dissociation of ClO_2^- is as follows:

	ClO_2^-	+ H_2O	$\rightleftharpoons$	$HClO_2$	+ OH^-
Initial	2.40			0	0
Change	$-x$			$+x$	$+x$
Equil.	$2.40-x$			x	x

$$K_b = \frac{[HClO_2][OH^-]}{[ClO_2^-]}$$

$$9.1\times10^{-13} = \frac{x \cdot x}{2.40 - \cancel{x}}$$

Solving the above simplified expression for x, we get $x = [OH^-] = 1.478\times10^{-6}$.

$$pH = 14 - pOH = 14 - \left[-\log\left(1.478\times10^{-6}\right)\right] = 8.2$$

CHAPTER 17

ADDITIONAL ASPECTS OF ACID–BASE EQUILIBRIA

PRACTICE EXAMPLES

1A **(D)** Organize the solution around the balanced chemical equation, as we have done before.

Equation: $\quad HF(aq) + H_2O(l) \rightleftharpoons H_3O^+(aq) + F^-(aq)$

Initial: $\qquad$ 0.500 M $\qquad$ – $\qquad \approx 0$ M $\qquad$ 0 M

Changes: $\qquad -x$ M $\qquad$ – $\qquad +x$ M $\qquad +x$ M

Equil: $\qquad (0.500 - x)$ M $\quad$ – $\qquad x$ M $\qquad x$ M

$$K_a = \frac{[H_3O^+][F^-]}{[HF]} = \frac{(x)(x)}{0.500 - x} = 6.6 \times 10^{-4} \approx \frac{x^2}{0.500} \qquad \text{assuming } x \ll 0.500$$

$$x = \sqrt{0.500 \times 6.6 \times 10^{-4}} = 0.018\,M \qquad \text{One further cycle of approximations gives:}$$

$$x = \sqrt{(0.500 - 0.018) \times 6.6 \times 10^{-4}} = 0.018\,M = [H_3O^+]$$

Thus, $[HF] = 0.500\,M - 0.018\,M = 0.482\,M$

Recognize that 0.100 M HCl means $\left[H_3O^+\right]_{\text{initial}} = 0.100\,M$, since HCl is a strong acid.

Equation: $\quad HF(aq) + H_2O(l) \rightleftharpoons H_3O^+(aq) + F^-(aq)$

Initial: $\qquad$ 0.500 M $\qquad$ – $\qquad$ 0.100 M $\qquad$ 0 M

Changes: $\qquad -x$ M $\qquad$ – $\qquad +x$ M $\qquad +x$ M

Equil: $\qquad (0.500 - x)$ M $\quad$ – $\qquad (0.100 + x)$ M $\quad x$ M

$$K_a = \frac{[H_3O^+][F^-]}{[HF]} = \frac{(x)(0.100 + x)}{0.500 - x} = 6.6 \times 10^{-4} \approx \frac{0.100\,x}{0.500} \quad \text{assuming } x \ll 0.100$$

$$x = \frac{6.6 \times 10^{-4} \times 0.500}{0.100} = 3.3 \times 10^{-3}\,M = \left[F^-\right] \qquad \text{The assumption is valid.}$$

$[HF] = 0.500\,M - 0.003\,M = 0.497\,M$

$[H_3O^+] = 0.100\,M + x = 0.100\,M + 0.003\,M = 0.103\,M$

1B **(M)** From Example 17-6 *in the text*, we know that $[H_3O^+] = [CH_3COO^-] = 1.3 \times 10^{-3}$ M in 0.100 M CH_3COOH. We base our calculation, as usual, on the balanced chemical equation. The concentration of H_3O^+ from the added HCl is represented by x.

Equation: $CH_3COOH(aq) + H_2O(l) \rightleftharpoons H_3O^+(aq) + CH_3COO^-(aq)$

Initial:	0.100 M	–	≈ 0 M	0 M
Changes:	– 0.00010 M	–	+0.00010 M	+0.00010 M
From HCl:			$+x$ M	
Equil:	0.100 M	–	$(0.00010 + x)$ M	0.00010 M

$$K_a = \frac{[H_3O^+][CH_3COO^-]}{[CH_3COOH]} = \frac{(0.00010 + x)0.00010}{0.100} = 1.8 \times 10^{-5}$$

$$0.00010 + x = \frac{1.8 \times 10^{-5} \times 0.100}{0.00010} = 0.018 \text{ M} \qquad x = 0.018 \text{ M} - 0.00010 \text{ M} = 0.018 \text{ M}$$

$$V_{12\,\text{M HCl}} = 1.00 \text{ L} \times \frac{0.018 \text{ mol } H_3O^+}{1 \text{ L}} \times \frac{1 \text{ mol HCl}}{1 \text{ mol } H_3O^+} \times \frac{1 \text{ L soln}}{12 \text{ mol HCl}} \times \frac{1000 \text{ mL}}{1 \text{ L}} \times \frac{1 \text{ drop}}{0.050 \text{ mL}} = 30. \text{ drops}$$

Since 30. drops corresponds to 1.5 mL of 12 M solution, we see that the volume of solution does indeed remain approximately 1.00 L after addition of the 12 M HCl.

2A **(M)** We again organize the solution around the balanced chemical equation.

Equation: $HCOOH(aq) + H_2O(l) \rightleftharpoons HCOO^-(aq) + H_3O^+(aq)$

Initial:	0.100 M	–	0.150 M	≈ 0 M
Changes:	$-x$ M	–	$+x$ M	$+x$ M
Equil:	$(0.100 - x)$ M	–	$(0.150 + x)$ M	x M

$$K_a = \frac{[HCOO^-][H_3O^+]}{[HCOOH]} = \frac{(0.150 + x)(x)}{0.100 - x} = 1.8 \times 10^{-4} \approx \frac{0.150\,x}{0.100} \qquad \text{assuming } x \ll 0.100$$

$$x = \frac{0.100 \times 1.8 \times 10^{-4}}{0.150} = 1.2 \times 10^{-4} \text{ M} = [H_3O^+], \; x \ll 0.100, \text{ thus our assumption is valid}$$

$$[HCOO^-] = 0.150 \text{ M} + 0.00012 \text{ M} = 0.150 \text{ M}$$

2B **(M)** This time, a solid sample of a weak base is being added to a solution of its conjugate acid. We let x represent the concentration of acetate ion from the added sodium acetate. Notice that sodium acetate is a strong electrolyte, thus, it completely dissociates in aqueous solution. $[H_3O^+] = 10^{-pH} = 10^{-5.00} = 1.0 \times 10^{-5}\,M = 0.000010\,M$

Equation: $CH_3COOH(aq) + H_2O(l) \rightleftharpoons CH_3COO^-(aq) + H_3O^+(aq)$

Initial: $0.100\,M$ – $0\,M$ $\approx 0\,M$

Changes: $-0.000010\,M$ – $+0.000010\,M$ $+0.000010\,M$

From NaAc: $+x\,M$

Equil: $0.100\,M$ – $(0.000010+x)\,M$ $0.000010\,M$

$$K_a = \frac{[H_3O^+][CH_3COO^-]}{[CH_3COOH]} = \frac{0.000010(0.000010+x)}{0.100} = 1.8 \times 10^{-5}$$

$$0.000010 + x = \frac{1.8 \times 10^{-5} \times 0.100}{0.000010} = 0.18\,M \qquad x = 0.18\,M - 0.000010\,M = 0.18\,M$$

$$\text{mass of NaCH}_3\text{COO} = 1.00\,L \times \frac{0.18\,\text{mol CH}_3\text{COO}^-}{1\,L} \times \frac{1\,\text{mol NaCH}_3\text{COO}}{1\,\text{mol CH}_3\text{COO}^-} \times \frac{82.03\,\text{g NaCH}_3\text{COO}}{1\,\text{mol NaCH}_3\text{COO}}$$

$$= 15\,\text{g NaCH}_3\text{COO}$$

3A **(M)** A strong acid dissociates essentially completely, and effectively is a source of H_3O^+. $NaCH_3COO$ also dissociates completely in solution. The hydronium ion and the acetate ion react to form acetic acid: $H_3O^+(aq) + CH_3COO^-(aq) \rightleftharpoons CH_3COOH(aq) + H_2O(l)$

All that is necessary to form a buffer is to have approximately equal amounts of a weak acid and its conjugate base together in solution. This will be achieved if we add an amount of HCl equal to approximately half the original amount of acetate ion.

3B **(M)** HCl dissociates essentially completely in water and serves as a source of hydronium ion. This reacts with ammonia to form ammonium ion:

$$NH_3(aq) + H_3O^+(aq) \rightleftharpoons NH_4^+(aq) + H_2O(l).$$

Because a buffer contains approximately equal amounts of a weak base (NH_3) and its conjugate acid (NH_4^+), to prepare a buffer we simply add an amount of HCl equal to approximately half the amount of $NH_3(aq)$ initially present.

4A **(M)** We first find the formate ion concentration, remembering that $NaCOO$ is a strong electrolyte, existing in solution as $Na^+(aq)$ and $HCOO^-(aq)$.

$$[HCOO^-] = \frac{23.1\,g\,NaCOO}{500.0\,mL\,soln} \times \frac{1000\,mL}{1L} \times \frac{1\,mol\,NaCOO}{68.01\,g\,NaCOO} \times \frac{1\,mol\,HCOO^-}{1\,mol\,NaCOO} = 0.679\,M$$

As usual, the solution to the problem is organized around the balanced chemical equation.

Equation: $HCOOH(aq) + H_2O(l) \rightleftharpoons HCOO^-(aq) + H_3O^+(aq)$

Initial: $0.432\,M$ — $0.679\,M$ $\approx 0\,M$

Changes: $-x\,M$ — $+x\,M$ $+x\,M$

Equil: $(0.432-x)\,M$ — $(0.679+x)\,M$ $x\,M$

$$K_a = \frac{[H_3O^+][HCOO^-]}{[HCOOH]} = \frac{x(0.679+x)}{0.432-x} = 1.8\times10^{-4} \approx \frac{0.679x}{0.432} \qquad x = \frac{0.432\times1.8\times10^{-4}}{0.679}$$

This gives $[H_3O^+] = 1.14\times10^{-4}$ M. The assumption that $x \ll 0.432$ is clearly correct.

$$pH = -\log[H_3O^+] = -\log(1.14\times10^{-4}) = 3.94 \approx 3.9$$

4B **(M)** The concentrations of the components in the 100.0 mL of buffer solution are found via the dilution factor. Remember that $NaC_2H_3O_2$ is a strong electrolyte, existing in solution as

$Na^+(aq)$ and $CH_3COO^-(aq)$.

$$[CH_3COOH] = 0.200\,M \times \frac{63.0\,mL}{100.0\,mL} = 0.126\,M \quad [CH_3COO^-] = 0.200\,M \times \frac{37.0\,mL}{100.0\,mL} = 0.0740\,M$$

As usual, the solution to the problem is organized around the balanced chemical equation.

Equation: $CH_3COOH(aq) + H_2O(l) \rightleftharpoons CH_3COO^-(aq) + H_3O^+(aq)$

Initial: $0.126\,M$ — $0.0740\,M$ $\approx 0\,M$

Changes: $-x\,M$ — $+x\,M$ $+x\,M$

Equil: $(0.126-x)\,M$ — $(0.0740+x)\,M$ $x\,M$

$$K_a = \frac{[H_3O^+][CH_3COO^-]}{[CH_3COOH]} = \frac{x(0.0740+x)}{0.126-x} = 1.8\times10^{-5} \approx \frac{0.0740x}{0.126}$$

$$x = \frac{1.8\times10^{-5}\times0.126}{0.0740} = 3.1\times10^{-5} \; [H_3O^+] = 3.1\times10^{-5}\,M;$$

$$pH = -\log[H_3O^+] = -\log 3.1\times10^{-5} = 4.51$$

Note that the assumption is valid: $x \ll 0.0740 < 0.126$.
Thus, x is neglected when added or subtracted

5A **(M)** We know the initial concentration of NH_3 in the buffer solution and can use the pH to find the equilibrium $[OH^-]$. The rest of the solution is organized around the balanced chemical equation. Our first goal is to determine the initial concentration of NH_4^+.

$$pOH = 14.00 - pH = 14.00 - 9.00 = 5.00 \qquad [OH^-] = 10^{-pOH} = 10^{-5.00} = 1.0 \times 10^{-5} \text{ M}$$

Equation:	$NH_3(aq)$	$+$	$H_2O(l)$	$\rightleftharpoons$	$NH_4^+(aq)$	$+$	$OH^-(aq)$
Initial:	0.35 M		–		x M		≈ 0 M
Changes:	-1.0×10^{-5} M		–		$+1.0 \times 10^{-5}$ M		$+1.0 \times 10^{-5}$ M
Equil:	$(0.35 - 1.0 \times 10^{-5})$M		–		$(x + 1.0 \times 10^{-5})$M		1.0×10^{-5} M

$$K_b = \frac{[NH_4^+][OH^-]}{[NH_3]} = 1.8 \times 10^{-5} = \frac{(x + 1.0 \times 10^{-5})(1.0 \times 10^{-5})}{0.35 - 1.0 \times 10^{-5}} = \frac{1.0 \times 10^{-5} \times x}{0.35}$$

Assume $x \gg 1.0 \times 10^{-5}$ $\qquad x = \dfrac{0.35 \times 1.8 \times 10^{-5}}{1.0 \times 10^{-5}} = 0.63$ M $=$ initial NH_4^+ concentration

$$\text{mass}(NH_4)_2 SO_4 = 0.500 \text{ L} \times \frac{0.63 \text{ mol } NH_4^+}{1 \text{ L soln}} \times \frac{1 \text{ mol } (NH_4)_2 SO_4}{2 \text{ mol } NH_4^+} \times \frac{132.1 \text{ g } (NH_4)_2 SO_4}{1 \text{ mol } (NH_4)_2 SO_4}$$

Mass of $(NH_4)_2SO_4 = 21$ g

5B **(M)** The solution is composed of 33.05 g $NaCH_3COO \cdot 3 H_2O$ dissolved in 300.0 mL of 0.250 M HCl. $NaCH_3COO \cdot 3 H_2O$, a strong electrolyte, exists in solution as $Na^+(aq)$ and $CH_3COO^-(aq)$ ions. First we calculate the number of moles of $NaCH_3COO \cdot 3 H_2O$, which, based on the 1:1 stoichiometry, is also equal to the number of moles of CH_3COO^- that are released into solution. From this we can calculate the initial $[CH_3COO^-]$ assuming the solution's volume remains at 300. mL.

moles of $NaCH_3COO \cdot 3 H_2O$ (and moles of CH_3COO^-)

$$= \frac{33.05 \text{ g } NaCH_3COO \cdot 3 H_2O}{\dfrac{1 \text{ mole } NaCH_3COO \cdot 3 H_2O}{136.08 \text{ g } NaCH_3COO \cdot 3 H_2O}} = 0.243 \text{ moles } NaCH_3COO \cdot 3 H_2O = \text{moles } CH_3COO^-$$

$$[CH_3COO^-] = \frac{0.243 \text{ mol } C_2H_3O_2^-}{0.300 \text{ L soln}} = 0.810 \text{ M}$$

However, we must recognize that the H_3O^+ produced by the ionization of HCl will react with some of the CH_3COO^- to produce CH_3COOH. We can organize this information as follows.

Equation: $\quad\quad CH_3COOH(aq) + H_2O(l) \rightleftharpoons \quad\quad CH_3COO^-(aq) \quad\quad + H_3O^+(aq)$

Initial:	0 M	–	0.810 M	0.250 M
Form HAc:	+0.250 M	–	–0.250 M	–0.250 M
	0.250 M	–	0.560 M	≈ 0 M
Changes:	–x M	–	+x M	+x M
Equil:	(0.250 – x) M	–	(0.560 + x) M	+x M

$$K_a = \frac{[H_3O^+][CH_3COO^-]}{[CH_3COOH]} = \frac{x(0.560 + x)}{0.250 - x} = 1.8 \times 10^{-5} \approx \frac{0.560\,x}{0.250}$$

$$x = \frac{1.8 \times 10^{-5} \times 0.250}{0.560} = 8.0 \times 10^{-6} \; [H_3O^+] = 8.0 \times 10^{-6} \text{ M}$$

(The approximation was valid since $x \ll$ both 0.250 and 0.560)

$$pH = -\log[H_3O^+] = -\log 8.0 \times 10^{-6} = 5.09 \approx 5.1$$

6A **(D)**

(a) For formic acid, $pK_a = -\log(1.8 \times 10^{-4}) = 3.74$. The Henderson-Hasselbalch equation provides the pH of the original buffer solution:

$$pH = pK_a + \log\frac{[HCOO^-]}{[HCOOH]} = 3.74 + \log\frac{0.350}{0.550} = 3.54$$

(b) The added acid can be considered completely reacted with the formate ion to produce formic acid. Each mole/L of added acid consumes 1 M of formate ion and forms 1 M of formic acid: $HCOO^-(aq) + H_3O^+(aq) \longrightarrow HCOOH(aq) + H_2O(l)$.

$K_{neut} = K_b/K_w \approx 5600$. Thus, $\left[HCOO^-\right] = 0.350 \text{ M} - 0.0050 \text{ M} = 0.345 \text{ M}$

and $[HCOOH] = 0.550 \text{ M} + 0.0050 \text{ M} = 0.555 \text{ M}$. By using the Henderson-Hasselbalch equation

$$pH = pK_a + \log\frac{[HCOO^-]}{[HCOOH]} = 3.74 + \log\frac{0.345}{0.555} = 3.53$$

(c) Added base reacts completely with formic acid producing, an equivalent amount of formate ion. This continues until all of the formic acid is consumed. Each 1 mole of added base consumes 1 mol of formic acid and forms 1 mol of formate ion: $HCOOH + OH^- \longrightarrow HCOO^- + H_2O$. $K_{neut} = K_a/K_w \approx 1.8 \times 10^{10}$.

Thus, $\left[HCOO^-\right] = 0.350 \text{ M} + 0.0050 \text{ M} = 0.355 \text{ M}$

$[HCOOH] = (0.550 - 0.0050)\text{ M} = 0.545 \text{ M}$. With the Henderson-Hasselbalch equation

we find $pH = pK_a + \log\frac{\left[HCOO^-\right]}{[HCOOH]} = 3.74 + \log\frac{0.355}{0.545} = 3.55$

6B **(D)** The buffer cited has the same concentration as weak acid and its anion, as does the buffer of Example 17-6. Our goal is to reach $pH = 5.03$ or

$$\left[H_3O^+\right] = 10^{-pH} = 10^{-5.03} = 9.3 \times 10^{-6} \text{ M}.$$ Adding strong acid $\left(H_3O^+\right)$, of course, produces $HC_2H_3O_2$ at the expense of $C_2H_3O_2^-$. Thus, adding H^+ drives the reaction to the left. Again, we use the data around the balanced chemical equation.

Equation: $CH_3COOH\,(aq)\; +\; H_2O(l)\; \rightleftharpoons\; CH_3COO^-\,(aq)\; +\; H_3O^+\,(aq)$

Initial:	0.250 M	–	0.560 M	8.0×10^{-6} M
Add acid:				$+y$ M
Form HAc:	$+y$ M	–	$-y$ M	$-y$ M
	$(0.250 + y)$ M	–	$(0.560 - y)$ M	≈ 0 M
Changes:	$-x$ M	–	$+x$ M	$+x$ M
Equil:	$(0.250 + y - x)$ M	–	$(0.560 - y + x)$ M	9.3×10^{-6} M

$$K_a = \frac{[H_3O^+][CH_3COO^-]}{[CH_3COOH]} = \frac{9.3 \times 10^{-6}(0.560 - y + x)}{0.250 + y - x} = 1.8 \times 10^{-5} \approx \frac{9.3 \times 10^{-6}(0.560 - y)}{0.250 + y}$$

(Assume that x is negligible compared to y)

$$\frac{1.8 \times 10^{-5} \times (0.250 + y)}{9.3 \times 10^{-6}} = 0.484 + 1.94\ y = 0.560 - y \qquad y = \frac{0.560 - 0.484}{1.94 + 1.00} = 0.026 \text{ M}$$

Notice that our assumption is valid: $x \ll 0.250 + y\ (= 0.276) < 0.560 - y\ (= 0.534)$.

$$V_{HNO_3} = 300.0 \text{ mL buffer} \times \frac{0.026 \text{ mmol } H_3O^+}{1 \text{ mL buffer}} \times \frac{1 \text{ mL } HNO_3\,(aq)}{6.0 \text{ mmol } H_3O^+} = 1.3 \text{ mL of } 6.0 \text{ M } HNO_3$$

Instead of the algebraic solution, we could have used the Henderson-Hasselbalch equation, since the final pH falls within one pH unit of the pK_a of acetic acid. We let z indicate the increase in $[CH_3COOH]$, and also the decrease in $\left[CH_3COO^-\right]$

$$pH = pK_a + \log\frac{\left[CH_3COO^-\right]}{[CH_3COOH]} = 4.74 + \log\frac{0.560 - z}{0.250 + z} = 5.03 \qquad \frac{0.560 - z}{0.250 + z} = 10^{5.03 - 4.74} = 1.95$$

$$0.560 - z = 1.95\,(0.250 + z) = 0.488 + 1.95\ z \qquad z = \frac{0.560 - 0.488}{1.95 + 1.00} = 0.024 \text{ M}$$

This is, and should be, almost exactly the same as the value of y we obtained by the I.C.E. table method. The slight difference is due to imprecision arising from rounding errors.

7A **(D)**

(a) The initial pH is the pH of 0.150 M HCl, which we obtain from $\left[H_3O^+\right]$ of that strong acid solution.

$$[H_3O^+] = \frac{0.150 \text{ mol HCl}}{1 \text{ L soln}} \times \frac{1 \text{ mol } H_3O^+}{1 \text{ mol HCl}} = 0.150 \text{ M},$$

$$pH = -\log[H_3O^+] = -\log(0.150) = 0.824$$

(b) To determine $\left[H_3O^+\right]$ and then pH at the 50.0% point, we need the volume of the solution and the amount of H_3O^+ left unreacted. First we calculate the amount of hydronium ion present and then the volume of base solution needed for its complete neutralization.

$$\text{amount } H_3O^+ = 25.00 \text{ mL} \times \frac{0.150 \text{ mmol HCl}}{1 \text{ mL soln}} \times \frac{1 \text{ mmol } H_3O^+}{1 \text{ mmol HCl}} = 3.75 \text{ mmol } H_3O^+$$

$$V_{\text{acid}} = 3.75 \text{ mmol } H_3O^+ \times \frac{1 \text{ mmol } OH^-}{1 \text{ mmol } H_3O^+} \times \frac{1 \text{ mmol NaOH}}{1 \text{ mmol } OH^-} \times \frac{1 \text{ mL titrant}}{0.250 \text{ mmol NaOH}}$$

$$= 15.0 \text{ mL titrant}$$

At the 50.0% point, half of the H_3O^+ (1.88 mmol H_3O^+) will remain unreacted and only half (7.50 mL titrant) of the titrant solution will be added. From this information, and the original 25.00-mL volume of the solution, we calculate $\left[H_3O^+\right]$ and then pH.

$$\left[H_3O^+\right] = \frac{1.88 \text{ mmol } H_3O^+ \text{ left}}{25.00 \text{ mL original} + 7.50 \text{ mL titrant}} = 0.0578 \text{ M}$$

$$pH = -\log(0.0578) = 1.238$$

(c) Since this is the titration of a strong acid by a strong base, at the equivalence point, the pH = 7.00. This is because the only ions of appreciable concentration in the equivalence point solution are $Na^+(aq)$ and $Cl^-(aq)$, and neither of these species undergoes detectable hydrolysis reactions.

(d) Beyond the equivalence point, the solution pH is determined almost entirely by the concentration of excess $OH^-(aq)$ ions. The volume of the solution is $40.00 \text{ mL} + 1.00 \text{ mL} = 41.00 \text{ mL}$. The amount of hydroxide ion in the excess titrant is calculated and used to determine $\left[OH^-\right]$, from which pH is computed.

$$\text{amount of } OH^- = 1.00 \text{ mL} \times \frac{0.250 \text{ mmol NaOH}}{1 \text{ mL}} = 0.250 \text{ mmol } OH^-$$

$$\left[OH^-\right] = \frac{0.250 \text{ mmol } OH^-}{41.00 \text{ mL}} = 0.006098 \text{ M}$$

$$pOH = -\log(0.006098) = 2.215; \quad pH = 14.00 - 2.215 = 11.785$$

7B **(D)**

(a) The initial pH is simply the pH of $0.00812 \text{ M Ba(OH)}_2$, which we obtain from $\left[OH^-\right]$ for the solution.

$$\left[OH^-\right] = \frac{0.00812 \text{ mol Ba(OH)}_2}{1 \text{ L soln}} \times \frac{2 \text{ mol } OH^-}{1 \text{ mol Ba(OH)}_2} = 0.01624 \text{ M}$$

$$pOH = -\log\left[OH^-\right] = -\log(0.0162) = 1.790; \quad pH = 14.00 - pOH = 14.00 - 1.790 = 12.21$$

(b) To determine $\left[OH^-\right]$ and then pH at the 50.0% point, we need the volume of the solution and the amount of OH^- unreacted. First we calculate the amount of hydroxide ion present and then the volume of acid solution needed for its complete neutralization.

$$\text{amount } OH^- = 50.00\,mL \times \frac{0.00812\,mmol\,Ba(OH)_2}{1\,mL\,soln} \times \frac{2\,mmol\,OH^-}{1\,mmol\,Ba(OH)_2} = 0.812\,mmol\,OH^-$$

$$V_{acid} = 0.812\,mmol\,OH^- \times \frac{1\,mmol\,H_3O^+}{1\,mmol\,OH^-} \times \frac{1\,mmol\,HCl}{1\,mmol\,H_3O^+} \times \frac{1\,mL\,titrant}{0.0250\,mmol\,HCl} = 32.48\,mL\,titrant$$

At the 50.0 % point, half (0.406 mmol OH^-) will remain unreacted and only half (16.24 mL titrant) of the titrant solution will be added. From this information, and the original 50.00 mL volume of the solution, we calculate $\left[OH^-\right]$ and then pH.

$$\left[OH^-\right] = \frac{0.406\,mmol\,OH^-\,left}{50.00\,mL\,original + 16.24\,mL\,titrant} = 0.00613\,M$$

$$pOH = -\log(0.00613) = 2.213; \quad pH = 14.00 - pOH = 11.79$$

(c) Since this is the titration of a strong base by a strong acid, at the equivalence point, pH = 7.00. The solution at this point is neutral because the dominant ionic species in solution, namely $Ba^{2+}(aq)$ and $Cl^-(aq)$, do not react with water to a detectable extent.

8A (D)

(a) Initial pH is just that of 0.150 M HF ($pK_a = -\log(6.6 \times 10^{-4}) = 3.18$).

[Initial solution contains $20.00\,mL \times \frac{0.150\,mmol\,HF}{1\,mL} = 3.00\,mmol\,HF$]

Equation: $HF(aq)$ + $H_2O(l)$ $\rightleftharpoons$ $H_3O^+(aq)$ + $F^-(aq)$

Initial:	0.150 M	–	$\approx 0\,M$	0 M
Changes:	$-x\,M$	–	$+x\,M$	$+x\,M$
Equil:	$(0.150-x)\,M$	–	$x\,M$	$x\,M$

$$K_a = \frac{[H_3O^+][F^-]}{[HF]} = \frac{x \cdot x}{0.150-x} \approx \frac{x^2}{0.150} = 6.6 \times 10^{-4}$$

$$x = \sqrt{0.150 \times 6.6 \times 10^{-4}} = 9.9 \times 10^{-3}$$

$x > 0.05(0.150)$. The assumption is invalid. After a second cycle of approximation, $\left[H_3O^+\right] = 9.6 \times 10^{-3}\,M$; $pH = -\log(9.6 \times 10^{-3}) = 2.02$

(b) When the titration is 25.0% complete, there are $(0.25 \times 3.00=)$ 0.75 mmol F^- for every 3.00 mmol HF that were present initially. Therefore, $(3.00 - 0.75 =)$ 2.25 mmol HF remain untitrated. We designate the solution volume (the volume holding these 3.00 mmol total) as V and use the Henderson-Hasselbalch equation to find the pH.

$$\text{pH} = \text{p}K_a + \log\frac{\left[F^-\right]}{\left[HF\right]} = 3.18 + \log\frac{0.75 \text{ mmol}/V}{2.25 \text{ mmol}/V} = 2.70$$

(c) At the midpoint of the titration of a weak base, $\text{pH} = \text{p}K_a = 3.18$.

(d) At the endpoint of the titration, the pH of the solution is determined by the conjugate base hydrolysis reaction. We calculate the amount of anion and the volume of solution in order to calculate its initial concentration.

$$\text{amount } F^- = 20.00 \text{ mL} \times \frac{0.150 \text{ mmol HF}}{1 \text{ mL soln}} \times \frac{1 \text{ mmol } F^-}{1 \text{ mmol HF}} = 3.00 \text{ mmol } F^-$$

$$\text{volume titrant} = 3.00 \text{ mmol HF} \times \frac{1 \text{mmol } OH^-}{1 \text{mmol HF}} \times \frac{1 \text{ mL titrant}}{0.250 \text{ mmol } OH^-} = 12.0 \text{ mL titrant}$$

$$\left[F^-\right] = \frac{3.00 \text{ mmol } F^-}{20.00 \text{ mL original volume} + 12.0 \text{ mL titrant}} = 0.0938 \text{ M}$$

We organize the solution of the hydrolysis problem around its balanced equation.

Equation: $\quad F^-(aq) \;+\; H_2O(l) \;\rightleftharpoons\; HF(aq) \;+\; OH^-(aq)$

Initial: $\qquad$ 0.0938M $\qquad$ – $\qquad\qquad$ 0 M $\qquad \approx 0$ M

Changes: $\qquad -x$ M $\qquad$ – $\qquad\qquad +x$ M $\qquad +x$ M

Equil: $\qquad (0.0938-x)$ M $\quad$ – $\qquad\qquad x$ M $\qquad\quad x$ M

$$K_b = \frac{\left[HF\right]\left[OH^-\right]}{\left[F^-\right]} = \frac{K_w}{K_a} = \frac{1.0 \times 10^{-14}}{6.6 \times 10^{-4}} = 1.5 \times 10^{-11} = \frac{x \cdot x}{0.0938 - x} \approx \frac{x^2}{0.0938}$$

$x = \sqrt{0.0934 \times 1.5 \times 10^{-11}} = 1.2 \times 10^{-6} = \left[OH^-\right] = 1.2 \times 10^{-6}$ M

The assumption is valid $(x \ll 0.0934)$.

$\text{pOH} = -\log\left(1.2 \times 10^{-6}\right) = 5.92; \quad \text{pH} = 14.00 - \text{pOH} = 14.00 - 5.92 = 8.08$

8B **(D)**

(a) The initial pH is simply that of 0.106 M NH_3.

Equation: $NH_3(aq) \;+\; H_2O(l) \rightleftharpoons NH_4^+(aq) \;+\; OH^-(aq)$

Initial: $\qquad$ 0.106 M $\qquad$ – $\qquad\qquad$ 0 M $\qquad \approx 0$ M

Changes: $\qquad -x$ M $\qquad$ – $\qquad\qquad +x$ M $\qquad +x$ M

Equil: $\qquad (0.106-x)$ M $\quad$ – $\qquad\qquad x$ M $\qquad\quad x$ M

$$K_b = \frac{\left[NH_4^+\right]\left[OH^-\right]}{\left[NH_3\right]} = \frac{x \cdot x}{0.106 - x} \approx \frac{x^2}{0.106} = 1.8 \times 10^{-5}$$

$x = \sqrt{0.106 \times 1.8 \times 10^{-4}} = 1.4 \times 10^{-3} \; \left[OH^-\right] = 1.4 \times 10^{-3}$ M

The assumption is valid ($x \ll 0.106$).

$$\text{pOH} = -\log(0.0014) = 2.85 \quad \text{pH} = 14.00 - \text{pOH} = 14.00 - 2.85 = 11.15$$

(b) When the titration is 25.0% complete, there are 25.0 mmol NH_4^+ for every 100.0 mmol of NH_3 that were present initially (i.e., there are 1.33 mmol of NH_4^+ in solution), 3.98 mmol NH_3 remain untitrated. We designate the solution volume (the volume holding these 5.30 mmol total) as V and use the basic version of the Henderson-Hasselbalch equation to find the pH.

$$\text{pOH} = pK_b + \log \frac{\left[\text{NH}_4^+\right]}{\left[\text{NH}_3\right]} = 4.74 + \log \frac{\dfrac{1.33 \text{ mmol}}{V}}{\dfrac{3.98 \text{ mmol}}{V}} = 4.26$$

$$\text{pH} = 14.00 - 4.26 = 9.74$$

(c) At the midpoint of the titration of a weak base, $\text{pOH} = pK_b = 4.74$ and $\text{pH} = 9.26$

(d) At the endpoint of the titration, the pH is determined by the conjugate acid hydrolysis reaction. We calculate the amount of that cation and the volume of the solution in order to determine its initial concentration.

$$\text{amount } \text{NH}_4^+ = 50.00 \text{ mL} \times \frac{0.106 \text{ mmol } \text{NH}_3}{1 \text{ mL soln}} \times \frac{1 \text{ mmol } \text{NH}_4^+}{1 \text{ mmol } \text{NH}_3}$$

$$\text{amount } \text{NH}_4^+ = 5.30 \text{ mmol } \text{NH}_4^+$$

$$V_{\text{titrant}} = 5.30 \text{ mmol } \text{NH}_3 \times \frac{1 \text{ mmol } \text{H}_3\text{O}^+}{1 \text{ mmol } \text{NH}_3} \times \frac{1 \text{ mL titrant}}{0.225 \text{ mmol } \text{H}_3\text{O}^+} = 23.6 \text{ mL titrant}$$

$$\left[\text{NH}_4^+\right] = \frac{5.30 \text{ mmol } \text{NH}_4^+}{50.00 \text{ mL original volume} + 23.6 \text{ mL titrant}} = 0.0720 \text{ M}$$

We organize the solution of the hydrolysis problem around its balanced chemical equation.

Equation: $\text{NH}_4^+(\text{aq}) + \text{H}_2\text{O}(\text{l}) \rightleftharpoons \text{NH}_3(\text{aq}) + \text{H}_3\text{O}^+(\text{aq})$

Initial:	0.0720 M	–	0 M	≈ 0 M
Changes:	$-x$ M	–	$+x$ M	$+x$ M
Equil:	$(0.0720 - x)$ M	–	x M	x M

$$K_b = \frac{\left[\text{NH}_3\right]\left[\text{H}_3\text{O}^+\right]}{\left[\text{NH}_4^+\right]} = \frac{K_w}{K_b} = \frac{1.0 \times 10^{-14}}{1.8 \times 10^{-5}} = 5.6 \times 10^{-10} = \frac{x \cdot x}{0.0720 - x} \approx \frac{x^2}{0.0720}$$

$$x = \sqrt{0.0720 \times 5.6 \times 10^{-10}} = 6.3 \times 10^{-6} \, [\text{H}_3\text{O}^+] = 6.3 \times 10^{-6} \text{ M}$$

The assumption is valid ($x \ll 0.0720$).

$$\text{pH} = -\log(6.3 \times 10^{-6}) = 5.20$$

9A **(M)** The acidity of the solution is principally the result of the hydrolysis of the carbonate ion, which is considered first.

Equation: $\quad CO_3^{2-}(aq) + H_2O(l) \rightleftharpoons HCO_3^-(aq) + OH^-(aq)$

Initial:	1.0 M	0 M	≈ 0 M
Changes:	$-x$ M	$+x$ M	$+x$ M
Equil:	$(1.0-x)$ M	x M	x M

$$K_b = \frac{K_w}{K_a(HCO_3^-)} = \frac{1.0 \times 10^{-14}}{4.7 \times 10^{-11}} = 2.1 \times 10^{-4} = \frac{[HCO_3^-][OH^-]}{[CO_3^{2-}]} = \frac{x \cdot x}{1.0-x} \approx \frac{x^2}{1.0}$$

$x = \sqrt{1.0 \times 2.1 \times 10^{-4}} = 0.015$ $[OH^-] = 1.5 \times 10^{-2}$ M. The assumption is valid $(x \ll 1.0M)$.

Now we consider the hydrolysis of the bicarbonate ion.

Equation: $\quad HCO_3^-(aq) + H_2O(l) \rightleftharpoons H_2CO_3(aq) + OH^-(aq)$

Initial:	0.015 M	0 M	0.015 M
Changes:	$-y$ M	$+y$ M	$+y$ M
Equil:	$(0.015-y)$ M	y M	$(0.015+y)$ M

$$K_b = \frac{K_w}{K_a(H_2CO_3)} = \frac{1.0 \times 10^{-14}}{4.4 \times 10^{-7}} = 2.3 \times 10^{-8} = \frac{[H_2CO_3][OH^-]}{[HCO_3^-]} = \frac{y(0.015+y)}{0.015-x} \approx \frac{0.015y}{0.015} = y$$

The assumption is valid $(y \ll 0.015)$ and $[H_2CO_3] = 2.3 \times 10^{-8}$ M. Clearly, the second hydrolysis makes a negligible contribution to the acidity of the solution. For the entire solution, then

$$pOH = -\log[OH^-] = -\log(0.015) = 1.82 \qquad pH = 14.00 - 1.82 = 12.18$$

9B **(M)** The acidity of the solution is principally the result of hydrolysis of the sulfite ion.

Equation: $\quad SO_3^{2-}(aq) + H_2O(l) \rightleftharpoons HSO_3^-(aq) + OH^-(aq)$

Initial:	0.500 M	0 M	≈ 0 M
Changes:	$-x$ M	$+x$ M	$+x$ M
Equil:	$(0.500-x)$ M	x M	x M

$$K_b = \frac{K_w}{K_a HSO_3^-} = \frac{1.0 \times 10^{-14}}{6.2 \times 10^{-8}} = 1.6 \times 10^{-7} = \frac{[HSO_3^-][OH^-]}{[SO_3^{2-}]} = \frac{x \cdot x}{0.500-x} \approx \frac{x^2}{0.500}$$

$x = \sqrt{0.500 \times 1.6 \times 10^{-7}} = 2.8 \times 10^{-4}$ $[OH^-] = 2.8 \times 10^{-4}$ M

The assumption is valid $(x \ll 0.500)$.

Next we consider the hydrolysis of the bisulfite ion.

Equation: $HSO_3^-(aq) + H_2O(l) \rightleftharpoons H_2SO_3(aq) + OH^-(aq)$

Initial:	0.00028 M	–	0 M	0.00028 M
Changes:	$-y$ M	–	$+y$ M	$+y$ M
Equil:	$(0.00028 - y)$ M	–	y M	$(0.00028 + y)$ M

$$K_b = \frac{K_w}{K_{a_1}(H_2SO_3)} = \frac{1.0\times10^{-14}}{1.3\times10^{-2}} = 7.7\times10^{-13}$$

$$K_b = 7.7\times10^{-13} = \frac{[H_2SO_3][OH^-]}{[HSO_3^-]} = \frac{y(0.00028 + y)}{0.00028 - y} \approx \frac{0.00028\,y}{0.00028} = y$$

The assumption is valid ($y \ll 0.00028$) and $y = [H_2SO_3] = 7.7\times10^{-13}$ M.

Clearly, the second hydrolysis makes a negligible contribution to the acidity of the solution. For the entire solution, then

$$pOH = -\log[OH^-] = -\log(0.00028) = 3.55 \qquad pH = 14.00 - 3.55 = 10.45$$

INTEGRATIVE EXAMPLE

A **(D)**

From the given information, the following can be calculated:

pH of the solution = 2.716   therefore, $[H^+] = 1.92\times10^{-3}$

pH at the halfway point = pK_a
pH = 4.602 = pK_a
$pK_a = -\log K_a$ therefore $K_a = 2.50 \times 10^{-5}$

$$FP = 0\,°C + \Delta T_f$$
$$\Delta T_f = -i \times K_f \times m$$
$$\Delta T_f = -1 \times 1.86\,°C/m \times \text{molality}$$

$$\text{molality} = \frac{\#\,\text{moles solute}}{\text{kg solvent}} = \frac{\#\,\text{moles solute}}{0.500\,\text{kg} - 0.00750\,\text{kg}}$$

To determine the number of moles of solute, convert 7.50 g of unknown acid to moles by using its molar mass. The molar mass can be calculated as follows:

pH = 2.716 $[H^+] = 1.92 \times 10^{-3}$

HA $\rightleftharpoons$ H^+ + A^-

	$\dfrac{7.50\ \text{g}}{\text{MM}}\Big/ 0.500\ \text{L}$		0	0
Initial				
Change	$-x$		x	x
Equilibrium	$\left(\dfrac{7.50\ \text{g}}{\text{MM}}\Big/ 0.500\ \text{L}\right)-1.92\times10^{-3}$		1.92×10^{-3}	1.92×10^{-3}

$$2.50\times10^{-5}=\frac{[\text{H}^+][\text{A}^-]}{[\text{HA}]}$$

$$2.50\times10^{-5}=\frac{(1.92\times10^{-3})^2}{\left(\dfrac{7.50\big/\text{MM}}{0.500}-1.92\times10^{-3}\right)}$$

$$\text{MM}=100.4\ \text{g/mol}$$

$$\#\,\text{moles of solute}=7.50\ \text{g}\times\frac{1\ \text{mol}}{100.4\ \text{g}}=0.0747\ \text{mol}$$

$$\text{Molality}=\frac{\#\ \text{moles solute}}{\text{kg solvent}}=\frac{0.0747\ \text{mol}}{0.500\ \text{kg}-0.00750\ \text{kg}}=0.152\ m$$

$$\Delta T_{\text{f}}=-\text{i}\times K_{\text{f}}\times m$$
$$\Delta T_{\text{f}}=-1\times1.86\ ^\circ\text{C/m}\times0.152\ m$$
$$\Delta T_{\text{f}}=-0.283\ ^\circ\text{C}$$
$$\text{FP}=0\ ^\circ\text{C}+\Delta T_{\text{f}}=0\ ^\circ\text{C}-0.283\ ^\circ\text{C}=-0.283\ ^\circ\text{C}$$

B (M) By looking at the titration curve provided, one can deduce that the titrant was a strong acid. The pH before titrant was added was basic, which means that the substance that was titrated was a base. The pH at the end of the titration after excess titrant was added was acidic, which means that the titrant was an acid.

Based on the titration curve provided, the equivalence point is at approximately 50 mL of titrant added. At the halfway point, of approximately 25 mL, the pH = pK_a. A pH ~8 is obtained by extrapolation a the halfway point.

$$\text{pH}=8=pK_a$$
$$K_a=10^{-8}=1\times10^{-8}$$
$$K_b=K_w/K_a=1\times10^{-6}$$

~50 mL of 0.2 M strong acid (1×10^{-2} mol) was needed to reach the equivalence point. This means that the unknown contained 1×10^{-2} mol of weak base. The molar mass of the unknown can be determined as follows:

$$\frac{0.800\ \text{g}}{1\times10^{-2}\ \text{mol}}=80\ \text{g/mol}$$

EXERCISES

The Common-Ion Effect

1. **(M)**

(a) Note that HI is a strong acid and thus the initial $\left[H_3O^+\right] = \left[HI\right] = 0.0892M$

Equation:	CH_3CH_2COOH + H_2O	$\rightleftharpoons$	$CH_2CH_2COO^-$ +	H_3O^+
Initial:	0.275 M	–	0M	0.0892M
Changes:	$-x$ M	–	$+x$ M	$+x$ M
Equil:	$(0.275 - x)$ M	–	x M	$(0.0892 + x)$ M

$$K_a = \frac{\left[CH_2CH_2COO^-\right]\left[H_3O^+\right]}{\left[CH_3CH_2COOH\right]} = 1.3 \times 10^{-5} = \frac{x(0.0892 + x)}{0.275 - x} \approx \frac{0.0892x}{0.275}$$

$$x = 4.0 \times 10^{-5} \text{ M}$$

The assumption that $x \ll 0.0892$ M is correct. $\left[H_3O^+\right] = 0.0892M$

(b) $\left[OH^-\right] = \dfrac{K_w}{\left[H_3O^+\right]} = \dfrac{1.0 \times 10^{-14}}{0.0892} = 1.1 \times 10^{-13}$ M

(c) $\left[CH_2CH_2COO^-\right] = x = 4.0 \times 10^{-5}$ M

(d) $\left[I^-\right] = \left[HI\right]_{intial} = 0.0892$ M

3. **(M)**

(a) We first determine the pH of 0.100 M HNO_2.

Equation	$HNO_2(aq)$ +	$H_2O(l)$	$\rightleftharpoons$	$NO_2^-(aq)$ +	$H_3O^+(aq)$
Initial:	0.100 M	–		0M	≈ 0M
Changes:	$-x$ M	–		$+x$ M	$+x$ M
Equil:	$(0.100 - x)$ M	–		x M	x M

$$K_a = \frac{\left[NO_2^-\right]\left[H_3O^+\right]}{\left[HNO_2\right]} = 7.2 \times 10^{-4} = \frac{x^2}{0.100 - x}$$

Via the quadratic equation roots formula or via successive approximations,

$$x = 8.1 \times 10^{-3} \quad \left[H_3O^+\right] = 8.1 \times 10^{-3} \text{ M}$$

Thus $pH = -\log(8.1 \times 10^{-3}) = 2.09$

When 0.100 mol $NaNO_2$ is added to 1.00 L of a 0.100 M HNO_2, a solution with $\left[NO_2^-\right] = 0.100 M = \left[HNO_2\right]$ is produced. The answer obtained with the Henderson-Hasselbalch equation, is $pH = pK_a = -\log\left(7.2 \times 10^{-4}\right) = 3.14$. Thus, the addition has caused a pH change of 1.05 units.

(b) $NaNO_3$ contributes nitrate ion, NO_3^-, to the solution. Since, however, there is no molecular $HNO_3(aq)$ in equilibrium with hydrogen and nitrate ions, there is no equilibrium to be shifted by the addition of nitrate ions. The $[H_3O^+]$ and the pH are thus unaffected by the addition of $NaNO_3$ to a solution of nitric acid. The pH changes are not the same because there is an equilibrium system to be shifted in the first solution, whereas there is no equilibrium, just a change in total ionic strength, for the second solution.

5. **(M)**

(a) The strong acid HCl suppresses the ionization of the weak acid HOCl to such an extent that a negligible concentration of H_3O^+ is contributed to the solution by HOCl. Thus, $\left[H_3O^+\right] = \left[HCl\right]_{initial} = 0.035$ M

(b) This is a buffer solution. Consequently, we can use the Henderson-Hasselbalch equation to determine its pH. $pK_a = -\log\left(7.2 \times 10^{-4}\right) = 3.14;$

$$pH = pK_a + \log\frac{\left[NO_2^-\right]}{\left[HNO_2\right]} = 3.14 + \log\frac{0.100\,M}{0.0550\,M} = 3.40$$

$$\left[H_3O^+\right] = 10^{-3.40} = 4.0 \times 10^{-4}\ M$$

(c) As summarized below, the components react to form a buffer solution.

Equation: $H_3O^+\left(aq,\ from\ HCl\right) + CH_3COO^-\left(aq,\ from\ NaCH_3COO\right) \rightarrow CH_3COOH\left(aq\right) + H_2O(l)$

initially present:	0.0525 M	0.0768 M	0 M	–
changes:	−0.0525 M	−0.0525 M	+0.0525 M	–
after reaction:	≈ 0M	0.0243 M	0.0525 M	–

Now the Henderson-Hasselbalch equation can be used to find the pH.

$$pK_a = -\log\left(1.8 \times 10^{-5}\right) = 4.74$$

$$pH = pK_a + \log\frac{\left[CH_3COO^-\right]}{\left[CH_3COO\right]} = 4.74 + \log\frac{0.0243\,M}{0.0525\,M} = 4.41$$

$$\left[H_3O^+\right] = 10^{-4.41} = 3.9 \times 10^{-5}\ M$$

Buffer Solutions

7. **(M)** $[H_3O^+] = 10^{-4.06} = 8.7 \times 10^{-5} M$. We let $S = [HCOO^-]_{int}$

Equation: $\quad HCOOH(aq) + H_2O(l) \rightleftharpoons HCOO^-(aq) + H_3O^+(aq)$

Initial: $\qquad 0.366\,M \qquad\qquad - \qquad\qquad S\,M \qquad\qquad \approx 0\,M$

Changes: $\quad -8.7 \times 10^{-5}\,M \qquad - \qquad +8.7 \times 10^{-5}\,M \qquad +8.7 \times 10^{-5}\,M$

Equil: $\qquad 0.366\,M \qquad\qquad - \qquad (S + 8.7 \times 10^{-5})\,M \quad 8.7 \times 10^{-5}\,M$

$$K_a = \frac{[H_3O^+][HCOO^-]}{[HCOOH]} = 1.8 \times 10^{-4} = \frac{(S + 8.7 \times 10^{-5})8.7 \times 10^{-5}}{0.366} \approx \frac{8.7 \times 10^{-5}\,S}{0.366}; \quad S = 0.76$$

To determine S, we assumed $S \gg 8.7 \times 10^{-5}$, which is clearly a valid assumption. Or, we could have used the Henderson-Hasselbalch equation (see below).

$$pK_a = -\log(1.8 \times 10^{-4}) = 3.74$$

$$4.06 = 3.74 + \log\frac{[HCOO^-]}{[HCOOH]}; \quad \frac{[HCOO^-]}{[HCOOH]} = 2.1; \quad [HCOO^-] = 2.1 \times 0.366 = 0.77\,M$$

The difference in the two answers is due simply to rounding.

9. **(M)**

(a) Equation: $\quad C_6H_5COOH(aq) + H_2O(l) \rightleftharpoons C_6H_5COO^-(aq) + H_3O^+(aq)$

Initial: $\qquad 0.012\,M \qquad\qquad - \qquad\qquad 0.033\,M \qquad\qquad \approx 0\,M$

Changes: $\qquad -x\,M \qquad\qquad - \qquad\qquad +x\,M \qquad\qquad +x\,M$

Equil: $\qquad (0.012 - x)\,M \qquad - \qquad (0.033 + x)\,M \qquad x\,M$

$$K_a = \frac{[H_3O^+][C_6H_5COO^-]}{[C_6H_5COOH]} = 6.3 \times 10^{-5} = \frac{x(0.033 + x)}{0.012 - x} \approx \frac{0.033x}{0.012} \quad x = 2.3 \times 10^{-5}$$

To determine the value of x, we assumed $x \ll 0.012$, which is an assumption that clearly is correct. $[H_3O^+] = 2.3 \times 10^{-5}\,M \qquad pH = -\log(2.3 \times 10^{-5}) = 4.64$

(b) Equation: $\quad NH_3(aq) + H_2O(l) \rightleftharpoons NH_4^+(aq) + OH^-(aq)$

Initial: $\qquad 0.408\,M \qquad\qquad - \qquad\qquad 0.153\,M \qquad\qquad \approx 0\,M$

Changes: $\qquad -x\,M \qquad\qquad - \qquad\qquad +x\,M \qquad\qquad +x\,M$

Equil: $\qquad (0.408 - x)\,M \qquad - \qquad (0.153 + x)\,M \qquad x\,M$

$$K_b = \frac{[NH_4^+][OH^-]}{[NH_3]} = 1.8 \times 10^{-5} = \frac{x(0.153 + x)}{0.408 - x} \approx \frac{0.153x}{0.408} \quad x = 4.8 \times 10^{-5}$$

To determine the value of x, we assumed $x \ll 0.153$, which clearly is a valid assumption.

$$[OH^-] = 4.8 \times 10^{-5}\,M; \quad pOH = -\log(4.8 \times 10^{-5}) = 4.32; \quad pH = 14.00 - 4.32 = 9.68$$

11. (M)

(a) 0.100 M NaCl is not a buffer solution. Neither ion reacts with water to a detectable extent.

(b) 0.100 M NaCl—0.100 M NH_4Cl is not a buffer solution. Although a weak acid, NH_4^+, is present, its conjugate base, NH_3, is not.

(c) 0.100 M CH_3NH_2 and 0.150 M $CH_3NH_3^+Cl^-$ is a buffer solution. Both the weak base, CH_3NH_2, and its conjugate acid, $CH_3NH_3^+$, are present in approximately equal concentrations.

(d) 0.100 M HCl—0.050 M $NaNO_2$ is not a buffer solution. All the NO_2^- has converted to HNO_2 and thus the solution is a mixture of a strong acid and a weak acid.

(e) 0.100 M HCl—0.200 M $NaCH_3COO$ is a buffer solution. All of the HCl reacts with half of the CH_3COO^- to form a solution with 0.100 M CH_3COOH, a weak acid, and 0.100 M CH_3COO^-, its conjugate base.

(f) 0.100 M CH_3COOH and 0.125 M $NaCH_3CH_2COO$ is not a buffer in the strict sense because it does not contain a weak acid and its conjugate base, but rather the conjugate base of another weak acid. These two weak acids (acetic, $K_a = 1.8 \times 10^{-5}$, and propionic, $K_a = 1.35 \times 10^{-5}$) have approximately the same strength, however, this solution would resist changes in its pH on the addition of strong acid or strong base, consequently, it could be argued that this system should also be called a buffer.

13. (M) moles of solute $= 1.15 \text{ mg} \times \dfrac{1 \text{ g}}{1000 \text{ mg}} \times \dfrac{1 \text{ mol } C_6H_5NH_3^+Cl^-}{129.6 \text{ g}} \times \dfrac{1 \text{ mol } C_6H_5NH_3^+}{1 \text{ mol } C_6H_5NH_3^+Cl^-}$

$$= 8.87 \times 10^{-6} \text{ mol } C_6H_5NH_3^+$$

$$\left[C_6H_5NH_3^+\right] = \dfrac{8.87 \times 10^{-6} \text{ mol } C_6H_5NH_3^+}{3.18 \text{ L soln}} = 2.79 \times 10^{-6} \text{ M}$$

Equation: $\quad C_6H_5NH_2 \text{ (aq)} + H_2O(l) \rightleftharpoons C_6H_5NH_3^+ \text{ (aq)} + OH^- \text{ (aq)}$

Initial:	0.105 M	–	2.79×10^{-6} M	≈ 0 M
Changes:	$-x$ M	–	$+x$ M	$+x$ M
Equil:	$(0.105 - x)$M	–	$(2.79 \times 10^{-6} + x)$M	x M

$$K_b = \dfrac{\left[C_6H_5NH_3^+\right]\left[OH^-\right]}{\left[C_6H_5NH_2\right]} = 7.4 \times 10^{-10} = \dfrac{\left(2.79 \times 10^{-6} + x\right)x}{0.105 - x}$$

$7.4 \times 10^{-10}(0.105 - x) = \left(2.79 \times 10^{-6} + x\right)x; \quad 7.8 \times 10^{-11} - 7.4 \times 10^{-10}x = 2.79 \times 10^{-6}x + x^2$

$x^2 + \left(2.79 \times 10^{-6} + 7.4 \times 10^{-10}\right)x - 7.8 \times 10^{-11} = 0; \quad x^2 + 2.79 \times 10^{-6}x - 7.8 \times 10^{-11} = 0$

$$x = \dfrac{-b \pm \sqrt{b^2 - 4ac}}{2a} = \dfrac{-2.79 \times 10^{-6} \pm \sqrt{7.78 \times 10^{-12} + 3.1 \times 10^{-10}}}{2} = 7.5 \times 10^{-6}$$

$pOH = -\log\left(7.5 \times 10^{-6}\right) = 5.12 \qquad pH = 14.00 - 5.12 = 8.88$

15. **(M)**

(a) First use the Henderson-Hasselbalch equation. $pK_b = -\log(1.8 \times 10^{-5}) = 4.74$,

$pK_a = 14.00 - 4.74 = 9.26$ to determine $\left[NH_4^+\right]$ in the buffer solution.

$$pH = 9.45 = pK_a + \log\frac{[NH_3]}{\left[NH_4^+\right]} = 9.26 + \log\frac{[NH_3]}{\left[NH_4^+\right]}; \quad \log\frac{[NH_3]}{\left[NH_4^+\right]} = 9.45 - 9.26 = +0.19$$

$$\frac{[NH_3]}{\left[NH_4^+\right]} = 10^{0.19} = 1.5\underline{5} \quad \left[NH_4^+\right] = \frac{[NH_3]}{1.5\underline{5}} = \frac{0.258M}{1.5\underline{5}} = 0.17M$$

We now assume that the volume of the solution does not change significantly when the solid is added.

$$\text{mass}(NH_4)_2SO_4 = 425 \text{ mL} \times \frac{1 \text{ L soln}}{1000 \text{ mL}} \times \frac{0.17 \text{ mol } NH_4^+}{1 \text{ L soln}} \times \frac{1 \text{ mol}(NH_4)_2SO_4}{2 \text{ mol } NH_4^+}$$

$$\times \frac{132.1 \text{ g}(NH_4)_2SO_4}{1 \text{ mol}(NH_4)_2SO_4} = 4.8 \text{ g }(NH_4)_2SO_4$$

(b) We can use the Henderson-Hasselbalch equation to determine the ratio of concentrations of cation and weak base in the altered solution.

$$pH = 9.30 = pK_a + \log\frac{[NH_3]}{\left[NH_4^+\right]} = 9.26 + \log\frac{[NH_3]}{\left[NH_4^+\right]} \quad \log\frac{[NH_3]}{\left[NH_4^+\right]} = 9.30 - 9.26 = +0.04$$

$$\frac{[NH_3]}{\left[NH_4^+\right]} = 10^{0.04} = 1.1 = \frac{0.258}{0.17 \text{ M} + x \text{ M}} \qquad 0.19 + 1.1\,x = 0.258 \quad x = 0.062$$

The reason we decided to add x to the denominator follows. (Notice we cannot remove a component.) A pH of 9.30 is more acidic than a pH of 9.45 and therefore the conjugate acid's (NH_4^+) concentration must increase. Additionally, mathematics tells us that for the concentration ratio to decrease from 1.5$\underline{5}$ to 1.1, its denominator must increase. We solve this expression for x to find a value of 0.062 M. We need to add NH_4^+ to increase its concentration by 0.062 M in 100 mL of solution.

$$(NH_4)_2SO_4 \text{ mass} = 0.100 \text{ L} \times \frac{0.062 \text{ mol } NH_4^+}{1 \text{ L}} \times \frac{1 \text{ mol}(NH_4)_2SO_4}{2 \text{ mol } NH_4^+} \times \frac{132.1\text{g }(NH_4)_2SO_4}{1 \text{ mol }(NH_4)_2SO_4}$$

$$= 0.41\text{g }(NH_4)_2SO_4 \qquad \text{Hence, we need to add} \approx 0.4 \text{ g}$$

17. **(M)** The added HCl will react with the ammonia, and the pH of the buffer solution will decrease. The original buffer solution has $[NH_3] = 0.258$ M and $[NH_4^+] = 0.17$ M. We first calculate the [HCl] in solution, reduced from 12 M because of dilution. [HCl]

added $= 12\,M \times \dfrac{0.55\,mL}{100.6\,mL} = 0.066\,M$ We then determine pK_a for ammonium ion:

$pK_b = -\log(1.8 \times 10^{-5}) = 4.74$ $pK_a = 14.00 - 4.74 = 9.26$

Equation:	$NH_3\,(aq)$	$+$	$H_3O^+\,(aq)$	$\rightleftharpoons$	$NH_4^+\,(aq)$	$+$	$H_2O\,(l)$
Buffer:	0.258 M		≈ 0 M		0.17 M		–
Added:			+0.066 M				
Changes:	−0.066 M		−0.066 M		+0.066 M		–
Final:	0.192 M		0 M		0.24 M		–

$$pH = pK_a + \log\frac{[NH_3]}{[NH_4^+]} = 9.26 + \log\frac{0.192}{0.24} = 9.16$$

19. **(M)** The pK_a's of the acids help us choose the one to be used in the buffer. It is the acid with a pK_a within 1.00 pH unit of 3.50 that will do the trick. $pK_a = 3.74$ for HCOOH, $pK_a = 4.74$ for CH_3COOH, and $pK_{a1} = 2.15$ for H_3PO_4. Thus, we choose HCOOH and NaCOO to prepare a buffer with pH = 3.50. The Henderson-Hasselbalch equation is used to determine the relative amount of each component present in the buffer solution.

$$pH = 3.50 = 3.74 + \log\frac{[HCOO^-]}{[HCOOH]} \qquad \log\frac{[HCOO^-]}{[HCOOH]} = 3.50 - 3.74 = -0.24$$

$$\frac{[HCOO^-]}{[HCOOH]} = 10^{-0.24} = 0.58$$

This ratio of concentrations is also the ratio of the number of moles of each component in the buffer solution, since both concentrations are a number of moles in a certain volume, and the volumes are the same (the two solutes are in the same solution). This ratio also is the ratio of the volumes of the two solutions, since both solutions being mixed contain the same concentration of solute. If we assume 100. mL of acid solution, $V_{acid} = 100.$ mL. Then the volume of salt solution is $V_{salt} = 0.58 \times 100.$ mL $= 58$ mL 0.100 M NaHCOO

21. **(M)**

(a) The pH of the buffer is determined via the Henderson-Hasselbalch equation.

$$pH = pK_a + \log\frac{[CH_3CH_2COO^-]}{[CH_3CH_2COOH]} = 4.89 + \log\frac{0.100M}{0.100M} = 4.89$$

The effective pH range is the same for every propionate buffer: from pH = 3.89 to pH = 5.89, one pH unit on either side of pK_a for propionic acid, which is 4.89.

(b) To each liter of 0.100 M $CH_3CH_2COOH - 0.100M$ $NaCH_3CH_2COO$ we can add 0.100 mol OH^- before all of the CH_3CH_2COOH is consumed, and we can add 0.100 mol H_3O^+ before all of the $CH_3CH_2COO^-$ is consumed. The buffer capacity thus is 100. millimoles (0.100 mol) of acid or base per liter of buffer solution.

23. **(M)**

(a) The pH of this buffer solution is determined with the Henderson-Hasselbalch equation.

$$pH = pK_a + \log\frac{\left[HCOO^-\right]}{\left[HCOOH\right]} = -\log\left(1.8\times10^{-4}\right) + \log\frac{8.5\,mmol/75.0\,mL}{15.5\,mmol/75.0\,mL}$$

$$= 3.74 - 0.26 = 3.48$$

[Note: the solution is not a good buffer, as $\left[HCOO^-\right] = 1.1\times10^{-1}$, which is only ~ 600 times K_a]

(b) Amount of added $OH^- = 0.25\,mmol\ Ba(OH)_2 \times \dfrac{2\,mmol\ OH^-}{1\,mmol\ Ba(OH)_2} = 0.50\,mmol\ OH^-$

The OH^- added reacts with the formic acid and produces formate ion.

Equation:	$HCOOH(aq)$	+	$OH^-(aq)$	$\rightleftharpoons$	$HCOO^-(aq)$	+	$H_2O(l)$
Buffer:	15.5 mmol		$\approx 0\,M$		8.5 mmol		–
Add base:			+0.50 mmol				
React:	−0.50 mmol		−0.50 mmol		+0.50 mmol		–
Final:	15.0 mmol		0 mmol		9.0 mmol		–

$$pH = pK_a + \log\frac{\left[HCOO^-\right]}{\left[HCOOH\right]} = -\log\left(1.8\times10^{-4}\right) + \log\frac{9.0\,mmol/75.0\,mL}{15.0\,mmol/75.0\,mL}$$

$$= 3.74 - 0.22 = 3.52$$

(c) Amount of added $H_3O^+ = 1.05\ mL\ acid \times \dfrac{12\ mmol\ HCl}{1\ mL\ acid} \times \dfrac{1\ mmol\ H_3O^+}{1\ mmol\ HCl} = 13\ mmol\ H_3O^+$

The H_3O^+ added reacts with the formate ion and produces formic acid.

Equation:	$HCOO^-(aq)$	+	$H_3O^+(aq)$	$\rightleftharpoons$	$HCOOH(aq)$	+	$H_2O(l)$
Buffer:	8.5 mmol		≈ 0 mmol		15.5 mmol		–
Add acid:			+13 mmol				
React:	−8.5 mmol		−8.5 mmol		+8.5 mmol		–
Final:	0 mmol		4.5 mmol		24.0 mmol		–

The buffer's capacity has been exceeded. The pH of the solution is determined by the excess strong acid present.

$$\left[H_3O^+\right] = \frac{4.5\,mmol}{75.0\,mL + 1.05\,mL} = 0.059\,M; \quad pH = -\log\left(0.059\right) = 1.23$$

472

25. **(D)**

(a) We use the Henderson-Hasselbalch equation to determine the pH of the solution. The total solution volume is

$$36.00\,\text{mL} + 64.00\,\text{mL} = 100.00\,\text{mL.} \quad pK_a = 14.00 - pK_b = 14.00 + \log\left(1.8\times10^{-5}\right) = 9.26$$

$$\left[NH_3\right] = \frac{36.00\,\text{mL}\times0.200\,\text{M NH}_3}{100.00\,\text{mL}} = \frac{7.20\,\text{mmol NH}_3}{100.0\,\text{mL}} = 0.0720\,\text{M}$$

$$\left[NH_4^+\right] = \frac{64.00\,\text{mL}\times0.200\,\text{M NH}_4^+}{100.00\,\text{mL}} = \frac{12.8\,\text{mmol NH}_4^+}{100.0\,\text{mL}} = 0.128\,\text{M}$$

$$pH = pK_a + \log\frac{[NH_3]}{[NH_4^+]} = 9.26 + \log\frac{0.0720\,\text{M}}{0.128\,\text{M}} = 9.01 \approx 9.00$$

(b) The solution has $\left[OH^-\right] = 10^{-4.99} = 1.0\times10^{-5}\,\text{M}$

The Henderson-Hasselbalch equation depends on the assumption that:

$$\left[NH_3\right] \gg 1.8\times10^{-5}\,\text{M} \ll \left[NH_4^+\right]$$

If the solution is diluted to 1.00 L, $\left[NH_3\right] = 7.20\times10^{-3}\,\text{M}$, and

$\left[NH_4^+\right] = 1.28\times10^{-2}\,\text{M}$. These concentrations are consistent with the assumption.

However, if the solution is diluted to 1000. L, $\left[NH_3\right] = 7.2\times10^{-6}\,\text{M}$, and $\left[NH_3\right] = 1.28\times10^{-5}\,\text{M}$, and these two concentrations are not consistent with the assumption. Thus, in 1000. L of solution, the given quantities of NH_3 and NH_4^+ will not produce a solution with pH = 9.00. With sufficient dilution, the solution will become indistinguishable from pure water (i.e.; its pH will equal 7.00).

(c) The 0.20 mL of added 1.00 M HCl does not significantly affect the volume of the solution, but it does add $0.20\,\text{mL}\times1.00\,\text{M HCl} = 0.20\,\text{mmol H}_3O^+$. This added H_3O^+ reacts with NH_3, decreasing its amount from 7.20 mmol NH_3 to 7.00 mmol NH_3, and increasing the amount of NH_4^+ from 12.8 mmol NH_4^+ to 13.0 mmol NH_4^+, as the reaction: $NH_3 + H_3O^+ \rightarrow NH_4^+ + H_2O$

$$pH = 9.26 + \log\frac{7.00\,\text{mmol NH}_3\,/\,100.20\,\text{mL}}{13.0\,\text{mmol NH}_4^+\,/\,100.20\,\text{mL}} = 8.99$$

(d) We see in the calculation of part (c) that the total volume of the solution does not affect the pOH of the solution, at least as long as the Henderson-Hasselbalch equation is obeyed. We let x represent the number of millimoles of H_3O^+ added, through 1.00 M HCl. This increases the amount of NH_4^+ and decreases the amount of NH_3, through the reaction $NH_3 + H_3O^+ \rightarrow NH_4^+ + H_2O$

$$pH = 8.90 = 9.26 + \log\frac{7.20 - x}{12.8 + x}; \quad \log\frac{7.20 - x}{12.8 + x} = 8.90 - 9.26 = -0.36$$

Inverting, we have:

$$\frac{12.8 + x}{7.20 - x} = 10^{0.36} = 2.29; \quad 12.8 + x = 2.29(7.20 - x) = 16.5 - 2.29x$$

$$x = \frac{16.5 - 12.8}{1.00 + 2.29} = 1.1 \text{ mmol } H_3O^+$$

$$\text{vol } 1.00\,M\, HCl = 1.1\,\text{mmol } H_3O^+ \times \frac{1\,\text{mmol HCl}}{1\,\text{mmol } H_3O^+} \times \frac{1\,\text{mL soln}}{1.00\,\text{mmol HCl}} = 1.1\,\text{mL } 1.00\,M\, HCl$$

Acid–Base Indicators

27. (E)

(a) The pH color change range is 1.00 pH unit on either side of pK_{HIn}. If the pH color change range is below $pH = 7.00$, the indicator changes color in acidic solution. If it is above $pH = 7.00$, the indicator changes color in alkaline solution. If $pH = 7.00$ falls within the pH color change range, the indicator changes color near the neutral point.

Indicator	K_{HIn}	pK_{HIn}	pH Color Change Range	Changes Color in?
Bromphenol blue	1.4×10^{-4}	3.85	2.9 (yellow) to 4.9 (blue)	acidic solution
Bromcresol green	2.1×10^{-5}	4.68	3.7 (yellow) to 5.7 (blue)	acidic solution
Bromthymol blue	7.9×10^{-8}	7.10	6.1 (yellow) to 8.1 (blue)	neutral solution
2,4-Dinitrophenol	1.3×10^{-4}	3.89	2.9 (colorless) to 4.9 (yellow)	acidic solution
Chlorophenol red	1.0×10^{-6}	6.00	5.0 (yellow) to 7.0 (red)	acidic solution
Thymolphthalein	1.0×10^{-10}	10.00	9.0 (colorless) to 11.0 (blue)	basic solution

(b) If bromcresol green is green, the pH is between 3.7 and 5.7, probably about $pH = 4.7$. If chlorophenol red is orange, the pH is between 5.0 and 7.0, probably about $pH = 6.0$.

29. (M)

(a) In an acid–base titration, the pH of the solution changes sharply at a definite pH that is known prior to titration. (This pH change occurs during the addition of a very small volume of titrant.) Determining the pH of a solution, on the other hand, is more difficult because the pH of the solution is not known precisely in advance. Since each indicator only serves to fix the pH over a quite small region, often less than 2.0 pH units, several indicators—carefully chosen to span the entire range of 14 pH units—must be employed to narrow the pH to ±1 pH unit or possibly lower.

(b) An indicator is, after all, a weak acid. Its addition to a solution will affect the acidity of that solution. Thus, one adds only enough indicator to show a color change and not enough to affect solution acidity.

31. (E)

(a) 0.10 M KOH is an alkaline solution and phenol red will display its basic color in such a solution; the solution will be red.

(b) 0.10 M CH_3COOH is an acidic solution, although that of a weak acid, and phenol red will display its acidic color in such a solution; the solution will be yellow.

(c) 0.10 M NH_4NO_3 is an acidic solution due to the hydrolysis of the ammonium ion. Phenol red will display its acidic color, that is, yellow, in this solution.

(d) 0.10 M HBr is an acidic solution, the aqueous solution of a strong acid. Phenol red will display its acidic color in this solution; the solution will be yellow.

(e) 0.10 M NaCN is an alkaline solution because of the hydrolysis of the cyanide ion. Phenol red will display its basic color, red, in this solution.

(f) An equimolar acetic acid–potassium acetate buffer has $pH = pK_a = 4.74$ for acetic acid. In this solution phenol red will display its acid color, namely, yellow.

33. (M) Moles of HCl = $C \times V$ = 0.04050 M × 0.01000 L = 4.050×10^{-4} moles
Moles of $Ba(OH)_2$ at endpoint = $C \times V$ = 0.01120 M × 0.01790 L = 2.005×10^{-4} moles.
Moles of HCl that react with $Ba(OH)_2$ = 2 × moles $Ba(OH)_2$
Moles of HCl in excess 4.050×10^{-4} moles – 4.010×10^{-4} moles = $4.0\underline{5} \times 10^{-6}$ moles
Total volume at the equivalence point = (10.00 mL + 17.90 mL) = 27.90 mL

$$[HCl]_{excess} = \frac{4.05 \times 10^{-6} \text{ mole HCl}}{0.02790 \text{ L}} = 1.45 \times 10^{-4} \text{ M}; pH = -\log(1.45 \times 10^{-4}) = 3.84$$

(a) The approximate pK_{HIn} = 3.84 (generally ± 1 pH unit)

(b) This is a relatively good indicator (with ≈ 1 % of the equivalence point volume), however, pK_{Hin} is not very close to the theoretical pH at the equivalence point (pH = 7.000) For very accurate work, a better indicator is needed (i.e., bromthymol blue (pK_{Hin} = 7.1). Note: 2,4-dinitrophenol works relatively well here because the pH near the equivalence point of a strong acid/strong base titration rises very sharply (≈ 6 pH units for an addition of only 2 drops (0.10 mL)).

Neutralization Reactions

35. **(E)** The reaction (to second equiv. pt.) is:

$$H_3PO_4(aq) + 2\,KOH(aq) \longrightarrow K_2HPO_4(aq) + 2\,H_2O(l).$$

The molarity of the H_3PO_4 solution is determined in the following manner.

$$H_3PO_4 \text{ molarity} = \frac{31.15\,\text{mL KOH soln} \times \dfrac{0.2420\,\text{mmol KOH}}{1\,\text{mL KOH soln}} \times \dfrac{1\,\text{mmol}\,H_3PO_4}{2\,\text{mmol KOH}}}{25.00\,\text{mL}\,H_3PO_4\,\text{soln}} = 0.1508\,M$$

37. **(M)** Here we must determine the amount of H_3O^+ or OH^- in each solution, and the amount of excess reagent.

$$\text{amount}\ H_3O^+ = 50.00\,\text{mL} \times \frac{0.0150\,\text{mmol}\,H_2SO_4}{1\,\text{mL soln}} \times \frac{2\,\text{mmol}\,H_3O^+}{1\,\text{mmol}\,H_2SO_4} = 1.50\,\text{mmol}\,H_3O^+$$

(assuming complete ionization of H_2SO_4 and HSO_4^- in the presence of OH^-)

$$\text{amount}\ OH^- = 50.00\,\text{mL} \times \frac{0.0385\,\text{mmol NaOH}}{1\,\text{mL soln}} \times \frac{1\,\text{mmol}\,OH^-}{1\,\text{mmol NaOH}} = 1.93\,\text{mmol}\,OH^-$$

Result: Titration reaction : $OH^-(aq)\ +\ H_3O^+(aq) \rightleftharpoons 2\,H_2O(l)$

Initial amounts :   1.93 mmol   1.50 mmol

After reaction :   0.43 mmol ≈ 0 mmol

$$\left[OH^-\right] = \frac{0.43\,\text{mmol}\,OH^-}{100.0\,\text{mL soln}} = 4.3 \times 10^{-3}\,M$$

$$pOH = -\log\left(4.3 \times 10^{-3}\right) = 2.37 \qquad pH = 14.00 - 2.37 = 11.63$$

Titration Curves

39. **(M)** First we calculate the amount of HCl. The relevant titration reaction is

$$HCl(aq)\ +\ KOH(aq)\ \rightarrow\ KCl(aq)\ +\ H_2O(l)$$

$$\text{amount HCl} = 25.00\,\text{mL} \times \frac{0.160\,\text{mmol HCl}}{1\,\text{mL soln}} = 4.00\,\text{mmol HCl} = 4.00\,\text{mmol}\,H_3O^+ \text{present}$$

Then, in each case, we calculate the amount of OH^- that has been added, determine which ion, $OH^-(aq)$ or $H_3O^+(aq)$, is in excess, compute the concentration of that ion, and determine the pH.

(a) amount $OH^- = 10.00 \text{ mL} \times \dfrac{0.242 \text{ mmol } OH^-}{1 \text{ mL soln}} = 2.42 \text{ mmol } OH^-$; H_3O^+ is in excess.

$$[H_3O^+] = \dfrac{4.00 \text{ mmol } H_3O^+ - \left(2.42 \text{ mmol } OH^- \times \dfrac{1 \text{ mmol } H_3O^+}{1 \text{ mmol } OH^-}\right)}{25.00 \text{ mL originally} + 10.00 \text{ mL titrant}} = 0.0451 \text{ M}$$

$$pH = -\log(0.0451) = 1.346$$

(b) amount $OH^- = 15.00 \text{ mL} \times \dfrac{0.242 \text{ mmol } OH^-}{1 \text{ mL soln}} = 3.63 \text{ mmol } OH^-$; H_3O^+ is in excess.

$$[H_3O^+] = \dfrac{4.00 \text{ mmol } H_3O^+ - \left(3.63 \text{ mmol } OH^- \times \dfrac{1 \text{ mmol } H_3O^+}{1 \text{ mmol } OH^-}\right)}{25.00 \text{ mL originally} + 15.00 \text{ mL titrant}} = 0.00925 \text{ M}$$

$$pH = -\log(0.00925) = 2.034$$

41. **(M)** The titration reaction is $HNO_2(aq) + NaOH(aq) \rightarrow NaNO_2(aq) + H_2O(l)$

amount $HNO_2 = 25.00 \text{ mL} \times \dfrac{0.132 \text{ mmol } HNO_2}{1 \text{ mL soln}} = 3.30 \text{ mmol } HNO_2$

(a) The volume of the solution is $25.00 \text{ mL} + 10.00 \text{ mL} = 35.00 \text{ mL}$

amount $NaOH = 10.00 \text{ mL} \times \dfrac{0.116 \text{ mmol } NaOH}{1 \text{ mL soln}} = 1.16 \text{ mmol } NaOH$

1.16 mmol $NaNO_2$ are formed in this reaction and there is an excess of $(3.30 \text{ mmol } HNO_2 - 1.16 \text{ mmol } NaOH) = 2.14 \text{ mmol } HNO_2$. We can use the Henderson-Hasselbalch equation to determine the pH of the solution.

$$pK_a = -\log(7.2 \times 10^{-4}) = 3.14$$

$$pH = pK_a + \log\dfrac{[NO_2^-]}{[HNO_2]} = 3.14 + \log\dfrac{1.16 \text{ mmol } NO_2^-/35.00 \text{ mL}}{2.14 \text{ mmol } HNO_2/35.00 \text{ mL}} = 2.87$$

(b) The volume of the solution is $25.00 \text{ mL} + 20.00 \text{ mL} = 45.00 \text{ mL}$

amount $NaOH = 20.00 \text{ mL} \times \dfrac{0.116 \text{ mmol } NaOH}{1 \text{ mL soln}} = 2.32 \text{ mmol } NaOH$

2.32 mmol $NaNO_2$ are formed in this reaction and there is an excess of $(3.30 \text{ mmol } HNO_2 - 2.32 \text{ mmol } NaOH =) 0.98 \text{ mmol } HNO_2$.

$$pH = pK_a + \log\dfrac{[NO_2^-]}{[HNO_2]} = 3.14 + \log\dfrac{2.32 \text{ mmol } NO_2^-/45.00 \text{ mL}}{0.98 \text{ mmol } HNO_2/45.00 \text{ mL}} = 3.51$$

43. **(E)** The titration reaction is HA(aq) + NaOH(aq) → NaA(aq) + H$_2$O(l). The volume of titrant needed to reach the equivalence point is determined by the volume and concentration of the HA solution and the concentration of the NaOH solution, irrespective of whether HA is a strong acid or a weak acid. The volume of NaOH(aq) required to reach the equivalence point is

$$V_{NaOH} = 25.00 \text{ mL} \times \frac{0.100 \text{ mmol HA}}{1 \text{ mL}} \times \frac{1 \text{ mmol NaOH}}{1 \text{ mmol HA}} \times \frac{1 \text{ mL}}{0.100 \text{ mmol NaOH}} = 25.00 \text{ mL}$$

The pH at the equivalence point, however, is determined by the nature of the salt NaA, and especially by the nature of A$^-$. If HA is a strong acid, then A$^-$ is a very weak base that will not hydrolyze in water. Consequently, for the titration of a strong acid with NaOH, the solution will be neutral at the equivalence point. If HA is a weak acid, then A$^-$ is a weak base that will hydrolyze to produce OH$^-$ ions: A$^-$ + H$_2$O $\rightleftharpoons$ HA + OH$^-$. Therefore, for the titration of a weak acid with NaOH, the solution will be basic at the equivalence point.

45. **(D)**

    (a)   Initial $[OH^-] = 0.100$ M OH$^-$ $pOH = -\log(0.100) = 1.000$ $pH = 13.00$

Since this is the titration of a strong base with a strong acid, KI is the solute present at the equivalence point, and since KI is a neutral salt, the pH = 7.00. The titration reaction is:

$$KOH(aq) + HI(aq) \longrightarrow KI(aq) + H_2O(l)$$

$$V_{HI} = 25.0 \text{ mL KOH soln} \times \frac{0.100 \text{ mmol KOH}}{1 \text{ mL soln}} \times \frac{1 \text{ mmol HI}}{1 \text{ mmol KOH}} \times \frac{1 \text{ mL HI soln}}{0.200 \text{ mmol HI}}$$

$$= 12.5 \text{ mL HI soln}$$

Initial amount of KOH present = 25.0 mL KOH soln × 0.100 M = 2.50 mmol KOH
At the 40% titration point: 5.00 mL HI soln × 0.200 M HI = 1.00 mmol HI
excess KOH = 2.50 mmol KOH − 1.00 mmol HI = 1.50 mmol KOH

$$[OH^-] = \frac{1.50 \text{ mmol KOH}}{30.0 \text{ mL total}} \times \frac{1 \text{ mmol OH}^-}{1 \text{ mmol KOH}} = 0.0500 \text{ M} \quad pOH = -\log(0.0500) = 1.30$$
$$pH = 14.00 - 1.30 = 12.70$$

At the 80% titration point: 10.00 mL HI soln × 0.200 M HI = 2.00 mmol HI
excess KOH = 2.50 mmol KOH − 2.00 mmol HI = 0.50 mmol KOH

$$[OH^-] = \frac{0.50 \text{ mmol KOH}}{35.0 \text{ mL total}} \times \frac{1 \text{ mmol OH}^-}{1 \text{ mmol KOH}} = 0.0143 \text{ M} \quad pOH = -\log(0.0143) = 1.84$$
$$pH = 14.00 - 1.84 = 12.16$$

At the 110% titration point: 13.75 mL HI soln × 0.200 M HI = 2.75 mmol HI
excess HI = 2.75 mmol HI − 2.50 mmol HI = 0.25 mmol HI

$$[H_3O^+] = \frac{0.25 \text{ mmol HI}}{38.8 \text{ mL total}} \times \frac{1 \text{ mmol H}_3O^+}{1 \text{ mmol HI}} = 0.0064 \text{ M}; \quad pH = -\log(0.0064) = 2.19$$

Since the pH changes very rapidly at the equivalence point, from about $pH = 10$ to about $pH = 4$, most of the indicators in Figure 17-8 can be used. The main exceptions are alizarin yellow R, bromphenol blue, thymol blue (in its acid range), and methyl violet.

(b) *Initial pH*:

Since this is the titration of a weak base with a strong acid, NH_4Cl is the solute present at the equivalence point, and since NH_4^+ is a weakly acidic cation, the pH should be slightly lower than 7. The titration reaction is:

Equation: $\quad NH_3(aq) + H_2O(l) \rightleftharpoons NH_4^+(aq) + OH^-(aq)$

Initial:	$1.00\,M$	$-$	$0\,M$	$\approx 0\,M$
Changes:	$-x\,M$	$-$	$+x\,M$	$+x\,M$
Equil:	$(1.00-x)\,M$	$-$	$x\,M$	$x\,M$

$$K_b = \frac{\left[NH_4^+\right]\left[OH^-\right]}{\left[NH_3\right]} = 1.8\times10^{-5} = \frac{x^2}{1.00-x} \approx \frac{x^2}{1.00}$$

($x \ll 1.0$, thus the approximation is valid)

$$x = 4.2\times10^{-3}\left[OH^-\right] = 4.2\times10^{-3}\,M, \quad pOH = -\log\left(4.2\times10^{-3}\right) = 2.38,$$

$$pH = 14.00 - 2.38 = 11.62 = \text{initial pH}$$

Volume of titrant: $\quad NH_3 + HCl \longrightarrow NH_4Cl + H_2O$

$$V_{HCl} = 10.0\,mL \times \frac{1.00\,mmol\,NH_3}{1\,mL\,soln} \times \frac{1\,mmol\,HCl}{1\,mmol\,NH_3} \times \frac{1\,mL}{0.250\,mmol\,HCl} = 40.0\,mL$$

pH at equivalence point: The total solution volume at the equivalence point is $(10.0 + 40.0)\,mL = 50.0\,mL$

Also at the equivalence point, all of the NH_3 has reacted to form NH_4^+. It is this NH_4^+ that hydrolyzes to determine the pH of the solution.

$$\left[NH_4^+\right] = \frac{10.0\,mL \times \dfrac{1.00\,mmol\,NH_3}{1\,mL\,soln} \times \dfrac{1\,mmol\,NH_4^+}{1\,mmol\,NH_3}}{50.0\,mL\,\text{total solution}} = 0.200\,M$$

Equation: $\quad NH_4^+(aq) + H_2O(l) \rightleftharpoons NH_3(aq) + H_3O^+(aq)$

Initial:	$0.200\,M$	$-$	$0\,M$	$\approx 0\,M$
Changes:	$-x\,M$	$-$	$+x\,M$	$+x\,M$
Equil:	$(0.200-x)\,M$	$-$	$x\,M$	$x\,M$

$$K_a = \frac{K_w}{K_b} = \frac{1.0 \times 10^{-14}}{1.8 \times 10^{-5}} = \frac{[NH_3][H_3O^+]}{[NH_3]} = \frac{x^2}{0.200 - x} \approx \frac{x^2}{0.200}$$

($x \ll 0.200$, thus the approximation is valid)

$$x = 1.1 \times 10^{-5}; \quad [H_3O^+] = 1.1 \times 10^{-5}\,M; \quad pH = -\log(1.1 \times 10^{-5}) = 4.96$$

Of the indicators in Figure 17-8, one that has the pH of the equivalence point within its pH color change range is methyl red (yellow at $pH = 6.2$ and red at $pH = 4.5$); Bromcresol green would be another choice. At the 50% titration point, $[NH_3] = [NH_4^+]$ and $pOH = pK_b = 4.74$ $\quad pH = 14.00 - 4.74 = 9.26$

The titration curves for parts **(a)** and **(b)** follow.

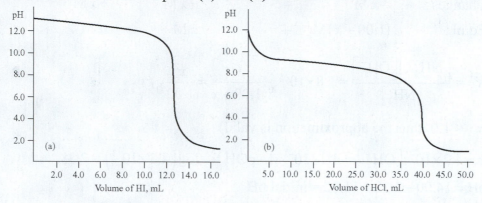

47. **(M)** A pH greater than 7.00 in the titration of a strong base with a strong acid means that the base is not completely titrated. A pH less than 7.00 means that excess acid has been added.

(a) We can determine $[OH^-]$ of the solution from the pH. $[OH^-]$ is also the quotient of the amount of hydroxide ion in excess divided by the volume of the solution: 20.00 mL base $+x$ mL added acid.

$$pOH = 14.00 - pH = 14.00 - 12.55 = 1.45 \quad [OH^-] = 10^{-pOH} = 10^{-1.45} = 0.035\,M$$

$$[OH^-] = \frac{\left(20.00\,mL\,base \times \dfrac{0.175\,mmol\,OH^-}{1\,mL\,base}\right) - \left(x\,mL\,acid \times \dfrac{0.200\,mmol\,H_3O^+}{1\,mL\,acid}\right)}{20.00\,mL + x\,mL} = 0.035\,M$$

$$3.50 - 0.200\,x = 0.70 + 0.035x; \quad 3.50 - 0.70 = 0.035\,x + 0.200\,x; \quad 2.80 = 0.235\,x$$

$$x = \frac{2.80}{0.235} = 11.9\,mL\,acid\,added.$$

(b) The set-up here is the same as for part (a).

$$pOH = 14.00 - pH = 14.00 - 10.80 = 3.20 \qquad \left[OH^-\right] = 10^{-pOH} = 10^{-3.20} = 0.00063\,M$$

$$[OH^-] = \frac{\left(20.00\,mL\,base \times \dfrac{0.175\,mmol\,OH^-}{1\,mL\,base}\right) - \left(x\,mL\,acid \times \dfrac{0.200\,mmol\,H_3O^+}{1\,mL\,acid}\right)}{20.00\,mL + x\,mL}$$

$$[OH^-] = 0.00063\,\frac{mmol}{mL} = 0.00063\,M$$

$$3.50 - 0.200x = 0.0126 + 0.00063x; \quad 3.50 - 0.0126 = 0.00063\,x + 0.200\,x; \quad 3.49 = 0.201\,x$$

$$x = \frac{3.49}{0.201} = 17.4\,mL\ \text{acid added. This is close to the equivalence point at } 17.5\,mL.$$

(c) Here the acid is in excess, so we reverse the set-up of part (a). We are just slightly beyond the equivalence point. This is close to the "mirror image" of part (b).

$$\left[H_3O^+\right] = 10^{-pH} = 10^{-4.25} = 0.000056\ M$$

$$[H_3O^+] = \frac{\left(x\,mL\,acid \times \dfrac{0.200\ mmol\ H_3O^+}{1\,mL\,acid}\right) - \left(20.00\ mL\ base \times \dfrac{0.175\ mmol\ OH^-}{1\,mL\,base}\right)}{20.00\ mL + x\ mL}$$

$$= 5.6 \times 10^{-5}\ M$$

$$0.200\,x - 3.50 = 0.0011 + 5.6 \times 10^{-5}\,x; \quad 3.50 + 0.0011 = -5.6 \times 10^{-5}\,x + 0.200\,x;$$

$$3.50 = 0.200\,x$$

$$x = \frac{3.50}{0.200} = 17.5\,mL\ \text{acid added, which is the equivalence point for this titration.}$$

49. **(D)** For each of the titrations, the pH at the half-equivalence point equals the pK_a of the acid.

The initial pH is that of 0.1000 M weak acid: $K_a = \dfrac{x^2}{0.1000 - x} \qquad x = [H_3O^+]$

x must be found using the quadratic formula roots equation unless the approximation is valid.

One method of determining if the approximation will be valid is to consider the ratio C_a/K_a. If the value of C_a/K_a is greater than 1000, the assumption should be valid, however, if the value of C_a/K_a is less than 1000, the quadratic should be solved exactly (i.e., the 5% rule will not be satisfied).

The pH at the equivalence point is that of 0.05000 M anion of the weak acid, for which the $\left[OH^-\right]$ is determined as follows.

$$K_b = \frac{K_w}{K_a} \approx \frac{x^2}{0.05000} \qquad x = \sqrt{\frac{K_w}{K_a} 0.05000} = [OH^-]$$

We can determine the pH at the quarter and three quarter of the equivalence point by using the Henderson-Hasselbalch equation (effectively ± 0.48 pH unit added to the pK_a).

And finally, when 0.100 mL of base has been added beyond the equivalence point, the pH is determined by the excess added base, as follows (for all three titrations).

$$\left[OH^-\right] = \frac{0.100\,mL \times \dfrac{0.1000\,mmol\,NaOH}{1\,mL\,NaOH\,soln} \times \dfrac{1\,mmol\,OH^-}{1\,mmol\,NaOH}}{20.1\ mL\ soln\ total} = 4.98 \times 10^{-4}\ M$$

$$pOH = -\log(4.98 \times 10^{-4}) = 3.303 \qquad pH = 14.000 - 3.303 = 10.697$$

(a) $C_a/K_a = 14.3$; thus, the approximation is <u>not</u> valid and the full quadratic equation must be solved.

Initial: From the roots equation $x = [H_3O^+] = 0.023\,M \qquad$ pH=1.63

Half equivalence point: pH $= pK_a = 2.15$

pH at quarter equivalence point $= 2.15 - 0.48 = 1.67$

pH at three quarter equivalence point $= 2.15 + 0.48 = 2.63$

Equiv: $x = \left[OH^-\right] = \sqrt{\dfrac{1.0 \times 10^{-14}}{7.0 \times 10^{-3}} \times 0.05000} = 2.7 \times 10^{-7} \qquad \begin{array}{l} pOH = 6.57 \\ pH = 14.00 - 6.57 = 7.43 \end{array}$

Indicator: bromthymol blue, yellow at pH $= 6.2$ and blue at pH $= 7.8$

(b) $C_a/K_a = 333$; thus, the approximation is <u>not</u> valid and the full quadratic equation must be solved.

Initial: From the roots equation $x = [H_3O^+] = 0.0053\,M \qquad$ pH=2.28

Half equivalence point: pH $= pK_a = 3.52$

pH at quarter equivalence point $= 3.52 - 0.48 = 3.04$

pH at three quarter equivalence point $= 3.52 + 0.48 = 4.00$

Equiv: $x = \left[OH^-\right] = \sqrt{\dfrac{1.0 \times 10^{-14}}{3.0 \times 10^{-4}} \times 0.05000} = 1.3 \times 10^{-6}$

pOH $= 5.89$

pH $= 14.00 - 5.89 = 8.11$

Indicator: thymol blue, yellow at pH $= 8.0$ and blue at pH $= 9.6$

(c) $C_a/K_a = 5 \times 10^6$; thus, the approximation <u>is</u> valid.

Initial: $[H_3O^+] = \sqrt{0.1000 \times 2.0 \times 10^{-8}} = 0.000045\ M$ pH $= 4.35$

pH at quarter equivalence point $= 4.35 - 0.48 = 3.87$

pH at three quarter equivalence point $= 4.35 + 0.48 = 4.83$

Half equivalence point: pH $= pK_a = 7.70$

$$\text{Equiv: } x = \left[OH^-\right] = \sqrt{\frac{1.0 \times 10^{-14}}{2.0 \times 10^{-8}} \times 0.0500} = 1.6 \times 10^{-4}$$

pOH = 3.80

pH = 14.00 − 3.80 = 10.20

Indicator: alizarin yellow R, yellow at pH = 10.0 and violet at pH = 12.0

The three titration curves are drawn with respect to the same axes in the diagram below.

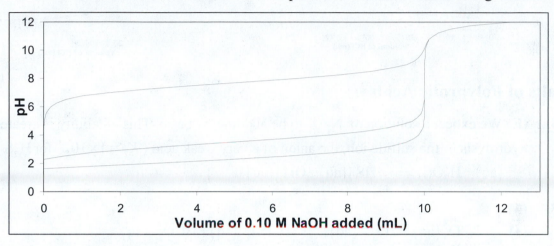

51. **(D)** 25.00 mL of 0.100 M NaOH is titrated with 0.100 M HCl

(i) Initial pOH for 0.100 M NaOH: $\left[OH^-\right]$ = 0.100 M, pOH = 1.000 or pH = 13.000

(ii) After addition of 24 mL: $[NaOH] = 0.100 \text{ M} \times \dfrac{25.00 \text{ mL}}{49.00 \text{ mL}} = 0.0510 \text{ M}$

$$[HCl] = 0.100 \text{ M} \times \frac{24.00 \text{ mL}}{49.00 \text{ mL}} = 0.0490 \text{ M}$$

NaOH is in excess by 0.0020 M = $\left[OH^-\right]$ pOH = 2.70

(iii) At the equivalence point (25.00 mL), the pOH should be 7.000 and pH = 7.000

(iv) After addition of 26 mL: $[NaOH] = 0.100\text{M} \times \dfrac{25.00}{51.00} = 0.0490 \text{ M}$

$$[HCl] = 0.100 \text{ M} \times \frac{26.00 \text{ mL}}{51.00 \text{ mL}} = 0.0510 \text{ M}$$

HCl is in excess by 0.0020 M = $\left[H_3O^+\right]$ pH = 2.70 or pOH = 11.30

(v) After addition of 33.00 mL HCl(xs) $[NaOH] = 0.100 \text{ M} \times \dfrac{25.00 \text{ mL}}{58.00 \text{ mL}} = 0.0431 \text{ M}$

$[HCl] = 0.100 \text{ M} \times \dfrac{33.00 \text{ mL}}{58.00 \text{ mL}} = 0.0569 \text{ M}$ $[HCl]_{excess}$ = 0.0138 M pH = 1.860, pOH = 12.140

The graphs look to be mirror images of one another. In reality, one must reflect about a horizontal line centered at pH or pOH = 7 to obtain the other curve.

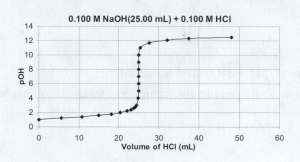
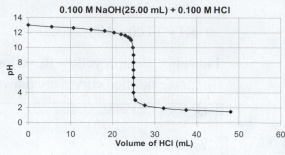

Salts of Polyprotic Acids

53. (E) We expect a solution of Na_2S to be alkaline, or basic. This alkalinity is created by the hydrolysis of the sulfide ion, the anion of a very weak acid ($K_2 = 1 \times 10^{-19}$ for H_2S).

$$S^{2-}(aq) + H_2O(l) \rightleftharpoons HS^-(aq) + OH^-(aq)$$

55. (M)

(a) $H_3PO_4(aq) + CO_3^{2-}(aq) \rightleftharpoons H_2PO_4^-(aq) + HCO_3^-(aq)$

$H_2PO_4^-(aq) + CO_3^{2-}(aq) \rightleftharpoons HPO_4^{2-}(aq) + HCO_3^-(aq)$

$HPO_4^{2-}(aq) + OH^-(aq) \rightleftharpoons PO_4^{3-}(aq) + H_2O(l)$

(b) The pH values of 1.00 M solutions of the three ions are;

1.0 M $OH^- \rightarrow$ pH = 14.00   1.0 M $CO_3^{2-} \rightarrow$ pH = 12.16   1.0 M $PO_4^{3-} \rightarrow$ pH = 13.15

Thus, we see that CO_3^{2-} is not a strong enough base to remove the third proton from H_3PO_4. As an alternative method of solving this problem, we can compute the equilibrium constants of the reactions of carbonate ion with H_3PO_4, $H_2PO_4^-$ and HPO_4^{2-}.

$$H_3PO_4 + CO_3^{2-} \rightleftharpoons H_2PO_4^- + HCO_3^- \quad K = \frac{K_{a1}\{H_3PO_4\}}{K_{a2}\{H_2CO_3\}} = \frac{7.1 \times 10^{-3}}{4.7 \times 10^{-11}} = 1.5 \times 10^8$$

$$H_2PO_4^- + CO_3^{2-} \rightleftharpoons HPO_4^{2-} + HCO_3^- \quad K = \frac{K_{a2}\{H_3PO_4\}}{K_{a2}\{H_2CO_3\}} = \frac{6.3 \times 10^{-8}}{4.7 \times 10^{-11}} = 1.3 \times 10^3$$

$$HPO_4^{2-} + CO_3^{2-} \rightleftharpoons PO_4^{3-} + HCO_3^- \quad K = \frac{K_{a3}\{H_3PO_4\}}{K_{a2}\{H_2CO_3\}} = \frac{4.2 \times 10^{-13}}{4.7 \times 10^{-11}} = 8.9 \times 10^{-3}$$

Since the equilibrium constant for the third reaction is much smaller than 1.00, we conclude that it proceeds to the right to only a negligible extent and thus is not a practical method of producing PO_4^{3-}. The other two reactions have large equilibrium constants, and products are expected to strongly predominate. They have the advantage of involving an inexpensive base and, even if they do not go to completion, they will be drawn to completion by reaction with OH^- in the last step of the process.

57. **(M)** Malonic acid has the formula $H_2C_3H_2O_4$ MM = 104.06 g/mol

$$\text{Moles of } H_2C_3H_2O_4 = 19.5 \text{ g} \times \frac{1 \text{ mol}}{104.06 \text{ g}} = 0.187 \text{ mol}$$

$$\text{Concentration of } H_2C_3H_2O_4 = \frac{\text{moles}}{V} = \frac{0.187 \text{ moles}}{0.250 \text{ L}} = 0.748 \text{ M}$$

The second proton that can dissociate has a negligible effect on pH (K_{a_2} is very small). Thus the pH is determined almost entirely by the first proton loss.

	$H_2A(aq)$	+	$H_2O(l)$	$\rightleftharpoons$	$HA^-(aq)$	+	$H_3O^+(aq)$
Initial	0.748 M		–		0 M		≈ 0 M
Change	$-x$		–		$+x$		$+x$
Equil	$0.748-x$		–		x		x

So, $x = \dfrac{x^2}{0.748-x} = K_a$; pH = 1.47, therefore, $[H_3O^+] = 0.034$ M $= x$,

$$K_{a_1} = \frac{(0.034)^2}{0.748-0.034} = 1.6 \times 10^{-3}$$

(1.5×10^{-3} in tables, difference owing to ionization of the second proton)

	$HA^-(aq)$	+	$H_2O(l)$	$\rightleftharpoons$	$A^{2-}(aq)$	+	$H_3O^+(aq)$
Initial	0.300 M		–		0 M		≈ 0 M
Change	$-x$		–		$+x$		$+x$
Equil	$0.300-x$		–		x		x

pH = 4.26, therefore, $[H_3O^+] = 5.5 \times 10^{-5}$ M $= x$, $K_{a_2} = \dfrac{(5.5\times10^{-5})^2}{0.300-5.5\times10^{-5}} = 1.0 \times 10^{-8}$

General Acid–Base Equilibria

59. **(E)**

(a) $Ba(OH)_2$ is a strong base.

$$pOH = 14.00 - 11.88 = 2.12 \quad [OH^-] = 10^{-2.12} = 0.0076 \text{ M}$$

$$\left[Ba(OH)_2\right]_{initial} = \frac{0.0076 \text{ mol } OH^-}{1 \text{ L}} \times \frac{1 \text{ mol } Ba(OH)_2}{2 \text{ mol } OH^-} = 0.0038 \text{ M}$$

(b) $pH = 4.52 = pK_a + \log\dfrac{\left[CH_3COO^-\right]}{\left[CH_3COOH\right]} = 4.74 + \log\dfrac{0.294 \text{ M}}{\left[CH_3COOH\right]}$

$$\log\frac{0.294 \text{ M}}{\left[CH_3COOH\right]} = 4.52 - 4.74 \qquad \frac{0.294 \text{ M}}{\left[CH_3COOH\right]} = 10^{-0.22} = 0.60$$

$$\left[CH_3COOH\right] = \frac{0.294 \text{ M}}{0.60} = 0.49 \text{ M}$$

61. (M)

(a) A solution can be prepared with equal concentrations of weak acid and conjugate base (it would be a buffer, with a buffer ratio of 1.00, where the pH = pK_a = 9.26). Clearly, this solution can be prepared, however, it would not have a pH of 6.07.

(b) These solutes can be added to the same solution, but the final solution will have an appreciable $[CH_3COOH]$ because of the reaction of $H_3O^+(aq)$ with $CH_3COO^-(aq)$

Equation:　　$H_3O^+(aq)$　+　$CH_3COO^-(aq)$　$\rightleftharpoons$　$CH_3COOH(aq) + H_2O(l)$

Initial	0.058 M	0.10 M	0 M　　–
Changes:	–0.058 M	–0.058 M	+0.058 M　　–
Equil:	≈0.000 M	0.04 M	0.058 M　　–

Of course, some H_3O^+ will exist in the final solution, but not equivalent to 0.058 M HI.

(c) Both 0.10 M KNO_2 and 0.25 M KNO_3 can exist together. Some hydrolysis of the $NO_2^-(aq)$ ion will occur, forming $HNO_2(aq)$ and $OH^-(aq)$.

(d) $Ba(OH)_2$ is a strong base and will react as much as possible with the weak conjugate acid NH_4^+, to form $NH_3(aq)$. We will end up with a solution of $BaCl_2(aq)$, $NH_3(aq)$, and unreacted $NH_4Cl(aq)$.

(e) This will be a benzoic acid–benzoate ion buffer solution. Since the two components have the same concentration, the buffer solution will have $pH = pK_a = -\log(6.3 \times 10^{-5}) = 4.20$. This solution can indeed exist.

(f) The first three components contain no ions that will hydrolyze. But CH_3COO^- is the anion of a weak acid and will hydrolyze to form a slightly basic solution. Since pH = 6.4 is an acidic solution, the solution described cannot exist.

INTEGRATIVE AND ADVANCED EXERCISES

63. (M)

(a) $NaHSO_4(aq) + NaOH(aq) \longrightarrow Na_2SO_4(aq) + H_2O(l)$

$HSO_4^-(aq) + OH^-(aq) \longrightarrow SO_4^{2-}(aq) + H_2O(l)$

(b) We first determine the mass of NaHSO₄.

$$\text{mass NaHSO}_4 = 36.56 \text{ mL} \times \frac{1 \text{ L}}{1000 \text{ mL}} \times \frac{0.225 \text{ mol NaOH}}{1 \text{ L}} \times \frac{1 \text{ mol NaHSO}_4}{1 \text{ mol NaOH}}$$

$$\times \frac{120.06 \text{ g NaHSO}_4}{1 \text{ mol NaHSO}_4} = 0.988 \text{ g NaHSO}_4$$

$$\% \text{ NaCl} = \frac{1.016 \text{ g sample} - 0.988 \text{ g NaHSO}_4}{1.016 \text{ g sample}} \times 100\% = 2.8\% \text{ NaCl}$$

(c) At the endpoint of this titration the solution is one of predominantly SO_4^{2-}, from which the pH is determined by hydrolysis. Since K_a for HSO_4^- is relatively large (1.1×10^{-2}), base hydrolysis of SO_4^{2-} should not occur to a very great extent. The pH of a neutralized solution should be very nearly 7, and most of the indicators represented in Figure 17-8 would be suitable. A more exact solution follows.

$$[SO_4^{2-}] = \frac{0.988 \text{ g } NaHSO_4}{0.03656 \text{ L}} \times \frac{1 \text{ mol } NaHSO_4}{120.06 \text{ g } NaHSO_4} \times \frac{1 \text{ mol } SO_4^{2-}}{1 \text{ mol } NaHSO_4} = 0.225 \text{ M}$$

Equation: SO_4^{2-} (aq) $+$ H_2O(l) $\rightleftharpoons$ HSO_4^- (aq) $+$ OH^- (aq)

Initial: 0.225 M $-$ 0 M ≈ 0 M

Changes: $-x$ M $-$ $+x$ M $+x$ M

Equil: $(0.225 - x)$ M $-$ x M x M

$$K_b = \frac{[HSO_4^-][OH^-]}{[SO_4^{2-}]} = \frac{K_w}{K_{a_2}} = \frac{1.0 \times 10^{-14}}{1.1 \times 10^{-2}} = 9.1 \times 10^{-13} = \frac{x \cdot x}{0.225 - x} \approx \frac{x^2}{0.225}$$

$$[OH^-] = \sqrt{9.1 \times 10^{-13} \times 0.225} = 4.5 \times 10^{-7} \text{ M}$$

(the approximation was valid since $x \ll 0.225$ M)

$$pOH = -\log(4.5 \times 10^{-7}) = 6.35 \qquad pH = 14.00 - 6.35 = 7.65$$

Thus, either bromthymol blue (pH color change range from pH = 6.1 to pH = 7.9) or phenol red (pH color change range from pH = 6.4 to pH = 8.0) would be a suitable indicator, since either changes color at pH = 7.65.

64. **(D)** The original solution contains

$$250.0 \text{ mL} \times \frac{0.100 \text{ mmol } CH_3CH_2COOH}{1 \text{ mL soln}} = 25.0 \text{ mmol } CH_3CH_2COOH$$

$pK_a = -\log(1.35 \times 10^{-5}) = 4.87$ We let V be the volume added to the solution, in mL.

(a) Since we add V mL of HCl solution (and each mL adds 1.00 mmol H_3O^+ to the solution), we have added V mmol H_3O^+ to the solution. Now the final $[H_3O^+] = 10^{-1.00} = 0.100$ M.

$$[H_3O^+] = \frac{V \text{ mmol } H_3O^+}{(250.0 + V) \text{ mL}} = 0.100 \text{ M}$$

$$V = 25.0 + 0.100\,V, \text{ therefore, } V = \frac{25.0}{0.900} = 27.8 \text{ mL added}$$

Now, we check our assumptions. The total solution volume is 250.0 mL + 27.8 mL = 277.8 mL There are 25.0 mmol CH_3CH_2COOH present before equilibrium is established,
and 27.8 mmol H_3O^+ also.

$$[CH_3COOH] = \frac{25.0 \text{ mmol}}{277.8 \text{ mL}} = 0.0900 \text{ M} \qquad [H_3O^+] = \frac{27.8 \text{ mmol}}{277.8 \text{ mL}} = 0.100 \text{ M}$$

Equation: $CH_3CH_2COOH(aq) + H_2O(l) \rightleftharpoons CH_3CH_2COO^-(aq) + H_3O^+(aq)$

Initial: 0.0900 M   —   0 M   0.100 M

Changes: $-x$ M   — $+x$ M $+x$ M

Equil: $(0.08999 - x)$M   — x M $(0.100 + x)$M

$$K_a = \frac{[H_3O^+][CH_3CH_2COO^-]}{[CH_3CH_2COOH]} = 1.35 \times 10^{-5} = \frac{x(0.100 + x)}{0.0900 - x} \approx \frac{0.100\ x}{0.0900} \quad x = 1.22 \times 10^{-5} M$$

The assumption used in solving this equilibrium situation, that $x \ll 0.0900$ clearly is correct. In addition, the tacit assumption that virtually all of the H_3O^+ comes from the HCl also is correct.

(b) Let V_{soln} be the volume of 1.00 M $NaCH_3CH_2COO$ required. Since the pH desired is within 1.00 pH unit of the pK_a, we can use the Henderson-Hasselbalch equation to find the required buffer ratio.

$$pH = pK_a + \frac{[A^-]}{[HA]} = pK_a + \log \frac{n_{A^-}/V}{n_{HA}/V} = pK_a + \log \frac{n_{A^-}}{n_{HA}} = 4.00 = 4.87 + \log \frac{n_{A^-}}{n_{HA}}$$

$$\log \frac{n_{A^-}}{n_{HA}} = -0.87 \qquad \frac{n_{A^-}}{n_{HA}} = 10^{-0.87} = 0.13 = \frac{n_{A^-}}{25.00} \qquad n_{A^-} = 3.3$$

$$V = 3.3 \text{ mmol } CH_3CH_2COO^- \times \frac{1 \text{ mmol } NaCH_3CH_2COO}{1 \text{ mmol } CH_3CH_2COO^-} \times \frac{1 \text{ mL soln}}{1.00 \text{ mmol } NaCH_3CH_2COO}$$

$$= 3.3 \text{ mL added}$$

We have assumed that all of the $CH_3CH_2COO^-$ is obtained from the $NaCH_3CH_2COO$ solution, since the addition of that ion in the solution should suppress the ionization of CH_3CH_2COOH.

(c) We let V be the final volume of the solution.

Equation: $CH_3CH_2COOH(aq) + H_2O(l) \rightleftharpoons CH_3CH_2COO^-(aq) + H_3O^+(aq)$

Initial: $25.0/V$ —   0 M $\approx$ 0 M

Changes: $-x/V$ M   — $+x/V$ M $+x/V$ M

Equil: $(25.0/V - x/V)$ M   — x/V M x/V M

$$K_a = \frac{[H_3O^+][CH_3CH_2COO^-]}{[CH_3CH_2COOH]} = 1.35 \times 10^{-5} = \frac{(x/V)^2}{25.0/V - x/V} \approx \frac{x^2/V}{25.0}$$

When $V = 250.0$ mL, $x = 0.29$ mmol H_3O^+ $[H_3O^+] = \frac{0.29 \text{ mmol}}{250.0 \text{ mL}} = 1.2 \times 10^{-3}$ M

$pH = -\log(1.2 \times 10^{-3}) = 2.92$

An increase of 0.15 pH unit gives pH = 2.92 + 0.15 = 3.07

$[H_3O^+] = 10^{-3.07} = 8.5 \times 10^{-4}$ M   This is the value of x/V. Now solve for V.

$$1.3 \times 10^{-5} = \frac{(8.5 \times 10^{-4})^2}{25.0/V - 8.5 \times 10^{-4}} \qquad 25.0/V - 8.5 \times 10^{-4} = \frac{(8.5 \times 10^{-4})^2}{1.3 \times 10^{-5}} = 0.056$$

$$25.0/V = 0.056 + 0.00085 = 0.057 \qquad V = \frac{25.0}{0.057} = 4.4 \times 10^2 \text{ mL}$$

On the other hand, if we had used $[H_3O^+] = 1.16 \times 10^{-3}$ M (rather than 1.2×10^{-3} M), we would obtain $V = 4.8 \times 10^2$ mL. The answer to the problem thus is sensitive to the last significant figure that is retained. We obtain $V = 4.6 \times 10^2$ mL, requiring the addition of 2.1×10^2 mL of H_2O.

Another possibility is to recognize that $[H_3O^+] = \sqrt{K_a C_a}$ for a weak acid with ionization constant K_a and initial concentration C_a if the approximation is valid. If C_a is changed to $C_a/2$, $[H_3O^+] = \sqrt{K_a/C_a} \times \sqrt{2}/2$. Since, pH $= -\log[H_3O^+]$, the change in pH given by: ΔpH $= -\log\sqrt{2} = -\log 2^{1/2} = -0.5 \log 2 = -0.5 \times 0.30103 = -0.15$. This corresponds to doubling the solution volume, that is, to adding 250 mL water. Diluted by half with water, $[H_3O^+]$ goes down and pH rises, then $(\text{pH}_1 - \text{pH}_2) < 0$.

66. (E)

(a) At a pH = 2.00, in Figure 17-8 the pH is changing gradually with added NaOH. There would be no sudden change in color with the addition of a small volume of NaOH.

(b) At pH = 2.0 in Figure 17-8, approximately 20.5 mL have been added. Since, equivalence required the addition of 25.0 mL, there are 4.5 mL left to add.

Therefore,

$$\% \text{ HCl unneutralized} = \frac{4.5}{25.0} \times 100\% = 18\%$$

69. (M)

(a) We concentrate on the ratio of concentrations of which the logarithm is taken.

$$\frac{[\text{conjugate base}]_{eq}}{[\text{weak acid}]_{eq}} = \frac{\text{equil. amount conj. base}}{\text{equil. amount weak acid}} = \frac{f \times \text{initial amt. weak acid}}{(1-f) \times \text{initial amt. weak acid}} = \frac{f}{1-f}$$

The first transformation is the result of realizing that the volume in which the weak acid and its conjugate base are dissolved is the same volume, and therefore the ratio of equilibrium amounts is the same as the ratio of concentrations. The second transformation is the result of realizing, for instance, that if 0.40 of the weak acid has been titrated, 0.40 of the original amount of weak acid now is in the form of its conjugate base, and 0.60 of that amount remains as weak acid. Equation 17.2 then is: pH $= \text{p}K_a + \log(f/(1-f))$

(b) We use the equation just derived. $\text{pH} = 10.00 + \log\dfrac{0.27}{(1 - 0.23)} = 9.56$

70. **(M)**

(a) $pH = pK_{a2} + \log\dfrac{[HPO_4^{2-}]}{[H_2PO_4^-]} = 7.20 + \log\dfrac{[HPO_4^{2-}]}{[H_2PO_4^-]} = 7.40 \qquad \dfrac{[HPO_4^{2-}]}{[H_2PO_4^-]} = 10^{+0.20} = 1.6$

(b) In order for the solution to be isotonic, it must have the same concentration of ions as does the isotonic NaCl solution.

$$[ions] = \dfrac{9.2\ g\ NaCl}{1\ L\ soln} \times \dfrac{1\ mol\ NaCl}{58.44\ g\ NaCl} \times \dfrac{2\ mol\ ions}{1\ mol\ NaCl} = 0.31\ M$$

Thus, 1.00 L of the buffer must contain 0.31 moles of ions. The two solutes that are used to formulate the buffer both ionize: KH_2PO_4 produces 2 mol of ions (K^+ and $H_2PO_4^-$) per mole of solute, while $Na_2HPO_4 \cdot 12\ H_2O$ produces 3 mol of ions (2 Na^+ and HPO_4^{2-}) per mole of solute. We let x = amount of KH_2PO_4 and y = amount of $Na_2HPO_4 \cdot 12\ H_2O$.

$$\dfrac{y}{x} = 1.6 \quad \text{or} \quad y = 1.6x \qquad 2x + 3y = 0.31 = 2x + 3(1.6x) = 6.8x$$

$$x = \dfrac{0.31}{6.8} = 0.046\ mol\ KH_2PO_4 \qquad y = 1.6 \times 0.046 = 0.074\ mol\ Na_2HPO_4 \cdot 12\ H_2O$$

$$mass\ KH_2PO_4 = 0.046\ mol\ KH_2PO_4 \times \dfrac{136.08\ g\ KH_2PO_4}{1\ mol\ KH_2PO_4} = 6.3\ g\ KH_2PO_4$$

$$mass\ of\ Na_2HPO_4 \cdot 12H_2O = 0.074\ mol\ Na_2HPO_4 \cdot 12H_2O \times \dfrac{358.1\ g\ Na_2HPO_4 \cdot 12H_2O}{1\ mol\ Na_2HPO_4 \cdot 12H_2O}$$

$$= 26\ g\ Na_2HPO_4 \cdot 12H_2O$$

71. **(M)** A solution of NH_4Cl should be acidic, hence, we should add an alkaline solution to make it pH neutral. We base our calculation on the ionization equation for $NH_3(aq)$, and assume that little $NH_4^+(aq)$ is transformed to $NH_3(aq)$ because of the inhibition of that reaction by the added $NH_3(aq)$, and because the added volume of $NH_3(aq)$ does not significantly alter the V_{total}.

Equation: $\qquad NH_3(aq) + H_2O(l) \rightleftharpoons NH_4^+(aq) \quad + \quad OH^-(aq)$

Initial: $\qquad\qquad x M \qquad\qquad\qquad\qquad 0.500\ M \qquad\quad 1.0 \times 10^{-7} M$

$$K_b = \dfrac{[NH_4^+][OH^-]}{[NH_3]} = \dfrac{(0.500\ M)(1.0 \times 10^{-7}\ M)}{x\ M} = 1.8 \times 10^{-5}$$

$$x\ M = \dfrac{(0.500\ M)(1.0 \times 10^{-7}\ M)}{1.8 \times 10^{-5}\ M} = 2.8 \times 10^{-3}\ M = [NH_3]$$

$$V = 500\ mL \times \dfrac{2.8 \times 10^{-3} mol\ NH_3}{1\ L\ final\ soln} \times \dfrac{1\ L\ conc.\ soln}{10.0\ mol\ NH_3} \times \dfrac{1\ drop}{0.05\ mL} = 2.8\ drops \approx 3\ drops$$

73. (M)

(a) Equation (1) is the reverse of the equation for the self-ionization of water. Thus, its equilibrium constant is simply the inverse of K_w.

$$K = \frac{1}{K_w} = \frac{1}{1.0 \times 10^{-14}} = 1.00 \times 10^{14}$$

Equation (2) is the reverse of the hydrolysis reaction for NH^{4+}. Thus, its equilibrium constant is simply the inverse of the acid ionization constant for NH^{4+}, $K_a = 5.6 \times 10 - 10$

$$K' = \frac{1}{K_a} = \frac{1}{5.6 \times 10^{-10}} = 1.8 \times 10^9$$

(b) The extremely large size of each equilibrium constant indicates that each reaction goes essentially to completion. In fact, a general rule of thumb suggests that a reaction is considered essentially complete if $K > 1000$ for the reaction.

77. (D)

(a) The *initial pH* is that of a solution of HCO_3^-(aq). This is an anion that can ionize to CO_3^{2-}(aq) or be hydrolyzed to H_2CO_3(aq). Thus,

$$pH = \tfrac{1}{2}(pK_{a_1} + pK_{a_2}) = \tfrac{1}{2}(6.35 + 10.33) = 8.34$$

The *pH at the equivalence point* is that of 0.500 M H_2CO_3(aq). All of the NaOH(aq) has been neutralized, as well as all of the HCO_3^-(aq), by the added HCl(aq).

Equation: $\quad H_2CO_3(aq) + H_2O(l) \rightleftharpoons HCO_3^-(aq) + H_3O^+(aq)$

Initial:	0.500 M	—	0 M	≈ 0 M
Changes:	$-x$ M	—	$+x$ M	$+x$ M
Equil:	$(0.500 - x)$ M	—	x M	x M

$$K_b = \frac{[HCO_3^-][H_3O^+]}{[H_2CO_3]} = 4.43 \times 10^{-7} = \frac{x \cdot x}{0.500 - x} \approx \frac{x^2}{0.500}$$

($c_a/K_a = 1.1 \times 10^6$; thus, the approximation is valid)

$$[H_3O^+] = \sqrt{0.500 \times 4.4 \times 10^{-7}} = 4.7 \times 10^{-4} M \qquad pH = 3.33$$

During the course of the titration, the pH is determined by the Henderson-Hasselbalch equation, with the numerator being the percent of bicarbonate ion remaining, and as the denominator being the percent of bicarbonate ion that has been transformed to H_2CO_3.

90% titrated: $pH = 6.35 + \log\dfrac{10\%}{90\%} = 5.40$ 10% titrated: $pH = 6.35 + \log\dfrac{90\%}{10\%} = 7.30$

95% titrated: $pH = 6.35 + \log\dfrac{5\%}{95\%} = 5.07$ 5% titrated: $pH = 6.35 + \log\dfrac{95\%}{5\%} = 7.63$

After the equivalence point, the pH is determined by the excess H_3O^+.

For 0.5 mL excess of HCl: $[H_3O^+] = \dfrac{0.50 \text{ mL} \times 1.00 \text{ M}}{20.5 \text{ mL}} = 0.024$ $\qquad pH = 1.61$

For 2.00 mL excess of HCl: $[H_3O^+] = \dfrac{2.00 \text{ mL} \times 1.00 \text{ M}}{22.0 \text{ mL}} = 0.091 \text{ M}$ $\qquad$ pH $= 1.04$

The titration curve derived from these data is sketched below.

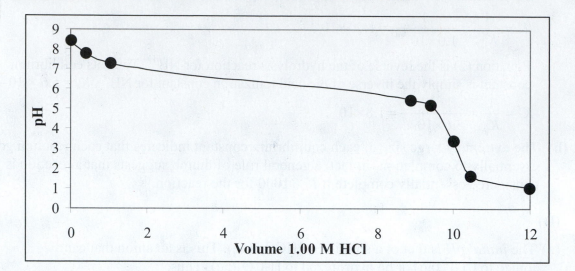

(b) In the titration of Na_2CO_3 with HCl, the pH at the first equivalence point is that of a solution of HCO_3^-(aq). This is an anion that can be ionized to CO_3^{2-}(aq) or hydrolyzed to H_2CO_3(aq). Thus, pH $= \frac{1}{2} (pK_{a_1} + pK_{a_2}) = \frac{1}{2} (6.35 + 10.33) = 8.34$. The *initial pH* is that of 1.000 M CO_3^{2-}(aq), which is determined by considering the hydrolysis of the anion.

Equation:	CO_3^{2-} (aq)	+	H_2O(l)	$\rightleftharpoons$	HCO_3^- (aq)	+	OH^- (aq)
Initial:	1.000 M		–		0 M		≈ 0 M
Changes:	$-x$ M		–		$+x$ M		$+x$ M
Equil:	$(1.000 - x)$ M		–		x M		x M

$$K_b = \frac{[HCO_3^-][OH^-]}{[CO_3^{2-}]} = \frac{K_w}{K_{a_2}} = \frac{1.0 \times 10^{-14}}{4.7 \times 10^{-11}} = 2.1 \times 10^{-4} = \frac{x \cdot x}{1.000 - x} \approx \frac{x^2}{1.000}$$

$[OH^-] = \sqrt{1.000 \times 2.1 \times 10^{-4}} = 1.4 \times 10^{-2}$ M $\qquad$ pOH $= 1.85$ $\qquad$ pH $= 12.15$

Before the first equivalence point is reached, the pH is determined by the Henderson-Hasselbalch equation, modified as in Exercise 69, but using as the numerator the percent of carbonate ion remaining, and as the denominator the percent of carbonate ion that has been transformed to HCO_3^-.

90% titrated: pH $= 10.33 + \log \dfrac{10\%}{90\%} = 9.38$ $\qquad$ 10% titrated: pH $= 10.33 + \log \dfrac{90\%}{10\%} = 11.28$

95% titrated: pH $= 10.33 + \log \dfrac{5\%}{95\%} = 9.05$ $\qquad$ 5% titrated: pH $= 10.33 + \log \dfrac{95\%}{5\%} = 11.61$

During the course of the second step of the titration, the values of pH are precisely as they are for the titration of $NaHCO_3$, except the titrant volume is 10.00 mL more (the volume needed to reach the first equivalence point). The solution *at the second equivalence* point is 0.333 M H_2CO_3, for which the set-up is similar to that for 0.500 M H_2CO_3.

$$[H_3O^+] = \sqrt{0.333 \times 4.4 \times 10^{-7}} = 3.8 \times 10^{-4} \text{ M} \qquad pH = 3.42$$

After the equivalence point, the pH is determined by the excess H_3O^+.

For 0.5 mL excess of HCl: $[H_3O^+] = \dfrac{0.50 \text{ mL} \times 1.00 \text{ M}}{30.5 \text{ mL}} = 0.0164 \text{ M} \qquad pH = 1.8$

For 2.00 mL excess of HCl: $[H_3O^+] = \dfrac{2.00 \text{ mL} \times 1.00 \text{ M}}{32.0 \text{ mL}} = 0.0625 \text{ M} \qquad pH = 1.20$

The titration curve from these data is sketched below.

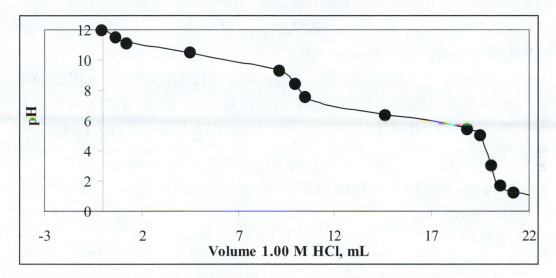

(c) $V_{HCl} = 1.00 \text{ g NaHCO}_3 \times \dfrac{1 \text{ mol NaHCO}_3}{84.01 \text{ g NaHCO}_3} \times \dfrac{1 \text{ mol HCl}}{1 \text{ mol NaHCO}_3} \times \dfrac{1000 \text{ mL}}{0.100 \text{ mol HCl}}$

  $= 119 \text{ mL } 0.100 \text{ M HCl}$

(d) $V_{HCl} = 1.00 \text{ g Na}_2CO_3 \times \dfrac{1 \text{ mol Na}_2CO_3}{105.99 \text{ g Na}_2CO_3} \times \dfrac{2 \text{ mol HCl}}{1 \text{ mol Na}_2CO_3} \times \dfrac{1000 \text{ mL}}{0.100 \text{ mol HCl}}$

  $= 189 \text{ mL } 0.100 \text{ M HCl}$

(e) The phenolphthalein endpoint occurs at pH = 8.00 and signifies that the NaOH has been neutralized, and that Na_2CO_3 has been half neutralized. The methyl orange endpoint occurs at about pH = 3.3 and is the result of the second equivalence point of Na_2CO_3. The mass of Na_2CO_3 can be determined as follows:

$$\text{mass Na}_2CO_3 = 0.78 \text{ mL} \times \dfrac{1 \text{ L}}{1000 \text{ mL}} \times \dfrac{0.1000 \text{ mol HCl}}{1 \text{ L soln}} \times \dfrac{1 \text{ mol HCO}_3^-}{1 \text{ mol HCl}} \dfrac{1 \text{ mol Na}_2CO_3}{1 \text{ mol HCO}_3^-}$$

$$\times \dfrac{105.99 \text{ g Na}_2CO_3}{1 \text{ mol Na}_2CO_3} = 0.0083 \text{ g Na}_2CO_3$$

$$\% \text{ Na}_2CO_3 = \dfrac{0.0083 \text{g}}{0.1000 \text{g}} \times 100 = 8.3\% \text{ Na}_2CO_3 \text{ by mass}$$

81. **(D)**

(a) A buffer solution is able to react with small amounts of added acid or base. When strong acid is added, it reacts with formate ion.

$$HCOO^-(aq) + H_3O^+(aq) \longrightarrow HCOOH(aq) + H_2O$$

Added strong base reacts with acetic acid.

$$CH_3COOH(aq) + OH^-(aq) \longrightarrow H_2O(l) + CH_3COO^-(aq)$$

Therefore neither added strong acid nor added strong base alters the pH of the solution very much. Mixtures of this type are referred to as buffer solutions.

(b) We begin with the two ionization reactions.

$$HCOOH(aq) + H_2O(l) \rightleftharpoons HCOO^-(aq) + H_3O^+(aq)$$

$$CH_3COOH(aq) + H_2O(l) \rightleftharpoons CH_3COO^-(aq) + H_3O^+(aq)$$

$[Na^+] = 0.250 \quad [OH^-] \approx 0 \qquad [H_3O^+] = x \quad [CH_3COO^-] = y \quad [HCOO^-] = z$

$0.150 = [CH_3COOH] + [CH_3COO^-] \quad [CH_3COOH] = 0.150 - [CH_3COO^-] = 0.150 - y$

$0.250 = [HCOOH] + [HCOO^-] \qquad [HCOOH] = 0.250 - [HCOO^-] = 0.250 - z$

$[Na^+] + [H_3O^+] = [CH_3COO^-] + [HCOO^-] + [OH^-]$ (electroneutrality)

$$0.250 + x = [CH_3COO^-] + [HCOO^-] = y + z \tag{1}$$

$$\frac{[H_3O^+][CH_3COO^-]}{[CH_3COOH]} = K_A = 1.8 \times 10^{-5} = \frac{x \cdot y}{0.150 - y} \tag{2}$$

$$\frac{[H_3O^+][HCOO^-]}{[HCOOH]} = K_F = 1.8 \times 10^{-4} = \frac{x \cdot z}{0.250 - z} \tag{3}$$

There now are three equations—(1), (2), and (3)—in three unknowns—x, y, and z. We solve equations (2) and (3), respectively, for y and z in terms of x.

$$0.150\,K_A - y\,K_A = xy \qquad y = \frac{0.150\,K_A}{K_A + x}$$

$$0.250\,K_F - z\,K_F = xz \qquad z = \frac{0.250\,K_F}{K_F + x}$$

Then we substitute these expressions into equation (1) and solve for x.

$$0.250 + x = \frac{0.150\,K_A}{K_A + x} + \frac{0.250\,K_F}{K_F + x} \approx 0.250 \qquad \text{since } x << 0.250$$

$$0.250\,(K_F + x)(K_A + x) = 0.150\,K_A\,(K_F + x) + 0.250\,K_F\,(K_A + x)$$

$$K_A K_F + (K_A + K_F)x + x^2 = 1.60\,K_A K_F + x(0.600\,K_A + 1.00\,K_F)$$

$$x^2 + 0.400\,K_A x - 0.600\,K_A K_F = 0 = x^2 + 7.2 \times 10^{-6}\,x - 1.9 \times 10^{-9}$$

$$x = \frac{-7.2 \times 10^{-6} \pm \sqrt{5.2 \times 10^{-11} + 7.6 \times 10^{-9}}}{2} = 4.0 \times 10^{-5}\ M = [H_3O^+]$$

$pH = 4.40$

(c) Adding 1.00 L of 0.100 M HCl to 1.00 L of buffer of course dilutes the concentrations of all components by a factor of 2. Thus, $[Na^+]=0.125$ M; total acetate concentration $=$ 0.0750 M; total formate concentration $=0.125$ M. Also, a new ion is added to the solution, namely, $[Cl^-]=0.0500$ M.

$$[Na^+]=0.125 \quad [OH^-]\approx 0 \quad [H_3O^+]=x \quad [Cl^-]=0.0500\text{ M} \quad [CH_3COO^-]=y \quad [HCOO^-]=z$$

$$0.0750=[CH_3COOH]+[CH_3COO^-] \quad [CH_3COOH]=0.0750-[CH_3COO^-]=0.0750-y$$
$$0.125=[HCOOH]+[HCOO^-] \quad\quad [HCOOH]=0.125-[HCOO^-]=0.125-z$$

$$[Na^+]+[H_3O^+]=[CH_3COO^-]+[OH^-]+[HCOO^-]+[Cl^-]\ (\text{electroneutrality})$$

$$0.125+x=[CH_3COO^-]+[HCOO^-]+[Cl^-]=y+z+0+0.0500 \quad\quad 0.075+x=y+z$$

$$\frac{[H_3O^+][CH_3COO^-]}{[CH_3COOH]}=K_A=1.8\times10^{-5}=\frac{x\cdot y}{0.0750-y}$$

$$\frac{[H_3O^+][HCOO^-]}{[HCOOH]}=K_F=1.8\times10^{-4}=\frac{x\cdot z}{0.125-z}$$

Again, we solve the last two equations for y and z in terms of x.

$$0.0750\,K_A-y\,K_A=xy \quad\quad y=\frac{0.0750\,K_A}{K_A+x}$$

$$0.125\,K_F-z\,K_F=xz \quad\quad z=\frac{0.125\,K_F}{K_F+x}$$

Then we substitute these expressions into equation (1) and solve for x.

$$0.0750+x=\frac{0.0750\,K_A}{K_A+x}+\frac{0.125\,K_F}{K_F+x}\approx0.0750 \quad \text{since } x\ll0.0750$$

$$0.0750\,(K_F+x)(K_A+x)=0.0750\,K_A\,(K_F+x)+0.125\,K_F\,(K_A+x)$$

$$K_AK_F+(K_A+K_F)x+x^2=2.67\,K_AK_F+x(1.67\,K_F+1.00\,K_A)$$

$$x^2-0.67\,K_Fx-1.67\,K_AK_F=0=x^2-1.2\times10^{-4}x-5.4\times10^{-9}$$

$$x=\frac{1.2\times10^{-4}\pm\sqrt{1.4\times10^{-8}+2.2\times10^{-8}}}{2}=1.55\times10^{-4}\text{ M}=[H_3O^+]$$

$$pH=3.81\approx3.8$$

As expected, the addition of HCl(aq), a strong acid, caused the pH to drop. The decrease in pH was relatively small, nonetheless, because the H_3O^+(aq) was converted to the much weaker acid HCOOH via the neutralization reaction:

$$HCOO^-(aq)+H_3O^+(aq)\rightarrow HCOOH(aq)+H_2O(l)\quad(\text{buffering action})$$

82. **(M)** First we find the pH at the equivalence point:

$$CH_3CH(OH)COOH(aq) + OH^-(aq) \Longrightarrow CH_3CH(OH)COO^-(aq) + H_2O(l)$$

$$\text{1 mmol} \qquad\qquad \text{1 mmol} \qquad\qquad \text{1 mmol}$$

The concentration of the salt is $1 \times 10^{-3} \text{mol}/0.1 \text{ L} = 0.01 \text{ M}$

The lactate anion undergoes hydrolysis thus:

$$CH_3CH(OH)COO^-(aq) + H_2O(l) \Longrightarrow CH_3CH(OH)COOH(aq) + OH^-(aq)$$

Initial	0.01 M	–	0 M	≈ 0 M
Change	$-x$	–	$+x$	$+x$
Equilibrium	$(0.01-x)$ M	–	x	x

Where x is the [hydrolyzed lactate ion], as well as that of the $[OH^-]$ produced by hydrolysis

$$K \text{ for the above reaction} = \frac{[CH_3CH(OH)COOH][OH^-]}{[CH_3CH(OH)COO^-]} = \frac{K_w}{K_a} = \frac{1.0 \times 10^{-14}}{10^{-3.86}} = 7.2 \times 10^{-11}$$

so $\dfrac{x^2}{0.01-x} \approx \dfrac{x^2}{0.01} = 7.2 \times 10^{-11}$ and $x = 8.49 \times 10^{-7}$; $[OH^-] = 8.49 \times 10^{-7}$ M

$x \ll 0.01$, thus, the assumption is valid

$$pOH = -\log(8.49 \times 10^{-7}) = 6.07 \qquad pH = 14 - pOH = 14.00 - 6.07 = 7.93$$

(a) Bromthymol blue or phenol red would be good indicators for this titration since they change color over this pH range.

(b) The $H_2PO_4^-/HPO_4^{2-}$ system would be suitable because the pK_a for the acid (namely, $H_2PO_4^-$) is close to the equivalence point pH of 7.93. An acetate buffer would be too acidic, an ammonia buffer too basic.

(c) $H_2PO_4^- + H_2O \Longrightarrow H_3O^+ + HPO_4^{2-} \quad K_{a2} = 6.3 \times 10^{-8}$

Solving for $\dfrac{[HPO_4^{2-}]}{[H_2PO_4^-]} = \dfrac{K_{a2}}{[H_3O^+]} = \dfrac{6.3 \times 10^{-8}}{10^{-7.93}} = 5.4$ (buffer ratio required)

83. **(M)**

$$H_2O_2(aq) + H_2O(l) \Longrightarrow H_3O^+(aq) + HO_2^-(aq) \quad K_a = \frac{[H_3O^+][HO_2^-]}{[H_2O_2]}$$

Data taken from experiments 1 and 2:

$[H_2O_2] + [HO_2^-] = 0.259 \text{ M} \qquad (6.78) \times (0.00357 \text{ M}) + [HO_2^-] = 0.259 \text{ M}$

$[HO_2^-] = 0.235 \text{ M} \qquad [H_2O_2] = (6.78)(0.00357 \text{ M}) = 0.0242 \text{ M}$

$[H_3O^+] = 10^{-(pK_w - pOH)}$

$-\log(0.250 - 0.235) \text{ M NaOH} = -\log(0.015 \text{ M NaOH}) = 1.82\underline{4}$

$[H_3O^+] = 10^{-(14.94 - 1.82\underline{4})} = 7.7 \times 10^{-14}$

$$K_a = \frac{[H_3O^+][HO_2^-]}{[H_2O_2]} = \frac{(7.7\times10^{-14})(0.235)}{(0.0242)} = 7.4\times10^{-13}$$

$$pK_a = 12.14$$

From data taken from experiments 1 and 3:

$[H_2O_2] + [HO_2^-] = 0.123$ M $\qquad (6.78)(0.00198$ M$) + [HO_2^-] = 0.123$ M

$[HO_2^-] = 0.109\underline{6}$ M

$[H_2O_2] = (6.78)(0.00198$ M$) = 0.0134$ M

$[H_3O^+] = 10^{-(pK_w - pOH)}, \quad pOH = -\log[(0.125 - 0.109\underline{6})$ M NaOH$] = 1.81$

$[H_3O^+] = 10^{-(14.94 - 1.81)} = 7.41\times10^{-14}$

$$K_a = \frac{[H_3O^+][HO_2^-]}{[H_2O_2]} = \frac{(7.41\times10^{-14})(0.1096)}{(0.0134)} = 6.06\times10^{-13}$$

$$pK_a = -\log(6.06\times10^{-13}) = 12.22$$

Average value for $pK_a = 12.17$

84. **(D)** Let's consider some of the important processes occurring in the solution.

(1) $HPO_4^{2-}(aq) + H_2O(l) \rightleftharpoons H_2PO_4^-(aq) + OH^-(aq) \quad K_{(1)} = K_{a_2} = 4.2\times10^{-13}$

(2) $HPO_4^{2-}(aq) + H_2O(l) \rightleftharpoons H_3O^+(aq) + PO_4^{3-}(aq) \quad K_{(2)} = K_b = \dfrac{1.0\times10^{-14}}{6.3\times10^{-8}} = 1.6\times10^{-7}$

(3) $NH_4^+(aq) + H_2O(l) \rightleftharpoons H_3O^+(aq) + NH_3(aq) \quad K_{(3)} = K_a = \dfrac{1.0\times10^{-14}}{1.8\times10^{-5}} = 5.6\times10^{-10}$

(4) $NH_4^+(aq) + OH^-(aq) \rightleftharpoons H_2O(l) + NH_3(aq) \quad K_{(4)} = \dfrac{1}{1.8\times10^{-5}} = 5.6\times10^4$

May have interaction between NH_4^+ and OH^- formed from the hydrolysis of HPO_4^{2-}.

	$NH_4^+(aq)$	$+$	$HPO_4^{2-}(aq)$	$\rightleftharpoons$	$H_2PO_4^-(aq)$	$+$	$NH_3(aq)$
Initial	0.10M		0.10M		0 M		0 M
Change	$-x$		$-x$		$+x$		$+x$
Equil.	$0.100 - x$		$0.100 - x$		x		x

(where x is the molar concentration of NH_4^+ that hydrolyzes)

$K = K_{(2)} \times K_{(4)} = (1.6\times10^{-7}) \times (5.6\times10^4) = 9.0\times10^{-3}$

$$K = \frac{[H_2PO_4^-][NH_3]}{[NH_4^+][HPO_4^{2-}]} = \frac{[x]^2}{[0.10 - x]^2} = 9.0\times10^{-3} \text{ and } x = 8.7\times10^{-3}$$

Finding the pH of this buffer system:

$$pH = pK_a + \log\frac{[HPO_4^{2-}]}{[H_2PO_4^-]} = -\log(6.3\times10^{-8}) + \log\left(\frac{0.100 - 8.7\times10^{-3}M}{8.7\times10^{-3}M}\right) = 8.2$$

As expected, we get the same result using the NH_3/NH_4^+ buffer system.

89. **(D)**

(a) We start by writing the equilibrium expression for all reactions:

$$K_1 = \frac{[CO_2(aq)]}{[CO_2(g)]} \qquad\qquad K_2 = [Ca^{2+}][CO_3^{2-}]$$

$$K_3 = \frac{[HCO_3^-]}{[H_3O^+][CO_3^{2-}]} \qquad\qquad K_4 = \frac{[CO_2(aq)]}{[HCO_3^-][H_3O^+]}$$

First, we have to express $[H_3O^+]$ using the available expressions:

$$[H_3O^+] = \frac{[HCO_3^-]}{K_3[CO_3^{2-}]}$$

$$[H_3O^+] = \frac{[CO_2(aq)]}{K_4[HCO_3^-]}$$

$$[H_3O^+]^2 = \frac{[HCO_3^-]}{K_3[CO_3^{2-}]} \times \frac{[CO_2(aq)]}{K_4[HCO_3^-]} = \frac{[CO_2(aq)]}{K_3\cdot K_4[CO_3^{2-}]}$$

From the expression for K_1, we know that $[CO_2(aq)] = K_1[CO_2(g)]$. Therefore,

$$[H_3O^+]^2 = \frac{K_1[CO_2(g)]}{K_3\cdot K_4[CO_3^{2-}]}$$

Now, the expression for K_2 can be plugged into the above expression as follows:

$$[H_3O^+]^2 = \frac{K_1[CO_2(g)]}{K_3\cdot K_4[CO_3^{2-}]} = \frac{K_1[CO_2(g)][Ca^{2+}]}{K_2\cdot K_3\cdot K_4}$$

$$[H_3O^+] = \sqrt{\frac{K_1[CO_2(g)][Ca^{2+}]}{K_2\cdot K_3\cdot K_4}}$$

(b) The K values for the reactions are as follows:

$K_1 = 0.8317$ (given in the problem)

$K_2 = 2.8 \times 10^{-9}$ (given in the problem)

$K_3 = 1/K_a$ of $HCO_3^- = 1/(4.7 \times 10^{-11}) = 2.13 \times 10^{10}$

$K_4 = 1/K_a$ of $H_2CO_3 = 1/(4.4 \times 10^{-7}) = 2.27 \times 10^6$

$$[CO_2(g)] = \frac{280 \times 10^{-6} \text{ L } CO_2}{\text{L air}} \times \frac{1 \text{ mol } CO_2}{24.45 \text{ L } CO_2} = 1.145 \times 10^{-5} \frac{\text{mol } CO_2}{\text{L air}}$$

$$[H_3O^+] = \sqrt{\frac{K_1 [CO_2(g)][Ca^{2+}]}{K_2 \cdot K_3 \cdot K_4}} = \sqrt{\frac{(0.8317)(1.145 \times 10^{-5})(10.24 \times 10^{-3})}{(2.8 \times 10^{-9})(2.13 \times 10^{10})(2.27 \times 10^6)}}$$

$$= 2.684 \times 10^{-8} \text{ M}$$

$$pH = -\log(2.684 \times 10^{-8}) = 7.57$$

91. **(M)**

(a) We note that the pH is 5.0. Therefore, $[H_3O^+] = 10^{-5.0} = 1.0 \times 10^{-5}$ M

$$\beta = \frac{C \cdot K_a \cdot [H_3O^+]}{(K_a + [H_3O^+])^2} = \frac{(2.0 \times 10^{-2})(1.8 \times 10^{-5})(1.0 \times 10^{-5})}{(1.8 \times 10^{-5} + 1.0 \times 10^{-5})^2} = 4.6 \times 10^{-3}$$

(b) $\beta \approx -dC_A / d(pH) \Rightarrow d(pH) \approx -dC_A / \beta$

$$d(pH) = -(1.0 \times 10^{-3}) / 4.6 \times 10^{-3} = -0.22$$

$$pH = 5 - 0.22 = 4.78$$

(c) At an acetic acid concentration of 0.1 M, C is also 0.1, because C is the total concentration of the acetic acid and acetate. The maximum buffer index β (2.50×10^{-2}) happens at a pH of 4.75, where $[HAc] = [Ac^-]$. The minima are located at pH values of 8.87 (which correspond to pH of a solution of 0.1 M acetic acid and 0.1 M acetate, respectively).

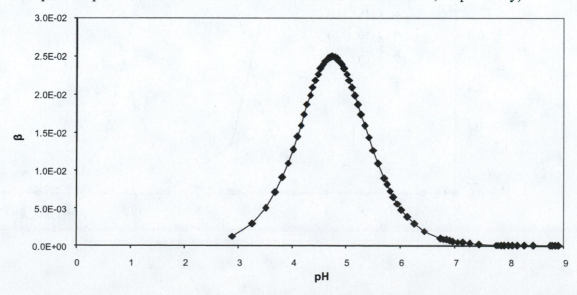

FEATURE PROBLEMS

92. **(D)**

(a) The two curves cross the point at which half of the total acetate is present as acetic acid and half is present as acetate ion. This is the half equivalence point in a titration, where $pH = pK_a = 4.74$.

(b) For carbonic acid, there are three carbonate containing species: "H_2CO_3" which predominates at low pH, HCO_3^-, and CO_3^{2-}, which predominates in alkaline solution. The points of intersection should occur at the half-equivalence points in each step-wise titration: at $pH = pK_{a_1} = -\log(4.4 \times 10^{-7}) = 6.36$ and at

$pH = pK_{a_2} = -\log(4.7 \times 10^{-11}) = 10.33$. The following graph was computer-calculated (and then drawn) from these equations. f in each instance represents the fraction of the species whose formula is in parentheses.

$$\frac{1}{f(H_2A)} = 1 + \frac{K_1}{[H^+]} + \frac{K_1 K_2}{[H^+]^2}$$

$$\frac{1}{f(HA^-)} = \frac{[H^+]}{K_1} + 1 + \frac{K_2}{[H^+]}$$

$$\frac{1}{f(A^{2-})} = \frac{[H^+]^2}{K_1 K_2} + \frac{[H^+]}{K_2} + 1$$

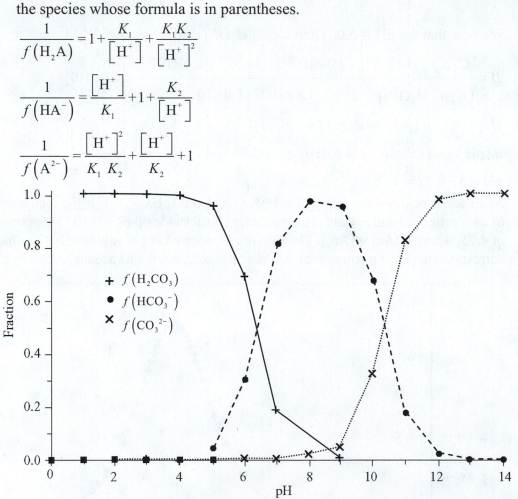

(c) For phosphoric acid, there are four phosphate containing species: H_3PO_4 under acidic conditions, $H_2PO_4^-$, HPO_4^{2-}, and PO_4^{3-}, which predominates in alkaline solution. The points of intersection should occur at $pH = pK_{a_1} = -\log(7.1 \times 10^{-3}) = 2.15$, $pH = pK_{a_2} = -\log(6.3 \times 10^{-8}) = 7.20$, and $pH = pK_{a_3} = -\log(4.2 \times 10^{-13}) = 12.38$, a quite alkaline solution. The graph that follows was computer-calculated and drawn.

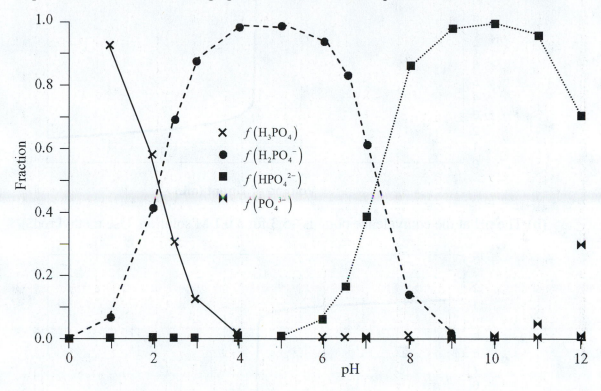

SELF-ASSESSMENT EXERCISES

98. (E)

(a) $HCOOH + OH^- \rightarrow HCOO^- + H_2O$

$HCOO^- + H_3O^+ \rightarrow HCOOH + H_2O$

(b) $C_6H_5NH_3^+ + OH^- \rightarrow C_6H_5NH_2 + H_2O$

$C_6H_5NH_2 + H_3O^+ \rightarrow C_6H_5NH_3^+ + H_2O$

(c) $H_2PO_4^- + OH^- \rightarrow HPO_4^{2-} + H_2O$

$HPO_4^{2-} + H_3O^+ \rightarrow H_2PO_4^- + H_2O$

99. **(M)**
(a) The pH at theequivalence point is 7. Use bromthymol blue.

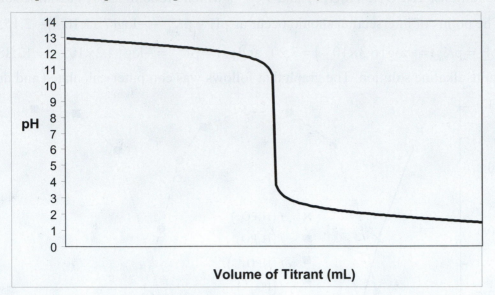

(b) The pH at the equivalence point is ~5.3 for a 0.1 M solution. Use methyl red.

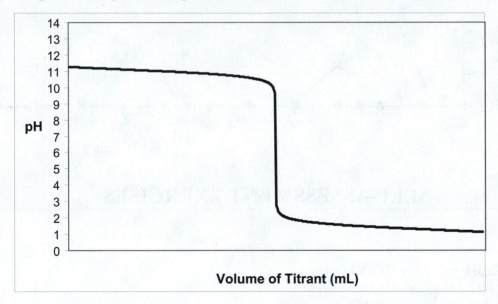

(c) The pH at the equivalence point is ~8.7 for a 0.1 M solution. Use phenolphthalein, because it just begins to get from clear to pink around the equivalence point.

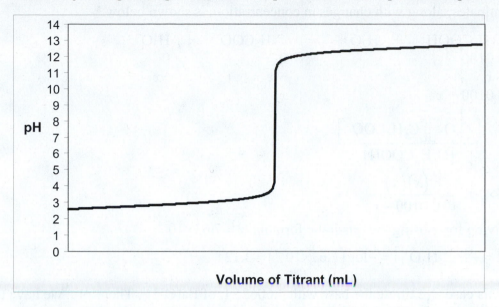

(d) The pH for the first equivalence point ($NaH_2PO_4^-$ to $Na_2HPO_4^{2-}$) for a 0.1 M solution is right around ~7, so use bromthymol blue.

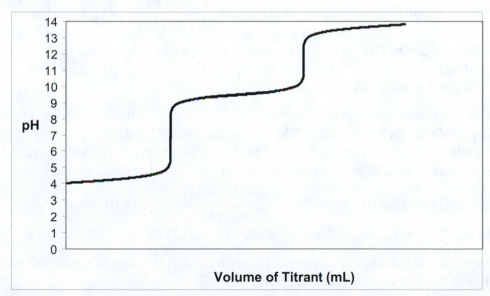

100. (D)

(a) This is the initial equilibrium before any base has reacted with the acid. The reaction that dominates, along with changes in concentration, is shown below:

C_6H_5COOH	$+ H_2O$	$\rightleftharpoons$	$C_6H_5COO^-$	$+ H_3O^+$
0.0100			0	0
$-x$			$+x$	$+x$
$0.0100 - x$			x	x

$$K_a = \frac{\left[H_3O^+\right]\left[C_6H_5COO^-\right]}{\left[C_6H_5COOH\right]}$$

$$6.3 \times 10^{-5} = \frac{(x)(x)}{0.0100 - x}$$

Solving for x using the quadratic formula, $x = 7.63 \times 10^{-4}$.

$$pH = -\log\left[H_3O^+\right] = -\log\left(7.63 \times 10^{-4}\right) = 3.12$$

(b) In this case, we titrate the base with 0.00625 L of $Ba(OH)_2$. Therefore, we have to calculate the final moles of the base and the total volume to determine the concentration.

$$\text{mol } C_6H_5COOH = 0.02500 \text{ L} \times 0.0100 \text{ M } C_6H_5COOH = 2.5 \times 10^{-4} \text{ mol}$$

$$\text{mol } OH^- = 0.00625 \text{ L} \times 0.0100 \text{ M } Ba(OH)_2 \times \frac{2 \text{ mol } OH^-}{1 \text{ mol } Ba(OH)_2} = 1.25 \times 10^{-4} \text{ mol}$$

Since the amount of OH^- is half of the initial amount of C_6H_5COOH, the moles of C_6H_5COOH and $C_6H_5COO^-$ are equal. Therefore, $K_a = [H_3O^+]$, and $pH = -\log(6.3 \times 10^{-5}) = 4.20$.

(c) At the equivalence point, there is initially no C_6H_5COOH. The equilibrium is dominated by $C_6H_5COO^-$ hydrolyzing water. The concentration of $C_6H_5COO^-$ is:
Total moles of $C_6H_5COOH = 2.5 \times 10^{-4}$ mol as shown previously. At the equivalence point, the moles of acid equal moles of OH^-.

$$\text{mol } Ba(OH)_2 = 2.5 \times 10^{-4} \text{ mol } OH^- \times \frac{1 \text{ mol } Ba(OH)_2}{2 \text{ mol } OH^-} = 1.25 \times 10^{-4} \text{ mol}$$

Vol of $Ba(OH)_2 = 1.25 \times 10^{-4}$ mol / 0.0100 M = 0.0125 L

Total volume of the solution at the equivalence point is the sum of the initial volume plus the volume of $Ba(OH)_2$ added. That is,

$V_{TOT} = 0.02500$ L $+ 0.0125$ L $= 0.0375$ L

Therefore, the concentration of $C_6H_5COO^- = 2.5 \times 10^{-4}/0.0375$ L $= 0.00667$ M.

$C_6H_5COO^-$	$+ H_2O$	$\rightleftharpoons$	C_6H_5COOH	$+ OH^-$
0.00667			0	0
$-x$			$+x$	$+x$
$0.00667 - x$			x	x

Since this is a base reaction, $K_b = K_w/K_a = (1.00\times10^{-4})/6.3\times10^{-5} = 1.587\times10^{-10}$.

$$K_b = \frac{[OH^-][C_6H_5COOH]}{[C_6H_5COO^-]}$$

$$1.59\times10^{-10} = \frac{(x)(x)}{0.00667 - x}$$

solving for x (by simplifying the formula above) yields $x = 1.03\times10^{-6}$ M.

$$pH = 14 - pOH = 14 - \left[-\log(1.03\times10^{-6})\right] = 14 - 6.00 = 8.00$$

(d) In this part, we have an excess of a strong base. As such, we have to determine how much excess base there is and what is the final volume of the solution.

$$mol(OH^-) = 0.01500 \text{ L} \times 0.0100 \frac{\text{mol Ba}(OH)_2}{\text{L}} \times \frac{2 \text{ mol OH}^-}{1 \text{ mol Ba}(OH)_2}$$

$$= 3.000\times10^{-4} \text{ mol OH}^-$$

$$\text{Excess mol OH}^- = \text{mol HC}_7H_5O_2 - \text{mol OH}^- = 3.000\times10^{-4} - 2.500\times10^{-4}$$

$$= 5.000\times10^{-5} \text{ mol}$$

$$[OH^-] = \frac{5.0\times10^{-5} \text{ mol}}{(0.02500 \text{ L} + 0.01500 \text{ L})} = 0.00125 \text{ M}$$

$$pH = 14 - pOH = 14 - \left[-\log(0.00125)\right] = 14 - 2.903 = 11.1$$

101. (E) The answer is (c); because of the common ion-effect, the presence of $HCOO^-$ will repress ionization of formic acid.

102. (E) The answer is (d), because $NaHCO_3$ is a weak base and will react with protons in water, shifting the formic acid ionization equilibrium to the right.

103. (E) The answer is (b), raise the pH. NH_4^+ is an acid, and to be converted to its conjugate base, it must react with a base to abstract its proton.

104. (E) The answer is (b), because at that point, the number of moles of weak base remaining is the same as its conjugate acid, and the equilibrium expression simplifies to $K_a = [H_3O^+]$.

105. (M) The base, $C_2H_5NH_2$, reacts with $HClO_4$. The reaction is:

$$C_2H_5NH_2 + HClO_4 \rightarrow C_2H_5NH_3^+ + ClO_4^-$$

Assuming a volume of 1 L for each solution,

$$[C_2H_5NH_2] = \frac{1.49 \text{ mol} - 1.001 \text{ mol}}{2 \text{ L}} = \frac{0.489 \text{ mol}}{2 \text{ L}} = 0.2445 \text{ M}$$

$$\left[C_2H_5NH_3^+\right] = \frac{1.001\ \text{mol}}{2\ \text{L}} = 0.5005\ \text{M}$$

$$4.3\times10^{-4} = \frac{\left[C_2H_5NH_3^+\right]\left[OH^-\right]}{\left[C_2H_5NH_2\right]} = \frac{(0.5005 + x)(x)}{(0.2445 - x)}$$

$$x = 2.10\times10^{-4}$$

$$\text{pOH} = -\log\left(2.10\times10^{-4}\right) = 3.68$$

$$\text{pH} = 14 - 3.68 = 10.32$$

106. **(D)** We assume that all of $Ca(HSe)_2$ dissociates in water. The concentration of HSe^- is therefore:

$$0.5\ \text{M Ca}\left(HSe\right)_2 \times \frac{2\ \text{mol HSe}^-}{1\ \text{mol Ca}\left(HSe\right)_2} = 1.0\ \text{M HSe}^-$$

We note that HSe^- is amphoteric; that is, it can act either as an acid or a base. The acid reaction of HSe^- and the concentration of $[H_3O^+]$ generated, are as follows:

$$HSe^- + H_2O \rightleftharpoons Se^{2-} + H_3O^+$$

$$1.00\times10^{-11} = \frac{\left[Se^{2-}\right]\left[H_3O^+\right]}{\left[HSe^-\right]} = \frac{(x)(x)}{(1.00 - x)}$$

$$x = \sqrt{1.00\times10^{-11}} = 3.16\times10^{-6}$$

The basic reaction of HSe^- and the concentration of $[OH^-]$ generated, are as follows:

$$HSe^- + H_2O \rightleftharpoons H_2Se + OH^-$$

$$K_b = 1.00 \times 10^{-14}/1.3 \times 10^{-4} = 7.69 \times 10^{-11}$$

$$7.69\times10^{-11} = \frac{\left[H_2Se\right]\left[OH^-\right]}{\left[HSe^-\right]} = \frac{(x)(x)}{(1.00 - x)}$$

$$x = \sqrt{7.69\times10^{-11}} = 8.77\times10^{-6}$$

Therefore, we have $[H_3O^+] = 3.16\times10^{-6}$ and $[OH^-] = 8.77 \times 10^{-6}$. Since these two react to give H_2O, the result is $8.77 \times 10^{-6} - 3.16\times10^{-6} = 5.61 \times 10^{-6}$ M $[OH^-]$. The pH of the solution is:

$$\text{pH} = 14 - \text{pOH} = 14 - 5.25 = 8.75$$

107. **(E)** The answer is (a). The solution system described is a buffer, and will resist large changes in pH. Adding KOH should raise the pH slightly.

108. (E) The answer is (b), because HSO_3^- is a much stronger acid ($K_a = 1.3 \times 10^{-2}$) than $H_2PO_4^-$ ($K_a = 6.3 \times 10^{-8}$).

109. (E) The answer is (b). The pK_a of the acid is 9, which puts it squarely in the middle of the 8–10 pH range for the equivalence point.

110. (E)

(a) $NaHCO_3$ titrated with NaOH: pH > 7, because HCO_3^- is itself slightly basic, and is being titrated with NaOH to yield CO_3^{2-} at the equivalence point, which is even more basic.

(b) HCl titrated with NH_3: pH < 7, because the resulting NH_4^+ at the equivalence point is acidic.

(c) KOH titrated with HI: pH = 7, because a strong base is being titrated by a strong acid, and the resulting anions and cations are all non-basic and non-acidic.

CHAPTER 18
SOLUBILITY AND COMPLEX-ION EQUILIBRIA
PRACTICE EXAMPLES

1A **(E)** In each case, we first write the balanced equation for the solubility equilibrium and then the equilibrium constant expression for that equilibrium, the K_{sp} expression:

(a) $MgCO_3(s) \rightleftharpoons Mg^{2+}(aq) + CO_3^{2-}(aq)$ $\qquad K_{sp} = [Mg^{2+}][CO_3^{2-}]$

(b) $Ag_3PO_4(s) \rightleftharpoons 3Ag^+(aq) + PO_4^{3-}(aq)$ $\qquad K_{sp} = [Ag^+]^3[PO_4^{3-}]$

1B **(E)**

(a) Provided the $[OH^-]$ is not too high, the hydrogen phosphate ion is not expected to ionize in aqueous solution to a significant extent because of the quite small values for the second and third ionization constants of phosphoric acid.
$$CaHPO_4(s) \rightleftharpoons Ca^{2+}(aq) + HPO_4^{2-}(aq)$$

(b) The solubility product constant is written in the manner of a K_c expression:
$$K_{sp} = [Ca^{2+}][HPO_4^{2-}] = 1. \times 10^{-7}$$

2A **(M)** We calculate the solubility of silver cyanate, s, as a molarity. We then use the solubility equation to (1) relate the concentrations of the ions and (2) write the K_{sp} expression.

$$s = \frac{7 \text{ mg AgOCN}}{100 \text{ mL}} \times \frac{1000 \text{ mL}}{1 \text{ L}} \times \frac{1 \text{ g}}{1000 \text{ mg}} \times \frac{1 \text{ mol AgOCN}}{149.9 \text{ g AgOCN}} = 5 \times 10^{-4} \text{ mol/L}$$

Equation : $\quad AgOCN(s) \rightleftharpoons Ag^+(aq) + OCN^-(aq)$

Solubility Product : $\qquad\qquad\qquad s \qquad\qquad s$

$$K_{sp} = [Ag^+][OCN^-] = (s) \times (s) = s^2 = (5 \times 10^{-4})^2 = 3 \times 10^{-7}$$

2B **(E)** We calculate the solubility of lithium phosphate, s, as a molarity. We then use the solubility equation to (1) relate the concentrations of the ions and (2) write the K_{sp} expression.

$$s = \frac{0.034 \text{ g Li}_3PO_4}{100 \text{ mL soln}} \times \frac{1000 \text{ mL}}{1 \text{ L}} \times \frac{1 \text{ mol Li}_3PO_4}{115.79 \text{ g Li}_3PO_4} = 0.0029 \text{ mol/L}$$

Equation: $Li_3PO_4(s) \rightleftharpoons 3Li^+(aq) + PO_4^{3-}(aq)$

Solubility Product: $\qquad\qquad (3s)^3 \qquad\quad s$

$$K_{sp} = [Li^+]^3[PO_4^{3-}] = (3s)^3 \cdot (s) = 27s^4 = 27(0.0029)^4 = 1.9 \times 10^{-9}$$

3A **(E)** We use the solubility equilibrium to write the K_{sp} expression, which we then solve to obtain the molar solubility, s, of $Cu_3(AsO_4)_2$.

$$Cu_3(AsO_4)_2(s) \rightleftharpoons 3\ Cu^{2+}(aq) + 2\ AsO_4^-(aq)$$

$$K_{sp} = \left[Cu^{2+}\right]^3\left[AsO_4^-\right]^2 = (3s)^3(2s)^2 = 108s^5 = 7.6\times10^{-36}$$

Solubility: $\quad s = \sqrt[5]{\dfrac{7.6\times10^{-36}}{108}} = 3.7\times10^{-8}\,M$

3B **(E)** First we determine the solubility of $BaSO_4$, and then find the mass dissolved.

$$BaSO_4(aq) \rightleftharpoons Ba^{2+}(aq) + SO_4^{2-}(aq) \qquad\qquad K_{sp} = \left[Ba^{2+}\right]\left[SO_4^{2-}\right] = s^2$$

The last relationship is true because $\left[Ba^{2+}\right] = \left[SO_4^{2-}\right]$ in a solution produced by dissolving $BaSO_4$ in pure water. Thus, $s = \sqrt{K_{sp}} = \sqrt{1.1\times10^{-10}} = 1.05\times10^{-5}\,M$.

$$\text{mass } BaSO_4 = 225\ mL \times \frac{1.05\times10^{-5}\,mmol\ BaSO_4}{1\ mL\ sat'd\ soln} \times \frac{233.39\,mg\ BaSO_4}{1\,mmol\ BaSO_4} = 0.55\,mg\ BaSO_4$$

4A **(M)** For PbI_2, $K_{sp} = \left[Pb^{2+}\right]\left[I^-\right]^2 = 7.1\times10^{-9}$. The solubility equilibrium is the basis of the calculation.

Equation:	$PbI_2(s)$	$\rightleftharpoons$	$Pb^{2+}(aq)$	$+$	$2I^-(aq)$
Initial:	—		0.10 M		0 M
Changes:	—		$+s$ M		$+2s$ M
Equil:	—		$(0.10 + s)$ M		$2s$ M

$$K_{sp} = \left[Pb^{2+}\right]\left[I^-\right]^2 = 7.1\times10^{-9} = (0.10+s)(2s)^2 \approx 0.40\,s^2 \qquad s = \sqrt{\frac{7.1\times10^{-9}}{0.40}} = 1.3\times10^{-4}\,M$$

(assumption $0.10 \gg s$ is valid)

This value of s is the solubility of PbI_2 in $0.10\,M\ Pb(NO_3)_2(aq)$.

4B **(E)** We find pOH from the given pH:

pOH $= 14.00 - 8.20 = 5.80$; $[OH^-] = 10^{-pOH} = 10^{-5.80} = 1.6\times10^{-6}\,M$.

We assume that pOH remains constant, and use the K_{sp} expression for $Fe(OH_3)$.

$$K_{sp} = \left[Fe^{3+}\right]\left[OH^-\right]^3 = 4\times10^{-38} = \left[Fe^{3+}\right]\left(1.6\times10^{-6}\right)^3 \quad \left[Fe^{3+}\right] = \frac{4\times10^{-38}}{\left(1.6\times10^{-6}\right)^3} = 9.8\times10^{-21}\,M$$

Therefore, the molar solubility of $Fe(OH)_3$ is 9.8×10^{-21} M.

The dissolved $Fe(OH)_3$ does not significantly affect $\left[OH^-\right]$.

5A **(M)** First determine $\left[I^-\right]$ as altered by dilution. We then compute Q_{sp} and compare it with K_{sp}.

$$\left[I^-\right] = \frac{3\,\text{drops} \times \dfrac{0.05\,\text{mL}}{1\,\text{drop}} \times \dfrac{0.20\,\text{mmol KI}}{1\,\text{mL}} \times \dfrac{1\,\text{mmol I}^-}{1\,\text{mmol KI}}}{100.0\,\text{mL soln}} = 3\times10^{-4}\,\text{M}$$

$$Q_{sp} = \left[Ag^+\right]\left[I^-\right] = (0.010)(3\times10^{-4}) = 3\times10^{-6}$$

$$Q_{sp} > 8.5\times10^{-17} = K_{sp} \qquad \text{Thus, precipitation should occur.}$$

5B **(M)** We first use the solubility product constant expression for PbI_2 to determine the $\left[I^-\right]$ needed in solution to just form a precipitate when $\left[Pb^{2+}\right] = 0.010$ M. We assume that the volume of solution added is small and that $\left[Pb^{2+}\right]$ remains at 0.010 M throughout.

$$K_{sp} = \left[Pb^{2+}\right]\left[I^-\right]^2 = 7.1\times10^{-9} = (0.010)\left[I^-\right]^2 \qquad \left[I^-\right] = \sqrt{\frac{7.1\times10^{-9}}{0.010}} = 8.4\times10^{-4}\,\text{M}$$

We determine the volume of 0.20 M KI needed.

$$\text{volume of KI(aq)} = 100.0\,\text{mL} \times \frac{8.4\times10^{-4}\,\text{mmol I}^-}{1\,\text{mL}} \times \frac{1\,\text{mmol KI}}{1\,\text{mmol I}^-} \times \frac{1\,\text{mL KI(aq)}}{0.20\,\text{mmol KI}} \times \frac{1\,\text{drop}}{0.050\,\text{mL}}$$

$$= 8.4\,\text{drops} = 9\,\text{drops}$$

Since one additional drop is needed, 10 drops will be required. This is an insignificant volume compared to the original solution, so $\left[Pb^{2+}\right]$ remains constant.

6A **(E)** Here we must find the maximum $\left[Ca^{2+}\right]$ that can coexist with $\left[OH^-\right] = 0.040$ M.

$$K_{sp} = 5.5\times10^{-6} = \left[Ca^{2+}\right]\left[OH^-\right]^2 = \left[Ca^{2+}\right](0.040)^2; \qquad \left[Ca^{2+}\right] = \frac{5.5\times10^{-6}}{(0.040)^2} = 3.4\times10^{-3}\,\text{M}$$

For precipitation to be considered complete, $\left[Ca^{2+}\right]$ should be less than 0.1% of its original value. 3.4×10^{-3} M is 34% of 0.010 M and therefore precipitation of $Ca(OH)_2$ is not complete under these conditions.

6B **(E)** We begin by finding $\left[Mg^{2+}\right]$ that corresponds to $1\,\mu g\ Mg^{2+}$ / L.

$$\left[Mg^{2+}\right] = \frac{1\,\mu g\ Mg^{2+}}{1\,\text{L soln}} \times \frac{1\,\text{g}}{10^6\,\mu g} \times \frac{1\,\text{mol Mg}^{2+}}{24.3\,\text{g Mg}^{2+}} = 4\times10^{-8}\,\text{M}$$

Now we use the K_{sp} expression for $Mg(OH)_2$ to determine $\left[OH^-\right]$.

$$K_{sp} = 1.8\times10^{-11} = \left[Mg^{2+}\right]\left[OH^-\right]^2 = (4\times10^{-8})\left[OH^-\right]^2 \qquad \left[OH^-\right] = \sqrt{\frac{1.8\times10^{-11}}{4\times10^{-8}}} = 0.02\,\text{M}$$

7A **(M)** Let us first determine $[Ag^+]$ when $AgCl(s)$ just begins to precipitate. At this point, Q_{sp} and K_{sp} are equal.

$$K_{sp} = 1.8 \times 10^{-10} = [Ag^+][Cl^-] = Q_{sp} = [Ag^+] \times 0.115 M \qquad [Ag^+] = \frac{1.8 \times 10^{-10}}{0.115} = 1.6 \times 10^{-9} M$$

Now let us determine the maximum $[Br^-]$ that can coexist with this $[Ag^+]$.

$$K_{sp} = 5.0 \times 10^{-13} = [Ag^+][Br^-] = 1.6 \times 10^{-9} M \times [Br^-]; \qquad [Br^-] = \frac{5.0 \times 10^{-13}}{1.6 \times 10^{-9}} = 3.1 \times 10^{-4} M$$

The remaining bromide ion has precipitated as $AgBr(s)$ with the addition of $AgNO_3(aq)$.

$$\text{Percent of } Br^- \text{ remaining} = \frac{[Br^-]_{final}}{[Br^-]_{initial}} \times 100\% = \frac{3.1 \times 10^{-4} M}{0.264 M} \times 100\% = 0.12\%$$

7B **(M)** Since the ions have the same charge and the same concentrations, we look for two K_{sp} values for the salt with the same anion that are as far apart as possible. The K_{sp} values for the carbonates are very close, while those for the sulfates and fluorides are quite different. However, the difference in the K_{sp} values is greatest for the chromates; K_{sp} for $BaCrO_4 (1.2 \times 10^{-10})$ is so much smaller than K_{sp} for $SrCrO_4 (2.2 \times 10^{-5})$, $BaCrO_4$ will precipitate first and $SrCrO_4$ will begin to precipitate when $[CrO_4^{2-}]$ has the value:

$$[CrO_4^{2-}] = \frac{K_{sp}}{[Sr^{2+}]} = \frac{2.2 \times 10^{-5}}{0.10} = 2.2 \times 10^{-4} M.$$

At this point $[Ba^{2+}]$ is found as follows.

$$[Ba^{2+}] = \frac{K_{sp}}{[CrO_4^{2-}]} = \frac{1.2 \times 10^{-10}}{2.2 \times 10^{-4}} = 5.5 \times 10^{-7} M;$$

$[Ba^{2+}]$ has dropped to 0.00055% of its initial value and therefore is considered to be completely precipitated, before $SrCrO_4$ begins to precipitate. The two ions are thus effectively separated as chromates. The best precipitating agent is a group 1 chromate salt.

8A **(M)** First determine $[OH^-]$ resulting from the hydrolysis of acetate ion.

Equation:	$CH_3COO^-(aq)$	$+ H_2O(l)$	$\rightleftharpoons$	$CH_3COOH(aq)$	$+ OH^-(aq)$
Initial:	0.10 M	—		0 M	$\approx 0 M$
Changes:	$-x M$	—		$+x M$	$+x M$
Equil:	$(0.10-x) M$	—		$x M$	$x M$

$$K_b = \frac{1.0 \times 10^{-14}}{1.8 \times 10^{-5}} = 5.6 \times 10^{-10} = \frac{[CH_3COOH][OH^-]}{[CH_3COO^-]} = \frac{x \cdot x}{0.10 - x} \approx \frac{x^2}{0.10}$$

$x = [OH^-] = \sqrt{0.10 \times 5.6 \times 10^{-10}} = 7.5 \times 10^{-6} \, M$ (the assumption $x \ll 0.10$ was valid)

Now compute the value of the ion product in this solution and compare it with the value of K_{sp} for $Mg(OH)_2$.

$$Q_{sp} = [Mg^{2+}][OH^-]^2 = (0.010 M)(7.5 \times 10^{-6} M)^2 = 5.6 \times 10^{-13} < 1.8 \times 10^{-11} = K_{sp}[Mg(OH)_2]$$

Because Q_{sp} is smaller than K_{sp}, this solution is unsaturated and precipitation of $Mg(OH)_2(s)$ will not occur.

8B **(M)** Here we can use the Henderson–Hasselbalch equation to determine the pH of the buffer.

$$pH = pK_a + \log \frac{[CH_3COO^-]}{[CH_3COOH]} = -\log(1.8 \times 10^{-5}) + \log \frac{0.250 \, M}{0.150 \, M} = 4.74 + 0.22 = 4.96$$

$pOH = 14.00 - pH = 14.00 - 4.96 = 9.04$ $[OH^-] = 10^{-pOH} = 10^{-9.04} = 9.1 \times 10^{-10} \, M$

Now we determine Q_{sp} to see if precipitation will occur.

$$Q_{sp} = [Fe^{3+}][OH^-]^3 = (0.013 \, M)(9.1 \times 10^{-10})^3 = 9.8 \times 10^{-30}$$

$Q_{sp} > 4 \times 10^{-38} = K_{sp}$; Thus, $Fe(OH)_3$ precipitation should occur.

9A **(M)** Determine $[OH^-]$, and then the pH necessary to prevent the precipitation of $Mn(OH)_2$.

$$K_{sp} = 1.9 \times 10^{-13} = [Mn^{2+}][OH^-]^2 = (0.0050 M)[OH^-]^2$$

$$[OH^-] = \sqrt{\frac{1.9 \times 10^{-13}}{0.0050}} = 6.2 \times 10^{-6} \, M$$

$pOH = -\log(6.2 \times 10^{-6}) = 5.21$ $pH = 14.00 - 5.21 = 8.79$

We will use this pH in the Henderson–Hasselbalch equation to determine $[NH_4^+]$.

$pK_b = 4.74$ for NH_3.

$$pH = pK_a + \log \frac{[NH_3]}{[NH_4^+]} = 8.79 = (14.00 - 4.74) + \log \frac{0.025 \, M}{[NH_4^+]}$$

$$\log \frac{0.025 \, M}{[NH_4^+]} = 8.79 - (14.00 - 4.74) = -0.47 \qquad \frac{0.025}{[NH_4^+]} = 10^{-0.47} = 0.34$$

$$[NH_4^+] = \frac{0.025}{0.34} = 0.074 M$$

9B **(M)** First we must calculate the $[H_3O^+]$ in the buffer solution that is being employed to dissolve the magnesium hydroxide:

$$NH_3(aq) + H_2O(l) \rightleftharpoons NH_4^+(aq) + OH^-(aq) \; ; \; K_b = 1.8 \times 10^{-5}$$

$$K_b = \frac{[NH_4^+][OH^-]}{[NH_3]} = \frac{[0.100M][OH^-]}{[0.250M]} = 1.8 \times 10^{-5}$$

$$[OH^-] = 4.5 \times 10^{-5} \, M \; ; \; [H_3O^+] = \frac{1.00 \times 10^{-14} \, M^2}{4.5 \times 10^{-5} \, M} = 2.2_2 \times 10^{-10} \, M$$

Now we can employ Equation 18.4 to calculate the molar solubility of $Mg(OH)_2$ in the buffer solution; molar solubility $Mg(OH)_2 = [Mg^{2+}]_{equil}$

$$Mg(OH)_2(s) + 2 \, H_3O^+(aq) \xrightarrow{K=1.8 \times 10^{17}} Mg^{2+}(aq) + 4H_2O(l) \; ;$$

Equilibrium $\qquad - \qquad\qquad 2.2 \times 10^{-10} \qquad\qquad\qquad\quad x \qquad\qquad -$

$$K = \frac{[Mg^{2+}]}{[H_3O^+]^2} = \frac{x}{[2.2 \times 10^{-10} \, M]^2} = 1.8 \times 10^{17} \qquad x = 8.7 \times 10^{-3} \, M = [Mg^{2+}]_{equil}$$

So, the molar solubility for $Mg(OH)_2 = 8.7 \times 10^{-3} M$.

10A **(M)**

(a) In solution are $Cu^{2+}(aq)$, $SO_4^{2-}(aq)$, $Na^+(aq)$, and $OH^-(aq)$.

$$Cu^{2+}(aq) + 2\,OH^-(aq) \rightarrow Cu(OH)_2(s)$$

(b) In the solution above, $Cu(OH)_2(s)$ is $Cu^{2+}(aq)$:

$$Cu(OH)_2(s) \rightleftharpoons Cu^{2+}(aq) + 2\,OH^-(aq)$$

This $Cu^{2+}(aq)$ reacts with the added $NH_3(aq)$:

$$Cu^{2+}(aq) + 4\,NH_3(aq) \rightleftharpoons [Cu(NH_3)_4]^{2+}(aq)$$

The overall result is: $Cu(OH)_2(s) + 4\,NH_3(aq) \rightleftharpoons [Cu(NH_3)_4]^{2+}(aq) + 2\,OH^-(aq)$

(c) $HNO_3(aq)$ (a strong acid), forms $H_3O^+(aq)$, which reacts with $OH^-(aq)$ and $NH_3(aq)$.

$$OH^-(aq) + H_3O^+(aq) \rightarrow 2\,H_2O(l); \qquad NH_3(aq) + H_3O^+(aq) \rightarrow NH_4^+(aq) + H_2O(l)$$

As $NH_3(aq)$ is consumed, the reaction below shifts to the left.

$$Cu(OH)_2(s) + 4\,NH_3(aq) \rightleftharpoons [Cu(NH_3)_4]^{2+}(aq) + 2\,OH^-(aq)$$

But as $OH^-(aq)$ is consumed, the dissociation reactions shift to the side with the dissolved ions: $Cu(OH)_2(s) \rightleftharpoons Cu^{2+}(aq) + 2\,OH^-(aq)$

The species in solution at the end of all this are $Cu^{2+}(aq)$, $NO_3^-(aq)$, $NH_4^+(aq)$, excess $H_3O^+(aq)$, $Na^+(aq)$, and $SO_4^{2-}(aq)$ (probably $HSO_4^-(aq)$ as well).

10B (M)

(a)   In solution are $Zn^{2+}(aq)$, $SO_4^{2-}(aq)$, and $NH_3(aq)$,

$$Zn^{2+}(aq)+4NH_3(aq)\rightleftharpoons\left[Zn(NH_3)_4\right]^{2+}(aq)$$

(b) $HNO_3(aq)$, a strong acid, produces $H_3O^+(aq)$, which reacts with $NH_3(aq)$.

$NH_3(aq)+H_3O^+(aq)\rightarrow NH_4^+(aq)+H_2O(l)$ As $NH_3(aq)$ is consumed, the tetrammine complex ion is destroyed.

$$\left[Zn(NH_3)_4\right]^{2+}(aq)+4H_3O^+(aq)\rightleftharpoons\left[Zn(H_2O)_4\right]^{2+}(aq)+4NH_4^+(aq)$$

(c)   NaOH(aq) is a strong base that produces $OH^-(aq)$, forming a hydroxide precipitate.

$$\left[Zn(H_2O)_4\right]^{2+}(aq)+2OH^-(aq)\rightleftharpoons Zn(OH)_2(s)+4H_2O(l)$$

Another possibility is a reversal of the reaction of part **(b)**.

$$\left[Zn(H_2O)_4\right]^{2+}(aq)+4NH_4^+(aq)+4OH^-(aq)\rightleftharpoons\left[Zn(NH_3)_4\right]^{2+}(aq)+8H_2O(l)$$

(d)   The precipitate dissolves in excess base.

$$Zn(OH)_2(s)+2OH^-(aq)\rightleftharpoons\left[Zn(OH)_4\right]^{2-}(aq)$$

11A (M) We first determine $[Ag^+]$ in a solution that is 0.100 M $Ag^+(aq)$ (from $AgNO_3$) and 0.225 M $NH_3(aq)$. Because of the large value of $K_f=1.6\times10^7$, we start by having the reagents form as much complex ion as possible, and approach equilibrium from this point.

Equation:	$Ag^+(aq)$ +	$2NH_3(aq)$	$\rightleftharpoons$	$\left[Ag(NH_3)_2\right]^+(aq)$
In soln	0.100 M	0.225 M		0 M
Form complex	–0.100 M	–0.200 M		+0.100 M
Initial	0 M	0.025 M		0.100 M
Changes	+x M	+2x M		–x M
Equil	x M	(0.025 + x) M		(0.100 – x) M

$$K_f=1.6\times10^7=\frac{\left[\left[Ag(NH_3)_2\right]^+\right]}{\left[Ag^+\right]\left[NH_3\right]^2}=\frac{0.100-x}{x(0.025+2x)^2}\approx\frac{0.100}{x(0.025)^2}$$

$$x=\frac{0.100}{(0.025)^2\,1.6\times10^7}=1.0\times10^{-5}\,M=\left[Ag^+\right]=\text{concentration of free silver ion}$$

($x\ll0.025$ M, so the approximation was valid)

The $\left[Cl^-\right]$ is diluted: $\left[Cl^-\right]_{final}=\left[Cl^-\right]_{initial}\times\dfrac{1.00\,mL_{initial}}{1,500\,mL_{final}}=3.50\,M\div1500=0.00233\,M$

Finally we compare Q_{sp} with K_{sp} to determine if precipitation of AgCl(s) will occur.

$$Q_{sp}=\left[Ag^+\right]\left[Cl^-\right]=(1.0\times10^{-5}\,M)(0.00233\,M)=2.3\times10^{-8}>1.8\times10^{-10}=K_{sp}$$

Because the value of the Q_{sp} is larger than the value of the K_{sp}, precipitation of AgCl(s) should occur.

11B **(M)** We organize the solution around the balanced equation of the formation reaction.

Equation: $\quad$ $Pb^{2+}(aq)$ $\quad + \quad$ $EDTA^{4-}(aq)$ $\rightleftharpoons$ $[PbEDTA]^{2-}(aq)$

Initial	0.100 M	0.250 M	0 M
Form Complex:	–0.100 M	(0.250 – 0.100) M	0.100 M
Equil	x M	$(0.150 + x)$ M	$(0.100 - x)$ M

$$K_f = \frac{\left[[PbEDTA]^{2-}\right]}{\left[Pb^{2+}\right]\left[EDTA^{4-}\right]} = 2\times10^{18} = \frac{0.100 - x}{x(0.150 + x)} \approx \frac{0.100}{0.150x}$$

$$x = \frac{0.100}{0.150 \times 2\times10^{18}} = 3\times10^{-19} \text{ M} \qquad (x \ll 0.100 \text{ M, thus the approximation was valid.})$$

We calculate Q_{sp} and compare it to K_{sp} to determine if precipitation will occur.

$$Q_{sp} = \left[Pb^{2+}\right]\left[I^-\right]^2 = \left(3\times10^{-19}\text{M}\right)\left(0.10\text{M}\right)^2 = 3\times10^{-21}.$$

$Q_{sp} < 7.1\times10^{-9} = K_{sp}$ $\quad$ Thus precipitation will not occur.

12A **(M)** We first determine the maximum concentration of free Ag^+.

$$K_{sp} = \left[Ag^+\right]\left[Cl^-\right] = 1.8\times10^{-10} \qquad \left[Ag^+\right] = \frac{1.8\times10^{-10}}{0.0075} = 2.4\times10^{-8} \text{ M.}$$

This is so small that we assume that all the Ag^+ in solution is present as complex ion:

$\left[\left[Ag(NH_3)_2\right]^+\right] = 0.13$ M. We use K_f to determine the concentration of free NH_3.

$$K_f = \frac{\left[\left[Ag(NH_3)_2\right]^+\right]}{\left[Ag^+\right]\left[NH_3\right]^2} = 1.6\times10^7 = \frac{0.13\text{M}}{2.4\times10^{-8}\left[NH_3\right]^2}$$

$$\left[NH_3\right] = \sqrt{\frac{0.13}{2.4\times10^{-8}\times1.6\times10^7}} = 0.58 \text{ M.}$$

If we combine this with the ammonia present in the complex ion, the total ammonia concentration is $0.58\text{M} + (2\times0.13\text{M}) = 0.84\text{M}$. Thus, the <u>minimum</u> concentration of ammonia necessary to keep AgCl(s) from forming is 0.84 M.

12B **(M)** We use the solubility product constant expression to determine the maximum $[Ag^+]$ that can be present in 0.010 M Cl^- without precipitation occurring.

$$K_{sp} = 1.8 \times 10^{-10} = \left[Ag^+\right]\left[Cl^-\right] = \left[Ag^+\right](0.010\,M) \qquad \left[Ag^+\right] = \frac{1.8 \times 10^{-10}}{0.010} = 1.8 \times 10^{-8}\,M$$

This is also the concentration of free silver ion in the K_f expression. Because of the large value of K_f, practically all of the silver ion in solution is present as the complex ion, $[Ag(S_2O_3)_2]^{3-}$. We solve the expression for $[S_2O_3^{2-}]$ and then add the $[S_2O_3^{2-}]$ "tied up" in the complex ion.

$$K_f = 1.7 \times 10^{13} = \frac{\left[\left[Ag(S_2O_3)_2\right]^{3-}\right]}{\left[Ag^+\right]\left[S_2O_3^{2-}\right]^2} = \frac{0.10\,M}{1.8 \times 10^{-8}\,M\,\left[S_2O_3^{2-}\right]^2}$$

$$\left[S_2O_3^{2-}\right] = \sqrt{\frac{0.10}{1.8 \times 10^{-8} \times 1.7 \times 10^{13}}} = 5.7 \times 10^{-4}\,M = \text{concentration of free } S_2O_3^{2-}$$

$$\text{total } \left[S_2O_3^{2-}\right] = 5.7 \times 10^{-4}\,M + 0.10\,M\left[Ag(S_2O_3)_2\right]^{3-} \times \frac{2\,mol\,S_2O_3^{2-}}{1\,mol\left[Ag(S_2O_3)_2\right]^{3-}}$$

$$= 0.20\,M + 0.00057\,M = 0.20\,M$$

13A **(M)** We must combine the two equilibrium expressions, for K_f and for K_{sp}, to find $K_{overall}$.

$$Fe(OH)_3\,(s) \rightleftharpoons Fe^{3+}\,(aq) + 3\,OH^-\,(aq) \qquad\qquad K_{sp} = 4 \times 10^{-38}$$

$$Fe^{3+}\,(aq) + 3\,C_2O_4^{2-}\,(aq) \rightleftharpoons \left[Fe(C_2O_4)_3\right]^{3-}\,(aq) \qquad K_f = 2 \times 10^{20}$$

$$Fe(OH)_3\,(s) + 3\,C_2O_4^{2-}\,(aq) \rightleftharpoons \left[Fe(C_2O_4)_3\right]^{3-}\,(aq) + 3\,OH^-\,(aq) \quad K_{overall} = 8 \times 10^{-18}$$

Initial	0.100 M	0 M	$\approx 0\,M$
Changes	$-3x$ M	$+x$ M	$+3x$ M
Equil	$(0.100 - 3x)$ M	x M	$3x$ M

$$K_{overall} = \frac{\left[\left[Fe(C_2O_4)_3\right]^{3-}\right]\left[OH^-\right]^3}{\left[C_2O_4^{2-}\right]^3} = \frac{(x)(3x)^3}{(0.100 - 3x)^3} = 8 \times 10^{-18} \approx \frac{27x^4}{(0.100)^3}$$

($3x \ll 0.100$ M, thus the approximation was valid.)

$$x = \sqrt[4]{\frac{(0.100)^3\,8 \times 10^{-18}}{27}} = 4 \times 10^{-6}\,M \quad \text{The assumption is valid.}$$

Thus the solubility of $Fe(OH)_3$ in 0.100 M $C_2O_4^{2-}$ is $4 \times 10^{-6}\,M$.

13B **(M)** In Example 18-13 we saw that the expression for the solubility, s, of a silver halide in an aqueous ammonia solution, where $[NH_3]$ is the concentration of aqueous ammonia, is given by:

$$K_{sp} \times K_f = \left(\frac{s}{[NH_3] - 2s} \right)^2 \quad or \quad \sqrt{K_{sp} \times K_f} = \frac{s}{[NH_3] - 2s}$$

For all scenarios, the $[NH_3]$ stays fixed at 0.1000 M and K_f is always 1.6×10^7. We see that s will decrease as does K_{sp}. The relevant values are:

$$K_{sp}(AgCl) = 1.8 \times 10^{-10}, \; K_{sp}(AgBr) = 5.0 \times 10^{-13}, \; K_{sp}(AgI) = 8.5 \times 10^{-17}.$$

Thus, the order of decreasing solubility must be: AgI > AgBr > AgCl.

14A **(M)** For FeS, we know that $K_{spa} = 6 \times 10^2$; for Ag_2S, $K_{spa} = 6 \times 10^{-30}$.

We compute the value of Q_{spa} in each case, with $[H_2S] = 0.10 \, M$ and $[H_3O^+] = 0.30 \, M$.

For FeS, $Q_{spa} = \dfrac{0.020 \times 0.10}{(0.30)^2} = 0.022 < 6 \times 10^2 = K_{spa}$

Thus, precipitation of FeS should not occur.

For Ag_2S, $Q_{spa} = \dfrac{(0.010)^2 \times 0.10}{(0.30)^2} = 1.1 \times 10^{-4}$

$Q_{spa} > 6 \times 10^{-30} = K_{spa}$; thus, precipitation of Ag_2S should occur.

14B **(M)** The $[H_3O^+]$ needed to just form a precipitate can be obtained by direct substitution of the provided concentrations into the K_{spa} expression. When that expression is satisfied, a precipitate will just form.

$$K_{spa} = \frac{[Fe^{2+}][H_2S]}{[H_3O^+]^2} = 6 \times 10^2 = \frac{(0.015 \, M \, Fe^{2+})(0.10 \, M \, H_2S)}{[H_3O^+]^2}, \; [H_3O^+] = \sqrt{\frac{0.015 \times 0.10}{6 \times 10^2}} = 0.002 \, M$$

$$pH = -\log[H_3O^+] = -\log(0.002) = 2.7$$

INTEGRATIVE EXAMPLES

A **(D)** To determine the amount of $Ca(NO_3)_2$ needed, one has to calculate the amount of Ca^{2+} that will result in only 1.00×10^{-12} M of PO_4^{3-}

$$Ca_3(PO_4)_2 \rightleftharpoons 3\,Ca^{2+} + 2\,PO_4^{3-}$$

$$K_{sp} = (3s)^3 (2s)^2$$

Using the common–ion effect, we will try to determine what concentration of Ca^{2+} ions forces the equilibrium in the lake to have only 1.00×10^{-12} M of phosphate, noting that $(2s)$ is the equilibrium concentration of phosphate.

$$1.30 \times 10^{-32} = \left[Ca^{2+}\right]^3 \left(1.00 \times 10^{-12}\right)^2$$

Solving for $[Ca^{2+}]$ yields a concentration of 0.00235 M.

The volume of the lake: $V = 300 \text{ m} \times 150 \text{ m} \times 5 \text{ m} = 225000 \text{ m}^3$ or 2.25×10^8 L.

Mass of $Ca(NO_3)_2$ is determined as follows:

$$\text{mass Ca}(NO_3)_2 = 2.25 \times 10^8 \text{ L} \times \frac{0.00235 \text{ mol Ca}^{2+}}{\text{L}} \times \frac{1 \text{ mol Ca}(NO_3)_2}{1 \text{ mol Ca}^{2+}} \times \frac{164.1 \text{ g Ca}(NO_3)_2}{1 \text{ mol Ca}(NO_3)_2}$$

$$= 87 \times 10^6 \text{ g Ca}(NO_3)_2$$

B **(M)** The reaction of $AgNO_3$ and Na_2SO_4 is as follows:

$$2AgNO_3 + Na_2SO_4 \rightarrow 2NaNO_3 + Ag_2SO_4$$

$$\text{mol Ag}_2SO_4 = (0.350 \text{ L} \times 0.200 \text{ M}) \text{ AgNO}_3 \times \frac{1 \text{ mol Ag}_2SO_4}{2 \text{ mol AgNO}_3} = 3.5 \times 10^{-2} \text{ mol}$$

$$\text{mol NaNO}_3 = (0.250 \text{ L} \times 0.240 \text{ M}) \text{ Na}_2SO_4 \times \frac{2 \text{ mol NaNO}_3}{1 \text{ mol Na}_2SO_4} = 0.12 \text{ mol}$$

Ag_2SO_4 is the precipitate. Since it is also the limiting reagent, there are 3.5×10^{-2} moles of Ag_2SO_4 produced.

The reaction of Ag_2SO_4 with $Na_2S_2O_3$ is as follows:

$$Na_2S_2O_3 + Ag_2SO_4 \rightarrow Na_2SO_4 + Ag_2S_2O_3$$

$$\text{mol Na}_2S_2O_3 = (0.400 \text{ L} \times 0.500 \text{ M}) \text{ Na}_2S_2O_3 = 0.200 \text{ mol}$$

Ag_2SO_4 is the limiting reagent, so no Ag_2SO_4 precipitate is left.

EXERCISES

K_{sp} and Solubility

1. **(E)**

    (a) $Ag_2SO_4(s) \rightleftharpoons 2\,Ag^+(aq) + SO_4{}^{2-}(aq)$ $\qquad$ $K_{sp} = \left[Ag^+\right]^2\left[SO_4{}^{2-}\right]$

    (b) $Ra(IO_3)_2(s) \rightleftharpoons Ra^{2+}(aq) + 2\,IO_3{}^-(aq)$ $\qquad$ $K_{sp} = \left[Ra^{2+}\right]\left[IO_3{}^-\right]^2$

    (c) $Ni_3(PO_4)_2(s) \rightleftharpoons 3\,Ni^{2+}(aq) + 2\,PO_4{}^{3-}(aq)$ $\quad$ $K_{sp} = \left[Ni^{2+}\right]^3\left[PO_4{}^{3-}\right]^2$

    (d) $PuO_2CO_3(s) \rightleftharpoons PuO_2{}^{2+}(aq) + CO_3{}^{2-}(aq)$ $\quad$ $K_{sp} = \left[PuO_2{}^{2+}\right]\left[CO_3{}^{2-}\right]$

3. **(E)**

    (a) $CrF_3(s) \rightleftharpoons Cr^{3+}(aq) + 3\,F^-(aq)$ $\qquad\qquad$ $K_{sp} = \left[Cr^{3+}\right]\left[F^-\right]^3 = 6.6 \times 10^{-11}$

    (b) $Au_2(C_2O_4)_3(s) \rightleftharpoons 2\,Au^{3+}(aq) + 3\,C_2O_4{}^{2-}(aq)$ $\quad$ $K_{sp} = \left[Au^{3+}\right]^2\left[C_2O_4{}^{2-}\right]^3 = 1 \times 10^{-10}$

    (c) $Cd_3(PO_4)_2(s) \rightleftharpoons 3\,Cd^{2+}(aq) + 2\,PO_4{}^{3-}(aq)$ $\quad$ $K_{sp} = \left[Cd^{2+}\right]^3\left[PO_4{}^{3-}\right]^2 = 2.1 \times 10^{-33}$

    (d) $SrF_2(s) \rightleftharpoons Sr^{2+}(aq) + 2\,F^-(aq)$ $\qquad\qquad$ $K_{sp} = \left[Sr^{2+}\right]\left[F^-\right]^2 = 2.5 \times 10^{-9}$

5. **(E)** We use the value of K_{sp} for each compound in determining the molar solubility in a saturated solution. In each case, s represents the molar solubility of the compound.

    $AgCN$ $\qquad$ $K_{sp} = \left[Ag^+\right]\left[CN^-\right] = (s)(s) = s^2 = 1.2 \times 10^{-16}$ $\qquad$ $s = 1.1 \times 10^{-8}\,M$

    $AgIO_3$ $\qquad$ $K_{sp} = \left[Ag^+\right]\left[IO_3{}^-\right] = (s)(s) = s^2 = 3.0 \times 10^{-8}$ $\qquad$ $s = 1.7 \times 10^{-4}\,M$

    AgI $\qquad$ $K_{sp} = \left[Ag^+\right]\left[I^-\right] = (s)(s) = s^2 = 8.5 \times 10^{-17}$ $\qquad$ $s = 9.2 \times 10^{-9}\,M$

    $AgNO_2$ $\qquad$ $K_{sp} = \left[Ag^+\right]\left[NO_2{}^-\right] = (s)(s) = s^2 = 6.0 \times 10^{-4}$ $\qquad$ $s = 2.4 \times 10^{-2}\,M$

    Ag_2SO_4 $\qquad$ $K_{sp} = \left[Ag^+\right]^2\left[SO_4^{2-}\right] = (2s)^2(s) = 4s^3 = 1.4 \times 10^{-5}$ $\quad$ $s = 1.5 \times 10^{-2}\,M$

    Thus, in order of increasing molar solubility, from smallest to largest:

    $AgI < AgCN < AgIO_3 < Ag_2SO_4 < AgNO_2$

7. **(M)** We determine $[F^-]$ in saturated CaF_2, and from that value the concentration of F^- in ppm.

    For CaF_2 $\quad$ $K_{sp} = \left[Ca^{2+}\right]\left[F^-\right]^2 = (s)(2s)^2 = 4s^3 = 5.3 \times 10^{-9}$ $\qquad$ $s = 1.1 \times 10^{-3}\,M$

519

The solubility in ppm is the number of grams of CaF_2 in 10^6 g of solution. We assume a solution density of 1.00 g/mL.

$$\text{mass of } F^- = 10^6 \text{ g soln} \times \frac{1 \text{ mL}}{1.00 \text{ g soln}} \times \frac{1 \text{ L soln}}{1000 \text{ mL}} \times \frac{1.1 \times 10^{-3} \text{ mol CaF}_2}{1 \text{ L soln}}$$

$$\times \frac{2 \text{ mol F}^-}{1 \text{ mol CaF}_2} \times \frac{19.0 \text{ g F}^-}{1 \text{ mol F}^-} = 42 \text{ g F}^-$$

This is 42 times more concentrated than the optimum concentration of fluoride ion for fluoridation. CaF_2 is, in fact, more soluble than is necessary. Its uncontrolled use might lead to excessive F^- in solution.

9. **(M)** We first assume that the volume of the solution does not change appreciably when its temperature is lowered. Then we determine the mass of $Mg(CH_3(CH_2)_{14}COO)_2$ dissolved in each solution, recognizing that the molar solubility of $Mg(CH_3(CH_2)_{14}COO)_2$ equals the cube root of one fourth of its solubility product constant, since it is the only solute in the solution.

$$K_{sp} = 4s^3 \qquad\qquad s = \sqrt[3]{K_{sp}/4}$$

At 50°C: $s = \sqrt[3]{4.8 \times 10^{-12}/4} = 1.1 \times 10^{-4} \text{ M}$; At 25°C: $s = \sqrt[3]{3.3 \times 10^{-12}/4} = 9.4 \times 10^{-5} \text{ M}$

amount of $Mg(CH_3(CH_2)_{14}COO)_2$ (50°C)

$$= 0.965 \text{ L} \times \frac{1.1 \times 10^{-4} \text{ mol Mg}(CH_3(CH_2)_{14}COO)_2}{1 \text{ L soln}} = 1.1 \times 10^{-4} \text{ mol}$$

amount of $Mg(CH_3(CH_2)_{14}COO)_2$ (25°C)

$$= 0.965 \text{ L} \times \frac{9.4 \times 10^{-5} \text{ mol Mg}(CH_3(CH_2)_{14}COO)_2}{1 \text{ L soln}} = 0.91 \times 10^{-4} \text{ mol}$$

mass of $Mg(CH_3(CH_2)_{14}COO)_2$ precipitated:

$$= (1.1 - 0.91) \times 10^{-4} \text{ mol} \times \frac{535.15 \text{ g Mg}(CH_3(CH_2)_{14}COO)_2}{1 \text{ mol Mg}(CH_3(CH_2)_{14}COO)_2} \times \frac{1000 \text{ mg}}{1 \text{ g}} = 11 \text{ mg}$$

11. **(M)** First we determine $\left[I^-\right]$ in the saturated solution.

$$K_{sp} = \left[Pb^{2+}\right]\left[I^-\right]^2 = 7.1 \times 10^{-9} = (s)(2s)^2 = 4s^3 \qquad\qquad s = 1.2 \times 10^{-3} \text{ M}$$

The $AgNO_3$ reacts with the I^- in this saturated solution in the titration.

$Ag^+(aq) + I^-(aq) \rightarrow AgI(s)$ We determine the amount of Ag^+ needed for this titration, and then $\left[AgNO_3\right]$ in the titrant.

$$\text{moles Ag}^+ = 0.02500\,\text{L} \times \frac{1.2 \times 10^{-3}\,\text{mol PbI}_2}{1\,\text{L soln}} \times \frac{2\,\text{mol I}^-}{1\,\text{mol PbI}_2} \times \frac{1\,\text{mol Ag}^+}{1\,\text{mol I}^-} = 6.0 \times 10^{-5}\,\text{mol Ag}^+$$

$$\text{AgNO}_3 \text{ molarity} = \frac{6.0 \times 10^{-5}\,\text{mol Ag}^+}{0.0133\,\text{L soln}} \times \frac{1\,\text{mol AgNO}_3}{1\,\text{mol Ag}^+} = 4.5 \times 10^{-3}\,\text{M}$$

13. **(M)** We first use the ideal gas law to determine the moles of H_2S gas used.

$$n = \frac{PV}{RT} = \frac{\left(748\,\text{mmHg} \times \dfrac{1\,\text{atm}}{760\,\text{mmHg}}\right) \times \left(30.4\,\text{mL} \times \dfrac{1\,\text{L}}{1000\,\text{mL}}\right)}{0.08206\,\text{L atm mol}^{-1}\,\text{K}^{-1} \times (23+273)\,\text{K}} = 1.23 \times 10^{-3}\,\text{moles}$$

If we assume that all the H_2S is consumed in forming Ag_2S, we can compute the $[Ag^+]$ in the $AgBrO_3$ solution. This assumption is valid if the equilibrium constant for the cited reaction is large, which is the case, as shown below:

$$2Ag^+(aq) + HS^-(aq) + OH^-(aq) \rightleftharpoons Ag_2S(s) + H_2O(l) \qquad K_{a_2}/K_{sp} = \frac{1.0 \times 10^{-19}}{2.6 \times 10^{-51}} = 3.8 \times 10^{31}$$

$$H_2S(aq) + H_2O(l) \rightleftharpoons HS^-(aq) + H_3O^+(aq) \qquad K_1 = 1.0 \times 10^{-7}$$

$$2H_2O(l) \rightleftharpoons H_3O^+(aq) + OH^-(aq) \qquad K_w = 1.0 \times 10^{-14}$$

$$\overline{2Ag^+(aq) + H_2S(aq) + 2H_2O(l) \rightleftharpoons Ag_2S(s) + 2H_3O^+(aq)}$$

$$K_{overall} = (K_{a_2}/K_{sp})(K_1)(K_w) = (3.8 \times 10^{31})(1.0 \times 10^{-7})(1.0 \times 10^{-14}) = 3.8 \times 10^{10}$$

$$[Ag^+] = \frac{1.23 \times 10^{-3}\,\text{mol H}_2S}{338\,\text{mL soln}} \times \frac{1000\,\text{mL}}{1\,\text{L soln}} \times \frac{2\,\text{mol Ag}^+}{1\,\text{mol H}_2S} = 7.28 \times 10^{-3}\,\text{M}$$

Then, for $AgBrO_3$, $K_{sp} = [Ag^+][BrO_3^-] = (7.28 \times 10^{-3})^2 = 5.30 \times 10^{-5}$

The Common-Ion Effect

15. **(E)** We let s = molar solubility of $Mg(OH)_2$ in moles solute per liter of solution.

   (a) $K_{sp} = [Mg^{2+}][OH^-]^2 = (s)(2s)^2 = 4s^3 = 1.8 \times 10^{-11} \qquad s = 1.7 \times 10^{-4}\,\text{M}$

   (b) Equation:

	$Mg(OH)_2(s)$	$\rightleftharpoons$	$Mg^{2+}(aq)$	+	$2OH^-(aq)$
Initial:	–		0.0862 M		≈ 0 M
Changes:	–		$+s$ M		$+2s$ M
Equil:	–		$(0.0862+s)$ M		$2s$ M

$$K_{sp} = (0.0862+s) \times (2s)^2 = 1.8 \times 10^{-11} \approx (0.0862) \times (2s)^2 = 0.34\,s^2 \quad s = 7.3 \times 10^{-6}\,\text{M}$$

($s \ll 0.0802$ M, thus, the approximation was valid.)

(c) $\left[OH^-\right] = \left[KOH\right] = 0.0355 \text{ M}$

Equation :	$Mg(OH)_2 (s)$	$\rightleftharpoons$	$Mg^{2+} (aq)$	$+$	$2OH^- (aq)$
Initial :			$0\,M$		$0.0355\,M$
Changes :		$-$	$+s\,M$		$+2s\,M$
Equil :		$-$	$s\,M$		$(0.0355 + 2s)\,M$

$$K_{sp} = (s)(0.0355 + 2s)^2 = 1.8 \times 10^{-11} \approx (s)(0.0355)^2 = 0.0013\,s \qquad s = 1.4 \times 10^{-8} \text{ M}$$

17. **(E)** The presence of KI in a solution produces a significant $\left[I^-\right]$ in the solution. Not as much AgI can dissolve in such a solution as in pure water, since the ion product, $\left[Ag^+\right]\left[I^-\right]$, cannot exceed the value of K_{sp} (i.e., the I^- from the KI that dissolves represses the dissociation of AgI(s)). In similar fashion, $AgNO_3$ produces a significant $\left[Ag^+\right]$ in solution, again influencing the value of the ion product; not as much AgI can dissolve as in pure water.

19. **(E)** Equation:

	$Ag_2SO_4 (s)$	$\rightleftharpoons$	$2Ag^+ (aq)$	$+$	$SO_4^{2-} (aq)$
Original:	—		$0\,M$		$0.150\,M$
Add solid:	—		$+2x\,M$		$+x\,M$
Equil:	—		$2x\,M$		$(0.150 + x)\,M$

$$2x = \left[Ag^+\right] = 9.7 \times 10^{-3}\,M; \quad x = 0.00485\,M$$

$$K_{sp} = \left[Ag^+\right]^2 \left[SO_4^{2-}\right] = (2x)^2 (0.150 + x) = (9.7 \times 10^{-3})^2 (0.150 + 0.00485) = 1.5 \times 10^{-5}$$

21. **(M)** For $PbI_2, K_{sp} = 7.1 \times 10^{-9} = \left[Pb^{2+}\right]\left[I^-\right]^2$

In a solution where 1.5×10^{-4} mol PbI_2 is dissolved, $\left[Pb^{2+}\right] = 1.5 \times 10^{-4}$ M, and $\left[I^-\right] = 2\left[Pb^{2+}\right] = 3.0 \times 10^{-4}$ M

Equation:

	$PbI_2 (s)$	$\rightleftharpoons$	$Pb^{2+} (aq)$	$+$	$2I^- (aq)$
Initial:	—		1.5×10^{-4} M		3.0×10^{-4} M
Add lead(II):	—		$+x\,M$		
Equil:	—		$(0.00015 + x)$ M		0.00030 M

$$K_{sp} = 7.1 \times 10^{-9} = (0.00015 + x)(0.00030)^2 ; (0.00015 + x) = 0.079 ; x = 0.079 \text{ M} = \left[Pb^{2+}\right]$$

23. **(D)** For Ag_2CrO_4, $K_{sp} = 1.1 \times 10^{-12} = \left[Ag^+\right]^2 \left[CrO_4^{2-}\right]$

In a 5.0×10^{-8} M solution of Ag_2CrO_4, $\left[CrO_4^{2-}\right] = 5.0 \times 10^{-8}$ M and $\left[Ag^+\right] = 1.0 \times 10^{-7}$ M

Equation: $\qquad Ag_2CrO_4(s) \rightleftharpoons 2Ag^+(aq) + CrO_4^{2-}(aq)$

Initial: $\qquad$ — $\qquad\qquad\qquad 1.0 \times 10^{-7}$ M $\quad 5.0 \times 10^{-8}$ M
Add chromate: $\quad$ — $\qquad\qquad\qquad\qquad\qquad\qquad +x$ M
Equil: $\qquad$ — $\qquad\qquad\qquad 1.0 \times 10^{-7}$ M $\quad (5.0 \times 10^{-8} + x)$M

$K_{sp} = 1.1 \times 10^{-12} = (1.0 \times 10^{-7})^2 (5.0 \times 10^{-8} + x); \quad (5.0 \times 10^{-8} + x) = 1.1 \times 10^2 = \left[CrO_4^{2-}\right]$.

This is an impossibly high concentration to reach. Thus, we cannot lower the solubility of Ag_2CrO_4 to 5.0×10^{-8} M with CrO_4^{2-} as the common ion. Let's consider using Ag^+ as the common ion.

Equation: $\qquad Ag_2CrO_4(s) \rightleftharpoons 2Ag^+(aq) + CrO_4^{2-}(aq)$

Initial: $\qquad$ — $\qquad\qquad\qquad 1.0 \times 10^{-7}$ M $\quad 5.0 \times 10^{-8}$ M
Add silver(I) ion: $\quad$ — $\qquad\qquad +x$ M
Equil: $\qquad$ — $\qquad\qquad (1.0 \times 10^{-7} + x)$ M $\quad 5.0 \times 10^{-8}$ M

$K_{sp} = 1.1 \times 10^{-12} = (1.0 + x)^2 (5.0 \times 10^{-8}) \qquad (1.0 \times 10^{-7} + x) = \sqrt{\dfrac{1.1 \times 10^{-12}}{5.0 \times 10^{-8}}} = 4.7 \times 10^{-3}$

$x = 4.7 \times 10^{-3} - 1.0 \times 10^{-7} = 4.7 \times 10^{-3}$ M $= \left[I^-\right]$; this is an easy-to-reach concentration. Thus, the solubility can be lowered to 5.0×10^{-8} M by carefully adding $Ag^+(aq)$.

25. **(E)** $\left[Ca^{2+}\right] = \dfrac{115 \text{ g } Ca^{2+}}{10^6 \text{ g soln}} \times \dfrac{1 \text{ mol } Ca^{2+}}{40.08 \text{ g } Ca^{2+}} \times \dfrac{1000 \text{ g soln}}{1 \text{ L soln}} = 2.87 \times 10^{-3}$ M

$\left[Ca^{2+}\right]\left[F^-\right]^2 = K_{sp} = 5.3 \times 10^{-9} = (2.87 \times 10^{-3})\left[F^-\right]^2 \qquad \left[F^-\right] = 1.4 \times 10^{-3}$ M

ppm $F^- = \dfrac{1.4 \times 10^{-3} \text{ mol } F^-}{1 \text{ L soln}} \times \dfrac{19.00 \text{ g } F^-}{1 \text{ mol } F^-} \times \dfrac{1 \text{ L soln}}{1000 \text{ g}} \times 10^6 \text{ g soln} = 27$ ppm

Criteria for Precipitation from Solution

27. **(E)** We first determine $\left[Mg^{2+}\right]$, and then the value of Q_{sp} in order to compare it to the value of K_{sp}. We express molarity in millimoles per milliliter, entirely equivalent to moles per liter.

$[Mg^{2+}] = \dfrac{22.5 \text{ mg } MgCl_2}{325 \text{ mL soln}} \times \dfrac{1 \text{ mmol } MgCl_2 \cdot 6H_2O}{203.3 \text{ mg } MgCl_2 \cdot 6H_2O} \times \dfrac{1 \text{ mmol } Mg^{2+}}{1 \text{ mmol } MgCl_2} = 3.41 \times 10^{-4}$ M

$Q_{sp} = [Mg^{2+}][F^-]^2 = (3.41 \times 10^{-4})(0.035)^2 = 4.2 \times 10^{-7} > 3.7 \times 10^{-8} = K_{sp}$

Thus, precipitation of $MgF_2(s)$ should occur from this solution.

29. **(E)** We determine the $\left[OH^-\right]$ needed to just initiate precipitation of $Cd(OH)_2$.

$$K_{sp} = \left[Cd^{2+}\right]\left[OH^-\right]^2 = 2.5 \times 10^{-14} = (0.0055 \text{ M})\left[OH^-\right]^2 \qquad \left[OH^-\right] = \sqrt{\frac{2.5 \times 10^{-14}}{0.0055}} = 2.1 \times 10^{-6} \text{ M}$$

$$pOH = -\log\left(2.1 \times 10^{-6}\right) = 5.68 \qquad pH = 14.00 - 5.68 = 8.32$$

Thus, $Cd(OH)_2$ will precipitate from a solution with $pH > 8.32$.

31. **(D)**

(a) First we determine $\left[Cl^-\right]$ due to the added NaCl.

$$\left[Cl^-\right] = \frac{0.10 \text{ mg NaCl}}{1.0 \text{ L soln}} \times \frac{1 \text{ g}}{1000 \text{ mg}} \times \frac{1 \text{ mol NaCl}}{58.4 \text{ g NaCl}} \times \frac{1 \text{ mol Cl}^-}{1 \text{ mol NaCl}} = 1.7 \times 10^{-6} \text{ M}$$

Then we determine the value of the ion product and compare it to the solubility product constant value.

$$Q_{sp} = \left[Ag^+\right]\left[Cl^-\right] = (0.10)\left(1.7 \times 10^{-6}\right) = 1.7 \times 10^{-7} > 1.8 \times 10^{-10} = K_{sp} \text{ for AgCl}$$

Thus, precipitation of AgCl(s) should occur.

(b) The KBr(aq) is diluted on mixing, but the $\left[Ag^+\right]$ and $\left[Cl^-\right]$ are barely affected by dilution.

$$\left[Br^-\right] = 0.10 \text{ M} \times \frac{0.05 \text{ mL}}{0.05 \text{ mL} + 250 \text{ mL}} = 2 \times 10^{-5} \text{ M}$$

Now we determine $\left[Ag^+\right]$ in a saturated AgCl solution.

$$K_{sp} = \left[Ag^+\right]\left[Cl^-\right] = (s)(s) = s^2 = 1.8 \times 10^{-10} \qquad s = 1.3 \times 10^{-5} \text{ M}$$

Then we determine the value of the ion product for AgBr and compare it to the solubility product constant value.

$$Q_{sp} = \left[Ag^+\right]\left[Br\right] = \left(1.3 \times 10^{-5}\right)\left(2 \times 10^{-5}\right) = 3 \times 10^{-10} > 5.0 \times 10^{-13} = K_{sp} \text{ for AgBr}$$

Thus, precipitation of AgBr(s) should occur.

(c) The hydroxide ion is diluted by mixing the two solutions.

$$\left[OH^-\right] = 0.0150 \text{ M} \times \frac{0.05 \text{ mL}}{0.05 \text{ mL} + 3000 \text{ mL}} = 2.5 \times 10^{-7} \text{ M}$$

But the $\left[Mg^{2+}\right]$ does not change significantly.

$$\left[Mg^{2+}\right] = \frac{2.0 \text{ mg Mg}^{2+}}{1.0 \text{ L soln}} \times \frac{1 \text{ g}}{1000 \text{ mg}} \times \frac{1 \text{ mol Mg}^{2+}}{24.3 \text{ g Mg}} = 8.2 \times 10^{-5} \text{ M}$$

Then we determine the value of the ion product and compare it to the solubility product constant value.

$$Q_{sp} = \left[Mg^{2+}\right]\left[OH^-\right]^2 = \left(2.5\times10^{-7}\right)^2\left(8.2\times10^{-5}\right) = 5.1\times10^{-18}$$

$Q_{sp} < 1.8\times10^{-11} = K_{sp}$ for $Mg(OH)_2$ Thus, no precipitate forms.

33. **(D)** First we must calculate the initial $[H_2C_2O_4]$ upon dissolution of the solid acid:

$$[H_2C_2O_4]_{initial} = 1.50 \text{ g } H_2C_2O_4 \times \frac{1 \text{ mol } H_2C_2O_4}{90.036 \text{ g } H_2C_2O_4} \times \frac{1}{0.200 \text{ L}} = 0.0833 \text{ M}$$

(We assume the volume of the solution stays at 0.200 L.)
Next we need to determine the $[C_2O_4^{2-}]$ in solution after the hydrolysis reaction between oxalic acid and water reaches equilibrium. To accomplish this we will need to solve two ICE tables:

Table 1:	$H_2C_2O_4(aq)$	$+ H_2O(l)$	$\xrightarrow{K_{a1}=5.2\times10^{-2}}$	$HC_2O_4^-(aq)$	$+ H_3O^+(aq)$
Initial:	0.0833 M	—		0 M	0 M
Change:	$-x$	—		$+x$ M	$+x$ M
Equilibrium:	$0.0833 - x$ M	—		x M	x M

Since $C_a/K_a = 1.6$, the approximation cannot be used, and thus the full quadratic equation must be solved: $x^2/(0.0833 - x) = 5.2\times10^{-2}$; $x^2 + 5.2\times10^{-2}x - 4.3\underline{3}\times10^{-3}$

$$x = \frac{-5.2\times10^{-2} \pm \sqrt{2.7\times10^{-3} + 0.0173}}{2} \qquad x = 0.045 \text{ M} = [HC_2O_4^-] \approx [H_3O^+]$$

Now we can solve the ICE table for the second proton loss:

Table 2:	$HC_2O_4^-(aq)$	$+ H_2O(l)$	$\xrightarrow{K_{a1}=5.4\times10^{-5}}$	$C_2O_4^{2-}(aq)$	$+ H_3O^+(aq)$
Initial:	0.045 M	—		0 M	≈ 0.045 M
Change:	$-y$	—		$+y$ M	$+y$ M
Equilibrium:	$(0.045 - y)$ M	—		y M	$\approx (0.045 + y)$ M

Since $C_a/K_a = 833$, the approximation may not be valid and we yet again should solve the full quadratic equation:

$$\frac{y\times(0.045+y)}{(0.045-y)} = 5.4\times10^{-5}; \qquad y^2 + 0.045y = 2.4\underline{3}\times10^{-6} - 5.4\times10^{-5}y$$

$$y = \frac{-0.045 \pm \sqrt{2.0\underline{3}\times10^{-3} + 9.7\underline{2}\times10^{-6}}}{2} \qquad y = 8.2\times10^{-5} \text{ M} = [C_2O_4^{2-}]$$

Now we can calculate the Q_{sp} for the calcium oxalate system:
$Q_{sp} = [Ca^{2+}]_{initial} \times [C_2O_4^{2-}]_{initial} = (0.150)(8.2\times10^{-5}) = 1.2\times10^{-5} > 1.3\times10^{-9}$ (K_{sp} for CaC_2O_4)
Thus, CaC_2O_4 should precipitate from this solution.

Completeness of Precipitation

35. **(M)** First determine that a precipitate forms. The solutions mutually dilute each other.

$$\left[CrO_4^{2-}\right] = 0.350 \text{ M} \times \frac{200.0 \text{ mL}}{200.0 \text{ mL} + 200.0 \text{ mL}} = 0.175 \text{ M}$$

$$\left[Ag^+\right] = 0.0100 \text{ M} \times \frac{200.0 \text{ mL}}{200.0 \text{ mL} + 200.0 \text{ mL}} = 0.00500 \text{ M}$$

We determine the value of the ion product and compare it to the solubility product constant value.

$$Q_{sp} = \left[Ag^+\right]^2\left[CrO_4^{2-}\right] = (0.00500)^2(0.175) = 4.4 \times 10^{-6} > 1.1 \times 10^{-12} = K_{sp} \text{ for } Ag_2CrO_4$$

Ag_2CrO_4 should precipitate.

Now, we assume that as much solid forms as possible, and then we approach equilibrium by dissolving that solid in a solution that contains the ion in excess.

Equation:	$Ag_2CrO_4\,(s)$	$\xrightarrow{1.1 \times 10^{-12}}$	$2Ag^+\,(aq)$	$+$	$CrO_4^{2-}\,(aq)$
Orig. soln :	–		0.00500 M		0.175 M
Form solid :	–		−0.00500 M		−0.00250 M
Not at equilibrium	–		0 M		0.173 M
Changes :	–		+2x M		+x M
Equil :	–		2x M		$(0.173 + x)$ M

$$K_{sp} = \left[Ag^+\right]^2\left[CrO_4^{2-}\right] = 1.1 \times 10^{-12} = (2x)^2(0.173 + x) \approx (4x^2)(0.173)$$

$$x = \sqrt{1.1 \times 10^{-12}/(4 \times 0.173)} = 1.3 \times 10^{-6} \text{ M} \qquad \left[Ag^+\right] = 2x = 2.6 \times 10^{-6} \text{ M}$$

$$\% \text{ Ag}^+ \text{ unprecipitated} = \frac{2.6 \times 10^{-6} \text{ M final}}{0.00500 \text{ M initial}} \times 100\% = 0.052\% \text{ unprecipitated}$$

37. **(M)** We first use the solubility product constant expression to determine $\left[Pb^{2+}\right]$ in a solution with 0.100 M Cl^-.

$$K_{sp} = \left[Pb^{2+}\right]\left[Cl^-\right]^2 = 1.6 \times 10^{-5} = \left[Pb^{2+}\right](0.100)^2 \qquad \left[Pb^{2+}\right] = \frac{1.6 \times 10^{-5}}{(0.100)^2} = 1.6 \times 10^{-3} \text{ M}$$

Thus, % unprecipitated $= \dfrac{1.6 \times 10^{-3} \text{ M}}{0.065 \text{ M}} \times 100\% = 2.5\%$

Now, we want to determine what $[Cl^-]$ must be maintained to keep $[Pb^{2+}]_{final} = 1\%$;

$[Pb^{2+}]_{initial} = 0.010 \times 0.065 \text{ M} = 6.5 \times 10^{-4} \text{ M}$

$$K_{sp} = \left[Pb^{2+}\right]\left[Cl^-\right]^2 = 1.6 \times 10^{-5} = \left(6.5 \times 10^{-4}\right)\left[Cl^-\right]^2 \qquad \left[Cl^-\right]\sqrt{\frac{6 \times 10^{-5}}{6.5 \times 10^{-4}}} = 0.16 \text{ M}$$

Fractional Precipitation

39. **(M)** First, assemble all of the data. K_{sp} for $Ca(OH)_2 = 5.5 \times 10^{-6}$, K_{sp} for $Mg(OH)_2 = 1.8 \times 10^{-11}$

$$[Ca^{2+}] = \frac{440 \text{ g } Ca^{2+}}{1000 \text{ kg seawater}} \times \frac{1 \text{ mol } Ca^{2+}}{40.078 \text{ g } Ca^{2+}} \times \frac{1 \text{ kg seawater}}{1000 \text{ g seawater}} \times \frac{1.03 \text{ kg seawater}}{1 \text{ L seawater}} = 0.0113 \text{ M}$$

$[Mg^{2+}] = 0.059$ M, obtained from Example 18-6. $[OH^-] = 0.0020$ M (maintained)

(a) $Q_{sp} = [Ca^{2+}] \times [OH^-]^2 = (0.0113)(0.0020)^2 = 4.5 \times 10^{-8}$ $\quad Q_{sp} < K_{sp} \therefore$ no precipitate forms.

(b) For the separation to be complete, $\gg 99.9\%$ of the Mg^{2+} must be removed before Ca^{2+} begins to precipitate. We have already shown that Ca^{2+} will not precipitate if the $[OH^-] = 0.0020$ M and is maintained at this level. Let us determine how much of the 0.059 M Mg^{2+} will still be in solution when $[OH^-] = 0.0020$ M.
$K_{sp} = [Mg^{2+}] \times [OH^-]^2 = (x)(0.0020)^2 = 1.8 \times 10^{-11}$ $\qquad x = 4.5 \times 10^{-6}$ M

The percent Mg^{2+} ion left in solution $= \dfrac{4.5 \times 10^{-6}}{0.059} \times 100\% = 0.0076 \%$

This means that $100\% - 0.0076\%$ Mg $= 99.992\%$ has precipitated.
Clearly, the magnesium ion has been separated from the calcium ion (i.e., $\gg 99.9\%$ of the Mg^{2+} ions have precipitated and virtually all of the Ca^{2+} ions are still in solution.)

41. **(M)**

(a) Here we need to determine $\left[I^- \right]$ when AgI just begins to precipitate, and $\left[I^- \right]$ when PbI_2 just begins to precipitate.

$$K_{sp} = \left[Ag^+ \right]\left[I^- \right] = 8.5 \times 10^{-17} = (0.10)\left[I^- \right] \qquad \left[I^- \right] = 8.5 \times 10^{-16} \text{ M}$$

$$K_{sp} = \left[Pb^{2+} \right]\left[I^- \right]^2 = 7.1 \times 10^{-9} = (0.10)\left[I^- \right]^2 \qquad \left[I^- \right] = \sqrt{\frac{7.1 \times 10^{-9}}{0.10}} = 2.7 \times 10^{-4} \text{ M}$$

Since 8.5×10^{-16} M is less than 2.7×10^{-4} M, AgI will precipitate before PbI_2.

(b) $\left[I^- \right]$ must be equal to 2.7×10^{-4} M before the second cation, Pb^{2+}, begins to precipitate.

(c) $K_{sp} = \left[Ag^+ \right]\left[I^- \right] = 8.5 \times 10^{-17} = \left[Ag^+ \right]\left(2.7 \times 10^{-4}\right) \quad \left[Ag^+ \right] = 3.1 \times 10^{-13} \text{ M}$

(d) Since $\left[Ag^+ \right]$ has decreased to much less than 0.1% of its initial value before any PbI_2 begins to precipitate, we conclude that Ag^+ and Pb^{2+} can be separated by precipitation with iodide ion.

43. **(M)** First, let's assemble all of the data. K_{sp} for AgCl $= 1.8 \times 10^{-10}$ K_{sp} for AgI $= 8.5 \times 10^{-17}$

$[Ag^+] = 2.00$ M $[Cl^-] = 0.0100$ M $[I^-] = 0.250$ M

(a) AgI(s) will be the first to precipitate by virtue of the fact that the K_{sp} value for AgI is about 2 million times smaller than that for AgCl.

(b) AgCl(s) will begin to precipitate when the Q_{sp} for AgCl(s) $> K_{sp}$ for AgCl(s). The concentration of Ag^+ required is: $K_{sp} = [Ag^+][Cl^-] = 1.8 \times 10^{-10} = (0.0100) \times (x)$
$x = 1.8 \times 10^{-8}$ M
Using this data, we can determine the remaining concentration of I^- using the K_{sp}.
$K_{sp} = [Ag^+][I^-] = 8.5 \times 10^{-17} = (x) \times (1.8 \times 10^{-8})$ $x = 4.7 \times 10^{-9}$ M

(c) In part (b) we saw that the $[I^-]$ drops from 0.250 M $\rightarrow 4.7 \times 10^{-9}$ M. Only a small percentage of the ion remains in solution. $\dfrac{4.7 \times 10^{-9}}{0.250} \times 100\% = 0.0000019\%$

This means that 99.999998% of the I^- ion has been precipitated before any of the Cl^- ion has precipitated. Clearly, the fractional separation of Cl^- from I^- is feasible.

Solubility and pH

45. **(E)** In each case we indicate whether the compound is more soluble in water. We write the net ionic equation for the reaction in which the solid dissolves in acid. Substances are more soluble in acid if either (1) an acid-base reaction occurs or (2) a gas is produced, since escape of the gas from the reaction mixture causes the reaction to shift to the right.

Same: KCl (K^+ and Cl^- do not react appreciably with H_2O)

Acid: $MgCO_3(s) + 2H^+(aq) \longrightarrow Mg^{2+}(aq) + H_2O(l) + CO_2(g)$

Acid: $FeS(s) + 2H^+(aq) \longrightarrow Fe^{2+}(aq) + H_2S(g)$

Acid: $Ca(OH)_2(s) + 2H^+(aq) \longrightarrow Ca^{2+}(aq) + 2H_2O(l)$

Water: C_6H_5COOH is less soluble in acid, because of the H_3O^+ common ion.

47. **(E)** We determine $\left[Mg^{2+}\right]$ in the solution.

$$\left[Mg^{2+}\right] = \frac{0.65 \text{ g } Mg(OH)_2}{1 \text{ L soln}} \times \frac{1 \text{ mol } Mg(OH)_2}{58.3 \text{ g } Mg(OH)_2} \times \frac{1 \text{ mol } Mg^{2+}}{1 \text{ mol } Mg(OH)_2} = 0.011 \text{M}$$

Then we determine $\left[OH^-\right]$ in the solution, and its pH.

$$K_{sp} = \left[Mg^{2+}\right]\left[OH^-\right]^2 = 1.8 \times 10^{-11} = (0.011)\left[OH^-\right]^2 ; \quad \left[OH^-\right] = \sqrt{\frac{1.8 \times 10^{-11}}{0.011}} = 4.0 \times 10^{-5} \text{ M}$$

$$pOH = -\log(4.0 \times 10^{-5}) = 4.40 \qquad pH = 14.00 - 4.40 = 9.60$$

49. **(M)**

(a) Here we calculate $\left[OH^-\right]$ needed for precipitation.

$$K_{sp} = [Al^{3+}][OH^-]^3 = 1.3\times10^{-33} = (0.075 \text{ M})[OH^-]^3$$

$$[OH^-] = \sqrt[3]{\frac{1.3\times10^{-33}}{0.075}} = 2.6\times10^{-11} \quad pOH = -\log\left(2.6\times10^{-11}\right) = 10.59$$

$$pH = 14.00 - 10.59 = 3.41$$

(b) We can use the Henderson–Hasselbalch equation to determine $\left[CH_3COO^-\right]$.

$$pH = 3.41 = pK_a + \log\frac{\left[CH_3COO^-\right]}{\left[CH_3COOH\right]} = 4.74 + \log\frac{\left[CH_3COO^-\right]}{1.00 \text{ M}}$$

$$\log\frac{\left[CH_3COO^-\right]}{1.00 \text{ M}} = 3.41 - 4.74 = -1.33; \quad \frac{\left[CH_3COO^-\right]}{1.00 \text{ M}} = 10^{-1.33} = 0.047; \quad \left[CH_3COO^-\right] = 0.047 \text{ M}$$

This situation does not quite obey the guideline that the ratio of concentrations must fall in the range 0.10 to 10.0, but the resulting error is a small one in this circumstance.

$$\text{mass NaCH}_3\text{COO} = 0.2500 \text{ L} \times \frac{0.047 \text{ mol CH}_3\text{COO}^-}{1 \text{ L soln}} \times \frac{1 \text{ mol NaCH}_3\text{COO}}{1 \text{ mol CH}_3\text{COO}^-} \times \frac{82.03 \text{ g NaCH}_3\text{COO}}{1 \text{ mol NaCH}_3\text{COO}}$$

$$= 0.96 \text{ g NaCH}_3\text{COO}$$

Complex-Ion Equilibria

51. **(E)** Lead(II) ion forms a complex ion with chloride ion. It forms no such complex ion with nitrate ion. The formation of this complex ion decreases the concentrations of free $Pb^{2+}(aq)$ and free $Cl^-(aq)$. Thus, $PbCl_2$ will dissolve in the HCl(aq) up until the value of the solubility product is exceeded. $Pb^{2+}(aq) + 3Cl^-(aq) \rightleftharpoons [PbCl_3]^-(aq)$

53. **(E)** We substitute the given concentrations directly into the K_f expression to calculate K_f.

$$K_f = \frac{[[Cu(CN)_4^{3-}]]}{[Cu^+][CN^-]^4} = \frac{0.0500}{(6.1\times10^{-32})(0.80)^4} = 2.0\times10^{30}$$

55. **(M)** We first find the concentration of free metal ion. Then we determine the value of Q_{sp} for the precipitation reaction, and compare its value with the value of K_{sp} to determine whether precipitation should occur.

Equation:	$Ag^+(aq)+$	$2\,S_2O_3^{2-}(aq)$	$\rightleftharpoons$	$\left[Ag(S_2O_3)_2\right]^{3-}(aq)$
Initial:	0 M	0.76 M		0.048 M
Changes:	$+x$ M	$+2x$ M		$-x$ M
Equil:	x M	$(0.76+2x)$ M		$(0.048-x)$ M

$$K_f = \frac{\left[\left[Ag(S_2O_3)_2\right]^{3-}\right]}{\left[Ag^+\right]\left[S_2O_3^{2-}\right]^2} = 1.7\times10^{13} = \frac{0.048-x}{x(0.76+2x)^2} \approx \frac{0.048}{(0.76)^2 x}; x = 4.9\times10^{-15}\,M = \left[Ag^+\right]$$

($x \ll 0.048$ M, thus the approximation was valid.)

$$Q_{sp} = \left[Ag^+\right]\left[I^-\right] = (4.9\times10^{-15})(2.0) = 9.8\times10^{-15} > 8.5\times10^{-17} = K_{sp}.$$

Because $Q_{sp} > K_{sp}$, precipitation of AgI(s) should occur.

57. **(M)** We first compute the free $\left[Ag^+\right]$ in the original solution. The size of the complex ion formation equilibrium constant indicates that the reaction lies far to the right, so we form as much complex ion as possible stoichiometrically.

Equation:	$Ag^+(aq)+$	$2NH_3(aq)$	$\rightleftharpoons$	$\left[Ag(NH_3)_2\right]^+(aq)$
In soln:	0.10 M	1.00 M		0 M
Form complex:	-0.10 M	-0.20 M		$+0.10$ M
	0 M	0.80 M		0.10 M
Changes:	$+x$ M	$+2x$ M		$-x$ M
Equil:	x M	$(0.80+2x)$ M		$(0.10-x)$ M

$$K_f = 1.6\times10^7 = \frac{\left[\left[Ag(NH_3)_2\right]^+\right]}{\left[Ag^+\right]\left[NH_3\right]^2} = \frac{0.10-x}{x(0.80+2x)^2} \approx \frac{0.10}{x(0.80)^2} \quad x = \frac{0.10}{1.6\times10^7 (0.80)^2} = 9.8\times10^{-9}\,M.$$

($x \ll 0.80$ M, thus the approximation was valid.)

Thus, $\left[Ag^+\right] = 9.8\times10^{-9}$ M. We next determine the $\left[I^-\right]$ that can coexist in this solution without precipitation.

$$K_{sp} = \left[Ag^+\right]\left[I^-\right] = 8.5\times10^{-17} = (9.8\times10^{-9})\left[I^-\right]; \quad \left[I^-\right] = \frac{8.5\times10^{-17}}{9.8\times10^{-9}} = 8.7\times10^{-9}\,M$$

Finally, we determine the mass of KI needed to produce this $\left[I^-\right]$

$$\text{mass KI} = 1.00\ \text{L soln}\times\frac{8.7\times10^{-9}\,\text{mol I}^-}{1\ \text{L soln}}\times\frac{1\,\text{mol KI}}{1\,\text{mol I}^-}\times\frac{166.0\ \text{g KI}}{1\,\text{mol KI}} = 1.4\times10^{-6}\ \text{g KI}$$

Precipitation and Solubilities of Metal Sulfides

59. **(M)** We know that $K_{spa} = 3 \times 10^7$ for MnS and $K_{spa} = 6 \times 10^2$ for FeS. The metal sulfide will begin to precipitate when $Q_{spa} = K_{spa}$. Let us determine the $[H_3O^+]$ just necessary to form each precipitate. We assume that the solution is saturated with H_2S, $[H_2S] = 0.10$ M.

$$K_{spa} = \frac{[M^{2+}][H_2S]}{[H_3O^+]^2} \quad [H_3O^+] = \sqrt{\frac{[M^{2+}][H_2S]}{K_{spa}}} = \sqrt{\frac{(0.10)(0.10)}{3 \times 10^7}} = 1.8 \times 10^{-5} \text{ M for MnS}$$

$$[H_3O^+] = \sqrt{\frac{(0.10)(0.10)}{6 \times 10^2}} = 4.1 \times 10^{-3} \text{ M for FeS}$$

Thus, if the $[H_3O^+]$ is maintained just a bit higher than 1.8×10^{-5} M, FeS will precipitate and $Mn^{2+}(aq)$ will remain in solution. To determine if the separation is complete, we see whether $[Fe^{2+}]$ has decreased to 0.1% or less of its original value when the solution is held at the aforementioned acidity. Let $[H_3O^+] = 2.0 \times 10^{-5}$ M and calculate $[Fe^{2+}]$.

$$K_{spa} = \frac{[Fe^{2+}][H_2S]}{[H_3O^+]^2} = 6 \times 10^2 = \frac{[Fe^{2+}](0.10)}{(2.0 \times 10^{-5})^2}; \quad [Fe^{2+}] = \frac{(6 \times 10^2)(2.0 \times 10^{-5})^2}{0.10} = 2.4 \times 10^{-6} \text{ M}$$

$$\% Fe^{2+}(aq) \text{ remaining} = \frac{2.4 \times 10^{-6} \text{ M}}{0.10 \text{ M}} \times 100\% = 0.0024\% \quad \therefore \text{ Separation is complete.}$$

61. **(M)**

(a) We can calculate $[H_3O^+]$ in the buffer with the Henderson–Hasselbalch equation.

$$pH = pK_a + \log \frac{[CH_3COO^-]}{[CH_3COOH]} = 4.74 + \log \frac{0.15 \text{ M}}{0.25 \text{ M}} = 4.52 \quad [H_3O^+] = 10^{-4.52} = 3.0 \times 10^{-5} \text{ M}$$

We use this information to calculate a value of Q_{spa} for MnS in this solution and then comparison of Q_{spa} with K_{spa} will allow us to decide if a precipitate will form.

$$Q_{spa} = \frac{[Mn^{2+}][H_2S]}{[H_3O^+]^2} = \frac{(0.15)(0.10)}{(3.0 \times 10^{-5})^2} = 1.7 \times 10^7 < 3 \times 10^7 = K_{spa} \text{ for MnS}$$

Thus, precipitation of MnS(s) will not occur.

(b) We need to change $[H_3O^+]$ so that

$$Q_{spa} = 3 \times 10^7 = \frac{(0.15)(0.10)}{[H_3O^+]^2}; \quad [H_3O^+] = \sqrt{\frac{(0.15)(0.10)}{3 \times 10^7}} \quad [H_3O^+] = 2.2 \times 10^{-5} \text{ M} \quad pH = 4.66$$

This is a more basic solution, which we can produce by increasing the basic component of the buffer solution, namely, the acetate ion. We can find out the necessary acetate ion concentration with the Henderson–Hasselbalch equation.

$$pH = pK_a + \log\frac{[CH_3COO^-]}{[CH_3COOH]} = 4.66 = 4.74 + \log\frac{[CH_3COO^-]}{0.25\ M}$$

$$\log\frac{[CH_3COO^-]}{0.25\ M} = 4.66 - 4.74 = -0.08$$

$$\frac{[CH_3COO^-]}{0.25\ M} = 10^{-0.08} = 0.83 \quad [CH_3COO^-] = 0.83 \times 0.25\ M = 0.21\ M$$

Qualitative Cation Analysis

63. **(E)** The purpose of adding hot water is to separate Pb^{2+} from $AgCl$ and Hg_2Cl_2. Thus, the most important consequence would be the absence of a valid test for the presence or absence of Pb^{2+}. In addition, if we add NH_3 first, $PbCl_2$ may form $Pb(OH)_2$. If $Pb(OH)_2$ does form, it will be present with Hg_2Cl_2 in the solid, although $Pb(OH)_2$ will not darken with added NH_3. Thus, we might falsely conclude that Ag^+ is present.

65. **(E)**

(a) Ag^+ and/or Hg_2^{2+} are probably present. Both of these cations form chloride precipitates from acidic solutions of chloride ion.

(b) We cannot tell whether Mg^{2+} is present or not. Both MgS and $MgCl_2$ are water soluble.

(c) Pb^{2+} possibly is absent; it is the only cation of those given which forms a precipitate in an acidic solution that is treated with H_2S, and no sulfide precipitate was formed.

(d) We cannot tell whether Fe^{2+} is present. FeS will not precipitate from an acidic solution that is treated with H_2S; the solution must be alkaline for a FeS precipitate to form.

(a) and (c) are the valid conclusions.

INTEGRATIVE AND ADVANCED EXERCISES

67. **(M)** We determine s, the solubility of $CaSO_4$ in a saturated solution, and then the concentration of $CaSO_4$ in ppm in this saturated solution, assuming that the solution's density is 1.00 g/mL.

$$K_{sp} = [Ca^{2+}][SO_4^{2-}] = (s)(s) = s^2 = 9.1\times10^{-6} \qquad s = 3.0\times10^{-3}\ M$$

$$ppm\ CaSO_4 = 10^6 g\ soln \times \frac{1\ mL}{1.00\ g} \times \frac{1\ L\ soln}{1000\ mL} \times \frac{0.0030\ mol\ CaSO_4}{1\ L\ soln} \times \frac{136.1\ g\ CaSO_4}{1\ mol\ CaSO_4} = 4.1\times10^2\ ppm$$

Now we determine the volume of solution remaining after we evaporate the 131 ppm $CaSO_4$ down to a saturated solution (assuming that both solutions have a density of 1.00 g/mL.)

$$\text{volume sat'd soln} = 131 \text{ g } CaSO_4 \times \frac{10^6 \text{ g sat'd soln}}{4.1 \times 10^2 \text{ g } CaSO_4} \times \frac{1 \text{ mL}}{1.00 \text{ g}} = 3.2 \times 10^5 \text{ mL}$$

Thus, we must evaporate 6.8×10^5 mL of the original 1.000×10^6 mL of solution, or 68% of the water sample.

69. **(M)** The solutions mutually dilute each other and, because the volumes are equal, the concentrations are halved in the final solution: $[Ca^{2+}]$ = 0.00625 M, $[SO_4^{2-}]$ = 0.00760 M. We cannot assume that either concentration remains constant during the precipitation. Instead, we assume that precipitation proceeds until all of one reagent is used up. Equilibrium is reached from that point.

Equation:	$CaSO_4(s) \rightleftharpoons$	$Ca^{2+}(aq)$	+	$SO_4^{2-}(aq)$	$K_{sp} = 9.1 \times 10^{-6}$
In soln	–	0.00625 M		0.00760 M	$K_{sp} = [Ca^{2+}][SO_4^{2-}]$
Form ppt	–	– 0.00625 M		– 0.00625 M	$K_{sp} = (x)(0.00135 + x)$
Not at equil	–	0 M		0.00135 M	$K_{sp} \approx 0.00135\, x$
Changes	–	+ x M		+ x M	$x = \dfrac{9.1 \times 10^{-6}}{0.00135} = 6.7 \times 10^{-3} \text{ M}$
Equil:	–	x M		(0.00135 + x) M	{not a reasonable assumption!}

Solving the quadratic $0 = x^2 + (1.35 \times 10^{-2})x - 9.1 \times 10^{-6}$ yields $x = 2.4 \times 10^{-3}$.

$$\text{\% unprecipitated } Ca^{2+} = \frac{2.4 \times 10^{-3} \text{ M}_{final}}{0.00625 \text{ M}_{initial}} \times 100\% = 38\% \text{ unprecipitated}$$

71. **(M)** The pH of the buffer establishes $[H_3O^+] = 10^{-pH} = 10^{-3.00} = 1.0 \times 10^{-3}$ M.

Now we combine the two equilibrium expressions, and solve the resulting expression for the molar solubility of $Pb(N_3)_2$ in the buffer solution.

$Pb(N_3)_2(s) \rightleftharpoons Pb^{2+}(aq) + 2 N_3^-(aq)$ $\qquad\qquad K_{sp} = 2.5 \times 10^{-9}$

$\underline{2 H_3O^+(aq) + 2 N_3^-(aq) \rightleftharpoons 2 HN_3(aq) + 2 H_2O(l)} \quad 1/K_a^2 = \dfrac{1}{(1.9 \times 10^{-5})^2} = 2.8 \times 10^9$

$Pb(N_3)_2(s) + 2 H_3O^+(aq) \rightleftharpoons Pb^{2+}(aq) + 2 HN_3(aq) + 2 H_2O(l) \quad K = K_{sp} \times (1/K_a^2) = 7.0$

$$\text{Pb(N}_3)_2(s) + 2\,\text{H}_3\text{O}^+(aq) \rightleftharpoons \text{Pb}^{2+}(aq) \;+\; 2\,\text{HN}_3(aq) \;+\; 2\,\text{H}_2\text{O}\,(l)$$

Buffer:	–	0.0010 M	0 M	0 M	–
Dissolving:	–	(buffer)	$+x$ M	$+2x$ M	–
Equilibrium:	–	0.0010 M	x M	$2x$ M	–

$$K = \frac{[\text{Pb}^{2+}][\text{HN}_3]^2}{[\text{H}_3\text{O}^+]^2} = 7.0 = \frac{(x)\,(2x)^2}{(0.0010)^2} \qquad 4x^3 = 7.0\,(0.0010)^2 \qquad x = \sqrt[3]{\frac{7.0(0.0010)^2}{4}} = 0.012 \text{ M}$$

Thus, the molar solubility of Pb(N$_3$)$_2$ in a pH = 3.00 buffer is 0.012 M.

74. We use the Henderson-Hasselbalch equation to determine $[\text{H}_3\text{O}^+]$ in this solution, and then use the K_{sp} expression for MnS to determine $[\text{Mn}^{2+}]$ that can exist in this solution without precipitation occurring.

$$\text{pH} = pK_a + \log\frac{[\text{CH}_3\text{COO}^-]}{[\text{CH}_3\text{COOH}]} = 4.74 + \log\frac{0.500 \text{ M}}{0.100 \text{ M}} = 4.74 + 0.70 = 5.44$$

$$[\text{H}_3\text{O}^+] = 10^{-5.44} = 3.6 \times 10^{-6} \text{ M}$$

$$\text{MnS}(s) + 2\,\text{H}_3\text{O}^+(aq) \rightleftharpoons \text{Mn}^{2+}(aq) + \text{H}_2\text{S}(aq) + 2\,\text{H}_2\text{O}(l) \qquad K_{spa} = 3 \times 10^7$$

Note that $[\text{H}_2\text{S}] = [\text{Mn}^{2+}] = s$, the molar solubility of MnS.

$$K_{spa} = \frac{[\text{Mn}^{2+}][\text{H}_2\text{S}]}{[\text{H}_3\text{O}^+]^2} = 3 \times 10^7 = \frac{s^2}{(3.6 \times 10^{-6}\,\text{M})^2} \qquad s = 0.02 \text{ M}$$

$$\text{mass MnS/L} = \frac{0.02 \text{ mol Mn}^{2+}}{1 \text{ L soln}} \cdot \frac{1 \text{ mol MnS}}{1 \text{ mol Mn}^{2+}} \times \frac{87 \text{ g MnS}}{1 \text{ mol MnS}} = 2 \text{ g MnS/L}$$

77. **(M)**

(a)

$$\text{BaSO}_4(s) \rightleftharpoons \text{Ba}^{2+}(aq) + \text{SO}_4^{2-}(aq) \qquad\qquad K_{sp} = 1.1 \times 10^{-10}$$

$$\underline{\text{Ba}^{2+}(aq) + \text{CO}_3^{2-}(aq) \rightleftharpoons \text{BaCO}_3(s) \qquad\qquad \frac{1}{K_{sp}} = \frac{1}{5.1 \times 10^{-9}}}$$

$$\text{Sum} \quad \text{BaSO}_4(s) + \text{CO}_3^{2-}(aq) \rightleftharpoons \text{BaCO}_3(s) + \text{SO}_4^{2-}(aq) \quad K_{overall} = \frac{1.1 \times 10^{-10}}{5.1 \times 10^{-9}} = 0.0216$$

Initial	–	3 M	–	≈ 0
Equil.	–	$(3-x)$M	–	x

(where x is the carbonate used up in the reaction)

$$K = \frac{[\text{SO}_4^{2-}]}{[\text{CO}_3^{2-}]} = \frac{x}{3-x} = 0.0216 \text{ and } x = 0.063 \quad \text{Since } 0.063 \text{ M} > 0.050 \text{ M, the response is yes.}$$

(b)

$$2\,AgCl(s) \rightleftharpoons 2\,Ag^+(aq) + 2\,Cl^-(aq) \qquad K_{sp} = (1.8 \times 10^{-10})^2$$

$$2\,Ag^+(aq) + CO_3^{2-}(aq) \rightleftharpoons Ag_2CO_3(s) \qquad 1/K_{sp} = \dfrac{1}{8.5 \times 10^{-12}}$$

Sum $\quad 2\,AgCl(s) + CO_3^{2-}(aq) \rightleftharpoons Ag_2CO_3(s) + 2\,Cl^-(aq) \qquad K_{overall} = \dfrac{(1.8 \times 10^{-10})^2}{8.5 \times 10^{-12}} = 3.8 \times 10^{-9}$

Initial	–	3 M	–	≈ 0 M
Equil.	–	$(3-x)$ M	–	$2x$

(where x is the carbonate used up in the reaction)

$$K_{overall} = \frac{[Cl^-]^2}{[CO_3^{2-}]} = \frac{(2x)^2}{3-x} = 3.8 \times 10^{-9} \text{ and } x = 5.3 \times 10^{-5} \text{ M}$$

Since $2x$, $(2(5.35 \times 10^{-5} \text{M}))$, $\ll 0.050$ M, the response is no.

(c)

$$MgF_2(s) \rightleftharpoons Mg^{2+}(aq) + 2\,F^-(aq) \qquad K_{sp} = 3.7 \times 10^{-8}$$

$$Mg^{2+}(aq) + CO_3^{2-}(aq) \rightleftharpoons MgCO_3(s) \qquad 1/K_{sp} = \dfrac{1}{3.5 \times 10^{-8}}$$

sum $\quad MgF_2(s) + CO_3^{2-}(aq) \rightleftharpoons MgCO_3(s) + 2\,F^-(aq) \qquad K_{overall} = \dfrac{3.7 \times 10^{-8}}{3.5 \times 10^{-8}} = 1.1$

initial	–	3 M	–	0 M
equil.	–	$(3-x)$ M	–	$2x$

(where x is the carbonate used up in the reaction)

$$K_{overall} = \frac{[F^-]^2}{[CO_3^{2-}]} = \frac{(2x)^2}{3-x} = 1.1 \text{ and } x = 0.769 \text{M}$$

Since $2x$, $(2(0.769\text{M})) > 0.050$ M, the response is yes.

80. (M) We combine the solubility product expression for AgCN(s) with the formation expression for $[Ag(NH_3)_2]^+(aq)$.

Solubility: $\qquad\qquad AgCN(s) \rightleftharpoons Ag^+(aq) + CN^-(aq) \qquad K_{sp} = ?$

Formation: $\quad Ag^+(aq) + 2\,NH_3(aq) \rightleftharpoons [Ag(NH_3)_2]^+(aq) \qquad K_f = 1.6 \times 10^{+7}$

Net reaction: $\quad AgCN(s) + 2\,NH_3(aq) \rightleftharpoons [Ag(NH_3)_2]^+ + CN^-(aq) \; K_{overall} = K_{sp} \times K_f$

$$K_{overall} = \frac{[[Ag(NH_3)_2]^+][CN^-]}{[NH_3]^2} = \frac{(8.8 \times 10^{-6})^2}{(0.200)^2} = 1.9 \times 10^{-9} = K_{sp} \times 1.6 \times 10^{+7}$$

$$K_{sp} = \frac{1.9 \times 10^{-9}}{1.6 \times 10^{+7}} = 1.2 \times 10^{-16}$$

Because of the extremely low solubility of AgCN in the solution, we assumed that $[NH_3]$ was not altered by the formation of the complex ion.

82. (M) We first determine $[Pb^{2+}]$ in this solution, which has $[Cl^-] = 0.10$ M.

$$K_{sp} = 1.6 \times 10^{-5} = [Pb^{2+}][Cl^-]^2 \qquad [Pb^{2+}] = \frac{K_{sp}}{[Cl^-]^2} = \frac{1.6 \times 10^{-5}}{(0.10)^2} = 1.6 \times 10^{-3}\,M$$

Then we use this value of $[Pb^{2+}]$ in the K_f expression to determine $[[PbCl_3]^-]$.

$$K_f = \frac{[[PbCl_3]^-]}{[Pb^{2+}][Cl^-]^3} = 24 = \frac{[[PbCl_3]^-]}{(1.6 \times 10^{-3})(0.10)^3} = \frac{[[PbCl_3]^-]}{1.6 \times 10^{-6}}$$

$$[[PbCl_3]^-] = 24 \times 1.6 \times 10^{-6} = 3.8 \times 10^{-5}\,M$$

The solubility of $PbCl_2$ in 0.10 M HCl is the following sum.

$$\text{solubility} = [Pb^{2+}] + [[PbCl_3]^-] = 1.6 \times 10^{-3}\,M + 3.8 \times 10^{-5}\,M = 1.6 \times 10^{-3}\,M$$

83. (M) Two of the three relationships needed to answer this question are the two solubility product expressions. $\quad 1.6 \times 10^{-8} = [Pb^{2+}][SO_4^{2-}] \quad 4.0 \times 10^{-7} = [Pb^{2+}][S_2O_3^{2-}]$

The third expression required is the electroneutrality equation, which states that the total positive charge concentration must equal the total negative charge concentration: $[Pb^{2+}] = [SO_4^{2-}] + [S_2O_3^{2-}]$, provided $[H_3O^+] = [OH^-]$.

Or, put another way, there is one Pb^{2+} ion in solution for each SO_4^{2-} ion *and* for each $S_2O_3^{2-}$ ion. We solve each of the first two expressions for the concentration of each anion, substitute these expressions into the electroneutrality expression, and then solve for $[Pb^{2+}]$.

$$[SO_4^{2-}] = \frac{1.6 \times 10^{-8}}{[Pb^{2+}]} \qquad [S_2O_3^{2-}] = \frac{4.0 \times 10^{-7}}{[Pb^{2+}]} \qquad [Pb^{2+}] = \frac{1.6 \times 10^{-8}}{[Pb^{2+}]} + \frac{4.0 \times 10^{-7}}{[Pb^{2+}]}$$

$$[Pb^{2+}]^2 = 1.6 \times 10^{-8} + 4.0 \times 10^{-7} = 4.2 \times 10^{-7} \qquad [Pb^{2+}] = \sqrt{4.2 \times 10^{-7}} = 6.5 \times 10^{-4}\,M$$

85. (D) (a) First let us determine if there is sufficient Ag_2SO_4 to produce a saturated solution, in which $[Ag^+] = 2\,[SO_4^{2-}]$.

$$K_{sp} = 1.4 \times 10^{-5} = [Ag^+]^2[SO_4^{2-}] = 4[SO_4^{2-}]^3 \qquad [SO_4^{2-}] = \sqrt[3]{\frac{1.4 \times 10^{-5}}{4}} = 0.015\,M$$

$$[SO_4^{2-}] = \frac{2.50\,g\,Ag_2SO_4}{0.150\,L} \times \frac{1\,mol\,Ag_2SO_4}{311.8\,g\,Ag_2SO_4} \times \frac{1\,mol\,SO_4^{2-}}{1\,mol\,Ag_2SO_4} = 0.0535\,M$$

Thus, there is more than enough Ag_2SO_4 present to form a saturated solution. Let us now see if AgCl or $BaSO_4$ will precipitate under these circumstances. $[SO_4^{2-}] = 0.015$ M and $[Ag^+] = 0.030$ M.

$Q = [Ag^+][Cl^-] = 0.030 \times 0.050 = 1.5 \times 10^{-3} > 1.8 \times 10^{-10} = K_{sp}$, thus AgCl should precipitate.

$Q = [Ba^{2+}][SO_4^{2-}] = 0.025 \times 0.015 = 3.8 \times 10^{-4} > 1.1 \times 10^{-10} = K_{sp}$, thus $BaSO_4$ should precipitate.

Thus, the net ionic equation for the reaction that will occur is as follows.

$$Ag_2SO_4(s) + Ba^{2+}(aq) + 2\,Cl^-(aq) \longrightarrow BaSO_4(s) + 2\,AgCl(s)$$

(b) Let us first determine if any $Ag_2SO_4(s)$ remains or if it is all converted to $BaSO_4(s)$ and $AgCl(s)$. Thus, we have to solve a limiting reagent problem.

$$\text{amount } Ag_2SO_4 = 2.50 \text{ g } Ag_2SO_4 \times \frac{1 \text{ mol } Ag_2SO_4}{311.8 \text{ g } Ag_2SO_4} = 8.02 \times 10^{-3} \text{ mol } Ag_2SO_4$$

$$\text{amount } BaCl_2 = 0.150 \text{ L} \times \frac{0.025 \text{ mol } BaCl_2}{1 \text{ L soln}} = 3.75 \times 10^{-3} \text{ mol } BaCl_2$$

Since the two reactants combine in a 1 mole to 1 mole stoichiometric ratio, $BaCl_2$ is the limiting reagent. Since there must be some Ag_2SO_4 present and because Ag_2SO_4 is so much more soluble than either $BaSO_4$ or $AgCl$, we assume that $[Ag^+]$ and $[SO_4^{2-}]$ are determined by the solubility of Ag_2SO_4. They will have the same values as in a saturated solution of Ag_2SO_4.

$$[SO_4^{2-}] = 0.015 \text{ M} \qquad\qquad [Ag^+] = 0.030 \text{ M}$$

We use these values and the appropriate K_{sp} values to determine $[Ba^{2+}]$ and $[Cl^-]$.

$$[Ba^{2+}] = \frac{1.1 \times 10^{-10}}{0.015} = 7.3 \times 10^{-9} \text{ M} \qquad\qquad [Cl^-] = \frac{1.8 \times 10^{-10}}{0.030} = 6.0 \times 10^{-9} \text{ M}$$

Since $BaCl_2$ is the limiting reagent, we can use its amount to determine the masses of $BaSO_4$ and $AgCl$.

$$\text{mass } BaSO_4 = 0.00375 \text{ mol } BaCl_2 \times \frac{1 \text{ mol } BaSO_4}{1 \text{ mol } BaCl_2} \times \frac{233.4 \text{ g } BaSO_4}{1 \text{ mol } BaSO_4} = 0.875 \text{ g } BaSO_4$$

$$\text{mass } AgCl = 0.00375 \text{ mol } BaCl_2 \times \frac{2 \text{ mol } AgCl}{1 \text{ mol } BaCl_2} \times \frac{143.3 \text{ g } AgCl}{1 \text{ mol } AgCl} = 1.07 \text{ g } AgCl$$

The mass of unreacted Ag_2SO_4 is determined from the initial amount and the amount that reacts with $BaCl_2$.

$$\text{mass } Ag_2SO_4 = \left(0.00802 \text{ mol } Ag_2SO_4 - 0.00375 \text{ mol } BaCl_2 \times \frac{1 \text{ mol } Ag_2SO_4}{1 \text{ mol } BaCl_2} \right) \times \frac{311.8 \text{ g } Ag_2SO_4}{1 \text{ mol } Ag_2SO_4}$$

$$\text{mass } Ag_2SO_4 = 1.33 \text{ g } Ag_2SO_4 \text{ unreacted}$$

Of course, there is some Ag_2SO_4 dissolved in solution. We compute its mass.

$$\text{mass dissolved } Ag_2SO_4 = 0.150 \text{ L} \times \frac{0.0150 \text{ mol } Ag_2SO_4}{1 \text{ L soln}} \times \frac{311.8 \text{ g } Ag_2SO_4}{1 \text{ mol } Ag_2SO_4}$$

$$= 0.702 \text{ g dissolved}$$

$$\text{mass } Ag_2SO_4(s) = 1.33 \text{ g } - 0.702 \text{ g} = 0.63 \text{ g } Ag_2SO_4(s)$$

FEATURE PROBLEMS

87. **(M)** $\left[Ca^{2+}\right]=\left[SO_4^{2-}\right]$ in the saturated solution. Let us first determine the amount of H_3O^+ in the 100.0 mL diluted effluent. $H_3O^+(aq)+NaOH(aq)\longrightarrow 2H_2O(l)+Na^+(aq)$

$$\text{mmol } H_3O^+ = 100.0 \text{ mL} \times \frac{8.25 \text{ mL base}}{10.00 \text{ mL sample}} \times \frac{0.0105 \text{ mmol NaOH}}{1 \text{ mL base}} \times \frac{1 \text{ mmol } H_3O^+}{1 \text{ mmol NaOH}}$$

$$= 0.866 \text{ mmol } H_3O^+(aq)$$

Now we determine $\left[Ca^{2+}\right]$ in the original 25.00 mL sample, remembering that $2H_3O^+$ were produced for each Ca^{2+}.

$$\left[Ca^{2+}\right] = \frac{0.866 \text{ mmol } H_3O^+(aq) \times \dfrac{1 \text{ mmol } Ca^{2+}}{2 \text{ mmol } H_3O^+}}{25.00 \text{ mL}} = 0.0173 \text{ M}$$

$$K_{sp} = \left[Ca^{2+}\right]\left[SO_4^{2-}\right] = (0.0173)^2 = 3.0\times10^{-4}; \text{ the } K_{sp} \text{ for } CaSO_4 \text{ is } 9.1\times10^{-6} \text{ in Appendix D.}$$

89. **(D)**

(a) We need to calculate the $\left[Mg^{2+}\right]$ in a solution that is saturated with $Mg(OH)_2$.

$$K_{sp} = 1.8\times10^{-11} = \left[Mg^{2+}\right]\left[OH^-\right]^2 = (s)(2s)^2 = 4s^3$$

$$s = \sqrt[3]{\frac{1.8\times10^{-11}}{4}} = 1.7\times10^{-4} \text{ M} = \left[Mg^{2+}\right]$$

(b) Even though water has been added to the original solution, it remains saturated (it is in equilibrium with the undissolved solid $Mg(OH)_2$). $\left[Mg^{2+}\right] = 1.7\times10^{-4} \text{M}$.

(c) Although HCl(aq) reacts with OH^-, it will not react with Mg^{2+}. The solution is simply a more dilute solution of Mg^{2+}.

$$\left[Mg^{2+}\right] = 1.7\times10^{-4} \text{ M} \times \frac{100.0 \text{ mL initial volume}}{(100.0+500.) \text{ mL final volume}} = 2.8\times10^{-5} \text{ M}$$

(d) In this instance, we have a dual dilution to a 275.0 mL total volume, followed by a common-ion scenario.

$$\text{initial } \left[Mg^{2+}\right] = \frac{\left(25.00 \text{ mL} \times \dfrac{1.7\times10^{-4} \text{ mmol } Mg^{2+}}{1 \text{ mL}}\right) + \left(250.0 \text{ mL} \times \dfrac{0.065 \text{ mmol } Mg^{2+}}{1 \text{ mL}}\right)}{275.0 \text{ mL total volume}}$$

$$= 0.059 \text{ M}$$

$$\text{initial } \left[OH^-\right] = \frac{25.00 \text{ mL} \times \dfrac{1.7\times10^{-4} \text{ mmol } Mg^{2+}}{1 \text{ mL}} \times \dfrac{2 \text{ mmol } OH^-}{1 \text{ mmol } Mg^{2+}}}{275.0 \text{ mL total volume}} = 3.1\times10^{-5} \text{ M}$$

Let's see if precipitation occurs.

$$Q_{sp} = \left[Mg^{2+} \right]\left[OH^- \right]^2 = (0.059)\left(3.1 \times 10^{-5}\right)^2 = 5.7 \times 10^{-11} > 1.8 \times 10^{-11} = K_{sp}$$

Thus, precipitation does occur, but very little precipitate forms. If $\left[OH^- \right]$ goes down by 1.4×10^{-5} M (which means that $\left[Mg^{2+} \right]$ drops by 0.7×10^{-5} M), then $\left[OH^- \right] = 1.7 \times 10^{-5}$ M and

$$\left[Mg^{2+} \right] = \left(0.059 \text{ M} - 0.7 \times 10^{-5} \text{ M} =\right) 0.059 \text{ M}, \text{ then } Q_{sp} < K_{sp} \text{ and precipitation will stop. Thus, } \left[Mg^{2+} \right] = 0.059 \text{ M}.$$

(e) Again we have a dual dilution, now to a 200.0 mL final volume, followed by a common-ion scenario.

$$\text{initial } \left[Mg^{2+} \right] = \frac{50.00 \text{ mL} \times \dfrac{1.7 \times 10^{-4} \text{ mmol } Mg^{2+}}{1 \text{ mL}}}{200.0 \text{ mL total volume}} = 4.3 \times 10^{-5} \text{ M}$$

$$\text{initial } \left[OH^- \right] = 0.150 \text{ M} \times \frac{150.0 \text{ mL initial volume}}{200.0 \text{ mL total volume}} = 0.113 \text{ M}$$

Now it is evident that precipitation will occur. Next we determine the $\left[Mg^{2+} \right]$ that can exist in solution with 0.113 M OH^-. It is clear that $\left[Mg^{2+} \right]$ will drop dramatically to satisfy the K_{sp} expression but the larger value of $\left[OH^- \right]$ will scarcely be affected.

$$K_{sp} = \left[Mg^{2+} \right]\left[OH^- \right]^2 = 1.8 \times 10^{-11} = \left[Mg^{2+} \right](0.0113 \text{ M})^2$$

$$\left[Mg^{2+} \right] = \frac{1.8 \times 10^{-11}}{(0.113)^2} = 1.4 \times 10^{-9} \text{ M}$$

SELF-ASSESSMENT EXERCISES

93. **(E)** The answer is (d). See the reasoning below.
(a) Wrong, because the stoichiometry is wrong
(b) Wrong, because $K_{sp} = [Pb^{2+}] \cdot [I^-]^2$
(c) Wrong, because $[Pb^{2+}] = K_{sp}/[I^-]^2$
(d) Correct because of the $[Pb^{2+}]:2[I^-]$ stoichiometry

94. **(E)** The answer is (a). $BaSO_4(s) \rightleftharpoons Ba^{2+} + SO_4^{2-}$. If Na_2SO_4 is added, the common–ion effect forces the equilibrium to the left and reduces $[Ba^{2+}]$.

95. **(E)** The answer is (c). Choices (b) and (d) reduce solubility because of the common–ion effect. In the case of choice (c), the diverse non-common–ion effect or the "salt effect" causes more Ag_2CrO_4 to dissolve.

96. **(E)** The answer is (b). The sulfate salt of Cu is soluble, whereas the Pb salt is insoluble.

97. **(M)** The answers are (c) and (d). Adding NH_3 causes the solution to become basic (adding OH^-). Mg, Fe, Cu, and Al all have insoluble hydroxides. However, only Cu can form a complex ion with NH_3, which is soluble. In the case of $(NH_4)_2SO_4$, it is slightly acidic and dissolved readily in a base.

98. **(M)** The answer is (a). $CaCO_3$ is slightly basic, so it is more soluble in an acid. The only option for an acid given is NH_4Cl.

99. **(M)** The answer is (c). Referring to Figure 18-7, it is seen that ammonia is added to an aqueous H_2S solution to precipitate more metal ions. Since ammonia is a base, increasing the pH should cause more precipitation.

100. **(M)**
(a) $H_2C_2O_4$ is a moderately strong acid, so it is more soluble in a basic solution.
(b) $MgCO_3$ is slightly basic, so it is more soluble in an acidic solution.
(c) CdS is more soluble in acidic solutions, but the solubility is still so small that it is essentially insoluble even in acidic solutions.
(d) KCl is a neutral salt, and therefore its solubility is independent of pH.
(e) $NaNO_3$ is a neutral salt, and therefore its solubility is independent of pH.
(f) $Ca(OH)_2$, a strong base, is more soluble in an acidic solution.

101. **(E)** The answer is NH_3. NaOH(aq) precipitates both, and HCl(aq) precipitates neither. $Mg(OH)_2$ precipitates from an NH_3(aq) solution but forms the soluble complex $Cu(NH_3)_4(OH)_2$.

102. **(D)** $Al(OH)_3$ will precipitate. To demonstrate this, the pH of the acetate buffer needs to be determined first, from which the OH^- can be determined. The OH^- concentration can be used to calculate Q_{sp}, which can then be compared to K_{sp} to see if any $Al(OH)_3$ will precipitate. This is shown below:

$$pH = pK_a + \log\frac{\left[CH_3COO^-\right]}{\left[CH_3COOH\right]} = -\log\left(1.8\times10^{-5}\right) + \log\frac{0.35\ M}{0.45\ M} = 4.65$$

$$[OH^-] = 10^{-(14-pH)} = 4.467\times10^{-10}\ M$$

$$Al(OH)_3 \rightleftharpoons Al^{3+} + 3\,OH^-$$

$$Q_{sp} = (s)(3s)^3$$

$$Q_{sp} = (0.275)(4.467 \times 10^{-10})^3 = 2.45 \times 10^{-29}$$

$Q_{sp} > K_{sp}$, therefore there will be precipitation.

103. **(E)** The answer is (b). Based on solubility rules, $Cu_3(PO_4)_2$ is the only species that is sparingly soluble in water.

104. **(M)** The answer is (d). The abbreviated work shown for each part calculates the molar solubility (s) for all the salts. They all follow the basic outlined below:

$$M_xA_y(s) \rightleftharpoons xM + yA(aq)$$

$$K_{sp} = (x \cdot s)^x \cdot (y \cdot s)^y$$

and we solve for s.

(a) $MgF_2(s) \rightleftharpoons Mg^{2+}(aq) + 2\,F^-(aq)$

$$3.7 \times 10^{-8} = s \cdot (2s)^2 = 4s^3$$

$$s = 2.1 \times 10^{-3}\ M$$

(c) $Mg_3(PO_4)_2(s) \rightleftharpoons 3\,Mg^{2+}(aq) + 2\,PO_4^{\,3-}(aq)$

$$1 \times 10^{-25} = (3s)^3 \cdot (2s)^2 = 108\,s^5$$

$$s = 3.9 \times 10^{-6}\ M$$

(b) $MgCO_3(s) \rightleftharpoons Mg^{2+}(aq) + CO_3^{2-}(aq)$

$$3.5 \times 10^{-8} = s \cdot s$$

$$s = 1.9 \times 10^{-4}\ M$$

(d) $Li_3PO_4(s) \rightleftharpoons 3\,Li^+(aq) + PO_4^{\,3-}(aq)$

$$3.2 \times 10^{-9} = (3s)^3 \cdot s = 27\,s^4$$

$$s = 3.3 \times 10^{-3}\ M$$

105. **(E)** The answer is (b). This is due to the "salt effect." The more moles of salt there are available, the greater the solubility. For (b), there are 0.300 moles of ions ($3 \times 0.100\ M$ $Na_2S_2O_3$).

106. **(E)** It will decrease the amount of precipitate. Since all salts of NO_3^- are highly soluble and Ag^+ and Hg^{2+} salts of I^- are not, anything that forms a soluble complex ion with Ag^+ and Hg^{2+} will reduce the amount of those ions available for precipitation with I^- and therefore will reduce the amount of precipitate.

107. **(M)** No precipitate will form. To demonstrate this, one has to calculate $[Ag^+]$, and then, using $[I^-]$, determine the Q_{sp} and compare it to K_{sp} of AgI.

Since K_f is for the formation of the complex ion $Ag(CN)_2^-$, its inverse is for the dissociation of $Ag(CN)_2^-$ to Ag^+ and CN^-.

$$K_{dis} = 1/K_f = 1/5.6 \times 10^{18} = 1.786 \times 10^{-19}$$

The dissociation of $Ag(CN)_2^-$ is as follows:

$$Ag(CN)_2^-(aq) \rightleftharpoons Ag^+ + 2CN^-$$

$$K_{dis} = \frac{x(1.05 + x)}{(0.012 - x)} = 1.786 \times 10^{-19}$$

We can simplify the calculations by noting that x is very small in relation to $[Ag^+]$ and $[CN^-]$. Therefore, $x = 1.70 \times 10^{-19}$ M.

Dissociation of AgI is as follows:

$$AgI \rightleftharpoons Ag^+ + I^-$$

$$Q_{sp} = (1.7 \times 10^{-19})(2.0) = 3.4 \times 10^{-19}$$

Since $Q_{sp} < K_{sp}$, no precipitate will form.

108. **(M)** In both cases, the dissolution reaction is similar to reaction (18.5), for which $K = K_f \times K_{sp}$. This is an exceedingly small quantity for (a) but large for (b). $CuCO_3$ is soluble in NH_3(aq) and CuS(s) is not.

CHAPTER 19

ELECTROCHEMISTRY

PRACTICE EXAMPLES

1A **(E)** The conventions state that the anode material is written first, and the cathode material is written last.

Anode, oxidation: $Sc(s) \rightarrow Sc^{3+}(aq) + 3e^-$

Cathode, reduction: $\{Ag^+(aq) + e^- \rightarrow Ag(s)\} \times 3$

Net reaction $Sc(s) + 3Ag^+(aq) \rightarrow Sc^{3+}(aq) + 3Ag(s)$

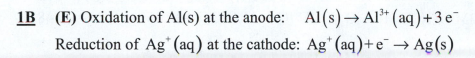

1B **(E)** Oxidation of Al(s) at the anode: $Al(s) \rightarrow Al^{3+}(aq) + 3e^-$

Reduction of $Ag^+(aq)$ at the cathode: $Ag^+(aq) + e^- \rightarrow Ag(s)$

Overall reaction in cell: $Al(s) + 3Ag^+(aq) \rightarrow Al^{3+}(aq) + 3Ag(s)$

Diagram: $Al(s)|Al^{3+}(aq)||Ag^+(aq)|Ag(s)$

2A **(E)** Anode, oxidation: $Sn(s) \rightarrow Sn^{2+}(aq) + 2e^-$

Cathode, reduction: $\{Ag^+(aq) + 1e^- \rightarrow Ag(s)\} \times 2$

Overall reaction in cell: $Sn(s) + 2Ag^+(aq) \rightarrow Sn^{2+}(aq) + 2Ag(s)$

2B **(E)** Anode, oxidation: $\{In(s) \rightarrow In^{3+}(aq) + 3e^-\} \times 2$

Cathode, reduction: $\{Cd^{2+}(aq) + 2e^- \rightarrow Cd(s)\} \times 3$

Overall reaction in cell: $2In(s) + 3Cd^{2+}(aq) \rightarrow 2In^{3+}(aq) + 3Cd(s)$

3A **(M)** We obtain the two balanced half-equations and the half-cell potentials from Table 19.1.

Oxidation: $\{Fe^{2+}(aq) \rightarrow Fe^{3+}(aq) + e^-\} \times 2$ $\qquad -E^\circ = -0.771V$

Reduction: $Cl_2(g) + 2e^- \rightarrow 2Cl^-(aq)$ $\qquad E^\circ = +1.358V$

Net: $2Fe^{2+}(aq) + Cl_2(g) \rightarrow 2Fe^{3+}(aq) + 2Cl^-(aq)$; $E^\circ_{cell} = +1.358V - 0.771V = +0.587V$

3B **(M)** Since we need to refer to Table 19.1, in any event, it is perhaps a bit easier to locate the two balanced half-equations in the table. There is only one half-equation involving both $Fe^{2+}(aq)$ and $Fe^{3+}(aq)$ ions. It is reversed and written as an oxidation below. The half-equation involving $MnO_4^-(aq)$ is also written below. [Actually, we need to know that in acidic solution $Mn^{2+}(aq)$ is the principal reduction product of $MnO_4^-(aq)$.]

Oxidation: $\{Fe^{2+}(aq) \rightarrow Fe^{3+}(aq) + e^-\} \times 5$ $\qquad$ $-E^\circ = -0.771V$

Reduction: $MnO_4^-(aq) + 8\,H^+(aq) + 5\,e^- \rightarrow Mn^{2+}(aq) + 4\,H_2O(l)$ $\qquad$ $E^\circ = +1.51V$

Net: $MnO_4^-(aq) + 5\,Fe^{2+}(aq) + 8\,H^+(aq) \rightarrow Mn^{2+}(aq) + 5\,Fe^{3+}(aq) + 4\,H_2O(l)$

$E^\circ_{cell} = +1.51\,V - 0.771\,V = +0.74\,V$

4A **(M)** We write down the oxidation half-equation following the method of Chapter 5, and obtain the reduction half-equation from Table 19.1, along with the reduction half-cell potential.

Oxidation: $\{H_2C_2O_4(aq) \longrightarrow 2\,CO_2(aq) + 2\,H^+(aq) + 2\,e^-\} \times 3$ $\qquad$ $-E^\circ\{CO_2/H_2C_2O_4\}$

Reduction: $Cr_2O_7^{2-}(aq) + 14\,H^+(aq) + 6\,e^- \rightarrow 2\,Cr^{3+}(aq) + 7\,H_2O(l)$ $\qquad$ $E^\circ = +1.33\,V$

Net: $Cr_2O_7^{2-}(aq) + 8\,H^+(aq) + 3\,H_2C_2O_4(aq) \rightarrow 2\,Cr^{3+}(aq) + 7\,H_2O(l) + 6\,CO_2(g)$

$E^\circ_{cell} = +1.81\,V = +1.33\,V - E^\circ\{CO_2/H_2C_2O_4\}; \quad E^\circ\{CO_2/H_2C_2O_4\} = 1.33\,V - 1.81\,V = -0.48\,V$

4B **(M)** The 2nd half-reaction must have $O_2(g)$ as reactant and $H_2O(l)$ as product.

Oxidation: $\{Cr^{2+} + (aq) \longrightarrow Cr^{3+}(aq) + e^-\} \times 4$ $\qquad$ $-E^\circ\{Cr^{3+}/Cr^{2+}\}$

Reduction: $O_2(g) + 4\,H^+(aq) + 4\,e^- \rightarrow 2\,H_2O(l)$ $\qquad$ $E^\circ = +1.229V$

Net: $O_2(g) + 4\,H^+(aq) + 4\,Cr^{2+}(aq) \rightarrow 2\,H_2O(l) + 4\,Cr^{3+}(aq)$

$E^\circ_{cell} = +1.653V = +1.229V - E^\circ\{Cr^{3+}/Cr^{2+}\}; \quad E^\circ\{Cr^{3+}/Cr^{2+}\} = 1.229V - 1.653V = -0.424V$

5A **(M)** First we write down the two half-equations, obtain the half-cell potential for each, and then calculate E°_{cell}. From that value, we determine $\Delta_r G^\circ$

Oxidation: $\{Al(s) \rightarrow Al^{3+}(aq) + 3\,e^-\} \times 2$ $\qquad$ $-E^\circ = +1.676V$

Reduction: $\{Br_2(l) + 2\,e^- \rightarrow 2\,Br^-(aq)\} \times 3$ $\qquad$ $E^\circ = +1.065V$

Net: $2\,Al(s) + 3\,Br_2(l) \rightarrow 2\,Al^{3+}(aq) + 6\,Br^-(aq)$ $\quad E^\circ_{cell} = 1.676\,V + 1.065\,V = 2.741\,V$

$\Delta_r G^\circ = -zFE^\circ_{cell} = -6\text{ mol e}^- \times \dfrac{96,485\text{ C}}{1\text{ mol e}^-} \times 2.741\text{ V} = -1.587 \times 10^6\text{ J} = -1587\text{ kJ}$

5B **(M)** First we write down the two half-equations, one of which is the reduction equation from the previous example. The other is the oxidation that occurs in the standard hydrogen electrode.

Oxidation: $2\,H_2(g) \rightarrow 4\,H^+(aq) + 4\,e^-$

Reduction: $O_2(g) + 4\,H^+(aq) + 4\,e^- \rightarrow 2\,H_2O(l)$

Net: $\qquad 2\,H_2(g) + O_2(g) \rightarrow 2\,H_2O(l)$ $\qquad z = 4$ in this reaction.

This net reaction is simply twice the formation reaction for $H_2O(l)$ and, therefore,

$$\Delta_r G° = 2\Delta_f G° \left[H_2O(l) \right] = 2 \times \left(-237.1 \text{ kJ mol}^{-1} \right) = -474.2 \times 10^3 \text{ J mol}^{-1} = -zFE°_{cell}$$

$$E°_{cell} = \frac{-\Delta_r G°}{zF} = \frac{-\left(-474.2 \times 10^3 \text{ J mol}^{-1} \right)}{4 \times \dfrac{96{,}485 \text{ C}}{\text{mol}}} = +1.229 \text{ V} = E°, \text{ as we might expect.}$$

6A **(M)** Cu(s) will displace metal ions of a metal less active than copper. Silver ion is one example.

Oxidation: $Cu(s) \rightarrow Cu^{2+}(aq) + 2e^-$ $\quad -E° = -0.340\text{V}$ (from Table 19.1)

Reduction: $\{Ag^+(aq) + e^- \rightarrow Ag(s)\} \times 2$ $\quad E° = +0.800\text{V}$

Net: $2\,Ag^+(aq) + Cu(s) \rightarrow Cu^{2+}(aq) + 2\,Ag(s)$ $\quad E°_{cell} = -0.340 \text{ V} + 0.800 \text{ V} = +0.460 \text{ V}$

6B **(M)** We determine the value for the hypothetical reaction's cell potential.

Oxidation: $\{Na(s) \rightarrow Na^+(aq) + e^-\} \times 2$ $\quad -E° = +2.713\text{ V}$

Reduction: $Mg^{2+}(aq) + 2e^- \rightarrow Mg(s)$ $\quad E° = -2.356 \text{ V}$

Net: $2\,Na(s) + Mg^{2+}(aq) \rightarrow 2\,Na^+(aq) + Mg(s)$ $\quad E°_{cell} = 2.713 \text{ V} - 2.356 \text{ V} = +0.357 \text{ V}$

The method is not feasible because another reaction occurs that has a more positive cell potential, i.e., Na(s) reacts with water to form $H_2(g)$ and NaOH(aq):

Oxidation: $\{Na(s) \rightarrow Na^+(aq) + e^-\} \times 2$ $\quad -E° = +2.713\text{ V}$

Reduction: $2\,H_2O + 2e^- \rightarrow H_2(g) + 2\,OH^-(aq)$ $\quad E° = -0.828\text{ V}$

$E°_{cell} = 2.713 \text{ V} - 0.828 \text{ V} = +1.885 \text{ V}.$

7A **(M)** The oxidation is that of SO_4^{2-} to $S_2O_8^{2-}$, the reduction is that of O_2 to H_2O.

Oxidation: $\{2\,SO_4^{2-}(aq) \rightarrow S_2O_8^{2-}(aq) + 2e^-\} \times 2$ $\quad -E° = -2.01 \text{ V}$

Reduction: $O_2(g) + 4\,H^+(aq) + 4e^- \rightarrow 2\,H_2O(l)$ $\quad E° = +1.229 \text{ V}$

Net: $\quad O_2(g) + 4\,H^+(aq) + 2\,SO_4^{2-}(aq) \rightarrow S_2O_8^{2-}(aq) + 2\,H_2O(l)$ $\quad E°_{cell} = -0.78 \text{ V}$

Because the standard cell potential is negative, we conclude that this cell reaction is nonspontaneous under standard conditions. This would not be a feasible method of producing peroxodisulfate ion.

7B **(M)**

(1) The oxidation is that of $Sn^{2+}(aq)$ to $Sn^{4+}(aq)$; the reduction is that of O_2 to H_2O.

Oxidation: $\{Sn^{2+}(aq) \rightarrow Sn^{4+}(aq) + 2e^-\} \times 2$ $\quad -E° = -0.154 \text{ V}$

Reduction: $O_2(g) + 4\,H^+(aq) + 4e^- \rightarrow 2\,H_2O(l)$ $\quad E° = +1.229 \text{ V}$

Net: $\quad O_2(g) + 4\,H^+(aq) + 2\,Sn^{2+}(aq) \rightarrow 2\,Sn^{4+}(aq) + 2\,H_2O(l)$ $\quad E°_{cell} = +1.075 \text{ V}$

Since the standard cell potential is positive, this cell reaction is spontaneous under standard conditions.

(2) The oxidation is that of Sn(s) to $Sn^{2+}(aq)$; the reduction is still that of O_2 to H_2O.

Oxidation: $\{Sn(s) \rightarrow Sn^{2+}(aq) + 2\,e^-\} \times 2$ $-E^\circ = +0.137$ V

Reduction: $O_2(g) + 4\,H^+(aq) + 4\,e^- \rightarrow 2\,H_2O(l)$ $E^\circ = +1.229$ V

Net: $O_2(g) + 4\,H^+(aq) + 2\,Sn(s) \rightarrow 2\,Sn^{2+}(aq) + 2\,H_2O(l)$

$E^\circ_{cell} = 0.137\ V + 1.229\ V = +1.366$ V

The standard cell potential for this reaction is more positive than that for situation (1). Thus, reaction (2) should occur preferentially. Also, if $Sn^{4+}(aq)$ is formed, it should react with Sn(s) to form $Sn^{2+}(aq)$.

Oxidation: $Sn(s) \rightarrow Sn^{2+}(aq) + 2\,e^-$ $-E^\circ = +0.137$ V

Reduction: $Sn^{4+}(aq) + 2\,e^- \rightarrow Sn^{2+}(aq)$ $E^\circ = +0.154$ V

Net: $Sn^{4+}(aq) + Sn(s) \rightarrow 2\,Sn^{2+}(aq)$ $E^\circ_{cell} = +0.137\ V + 0.154\ V = +0.291$ V

8A **(M)** For the reaction $2\,Al(s) + 3\,Cu^{2+}(aq) \rightarrow 2\,Al^{3+}(aq) + 3\,Cu(s)$ we know $z = 6$ and

$E^\circ_{cell} = +2.013$ V. We calculate the value of K_{eq}.

$E^\circ_{cell} = \dfrac{0.0257\ V}{z} \ln K; \quad \ln K = \dfrac{zE^\circ_{cell}}{0.0257\ V} = \dfrac{6 \times (+2.013\ V)}{0.0257\ V} = 470; \quad K = e^{470} = 10^{204}$

The huge size of the equilibrium constant indicates that this reaction indeed will go to essentially 100% to completion.

8B **(M)** We first determine the value of E°_{cell} from the half-cell potentials.

Oxidation: $Sn(s) \rightarrow Sn^{2+}(aq) + 2\,e^-$ $-E^\circ = +0.137$ V

Reduction: $Pb^{2+}(aq) + 2\,e^- \rightarrow Pb(s)$ $E^\circ = -0.125$ V

Net: $Pb^{2+}(aq) + Sn(s) \rightarrow Pb(s) + Sn^{2+}(aq)$ $E^\circ_{cell} = +0.137\ V - 0.125\ V = +0.012$ V

$E^\circ_{cell} = \dfrac{0.0257\ V}{z} \ln K \quad \ln K = \dfrac{zE^\circ_{cell}}{0.0257\ V} = \dfrac{2 \times (+0.012\ V)}{0.0257\ V} = 0.93 \quad K = e^{0.93} = 2.5$

The equilibrium constant's small size $(0.001 < K < 1000)$ indicates that this reaction will not go to completion.

9A **(M)** We first need to determine the standard cell voltage and the cell reaction.

Oxidation: $\{Al(s) \rightarrow Al^{3+}(aq) + 3\,e^-\} \times 2$ $-E^\circ = +1.676$ V

Reduction: $\{Sn^{4+}(aq) + 2\,e^- \rightarrow Sn^{2+}(aq)\} \times 3$ $E^\circ = +0.154$ V

Net: $2\,Al(s) + 3\,Sn^{4+}(aq) \rightarrow 2\,Al^{3+}(aq) + 3\,Sn^{2+}(aq)$ $E^\circ_{cell} = +1.676\ V + 0.154\ V = +1.830$ V

Note that $z = 6$. We now set up and substitute into the Nernst equation.

$$E_{cell} = E_{cell}^{\circ} - \frac{0.0257\ V}{z} \ln \frac{\left[Al^{3+}\right]^2 \left[Sn^{2+}\right]^3}{\left[Sn^{4+}\right]^3} = +1.830\ V - \frac{0.0257\ V}{6} \ln \frac{(0.36)^2 (0.54)^3}{(0.086)^3}$$

$$= +1.830\ V - 0.0149\ V = +1.815\ V$$

9B **(M)** We first need to determine the standard cell voltage and the cell reaction.

Oxidation: $2\ Cl^- (1.0\ M) \rightarrow Cl_2 (1\ bar) + 2\ e^- \qquad -E^{\circ} = -1.358\ V$

Reduction: $PbO_2 (s) + 4 H^+ (aq) + 2 e^- \rightarrow Pb^{2+} (aq) + 2 H_2O(l) \qquad E^{\circ} = +1.455\ V$

Net: $PbO_2 (s) + 4 H^+ (0.10\ M) + 2 Cl^- (1.0\ M) \rightarrow Cl_2 (1\ atm) + Pb^{2+} (0.050\ M) + 2 H_2O(l)$

$E_{cell}^{\circ} = -1.358\ V + 1.455\ V = +0.097\ V$ Note that $z = 2$. Substitute values into the Nernst equation.

$$E_{cell} = E_{cell}^{\circ} - \frac{0.0257\ V}{z} \ln \frac{P\{Cl_2\}\left[Pb^{2+}\right]}{\left[H^+\right]^4 \left[Cl^-\right]^2} = +0.097\ V - \frac{0.0257\ V}{2} \ln \frac{(1.0)(0.050)}{(0.10)^4 (1.0)^2}$$

$$= +0.097\ V - 0.080\ V = +0.017\ V$$

10A **(M)** The cell reaction is $2\ Fe^{3+} (0.35\ M) + Cu(s) \rightarrow 2\ Fe^{2+} (0.25\ M) + Cu^{2+} (0.15\ M)$ with $z = 2$ and $E_{cell}^{\circ} = -0.337\ V + 0.771\ V = 0.434\ V$ Next, substitute this voltage and the concentrations into the Nernst equation.

$$E_{cell} = E_{cell}^{\circ} - \frac{0.0257\ V}{z} \ln \frac{\left[Fe^{2+}\right]^2 \left[Cu^{2+}\right]}{\left[Fe^{3+}\right]^2} = 0.434\ V - \frac{0.0257\ V}{2} \ln \frac{(0.25)^2 (0.15)}{(0.35)^2} = 0.434 + 0.033$$

$E_{cell} = +0.467\ V$ Thus the reaction is spontaneous under standard conditions as written.

10B **(M)** The reaction is not spontaneous under standard conditions in either direction when $E_{cell} = 0.000\ V$. We use the standard cell potential from Example 19-10.

$$E_{cell} = E_{cell}^{\circ} - \frac{0.0257\ V}{z} \ln \frac{\left[Ag^+\right]^2}{\left[Hg^{2+}\right]}; \qquad 0.000\ V = -0.054\ V - \frac{0.0257\ V}{2} \ln \frac{\left[Ag^+\right]^2}{\left[Hg^{2+}\right]}$$

$$\ln \frac{\left[Ag^+\right]^2}{\left[Hg^{2+}\right]} = \frac{-0.054\ V \times 2}{0.0257\ V} = -1.82; \qquad \frac{\left[Ag^+\right]^2}{\left[Hg^{2+}\right]} = 10^{-1.82} = 0.0150$$

11A **(M)** In this concentration cell $E_{cell}^{\circ} = 0.000\ V$ because the same reaction occurs at anode and cathode, only the concentrations of the ions differ. $\left[Ag^+\right] = 0.100\ M$ in the cathode compartment. The anode compartment contains a saturated solution of AgCl(aq).

$$K_{sp} = 1.8 \times 10^{-10} = \left[Ag^+\right]\left[Cl^-\right] = s^2; \quad s = \sqrt{1.8 \times 10^{-10}} = 1.3 \times 10^{-5} \text{ M}$$

Now we apply the Nernst equation. The cell reaction is

$$Ag^+ \left(0.100 \text{ M}\right) \rightarrow Ag^+ \left(1.3 \times 10^{-5} \text{ M}\right)$$

$$E_{cell} = 0.000 \text{ V} - \frac{0.0257 \text{ V}}{1} \ln \frac{1.3 \times 10^{-5} \text{ M}}{0.100 \text{ M}} = +0.23 \text{ V}$$

11B **(D)** Because the electrodes in this cell are identical, the standard electrode potentials are numerically equal and subtracting one from the other leads to the value $E^\circ_{cell} = 0.000 \text{ V}$. However, because the ion concentrations differ, there is a potential difference between the two half cells (non-zero nonstandard voltage for the cell). $\left[Pb^{2+}\right] = 0.100 \text{ M}$ in the cathode compartment, while the anode compartment contains a saturated solution of PbI_2. We use the Nernst equation (with $n = 2$) to determine $\left[Pb^{2+}\right]$ in the saturated solution.

$$E_{cell} = +0.0567 \text{ V} = 0.000 \text{ V} - \frac{0.0257 \text{ V}}{2} \ln \frac{x \text{ M}}{0.100 \text{ M}}; \quad \ln \frac{x \text{ M}}{0.100 \text{ M}} = \frac{2 \times 0.0567 \text{ V}}{-0.0257 \text{ V}} = -1.92$$

$$\frac{x \text{ M}}{0.100 \text{ M}} = e^{-4.41} = 0.012; \quad \left[Pb^{2+}\right]_{anode} = x \text{ M} = 0.012 \times 0.100 \text{ M} = 0.0012 \text{ M};$$

$$\left[I^-\right] = 2 \times 0.0012 \text{ M} = 0.0024 \text{ M}$$

$$K_{sp} = \left[Pb^{2+}\right]\left[I^-\right]^2 = (0.0012)(0.0024)^2 = 6.9 \times 10^{-9} \text{ compared with } 7.1 \times 10^{-9} \text{ in Appendix D}$$

12A **(M)** From Table 19.1 we choose one oxidations and one reductions reaction so as to get the least negative cell voltage. This will be the most likely pair of ½-reactions to occur.

Oxidation: $\quad 2I^- (aq) \rightarrow I_2 (s) + 2 e^- \qquad\qquad -E^\circ = -0.535 \text{ V}$

$\qquad\qquad 2 H_2O(l) \rightarrow O_2 (g) + 4 H^+ (aq) + 4 e^- \qquad -E^\circ = -1.229 \text{ V}$

Reduction: $\quad K^+ (aq) + e^- \rightarrow K(s) \qquad\qquad\qquad E^\circ = -2.924 \text{ V}$

$\qquad\qquad 2 H_2O(l) + 2 e^- \rightarrow H_2 (g) + 2 OH^- (aq) \qquad E^\circ = -0.828 \text{ V}$

The least negative standard cell potential $(-0.535 \text{ V} - 0.828 \text{ V} = -1.363 \text{ V})$ occurs when $I_2(s)$ is produced by oxidation at the anode, and $H_2 (g)$ is produced by reduction at the cathode.

12B **(M)** We obtain from Table 19.1 all the possible oxidations and reductions and choose one of each to get the least negative cell voltage. That pair is the most likely pair of half-reactions to occur.

Oxidation: $\quad 2 H_2O(l) \rightarrow O_2 (g) + 4 H^+ (aq) + 4 e^- \qquad -E^\circ = -1.229 \text{V}$

$\qquad\qquad Ag(s) \rightarrow Ag^+ (aq) + e^- \qquad\qquad\qquad -E^\circ = -0.800 \text{V}$

[We cannot further oxidize NO_3^- (aq) or Ag^+ (aq).]

Reduction: Ag^+ (aq) $+ e^- \rightarrow Ag(s)$ $\qquad\qquad E° = +0.800 \text{V}$

$\qquad\qquad 2\,H_2O(l) + 2e^- \rightarrow H_2(g) + 2\,OH^-$ (aq) $\qquad E° = -0.828 \text{V}$

Thus, we expect to form silver metal at the cathode and Ag^+ (aq) at the anode.

13A (M) The half-cell equation is Cu^{2+} (aq) $+ 2e^- \rightarrow Cu(s)$, indicating that two moles of electrons are required for each mole of copper deposited. Current is measured in amperes, or coulombs per second. We convert the mass of copper to coulombs of electrons needed for the reduction and the time in hours to seconds.

$$\text{Current} = \frac{12.3\ \text{g Cu} \times \dfrac{1\ \text{mol Cu}}{63.6463\ \text{g Cu}} \times \dfrac{2\ \text{mol e}^-}{1\ \text{mol Cu}} \times \dfrac{96{,}485\ \text{C}}{1\ \text{mol e}^-}}{5.50\ \text{h} \times \dfrac{60\ \text{min}}{1\text{h}} \times \dfrac{60\ \text{s}}{1\ \text{min}}} = \frac{3.735 \times 10^4\ \text{C}}{1.98 \times 10^4\ \text{s}} = 1.89\ \text{amperes}$$

13B (D) We first determine the moles of O_2 (g) produced with the ideal gas equation.

$$\text{moles } O_2(g) = \frac{\left(738\ \text{mmHg} \times \dfrac{1.01325\ \text{atm}}{760\ \text{mmHg}} \right) \times 2.62\ \text{L}}{0.08314\ \text{L atm mol}^{-1}\ \text{K}^{-1} \times (26.2 + 273.15)\ \text{K}} = 0.104\ \text{mol } O_2$$

Then we determine the time needed to produce this amount of O_2.

$$\text{elapsed time} = 0.104\ \text{mol } O_2 \times \frac{4\ \text{mol e}^-}{1\ \text{mol } O_2} \times \frac{96{,}485\ \text{C}}{1\text{mol e}^-} \times \frac{1\text{s}}{2.13\,\text{C}} \times \frac{1\text{h}}{3600\,\text{s}} = 5.23\,\text{h}$$

INTEGRATIVE EXAMPLES

A (D) In this problem we are asked to determine $E°$ for the reduction of $CO_2(g)$ to $C_3H_8(g)$ in an acidic solution. We proceed by first determining $\Delta_r G°$ for the reaction using tabulated values for $\Delta_f G°$ in Appendix D. Next, $E°_{cell}$ for the reaction can be determined using $\Delta G° = -zFE°_{cell}$. Given reaction can be separated into reduction and oxidation. Since we are in acidic medium, the reduction half-cell potential can be found in Table 19.1. Lastly, the oxidation half-cell potential can be calculated using

$E°_{cell} = E°(\text{reduction half-cell}) - E°(\text{oxidation half-cell})$.

Stepwise approach

First determine $\Delta_r G°$ for the reaction using tabulated values for $\Delta_f G°$ in Appendix D:

$$\begin{array}{ccccccc} C_3H_8(g) & + & 5O_2(g) & \rightarrow & 3CO_2(g) & + & 4H_2O(l) \\ \Delta_f G° \quad -23.3\ \text{kJ/mol} & & 0\ \text{kJ/mol} & & -394.4\ \text{kJ/mol} & & -237.1\ \text{kJ/mol} \end{array}$$

$\Delta_r G° = 3 \times \Delta_f G°(CO_2(g)) + 4 \times \Delta_f G°(H_2O(l)) - [\Delta_f G°(C_3H_8(g)) + 5 \times \Delta_f G°(O_2(g))]$

$\Delta_r G° = [3 \times (-394.4) + 4 \times (-237.1) - (-23.3 + 5 \times 0)]\ \text{kJ/mol}$

$\Delta_r G° = -2108\ \text{kJ/mol}$

In order to calculate $E°_{cell}$ for the reaction using $\Delta_r G° = -zFE°_{cell}$, z must be first determined. We proceed by separating the given reaction into oxidation and reduction:

Reduction: $5 \times \{O_2(g) + 4\,H^+(aq) + 4\,e^- \rightarrow 2\,H_2O(l)\}$ $E° = +1.229$ V

Oxidation: $C_3H_8(g) + 6\,H_2O(l) \rightarrow 3\,CO_2(g) + 20\,H^+ + 20\,e^-$ $E° = x$V

Overall: $C_3H_8(g) + 5O_2(g) \rightarrow 3\,CO_2(g) + 4\,H_2O(l)$ $E° = +\,1.229$ V$-x$V

Since $z = 20$, $E°_{cell}$ can now be calculated using $\Delta_r G° = -zFE°_{cell}$:

$$\Delta_r G° = -zFE°_{cell}$$

$$-2108 \times 1000 \text{ J/mol} = -20 \times \frac{96,485 \text{ C}}{1 \text{ mol}} \times E°_{cell}$$

$$E°_{cell} = \frac{-2108 \times 1000}{-20 \times 96485}\text{V} = 1.092 \text{ V}$$

Finally, $E°$(reduction half-cell) can be calculated using

$E°_{cell} = E°$(reduction half-cell) $- E°$(oxidation half-cell) :

1.092 V $= 1.229$ V $- E°$(oxidation half-cell) V

$E°$(oxidation half-cell) $= 1.229$ V $- 1.092$ V $= 0.137$ V

Therefore, $E°$ for the reduction of $CO_2(g)$ to $C_3H_8(g)$ in an acidic medium is 0.137 V.

B   **(D)** This is a multi component problem dealing with a flow battery in which oxidation occurs at an aluminum anode and reduction at a carbon-air cathode. Al^{3+} produced at the anode is complexed with OH^- anions from $NaOH(aq)$ to form $[Al(OH)_4]^-$.

Stepwise approach:

Part (a): The flow battery consists of aluminum anode where oxidation occurs and the formed Al^{3+} cations are complexes with OH^- anions to form $[Al(OH)_4]^-$. The plausible half-reaction for the oxidation is:

Oxidation: $Al(s) + 4\,OH^-(aq) \rightarrow [Al(OH)_4]^- + 3\,e^-$

The cathode, on the other hand consists of carbon and air. The plausible half-reaction for the reduction involves the conversion of O_2 and water to form OH^- anions (basic medium):

Reduction: $O_2(g) + 2\,H_2O(l) + 4\,e^- \rightarrow 4\,OH^-(aq)$

Combining the oxidation and reduction half-reactions we obtain overall reaction for the process:

Oxidation: $\{Al(s) + 4\,OH^-(aq) \rightarrow [Al(OH)_4]^- + 3\,e^-\} \times 4$

Reduction: $\{O_2(g) + 2\,H_2O(l) + 4\,e^- \rightarrow 4\,OH^-(aq)\} \times 3$

Overall: $4\,Al(s) + 4\,OH^-(aq) + 3\,O_2(g) + 6\,H_2O(l) \rightarrow 4[Al(OH)_4]^-(aq)$

Part(b): In order to find $E°$ for the reduction, use the known value for $E°_{cell}$ as well as E° for the reduction half-reaction from Table 19.1:

$E°_{cell} = E°(\text{reduction half-cell}) - E°(\text{oxidation half-cell})$

$E°_{cell} = +0.401 \text{ V} - E°(\text{oxidation half-cell}) = +2.73 \text{ V}$

$E°(\text{oxidation half-cell}) = +0.401 \text{ V} - 2.73 \text{ V} = -2.329 \text{ V}$

Part (c): From the given value for $E°_{cell}$ (+2.73V) first calculate $\Delta_r G°$ using $\Delta_r G° = -zFE°_{cell}$ (notice that $z = 12$ from part (a) above):

$$\Delta_r G° = -zFE°_{cell} = -12 \text{ mol e}^- \times \frac{96,485 \text{ C}}{1 \text{ mol e}^-} \times 2.73 \text{ V}$$

$\Delta_r G° = -3161 \text{ kJ/mol}$

Given the overall reaction (part (a)) and $\Delta_f G°$ for OH⁻(aq) anions and $H_2O(l)$, we can calculate the Gibbs energy of formation of the aluminate ion, $[Al(OH)_4]^-$:

Overall reaction: $4 \text{ Al(s)} + 4 \text{ OH}^-(\text{aq}) + 3 \text{ O}_2(\text{g}) + 6 \text{ H}_2\text{O(l)} \rightarrow 4[\text{Al(OH)}_4]^-(\text{aq})$

$\Delta_f G°$ 0 kJ/mol   −157 kJ/mol   0 kJ/mol   −237.2 kJ/mol x

$\Delta_r G° = \{4 \times x - [4 \times 0 + 4 \times (-157) + 3 \times 0 + 6 \times (-237.2)]\} \text{ kJ/mol} = -3161 \text{ kJ/mol}$

$4x = -3161 - 2051.2 = -5212.2 \text{ kJ/mol}$

$x = -1303 \text{ kJ/mol}$

Therefore, $\Delta_f G°([\text{Al(OH)}_4]^-) = -1303 \text{ kJ/mol}$

Part (d): First calculate the number of moles of electrons:

$$\text{number of mol e}^- = \text{current(C/s)} \times \text{time(s)} \times \frac{1 \text{ mol e}^-}{96,485 \text{ C}}$$

$$\text{number of mol e}^- = 4.00 \text{ h} \times \frac{60 \text{ min}}{1 \text{ h}} \times \frac{60 \text{ s}}{1 \text{ min}} \times 10.0 \frac{\text{C}}{\text{s}} \times \frac{1 \text{ mol e}^-}{96,485 \text{ C}}$$

number of mol e⁻ = 1.49 mol e⁻

Now, use the oxidation half-reaction to determine the mass of Al(s) consumed:

$$\text{mass (Al)} = 1.49 \text{ mol e}^- \times \frac{1 \text{ mol Al}}{3 \text{ mol e}^-} \times \frac{26.9815 \text{ g Al}}{1 \text{ mol Al}} = 13.4 \text{ g}$$

EXERCISES

Standard Electrode Potential

1. **(E)**

   (a) If the metal dissolves in HNO_3, it has a reduction potential that is smaller than $E^\circ\{NO_3^-(aq)/NO(g)\} = 0.956$ V. If it also does not dissolve in HCl, it has a reduction potential that is larger than $E^\circ\{H^+(aq)/H_2(g)\} = 0.000$ V. If it displaces $Ag^+(aq)$ from solution, then it has a reduction potential that is smaller than $E^\circ\{Ag^+(aq)/Ag(s)\} = 0.800$ V. But if it does not displace $Cu^{2+}(aq)$ from solution, then its reduction potential is larger than $E^\circ\{Cu^{2+}(aq)/Cu(s)\} = 0.340$ V   Thus, 0.340 V $< E^\circ < 0.800$ V

   (b) If the metal dissolves in HCl, it has a reduction potential that is smaller than $E^\circ\{H^+(aq)/H_2(g)\} = 0.000$ V. If it does not displace $Zn^{2+}(aq)$ from solution, its reduction potential is larger than $E^\circ\{Zn^{2+}(aq)/Zn(s)\} = -0.763$ V. If it also does not displace $Fe^{2+}(aq)$ from solution, its reduction potential is larger than $E^\circ\{Fe^{2+}(aq)/Fe(s)\} = -0.440$ V. -0.440 V $< E^\circ < 0.000$ V

3. **(M)** We separate the given equation into its two half-equations. One of them is the reduction of nitrate ion in acidic solution, whose standard half-cell potential we retrieve from Table 19.1 and use to solve the problem.

   Oxidation: $\{Pt(s) + 4\,Cl^-(aq) \longrightarrow [PtCl_4]^{2-}(aq) + 2\,e^-\} \times 3$; $-E^\circ\{[PtCl_4]^{2-}(aq)/Pt(s)\}$

   Reduction: $\{NO_3^-(aq) + 4\,H^+(aq) + 3\,e^- \rightarrow NO(g) + 2\,H_2O(l)\} \times 2$; $E^\circ = +0.956$ V

   Net: $3\,Pt(s) + 2\,NO_3^-(aq) + 8\,H^+(aq) + 12\,Cl^-(aq) \rightarrow 3[PtCl_4]^{2-}(aq) + 2\,NO(g) + 6\,H_2O(l)$

   $E^\circ_{cell} = 0.201$ V $= +0.956$ V $- E^\circ\{[PtCl_4]^{2-}(aq)/Pt(s)\}$

   $E^\circ\{[PtCl_4]^{2-}(aq)/Pt(s)\} = 0.956$ V $- 0.201$ V $= +0.755$ V

5. **(M)** We divide the net cell equation into two half-equations.

   Oxidation: $\{Al(s) + 4\,OH^-(aq) \rightarrow [Al(OH)_4]^-(aq) + 3\,e^-\} \times 4$; $-E^\circ\{[Al(OH)_4]^-(aq)/Al(s)\}$

   Reduction: $\{O_2(g) + 2\,H_2O(l) + 4\,e^- \rightarrow 4\,OH^-(aq)\} \times 3$; $E^\circ = +0.401$ V

   Net: $4\,Al(s) + 3\,O_2(g) + 6\,H_2O(l) + 4\,OH^-(aq) \rightarrow 4[Al(OH)_4]^-(aq)$ $E^\circ_{cell} = 2.71$ V

   $E^\circ_{cell} = 2.71$ V $= +0.401$ V $- E^\circ\{[Al(OH)_4]^-(aq)/Al(s)\}$

   $E^\circ\{[Al(OH)_4]^-(aq)/Al(s)\} = 0.401$ V $- 2.71$ V $= -2.31$ V

7. **(M)**

(a) We need standard reduction potentials for the given half-reactions from Table 19.1:

$Ag^+(aq) + e^- \rightarrow Ag(s)$ $E° = +0.800$ V

$Zn^{2+}(aq) + 2\,e^- \rightarrow Zn(s)$ $E° = -0.763$ V

$Cu^{2+}(aq) + 2\,e^- \rightarrow Cu(s)$ $E° = +0.340$ V

$Al^{3+}(aq) + 3\,e^- \rightarrow Al(s)$ $E° = -1.676$ V

Therefore, the largest positive cell potential will be obtained for the reaction involving the oxidation of Al(s) to Al^{3+}(aq) and the reduction of Ag^+(aq) to Ag(s):

$Al(s) + 3\,Ag^+(aq) \rightarrow 3\,Ag(s) + Al^{3+}(aq)$ $E°_{cell} = +1.676$ V $+ 0.800$ V $= 2.476$ V

Ag is the anode and Al is the cathode.

(b) Reverse to the above, the cell with the smallest positive cell potential will be obtained for the reaction involving the oxidation of Zn(s) to Zn^{2+}(aq) (anode) and the reduction of Cu^{+2}(aq) to Cu(s) (cathode):

$Zn(s) + Cu^{2+}(aq) \rightarrow Cu(s) + Zn^{2+}(aq)$ $E°_{cell} = 0.763$ V $+ 0.340$ V $= 1.103$ V

Predicting Oxidation-Reduction Reactions

9. **(E)** **(a)** Ni^{2+}, **(b)** Cd.

11. **(M)**

(a) Oxidation: $Sn(s) \rightarrow Sn^{2+}(aq) + 2\,e^-$ $-E° = +0.137$ V

Reduction: $Pb^{2+}(aq) + 2\,e^- \rightarrow Pb(s)$ $E° = -0.125$ V

Net: $Sn(s) + Pb^{2+}(aq) \rightarrow Sn^{2+}(aq) + Pb(s)$ $E°_{cell} = +0.012$ V   Spontaneous

(b) Oxidation: $2\,I^-(aq) \rightarrow I_2(s) + 2\,e^-$ $-E° = -0.535$ V

Reduction: $Cu^{2+}(aq) + 2\,e^- \rightarrow Cu(s)$ $E° = +0.340$ V

Net: $2\,I^-(aq) + Cu^{2+}(aq) \rightarrow Cu(s) + I_2(s)$ $E°_{cell} = -0.195$ V   Nonspontaneous

(c) Oxidation: $\{2\,H_2O(l) \rightarrow O_2(g) + 4\,H^+(aq) + 4\,e^-\} \times 3$ $-E° = -1.229$ V

Reduction: $\{NO_3^-(aq) + 4\,H^+(aq) + 3\,e^- \rightarrow NO(g) + 2\,H_2O(l)\} \times 4$ $E° = +0.956$ V

Net: $4\,NO_3^-(aq) + 4\,H^+(aq) \rightarrow 3\,O_2(g) + 4\,NO(g) + 2\,H_2O(l)$ $E°_{cell} = -0.273$ V

Nonspontaneous

(d) Oxidation: $Cl^-(aq) + 2\,OH^-(aq) \rightarrow OCl^-(aq) + H_2O(l) + 2\,e^-$ $-E° = -0.890$ V

Reduction: $O_3(g) + H_2O(l) + 2\,e^- \rightarrow O_2(g) + 2\,OH^-(aq)$ $E° = +1.246$ V

Net: $Cl^-(aq) + O_3(g) \rightarrow OCl^-(aq) + O_2(g)$ (basic solution) $E°_{cell} = +0.356$ V

Spontaneous

13. **(M)**

(a) Oxidation: $Mg(s) \rightarrow Mg^{2+}(aq) + 2\,e^-$ $-E^\circ = +2.356\,V$

Reduction: $Pb^{2+}(aq) + 2\,e^- \rightarrow Pb(s)$ $E^\circ = -0.125\,V$

Net: $Mg(s) + Pb^{2+}(aq) \rightarrow Mg^{2+}(aq) + Pb(s)$ $E^\circ_{cell} = +2.231\,V$

This reaction occurs to a significant extent.

(b) Oxidation: $Sn(s) \rightarrow Sn^{2+}(aq) + 2\,e^-$ $-E^\circ = +0.137\,V$

Reduction: $2\,H^+(aq) \rightarrow H_2(g)$ $E^\circ = 0.000\,V$

Net: $Sn(s) + 2\,H^+(aq) \rightarrow Sn^{2+}(aq) + H_2(g)$ $E^\circ_{cell} = +0.137\,V$

This reaction will occur to a significant extent.

(c) Oxidation: $Sn^{2+}(aq) \rightarrow Sn^{4+}(aq) + 2\,e^-$ $-E^\circ = -0.154\,V$

Reduction: $SO_4^{2-}(aq) + 4\,H^+(aq) + 2\,e^- \rightarrow SO_2(g) + 2\,H_2O(l)$ $E^\circ = +0.17\,V$

Net: $Sn^{2+}(aq) + SO_4^{2-}(aq) + 4\,H^+(aq) \rightarrow Sn^{4+}(aq) + SO_2(g) + 2\,H_2O(l)$ $E^\circ_{cell} = +0.02\,V$

This reaction will occur, but not to a large extent.

(d) Oxidation: $\{H_2O_2(aq) \rightarrow O_2(g) + 2\,H^+(aq) + 2\,e^-\} \times 5$ $-E^\circ = -0.695\,V$

Reduction: $\{MnO_4^-(aq) + 8\,H^+(aq) + 5\,e^- \rightarrow Mn^{2+}(aq) + 4\,H_2O(l)\} \times 2$ $E^\circ = +1.51\,V$

Net: $5\,H_2O_2(aq) + 2\,MnO_4^-(aq) + 6\,H^+(aq) \rightarrow 5\,O_2(g) + 2\,Mn^{2+}(aq) + 8\,H_2O(l)$ $E^\circ_{cell} = +0.82\,V$

This reaction will occur to a significant extent.

(e) Oxidation: $2\,Br^-(aq) \rightarrow Br_2(aq) + 2\,e^-$ $-E^\circ = -1.065\,V$

Reduction: $I_2(s) + 2\,e^- \rightarrow 2\,I^-(aq)$ $E^\circ = +0.535\,V$

Net: $2\,Br^-(aq) + I_2(s) \rightarrow Br_2(aq) + 2\,I^-(aq)$ $E^\circ_{cell} = -0.530\,V$

This reaction will not occur to a significant extent.

15. **(M)** If E°_{cell} is positive, the reaction will occur. For the reduction of $Cr_2O_7^{2-}$ to $Cr^{3+}(aq)$:

$Cr_2O_7^{2-}(aq) + 14\,H^+(aq) + 6\,e^- \rightarrow 2\,Cr^{3+}(aq) + 7\,H_2O(l)$ $E^\circ = +1.33\,V$

If the oxidation has $-E^\circ$ smaller (more negative) than $-1.33\,V$, the oxidation will not occur.

(a) $Sn^{2+}(aq) \rightarrow Sn^{4+}(aq) + 2\,e^-$ $-E^\circ = -0.154\,V$

Hence, $Sn^{2+}(aq)$ can be oxidized to $Sn^{4+}(aq)$ by $Cr_2O_7^{2-}(aq)$.

(b) $I_2(s) + 6\,H_2O(l) \rightarrow 2\,IO_3^-(aq) + 12\,H^+(aq) + 10\,e^-$ $-E^\circ = -1.20\,V$

$I_2(s)$ can be oxidized to $IO_3^-(aq)$ by $Cr_2O_7^{2-}(aq)$.

(c) $Mn^{2+}(aq) + 4\,H_2O(l) \rightarrow MnO_4^-(aq) + 8\,H^+(aq) + 5\,e^-$ $-E^\circ = -1.51\,V$

$Mn^{2+}(aq)$ cannot be oxidized to $MnO_4^-(aq)$ by $Cr_2O_7^{2-}(aq)$.

17. (M)

(a) Oxidation: $\{Ag(s) \rightarrow Ag^+(aq)+e^-\} \times 3$ $\qquad\qquad -E° = -0.800\ V$

Reduction: $NO_3^-(aq)+4\ H^+(aq)+3\ e^- \rightarrow NO(g)+2\ H_2O(l)$ $\qquad E° = +0.956\ V$

Net: $3\ Ag(s)+NO_3^-(aq)+4\ H^+(aq) \rightarrow 3\ Ag^+(aq)+NO(g)+2\ H_2O(l)$ $\quad E°_{cell} = +0.156\ V$

$Ag(s)$ reacts with $HNO_3(aq)$ to form a solution of $AgNO_3(aq)$.

(b) Oxidation: $Zn(s) \rightarrow Zn^{2+}(aq)+2\ e^-$ $\qquad\qquad -E° = +0.763\ V$

Reduction: $2\ H^+(aq)+2\ e^- \rightarrow H_2(g)$ $\qquad\qquad E° = 0.000\ V$

Net: $Zn(s)+2\ H^+(aq) \rightarrow Zn^{2+}(aq)+H_2(g)$ $\qquad E°_{cell} = +0.763\ V$

$Zn(s)$ reacts with $HI(aq)$ to form a solution of $ZnI_2(aq)$.

(c) Oxidation: $Au(s) \rightarrow Au^{3+}(aq)+3\ e^-$ $\qquad\qquad -E° = -1.52\ V$

Reduction: $NO_3^-(aq)+4\ H^+(aq)+3\ e^- \rightarrow NO(g)+2\ H_2O(l)$ $\qquad E° = +0.956\ V$

Net: $Au(s)+NO_3^-(aq)+4\ H^+(aq) \rightarrow Au^{3+}(aq)+NO(g)+2\ H_2O(l); E°_{cell} = -0.56\ V$

$Au(s)$ does not react with $1.00\ M\ HNO_3(aq)$.

Galvanic Cells

19. (M)

(a) Oxidation: $\{Al(s) \rightarrow Al^{3+}(aq)+3\ e^-\} \times 2$ $\qquad\qquad -E° = +1.676\ V$

Reduction: $\{Sn^{2+}(aq)+2\ e^- \rightarrow Sn(s)\} \times 3$ $\qquad\qquad E° = -0.137\ V$

Net: $\quad 2\ Al(s)+3\ Sn^{2+}(aq) \rightarrow 2\ Al^{3+}(aq)+3\ Sn(s)$ $\qquad E°_{cell} = +1.539\ V$

(b) Oxidation: $Fe^{2+}(aq) \rightarrow Fe^{3+}(aq)+e^-$ $\qquad\qquad -E° = -0.771\ V$

Reduction: $Ag^+(aq)+e^- \rightarrow Ag(s)$ $\qquad\qquad E° = +0.800\ V$

Net: $\quad Fe^{2+}(aq)+Ag^+(aq) \rightarrow Fe^{3+}(aq)+Ag(s)$ $\qquad E°_{cell} = +0.029\ V$

(c) Oxidation: $\{Cr(s) \rightarrow Cr^{2+}(aq)+2\ e^-\} \times 3$ $\qquad\qquad -E° = +0.90V$

Reduction: $\{Au^{3+}(aq)+3\ e^- \rightarrow Au(s)\} \times 2$ $\qquad\qquad E° = 1.52V$

Net: $\quad 3\ Cr(s)+2\ Au^{3+}(aq) \rightarrow 3\ Cr^{2+}(aq)+2\ Au(s)$ $\qquad E°_{cell} = 2.42\ V$

(d) Oxidation: $2\ H_2O(l) \rightarrow O_2(g)+4\ H^+(aq)+4\ e^-$ $\qquad\qquad -E° = -1.229\ V$

Reduction: $O_2(g)+2\ H_2O(l)+4\ e^- \rightarrow 4\ OH^-(aq)$ $\qquad\qquad E° = +0.401\ V$

Net: $\quad H_2O(l) \rightarrow H^+(aq)+OH^-(aq)$ $\qquad\qquad E°_{cell} = -0.828\ V$

23. **(M)** In each case, we determine whether E°_{cell} is greater than zero; if so, the reaction will occur.

(a) Oxidation: $Ag(s) \rightarrow Ag^+(aq) + e^-$ $\qquad\qquad -E^\circ = -0.800$ V

Reduction: $Fe^{3+}(aq) + e^- \rightarrow Fe^{2+}(aq)$ $\qquad\qquad E^\circ = +0.771$ V

Net: $Ag(s) + Fe^{3+}(aq) \rightarrow Ag^+(aq) + Fe^{2+}(aq)$ $\qquad E^\circ_{cell} = -0.029$ V

The reaction is not spontaneous under standard conditions as written.

(b) Oxidation: $Sn(s) \rightarrow Sn^{2+}(aq) + 2\,e^-$ $\qquad\qquad -E^\circ = +0.137$ V

Reduction: $Sn^{4+}(aq) + 2\,e^- \rightarrow Sn^{+2}(aq)$ $\qquad\qquad E^\circ = +0.154$ V

Net: $Sn(s) + Sn^{4+}(aq) \rightarrow 2\,Sn^{2+}(aq)$ $\qquad E^\circ_{cell} = +0.291$ V

The reaction is spontaneous under standard conditions as written.

(c) Oxidation: $2\,Br^-(aq) \rightarrow Br_2(l) + 2\,e^-$ $\qquad\qquad -E^\circ = -1.065$ V

Reduction: $2\,Hg^{2+}(aq) + 2\,e^- \rightarrow Hg_2^+(aq)$ $\qquad\qquad E^\circ = +0.630$ V

Net: $2\,Br^-(aq) + 2\,Hg^{2+}(aq) \rightarrow Br_2(l) + Hg_2^+$ $\qquad E^\circ_{cell} = -0.435$ V

The reaction is not spontaneous under standard conditions as written.

(d) Oxidation: $\{NO_3^-(aq) + 2\,H^+(aq) + e^- \rightarrow NO_2(g) + H_2O(l)\} \times 2$ $\qquad -E^\circ = -2.326$ V

Reduction: $Zn(s) \rightarrow Zn^{2+}(aq) + 2e^-$ $\qquad\qquad E^\circ = +1.563$ V

Net: $2\,NO_3^-(aq) + 4\,H^+(aq) + Zn(s) \rightarrow 2\,NO_2(g) + 2\,H_2O(l) + Zn^{2+}(aq)$ $\quad E^\circ_{cell} = -0.435$ V

The reaction is not spontaneous under standard conditions as written.

25. **(M)**

(a) Oxidation: $\{Al(s) \rightarrow Al^{3+}(aq) + 3e^-\} \times 2$ $\qquad\qquad -E^\circ = +1.676$ V

Reduction: $\{Cu^{2+}(aq) + 2e^- \rightarrow Cu(s)\} \times 3$ $\qquad\qquad E^\circ = +0.337$ V

Net: $2\,Al(s) + 3\,Cu^{2+}(aq) \rightarrow 2\,Al^{3+}(aq) + 3\,Cu(s)$ $\qquad E^\circ_{cell} = +2.013$ V

$\Delta_r G^\circ = -zFE^\circ_{cell} = -(6)(96,485\ C/mol)(2.013\ V)$

$\Delta_r G^\circ = -1.165 \times 10^6$ J mol^{-1} = -1.165×10^3 kJ mol^{-1}

(b) Oxidation: $\{2\,I^-(aq) \rightarrow I_2(s) + 2e^-\} \times 2$ $\qquad\qquad -E^\circ = -0.535$ V

Reduction: $O_2(g) + 4\,H^+(aq) + 4e^- \rightarrow 2\,H_2O(l)$ $\qquad\qquad E^\circ = +1.229$ V

Net: $4\,I^-(aq) + O_2(g) + 4\,H^+(aq) \rightarrow 2\,I_2(s) + 2\,H_2O(l)$ $\quad E^\circ_{cell} = +0.694$ V

$\Delta_r G^\circ = -zFE^\circ_{cell} = -(4)(96,485\ C/mol)(0.694\ V) = -2.68 \times 10^5$ J mol^{-1} = -268 kJ mol^{-1}

(c) Oxidation: $\{Ag(s) \rightarrow Ag^+(aq) + e^-\} \times 6$ $\qquad\qquad -E^\circ = -0.800$ V

Reduction: $Cr_2O_7^{2-}(aq) + 14\,H^+(aq) + 6\,e^- \rightarrow 2\,Cr^{3+}(aq) + 7\,H_2O(l)$ $\quad E^\circ = +1.33$ V

Net: $6\,Ag(s) + Cr_2O_7^{2-}(aq) + 14\,H^+(aq) \rightarrow 6\,Ag^+(aq) + 2\,Cr^{3+}(aq) + 7\,H_2O(l)$

$E^\circ_{cell} = -0.800\ V + 1.33\ V = +0.53\ V$

$\Delta_r G^\circ = -zFE^\circ_{cell} = -(6)(96,485\ C/mol)(0.53\ V) = -3.1 \times 10^5$ J mol^{-1} = -3.1×10^2 kJ mol^{-1}

27. **(M)** First calculate $E°_{cell}$ from standard electrode reduction potentials (Table 19.1). Then use $\Delta_r G° = -zFE°_{cell} = -RT \ln K$ to determine $\Delta_r G°$ and K.

(a)

Oxidation: $\{Ce^{3+}(aq) \rightarrow Ce^{4+}(aq) + e^-\} \times 5$	$-E° = -1.61$ V
Reduction: $MnO_4^-(aq) + 8H^+(aq) + 5e^- \rightarrow Mn^{2+}(aq) + 4H_2O(l)$	$E° = +1.51$ V

Net: $MnO_4^-(aq) + 8H^+(aq) + 5Ce^{3+}(aq) \rightarrow Mn^{2+}(aq) + 4H_2O(l) + 5Ce^{4+}(aq)$ $E°_{cell} = -0.100$ V

(b) $\Delta_r G° = -zFE°_{cell} = -RT \ln K$;

$$\Delta_r G° = -5 \times 96,485 \frac{C}{mol} \times (-0.100 \text{ V}) = 48.24 \text{ kJ mol}^{-1}$$

(c) $\Delta_r G° = -RT \ln K \Rightarrow 48.24 \times 1000 \text{ J mol}^{-1} = -8.314 \text{ J K}^{-1}\text{mol}^{-1} \times 298.15 \text{ K } \ln K$

$\ln K = -19.46 \Rightarrow K = e^{-19.46} = 3.5 \times 10^{-9}$

(d) Since K is very small the reaction will not go to completion.

29. **(M)**

(a) A negative value of $E°_{cell}$ (-0.0050 V) indicates that $\Delta_r G° = -zFE°_{cell}$ is positive which in turn indicates that K is less than one $(K < 1.00)$; $\Delta_r G° = -RT \ln K$.

$$K = \frac{\left[Cu^{2+}\right]^2 \left[Sn^{2+}\right]}{\left[Cu^+\right]^2 \left[Sn^{4+}\right]}$$

Thus, when all concentrations are the same, the ion product, Q, equals 1.00. From the negative standard cell potential, it is clear that K must be (slightly) less than one. Therefore, all the concentrations cannot be 0.500 M at the same time.

(b) In order to establish equilibrium, that is, to have the ion product become less than 1.00, and equal the equilibrium constant, the concentrations of the products must decrease and those of the reactants must increase. A net reaction to the left (towards the reactants) will occur.

31. **(M)** Cell reaction: $Zn(s) + Ag_2O(s) \rightarrow ZnO(s) + 2Ag(s)$. We assume that the cell operates at 298 K.

$\Delta_r G° = \Delta_f G°\left[ZnO(s)\right] + 2\Delta_f G°\left[Ag(s)\right] - \Delta_f G°\left[Zn(s)\right] - \Delta_f G°\left[Ag_2O(s)\right]$

$= -318.3 \text{ kJ/mol} + 2(0.00 \text{ kJ/mol}) - 0.00 \text{ kJ/mol} - (-11.20 \text{ kJ/mol})$

$= -307.1 \text{ kJ/mol} = -zFE°_{cell}$

$$E°_{cell} = -\frac{\Delta_r G°}{zF} = -\frac{-307.1 \times 10^3 \text{ J/mol}}{2 \times 96,485 \text{ C/mol}} = 1.591 \text{ V}$$

33. **(D)** From the data provided we can construct the following Latimer diagram.

$$\text{IrO}_2 \xrightarrow{\;0.223 \text{ V}\;} \text{Ir}^{3+} \xrightarrow{\;1.156 \text{ V}\;} \text{Ir} \quad \text{(Acidic conditions)}$$
$$\text{(IV)} \qquad\qquad \text{(III)} \qquad\qquad \text{(0)}$$

Latimer diagrams are used to calculate the standard potentials of non-adjacent half-cell couples. Our objective in this question is to calculate the voltage differential between IrO_2 and iridium metal (Ir), which are separated in the diagram by Ir^{3+}. The process basically involves adding two half-reactions to obtain a third half-reaction. The potentials for the two half-reactions cannot, however, simply be added to get the target half-cell voltage because the electrons are not cancelled in the process of adding the two half-reactions. Instead, to find $E^\circ_{1/2 \text{ cell}}$ for the target half-reaction, we must use free energy changes, which are additive. To begin, we will balance the relevant half-reactions in acidic solution:

$$4\,\text{H}^+(\text{aq}) + \text{IrO}_2(\text{s}) + \text{e}^- \;\rightarrow\; \text{Ir}^{3+}(\text{aq}) + 2\,\text{H}_2\text{O}(\text{l}) \qquad E^\circ_{1/2\text{red(a)}} = 0.223 \text{ V}$$
$$\text{Ir}^{3+}(\text{aq}) + 3\,\text{e}^- \rightarrow \text{Ir}(\text{s}) \qquad E^\circ_{1/2\text{red(b)}} = 1.156 \text{V}$$

$$4\,\text{H}^+(\text{aq}) + \text{IrO}_2(\text{s}) + 4\,\text{e}^- \;\rightarrow\; 2\,\text{H}_2\text{O}(\text{l}) + \text{Ir}(\text{s}) \qquad E^\circ_{1/2\text{red(c)}} = ?$$

$E^\circ_{1/2\text{red(c)}} \neq E^\circ_{1/2\text{red(a)}} + E^\circ_{1/2\text{red(b)}}$ but $\Delta_r G^\circ_{(a)} + \Delta_r G^\circ_{(b)} = \Delta_r G^\circ_{(c)}$ and $\Delta_r G^\circ = -zFE^\circ$

$$-4F(E^\circ_{1/2\text{red(c)}}) = -1F(E^\circ_{1/2\text{red(a)}}) + -3F(E^\circ_{1/2\text{red(b)}})$$
$$-4F(E^\circ_{1/2\text{red(c)}}) = -1F(0.223) + -3F(1.156)$$
$$E^\circ_{1/2\text{red(c)}} = \frac{-1F(0.223) + -3F(1.156)}{-4F} = \frac{-1(0.223) + -3(1.156)}{-4} = 0.923 \text{ V}$$

In other words, $E^\circ_{(c)}$ is the weighted average of $E^\circ_{(a)}$ and $E^\circ_{(b)}$

Concentration Dependence of E_{cell}—the Nernst Equation

35. **(M)** In this problem we are asked to determine the concentration of $[\text{Ag}^+]$ ions in electrochemical cell that is under nonstandard conditions. We proceed by first determining E°_{cell}. Using the Nerst equation and the known value of E, we can then calculate the concentration of $[\text{Ag}^+]$.

Oxidation: $\text{Zn}(\text{s}) \rightarrow \text{Zn}^{2+}(\text{aq}) + 2\,\text{e}^- \qquad\qquad -E^\circ = +0.763 \text{ V}$

Reduction: $\{\text{Ag}^+(\text{aq}) + \text{e}^- \rightarrow \text{Ag}(\text{s})\} \times 2 \qquad\qquad E^\circ = +0.800 \text{ V}$

Net: $\text{Zn}(\text{s}) + 2\,\text{Ag}^+(\text{aq}) \rightarrow \text{Zn}^{2+}(\text{aq}) + 2\,\text{Ag}(\text{s}) \qquad E^\circ_{\text{cell}} = +1.563 \text{ V}$

Use the Nerst equation and the known value of E to solve for $[\text{Ag}]^+$:

$$E = E^\circ_{\text{cell}} - \frac{0.0257 \text{ V}}{z} \ln \frac{[\text{Zn}^{2+}]}{[\text{Ag}^+]^2} = +1.563 \text{ V} - \frac{0.0257 \text{ V}}{2} \ln \frac{1.00}{x^2} = +1.250 \text{ V}$$

$$\ln \frac{1.00}{x^2} = \frac{-2 \times (1.250 - 1.563) \text{ V}}{0.0257 \text{ V}} = 10.6; \quad x = \sqrt{2.5 \times 10^{-11}} = 5 \times 10^{-6}$$

Therefore, $[\text{Ag}^+] = 5 \times 10^{-6}$ M

37. **(M)** We first calculate $E°_{cell}$ for each reaction and then use the Nernst equation to calculate E_{cell}.

(a) Oxidation: $\{Al(s) \rightarrow Al^{3+}(0.18\ M) + 3\ e^-\} \times 2$ $-E° = +1.676\ V$

Reduction: $\{Fe^{2+}(0.85\ M) + 2\ e^- \rightarrow Fe(s)\} \times 3$ $E° = -0.440\ V$

Net: $2\ Al(s) + 3\ Fe^{2+}(0.85\ M) \rightarrow 2\ Al^{3+}(0.18\ M) + 3\ Fe(s)$ $E°_{cell} = +1.236\ V$

$$E_{cell} = E°_{cell} - \frac{0.0257\ V}{z} \ln \frac{\left[Al^{3+}\right]^2}{\left[Fe^{2+}\right]^3} = 1.236\ V - \frac{0.0257\ V}{6} \ln \frac{(0.18)^2}{(0.85)^3} = 1.249\ V$$

(b) Oxidation: $\{Ag(s) \rightarrow Ag^+(0.34\ M) + e^-\} \times 2$ $-E° = -0.800\ V$

Reduction: $Cl_2(0.55\ bar) + 2\ e^- \rightarrow 2\ Cl^-(0.098\ M)$ $E° = +1.358\ V$

Net: $Cl_2(0.55\ bar) + 2\ Ag(s) \rightarrow 2\ Cl^-(0.098\ M) + 2\ Ag^+(0.34\ M);\ E°_{cell} = +0.558\ V$

$$E_{cell} = E°_{cell} - \frac{0.0257\ V}{z} \ln \frac{\left[Cl^-\right]^2 \left[Ag^+\right]^2}{P\{Cl_2(g)\}} = +0.558\ V - \frac{0.0257\ V}{2} \ln \frac{(0.34)^2 (0.098)^2}{0.55} = +0.638\ V$$

39. **(M)** All these observations can be understood in terms of the procedure we use to balance half-equations: the ion—electron method.

(a) The reactions for which E depends on pH are those that contain either $H^+(aq)$ or $OH^-(aq)$ in the balanced half-equation. These reactions involve oxoacids and oxoanions whose central atom changes oxidation state.

(b) $H^+(aq)$ will inevitably be on the left side of the reduction of an oxoanion because reduction is accompanied by not only a decrease in oxidation state, but also by the loss of oxygen atoms, as in $ClO_3^- \rightarrow ClO_2^-$, $SO_4^{2-} \rightarrow SO_2$, and $NO_3^- \rightarrow NO$. These oxygen atoms appear on the right-hand side as H_2O molecules. The hydrogens that are added to the right-hand side with the water molecules are then balanced with $H^+(aq)$ on the left-hand side.

(c) If a half-reaction with $H^+(aq)$ ions present is transferred to basic solution, it may be re-balanced by adding to each side $OH^-(aq)$ ions equal in number to the $H^+(aq)$ originally present. This results in $H_2O(l)$ on the side that had $H^+(aq)$ ions (the left side in this case) and $OH^-(aq)$ ions on the other side (the right side.)

41. **(M)** Oxidation: $Zn(s) \rightarrow Zn^{2+}(aq) + 2\ e^-$ $-E° = +0.763\ V$

Reduction: $Cu^{2+}(aq) + 2\ e^- \rightarrow Cu(s)$ $E° = +0.337\ V$

Net: $Zn(s) + Cu^{2+}(aq) \rightarrow Cu(s) + Zn^{2+}(aq)$ $E°_{cell} = +1.100\ V$

(a) We set $E = 0.000$ V, $\left[Zn^{2+} \right] = 1.00$ M, and solve for $\left[Cu^{2+} \right]$ in the Nernst equation.

$$E_{cell} = E^{\circ}{}_{cell} - \frac{0.0257 \text{ V}}{2} \ln \frac{\left[Zn^{2+} \right]}{\left[Cu^{2+} \right]} ; \quad 0.000 = 1.100 - 0.0129 \text{ V} \ln \frac{1.0 \text{ M}}{\left[Cu^{2+} \right]}$$

$$\ln \frac{1.0 \text{ M}}{\left[Cu^{2+} \right]} = \frac{0.000 - 1.100}{-0.0129 \text{ V}} = 85.3; \quad \left[Cu^{2+} \right] = e^{-85.3} = 9 \times 10^{-38} \text{ M}$$

(b) If we work the problem the other way, by assuming initial concentrations of $\left[Cu^{2+} \right]_{initial} = 1.0 \text{ M}$ and $\left[Zn^{2+} \right]_{initial} = 0.0 \text{ M}$, we obtain $\left[Cu^{2+} \right]_{final} = 6 \times 10^{-38} \text{ M}$ and $\left[Zn^{2+} \right]_{final} = 1.0 \text{ M}$. Thus, we would conclude that this reaction goes to completion.

43. (M)

(a) The two half-equations and the cell equation are given below. $E^{\circ}_{cell} = 0.000$ V

Oxidation: $H_2 (g) \rightarrow 2 H^+ (0.65 \text{ M KOH}) + 2 e^-$

Reduction: $2 H^+ (1.0 \text{ M}) + 2 e^- \rightarrow H_2 (g)$

Net: $2 H^+ (1.0 \text{ M}) \rightarrow 2 H^+ (0.65 \text{ M KOH})$

$$\left[H^+ \right]_{base} = \frac{K_w}{\left[OH^- \right]} = \frac{1.00 \times 10^{-14}}{0.65} = 1.5 \times 10^{-14} \text{ M}$$

$$E_{cell} = E^{\circ}{}_{cell} - \frac{0.0257 \text{ V}}{2} \ln \frac{\left[H^+ \right]_{base}^2}{\left[H^+ \right]_{acid}^2} = 0.000 \text{ V} - \frac{0.0257 \text{ V}}{2} \ln \frac{\left(1.5 \times 10^{-14} \right)^2}{\left(1.0 \right)^2} = +0.818 \text{ V}$$

(b) For the reduction of $H_2O(l)$ to $H_2 (g)$ in basic solution, $2 H_2O(l) + 2 e^- \rightarrow 2 H_2 (g) + 2 OH^- (aq)$, $E^{\circ} = -0.828$ V. This reduction is the reverse of the reaction that occurs in the anode of the cell described, with one small difference: in the standard half-cell, $[OH^-] = 1.00$ M, while in the anode half-cell in the case at hand, $[OH^-] = 0.65$ M. Or, viewed in another way, in 1.00 M KOH, $[H^+]$ is smaller still than in 0.65 M KOH. The forward reaction (dilution of H^+) should be even more spontaneous, (i.e. a more positive voltage will be created), with 1.00 M KOH than with 0.65 M KOH. We expect that E°_{cell} (1.000 M NaOH) should be a little larger than E_{cell} (0.65 M NaOH), which, is in fact, the case.

45. (M) First we need to find $\left[Mg^{2+} \right]$ in a saturated solution of $Mg_3(PO_4)_2$.

$Mg_3(PO_4)_2 (s) \rightarrow 3 Mg^{2+} (aq) + 2 PO_4{}^{3-} (aq)$

$$K_{sp} = \left[mg^{2+} \right]^3 \left[PO_4^{3-} \right]^2 = (3s)^3 (2s)^2 = 108s^5 = 1.0 \times 10^{-25} \quad s = \sqrt[5]{\frac{1.0 \times 10^{-25}}{108}} = 3.9 \times 10^{-6} \text{ M}$$

The cell diagrammed is a concentration cell, for which $E°_{cell} = 0.000$ V, $z = 2$, and

$$\left[Mg^{2+}\right]_{anode} = 3s = 3 \times 3.9 \times 10^{-6} \text{ M} = 1.17 \times 10^{-5} \text{ M}$$

Cell reaction: $Mg(s) + Mg^{2+}(0.125 \text{ M}) \rightarrow Mg(s) + Mg^{2+}(1.17 \times 10^{-5}\text{M})$

$$E_{cell} = E°_{cell} - \frac{0.0257}{2}\ln\frac{1.17 \times 10^{-5}\text{ M}}{0.125 \text{ M}} = 0.000 + 0.119 \text{ V} = 0.119 \text{ V}$$

47. (D)

(a) Oxidation: $Sn(s) \rightarrow Sn^{2+}(0.075 \text{ M}) + 2 e^-$ $\qquad -E° = +0.137$ V

Reduction: $Pb^{2+}(0.600 \text{ M}) + 2 e^- \rightarrow Pb(s)$ $\qquad E° = -0.125$ V

Net: $Sn(s) + Pb^{2+}(0.600 \text{ M}) \rightarrow Pb(s) + Sn^{2+}(0.075 \text{ M}); E°_{cell} = +0.012$ V

$$E_{cell} = E°_{cell} - \frac{0.0257 \text{ V}}{2}\ln\frac{\left[Sn^{2+}\right]}{\left[Pb^{2+}\right]} = 0.012 \text{ V} - 0.0129 \text{ V} \ln\frac{0.075}{0.600} = 0.012 \text{ V} + 0.027 \text{ V} = 0.039 \text{ V}$$

(b) As the reaction proceeds, $[Sn^{2+}]$ increases while $[Pb^{2+}]$ decreases. These changes cause the driving force behind the reaction to steadily decrease with the passage of time. This decline in driving force is manifested as a decrease in E_{cell} with time.

(c) When $\left[Pb^{2+}\right] = 0.500 \text{ M} = 0.600 \text{ M} - 0.100 \text{ M}$, $\left[Sn^{2+}\right] = 0.075 \text{ M} + 0.100 \text{ M}$, because

the stoichiometry of the reaction is 1:1 for Sn^{2+} and Pb^{2+}.

$$E_{cell} = E°_{cell} - \frac{0.0257 \text{ V}}{2}\ln\frac{\left[Sn^{2+}\right]}{\left[Pb^{2+}\right]} = 0.012 \text{ V} - 0.0129 \text{ V} \ln\frac{0.175}{0.500} = (0.012 + 0.013)\text{V} = 0.025 \text{ V}$$

(d)

Reaction:	$Sn(s) +$	$Pb^{2+}(aq)$	$\rightarrow$	$Pb(s)$	$+$	$Sn^{2+}(aq)$
Initial:	—	0.600 M		—		0.075 M
Changes:	—	$-x$ M		—		$+x$ M
Final:	—	$(0.600 - x)$ M		—		$(0.075 + x)$ M

$$E_{cell} = E°_{cell} - \frac{0.0257 \text{ V}}{2}\ln\frac{\left[Sn^{2+}\right]}{\left[Pb^{2+}\right]}, \text{ so } 0.020 \text{ V} = 0.012 \text{ V} - 0.0129 \text{ V} \ln\frac{0.075 + x}{0.600 - x}$$

$$\ln\frac{0.075 + x}{0.600 - x} = \frac{E_{cell} - 0.012 \text{ V}}{-0.0129 \text{ V}} = \frac{0.020 - 0.012}{-0.0129 \text{ V}} = -0.27; \quad \frac{0.075 + x}{0.600 - x} = e^{-0.62} = 0.54$$

$$0.075 + x = 0.54(0.600 - x) = 0.324 - 0.54x; \quad x = \frac{0.324 - 0.075}{1.54} = 0.162$$

$$\left[Sn^{2+}\right] = (0.075 + 0.162)\text{ M} = 0.237 \text{ M}$$

(e) Here we use the expression developed in part (d).

$$\ln \frac{0.075 + x}{0.600 - x} = \frac{E_{cell} - 0.012\ V}{-0.0129\ V} = \frac{(0.000 - 0.012)\ V}{-0.0129\ V} = +0.93$$

$$\frac{0.075 + x}{0.600 - x} = e^{+0.93} = 2.6; \quad 0.075 + x = 2.6(0.600 - x) = 1.6 - 2.6x$$

$$x = \frac{1.6 - 0.075}{3.6} = 0.42\ M$$

$$\left[Sn^{2+}\right] = (0.075 + 0.42)M = 0.50\ M; \qquad \left[Pb^{2+}\right] = (0.600 - 0.42)M = 0.18\ M$$

49. **(M)** First we will need to come up with a balanced equation for the overall redox reaction. Clearly, the reaction must involve the oxidation of $Cl^-(aq)$ and the reduction of $Cr_2O_7^{2-}(aq)$:

$$14\ H^+(aq) + Cr_2O_7^{2-}(aq) + 6\ e^- \rightarrow 2\ Cr^{3+}(aq) + 7\ H_2O(l) \qquad\qquad E°_{red} = 1.33\ V$$
$$\{Cl^-(aq) \rightarrow 1/2\ Cl_2(g) + 1\ e^-\} \times 6 \qquad\qquad E°_{ox} = -1.358\ V$$

$$\overline{14\ H^+(aq) + Cr_2O_7^{2-}(aq) + 6\ Cl^-(aq) \rightarrow 2\ Cr^{3+}(aq) + 7\ H_2O(l) + 3\ Cl_2(g)\quad E°_{cell} = -0.03\ V}$$

A <u>negative</u> cell potential means, the oxidation of $Cl^-(aq)$ to $Cl_2(g)$ by $Cr_2O_7^{2-}(aq)$ at standard conditions will <u>not</u> occur spontaneously. We could obtain some $Cl_2(g)$ from this reaction by driving it to the product side with an external voltage. In other words, the reverse reaction is the spontaneous reaction at standard conditions and if we want to produce some $Cl_2(g)$ from the system, we must push the non-spontaneous reaction in its forward direction with an external voltage, (i.e., a DC power source). Since $E°_{cell}$ is only slightly negative, we could also drive the reaction by removing products as they are formed and replenishing reactants as they are consumed.

Batteries and Fuel Cells

51. **(M)**

(a) The cell diagram begins with the anode and ends with the cathode.

Cell diagram: $Cr(s)|Cr^{2+}(aq), Cr^{3+}(aq)\|Fe^{2+}(aq), Fe^{3+}(aq)|Fe(s)$

(b) Oxidation: $Cr^{2+}(aq) \rightarrow Cr^{3+}(aq) + e^- \qquad\qquad -E° = +0.424\ V$

Reduction: $Fe^{3+}(aq) + e^- \rightarrow Fe^{2+}(aq) \qquad\qquad E° = +0.771\ V$

Net: $\ Cr^{2+}(aq) + Fe^{3+}(aq) \rightarrow Cr^{3+}(aq) + Fe^{2+}(aq) \qquad E°_{cell} = +1.195\ V$

53. **(M)**

(a) Cell reaction: $2\ H_2(g) + O_2(g) \rightarrow 2\ H_2O(l)$

$$\Delta_r G° = 2\Delta_f G°\left[H_2O(l)\right] = 2(-237.1\ kJ/mol) = -474.2\ kJ/mol$$

$$E°_{cell} = -\frac{\Delta_r G°}{zF} = -\frac{-474.2 \times 10^3\ J/mol}{4 \times 96,485\ C/mol} = 1.229\ V$$

(b) Anode, oxidation: $\{Zn(s) \rightarrow Zn^{2+}(aq) + 2\,e^-\} \times 2$ $\qquad -E° = +0.763 \text{ V}$

Cathode, reduction: $O_2(g) + 4\,H^+(aq) + 4\,e^- \rightarrow 2\,H_2O(l)$ $\qquad E° = +1.229 \text{ V}$

Net: $2\,Zn(s) + O_2(g) + 4\,H^+(aq) \rightarrow 2\,Zn^{2+}(aq) + 2\,H_2O(l)$ $\quad E°_{cell} = +1.992 \text{ V}$

(c) Anode, oxidation: $Mg(s) \rightarrow Mg^{2+}(aq) + 2\,e^-$ $\qquad -E° = +2.356 \text{ V}$

Cathode, reduction: $I_2(s) + 2\,e^- \rightarrow 2\,I^-(aq)$ $\qquad E° = +0.535 \text{ V}$

Net: $Mg(s) + I_2(s) \rightarrow Mg^{2+}(aq) + 2\,I^-(aq)$ $\qquad E°_{cell} = +2.891 \text{ V}$

55. **(M)** Aluminum-Air Battery: $2\,Al(s) + 3/2\,O_2(g) \rightarrow Al_2O_3(s)$
Zinc-Air Battery: $\qquad Zn(s) + \frac{1}{2}\,O_2(g) \rightarrow ZnO(s)$
Iron-Air Battery $\qquad Fe(s) + \frac{1}{2}\,O_2(g) \rightarrow FeO(s)$
Calculate the quantity of charge transferred when 1.00 g of metal is consumed in each cell.

Aluminum-Air Cell:

$$1.00 \text{ g Al(s)} \times \frac{1 \text{ mol Al(s)}}{26.98 \text{ g Al(s)}} \times \frac{3 \text{ mol e}^-}{1 \text{ mol Al(s)}} \times \frac{96,485\,C}{1 \text{ mol e}^-} = 1.07 \times 10^4 \text{ C}$$

Zinc-Air Cell:

$$1.00 \text{ g Zn(s)} \times \frac{1 \text{ mol Zn(s)}}{65.38 \text{ g Zn(s)}} \times \frac{2 \text{ mol e}^-}{1 \text{ mol Zn(s)}} \times \frac{96,485 \text{ C}}{1 \text{ mol e}^-} = 2.95 \times 10^3 \text{ C}$$

Iron-Air Cell:

$$1.00 \text{ g Fe(s)} \times \frac{1 \text{ mol Fe(s)}}{55.845 \text{ g Fe(s)}} \times \frac{2 \text{ mol e}^-}{1 \text{ mol Fe(s)}} \times \frac{96,485 \text{ C}}{1 \text{ mol e}^-} = 3.46 \times 10^3 \text{ C}$$

As expected, aluminum has the greatest quantity of charge transferred per unit mass (1.00 g) of metal oxidized. This is because aluminum has the smallest molar mass and forms the most highly charged cation (3+ for aluminum vs 2+ for Zn and Fe).

57. **(M)**
Oxidation (anode): $Li(s) \rightarrow Li^+(aq) + e^-$ $\qquad\qquad -E° = +3.040 \text{ V}$
Reduction (cathode): $MnO_2(s) + 2\,H_2O(l) + e^- \rightarrow Mn(OH)_3(s) + OH^-(aq)$ $\quad E° = -0.20 \text{ V}$

Net: $MnO_2(s) + 2\,H_2O(l) + Li(s) \rightarrow Mn(OH)_3(s) + OH^-(aq) + Li^+(aq)$ $\quad E°_{cell} = 2.84 \text{ V}$
Cell diagram: $Li(s), Li^+(aq) | KOH(satd) | MnO_2(s), Mn(OH)_3(s)|$

Electrochemical Mechanism of Corrosion

59. **(M)**

(a) Because copper is a less active metal than is iron (i.e. a weaker reducing agent), this situation would be similar to that of an iron or steel can plated with a tin coating that has been scratched. Oxidation of iron metal to Fe^{2+}(aq) should be enhanced in the body of the nail (blue precipitate), and hydroxide ion should be produced in the vicinity of the copper wire (pink color), which serves as the cathode.

(b) Because a scratch tears the iron and exposes "fresh" metal, it is more susceptible to corrosion. We expect blue precipitate in the vicinity of the scratch.

(c) Zinc should protect the iron nail from corrosion. There should be almost no blue precipitate; the zinc corrodes instead. The pink color of OH^- should continue to form.

61. **(M)** During the process of corrosion, the metal that corrodes loses electrons. Thus, the metal in these instances behaves as an anode and, hence, can be viewed as bearing a latent negative polarity. One way in which we could retard oxidation of the metal would be to convert it into a cathode. Once transformed into a cathode, the metal would develop a positive charge and no longer release electrons (or oxidize). This change in polarity can be accomplished by hooking up the metal to an inert electrode in the ground and then applying a voltage across the two metals in such a way that the inert electrode becomes the anode and the metal that needs protecting becomes the cathode. This way, any oxidation that occurs will take place at the negatively charged inert electrode rather than the positively charged metal electrode.

Electrolysis Reactions

63. **(M)** Here we start by calculating the total amount of charge passed and the number of moles of electrons transferred.

$$\text{mol e}^- = 75 \text{ min} \times \frac{60 \text{ s}}{1 \text{ min}} \times \frac{2.15 \text{ C}}{1 \text{ s}} \times \frac{1 \text{ mol e}^-}{96,485 \text{ C}} = 0.10 \text{ mol e}^-$$

(a) $\text{mass Zn} = 0.10 \text{ mol e}^- \times \dfrac{1 \text{ mol Zn}^{2+}}{2 \text{ mol e}^-} \times \dfrac{1 \text{ mol Zn}}{1 \text{ mol Zn}^{2+}} \times \dfrac{65.38 \text{ g Zn}}{1 \text{ mol Zn}} = 3.3 \text{ g Zn}$

(b) $\text{mass Al} = 0.10 \text{ mol e}^- \times \dfrac{1 \text{ mol Al}^{3+}}{3 \text{ mol e}^-} \times \dfrac{1 \text{ mol Al}}{1 \text{ mol Al}^{3+}} \times \dfrac{26.9815 \text{ g Al}}{1 \text{ mol Al}} = 0.90 \text{ g Al}$

(c) $\text{mass Ag} = 0.10 \text{ mol e}^- \times \dfrac{1 \text{ mol Ag}^+}{1 \text{ mol e}^-} \times \dfrac{1 \text{ mol Ag}}{1 \text{ mol Ag}^+} \times \dfrac{107.8682 \text{ g Ag}}{1 \text{ mol Ag}} = 11 \text{ g Ag}$

(d) $\text{mass Ni} = 0.10 \text{ mol e}^- \times \dfrac{1 \text{ mol Ni}^{2+}}{2 \text{ mol e}^-} \times \dfrac{1 \text{ mol Ni}}{1 \text{ mol Ni}^{2+}} \times \dfrac{58.6934 \text{ g Ni}}{1 \text{ mol Ni}} = 2.9 \text{ g Ni}$

65. **(M)** Here we must determine the standard cell voltage of each chemical reaction. Those chemical reactions that have a negative voltage are the ones that require electrolysis.

(a) Oxidation: $2\,H_2O(l) \rightarrow 4\,H^+(aq) + O_2(g) + 4\,e^-$ $\qquad -E° = -1.229\ V$

Reduction: $\{2\,H^+(aq) + 2\,e^- \rightarrow H_2(g)\} \times 2$ $\qquad E° = 0.000\ V$

Net: $2\,H_2O(l) \rightarrow 2\,H_2(g) + O_2(g)$ $\qquad E°_{cell} = -1.229\ V$

This reaction requires electrolysis, with an applied voltage greater than $+1.229\,V$.

(b) Oxidation: $Zn(s) \rightarrow Zn^{2+}(aq) + 2\,e^-$ $\qquad -E° = +0.763\ V$

Reduction: $Fe^{2+}(aq) + 2\,e^- \rightarrow Fe(s)$ $\qquad E° = -0.440\ V$

Net: $Zn(s) + Fe^{2+}(aq) \rightarrow Fe(s) + Zn^{2+}(aq)$ $\qquad E°_{cell} = +0.323\ V$

This is a spontaneous reaction under standard conditions.

(c) Oxidation: $\{Fe^{2+}(aq) \rightarrow Fe^{3+}(aq) + e^-\} \times 2$ $\qquad -E° = -0.771\ V$

Reduction: $I_2(s) + 2\,e^- \rightarrow 2\,I^-(aq)$ $\qquad E° = +0.535\ V$

Net: $2\,Fe^{2+}(aq) + I_2(s) \rightarrow 2\,Fe^{3+}(aq) + 2\,I^-(aq)$ $\qquad E°_{cell} = -0.236\ V$

This reaction requires electrolysis, with an applied voltage greater than $+0.236\ V$.

(d) Oxidation: $Cu(s) \rightarrow Cu^{2+}(aq) + 2\,e^-$ $\qquad -E° = -0.337\ V$

Reduction: $\{Sn^{4+}(aq) + 2\,e^- \rightarrow Sn^{2+}(aq)\} \times 2$ $\qquad E° = +0.154\ V$

Net: $Cu(s) + Sn^{4+}(aq) \rightarrow Cu^{2+}(aq) + Sn^{2+}(aq)$ $\quad E°_{cell} = -0.183\ V$

This reaction requires electrolysis, with an applied voltage greater than $+0.183\ V$.

67. **(M)**

(a) The two gases that are produced are $H_2(g)$ and $O_2(g)$.

(b) At the anode: $2\,H_2O(l) \rightarrow 4\,H^+(aq) + O_2(g) + 4\,e^-$ $\qquad -E° = -1.229\ V$

At the cathode: $\{2\,H^+(aq) + 2\,e^- \rightarrow H_2(g)\} \times 2$ $\qquad E° = 0.000\ V$

Net cell reaction: $2\,H_2O(l) \rightarrow 2\,H_2(g) + O_2(g)$ $\qquad E°_{cell} = -1.229\ V$

69. **(M)**

(a) $Zn^{2+}(aq) + 2\,e^- \rightarrow Zn(s)$

$$\text{mass of Zn} = 42.5\ min \times \frac{60\ s}{1\ min} \times \frac{1.87\ C}{1\ s} \times \frac{1\ mol\ e^-}{96{,}485\ C} \times \frac{1\ mol\ Zn}{2\ mol\ e^-} \times \frac{65.38\ g\ Zn}{1\ mol\ Zn} = 1.62\ g\ Zn$$

(b) $2\,I^-(aq) \rightarrow I_2(s) + 2\,e^-$

$$\text{time needed} = 2.79\ g\ I_2 \times \frac{1\ mol\ I_2}{253.8090\ g\ I_2} \times \frac{2\ mol\ e^-}{1\ mol\ I_2} \times \frac{96{,}485\ C}{1\ mol\ e^-} \times \frac{1\ s}{1.75\ C} \times \frac{1\ min}{60\ s} = 20.2\ min$$

71. **(M)**

(a) $\text{charge} = 1.206 \text{ g Ag} \times \dfrac{1 \text{ mol Ag}}{107.8682 \text{ g Ag}} \times \dfrac{1 \text{ mol e}^-}{1 \text{ mol Ag}} \times \dfrac{96,485 \text{ C}}{1 \text{ mol e}^-} = 1079 \text{ C}$

(b) $\text{current} = \dfrac{1079 \text{ C}}{1412 \text{ s}} = 0.7642 \text{ A}$

73. **(D)**

(a) The electrochemical reaction is:

Anode (oxidation): $\{\text{Ag(s)} \rightarrow \text{Ag}^+(\text{aq}) + \text{e}^-(\text{aq})\} \times 2$ $-E° = -0.800 \text{ V}$

Cathode (reduction): $\text{Cu}^{2+}(\text{aq}) + 2 \text{ e}^- \rightarrow \text{Cu(s)}$ $E° = +0.340 \text{ V}$

Net: $2 \text{ Ag(s)} + \text{Cu}^{2+}(\text{aq}) \rightarrow \text{Cu(s)} + 2 \text{ Ag}^+(\text{aq})$ $E°_{\text{cell}} = -0.46 \text{ V}$

Therefore, copper should plate out first.

(b) $\text{current} = \dfrac{\text{charge}}{2.50 \text{ h}} \times \dfrac{1 \text{ h}}{3600 \text{ s}} = 0.75 \text{ A} \Rightarrow \text{charge} = 6750 \text{ C}$

$\text{mass} = 6750 \text{ C} \times \dfrac{1 \text{ mole}^-}{96,485 \text{ C}} \times \dfrac{1 \text{ mol Cu}}{2 \text{ mol e}^-} \times \dfrac{63.5463 \text{ g Cu}}{1 \text{ mol Cu}} = 2.22 \text{ g Cu}$

(c) The total mass of the metal is 3.50 g out of which 2.22 g is copper. Therefore, the mass of silver in the sample is 3.50 g – 2.22 g = 1.28 g or (1.28/3.50) × 100 = 37%.

INTEGRATIVE AND ADVANCED EXERCISES

75. **(M)** Oxidation: $\text{V}^{3+} + \text{H}_2\text{O} \longrightarrow \text{VO}^{2+} + 2 \text{ H}^+ + \text{e}^-$ $-E°_a$

Reduction: $\text{Ag}^+ + \text{e}^- \longrightarrow \text{Ag(s)}$ $E° = +0.800 \text{ V}$

Net: $\text{V}^{3+} + \text{H}_2\text{O} + \text{Ag}^+ \longrightarrow \text{VO}^{2+} + 2 \text{ H}^+ + \text{Ag(s)}$ $E°_{\text{cell}} = 0.439 \text{ V}$

$0.439 \text{ V} = -E°_a + 0.800 \text{ V}$ $E°_a = 0.800 \text{ V} - 0.439 \text{ V} = 0.361 \text{ V}$

Oxidation: $\text{V}^{2+} \longrightarrow \text{V}^{3+} + \text{e}^-$ $-E°_b$

Reduction: $\text{VO}^{2+} + 2 \text{ H}^+ + \text{e}^- \longrightarrow \text{V}^{3+} + \text{H}_2\text{O}$ $E° = +0.361 \text{ V}$

Net: $\text{V}^{2+} + \text{VO}^{2+} + 2 \text{ H}^+ \longrightarrow 2 \text{ V}^{3+} + \text{H}_2\text{O}$ $E°_{\text{cell}} = +0.616 \text{ V}$

$0.616 \text{ V} = -E°_b + 0.361 \text{ V}$ $E°_b = 0.361 \text{ V} - 0.616 \text{ V} = -0.255 \text{ V}$

Thus, for the cited reaction: $\text{V}^{3+} + \text{e}^- \longrightarrow \text{V}^{2+}$ $E° = -0.255 \text{ V}$

77. **(M)** The cell reaction is $2 \text{ Cl}^-(\text{aq}) + 2 \text{ H}_2\text{O(l)} \longrightarrow 2 \text{ OH}^-(\text{aq}) + \text{H}_2(\text{g}) + \text{Cl}_2(\text{g})$

We first determine the charge transferred per 1000 kg Cl_2.

$\text{charge} = 1000 \text{ kg Cl}_2 \times \dfrac{1000 \text{ g}}{1 \text{ kg}} \times \dfrac{1 \text{ mol Cl}_2}{70.8920 \text{ g Cl}_2} \times \dfrac{2 \text{ mol e}^-}{1 \text{ mol Cl}_2} \times \dfrac{96,485 \text{ C}}{1 \text{ mol e}^-} = 2.72 \times 10^9 \text{ C}$

(a) $\text{energy} = 3.45\text{ V} \times 2.72 \times 10^{9}\text{ C} \times \dfrac{1\text{ J}}{1\text{ V} \cdot \text{C}} \times \dfrac{1\text{ kJ}}{1000\text{ J}} = 9.38 \times 10^{6}\text{ kJ}$

(b) $\text{energy} = 9.38 \times 10^{9}\text{ J} \times \dfrac{1\text{ W} \cdot \text{s}}{1\text{ J}} \times \dfrac{1\text{ h}}{3600\text{ s}} \times \dfrac{1\text{ kWh}}{1000\text{ W} \cdot \text{h}} = 2.61 \times 10^{3}\text{ kWh}$

80. **(D)** We first note that we are dealing with a concentration cell, one in which the standard oxidation reaction is the reverse of the standard reduction reaction, and consequently its standard cell potential is precisely zero volts. For this cell, the Nernst equation permits us to determine the ratio of the two silver ion concentrations.

$E_{\text{cell}} = 0.000\text{ V} - \dfrac{0.0257\text{ V}}{1}\ln\dfrac{[\text{Ag}^{+}(\text{satd AgI})]}{[\text{Ag}^{+}(\text{satd AgCl, }x\text{ M Cl}^{-})]} = 0.0860\text{ V}$

$\ln\dfrac{[\text{Ag}^{+}(\text{satd AgI})]}{[\text{Ag}^{+}(\text{satd AgCl, }x\text{ M Cl}^{-})]} = \dfrac{-0.0860\text{ V}}{0.0257\text{ V}} = -3.35 \quad \dfrac{[\text{Ag}^{+}(\text{satd AgI})]}{[\text{Ag}^{+}(\text{satd AgCl, }x\text{ M Cl}^{-})]} = e^{-3.35} = 0.035$

We can determine the numerator concentration from the solubility product expression for $\text{AgI}(s)$

$K_{\text{sp}} = [\text{Ag}^{+}][\text{I}^{-}] = 8.5 \times 10^{-17} = s^{2} \qquad s = \sqrt{8.5 \times 10^{-17}} = 9.2 \times 10^{-9}\text{ M}$

This permits the determination of the concentration in the denominator.

$[\text{Ag}^{+}(\text{satd AgCl, }x\text{ M Cl}^{-})] = \dfrac{9.2 \times 10^{-9}}{0.035} = 2.6 \times 10^{-7}\text{ M}$

We now can determine the value of x. Note: Cl^{-} arises from two sources, one being the dissolved AgCl.

$K_{\text{sp}} = [\text{Ag}^{+}][\text{Cl}^{-}] = 1.8 \times 10^{-10} = (2.6 \times 10^{-7})(2.6 \times 10^{-7} + x) = 6.8 \times 10^{-14} + 2.6 \times 10^{-7}\,x$

$x = \dfrac{1.8 \times 10^{-10} - 6.8 \times 10^{-14}}{2.6 \times 10^{-7}} = 6.9 \times 10^{-4}\text{ M} = [\text{Cl}^{-}]$

82. **(M)** In this problem we are asked to determine ΔG_{f}° for $N_2H_4(\text{aq})$ using the electrochemical data for hydrazine fuel cell. We first determine the value of ΔG° for the cell reaction, a reaction in which $z = 4$. $\Delta_f G^{\circ}$ can then be determined using data in Appendix D.

Stewise approach:

Calculate ΔG° for the cell reaction (n=4):

$\Delta_r G^{\circ} = -zFE^{\circ}_{\text{cell}} - 4\text{ mol e}^{-} \times \dfrac{96{,}485\text{ C}}{1\text{ mol}} \times 1.559\text{ V} = -6.017 \times 10^{5}\text{ J mol}^{-1} = -601.7\text{ kJ mol}^{-1}$

Using the data in Appendix D, determine $\Delta_f G^{\circ}$ for hydrazine (N_2H_4):

$-601.7\text{ kJ/mol} = \Delta_f G^{\circ}[\text{N}_2(\text{g})] + 2\,\Delta_f G^{\circ}[\text{H}_2\text{O(l)}] - \Delta_f G^{\circ}[\text{N}_2\text{H}_4(\text{aq})] - \Delta_f G^{\circ}[\text{O}_2(\text{g})]$

$= 0.00\text{ kJ/mol} + 2 \times (-237.2\text{ kJ/mol}) - \Delta_f G^{\circ}[\text{N}_2\text{H}_4(\text{aq})] - 0.00\text{ kJ/mol}$

$\Delta_f G^{\circ}[\text{N}_2\text{H}_4(\text{aq})] = 2 \times (-237.2\text{ kJ/mol}) + 601.7\text{ kJ/mol} = +127.3\text{ kJ/mol}$

87. (M) We first determine the change in the amount of M^{2+} ion in each compartment.

$$\text{Change in amount of } M^{2+} = 10.00 \text{ h} \times \frac{3600 \text{ s}}{1 \text{ h}} \times \frac{0.500 \text{ C}}{1 \text{ s}} \times \frac{1 \text{ mol e}^-}{96,485 \text{ C}} \times \frac{1 \text{ mol } M^{2+}}{2 \text{ mol e}^-} = 0.0933 \text{ mol } M^{2+}$$

This change in amount is the increase in the amount of Cu^{2+} and the decrease in the amount of Zn^{2+}. We can also calculate the change in each of the concentrations.

$$\Delta[Cu^{2+}] = \frac{+0.0933 \text{ mol } Cu^{2+}}{0.1000 \text{ L}} = +0.933 \text{ M} \qquad \Delta[Zn^{2+}] = \frac{-0.0933 \text{ mol } Zn^{2+}}{0.1000 \text{ L}} = -0.933 \text{ M}$$

Then the concentrations of the two ions are determined.

$$[Cu^{2+}] = 1.000 \text{ M} + 0.933 \text{ M} = 1.933 \text{ M} \qquad [Zn^{2+}] = 1.000 \text{ M} - 0.933 \text{ M} = 0.067 \text{ M}$$

Now we run the cell as a voltaic cell, first determining the value of $E°_{cell}$.

Oxidation: $Zn(s) \longrightarrow Zn^{2+}(aq) + 2 e^- \qquad -E° = +0.763 \text{ V}$

Reduction: $Cu^{2+}(aq) + 2 e^- \quad Cu(s) \qquad\qquad E° = +0.340 \text{ V}$

Net: $Zn(s) + Cu^{2+}(aq) \longrightarrow Cu(s) \qquad E°_{cell} = +1.103 \text{ V}$

Then we use the Nernst equation to determine the voltage of this cell.

$$E_{cell} = E°_{cell} - \frac{0.0257 \text{ V}}{2} \ln\frac{[Zn^{2+}]}{[Cu^{2+}]} = 1.103 - \frac{0.0257 \text{ V}}{2} \ln\frac{0.067 \text{ M}}{1.933 \text{ M}} = (1.103 + 0.043) \text{ V} = 1.146 \text{ V}$$

90. (M) We assume that the $Pb^{2+}(aq)$ is "ignored" by the silver electrode, that is, the silver electrode detects only silver ion in solution.

Oxidation : $H_2(g) \longrightarrow 2 H^+(aq) + 2 e^- \qquad -E° = 0.000 \text{ V}$

Reduction: $\{Ag^+(aq) + e^- \longrightarrow Ag(s)\} \times 2 \qquad E° = 0.800 \text{ V}$

Net: $H_2(g) + 2 Ag^+ \longrightarrow 2 H^+(aq) + 2 Ag(s) \quad E°_{cell} = 0.800 \text{ V}$

$$E_{cell} = E°_{cell} - \frac{0.0257 \text{ V}}{2} \ln\frac{[H^+]^2}{[Ag^+]^2} \quad 0.503 \text{ V} = 0.800 \text{ V} - \frac{0.0257 \text{ V}}{2} \ln\frac{1.00^2}{[Ag^+]^2}$$

$$\log\frac{1.00^2}{[Ag^+]^2} = \frac{2(0.800 - 0.503)}{0.0257 \text{ V}} = 23.1 \qquad \frac{1.00^2}{[Ag^+]^2} = e^{+23.1} = 1.0\times10^{10}$$

$$[Ag^+]^2 = 1.0\times10^{-10} \text{ M}^2 \Rightarrow [Ag^+] = 1.0\times10^{-5} \text{ M}$$

$$\text{mass Ag} = 0.500 \text{ L} \times \frac{1.0\times10^{-5} \text{ mol } Ag^+}{1 \text{ L soln}} \times \frac{1 \text{ mol Ag}}{1 \text{ mol } Ag^+} \times \frac{107.87 \text{ g Ag}}{1 \text{ mol Ag}} = 5.4\times10^{-4} \text{ g Ag}$$

$$\% \text{ Ag} = \frac{5.4\times10^{-4} \text{ g Ag}}{1.050 \text{ g sample}} \times 100 \% = 0.051\% \text{ Ag (by mass)}$$

92. **(M)** First we determine the molar solubility of AgBr in 1 M NH_3 .

Sum	$AgBr(s) + 2\,NH_3(aq) \;\rightleftharpoons\; Ag(NH_3)_2^{+}(aq) + Br^-(aq)$			$K = K_{sp} \times K_f = 8.0 \times 10^{-6}$
Initial	1.00 M	0 M	0 M	
Equil.	$1.00 - 2s$	s	s	{s = AgBr molar solubility}

$$K = \frac{[Ag(NH_3)_2^{+}][Br^-]}{[NH_3(aq)]} = \frac{s^2}{(1-2s)^2} = 8.0 \times 10^{-6} \qquad s = 2.8\underline{1} \times 10^{-3}\ \text{M (also } [Br^-]\text{)}$$

So $[Ag^+] = \dfrac{K_{sp}}{[Br^-]} = \dfrac{5.0 \times 10^{-13}}{2.8\underline{1} \times 10^{-3}} = 1.8 \times 10^{-10}\,\text{M}$

$$AgBr(s) \rightleftharpoons Ag^+ + Br^- \qquad\qquad K_{sp} = 5.0 \times 10^{-13}$$

$$\underline{Ag^+ + 2\,NH_3 \rightleftharpoons Ag(NH_3)_2^{+} \qquad\qquad K_f = 1.6 \times 10^7}$$

(sum) $AgBr(s) + 2\,NH_3(aq) \rightleftharpoons Ag(NH_3)_2^{+} + Br^- \qquad K = K_{sp} \times K_f = 8.0 \times 10^{-6}$

Now, let's construct the cell, guessing that the standard hydrogen electrode is the anode.

Oxidation:	$H_2(g) \longrightarrow 2\,H^+ + 2\,e^-$	$E^\circ = -0.000$ V
Reduction:	{$Ag^+(aq) + e^- \longrightarrow Ag(s)$ }$\times 2$	$E^\circ = +0.800$ V
Net:	$2\,Ag^+(aq) + H_2(g) \longrightarrow 2\,Ag(s) + 2\,H^+(aq)$	$E^\circ_{cell} = +0.800$ V

From the Nernst equation:

$$E = E^\circ - \frac{0.0257\ \text{V}}{z}\ln Q = E^\circ - \frac{0.0257\ \text{V}}{z}\ln\frac{[H^+]^2}{[Ag^+]^2} = 0.800\ \text{V} - \frac{0.0257\ \text{V}}{2}\ln\frac{1^2}{(1.7\underline{8}\times10^{-10})^2}$$

and $E = 0.22\underline{3}$ V. Since the voltage is positive, our guess is correct and the standard hydrogen electrode is the anode (oxidizing el*ectrode*).

96. **(M)**

(a) The metal has to have a reduction potential more negative than –0.691 V, so that its oxidation can reverse the tarnishing reaction's –0.691 V reduction potential. Aluminum is a good candidate, because it is inexpensive, readily available, will not react with water and has an E° of –1.676 V. Zinc is also a possibility with an E° of –0.763 V, but we don't choose it because there may be an overpotential associated with the tarnishing reaction.

(b)

Oxidation:	{$Al(s) \longrightarrow Al^{3+}(aq) + 3\,e^-$}$\times 2$
Reduction:	{$Ag_2S(s) + 2\,e^- \longrightarrow Ag(s) + S^{2-}(aq)$ }$\times 3$
Net:	$2\,Al(s) + 3\,Ag_2S(s) \longrightarrow 6\,Ag(s) + 2\,Al^{3+}(aq) + 3\,S^{2-}(aq)$

(c) The dissolved $NaHCO_3(s)$ serves as an electrolyte. It would also enhance the electrical contact between the two objects.

(d) There are several chemicals involved: Al, H_2O, and $NaHCO_3$. Although the aluminum plate will be consumed very slowly because silver tarnish is very thin, it will, nonetheless, eventually erode away. We should be able to detect loss of mass after many uses.

99. **(D)** Recall that under non-standard conditions $\Delta_r G = \Delta_r G° + RT \ln K$. Substituting $\Delta_r G° = \Delta_r H° - T\Delta_r S°$ and $\Delta_r G = -zFE_{cell}$ one obtains:

$$-zFE_{cell} = \Delta_r H° - T\Delta_r S° + RT \ln Q$$

For two different temperatures (T_1 and T_2) we can write:

$$-zFE_{cell}(T_1) = \Delta_r H° - T_1\Delta_r S° + RT_1 \ln Q$$

$$-zFE_{cell}(T_2) = \Delta_r H° - T_2\Delta_r S° + RT_2 \ln Q$$

$$-E_{cell}(T_1) + E_{cell}(T_2) = \frac{\Delta_r H° - T_1\Delta_r S° + RT_1 \ln Q - \Delta_r H° + T_2\Delta_r S° - RT_2 \ln Q}{zF}$$

$$-E_{cell}(T_1) + E_{cell}(T_2) = \frac{-T_1\Delta_r S° + RT_1 \ln Q + T_2\Delta_r S° - RT_2 \ln Q}{zF}$$

$$E_{cell}(T_1) - E_{cell}(T_2) = \frac{T_1\Delta_r S° - RT_1 \ln Q}{zF} - \frac{T_2\Delta_r S° + RT_2 \ln Q}{zF}$$

$$E_{cell}(T_1) - E_{cell}(T_2) = T_1\left(\frac{\Delta_r S° - R \ln Q}{zF}\right) - T_2\left(\frac{\Delta_r S° - R \ln Q}{zF}\right)$$

$$E_{cell}(T_1) - E_{cell}(T_2) = (T_1 - T_2)\left(\frac{\Delta_r S° - R \ln Q}{zF}\right)$$

The value of Q at 25 °C can be calculated from $E°_{cell}$ and E_{cell}. First calculate $E°_{cell}$:

Oxidation:	$Cu(s) \rightarrow Cu^{2+}(aq) + 2\ e^-$	$-E° = -0.340$ V
Reduction:	$2\ Fe^{3+}(aq) + 2\ e^- \rightarrow 2\ Fe^{2+}(aq)$	$-E° = +0.771$ V
Overall:	$Cu(s) + 2\ Fe^{3+} \rightarrow Cu^{2+}(aq) + 2\ Fe^{2+}$	$E°_{cell} = 0.431$ V

$$0.370 = 0.431 - \frac{0.0257\ V}{2} \ln Q \Rightarrow 0.0129\ V \ln Q = 0.431 - 0.370 = 0.061$$

$$\ln Q = 4.73 \Rightarrow Q = e^{4.73} = 83.9$$

Now, use the above derived equation and solve for $\Delta_r S°$:

$$0.394 - 0.370 = (50 - 25)\left(\frac{\Delta_r S° - 8.314 \times \ln 83.9}{2 \times 96,485}\right)$$

$$0.024 = 25 \times \left(\frac{\Delta_r S° - 3.96}{192,970}\right) \Rightarrow \Delta_r S° - 3.96 = 185.3 \Rightarrow \Delta_r S° = 189.2\ J\ K^{-1}$$

Since $\Delta_r G° = \Delta_r H° - T\Delta_r S° = -zFE°_{cell}$ we can calculate $\Delta_r G°$, K (at 25 °C) and $\Delta_r H°$:

$$\Delta_r G° = -zFE°_{cell} = -2 \times 96,485 \times 0.431 = -83.2 \text{ kJ}$$

$$\Delta_r G° = -RT \ln K \Rightarrow -83.2 \times 1000 = -8.314 \times 298.1 \times \ln K$$

$$\ln K = 33.56 \Rightarrow K = e^{33.56} = 3.77 \times 10^{14}$$

$$-83.2 \text{ kJ} = \Delta_r H° - 298.15 \times \frac{189.2}{1000} \text{ kJ} \Rightarrow \Delta_r H° = -83.2 + 56.4 = -26.8 \text{ kJ}$$

Since we have $\Delta_r H°$ and $\Delta_r S°$ we can calculate the value of the equilibrium constant at 50 °C:

$$\Delta_r G° = \Delta_r H° - T\Delta_r S° = -26.8 \text{ kJ} - (273.15 + 50)\text{K} \times \frac{189.2}{1000} \text{ kJ K}^{-1} = -87.9 \text{ kJ}$$

$$-87.9 \times 1000 \text{ J} = -8.314 \text{ J K}^{-1} \text{ mol}^{-1} \times (273.15 + 50)\text{K} \times \ln K$$

$$\ln K = 32.7 \Rightarrow K = e^{32.7} = 1.59 \times 10^{14}$$

Choose the values for the concentrations of Fe^{2+}, Cu^{2+} and Fe^{3+} that will give the value of the above calculated Q. For example:

$$Q = \frac{\left[Fe^{2+}\right]^2 \left[Cu^{2+}\right]}{\left[Fe^{3+}\right]^2} = 1.61$$

$$\frac{0.1^2 \times 1.61}{0.1^2} = 1.61$$

Determine the equilibrium concentrations at 50 °C. Notice that since $Q < K$, a net change occurs from left to right (the direction of the forward reaction):

$$Cu(s) + 2 Fe^{3+} \rightarrow Cu^{2+}(aq) + 2 Fe^{2+}$$

Initial:	0.1	1.61	0.1
Change:	0.1 − x	1.61 + x	0.1 + x

$$K = \frac{\left[Fe^{2+}\right]_{eq}^2 \left[Cu^{2+}\right]_{eq}}{\left[Fe^{3+}\right]_{eq}^2} = 1.59 \times 10^{14} = \frac{(0.1+x)^2 \times (1.61+x)}{(0.1-x)^2}$$

Obviously, the reaction is almost completely shifted towards products. First assume that the reaction goes to completion, and then let the equilibrium be shifted towards reactants:

$$Cu(s) + 2 Fe^{3+} \rightarrow Cu^{2+}(aq) + 2 Fe^{2+}$$

Initial:	0.1	1.61	0.1
Final	0	1.71	0.2
Equilibrium	0 + x	1.71 − x	0.2 − x

$$K = \frac{\left[Fe^{2+}\right]^2_{eq}\left[Cu^{2+}\right]_{eq}}{\left[Fe^{3+}\right]^2_{eq}} = 1.59\times10^{14} = \frac{(0.2-x)^2\times(1.71-x)}{(x)^2} \approx \frac{0.2^2\times1.71}{x^2}$$

$$x^2 \approx \frac{0.2^2\times1.71}{1.59\times10^{14}} \approx 4.3\times10^{-16}$$

$$x \approx 2.1\times10^{-8}\,M$$

Therefore, $[Cu^{2+}] \approx 1.7$ M, $[Fe^{2+}] \approx 0.2$ M and $[Fe^{3+}] \approx 2.1 \times 10^{-8}$ M

100. **(D)** This problem can be solved by utilizing the relationship between $\Delta_r G^\circ$ and E°_{cell} $(\Delta_r G^\circ = -zFE^\circ_{cell})$:

Consider a hypothetical set of the following reactions:

$$A + ne^- \rightarrow A^{n-} \quad E^\circ_1 \text{ and } \Delta_r G^\circ_1$$

$$B + me^- \rightarrow B^{m-} \quad E^\circ_2 \text{ and } \Delta_r G^\circ_2$$

Overall: $A + B + ne^- + me^- \rightarrow A^{n-} + B^{m-} \quad E^\circ_r = ? \text{ and } \Delta_r G^\circ_{rxn} = \Delta_r G^\circ_1 + \Delta_r G^\circ_2$

$$\Delta_r G^\circ = \Delta_r G^\circ_1 + \Delta_r G^\circ_2 = -nFE^\circ_1 - mFE^\circ_2$$

$$-(n+m)FE^\circ = -nFE^\circ_1 - mFE^\circ_2$$

$$(n+m)FE^\circ = nFE^\circ_1 + mFE^\circ_2 \Rightarrow E^\circ = \frac{nFE^\circ_1 + mFE^\circ_2}{n+m}$$

Therefore, for n-sets of half-reactions:

$$E^\circ = \frac{\sum n_i E^\circ_i}{\sum n_i}$$

The E° for the given half-reaction can be determined by combining four half-reactions:

$$H_6IO_6 + H^+ + 2\,e^- \rightarrow IO_3^- + 3\,H_2O \Rightarrow E^\circ = 1.60 \text{ V}$$

$$IO_3^- + 6\,H^+ + 5\,e^- \rightarrow \frac{1}{2}I_2 + 3\,H_2O \Rightarrow E^\circ = 1.19 \text{ V}$$

$$I_2 + 2\,H_2O \rightarrow 2\,HIO + 2\,H^+ + 2\,e^- \Rightarrow E^\circ = -1.45 \text{ V}$$

$$2\,I^- \rightarrow I_2 + 2\,e^- \Rightarrow E^\circ = -0.535 \text{ V}$$

Overall :

$$H_6IO_6 + H^+ + 2\,e^- + IO_3^- + 6\,H^+ + 5\,e^- + I_2 + 2\,H_2O + 2\,I^- \rightarrow$$

$$IO_3^- + 3\,H_2O + \frac{1}{2}I_2 + 3\,H_2O + 2\,HIO + 2\,H^+ + 2\,e^- + I_2 + 2\,e^-$$

$$H_6IO_6 + 5\,H^+ + 2\,I^- + 3\,e^- \rightarrow \frac{1}{2}I_2 + 4\,H_2O + 2\,HIO$$

$$E^\circ = \frac{1.60\times3 + 1.19\times5 - 1.45\times2 - 0.535\times2}{2+5-2-2} = 2.26 \text{ V}$$

FEATURE PROBLEMS

<u>**102.**</u> **(D)**

(a) **(1)** anode: $Na(s) \rightarrow Na^+(\text{in ethylamine}) + e^-$

cathode: $Na^+(\text{in ethylamine}) \rightarrow Na(\text{amalgam}, 0.206\ \%)$

Net: $Na(s) \rightarrow Na(\text{amalgam}, 0.206\%)$

(2) anode: $2\ Na(\text{amalgam}, 0.206\%) \rightarrow 2\ Na^+(1\ M) + 2\ e^-$

cathode: $2\ H^+(aq, 1\ M) + 2\ e^- \rightarrow H_2(g, 1\ atm)$

Net: $2\ Na(\text{amalgam}, 0.206\%) + 2\ H^+(aq, 1\ M) \rightarrow 2\ Na^+(1\ M) + H_2(g, 1\ atm)$

(b) **(1)** $\Delta_r G = -1 \times \dfrac{96,485\ C}{1\ mol} \times 0.8453\ V = -8.156 \times 10^4\ J\ mol^{-1}$ or $-81.56\ kJ\ mol^{-1}$

(2) $\Delta_r G = -2 \times \dfrac{96,485\ C}{1\ mol} \times 1.8673\ V = -36.033 \times 10^4\ J\ mol^{-1}$ or $-360.33\ kJ\ mol^{-1}$

(c) **(1)** $2\ Na(s) \rightarrow 2\ Na(\text{amalgam}, 0.206\%)$ $\qquad \Delta_r G_1 = -2 \times 8.156 \times 10^4\ J\ mol^{-1}$

(2) $2\ Na\ (\text{amalg}, 0.206\%) + 2\ H^+(aq, 1\ M) \rightarrow 2\ Na^+(1\ M) + H_2(g, 1\ atm)$

$\Delta_r G_2 = -36.033 \times 10^4\ J\ mol^{-1}$

Overall: $2\ Na(s) + 2\ H^+(aq) \rightarrow 2\ Na^+(1\ M) + H_2(g, 1\ atm)$

$\Delta_r G = \Delta_r G_1 + \Delta_r G_2 = -16.312 \times 10^4\ J\ mol^{-1} - 36.033 \times 10^4\ J\ mol^{-1} = -52.345 \times 10^4\ J\ mol^{-1}$
or $-523.45\ kJ\ mol^{-1}$

Since standard conditions are implied in the overall reaction, $\Delta_r G = \Delta_r G°$.

(d) $E°_{\text{cell}} = -\dfrac{-52.345 \times 10^4\ J\ mol^{-1}}{2 \times \dfrac{96,485\ C}{1\ mol} \times \dfrac{1\ J\ C^{-1}}{1\ V}} = E°\{H^+(1\ M)/H_2(1\ atm)\} - E°\{Na^+(1\ M)/Na(s)\}$

$E°\{Na^+(1\ M)/Na(s)\} = -2.713\ V$. This is precisely the same as the value in Appendix D.

104. (D)

(a) The capacitance of the cell membrane is given by the following equation,

$$C = \frac{\varepsilon_0 \varepsilon A}{l}$$

where $\varepsilon_0 \varepsilon = 3 \times 8.854 \times 10^{-12}$ C^2 N^{-1} m^{-2};
$A = 1 \times 10^{-6}$ cm^2; and $l = 1 \times 10^{-6}$ cm.
Together with the factors necessary to convert
from cm to m and from cm^2 to m^2, these data yield

$$C = \frac{(3)\left(8.854\times10^{-12}\ \frac{C^2}{N^1\ m^2}\right)(1\times10^{-6}\ cm^2)\left(\frac{1\ m}{100\ cm}\right)^2}{(1\times10^{-6}cm)\left(\frac{1\ m}{100\ cm}\right)} = 2.66 \times 10^{-13}\ \frac{C^2}{N\ m}$$

$$C = \left(2.66\times10^{-13}\ \frac{C^2}{N\ m}\right)\left(\frac{1\ F}{1\ \frac{C^2}{N\ m}}\right) = 2.66\times10^{-13}\ F$$

(b) Since the capacitance C is the charge in coulombs per volt, the charge on the membrane, Q, is given by the product of the capacitance and the potential across the cell membrane.

$$Q = 2.66 \times 10^{-13}\ \frac{C}{V} \times 0.085\ V = 2.26 \times 10^{-14}\ C$$

(c) The number of K^+ ions required to produce this charge is

$$\frac{Q}{e} = \frac{2.26\times10^{-14}\ C}{1.602\times10^{-19}\ C/ion} = 1.41\times10^5\ K^+\ \text{ions}$$

(d) The number of K^+ ions in a typical cell is

$$\left(6.022\times10^{23}\ \frac{\text{ions}}{\text{mol}}\right)\left(155\times10^{-3}\ \frac{\text{mol}}{L}\right)\left(\frac{1\ L}{1000\ cm^3}\right)(1\times10^{-8}\ cm^3) = 9.3\times10^{11}\ \text{ions}$$

(e) The fraction of the ions involved in establishing the charge on the cell membrane is

$$\frac{1.4\times10^5\ \text{ions}}{9.3\times10^{11}\ \text{ions}} = 1.5\times10^{-7}\ (\sim 0.000015\ \%)$$

Thus, the concentration of K^+ ions in the cell remains constant at 155 mM.

105. (M) Reactions with a positive cell potential are reactions for which $\Delta_r G° < 0$, or reactions for which $K > 1$. $\Delta_r S°$, $\Delta_r H°$ and $\Delta_r U°$ cannot be used alone to determine whether a particular electrochemical reaction will have a positive or negative value.

106. (M) The half-reactions for the first cell are:

Anode (oxidation): $X(s) \rightarrow X^+(aq) + e^-$ $-E°_{X^+/X}$

Cathode (reduction): $2H^+(aq) + 2e^- \rightarrow H_2(g)$ $E° = 0$ V

Since the electrons are flowing from metal X to the standard hydrogen electrode, $E°_{X^+/X} < 0$ V.

The half-reactions for the second cell are:

Anode (oxidation): $X(s) \rightarrow X^+(aq) + e^-$ $-E^\circ_{X^+/X}$

Cathode (reduction): $Y^{2+} + 2e^- \rightarrow Y(s)$ $E^\circ_{Y^{2+}/Y}$

Since the electrons are flowing from metal X to metal Y, $-E^\circ_{X^+/X} + E^\circ_{Y^{2+}/Y} > 0$.

From the first cell we know that $E^\circ_{X^+/X} < 0$ V. Therefore, $E^\circ_{X^+/X} > E^\circ_{Y^{2+}/Y}$.

107. (M) The standard reduction potential of the $Fe^{2+}(aq)/Fe(s)$ couple can be determined from:

$Fe^{2+}(aq) \rightarrow Fe^{3+}(aq) + e^- \quad E^\circ = -0.771$ V

$Fe^{3+}(aq) + 3e^- \rightarrow Fe(s) \quad E^\circ = -0.04$ V

Overall:

$Fe^{2+}(aq) + 2e^- \rightarrow Fe(s)$

We proceed similarly to the solution for 100:

$$E^\circ = \frac{\sum n_i E^\circ_i}{\sum n_i} = \frac{-0.771 \times (1) - 0.04 \times 3}{3-1} = -0.445 \text{ V}$$

SELF-ASSESSMENT EXERCISES

114. (M)

(a)   False. The cathode is the positive electrode in a voltaic cell and negative in electrolytic cell.

(b)   False. The function of the salt bridge is to permit the migration of the ions not electrons.

(c)   True. The anode is the negative electrode in a voltaic cell.

(d)   True.

(e)   True. Reduction always occurs at the cathode of an electrochemical cell. Because of the removal of electrons by the reduction half-reaction, the cathode of a voltaic cell is positive. Because of the electrons forced onto it, the cathode of an electrolytic cell is negative. For both types, the cathode is the electrode at which electrons enter the cell.

(f)   False. Reversing the direction of the electron flow changes the voltaic cell into an electrolytic cell.

(g)   True. The cell reaction is an oxidation-reduction reaction.

115. (M) The correct answer is (b), $Hg^{2+}(aq)$ is more readily reduced than $H^+(aq)$.

116. **(M)** Under non-standard conditions, apply the Nernst equation to calculate E_{cell}:

$$E_{cell} = E°_{cell} - \frac{0.0257 \text{ V}}{z} \ln Q$$

$$E_{cell} = 0.66 \text{ V} - \frac{0.0257 \text{ V}}{2} \ln \frac{0.10}{0.01} = 0.63 \text{ V}$$

The correct answer is (d).

117. **(E)** **(c)** The displacement of Ni(s) from the solution will proceed to a considerable extent, but the reaction will not go to completion.

118. **(E)** The gas evolved at the anode when K_2SO_4(aq) is electrolyzed between Pt electrodes is most likely oxygen.

119. **(M)** The electrochemical reaction in the cell is:

Anode (oxidation): $\{Al(s) \rightarrow Al^{3+}(aq) + 3 e^-\} \times 2$

Cathode (reduction): $\{H_2(g) + 2 e^- \rightarrow 2 H^+(aq)\} \times 3$

Overall: $2 Al(s) + 3 H_2(g) \rightarrow 2 Al^{3+}(aq) + 6 H^+(aq)$

$$4.5 \text{ g Al} \times \frac{1 \text{ mol Al}}{26.98 \text{ g Al}} \times \frac{3 \text{ mol } H_2}{2 \text{ mol Al}} = 0.250 \text{ mol } H_2$$

$$0.250 \text{ mol } H_2 \times \frac{22.4 \text{ L } H_2}{1 \text{ mol } H_2} = 5.6 \text{ L } H_2$$

120. **(E)** The correct answer is (a) $\Delta_r G$.

121. **(M)**

Anode (oxidation): $\{Zn(s) \rightarrow Zn^{2+}(aq) + 2 e^-\} \times 3$ $-E° = 0.763 \text{ V}$

Cathode (reduction): $\{NO_3^-(aq) + 4 H^+(aq) + 3 e^- \rightarrow NO(g) + 2 H_2O(l)\} \times 2$ $E° = +0.956 \text{ V}$

Overall: $3 Zn(s) + 2 NO_3^-(aq) + 8 H^+(aq) \rightarrow 3 Zn^{2+}(aq) + 2 NO(g) + 4 H_2O(l)$ $E°_{cell} = 1.719 \text{ V}$

Cell diagram: $Zn(s) | Zn^{2+}(1 \text{ M}) \| H^+(1 \text{ M}), NO_3^-(1 \text{ M}) | NO(g, 1 \text{ atm}) | Pt(s)$

122. **(M)** Apply the Nernst equation:

$$E_{cell} = E°_{cell} - \frac{0.0257 \text{ V}}{z} \ln Q$$

$$0.108 \text{ V} = 0 \text{ V} - \frac{0.0257 \text{ V}}{2} \ln x^2 \Rightarrow \ln x^2 = -8.40$$

$$x^2 = e^{-8.40} \Rightarrow x = 0.0150$$

$$pH = -\log(0.0150) = 1.82$$

123. (M)

(a) Since we are given E°_{cell}, we can calculate K for the given reaction:

$$E^{\circ}_{cell} = \frac{RT}{zF} \ln K$$

$$-0.0050\ \text{V} = \frac{8.314\ \text{J K}^{-1}\text{mol}^{-1} \times 298\ \text{K}}{2 \times 96,485\ \text{C mol}^{-1}} \ln K \Rightarrow \ln K = -0.389$$

$$K = e^{-0.389} = 0.68$$

Since for the given conditions $Q = 1$, the system is not at equilibrium.

(b) Because $Q > K$, a net reaction occurs to the left.

124. (M)

(a) $Fe(s) + Cu^{2+}(1\ \text{M}) \rightarrow Fe^{2+}(1\ \text{M}) + Cu(s)$, $E^{\circ}_{cell} = -0.780\ \text{V}$, electrons flow from B to A

(b) $Sn^{2+}(1\ \text{M}) + 2\ Ag^{+}(1\ \text{M}) \rightarrow Sn^{4+}(aq) + 2\ Ag(s)$, $E^{\circ}_{cell} = +0.646\ \text{V}$, electrons flow from A to B.

(c) $Zn(s) + Fe^{2+}(0.0010\ \text{M}) \rightarrow Zn^{2+}(0.10\ \text{M}) + Fe(s)$, $E^{\circ}_{cell} = +0.264\ \text{V}$, electrons flow from A to B.

125. (M)

(a) $Cl_2(g)$ at anode and $Cu(s)$ at cathode.

(b) $O_2(g)$ at anode and $H_2(g)$ and $OH^-(aq)$ at cathode.

(c) $Cl_2(g)$ at anode and $Ba(l)$ at cathode.

(d) $O_2(g)$ at anode and $H_2(g)$ and $OH^-(aq)$ at cathode.

CHAPTER 20
CHEMICAL KINETICS

PRACTICE EXAMPLES

1A   **(E)** The rate of consumption for a reactant is expressed as the negative of the change in molarity divided by the time interval. The rate of reaction is expressed as the rate of consumption of a reactant or production of a product divided by its stoichiometric coefficient.

$$\text{rate of consumption of A} = \frac{-\Delta[A]}{\Delta t} = \frac{-(0.3187 \text{ M} - 0.3629 \text{ M})}{8.25 \text{ min}} \times \frac{1 \text{ min}}{60 \text{ s}} = 8.93 \times 10^{-5} \text{ M s}^{-1}$$

$$\text{rate of reaction} = \text{rate of consumption of A} \div 2 = \frac{8.93 \times 10^{-5} \text{ M s}^{-1}}{2} = 4.46 \times 10^{-5} \text{ M s}^{-1}$$

1B   **(E)** We use the rate of reaction of A to determine the rate of formation of B, noting from the balanced equation that 3 moles of B form (+3 moles B) when 2 moles of A react (−2 moles A). (Recall that "M" means "moles per liter.")

$$\text{rate of B formation} = \frac{0.5522 \text{ M A} - 0.5684 \text{ M A}}{2.50 \text{ min} \times \dfrac{60 \text{ s}}{1 \text{ min}}} \times \frac{+3 \text{ moles B}}{-2 \text{ moles A}} = 1.62 \times 10^{-4} \text{ M s}^{-1}$$

2A   **(M)**

     (a)   The 2400-s tangent line intersects the 1200-s vertical line at 0.75 M and reaches 0 M at 3500 s. The slope of that tangent line is thus

$$\text{slope} = \frac{0 \text{ M} - 0.75 \text{ M}}{3500 \text{ s} - 1200 \text{ s}} = -3.3 \times 10^{-4} \text{ M s}^{-1} = - \text{ instantaneous rate of reaction}$$

         The instantaneous rate of reaction $= 3.3 \times 10^{-4} \text{ M s}^{-1}$.

     (b)   At 2400 s, $[H_2O_2] = 0.39$ M. At 2450 s, $[H_2O_2] = 0.39$ M $+$ rate $\times \Delta t$

$$\text{At 2450 s,} [H_2O_2] = 0.39 \text{ M} + \left[-3.3 \times 10^{-4} \text{ mol } H_2O_2 \text{ L}^{-1}\text{s}^{-1} \times 50 \text{s} \right]$$

$$= 0.39 \text{ M} - 0.017 \text{ M} = 0.37 \text{ M}$$

2B   **(M)** With only the data of Table 20.2 we can use only the reaction rate during the first 400 s, $-\Delta[H_2O_2]/\Delta t = 15.0 \times 10^{-4} \text{ M s}^{-1}$, and the initial concentration, $[H_2O_2]_o = 2.32$ M. We calculate the change in $[H_2O_2]$ and add it to $[H_2O_2]_o$ to determine $[H_2O_2]_{100}$.

$$\Delta[H_2O_2] = \text{ rate of reaction of } H_2O_2 \times \Delta t = -15.0 \times 10^{-4} \text{ M s}^{-1} \times 100 \text{ s} = -0.15 \text{ M}$$

$$[H_2O_2]_{100} = [H_2O_2]_o + \Delta[H_2O_2] = 2.32 \text{ M} + (-0.15 \text{ M}) = 2.17 \text{ M}$$

This value differs from the value of 2.15 M determined in *text* Example 20-2b because the *text* used the initial rate of reaction $(17.1 \times 10^{-4} \text{ M s}^{-1})$, which is a bit faster than the average rate over the first 400 seconds.

3A **(M)** We write the equation for each rate, divide them into each other, and solve for n.

$$R_1 = k \times [N_2O_5]_1^n = 5.45 \times 10^{-5} \text{ M s}^{-1} = k(3.15 \text{ M})^n$$

$$R_2 = k \times [N_2O_5]_2^n = 1.35 \times 10^{-5} \text{ M s}^{-1} = k(0.78 \text{ M})^n$$

$$\frac{R_1}{R_2} = \frac{5.45 \times 10^{-5} \text{ M s}^{-1}}{1.35 \times 10^{-5} \text{ M s}^{-1}} = 4.04 = \frac{k \times [N_2O_5]_1^n}{k \times [N_2O_5]_2^n} = \frac{k(3.15 \text{ M})^n}{k(0.78 \text{ M})^n} = \left(\frac{3.15}{0.78}\right)^n = (4.0\underline{4})^n$$

We kept an extra significant figure (4) to emphasize that the value of $n = 1$. Thus, the reaction is first-order in N_2O_5.

3B **(E)** For the reaction, we know that $\text{rate} = k[HgCl_2]^1[C_2O_4^{2-}]^2$. Here we will compare Expt. 4 to Expt. 1 to find the rate.

$$\frac{\text{rate}_4}{\text{rate}_1} = \frac{k[HgCl_2]^1[C_2O_4^{2-}]^2}{k[HgCl_2]^1[C_2O_4^{2-}]^2} = \frac{0.025 \text{ M} \times (0.045 \text{ M})^2}{0.105 \text{ M} \times (0.150 \text{ M})^2} = 0.0214 = \frac{\text{rate}_4}{1.8 \times 10^{-5} \text{ M min}^{-1}}$$

The desired rate is $\text{rate}_4 = 0.0214 \times 1.8 \times 10^{-5} \text{ M min}^{-1} = 3.9 \times 10^{-7} \text{ M min}^{-1}$.

4A **(E)** We place the initial concentrations and the initial rates into the rate law and solve for k.

$$\text{rate} = k[A]^2[B] = 4.78 \times 10^{-2} \text{ M s}^{-1} = k(1.12 \text{ M})^2(0.87 \text{ M})$$

$$k = \frac{4.78 \times 10^{-2} \text{ M s}^{-1}}{(1.12 \text{ M})^2 \, 0.87 \text{ M}} = 4.4 \times 10^{-2} \text{ M}^{-2} \text{ s}^{-1}$$

4B **(E)** We know that $\text{rate} = k[HgCl_2]^1[C_2O_4^{2-}]^2$ and $k = 7.6 \times 10^{-3} \text{ M}^{-2}\text{min}^{-1}$.

Thus, insertion of the starting concentrations and the k value into the rate law yields:

$$\text{Rate} = 7.6 \times 10^{-3} \text{ M}^{-2} \text{ min}^{-1}(0.050 \text{ M})^1(0.025 \text{ M})^2 = 2.4 \times 10^{-7} \text{ M min}^{-1}$$

5A **(E)** Here we substitute directly into the integrated rate law equation.

$$\ln[A]_t = -kt + \ln[A]_o = -3.02 \times 10^{-3} \text{ s}^{-1} \times 325 \text{ s} + \ln(2.80) = -0.982 + 1.030 = 0.048$$

$$[A]_t = e^{0.048} = 1.0 \text{ M}$$

5B **(M)** This time we substitute the provided values into text Equation 20.13.

$$\ln\frac{[H_2O_2]_t}{[H_2O_2]_o} = -kt = -k \times 600 \text{ s} = \ln\frac{1.49 \text{ M}}{2.32 \text{ M}} = -0.443 \qquad k = \frac{-0.443}{-600 \text{ s}} = 7.38 \times 10^{-4} \text{ s}^{-1}$$

Now we choose $[H_2O_2]_o = 1.49 \text{ M}$, $[H_2O_2]_t = 0.62$, $\qquad t = 1800 \text{ s} - 600 \text{ s} = 1200 \text{ s}$

$$\ln\frac{[H_2O_2]_t}{[H_2O_2]_o} = -kt = -k \times 1200 \text{ s} = \ln\frac{0.62 \text{ M}}{1.49 \text{ M}} = -0.88 \qquad k = \frac{-0.88}{-1200 \text{ s}} = 7.3 \times 10^{-4} \text{ s}^{-1}$$

These two values agree within the limits of the experimental error and thus, the reaction is first-order in $[H_2O_2]$.

6A (M) We can use the integrated rate equation to find the ratio of the final and initial concentrations. This ratio equals the fraction of the initial concentration that remains at time t.

$$\ln\frac{[A]_t}{[A]_o} = -kt = -2.95\times10^{-3}\ s^{-1}\times150\ s = -0.443$$

$$\frac{[A]_t}{[A]_o} = e^{-0.443} = 0.642; \quad 64.2\% \text{ of } [A]_o \text{ remains.}$$

6B (M) After two-thirds of the sample has decomposed, one-third of the sample remains. Thus $[H_2O_2]_t = [H_2O_2]_o \div 3$, and we have

$$\ln\frac{[H_2O_2]_t}{[H_2O_2]_o} = -kt = \ln\frac{[H_2O_2]_o \div 3}{[H_2O_2]_o} = \ln(1/3) = -1.099 = -7.30\times10^{-4}\ s^{-1}t$$

$$t = \frac{-1.099}{-7.30\times10^{-4}\ s^{-1}} = 1.51\times10^3\ s\times\frac{1\ min}{60\ s} = 25.1\ min$$

7A (M) At the end of one half-life the pressure of DTBP will have been halved, to 400 mmHg. At the end of another half-life, at 160 min, the pressure of DTBP will have halved again, to 200 mmHg. Thus, the pressure of DTBP at 125 min will be intermediate between the pressure at 80.0 min (400 mmHg) and that at 160 min (200 mmHg). To obtain an exact answer, first we determine the value of the rate constant from the half-life.

$$k = \frac{0.693}{t_{1/2}} = \frac{0.693}{80.0\ min} = 0.00866\ min^{-1}$$

$$\ln\frac{(P_{DTBP})_t}{(P_{DTBP})_o} = -kt = -0.00866\ min^{-1}\times125\ min = -1.08$$

$$\frac{(P_{DTBP})_t}{(P_{DTBP})_o} = e^{-1.08} = 0.340$$

$$(P_{DTBP})_t = 0.340\times(P_{DTBP})_o = 0.340\times800\ mmHg = 272\ mmHg$$

7B (M)

(a) We use partial pressures in place of concentrations in the integrated first-order rate equation. Notice first that more than 30 half-lives have elapsed, and thus the ethylene oxide pressure has declined to at most $(0.5)^{30} = 9\times10^{-10}$ of its initial value.

$$\ln\frac{P_{30}}{P_o} = -kt = -2.05\times10^{-4}\ s^{-1}\times30.0\ h\times\frac{3600\ s}{1\ h} = -22.1 \quad \frac{P_{30}}{P_o} = e^{-22.1} = 2.4\times10^{-10}$$

$$P_{30} = 2.4\times10^{-10}\times P_o = 2.4\times10^{-10}\times782\ mmHg = 1.9\times10^{-7}\ mmHg$$

(b) $P_{\text{ethylene oxide}}$ initially 782 mmHg $\rightarrow$ 1.9 × 10^{-7} mmHg ($\sim$ 0). Essentially all of the ethylene oxide is converted to CH$_4$ and CO. Since pressure is proportional to moles, the final pressure will be twice the initial pressure (1 mole gas $\rightarrow$ 2 moles gas; 782 mmHg $\rightarrow$ 1564 mmHg). The final pressure will be 1.56 × 10^3 mmHg.

8A **(D)** We first begin by looking for a constant rate, indicative of a zero-order reaction. If the rate is constant, the concentration will decrease by the same quantity during the same time period. If we choose a 25-s time period, we note that the concentration decreases $(0.88\ \text{M} - 0.74\ \text{M} =)\ 0.14\ \text{M}$ during the first 25 s, $(0.74\ \text{M} - 0.62\ \text{M} =)\ 0.12\ \text{M}$ during the second 25 s, $(0.62\ \text{M} - 0.52\ \text{M} =)\ 0.10\ \text{M}$ during the third 25 s, and $(0.52\ \text{M} - 0.44\ \text{M} =)\ 0.08\ \text{M}$ during the fourth 25-s period. This is hardly a constant rate and we thus conclude that the reaction is not zero-order.

We next look for a constant half-life, indicative of a first-order reaction. The initial concentration of 0.88 M decreases to one half of that value, 0.44 M, during the first 100 s, indicating a 100-s half-life. The concentration halves again to 0.22 M in the second 100 s, another 100-s half-life. Finally, we note that the concentration halves also from 0.62 M at 50 s to 0.31 M at 150 s, yet another 100-s half-life. The rate is established as first-order.

The rate constant is $k = \dfrac{0.693}{t_{1/2}} = \dfrac{0.693}{100\ \text{s}} = 6.93 \times 10^{-3}\ \text{s}^{-1}$.

That the reaction is first-order is made apparent by the fact that the ln[B] vs time plot is a straight line with slope $= -k$ ($k = 6.85 \times 10^{-3}\ \text{s}^{-1}$).

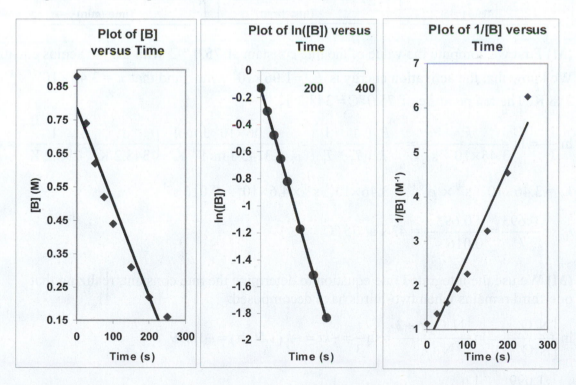

8B **(D)** We plot the data in three ways to determine the order. (1) A plot of [A] vs. time is linear if the reaction is zero-order. (2) A plot of ln [A] vs. time will be linear if the reaction is first-order. (3) A plot of 1/[A] vs. time will be linear if the reaction is second-order. It is obvious from the plots below that the reaction is zero-order. The negative of the slope of the line equals $k = -(0.083\ \text{M} - 0.250\ \text{M}) \div 18.00\ \text{min} = 9.28 \times 10^{-3}\ \text{M/min}$ ($k = 9.30 \times 10^{-3}\ \text{M/min}$ using a graphical approach).

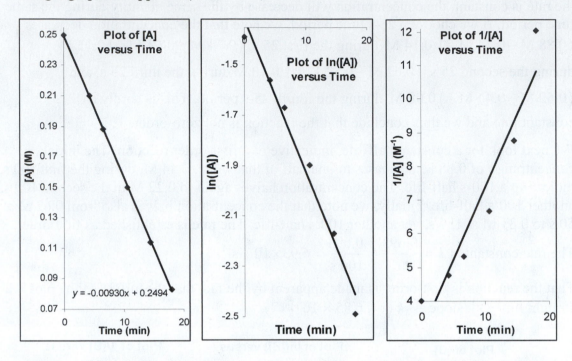

9A **(M)** First we compute the value of the rate constant at $75.0\ ^\circ\text{C}$ with the Arrhenius equation. We know that the activation energy is $E_\text{a} = 1.06 \times 10^5$ J/mol, and that $k = 3.46 \times 10^{-5}\ \text{s}^{-1}$ at 298 K. The temperature of $75.0\ ^\circ\text{C} = 348.2$ K.

$$\ln\frac{k_2}{k_1} = \ln\frac{k_2}{3.46 \times 10^{-5}\ \text{s}^{-1}} = -\frac{E_\text{a}}{R}\left(\frac{1}{T_2} - \frac{1}{T_1}\right) = -\frac{1.06 \times 10^5\ \text{J/mol}}{8.3145\ \text{J mol}^{-1}\ \text{K}^{-1}}\left(\frac{1}{348.2\ \text{K}} - \frac{1}{298.2\ \text{K}}\right) = 6.14$$

$$k_2 = 3.46 \times 10^{-5}\ \text{s}^{-1} \times e^{+6.14} = 3.46 \times 10^{-5}\ \text{s}^{-1} \times 4.6 \times 10^2 = 0.016\ \text{s}^{-1}$$

$$t_{1/2} = \frac{0.693}{k} = \frac{0.693}{0.016\ \text{s}^{-1}} = 43\ \text{s at } 75\ ^\circ\text{C}$$

9B **(M)** We use the integrated rate equation to determine the rate constant, realizing that one-third remains when two-thirds have decomposed.

$$\ln\frac{[\text{N}_2\text{O}_5]_t}{[\text{N}_2\text{O}_5]_\text{o}} = \ln\frac{[\text{N}_2\text{O}_5]_\text{o} \div 3}{[\text{N}_2\text{O}_5]_\text{o}} = \ln\frac{1}{3} = -kt = -k(1.50\ \text{h}) = -1.099$$

$$k = \frac{1.099}{1.50\ \text{h}} \times \frac{1\ \text{h}}{3600\ \text{s}} = 2.03 \times 10^{-4}\ \text{s}^{-1}$$

Now use the Arrhenius equation to determine the temperature at which the rate constant is $2.04 \times 10^{-4} \text{ s}^{-1}$.

$$\ln \frac{k_2}{k_1} = \ln \frac{2.04 \times 10^{-4} \text{ s}^{-1}}{3.46 \times 10^{-5} \text{ s}^{-1}} = 1.77 = -\frac{E_a}{R}\left(\frac{1}{T_2} - \frac{1}{T_1}\right) = -\frac{1.06 \times 10^5 \text{ J/mol}}{8.3145 \text{ J mol}^{-1} \text{ K}^{-1}}\left(\frac{1}{T_2} - \frac{1}{298 \text{ K}}\right)$$

$$\frac{1}{T_2} = \frac{1}{298 \text{ K}} + \frac{1.77 \times 8.3145 \text{ K}^{-1}}{-1.06 \times 10^5} = 3.22 \times 10^{-3} \text{ K}^{-1} \qquad T_2 = 311 \text{ K}$$

10A **(M)** The two steps of the mechanism must add, in a Hess's law fashion, to produce the overall reaction.

Overall reaction: $CO + NO_2 \longrightarrow CO_2 + NO$ or $CO + NO_2 \longrightarrow CO_2 + NO$

Second step: $-\left(NO_3 + CO \longrightarrow NO_2 + CO_2\right)$ or $+\left(NO_2 + CO_2 \longrightarrow NO_3 + CO\right)$

First step: $\qquad\qquad\qquad\qquad\qquad\qquad\qquad 2\, NO_2 \longrightarrow NO + NO_3$

If the first step is the slow step, then it will be the rate-determining step, and the rate of that step will be the rate of the reaction, namely, rate of reaction $= k_1[NO_2]^2$.

10B **(M)**

(1) The steps of the mechanism must add, in a Hess's law fashion, to produce the overall reaction. This is done below. The two intermediates, $NO_2F_2(g)$ and $F(g)$, are each produced in one step and consumed in the next one.

Fast: $\quad NO_2(g) + F_2(g) \rightleftharpoons NO_2F_2(g)$

Slow: $\quad NO_2F_2(g) \rightarrow NO_2F(g) + F(g)$

Fast: $\quad F(g) + NO_2(g) \rightarrow NO_2F(g)$

Net: $\quad 2\, NO_2(g) + F_2(g) \rightarrow 2\, NO_2F(g)$

(2) The proposed mechanism must agree with the rate law. We expect the rate-determining step to determine the reaction rate: Rate $= k_3[NO_2F_2]$. To eliminate $[NO_2F_2]$, we recognize that the first elementary reaction is very fast and will have the same rate forward as reverse: $R_f = k_1[NO_2][F_2] = k_2[NO_2F_2] = R_r$. We solve for the concentration of intermediate: $[NO_2F_2] = k_1[NO_2][F_2]/k_2$. We now substitute this expression for $[NO_2F_2]$ into the rate equation: Rate $= (k_1 k_3/k_2)[NO_2][F_2]$. Thus the predicted rate law agrees with the experimental rate law.

INTEGRATIVE EXAMPLE

A **(M)**

(a) The time required for the fixed (c) process of souring is three times as long at 3 °C refrigerator temperature (276 K) as at 20 °C room temperature (293 K).

$$\ln\frac{c/t_2}{c/t_1} = \ln\frac{t_1}{t_2} = \ln\frac{64\text{ h}}{3\times 64\text{ h}} = -1.10 = -\frac{E_a}{R}\left(\frac{1}{T_2}-\frac{1}{T_1}\right) = -\frac{E_a}{R}\left(\frac{1}{276\text{ K}}-\frac{1}{293\text{ K}}\right) = -\frac{E_a}{R}(2.10\times10^{-4})$$

$$E_a = \frac{1.10\,R}{2.10\times10^{-4}\text{ K}^{-1}} = \frac{1.10\times8.3145\text{ J mol}^{-1}\text{ K}^{-1}}{2.10\times10^{-4}\text{ K}^{-1}} = 4.4\times10^4\text{ J/mol} = 44\text{ kJ/mol}$$

(b) Use the E_a determined in part (a) to calculate the souring time at 40 °C = 313 K.

$$\ln\frac{t_1}{t_2} = -\frac{E_a}{R}\left(\frac{1}{T_2}-\frac{1}{T_1}\right) = -\frac{4.4\times10^4\text{ J/mol}}{8.3145\text{ J mol}^{-1}\text{ K}^{-1}}\left(\frac{1}{293\text{ K}}-\frac{1}{313\text{ K}}\right) = -1.15 = \ln\frac{t_1}{64\text{ h}}$$

$$\frac{t_1}{64\text{ h}} = e^{-1.15} = 0.317 \quad t_1 = 0.317\times64\text{ h} = 20.\text{ h}$$

B **(M)** The species A^* is a reactive intermediate. Let's deal with this species by using a steady state approximation.

$d[A^*]/dt = 0 = k_1[A]^2 - k_{-1}[A^*][A] - k_2[A^*]$. Solve for $[A^*]$. $k_{-1}[A^*][A] + k_2[A^*] = k_1[A]^2$

$$[A^*] = \frac{k_1[A]^2}{k_{-1}[A]+k_2} \quad \text{The rate of reaction is: Rate} = k_2[A^*] = \frac{k_2k_1[A]^2}{k_{-1}[A]+k_2}$$

At low pressures ($[A]\sim 0$ and hence $k_2\gg k_{-1}[A]$), the denominator becomes $\sim k_2$ and the rate law is

$$\text{Rate} = \frac{k_2k_1[A]^2}{k_2} = k_1[A]^2 \text{ Second-order with respect to } [A]$$

At high pressures ($[A]$ is large and $k_{-1}[A]\gg K_2$), the denominator becomes $\sim k_{-1}[A]$ and the rate law is

$$\text{Rate} = \frac{k_2k_1[A]^2}{k_{-1}[A]} = \frac{k_2k_1[A]}{k_{-1}} \text{ First-order with respect to } [A]$$

EXERCISES

Rates of Reactions

1. **(M)** $2A + B \rightarrow C + 3D$ $\qquad -\dfrac{\Delta[A]}{\Delta t} = 6.2\times10^{-4}\text{ M s}^{-1}$

(a) $\text{Rate} = -\dfrac{1}{2}\dfrac{\Delta[A]}{\Delta t} = 1/2(6.2\times10^{-4}\text{ M s}^{-1}) = 3.1\times10^{-4}\text{ M s}^{-1}$

(b) Rate of disappearance of B $= -\dfrac{1}{2}\dfrac{\Delta[A]}{\Delta t} = 1/2(6.2\times10^{-4}\text{ M s}^{-1}) = 3.1\times10^{-4}\text{ M s}^{-1}$

(c) Rate of appearance of D $= -\dfrac{3}{2}\dfrac{\Delta[A]}{\Delta t} = 3(6.2\times10^{-4}\text{ M s}^{-1}) = 9.3\times10^{-4}\text{ M s}^{-1}$

3. **(E)** Rate $= -\dfrac{\Delta[A]}{\Delta t} = -\dfrac{(0.474\ M - 0.485\ M)}{82.4\ s - 71.5\ s} = 1.0 \times 10^{-3}\ M\ s^{-1}$

5. **(M)**

(a) $[A] = [A]_i + \Delta[A] = 0.588\ M - 0.013\ M = 0.575\ M$

(b) $\Delta[A] = 0.565\ M - 0.588\ M = -0.023\ M$

$\Delta t = \Delta[A]\dfrac{\Delta t}{\Delta[A]} = \dfrac{-0.023\ M}{-2.2 \times 10^{-2}\ M/min} = 1.0\ min$

time $= t + \Delta t = (4.40 + 1.0)\ min = 5.4\ min$

7. **(M)**

(a) Rate $= \dfrac{-\Delta[A]}{\Delta t} = \dfrac{\Delta[C]}{2\Delta t} = 1.76 \times 10^{-5}\ M\ s^{-1}$

$\dfrac{\Delta[C]}{\Delta t} = 2 \times 1.76 \times 10^{-5}\ M\ s^{-1} = 3.52 \times 10^{-5}\ M/s$

(b) $\dfrac{\Delta[A]}{\Delta t} = -\dfrac{\Delta[C]}{2\Delta t} = -1.76 \times 10^{-5}\ M\ s^{-1}$ Assume this rate is constant.

$[A] = 0.3580\ M + \left(-1.76 \times 10^{-5}\ M\ s^{-1} \times 1.00\ min \times \dfrac{60\ s}{1\ min}\right) = 0.357\ M$

(c) $\dfrac{\Delta[A]}{\Delta t} = -1.76 \times 10^{-5}\ M\ s^{-1}$

$\Delta t = \dfrac{\Delta[A]}{-1.76 \times 10^{-5}\ M/s} = \dfrac{0.3500\ M - 0.3580\ M}{-1.76 \times 10^{-5}\ M/s} = 4.5 \times 10^{2}\ s$

9. **(M)** Notice that, for every 1000 mmHg drop in the pressure of A(g), there will be a corresponding 2000 mmHg rise in the pressure of B(g) plus a 1000 mmHg rise in the pressure of C(g).

(a) We set up the calculation with three lines of information below the balanced equation: (1) the initial conditions, (2) the changes that occur, which are related to each other by reaction stoichiometry, and (3) the final conditions, which simply are initial conditions + changes.

	A(g)	→	2B(g)	+	C(g)
Initial	1000. mmHg		0. mmHg		0. mmHg
Changes	−1000. mmHg		+2000. mmHg		+1000. mmHg
Final	0. mmHg		2000. MmHg		1000. mmHg

Total final pressure $= 0.\ mmHg + 2000.\ mmHg + 1000.\ mmHg = 3000.\ mmHg$

(b)

	A(g)	→	2B(g)	+	C(g)
Initial	1000. mmHg		0. mmHg		0. mmHg
Changes	−200. mmHg		+400. mmHg		+200. mmHg
Final	800 mmHg		400. mmHg		200. mmHg

Total pressure = 800. mmHg + 400. mmHg + 200. mmHg = 1400. mmHg

Method of Initial Rates

11. **(M)**

(a) From Expt. 1 to Expt. 3, [A] is doubled, while [B] remains fixed. This causes the rate

to increases by a factor of $\dfrac{6.75 \times 10^{-4} \text{ M s}^{-1}}{3.35 \times 10^{-4} \text{ M s}^{-1}} = 2.01 \approx 2$.

Thus, the reaction is first-order with respect to A.

From Expt. 1 to Expt. 2, [B] doubles, while [A] remains fixed. This causes the rate to

increases by a factor of $\dfrac{1.35 \times 10^{-3} \text{ M s}^{-1}}{3.35 \times 10^{-4} \text{ M s}^{-1}} = 4.03 \approx 4$.

Thus, the reaction is second-order with respect to B.

(b) Overall reaction order = order with respect to A + order with respect to B = 1 + 2 = 3. The reaction is third-order overall.

(c) $\text{Rate} = 3.35 \times 10^{-4} \text{ M s}^{-1} = k(0.185 \text{ M})(0.133 \text{ M})^2$

$k = \dfrac{3.35 \times 10^{-4} \text{ M s}^{-1}}{(0.185 \text{ M})(0.133 \text{ M})^2} = 0.102 \text{ M}^{-2} \text{ s}^{-1}$

13. **(M)** From Experiment 1 to 2, [NO] remains constant while [Cl$_2$] is doubled. At the same time the initial rate of reaction is found to double. Thus, the reaction is first-order with respect to [Cl$_2$], since dividing reaction 2 by reaction 1 gives $2 = 2^x$ when $x = 1$. From Experiment 1 to 3, [Cl$_2$] remains constant, while [NO] is doubled, resulting in a quadrupling of the initial rate of reaction. Thus, the reaction must be second-order in [NO], since dividing reaction 3 by reaction 1 gives $4 = 2^x$ when $x = 2$. Overall the reaction is third-order: Rate = k [NO]^2[Cl$_2$]. The rate constant may be calculated from any one of the experiments. Using data from Exp. 1,

$k = \dfrac{\text{Rate}}{[\text{NO}]^2[\text{Cl}_2]} = \dfrac{2.27 \times 10^{-5} \text{ M s}^{-1}}{(0.0125 \text{ M})^2(0.0255 \text{ M})} = 5.70 \text{ M}^{-2} \text{ s}^{-1}$

First-Order Reactions

15. **(E)**

    (a)   TRUE   The rate of the reaction does decrease as more and more of B and C are formed, but not because more and more of B and C are formed. Rather, the rate decreases because the concentration of A must decrease to form more and more of B and C.

    (b)   FALSE   The time required for one half of substance A to react—the half-life—is independent of the quantity of A present.

17. **(M)**

    (a)   Since the half-life is 180 s, after 900 s five half-lives have elapsed, and the original quantity of A has been cut in half five times.

final quantity of $A = (0.5)^5 \times$ initial quantity of $A = 0.03125 \times$ initial quantity of A

About 3.13% of the original quantity of A remains unreacted after 900 s.

or

More generally, we would calculate the value of the rate constant, k, using

$$k = \frac{\ln 2}{t_{1/2}} = \frac{0.693}{180 \text{ s}} = 0.00385 \text{ s}^{-1} \quad \text{Now ln (\% unreacted)} = -kt = -0.00385 \text{ s}^{-1} \times (900\text{s})$$

$$= -3.46\underline{5}$$

(% unreacted) = $0.0313 \times 100\% = 3.13\%$ of the original quantity.

    (b)   Rate $= k[A] = 0.00385 \text{ s}^{-1} \times 0.50 \text{ M} = 0.00193 \text{ M/s}$

19. **(M)**

    (a)   The mass of A has decreased to one fourth of its original value, from 1.60 g to 0.40 g. Since $\frac{1}{4} = \frac{1}{2} \times \frac{1}{2}$, we see that two half-lives have elapsed.

Thus, $2 \times t_{1/2} = 38$ min, or $t_{1/2} = 19$ min.

    (b) $k = 0.693 / t_{1/2} = \dfrac{0.693}{19 \text{ min}} = 0.036 \text{ min}^{-1}$ $\ln \dfrac{[A]_t}{[A]_o} = -kt = -0.036 \text{ min}^{-1} \times 60 \text{ min} = -2.2$

$$\frac{[A]_t}{[A]_o} = e^{-2.2} = 0.1\underline{1} \quad \text{or} \quad [A]_t = [A]_o \, e^{-kt} = 1.60 \text{ g A} \times 0.1\underline{1} = 0.1\underline{8} \text{ g A}$$

21. **(M)** We determine the value of the first-order rate constant and from that we can calculate the half-life. If the reactant is 99% decomposed in 137 min, then only 1% (0.010) of the initial concentration remains.

$$\ln \frac{[A]_t}{[A]_o} = -kt = \ln \frac{0.010}{1.000} = -4.61 = -k \times 137 \text{min} \qquad k = \frac{-4.61}{-137 \text{ min}} = 0.0336 \text{ min}^{-1}$$

$$t_{1/2} = \frac{0.0693}{k} = \frac{0.693}{0.0336 \text{ min}^{-1}} = 20.6 \text{ min}$$

23. **(D)**

(a) $\ln\left(\dfrac{\dfrac{35}{100}[A]_o}{[A]_o}\right) = \ln(0.35) = -kt = (-4.81 \times 10^{-3}\text{ min}^{-1})t$ $t = 218$ min.

Note: We did not need to know the initial concentration of acetoacetic acid to answer the question.

(b) Let's assume that the reaction takes place in a 1.00 L container.

$$10.0\text{ g acetoacetic acid} \times \dfrac{1\text{ mol acetoacetic acid}}{102.090\text{ g acetoacetic acid}} = 0.09795\text{ mol acetoacetic acid.}$$

After 575 min. (~ 4 half lives, hence, we expect ~ 6.25% remains as a rough approximation), use integrated form of the rate law to find $[A]_t = 575$ min.

$$\ln\left(\dfrac{[A]_t}{[A]_o}\right) = -kt = (-4.81 \times 10^{-3}\text{ min}^{-1})(575\text{ min}) = -2.766$$

$$\dfrac{[A]_t}{[A]_o} = e^{-2.766} = 0.06293 \text{ (~ 6.3\% remains)} \qquad \dfrac{[A]_t}{0.09795\text{ moles}} = 0.063 \ [A]_t = 6.2 \times 10^{-3}$$

moles.

$[A]_{reacted} = [A]_o - [A]_t = (0.098 - 6.2 \times 10^{-3})$ moles $= 0.092$ moles acetoacetic acid. The stoichiometry is such that for every mole of acetoacetic acid consumed, one mole of CO_2 forms. Hence, we need to determine the volume of 0.0918 moles CO_2 at 24.5 °C (297.65 K) and 748 torr (0.984 atm) by using the Ideal Gas law.

$$V = \dfrac{nRT}{P} = \dfrac{0.0918\text{ mol}\left(0.08206\dfrac{\text{L atm}}{\text{K mol}}\right)297.65\text{ K}}{0.984\text{ atm}} = 2.3\text{ L }CO_2$$

25. **(D)**

(a) If the reaction is first-order, we will obtain the same value of the rate constant from several sets of data.

$$\ln\dfrac{[A]_t}{[A]_o} = -kt = \ln\dfrac{0.497\text{ M}}{0.600\text{ M}} = -k \times 100\text{ s} = -0.188, \quad k = \dfrac{0.188}{100\text{ s}} = 1.88 \times 10^{-3}\text{ s}^{-1}$$

$$\ln\dfrac{[A]_t}{[A]_o} = -kt = \ln\dfrac{0.344\text{ M}}{0.600\text{ M}} = -k \times 300\text{ s} = -0.556, \quad k = \dfrac{0.556}{300\text{ s}} = 1.85 \times 10^{-3}\text{ s}^{-1}$$

$$\ln\dfrac{[A]_t}{[A]_o} = -kt = \ln\dfrac{0.285\text{ M}}{0.600\text{ M}} = -k \times 400\text{ s} = -0.744, \quad k = \dfrac{0.744}{400\text{ s}} = 1.86 \times 10^{-3}\text{ s}^{-1}$$

$$\ln\dfrac{[A]_t}{[A]_o} = -kt = \ln\dfrac{0.198\text{ M}}{0.600\text{ M}} = -k \times 600\text{ s} = -1.109, \quad k = \dfrac{1.109}{600\text{ s}} = 1.85 \times 10^{-3}\text{ s}^{-1}$$

$$\ln\dfrac{[A]_t}{[A]_o} = -kt = \ln\dfrac{0.094\text{ M}}{0.600\text{ M}} = -k \times 1000\text{ s} = -1.854, \quad k = \dfrac{1.854}{1000\text{ s}} = 1.85 \times 10^{-3}\text{ s}^{-1}$$

The virtual constancy of the rate constant throughout the time of the reaction confirms that the reaction is first-order.

(b) For this part, we assume that the rate constant equals the average of the values obtained in part (a).

$$k = \frac{1.88 + 1.85 + 1.86 + 1.85}{4} \times 10^{-3} \text{ s}^{-1} = 1.86 \times 10^{-3} \text{ s}^{-1}$$

(c) We use the integrated first-order rate equation:

$$[A]_{750} = [A]_0 \exp(-kt) = 0.600 \text{ M} \exp(-1.86 \times 10^{-3} \text{ s}^{-1} \times 750 \text{ s})$$

$$[A]_{750} = 0.600 \text{ M e}^{-1.40} = 0.148 \text{ M}$$

Reactions of Various Orders

27. (M)

(a) Set II is data from a zero-order reaction. We know this because the rate of set II is constant.

0.25 M/25 s = 0.010 M s⁻¹. Zero-order reactions have constant rates of reaction.

(b) A first-order reaction has a constant half-life. In set I, the first half-life is slightly less than 75 sec, since the concentration decreases by slightly more than half (from 1.00 M to 0.47 M) in 75 s. Again, from 75 s to 150 s the concentration decreases from 0.47 M to 0.22 M, again by slightly more than half, in a time of 75 s. Finally, two half-lives should see the concentration decrease to one-fourth of its initial value. This, in fact, is what we see. From 100 s to 250 s, 150 s of elapsed time, the concentration decreases from 0.37 M to 0.08 M, i.e., to slightly less than one-fourth of its initial value. Notice that we cannot make the same statement of constancy of half-life for set III. The first half-life is 100 s, but it takes more than 150 s (from 100 s to 250 s) for [A] to again decrease by half.

(c) For a second-order reaction, $1/[A]_t - 1/[A]_o = kt$. For the initial 100 s in set III, we have

$$\frac{1}{0.50 \text{ M}} - \frac{1}{1.00 \text{ M}} = 1.0 \text{ L mol}^{-1} = k\,100 \text{ s}, \quad k = 0.010 \text{ L mol}^{-1} \text{ s}^{-1}$$

For the initial 200 s, we have

$$\frac{1}{0.33 \text{ M}} - \frac{1}{1.00 \text{ M}} = 2.0 \text{ L mol}^{-1} = k\,200 \text{ s}, \quad k = 0.010 \text{ L mol}^{-1} \text{ s}^{-1}$$

Since we obtain the same value of the rate constant using the equation for second-order kinetics, set III must be second-order.

29. (M) Set I is the data for a first-order reaction; we can analyze those items of data to determine the half-life. In the first 75 s, the concentration decreases by a bit more than half. This implies a half-life slightly less than 75 s, perhaps 70 s. This is consistent with the other time periods noted in the answer to Review Question 18 (b) and also to the fact that in the 150-s interval from 50 s to 200 s, the concentration decreases from 0.61 M to 0.14 M, which is a bit more than a factor-of-four decrease. The factor-of-four decrease, to one-fourth of the initial value, is what we would expect for two successive half-lives. We can determine the half-life more accurately, by obtaining a value of k from the relation $\ln\left([A]_t / [A]_o\right) = -kt$ followed by $t_{1/2} = 0.693 / k$ For instance, $\ln(0.78/1.00) = -k\,(25\text{ s})$; $k = 9.9\underline{4} \times 10^{-3}\text{ s}^{-1}$. Thus, $t_{1/2} = 0.693/9.9\underline{4} \times 10^{-3}\text{ s}^{-1} = 70\text{ s}$.

31. **(M)** The approximate rate at 75 s can be taken as the rate over the time period from 50 s to 100 s.

(a) $\text{Rate}_{\text{II}} = -\dfrac{\Delta[A]}{\Delta t} = -\dfrac{0.00\text{ M} - 0.50\text{ M}}{100\text{ s} - 50\text{ s}} = 0.010\text{ M s}^{-1}$

(b) $\text{Rate}_{\text{I}} = -\dfrac{\Delta[A]}{\Delta t} = -\dfrac{0.37\text{ M} - 0.61\text{ M}}{100\text{ s} - 50\text{ s}} = 0.0048\text{ M s}^{-1}$

(c) $\text{Rate}_{\text{III}} = -\dfrac{\Delta[A]}{\Delta t} = -\dfrac{0.50\text{ M} - 0.67\text{ M}}{100\text{ s} - 50\text{ s}} = 0.0034\text{ M s}^{-1}$

Alternatively we can use [A] at 75 s (the values given in the table) in the relationship $\text{Rate} = k[A]^m$, where $m = 0,\ 1,$ or 2.

(a) $\text{Rate}_{\text{II}} = 0.010\text{ M s}^{-1} \times (0.25\text{ mol/L})^0 = 0.010\text{ M s}^{-1}$

(b) Since $t_{1/2} = 70\text{ s}$, $k = 0.693 / 70\text{ s} = 0.0099\text{ s}^{-1}$
$\text{Rate}_{\text{I}} = 0.0099\,s^{-1} \times (0.47\,\text{mol/L})^1 = 0.0047\,\text{M s}^{-1}$

(c) $\text{Rate}_{\text{III}} = 0.010\text{ L mol}^{-1}\text{ s}^{-1} \times (0.57\text{ mol/L})^2 = 0.0032\text{ M s}^{-1}$

33. **(E)** Substitute the given values into the rate equation to obtain the rate of reaction.
$\text{Rate} = k[A]^2[B]^0 = \left(0.0103\text{ M}^{-1}\text{min}^{-1}\right)(0.116\text{ M})^2(3.83\text{ M})^0 = 1.39 \times 10^{-4}\text{ M / min}$

35. **(M)** For reaction: $HI(g) \rightarrow 1/2\ H_2(g) + 1/2\ I_2(g)$ (700 K)

Time (s)	[HI] (M)	ln[HI]	1/[HI](M^{-1})
0	1.00	0	1.00
100	0.90	−0.105	1.11
200	0.81	−0.211	1.23$\underline{5}$
300	0.74	−0.301	1.35
400	0.68	−0.386	1.47

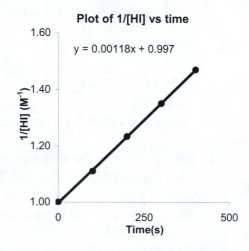

Plot of 1/[HI] vs time

y = 0.00118x + 0.997

From data above, a plot of 1/[HI] vs. t yields a straight line. The reaction is second-order in HI at 700 K. $\text{Rate} = k[HI]^2$. The slope of the line $= k = 0.00118\text{ M}^{-1}\text{s}^{-1}$

37. **(M)**

(a) Plot [A] vs t, ln[A] vs t, and 1/[A] vs t and see which yields a straight line.

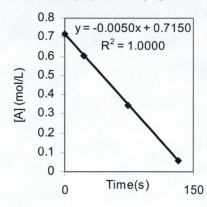

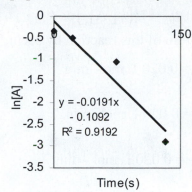

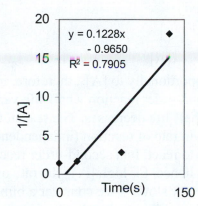

Clearly we can see that the reaction is zero-order in reactant A with a rate constant of 5.0×10^{-3}.

(b) The half-life of this reaction is the time needed for one half of the initial [A] to react.

Thus, $\Delta[A] = 0.715\ M \div 2 = 0.358\ M$ and $t_{1/2} = \dfrac{0.358\ M}{5.0 \times 10^{-3}\ M/s} = 72$ s.

39. **(E)**

(a) initial rate $= -\dfrac{\Delta[A]}{\Delta t} = -\dfrac{1.490\ M - 1.512\ M}{1.0\ min - 0.0\ min} = +0.022\ M/min$

initial rate $= -\dfrac{\Delta[A]}{\Delta t} = -\dfrac{2.935\ M - 3.024\ M}{1.0\ min - 0.0\ min} = +0.089\ M/min$

(b) When the initial concentration is doubled $(\times 2.0)$, from 1.512 M to 3.024 M, the initial rate quadruples $(\times 4.0)$. Thus, the reaction is second-order in A (since $2.0^x = 4.0$ when $x = 2$).

41. **(M)** The half-life of the reaction depends on the concentration of A and, thus, this reaction cannot be first-order. For a second-order reaction, the half-life varies inversely with the reaction rate: $t_{1/2} = 1/(k[A]_o)$ or $k = 1/(t_{1/2}[A]_o)$. Let us attempt to verify the second-order nature of this reaction by seeing if the rate constant is fixed.

$$k = \frac{1}{1.00 \text{ M} \times 50 \text{ min}} = 0.020 \text{ L mol}^{-1}\text{min}^{-1}$$

$$k = \frac{1}{2.00 \text{ M} \times 25 \text{ min}} = 0.020 \text{ L mol}^{-1}\text{min}^{-1}$$

$$k = \frac{1}{0.50 \text{ M} \times 100 \text{ min}} = 0.020 \text{ L mol}^{-1} \text{ min}^{-1}$$

The constancy of the rate constant demonstrates that this reaction indeed is second-order. The rate equation is $\text{Rate} = k[A]^2$ and $k = 0.020 \text{ L mol}^{-1}\text{min}^{-1}$.

43. **(M)** Zero-order: $t_{1/2} = \dfrac{[A]_o}{2k}$ Second-order: $t_{1/2} = \dfrac{1}{k[A]_o}$

A zero-order reaction has a half life that varies proportionally to $[A]_o$, therefore, increasing $[A]_o$ increases the half-life for the reaction. A second-order reaction's half-life varies inversely proportional to $[A]_o$, that is, as $[A]_o$ increases, the half-life decreases. The reason for the difference is that a zero-order reaction has a constant rate of reaction (independent of $[A]_o$). The larger the value of $[A]_o$, the longer it will take to react. In a second-order reaction, the rate of reaction increases as the square of the $[A]_o$, hence, for high $[A]_o$, the rate of reaction is large and for very low $[A]_o$, the rate of reaction is very slow. If we consider a bimolecular elementary reaction, we can easily see that a reaction will not take place unless two molecules of reactants collide. This is more likely when the $[A]_o$ is large than when it is small.

Collision Theory; Activation Energy

45. **(M)**

(a) The rate of a reaction depends on at least two factors other than the frequency of collisions. The first of these is whether each collision possesses sufficient energy to get over the energy barrier to products. This depends on the activation energy of the reaction; the higher it is, the smaller will be the fraction of successful collisions. The second factor is whether the molecules in a given collision are properly oriented for a successful reaction. The more complex the molecules are, or the more freedom of motion the molecules have, the smaller will be the fraction of collisions that are correctly oriented.

(b) Although the collision frequency increases relatively slowly with temperature, the fraction of those collisions that have sufficient energy to overcome the activation energy increases much more rapidly. Therefore, the rate of reaction will increase dramatically with temperature.

(c) The addition of a catalyst has the net effect of decreasing the activation energy of the overall reaction, by enabling an alternative mechanism. The lower activation energy of the alternative mechanism, (compared to the uncatalyzed mechanism), means that a larger fraction of molecules have sufficient energy to react. Thus the rate increases, even though the temperature does not.

47. (M)
(a) The products are 21 kJ/mol closer in energy to the energy activated complex than are the reactants. Thus, the activation energy for the reverse reaction is 84 kJ/mol − 21 kJ/mol = 63 kJ/mol.

(b) The reaction profile for the reaction in Figure 20-10 is sketched below.

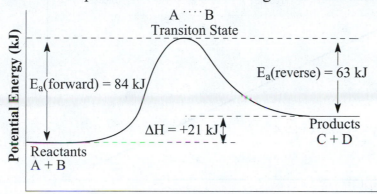

Progress of Reaction

49. (E)
(a) There are two intermediates (B and C).

(b) There are three transition states (peaks/maxima) in the energy diagram.

(c) The fastest step has the smallest E_a, hence, step 3 is the fastest step in the reaction with step 2 a close second.

(d) Reactant A (step 1) has the highest E_a, and therefore the slowest smallest constant

(e) Endothermic; energy is needed to go from A to B.

(f) Exothermic; energy is released moving from A to D.

Effect of Temperature on Rates of Reaction

51. (M)
$$\ln\frac{k_2}{k_1} = -\frac{E_a}{R}\left(\frac{1}{T_2} - \frac{1}{T_1}\right) = \ln\frac{2.8\times10^{-2}\ \text{L mol}^{-1}\ \text{s}^{-1}}{5.4\times10^{-4}\ \text{L mol}^{-1}\ \text{s}^{-1}} = -\frac{E_a}{R}\left(\frac{1}{683\ \text{K}} - \frac{1}{599\ \text{K}}\right)$$

$$3.95R = E_a \times 2.05\times10^{-4}$$

$$E_a = \frac{3.95\ R}{2.05\times10^{-4}} = 1.93\times10^4\ \text{K}^{-1}\times8.3145\ \text{J mol}^{-1}\ \text{K}^{-1} = 1.60\times10^5\ \text{J / mol} = 160\ \text{kJ / mol}$$

53. **(D)**

    (a) First we need to compute values of $\ln k$ and $1/T$. Then we plot the graph of $\ln k$ versus 1/T.

$T, °C$	0 °C	10 °C	20 °C	30 °C
T, K	273 K	283 K	293 K	303 K
$1/T$, K^{-1}	0.00366	0.00353	0.00341	0.00330
k, s^{-1}	5.6×10^{-6}	3.2×10^{-5}	1.6×10^{-4}	7.6×10^{-4}
$\ln k$	-12.09	-10.35	-8.74	-7.18

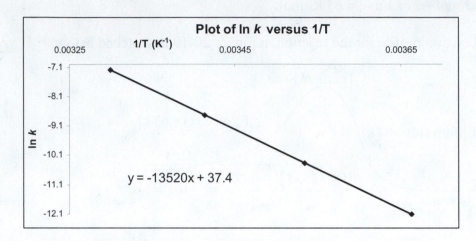

Plot of ln *k* versus 1/T

$y = -13520x + 37.4$

    (b) The slope $= -E_a/R$.

$$E_a = -R \times \text{slope} = -8.3145 \frac{J}{\text{mol K}} \times -1.35_2 \times 10^4 \text{ K} \times \frac{1 \text{ kJ}}{1000 \text{ J}} = 112 \frac{\text{kJ}}{\text{mol}}$$

    (c) We apply the Arrhenius equation, with $k = 5.6 \times 10^{-6}$ s^{-1} at 0 °C (273 K), $k = ?$ at 40°C (313 K), and $E_a = 113 \times 10^3$ J/mol.

$$\ln \frac{k}{5.6 \times 10^{-6} \text{ s}^{-1}} = -\frac{E_a}{R}\left(\frac{1}{T_2} - \frac{1}{T_1}\right) = -\frac{112 \times 10^3 \text{ J/mol}}{8.3145 \text{ J mol}^{-1} \text{ K}^{-1}}\left(\frac{1}{313 \text{ K}} - \frac{1}{273 \text{ K}}\right) = 6.30\underline{6}$$

$$e^{6.30\underline{6}} = 548 = \frac{k}{5.6 \times 10^{-6} \text{ s}^{-1}} \qquad k = 548 \times 5.6 \times 10^{-6} \text{ s}^{-1} = 3.0\underline{7} \times 10^{-3} \text{ s}^{-1}$$

$$t_{1/2} = \frac{0.693}{k} = \frac{0.693}{3.0\underline{7} \times 10^{-3} \text{ s}^{-1}} = 2.3 \times 10^2 \text{ s}$$

55. **(M)** The half-life of a first-order reaction is inversely proportional to its rate constant: $k = 0.693/t_{1/2}$. Thus we can apply a modified version of the Arrhenius equation to find E_a.

    (a)
$$\ln \frac{k_2}{k_1} = \ln \frac{(t_{1/2})_1}{(t_{1/2})_2} = -\frac{E_a}{R}\left(\frac{1}{T_2} - \frac{1}{T_1}\right) = \ln \frac{46.2 \text{ min}}{2.6 \text{ min}} = -\frac{E_a}{R}\left(\frac{1}{(102 + 273) \text{ K}} - \frac{1}{298 \text{ K}}\right)$$

$$2.88 = \frac{E_a}{R} 6.89 \times 10^{-4} \qquad E_a = \frac{2.88 \times 8.3145 \text{ J mol}^{-1} \text{ K}^{-1}}{6.89 \times 10^{-4} \text{ K}^{-1}} \times \frac{1 \text{ kJ}}{1000 \text{ J}} = 34.8 \text{ kJ/mol}$$

(b) $\ln\dfrac{10.0 \text{ min}}{46.2 \text{ min}} = -\dfrac{34.8\times10^3 \text{ J/mol}}{8.3145 \text{ J mol}^{-1}\text{ K}^{-1}}\left(\dfrac{1}{298}-\dfrac{1}{T}\right) = -1.53 = -4.19\times10^3\left(\dfrac{1}{298}-\dfrac{1}{T}\right)$

$\left(\dfrac{1}{298}-\dfrac{1}{T}\right) = \dfrac{-1.53}{-4.19\times10^3} = -3.65\times10^{-4}$ $\dfrac{1}{T} = 2.99\times10^{-3}$ $T = 334 \text{ K} = 61 \text{ °C}$

57. **(M)**

(a) It is the change in the value of the rate constant that causes the reaction to go faster. Let k_1 be the rate constant at room temperature, 20 °C (293 K). Then, ten degrees higher (30 °C or 303 K), the rate constant $k_2 = 2\times k_1$.

$\ln\dfrac{k_2}{k_1} = \ln\dfrac{2\times k_1}{k_1} = 0.693 = -\dfrac{E_a}{R}\left(\dfrac{1}{T_2}-\dfrac{1}{T_1}\right) = -\dfrac{E_a}{R}\left(\dfrac{1}{303}-\dfrac{1}{293 \text{ K}}\right) = 1.13\times10^{-4} \text{ K}^{-1}\dfrac{E_a}{R}$

$E_a = \dfrac{0.693\times8.3145 \text{ J mol}^{-1}\text{ K}^{-1}}{1.13\times10^{-4} \text{ K}^{-1}} = 5.1\times10^4 \text{ J / mol} = 51 \text{ kJ / mol}$

(b) Since the activation energy for the depicted reaction (i.e., $N_2O + NO \rightarrow N_2 + NO_2$) is 209 kJ/mol, we would not expect this reaction to follow the rule of thumb.

Catalysis

59. **(E)**

(a) Although a catalyst is *recovered unchanged from the reaction mixture*, it does "take part in the reaction." Some catalysts actually slow down the rate of a reaction. Usually, however, these negative catalysts are called inhibitors.

(b) The function of a catalyst is to *change the mechanism of a reaction*. The new mechanism is one that has a different (lower) activation energy (and frequently a different A value), than the original reaction.

61. **(E)** Both platinum and an enzyme have a metal center that acts as the active site. Generally speaking, platinum is not dissolved in the reaction solution (heterogeneous), whereas enzymes are generally soluble in the reaction media (homogeneous). The most important difference, however, is one of specificity. Platinum is rather nonspecific, catalyzing many different reactions. An enzyme, however, is quite specific, usually catalyzing only one reaction rather than all reactions of a given class.

63. **(E)** For the straight-line graph of Rate versus [Enzyme], an excess of substrate must be present.

Reaction Mechanisms

65. **(E)** The molecularity of an elementary process is the number of reactant molecules in the process. This molecularity is equal to the order of the overall reaction only if the elementary process in question is the slowest and, thus, the rate-determining step of the overall reaction. In addition, the elementary process in question should be the only elementary step that influences the rate of the reaction.

67. **(M)** The three elementary steps must sum to give the overall reaction. That is, the overall reaction is the sum of step 1 + step 2 + step 3. Hence, step 2 = overall reaction −step 1 −step 3. Note that all species in the equations below are gases.

Overall: $2\,NO + 2\,H_2 \rightarrow N_2 + 2\,H_2O$ $\qquad$ $2\,NO + 2\,H_2 \rightarrow N_2 + 2\,H_2O$

−First: $\qquad -\left(2\,NO \rightleftharpoons N_2O_2\right)$ $\qquad$ $N_2O_2 \rightleftharpoons 2\,NO$

−Third $\qquad -\left(N_2O + H_2 \rightarrow N_2 + H_2O\right)$ or $N_2 + H_2O \rightarrow N_2O + H_2$

The result is the second step, which is slow: $\qquad H_2 + N_2O_2 \rightarrow H_2O + N_2O$

The rate of this rate-determining step is: $\quad \text{Rate} = k_2\left[H_2\right]\left[N_2O_2\right]$

Since N_2O_2 does not appear in the overall reaction, we need to replace its concentration with the concentrations of species that do appear in the overall reaction. To do this, recall that the first step is rapid, with the forward reaction occurring at the same rate as the reverse reaction. $k_1\left[NO\right]^2 = $ forward rate $=$ reverse rate $= k_{-1}\left[N_2O_2\right]$. This expression is solved for $\left[N_2O_2\right]$, which then is substituted into the rate equation for the overall reaction.

$$\left[N_2O_2\right] = \frac{k_1\left[NO\right]^2}{k_{-1}} \qquad\qquad \text{Rate} = \frac{k_2 k_1}{k_{-1}}\left[H_2\right]\left[NO\right]^2$$

The reaction is first-order in $\left[H_2\right]$ and second-order in $[NO]$. This result conforms to the experimentally determined reaction order.

69. **(M)** Proposed mechanism: $\qquad Cl_2(g) \underset{k_{-1}}{\overset{k_1}{\rightleftharpoons}} 2\,Cl(g)$ $\qquad\qquad$ Observed rate law:

$$\frac{2\,Cl(g) + 2\,NO(g) \overset{k_2}{\longrightarrow} 2\,NOCl(g)}{Cl_2(g) + 2NO(g) \rightarrow 2\,NOCl(g)} \qquad \text{Rate} = k[Cl_2][NO]^2$$

The first step is a fast equilibrium reaction and step 2 is slow. Thus, the predicted rate law is Rate $= k_2[Cl]^2[NO]^2$ In the first step, set the rate in the forward direction for the equilibrium equal to the rate in the reverse direction. Then express $[Cl]^2$ in terms of k_1, k_{-1} and $[Cl_2]$. This mechanism is almost certainly not correct because it involves a tetra molecular second step.

Rate$_{\text{forward}}$ = Rate$_{\text{reverse}}$ $\qquad$ Use: Rate$_{\text{forward}} = k_1[Cl_2]$ and Rate$_{\text{reverse}} = k_{-1}[Cl]^2$
From this we see: $k_1[Cl_2] = k_{-1}[Cl]^2$. Rearranging (solving for $[Cl]^2$)

$[Cl]^2 = \dfrac{k_1[Cl_2]}{k_{-1}}$ Substitute into Rate $= k_2[Cl]^2[NO]^2 = k_2 \dfrac{k_1[Cl_2]}{k_{-1}}[NO]^2 = k_{\text{obs}}[Cl_2][NO_2]^2$

There is another plausible mechanism. $Cl_2(g) + NO(g) \underset{k_{-1}}{\overset{k_1}{\rightleftharpoons}} NOCl(g) + Cl(g)$

$$Cl(g) + NO(g) \underset{k_{-1}}{\overset{k_1}{\rightleftharpoons}} NOCl(g)$$

$$\overline{Cl_2(g) + 2NO(g) \rightarrow 2\ NOCl(g)}$$

$Rate_{forward} = Rate_{reverse}$ Use: $Rate_{forward} = k_1[Cl_2][NO]$ and $Rate_{reverse} = k_{-1}[Cl][NOCl]$

From this we see: $k_1[Cl_2][NO] = k_{-1}[Cl][NOCl]$. Rearranging (solving for [Cl])

$[Cl] = \dfrac{k_1[Cl_2][NO]}{k_{-1}[NOCl]}$ Substitute into Rate $= k_2[Cl][NO] = \dfrac{k_2 k_1[Cl_2][NO]^2}{k_{-1}[NOCl]}$

If [NOCl], the product is assumed to be constant (~ 0 M using method of initial rates), then

$\dfrac{k_2 k_1}{k_{-1}[NOCl]} = \text{constant} = k_{obs}$ Hence, the predicted rate law is $k_{obs}[Cl_2][NO]^2$ which agrees

with the experimental rate law. Since the predicted rate law agrees with the experimental rate law, both this and the previous mechanism are plausible, however, the first is dismissed as it has a tetramolecular elementary reaction (extremely unlikely to have four molecules simultaneously collide).

71. **(M)**

$S_1 + S_2 \underset{k_{-1}}{\overset{k_1}{\rightleftharpoons}} (S_1 : S_2)^*$ (fast)

$(S_1 : S_2)^* \overset{k_2}{\longrightarrow} S_1 : S_2$ (slow)

The first step is a fast equilibrium, so the rate of the forward reaction is equal to the rate of the reverse reaction.

$k_1[S_1][S_2] = k_{-1}\left[(S_1 : S_2)\right]^*$

$\left[(S_1 : S_2)\right]^* = \dfrac{k_1}{k_{-1}}[S_1][S_2]$

$\dfrac{d[S_1 : S_2]}{dt} = k_2\left[(S_1 : S_2)\right]^* = \dfrac{k_2 \cdot k_1[S_1][S_2]}{k_{-1}}$

INTEGRATIVE AND ADVANCED EXERCISES

74. (M)

(a) The concentration vs. time graph is not linear. Thus, the reaction is obviously not zero-order (the rate is not constant with time). A quick look at various half lives for this reaction shows the ~2.37 min (1.000 M to 0.5 M), ~2.32 min (0.800 M to 0.400 M), and ~2.38 min(0.400 M to 0.200 M). Since the half-life is constant, the reaction is probably first-order.

(b) average $t_{1/2} = \dfrac{(2.37 + 2.32 + 2.38)}{3} = 2.36$ min $\quad k = \dfrac{0.693}{t_{1/2}} = \dfrac{0.693}{2.36\ \text{min}} = 0.294\,\text{min}^{-1}$

or perhaps better expressed as $k = 0.29\ \text{min}^{-1}$ due to imprecision.

(c) When $t = 3.5$ min, $[A] = 0.352$ M.

Then, rate $= k[A] = 0.294\ \text{min}^{-1} \times 0.352\ \text{M} = 0.103$ M/min.

(d) Slope $= \dfrac{\Delta[A]}{\Delta t} = -\text{Rate} = \dfrac{0.1480\ \text{M} - 0.339\ \text{M}}{6.00\ \text{min} - 3.00\ \text{min}} = -0.0637\ \text{M / min}$ Rate $= 0.064$ M/min.

(e) Rate $= k[A] = 0.294\ \text{min}^{-1} \times 1.000\ \text{M} = 0.294$ M/min.

75. (M) The reaction being investigated is:

$$2\ MnO_4^-(aq) + 5\ H_2O_2(aq) + 6\ H^+(aq) \longrightarrow 2\ Mn^{2+}(aq) + 8\ H_2O(l) + 5\ O_2(g)$$

We use the stoichiometric coefficients in this balanced reaction to determine $[H_2O_2]$.

$$[H_2O_2] = \dfrac{37.1\ \text{mL titrant} \times \dfrac{0.1000\ \text{mmol } MnO_4^-}{1\ \text{mL titrant}} \times \dfrac{5\ \text{mmol } H_2O_2}{2\ \text{mmol } MnO_4^-}}{5.00\ \text{mL}} = 1.86\ \text{M}$$

78. (M) We know that rate has the units of M/s, and also that concentration has the units of M. The generalized rate equation is Rate $= k[A]_o$. In terms of units, this becomes

M/s = {units of k} M_o. Therefore {Units of k} $= \dfrac{\text{M/s}}{M_o} = M_{1-o}\ s^{-1}$

82. **(M)** For this first-order reaction $\ln\dfrac{P_t}{P_o} = -kt$ Elapsed time is computed as: $t = -\dfrac{1}{k}\ln\dfrac{P_t}{P_o}$

We first determine the pressure of DTBP when the total pressure equals 2100 mmHg.

Reaction: $\quad C_8H_{18}O_2(g) \longrightarrow 2C_3H_6O(g) + C_2H_6(g)$ [Equation 15.17]

Initial: $\quad\quad$ 800.0 mmHg

Changes: $\quad -x$ mmHg $\quad\quad\quad +2x$ mmHg $\quad\quad +x$ mmHg

Final: $\quad\quad (800.0-x)$ mmHg $\quad\quad 2x$ mmHg $\quad\quad x$ mmHg

Total pressure $= (800.0-x) + 2x + x = 800.0 + 2x = 2100.$

$x = 650.$ mmHg $\quad P\{C_8H_{18}O_2(g)\} = 800.$ mmHg $- 650.$ mmHg $= 150.$ mmHg

$t = -\dfrac{1}{k}\ln\dfrac{P_t}{P_o} = -\dfrac{1}{8.7\times10^{-3}\ \text{min}^{-1}}\ln\dfrac{150.\ \text{mmHg}}{800.\ \text{mmHg}} = 19_2\ \text{min} = 1.9\times10^2\ \text{min}$

87. **(M)**

$\dfrac{\Delta CCl_3}{\Delta t} = \text{rate}_{\text{formation}} + \text{rate}_{\text{disappearance}} = 0 \quad\quad$ so $\text{rate}_{\text{formation}} = \text{rate}_{\text{decomposition}}$

$k_2[Cl(g)][CHCl_3] = k_3[CCl_3][[Cl(g)]$ and, simplifying, $[CCl_3] = \dfrac{k_2}{k_3}[CHCl_3]$

since $\text{rate} = k_3[CCl_3][Cl(g)] = k_3\left(\dfrac{k_2}{k_3}[CHCl_3]\right)[Cl(g)] = k_2[CHCl_3][Cl(g)]$

We know: $[Cl(g)] = \left(\dfrac{k_1}{k_{-1}}[Cl_2(g)]\right)^{1/2}$ then $\text{rate}_{\text{overall}} = k_2[CHCl_3]\left(\dfrac{k_1}{k_{-1}}[Cl_2(g)]\right)^{1/2}$

and the rate constant k will be: $k = k_2\left(\dfrac{k_1}{k_{-1}}\right)^{1/2} = (1.3\times10^{-2})\left(\dfrac{4.8\times10^3}{3.6\times10^3}\right)^{1/2} = 0.015$

90. **(D)** Let 250-2x equal the partial pressure of CO(g) and x be the partial pressure of CO$_2$(g).

$2CO \rightarrow CO_2 + C(s) \quad\quad P_{tot} = P_{CO} + P_{CO2} = 250 - 2x + x = 250 - x$

$250 - 2x \quad\quad x \quad\quad -$

P_{tot} [torr]	Time [sec]	P_{CO2}	P_{CO} [torr]
250	0	0	250
238	398	12	226
224	1002	26	198
210	1801	40	170

The plots that follow show that the reaction appears to obey a second-order rate law.
Rate = k[CO]^2

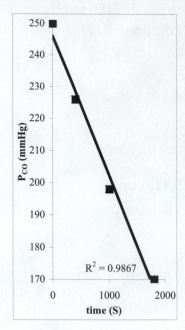 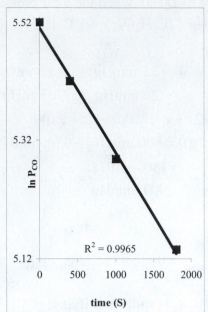 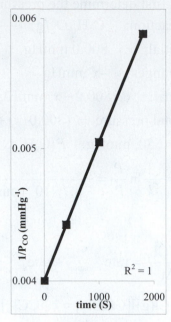

ZERO ORDER PLOT		1st ORDER PLOT		2nd ORDER PLOT		(Best correlation coefficient)

T	P_{CO}
0	250
398	226
1002	198
1801	170

T	lnP_{CO}
0	5.521461
398	5.420535
1002	5.288267
1801	5.135798

T	1/CO
0	0.004
398	0.004425
1002	0.005051
1801	0.005882

91. **(D)** Let 100-4x equal the partial pressure of PH$_3$(g), x be the partial pressure of P$_4$(g)and 6x be the partial pressure of H$_2$ (g)

$$4 PH_3(g) \rightarrow P_4(g) + 6 H_2(g)$$
$$100 - 4x \qquad x \qquad 6x$$

$$P_{tot} = P_{PH_3} + P_{P_4} + P_{H_2} = 100 - 4x + x + 6x = 100 + 3x$$

P_{tot} [torr]	Time [sec]	P_{P_4} [torr]	P_{PH_3} [torr]
100	0	0	100
150	40	50/3	100 − (4)(50/3)
167	80	67/3	100 − (4)(67/3)
172	120	72/3	100 − (4)(72/3)

The plots to follow show that the reaction appears to obey a first-order rate law.
Rate = k[PH₃]

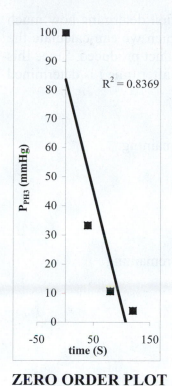

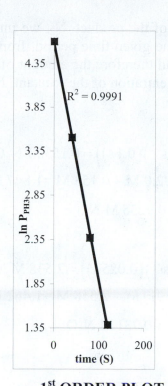

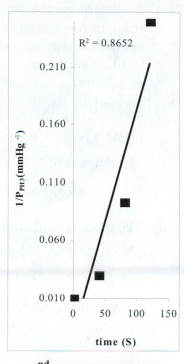

ZERO ORDER PLOT	
T	P_{PH3}
0	100
40	33.3
80	10.7
120	4

1ˢᵗ ORDER PLOT	
T	$\ln P_{PH3}$
0	4.61
40	3.51
80	2.37
120	1.39

2ⁿᵈ ORDER PLOT	
T	$1/P_{PH3}$
0	0.010
40	0.030
80	0.093
120	0.250

94. (M)

(a) The first elementary step $HBr + O_2 \xrightarrow{k1} HOOBr$ is rate-determining if the reaction obeys reaction rate = k [HBr][O₂] since the rate of this step is identical to that of the experimental rate law.

(b) No, mechanisms cannot be shown to be absolutely correct, only consistent with experimental observations.

(c) Yes; the sum of the elementary steps (3 HBr + O₂ → HOBr + Br₂ + H₂O) is not consistent with the overall stoichiometry (since HOBr is not detected as a product) of the reaction and therefore cannot be considered a valid mechanism.

95. (M)

(a) Both reactions are first-order, because the units on the rate constants are s^{-1}.

(b) k_2 is the slow reaction.

(c) To determine the concentration of the product, N_2, we must first determine how much reactant remains at the end of the given time period, from which we can calculate the amount of reactant consumed and therefore the amount of product produced. Since this is a first-order reaction, the concentration of the reactant, N_2O after time t is determined as follows:

$$[A]_t = [A]_o \, e^{-kt}$$

$$[N_2O]_{0.1} = (2.0 \text{ M}) \cdot \exp\left(-(25.7 \text{ s}^{-1})(0.1 \text{ s})\right) = 0.153 \text{ M } N_2O \text{ remaining}$$

The amount of N_2O consumed $= 2.0$ M $- 0.153$ M $= 1.847$ M

$$[N_2] = 1.847 \text{ M NO} \times \frac{1 \text{ M } N_2}{2 \text{ M NO}} = 0.9235 \text{ M } N_2$$

(d) The process is identical to step (c).

$$[N_2O]_{0.1} = (4.0 \text{ M}) \cdot \exp\left(-(18.2 \text{ s}^{-1})(0.025 \text{ s})\right) = 2.538 \text{ M } N_2O \text{ remaining}$$

The amount of N_2O consumed $= 4.0$ M $- 2.538$ M $= 1.462$ M

$$[N_2O] = 1.462 \text{ M NO} \times \frac{1 \text{ M } N_2O}{2 \text{ M NO}} = 0.731 \text{ M } N_2O$$

FEATURE PROBLEMS

96. (D)

(a) To determine the order of the reaction, we need $[C_6H_5N_2Cl]$ at each time. To determine this value, note that 58.3 mL $N_2 (g)$ evolved corresponds to total depletion of $C_6H_5N_2Cl$, to $[C_6H_5N_2Cl] = 0.000$ M.

Thus, at any point in time, $[C_6H_5N_2Cl] = 0.071 \text{ M} - \left(\text{volume } N_2(g) \times \dfrac{0.071 \text{ M } C_6H_5N_2Cl}{58.3 \text{ mL } N_2(g)} \right)$

Consider 21 min: $[C_6H_5N_2Cl] = 0.071 \text{ M} - \left(44.3 \text{ mL } N_2 \times \dfrac{0.071 \text{ M } C_6H_5N_2Cl}{58.3 \text{ mL } N_2(g)} \right) = 0.017 \text{ M}$

The numbers in the following table are determined with this method.

Time, min	0	3	6	9	12	15	18	21	24	27	30	∞
V_{N_2}, mL	0	10.8	19.3	26.3	32.4	37.3	41.3	44.3	46.5	48.4	50.4	58.3
$[C_6H_5N_2Cl]$, mM	71	58	47	39	32	26	21	17	14	12	10	0

[The concentration is given in thousandths of a mole per liter (mM).]

602

(b)

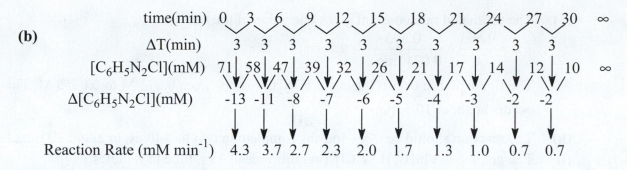

time(min)	0		3		6		9		12		15		18		21		24		27		30		∞

ΔT(min) 3 3 3 3 3 3 3 3 3 3

$[C_6H_5N_2Cl]$(mM) 71 58 47 39 32 26 21 17 14 12 10 ∞

$\Delta[C_6H_5N_2Cl]$(mM) -13 -11 -8 -7 -6 -5 -4 -3 -2 -2

Reaction Rate (mM min^{-1}) 4.3 3.7 2.7 2.3 2.0 1.7 1.3 1.0 0.7 0.7

(c) The two graphs are drawn on the same axes.

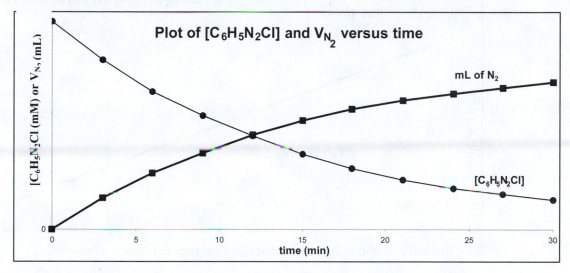

Plot of $[C_6H_5N_2Cl]$ and V_{N_2} versus time

(d) The rate of the reaction at $t = 21$ min is the slope of the tangent line to the $[C_6H_5N_2Cl]$ curve. The tangent line intercepts the vertical axis at about $[C_6H_5N_2Cl] = 39$ mM and the horizontal axis at about 37 min

$$\text{Reaction rate} = \frac{39 \times 10^{-3} \text{ M}}{37 \text{ min}} = 1.0_5 \times 10^{-3} \text{ M min}^{-1} = 1.1 \times 10^{-3} \text{ M min}^{-1}$$

The agreement with the reported value is very good.

(e) The initial rate is the slope of the tangent line to the $[C_6H_5N_2Cl]$ curve at $t = 0$. The intercept with the vertical axis is 71 mM, of course. That with the horizontal axis is about 13 min.

$$\text{Rate} = \frac{71 \times 10^{-3} \text{ M}}{13 \text{ min}} = 5.5 \times 10^{-3} \text{ M min}^{-1}$$

(f) The first-order rate law is Rate $= k[C_6H_5N_2Cl]$, which we solve for k:

$$k = \frac{\text{Rate}}{[C_6H_5N_2Cl]} \qquad k_0 = \frac{5.5 \times 10^{-3} \text{ M min}^{-1}}{71 \times 10^{-3} \text{ M}} = 0.077 \text{ min}^{-1}$$

$$k_{21} = \frac{1.1 \times 10^{-3} \text{ M min}^{-1}}{17 \times 10^{-3} \text{ M}} = 0.065 \text{ min}^{-1}$$

An average value would be a reasonable estimate: $k_{avg} = 0.071 \text{ min}^{-1}$

(g) The estimated rate constant gives one value of the half-life:

$$t_{1/2} = \frac{0.693}{k} = \frac{0.693}{0.071 \text{ min}^{-1}} = 9.8 \text{ min}$$

The first half-life occurs when $\left[C_6H_5N_2Cl\right]$ drops from 0.071 M to 0.0355 M. This occurs at about 10.5 min.

(h) The reaction should be three-fourths complete in two half-lives, or about 20 minutes.

(i) The graph plots $\ln\left[C_6H_5N_2Cl\right]$ (in millimoles/L) vs. time in minutes.

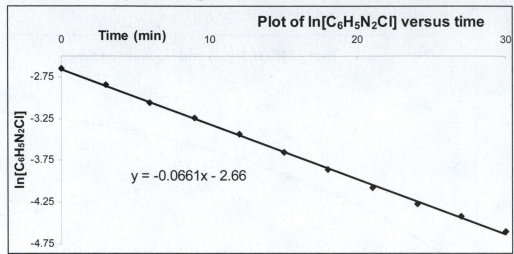

The linearity of the graph demonstrates that the reaction is first-order.

(j) $k = - \text{slope} = -\left(-6.61 \times 10^{-2}\right) \text{min}^{-1} = 0.0661 \text{ min}^{-1}$

$$t_{1/2} = \frac{0.693}{0.0661 \text{ min}^{-1}} = 10.5 \text{ min, in good agreement with our previously determined values.}$$

97. **(D)**

(a) In Experiments 1 & 2, [KI] is the same (0.20 M), while $\left[(NH_4)_2 S_2O_8\right]$ is halved, from 0.20 M to 0.10 M. As a consequence, the time to produce a color change doubles (i.e., the rate is halved). This indicates that reaction (a) is first-order in $S_2O_8{}^{2-}$. Experiments 2 and 3 produce a similar conclusion. In Experiments 4 and 5, $\left[(NH_4)_2 S_2O_8\right]$ is the same (0.20 M) while [KI] is halved, from 0.10 to 0.050 M. As a consequence, the time to produce a color change nearly doubles, that is, the rate is halved. This indicates that reaction (a) is also first-order in I^-. Reaction (a) is (1 + 1) second-order overall.

(b) The blue color appears when all the $S_2O_3^{2-}$ has been consumed, for only then does reaction (b) cease. The same amount of $S_2O_3^{2-}$ is placed in each reaction mixture.

$$\text{amount } S_2O_3^{2-} = 10.0 \text{ mL} \times \frac{1 \text{ L}}{1000 \text{ mL}} \times \frac{0.010 \text{ mol Na}_2S_2O_3}{1 \text{ L}} \times \frac{1 \text{ mol } S_2O_3^{2-}}{1 \text{ mol Na}_2S_2O_3} = 1.0 \times 10^{-4} \text{mol}$$

Through stoichiometry, we determine the amount of each reactant that reacts before this amount of $S_2O_3^{2-}$ will be consumed.

$$\text{amount } S_2O_8^{2-} = 1.0 \times 10^{-4} \text{ mol } S_2O_3^{2-} \times \frac{1 \text{ mol } I_3^-}{2 \text{ mol } S_2O_3^{2-}} \times \frac{1 \text{ mol } S_2O_8^{2-}}{1 \text{ mol } I_3^-}$$

$$= 5.0 \times 10^{-5} \text{ mol } S_2O_8^{2-}$$

$$\text{amount } I^- = 5.0 \times 10^{-5} \text{ mol } S_2O_8^{2-} \times \frac{2 \text{ mol } I^-}{1 \text{ mol } S_2O_8^{2-}} = 1.0 \times 10^{-4} \text{ mol } I^-$$

Note that we do not use "3 mol I^-" from equation (a) since one mole has not been oxidized; it simply complexes with the product I_2. The total volume of each solution is $(25.0 \text{ mL} + 25.0 \text{ mL} + 10.0 \text{ mL} + 5.0 \text{ mL} =) 65.0 \text{ mL}$, or 0.0650 L.

The amount of $S_2O_8^{2-}$ that reacts in each case is 5.0×10^{-5} mol and thus

$$\Delta\left[S_2O_8^{2-}\right] = \frac{-5.0 \times 10^{-5} \text{ mol}}{0.0650 \text{ L}} = -7.7 \times 10^{-4} \text{ M}$$

$$\text{Thus, Rate}_1 = \frac{-\Delta\left[S_2O_8^{2-}\right]}{\Delta t} = \frac{+7.7 \times 10^{-4} \text{ M}}{21 \text{ s}} = 3.7 \times 10^{-5} \text{ M s}^{-1}$$

(c) For Experiment 2, $\text{Rate}_2 = \dfrac{-\Delta\left[S_2O_8^{2-}\right]}{\Delta t} = \dfrac{+7.7 \times 10^{-4} \text{ M}}{42 \text{ s}} = 1.8 \times 10^{-5} \text{ M s}^{-1}$

To determine the value of k, we need initial concentrations, as altered by dilution.

$$\left[S_2O_8^{2-}\right]_1 = 0.20 \text{ M} \times \frac{25.0 \text{ mL}}{65.0 \text{ mL total}} = 0.077 \text{ M} \qquad \left[I^-\right]_1 = 0.20 \text{ M} \times \frac{25.0 \text{ mL}}{65.0 \text{ mL}} = 0.077 \text{ M}$$

$$\text{Rate}_1 = 3.7 \times 10^{-5} \text{ M s}^{-1} = k\left[S_2O_8^{2-}\right]^1\left[I^-\right]^1 = k(0.077 \text{ M})^1 (0.077 \text{ M})^1$$

$$k = \frac{3.7 \times 10^{-5} \text{ M s}^{-1}}{0.077 \text{ M} \times 0.077 \text{ M}} = 6.2 \times 10^{-3} \text{ M}^{-1} \text{ s}^{-1}$$

$$\left[S_2O_8^{2-}\right]_2 = 0.10 \text{ M} \times \frac{25.0 \text{ mL}}{65.0 \text{ mL total}} = 0.038 \text{ M} \qquad \left[I^-\right]_2 = 0.20 \text{ M} \times \frac{25.0 \text{ mL}}{65.0 \text{ mL}} = 0.077 \text{ M}$$

$$\text{Rate}_2 = 1.8 \times 10^{-5} \text{ M s}^{-1} = k\left[S_2O_8^{2-}\right]^1\left[I^-\right]^1 = k(0.038 \text{ M})^1 (0.077 \text{ M})^1$$

$$k = \frac{1.8 \times 10^{-5} \text{ M s}^{-1}}{0.038 \text{ M} \times 0.077 \text{ M}} = 6.2 \times 10^{-3} \text{ M}^{-1} \text{ s}^{-1}$$

(d) First we determine concentrations for Experiment 4.

$$\left[S_2O_8^{2-}\right]_4 = 0.20 \text{ M} \times \frac{25.0 \text{ mL}}{65.0 \text{ mL total}} = 0.077 \text{ M} \qquad \left[I^-\right]_4 = 0.10 \times \frac{25.0 \text{ mL}}{65.0 \text{ mL}} = 0.038 \text{ M}$$

We have two expressions for Rate; let us equate them and solve for the rate constant.

$$\text{Rate}_4 = \frac{-\Delta\left[S_2O_8^{2-}\right]}{\Delta t} = \frac{+7.7\times10^{-4} \text{ M}}{\Delta t} = k\left[S_2O_8^{2-}\right]_4^1\left[I^-\right]_4^1 = k(0.077 \text{ M})(0.038 \text{ M})$$

$$k = \frac{7.7\times10^{-4} \text{ M}}{\Delta t \times 0.077 \text{ M} \times 0.038 \text{ M}} = \frac{0.26 \text{ M}^{-1}}{\Delta t} \qquad k_3 = \frac{0.26 \text{ M}^{-1}}{189 \text{ s}} = 0.0014 \text{ M}^{-1}\text{ s}^{-1}$$

$$k_{13} = \frac{0.26 \text{ M}^{-1}}{88 \text{ s}} = 0.0030 \text{ M}^{-1}\text{ s}^{-1} \qquad k_{24} = \frac{0.26 \text{ M}^{-1}}{42 \text{ s}} = 0.0062 \text{ M}^{-1}\text{ s}^{-1}$$

$$k_{33} = \frac{0.26 \text{ M}^{-1}}{21 \text{ s}} = 0.012 \text{ M}^{-1}\text{ s}^{-1}$$

(e) We plot $\ln k$ vs. $1/T$ The slope of the line $= -E_a/R$.

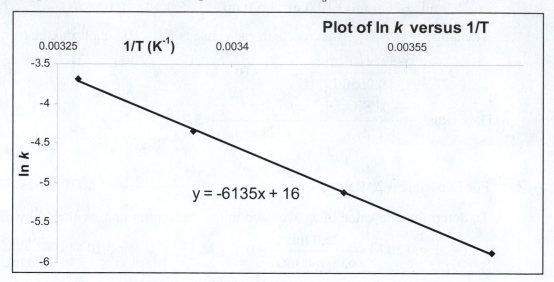

Plot of ln k versus 1/T

$$E_a = +6135 \text{ K} \times 8.3145 \text{ J mol}^{-1}\text{ K}^{-1} = 51.0\times10^3 \text{ J/mol} = 51.0 \text{ kJ/mol}$$

The scatter of the data permits only a two significant figure result: 51 kJ/mol

(f) For the mechanism to agree with the reaction stoichiometry, the steps of the mechanism must sum to the overall reaction, in the manner of Hess's law.

(slow) $I^- + S_2O_8^{2-} \rightarrow IS_2O_8^{3-}$

(fast) $IS_2O_8^{3-} \rightarrow 2\ SO_4^{2-} + I^+$

(fast) $I^+ + I^- \rightarrow I_2$

(fast) $I_2 + I^- \rightarrow I_3^-$

(net) $3\ I^- + S_2O_8^{2-} \rightarrow 2\ SO_4^{2-} + I_3^-$

Each of the intermediates cancels: $IS_2O_8^{3-}$ is produced in the first step and consumed in the second, I^+ is produced in the second step and consumed in the third, I_2 is produced in the third step and consumed in the fourth. The mechanism is consistent with the stoichiometry. The rate of the slow step of the mechanism is

$$\text{Rate}_1 = k_1\left[S_2O_8^{2-}\right]^1\left[I^-\right]^1$$

This is exactly the same as the experimental rate law. It is reasonable that the first step be slow since it involves two negatively charged species coming together. We know that like charges repel, and thus this should not be an easy or rapid process.

SELF-ASSESSMENT EXERCISES

101. **(E)** The answer is (c). The rate constant k is only dependent on temperature, not on the concentration of the reactants

102. **(E)** The answers are (b) and (e). Because half-life is 75 seconds, the quantity of reactant left at two half-lives ($75 + 75 = 150$) equals one-half of the level at 75 seconds. Also, if the initial concentration is doubled, after one half-life the remaining concentration would have to be twice as much as the original concentration.

103. **(E)** The answer is (a). Half-life $t_{1/2} = 13.9$ min, $k = \ln 2/t_{1/2} = 0.050$ min^{-1}. Rate of a first-order reaction is as follows:

$$\frac{d[A]}{dt} = k[A] = \left(0.050 \text{ min}^{-1}\right)\left(0.40 \text{ M}\right) = 0.020 \text{ M min}^{-1}$$

104. **(E)** The answer is (d). A second-order reaction is expressed as follows:

$$\frac{d[A]}{dt} = k[A]^2$$

If the rate of the reaction when $[A] = 0.50$ is $k(0.50)^2 = k(0.25)$. If $[A] = 0.25$ M, then the rate is $k(0.0625)$, which is ¼ of the rate at $[A] = 0.50$.

105. **(M)** The answer is (b). Going to slightly higher temperatures broadens the molecular speed distribution, which in turn increases the fraction of molecules at the high kinetic energy range (which are those sufficiently energetic to make a reaction happen).

106. **(E)** The answer is (c). Since the reaction at hand is described as an elementary one, the rate of the reaction is $k[A][B]$.

107. **(E)** We note that from the given data, the half-life of the reaction is 100 seconds (at t = 0, $[A] = 0.88$ M/s, whereas at t = 100, $[A] = 0.44$ M/s). Therefore, the rate constant k is: $k = \ln 2/100$ s $= 0.00693$ s^{-1}. We can now calculate instantaneous rate of the reaction: $d[A]/dt = (0.00693 \text{ s}^{-1})(0.44 \text{ M}) = 3.0 \times 10^{-3}$ M·s^{-1}

108. (M)

    (a) For a first-order reaction,

$$k = \frac{\ln 2}{t_{1/2}} = \frac{0.693}{30} = 0.0231 \text{ min}^{-1}$$

$$\ln[A]_t - \ln[A]_o = -kt$$

$$t = \frac{\ln[A]_t - \ln[A]_o}{-k} = \frac{\ln(0.25) - 0}{-0.0231 \text{ min}^{-1}} = 60.0 \text{ min}$$

    (b) For a zero-order reaction,

$$k = \frac{0.5}{t_{1/2}} = \frac{0.5}{30} = 0.0167 \text{ min}^{-1}$$

$$[A]_t - [A]_o = -kt$$

$$t = \frac{[A]_t - [A]_o}{-k} = \frac{0.25 - 1.00}{-0.0167 \text{ min}^{-1}} = 45.0 \text{ min}$$

109. (M) The reaction is second-order, because the half-life doubles with each successive half-life period.

110. (M)

    (a) The initial rate = $\Delta M / \Delta t$ = (1.204 M – 1.180 M)/(1.0 min) = 0.024 M/min

    (b) In experiment 2, the initial concentration is twice that of experiment A. For a second-order reaction:

$$\text{Rate} = k \,[A_{exp\,2}]^2 = k\,[2 \times A_{exp\,1}]^2 = 4\,k\,[A_{exp\,1}]^2$$

    This means that if the reaction is second order, its initial rate of experiment 2 will be 4 times that of experiment 1 (that is, 4 times as many moles of A will be consumed in a given amount of time). The initial rate is 4 × 0.024 M/min = 0.096 M/s. Therefore, at 1 minute, [A] = 2.408 – 0.0960 = 2.312 M.

    (c) The half-life of the reaction, obtained from experiment 1, is 35 minutes. If the reaction is first-order, then k = ln 2/35 min = 0.0198 min^{-1}.

    For a first-order reaction,

$$[A]_t = [A]_o\, e^{-kt}$$

$$[A]_{35\,min} = (2.408)\exp\left(-0.0198 \text{ min}^{-1} \cdot 30 \text{ min}\right) = 1.33 \text{ M}$$

111. (D) The overall stoichiometry of the reaction is determined by adding the two reactions with each other: $A + 2B \rightleftharpoons C + D$

(a) Since I is made slowly but is used very quickly, its rate of formation is essentially zero. The amount of I at any given time during the reaction can be expressed as follows:

$$\frac{d[I]}{dt} = 0 = k_1[A][B] - k_2[B][I]$$

$$[I] = \frac{k_1}{k_2}[A]$$

Using the above expression for [I], we can now determine the overall reaction rate law:

$$\frac{d[C]}{dt} = k_2[I][B] = k_2 \cdot \frac{k_1}{k_2}[A] \cdot [B] = k_1[A][B]$$

(b) Adding the two reactions given, we still get the same overall stoichiometry as part (a). However, with the given proposed reaction mechanisms, the rate law for the product(s) is given as follows:

$$\frac{d[B_2]}{dt} = k_1[B]^2 - k_{-1}[B_2] - k_2[A][B] = 0$$

$$[B_2] = \frac{k_1[B]^2}{k_{-1} + k_2[A]}$$

Therefore,

$$\frac{d[C]}{dt} = k_2[A][B_2] = \frac{k_2 k_1[A][B]^2}{k_{-1} + k_2[A]}$$

which does not agree with the observed reaction rate law.

112. (M) The answer is (b), first-order, because only in a first-order reaction is the half-life independent of the concentration of the reacting species.

113. (E) The answer is (a), zero-order, because in a zero-order reaction the relationship between concentration and time is: $[A]_t = kt + [A]_o$

114. (E) The answer is (c), remain the same. This is because for a zero-order reaction, $d[A]/dt = k[A]^0 = k$. Therefore, the reaction rate is independent of the concentration of the reactant.

CHAPTER 21
CHEMISTRY OF THE MAIN-GROUP ELEMENTS I:
GROUPS 1, 2, 13, AND 14

PRACTICE EXAMPLES

1A **(E)** From Figure 21-2, the route from sodium chloride to sodium nitrate begins with electrolysis of NaCl(aq) to form NaOH(aq).

$2 NaCl(aq) + 2 H_2O(l) \xrightarrow{electrolysis} 2 NaOH(aq) + H_2(g) + Cl_2(g)$ followed by addition of

$NO_2(g)$ to NaOH(aq). $2 NaOH(aq) + 3 NO_2(g) \rightarrow 2 NaNO_3(aq) + NO(g) + H_2O(l)$

1B **(E)** From Figure 21-2, we see that the route from sodium chloride to sodium thiosulfate begins with the electrolysis of NaCl(aq) to produce NaOH(aq),

$2 NaCl(aq) + 2 H_2O(l) \xrightarrow{electrolysis} 2 NaOH(aq) + H_2(g) + Cl_2(g)$

and continues through the reaction of $SO_2(g)$ with the NaOH(aq) in an acid-base reaction

$[SO_2(g)$ is an acid anhydride] to produce

$Na_2SO_3(aq):\quad 2 NaOH(aq) + SO_2(g) \rightarrow Na_2SO_3(aq) + H_2O(l)$ and (3) concludes with

the addition of S to the boiling solution: $Na_2SO_3(aq) + S(s) \xrightarrow{boil} Na_2S_2O_3(aq)$.

2A **(M)** The first reaction indicates that 0.1 mol of $NaNO_2$ reacts with 0.3 mol of Na to yield 0.2 mol of compound X and 0.05 mol of N_2. Since this reaction liberates nitrogen gas, it is very likely that compound X is an oxide of sodium. The two possibilities include Na_2O and Na_2O2. Na_2O_2 will not react with oxygen, but Na_2O will. Therefore, compound X is most likely Na_2O and compound Y is Na_2O_2. The balanced chemical equations for two processes are:

$$NaNO_2(s) + 3 Na(s) \rightarrow 2 Na_2O(s) + \frac{1}{2} N_2(g)$$

$$Na_2O(s) + \frac{1}{2} O_2(g) \rightarrow Na_2O_2(s)$$

2B **(M)** Since compound X is used in plaster of Paris it must contain calcium. Calcium reacts with carbon to form calcium carbide, according to the following chemical equation:
$Ca(s) + 2C(s) \rightarrow CaC_2(s)$

Furthermore, compound X or calcium carbide reacts with nitrogen gas to form compound Y or calcium cyanamide:
$CaC_2(s) + N_2(g) \rightarrow CaCN_2(s) + C(s)$

$CN_2{}^{2-}$ anion is isoelectronic with CO_2 and it therefore contains 16 electrons. The structure of this anion is:

$$\overset{\bullet\bullet}{\underset{\bullet\bullet}{N}} = C = \overset{\bullet\bullet}{\underset{\bullet\bullet}{N}}$$

3A **(M)** The first two reactions, are those from Example 21-3, used to produce B_2O_3.

$$Na_2B_4O_7 \cdot 10\,H_2O(s) + H_2SO_4(l) \longrightarrow 4\,B(OH)_3(s) + Na_2SO_4(s) + 5\,H_2O(l)$$

$$2\,B(OH)_3(s) \overset{\Delta}{\longrightarrow} B_2O_3(s) + 3\,H_2O(g)$$

The next reaction is conversion to BCl_3 with heat, carbon, and chlorine.

$$2\,B_2O_3(s) + 3\,C(s) + 6\,Cl_2(g) \overset{\Delta}{\longrightarrow} 4\,BCl_3(g) + 3\,CO_2(g)$$

$LiAlH_4$ is used as a reducing agent to produce diborane.

$$4\,BCl_3(g) + 3\,LiAlH_4(s) \rightarrow 2\,B_2H_6(g) + 3\,LiCl(s) + 3\,AlCl_3(s)$$

3B **(M)** $Na_2B_4O_7\,10\,H_2O(s) + H_2SO_4(aq) \rightarrow 4\,B(OH)_3(s) + Na_2SO_4(aq) + 5\,H_2O(l)$

$2\,B(OH)_3(s) \overset{\Delta}{\longrightarrow} B_2O_3(s) + 3\,H_2O(l)$

$B_2O_3(s) + 3\,CaF_2(s) + 3\,H_2SO_4(l) \overset{\Delta}{\longrightarrow} 2\,BF_3(g) + 3\,CaSO_4(s) + 3\,H_2O(g)$

INTEGRATIVE EXAMPLES

A **(M)** NaCN and $Al(NO_3)_3$ dissociate in water according to the following equations:

$NaCN(aq) \rightarrow Na^+(aq) + CN^-(aq)$

$Al(NO_3)_3(aq) \rightarrow Al^{3+}(aq) + 3\,NO_3^-(aq)$

NaCN is a salt of strong base (NaOH) and weak acid (HCN) and its solution is therefore basic.

We proceed by first determining the $[OH]^-$ concentration in a solution which is 1.0M in NaCN:

$$CN^- + H_2O \rightleftharpoons HCN + OH^-$$

1.0	/	/
$1.0 - x$	x	x

$$K_b = \frac{K_w}{K_a(HCN)} = \frac{1.0 \times 10^{-14}}{6.2 \times 10^{-10}} = 1.6 \times 10^{-5}$$

$$K_b = \frac{[HCN][OH^-]}{[CN^-]} = \frac{x \times x}{1.0 - x} = \frac{x^2}{1.0 - x} = 1.6 \times 10^{-5}$$

Assume, $x \ll 1$ and $1.0 - x \approx 1.0$

$$x^2 = 1.6 \times 10^{-5} \Rightarrow x = \sqrt{1.6 \times 10^{-5}} = 4.0 \times 10^{-3}$$

$$[OH^-] = 4.0 \times 10^{-3}$$

On mixing, $[OH^-] = 2.0 \times 10^{-3}$ M and $[Al^{3+}] = 0.50$ M. The precipitation will occur according to the following equation:

$$Al^{3+}(aq) + 3\,OH^-(aq) \rightleftharpoons Al(OH)_3(s)$$

The product $[Al^{3+}][OH^-]^3 = 0.50 \times (2.0 \times 10^{-3})^3 = 4.0 \times 10^{-9}$ is much greater than K_{sp} for $Al(OH)_3$ (1.3×10^{-33}) and therefore precipitation will occur.

B **(M)** When $BeCl_2 \cdot 4H_2O$ is heated, it decomposes to $Be(OH)_2(s)$, $H_2O(g)$, and $HCl(g)$, as discussed in Section 21-3. $BeCl_2 \cdot 4H_2O$ comprises $[Be(H_2O)_4]^{2+}$ and Cl^- ions. Because of the high polarizing power of Be^{2+}, it is difficult to remove the coordinated water molecules by heating the solid and the acidity of the coordinated H_2O molecules is enhanced. When $BeCl_2 \cdot 4H_2O$ is dissolved in water, $[Be(H_2O)_4]^{2+}$ ions react with water, and $[Be(H_2O)_3(OH)]^+$ and H_3O^+ ions are produced. Hence, a solution of $BeCl_2 \cdot 4H_2O$ is expected to be acidic.

When $CaCl_2 \cdot 6H_2O$ is heated, $CaCl_2(s)$ and $H_2O(g)$ are produced. Because the charge density and polarizing power of Ca^{2+} is much less than that of Be^{2+}, it is much easier to drive off the coordinated water molecules by heating the solid. When $CaCl_2 \cdot 6H_2O$ is dissolved in water, $Ca^{2+}(aq)$ and $Cl^-(aq)$ are produced. However, the charge density of the Ca^{2+} ion is too low to affect the acidity of water molecules in the hydration sphere of the Ca^{2+} ion. Hence, a solution of $CaCl_2 \cdot 6H_2O$ has a neutral pH.

EXERCISES

Group 1: The Alkali Metals

1. **(E)** **(a)** $2\,Cs(s) + Cl_2(g) \longrightarrow 2\,CsCl(s)$

(b) $2\,Na(s) + O_2(g) \longrightarrow Na_2O_2(s)$

(c) $Li_2CO_3(s) \xrightarrow{\Delta} Li_2O(s) + CO_2(g)$

(d) $Na_2SO_4(s) + 4\,C(s) \rightarrow Na_2S(s) + 4\,CO(g)$

(e) $K(s) + O_2(g) \longrightarrow KO_2(s)$

3. **(E)** Both LiCl and KCl are soluble in water, but Li_3PO_4 is not very soluble. Hence the addition of $K_3PO_4(aq)$ to a solution of the white solid will produce a precipitate if the white solid is LiCl, but no precipitate if the white solid is KCl. The best method is a flame test; lithium gives a red color to a flame, while the potassium flame test is violet.

5. **(M)** First we note that sodium carbonate ionizes virtually completely when dissolved in H_2O and thus is described as highly soluble in water. This is made evident by the fact that there is no K_{sp} value for Na_2CO_3. By contrast, the existence of K_{sp} values for $MgCO_3$ and Li_2CO_3 shows that these compounds have lower solubilities in water than Na_2CO_3. To decide whether $MgCO_3$ is more or less soluble than Li_2CO_3, we must calculate the molar solubility for each salt and then compare the two values. Clearly, the salt that has the larger molar solubility will be more soluble in water. The molar solubilities can be found using the respective K_{sp} expressions for the two salts.

1. $MgCO_3 \xrightleftharpoons{K_{sp} = 3.5 \times 10^{-8}} Mg^{2+}(aq) + CO_3^{2-}(aq)$ Let "s" = molar solubility of

    $MgCO_3$ excess $- s$ $\qquad\qquad\qquad\quad s \qquad\quad s \qquad$ $MgCO_3$

    $s^2 = 3.5 \times 10^{-8} \qquad s = 1.9 \times 10^{-4}$ M

2. $Li_2CO_3 \xrightleftharpoons{K_{sp} = 2.5 \times 10^{-2}} 2\, Li^+(aq) + CO_3^{2-}(aq) \qquad$ Let "s" = molar solubility of

    Li_2CO_3 excess $- s$ $\qquad\qquad\qquad\quad 2s \qquad\quad s \qquad$ Li_2CO_3

    $(2s)^2 \times (s) = 2.5 \times 10^{-8} \quad 4s^3 = 2.5 \times 10^{-8} \quad s = \sqrt[3]{\dfrac{2.5 \times 10^{-8}}{4}} = 0.18$ M

We can conclude that Li_2CO_3 is more soluble than $MgCO_3$. Thus, the expected order of increasing solubility in water is: $\quad MgCO_3 < Li_2CO_3 < Na_2CO_3$.

7. (M)
    (a) $H_2(g)$ and $Cl_2(g)$ are produced during the electrolysis of $KCl(aq)$.
    The electrode reactions are:
    Anode, oxidation: $\qquad 2\, Cl^-(aq) \rightarrow Cl_2(g) + 2\, e^-$
    Cathode, reduction: $\qquad 2\, H_2O(l) + 2\, e^- \rightarrow H_2(g) + 2\, OH^-(aq)$
    We can compute the amount of OH^- produced at the cathode.

    $$mol\ OH^- = 3.50\ min \times \frac{60\ s}{1\ min} \times \frac{0.910\ C}{1 s} \times \frac{1\ mol\ e^-}{96485\ C} \times \frac{2\ mol\ OH^-}{2\ mol\ e^-} = 1.98 \times 10^{-3}\ mol$$

    Now, we can calculate the $[OH^-]$ and, from that, the pH of the solution:

    $$[OH^-] = \frac{1.98 \times 10^{-3}\ mol\ OH^-}{1.26\ L\ soln} = 1.57 \times 10^{-3}\ M$$

    $pOH = -\log(1.57 \times 10^{-3}) = 2.80$
    $pH = 14.000 - 2.80 = 11.2$

    (b)   As long as KCl is in excess and the volume of the solution is nearly constant, the solution pH only depends on the number of electrons transferred.

9. (M)
    (a)   We first compute the mass of $NaHCO_3$ that should be produced from 1.00 ton NaCl, assuming that all of the Na in the NaCl ends up in the $NaHCO_3$. We use the unit, ton-mole, to simplify the calculations.

    $$mass\ NaHCO_3 = 1.00\ ton\ NaCl \times \frac{1\ ton\text{-}mol\ NaCl}{58.4\ ton\ NaCl} \times \frac{1\ ton\text{-}mol\ Na}{1\ ton\text{-}mol\ NaCl}$$

    $$\times \frac{1\ tol\text{-}mol\ NaHCO_3}{1\ ton\text{-}mol\ Na} \times \frac{84.0\ ton\ NaHCO_3}{1\ ton\ mol\ NaHCO_3} = 1.44\ ton\ NaHCO_3$$

    $$\%\ yield = \frac{1.03\ ton\ NaHCO_3\ produced}{1.44\ ton\ NaHCO_3\ expected} \times 100\% = 71.5\%\ yield$$

(b) NH_3 is used in the principal step of the Solvay process to produce a solution in which $NaHCO_3$ is formed and from which it will precipitate. The filtrate contains NH_4Cl, from which NH_3 is recovered by treatment with $Ca(OH)_2$. Thus, NH_3 is simply used during the Solvay process to produce the proper conditions for the desired reactions. Any net consumption of NH_3 is the result of unavoidable losses during production.

11. **(M)** Use $\Delta_f G°$ values to calculate $\Delta_r G°$ for the reaction and then the equilibrium constant K at 298 K.

$$Na_2O_2(s) \rightleftharpoons Na_2O(s) + \frac{1}{2}O_2(g)$$

$\Delta_f G°(\text{kJ mol}^{-1})\ -449.63\ -379.09$

$\Delta_r G° = -379.09 - (-449.63) = 70.54 \text{ kJ mol}^{-1}$

$\Delta_r G° = -RT \ln K$

$-8.314 \text{ JK}^{-1}\text{mol}^{-1} \times 298 \text{ K} \times \ln K = 70.54 \times 1000 \text{ Jmol}^{-1}$

$$\ln K = \frac{70.54 \times 1000 \text{ Jmol}^{-1}}{-8.314 \text{ JK}^{-1}\text{mol}^{-1} \times 298 \text{ K}} = -28.47 \Rightarrow K = e^{-28.47} = 4.32 \times 10^{-13}$$

Since $K = p_{O_2}^{\frac{1}{2}} \Rightarrow p_{O_2} = K^2 = 1.87 \times 10^{-25}$. The equilibrium constant and partial pressure of oxygen are both very small at 298 K. Therefore, $Na_2O_2(s)$ is thermodynamically stable with respect to $Na_2O(s)$ and $O_2(g)$ at 298 K.

Group 2: The Alkaline Earth Metals

13. **(M)** $CaO \xleftarrow{\Delta} CaCO_3 \xleftarrow{CO_2} Ca(OH)_2 \xrightarrow{HCl} CaCl_2 \xrightarrow{\text{electrolysis}} Ca$

$CaHPO_4 \xleftarrow{H_3PO_4} \quad \downarrow{H_2SO_4} \quad CaSO_4$

The reactions are as follows. $\quad Ca(OH)_2(s) + 2\,HCl(aq) \rightarrow CaCl_2(aq) + 2\,H_2O(l)$

$CaCl_2(l) \xrightarrow{\Delta,\text{electrolysis}} Ca(l) + Cl_2(g) \quad Ca(OH)_2(s) + CO_2(g) \rightarrow CaCO_3(s) + H_2O(g)$

$CaCO_3(s) \xrightarrow{\Delta} CaO(s) + CO_2(g) \quad Ca(OH)_2(s) + H_2SO_4(aq) \rightarrow CaSO_4(s) + 2\,H_2O(l)$

$Ca(OH)_2(s) + H_3PO_4(aq) \rightarrow CaHPO_4(aq) + 2\,H_2O(l)$. Actually $CaO(s)$ is the industrial starting material from which $Ca(OH)_2$ is made. $CaO(s) + H_2O(l) \longrightarrow Ca(OH)_2(s)$

15. **(M)** The reactions involved are:

$Mg^{2+}(aq) + Ca^{2+}(aq) + 2\,OH^-(aq) \rightarrow Mg(OH)_2(s) + Ca^{2+}(aq)$

$Mg(OH)_2(s) + 2\,H^+(aq) + 2\,Cl^-(aq) \rightarrow Mg^{2+}(aq) + 2\,H_2O(l) + 2\,Cl^-(aq)$

$Mg^{2+}(aq) + 2\,Cl^-(aq) \xrightarrow{\Delta} MgCl_2(s)$

$MgCl_2(s) \text{— (electrolysis)} \longrightarrow Mg(l) + Cl_2(g)$

$\underline{Mg(l) \rightarrow Mg(s)}$

$Mg^{2+} + 2\,Cl^-(aq) \rightarrow Mg(s) + Cl_2(g)$ (Overall reaction)

As can be seen, the process does not violate the principle of conservation of charge.

17. **(M)** **(a)** $\qquad BeF_2(s) + Mg(s) \xrightarrow{\Delta} Be(s) + MgF_2(s)$

(b) $\quad Ba(s) + Br_2(l) \longrightarrow BaBr_2(s)$

(c) $\quad UO_2(s) + 2\,Ca(s) \longrightarrow U(s) + 2\,CaO(s)$

(d) $\quad MgCO_3 \cdot CaCO_3(s) \xrightarrow{\Delta} MgO(s) + CaO(s) + 2\,CO_2(g)$

(e) $\quad 2\,H_3PO_4(aq) + 3\,CaO(s) \longrightarrow Ca_3(PO_4)_2(s) + 3\,H_2O(l)$

19. **(D)** Let us compute the value of the equilibrium constant for each reaction by combining the two solubility product constants. Large values of equilibrium constants indicate that the reaction is displaced far to the right. Values of K that are much smaller than 1 indicate that the reaction is displaced far to the left.

(a) $BaSO_4(s) \rightleftharpoons Ba^{2+}(aq) + SO_4^{2-}(aq)$ $\qquad\qquad K_{sp} = 1.1 \times 10^{-10}$

$\underline{Ba^{2+}(aq) + CO_3^{2-}(aq) \rightleftharpoons BaCO_3(s) \qquad\qquad 1/K_{sp} = 1/(5.1 \times 10^{-9})}$

$BaSO_4(s) + CO_3^{2-}(aq) \rightleftharpoons BaCO_3(s) + SO_4^{2-}(aq) \quad K = \dfrac{1.1 \times 10^{-10}}{5.1 \times 10^{-9}} = 2.2 \times 10^{-2}$

Thus, the equilibrium lies slightly to the left.

(b) $Mg_3(PO_4)_2(s) \rightleftharpoons 3\,Mg^{2+}(aq) + 2\,PO_4^{3-}(aq) \qquad K_{sp} = 2.1 \times 10^{-25}$

$\underline{3\{Mg^{2+}(aq) + CO_3^{2-}(aq) \rightleftharpoons MgCO_3(s)\} \qquad 1/(K_{sp})^3 = 1/(3.5 \times 10^{-8})^3}$

$Mg_3(PO_4)_2(s) + 3\,CO_3^{2-}(aq) \rightleftharpoons 3\,MgCO_3(s) + 2\,PO_4^{3-}(aq)$

$K = \dfrac{2.1 \times 10^{-25}}{(3.5 \times 10^{-8})^3} = 4.9 \times 10^{-3}$ Thus, the equilibrium lies to the left.

(c) $Ca(OH)_2(s) \rightleftharpoons Ca^{2+}(aq) + 2\,OH^-(aq) \qquad\qquad K_{sp} = 5.5 \times 10^{-6}$

$\underline{Ca^{2+}(aq) + 2\,F^-(aq) \rightleftharpoons CaF_2(s) \qquad\qquad 1/K_{sp} = 1/(5.3 \times 10^{-9})}$

$Ca(OH)_2(s) + 2\,F^-(aq) \rightleftharpoons CaF_2(s) + 2\,OH^-(aq) \quad K = \dfrac{5.5 \times 10^{-6}}{5.3 \times 10^{-9}} = 1.0 \times 10^3$

Thus, the equilibrium lies to the right.

21. **(M)** The SO_4^{2-} ion is a large polarizable ion. A cation with a high polarizing power will polarize the SO_4^{2-} ion and kinetically assist the decomposition to SO_3. Because Be^{2+} has the largest charge density of the group 2 cations, we expect Be^{2+} to be the most polarizing and thus, $BeSO_4$ will be the least stable with respect to decomposition.

Group 13: The Boron Family

23. **(M)**

(a) B_4H_{10} contains a total of $4\times3+10\times1 = 22$ valence electrons or 11 pairs. Ten of these pairs could be allocated to form 10 B—H bonds, leaving but one pair to bond the four B atoms together, which is clearly an electron deficient situation.

(b) In our analysis in part (a), we noted that the four B atoms had but one electron pair to bond them together. To bond these four atoms into a chain requires three electron pairs. Since each electron pair in a bridging bond replaces two "normal" bonds, there must be at least two bridging bonds in the B_4H_{10} molecules. By analogy with B_2H_6, we might write the structure below left. But this structure uses only a total of 20 electrons. (The bridge bonds are shown as dots, normal bonds—electron pairs—as dashes.) In the structure at right below, we have retained some of the form of B_2H_6, and produced a compound with the formula B_4H_{10} and 11 electron pairs. (The experimentally determined structure of B_4H_{10} consists of a four-membered ring of alternating B and H atoms, held together by bridging bonds. Two of the B atoms have two H atoms bonded to each of them by normal covalent bonds. The other two B atoms have one H atom covalently bonded to each. One final B—B bond joins these last two B atoms, across the diameter of the ring.). See the diagram that follows:

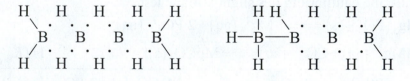

(c) C_4H_{10} contains a total of $4\times4+10\times1 = 26$ valence electrons or 13 pairs. A plausible Lewis structure follows. (Note that each atom possess an octet of electrons.)

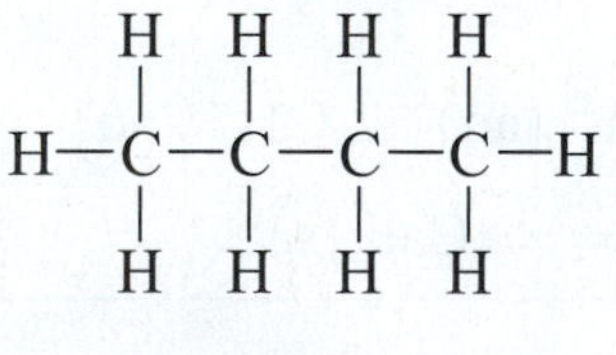

$$\begin{array}{ccccc} & H & H & H & H \\ & | & | & | & | \\ H- & C- & C- & C- & C-H \\ & | & | & | & | \\ & H & H & H & H \end{array}$$

25. **(M)**

(a) $2\,BBr_3(l) + 3\,H_2(g) \longrightarrow 2\,B(s) + 6\,HBr(g)$

(b) i) $B_2O_3(s) + 3\,C(s) \xrightarrow{\Delta} 3\,CO(g) + 2\,B(s)$

ii) $2\,B(s) + 3\,F_2(g) \xrightarrow{\Delta} 2\,BF_3(g)$

(c) $2\,B(s) + 3\,N_2O(g) \xrightarrow{\Delta} 3\,N_2(g) + B_2O_3(s)$

27. **(M) (a)** $2\ Al(s) + 6\ HCl(aq) \rightarrow 2\ AlCl_3(aq) + 3\ H_2(g)$

(b) $2\ NaOH(aq) + 2\ Al(s) + 6\ H_2O(l) \rightarrow 2\ Na^+(aq) + 2\left[Al(OH)_4\right]^-(aq) + 3\ H_2(g)$

(c) Oxidation: $\{Al(s) \longrightarrow Al^{3+}(aq) + 3\ e^-\}$ $\qquad \times 2$

Reduction: $\{SO_4^{\ 2-}(aq) + 4\ H^+(aq) + 2e^- \longrightarrow SO_2(g) + 2\ H_2O(l)\}$ $\qquad \times 3$

Net: $2\ Al(s) + 3\ SO_4^{\ 2-}(aq) + 12\ H^+(aq) \longrightarrow 2\ Al^{3+}(aq) + 3\ SO_2(aq) + 6\ H_2O(l)$

29. **(M)** One method of analyzing this reaction is to envision the HCO_3^- ion as a combination of CO_2 and OH^-. Then the OH^- reacts with Al^{3+} and forms $Al(OH)_3$. This method of envisioning HCO_3^- does have its basis in reality. After all,

$$H_2CO_3\,(= H_2O + CO_2) + OH^- \longrightarrow HCO_3^- + H_2O$$
$$Al^{3+}(aq) + 3\ HCO_3^-(aq) \longrightarrow Al(OH)_3(s) + 3\ CO_2(g)$$

Another method is to consider the reaction as, first, the hydrolysis of hydrated aluminum ion to produce $Al(OH)_3(s)$ and an acidic solution, followed by the reaction of the acid with bicarbonate ion.

$$\left[Al(H_2O)_6\right]^{3+}(aq) + 3\ H_2O(l) \longrightarrow Al(OH)_3(H_2O)_3(s) + 3\ H_3O^+(aq)$$

$$3\ H_3O^+(aq) + 3\ HCO_3^-(aq) \longrightarrow 6\ H_2O(l) + 3\ CO_2(g)$$

This gives the same net reaction:

$$\left[Al(H_2O)_6\right]^{3+}(aq) + 3\ HCO_3^-(aq) \longrightarrow Al(OH)_3(H_2O)_3(s) + 3\ CO_2(g) + 3\ H_2O(l)$$

31. **(M)** Aluminum and its oxide are soluble in both acid and base.

$$2\ Al(s) + 6\ H^+(aq) \longrightarrow 2\ Al^{3+}(aq) + 3\ H_2(g)$$
$$Al_2O_3(s) + 6\ H^+(aq) \longrightarrow 2\ Al^{3+}(aq) + 3\ H_2O(l)$$
$$2\ Al(s) + 2\ OH^-(aq) + 6\ H_2O(l) \longrightarrow 2\left[Al(OH)_4\right]^-(aq) + 3\ H_2(g)$$
$$Al_2O_3(s) + 2\ OH^-(aq) + 3\ H_2O(l) \longrightarrow 2\left[Al(OH)_4\right]^-(aq)$$

$Al(s)$ is resistant to corrosion only over the pH range 4.5 to 8.5. Thus, aluminum is inert only when the medium to which it is exposed is neither highly acidic nor highly basic.

33. **(M)** $CO_2(g)$ is, of course, the anhydride of an acid. The reaction here is an acid-base reaction.

$$\left[Al(OH)_4\right]^-(aq) + CO_2(aq) \longrightarrow Al(OH)_3(s) + HCO_3^-(aq)$$

$HCl(aq)$, being a strong acid, can't be used because it will dissolve the $Al(OH)_3(s)$.

35. (M) $2 \text{ KOH(aq)} + 2 \text{ Al(s)} + 6 \text{ H}_2\text{O(l)} \rightarrow 2 \text{ K[Al(OH)}_4\text{](aq)} + 3 \text{ H}_2\text{(g)}$

$2 \text{ K[Al(OH)}_4\text{](aq)} + 4 \text{ H}_2\text{SO}_4\text{(aq)} \rightarrow \text{K}_2\text{SO}_4\text{(aq)} + \text{Al}_2\text{(SO}_4\text{)}_3\text{(aq)} + 8 \text{ H}_2\text{O(l)}$
— crystallize→ $2 \text{ KAl(SO}_4\text{)}_2\text{(s)}$

37. (M) The chemical reaction for the disproportionation of $\text{FB(BF}_2\text{)}_2$ is
$3 \text{ FB(BF}_2\text{)}_2 \rightarrow \text{BF}_3 + \text{B}_8\text{F}_{12}$
The structure of B_8F_{12} is

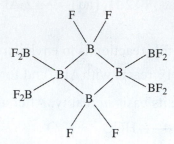

Group 14: The Carbon Family

39. (E) In the sense that diamonds react imperceptibly slowly at room temperature (either with oxygen to form carbon dioxide, or in its transformation to the more stable graphite), it is essentially true that "diamonds last forever." However, at elevated temperatures, diamond will burn to form $\text{CO}_2\text{(g)}$ and thus the statement is false. Also, the transformation C(diamond) → C(graphite) might occur more rapidly under other conditions. Eventually, of course, the conversion to graphite occurs.

41. (M)
(a) $3 \text{ SiO}_2\text{(s)} + 4 \text{ Al(s)} \xrightarrow{\Delta} 2 \text{ Al}_2\text{O}_3\text{(s)} + 3 \text{ Si(s)}$

(b) $\text{K}_2\text{CO}_3\text{(s)} + \text{SiO}_2\text{(s)} \xrightarrow{\Delta} \text{CO}_2\text{(g)} + \text{K}_2\text{SiO}_3\text{(s)}$

(c) $\text{Al}_4\text{C}_3\text{(s)} + 12 \text{ H}_2\text{O(l)} \longrightarrow 3 \text{ CH}_4\text{(g)} + 4\text{Al(OH)}_3\text{(s)}$

43. (M) A silane is a silicon-hydrogen compound, with the general formula $\text{Si}_n\text{H}_{2n+2}$. A silanol is a compound in which one or more of the hydrogens of silane is replaced by an —OH group. Then, the general formula becomes $\text{Si}_n\text{H}_{2n+1}\left(\text{OH}\right)$. In both of these classes of compounds, the number of silicon atoms, n, ranges from 1 to 6. Silicones are produced when silanols condense into chains, with the elimination of a water molecule between every two silanol molecules.

$\text{HO—Si}_n\text{H}_{2n}\text{—OH} + \text{HO—Si}_n\text{H}_{2n}\text{—OH} \longrightarrow \text{HO—Si}_n\text{H}_{2n}\text{—O—Si}_n\text{H}_{2n}\text{—OH} + \text{H}_2\text{O}$

45. (M)
(1) $2 \text{ CH}_4\left(\text{g}\right) + \text{S}_8\left(\text{g}\right) \longrightarrow 2 \text{ CS}_2\left(\text{g}\right) + 4 \text{ H}_2\text{S}\left(\text{g}\right)$

(2) $\text{CS}_2\left(\text{g}\right) + 3 \text{ Cl}_2\left(\text{g}\right) \longrightarrow \text{CCl}_4\left(\text{l}\right) + \text{S}_2\text{Cl}_2\left(\text{l}\right)$

(3) $4 \text{ CS}_2\left(\text{g}\right) + 8 \text{ S}_2\text{Cl}_2\left(\text{g}\right) \longrightarrow 4 \text{ CCl}_4\left(\text{l}\right) + 3 \text{ S}_8\left(\text{s}\right)$

47. **(D)** Muscovite or white mica has the formula $KAl_2(OH)_2(AlSi_3O_{10})$. Since they are not segregated into O_2 units in the formula, all of the oxygen atoms in the mineral must be in the −2 oxidation state. Potassium is obviously in the +1 oxidation state, as are the hydrogen atoms in the hydroxyl groups. Up to this point, we have -24 from the twelve oxygen atoms and +3 from the potassium and hydrogen atoms for a net number of -21 for the oxidation state. We still have three aluminum atoms and three silicon atoms to account for. In oxygen-rich salts such as mica, we would expect that the silicon and the aluminum atoms would be in their highest possible oxidation states, namely +4 and +3, respectively. Since the salt is neutral, the oxidation numbers for the silicon and aluminum atoms must add up to +21. This is precisely the total that is obtained if the silicon and aluminum atoms are in their highest possible oxidation states: $(3 \times (+3) + 3 \times (+4) = +21)$. Consequently, the empirical formula for white muscovite is consistent with the expected oxidation state for each element present.

49. **(M)**

(a) $PbO(s) + 2\ HNO_3(aq) \longrightarrow Pb(NO_3)_2(s) + H_2O(l)$

(b) $SnCO_3(s) \xrightarrow{\Delta} SnO(s) + CO_2(g)$

(c) $PbO(s) + C(s) \xrightarrow{\Delta} Pb(l) + CO(g)$

(d) $2\ Fe^{3+}(aq) + Sn^{2+}(aq) \longrightarrow 2\ Fe^{2+}(aq) + Sn^{4+}(aq)$

(e) $2\ PbS(s) + 3\ O_2(g) \xrightarrow{\Delta} 2\ PbO(s) + 2\ SO_2(g)$

$2\ SO_2(g) + O_2(g) \longrightarrow 2\ SO_3(g)$

$SO_3(g) + PbO(s) \longrightarrow PbSO_4(s)$

Or perhaps simply: $PbS(s) + 2O_2(g) \longrightarrow PbSO_4(s)$

Yet a third possibility:

$PbO(s) + SO_2(s) \longrightarrow PbSO_3(s)$, followed by $2\ PbSO_3(s) + O_2(s) \longrightarrow 2PbSO_4(s)$

51. **(M)** We start by using the Nernst equation to determine whether the cell voltage still is positive when the reaction has gone to completion.

(a) Oxidation: $Fe^{2+}(aq) \rightarrow Fe^{3+}(aq) + e^-\}\ \times 2$ $\qquad\qquad -E° = -0.771\ V$

Reduction: $PbO_2(s) + 4\ H^+(aq) + 2\ e^- \longrightarrow Pb^{2+}(aq) + 2\ H_2O(l)\ \ E° = +1.455\ V$

Net: $2\ Fe^{2+}(aq) + PbO_2(s) + 4\ H^+(aq) \longrightarrow 2\ Fe^{3+}(aq) + Pb^{2+}(aq) + 2\ H_2O(l)$

$E°_{cell} = -0.771\ V + 1.455\ V = +0.684\ V$

In this case, when the reaction has gone to completion,

$\left[Fe^{2+}\right] = 0.001\ M, \left[Fe^{3+}\right] = 0.999\ M$, and $\left[Pb^{2+}\right] = 0.500\ M$.

$E_{cell} = E°_{cell} - \dfrac{0.0257}{2} \ln \dfrac{[Fe^{3+}]^2[Pb^{2+}]}{[Fe^{2+}]^2} = 0.684\ V - \dfrac{0.0257}{2} \ln \dfrac{[0.999]^2[0.500]}{[0.001]^2} = 0.515\ V.$

Thus, this reaction will go to completion.

(b) Oxidation: $2\,SO_4^{2-}(aq) \longrightarrow S_2O_8^{2-}(aq) + 2\,e^-$ $\qquad\qquad -E^\circ = -2.01\,V$

Reduction: $PbO_2(s) + 4\,H^+(aq) + 2\,e^- \longrightarrow Pb^{2+}(aq) + 2\,H_2O(l)$ $\quad E^\circ = +1.455\,V$

$\overline{2\,SO_4^{2-}(aq) + PbO_2(s) + 4\,H^+(aq) \longrightarrow S_2O_8^{2-}(aq) + Pb^{2+}(aq) + 2\,H_2O(l)}$

$E^\circ_{cell} = -2.01 + 1.455 = -0.56\,V$ This reaction is not even spontaneous initially.

(c) Oxidation: $\{Mn^{2+}(1\times10^{-4}M) + 4\,H_2O(l) \longrightarrow MnO_4^-(aq) + 8\,H^+(aq) + 5\,e^-\}\times 2;\ -E^\circ = -1.51\,V$

Reduction: $\{PbO_2(s) + 4\,H^+(aq) + 2\,e^- \longrightarrow Pb^{2+}(aq) + 2\,H_2O(l)\}\times 5$ $\qquad E^\circ = +1.455\,V$

Net: $2\,Mn^{2+}(1\times10^{-4}M) + 5\,PbO_2(s) + 4\,H^+ \longrightarrow 2\,MnO_4^-(aq) + 5\,Pb^{2+}(aq) + 2\,H_2O(l)$

$E^\circ_{cell} = -1.51 + 1.455 = -0.06\,V$. The standard cell potential indicates that this reaction is not spontaneous when all concentrations are 1 M. Since the concentration of a reactant (Mn^{2+}) is lower than 1.00 M, this reaction is even less spontaneous than the standard cell potential indicates.

53. **(E)** As we move down a group, the lower oxidation state is generally favored. Thus, we expect $SnCl_2$, with tin in the +2 oxidation state, to be the product.

INTEGRATIVE AND ADVANCED EXERCISES

58. **(M)** The triiodide ion is linear (AX_2E_3). The Li^+ ion has a high charge density and significant polarizing power. The Li^+ ion will polarize the I_3^- to a significant extent and presumably assists the decomposition of I_3^- to I_2 and I^-.

59. **(M)** Use $\Delta_f G^\circ$ values to calculate $\Delta_r G^\circ$ for the reaction and then K. Because $K = P_{O_2}^{\frac{1}{2}}$, the value of K can then be used to calculate P_{O_2}.

$$Li_2O_2(s) \rightleftharpoons Li_2O(s) + \frac{1}{2}O_2(g)$$

$\Delta_f G^\circ\,[kJ/mol] \quad -419.02 \quad -466.40$

$\Delta_r G^\circ = -466.40 - (-419.02) = -47.38\,kJ\,mol^{-1}$

$\Delta_r G^\circ = -RT \ln K$

$-8.314\,J\,K^{-1}mol^{-1} \times 1000\,K \times \ln K = -47.38 \times 1000\,J\,mol^{-1}$

$\ln K = \dfrac{-47.38 \times 1000\,J\,mol^{-1}}{-8.314\,J\,K^{-1}mol^{-1} \times 1000\,K} = 5.70 \Rightarrow K = e^{5.70} = 298$

$K = P_{O_2}^{\frac{1}{2}} \Rightarrow P_{O_2} = K^2 = 298^2 = 8.8\times10^4$

62. **(D)**

(a) The mass of CaO(s) that would be produced can be determined with information from the balanced chemical equation for the oxidation of Ca(s).

$$2\ Ca(s) + O_2(g) \longrightarrow 2\ CaO(s)$$

$$mass\ CaO = 0.250\ g\ Ca \times \frac{1\ mol\ Ca}{40.0784\ g\ Ca} \times \frac{2\ mol\ CaO}{2\ mol\ Ca} \times \frac{56.0774\ g\ CaO}{1\ mol\ CaO}$$

$$= 0.350\ g\ CaO$$

(b) If the mass of product formed from the starting Ca differs from the 0.350 g CaO predicted, then the product could be a mixture of calcium nitride and calcium oxide. This scenario is not implausible since molecular nitrogen is readily available in the atmosphere. $3\ Ca(s) + N_2(g) \longrightarrow Ca_3N_2(s)$

$$mass\ Ca_3N_2 = 0.250\ g\ Ca \times \frac{1\ mol\ Ca}{40.0784\ g\ Ca} \times \frac{1\ mol\ Ca_3N_2}{3\ mol\ Ca} \times \frac{148.2480\ g\ Ca_3N_2}{1\ mol\ Ca_3N_2}$$

$$= 0.308\ g\ Ca_3N_2$$

We can use a technique similar to determining the percent abundance of an isotope to find the mass fraction of CaO in the product. Let f = the mass fraction of CaO in the product. Then, $(1.000 - f)$ is the mass fraction of Ca_3N_2 in the product. Finally,
product mass = mass CaO $\times f$ + mass $Ca_3N_2 \times (1.000 - f)$
0.325 g product = $0.350 \times f + 0.308 \times (1.000 - f)] = 0.308 + f(0.350 - 0.308)$

$$f = \frac{0.325 - 0.308}{0.350 - 0.308} = 0.40$$

Therefore, the product is 40% by mass CaO.

63. **(M)** Reaction (21.4) is $KCl(l) + Na(l) \xrightarrow{850\,°C} NaCl(l) + K(g)$ It is practical when the reactant element is a liquid and the product element is a gas at the same temperature.

(a) Since Li has a higher boiling point (1347 °C) than K (773.9 °C), a reaction similar to (21.4) is not a feasible way of producing Li metal from LiCl.

(b) On one hand, Cs has a lower boiling point (678.5 °C) than K (773.9 °C), and thus a reaction similar to (21.4) is a feasible method of producing Cs metal from CsCl. However, the ionization energy of Na is considerably larger than that of Cs, making it difficult to transfer an electron from Na to Cs^+.

64. **(M)**

(a) $2\ Al(s) + Fe_2O_3(s) \rightarrow 2\ Fe(s) + Al_2O_3(s)$ $\qquad$ $\Delta_r H° = -852\ kJ\ mol^{-1}$

(b) $4\ Al(s) + 3\ MnO_2(s) \rightarrow 2\ Al_2O_3(s) + 3\ Mn(s)$ $\qquad$ $\Delta_r H° = -1792\ kJ\ mol^{-1}$

(c) $2\ Al(s) + 3\ MgO(s) \rightarrow Al_2O_3(s) + 3\ Mg(s)$ $\qquad$ $\Delta_r H° = +129\ kJ\ mol^{-1}$

66. (M) The partial pressure of H_2 is 748 mmHg $-$ 21 mmHg $=$ 727 mmHg

$$\text{amount } H_2 = \frac{PV}{RT} = \frac{\left(727 \text{ mm Hg} \times (1 \text{ atm}/760 \text{ mmHg})\right) \times 0.104 \text{ L}}{0.08206 \text{ L atm K}^{-1} \text{mol}^{-1} \times 296 \text{ K}} = 4.10 \times 10^{-3} \text{ mol } H_2$$

The electrode reactions are the following.

Anode: $2 Cl^-(aq) \longrightarrow Cl_2(g) + 2 e^-$ Cathode: $2 H_2O(l) + 2 e^- \longrightarrow 2 OH^-(aq) + H_2(g)$

Thus 2 mol $OH^-(aq)$ are produced per mole of $H_2(g)$; 8.20×10^{-3} mol $OH^-(aq)$ are produced.

$$[OH^-] = \frac{8.20 \times 10^{-3} \text{ mol } OH^-}{0.250 \text{ L soln}} = 3.28 \times 10^{-2} \text{ M} = 0.0328 \text{ M}$$

Then we compute the ion product and compare its value to the value of K_{sp} for $Mg(OH)_2$.

$$Q_{sp} = [Mg^{2+}][OH^-]^2 = (0.220)(0.0328)^2 = 2.37 \times 10^{-4} > 1.8 \times 10^{-11} = K_{sp} \text{ for } Mg(OH)_2$$

Thus, $Mg(OH)_2$ should precipitate.

67. (D)

(a) We determine the amount of Ca^{2+} associated with each anion in 10^6 g of the water.

$$\text{amount } Ca^{2+} (SO_4^{2-}) = 56.9 \text{ g } SO_4^{2-} \times \frac{1 \text{ mol } SO_4^{2-}}{96.06 \text{ g } SO_4^{2-}} \times \frac{1 \text{ mol } Ca^{2+}}{1 \text{ mol } SO_4^{2-}} = 0.592 \text{ mol}$$

$$\text{amount } Ca^{2+} (HCO_3^-) = 176 \text{ g } HCO_3^- \times \frac{1 \text{ mol } HCO_3^-}{61.02 \text{ g } HCO_3^-} \times \frac{1 \text{ mol } Ca^{2+}}{2 \text{ mol } HCO_3^-} = 1.44 \text{ mol}$$

Then we determine the total mass of Ca^{2+}, numerically equal to the ppm Ca^{2+}.

$$\text{mass } Ca^{2+} = (0.592 + 1.44) \text{ mol } Ca^{2+} \times \frac{40.08 \text{ g } Ca^{2+}}{1 \text{ mol } Ca^{2+}} = 81.4 \text{ g } Ca^{2+} \longrightarrow 81.4 \text{ ppm } Ca^{2+}$$

(b) The reactions for the removal of $HCO_3^-(aq)$ begin with the formation of hydroxide ion resulting from dissolving $CaO(s)$. Hydroxide ion reacts with bicarbonate ion to form carbonate ion, which then combines with calcium ion to form the $CaCO_3(s)$ precipitate.

$$CaO(s) + H_2O(l) \longrightarrow Ca^{2+}(aq) + 2 OH^-(aq)$$

$$OH^-(aq) + HCO_3^-(aq) \longrightarrow H_2O(l) + CO_3^{2-}(aq)$$

$$Ca^{2+}(aq) + CO_3^{2-}(aq) \longrightarrow CaCO_3(s)$$

For each 1.000×10^6 g of water, we need to remove 176 g $HCO_3^-(aq)$ with the added $CaO(s)$.

$$\text{mass } CaO = 602 \times 10^3 \text{ g water} \times \frac{176 \text{ g } HCO_3^-}{1.000 \times 10^6 \text{ g water}} \times \frac{1 \text{ mol } HCO_3^-}{61.02 \text{ g } HCO_3^-} \times \frac{1 \text{ mol } OH^-}{1 \text{ mol } HCO_3^-}$$

$$\times \frac{1 \text{ mol } CaO}{2 \text{ mol } OH^-} \times \frac{56.08 \text{ g } CaO}{1 \text{ mol } CaO} = 48.7 \text{ g } CaO$$

(c) Here we determine the total amount of Ca^{2+} in 602 kg of water, <u>and</u> that added as CaO.

$$Ca^{2+} = \left(\frac{602 \text{ kg}}{1\times10^6 \text{ kg}}\right)(0.592+1.44) \text{ mol } Ca^{2+} + \left(48.7 \text{ g CaO}\times\frac{1 \text{ mol CaO}}{56.08 \text{ g CaO}}\times\frac{1 \text{ mol } Ca^{2+}}{1 \text{ mol CaO}}\right) = 2.09 \text{ mol } Ca^{2+}$$

The amount of Ca^{2+} that has precipitated equals the amount of HCO_3^- in solution, since each mole of HCO_3^- is transformed into 1 mole of CO_3^{2-}, which reacts with and then precipitates one mole of Ca^{2+}. Thus the amount of Ca^{2+} that has precipitated is $(0.602)(1.44) = 0.867$ mol Ca^{2+} as $CaCO_3(s)$. Then we can determine the concentration of Ca^{2+} remaining in solution.

$$[Ca^{2+}] = \frac{2.09 \text{ mol } Ca^{2+} \text{ total} - 0.867 \text{ mol } Ca^{2+}}{10^6 \text{ g water}}\times\frac{10^3 \text{ g water}}{1 \text{ L water}} = 2.03\times10^{-3} \text{ M}$$

To consider the Ca^{2+} "removed", its concentration should be decreased to 0.1% (0.001) of its initial value, or 2.03×10^{-6} M. We use the K_{sp} expression for $CaCO_3$ to determine the needed $[CO_3^{2-}]$

$$K_{sp} = [Ca^{2+}][CO_3^{2-}] = 2.8\times10^{-9} = 1.46\times10^{-6} \text{ M}$$

$$[CO_3^{2-}] = \frac{2.8\times10^{-9}}{2.03\times10^{-6}} = 0.0014 \text{ M}$$

This carbonate ion concentration can readily be achieved by adding solid Na_2CO_3.

(d) The amount of CO_3^{2-} needed is that which ends up in the precipitate plus that needed to attain the 0.0014 M concentration in the 602 kg = 602 L of water.

$$n_{CO_3^{-2}} = 602 \text{ L}\times\left(\left(\frac{0.00203 \text{ mol } Ca^{2+}}{1 \text{ L water}}\times\frac{1 \text{ mol } CO_3^{2-}}{1 \text{ mol } Ca^{2+}}\right) + \frac{0.0014 \text{ mol } CO_3^{2-}}{1 \text{ L water}}\right) = 2.1 \text{ mol } CO_3^{2-}$$

This CO_3^{2-} comes from the added $Na_2CO_3(s)$.

$$\text{mass } Na_2CO_3 = 2.0 \text{ mol } CO_3^{2-}\times\frac{1 \text{ mol } Na_2CO_3}{1 \text{ mol } CO_3^{2-}}\times\frac{105.99 \text{ g } Na_2CO_3}{1 \text{ mol } Na_2CO_3} = 2.2\times10^2 \text{ g } Na_2CO_3$$

68. (M)

(a) In the cell reaction, three moles of electrons are required to reduce each mole of Al^{3+}.

$$\text{mass Al} = 8.00 \text{ h}\times\frac{3600 \text{ s}}{1 \text{ h}}\times\frac{1.00\times10^5 \text{ C}}{1 \text{ s}}\times\frac{1 \text{ mol } e^-}{96,485 \text{ C}}\times\frac{1 \text{ mol Al}}{3 \text{ mol } e^-}\times\frac{26.98 \text{ g Al}}{1 \text{ mol Al}} = 2.68\times10^5 \text{ g Al}$$

The 38% efficiency is not considered in this calculation. All of the electrons produced must pass through the electrolytic cell; they simply require a higher than optimum voltage to do so, leading to resistance heating (which consumes some of the electrical energy).

(b) $$\text{total energy} = 4.5 \text{ V}\times 8.00 \text{ h}\times\frac{3600 \text{ s}}{1 \text{ h}}\times\frac{1.00\times10^5 \text{ C}}{1 \text{ s}}\times\frac{1 \text{ J}}{1 \text{ V}\cdot\text{C}}\times\frac{1 \text{ kJ}}{1000 \text{ J}} = 1.3\times10^7 \text{ kJ}$$

$$\text{mass coal} = 1.3\times10^7 \text{ kJ electricity}\times\frac{1 \text{ kJ heat}}{0.35 \text{ kJ electricity}}\times\frac{1 \text{ g C}}{32.8 \text{ kJ}}\times\frac{1 \text{ g coal}}{0.85 \text{ g C}}\times\frac{1 \text{ metric ton}}{10^6 \text{g}}$$

mass coal = 1.3 metric tons of coal

69. (M) We begin by rewriting Equation 21.25 for the electrolysis of $Al_2O_3(s)$ with $z = 12$ e⁻.

$$3\ C(s) + 2\ Al_2O_3(s) \rightarrow 4\ Al(s) + 3\ CO_2(g)$$

$$\Delta_r G° = 4(0\ kJ\ mol^{-1}) + 3(-394\ kJ\ mol^{-1}) - [3(0\ kJ\ mol^{-1}) + 2(-1520\ kJ\ mol^{-1})]$$
$$= +1858\ kJ\ mol^{-1}$$

$$\Delta_r G° = 1.858 \times 10^6\ J = -zFE° = -12(96,485\ C/mol)E°$$

$E° = -1.605$ V Note: this is just an estimate because $\Delta G°$ values are at 298 K, whereas the reaction occurs at a temperature that is much higher than 298 K.

If the oxidation of C(s) to $CO_2(g)$ did not occur, then the cell reaction would just be the reverse of the formation reaction of Al_2O_3 with n = 6 e⁻.

$$E° = \Delta E° = \frac{-\Delta G°}{zF} = \frac{1.520 \times 10^6\ J}{6\ mol\ e^- \times 96485\ C/mol\ e^-} = -2.626\ V$$

70. (M) It would not be unreasonable to predict that Raoult's law holds under these circumstances. This is predicated on the assumption, of course, that $Pb(NO_3)_2$ is completely ionized in aqueous solution.

$$P_{water} = x_{water}\ P°_{water} \qquad x_{water} = \frac{P_{water}}{P°_{water}} = 97\% = 0.97$$

Thus, there are 97 mol H_2O in every 100 mol solution. The remaining 3 moles are 1 mol Pb^{2+} and 2 mol NO_3^-. Thus there is 1 mol $Pb(NO_3)_2$ for every 97 mol H_2O. Compute the mass of $Pb(NO_3)_2$ in 100 g H_2O.

$$mass_{Pb(NO_3)_2} = 100.0\ g\ H_2O \times \frac{1\ mol\ H_2O}{18.02\ g\ H_2O} \times \frac{1\ mol\ Pb(NO_3)_2}{97\ mol\ H_2O} \times \frac{331.21\ g\ Pb(NO_3)_2}{1\ mol\ Pb(NO_3)_2} = 19\ g\ Pb(NO_3)_2$$

If we did not assume complete ionization of $Pb(NO_3)_2$, we would obtain 3 moles of un-ionized $Pb(NO_3)_2$ in solution for every 97 moles of H_2O. Then there would be 57 g $Pb(NO_3)_2$ dissolved in 100 g H_2O. A handbook gives the solubility as 56 g $Pb(NO_3)_2/100$ g H_2O, indicating only partial dissociation or, more probably, extensive re-association into ion pairs, triplets, quadruplets, etc.

73. (D) (a) 1.00 M NH_4Cl is a somewhat acidic solution due to hydrolysis of NH_4^+(aq); $MgCO_3$ should be most soluble in this solution. The remaining two solutions are buffer solutions, and the 0.100 M NH_3–1.00 M NH_4Cl buffer is more acidic of the two. In this solution, $MgCO_3$ has intermediate solubility. $MgCO_3$ is least soluble in the most alkaline solution, namely, 1.00 M NH_3–1.00 M NH_4Cl.

$$MgCO_3(s) \rightleftharpoons Mg^{2+}(aq) + CO_3^{2-}(aq) \qquad K_{sp} = 3.5 \times 10^{-8}$$

$$NH_4^+(aq) + OH^-(aq) \rightleftharpoons NH_3(aq) + H_2O(l) \qquad 1/K_b = 1/1.8 \times 10^{-5}$$

$$CO_3^{2-}(aq) + H_3O^+(aq) \rightleftharpoons HCO_3^-(aq) + H_2O(l) \qquad 1/K_2 = 1/4.7 \times 10^{-11}$$

$$2\ H_2O(l) \rightleftharpoons H_3O^+(aq) + OH^-(aq) \qquad K_w = 1.0 \times 10^{-14}$$

$$\overline{MgCO_3(s) + NH_4^+(aq) \rightleftharpoons Mg^{2+} + HCO_3^-(aq) + NH_3(aq)}$$

$$K = \frac{K_{sp} \times K_w}{K_b \times K_2} = \frac{3.5 \times 10^{-8} \times 1.0 \times 10^{-14}}{1.8 \times 10^{-5} \times 4.7 \times 10^{-11}} = 4.1 \times 10^{-7}$$

In 1.00 M NH₄Cl

Reaction:	MgCO$_3$(s)	+	NH$_4^+$ (aq)	$\rightleftharpoons$	Mg^{2+}(aq)	+	HCO$_3^-$ (aq)	+	NH$_3$(aq)
Initial:	–		1.00 M		0 M		0 M		0 M
Changes:	–		$-x$ M		$+x$ M		$+x$ M		$+x$ M
Equil:	–		$(1.00 - x)$M		x M		x M		x M

$$K = \frac{[\text{Mg}^{2+}][\text{HCO}_3^-][\text{NH}_3]}{[\text{NH}_4^+]} = \frac{x \cdot x \cdot x}{1.00 - x} \approx \frac{x^3}{1.00}$$

$$x = \sqrt[3]{4.1 \times 10^{-7}} = 7.4 \times 10^{-3} \text{ M} = \text{ molar solubility of MgCO}_3 \text{ in 1.00 M NH}_4\text{Cl}$$

($x \ll 1.00$ M, thus the approximation was valid)

(b) In 1.00 M NH₃–1.00 M NH₄Cl

Reaction:	MgCO$_3$(s)	+	NH$_4^+$(aq)	$\rightleftharpoons$	Mg^{2+}(aq)	+	HCO$_3^-$ (aq)	+	NH$_3$(aq)
Initial:	–		1.00 M		0 M		0 M		1.00 M
Changes:	–		$-x$ M		$+x$ M		$+x$ M		$+x$ M
Equil:	–		$(1.00 - x)$M		x M		$+x$M		$(1.00 + x)$M

$$K = \frac{[\text{Mg}^{2+}][\text{HCO}_3^-][\text{NH}_3]}{[\text{NH}_4^+]} = \frac{x \cdot x \cdot (1.00 + x)}{1.00 - x} \approx \frac{x^2 \cdot 1.00}{1.00}$$

$$x = \sqrt{4.1 \times 10^{-7}} = 6.4 \times 10^{-4} \text{ M} = \text{molar solubility of MgCO}_3 \text{ in 1.00 M NH}_3 - 1.00 \text{ M NH}_4\text{Cl}$$

($x \ll 1.00$ M, thus the approximation was valid)

(c) In 0.100 M NH₃–1.00 M NH₄Cl

Reaction:	MgCO$_3$(s)	+	NH$_4^+$(aq)	$\rightleftharpoons$	Mg^{2+}(aq)	+	HCO$_3^-$ (aq)	+	NH$_3$(aq)
Initial:	–		1.00 M		0 M		0 M		0.100 M
Changes:	–		$-x$ M		$+x$ M		$+x$ M		$+x$ M
Equil:	–		$(1.00 - x)$M		xM		$+x$M		$(0.100 + x)$M

$$K = \frac{[\text{Mg}^{2+}][\text{HCO}_3^-][\text{NH}_3]}{[\text{NH}_4^+]} = \frac{x \cdot x \cdot (0.100 + x)}{1.00 - x} \approx \frac{x^2 \times 0.100}{1.00}$$

$$x = \sqrt{\frac{4.1 \times 10^{-7}}{0.100}} = 2.0 \times 10^{-3} \text{ M} = \text{solubility of MgCO}_3 \text{ in 0.100 M NH}_3 - 1.00 \text{ M NH}_4\text{Cl}$$

(x is 2 % of 0.100 M, so the approximation was valid)

76. **(M)** We would expect MgO(s) to have a larger value of lattice energy than MgS(s) because of the smaller interionic distance in MgO(s).

Lattice energy of MgO:

Enthalpy of formation:	$Mg(s) + O_2(g) \longrightarrow MgO(s)$	$\Delta_f H° = -6.02 \times 10^2$ kJ mol^{-1}
Sublimation of Mg(s):	$Mg(s) \longrightarrow Mg(g)$	$\Delta_{sub} H° = +146$ kJ mol^{-1}
Ionization of Mg(g)	$Mg(g) \longrightarrow Mg^+(g) + e^-$	$\Delta I_1 = +737.7$ kJ mol^{-1}
Ionization of Mg(g):	$Mg^+(g) \longrightarrow Mg^{2+}(g) + e^-$	$\Delta I_2 = +1451$ kJ mol^{-1}
$\frac{1}{2}$ Dissociation O$_2$(g):	$\frac{1}{2} O_2(g) \longrightarrow O(g)$	$DE = \frac{1}{2} \times 497.4 = 248.7$ kJ mol^{-1}
O(g) electron affinity:	$O(g) + e^- \longrightarrow O^-(g)$	$EA_1 = -141$ kJ mol^{-1}
O(g) electron affinity:	$O^-(g) + e^- \longrightarrow O^{2-}(g)$	$EA_2 = +744$ kJ mol^{-1}
Lattice energy:	$Mg^{2+}(g) + O^{2-}(g) \rightarrow MgO(s)$	$L.E = -3789$ kJ mol^{-1} (see below)

L.E = –602 kJ mol^{-1} – 146 kJ mol^{-1} – 737.7 kJ mol^{-1} – 1451 kJ mol^{-1} – 248.7 kJ mol^{-1} + 141 kJ mol^{-1} – 744 kJ mol^{-1} = –3789 kJ mol^{-1}

Lattice energy of MgS:

Enthalpy of formation:	$Mg(s) + S(g) \longrightarrow MgS(s)$	$\Delta_f H° = -3.46 \times 10^2$ kJ mol^{-1}
Sublimation of Mg(s):	$Mg(s) \longrightarrow Mg(g)$	$\Delta_{sub} H = +146$ kJ mol^{-1}
Ionization of Mg(g)	$Mg(g) \longrightarrow Mg^+(g) + e^-$	$\Delta I_1 = +737.7$ kJ mol^{-1}
Ionization of Mg(g):	$Mg^+(g) \longrightarrow Mg^{2+}(g) + e^-$	$\Delta I_2 = +1451$ kJ mol^{-1}
$\frac{1}{2}$ Dissociation of S$_2$(g):	$\frac{1}{2} S_{rhombic}(g) \longrightarrow S(g)$	$DE = \frac{1}{2} \times 557.6 = 278.8$ kJ mol^{-1}
S(g) electron affinity:	$S(g) + e^- \longrightarrow S^-(g)$	$EA_1 = -200.4$ kJ mol^{-1}
S(g) electron affinity:	$S^-(g) + e^- \longrightarrow S^{2-}(g)$	$EA_2 = +456$ kJ mol^{-1}
Lattice energy:	$Mg^{2+}(g) + S^{2-}(g) \rightarrow MgS(s)$	$L.E = -3215$ kJ mol^{-1}

L.E = –346 kJ mol^{-1} – 146 kJ mol^{-1} – 737.7 kJ mol^{-1} – 1451 kJ mol^{-1} – 278.8 kJ mol^{-1} + 200.4 kJ mol^{-1} – 456 kJ mol^{-1} = –3215 kJ mol^{-1}

77. **(M)** In one unit cell, there are $\frac{1}{8} \times 8 + \frac{1}{2} \times 6 = 4$ fulleride ions and $\frac{1}{4} \times 12 + 1 + 8 = 12$ alkali metal ions. The ratio of cations to anions is 12:4 = 3:1, and so the fulleride ion is C_{60}^{3-} and the empirical formula is M_3C_{60}.

FEATURE PROBLEMS

78. **(D)** $Li(s) \rightarrow Li(g)$ 159.4 kJ mol^{-1}; (Using $\Delta_r H° = \Sigma\Delta_f H°_{products} - \Sigma\Delta_f H°_{reactants}$ in Appendix D)
$Li(g) \rightarrow Li^+(g) + e^-$ 520.2 kJ mol^{-1}; (Data given in Table 21.2, Chapter 21)
$\underline{Li^+(g) \rightarrow Li^+(aq) \quad\quad -506 \text{ kJ mol}^{-1};}$ (Provided in the question)
$Li(s) \rightarrow Li^+(aq) \quad\quad + e^- \quad 174 \text{ kJ mol}^{-1}$

$1/2\ H_2(g) \rightarrow H(g)$ 218.0 kJ mol^{-1} (Using $\Delta_r H° = \Sigma\Delta_f H°_{products} - \Sigma\Delta_f H°_{reactants}$ in Appendix D)
$H(g) \rightarrow H^+(g) + e^-$ 1312 kJ mol^{-1} (Use $R_H(N_A)$Bohr theory = 2.179×10^{-18} J(6.022×10^{23}))
$\underline{H^+(g) \rightarrow H^+(aq) \quad\quad -1079 \text{ kJ mol}^{-1}}$ (Provided in the question)
$1/2\ H_2(g) \rightarrow H^+(aq) + e^-$ 451 kJ mol^{-1}

(a) $Li(s) + H^+(aq) \rightarrow Li^+(aq) + 1/2\ H_2(g)$
$\Delta_r H° = 174 \text{ kJ mol}^{-1} - 451 \text{ kJ mol}^{-1} = -277 \text{ kJ mol}^{-1} \approx \Delta_r G°$

$$E_{ox}° = \frac{\Delta_r G°}{-zF} = \frac{-277 \times 10^3 \text{ J mol}^{-1}}{-1\left(96{,}485 \dfrac{C}{mol}\right)} = 2.87 \frac{J}{C} = 2.87 \text{ V}$$

In this reaction, Li(s) is being oxidized to Li$^+$(aq). If we wish to compare this to the reduction potential for Li$^+$(aq) being reduced to Li(s), we would have to reverse the reaction, which would result in the same answer only as a negative value. Alternatively, we can change the sign of the oxidation potential (Li $\rightarrow$ Li$^+$) to -2.87 V for the reduction potential for Li$^+ \rightarrow$ Li.

(it apears as -3.040 V in Appendix D; here, we see much better agreement.)

(b) $\Delta_r S = \Sigma S°_{products} - \Sigma S°_{reactants}$

$$= [1\left(13.4 \frac{J}{K \text{ mol}}\right) + 0.5\left(130.7 \frac{J}{K \text{ mol}}\right)] -$$

$$[1\left(29.12 \frac{J}{K \text{ mol}}\right) + 1\left(0 \frac{J}{K \text{ mol}}\right)] = 49.6 \frac{J}{K \text{ mol}}$$

$$\Delta_r G° = \Delta_r H° - T\Delta_r S° = -277 \text{ kJ mol}^{-1} - 298.15 \text{ K}\left(49.6 \frac{J}{K \text{ mol}}\right) \times \frac{1 \text{ kJ}}{1000 \text{ J}} =$$

$$-292 \times 10^3 \text{ J mol}^{-1}$$

$$E_{ox}° = \frac{\Delta_r G°}{-zF} = \frac{-292 \times 10^3 \text{ J mol}^{-1}}{-1\left(96{,}485 \dfrac{C}{mol}\right)} = 3.03 \frac{J}{C} = 3.03 \text{ V}$$

As mentioned in part (a) of this question, we have calculated the oxidation potential for the half reaction $Li(s) \rightarrow Li^+(aq) + e^-$. The reduction potential is the reverse half-reaction and has a potential of -3.03 V. This is in excellent agreement with -3.040 V given in Appendix D.

SELF-ASSESSMENT EXERCISES

83. **(E) (c)** MgO The reason for MgO having the highest melting point is that MgO is ionic with a +2 on magnesium and a –2 on the oxygen.

84. **(E) (f)** PbO_2 Lead(IV) easily undergoes reduction to lead(II). Much more so than the reduction of the other elements, e.g., Li^+, Mg^{2+}, Al^{3+}, C^{4+}, and Sn^{4+}.

85. **(M) (a)** $BCl_3NH_3(g)$ (an adduct); **(b)** $KO_2(s)$; **(c)** $Li_2O(s)$; **(d)** $Ba(OH)_2(aq)$ and $H_2O_2(aq)$; $H_2O_2(aq)$ slowly disproportionates into $H_2O(l)$ and $O_2(g)$.

86. **(M)** The thermite reaction is evidence that aluminum will readily extract oxygen from Fe_2O_3. Aluminum can be used for making products that last and for structural purposes, because aluminum develops a coating of Al_2O_3 that protects the metal beneath it.

87. **(M) (b)** $Ca(s)$ and $CaH_2(s)$

88. **(M)**

(a) $Li_2CO_3(s) \xrightarrow{\Delta} Li_2O(s) + CO_2(g)$

(b) $CaCO_3(s) + 2\,HCl(aq) \longrightarrow CaCl_2(aq) + H_2O(l) + CO_2(g)$

(c) $2\,Al(s) + 2\,NaOH(aq) + 6H_2O(l) \longrightarrow 2\,Na[Al(OH)_4](aq) + 2\,H_2(g)$

(d) $BaO(s) + H_2O(l) \longrightarrow Ba(OH)_2(aq,\ limited\ solubility)$

(e) $2\,Na_2O_2(s) + 2\,CO_2(g) \longrightarrow 2\,Na_2CO_3(s) + O_2(g)$

89. **(M)**

(a) $MgCO_3(s) + 2\,HCl(aq) \longrightarrow MgCl_2(aq) + H_2O(l) + CO_2(g)$

(b) $2\,Na(s) + 2\,H_2O(l) \longrightarrow 2\,NaOH(aq) + H_2(g)$ followed by the reaction in Exercise 82(c)

(c) $2\,NaCl(s) + H_2SO_4(concd.,\ aq) \xrightarrow{\Delta} Na_2SO_4(s) + 2\,HCl(g)$

90. **(M)**

(a) $K_2CO_3(aq) + Ba(OH)_2(aq) \longrightarrow BaCO_3(s) + 2\,KOH(aq)$

(b) $Mg(HCO_3)_2(aq) \xrightarrow{\Delta} MgCO_3(s) + H_2O(l) + CO_2(g)$

(c) $SnO(s) + C(s) \xrightarrow{\Delta} Sn(l) + CO(g)$

(d) $CaF_2(s) + H_2SO_4(concd.\ aq) \xrightarrow{\Delta} CaSO_4(s) + 2\,HF(g)$

(e) $NaHCO_3(s) + HCl(aq) \longrightarrow NaCl(aq) + H_2O(l) + CO_2(g)$

(f) $PbO_2(s) + 4\,HBr(aq) \longrightarrow PbBr_2(s) + Br_2(l) + 2\,H_2O(l)$

(g) $SiF_4(g) + 4\,Na(l) \longrightarrow Si(s) + 4\,NaF(s)$

91. **(M)** $CaSO_4 \cdot 2\,H_2O(s) + (NH_4)_2CO_3(aq) \longrightarrow CaCO_3(s) + (NH_4)_2SO_4(aq) + 2\,H_2O(l)$.
The reaction proceeds to the right because K_{sp} for $CaCO_3 < K_{sp}$ for $CaSO_4$

92. **(M)**
(a) $2\,B(OH)_3(s) \longrightarrow B_2O_3(s) + 3\,H_2O(g)$
(b) no reaction
(c) $CaSO_4 \cdot 2\,H_2O(s)[\text{gypsum}] \overset{\Delta}{\longrightarrow} CaSO_4 \times \dfrac{1}{2}H_2O(s)[\text{plaster of Paris}] + \dfrac{3}{2}H_2O(g)$

93. **(M)**
(a) $Pb(NO_3)_2(aq) + 2\,NaHCO_3(aq) \longrightarrow PbCO_3(s) + 2\,NaNO_3(aq) + H_2O(l) + CO_2(g)$
In $NaHCO_3(aq)$ the concentration of $CO_3^{2-}(aq)$ is high enough that K_{sp} of $PbCO_3$ is exceeded.
(b) $Li_2O(s) + (NH_4)_2CO_3(aq) \longrightarrow Li_2CO_3(s) + 2\,NH_3(g) + H_2O(l)$. $Li_2O(s)$ is the anhydride of the strong base LiOH, which reacts with acidic $NH_4^+(aq)$ to liberate $NH_3(g)$.
$Li_2CO_3(s)$ is only slightly soluble, so most of it precipitates.
(c) $H_2SO_4(aq) + BaO_2(aq) \longrightarrow H_2O_2(aq) + BaSO_4(s)$. The forward reaction is favored by the formation of a precipitate, $BaSO_4(s)$.
(d) $2\,PbO(s) + Ca(OCl)_2(aq) \longrightarrow CaCl_2(aq) + 2\,PbO_2(s)$. Hypochlorite ion oxidizes lead(II) oxide to lead(IV) oxide.

94. **(M)**
(a) Stalactites are primarily $CaCO_3(s)$.
(b) Gypsum is $CaSO_4 \cdot 2\,H_2O(s)$.
(c) Suspension of $BaSO_4(s)$ in water.
(d) $Al_2O_3(s)$ with Fe^{3+} and Ti^{4+} replacing some Al^{3+} in the crystal structure.

95. **(M)** Dolomite, molar mass of 184.4 g/mol yields 2 mol $CO_2(g)$ per mol of dolomite on decomposition. The 5.00×10^3 kg sample yields 5.42×10^4 mol CO_2 which, under the stated conditions occupies a volume of 1.27×10^3 m^3.

CHAPTER 22

CHEMISTRY OF THE MAIN GROUP ELEMENTS II: GROUPS 18, 17, 16, 15, AND HYDROGEN

PRACTICE EXAMPLES

1A **(D)** This question involves calculating $E°$ for the reduction half-reaction:

$$ClO_3^-(aq) + 3\,H_2O(l) + 6\,e^- \rightarrow Cl^-(aq) + 6\,OH^-(aq)$$

Here we will consider just one of the several approaches available to solve this problem. The four half-reactions (and their associated $E°$ values) that are used in this method to come up with the "missing $E°$ value" are given below. (Note: the $E°$ for the first reaction was determined in Example 22-1)

1) $ClO_3^-(aq) + H_2O(l) + 2\,e^- \rightarrow ClO_2^-(aq) + 2\,OH^-(aq)$ $E° = 0.295$ V $\Delta_rG°_1 = -2FE°$

2) $ClO_2^-(aq) + H_2O(l) + 2\,e^- \rightarrow OCl^-(aq) + 2\,OH^-(aq)$ $E° = 0.681$ V $\Delta_rG°_2 = -2FE°$

3) $OCl^-(aq) + H_2O(l) + 1\,e^- \rightarrow 1/2\,Cl_2(aq) + 2\,OH^-(aq)$ $E° = 0.421$ V $\Delta_rG°_3 = -1FE°$

4) $1/2\,Cl_2(aq) + 1\,e^- \rightarrow Cl^-(aq)$ $E° = 1.358$ V $\Delta_rG°_4 = -1FE°$

Although the reactions themselves may be added to obtain the desired equation, the $E°$ for this equation is not the sum of the $E°$ values for the above four reactions. The $E°$ value for the desired equation is actually the weighted average of the $E°$ values for reactions (1) to (4). It can be calculated by summing up the free energy changes for the four reactions. (The standard voltages for half-reactions of the same type are not additive. The $\Delta_rG°$ values for these reactions can, however, be summed together.)

When (1), (2), (3) and (4) are added together we obtain

1)	$ClO_3^-(aq) + H_2O(l) + 2\,e^- \rightarrow ClO_2^-(aq) + 2\,OH^-(aq)$	$E°_1 = 0.295$ V
+ 2)	$ClO_2^-(aq) + H_2O(l) + 2\,e^- \rightarrow OCl^-(aq) + 2\,OH^-(aq)$	$E°_2 = 0.681$ V
+ 3)	$OCl^-(aq) + H_2O(l) + 1\,e^- \rightarrow 1/2\,Cl_2(aq) + 2\,OH^-(aq)$	$E°_3 = 0.421$ V
+ 4)	$1/2\,Cl_2(aq) + 1\,e^- \rightarrow Cl^-(aq)$	$E°_4 = 1.358$ V
	$ClO_3^-(aq) + 3\,H_2O(l) + 6\,e^- \rightarrow Cl^-(aq) + 6\,OH^-(aq)$	$E°_5 = ?$

and $\Delta G°_5 = \Delta G°_1 + \Delta G°_2 + \Delta G°_3 + \Delta G°_4$

so, $-6FE°_5 = -2F(0.295\text{ V}) + -2F(0.681\text{V}) + -1F(0.421\text{ V}) + -1F(1.358\text{ V})$

Hence, $E°_5 = \dfrac{-2F(0.295\text{ V}) + -2F(0.681\text{V}) + -1F(0.421\text{ V}) + -1F(1.358\text{ V})}{-6\,F} = 0.622$ V

1B **(D)** This question involves calculating $E°$ for the reduction half-reaction:

$$2\ ClO_3^-(aq) + 12\ H^+(aq) + 10\ e^- \rightarrow Cl_2(aq) + 6\ H_2O(l)$$

The three half-reactions (and their associated $E°$ values) are given below:

(1) $\quad ClO_3^-(aq) + 3\ H^+(aq) + 2\ e^- \rightarrow HClO_2(aq) + H_2O(l)\quad E° = 1.181\ V$

(2) $\quad HClO_2(aq) + 2\ H^+(aq) + 2\ e^- \rightarrow HClO(aq) + H_2O(l)\quad E° = 1.645\ V$

(3) $\quad 2\ HClO\ (aq) + 2\ H^+(aq) + 2\ e^- \rightarrow Cl_2(aq) + 2\ H_2O(l)\quad E° = 1.611\ V$
$\quad$ (remember that $\Delta_r G° = -zFE°$)

Although the reactions themselves can be added to obtain the desired equation, the $E°$ for this equation is not the sum of the $E°$ values for the above three reactions. The $E°$ for the desired equation is actually the weighted average of the $E°$ values for reactions (1) to (3). It can be obtained by summing up the free energy changes for the three reactions. (For reactions of the same type, standard voltages are not additive; $\Delta G°$ values are additive, however.) When (1) and (2) (each multiplied by two) are added to (3) we obtain:

$2 \times$ (1) $\ 2\ ClO_3^-(aq) + 6\ H^+(aq) + 4\ e^- \rightarrow 2\ HClO_2(aq) + 2\ H_2O(l)\qquad E°_1 = 1.181\ V$
$2 \times$ (2) $\ 2\ HClO_2(aq) + 4\ H^+(aq) + 4\ e^- \rightarrow 2\ HClO(aq) + 2\ H_2O(l)\qquad E°_2 = 1.645\ V$
(3) $\qquad 2\ HClO\ (aq) + 2\ H^+(aq) + 2\ e^- \rightarrow Cl_2(aq) + 2\ H_2O(l)\qquad E°_3 = 1.611\ V$

$$2\ ClO_3^-(aq) + 12\ H^+(aq) + 10\ e^- \rightarrow Cl_2(aq) + 6\ H_2O(l)\qquad E°_4 = ?$$

and $\quad \Delta G°_4 = \Delta G°_1 + \Delta G°_2 + \Delta G°_3$

so, $\quad -10FE°_4 = -4F(1.181\ V) + -4F(1.645\ V) + -2F(1.611\ V)$

Hence, $E°_4 = \dfrac{-4F(1.181\ V) + -4F(1.645\ V) + -2F(1.611\ V)}{-10\ F} = 1.453\ V$

2A **(M)** The dissociation reaction is the reverse of the formation reaction, and thus $\Delta_r G°$ for the dissociation reaction is the negative of $\Delta_f G°$

$$HF(g) \longrightarrow \tfrac{1}{2}H_2(g) + \tfrac{1}{2}F_2(g)\qquad \Delta_r G° = -\Delta_f G° = -(-273.2\ kJ/mol)$$

We know that

$$\Delta_r G° = -RT \ln K_p \qquad +273.2 \times 10^3\ J/mol = -8.3145\ J\ mol^{-1}\ K^{-1} \times 298\ K \times \ln K_p$$

$$\ln K_p = \frac{273.2 \times 10^3\ J\ mol^{-1}}{-8.3145\ J\ mol^{-1}\ K^{-1} \times 298\ K} = -110\ K_p = e^{-110} = 1.7 \times 10^{-48} \approx 2 \times 10^{-48}$$

Virtually no dissociation of $HF(g)$ into its elements occurs.

2B **(M)** The dissociation reaction with all integer coefficients is twice the reverse of the formation reaction.

$$2\,HCl(g) \rightleftharpoons H_2(g) + Cl_2(g) \qquad \Delta_r G° = -2 \times (-95.30\ kJ/mol) = +190.6\ kJ/mol = -RT \ln K_p$$

$$\ln K_p = \frac{-\Delta_r G°}{RT} = \frac{-190.6 \times 10^3\ J/mol}{8.3145\ J\ mol^{-1}\ K^{-1} \times 298\ K} = -76.9 \qquad K_p = e^{-76.9} = 4 \times 10^{-34}$$

We assume an initial HCl(g) pressure of P atm, and calculate the final pressure of $Cl_2(g)$ and $H_2(g)$, x atm.

Reaction:	$2\,HCl(g)$	$\rightleftharpoons$	$H_2(g)$	$+$	$Cl_2(g)$
Initial:	P atm		0 atm		0 atm
Changes:	$-2x$ atm		$+x$ atm		$+x$ atm
Equil:	$(P - 2x)$ atm		x atm		x atm

$$K_p = \frac{P\{H_2(g)\}P\{Cl_2(g)\}}{P\{HCl(g)\}^2} = \frac{x \cdot x}{(P - 2x)^2} = \left(\frac{x}{P - 2x}\right)^2 \qquad \frac{x}{P - 2x} = \sqrt{4 \times 10^{-34}} = 2 \times 10^{-17}$$

$$x = 2 \times 10^{-17}(P - 2x) \approx 2 \times 10^{-17}\,P$$

$$\%\ decomposition = \frac{2x}{P} \times 100\% = 2 \times 2 \times 10^{-17} \times 100\% = 4 \times 10^{-15}\ \%\ decomposed$$

INTEGRATIVE EXAMPLES

A **(M)** The half-reactions are:

$NO_2^- + 2\,OH^- \rightarrow NO_3^- + H_2O + 2\,e^-$	$-E° = 0.04\ V$
$\{NO_2^- + H_2O + e^- \rightarrow NO + 2\,OH^-\} \times 2$	$E° = -0.46\ V$
$3\,NO_2^- + H_2O \rightarrow NO_3^- + 2\,NO + 2\,OH^-$	$E°_{cell} = -0.42\ V$

$E°$ for the first reaction (oxidation) can be calculated by using the approach in Example 22-1. The calculated value is very small, and so the $E°$ for the disproportionation reaction is negative. The disproportionation of NO_2^- to NO_3^- and NO is therefore not spontaneous.

B **(M)** The half-reactions are:

$HNO_2 + H_2O \rightarrow NO_3^- + 3\,H^+ + 2\,e^-$	$-E° = -0.934\ V$
$\{HNO_2 + H^+ + e^- \rightarrow NO + H_2O\} \times 2$	$E° = 0.996\ V$
$3\,HNO_2 \rightarrow NO_3^- + H^+ + H_2O + 2\,NO$	$E°_{cell} = 0.062\ V$

$E°$ for the first reaction (oxidation) can be calculated by using the approach in Example 22-1. $E°_{cell}$ for the reaction is positive and therefore disproportionation of HNO_2 to NO_3^- and NO is spontaneous under standard conditions.

EXERCISES

Periodic Trends in Bonding and Acid-Base Character of Oxides

1. **(E)** LiF, BeF_2, BF_3, CF_4, NF_3, OF_2. LiF is an ionic compound, BeF_2 is a network covalent compound, and the others are molecular covalent compounds.

3. **(E)** The metallic character of the elements increases as we move down a group, and so too does the basic character of the oxides. Bi_2O_3 is most basic, P_4O_6 is least basic (and is actually acidic), and Sb_4O_6 is amphoteric.

The Noble Gases

5. **(M)** First we use the ideal gas law to determine the amount in moles of argon.

$$n = \frac{PV}{RT} = \frac{145 \text{ atm} \times 55 \text{ L}}{0.08206 \text{ L atm mol}^{-1} \text{ K}^{-1} \times 299 \text{ K}} = 3.2\underline{5} \times 10^2 \text{ mol Ar}$$

$$\text{L air} = 3.2\underline{5} \times 10^2 \text{ mol Ar} \times \frac{22.711 \text{ L Ar STP}}{1 \text{ mol Ar}} \times \frac{100.000 \text{ L air}}{0.934 \text{ L Ar}} = 7.9 \times 10^5 \text{ L air}$$

7. **(M) (a)**

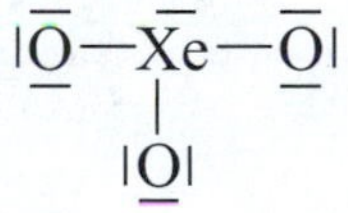

(The Lewis structures for XeO₃, XeO₄, and XeF₅⁺ appear here)

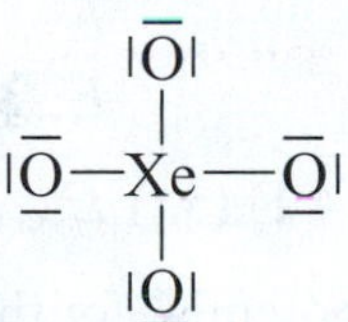

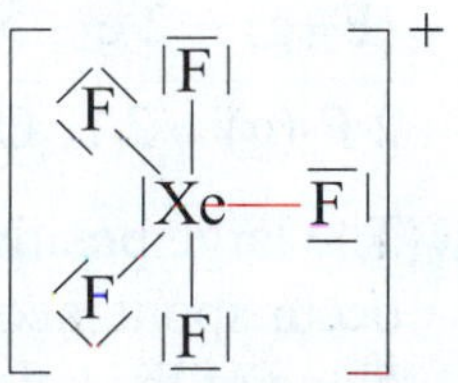

(a) The Lewis structure has three ligands and one lone pair on Xe. XeO_3 has a trigonal pyramidal shape.

(b) The Lewis structure has four ligands on Xe. XeO_4 has a tetrahedral shape.

(c) There are five ligands and one lone pair on Xe in XeF_5^+. Its shape is square pyramidal.

9. **(E)** $3 \text{ XeF}_4(aq) + 6 \text{ H}_2\text{O}(l) \rightarrow 2 \text{ Xe}(g) + 3/2 \text{ O}_2(g) + 12 \text{ HF}(g) + \text{XeO}_3(s)$

11. **(E)** For these noble gases, the bond energy of the noble gas fluorine is too small to offset the energy required to break the F—F bond.

The Halogens

13. **(M)** Iodide ion is slowly oxidized to iodine, which is yellow-brown in aqueous solution, by oxygen in the air.

Oxidation: $\{2\,I^-(aq) \longrightarrow I_2(aq) + 2\,e^-\} \times 2$ $\qquad -E° = -0.535\,V$

Reduction: $O_2(g) + 4\,H^+(aq) + 4\,e^- \longrightarrow 2\,H_2O(l)$ $\qquad E° = +1.229\,V$

Net: $\qquad 4\,I^-(aq) + O_2(g) + 4\,H^+(aq) \longrightarrow 2\,I_2(aq) + 2\,H_2O(l)$ $\qquad E°_{cell} = +0.694\,V$

Possibly followed by: $I_2(aq) + I^-(aq) \longrightarrow I_3^-(aq)$

15. **(M)** Displacement reactions involve one element displacing another element from solution. The element that dissolves in the solution is more *"active"* than the element supplanted from solution. Within the halogen group, the activity decreases from top to bottom. Thus, each halogen is able to displace the members of the group below it, but not those above it. For instance, molecular bromine can oxidize aqueous iodide ion but molecular iodine is incapable of oxidizing bromide ion:

$Br_2(aq) + 2\,I^-(aq) \rightarrow 2\,Br^-(aq) + I_2(aq)$ however, $I_2(aq) + 2\,Br^-(aq) \rightarrow$ NO RXN

The only halogen with sufficient oxidizing power to displace $O_2(g)$ from water is $F_2(g)$:

$2\,H_2O(l) \rightarrow 4\,H^+ + O_2(g) + 4\,e^-$ $\qquad E°_{ox} = -1.229\,V$

$\{F_2(g) + 2\,e^- \rightarrow 2\,F^-(aq)\} \times 2$ $\qquad E°_{red} = 2.866\,V$

$2\,F_2(g) + 2\,H_2O(l) \rightarrow 4\,H^+ + O_2(g) + 4\,F^-(aq)$ $\quad E°_{cell} = 1.637\,V$

The large positive standard reduction potential for this reaction indicates that the reaction will occur spontaneously, with products being strongly preferred under standard state conditions. None of the halogens reacts with water to form $H_2(g)$. In order to displace molecular hydrogen from water, one must add a strong reducing agent, such as sodium metal.

17. **(M)**

(a) $\qquad \text{mass } F_2 = 1\,km^3 \times \left(\dfrac{1000\,m}{1\,km} \times \dfrac{100\,cm}{1\,m}\right)^3 \times \dfrac{1.03\,g}{1\,cm^3} \times \dfrac{1\,lb}{454\,g} \times \dfrac{1\,ton}{2000\,lb} \times \dfrac{1\,g\,F^-}{1\,ton} \times \dfrac{37.996\,g\,F_2}{18.998\,g\,F^-}$

$\qquad\qquad = 2 \times 10^9\,g\,F_2 = 2 \times 10^6\,kg\,F_2$

(b) Bromine can be extracted by displacing it from solution with $Cl_2(g)$. Since there is no chemical oxidizing agent that is stronger than $F_2(g)$, this method of displacement would not work for $F_2(g)$. Even if there were a chemical oxidizing agent stronger than $F_2(g)$, it would displace $O_2(g)$ before it displaced $F_2(g)$. Obtaining $F_2(g)$ would require electrolysis of one of its molten salts, obtained from seawater evaporate.

19. **(M)** In order for the disproportionation reaction to occur under standard conditions, the $E°$ for the overall reaction must be greater than zero. To answer this question, we must refer to the Standard Electrode Potential Diagrams (also known as Latimer diagrams) provided in Figure 22-4 and the answer to Practice Example 22-1B.

(i)   Reduction half reaction (acidic solution)
$Cl_2(aq) + 2\ e^- \rightarrow 2\ Cl^-(aq)\quad E°_{red} = 1.358\ V$

(ii)   Oxidation half reaction (acidic solution)
$Cl_2(aq) + 6\ H_2O(l) \rightarrow 2\ ClO_3^-(aq) + 12\ H^+(aq) + 10\ e^-\quad E°_{ox} = -1.453\ V$

Combining (i) $\times$ 5 with (ii) $\times$ 1, we obtain the desired disproportionation reaction:
$6\ Cl_2(aq) + 6\ H_2O(l) \rightarrow 2\ ClO_3^-(aq) + 12\ H^+(aq) + 10\ Cl^-(aq)\qquad E°_{cell} = -0.095\ V$

Since the final cell voltage is negative, the disproportionation reaction will not occur spontaneously under standard conditions. Alternatively, we can calculate K by using $\ln K = -\Delta_r G°/RT$ and $\Delta_r G° = -zFE°$. This method gives a $K = 8.6 \times 10^{-17}$. Clearly, the reaction will not go to completion under standard conditions.

21. **(M)** First we must draw the Lewis structures for all of the species listed. Following this, we will deduce their electron-group geometries and molecular shapes following the VSEPR approach.

(a) BrF_3:   28 valence electrons
VSEPR class: AX_3E_2
Thus BrF_3 is T-shaped

(b) IF_5 42 valence electrons
VSEPR class: AX_5E
Thus IF_5 is square pyramidal

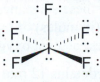

(c) Cl_3IF^- 36 valence electrons
VSEPR class: AX_4E_2
Thus Cl_3IF^- is square planar

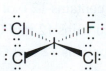

23. **(M)**
(a)   The half-reactions are:
Oxidation: $3\ I^- \rightarrow I_3^- + 2\ e^-$

Reduction: $I_2(aq) + 2\ e^- \rightarrow 2\ I^-(aq)$

Overall: $I_2(aq) + I^-(aq) \rightarrow I_3^-(aq)$

From the known value of the equilibrium constant at 25 °C, we can calculate $\Delta_r G°$ and consequently $E°$:

$$\Delta_r G° = -RT \ln K = -zFE°$$

$$-8.314 \text{ JK}^{-1}\text{mol}^{-1} \times 298.15 \text{ K} \ln K = -2 \times 96485 \text{ Cmol}^{-1} \times E°$$

$$E° = \frac{-8.314 \text{ JK}^{-1}\text{mol}^{-1} \times 298.15 \text{ K} \ln 7.7 \times 10^2}{-2 \times 96485 \text{ C mol}^{-1}} = 0.085 \text{ V}$$

(b) Set up the ICE table and solve for the equilibrium concentrations:

$$I_2(aq) + I^-(aq) \rightleftharpoons I_3^-(aq)$$

| 0.0010M | 0.0050M | 0 |
| 0.0010 – x | 0.0050 – x | x |

$$K = 7.7 \times 10^2 = \frac{x}{(0.0010 - x) \times (0.0050 - x)} = \frac{x}{5.0 \times 10^{-6} - 0.0060x + x^2}$$

$$7.7 \times 10^2 \times (5.0 \times 10^{-6} - 0.0060x + x^2) = x$$

$$7.7 \times 10^2 x^2 - 5.62x + 3.85 \times 10^{-3} = 0$$

Solving the quadratic equation gives $x = 7.65 \times 10^{-3}$ M.

Oxygen

25. **(E)(a)** $2 \text{ HgO(s)} \xrightarrow{\Delta} 2 \text{ Hg(l)} + O_2(g)$ **(b)** $2 \text{ KClO}_4(s) \xrightarrow{\Delta} 2 \text{ KClO}_3(s) + O_2(g)$

27. **(M)** We first write the formulas of the four substances: N_2O_4, Al_2O_3, P_4O_6, CO_2. The one constant in all these substances is oxygen. If we compare amounts of substance with the same amount (in moles) of oxygen, the one with the smallest mass of the other element will have the highest percent oxygen.

3 mol N_2O_4 contains 12 mol O and 6 mol N: $6 \times 14.0 = 84.0$ g N
4 mol Al_2O_3 contains 12 mol O and 8 mol Al: $8 \times 27.0 = 216$ g Al
2 mol P_4O_6 contains 12 mol O and 8 mol P: $8 \times 31.0 = 248$ g P
6 mol CO_2 contains 12 mol O and 6 mol C: $6 \times 12.0 = 72.0$ g C

Thus, of the oxides listed, CO_2 contains the largest percent oxygen by mass.

29. **(M)** Recall that fraction by volume and fraction by pressure are numerically equal. Additionally, one atmosphere pressure is equivalent to 760 mmHg. We combine these two facts.

$$P\{O_3\} = 760 \text{ mmHg} \times \frac{0.04 \text{ mmHg of } O_3}{10^6 \text{ mmHg of atmosphere}} = 3 \times 10^{-5} \text{ mmHg}$$

31. **(M)** The electrolysis reaction is $2\,H_2O(l) \xrightarrow{\text{electrolysis}} 2\,H_2(g) + O_2(g)$. In this reaction, 2 moles of $H_2(g)$ are produced for each mole of $O_2(g)$. By the law of combining volumes, we would expect the volume of hydrogen to be twice the volume of oxygen produced. (Actually the volumes are not exactly in the ratio of 2:1 because of the different solubilities of oxygen and hydrogen in water.)

33. **(D)** Since the pK_a for H_2O_2 had been provided to us, we can find the solution pH simply by solving an I.C.E. table for the hydrolysis of a 3.0 % H_2O_2 solution (by mass). Of course, in order to use this method, the mass percent must first be converted to molarity. We must assume that the density of the solution is 1.0 g mL^{-1}.

$$[H_2O_2] = \frac{3.0\ \text{g}\ H_2O_2}{100\ \text{g}\ \text{solution}} \times \frac{1\,\text{g solution}}{1\,\text{mL solution}} \times \frac{1\ \text{mol}\ H_2O_2}{34.015\ \text{g}\ H_2O_2} \times \frac{1000\,\text{mL}}{1\,\text{L}} = 0.88\ \text{M}$$

The pK_a for H_2O_2 is 11.75. The K_a for H_2O_2 is therefore $10^{-11.75}$ or 1.8×10^{-12}. By comparison with pure water, which has a K_a of 1.8×10^{-16} at 25 °C, one can see that H_2O_2 is indeed a stronger acid than water but the difference in acidity between the two is not that great. Consequently, we cannot ignore the contribution of protons of pure water when we work out the pH of the solution at equilibrium.

Reaction:

$$H_2O_2(aq) + H_2O(l) \underset{\phantom{K_a = 1.8 \times 10^{-12}}}{\overset{K_a = 1.8 \times 10^{-12}}{\rightleftharpoons}} H_3O^+(aq) + HO_2^-(aq)$$

	$H_2O_2(aq)$	$H_2O(l)$	$H_3O^+(aq)$	$HO_2^-(aq)$
Initial:	0.88 M	—	1.0×10^{-7} M	0 M
Change:	$-x$ M	—	$+x$ M	$+x$ M
Equilibrium:	$(0.88 - x)$ M ($\sim$0.88 M)	—	$(1.0 \times 10^{-7} + x)$M	x M

So, $1.8 \times 10^{-12} = \dfrac{x(x + 1.0 \times 10^{-7})}{\sim 0.88}$ $\qquad x^2 + 1.0 \times 10^{-7}x - 1.58 \times 10^{-12} = 0$

$$x = \frac{-1.0 \times 10^{-7} \pm \sqrt{1.0 \times 10^{-14} + 4(1.58 \times 10^{-12})}}{2}$$

The root that makes sense in this context is $x = 1.2 \times 10^{-6}$ M.
Thus, the $[H_3O^+] = 1.2 \times 10^{-6}$ M $+ 1.0 \times 10^{-7}$ M $= 1.3 \times 10^{-6}$ M
Consequently, the pH for the 3.0 % H_2O_2 solution (by mass) should be 5.89 (i.e., the solution is weakly acidic)

35. (E) $3 \, \overline{\overline{O}} = \overline{\overline{O}} \longrightarrow 2 \, \overline{\overline{O}} \cdot \cdot \cdot \overline{\overline{O}} \cdot \cdot \cdot \overline{\overline{O}}$

Bonds broken: $3 \times (O=O) = 3 \times 498$ kJ/mol $= 1494$ kJ/mol

Bonds formed: $4 \times (O \cdot \cdot \cdot O)$

$\Delta_r H° = +285$ kJ/mol = bonds broken − bonds formed = 1494 kJ / mol $- 4 \times (O \cdot \cdot \cdot O)$

$\quad 4 \times (O \cdot \cdot \cdot O) = 1494$ kJ/mol $- 285$ kJ/mol $= 1209$ kJ/mol

$\quad O \cdot \cdot \cdot O = 1209$ kJ/mol $+ 4 = 302$ kJ/mol

37. (E)

(a) H_2S, while polar, forms only weak hydrogen bonds. H_2O forms much stronger hydrogen bonds, leading to a higher boiling point and thus a liquid at room temperature.

(b) All electrons are paired in O_3, producing a diamagnetic molecule.

39. (M) Reactions that have K values greater than 1000 are considered to be essentially quantitative (i.e., they go virtually 100% to completion). So to answer this question we need only calculate the equilibrium constant for each reaction using the equation $E°_{cell} = (0.0257/z) \ln K$.

(a) $H_2O_2(aq) + 2 \, H^+(aq) + 2 \, e^- \rightleftharpoons 2 \, H_2O(l) \qquad\qquad E°_{red} = +1.763$ V

$\qquad\qquad 2 \, I^-(aq) \rightleftharpoons I_2(s) + 2 \, e^- \qquad\qquad\qquad E°_{ox} = -0.535$ V

$\rule{10cm}{0.4pt}$

$H_2O_2(aq) + 2 \, H^+(aq) + 2 \, I^-(aq) \rightleftharpoons I_2(s) + 2 \, H_2O(l) \quad E°_{cell} = +1.228$ V $(n = 2 \, e^-)$

$\ln K = \dfrac{1.228 \text{ V} \times 2}{0.0257 \text{ V}} = 95.56 \qquad K = 3.2 \times 10^{41}$

Therefore the reaction goes to completion (or very nearly so).

(b) $O_2(g) + 2 \, H_2O(l) + 4 \, e^- \rightleftharpoons 4 \, OH^-(aq) \qquad\qquad E°_{red} = +0.401$ V

$\quad 4 \, Cl^-(aq) \rightleftharpoons 2 \, Cl_2(g) + 4 \, e^- \qquad\qquad\qquad\quad E°_{ox} = -1.358$ V

$\rule{10cm}{0.4pt}$

$O_2(g) + 2 \, H_2O(l) + 4 \, Cl^-(aq) \rightleftharpoons 2 \, Cl_2(g) + 4 \, OH^-(aq) \; E°_{cell} = -0.957$ V

$\qquad\qquad\qquad\qquad\qquad\qquad\qquad\qquad\qquad\qquad\qquad (z = 4 \, e^-)$

$\ln K = \dfrac{-0.957 \text{ V} \times 4}{0.0257 \text{ V}} = -148.95 \qquad K = 2.1 \times 10^{-65}$

The extremely small value of K indicates that reactants are strongly preferred and thus, the reaction does not even come close to going to completion.

(c) $O_3(g) + 2 \, H^+(aq) + 2 \, e^- \rightleftharpoons O_2(g) + H_2O(l) \qquad\qquad E°_{red} = +2.075$ V

$\quad Pb^{2+}(aq) + 2 \, H_2O(l) \rightleftharpoons PbO_2(s) + 4 H^+(aq) + 2 \, e^- \qquad E°_{ox} = -1.455$ V

$\rule{10cm}{0.4pt}$

$O_3(g) + Pb^{2+}(aq) + H_2O(l) \rightleftharpoons PbO_2(s) + 2 \, H^+(aq) + O_2(g) \quad E°_{cell} = 0.620$ V

$\ln K = \dfrac{+0.62 \text{ V} \times 2}{0.0257 \text{ V}} = 48.25 \qquad K = 9.0 \times 10^{20}$

Therefore the reaction goes to completion (or very nearly so).

(d)
$$HO_2^-(aq) + H_2O(l) + 2\ e^- \rightleftharpoons 3\ OH^-(aq) \qquad\qquad E°_{red} = +0.878\ V$$
$$2\ Br^-(aq) \rightleftharpoons Br_2(l) + 2\ e^- \qquad\qquad E°_{ox} = -1.065\ V$$

$$HO_2^-(aq) + H_2O(l) + 2\ Br^-(aq) \rightleftharpoons Br_2(s) + 3\ OH^-(aq) \qquad E°_{cell} = -0.187\ V$$

$$\ln K = \frac{-0.187\ V \times 2}{0.0257\ V} = -14.55 \qquad\qquad K = 4.8 \times 10^{-7}$$

The extremely small value of K indicates the reaction heavily favors reactants at equilibrium and thus, the reaction does not even come close to going to completion.

41. **(M)** The reaction is $2\ KClO_3(s) \xrightarrow[\text{MnO}_2\,(s)]{\Delta} 2\ KCl(s) + 3\ O_2(g)$.

$$1.0\ g\ KClO_3 \times \frac{1\ mol}{122.549\ g} \times \frac{3\ mol\ O_2}{2\ mol\ KClO_3} \times \frac{31.998\ g\ O_2}{1\ mol\ O_2} = 0.392\ g\ O_2$$

$$pV = nRT \Rightarrow V = \frac{nRT}{p} = \frac{0.01224 \times 8.314\ J\ K^{-1}mol^{-1} \times 298.15\ K}{101\ kPa \times \dfrac{1\ atm}{101.325\ kPa}} = 30\ L$$

Sulfur

43. **(E)**
(a) MgS, magnesium sulfide	**(b)** H₂S, hydrogen sulfide
(c) Ca(HSO₃)₂, calcium hydrogen sulfite	**(d)** Na₂S₂O₃, sodium thiosulfate
(e) S₄N₄, tetrasulfur tetranitride	

(a) MgS, magnesium sulfide   **(b)** H_2S, hydrogen sulfide
(c) $Ca(HSO_3)_2$, calcium hydrogen sulfite   **(d)** $Na_2S_2O_3$, sodium thiosulfate
(e) S_4N_4, tetrasulfur tetranitride

45. **(M)**

(a) $FeS(s) + 2\ HCl(aq) \longrightarrow FeCl_2(aq) + H_2S(aq)$ MnS(s), ZnS(s), etc. also are possible.

(b) $CaSO_3(s) + 2\ HCl(aq) \longrightarrow CaCl_2(aq) + H_2O(l) + SO_2(g)$

(c) Oxidation: $SO_2(aq) + 2\ H_2O(l) \longrightarrow SO_4^{2-}(aq) + 4\ H^+(aq) + 2\ e^-$

　　　Reduction: $MnO_2(s) + 4\ H^+(aq) + 2\ e^- \longrightarrow Mn^{2+}(aq) + 2\ H_2O(l)$

　　　Net: $SO_2(aq) + MnO_2(s) \longrightarrow Mn^{2+}(aq) + SO_4^{2-}(aq)$

(d) Oxidation: $S_2O_3^{2-}(aq) + H_2O(l) \longrightarrow 2\ SO_2(g) + 2\ H^+(aq) + 4\ e^-$

　　　Reduction: $S_2O_3^{2-}(aq) + 6\ H^+(aq) + 4\ e^- \longrightarrow 2\ S(s) + 3\ H_2O(l)$

　　　Net: $S_2O_3^{2-}(aq) + 2\ H^+(aq) \longrightarrow S(s) + SO_2(g) + H_2O(l)$

47. **(M)** The decomposition of thiosulfate ion is more highly favored in an acidic solution. If the white solid is Na_2SO_4, there will be no reaction with strong acids such as HCl. By contrast, if the white solid is $Na_2S_2O_3$, $SO_2(g)$ will be liberated and a pale yellow precipitate of S(s,rhombic) will form upon addition of HCl(aq).

$$S_2O_3^{2-}(aq) + 2H^+(aq) \longrightarrow S(s) + SO_2(g) + H_2O(l)$$

Consequently, the solid can be identified by adding a strong mineral acid such as HCl(aq).

49. **(D)** $Mg^{2+}(aq)$ is only weakly polarizing, and it will not hydrolyze in aqueous medium. But $HSO_4^-(aq)$ will ionize further, $K_{a_2} = 1.1 \times 10^{-2}$ for $HSO_4^-(aq)$. We proceed by setting up the dissociation equilibrium for $[HSO_4^-(aq)]$ and then solving the quadratic equation to obtain $[H_3O^+(aq)]$:

$$Mg(HSO_4)_2(aq) \rightarrow Mg^{2+}(aq) + 2\,HSO_4^-(aq)$$

$$[HSO_4^-(aq)] = \frac{16.5\text{ g Mg}(HSO_4)_2}{250.0\text{ mL soln}} \times \frac{1000\text{ mL}}{1\text{ L soln}} \times \frac{1\text{ mol Mg}(HSO_4)_2}{218.43\text{ g Mg}(HSO_4)_2} \times \frac{2\text{ mol HSO}_4^-}{1\text{ mol Mg}(HSO_4)_2}$$

$$= 0.604\text{ M}$$

Reaction:	$HSO_4^-(aq)$	+	$H_2O(l)$	$\rightleftharpoons$	$SO_4^{2-}(aq)$	+	$H_3O+(aq)$
Initial:	0.604 M				0 M		0 M
Changes:	$-x$ M				$+x$ M		$+x$ M
Equil:	$(0.604 - x)$ M				x M		x M

Since the value of the equilibrium constant ($K = 0.011$) is moderate, the full quadratic equation must be solved (i.e., no assumptions can be made).

$$K_{a_2} = \frac{[H_3O^+][SO_4^{2-}]}{[HSO_4^-]} = 0.011 = \frac{x^2}{0.604 - x} \qquad x^2 = 0.00664 - 0.011x$$

$$x^2 + 0.011x - 0.00664 = 0$$

$$x = \frac{-b \pm \sqrt{b^2 - 4ac}}{2a} = \frac{-0.011 \pm \sqrt{1.2 \times 10^{-4} + 2.7 \times 10^{-2}}}{2} = 0.077 = [H_3O^+]$$

$$pH = -\log_{10}(0.077) = 1.1$$

51. **(M)** This question is concerned with assaying for the mass percent of iron in an ore. The assay in this instance involves the quantitative determination of the amount of metal in an ore by chemical analysis. The titration for iron in the sample does not occur directly but rather indirectly via the number of moles of $I_2(aq)$ produced from the reaction of $Fe^{3+}(aq)$ with excess $I^-(aq)$:

$$2\,Fe^{3+}(aq) + 2\,I^-(aq) \rightarrow 2\,Fe^{2+}(aq) + I_2(aq)$$

The number of moles of $I_2(aq)$ produced is then determined by titrating the iodide-treated sample with sodium thiosulfate. The balanced oxidation reaction that forms the basis for the titration is

$$I_2(aq) + 2\,S_2O_3^-(aq) \rightarrow 2\,I^-(aq) + S_4O_6^{2-}(aq)$$

The stoichiometric ratio is one $I_2(aq)$ reacting with two $S_2O_3^{2-}(aq)$ in this titration.

The number of moles of $I_2(aq)$ formed is therefore

$$0.01525 \text{ L } S_2O_3^-(aq) \times \frac{0.100 \text{ mol } S_2O_3^-(aq)}{1 \text{ L } S_2O_3^-(aq)} \times \frac{1 \text{ mol } I_2(aq)}{2 \text{ mol } S_2O_3^-(aq)} = 7.63 \times 10^{-4} \text{mol } I_2(aq)$$

Hence, the number of moles of $Fe^{3+}(aq)$ released when the sample is dissolved is

$$7.63 \times 10^{-4} \text{mol } I_2(aq) \times \frac{2 \text{ mol } Fe^{3+}(aq)}{1 \text{ mol } I_2(aq)} = 1.53 \times 10^{-3} \text{mol } Fe^{3+}(aq)$$

Consequently, the mass percent of iron in the ore is

$$1.53 \times 10^{-3} \text{mol } Fe^{3+}(aq) \times \frac{55.8452 \text{ g } Fe^{3+}(aq)}{1 \text{mol } Fe^{3+}(aq)} \times \frac{1}{1.500 \text{ g of Fe ore}} \times 100\% = 5.70\%$$

53. **(E)**
    (a) $Mg = +2$, $O = -2$, $S = -(+2 - 3 \times 2) = +4$;
    (b) $Cl = -4$, $S = +4$;
    (c) $Mg = +2$, $O = -2$, $S = -(+2 - 4 \times 2) = +6$;

    (d) $I = -1$, $S = -\dfrac{1}{2} \times (-2) = +1$;

    (e) $N = -3$, $S = -1/4 \times (-3 \times 4) = +3$.

Nitrogen Family

55. **(M)**
    (a)   The Haber-Bosch process is the principal artificial method of fixing atmospheric nitrogen. $N_2(g) + 3 H_2(g) \rightleftharpoons 2 NH_3(g)$

    (b)   The first step of the Ostwald process: $4 NH_3(g) + 5 O_2(g) \xrightarrow{850\,°C,\, Pt} 4 NO(g) + 6 H_2O(g)$

    (c)   The second and third steps of the process: $2 NO(g) + O_2(g) \rightarrow 2 NO_2(g)$
$$3 NO_2(g) + H_2O(l) \rightarrow 2 HNO_3(aq) + NO(g)$$

57. **(D)** Begin with the chemical formulas of the species involved:

$$Na_2CO_3(aq) + O_2(g) + NO(g) \rightarrow NaNO_2(aq)$$

Oxygen is reduced and nitrogen (in NO) is oxidized. We use the ion-electron method.

Two couples:	$NO(g) \rightarrow NO_2^-(aq)$ and	$O_2 \rightarrow$ products
Balance oxygens:	$H_2O + NO \rightarrow NO_2^-$	$O_2 \rightarrow 2 H_2O$
Balance hydrogens:	$H_2O + NO \rightarrow NO_2^- + 2 H^+$	$4 H^+ + O_2 \rightarrow 2 H_2O$
Balance charge:	$H_2O + NO \rightarrow NO_2^- + 2 H^+ + e^-$	$4 e^- + 4 H^+ + O_2 \rightarrow 2 H_2O$
Combine:	$4 H_2O + 4 NO + 4 H^+ + O_2 \rightarrow 2 H_2O + 4NO_2^- + 8 H^+$	

Simplify: $2 H_2O + 4 NO + O_2 \rightarrow 4NO_2^- + 4 H^+$

Add spectator ions: $4 Na^+ + 2CO_3^{2-} \rightarrow 4 Na^+ + 2CO_3^{2-}$

$2 Na_2CO_3 + 2 H_2O + 4 NO + O_2 \rightarrow 4 NaNO_2 + 2 H_2O + 2 CO_2$

simplify: $2 Na_2CO_3 (aq) + 4 NO(g) + O_2 (g) \rightarrow 4 NaNO_2 (aq) + 2 CO_2 (g)$

59. **(E)** $75 \times 10^9 \text{ gal} \times \dfrac{15 \text{ miles}}{1 \text{ gal}} \times \dfrac{5 \text{ g}}{1 \text{ mile}} \times \dfrac{1 \text{ kg}}{1000 \text{ g}} = 6 \times 10^9$ kg of nitrogen oxides released.

61. **(M)**

(a) $2 NO_2 (g) \rightleftharpoons N_2O_4 (g)$

(b) i) $HNO_2(aq) + N_2H_5^+(aq) \rightarrow HN_3(aq) + 2H_2O(l) + H^+(aq)$

ii) $HN_3(aq) + HNO_2(aq) \rightarrow N_2(g) + H_2O(l) + N_2O(g)$

(c) $H_3PO_4(aq) + 2 NH_3(aq) \rightarrow (NH_4)_2HPO_4(aq)$

63. **(M)**

(a)	$CH_3(NH)NH_2$ has 20 valence electrons.	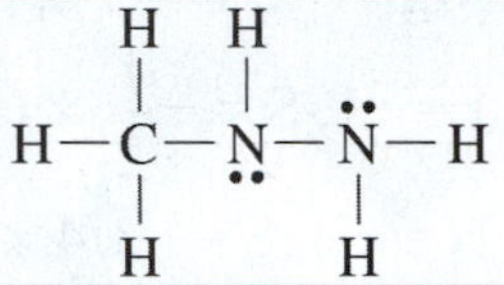
(b)	$(CH_3)_2NNH_2$ has 26 valence electrons.	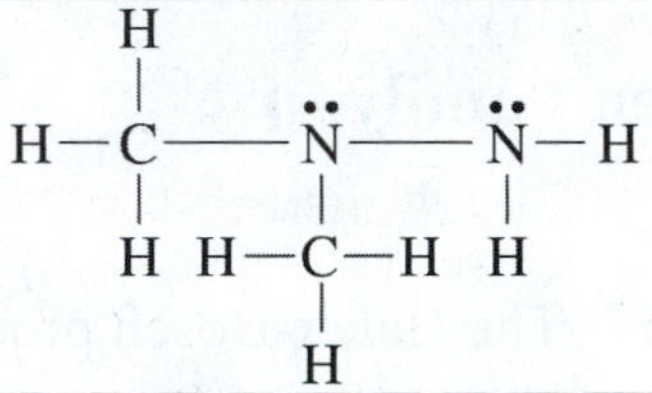
(c)	N_2O_4 has 34 valence electrons.	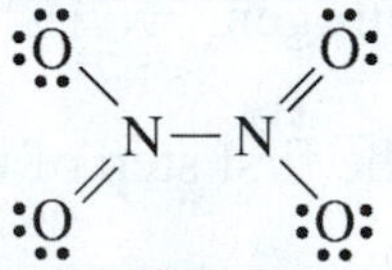
(d)	H_3PO_4 has 32 valence electrons.	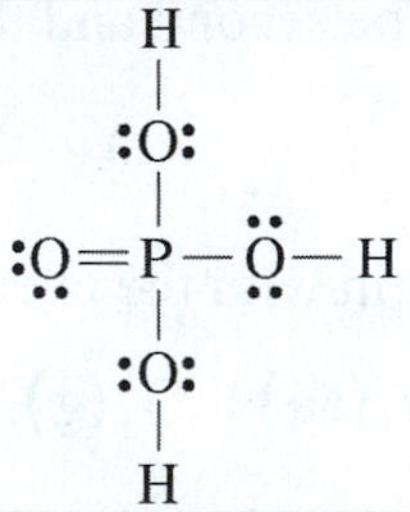
(e)	$ClNO_2$ has 24 valence electrons.	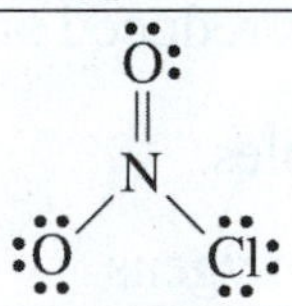

65. **(E)**

(a) HPO_4^{2-}, hydrogen phosphate ion

(b) $Ca_2P_2O_7$, calcium pyrophosphate or calcium diphosphate

(c) $H_6P_4O_{13}$, tetrapolyphosphoric acid

67. **(M)**

(i)	$2\ H^+(aq) + N_2O_4(aq) + 2\ e^- \rightarrow 2\ HNO_2(aq)$	$E°_1 = +1.065\ V$
(ii)	$2\ HNO_2(aq) + 2\ H^+(aq) + 2\ e^- \rightarrow 2\ NO(aq) + 2\ H_2O(l)$	$E°_2 = +0.996\ V$
(iii)	$N_2O_4(aq) + 4\ H^+(aq) + 4\ e^- \rightarrow 2\ NO(aq) + 2\ H_2O(l)$	$E°_3 = ?\ V$

Recall that $\Delta_r G° = -zFE°$ and that $\Delta_r G°$ values, not standard voltages are additive for reactions in which the number of electrons do not cancel out.

So, $-4\ FE°_3 = -2\ F(1.065\ V) + -2\ F(0.996\ V)$ $E°_3 = 1.031\ V$ (4 sig figs)

69. **(M)**

(a) The nitrogen atom cannot bond to five fluorine atoms because, as a second-row element, it cannot accommodate more than four electron pairs.

(b) The NF_3 molecule is trigonal pyramidal. The lone pair on the N atom causes the F—N—F bond angle to decrease from the ideal tetrahedral bond angle of 109° to 102.5°.

Hydrogen

71. **(M)** The four reactions of interest are: (Note: $\Delta_{comb}H° = \Sigma\Delta_f H°_{products} - \Sigma\Delta_f H°_{reactants}$)

$CH_4(g) + 2\ O_{2(g)} \rightarrow CO_2(g) + 2\ H_2O(l)$ $\Delta_{comb}H° = -890.3\ kJ\ mol^{-1}$
(Molar mass CH_4 = 16.0428 g mol^{-1})

$C_2H_6(g) + 7/2\ O_{2(g)} \rightarrow 2\ CO_2(g) + 3\ HO(l)$ $\Delta_{comb}H° = -1559.7\ kJ\ mol^{-1}$
(Molar mass C_2H_6 = 30.070 g mol^{-1})

$C_3H_8(g) + 5\ O_{2(g)} \rightarrow 3\ CO_2(g) + 4\ H_2O(l)$ $\Delta_{comb}H° = -2219.9\ kJ\ mol^{-1}$
(Molar mass C_3H_8 = 44.097 g mol^{-1})

$C_4H_{10}(g) + 13/2\ O_{2(g)} \rightarrow 4\ CO_2(g) + 5\ H_2O(l)$ $\Delta_{comb}H° = -2877.4\ kJ\ mol^{-1}$
(Molar mass C_4H_{10} = 58.123 g mol^{-1})

73. **(E)**

(a) $2\ Al(s) + 6\ HCl(aq) \rightarrow 2\ AlCl_3(aq) + 3\ H_2(g)$

(b) $3\ CO(g) + 7\ H_2(g) \rightarrow C_3H_8(g) + 3\ H_2O(g)$

(c) $MnO_2(s) + 2\ H_2(g) \xrightarrow{\Delta} Mn(s) + 2\ H_2O(g)$

75. **(M)** $CaH_2(s) + 2 H_2O(l) \rightarrow Ca(OH)_2(aq) + 2 H_2(g)$

$Ca(s) + 2 H_2O(l) \rightarrow Ca(OH)_2(aq) + H_2(g)$

$$\overline{2 Na(s) + 2 H_2O(l) \rightarrow 2 NaOH(aq) + H_2(g)}$$

(a) The reaction that produces the largest volume of $H_2(g)$ per liter of water also produces the largest amount of $H_2(g)$ per mole of water used. All three reactions use two moles of water and the reaction with $CaH_2(s)$ produces the most $H_2(g)$.

(b) We can compare three reactions that produce the same amount of hydrogen; the one that requires the smallest mass of solid produces the greatest amount of H_2 per gram of solid. The amount of hydrogen we will choose, to simplify matters, is 2 moles, which means that we compare 1 mol CaH_2 (42.09 g) with 2 mol Ca (80.16 g) and with 4 mol Na (91.96 g). Clearly CaH_2 produces the greatest amount of H_2 per gram of solid.

77. **(M)** Greatest mass percent hydrogen:
The atmosphere is mostly $N_2(g)$ and $O_2(g)$ with only a trace of hydrogen containing gas molecules. Seawater is $H_2O(l)$, natural gas is $CH_4(g)$ and ammonia is $NH_3(g)$. Each of these compounds has one non-hydrogen atom, and the non-hydrogen atoms have approximately the same mass ($\sim 14 \pm 2$ g mol^{-1}). Since CH_4 has the highest hydrogen atom to non-hydrogen atom ratio, this molecule has the greatest mass percent hydrogen.

79. **(M)** NH_2^- has 8 valence electrons (and is isoelectronic with H_2O). The first four MO's are fully occupied. Because the $2p_x$ orbital on N is strongly bonding in the bent configuration, as shown in Figure 22-27, the energy of the NH_2^- will be much lower for the bent configuration. (The same argument applies to H_2O.) On the basis of molecular orbital theory, we expect that NH_2^- will be bent.

INTEGRATIVE AND ADVANCED EXERCISES

81. **(M)** In step (1), oxygen is converted to H_2O by the reaction $H_2(g) + \frac{1}{2}O_2(g) \rightarrow H_2O(l)$.

Step (2) ensures that unreacted hydrogen from step (1) is also converted to $H_2O(l)$. The dehydrated zeolite has a very strong affinity for water molecules and thus, $H_2O(l)$ is removed from the gas mixture.

84. **(M)** $NaNO_2$ decolorizes acidic solution of $KMnO_4$ according to the following balanced chemical equation:
$2 KMnO_4 + 3 H_2SO_4 + 5 NaNO_2 \rightarrow 5 NaNO_3 + K_2SO_4 + 2 MnSO_4 + 3 H_2O$
$NaNO_3$ does not react with acidic solution of $KMnO_4$.

85. **(D)** First we balance the equation, then determine the number of millimoles of $NH_3(g)$ that are produced, and finally find the $[NO_3^-]$ in the original solution.

The skeleton half-equations: $NO_3^-(aq) \longrightarrow NH_3(g)$ $Zn(s) \longrightarrow Zn(OH)_4^{2-}(aq)$

Balance O's and H's: $NO_3^-(aq) \longrightarrow NH_3(g) + 3\,H_2O(l)$

$$NO_3^-(aq) + 9\,H^+(aq) \longrightarrow NH_3(g) + 3\,H_2O(l)$$

Balance charge and add $OH^-(aq)$ $NO_3^-(aq) + 6\,H_2O(l) + 8\,e^- \longrightarrow NH_3(g) + 9\,OH^-(aq)$

Add OH^- (ad)'s and then electrons: $Zn(s) + 4\,OH^-(aq) \longrightarrow Zn(OH)_4^{2-}(aq) + 2\,e^-$

Oxidation: $\{\,Zn(s) + 4\,OH^-(aq) \longrightarrow Zn(OH)_4^{2-}(aq) + 2\,e^-\,\}$ $\times 4$

Reduction: $NO_3^-(aq) + 6\,H_2O(l) + 8\,e^- \longrightarrow NH_3(g) + 9\,OH^-(aq)$

Net: $NO_3^-(aq) + 4\,Zn(s) + 6\,H_2O(l) + 7\,OH^-(aq) \longrightarrow Zn(OH)_4^{2-}(aq) + NH_3(g)$

The titration reactions are the following.

$HCl(aq) + NaOH(aq) \longrightarrow NaCl(aq) + H_2O(l)$ $NH_3(aq) + HCl(aq) \longrightarrow NH_4Cl(aq)$

$$\text{mmol excess HCl} = 32.10 \text{ mL} \times \frac{0.1000 \text{ mmol NaOH}}{1 \text{ mL soln}} \times \frac{1 \text{ mmol HCl}}{1 \text{ mmol NaOH}} = 3.210 \text{ mmol HCl}$$

$$\text{mmol HCl at start} = 50.00 \text{ mL} \times \frac{0.1500 \text{ mmol HCl}}{1 \text{ mL soln}} = 7.500 \text{ mmol HCl}$$

$$\text{mmol NH}_3 \text{ produced} = (7.500 - 3.210) \quad \text{mmol HCl} \times \frac{1 \text{ mmol NH}_3}{1 \text{ mmol HCl}} = 4.290 \text{ mmol NH}_3$$

$$[NO_3^-] = \frac{4.290 \text{ mmol NH}_3 \times \dfrac{1 \text{ mmol NO}_3^-}{1 \text{ mmol NH}_3}}{25.00 \text{ mL soln}} = 0.1716 \text{ M}$$

Notice that it was not necessary to balance the equation, since NO_3^- and NH_3 are the only nitrogen-containing species involved, and thus they must be in a one-to-one molar ratio.

86. **(M)** First we compute the root-mean-square speed of O at 1500 K.

$$u_{\text{rms}} = \sqrt{\frac{3RT}{M}} = \sqrt{\frac{3 \times 8.314 \text{ J mol}^{-1}\text{K}^{-1} \times 1500K}{0.0160 \text{ kg/mol}}} = 1.5 \times 10^3 \text{ m/s}$$

Kinetic Energy =

$$KE = \tfrac{1}{2}mu^2 = 0.5 \times \frac{0.0160 \text{ kg/mol}}{6.022 \times 10^{23} \text{ atoms/mol}} \times (1.5 \times 10^3 \text{ m/s})^2 = 3.0 \times 10^{-20} \text{ J/atom}$$

89. **(E)** The N center in the ammonium ion is the reducing agent, while Cl in the perchlorate anion is the oxidizing agent.

$$2\,NH_4ClO_4(s) \longrightarrow N_2(g) + 4\,H_2O(g) + Cl_2(g) + 2\,O_2(g)$$

92. **(M)** Each simple cubic unit cell has one Po atom at each of its eight corners, but each corner is shared among eight unit cells. Thus, there is a total of one Po atom per unit cell. The edge of that unit cell is 335 pm. From this information we obtain the density of polonium.

$$\text{density} = \frac{1 \text{ Po atom} \times \dfrac{1 \text{ mol Po}}{6.022 \times 10^{23} \text{ Po atoms}} \times \dfrac{209 \text{ g Po}}{1 \text{ mol Po}}}{\left(335 \text{ pm} \times \dfrac{1 \text{ m}}{10^{12} \text{ pm}} \times \dfrac{100 \text{ cm}}{1 \text{ m}}\right)^3} = 9.23 \text{ g/cm}^3$$

94. **(M)**

(a)

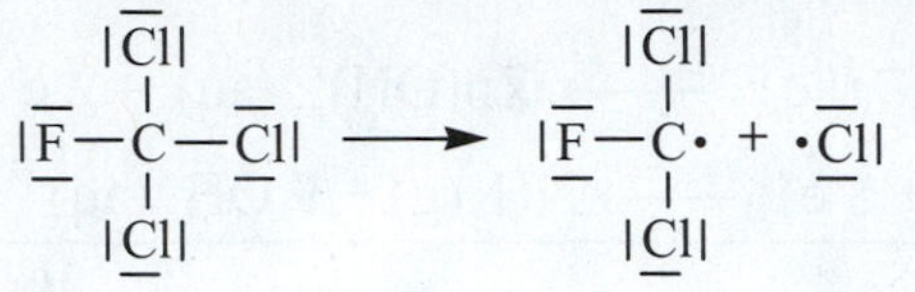

As shown in the Lewis structures above, the net result of this reaction is breaking a C—Cl bond. From Table 10-3, the energy of this bond is 339 kJ/mol, and this must be the energy of the photons involved in the reaction.

(b) $E = h\nu$ or $\nu = E/h$ on a molecular basis. Thus, we have the following.

$$\nu = \frac{339 \times 10^3 \text{ J/mol}}{6.626 \times 10^{-34} \text{ J} \cdot \text{s} \times 6.022 \times 10^{23} \text{ / mol}} = 8.50 \times 10^{14} \text{ s}^{-1}$$

$$\lambda = \frac{c}{\nu} = \frac{3.00 \times 10^8 \text{ m/s}}{8.50 \times 10^{14} \text{ / s}} = 3.53 \times 10^{-7} \text{ m} \times \frac{10^9 \text{ nm}}{1 \text{ m}} = 353 \text{ nm}$$

This radiation is in the near ultraviolet region of the electromagnetic spectrum.

95. **(M)**
(a) % P indicates the number of grams of P per 100 g of material, while % P_4O_{10} indicates the number of grams of P_4O_{10} per 100 g of material.

$$\text{mass P} = 1.000 \text{ g } P_4O_{10} \times \frac{1 \text{ mol } P_4O_{10}}{283.89 \text{ g } P_4O_{10}} \times \frac{4 \text{ mol P}}{1 \text{ mol } P_4O_{10}} \times \frac{30.974 \text{ g P}}{1 \text{ mol P}} = 0.436 \text{ g P}$$

Thus, multiplying the mass of P_4O_{10} by 0.436 will give the mass (or mass percent) of P. % BPL indicates the number of grams of $Ca_3(PO_4)_2$ per 100 g of material.

$$\text{mass } Ca_3(PO_4)_2 = 1.000 \text{ g } P_4O_{10} \times \frac{1 \text{ mol } P_4O_{10}}{283.89 \text{ g } P_4O_{10}} \times \frac{4 \text{ mol P}}{1 \text{ mol } P_4O_{10}}$$

$$\times \frac{1 \text{ mol } Ca_3(PO_4)_2}{2 \text{ mol P}} \times \frac{310.18 \text{ g } Ca_3(PO_4)_2}{1 \text{ mol } Ca_3(PO_4)_2} = 2.185 \text{ g } Ca_3(PO_4)_2$$

Thus, multiplying the mass of P_4O_{10} (283.88) by 2.185 will give the mass (or mass %) of BPL.

(b) A %BPL greater than 100% means that the material has a larger %P than does pure $Ca_3(PO_4)_2$.

(c) $\%P = \dfrac{6 \text{ mol P}}{1 \text{mol } 3Ca_3(PO_4)_2 \cdot CaF_2} \times \dfrac{1 \text{ mol } 3Ca_3(PO_4)_2 \cdot CaF_2}{1008.6 \text{ g } 3Ca_3(PO_4)_2 \cdot CaF_2} \times \dfrac{30.974 \text{ g P}}{1 \text{ mol P}} \times 100\%$

$= 18.43\% \text{ P}$

$\%P_4O_{10} = \dfrac{\%P}{0.436} = \dfrac{18.43}{0.436} = 42.3\% \text{ P}_4O_{10}$

$\%BPL = 2.185 \times \%P_4O_{10} = 2.185 \times 42.3\% \text{ P}_4O_{10} = 92.4\% \text{ BPL}$

99. **(M)** pH = 3.5 means $[H^+] = 10^{-3.5} = 3 \times 10^{-4}$ M. This is quite a dilute acidic solution, and we expect H_2SO_4 to be completely ionized under these circumstances.

mass $H_2SO_4 = 1.00 \times 10^3 \text{ L} \times \dfrac{3 \times 10^{-4} \text{ mol } H^+}{1 \text{ L}} \times \dfrac{1 \text{ mol } H_2SO_4}{2 \text{ mol } H^+} \times \dfrac{98.08 \text{ g } H_2SO_4}{1 \text{ mol } H_2SO_4} = 15 \text{ g } H_2 SO_4$

mass $Cl_2 = 1.00 \times 10^3 \text{ L} \times \dfrac{1000 \text{ cm}^3}{1 \text{ L}} \times \dfrac{1.03 \text{ g}}{1 \text{ cm}^3} \times \dfrac{70 \text{ g } Br_2}{10^6 \text{ g seawater}} \times \dfrac{1 \text{ mol } Br_2}{159.81 \text{ g } Br_2} \times \dfrac{1 \text{ mol } Cl_2}{1 \text{ mol } Br_2}$

$\times \dfrac{70.91 \text{ g } Cl_2}{1 \text{ mol } Cl_2} \times \dfrac{115 \text{ g } Cl_2 \text{used}}{100 \text{ g } Cl_2 \text{ needed}} = 37 \text{ g } Cl_2$

100. **(M)** $E = E^o - \dfrac{0.0592}{z} \log_{10} Q$;

assuming $E = 0$ for a process that is no longer spontaneous:

$\log_{10} Q = \dfrac{zE^\circ}{0.0592} = \dfrac{(16)(0.065)}{0.0592} = 17.\underline{6}$ $\qquad\qquad P_{SO_2} = 1 \times 10^{-6}$

$Q = 10^{17.\underline{6}} = 3.9 \times 10^{17} = \dfrac{(P_{SO_2})^8}{1^8 (H^+)^{16}} = \dfrac{(1 \times 10^{-6})^8}{1^8 (H^+)^{16}}$

Solving for $[H^+]$ yields a value of 8×10^{-5} M, which corresponds to a pH of 4.1

Thus, the solution is still acidic.

102. **(M)**

Oxidations: $XeF_4(g) + 3 H_2O \longrightarrow XeO_3(g) + 4 HF(aq) + 2 H^+(aq) + 2 e^-$

$\{2 H_2O(l) \longrightarrow O_2(g) + 4 H^+(aq) + 4 e^-\} \times \tfrac{3}{2}$

Reduction: $\{XeF_4(g) + 4 e^- \longrightarrow Xe(g) + 4 F^-(aq)\} \qquad \times 2$

Net: $\overline{3 XeF_4(g) + 6 H_2O(l) \longrightarrow 2 XeO_3(g) + 12 HF(aq) + \tfrac{3}{2} O_2(g) + 2 Xe(g)}$

The fact that O_2 is also produced indicates that there are two oxidation half-reactions. The production ratio of Xe and XeO_3 indicates the amount by which the other two half-reactions must be multiplied before they are added. Then the half-reaction for the production of O_2 is multiplied by $\tfrac{3}{2}$ to balance charge.

104. **(M)** The initial concentration of $Cl_2(aq) = \dfrac{6.4g}{70.91 \, g/mol} = 0.090M$

$$Cl_2(aq) \; + \; H_2O(l) \; \xrightleftharpoons{K_c = 4.4 \times 10^{-4}} \; HOCl(aq) + \; H^+(aq) \; + \; Cl^-(aq)$$

initial   0.090 M   0 M   0 M   0 M

equil.   (0.090-x) M x M x M x M

{where x is the molar solubility of Cl_2 in water}

$$K = \frac{[HOCl(aq)][\,H^+(aq)]\,[Cl^-(aq)]}{[Cl_2(aq)]} = \frac{x^3}{0.090\text{-}x} = 4.4 \times 10^{-4}$$

By using successive approximations, we find $x = 0.030$ M, so:

$[HOCl(aq)] = [H^+(aq)] \; = \; [Cl^-(aq)] = 0.030$ M and $[Cl_2(aq)] = 0.090 - 0.030 = 0.060$ M

105. **(M)**

$2e^- + N_2(g) + 4\,H^+(aq) + 2H_2O(l) \rightarrow 2\,NH_3OH^+(aq)$ $E° = -1.87$ V

$2\,NH_3OH^+(aq) + H^+(aq) + 2e^- \rightarrow N_2H_5^+(aq) + 2\,H_2O(l)$ $E° = 1.42$ V

$\underline{N_2H_5^+(aq) + 3\,H^+(aq) + 2e^- \rightarrow 2\,NH_4^+(aq) \hspace{3.5cm} E° = 1.275\ \text{V}}$

$N_2(g) + 8\,H^+(aq) + \; 6e^- \rightarrow 2\,NH_4^+(aq)$ $E° = 0.275$ V (see below)

$\Delta_r G° = \Delta G°_1 + \Delta G°_2 + \Delta G°_3 = -z_{tot} FE°_{tot} = -zFE°_1 - zFE°_2 - zFE°_3$

$$E°_{tot} = \frac{n_1 E°_1 + n_2 E°_2 + n_3 E°_3}{n_{tot}}$$

$E°_{tot} = [2(-1.87) + 2(1.42) + 2(1.275)]/6 = 0.275$ V

moving on...

$2\,HN_3(aq) \rightarrow 3\,N_2(g) + 2\,H^+(aq) + 2\,e^-$ $E° = +3.09$ V (oxidation)

$\underline{3\,N_2(g) + 24\,H^+(aq) + 18e^- \rightarrow 6\,NH_4^+(aq) \quad E° = 0.275\ \text{V (top equation multiplied by 3)}}$

$2\,HN_3(aq) \; + \; 22\,H^+(aq) \; + \; 16\,e^- \; \rightarrow \; 6\,NH_4^+(aq)$ [*sum*]

[divide by 2]

$HN_3(aq) + 11\,H^+(aq) + 8\,e^- \rightarrow 3\,NH_4^+(aq)$ $E° = 0.70$ V

since

$\Delta_r G° = \Delta G°_1 + \Delta G°_2 = -zFE°_{tot} = -zFE°_1 - zFE°_2$

$E°_{total} = [2(3.09) + 18(0.275)\,]/16 = 0.70$ V

107. **(E)** Both molecules are V shaped, with a lone pair of electrons on the central atom. For O_3, the structure is a hybrid of two equivalent structures. In each of these structures, the central atom has a formal charge of +1. The oxygen–oxygen bond order is between 1 and 2. Although many resonance structures can be drawn for SO_2, in the most important structure, the formal charge on the S atom is zero and the sulfur–oxygen bonds are double bonds.

109. **(M)** $\Delta_f H° = [\frac{3}{2}(498) - 3(36)] = 639$ kJ mol^{-1}

The formation reaction is very endothermic and thus energetically unfavorable. This result supports the observation that Xe(g) does not react directly with $O_2(g)$ to form $XeO_3(g)$. Because the reaction converts 2.5 moles of gas into 1 mole of gas, we expect $\Delta_f S° < 0$; thus, the reaction is also entropically unfavorable.

FEATURE PROBLEMS

110. **(D)** The goal here is to demonstrate that the three reactions result in the decomposition of water as the net reaction: Net: $2\,H_2O \rightarrow 2\,H_2 + O_2$ First balance each equation.

(1) $3\,FeCl_2 + 4\,H_2O \rightarrow Fe_3O_4 + HCl + H_2$ Balance by inspection. Notice that there are 3 Fe and 4 O on the right-hand side. Then balance Cl.

$3\,FeCl_2 + 4\,H_2O \rightarrow Fe_3O_4 + 6\,HCl + H_2$

(2) $Fe_3O_4 + HCl + Cl_2 \rightarrow FeCl_3 + H_2O + O_2$ Try the half-equation method.

$Cl_2 + 2\,e^- \rightarrow 2\,Cl^-$ But realize that $Fe_3O_4 = Fe_2O_3 \cdot FeO$.

$2\,FeO \rightarrow 2\,Fe^{3+} + O_2 + 6\,e^-$ Now combine the two half-equations.

$2\,FeO + 3\,Cl_2 \rightarrow 2\,Fe^{3+} + O_2 + 6\,Cl^-$

Add in $2\,Fe_2O_3 + 12\,H^+ \rightarrow 4\,Fe^{3+} + 6\,H_2O$

$2\,Fe_3O_4 + 3\,Cl_2 + 12\,H^+ \rightarrow 6\,Fe^{3+} + O_2 + 6\,Cl^- + 6\,H_2O$

And 12 Cl^- spectators: $2\,Fe_3O_4 + 3\,Cl_2 + 12\,HCl \rightarrow 6\,FeCl_3 + O_2 + 6\,H_2O$

(3) $FeCl_3 \rightarrow FeCl_2 + Cl_2$ by inspection $2\,FeCl_3 \rightarrow 2\,FeCl_2 + Cl_2$

One strategy is to consider each of the three equations and the net equation. Only equation (1) produces hydrogen. Thus, we must run it twice. Only equation (2) produces oxygen. Since only one mole of $O_2(g)$ is needed, we only have to run it once. Equation (3) can balance out the Cl_2 required by equation (2), but we have to run it three times to cancel all the $Cl_2(g)$.

$2\times(1)$ $6\,FeCl_2(s) + 8\,H_2O(l) \rightarrow 2\,Fe_3O_4(s) + 12\,HCl(l) + 2\,H_2(g)$

$1\times(2)$ $2\,Fe_3O_4(s) + 3\,Cl_2(g) + 12\,HCl(g) \rightarrow 6\,FeCl_3(s) + O_2(g) + 6\,H_2O(l)$

$3\times(3)$ $6\,FeCl_3(s) \rightarrow 6\,FeCl_2(s) + 3\,Cl_2(g)$

Net: $2\,H_2O(l) \rightarrow 2\,H_2(g) + O_2(g)$

112. (D)

$$ClO_3^- \xrightarrow{\text{(??)}} ClO_2 \xrightarrow{\text{(??)}} HClO_2$$

$$1.181 \text{ V}$$

In order to add ClO_2 to the Standard Electrode Potential Diagrams (also known as Latimer diagram) drawn above, we must calculate the voltages denoted by (?) and (??) . The equation associated with the reduction potential (?) is

(i) $2 H^+(aq) + ClO_3^-(aq) + 1 e^- \rightarrow ClO_2(g) + H_2O(l)$

The standard voltage for this half-reaction is given in Appendix D:

To finish up this problem, we just need to calculate the standard voltage (??) for the half-reaction (ii):

(ii) $H^+(aq) + ClO_2(aq) + 1 e^- \rightarrow HClO_2(g) \quad E^\circ = (??)$

To obtain the voltage for reaction (ii), we need to subtract reaction (i) from reaction (iii) below, which has been taken from Figure 22-2:

(iii) $3 H^+(aq) + ClO_3^-(aq) + 2 e^- \rightleftharpoons HClO_2(g) + H_2O(l) \quad E^\circ = 1.181 \text{ V}$

Thus,

$$\text{(iii)} \quad 3 H^+(aq) + ClO_3^-(aq) + 2 e^- \rightleftharpoons HClO_2(g) + H_2O(l) \quad E^\circ(\text{iii}) = 1.181 \text{ V}$$
$$-1\times \text{(i)} \quad 2 H^+(aq) + ClO_3^-(aq) + 1 e^- \rightarrow ClO_2(g) + H_2O(l) \quad E^\circ(\text{i}) = 1.175 \text{ V}$$

Net (ii) $H^+(aq) + ClO_2(g) + 1 e^- \rightarrow HClO_2(g) \quad\quad E^\circ(\text{ii}) = (??)$

Since reactions (i) and (iii) are both reduction half reactions, we cannot simply subtract the potential for (i) from the potential for (iii). Instead, we are forced to obtain the voltage for (ii) via the free energy changes for the three half reactions. Thus,

$$\Delta_r G^\circ \text{ (ii)} = \Delta_r G^\circ \text{ (iii)} - \Delta_r G^\circ \text{ (i)} = -1\ FE^\circ(\text{ii}) = -2\ F(1.181 \text{ V}) + 1\ F(1.175 \text{ V})$$

Dividing both sides by $-F$ gives

$$E^\circ \text{ (ii)} = 2(1.181 \text{ V}) - 1.175 \text{ V} \qquad\qquad \text{So, } E^\circ \text{ (??)} = 1.187 \text{ V.}$$

SELF-ASSESSMENT EXERCISES

120. (E) **(b)**

121. (E) **(d)**

122. (E) **(c)**

123. (E) **(a)**

124. (E) **(d)**

125. (E) **(b)**

126. (E) **(a)** and **(d)**

127. **(M)**

   (a) $Cl_2(g) + 2\,NaOH(aq) \rightarrow NaCl(aq) + NaOCl(aq) + H_2O(l)$

   (b) $2\,NaI(s) + 2\,H_2SO_4(concd\ aq) \rightarrow Na_2SO_4(a) + 2\,H_2O(g) + SO_2(g) + I_2(g)$

   (c) $Cl_2(g) + 2\,KI_3(aq) \rightarrow 2\,KCl(aq) + 3\,I_2(s)$

   (d) $3\,NaBr(s) + H_3PO_4(concd\ aq) \rightarrow Na_3PO_4 + 3\,HBr(g)$

   (e) $5\,HSO_3^-(aq) + 2\,MnO_4^-(aq) + H^+(aq) \rightarrow 5\,SO_4^{2-}(aq) + 2\,Mn^{2+}(aq) + 3\,H_2O(l)$

128. **(M)**

   (a) $2\,KClO_3(s) \rightarrow 2\,KCl(s) + 3\,O_2(g)$ a catalyst such as MnO_2 is required, electrolysis of H_2O is an alternate method.

   (b) $3\,Cu(s) + 8\,H+(aq) + 2\,NO_3^-(aq) \rightarrow 3\,Cu^{2+}(aq) + 4\,H_2O(l) + 2\,NO(g)$, difficult to control reaction so that $NO(g)$ is the only reduction product.

   (c) $Zn(s) + 2\,H^+(aq) \rightarrow Zn^{2+}(aq) + H_2(g)$, a number of other metals can be used.

   (d) $NH_4Cl(aq) + NaOH(aq) \rightarrow NaCl(aq) + H_2O(l) + NH_3(g)$, other ammonium salts and other bases work as well.

   (e) $CaCO_3(s) + 2\,HCl(aq) \rightarrow CaCl_2(aq) + H_2O(l) + CO_2(g)$, other carbonates and acids can be used.

129. **(M)**

   (a) $LiH(s) + H_2O(l) \rightarrow LiOH(aq) + H_2(g)$

   (b) $C(s) + H_2O(g) \rightarrow CO(g) + H_2(g)$

   (c) $3\,NO_2(g) + H_2O(l) \rightarrow 2\,HNO_3(aq) + NO(g)$

130. **(M)** $3\,Cl_2(g) + I^-(aq) + 3\,H_2O(l) \rightarrow 6\,Cl^-(aq) + IO_3^-(aq) + 6\,H^+(aq)$

         $Cl_2(g) + 2\,Br^-(aq) \rightarrow 2\,Cl^-(aq) + Br_2(aq)$ and Br_2 is extracted by CS_2.

131. **(M)**   One conversion pathway is ton $\rightarrow$ lb $\rightarrow$ kg $\rightarrow$ g $H_2SO_4 \rightarrow$ g S $\rightarrow$ mg S $\rightarrow$ L seawater $\rightarrow$ m^3 $\rightarrow$ km^3 seawater, leading to about 5.0 km^3 of seawater.

132. **(M)**   **(a)** AgAt, **(b)** sodium perxenate, **(c)** MgPo, **(d)** tellurous acid, **(e)** K_2SeSO_3, **(f)** potassium perastatate

133. **(D)**   Convert the $E°$ values to $\Delta_r G°$ values for the reduction of H_2SeO_3 to Se and from Se to H_2Se. Add those two $\Delta_r G°$ values to obtain $\Delta_r G°$ for the reduction of H_2SeO_3 to H_2Se. Convert this $\Delta_r G°$ value to $E°$, which proves to be 0.38 V.

134. **(M)**   **(a)** SO_3, **(b)** SO_2, **(c)** Cl_2O_7, **(d)** I_2O_5

135. **(M)**   **(a)** false, **(b)** true, **(c)** true.

CHAPTER 23
THE TRANSITION ELEMENTS
PRACTICE EXAMPLES

1A **(E) (a)** Cu_2O should form. $\qquad$ $2\,Cu_2S(s)+3\,O_2(g)\rightarrow 2\,Cu_2O(s)+2\,SO_2(g)$

$\quad$ **(b)** $\quad$ W(s) is the reduction product. $\quad$ $WO_3(s)+3\,H_2(g)\rightarrow W(s)+3\,H_2O(g)$

$\quad$ **(c)** $\quad$ Hg(l) forms. $\qquad\qquad\qquad$ $2\,HgO(s)\xrightarrow{\Delta}2\,Hg(l)+O_2(g)$

1B **(E) (a)** $SiO_2(s)$ is the oxidation product of Si.

$$3\,Si(s)+2\,Cr_2O_3(s)\xrightarrow{\Delta}3\,SiO_2(s)+4\,Cr(s)$$

$\quad$ **(b)** $\quad$ Roasting is simply heating in air. $\quad$ $2\,Co(OH)_3(s)\xrightarrow{\Delta,\,Air}Co_2O_3(s)+3\,H_2O(g)$

$\quad$ **(c)** $\quad$ $MnO_2(s)$ forms; (acidic solution). $\quad$ $Mn^{2+}(aq)+2\,H_2O(l)\rightarrow MnO_2(s)+4\,H^+(aq)+2\,e^-$

2A **(M)** We write and combine the half-equations for oxidation and reduction. If $E° > 0$, the reaction is spontaneous.

Oxidation: $\{V^{3+}(aq)+H_2O(l)\rightarrow VO^{2+}(aq)+2\,H^+(aq)+e^-\}\times 3$ $\qquad$ $E° = -0.337\,V$

Reduction: $NO_3^-(aq)+4\,H^+(aq)+3\,e^-\rightarrow NO(g)+2\,H_2O$ $\qquad\qquad$ $E° = +0.956\,V$

Net: $NO_3^-(aq)+3\,V^{3+}(aq)+H_2O(l)\rightarrow NO(g)+3\,VO^{2+}(aq)+2\,H^+(aq)$ $\quad$ $E°_{cell} = +0.619\,V$

Because the cell potential is positive, nitric acid can be used to oxidize $V^{3+}(aq)$ to $VO^{2+}(aq)$ under standard conditions.

2B **(M)** The reducing couple must have a half-cell potential of such a size and sign that a positive sum results when this half-cell potential is combined with $E°\{VO^{2+}(aq)|V^{2+}(aq)\} = 0.041\,V$ (this is the weighted average of the $VO^{2+}|V^{3+}$ and $V^{3+}|V^{2+}$ reduction potentials) *and* a negative sum must be produced when this half-cell potential is combined with $E°\{V^{2+}(aq)|V(s)\} = -1.13\,V$.

So, $-E°$ for the couple must be $> -0.041\,V$ and $< +1.13\,V$ (i.e., it cannot be more positive than 1.13V, nor more negative than $-0.041\,V$). Some possible reducing couples from Table 19-1 are: $-E°\{Cr^{3+}(aq)|Cr^{2+}(aq)\} = +0.42\,V$; $-E°\{Fe^{2+}(aq)|Fe(s)\} = +0.440\,V$ $-E°\{Zn^{2+}(aq)|Zn(s)\} = +0.763\,V$. Thus Fe(s), Cr^{2+}(aq), and Zn(s) will do the job.

INTEGRATIVE EXAMPLE

3A **(M)** $\quad$ Because the reduction potential for $PtCl_6^{2-}$ is more positive than that of V^{3+}, the following half-reactions occur spontaneously in the cell:

Oxidation: $[\,V^{2+}+e^-\rightarrow V^{3+}\,]\times 2$ $\qquad\qquad\qquad$ $E° = +0.255\,V$

Reduction: $PtCl_6^{2-}+2\,e^-\rightarrow PtCl_4^{2-}+2\,Cl^-$ $\qquad$ $E° = +0.68\,V$

Overall: $2\,V^{2+}+PtCl_6^{2-}\rightarrow 2\,V^{3+}+PtCl_4^{2-}+2\,Cl^-$ $\quad$ $E° = 0.94\,V$

The equilibrium constant for the overall reaction is $K = 10^{\frac{2(0.94)}{0.0592}} = 5.7 \times 10^{31}$. For the reverse reaction to be spontaneous, we need $Q > 5.7 \times 10^{31}$. Because $Q = [V^{3+}]^2 [PtCl_4^{2-}] [Cl^-]^2 / [V^{2+}] [PtCl_6^{2-}]$, the formation of $PtCl_6^{2-}$ is favored by using a very low concentration of V^{2+} and very high concentrations of V^{3+}, $PtCl_4^{2-}$, and Cl^-. In practical terms, though, the amount of $PtCl_6^{2-}$ that could be formed spontaneously would be small. A quick calculation shows that starting from $[V^{3+}]_o = [PtCl_4^{2-}]_o = [Cl^-]_o = 1.$ M and $[V^{2+}] = [PtCl_6^{2-}]_o = 0$, the equilibrium concentration of PtC_6^{2-} would be about 1.6×10^{-11} M. So, in practical terms, an external voltage source would be required to make a significant amount of $PtCl_6^{2-}$ from V^{3+}, $PtCl_4^{2-}$, and Cl^-.

3B **(M)** The disproportionation reaction is $3\ Ti^{2+} \rightarrow 2\ Ti^{3+} + Ti$ and $E° = -1.261$ V. Because $E° < 0$, the reaction is not spontaneous under standard conditions.

EXERCISES

Properties of the Transition Elements

1. **(E)**

(a) V [Ar]   4s: ↑↓   3d: ↑ | ↑ | ↑ | |

(b) Cr^{3+} [Ar]   4s: ☐   3d: ↑ | ↑ | ↑ | |

(c) Mn^{2+} [Ar]   4s: ☐   3d: ↑ | ↑ | ↑ | ↑ | ↑

(d) Fe^{2+} [Ar]   4s: ☐   3d: ↑↓ | ↑ | ↑ | ↑ | ↑

(e) Cu^{2+} [Ar]   4s: ☐   3d: ↑↓ | ↑↓ | ↑↓ | ↑↓ | ↑

(f) Ni^{2+} [Ar]   4s: ☐   3d: ↑↓ | ↑↓ | ↑↓ | ↑ | ↑

3. **(M)** A given main-group metal typically displays just one oxidation state, usually equal to its family number in the periodic table. Exceptions are elements such as Tl ($+1$ and $+3$), Pb ($+2$ and $+4$), and Sn ($+2$ and $+4$), for which the lower oxidation state represents a pair of s electrons not being ionized (a so-called "inert pair").

Main group metals do not form a wide variety of complex ions, with Al^{3+}, Sn^{2+}, Sn^{4+}, and Pb^{2+} being major exceptions. On the other hand, most transition metal ions form an extensive variety of complex ions. Most compounds of main group metals are colorless; exceptions occur when the anion is colored. On the other hand, many of the compounds of transition metal cations are colored, due to d-d electron transitions. Virtually every main-group metal cation has no unpaired electrons and hence is diamagnetic. On the other hand, many transition metals cations have one or more unpaired electrons and therefore are paramagnetic.

5. **(M)** When an electron and a proton are added to a main group element to create the element of next highest atomic number, it is the electron that influences the radius. The electron is added to the outermost shell (n value), which is farthest from the nucleus. Moreover, the added electron is well shielded from the nucleus and hence it is only weakly attracted to the nucleus. Thus, this electron billows out and, as a result, has a major influence on the size of the atom. However, when an electron and a proton are added to a transition metal atom to create the atom of next highest atomic number, the electron it is not added to the outermost shell. The electron is added to the d-orbitals, which are one principal quantum number lower than the outermost shell. Also, the electron in the same subshell as the added electron offers little shielding. Thus it has small effect on the size of the atom.

7. **(M)** Of the first transition series, manganese exhibits the greatest number of different oxidation states in its compounds, namely, every state from $+1$ to $+7$. One possible explanation might be its $3d^5 4s^2$ electron configuration. Removing one electron produces an electron configuration ($3d^5 4s^1$) with two half-filled subshells, removing two produces one with a half-filled and an empty subshell. Then there is no point of semistability until the remaining five d electrons are removed. These higher oxidation states all are stabilized by being present in oxides (MnO_2) or oxoanions (e.g., MnO_4^-).

9. **(M)** The greater ease of forming lanthanide cations compared to forming transition metal cations, is due to the larger size of lanthanide atoms. The valence (outer shell) electrons of these larger atoms are further from the nucleus, less strongly attracted to the positive charge of the nucleus in diffuse f-orbitals that are do not penetrate effectively and are very effectively shielded by the core electrons. As a result, they are removed much more readily.

Reactions of Transition Metals and Their Compounds

11. **(M)** **(a)** $TiCl_4(g) + 4\,Na(l) \xrightarrow{\Delta} Ti(s) + 4NaCl(l)$

(b) $Cr_2O_3(s) + 2Al(s) \xrightarrow{\Delta} 2\,Cr(l) + Al_2O_3(s)$

(c) $Ag(s) + HCl(aq) \rightarrow$ no reaction

(d) $K_2Cr_2O_7(aq) + 2KOH(aq) \rightarrow 2\,K_2CrO_4(aq) + H_2O(l)$

(e) $MnO_2(s) + 2\,C(s) \xrightarrow{\Delta} Mn(l) + 2CO(g)$

13. **(M) (a)** $Sc(OH)_3(s) + 3H^+(aq) \rightarrow Sc^{3+}(aq) + 3\,H_2O(l)$

(b) $3\,Fe^{2+}(aq) + MnO_4^-(aq) + 2\,H_2O(l) \rightarrow 3\,Fe^{3+}(aq) + MnO_2(s) + 4\,OH^-(aq)$

(c) $2\,KOH(l) + TiO_2(s) \xrightarrow{\Delta} K_2TiO_3(s) + H_2O(g)$

(d) $Cu(s) + 2\,H_2SO_4(conc, aq) \rightarrow CuSO_4(aq) + SO_2(g) + 2\,H_2O(l)$

15. **(M)** We write some of the following reactions as total equations rather than as net ionic equations so that the reagents used are indicated.

(a) $FeS(s) + 2\,HCl(aq) \rightarrow FeCl_2(aq) + H_2S(g)$

$4\,Fe^{2+}(aq) + O_2(g) + 4H^+(aq) \rightarrow 4Fe^{3+}(aq) + 2\,H_2O(l)$

$Fe^{3+}(aq) + 3\,OH^-(aq) \rightarrow Fe(OH)_3(s)$

(b) $BaCO_3(s) + 2\,HCl(aq) \rightarrow BaCl_2(aq) + H_2O(l) + CO_2(g)$

$2\,BaCl_2(aq) + K_2Cr_2O_7(aq) + 2\,NaOH(aq) \rightarrow 2\,BaCrO_4(s) + 2\,KCl(aq) + 2\,NaCl(aq) + H_2O(l)$

Extractive Metallurgy

17. **(E)** $HgS(s) + O_2(g) \xrightarrow{\Delta} Hg(l) + SO_2(g)$

$4\,HgS(s) + 4\,CaO(s) \xrightarrow{\Delta} 4\,Hg(l) + 3\,CaS(s) + CaSO_4(s)$

19. **(M)** The plot of $\Delta_r G°$ versus T will consist of three lines of increasing positive slope. The first line is joined to the second line at the melting point for Ca(s), while the second line is joined to the third at the boiling point for Ca(l).

$2\,Ca(s) + O_2(g) \rightarrow 2\,CaO(s)$ $\Delta_f H° = -1270.2\ kJ\ mol^{-1}$

$\Delta_r S° = 2(39.75\ J\ mol^{-1}\ K^{-1}) - [2 \times (41.42\ J\ mol^{-1}\ K^{-1}) + 205.1\ J\ mol^{-1}\ K^{-1}] = -208.4\ J\ mol^{-1}\ K^{-1}$

The graph should be similar to that for $2\,Mg(s) + O_2(g) \rightarrow 2\,MgO(s)$. We expect a positive slope with slight changes in the slope after the melting point $(839\ °C)$ and boiling point $(1484\ °C)$, mainly owing to changes in entropy. The plot will be below the $\Delta_r G°$ line for $2\,Mg(s) + O_2(g) \rightarrow 2\,MgO(s)$ at all temperatures.

Oxidation-Reduction

21. **(E)**

(a) $CuO(s) + 2 H^+(aq) + e^- \rightarrow Cu^+(aq) + H_2O$

(b) $FeO(s) + 2 H^+(aq) \rightarrow Fe^{3+}(aq) + H_2O + e^-$

23. **(M)**

(a) First we need the reduction potential for the couple $VO_2^+(aq)/V^{2+}(aq)$. We will use the half-cell addition method learned in Chapter 19.

$$VO_2^+(aq) + 2 H^+(aq) + e^- \rightarrow VO^{2+}(aq) + H_2O(l) \quad \Delta_rG^\circ = -1\,F(+1.000\text{ V})$$

$$VO^{2+}(aq) + 2 H^+(aq) + e^- \rightarrow V^{3+}(aq) + H_2O(l) \quad \Delta_rG^\circ = -1\,F(+0.337\text{ V})$$

$$V^{3+}(aq) + e^- \rightarrow V^{2+}(aq) \quad \Delta_rG^\circ = -1\,F(-0.255\text{ V})$$

$$VO_2^+(aq) + 4 H^+(aq) + 3\,e^- \rightarrow V^{2+}(aq) + 2 H_2O \quad \Delta_rG^\circ = -3\,FE^\circ$$

$$E^\circ = \frac{1.000\text{ V} + 0.337\text{ V} - 0.255\text{ V}}{3} = +0.361\text{ V}$$

We next analyze the oxidation-reduction reaction.

Oxidation: $\{2\ Br^-(aq) \rightarrow Br_2(l) + 2e^-\} \times 3 \qquad -E^\circ = -1.065\text{ V}$

Reduction: $\{VO_2^+(aq) + 4H^+(aq) + 3e^- \rightarrow V^{2+}(aq) + 2H_2O(l)\} \times 2 \quad E^\circ = +0.361\text{V}$

Net: $6\ Br^-(aq) + 2\ VO_2^+(aq) + 8\ H^+(aq) \rightarrow 3\ Br_2(l) + 2\ V^{2+}(aq) + 4\ H_2O$

$E^\circ_{cell} = -0.704\text{V}$

Thus, this reaction does not occur to a significant extent as written under standard conditions.

(b) Oxidation: $Fe^{2+}(aq) \rightarrow Fe^{3+}(aq) + e^- \qquad\qquad\qquad -E^\circ = -0.771\text{V}$

Reduction: $VO_2^+(aq) + 2\ H^+(aq) + e^- \rightarrow VO^{2+}(aq) + H_2O(l) \quad E^\circ = +1.000\text{V}$

Net: $\quad Fe^{2+}(aq) + VO_2^+(aq) + 2\ H^+(aq) \rightarrow Fe^{3+}(aq) + VO^{2+}(aq) + H_2O(l)$

$E^\circ_{cell} = +0.229\text{V}$

This reaction does occur to a significant extent under standard conditions.

(c) Oxidation: $H_2O_2 \rightarrow 2\ H^+(aq) + 2\ e^- + O_2(g) \qquad\qquad -E^\circ = -0.695\text{ V}$

Reduction: $MnO_2(s) + 4\ H^+(aq) + 2\ e^- \rightarrow Mn^{2+}(aq) + 2\ H_2O(l) \qquad E^\circ = +1.23\text{V}$

Net: $H_2O_2 + MnO_2(s) + 2\ H^+(aq) \rightarrow O_2(g) + Mn^{2+}(aq) + 2\ H_2O(l) \qquad E_{cell}^{\ \circ} = +0.54\text{V}$

Thus, this reaction does occur to a significant extent under standard conditions.

25. (M) The reducing couple that we seek must have a half-cell potential of such a size and sign that a positive sum results when this half-cell potential is combined with $E^\circ\{VO^{2+}(aq)|V^{3+}(aq)\} = 0.337\ V$ *and* a negative sum must be produced when this half-cell potential is combined with $E^\circ\{V^{3+}(aq)|V^{2+}(aq)\} = -0.255V$.

So, $-E^\circ$ for the couple must be$> -0.337\ V$ and $<+0.255\ V$ (i.e. it cannot be more positive than 0.255V, or more negative than −0.337 V)

Some possible reducing couples from Table 19.1 are:

$-E^\circ\{Sn^{2+}(aq)|Sn(s)\} = +0.137\ V$; $-E^\circ\{H^+(aq)|H_2(g)] = 0.000\ V$
$-E^\circ\{Pb^{2+}(aq)|Pb(s)] = +0.125\ V$; thus Pb(s), Sn(s), $H_2(g)$, to name but a few, will do the job.

27. (M) Table D-4 contains the following data: Cr^{3+}/Cr^{2+} reduction potential $= -0.424\ V$, $Cr_2O_7^{2-}/Cr^{3+}$ reduction potential $= 1.33\ V$ and Cr^{2+}/Cr reduction potential $= -0.90\ V$.

By using the additive nature of free energies and the fact that $\Delta G^\circ = -zFE^\circ$, we can determine the two unknown potentials and complete the diagram.

(i) $Cr_2O_7^{2-}/Cr^{2+}$: $E^\circ = \dfrac{3(1.33\ V) - 0.424\ V}{4} = 0.892\ V$

(ii) Cr^{3+}/Cr : $E^\circ = \dfrac{-0.424\ V + 2(-0.90)V}{3} = -0.74\ V$

Chromium and Chromium Compounds

29. (M) Orange dichromate ion is in equilibrium with yellow chromate ion in aqueous solution.
$$Cr_2O_7^{2-}(aq) + H_2O(l) \rightleftharpoons 2\ CrO_4^{2-}(aq) + 2\ H^+(aq)$$

The chromate ion in solution then reacts with lead(II) ion to form a precipitate of yellow lead(II) chromate. $Pb^{2+}(aq) + CrO_4^{2-}(aq) \rightleftharpoons PbCrO_4(s)$

$PbCrO_4(s)$ will form until $[H^+]$ from the first equilibrium increases to the appropriate level and both equilibria are simultaneously satisfied.

31. (D) Oxidation: $\{Zn(s) \rightarrow Zn^{2+}(aq) + 2\ e^-\} \times 3$
Reduction: $Cr_2O_7^{2-}(aq, orange) + 14\ H^+(aq) + 6\ e^- \rightarrow 2\ Cr^{3+}(aq, green) + 7\ H_2O(l)$

Net: $3\ Zn(s) + Cr_2O_7^{2-}(aq) + 14\ H^+(aq) \rightarrow 3\ Zn^{2+}(aq) + 2\ Cr^{3+}(aq) + 7\ H_2O(l)$

Oxidation: $Zn(s) \rightarrow Zn^{2+}(aq) + 2\ e^-$
Reduction: $\{Cr^{3+}(aq, green) + e^- \rightarrow Cr^{2+}(aq, blue)\} \times 2$

Net: $Zn(s) + 2\ Cr^{3+}(aq) \rightarrow Zn^{2+}(aq) + 2\ Cr^{2+}(aq)$

The green color is most likely due to a chloro complex of Cr^{3+}, such as $\left[Cr(H_2O)_4Cl_2\right]^+$.

Oxidation: $\{Cr^{2+}(aq, blue) \rightarrow Cr^{3+}(aq, green) + e^-\} \times 4$

Reduction: $O_2(g) + 4\,H^+(aq) + 4\,e^- \rightarrow 2\,H_2O(l)$

Net: $4\,Cr^{2+}(aq) + O_2(g) + 4\,H^+(aq) \rightarrow 4\,Cr^{3+}(aq) + 2\,H_2O(l)$

33. **(M)** Simple substitution into equation (23.19) yields $\left[Cr_2O_7{}^{2-}\right]$ in each case. The expression is readily solved for the desired concentration as follows:

$\left[Cr_2O_7{}^{2-}\right] = 3.2 \times 10^{14} \times \left[H^+\right]^2 \times \left[CrO_4{}^{2-}\right]^2$. In each case, we use the value of pH to determine $\left[H^+\right]$ $\left(\left[H^+\right] = 10^{-pH}\right)$.

(a) $\left[Cr_2O_7{}^{2-}\right] = 3.2 \times 10^{14} \times (10^{-7.12})^2 \times (0.20)^2 = 0.074$ M

(b) $\left[Cr_2O_7{}^{2-}\right] = 3.2 \times 10^{14} \times \left(10^{-9.15}\right)^2 \times (0.20)^2 = 6.4 \times 10^{-6}$ M

35. **(E)** Each mole of chromium metal plated out from a chrome plating bath (i.e., CrO_3 and H_2SO_4) requires six moles of electrons.

$$\text{mass Cr} = 1.00\ h \times \frac{3600\ s}{1\ hr} \times \frac{3.4\ C}{1\ s} \times \frac{1\ mol\ e^-}{96485\ C} \times \frac{1\ mol\ Cr}{6\ mol\ e^-} \times \frac{52.00\ g\ Cr}{1\ mol\ Cr} = 1.10\ g\ Cr$$

37. **(M)** Dichromate ion is the prevalent species in acidic solution. Oxoanions are better oxidizing agents in acidic solution because increasing the concentration of hydrogen ion favors formation of product. The half-equation is:

$Cr_2O_7^{2-}(aq) + 14\,H^+(aq) + 6\,e^- \longrightarrow 2\,Cr^{3+}(aq) + 7\,H_2O(l)$ Note that precipitation occurs most effectively in alkaline solution. In fact, adding an acid to a compound is often an effective way of dissolving a water-insoluble compound. Thus, we expect to see the form that predominates in alkaline solution to be the most effective precipitating agent. Notice also that $CrO_4{}^{2-}$ is smaller than is $Cr_2O_7{}^{2-}$, giving it a higher lattice energy in its compounds, which makes these compounds harder to dissolve.

The Iron Triad

39. **(M)** $4\,Fe^{2+}(aq) + O_2(g) + 4\,H^+ \rightarrow 4\,Fe^{3+}(aq) + 2\,H_2O(l)$ $\qquad E° = 0.44$ V

$[Fe^{2+}] = [Fe^{3+}]$; pH = 3.25 or $[H^+] = 5.6 \times 10^{-4}$ and $P_{O_2} = 0.20$ atm

$$E = E° - \frac{0.0592}{z}\log\left(\frac{[Fe^{3+}]^4}{[Fe^{2+}]^4[H^+]^4 P_{O_2}}\right) = 0.44\ V - \frac{0.0592}{4}\log\left(\frac{[Fe^{3+}]^4}{[Fe^{2+}]^4[10^{-3.25}]^4\,0.20}\right)$$

$E = 0.24$ V (spontaneous under these conditions)

41. (E) $Fe^{3+}(aq) + K_4[\overset{[II]}{Fe}(CN)_6](aq) \rightarrow K\overset{[III]}{Fe}[\overset{[II]}{Fe}(CN)_6](s) + 3\ K^+(aq)$

Alternate formulation: $4Fe^{3+} + 3[\overset{[II]}{Fe}(CN)_6]^{4-}(aq) + Fe_4[\overset{[II]}{Fe}(CN)_6]_3$

Group 11 Metals

43. (E) (a) $Cu^{2+}(aq) + H_2(g) \rightarrow Cu(s) + 2\ H^+(aq)$

(b) $Au^+(aq) + Fe^{2+}(aq) \rightarrow Au(s) + Fe^{3+}(aq)$

(c) $2\ Cu^{2+}(aq) + SO_2(g) + 2\ H_2O(l) \rightarrow 2\ Cu^+(aq) + SO_4^{2-}(aq) + 4\ H^+(aq)$

45. (M) The Integrative Example showed $K_c = 1.2 \times 10^6 = \dfrac{[Cu^{2+}]}{[Cu^+]^2}$ or

$[Cu^{2+}] = 1.2 \times 10^6 [Cu^+]^2$

(a) When $[Cu^+] = 0.20$ M, $[Cu^{2+}] = 1.2 \times 10^6 (0.20)^2 = 4.8 \times 10^4$ M. This is an

impossibly high concentration. Thus $[Cu^+] = 0.20$ M can never be achieved.

(b) When $[Cu^+] = 1.0 \times 10^{-10}$ M, $[Cu^{2+}] = 1.2 \times 10^6 (1.0 \times 10^{-10})^2 = 1.2 \times 10^{-14}$ M. This is an

entirely reasonable (even though small) concentration; $[Cu^+] = 1.0 \times 10^{-10}$ M can be

maintained in solution.

Group 12 Metals

47. (E) Given: Hg^{2+}/Hg reduction potential = 0.854 V and Hg_2^{2+}/Hg reduction potential = 0.796 V

Using the additive nature of free energies and the fact that $\Delta_r G° = -zFE°$, we can determine the

Hg^{2+}/Hg_2^{2+} potential as $\dfrac{2(0.854V) - 0.796V}{1} = 0.912$ V

49. (M) (a) Estimate K_p for $ZnO(s) + C(s) \rightarrow Zn(l) + CO(g)$ at 800 °C (Note: Zn(l) boils at 907 °C)

$\{2\ C(s) + O_2(g) \rightarrow 2\ CO(g)\} \times 1/2$ $\Delta_r G° \approx (-415\ kJ\ mol^{-1}) \times (1/2)$

$\{2\ ZnO(s) \rightarrow 2\ Zn(l) + O_2(g)\} \times 1/2$ $\Delta_r G° = (+485\ kJ\ mol^{-1}) \times (1/2)$

$ZnO(s) + C(s) \rightarrow Zn(l) + CO(g)$ $\Delta_r G° = 35\ kJ\ mol^{-1}$

Use $\Delta_r G° = -RT \ln K$ where $T = 800$ °C, $K = K_p$

$35\ kJ\ mol^{-1} = -(8.3145 \times 10^{-3}\ kJ\ mol^{-1}\ K^{-1})((273.15 + 800)\ K)(\ln K_p)$

$\ln K_p = -3.9$ or $K_p = 0.02$

(b) $K_p = P_{CO} = 0.02$ Hence, $P_{CO} = 0.02$ atm

51. **(M)** We must calculate the wavelength of light absorbed in order to promote an electron across each band gap.

First, a few relationships. $E_{mole} = N_A E_{photon}$ $\quad$ $E_{photon} = h\nu$ $\quad$ $c = \nu\lambda$ or $\nu = c/\lambda$

Then, some algebra. $E_{mole} = N_A E_{photon} = N_A h\nu = N_A hc/\lambda$ $\qquad$ or $\qquad$ $\lambda = N_A hc/E_{mole}$

For ZnO, $\lambda = \dfrac{6.022\times10^{23} \text{ mol}^{-1} \times 6.626\times10^{-34} \text{ J s} \times 2.998\times10^8 \text{ m s}^{-1}}{325\times10^3 \text{ J mol}^{-1}} \times \dfrac{10^9 \text{ nm}}{1 \text{ m}}$

$\qquad = \dfrac{1.196\times10^8 \text{ J mol}^{-1} \text{ nm}}{325\times10^3 \text{ J mol}^{-1}} = 368 \text{ nm}$ $\quad$ violet light

For CdS, $\lambda = \dfrac{1.196\times10^8 \text{ J mol}^{-1} \text{ nm}}{250\times10^3 \text{ J mol}^{-1}} = 479 \text{ nm}$ blue light

The blue light absorbed by CdS is subtracted from the white light incident on the surface of the solid. The remaining reflected light is yellow in this case. When the violet light is subtracted from the white light incident on the ZnO surface, the reflected light appears white.

INTEGRATIVE AND ADVANCED EXERCISES

53. **(M)** There are two reasons why Au is soluble in aqua regia while Ag is not. First, Ag^+, the oxidation product, forms a very insoluble chloride, AgCl, which probably adheres to the surface of the metal and prevents further reaction. $AuCl_3$ is not noted as being an insoluble chloride. Second, gold(III) forms a very stable complex ion with chloride ion, $[AuCl_4]^-$. This complex ion is much more stable than the corresponding dichloroargentate ion, $[AgCl_2]^-$.

57. **(M)** The noble gas formalism requires that the number of valence electrons possessed by the metal atom plus the number of sigma electrons be equal to the number of electrons in the succeeding noble gas atom.

(a) $Mo(CO)_6$; Mo has 42 electrons, 6 CO contribute another 12 for 54, which is the number of electrons in Xe

(b) $Os(CO)_5$; Os has 76 electrons, 5 CO contribute another 10 for 86, which is the number of electrons in Rn

(c) $Re(CO)_5^-$; Re anion has 76 electrons, 5 CO contribute another 10 for 86, which is the number of electrons in Rn

(d) All of the metal carbonyls have zero dipole moments and very weak intermolecular forces. Therefore one would expect these compounds to be liquids or low melting solids at room temperature. Since chromium carbonyls have more electrons than the other compounds it will be more polarizable and therefore have the strongest dispersion force leading to it being a solid while the other metal carbonyls are liquids. Another reason that the nickel and iron carbonyls are liquids and the chromium carbonyl is a solid is due to the differences in their shape. The nickel carbonyl $Ni(CO)_4$ being four-coordinate is tetrahedral while the iron carbonyl $Fe(CO)_5$ being penta-coordinate is trigonal bipyramidal. These two shapes do not lend themselves to forming crystals while the octahedral shape of chromium carbonyl, more spherical in nature, does fit better into a crystal lattice.

(e) This compound would be an ionic, salt-like material consisting of Na^+ and $V(CO)_6^-$ ions.

61. (M)

(a) Dissolution: $AgO(s) + 2\ H^+(aq) \longrightarrow Ag^{2+}(aq) + H_2O(l)$

Oxidation :	$2\ H_2O(l) \longrightarrow O_2(g) + 4\ H^+(aq) + 4\ e^-$		$-E^\circ = -1.229\ V$
Reduction :	$\{Ag^{2+}(aq) + e^- \longrightarrow Ag^+(aq)\}$	$\times 4$	$E^\circ = 1.98\ V$

Net : $\quad 2\ H_2O(l) + 4\ Ag^{2+}(aq) \longrightarrow O_2(g) + 4\ H^+(aq) + 4\ Ag^+(aq)$

(b) $E^\circ_{cell} = -1.229\ V + 1.98\ V = +0.75\ V \rightarrow$ spontaneous since $E^\circ_{cell} > 0$.

64. (D) The equations are first balanced with the ion-electron method. Oxalic acid is oxidized in each case.

Reaction 1:

Oxidation: $H_2C_2O_4(aq) \rightarrow 2\ H^+(aq) + 2e^- + CO_2(g)$

Reduction: $Fe_2O_3(s) + 6\ H^+(aq) + 2e^- \rightarrow 2\ Fe^{2+}(aq) + 3\ H_2O(l)$

Net: $H_2C_2O_4(aq) + Fe_2O_3(s) + 4\ H^+(aq) \rightarrow 2\ CO_2(g) + 2\ Fe^{2+}(aq) + 3\ H_2O(l)$

Reaction 2:

Oxidation: $\{H_2C_2O_4(aq) \rightarrow 2\ H^+(aq) + 2e^- + CO_2(g)\} \times 5$

Reduction: $\{MnO_4^-(aq) + 8\ H^+(aq) + 5e^- \rightarrow Mn^{2+}(aq) + 4\ H_2O(l)\} \times 2$

Net: $5\ H_2C_2O_4(aq) + 8\ H^+(aq) + 2\ MnO_4^-(aq) \rightarrow 2\ Mn^{2+}(aq) + 10\ CO_2(g) + 8\ H_2O(l)$

The mass of excess oxalic acid can be calculated from the second reaction:

$$\text{mass } H_2C_2O_4 \cdot 2\ H_2O = 0.03516\ L \times \frac{0.1000\ \text{mol } MnO_4^-}{1\ L\ \text{soln}} \times \frac{5\ \text{mol } H_2C_2O_4}{2\ \text{mol } MnO_4^-}$$

$$\times \frac{126.07\ g\ H_2C_2O_4 \cdot 2\ H_2O}{1\ \text{mol } H_2C_2O_4 \cdot 2\ H_2O} = 1.108\ g\ H_2C_2O_4 \cdot 2\ H_2O$$

Therefore, the mass of oxalic acid that reacted with Fe_2O_3 is 1.752 g – 1.108 g = 0.644 g. Finally, using reaction 1, we can calculate the mass and percent of Fe_2O_3 in the impure sample of hematite:

$$\text{mass } Fe_2O_3 = 0.644 \text{ g } H_2C_2O_4 \cdot 2\,H_2O \times \frac{1 \text{ mol } H_2C_2O_4 \cdot 2\,H_2O}{126.07 \text{ g } H_2C_2O_4 \cdot 2\,H_2O} \times \frac{1 \text{ mol } Fe_2O_3}{1 \text{ mol } H_2C_2O_4 \cdot 2\,H_2O}$$

$$\times \frac{159.69 \text{ g } Fe_2O_3}{1 \text{mol } Fe_2O_3} = 0.816 \text{ g } Fe_2O_3$$

$$\% Fe_2O_3 = \frac{0.816 \text{ g}}{0.960 \text{ g}} \times 100\% = 85.0\%$$

65. **(D)**

(a) We consider reaction with 1.00 millimoles of Fe^{2+}. In each case, we need the balanced chemical equation of the redox reaction.

$$Cr_2O_7^{2-}(aq) + 14\,H^+(aq) + 6\,Fe^{2+}(aq) \longrightarrow 2\,Cr^{3+}(aq) + 7\,H_2O(l) + 6\,Fe^{3+}(aq)$$

$$\begin{matrix} \text{titrant} \\ \text{volume} \end{matrix} = 1.00 \text{ mmol } Fe^{2+} \times \frac{1 \text{ mmol } Cr_2O_7^{2-}}{6 \text{ mmol } Fe^{2+}} \times \frac{1 \text{ mL solution}}{0.1000 \text{ mmol } Cr_2O_7^{2-}} = 1.67 \text{ mL soln}$$

$$MnO_4^-(aq) + 8\,H^+(aq) + 5\,Fe^{2+}(aq) \longrightarrow Mn^{2+}(aq) + 4\,H_2O + 5\,Fe^{3+}(aq)$$

$$\begin{matrix} \text{titrant} \\ \text{volume} \end{matrix} = 1.00 \text{ mmol } Fe^{3+} \times \frac{1 \text{ mmol } MnO_4^-}{5 \text{ mmol } Fe^{2+}} \times \frac{1 \text{ mL solution}}{0.1000 \text{ mmol } MnO_4^-} = 2.00 \text{ mL soln}$$

More of the 0.1000 M MnO_4^- solution would be required.

(b) We use the reaction of each with Fe^{2+} to find the equivalence between them.

$$V_{MnO_4^-} = 24.50 \text{ mL} \times \frac{0.1000 \text{ mmol } Cr_2O_7^{2-}}{1 \text{ mL soln}} \times \frac{6 \text{ mmol } Fe^{2+}}{1 \text{ mmol } Cr_2O_7^{2-}} \times \frac{1 \text{ mmol } MnO_4^-}{5 \text{ mmol } Fe^{2+}}$$

$$\times \frac{1 \text{ mL soln}}{0.1000 \text{ mmol } MnO_4^-} = 29.40 \text{ mL of } 0.1000 \text{ mmol } MnO_4^-$$

67. **(D)** The addition of the first 10.00 mL aliquot to $BaCl_2$ results in the precipitation of $BaCrO_4$. From this we are able to determine the amount of Cr in solution. The titration of the second 10.00 mL aliquot with Fe^{2+} results in the redox reaction between Fe^{2+}, MnO_4^- and $Cr_2O_7^{2-}$. Since we have the amount of Cr from the first titration we can determine the amount of $Cr_2O_7^{2-}$ was present and determine the amount of Fe^{2+} that was needed to reduce the dichromate from which we can determine the amount of Mn that was in solution.

Equation: $\quad MnO_4^-(aq) + 8\,H^+(aq) + 5\,Fe^{2+}(aq) \longrightarrow Mn^{2+}(aq) + 4\,H_2O + 5\,Fe^{3+}(aq)$

$$\text{mass of Cr} = 250.0 \text{ mL soln} \times \frac{0.549 \text{ g } BaCrO_4}{10.00 \text{ mL sample}} \times \frac{1 \text{ mol } BaCrO_4}{253.3 \text{ g } BaCrO_4} \times \frac{1 \text{ mol Cr}}{1 \text{ mol } BaCrO_4} \times \frac{52.00 \text{ g Cr}}{1 \text{ mol Cr}}$$

mass of Cr = 2.82 g Cr

equation: $\quad Cr_2O_7^{2-} + 14\,H^+ + 6\,Fe^{2+} \rightarrow 2Cr^{3+} + 7\,H_2O + 6\,Fe^{3+}$

volume of titrant to reduce $Cr_2O_7^{2-}$ = 2.82 g Cr $\times \dfrac{10\ \text{ml}}{250\ \text{ml}} \times \dfrac{1\ \text{mol}\ Cr^{3+}}{52.00\ \text{g}\ \ Cr} \times \dfrac{6\ \text{mol}\ Fe^{2+}}{1\ \text{mol}\ Cr_2O_7^{2-}}$

$\times \dfrac{1\ \text{mol}\ Cr_2O_7^{2-}}{2\ \text{mol}\ \ Cr^{3+}} \times \dfrac{1\ \text{L}\ \ \text{titrant}}{0.0750\ \text{mol}\ Fe^{2+}} \times \dfrac{1000\ \text{mL}}{1\ \text{L}} = 86.78$ ml Fe^{2+} titrant

Now the volume of titrant to reduce both MnO_4^- and $Cr_2O_7^{2-}$ must be $(15.95 + 86.78)$ mL

Volume of titrant $= 102.73$ mL.

Not just 15.95 mL, which must be the volume of titrant required for MnO_4^- *alone*.

mass Mn = 250.0 mL soln $\times \dfrac{15.95\ \ \text{mL}\ \ \text{titrant}}{10.00\ \ \text{mL}\ \ \text{sample}} \times \dfrac{0.0750\ \ \text{mmol}\ Fe^{2+}}{1\ \text{mL}\ \ \text{titrant}} \times \dfrac{1\ \ \text{mmol}\ \ MnO_4^-}{5\ \ \text{mmol}\ \ Fe^{2+}}$

$\times \dfrac{1\ \ \text{mmol}\ \ \text{Mn}}{1\ \ \text{mmol}\ \ MnO_4^-} \times \dfrac{54.94\ \ \text{mg}\ \ \text{Mn}}{1\ \ \text{mmol}\ \ \text{Mn}} \times \dfrac{1\ \text{g}\ \ \text{Mn}}{1000\ \ \text{mg}\ \ \text{Mn}} = 0.3286$ Mn

$\%Mn = \dfrac{0.3286\ \text{g Mn}}{10.000\ \text{g steel}} \times 100\% = 3.286\%\ \text{Mn}$ $\quad \%Cr = \dfrac{2.82\ \text{g Cr}}{10.000\ \text{g steel}} \times 100\% = 28.2\%\ \text{Cr}$

68. **(M) (a)** We use the technique from Chapter 3 to determine the empirical formula of palladium dimethylglyoximate.

$31.61\text{g Pd} \times \dfrac{1\ \text{mol Pd}}{106.42\ \text{g Pd}} = 0.2970\ \text{mol Pd} \div 0.2970 \rightarrow 1.000\ \text{mol Pd}$

$28.54\ \text{g C} \times \dfrac{1\ \text{mol C}}{12.011\ \text{g C}} = 2.376\ \text{mol C} \div 0.2970 \rightarrow 8.000\ \text{mol C}$

$4.19\ \text{g H} \times \dfrac{1\ \text{mol H}}{1.008\ \text{g H}} = 4.16\ \text{mol H} \div 0.2970 \rightarrow 14.01\ \text{mol H}$

$19.01\ \text{g O} \times \dfrac{1\ \text{mol O}}{15.999\ \text{g O}} = 1.188\ \text{mol O} \div 0.2970 \rightarrow 4.000\ \text{mol O}$

$16.64\text{g N} \times \dfrac{1\ \text{mol N}}{14.007\ \text{g N}} = 1.188\ \text{mol N} \div 0.2970 \rightarrow 4.000\ \text{mol N}$

The empirical formula of palladium dimethylglyoximate ($PdDmgH_2$) is $PdC_8H_{14}O_4N_4$, with an empirical molar mass of 336.64 g/mol.

(b) mass of Pd $= 250.0\,\text{mL} \times \dfrac{0.0784\,\text{g PdDmgH}_2}{10.00\,\text{mL}} \times \dfrac{1\,\text{mol PdDmgH}_2}{336.64\,\text{g PdDmgH}_2} \times \dfrac{1\ \text{mol Pd}}{1\,\text{mol PdDmgH}_2}$

$\times \dfrac{106.42\,\text{g Pd}}{1\,\text{mol Pd}} = 0.620\ \text{g Pd}$

% Pd in steel $= \dfrac{0.620\ \text{g Pd}}{16.312\ \text{g steel}} \times 100\% = 3.80\%$

71. **(M)**

(a)	(b)	(c)

$[Hg{-}Hg]^{+2}$

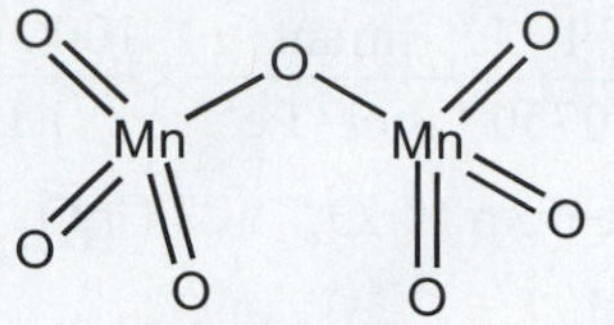

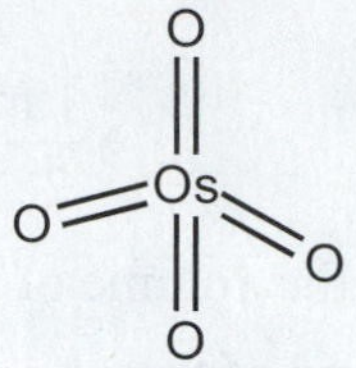

73. **(M)** ? Ti per unit cell $= \dfrac{1}{8}$ Ti per corner $\times$ 8 corners per cell $= 1$ Ti

? Ni per unit cell $= 1$

The empirical formula is NiTi. The % Ti by mass is $100 \times 47.88/(47.88 + 63.55) = 45\%$

FEATURE PROBLEMS

74. **(D)**

(a) If $\Delta n_{gas} = 0$, then $\Delta_r S^\circ \sim 0$ and $\Delta_r G^\circ$ is essentially independent of temperature

$(C(s) + O_2(g) \rightarrow CO_2(g))$

If $\Delta n_{gas} > 0$, then $\Delta_r S^\circ > 0$ and $\Delta_r G^\circ$ will become more negative with increasing temperature, hence the graph has a negative slope $(2\,C(s) + O_2(g) \rightarrow 2\,CO\,(g))$.

If $\Delta n_{gas} < 0$, then $\Delta_r S^\circ < 0$ and $\Delta_r G^\circ$ will become more positive with increasing temperature, hence the graph has a positive slope $(2\,CO(g) + O_2(g) \rightarrow 2\,CO_2\,(g))$.

(b) The additional blast furnace reaction, $C(s) + CO_2(g) \rightarrow 2\,CO(g)$, has

$\Delta_r H^\circ = [2 \times -110.5\ kJ] - [-393.5\ kJ + 0\ kJ] = 172.5\ kJ\ mol^{-1}$ and

$\Delta_r S^\circ = [2 \times 197.7\ J\ K^{-1}] - [1 \times 5.74\ J\ K^{-1} + 213.7\ J\ K^{-1}] = 176.0\ J\ K^{-1}\ mol^{-1}$

It can be obtained by adding reaction (b) to the reverse of reaction (c) (both appear in the provided figure)

$C(s) + O_2(g) \rightarrow CO_2(g)$ Reaction (b)

$2\,CO_2(s) \rightarrow 2\,CO(g) + O_2(g)$ Reverse of Reaction (c)

Net: $C(s) + CO_2(g) \rightarrow 2\,CO(g)$ (Additional Blast Furnace Reaction)

Consequently, the plot of $\Delta_r G^\circ$ for the *net* reaction as a function of temperature will be a straight line with a slope of $-[\{\Delta_r S^\circ\,(\text{Rxn b})\} - \{\Delta_r S^\circ\,(\text{Rxn c})\}]$ (in kJ mol^{-1} K^{-1}) and a y-intercept of $[\{\Delta_r H^\circ\,(\text{Rxn b})\} - \{\Delta_r H^\circ\,(\text{Rxn c})\}]$ (in kJ mol^{-1}).

Since $\Delta_r H^\circ$ (Rxn b) = -393.5 kJ mol^{-1}, $\Delta_r H^\circ$ (Rxn c) = -566 kJ/mol, $\Delta_r S^\circ$ (Rxn b) = 2.9 J mol^{-1} K^{-1} and $\Delta_r S^\circ$ (Rxn c) = -173.1 J mol^{-1} K^{-1}, the plot of $\Delta_r G^\circ$ vs. T for the reaction C(s) + CO$_2$(g) $\rightarrow$ 2 CO(g) will follow the equation $y = -0.176x + 172.5$

From the graph, we can see that the difference in $\Delta_r G^\circ$ (line b – line c at 1000 °C) is ~ -40 kJ mol^{-1}. K_p is readily calculated using this value of $\Delta_r G^\circ$.

$\Delta_r G^\circ = -RT \ln K_p = -40$ kJ = $-(8.3145 \times 10^{-3}$ kJ K^{-1} mol^{-1})(1273 K)(ln K_p)
lnK_p = 3.$\underline{8}$ Hence, $K_p = 4\underline{4}$

The equilibrium partial pressure for CO(g) is then determined by using the K_p expression:

$$K_p = \frac{(P_{CO})^2}{(P_{CO_2})} = 4\underline{4} = \frac{(P_{CO})^2}{(0.25 \text{ atm})} \quad \text{Hence, } (P_{CO})^2 = 1\underline{1} \text{ and } P_{CO} = 3.\underline{3} \text{ atm or 3 atm.}$$

Alternatively, we can determine the partial pressure for CO$_2$ at 1000 °C via the calculated $\Delta_r H^\circ$ and $\Delta_r S^\circ$ values for the reaction C(s) + CO$_2$(g) $\rightarrow$ 2 CO(g) to find $\Delta_r G^\circ$ at 1000 °C , and ultimately K_p with the relationship $\Delta_r G^\circ = -RT \ln K_p$ (here we are making the assumption that $\Delta_r H^\circ$ and $\Delta_r S^\circ$ are relatively constant over the temperature range 298 K to 1273 K). The calculated values of $\Delta_r H^\circ$ and $\Delta_r S^\circ$ (using Appendix D) are given below:

$\Delta_r H^\circ = [2 \times -110.5$ kJ$] - [-393.5$ kJ $+ 0$ kJ$] = 172.5$ kJ mol^{-1}
$\Delta_r S^\circ = [2 \times 197.7$ J K$^{-1}] - [1 \times 5.74$ J K$^{-1} + 213.7$ J K$^{-1}] = 176.0$ J K^{-1} mol^{-1}

To find $\Delta_r G^\circ$ at 1000°C, we simply plug $x = 1273$ K into the straight-line equation we developed above and solve for y ($\Delta_r G^\circ$).
So, $y = -0.176(1273$ K$) + 172.5$; $y = -51.5$ kJ mol^{-1}
Next we need to calculate the K_p for the reaction at 1000°C.

$\Delta_r G^\circ = -RT \ln K_p = -51.5$ kJ mol^{-1} = $-(8.3145 \times 10^{-3}$ kJ mol^{-1} K^{-1})(1273 K)(ln K_p)
lnK_p = 4.87. Hence, $K_p = 1.3 \times 10^2$
The equilibrium P$_{CO(g)}$ is then determined by using the K_p expression:

$$K_p = \frac{(P_{CO})^2}{(P_{CO_2})} = 130 = \frac{(P_{CO})^2}{(0.25 \text{ atm})} \quad \text{Hence, } (P_{CO})^2 = 32.\underline{5} \text{ atm and } P_{CO} = 5.7 \text{ atm}$$

SELF-ASSESSMENT EXERCISES

79. **(E)**

(a) Pig iron is impure iron (95% Fe) formed in the reduction of iron ore in a blast furnace.

(b) Ferromanganese is an iron-manganese alloy formed by the reduction of a mixture of iron and manganese oxides.

(c) Chromite ore, Fe(CrO$_2$)$_2$ is the principal chromium ore.

(d) Brass is an alloy of Zn and Cu with small amounts of Sn, Pb, and Fe.

(e) Aqua regia is a mixture of HCl(aq) and HNO_3(aq) that dissolves inactive metals by a combination of oxidation and complex ion formation.

(f) Blister copper, formed in the reduction of a mixture of Cu_2O(s) and Cu_2S(s), is impure Cu(s) containing SO_2(g).

(g) Stainless steel is an iron alloy with varying quantities of metals such as Cr, Mn, and Ni, and a small and carefully controlled percentage of carbon.

80. **(E)** (c), (f), and (g).

81. **(E)** (b)

82. **(E)** (d)

83. **(E)** (a)

84. **(E)** (d)

85. **(E)** (c) and (e)

86. **(M)** CrO_3(s), **(b)** potassium manganate, **(c)** chromium carbonyl, **(d)** $BaCr_2O_7$, **(e)** lanthanum(III) sulfate nonahydrate, **(f)** $Au(CN)_3 \cdot 3\,H_2O$.

87. **(M)**

(a) $2\,Fe_2S_3(s) + 3\,O_2(g) + 6\,H_2O(l) \rightarrow 4\,Fe(OH)_3(s) + 6\,S(s)$

(b) $2\,Mn^{2+}(aq) + 8\,H_2O(l) + 5\,S_2O_8^{2-}(aq) \rightarrow 2\,MnO_4^-(aq) + 16\,H^+(aq) + 10\,SO_4^{2-}(aq)$

(c) $4\,Ag(s) + 8\,CN^-(aq) + O_2(g) + 2\,H_2O(l) \rightarrow 4[Ag(CN_2]^-(aq) + 4\,OH^-(aq)$

88. **(M)** Atoms of Zn, Cd and Hg have configurations of $4s^2 3d^{10}$, $5s^2 4d^{10}$, and $6s^2 4f^{14} 5d^{10}$, respectively. The ns^2 electrons participate in bonding, and in this regard Zn, Cd, and Hg resemble the group 2 elements.

89. **(M)** HNO_3 is the oxidizing agent for oxidizing the metal to Au^{3+} but Au^{3+} must be stabilized in solution. In aqua regia, Cl^- ions combine with Au^{3+} to form $[AuCl_4]^-$, which is stable in solution.

90. **(M)** The Fe^{3+} ion forms a complex ion, $Fe(H_2O)_6^{3+}$, in aqueous solution. The complex behaves as a weak monoprotic acid in solution. See Equation (23.26).

CHAPTER 24
COMPLEX IONS AND COORDINATION COMPOUNDS
PRACTICE EXAMPLES

1A **(E)** There are two different kinds of ligands in this complex ion, I^- and CN^-. Both are monodentate ligands, that is, they each form only one bond to the central atom. Since there are five ligands in total for the complex ion, the coordination number is 5: C.N. = 5. Each CN^- ligand has a charge of -1, as does the I^- ligand. Thus, the O.S. must be such that:
$$O.S. + \left[(4+1) \times (-1) \right] = -3 = O.S. - 5. \text{ Therefore, O.S.} = +2.$$

1B **(E)** The ligands are all CN^-. Fe^{3+} is the central metal ion. The complex ion is $\left[Fe(CN)_6 \right]^{3-}$.

2A **(E)** There are six "Cl^-" ligands (chloride), each with a charge of $1-$. The platinum metal center has an oxidation state of $+4$. Thus, the complex ion is $\left[PtCl_6 \right]^{2-}$, and we need two K^+ to balance charge: $K_2 \left[PtCl_6 \right]$.

2B **(E)** The "SCN^-" ligand is the thiocyanato-S group, with a charge of $1-$, bonding to the central metal ion through the sulfur atom. The "NH_3" ligand is ammonia, a neutral ligand. There are five (penta) ammine ligands bonded to the metal. The oxidation state of the cobalt atom is $+3$. The complex ion is not negatively charged, so its name does not end with "-ate". The name of the compound is pentaamminethiocyanato-S-cobalt(III) chloride.

3A **(M)** The oxalato ligand must occupy two *cis-* positions. Either the two NH_3 or the two Cl^- ligands can be coplanar with the oxalate ligand, leaving the other two ligands axial. The other isomer has one NH_3 and one Cl^- ligand coplanar with the oxalate ligand. The structures are sketched below.

3B **(M)** We start out with the two pyridines, C_5H_5N, located *cis* to each other. With this assignment imposed, we can have the two Cl^- ligands *trans* and the two CO ligands *cis*, the two CO ligands *trans* and the two Cl^- ligands *cis*, or both the Cl^- ligands and the two CO ligands *cis*. If we now place the two pyridines *trans*, we can have either both other pairs *trans*, or both other pairs *cis*. There are five geometric isomers. They follow, in the order described.

4A **(M)** The F^- ligand is a weak field ligand. $\left[MnF_6\right]^{2-}$ is an octahedral complex. Mn^{4+} has three $3d$ electrons. The ligand field splitting diagram for $\left[MnF_6\right]^{2-}$ is sketched below. There are three unpaired electrons.

4B **(M)** Co^{2+} has seven $3d$ electrons. Cl^- is a weak field ligand. H_2O is a moderate field ligand. There are three unpaired electrons in each case. The number of unpaired electrons has no dependence on geometry for either metal ion complex.

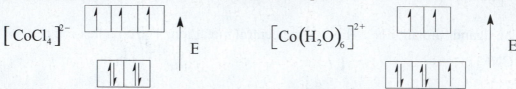

5A **(M)** CN^- is a strong field ligand. Co^{2+} has seven $3d$ electrons. In the absence of a crystal field, all five d orbitals have the same energy. Three of the seven d electrons in this case will be unpaired. We need an orbital splitting diagram in which there are three orbitals of the same energy at higher energy. This is the case with a tetrahedral orbital diagram.

⇅ ⇅ ↑ ↑ ↑   ↑ ↑ ↑   ↑ —

In absence of a

crystal field   ⇅ ⇅   ⇅ ⇅ ⇅

               Tetrahedral geometry   Octahedral geometry

Thus, $\left[Co(CN)_4\right]^{2-}$ must be tetrahedral (3 unpaired electrons) and not octahedral (1 unpaired electron) because the magnetic behavior of a tetrahedral arrangement would agree with the experimental observations (3 unpaired electrons).

5B **(M)** NH_3 is a strong field ligand. Cu^{2+} has nine $3d$ electrons. There is only one way to arrange nine electrons in five d-orbitals and that is to have four fully occupied orbitals (two electrons in each orbital), and one half-filled orbital. Thus, the complex ion must be paramagnetic to the extent of one unpaired electron, regardless of the geometry the ligands adopt around the central metal ion.

6A **(M)** We are certain that $\left[Co(H_2O)_6\right]^{2+}$ is octahedral with a moderate field ligand. Tetrahedral $\left[CoCl_4\right]^{2-}$ has a weak field ligand. The relative values of ligand field splitting for the same ligand are $\Delta_t = 0.44\,\Delta_o$. Thus, $\left[Co(H_2O)_6\right]^{2+}$ absorbs light of higher energy, blue or green light, leaving a light pink as the complementary color we observe. $\left[CoCl_4\right]^{2-}$ absorbs lower energy red light, leaving blue light to pass through and be seen.

6B **(M)** In order, the two complex ions are $\left[Fe(H_2O)_6\right]^{2+}$ and $\left[Fe(CN)_6\right]^{4-}$. We know that CN^- is a strong field ligand; it should give rise to a large value of Δ_o and absorb light of the shorter wavelength. We would expect the cyano complex to absorb blue or violet light and thus $K_4\left[Fe(CN)_6\right]\cdot 3\,H_2O$ should appear yellow. The compound $\left[Fe(H_2O)_6\right](NO_3)_2$, contains the weak field ligand H_2O and thus should be green. (The weak field would result in the absorption of light of long wavelength (namely, red light), which would leave green as the color we observe.)

INTEGRATIVE EXAMPLE

A **(M)**

(a) There is no reaction with $AgNO_3$ or en, so the compound must be *trans*-chloridobis(ethylenediamine)nitrito-*N*-cobalt(III) nitrite

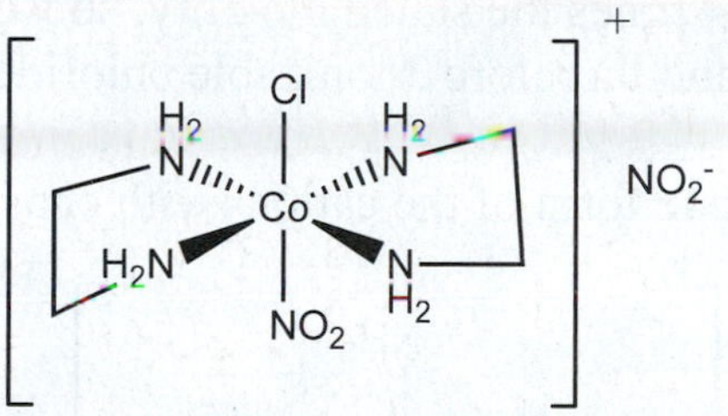

(b) If the compound reacts with $AgNO_3$, but not with en, it must be *trans*-bis(ethylenediamine)dinitrito-*N*-cobalt(III) chloride.

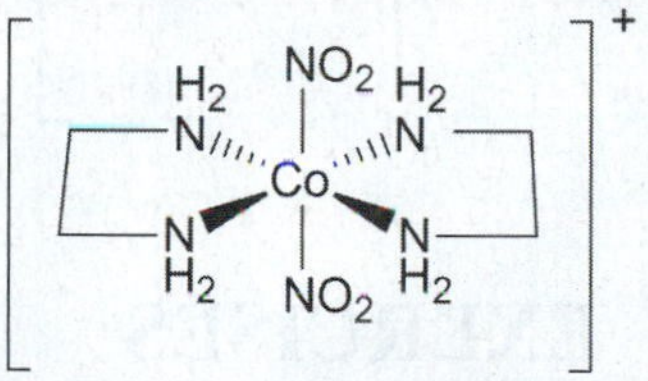

(c) If it reacts with $AgNO_3$ and en and is optically active, it must be *cis*-bis(ethylenediamine)dinitrito-*N*-cobalt(III) chloride.

669

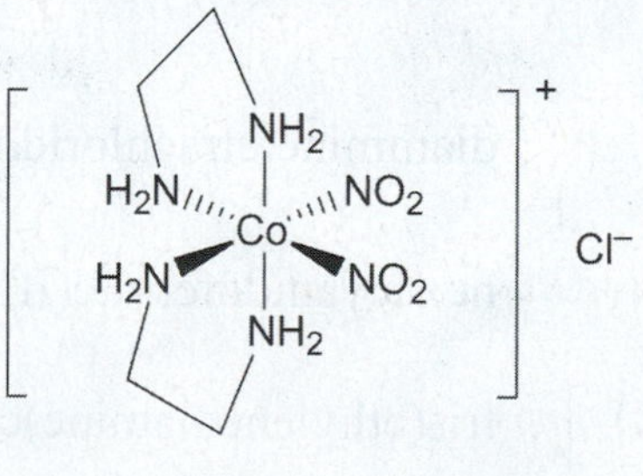

B   **(D)** We first need to compute the empirical formula of the complex compound:

$$46.2 \text{ g Pt} \times \frac{1 \text{ mol Pt}}{195.08 \text{ g Pt}} = 0.237 \text{ mol Pt} \Rightarrow \frac{0.237}{0.237} = 1 \text{ mol Pt}$$

$$33.6 \text{ g Cl} \times \frac{1 \text{ mol Cl}}{35.45 \text{ g Cl}} = 0.948 \text{ mol Cl} \Rightarrow \frac{0.948}{0.237} = 4 \text{ mol Cl}$$

$$16.6 \text{ g N} \times \frac{1 \text{ mol N}}{14.01 \text{ g N}} = 1.18 \text{ mol N} \Rightarrow \frac{1.18}{0.237} = 5 \text{ mol N}$$

$$3.6 \text{ g Pt} \times \frac{1 \text{ mol H}}{1.008 \text{ g H}} = 3.6 \text{ mol H} \Rightarrow \frac{3.6}{0.237} = 15 \text{ mol H}$$

The nitrogen ligand is NH_3, apparently, so the empirical formula is $PtCl_4(NH_3)_5$.

The effective molality of the solution is $m = \dfrac{\Delta T}{K_{fp}} = \dfrac{0.74\,°C}{1.86\,°C/(\text{mol kg}^{-1})} = 0.4 \text{ mol kg}^{-1}$.

The effective molality is 4 times the stated molality, so we have 4 particles produced per mole of Pt complex, and therefore 3 ionizable chloride ions. We can write this in the following way:

$[PtCl(NH_3)_5][Cl_3]$. Only one form of the cation (with charge 3+) shown below will exist.

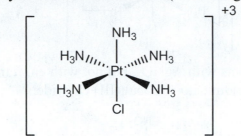

EXERCISES

Nomenclature

1.   **(E)**

    (a) $\left[CrCl_4(NH_3)_2 \right]^-$ diamminetetrachloridochromate(III) ion

    (b) $\left[Fe(CN)_6 \right]^{3-}$ hexacyanidoferrate(III) ion

    (c) $\left[Cr(en)_3 \right]_2 \left[Ni(CN)_4 \right]_3$.   tris(ethylenediamine)chromium(III) tetracyanidonickelate(II) ion

3.   **(M)**
    (a)   diamminesilver(I) chloride
    (b)   tetraamminediaquacopper(II) sulfate
    (c)   dichlorido(1,2-ethanediamine)platinum(II)

(d) pentaaquabromidochromium(III) ion
(e) rubidium tetrafluoridoargentate(III)
(f) sodium pentacyanidooxidonitrogenferrate(III)

Bonding and Structure in Complex Ions

5. **(E)** The Lewis structures are grouped together at the end.

(a) H_2O has $1 \times 2 + 6 = 8$ valence electrons, or 4 pairs.

(b) CH_3NH_2 has $4 + 3 \times 1 + 5 + 2 \times 1 = 14$ valence electrons, or 7 pairs.

(c) ONO^- has $2 \times 6 + 5 + 1 = 18$ valence electrons, or 9 pairs. The structure has a 1– formal charge on the oxygen that is singly bonded to N.

(d) SCN^- has $6 + 4 + 5 + 1 = 16$ valence electrons, or 8 pairs. This structure, appropriately, gives a 1– formal charge to N.

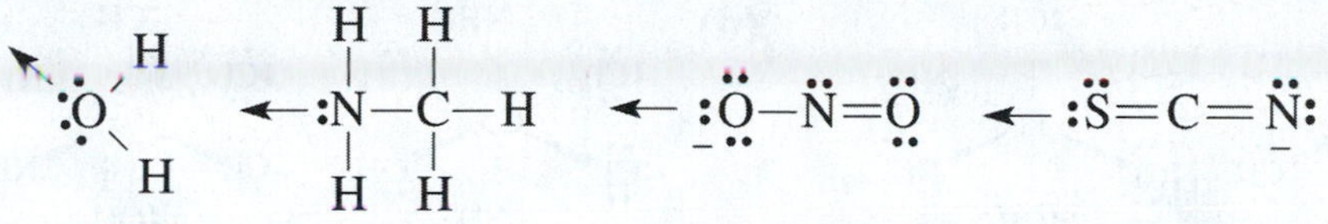

(a) **(b)** **(c)** **(d)**

7. **(M)**

(a)

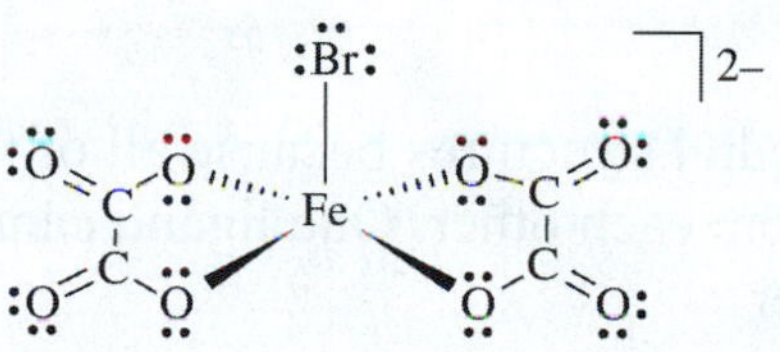

(b)

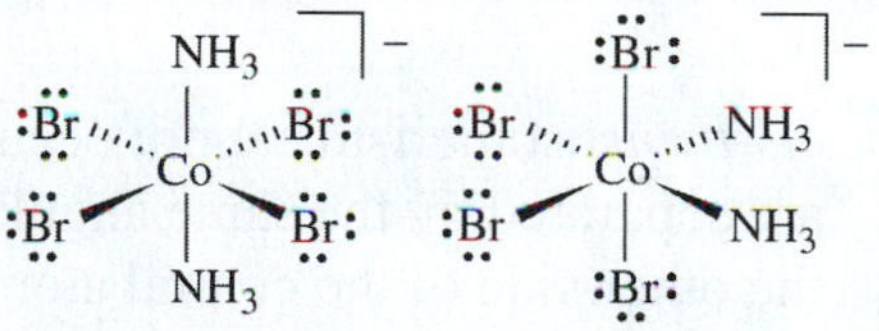

trans ***cis***

(c)

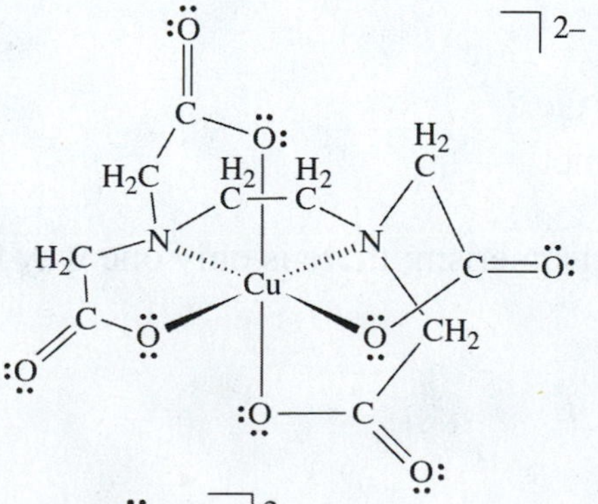

(d)

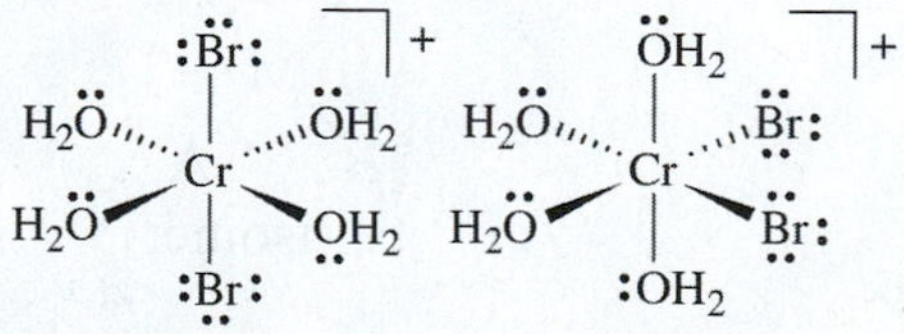

trans ***cis***

(e)

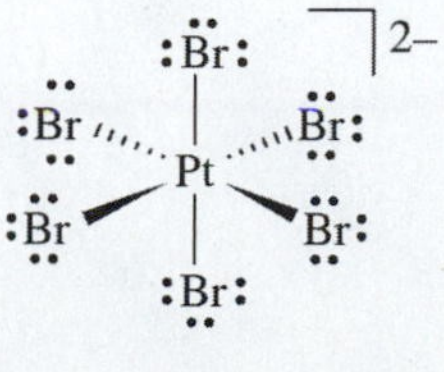

9. (M)

(a)

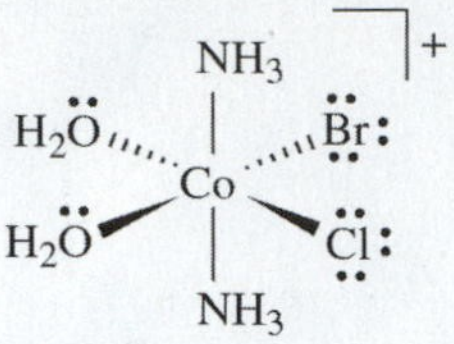

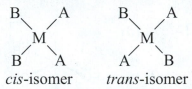

trans-NH₃ *cis*-H₂O   *trans*-NH₃ *trans*-H₂O   *cis*-NH₃ *trans*-H₂O   *cis*-NH₃ *cis*-H₂O

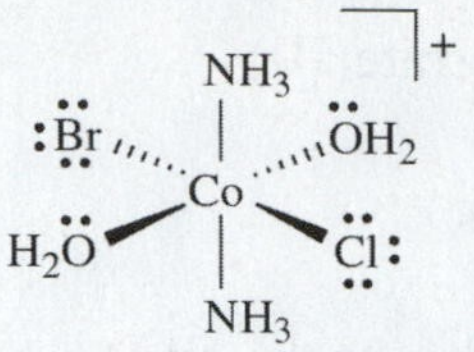

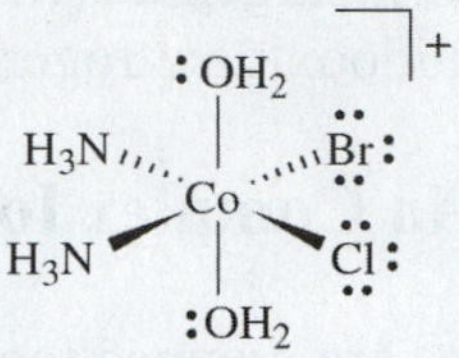

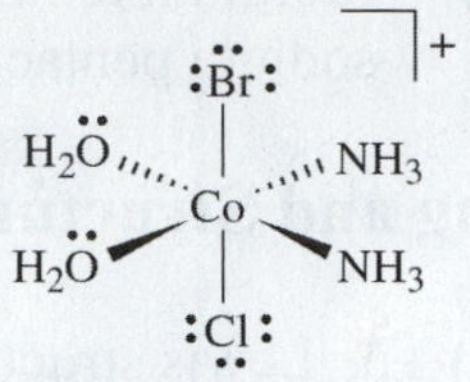

(b)

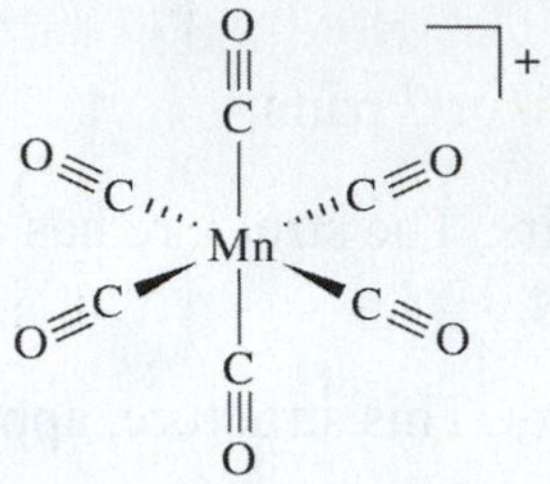

(c)

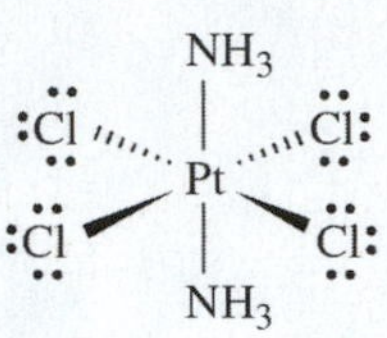

trans 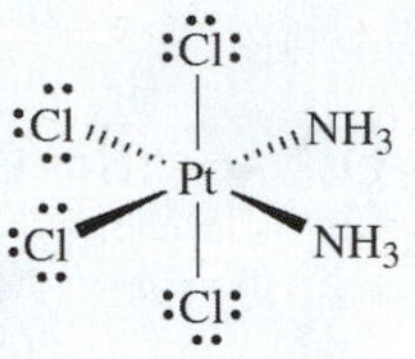   *cis*

(d)

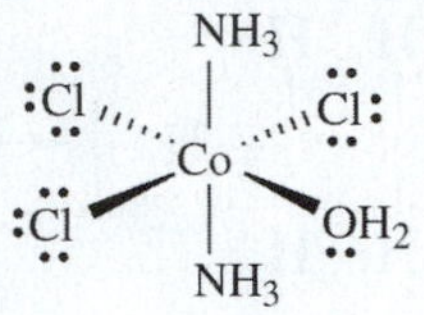

trans 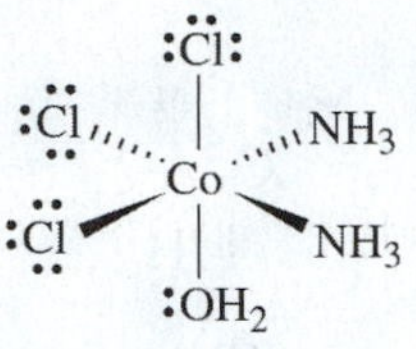   *cis*

Isomerism

11. (E)

(a) *cis-trans* isomerism cannot occur with tetrahedral structures because all of the ligands are separated by the same angular distance from each other. One ligand cannot be on the other side of the central atom from another.

(b) Square planar structures can show *cis-trans* isomerism. Examples are drawn following, with the *cis*-isomer drawn on the left, and the *trans*-isomer drawn on the right.

cis-isomer   *trans*-isomer

(c) Linear structures do not display *cis-trans* isomerism; there is only one way to bond the two ligands to the central atom.

13. **(M)**

(a) There are three different square planar isomers, with D, C, and B, respectively, trans to the A ligand. They are drawn below.

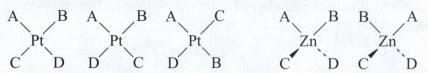

(b) Tetrahedral $[ZnABCD]^{2+}$ does display optical isomerism. The two optical isomers are drawn above.

15. **(M)** The *cis*-dichloridobis(ethylenediamine)cobalt(III) ion is optically active. The two optical isomers are drawn below. The *trans*-isomer is not optically active: the ion and its mirror image are superimposable.

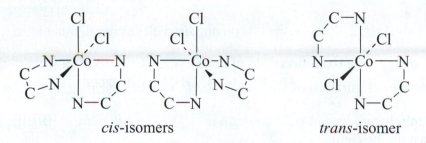

cis-isomers *trans*-isomer

Crystal Field Theory

17. **(E)** In crystal field theory, the five *d* orbitals of a central transition metal ion are split into two (or more) groups of different energies. The energy spacing between these groups often corresponds to the energy of a photon of visible light. Thus, the transition-metal complex will absorb light with energy corresponding to this spacing. If white light is incident on the complex ion, the light remaining after absorption will be missing some of its components. Thus, light of certain wavelengths (corresponding to the energies absorbed) will no longer be present in the formerly white light. The resulting light is colored. For example, if blue light is absorbed from white light, the remaining light will be yellow in color.

19. **(M)** We begin with the 7 electron *d*-orbital diagram for Co^{2+} [Ar]

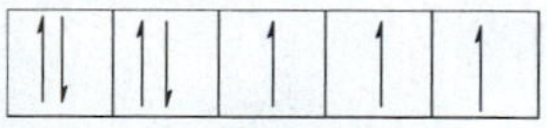

The strong field and weak field diagrams for octahedral complexes follow, along with the number of unpaired electrons in each case.

Strong Field e_g

(1 unpaired electron) t_{2g}

Weak Field e_g

(3 unpaired electrons) t_{2g}

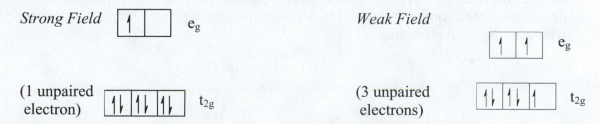

21. **(M)**

(a) Both of the central atoms have the same oxidation state, +3. We give the electron configuration of the central atom to the left, then the completed crystal field diagram in the center, and finally the number of unpaired electrons. The chlorido ligand is a weak field ligand in the spectrochemical series.

Mo^{3+} $[Kr] 4d^3$ *weak field*   ☐☐ e_g

       $[Kr]_{4d}$ ⬛⬛⬛☐☐   ⬆⬆⬆ t_{2g}

                               3 unpaired electrons; paramagnetic

The ethylenediamine ligand is a strong field ligand in the spectrochemical series.

Co^{3+} $[Ar] 3d^6$ *strong field*   ☐☐ e_g

$[Ar]$ ⬛⬛⬛⬛⬛⬛

                           ⬇⬆ ⬇⬆ ⬇⬆ t_{2g}

                    no unpaired electrons; diamagnetic

(b) In $[CoCl_4]^{2-}$ the oxidation state of cobalt is 2+. Chlorido is a weak field ligand. The electron configuration of Co^{2+} is $[Ar] 3d^7$ or $[Ar]$ ⬛⬛⬛⬛⬛

The tetrahedral ligand field diagram is   *weak field*   ⬆⬆⬆ t_{2g}
shown on the right.   ⬇⬆ ⬇⬆ e_g

                                         3 unpaired electrons

23. **(M)** The electron configuration of Ni^{2+} is $[Ar] 3d^8$ or $[Ar]$ ⬛⬛⬛⬛⬛

Ammonia is a strong field ligand. The ligand field diagrams follow, octahedral at left, tetrahedral in the center and square planar at right.

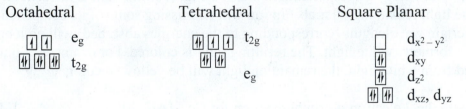

Octahedral   Tetrahedral   Square Planar

Since the octahedral and tetrahedral configurations have the same number of unpaired electrons (that is, 2 unpaired electrons), we cannot use magnetic behavior to determine whether the ammine complex of nickel(II) is octahedral or tetrahedral. But we can determine if the complex is square planar, since the square planar complex is diamagnetic with zero unpaired electrons.

Complex-Ion Equilibria

25. **(E)**

(a) $Zn(OH)_2(s) + 4\,NH_3(aq) \rightleftharpoons \left[Zn(NH_3)_4\right]^{2+}(aq) + 2\,OH^-(aq)$

(b) $Cu^{2+}(aq) + 2\,OH^-(aq) \rightleftharpoons Cu(OH)_2(s)$

The blue color is most likely due to the presence of some unreacted $[Cu(H_2O)_4]^{2+}$ (pale blue)

$Cu(OH)_2(s) + 4\,NH_3(aq) \rightleftharpoons \left[Cu(NH_3)_4\right]^{2+}(aq,\ \text{dark blue}) + 2\,OH^-(aq)$

$\left[Cu(NH_3)_4\right]^{2+}(aq) + 4\,H_3O^+(aq) \rightleftharpoons \left[Cu(H_2O)_4\right]^{2+}(aq) + 4\,NH_4^+(aq)$

27. **(M)** $[Cu(en)_3]^{2+}$ should have the largest overall K_f value. Generally, a complex ion containing polydentate ligands, such as 1,2-ethanediamine (en), will have a larger value for the formation constant than a complex ion containing only monodentate ligands. This phenomenon is known as the *chelate effect*. After one end of a polydentate ligand becomes attached to the central metal, the coordination of the remaining electron pairs is relatively easy because they are already in a close proximity (i.e., no migration from a distant point in the solution is required).

29. **(M)** First: $\left[Fe(H_2O)_6\right]^{3+}(aq) + en(aq) \rightleftharpoons [Fe(en)(H_2O)_4]^{3+}(aq) + 2H_2O(l)$ $K_1 = 10^{4.34}$

Second: $\left[Fe(en)(H_2O)_4\right]^{3+}(aq) + en(aq) \rightleftharpoons \left[Fe(en)_2(H_2O)^+\right]^{3+}(aq) + 2H_2O(l)$ $K_2 = 10^{3.31}$

Third: $\left[Fe(en)_2(H_2O)_2\right]^{3+}(aq) + en(aq) \rightleftharpoons \left[Fe(en)_3\right]^{3+}(aq) + 2H_2O(l)$ $K_3 = 10^{2.05}$

Net: $\left[Fe(H_2O)_6\right]^{3+}(aq) + 3\,en(aq) \rightleftharpoons \left[Fe(en)_3\right]^{3+}(aq) + 6H_2O(l)$ $K_f = K_1 \times K_2 \times K_3$

$\log K_f = 4.34 + 3.31 + 2.05 = 9.70$ $K_f = 10^{9.70} = 5.0 \times 10^9 = \beta_3$

31. **(M)**

(a) Aluminum(III) forms a stable (and soluble) hydroxo complex but not a stable ammine complex.

$\left[Al(H_2O)_3(OH)_3\right](s) + OH^-(aq) \rightleftharpoons \left[Al(H_2O)_2(OH)_4\right]^-(aq) + H_2O(l)$

(b) Although zinc(II) forms a soluble stable ammine complex ion, its formation constant is not sufficiently large to dissolve highly insoluble ZnS. However, it is sufficiently large to dissolve the moderately insoluble $ZnCO_3$. Said another way, ZnS does not produce sufficient $\left[Zn^{2+}\right]$ to permit the complex ion to form.

$ZnCO_3(s) + 4\,NH_3(aq) \rightleftharpoons \left[Zn(NH_3)_4\right]^{2+}(aq) + CO_3^{2-}(aq)$

(c) Chloride ion forms a stable complex ion with silver(I) ion, that dissolves the AgCl(s) that formed when $\left[Cl^- \right]$ is low.

$$AgCl(s) \rightleftharpoons Ag^+(aq) + Cl^-(aq) \qquad \text{and} \qquad Ag^+(aq) + 2\,Cl^-(aq) \rightleftharpoons \left[AgCl_2 \right]^-(aq)$$

$$\text{Overall:} \quad AgCl(s) + Cl^-(aq) \rightleftharpoons \left[AgCl_2 \right]^-(aq)$$

Acid-Base Properties

33. **(E)** $\left[Al(H_2O)_6 \right]^{3+}(aq)$ is capable of releasing H^+:

$$\left[Al(H_2O)_6 \right]^{3+}(aq) + H_2O(l) \rightleftharpoons \left[Al(H_2O)_5OH \right]^{2+}(aq) + H_3O^+(aq)$$

The value of its ionization constant ($pK_a = 5.01$) approximates that of acetic acid.

Applications

35. **(M)**

(a) Solubility: $\qquad\qquad\qquad\qquad AgBr(s) \xrightleftharpoons{K_{sp}=5.0\times10^{-13}} Ag^+(aq) + Br^-(aq)$

Cplx. Ion Formation: $\quad Ag^+(aq) + 2S_2O_3^{2-}(aq) \xrightleftharpoons{K_f=1.7\times10^{13}} \left[Ag(S_2O_3)_2 \right]^{3-}(aq)$

Net: $AgBr(s) + 2S_2O_3^{2-}(aq) \rightleftharpoons \left[Ag(S_2O_3)_2 \right]^{3-}(aq) + Br^-(aq) \qquad K_{overall} = K_{sp} \times K_f$

$K_{overall} = 5.0\times10^{-13} \times 1.7\times10^{13} = 8.5$

With a reasonably high $\left[S_2O_3^{2-} \right]$, this reaction will go essentially to completion.

(b) $NH_3(aq)$ cannot be used in the fixing of photographic film because of the relatively small value of K_f for $\left[Ag(NH_3)_2 \right]^+(aq)$, $K_f = 1.6\times10^7$. This would produce a value of $K = 8.0\times10^{-6}$ in the expression above, which is nowhere large enough to drive the reaction to completion.

The positive value of the standard cell potential indicates that this is a spontaneous reaction.

37. **(M)** To make the *cis* isomer, we must use ligands that show a strong tendency for directing incoming ligands to positions that are *trans* to themselves. I^- has a stronger tendency than does Cl^- or NH_3 for directing incoming ligands to the *trans* positions, and so it is beneficial to convert $K_2[PtCl_4]$ to $K_2[PtI_4]$ before replacing ligands around Pt with NH_3 molecules.

INTEGRATIVE AND ADVANCED EXERCISES

39. **(M)**
(a) cupric tetraammine ion   tetraamminecopper(II) ion $[Cu(NH_3)_4]^{2+}$
(b) dichloridotetraamminecobaltic chloride   tetraamminedichloridocobalt(III) chloride $[CoCl_2(NH_3)_4]Cl$
(c) platinic(IV) hexachlorido ion   hexachloridoplatinate(IV) ion $[PtCl_6]^{2-}$
(d) disodium copper tetrachloride   sodium tetrachloridocuprate(II) $Na_2[CuCl_4]$
(e) dipotassium antimony(III) pentachloride   potassium pentachloridoantimonate(III) $K_2[SbCl_5]$

40. **(E)** $[Pt(NH_3)_4][PtCl_4]$ tetraammineplatinum(II) tetrachloridoplatinate(II)

45. **(M)** The successive acid ionizations of a complex ion such as $[Fe(OH)_6]^{3+}$ are more nearly equal in magnitude than those of an uncharged polyprotic acid such as H_3PO_4 principally because the complex ion has a positive charge. The second proton is leaving a species which has one fewer positive charge but which is nonetheless positively charged. Since positive charges repel each other, successive ionizations should not show a great decrease in the magnitude of their ionization constants. In the case of polyprotic acids, on the other hand, successive protons are leaving a species whose negative charge is increasingly greater with each step. Since unlike charges attract each other, it becomes increasingly difficult to remove successive protons.

47. **(M)**

(a)
$$Fe(H_2O)_6^{3+}(aq) \;+\; H_2O(l) \xrightleftharpoons{K=9.0\times10^{-4}} Fe(H_2O)_5OH^{2+}(aq) \;+\; H_3O^+$$

Initial   0.100 M   –   0 M ≈ 0 M

Equil. $(0.100 - x)$ M   – x x

{where x is the molar quantity of $Fe(H_2O)_6^{3+}(aq)$ hydrolyzed}

$$K = \frac{[[Fe(H_2O)_5OH]^{2+}]\,[H_3O^+]}{[[Fe(H_2O)_6]^{3+}]} = \frac{x^2}{0.100 - x} = 9.0 \times 10^{-4}$$

Solving, we find $x = 9.5 \times 10^{-3}$ M, which is the $[H_3O^+]$ so:

$$pH = -\log(9.5 \times 10^{-3}) = 2.02$$

(b)
$$[Fe(H_2O)_6]^{3+}(aq) \;+\; H_2O(l) \xrightleftharpoons{K=9.0\times10^{-4}} [Fe(H_2O)_5OH]^{2+}(aq) \;+\; H_3O^+$$

initial   0.100 M   –   0 M   0.100 M

equil. $(0.100 - x)$ M   – x M $(0.100 + x)$ M

{where x is $[[Fe(H_2O)_6]^{3+}]$ reacting}

$$K = \frac{[[Fe(H_2O)_5OH]^{2+}]\,[H_3O^+]}{[[Fe(H_2O)_6]^{3+}]} = \frac{x\,(0.100 + x)}{(0.100 - x)} = 9.0 \times 10^{-4}$$

Solving, we find $x = 9.0 \times 10^{-4}$ M, which is the $[[Fe(H_2O)_5OH]^{2+}]$

(c) We simply substitute $[[Fe(H_2O)_5OH]^{2+} = 1.0 \times 10^{-6}$ M into the K_a expression with $[Fe(H_2O)_6]^{3+} = 0.100$ M and determine the concentration of H_3O^+

$$[H_3O^+] = \frac{K_a[[Fe(H_2O)_6]^{3+}]}{[[Fe(H_2O)_5OH]^{2+}]} = \frac{9.0 \times 10^{-4}(0.100 \text{ M})}{[1.0 \times 10^{-6} \text{M}]} = 90. \text{ M}$$

To maintain the concentration at this level requires an impossibly high concentration of H_3O^+

51. **(M)** If a 99% conversion to the chloro complex is achieved, which is the percent conversion necessary to produce a yellow color, $[[CuCl_4]^{2-}] = 0.99 \times 0.10$ M $= 0.099$ M, and $[[Cu(H_2O)_4]^{2+}] = 0.01 \times 0.10$ M $= 0.0010$ M. We substitute these values into the formation constant expression and solve for $[Cl^-]$.

$$K_f = \frac{[[CuCl_4]^{2-}]}{[[Cu(H_2O)_4]^{2+}][Cl^-]^4} = 4.2 \times 10^5 = \frac{0.099 \text{ M}}{0.0010 \text{ M }[Cl^-]^4}$$

$$[Cl^-] = \sqrt[4]{\frac{0.099}{0.0010 \times 4.2 \times 10^5}} = 0.12 \text{ M}$$

This is the final concentration of free chloride ion in the solution. If we wish to account for all chloride ion that has been added, we must include the chloride ion present in the complex ion.

total $[Cl^-] = 0.12$ M free $Cl^- + (4 \times 0.099$ M) bound $Cl^- = 0.52$ M

52. **(M)**

(a) Oxidation: $2H_2O(l) \longrightarrow O_2(g) + 4 H^+(aq) + 4 e^-$ $\qquad -E° = -1.229$ V

Reduction: $\{Co^{3+}((aq) + e^- \longrightarrow Co^{2+}(aq)\} \times 4$ $\qquad E° = +1.82$ V

Net: $4 Co^{3+}(aq) + 2 H_2O(l) \longrightarrow 4 Co^{2+}(aq) + 4 H^+(aq) + O_2(g)$ $E°_{cell} = +0.59$ V

(b) Reaction: $Co^{3+}(aq) + 6 NH_3(aq) \rightleftharpoons [Co(NH_3)_6]^{3+}(aq)$

Initial: $\quad$ 0.1 M $\qquad$ 0.10 M $\qquad\qquad$ 0 M

Changes: $- x$M $\qquad$ constant $\qquad\qquad$ $+ x$M

Equil: $(0.1 - x)$M $\quad$ 0.10 M $\qquad\qquad$ xM

$$K_f = \frac{[[Co(NH_3)_6]^{3+}]}{[Co^{3+}][NH_3]^6} = 4.5 \times 10^{33} = \frac{x}{(0.1-x)(0.10)^6} \qquad \frac{x}{0.1-x} = 4.5 \times 10^{27}$$

Thus $[[Co(NH_3)_6]^{3+}] = 0.1$ M because K is so large, and $[Co^{3+}] = \dfrac{1}{4.5 \times 10^{27}} = 2.2 \times 10^{-28}$ M

(c) The equilibrium and equilibrium constant for the reaction of NH_3 with water follows.

$NH_3(aq) + H_2O(l) \rightleftharpoons NH_4^+(aq) + OH^-(aq)$

$$K_b = 1.8 \times 10^{-5} = \frac{[NH_4^+][OH^-]}{[NH_3]} = \frac{[OH^-]^2}{0.10} \qquad [OH^-] = \sqrt{0.10(1.8 \times 10^{-5})} = 0.0013 \text{ M}$$

In determining the $[OH^-]$, we have noted that $[NH_4^+] = [OH^-]$ by stoichiometry, and also that $[NH_3] = 0.10$ M, as we assumed above.

$$[H_3O^+] = \frac{K_w}{[OH^-]} = \frac{1.0 \times 10^{-14}}{0.0013} = 7.7 \times 10^{-12} \text{ M}$$

We use the Nernst equation to determine the potential of reaction (24.12) at the conditions described.

$$E = E^\circ_{cell} - \frac{0.0257}{4} \ln \frac{[Co^{2+}]^4 \, [H^+]^4 \, P[O_2]}{[Co^{3+}]^4}$$

$$E = +0.59 - \frac{0.0257}{4} \ln \frac{(1 \times 10^{-4})^4 \, (7.7 \times 10^{-12})^4 \, 0.2}{(2.2 \times 10^{-28})^4}$$

$$= +0.59 \text{ V} - 0.732 \text{ V} = -0.142 \text{ V}$$

The negative cell potential indicates that the reaction indeed does not occur.

54. **(M)** We first determine $[Ag^+]$ in the cyanide solution.

$$K_f = \frac{[[Ag(CN)_2]^-]}{[Ag^+][CN^-]^2} = 5.6 \times 10^{18} = \frac{0.10}{[Ag^+] \, (0.10.)^2}$$

$$[Ag^+] = \frac{0.10}{5.6 \times 10^{18} \, (0.10)^2} = 1.8 \times 10^{-18}$$

The cell reaction is as follows. It has $E^\circ_{cell} = 0.000$ V; the same reaction occurs at both anode and cathode and thus the Nernstian voltage is influenced only by the Ag^+ concentration.

$$Ag^+ (0.10 \text{ M}) \longrightarrow Ag^+ (0.10 \text{ M} [Ag(CN)_2]^-, \, 0.10 \text{ M KCN})$$

$$E = E^\circ - \frac{0.0257}{1} \ln \frac{[Ag^+]_{CN}}{[Ag^+]} = 0.000 - 0.0257 \ln \frac{1.8 \times 10^{-18}}{0.10} = +0.99 \text{ V}$$

57. **(M)** The Cr^{3+} ion would have 3 unpaired electrons, each residing in a $3d$ orbital and would be sp^3d^2 hybridized. The hybrid orbitals would be hybrids of $4s$, $4p$, and $3d$ (or $4d$) orbitals. Each Cr–NH$_3$ coordinate covalent bond is a σ bond formed when a lone pair in an sp^3 orbital on N is directed toward an empty sp^3d^2 orbital on Cr^{3+}. The number of unpaired electrons predicted by valence bond theory would be the same as the number of unpaired electrons predicted by crystal field theory.

62. **(M)** The coordination compound is face-centered cubic, K^+ occupies tetrahedral holes, while $PtCl_6^{2-}$ occupies octahedral holes.

63. **(M)**

# NH$_3$	0	1	2	3
Formula:	K$_2$[PtCl$_6$]	K[PtCl$_5$(NH$_3$)]	PtCl$_4$(NH$_3$)$_2$	[PtCl$_3$(NH$_3$)$_3$]Cl
Total # ions	3	2	0	2
(per formula				
unit)				

# NH_3	4	5	6
Formula:	$[PtCl_2(NH_3)_4]Cl_2$	$[PtCl(NH_3)_5]Cl_3$	$[Pt(NH_3)_6]Cl_4$
Total # ions	3	4	5
(per formula			
unit)			

FEATURE PROBLEMS

64. **(M)**

(a) A trigonal prismatic structure predicts three geometric isomers for $[CoCl_2(NH_3)_4]^+$, which is one more than the actual number of geometric isomers found for this complex ion. All three geometric isomers arising from a trigonal prism are shown below.

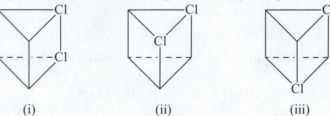

(i)	(ii)	(iii)

The fact that the trigonal prismatic structure does not afford the correct number of isomers is a clear indication that the ion actually adopts some other structural form (i.e., the theoretical model is contradicted by the experimental result). We know now of course, that this ion has an octahedral structure and as a result, it can exist only in *cis* and *trans* configurations.

(b) All attempts to produce optical isomers of $[Co(en)_3]^{3+}$ based upon a trigonal prismatic structure are shown below. The ethylenediamine ligand appears as an arc in diagrams below:

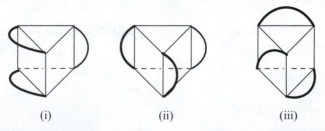

(i)	(ii)	(iii)

Only structure (iii), which has an ethylenediamine ligand connecting the diagonal corners of a face can give rise to optical isomers. Structure (iii) is highly unlikely, however, because the ethylenediamine ligand is simply too short to effectively span the diagonal distance across the face of the prism. Thus, barring any unusual stretching of the ethylenediamine ligand, a trigonal prismatic structure cannot account for the optical isomerism that was observed for $[Co(en)_3]^{3+}$.

SELF-ASSESSMENT EXERCISES

70. **(E)** (d)

71. **(E)** (e)

72. **(E)** (b)

73. **(E)** (a)

74. **(E)** (d)

75. **(E)** (c)

76. **(E)** (b)

77. **(M)**
 (a) pentaamminebromidocobalt(III)sulfate, no isomerism.
 (b) hexaamminechromium(III)hexacyanidocobaltate(III), no isomerism.
 (c) sodiumhexanitrito-N-cobaltate(III), no isomerism
 (d) tris(ethylenediamine)cobalt(III)chloride, two optical isomers.

78. **(M) (a)** $[Ag(CN)_2]^-$; **(b)** $[Pt(NH_3)_3(NO_2)]^+$; **(c)** $[CoCl(en)_2(H_2O)]^{2+}$; **(d)** $K_4[Cr(CN)_6]$

79. **(M)**

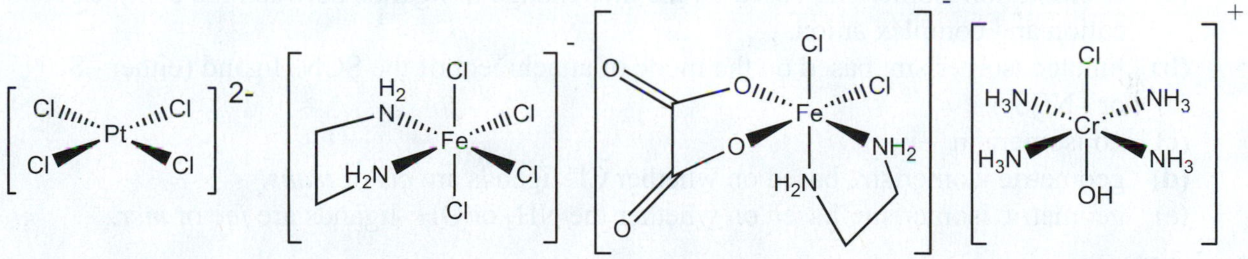

(a)   (b)   (c)   (d)

80. **(M)**
 (a) one structure only (all positions are equivalent for the H_2O ligand; NH_3 ligands attach at the remaining five sites).

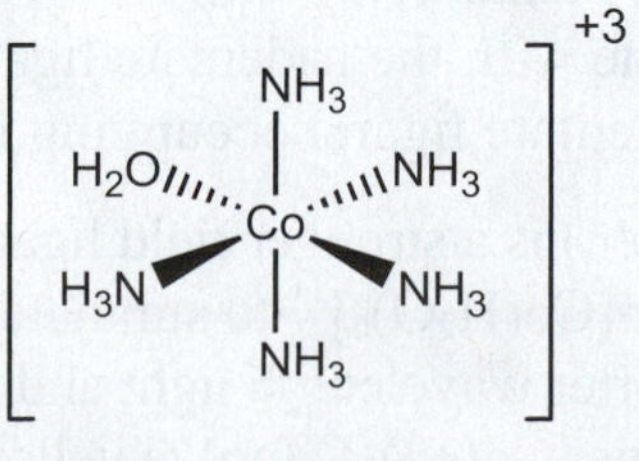

(b) two structures, *cis* and *trans*, based on the placement of H_2O.

$$\left[\begin{array}{c} NH_3 \\ H_2O\cdots\underset{H_3N}{\overset{}{Co}}\cdots NH_3 \\ NH_3 \end{array}\right]^{+3} \qquad \left[\begin{array}{c} NH_3 \\ H_2O\cdots\underset{H_2O}{\overset{}{Co}}\cdots NH_3 \\ NH_3 \end{array}\right]^{+3}$$

trans   *cis*

(c) two structures, *fac* and *mer*.

$$\left[\begin{array}{c} NH_3 \\ H_3N\cdots\underset{H_2O}{\overset{}{Co}}\cdots NH_3 \\ OH_2 \end{array}\right]^{+3} \qquad \left[\begin{array}{c} NH_3 \\ H_2O\cdots\underset{H_2O}{\overset{}{Co}}\cdots OH_2 \\ NH_3 \end{array}\right]^{+3}$$

fac   *mer*

(d) two structures, *cis* and *trans*, based on the placement of NH_3.

$$\left[\begin{array}{c} NH_3 \\ H_2O\cdots\underset{H_2O}{\overset{}{Co}}\cdots OH_2 \\ NH_3 \end{array}\right]^{+3} \qquad \left[\begin{array}{c} OH_2 \\ H_3N\cdots\underset{H_3N}{\overset{}{Co}}\cdots OH_2 \\ OH_2 \end{array}\right]^{+3}$$

trans   *cis*

81. **(M)**
   (a) coordination isomerism, based on the interchange of ligands between the complex cation and complex anion.
   (b) linkage isomerism, based on the mode of attachment of the SCN^- ligand (either $-SCN^-$ or $-NCS^-$).
   (c) no isomerism
   (d) geometric isomerism, based on whether Cl^- ligands are *cis* or *trans*.
   (e) geometric isomerism, based on whether the NH_3 or OH^- ligands are *fac* or *mer*.

82. **(M)**
   (a) geometric isomerism (*cis* and *trans*) and optical isomerism in the *cis* isomer.
   (b) geometric isomerism (*cis* and *trans*), optical isomerism in the *cis* isomer, and linkage isomerism in the thiocyanate ligand.
   (c) no isomers.
   (d) no isomers for this square-planar complex.
   (e) two geometric isomers, one with the tridentate ligand occupying meridional positions and the other with the tridentate ligand occupying facial positions.

83. **(M)** Because ethylenediamine (en) is a stronger field ligand than H_2O, more energy must be absorbed by $[Co(en)_3]^{3+}$ than by $[Co(H_2O)_6]^{3+}$ to simulate an electronic transition. This means that $[Co(en)_3]^{3+}$(aq) absorbs shorter wavelength light and transmits longer wavelength light than does $[Co(H_2O)_6]^{3+}$(aq). Thus $[Co(en)_3]^{3+}$(aq) is yellow and $[Co(H_2O)_6]^{3+}$(aq) is blue.

CHAPTER 25
NUCLEAR CHEMISTRY
PRACTICE EXAMPLES

1A **(E)** A β^- has a mass number of zero and an "atomic number" of -1. Emission of this electron has the effect of transforming a neutron into a proton. $^{241}_{94}Pu \rightarrow ^{241}_{95}Am + ^{0}_{-1}\beta$

1B **(E)** ^{58}Ni has a mass number of 58 and an atomic number of 28. A positron has a mass number of 0 and an effective atomic number of $+1$. Emission of a positron has the seeming effect of transforming a proton into a neutron. The parent nuclide must be copper-58.

$^{58}_{29}Cu \rightarrow ^{58}_{28}Ni + ^{0}_{+1}\beta$

2A **(E)** The sum of the mass numbers $(139 + 12 = ? + 147)$ tells us that the other product species has $A = 4$. The atomic number of La is 57, that of C is 6, and that of Eu is 63. The atomic number sum $(57 + 6 = ? + 63)$ indicates that the atomic number of this product species is zero. Therefore, four neutrons must have been emitted. $^{139}_{57}La + ^{12}_{6}C \rightarrow ^{147}_{63}Eu + 4^{0}_{1}n$

2B **(E)** An alpha particle is $^{4}_{2}He$ and a positron is $^{0}_{+1}\beta$. We note that the total mass number in the first equation is 125; the mass number of the additional product is 1. The total atomic number is 53; the atomic number of the additional product is 0; it is a neutron. $^{121}_{51}Sb + ^{4}_{2}He \rightarrow ^{124}_{53}I + ^{1}_{0}n$

In the second equation, the positron has a mass number of 0, meaning that the mass number of the product is 124. Because the atomic number of the positron is $+1$, that of the product is 52; it is $^{124}_{52}Te$.

$^{124}_{53}I \rightarrow ^{0}_{+1}\beta + ^{124}_{52}Te$

3A **(M) (a)** The decay constant is found from the 8.040-day half-life.

$$\lambda = \frac{0.693}{8.040 \text{ d}} = 0.0862 \text{ d}^{-1} \times \frac{1 \text{ d}}{24 \text{ h}} \times \frac{1 \text{ h}}{60 \text{ min}} \times \frac{1 \text{ min}}{60 \text{ s}} = 9.98 \times 10^{-7} \text{ s}^{-1}$$

(b) The number of ^{131}I atoms is used to find the activity.

$$\text{no.}^{131}I \text{ atoms} = 2.05 \text{ mg} \times \frac{1 \text{ g}}{1000 \text{ mg}} \times \frac{1 \text{ mol } ^{131}I}{131 \text{ g } ^{131}I} \times \frac{6.022 \times 10^{23} \text{ atoms}}{1 \text{ mol } ^{131}I}$$

$$= 9.42 \times 10^{18} \text{ atoms } ^{131}I$$

$$\text{activity} = \lambda N = 9.98 \times 10^{-7} \text{ s}^{-1} \times 9.42 \times 10^{18} \text{ atoms} = 9.40 \times 10^{12} \text{ disintegrations / second}$$

(c) We now determine the number of atoms remaining after 16 days. Because two half-lives elapse in 16 days, the number of atoms has been halved twice, to one-fourth (25%) the original number of atoms.

$$N_t = 0.25 \times N_0 = 0.25 \times 9.42 \times 10^{18} \text{ atoms} = 2.36 \times 10^{18} \text{ atoms}$$

(d) The rate after 14 days is determined by the number of atoms present on day 14.

$$\text{rate} = \lambda N_t = 9.98 \times 10^{-7} \text{ s}^{-1} \times 2.36 \times 10^{18} \text{ atoms} = 2.36 \times 10^{12} \text{ dis/s}$$

3B **(M)** First we determine the value of λ: $\lambda = \dfrac{0.693}{t_{1/2}} = \dfrac{0.693}{11.43 \text{ d}} = 0.0606 \text{ d}^{-1}$

Then we set $N_t = 1\% N_0 = 0.010 N_0$ in equation (25.12).

$$\ln \frac{N_t}{N_0} = -\lambda t = \ln \frac{0.010 N_0}{N_0} = \ln(0.010) = -4.61 = -(0.0608 \text{ d}^{-1})t$$

$$t = \frac{-4.61}{-0.0608 \text{ d}^{-1}} = 75.8 \text{ d}$$

4A **(M)** The half-life of ^{14}C is 5730 a and $\lambda = 1.21 \times 10^{-4} \text{ a}^{-1}$. The activity of ^{14}C when the object supposedly stopped growing was 15 dis/min per g C. We use equation (25.12) with activities (λN) in place of numbers of atoms (N).

$$\ln \frac{A_t}{A_0} = -\lambda t = \ln \frac{8.5 \text{ dis / min}}{15 \text{ dis / min}} = -(1.21 \times 10^{-4} \text{ a}^{-1})t = -0.56_8 \;;\; t = \frac{0.57}{1.21 \times 10^{-4} \text{ a}^{-1}} = 4.7 \times 10^3 \text{ a}$$

4B **(M)** The half-life of ^{14}C is 5730 a and $\lambda = 1.21 \times 10^{-4} \text{ a}^{-1}$. The activity of ^{14}C when the object supposedly stopped growing was 15 dis/min per g C. We use equation (25.12) with activities (λN) in place of numbers of atoms (N).

$$\ln \frac{A_t}{A_0} = -\lambda t = \ln \frac{A_t}{15 \text{ dis / min}} = -(1.21 \times 10^{-4} \text{ a}^{-1})(1100 \text{ a}) = -0.13$$

$$\frac{A_t}{15 \text{ dis/min}} = e^{-0.13} = 0.88 \;,\; A_t = 0.88 \times 15 \text{ dis/min} = 13 \text{ dis/min (per gram of C)}$$

5A **(M)** mass defect. $= 145.913053 \text{ u} \left(^{146}\text{Sm}\right) - 141.907719 \text{ u} \left(^{142}\text{Nd}\right) - 4.002603 \text{ u} \left(^4\text{He}\right) = 0.002731 \text{ u}$

Then, from the text, we have 931.5 MeV $= 1$ u $E = 0.002731 \text{ u} \times \dfrac{931.5 \text{ MeV}}{1 \text{ u}} = 2.544 \text{ MeV}$

5B **(M)** Unfortunately, we cannot use the result of Example 25–5 ($0.0045 \text{ u} = 4.2 \text{ MeV}$) because it is expressed to only two significant figures, and here we begin with four significant figures. But, we essentially work backwards through that calculation. The last conversion factor is from Table 2-1.

$$E = 5.590 \text{ MeV} \times \frac{1.602 \times 10^{-13} \text{ J}}{1 \text{ MeV}} = 8.955 \times 10^{-13} \text{ J} = mc^2 = m\left(2.9979 \times 10^8 \text{ m/s}\right)^2$$

$$m = \frac{8.955 \times 10^{-13} \text{ J}}{\left(2.9979 \times 10^8 \text{ m/s}\right)^2} \times \frac{1000 \text{ g}}{1 \text{ kg}} \times \frac{1.0073 \text{ u}}{1.673 \times 10^{-24} \text{ g}} = 0.005999 \text{ u}$$

Or we could use $\quad m = 5.590 \text{ MeV} \times \dfrac{1 \text{ u}}{931.5 \text{ MeV}} = 0.006001 \text{ u}$

6A **(E) (a)** ^{88}Sr has an even atomic number (38) and an even neutron number (50); its mass number (88) is not too far from the average mass (87.6) of Sr. It should be stable.

(b) ^{118}Cs has an odd atomic number (55) and a mass number (118) that is pretty far from the average mass of Cs (132.9). It should be radioactive.

(c) ^{30}S has an even atomic number (16) and an even neutron number (14); but its mass number (30) is too far from the average mass of S (32.1). It should be radioactive.

6B **(M)** We know that ^{19}F is stable, with approximately the same number of neutrons and protons: 9 protons, and 10 neutrons. Thus, nuclides of light elements with approximately the same number of neutrons and protons should be stable. In Practice Example 25–1 we saw that positron emission has the effect of transforming a proton into a neutron. β^- emission has the opposite effect, namely, the transformation of a neutron into a proton. The mass number does not change in either case. Now let us analyze our two nuclides.

^{17}F has 9 protons and 8 neutrons. Replacing a proton with a neutron would produce a more stable nuclide. Thus, we predict positron emission by ^{17}F to produce ^{17}O.

^{22}F has 9 protons and 13 neutrons. Replacing a neutron with a proton would produce a more stable nuclide. Thus, we predict β^- emission by ^{22}F to produce ^{22}Ne.

INTEGRATIVE EXAMPLE

A. (M) $\lambda = \dfrac{0.693}{1.26 \times 10^9 \text{ a}} = 5.50 \times 10^{-10} \text{ a}^{-1}$ Calculate the fraction of ^{40}K that remains after 1.5×10^9 a.

$$\ln \frac{N_t}{N_0} = -\lambda t = -5.50 \times 10^{-10} \text{ a}^{-1} \times 1.5 \times 10^9 \text{ a} = -0.83 \qquad \frac{N_t}{N_0} = 0.44$$

Thus, the fraction of ^{40}K that has decayed is $1.000 - 0.44 = 0.56$.

The fraction of the ^{40}K that has decayed into ^{40}Ar is $0.110 \times 0.56 = 0.062$.

This fraction is proportional to the mass of ^{40}Ar. Then the ratio of masses is determined.

$$\frac{\text{mass } ^{40}\text{Ar}}{\text{mass } ^{40}\text{K}} = \frac{0.062}{0.44} = 0.14$$

B. **(M) (a)** $Zr(s) + 6H_2O(l) \rightarrow ZrO_2(s) + 4\,H_3O^+(aq) + 4\,e^-$ 1.43 V

            $4\,H_2O(l) + 4\,e^- \rightarrow 2\,H_2(g) + 4\,OH^-(aq)$ -0.828 V

            $Zr(s) + 2\,H_2O(l) \rightarrow ZrO_2(s) + 2\,H_2(g)$ 0.602 V (spont)

            Yes, Zr can reduce water under standard conditions.

     (b) $E° = \dfrac{0.0592}{z} \ln K_{eq}$ $0.602\ \text{V} = \dfrac{0.0592}{4} \ln K_{eq}$ $K_{eq} = 4.67 \times 10^{40}$

     (c) $pH = 7$ Therefore, $[OH^-] = [H_3O^+] = 1.0 \times 10^{-7}$

         $E_{ox} = E°_{ox} - \dfrac{0.0257}{z} \ln Q = 1.43\ \text{V} - \dfrac{0.0257}{4} \ln (1.0 \times 10^{-7})^4 = 1.84\ \text{V}$

         $E_{red} = E°_{red} - \dfrac{0.0257}{z} \ln Q = -0.828\ \text{V} - \dfrac{0.0257}{4} \ln (1 \times 10^{-7})^4 = -0.414\ \text{V}$

         $E_{cell} = E_{ox} + E_{red} = 1.84 + (-0.414) = 1.43\ \text{V (spontaneous)}$

     (d)   Zr may be the culprit responsible for the $H_2(g)$ formation. In the Chernobyl accident, the reaction of carbon with superheated steam played a major role.
         Reaction: $H_2O(g) + C(s) \rightarrow CO_{(g)} + H_2(g)$

EXERCISES

Radioactive Processes

1.   **(E) (a)** $^{55}_{26}Fe \rightarrow {}^{55}_{27}Co + {}^{0}_{-1}e$

     (b) $^{238}_{92}U \rightarrow {}^{234}_{90}Th + {}^{4}_{2}He$

     (c) $^{222}_{86}Rn \rightarrow {}^{218}_{84}Po + {}^{4}_{2}He$; $^{218}_{84}Po \rightarrow {}^{214}_{82}Pb + {}^{4}_{2}He$

     (d) $^{64}_{29}Cu \rightarrow {}^{64}_{30}Zn + {}^{0}_{-1}e$; $^{64}_{30}Zn \rightarrow {}^{64}_{31}Ga + {}^{0}_{-1}e$

3.   **(E)** We would expect a neutron:proton ratio that is closer to 1:1 than that of ^{14}C. This would be achieved if the product were ^{14}N, which is the result of β^- decay: $^{14}_{6}C \rightarrow {}^{14}_{7}N + {}^{0}_{-1}e$.

Radioactive Decay Series

5.   **(M)** We first write conventional nuclear reactions for each step in the decay series.

$^{232}_{90}Th \rightarrow {}^{228}_{88}Ra + {}^{4}_{2}He$ $^{228}_{88}Ra \rightarrow {}^{228}_{89}Ac + {}^{0}_{-1}e$ $^{228}_{89}Ac \rightarrow {}^{228}_{90}Th + {}^{0}_{-1}e$

$^{228}_{90}Th \rightarrow {}^{224}_{88}Ra + {}^{4}_{2}He$ $^{224}_{88}Ra \rightarrow {}^{220}_{86}Rn + {}^{4}_{2}He$ $^{220}_{86}Rn \rightarrow {}^{216}_{84}Po + {}^{4}_{2}He$

Now for a branch in the series:

these two $\qquad$ ${}^{216}_{84}\text{Po} \rightarrow {}^{212}_{82}\text{Pb} + {}^{4}_{2}\text{He}$ $\qquad$ ${}^{212}_{82}\text{Pb} \rightarrow {}^{212}_{83}\text{Bi} + {}^{0}_{-1}\text{e}$

or these two $\qquad$ ${}^{216}_{84}\text{Po} \rightarrow {}^{216}_{85}\text{At} + {}^{0}_{-1}\text{e}$ $\qquad$ ${}^{216}_{85}\text{At} \rightarrow {}^{212}_{83}\text{Bi} + {}^{4}_{2}\text{He}$

And now a second branch:

these two $\qquad$ ${}^{212}_{83}\text{Bi} \rightarrow {}^{208}_{81}\text{Tl} + {}^{4}_{2}\text{He}$ $\qquad$ ${}^{208}_{81}\text{Tl} \rightarrow {}^{208}_{82}\text{Pb} + {}^{0}_{-1}\text{e}$

or these two $\qquad$ ${}^{212}_{83}\text{Bi} \rightarrow {}^{212}_{84}\text{Po} + {}^{0}_{-1}\text{e}$ $\qquad$ ${}^{212}_{84}\text{Po} \rightarrow {}^{208}_{82}\text{Pb} + {}^{4}_{2}\text{He}$

Both branches end at the isotope ${}^{208}_{82}\text{Pb}$. The graph, similar to Figure 25-2, is drawn below.

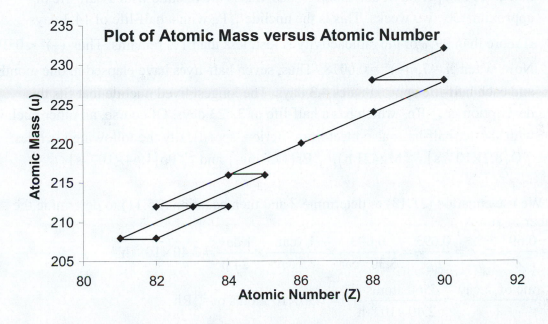

7. **(E)** In Figure 25–2, only the following mass numbers are represented: 206, 210, 214, 218, 222, 226, 230, 234, and 238. We see that these mass numbers are separated from each other by 4 units. The first of them, 206, equals $(4 \times 51) + 2$, that is $4n + 2$, where $n = 51$.

Nuclear Reactions

9. **(E)** **(a)** ${}^{160}_{74}\text{W} \rightarrow {}^{156}_{72}\text{Hf} + {}^{4}_{2}\text{He}$ $\qquad$ **(b)** ${}^{38}_{17}\text{Cl} \rightarrow {}^{38}_{18}\text{Ar} + {}^{0}_{-1}\beta$

 (c) ${}^{214}_{83}\text{Bi} \rightarrow {}^{214}_{84}\text{Po} + {}^{0}_{-1}\beta$ $\qquad$ **(d)** ${}^{32}_{17}\text{Cl} \rightarrow {}^{32}_{16}\text{S} + {}^{0}_{+1}\beta$

11. **(E)** **(a)** ${}^{7}_{3}\text{Li} + {}^{1}_{1}\text{H} \rightarrow {}^{8}_{4}\text{Be} + \gamma$ $\qquad$ **(b)** ${}^{9}_{4}\text{Be} + {}^{2}_{1}\text{H} \rightarrow {}^{10}_{5}\text{B} + {}^{1}_{0}\text{n}$

 (c) ${}^{14}_{7}\text{N} + {}^{1}_{0}\text{n} \rightarrow {}^{14}_{6}\text{C} + {}^{1}_{1}\text{H}$

13. **(E)** ${}^{238}_{92}\text{U} + {}^{2}_{1}\text{H} \rightarrow {}^{238}_{93}\text{Np} + 2{}^{1}_{0}\text{n}$; ${}^{238}_{93}\text{Np} \rightarrow 5{}^{4}_{2}\text{He} + {}^{218}_{83}\text{Bi}$

15. (M) $^{48}_{20}Ca + ^{249}_{98}Cf \rightarrow ^{249}_{118}Unk + ^{1}_{0}n + ^{1}_{0}n + ^{1}_{0}n$

17. (M) $^{58}_{26}Fe + ^{244}_{94}Pu \rightarrow ^{302}_{120}Unk$

Rate of Radioactive Decay

19. (M)

(a) related Since the decay constant is inversely to the half-life, the nuclide with the smallest half-life also has the largest value of its decay constant. This is the nuclide $^{214}_{84}Po$, with a half-life of 1.64×10^{-4} s.

(b) The nuclide that displays a 75% reduction in its radioactivity has passed through two half-lives in a period of one month. Thus, this is the nuclide with a half-life of approximately two weeks. This is the nuclide $^{32}_{15}P$, with a half-life of 14.3 days.

(c) If more than 99% of the radioactivity is lost, less than 1% remains. Thus $(\frac{1}{2})^n < 0.010$. Now, when $n = 7$, $(\frac{1}{2})^n = 0.0078$. Thus, seven half-lives have elapsed in one month, and each half-life approximates 4.3 days. The longest lived nuclide that fits this description is $^{222}_{86}Rn$, which has a half-life of 3.823 days. Of course, all other nuclides with shorter half-lives also meet this criterion, specifically the following nuclides: $^{13}_{8}O (8.7 \times 10^{-3} \text{ s})$, $^{28}_{12}Mg (21 \text{ h})$, $^{80}_{35}Br (17.6 \text{ min})$ and $^{214}_{84}Po (1.64 \times 10^{-4} \text{ s})$.

21. (M) We use equation (25.13) to determine λ and then equation (25.11) to determine the number of atoms.

$$t_{1/2} = \frac{0.693}{\lambda} \Rightarrow \lambda = \frac{0.693}{t_{1/2}} = \frac{0.693}{3.30 \text{ year}} \times \frac{1 \text{ year}}{365 \text{ days}} \times \frac{1 \text{ day}}{24 \text{ h}} = 2.40 \times 10^{-5} \text{ h}^{-1}$$

$$N = \frac{\text{rate of decay}}{\lambda} = \frac{4560 \text{ atoms h}^{-1}}{2.40 \times 10^{-5} \text{ h}^{-1}} = 1.90 \times 10^{8} \text{ atoms of } ^{101}_{45}Rh$$

23. (M) Let us use the first and the last values to determine the decay constant.

$$\ln \frac{R_t}{R_o} = -\lambda t = \ln \frac{138 \text{ cpm}}{1000 \text{ cpm}} = -\lambda 250 \text{ h} = -1.981 \qquad \lambda = \frac{1.981}{250 \text{ h}} = 0.00792 \text{ h}^{-1}$$

$$t_{1/2} = \frac{0.693}{\lambda} = \frac{0.693}{0.00792 \text{ h}^{-1}} = 87.5 \text{ h}$$

A slightly different value of $t_{1/2}$ may result from other combinations of R_o and R_t.

25. (M) $^{32}_{15}P$ half-life = 14.3 d. We need to determine the time necessary to get to the detectable limit, $\frac{1}{1000}$ of the initial value. Use $\lambda = \frac{0.693}{t_{1/2}} = \frac{0.693}{14.3 \text{ d}} = 0.0485 \text{ d}^{-1}$

$$\ln\left(\frac{1}{1000}\right) = -0.0485 \text{ d}^{-1}(t) \qquad t = 142 \text{ days}$$

Age Determinations with Radioisotopes

27. **(E)** Again we use equations (25.12) and (25.13) to determine the time elapsed. The initial rate of decay is about 15 dis/min. First we compute the decay constant.

$$\lambda = \frac{0.693}{5730 \text{ a}} = 1.21 \times 10^{-4} \text{ a}^{-1}$$

$$\ln \frac{10 \text{ dis / min}}{15 \text{ dis / min}} = -0.40_5 = -\lambda t; \qquad t = \frac{0.40_5}{1.21 \times 10^{-4} \text{ a}^{-1}} = 3.4 \times 10^3 \text{ a}$$

The object is a bit more than 3000 years old, and thus is probably not from the pyramid era, which occurred about 3000 B.C.

29. **(M)** First we determine the decay constant. $\lambda = \dfrac{0.693}{1.40 \times 10^{10} \text{ a}} = 4.99 \times 10^{-11} \text{ a}^{-1}$

Then we can determine the ratio of (N_t), the number of thorium atoms after 2.7×10^9 a, to (N_0), the initial number of thorium atoms:

$$\ln \frac{N_t}{N_0} = -kt = -\left(4.99 \times 10^{-11} \text{ a}^{-1}\right)\left(2.7 \times 10^9 \text{ a}\right) = -0.13 \qquad \frac{N_t}{N_0} = 0.88$$

Thus, for every mole of ^{232}Th present initially, after 2.7×10^9 a there are

0.88 mol ^{232}Th and 0.12 mol ^{208}Pb. From this information, we can compute the mass ratio.

$$\frac{0.12 \text{ mol } ^{208}\text{Pb}}{0.88 \text{ mol } ^{232}\text{Th}} \times \frac{1 \text{ mol } ^{232}\text{Th}}{232 \text{ g } ^{232}\text{Th}} \times \frac{208 \text{ g } ^{208}\text{Pb}}{1 \text{ mol } ^{208}\text{Pb}} = \frac{0.12 \text{ g } ^{208}\text{Pb}}{1 \text{ g } ^{232}\text{Th}}$$

31. **(M)** First convert argon-40 to the number of atoms/g in the sample. Next, convert % potassium to atoms/g in the sample. Finally, use equation (25.21) to determine the final answer 3.03×10^9 a.

Energetics of Nuclear Reactions

33. **(M)** The principal equation that we shall employ is $E = mc^2$, along with conversion factors.

(a) $E = 6.02 \times 10^{-23} \text{ g} \times \dfrac{1 \text{ kg}}{1000 \text{ g}} \times \left(3.00 \times 10^8 \text{ m/s}\right)^2 = 5.42 \times 10^{-9} \text{ kg m}^2 \text{ s}^{-2} = 5.42 \times 10^{-9} \text{ J}$

(b) $E = 4.0015 \text{ u} \times \dfrac{931.5 \text{ MeV}}{1 \text{ u}} = 3727 \text{ MeV}$

35. **(E)** The mass defect is the difference between the mass of the nuclide and the sum of the masses of its constituent particles. The binding energy is this mass defect expressed as an energy.

particle mass $\quad = 9p + 10n + 9e = 9(p + n + e) + n$

$\qquad\qquad\qquad = 9(1.0073 + 1.0087 + 0.0005486) \text{ u} + 1.0087 \text{ u} = 19.1576 \text{ u}$

mass defect $\quad = 19.1576 \text{ u} - 18.998403 \text{ u} = 0.1592 \text{ u}$

$$\text{binding energy per nucleon} = \frac{0.1592 \text{ u} \times \dfrac{931.5 \text{ MeV}}{1 \text{ u}}}{19 \text{ nucleons}} = 7.805 \text{ MeV/nucleon}$$

37. **(E)** mass defect $= (10.01294 \ u + 4.00260 \ u) - (13.00335 \ u + 1.00783 \ u) = 0.00436 \ u$

$$\text{energy} = 0.00436 \ u \times \frac{931.5 \ \text{MeV}}{1 \ u} = 4.06 \ \text{MeV}$$

39. **(E)** 1 neutron $\approx$ 1 amu $= 1.66 \times 10^{-27} \ \text{kg}$

$E = mc^2 = 1.66 \times 10^{-27} \ \text{kg} (2.998 \times 10^8 \ \text{m s}^{-1})^2 = 1.49 \times 10^{-10} \ \text{J}$ (1 neutron)

$1 \ \text{eV} = 1.602 \times 10^{-19} \ \text{J}$,

Hence, 1 neutron $= 1.49 \times 10^{-10} \ \text{J} \times \dfrac{1 \ \text{eV}}{1.602 \times 10^{-19} \ \text{J}} = 9.30 \times 10^8 \ \text{eV}$ or 930. MeV

$6.75 \times 10^6 \ \text{MeV} \times \dfrac{1 \ \text{neutron}}{930 \ \text{MeV}} = 7.26 \times 10^3 \ \text{neutrons}$

Nuclear Stability

41. **(E) (a)** We expect ^{20}Ne to be more stable than ^{22}Ne. A neutron-to-proton ratio of 1-to-1 is associated with stability for elements of low atomic number (with $Z \leq 20$).

(b) We expect ^{18}O to be more stable than ^{17}O. An even number of protons and an even number of neutrons are associated with a stable isotope.

(c) We expect ^{7}Li to be more stable than ^{6}Li. Both isotopes have an odd number of protons, but only ^{7}Li has an even number of neutrons.

43. **(M)** β^- emission has the effect of "converting" a neutron to a proton. β^+ emission, on the other hand, has the effect of "converting" a proton to a neutron.

(a) The most stable isotope of phosphorus is ^{31}P, with a neutron-to-proton ratio of close to 1-to-1 and an even number of neutrons. Thus, ^{29}P has "too few" neutrons, or too many protons. It should decay by β^+ emission. In contrast, ^{33}P has "too many" neutrons, or "too few" protons. Therefore, ^{33}P should decay by β^- emission.

(b) Based on the atomic mass of I (126.90447), we expect the isotopes of iodine to have mass numbers close to 127. This means that ^{120}I has "too few" neutrons and therefore should decay by β^+ emission, whereas ^{134}I has "too many" neutrons (or "too few" protons) and therefore should decay by β^- emission.

45. **(M)** A "doubly magic" nuclide is one in which the atomic number is a magic number (2, 8, 20, 28, 50, 82, 114) and the number of neutrons also is a magic number (2, 8, 20, 28, 50, 82, 126, 184). Nuclides that fit this description are given below.

Nuclide	^{4}He	^{16}O	^{40}Ca	^{56}Ni	^{208}Pb
No. of protons	2	8	20	28	82
No. of neutrons	2	8	20	28	126

Fission and Fusion

47. **(E)** We use the conversion factor between number of curies and mass of ^{131}I which was developed in the Integrative Example.

$$\text{no. g } ^{131}I = 170 \text{ curies} \times \frac{18.8 \text{ g } ^{131}I}{2.33 \times 10^6 \text{ curie}} = 1.37 \times 10^{-3} \text{ g} = 1.37 \text{ mg}$$

Effect of Radiation on Matter

49. **(E)** The term "rem" is an acronym for "radiation equivalent-man," and takes into account the quantity of biological damage done by a given dosage of radiation. On the other hand, the rad is the dosage that places 0.010 J of energy into each kilogram of irradiated matter. Thus, for living tissue, the rem provides a good idea of how much tissue damage a certain kind and quantity of radiation damage will do. But for nonliving materials, the rad is usually preferred, and indeed is often the only unit of utility.

51. **(M)** One reason why ^{90}Sr is hazardous is because strontium is in the same family of the periodic table as calcium, and hence often reacts in a similar fashion to calcium. The most likely place for calcium to be incorporated into the body is in bones, where it resides for a long time. Strontium is expected to behave in a similar fashion. Thus, it will be retained in the body for a long time. Bone is an especially dangerous place for a radioisotope to be present—even if it has low penetrating power, as do β^- rays—because blood cells are produced in bone marrow.

Applications of Radioisotopes

53. **(M)** Mix a small amount of tritium with the $H_2(g)$ and detect where the radioactivity appears with a Geiger counter.

55. **(M)** The recovered sample will be radioactive. When $NaCl(s)$ and $NaNO_3(s)$ are dissolved in solution, the ions (Na^+, Cl^-, and NO_3^-) are free to move throughout the solution. A given anion does not remain associated with a particular cation. Thus, all the anions and cations are shuffled and some of the radioactive ^{24}Na will end up in the crystallized $NaNO_3$.

INTEGRATIVE AND ADVANCED EXERCISES

59. **(M)** We use $\Delta H_f^\circ[CO_2(g)] = -393.51$ kJ/mol as the heat of combustion of 1 mole of carbon. In the text, the energy produced by the fission of 1.00 g ^{235}U is determined as 8.20×10^7 kJ.

$$\text{metric tons of coal required} = 1.00 \text{ kg } ^{235}U \times \frac{1000 \text{ g}}{1 \text{ kg}} \times \frac{8.20 \times 10^7 \text{kJ}}{1.00 \text{ g } ^{235}U} \times \frac{1 \text{mol C}}{393.5 \text{ kJ}} \times \frac{12.01 \text{ g C}}{1 \text{mol C}}$$

$$\times \frac{1.00 \text{ g coal}}{0.85 \text{ gC}} \times \frac{1 \text{ kg}}{1000 \text{g}} \times \frac{1 \text{ metric ton}}{1000 \text{ kg}} = 2.9 \times 10^3 \text{metric tons}$$

62. (D) First we find the decay constant. The activity (λN) is the product of the decay constant and the number of atoms.

$$\lambda = \frac{0.693}{29.1\ a} \times \frac{1\ a}{365.25\ d} \times \frac{1\ d}{24\ h} \times \frac{1\ h}{3600\ s} = 7.55 \times 10^{-10}\ s^{-1}$$

$$\text{radioactivity} = 1.00\ \text{mCi} \times \frac{1\ \text{Ci}}{1000\ \text{mCi}} \times \frac{3.7 \times 10^{10}\ \text{dis/s}}{1\ \text{Ci}} = 3.7 \times 10^{7}\ \text{dis/s}$$

$$N = \frac{\text{activity}}{\lambda} = \frac{3.7 \times 10^{7}\ \text{dis/s}}{7.55 \times 10^{-10}\ s^{-1}} = 4.9 \times 10^{16}\ ^{90}\text{Sr atoms} \times \frac{1\ \text{mol}\ ^{90}\text{Sr}}{6.022 \times 10^{23}\ \text{atoms}} \times \frac{90\ g\ ^{90}\text{Sr}}{1\ \text{mol}\ ^{90}\text{Sr}}$$

$$= 7.3 \times 10^{-6}\ g\ ^{90}\text{Sr} = 7.3\ \mu g\ ^{90}\text{Sr}$$

63. (D) $\text{decay rate} = 89.8\ \text{mCi} \times \frac{1\ \text{Ci}}{1000\ \text{mCi}} \times \frac{3.7 \times 10^{10}\ \text{dis/s}}{1\ \text{Ci}} = 3.3_2 \times 10^{9}\ \text{dis/s}$

$$N = 1.00\ \text{mg} \times \frac{1\ g}{1000\ \text{mg}} \times \frac{1\ \text{mol}\ ^{137}\text{Cs}}{137\ g\ ^{137}\text{Cs}} \times \frac{6.022 \times 10^{23}\ \text{atoms}}{1\ \text{mol}} = 4.40 \times 10^{18}\ ^{137}\text{Cs atoms}$$

$$\text{decay rate} = \lambda N \qquad \lambda = \frac{\text{decay rate}}{N} = \frac{3.3_2 \times 10^{9}\ \text{dis/s}}{4.40 \times 10^{18}\ \text{atoms}} = 7.5_5 \times 10^{-10}\ s^{-1}$$

$$t_{1/2} = \frac{0.693}{\lambda} = \frac{0.693}{7.5_5 \times 10^{-10}\ s^{-1}} \times \frac{1\ h}{3600\ s} \times \frac{1\ d}{24\ h} \times \frac{1\ a}{365.25\ d} = 29\ a$$

67. (D) Assume we have in our possession 100 g of the hydrogen/tritium mixture. This sample will afford us 95 g hydrogen and 5 g tritium.

$$\text{mol hydrogen} = \frac{95.00\ gH}{1.008\ g/mol} = 94.246\ \text{mol hydrogen}$$

$$\text{mol tritium} = \frac{5.00\ gH}{3.02\ g/mol} = 1.656\ \text{mol tritium}$$

$$\text{mole fraction tritium} = \frac{1.656\ \text{mol tritium}}{1.656\ \text{mol tritium} + 94.246\ \text{mol hydrogen}} = 1.72_7 \times 10^{-2}$$

$$\text{total moles of gas in mixture} = \frac{PV}{RT} = \frac{(1.05\ \text{atm})(4.65\ L)}{0.0821\dfrac{L\ atm}{mol\ K}(298.15\ K)} = 0.199_5\ \text{mol}$$

$$\text{mols of tritium} = (0.1995\ \text{mol})(1.727 \times 10^{-2}\ \text{mol tritium / mol mixture}) = 3.445 \times 10^{-3}\ \text{mol tritium}$$

$$\text{\# of tritium atoms (N)} = (2 \times 3.445 \times 10^{-3}\ \text{mol tritium})(6.022 \times 10^{23}\ \text{tritium atoms/mol}) = 4.15 \times 10^{21}\ \text{tritium atoms}$$

$$\text{rate} = \frac{0.693}{t_{1/2}} N = \frac{0.693}{12.3\ a \times \dfrac{365\ d}{1\ a} \times \dfrac{24\ h}{1\ d} \times \dfrac{60\ min}{1\ h} \times \dfrac{60\ s}{1\ min}} \times 4.15 \times 10^{21} = 7.42 \times 10^{12}\ \text{disintegrations/s}$$

7

$$\text{activity in curies} = \frac{7.42 \times 10^{12}\ \text{disintegrations/s}}{3.7 \times 10^{10}\ \dfrac{\text{disintegrations/s}}{\text{Ci}}} = 2.0 \times 10^{2}\ \text{Ci}$$

68. **(D)** $\text{energy} = 1.00 \times 10^3 \text{ cm}^3 \times \dfrac{2.5 \text{ g}}{1 \text{ cm}^3} \times \dfrac{0.006 \text{ g U}}{100.000 \text{ g shale}} \times \dfrac{1 \text{ mol U}}{238 \text{ g U}}$

$\times \dfrac{6.022 \times 10^{23} \text{ U atoms}}{1 \text{ mol U}} \times \dfrac{3.20 \times 10^{-11} \text{ J}}{1 \text{ U atom}} \times \dfrac{1 \text{ kJ}}{1000 \text{ J}} = 1.2 \times 10^7 \text{ kJ}$

FEATURE PROBLEMS

72. **(D)** First tabulate the isotope symbols, the mass of the isotope and its associated packing fraction.

Isotope Symbol	Mass of Isotope (u)	Packing Fraction
^{1}H	1.007825	0.007825
^{4}He	4.002603	0.000651
^{9}Be	9.012186	0.001354
^{12}C	12	0
^{16}O	15.994915	−0.000318
^{20}Ne	19.992440	−0.000378
^{24}Mg	23.985042	−0.000623
^{32}S	31.972074	−0.000873
^{40}Ar	39.962384	−0.000940
^{40}Ca	39.962589	−0.000935
^{48}Ti	47.947960	−0.001084
^{52}Cr	51.940513	−0.001144
^{56}Fe	55.934936	−0.001162
^{58}Ni	57.935342	−0.001115
^{64}Zn	63.929146	−0.001107
^{80}Se	79.916527	−0.001043
^{84}Kr	83.911503	−0.001054
^{90}Zr	89.904700	−0.001059
^{102}Ru	101.904348	−0.000938
^{114}Cd	113.903360	−0.000848
^{130}Te	129.906238	−0.000721
^{138}Ba	137.905000	−0.000688
^{142}Nd	141.907663	−0.000650
^{158}Gd	157.924178	−0.000480
^{166}Er	165.932060	−0.000409

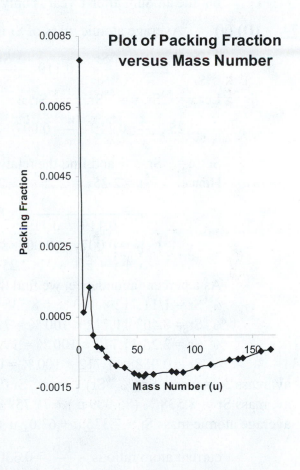

This graph and Fig. 25-6 are almost exactly the inverse of one another, with the maxima of one being the minima of the other. Actual nuclidic mass is often a number slightly less than the number of nucleons (mass number). This difference divided by the number of nucleons (packing fraction) is proportional to the negative of the mass defect per nucleon.

73. **(D)**

(a) The rate of decay depends on both the half-life and the number of radioactive atoms present. In the early stages of the decay chain, the larger number of radium-226, atoms multiplied by the very small decay constant is still larger than the product of the very small number of radon-222 atoms and its much larger decay constant. Only after some time has elapsed, does the rate of decay of radon-222 approach the rate at which it is formed from radium-226 and the amount of radon-222 reaches a maximum. Beyond this point, the rate of decay of radon-222 exceeds its rate of formation.

(b) $\dfrac{dD}{dt} = \lambda_p P - \lambda_d D = \lambda_p N_o e^{-\lambda_p t} - \lambda_d D$

(c) The number of radon-222 atoms at the proposed times are: 2.90×10^{15} atoms after 1 day; 1.26×10^{16} after 1 week; 1.75×10^{16} after 1 year; 1.68×10^{16} after one century; and 1.13×10^{16} after 1 millennium. The actual maximum comes after about 2 months, but the amount after 1 year is only slightly smaller.

74. **(D) (a)** Average atomic mass of Sr in the rock

$\dfrac{^{87}Sr}{^{86}Sr} = 2.25 \qquad \dfrac{^{86}Sr}{^{88}Sr} = 0.119 \qquad \dfrac{^{84}Sr}{^{88}Sr} = 0.007 \qquad$ Given: 15.5 ppm Sr

Let $x = {}^{86}Sr,\ y = {}^{88}Sr,\ z = {}^{87}Sr,\ w = {}^{84}Sr \qquad x + y + z + w = 15.5$ ppm

$\dfrac{z}{x} = 2.25,\quad \dfrac{x}{y} = 0.119,\quad \dfrac{w}{y} = 0.007$

Set $x = {}^{86}Sr = 1$ and find the relative atom ratio of the other

Hence, $\quad z = 2.25x = 2.25 \times 1 = 2.25$

$$y = \dfrac{x}{0.119} = \dfrac{1}{0.119} = 8.40\underline{3}$$

$$w = 0.007y = 0.007 \times 8.40\underline{3} = 0.058\underline{8}$$

$$x + y + z + w = 1 + 2.25 + 8.40\underline{3} + 0.058\underline{8} = 11.7\underline{12}$$

As a percent abundance, we find the following for the Sr in the sample.

$\%^{86}Sr = 1/11.7\underline{12} \times 100\% = 8.53\underline{8}\%$

$\%^{88}Sr = 8.40\underline{3}/11.7\underline{12} \times 100\% = 71.7\underline{5}\%$

$\%^{87}Sr = 2.25/11.7\underline{12} \times 100\% = 19.2\underline{1}\%$

$\%^{84}Sr = 0.058\underline{8}/11.7\underline{12} \times 100\% = 0.5\%$

av. mass Sr = mass^{86}Sr (% ^{86}Sr) + mass^{88}Sr (% ^{88}Sr) + mass^{87}Sr (% ^{87}Sr) + mass^{84}Sr (% ^{84}Sr)

av. mass Sr = 8.53$\underline{8}$ % (85.909 u) + 71.7$\underline{5}$ % (87.906 u)+ 19.2$\underline{1}$ % (86.909 u) + 0.5%(83.913 u)

average atomic mass Sr = 7.33$\underline{5}$ u + 63.0$\underline{7}$ u + 16.6$\underline{95}$ u + 0.4$\underline{2}$ u = 87.5 u

current atom ratio is $\dfrac{^{87}Rb}{^{85}Rb} = 0.330$

Set 1000 atoms for ^{85}Rb and 330 atoms ^{87}Rb or a total of 1330 atoms of Rb

Percent abundance of each isotope: ^{85}Rb = (1000/1330)×100% = 75.2 % ^{85}Rb

$\qquad\qquad\qquad\qquad\qquad\qquad\quad ^{87}$Rb = (1000/1330)×100% = 24.8 % ^{87}Rb

av. mass Rb = mass^{85}Rb (% ^{85}Rb) + mass^{87}Rb (% ^{87}Rb)

av. mass Rb = 75.2 % (84.912 u) + 24.8 % (86.909 u)

average atomic mass Rb = 63.8$\underline{5}$ u + 21.5$\underline{5}$ u = 85.4 u

(b) Original Rb in rock?

Need to convert atom ratio → isotope concentration in ppm.

^{85}Rb concentration in ppm

$$= \frac{1000 \text{ atoms } ^{85}\text{Rb}}{1330 \text{ atoms Rb}} \times \frac{1 \text{ atom Rb}}{85.4 \text{ u Rb}} \times \frac{84.912 \text{ u } ^{85}\text{Rb}}{1 \text{ atom } ^{85}\text{Rb}} \times 265.4 \text{ ppm Rb} = 198.4 \text{ ppm } ^{85}\text{Rb}$$

^{87}Rb concentration in ppm

$$= \frac{330 \text{ atoms } ^{87}\text{Rb}}{1330 \text{ atoms Rb}} \times \frac{1 \text{ atom Rb}}{85.4 \text{ u Rb}} \times \frac{86.909 \text{ u } ^{87}\text{Rb}}{1 \text{ atom } ^{87}\text{Rb}} \times 265.4 \text{ ppm Rb} = 67.0 \text{ ppm } ^{87}\text{Rb}$$

Currently 265.4 ppm (198.4 ppm ^{85}Rb + 67.0 ppm ^{87}Rb)

Recall earlier calculations showed: %^{86}Sr = 8.53$\underline{8}$ %;   %^{88}Sr = 71.7$\underline{5}$ %;
                  %^{87}Sr = 19.2$\underline{1}$ %;   %^{84}Sr = 0.5 %

Consider 100,000 atoms of Sr. Calculate the concentration (in ppm) of ^{86}Sr and ^{87}Sr.

^{86}Sr concentration in ppm

$$= \frac{8538 \text{ atoms } ^{86}\text{Sr}}{100,000 \text{ atoms Sr}} \times \frac{1 \text{ atom Sr}}{87.5 \text{ u Sr}} \times \frac{85.909 \text{ u } ^{86}\text{Sr}}{1 \text{ atom } ^{86}\text{Sr}} \times 15.5 \text{ ppm Sr} = 1.299 \text{ ppm } ^{86}\text{Sr}$$

^{87}Sr concentration in ppm

$$= \frac{19,210 \text{ atoms } ^{87}\text{Sr}}{100,000 \text{ atoms Sr}} \times \frac{1 \text{ atom Sr}}{87.5 \text{ u Sr}} \times \frac{86.909 \text{ u } ^{87}\text{Sr}}{1 \text{ atom } ^{87}\text{Sr}} \times 15.5 \text{ ppm Sr} = 2.957 \text{ ppm } ^{87}\text{Sr}$$

Currently: $\dfrac{^{87}\text{Sr}}{^{86}\text{Sr}} = 2.25 = \dfrac{19,210 \text{ atoms } ^{87}\text{Sr}}{8,538 \text{ atoms } ^{86}\text{Sr}}$

Originally: $\dfrac{^{87}\text{Sr}}{^{86}\text{Sr}} = 0.700$ or $^{87}\text{Sr} = {}^{86}\text{Sr} \times 0.700 = 8,538 \times 0.700 = 5,977$ atoms ^{87}Sr

Change in ^{87}Sr = 192$\underline{10}$ – 597$\underline{7}$ = 132$\underline{33}$ atoms ^{87}Sr (per 100,000 Sr atoms)
Currently, 19210 per 100,000 atoms is ^{87}Sr which represents 2.957 ppm.
A change of 13233 atoms represents (13233/19210)×2.957 ppm = 2.037 ppm ^{87}Sr

The source of ^{87}Sr is radioactive decay from ^{87}Rb (a 1:1 relation).

Change in the ^{87}Rb (through radioactive decay) = change in ^{87}Sr = 2.037 ppm

Isotope:	^{87}Rb	^{85}Rb	Total Rb
Current concentration	67.0 ppm	198.4 ppm	265.4 ppm
Change concentration	+2.037 ppm	—	+2.037 ppm
Original concentration	69.04 ppm	198.4 ppm	267.44 ppm

(c) $\% \, ^{87}\text{Rb decayed} = \left(\dfrac{2.037 \text{ ppm}}{69.04 \text{ ppm}} \right) \times 100\% = 2.95 \% \ (\% \, ^{87}\text{Rb remaining} = 97.05\%)$

(d) $\ln(0.9705) = -\lambda t \qquad \left(\lambda = \dfrac{0.693}{t_{1/2}} = \dfrac{0.693}{4.88 \times 10^{10} \text{ a}} = 1.420 \times 10^{-11} \text{ a}^{-1} \right)$

$\ln(0.9705) = -1.444 \times 10^{-11} \text{ a}^{-1} t; \qquad t = 2.11 \times 10^{9} \text{ years}$

SELF-ASSESSMENT EXERCISES

78. **(E)** (c)

79. **(E)** (b)

80. **(E)** (d)

81. **(E)** (c)

82. **(E)** (c)

83. **(E)** (d)

84. **(E)** (d)

85. **(M)**

(a) $^{214}_{88}\text{Ra} \rightarrow \, ^{210}_{86}\text{Rn} + \, ^{4}_{2}\text{He}$

(b) $^{205}_{85}\text{At} \rightarrow \, ^{205}_{84}\text{Po} + \, ^{0}_{+1}\text{b}$

(c) $^{212}_{87}\text{Fr} + \text{e}^{-} \rightarrow \, ^{212}_{86}\text{Rn}$

(d) $^{2}_{1}\text{H} + \, ^{2}_{1}\text{H} \rightarrow \, ^{3}_{2}\text{He} + \, ^{1}_{0}\text{n}$

(e) $^{241}_{95}\text{Am} + \, ^{4}_{2}\text{He} \rightarrow \, ^{243}_{97}\text{Bk} + 2 \, ^{1}_{0}\text{n}$

(f) $^{232}_{90}\text{Th} + \, ^{4}_{2}\text{He} \rightarrow \, ^{232}_{92}\text{U} + 4 \, ^{1}_{0}\text{n}$

86. **(M)** First use the equation $t_{1/2} = \dfrac{0.693}{\lambda}$ to determine λ from $t_{1/2}$:

$\lambda = \dfrac{0.693}{t_{1/2}} = \dfrac{0.693}{11.43 \text{d}} = 0.0606 \text{ d}^{-1}$

Then use the equation $\ln\left(\dfrac{N_t}{N_0} \right) = -\lambda t$ with $\dfrac{N_t}{N_0} = 0.01$ and solve for t:

$\ln(0.01) = -0.0606 t \Rightarrow t = \dfrac{-4.605}{-0.0606} = 76 \text{ days}$

87. **(M)** First use the equation $t_{1/2} = \dfrac{0.693}{\lambda}$ to determine λ from $t_{1/2}$:

$$\lambda = \frac{0.693}{t_{1/2}} = \frac{0.693}{87.9\text{d}} = 7.88 \times 10^{-3}\text{d}^{-1}$$

Then use the equation $\ln\left(\dfrac{N_t}{N_0}\right) = -\lambda t$ with $\dfrac{N_t}{N_0} = \dfrac{253}{1000}, \dfrac{104}{1000}$, and $\dfrac{52}{1000}$ and solve for t:

$$t_a = \frac{\ln\dfrac{253}{1000}}{7.88 \times 10^{-3}} = 174 \text{ days}$$

$$t_b = \frac{\ln\dfrac{104}{1000}}{7.88 \times 10^{-3}} = 287 \text{ days}$$

$$t_c = \frac{\ln\dfrac{52}{1000}}{7.88 \times 10^{-3}} = 375 \text{ days}$$

88. **(M)** (b)

89. **(M)** (d)

90. **(M)** (c)

91. **(M)** (a)

92. **(M)** (b)

CHAPTER 26
STRUCTURES OF ORGANIC COMPOUNDS
PRACTICE EXAMPLES

1A **(E)** We have shown only the C atoms and the bonds between them. Remember that there are four bonds to each C atom; the remaining bonds not shown are to H atoms. First we realize there is only one isomer with all six C atoms in one line. Then we draw the isomers with one 1-C branch. The isomers with two 1-C branches can have them both on the same atom or on different atoms. This accounts for all five isomers.

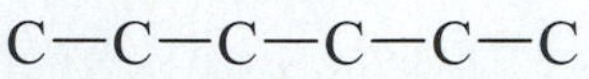

C—C—C—C—C—C

$CH_3CH_2CH_2CH_2CH_2CH_3$

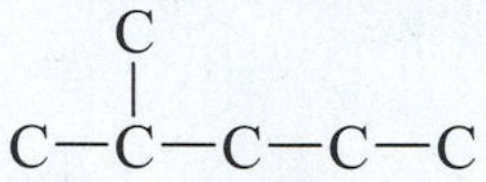

```
      C
      |
C—C—C—C—C
```

$(CH_3)_2CHCH_2CH_2CH_3$

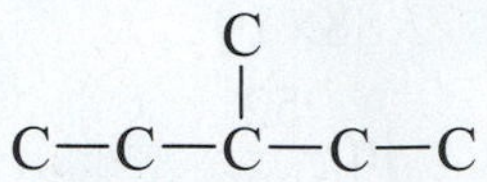

```
      C
      |
C—C—C—C—C
```

$CH_3CH_2CH(CH_3)CH_2CH_3$

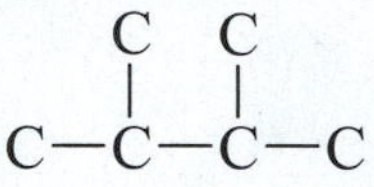

```
    C   C
    |   |
C—C—C—C
```

$(CH_3)_2CHCH(CH_3)_2$

```
      C
      |
C—C—C—C
      |
      C
```

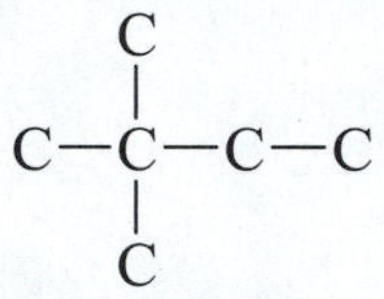

$(CH_3)_3CCH_2CH_3$

1B **(E)** We have shown only the C atoms and the bonds between them. Remember that there are four bonds to each C atom; the remaining bonds not shown are to H atoms.

C—C—C—C—C—C—C

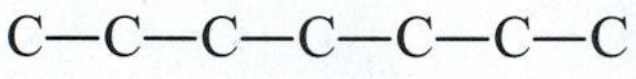

$CH_3CH_2CH_2CH_2CH_2CH_2CH_3$

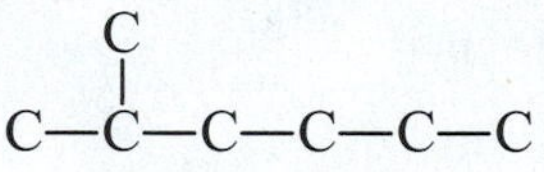

```
      C
      |
C—C—C—C—C
```

$(CH_3)_2CHCH_2CH_2CH_2CH_3$

```
      C
      |
C—C—C—C—C—C
```

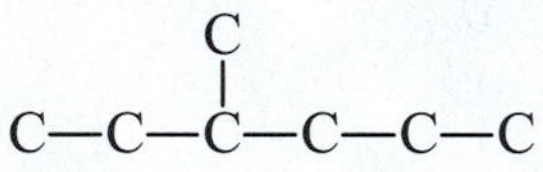

$CH_3CH_2CH(CH_3)CH_2CH_2CH_3$

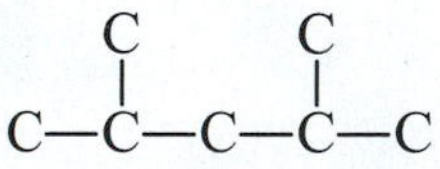

```
    C       C
    |       |
C—C—C—C—C
```

$(CH_3)_2CHCH_2CHC(CH_3)_2$

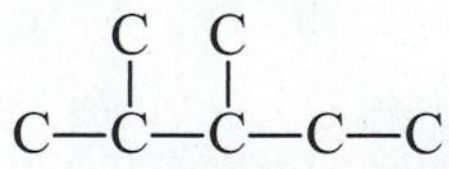

```
    C   C
    |   |
C—C—C—C—C
```

$(CH_3)_2CHCH(CH_3)CH_2CH_3$

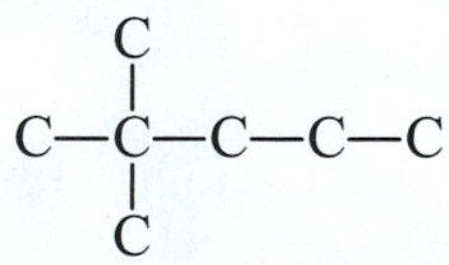

```
      C
      |
C—C—C—C—C
      |
      C
```

$(CH_3)_3CCH_2CH_2CH_3$

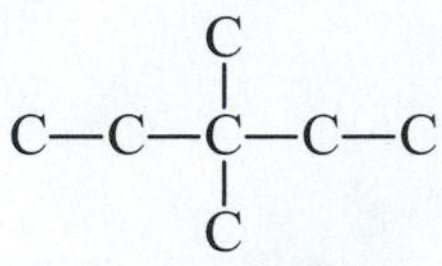

```
      C
      |
C—C—C—C—C
      |
      C
```

$CH_3CH_2C(CH_3)_2CH_2CH_3$

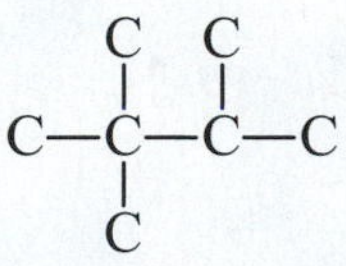

```
    C   C
    |   |
C—C—C—C
        |
        C
```

$(CH_3)_3CCH(CH_3)_2$

```
      C—C
      |
C—C—C—C—C
```

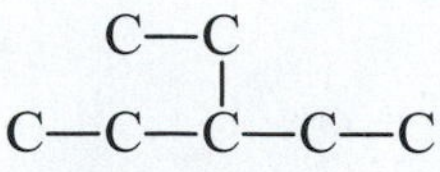

$CH(CH_2CH_3)_3$

2A **(M)**

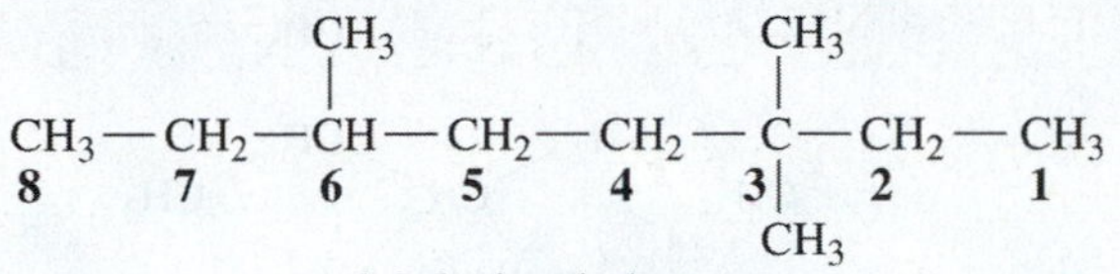

$$CH_3-\underset{8}{CH_2}-\underset{7}{CH_2}-\underset{6}{\overset{\overset{\displaystyle CH_3}{|}}{CH}}-\underset{5}{CH_2}-\underset{4}{CH_2}-\underset{3}{\overset{\overset{\displaystyle CH_3}{|}}{\underset{\underset{\displaystyle CH_3}{|}}{C}}}-\underset{2}{CH_2}-\underset{1}{CH_3}$$

3,3,6-trimethyloctane

Numbering starts from the right and goes left so that the substituents appear with the lowest numbers possible. This is 3,6,6-trimethyloctane.

2B **(M)**

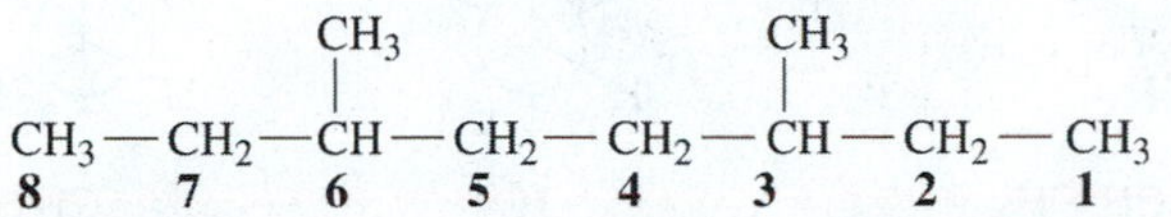

$$CH_3-\underset{8}{CH_2}-\underset{7}{CH_2}-\underset{6}{\overset{\overset{\displaystyle CH_3}{|}}{CH}}-\underset{5}{CH_2}-\underset{4}{CH_2}-\underset{3}{\overset{\overset{\displaystyle CH_3}{|}}{CH}}-\underset{2}{CH_2}-\underset{1}{CH_3}$$

This is a symmetrical molecule. Therefore it does not matter whether numbering starts from left or right. The compound's name is 3,6-dimethyloctane.

3A **(M)**

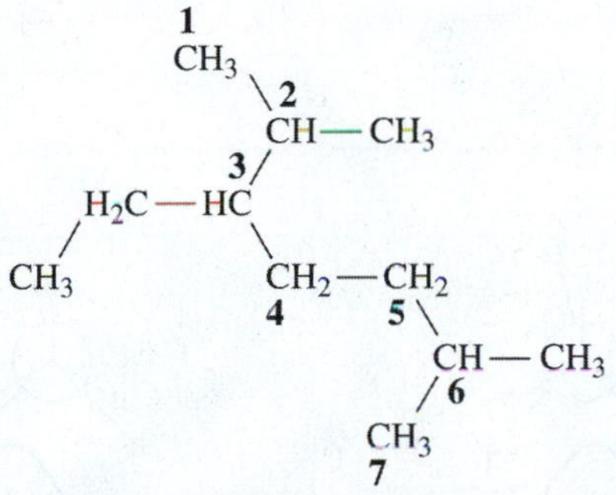

3-ethyl-2,6-dimethylheptane

3B **(M)**

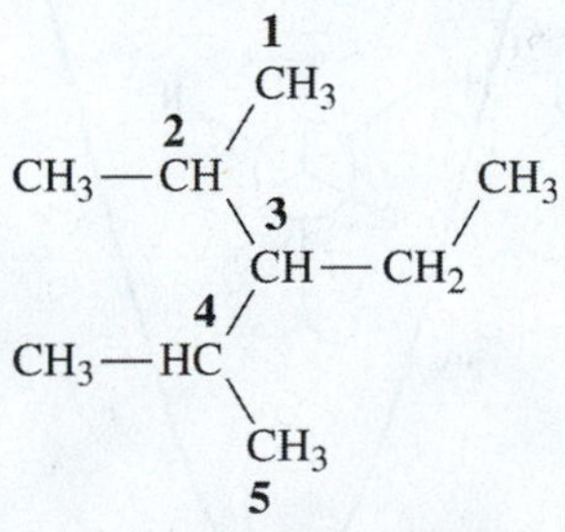

3-ethyl-2,4-dimethylpentane

4A **(M)** The structural diagram for 2-methylpentane is given below:

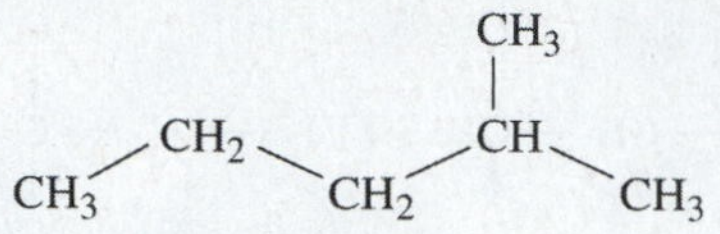

$$CH_3$$
$$|$$
$$CH_2 \qquad CH$$
$$CH_3 \qquad CH_2 \qquad CH_3$$

2-methylpentane

When the molecule is viewed along C1-C2 bond there are several possible staggered conformations:

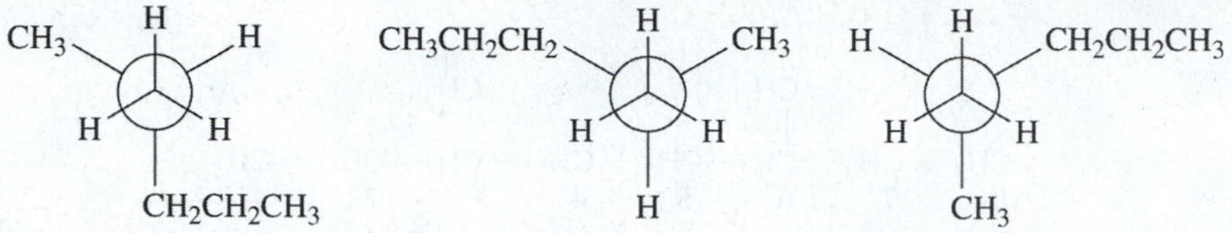

There are no gauche interactions in this molecule. Therefore, conformations (a), (b) and (c) all have the same energy.

4B **(D)** When 1-chloropropane molecule is viewed along the C1-C2 bond, several eclipsed and staggered conformations can be identified:

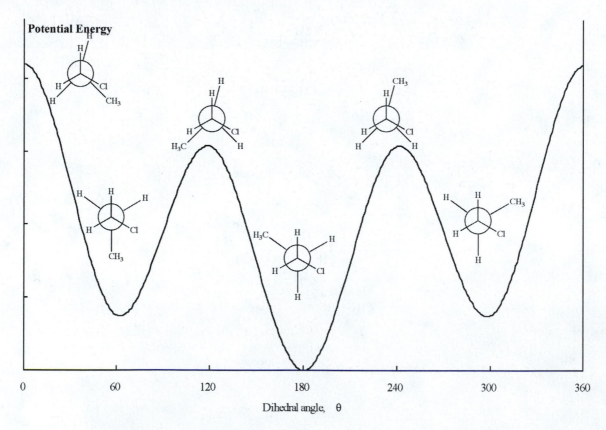

5A **(M)** We are dealing with a trans isomer, and so both methyl groups are adjacent to opposite face of the ring. The conformation of the lowest energy will be the one that has the methyl groups in the equatorial positions.

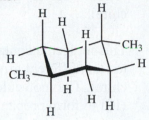

5B **(M)** For this molecule, it is more favorable to place tert-butyl group in the equatorial position.

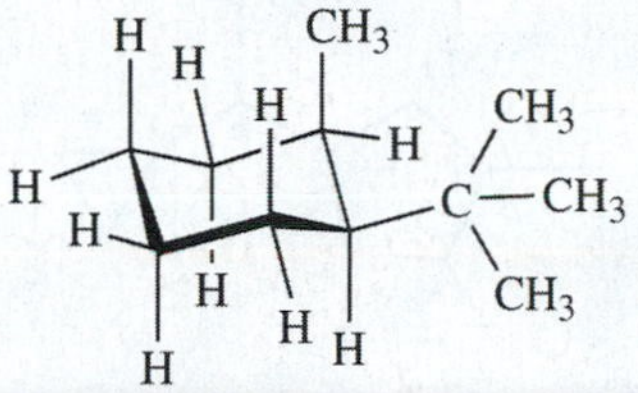

6A **(M)**

(a) All three carbon atoms in this molecule are attached to at least two groups of the same type; thus, the molecule is achiral

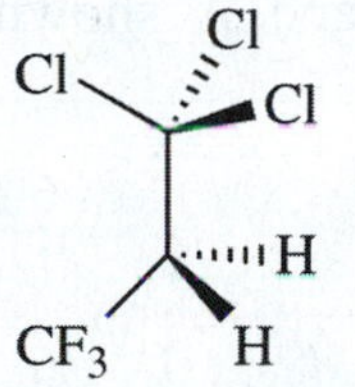

(b) This molecule contains a carbon atom that is bonded to four different groups; consequently, the molecule is chiral.

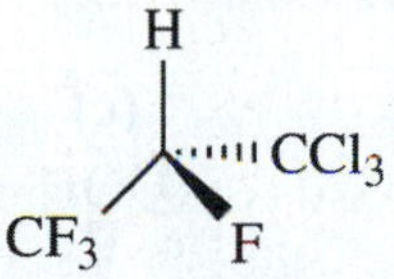

(c) This molecule contains a carbon atom that is attached to four different groups; consequently, the molecule is chiral.

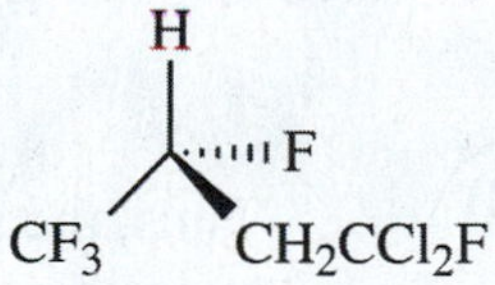

6B **(M)**

(a) None of the three carbon atoms in this alcohol are bonded to four different groups; consequently the molecule is achiral

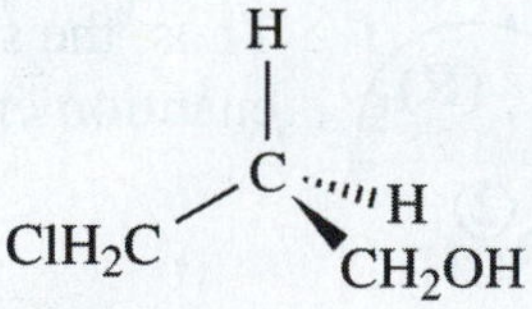

701

(b)

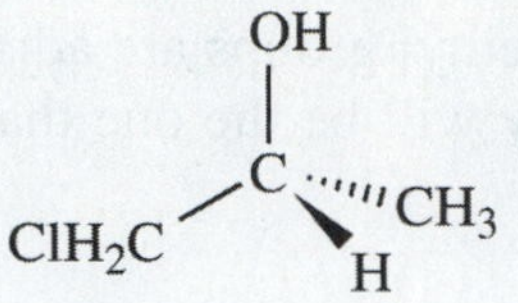

OH
|
ClH₂C—C·‴CH₃
|
H

This molecule contains a carbon atom that is bonded to four different groups; consequently, the molecule is chiral.

(c)

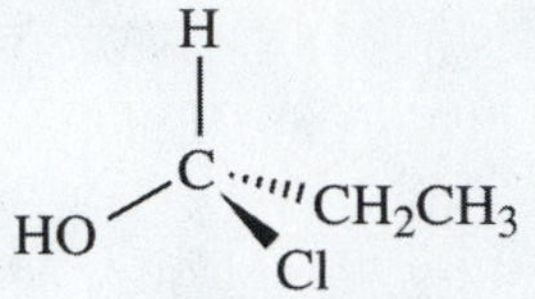

H
|
HO—C·‴CH₂CH₃
|
Cl

This molecule contains a carbon atom that is bonded to four different groups; consequently, the molecule is chiral. The molecule is also unstable and eliminates HCl to form propanal CH_3CH_2CHO.

7A **(M)** Chiral carbon atoms (*) are indicated below:

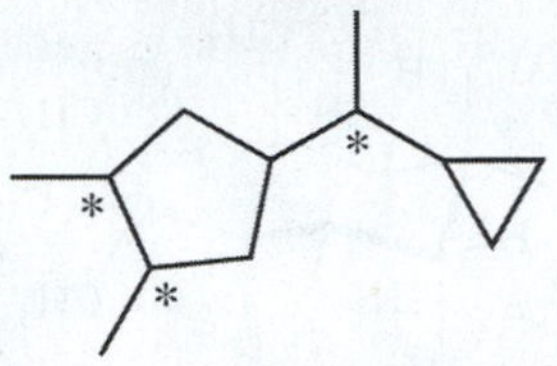

7B **(M)** The molecular structure of 1,1,3-trimethylcyclohexane is shown below and chiral carbon atoms indicated by "*":

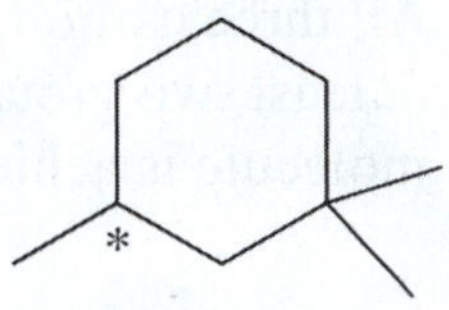

8A **(D)**

(a)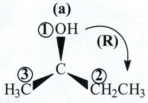

(b)

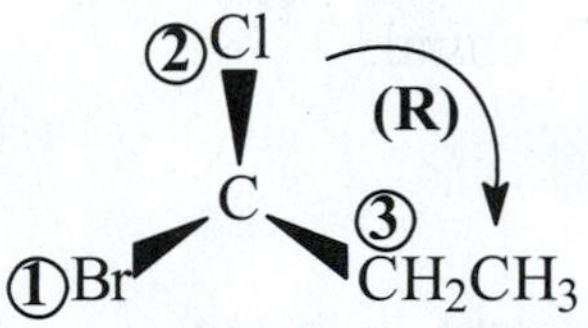

(c)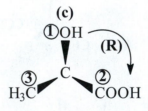

8B **(D)**

(a) 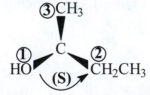 Thus, the structures are enantiomers.

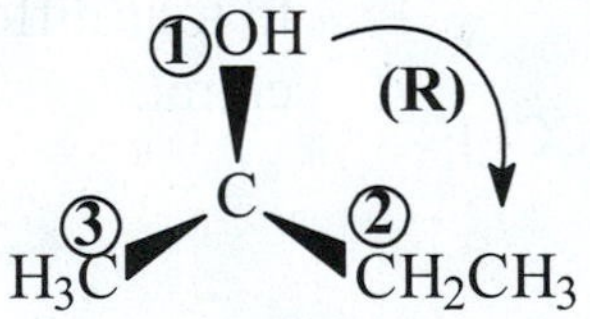

(b) 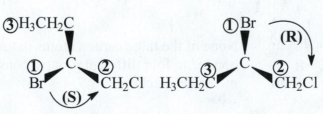 Thus, the structures are enantiomers.

702

9A **(M)** **(a)** This is the (E) stereoisomer

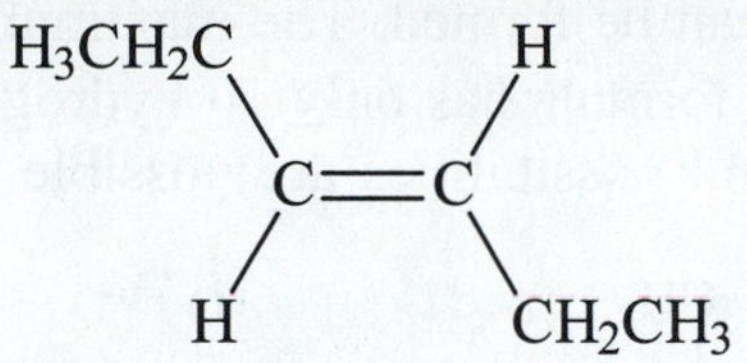

H₃CH₂C, H / C=C / H, CH₂CH₃

H_3CH_2C and H on one carbon; H and CH_2CH_3 on the other; C=C double bond.

(b) This is the (Z) stereoisomer

H_3C, CH_2Br / H, CH_3; C=C double bond.

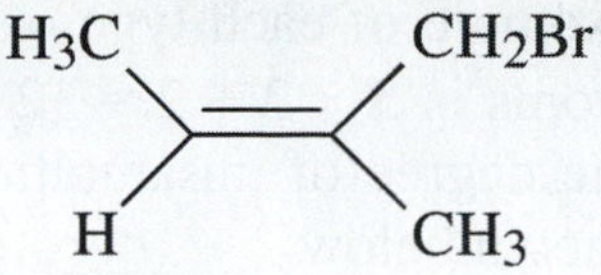

(c) This is the (Z) stereoisomer

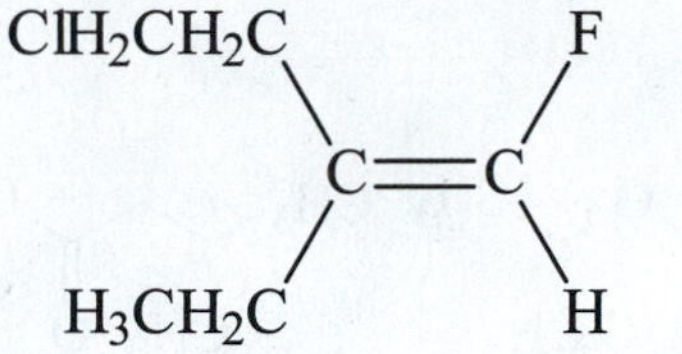

ClH_2CH_2C, F / H_3CH_2C, H; C=C double bond.

9B **(M)** **(a)**

(Z) stereoisomer

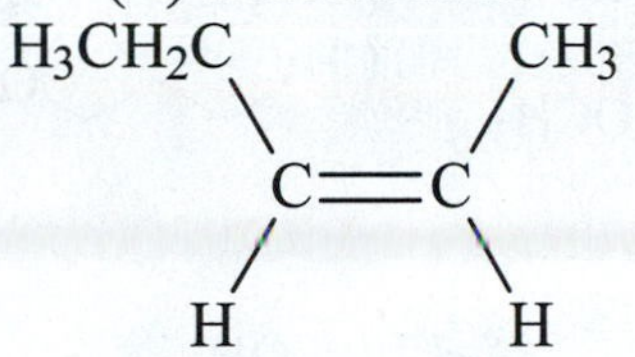

H_3CH_2C, CH_3 / H, H; C=C

(E) stereoisomer

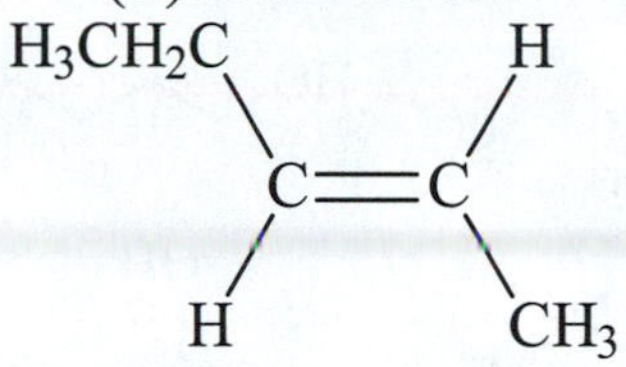

H_3CH_2C, H / H, CH_3; C=C

(b)

(Z) stereoisomer

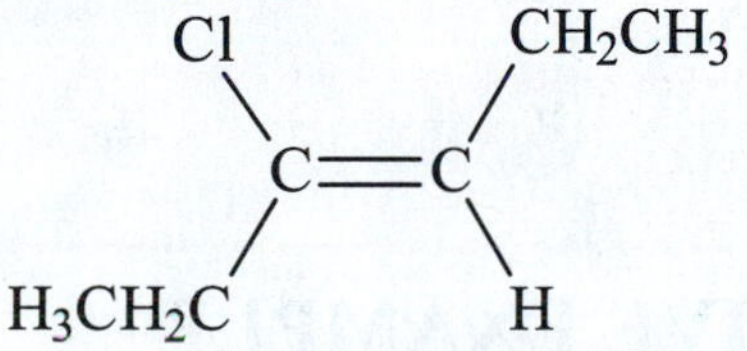

Cl, CH_2CH_3 / H_3CH_2C, H; C=C

(E) stereoisomer

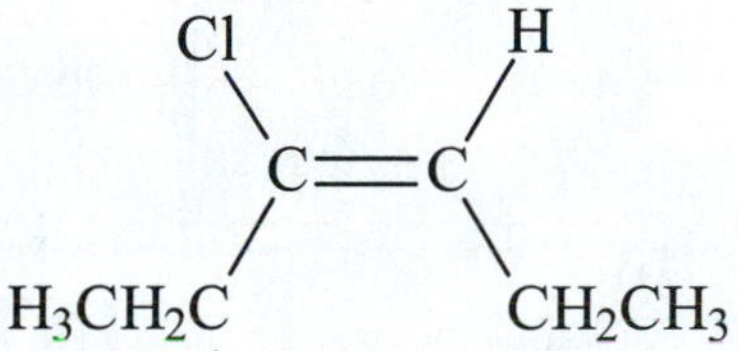

Cl, H / H_3CH_2C, CH_2CH_3; C=C

(c)

(Z) stereoisomer

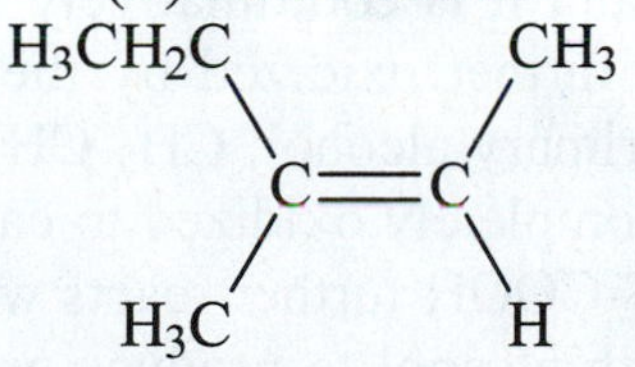

H_3CH_2C, CH_3 / H_3C, H; C=C

(E) stereoisomer

H_3CH_2C, H / H_3C, CH_3; C=C

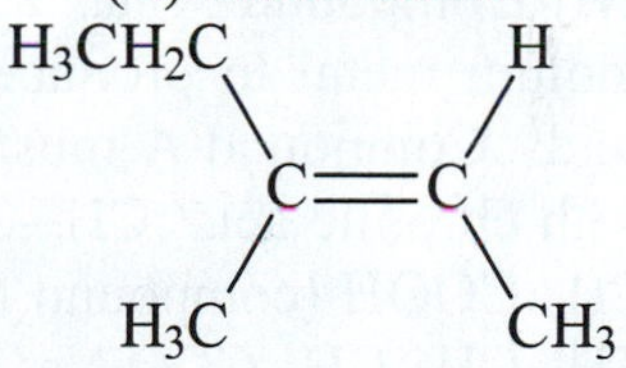

10A **(M)** We need to establish the degree of unsaturation in the molecule and then construct one example of each type of molecule that can be formed. The maximum number of hydrogen atoms is $5 \times 2 + 2 = 12$. The molecular formula has only 11 hydrogen atoms, so we know the degree of unsaturation is one. Several possibilities are possible:

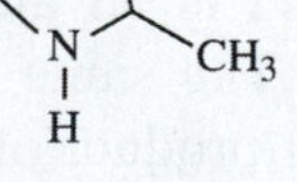

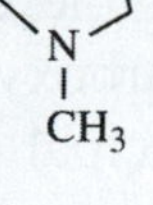

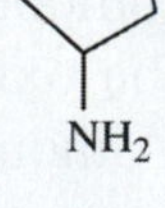

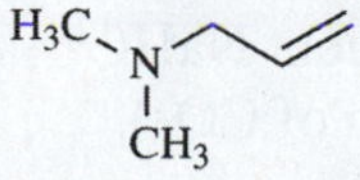

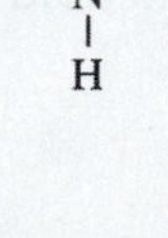

(Structures shown: a pyrrolidine ring with N–H and CH₃ substituent; a pyrrolidine ring with N–CH₃; a cyclopentane ring with NH₂; an H₃C–N(CH₃)–CH₂–CH=CH₂ chain; a piperidine ring with N–H)

10B **(D)** We need to establish the degree of unsaturation in the molecule and then construct one example of each type of molecule that can be formed. The maximum number of hydrogen atoms is $5 \times 2 + 2 = 12$. The molecular formula has only 10 hydrogen atoms, so we know the degree of unsaturation is one. Several possibilities are possible and some of them are shown below:

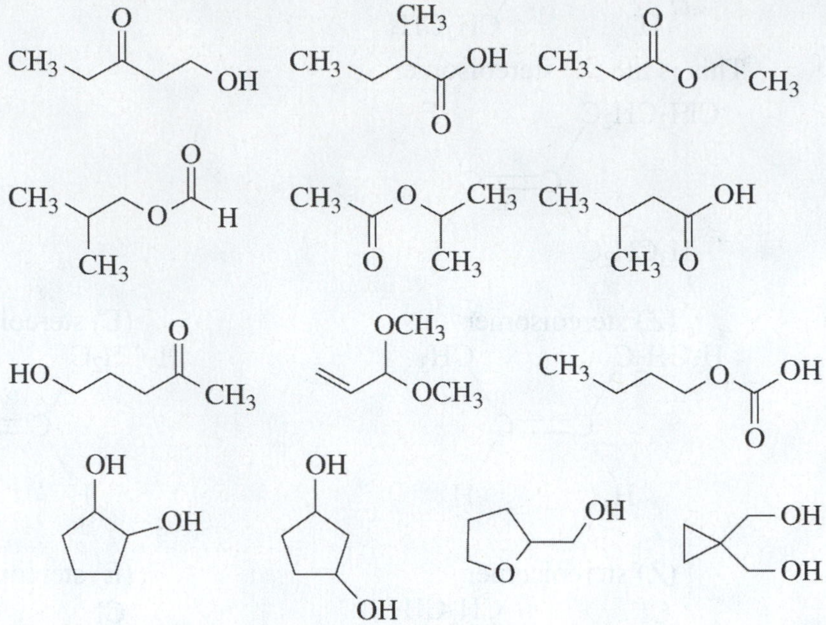

INTEGRATIVE EXAMPLE

A **(M)** Compound A has a formula C_3H_8O and it is coordinatively saturated. It reacts with sodium metal to produce gas and also is further oxidized by the treatment with chromic acid. Compound A must therefore be a primary alcohol, CH_3-CH_2-CH_2-OH. On treatment with chromic acid, CH_3-CH_2-CH_2-OH is completely oxidized to carboxylic acid, CH_3-CH_2-CH_2-COOH (compound B). CH_3-CH_2-CH_2-COOH further reacts with base Na_2CO_3 to yield CH_3-CH_2-CH_2-COO$^-$Na$^+$. It also reacts with ethanol to produce ester CH_3-CH_2-CH_2-COO-CH_2CH_3 (compound C).

B **(M)** Compounds (a) and (b) are both ethers. Ethers are relatively unreactive and the ether linkage is stable in the presence of most oxidizing and reducing agents, as well as dilute acids and alkalis. Compound (c) is unsaturated secondary alcohol. It will decolorize a Br_2/CCl_4 solution. Compound (d), on the other hand is an aldehyde. On treatment with chromic acid it will be oxidized to carboxylic acid. The same reaction would not happen with compound (c). Although the principal oxidation product of the unsaturated secondary alcohols with chromic acid will be a ketone, the reaction of chromic acid with the aldehyde will appear identical to that of the secondary alcohol (i.e., the chromic acid will change from a red solution to a green solution). A subsequent $NaHCO_3$ treatment would be needed to test for the presence of carboxylic acid (evolution of CO_2).

EXERCISES

Organic Structures

1. **(E)** In the following structural formulas, the hydrogen atoms are omitted for clarity. Remember that there are four bonds to each carbon atom. The missing bonds are C—H bonds.

(a)

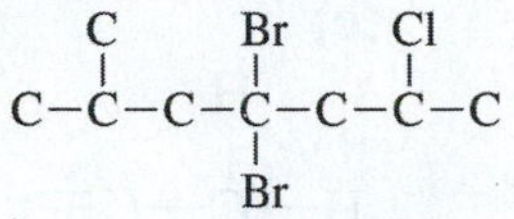

(b)

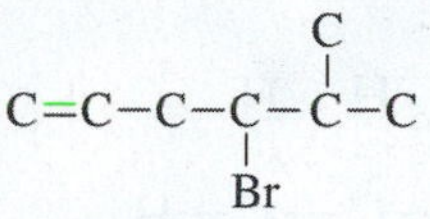

(c)

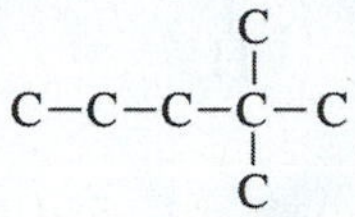

(d)

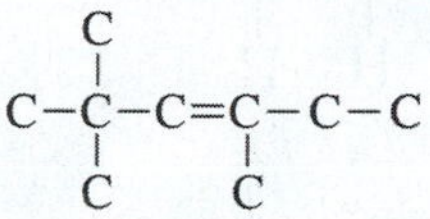

3. **(E)** (a)

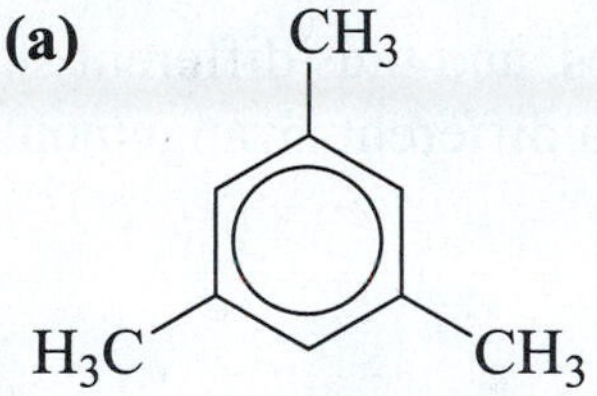

(b)

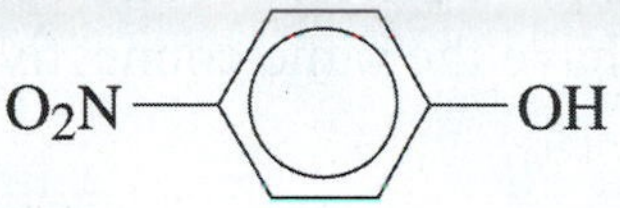

(c)

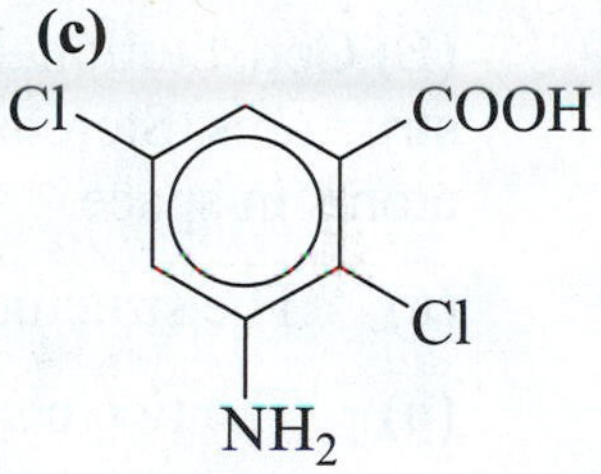

5. **(M)**

(a)

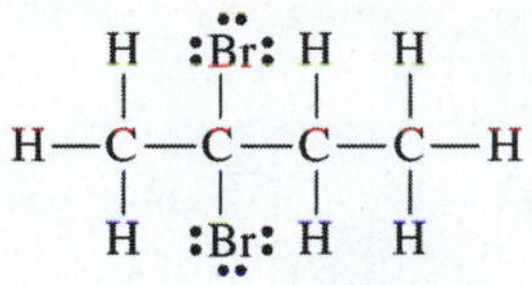

(b)

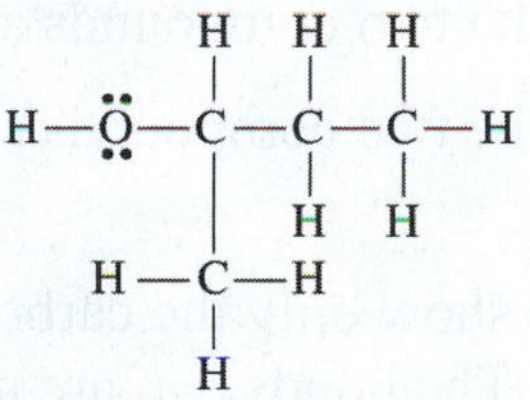

(c)

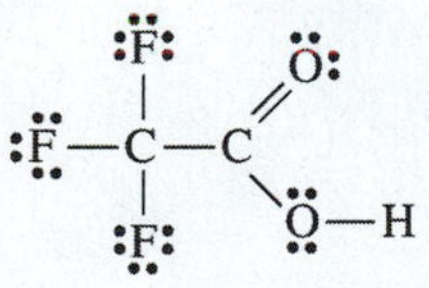

(d)

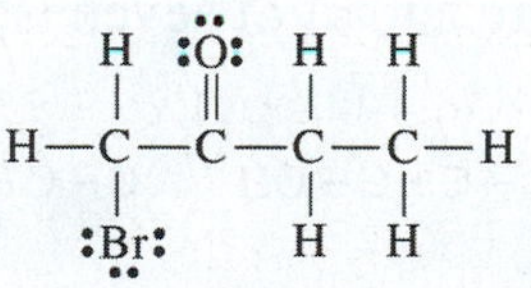

7. **(M)**

(a) Each carbon atom is sp^3 hybridized. All of the C – H bonds in the structure (drawn below) are sigma bonds, between the $1s$ orbital of H and the sp^3 orbital of C. The C–C bond is between sp^3 orbitals on each C atom.

(b) Both carbon atoms are sp^2 hybridized. All of the C – H bonds in the structure (below) are sigma bonds between the $1s$ orbital of H and the sp^2 orbital of C. The C–Cl bond is between the sp^2 orbital on C and the $3p$ orbital on Cl. The C = C double bond is composed of a sigma bond between the sp^2 orbitals on each C atom and a pi bond between the $2p_z$ orbitals on the two C atoms.

(c) The left-most C atom (in the structure drawn below) is sp^3 hybridized, and the C–H bonds to that C atom are between the sp^3 orbitals on C and the $1s$ orbital on H. The other two C atoms are sp hybridized. The right-hand C – H bond is between the sp orbital on C and the $1s$ orbital on H. The $C \equiv C$ triple bond is composed of one sigma bond formed by overlap of sp orbitals, one from each C atom, and two pi bonds, each formed by the overlap of two $2p$ orbitals, one from each C atom (that is a $2p_y - 2p_y$ overlap and a $2p_z - 2p_z$ overlap).

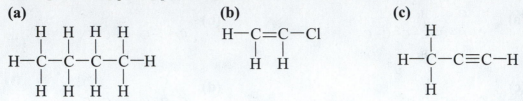

(a)

H H H H
| | | |
H—C—C—C—C—H
| | | |
H H H H

(b)

H—C=C—Cl
| |
H   H

(c)

    H
    |
H—C—C≡C—H
    |
    H

Isomers

9. **(E)** Constitutional isomers have different bond connectivities, and thus different skeletal structures. Stereoisomers have the same connectivities, but a different arrangement of the atoms in space.

(a) The structures are identical.

(b) The two compounds are constitutional isomers.

(c) The two compounds have no relationship. They have different molecular formulas.

(d) The two compounds are constitutional isomers.

(e) The two compounds are stereoisomers.

11. **(D)** We show only the carbon skeleton in each case. Remember that there are four bonds to carbon. The bonds that are not included in the following structures represent C—H bonds. There are a total of seven isomers of pentanol.

C–C–C–C–C–OH   C–C–C–Ċ–C   C–C–Ċ–C–C
                       | |
                       OH   OH

    C   C   C
    | | |
C–Ċ–C–C–OH   C–C–Ċ–C   C–C–Ċ–C
                         | |
                         OH   OH

    C
    |
C–Ċ–C–OH
    |
    C

13. **(M)**

(a)

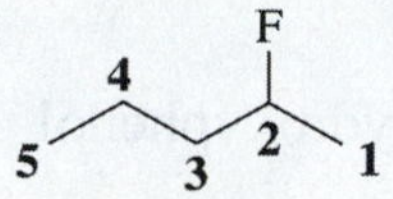

Carbon 2 is chiral.

(b)

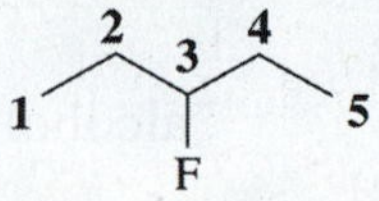

There are no chiral carbon atoms.

(c)

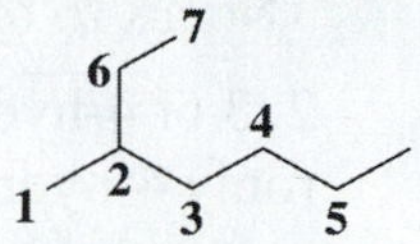

Carbon 2 is chiral.

15. **(M)** Chiral carbon atoms are indicated by *. Compound (b) does not contain any chiral carbon atoms.

(a)

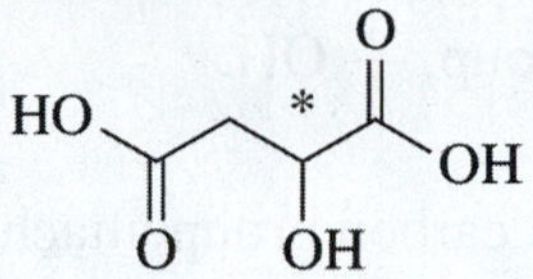

(b)

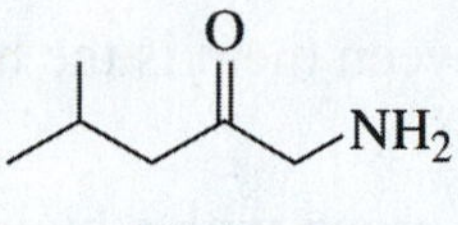

no chiral atoms

(c)

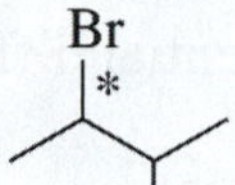

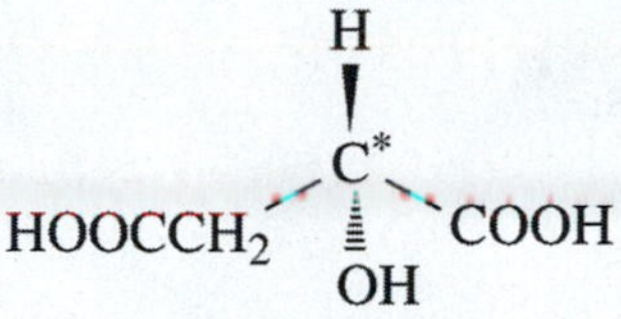

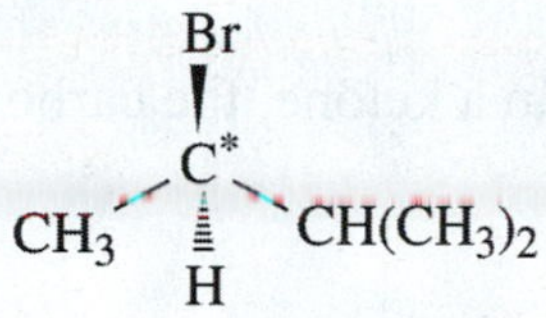

Functional Groups

17. **(E)**

(a)

Br
|
CH$_3$CHCH$_2$CH$_3$

2-Bromobutane

alkyl halide (bromide)

(b)

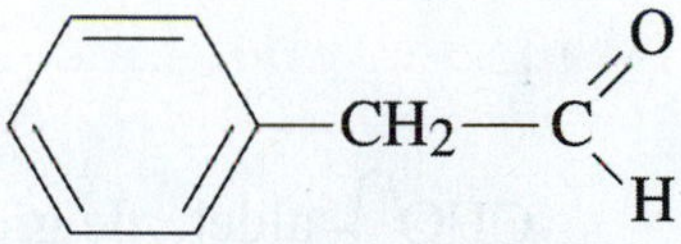

phenylacetaldehyde

aldehyde

(c)

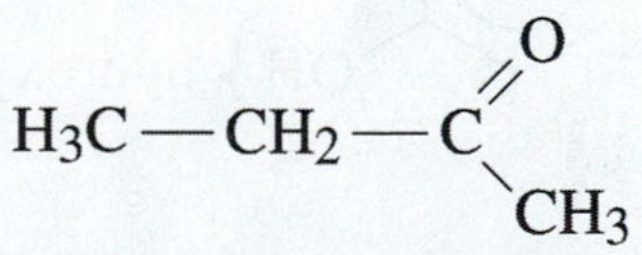

butan-2-one

ketone

(d)

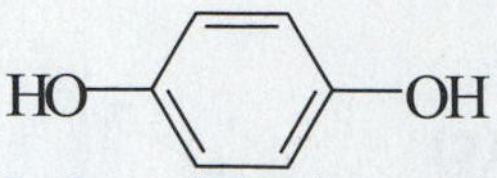

2, 3 or 4-hydroxyphenol   alcohol, specifically a hydroxyphenol
(only 4-hydroxyphenol shown)

19. **(M)**

(a)   A carbonyl group is $>C=O$, whereas a carboxyl group is $-C(O)-OH$

The essential difference between them is the hydroxyl group, $-OH$.

(b)   An aldehyde has a carbonyl group with a hydrogen and a carbon group attached.

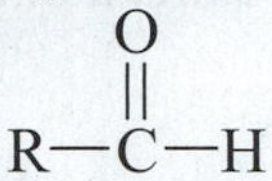

$$
\begin{array}{c}
O \\
\parallel \\
R-C-H
\end{array}
$$

In a ketone, the carbonyl group is attached to two carbons.

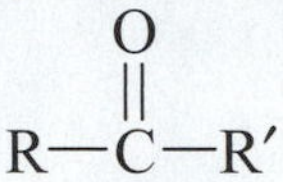

$$
\begin{array}{c}
O \\
\parallel \\
R-C-R'
\end{array}
$$

(c)   Acetic acid is $H_3C-\overset{\overset{\displaystyle O}{\parallel}}{C}-OH$, while an acetyl group is $H_3C-\overset{\overset{\displaystyle O}{\parallel}}{C}-$

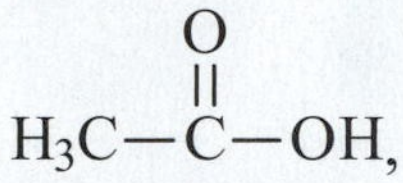

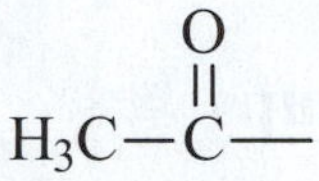

The essential difference is the presence of the $-OH$ group in acetic acid.

21. **(M)**   **(a)**

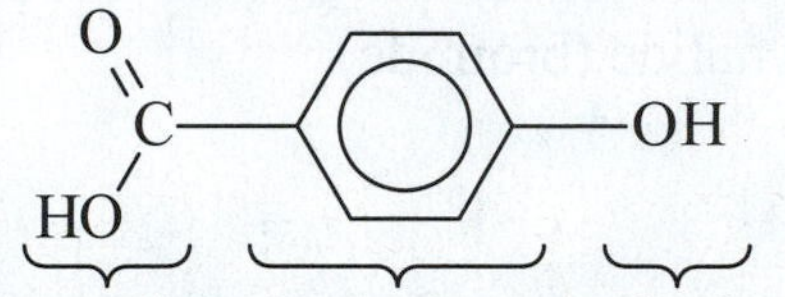

carboxylic   disubstituted   hydroxyl
acid group   aromatic ring   group

(b)

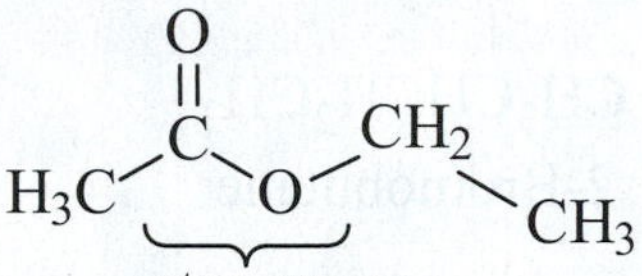

ester group

(c)

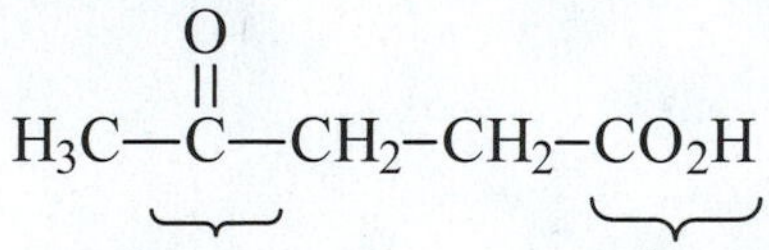

$$H_3C-\overset{\overset{\displaystyle O}{\parallel}}{C}-CH_2-CH_2-CO_2H$$

ketone   carboxylic acid
group   group

(d)

CHO } aldehyde group

disubstituted
aromatic ring

OH } hydroxyl group

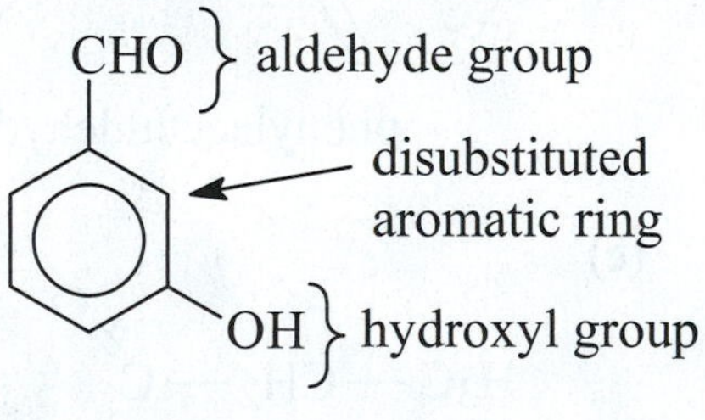

23.

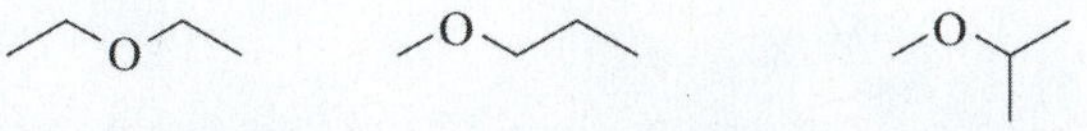

| IUPAC name: | ethoxyethane | 1-methoxypropane | 2-methoxypropane |
| common name: | diethyl ether | methyl propyl ether | isopropyl methyl ether |

25. **(M)**

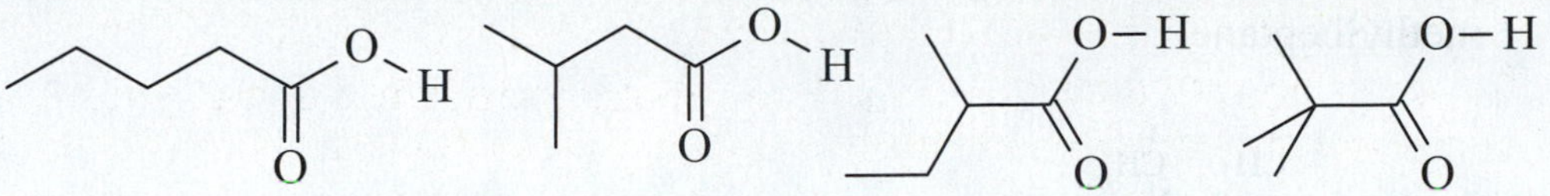

27. **(M)**

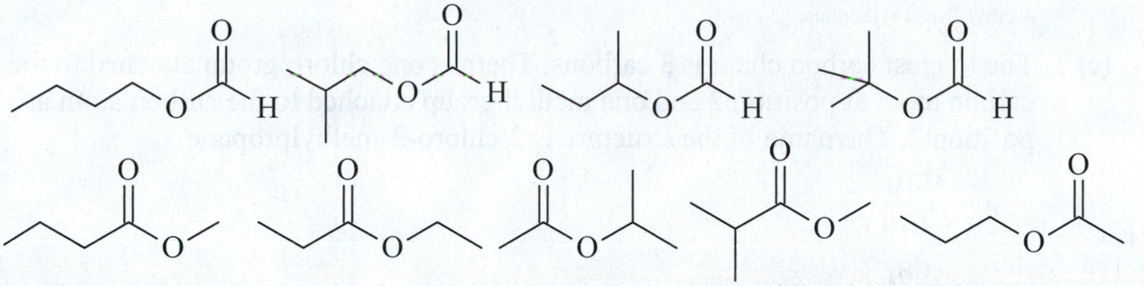

29. **(M)**

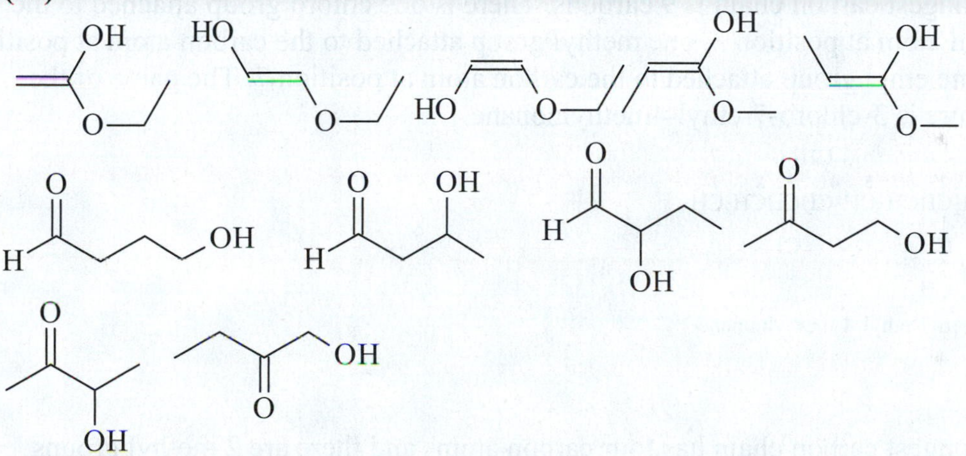

Nomenclature and Formulas

31. **(M)**
(a) The longest chain is eight carbons, and the three methyl groups are attached to carbon atoms at positions 2, 4, and 6. The name of the structure is 2,4,6-trimethyloctane.

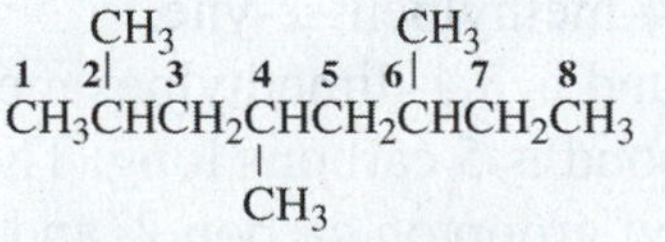

$$\underset{\text{2,4,6-trimethyloctane}}{\overset{\displaystyle \text{CH}_3 \qquad\qquad \text{CH}_3}{\underset{\overset{|}{\text{CH}_3}}{\overset{\text{1 \quad 2| \quad 3 \quad 4 \quad 5 \quad 6| \quad 7 \quad 8}}{\text{CH}_3\text{CHCH}_2\text{CHCH}_2\text{CHCH}_2\text{CH}_3}}}}$$

(b) The longest chain is seven carbons, the methyl group is attached to the carbon atom at position 2, and the ethyl group at position 4. The name of the structure is 4-ethyl-2-methylheptane.

$$\begin{array}{c}
CH_3 \\
| \\
CH_2 \quad\; ^1CH_3 \\
^7\;\;^6\;\;^5\;^4|\;\;^3\;\;^2| \\
CH_3CH_2CH_2CHCH_2CH \\
| \\
CH_3
\end{array}$$

4-ethyl-2-methylheptane

(c) The longest carbon chain is 3 carbons. There is one chloro group attached to the carbon atom at position 2 and one methyl group attached to the carbon atom at position 2. The name of the structure is 2-chloro-2-methylpropane.

$$\begin{array}{c}
^1CH_3 \\
^3\quad ^2| \\
CH_3-C-Cl \\
| \\
CH_3
\end{array}$$

2-chloro-2-methylpropane

(d) The longest carbon chain is 9 carbons. There is one chloro group attached to the carbon atom at position 3, one methyl group attached to the carbon atom at position 4, and one ethyl group attached to the carbon atom at position 7. The name of the structure is 3-chloro-7-ethyl-4methylnonane.

$$\begin{array}{c}
CH_3 \\
^9\;\;^8\;\;^7\;\;^6\;\;^5\;\;^4|\;\;^3\;\;^2\;\;^1 \\
CH_3CH_2CHCH_2CH_2CHCHCH_2CH_3 \\
| \qquad\qquad\quad | \\
CH_2 \qquad\qquad Cl \\
| \\
CH_3
\end{array}$$

3-chloro-7-ethyl-4-methylnonane

33. (M)

(a) The longest carbon chain has four carbon atoms and there are 2 methyl groups attached to carbon 2. This is 2,2-dimethylbutane.

(b) The longest chain is three carbons long, there is a double bond between carbons 1 and 2, and a methyl group attached to carbon 2. This is 2-methylpropene.

(c) Two methyl groups are attached to a three-carbon ring. This is 1,2-dimethylcyclopropane.

(d) The longest chain is 5 carbons long, there is a triple bond between carbons 2 and 3 and a methyl group attached to carbon 4. This is 4-methylpent-2-yne.

(e) The longest chain is 6 carbons long. This compound is 3,4-dimethylhexane

(f) The longest carbon chain containing the double bond is 5 carbons long. The double bond is between carbons 1 and 2. There is a propyl group on carbon 2, and carbons 3 and 4 each have one methyl group. This is 3,4-dimethyl-2-propylpent-1-ene.

35. **(M)**

(a) The name pentene is insufficient. pent-1-ene is $CH_2{=}CHCH_2CH_2CH_3$ and pent-2-ene is $CH_3CH{=}CHCH_2CH_3$.

(b) The name butanone is sufficient. There is only one four-carbon ketone.

(c) The name butyl alcohol is insufficient. There are numerous butanols. butan-1-ol is $HOCH_2CH_2CH_2CH_3$, butan-2-ol is$CH_3CH(OH)CH_2CH_3$, isobutyl alcohol is $HOCH_2CH(CH_3)_2$, and t-butyl alcohol is $(CH_3)_3COH$.

(d) The name methylaniline is insufficient. It specifies $CH_3{-}C_6H_4{-}NH_2$, with no relative locations for the $-NH_2$ and $-CH_3$ substituents.

(e) The name methylcyclopentane is sufficiently precise. It does not matter where on a five-carbon ring with only $C-C$ single bonds a methyl group is placed.

(f) Dibromobenzene is insufficient. It specifies $Br{—}C_6H_4{—}Br$, with no indication of the relative locations of the two bromo groups on the ring.

37. **(M)**

(a) 2,4,6-trinitrotoluene

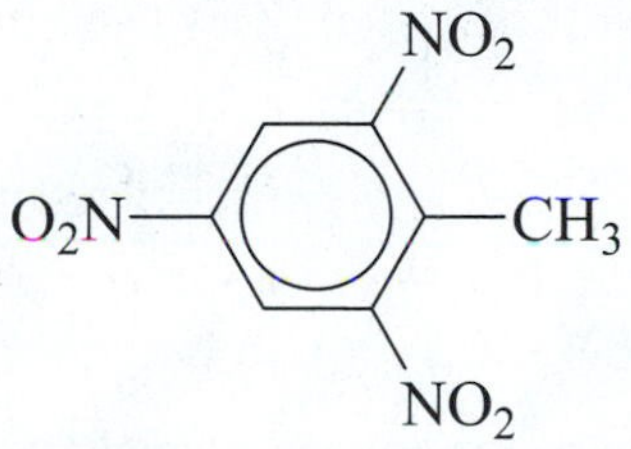

(b) methyl salicylate

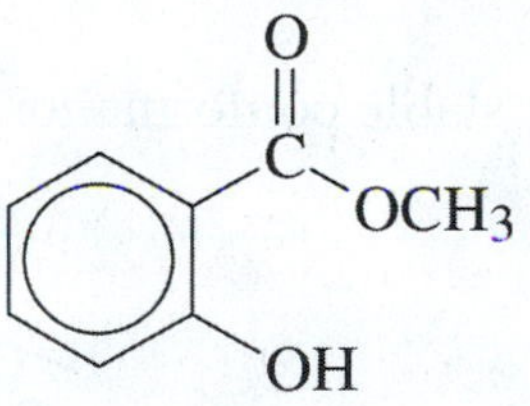

(c) 2-hydroxy-1,2,3-propanetricarboxylic acid: $HOOCCH_2C(OH)(COOH)CH_2COOH$

39. **(M)**

(a) $HN(CH_2CH_3)_2$

diethylamine

(b)

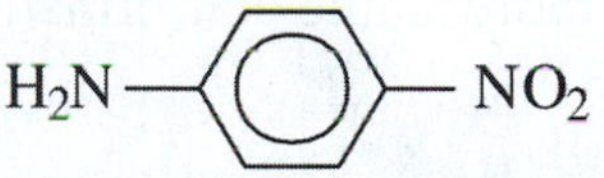

p-nitroaniline

(c)

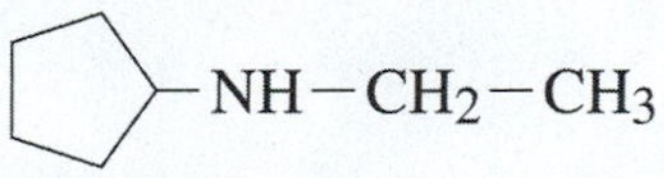

N-ethylcyclopentanamine

(d) $CH_3CH_2N(CH_3)CH_2CH_3$

N-ethyl-*N*-methylethanamine

Alkanes and Cycloalkanes

41. **(M)**

2-methylbutane

2,2-dimethylpropane

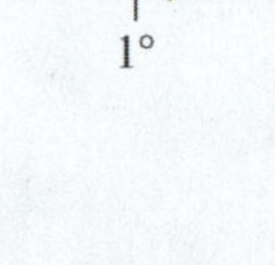

43. **(M)** Conformation (d) is lowest in energy. Conformation (a) is highest in energy.

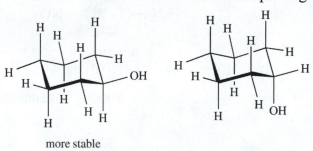

45. **(E)** The most stable conformation is:

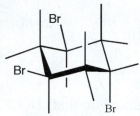

47. **(M)**

(a) The more stable conformation involves placing OH group in the equatorial position:

(b) The more stable configuration involves placing OH group in the axial position:

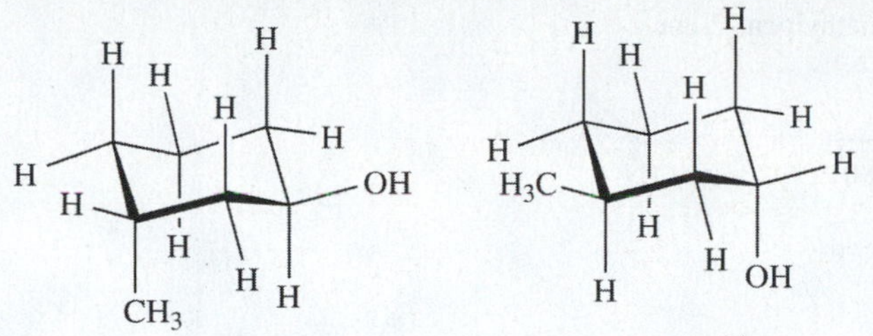

more stable

Alkenes

49. **(E)** In the case of ethene there are only two carbon atoms between which there can be a double bond. Thus, specifying the compound as ethene is unnecessary. In the case of propene, there can be a double bond only between the central carbon atom and a terminal carbon atom. Thus here also, specifying the compound as prop-1-ene is unnecessary. The case of butene is different, however, since but-1-ene, $CH_2 = CHCH_2CH_3$, is distinct from but-2-ene, $CH_3CH = CHCH_3$.

51. **(M)** The *E, Z* system is used to name highly substituted alkenes. The stereochemistry about the double bond is assigned *Z* if the two groups of higher priority at each end of the double bond are on the same side of the molecule. If the two groups of higher priority are on the opposite sides of the double bond, the configuration is denoted by *E*.

(a) *E* isomer

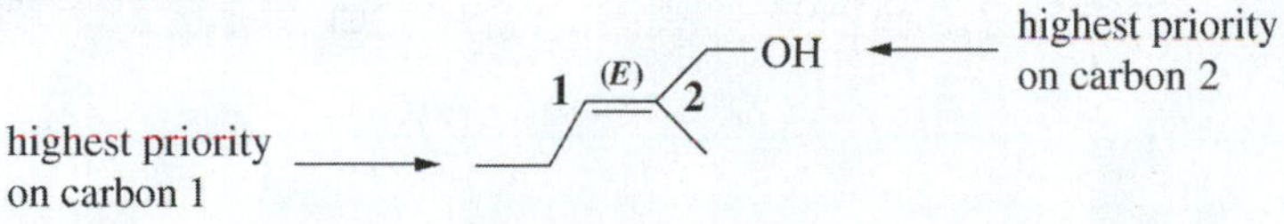

(b) *E* isomer

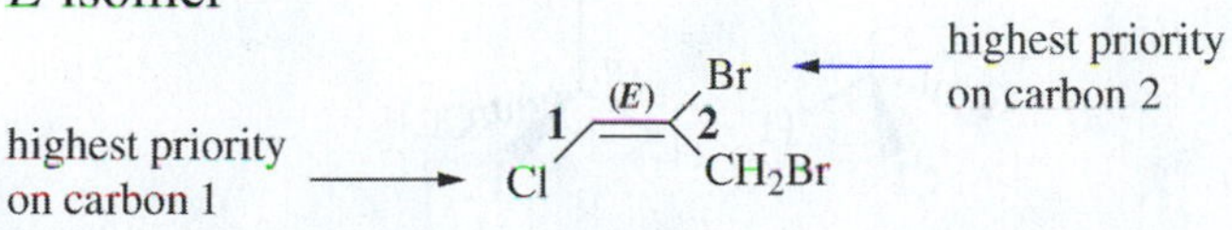

(c) *Z* isomer

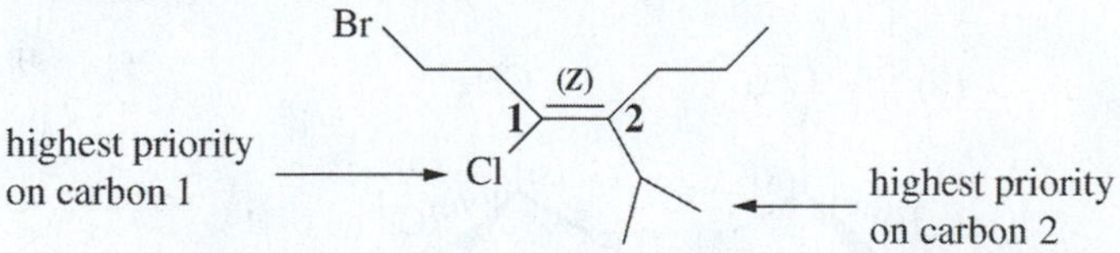

(d) *E* isomer

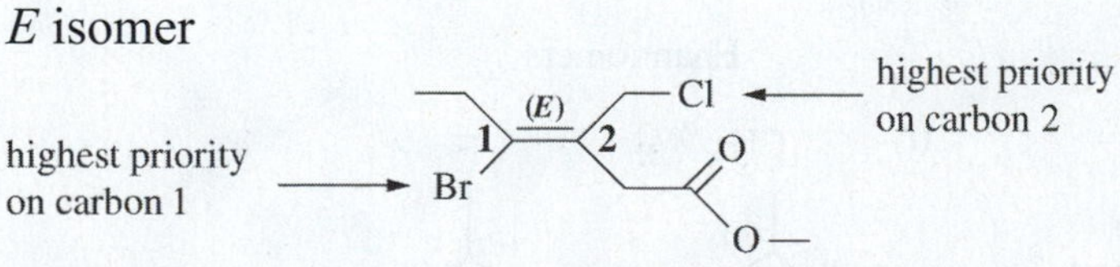

(e) *Z* isomer

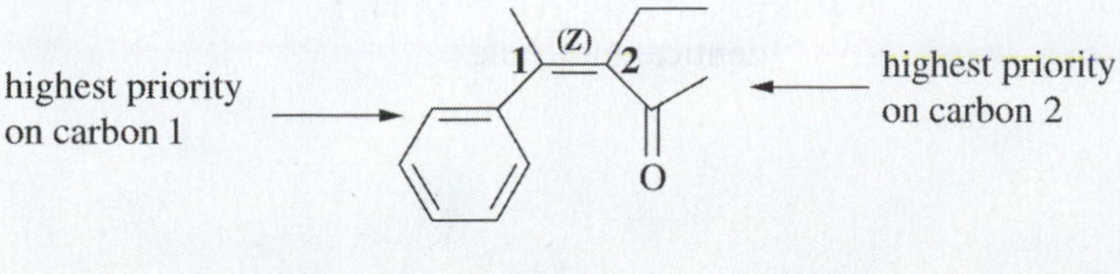

53. **(M)**

   (a) 2-chlorobut-2-ene   **(b)** 3-methylpent-2-ene

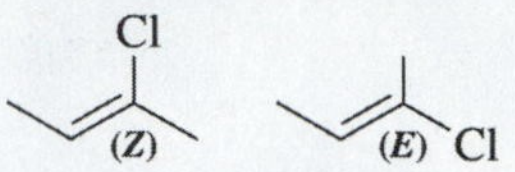

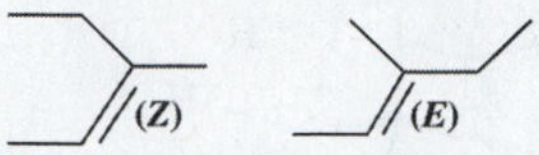

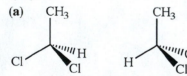

Aromatic Compounds

55. **(E)**

  (a)   phenylacetylene   **(b)**   meta-dichlorobenzene 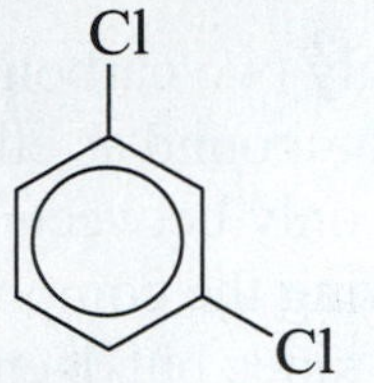   **(c)** 1,3,5-trihydroxybenzene

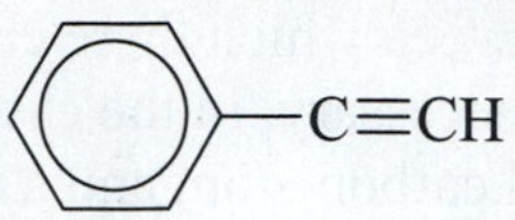

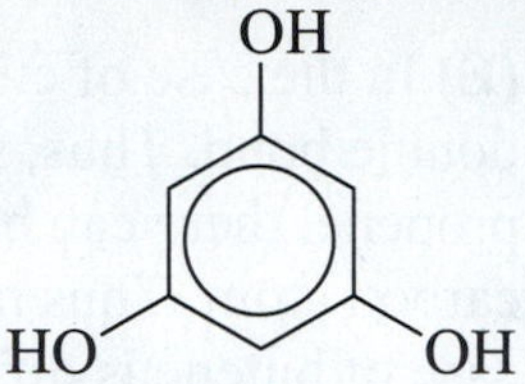

Organic Stereochemistry

57. **(E)** The four structural isomers are 1,2-dichloropropane, 1,3-dichloropropane, 2,2-dichloropropane and 1,1-dichloropropane.

  1,2-dichloropropane   1,3-dichloropropane    2,2-dichloropropane 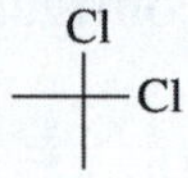   1,1-dichloropropane

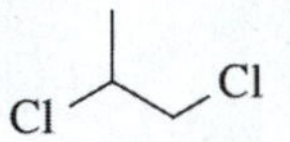

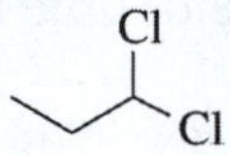

59. **(M)**

  (a)   **(b)**

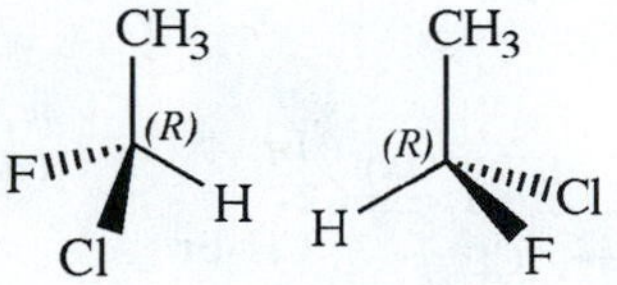

        Identical molecules   Identical molecules

  (c)   **(d)**

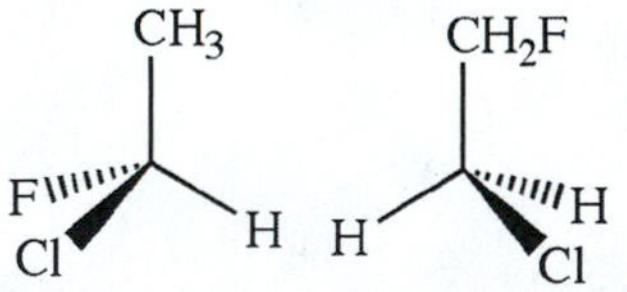

            Isomers   Enantiomers

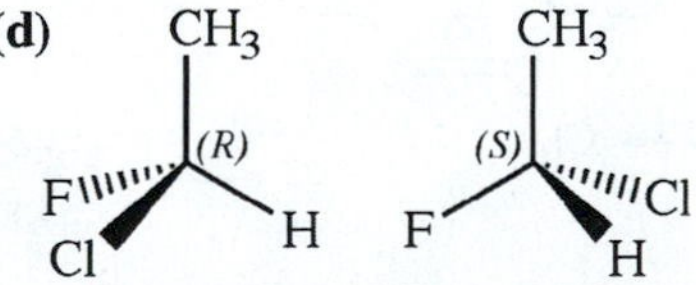

  (e)   **(f)**

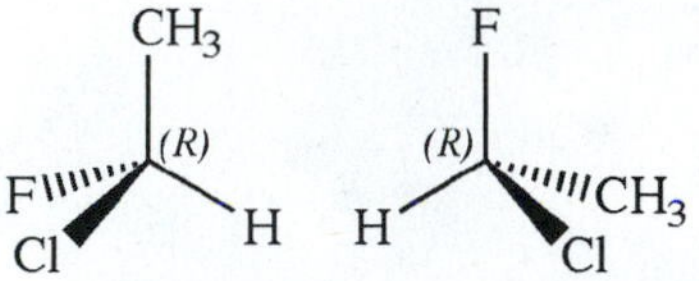

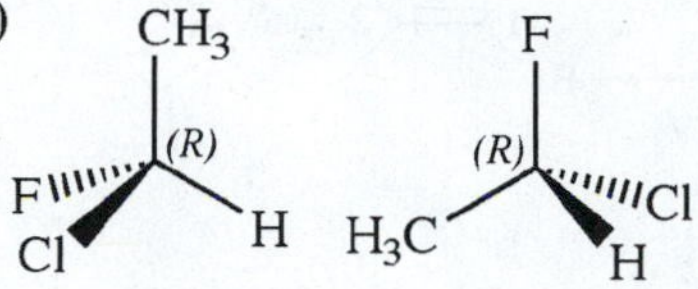

      Identical molecules   Identical molecules

61. **(M)**

(a) (*S*)-3-bromo-2-methylpentane

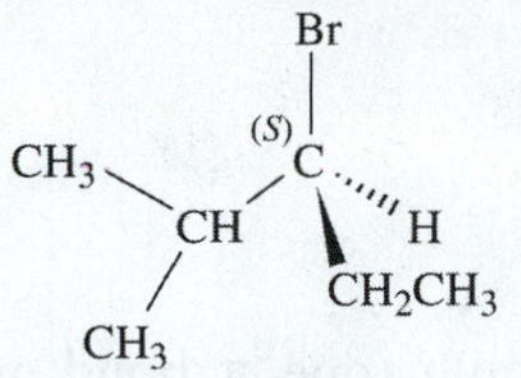

(b) (*S*)-1,2-dibromopentane

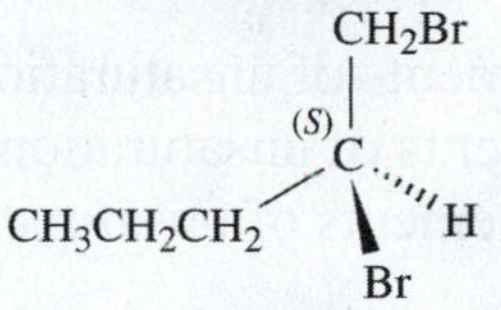

(c) (*R*)-3-(bromomethyl)-1-chloropentan-3-ol

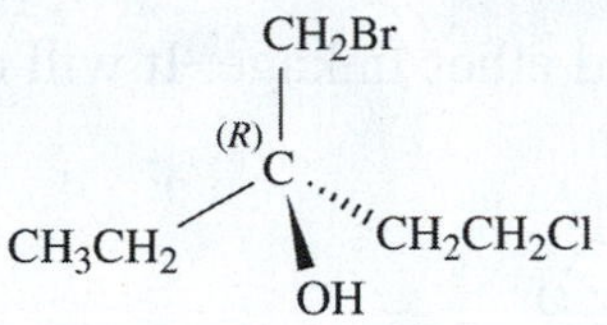

(d) (*S*)-1-bromopropan-2-ol

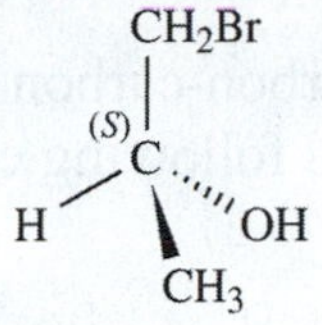

63. **(M)**

(a)

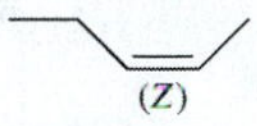

(*Z*)-pent-2-ene

(b)

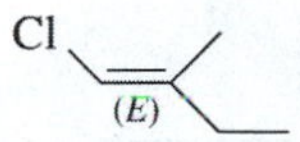

(*E*)-1-chloro-2-methylbut-1-ene

(c)

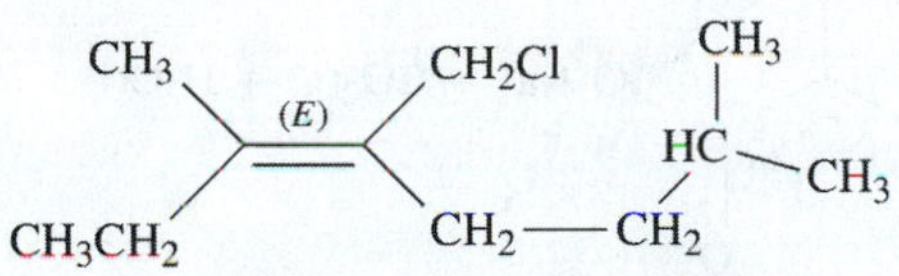

(*E*)-4-(chloromethyl)-3,7-dimethyloct-3-ene

(d)

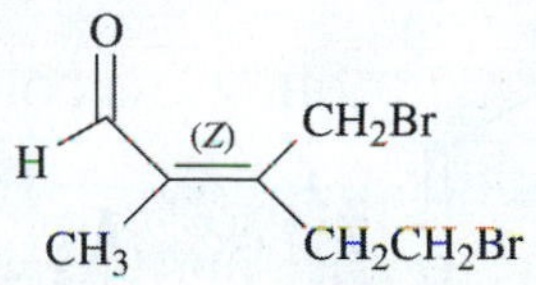

(*Z*)-5-bromo-3-(bromomethyl)-2-methylpent-2-enal

65. **(M)**

(a)

(*Z*)-1,3,5-tribromopent-2-ene

(b)

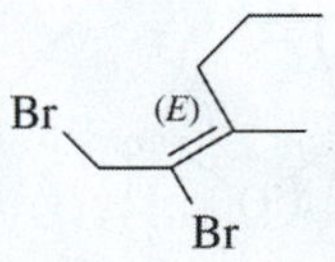

(*E*)-1,2-dibromo-3-methylhex-2-ene

(c)

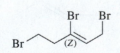

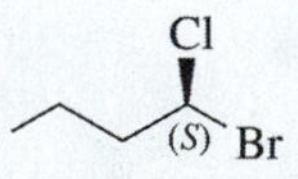

(*S*)-1-bromo-1-chlorobutane

(d)

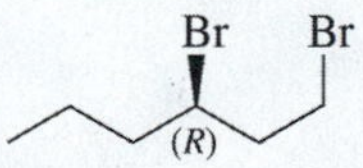

(*R*)-1,3-dibromohexane

(e)

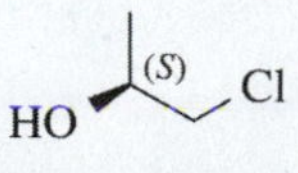

(*S*)-1-chloropropan-2-ol

Structures and Properties of Organic Compounds

67. **(E)**
    (a) $C_4H_{11}N$ has zero elements of unsaturation.
    (b) C_4H_6O has two elements of unsaturation.
    (c) $C_9H_{15}ClO$ has two elements of unsaturation.

69. **(M)** There are two elements of unsaturation in the molecule (one π bond and one ring structure). The molecular formula is $C_8H_{15}NO$.

71. **(M)** Compound 1 contains carbon-carbon double bond and ether linkage. It will decolorize bromine water according to the following chemical reaction:

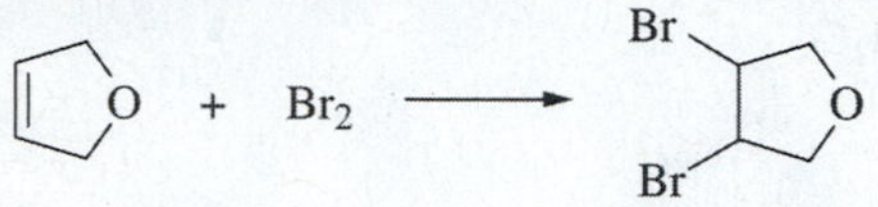

Compound 2 is a primary alcohol and is easily oxidized. Primary alcohols readily react with sodium metal to liberate hydrogen gas according to the following chemical reaction:

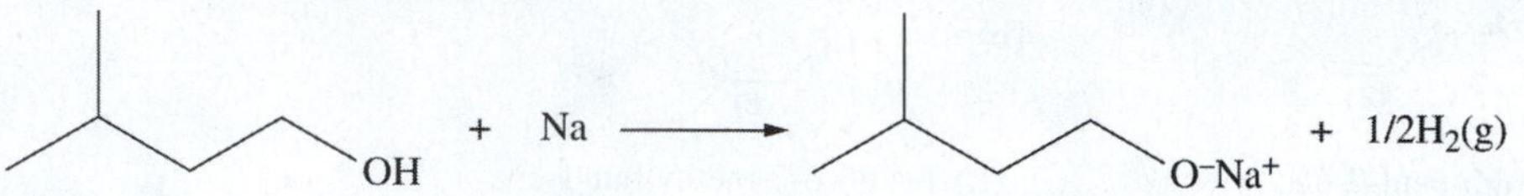

Compound 3 is a carboxylic acid. It reacts with $Na_2CO_3(aq)$ to generate $CO_2(g)$:

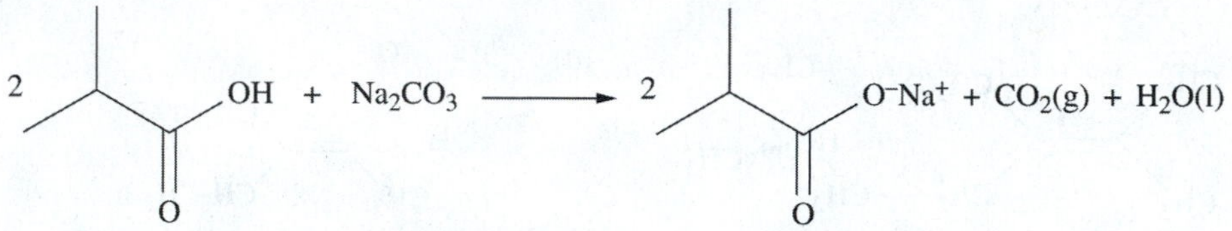

73. **(M)** Cyclic ethers do not contain OH groups. Possible structures for the compound with molecular formula C_4H_6O (two elements of unsaturation) are:

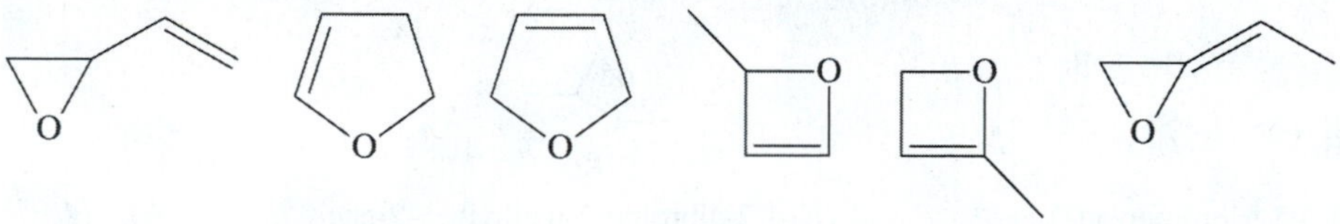

INTEGRATIVE AND ADVANCED EXERCISES

75. **(E)**

(a) cycloocta-1,5-diene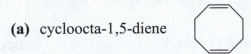

(b) 3,7,11-trimethyldodeca-2,6,10-trien-1-ol

$$HOCH_2CH = C(CH_3)CH_2CH_2CH = C(CH_3)CH_2CH_2CH = C(CH_3)_2$$

(c) 2,6-dimethylhept-5-enal

$$OHCCH(CH_3)CH_2CH_2CH = C(CH_3)_2$$

77. **(D)** In the structures below we have omitted the hydrogen atoms. Remember that there are four bonds to each C atom. The bonds that are missing are C—H bonds.

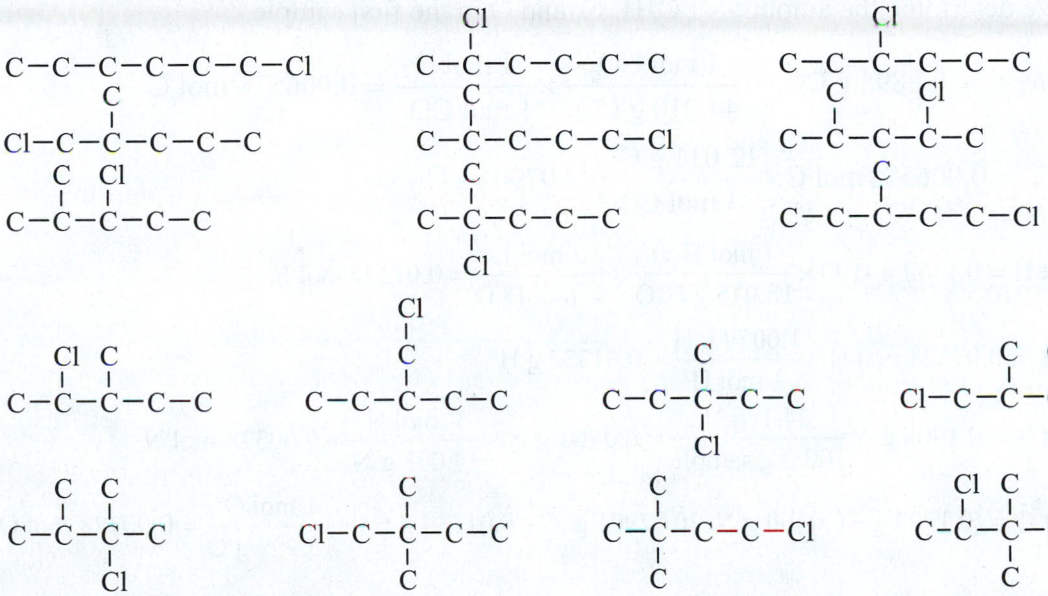

```
                         Cl                         Cl
                         |                          |
C—C—C—C—C—C—Cl      C—C—C—C—C—C            C—C—C—C—C—C
      C                  C                       C     Cl
      |                  |                       |     |
Cl—C—C—C—C—C        C—C—C—C—C—Cl           C—C—C—C—C
    C  Cl                C                       C
    |  |                 |                       |
C—C—C—C—C           C—C—C—C—C              C—C—C—C—C—Cl
                         |
                         Cl

                         Cl
                         |
    Cl C                 C                    C                 C  C
    |  |                 |                    |                 |  |
C—C—C—C—C           C—C—C—C—C            C—C—C—C—C         Cl—C—C—C—C
                                             |
                                             Cl

    C  C                 C                    C                 Cl C
    |  |                 |                    |                 |  |
C—C—C—C            Cl—C—C—C—C            C—C—C—C—Cl         C—C—C—C
    |                    |                    |                    |
    Cl                   C                    C                    C
```

79. **(D)**

(a) Oxdn : $\{Fe \longrightarrow Fe^{3+} + 3\,e^-\}$ $\qquad \times 2$

Redn : $\underline{C_6H_5NO_2 + 7\,H^+ + 6\,e^- \longrightarrow C_6H_5NH_3^+ + 2\,H_2O}$

$\qquad\qquad C_6H_5NO_2 + 7\,H^+ + 2\,Fe \longrightarrow C_6H_5NH_3^+ + 2\,H_2O + 2\,Fe^{3+}$

(b) Oxdn : $\{C_6H_5CH_2OH + H_2O \longrightarrow C_6H_5COOH + 4\,H^+ + 4\,e^-\} \times 3$

Redn : $\underline{\{Cr_2O_7^{2-} + 14\,H^+ + 6\,e^- \longrightarrow 2\,Cr^{3+} + 7\,H_2O\} \qquad \times 2}$

$\qquad 3\,C_6H_5CH_2OH + 2\,Cr_2O_7^{2-} + 16\,H^+ \longrightarrow 3\,C_6H_5COOH + 4\,Cr^{3+} + 11\,H_2O$

(c)   Oxdn: $\{CH_3CH{=}CH_2 + 2\ OH^- \longrightarrow CH_3CHOHCH_2OH + 2\ e^-\} \times 3$

Redn : $\underline{\{MnO_4^- + 2\,H_2O + 3\,e^- \longrightarrow MnO_2 + 4\,OH^-\}\qquad \times 2}$

$3\,CH_3CH{=}CH_2 + 2\,MnO_4^- + 4\,H_2O \longrightarrow 3\,CH_3CHOHCH_2OH + 2\,MnO_2 + 2\,OH^-$

81. **(D)** We suspect the compound contains C, H, N, and O. The %N by mass in the compound can be found using the ideal gas law:

$$\text{amount } N_2 = \frac{PV}{RT} = \frac{(735-9)\ \text{mmHg} \times \dfrac{1\,\text{atm}}{760\ \text{mmHg}} \times 0.0402\ \text{L}}{0.08206\ \text{L atm mol}^{-1}\,\text{K}^{-1} \times 298\ \text{K}} = 1.57 \times 10^{-3}\ \text{mol } N_2$$

$$\%\ N = \frac{1.57 \times 10^{-3}\ \text{mol } N_2 \times \dfrac{28.01\ \text{g } N_2}{1\ \text{mol } N_2}}{0.1825\ \text{g sample}} \times 100\% = 24.1\%\ N$$

Now we determine the amounts of C, H, N, and O in the first sample.

$$\text{amount } C = 0.2895\ \text{g } CO_2 \times \frac{1\ \text{mol } CO_2}{44.010\ \text{g } CO_2} \times \frac{1\ \text{mol C}}{1\ \text{mol } CO_2} = 0.006578\ \text{mol C}$$

$$\text{mass } C = 0.006578\ \text{mol C} \times \frac{12.011\ \text{g C}}{1\ \text{mol C}} = 0.07901\ \text{g C}$$

$$\text{amount } H = 0.1192\ \text{g } H_2O \times \frac{1\ \text{mol } H_2O}{18.015\ \text{g } H_2O} \times \frac{2\ \text{mol H}}{1\ \text{mol } H_2O} = 0.01323\ \text{mol H}$$

$$\text{mass } H = 0.01323\ \text{mol H} \times \frac{1.0079\ \text{g H}}{1\ \text{mol H}} = 0.01334\ \text{g H}$$

$$\text{amount } N = 0.1908\ \text{g } \times \frac{24.1\ \text{g N}}{100.0\ \text{g sample}} = 0.0460\ \text{g N} \times \frac{1\ \text{mol N}}{14.007\ \text{g N}} = 0.00328\ \text{mol N}$$

$$\text{amount } O = (0.1908\ \text{g} - 0.0460\ \text{g N} - 0.07901\ \text{g C} - 0.01334\ \text{g H}) \times \frac{1\ \text{mol O}}{15.999\ \text{g O}} = 0.00328\ \text{mol O}$$

From this information we determine the empirical formula of the compound.

$0.006578\ \text{mol C} \div 0.00328 \longrightarrow 2.01\ \text{mol C}$ $0.01323\ \text{mol H} \div 0.00328 \longrightarrow 4.03\ \text{mol H}$

$0.00328\ \text{mol N} \div 0.00328 \longrightarrow 1.00\ \text{mol N}$ $0.00328\ \text{mol O} \div 0.00328 \longrightarrow 1.00\ \text{mol O}$

The compound has an empirical formula of C_2H_4NO, which has a molar mass of 58.1 g/mol. We use the freezing point depression data to determine the number of moles of compound dissolved, and then its molar mass. First, the molality of the solution: $m = \dfrac{\Delta T_f}{K_f} = \dfrac{3.66\,°C - 5.50\,°C}{-5.12\,°C/m} = 0.359\ m$

amount solute $= 0.02600 \text{ kg benzene} \times \dfrac{0.359 \text{ mol solute}}{1 \text{ kg benzene}} = 0.00933 \text{ mol solute}$

$M = \dfrac{1.082 \text{ g}}{0.00933 \text{ mol}} = 116 \text{ g/mol}$ This molar mass is twice the empirical molar mass.

Thus, the molecular formula is twice the empirical formula. Namely, the molecular formula = $C_4H_8N_2O_2$

84. (D)

(1) The molar masses of the five compounds are the same to one significant figure:

butan-1-ol   diethyl ether   methyl propyl ether   butyraldehyde   propionic acid

74.12 g/mol   74.12 g/mol   74.12 g/mol   72.11 g/mol   74.08 g/mol

Since the freezing point depression is known to only one significant figure, the molar mass of the unknown can only be determined to one significant figure. Thus, the cited freezing point depression data are insufficiently precise to differentiate between the compounds.

(2) Propionic acid would produce an aqueous solution that would turn blue litmus red. The unknown compound cannot be propionic acid.

(3) Both alcohols and aldehydes are oxidized to carboxylic acids by aqueous $KMnO_4$. Ethers are not oxidized by $KMnO_4$. Thus, the unknown must be butan-1-ol or butyraldehyde.

The identity of the unknown can be established by treatment with a carboxylic acid. If the unknown is an alcohol, it will form a pleasant smelling ester upon heating with a carboxylic acid such as acetic acid. If the compound is an aldehyde, it will not react with the carboxylic acid under these conditions.

85. (M)

2,4-dimethylpenta-1,4-diene   2,3-dimethylpentane   1,2,4-tribromobenzene

methyl acetate 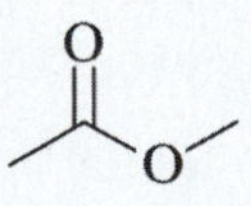   butan-2-one

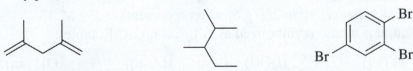

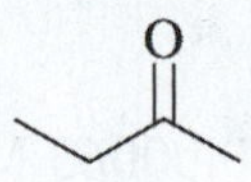

87. **(D)**

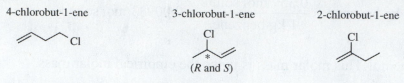

4-chlorobut-1-ene   3-chlorobut-1-ene   2-chlorobut-1-ene

(*R* and *S*)

(*Z*)-1-chlorobut-1-ene 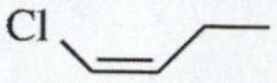   (*E*)-1-chlorobut-1-ene 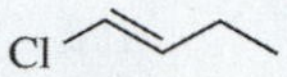   1-chloro-2-methylprop-1-ene

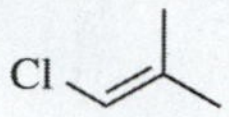

3-chloro-2-methylprop-1-ene 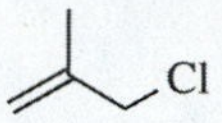   (*E*)-1-chlorobut-2-ene 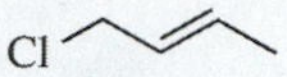   (*Z*)-1-chlorobut-2-ene

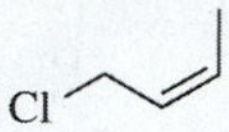

(*Z*)-2-chlorobut-2-ene 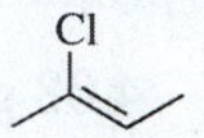   (*E*)-2-chlorobut-2-ene 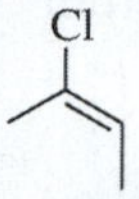   1-chloro-2-methylcyclopropane

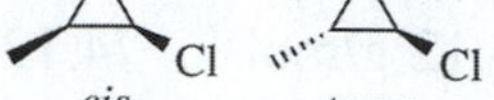

cis-   trans-

1-chloro-1-methylcyclopropane   (chloromethyl)cyclopropane 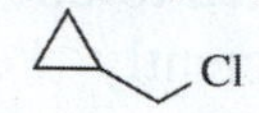   chlorocyclobutane

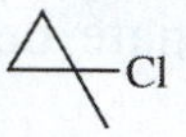

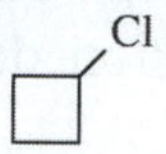

89. **(M)**

(a) ester, amine(tertiary), arene

(b) $C_1 = sp^2$, $C_2 = sp^3$, $C_3 = sp^3$, $C_4 = sp^3$, $N = sp^3$

(c) Carbons 2 and 4 are chiral.

91. **(M)**

(a) alcohol, amine(secondary), arene

(b) $C_1 = sp^2$, $C_2 = sp^3$, $C_3 = sp^3$, $C_4 = sp^3$, $N = sp^3$

(c) Carbons 2 and 3 are chiral.

(d) $C_{10}H_{15}NO$ has a molar mass of 165.24 g/mol.

Assume no volume change and
1 g water = 1 mL water, hence, 1g in 200 g of water represents 0.030 M.
Since the pH = 10.8, it is a base (symbolized as A⁻). Set up I.C.E. table.

Reaction	A⁻(aq)	+	H₂O(l)	⇔	HA(aq)	+	OH⁻(aq)
Initial	0.03$\underline{0}$ M		—		0 M		≈ 0 M
Change	-x		—		+x		+x
Equilibrium	0.03$\underline{0}$-x		—		x		x

pH = 10.8   hence pOH = 3.2 and [OH⁻] = 0.0006$\underline{3}$ M = x
$K_b = x^2/(0.3\underline{0}-x) = (0.0006\underline{3})^2/(0.03\underline{0}-0.00063) = 1.\underline{35} \times 10^{-5}$, $pK_b = 4.87$

93. **(M)**

(a)   (b)

95. **(D)** Cholesterol has eight chiral centers. The configuration of the carbon atom bonded to the OH group is S and the configuration of the double bond is Z.

SELF-ASSESSMENT EXERCISES

101. **(M)** The main carbon-carbon chain in isoheptane contains six carbon atoms. Furthermore, isoheptane contains isopropyl group, $CH(CH_3)_2$. The correct structure is structure (c).

102. **(M)** The molecular formula of cyclobutane is C_4H_8. The correct answer is (b).

103. **(M)** Hydrocarbons with four of more carbon atoms form isomers. The correct answer is (e). The three isomers of the compound with molecular formula C_5H_{12} are:

pentane   2-methylbutane   2,2-dimethylpropane

108. **(M)** The correct answer is (c).

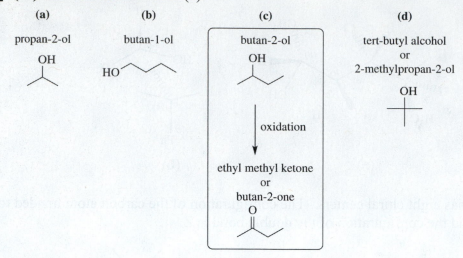

(a) propan-2-ol

(b) butan-1-ol

(c) butan-2-ol

↓ oxidation

ethyl methyl ketone
or
butan-2-one

(d) tert-butyl alcohol
or
2-methylpropan-2-ol

109. **(M)** C_4H_{10} has only two isomers (n-butane and 2-methylpropane), whereas C_4H_8 has six (but-1-ene, (*E*)-but-2-ene, (*Z*)-but-2-ene, 2-merthylpropene, cyclobutane, and methylcyclopropane.).

110. **(M)**

(a) C_6H_{12} is expected to have higher boiling point than C_2H_4.
(b) C_3H_7OH is expected to have higher solubility in water because of smaller hydrocarbon chain.
(c) Benzoic acid (C_6H_5COOH) has greater acidity than benzaldehyde.

111. **(M)**

(*E*)-3-benzyl-2,5-dichloro-4-methylhex-3-ene

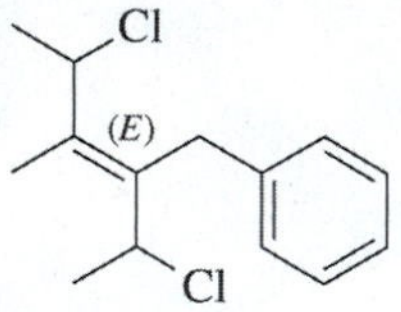

1-nitro-4-vinylbenzene

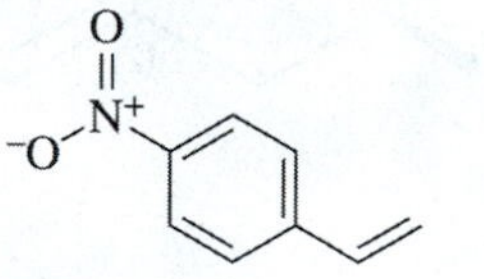

trans-1-(4-bromophenyl)-2-methylcyclohexane

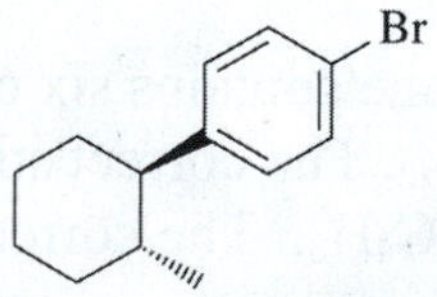

CHAPTER 27
REACTIONS OF ORGANIC COMPOUNDS
PRACTICE EXAMPLES

1A **(M)**

(a) Because hydrogen atom in benzene is replaced by a bromine atom, this reaction is a substitution reaction.

(b) This reaction involves only a change in the skeletal structure (constitution), and so the reaction is a rearrangement reaction.

(c) In this reaction, there is a reduction of C=O bond, i.e. the alcohol is adding across a ketone group. This is an example of an addition reaction.

1B **(M)**

(a) This in an example of an elimination reaction. HBr is being eliminated from bromobutane to yield but-1-ene.

(b) This is an example of a substitution reaction in which Cl group is replaced by $N(CH_2CH_3)_3$ group.

(c) This is an example of a substitution reaction in which bromine is being replaced by a butynyl group.

2A **(M)**

(a)

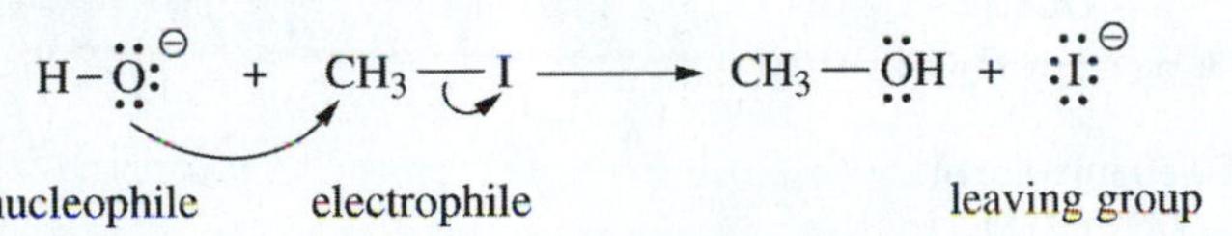

nucleophile   electrophile   leaving group

(b)

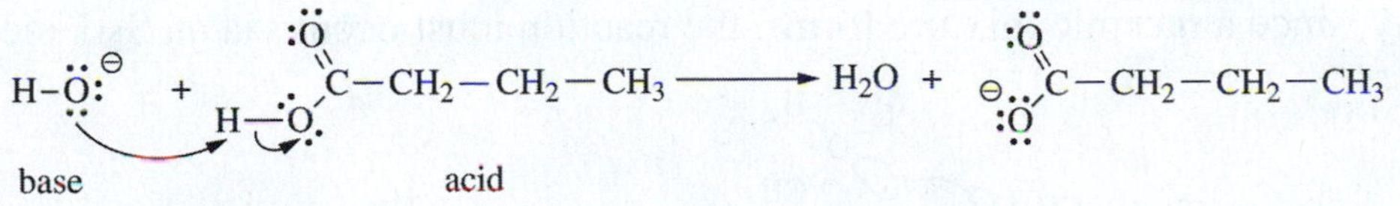

base   acid

2B **(M)**

(a)

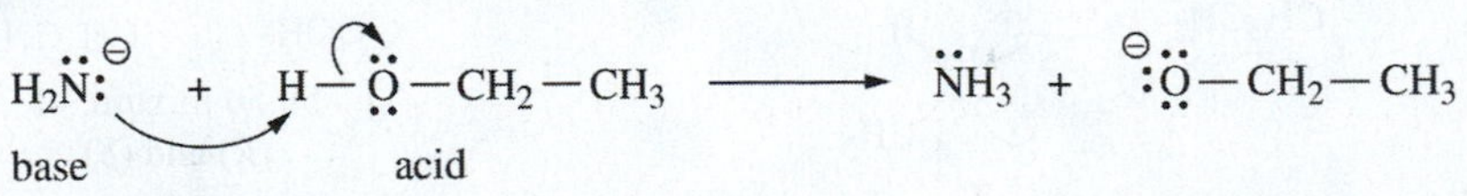

base   acid

(b)

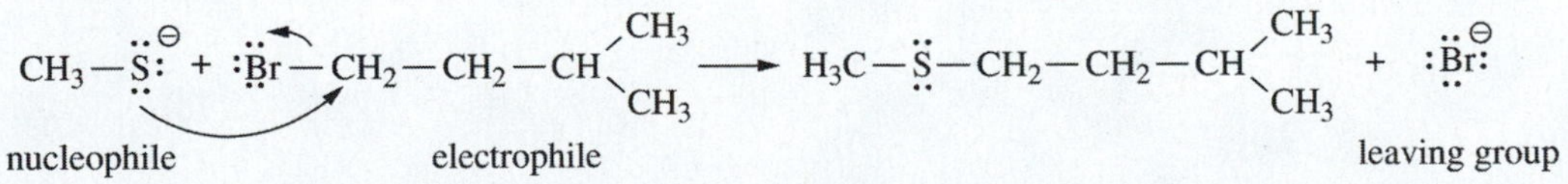

nucleophile   electrophile   leaving group

<u>3A</u> **(M)**

(a) $CH_3C \equiv C^- + CH_3Br$ — $S_N2 \rightarrow CH_3C \equiv CCH_3 + Br^-$

(b) $Cl^- + CH_3CH_2CN \rightarrow$ NO REACTION

The nucleophile in this reaction is Cl^- whereas the leaving group is the CN^-. The CN^- ion is a much stronger nucleophile than Cl^-, so the equilibrium will strongly favor the reactants. In other words, no reaction is expected.

(c) $CH_3NH_2 + (CH_3)_3CCl - S_N1 \rightarrow CH_3 + NH_2C(CH_3)_3 + Cl^-$

<u>3B</u> **(M)**

(a)

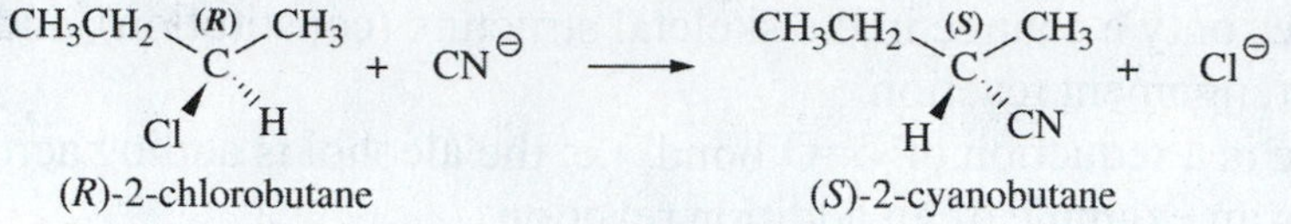

(R)-2-chlorobutane   (S)-2-cyanobutane

Because the configuration at the stereogenic carbon has undergone an inversion, we can conclude that the reaction has occurred via an S_N2 mechanism.

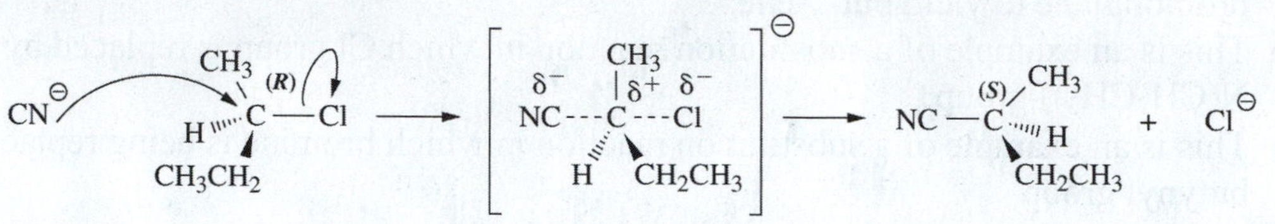

(b)

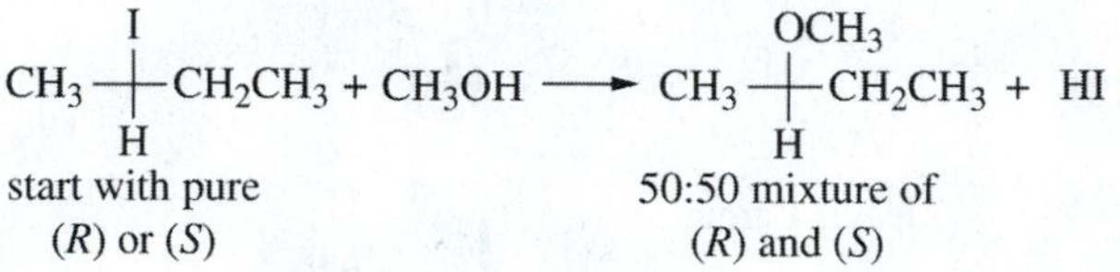

start with pure   50:50 mixture of
(R) or (S)   (R) and (S)

Clearly, since a racemic mixture forms, the reaction must occur via an S_N1 mechanism.

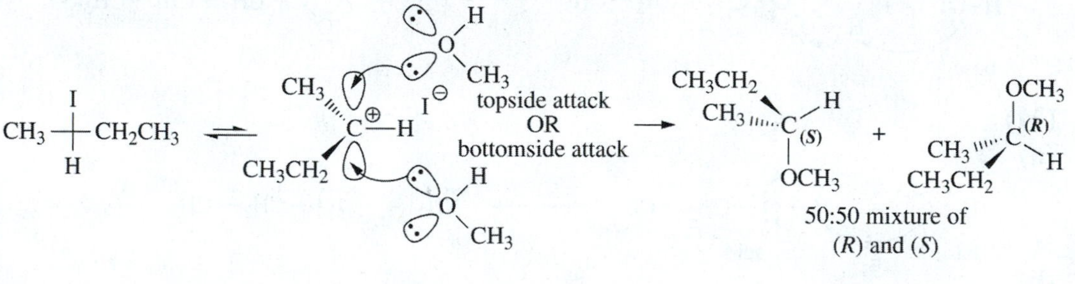

topside attack
OR
bottomside attack

50:50 mixture of
(R) and (S)

4A **(M)** SCH_3^- is a nucleophile and haloalkane is the electrophile. The nucleophile is charged and is therefore a strong nucleophile. The substitution reaction should occur via S_N2 mechanism.

4B **(M)** Water is a weak nucleophile. The reaction should proceed via S_N1 mechanism to give a mixture of products (R and S). Because the haloalkane is symmetrical only one compound will form.

5A **(M)** This is an S_N2 reaction leading to the alkyl sulfide product.

725

5B **(M)** Electrophile is a secondary haloalkane. The nucleophile is $(CH_3)_2CHO^-$, a relatively hindered strong base. Therefore, the reaction may proceed by $S_N1/E1$, although it may also proceed to a significant degree by S_N2 and E2.

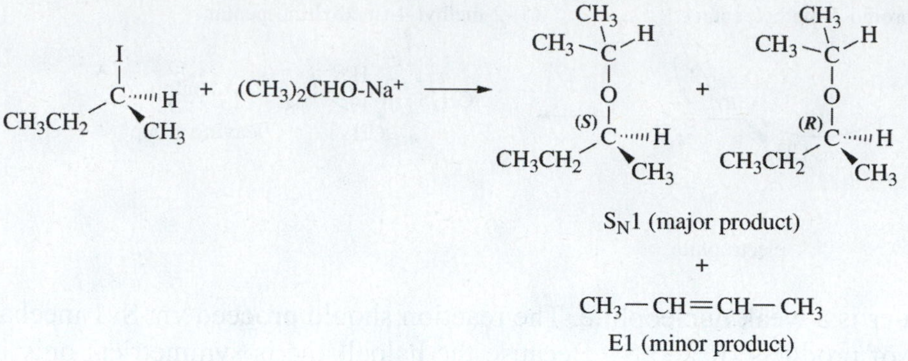

S_N1 (major product)

+

$CH_3-CH=CH-CH_3$
E1 (minor product)

6A **(M)**

(a) This reaction employs a strong oxidizing agent. The secondary alcohol will be oxidized to a ketone.

(R)-butan-2-ol   butan-2-one

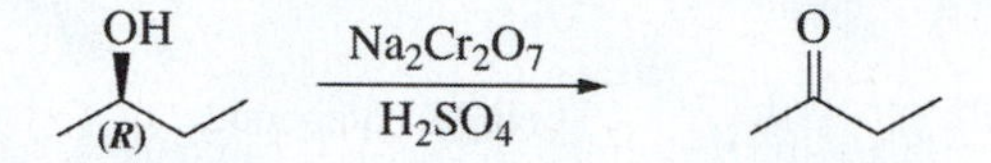

(b) Propan-1-ol is a primary alcohol and for the conditions specified, a dehydration reaction will occur. The product is prop-1-ene and it is formed by a concerted E2 reaction following protonation of the alcohol. The reaction mechanism is shown below:

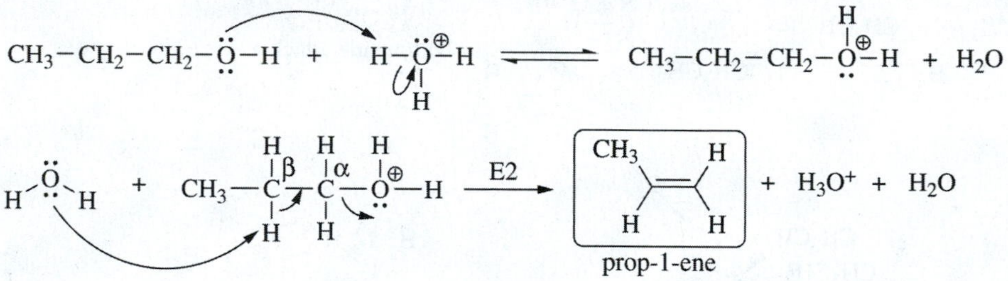

prop-1-ene

6B **(M)**

(a) Primary alcohols are readily converted to alkyl halides. The reactions proceeds via a S_N2 mechanism.

3-methylbutan-1-ol   1-iodo-3-methylbutane

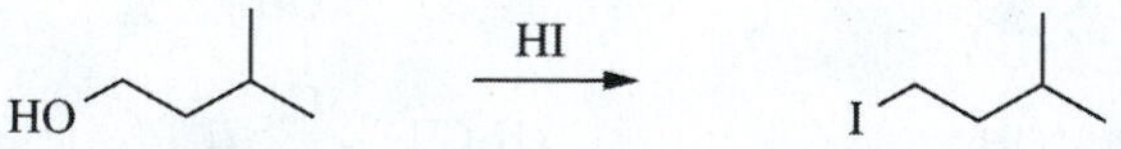

(b) In this reaction, an alcohol reacts with a sodium metal to produce sodium tert-butoxide and hydrogen gas.

2-methylpropan-2-ol

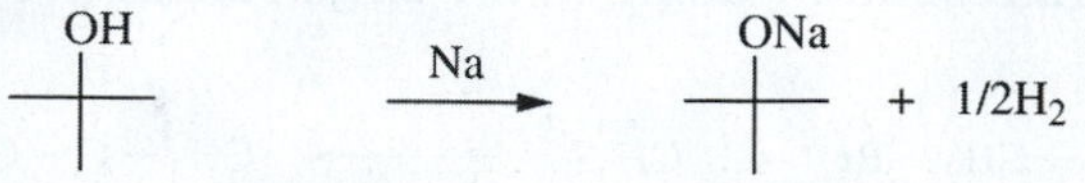

7A **(D)** In this reaction, three products are possible, the major one being 3-nitro-benzaldehyde.

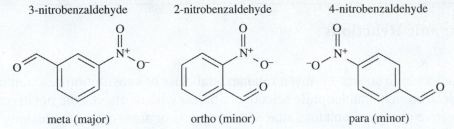

3-nitrobenzaldehyde	2-nitrobenzaldehyde	4-nitrobenzaldehyde
meta (major)	ortho (minor)	para (minor)

7B **(M)** Chlorine group is ortho and para directing. Several products are possible:

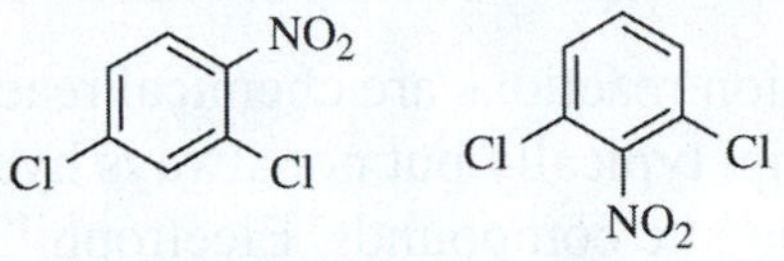

major   minor

INTEGRATIVE EXAMPLE

A. **(D)** Alkanes are relatively unreactive. However, they will undergo bromination reaction in the presence of light. Such reaction will generate bromo-cyclohexane in the first step. In the presence of a base, elimination will occur resulting in the formation of cyclohexene. Subsequent repetition of these two steps will genererate the desired product, 1,3-cyclohexadiene. Typically, in the second step, selective allylic bromination is achieved utilizing brominating agent, N-bromosuccinimide (NBS).

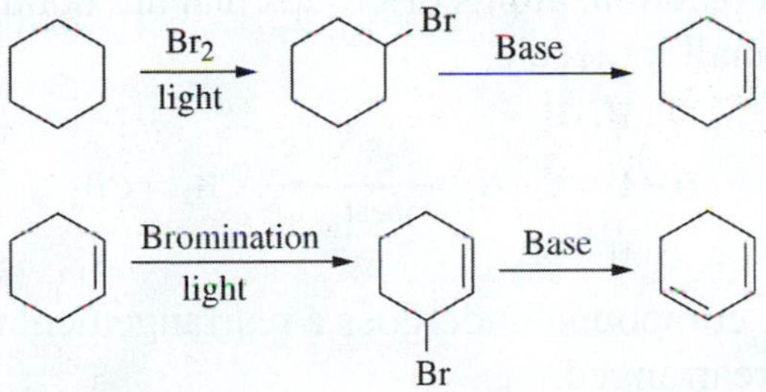

Brominating agent:

N-bromosuccinimide

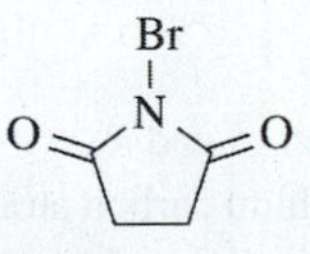

B. **(M)** SO_3H and NO_2 groups are meta directing, whereas bromine group is ortho and para directing. The given compound could therefore be synthesized using the following sequence of reactions:

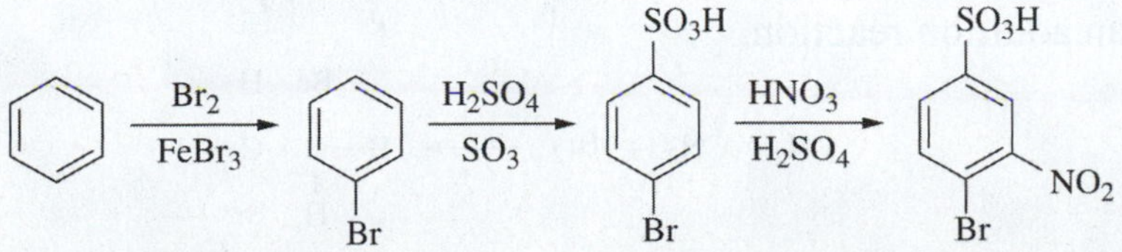

EXERCISES

Types of Organic Reactions

1. **(E)**

(a) Nucleophilic substitution is a fundamental class of substitution reaction in which an "electron rich" nucleophile selectively bonds with or attacks the positive or partially positive charge of an atom attached to a group or atom called the leaving group; the positive or partially positive atom is referred to as an electrophile.

$$\text{R-Br} + \text{OH}^- \rightarrow \text{R} - \text{OH} + \text{Br}^-$$

(b) Electrophilic substitution reactions are chemical reactions in which an electrophile displaces another group, typically but not always hydrogen. Electrophilic substitution is characteristic of aromatic compounds. Electrophilic aromatic substitution is an important way of introducing functional groups onto benzene rings.

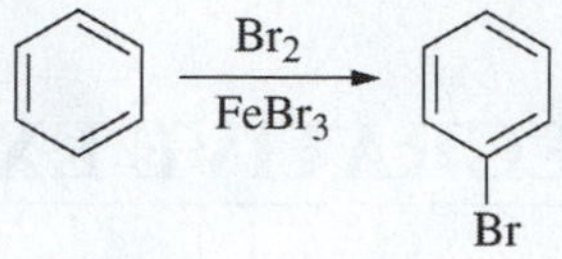

(c) Generally, in an addition reaction, a molecule adds across a double or triple bond in another molecule.

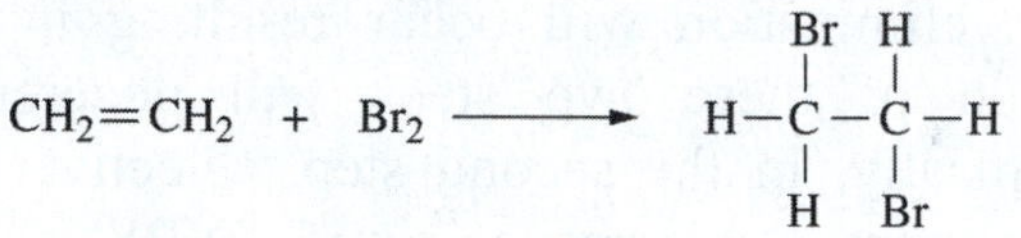

(d) In an elimination reaction, atoms or groups that are bonded to adjacent atoms are eliminated as a small molecule.

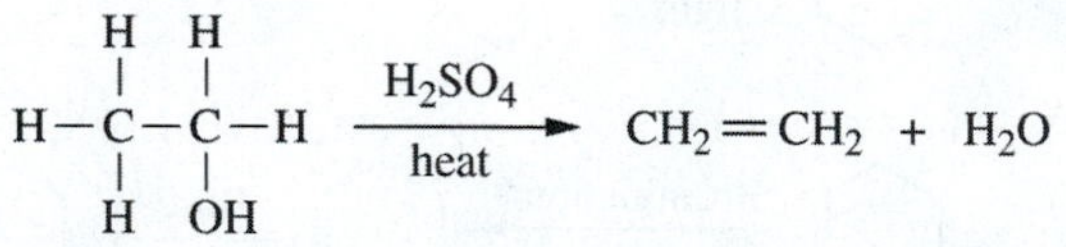

(e) When an organic compound undergoes a rearrangement reaction, the carbon skeleton of a molecule is rearranged.

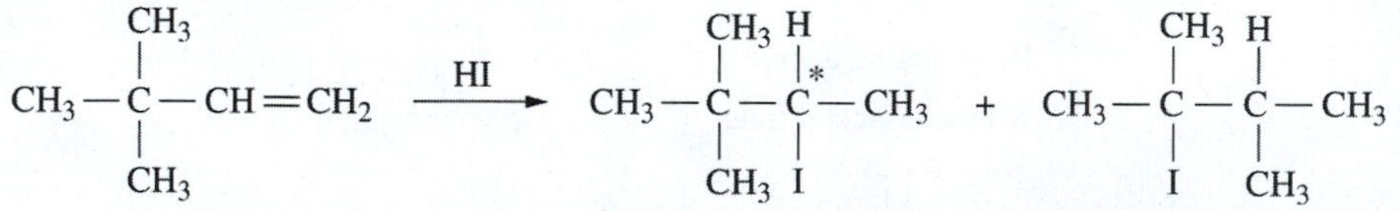

minor product (*S* and *R*) major product (*S* and *R*)

chiral carbon atom indicated by*

3. **(E)** **(a)** substitution reaction. **(b)** addition reaction. **(c)** substitution reaction.

5. **(M)**

(a) This is an addition reaction.

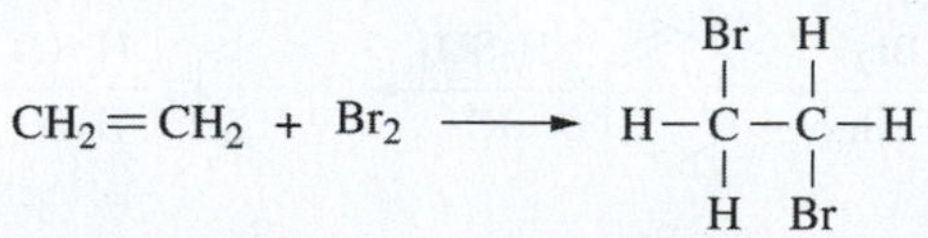

(b) This is an elimination reaction.

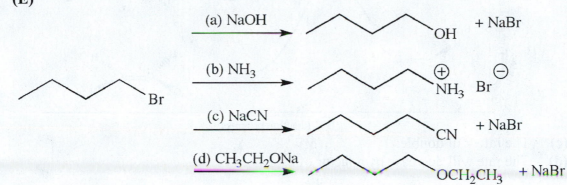

(c) This is a substitution reaction.

$$—Cl \; + \; NaOH \longrightarrow —OH \; + \; NaCl$$

Substitution and Elimination Reactions

7. **(E)**

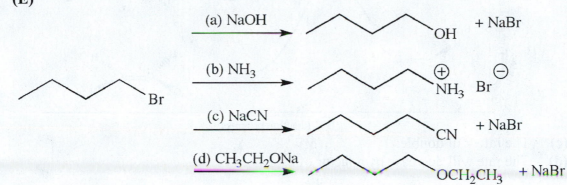

(a) NaOH → ... OH + NaBr

(b) NH₃ → ... NH₃⁺ Br⁻

(c) NaCN → ... CN + NaBr

(d) CH₃CH₂ONa → ... OCH₂CH₃ + NaBr

9. **(M)**

(a) rate = k[CH₃CH₂CH₂CH₂Br][NaOH]

(b)

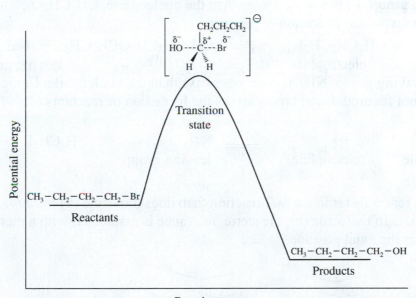

(c) Doubling the concentration of n-butyl bromide will also double the rate.

(d) Decreasing the concentration of NaOH by a factor of two will decrease the rate by a factor of 2.

11. **(M)**

    (a)   Rate = $k[CH_3CH_2CH_2Br][OH^-]$

    (b)

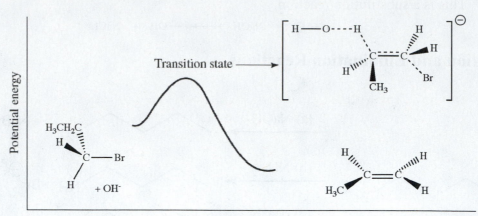

    (c)   The rate will double.

    (d)   The rate will decrease by a factor of 2.

13. **(M)** Equilibrium favors the formation of products if the leaving group is a weaker base than the nucleophile.

    (a)   The leaving group, I^-, is a weaker base than the nucleophile, $CH_3CH_2O^-$. Therefore, equilibrium favors the formation of products.

$$CH_3CH_2O^- + CH_3CH_2CH_2I \rightleftharpoons CH_3CH_2CH_2OCH_2CH_3 + NaI$$

    nucleophile   electrophile   leaving group

    (b)   Since the leaving group, NH_3, is a stronger base than the nucleophile, I^-, the forward reaction is not favored. Equilibrium favors the formation of reactants.

$$CH_3CH_2NH_3^+ + KI^- \rightleftharpoons NH_3 + CH_3CH_2I + K^+$$

    electrophile   nucleophile   leaving group

15. **(M)** Molecule (a) reacts faster in the S_N2 reaction than does molecule (b) because there is less steric hindrance. In molecule (b), the steric hindrance is associated with a methyl group being positioned in the axial position.

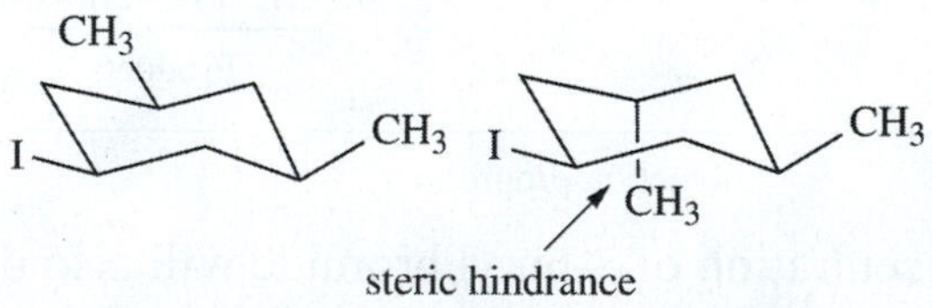

steric hindrance

17. **(M)** The reaction most likely proceeded via S_N1 mechanism resulting in racemization of the product (equal amounts of S and R isomers).

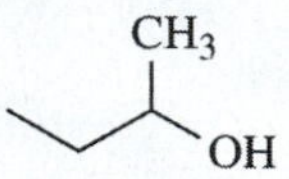

(*S* and *R*)

19. **(M)** The resulting product is an equimolar mixture of S and R isomers. The reaction proceeded via S_N1 mechanism.

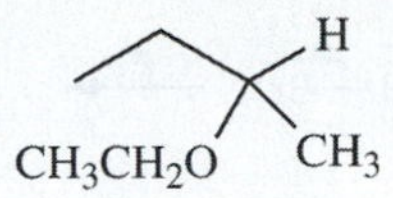

(S and R)

21. **(M)**

(a) In this reaction we are dealing with a primary alkyl halide and a strong but bulky base. The reaction is going to proceed via an E_2 mechanism. The product of the reaction is methylenecyclopentane.

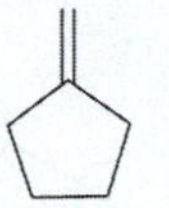

(b) In this reaction we are dealing with a secondary alkyl halide and a weak base which is also a good nucleophile. The reaction is most likely going to proceed via S_N2 mechanism.

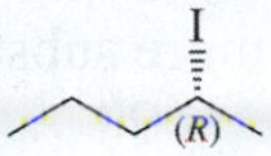

(c) A 2° substrate, a weak nucleophile, and a polar protic solvent favor substitution by S_N1. Substitution by S_N1 involves the formation of a carbocation in the rate-determining step, followed by nucleophilic attack. The major substitution products will be (R) and (S) $(CH_3)_3CCH(OCH_2CH_3)CH_3$. S_N1 is always accompanied by E1, so there will also be some elimination product, $CH_2 = CHC(CH_3)_3$. A methide shift could also occur leading to $(CH_3)_2C = C(CH_3)_2$.

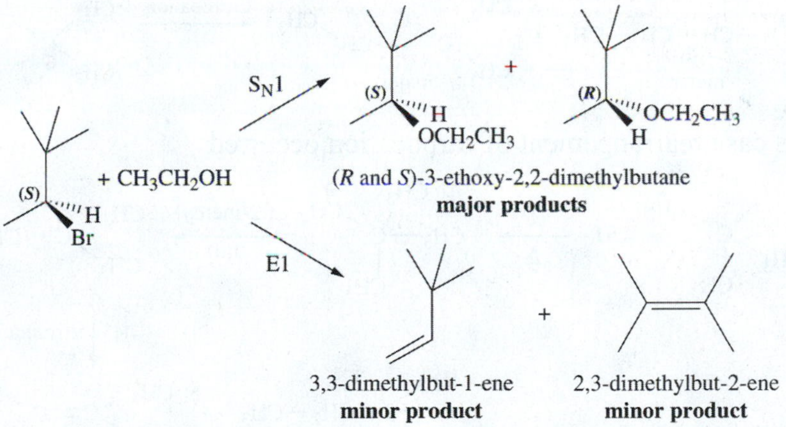

S_N1

(S) H OCH₂CH₃   +   (R) OCH₂CH₃ H

+ CH_3CH_2OH

(S) H Br

(R and S)-3-ethoxy-2,2-dimethylbutane
major products

E1

3,3-dimethylbut-1-ene
minor product

+

2,3-dimethylbut-2-ene
minor product

Alcohols and Alkenes

23. **(E)**

(a) $(CH_3)_2CHCH_2OH + HBr \longrightarrow (CH_3)_2CHCH_2Br + H_2O$

(b) $(CH_3)_3COH + K \longrightarrow (CH_3)_3CO^-K^+ + \dfrac{1}{2}H_2(g)$

(c) $(CH_3)_2CHOH \xrightarrow{[O]} (CH_3)_2CO$

(d) There is no reaction.

25. **(M)** We are converting alcohol to alkyl halide.

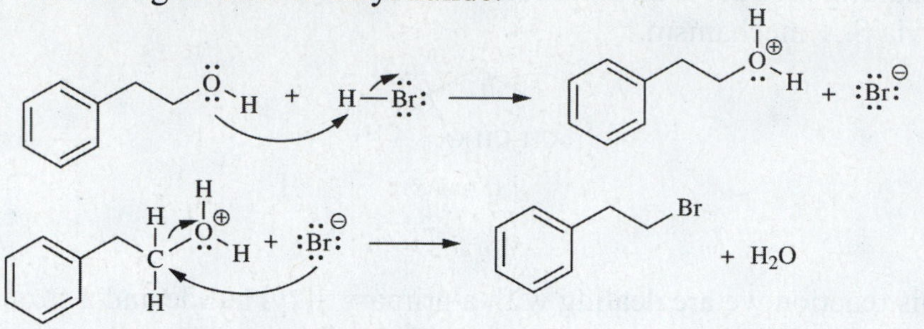

27. **(M)** If no rearrangement occurs the expected product would be 4,4-dimethyl-pent-2-ene

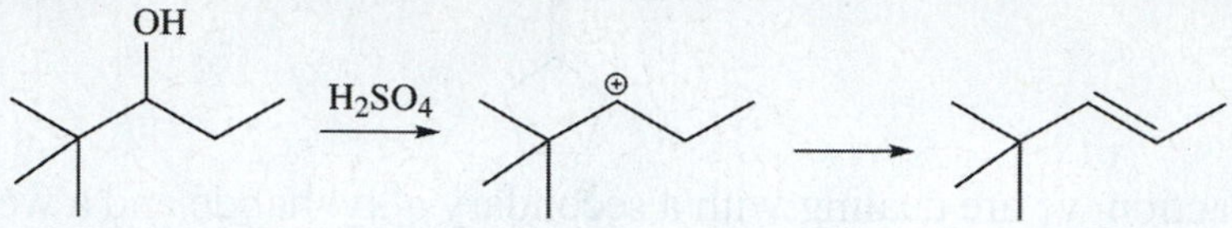

Since the main product is 2,3-dimethyl-2-pentene, the rearrangement of carbocation via 1.2-methyl shift occurred to afford more substituted alkene.

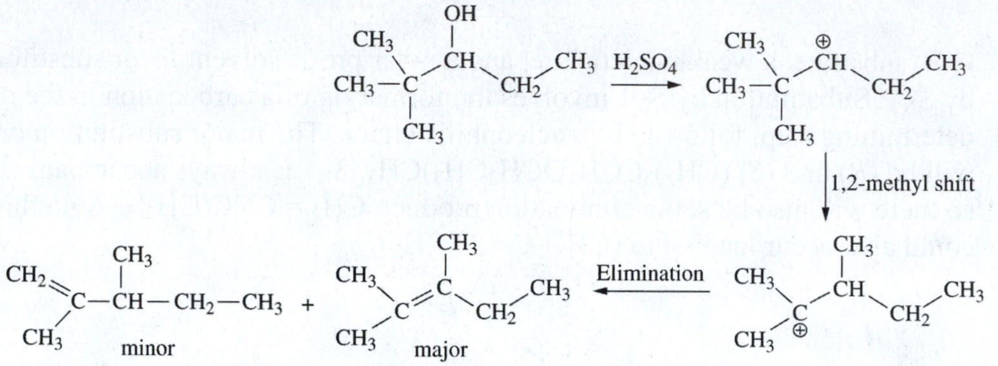

28. **(D)** In this case rearrangement of carbocation occurred

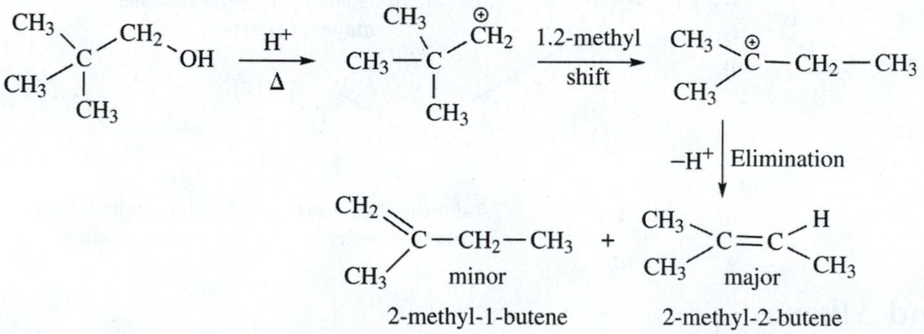

2-methyl-1-butene   2-methyl-2-butene

29. **(M)**

(a) $CH_3 - CH = CH_2 \xrightarrow{H_2 \atop Pt} CH_3 - CH_2 - CH_3$

(b)

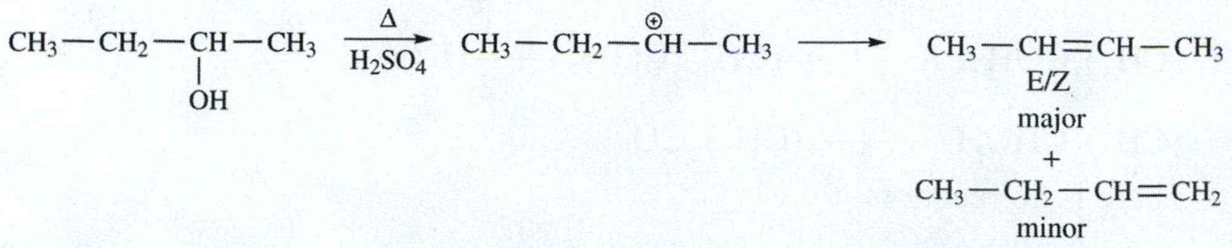

31. **(M)**

(a)

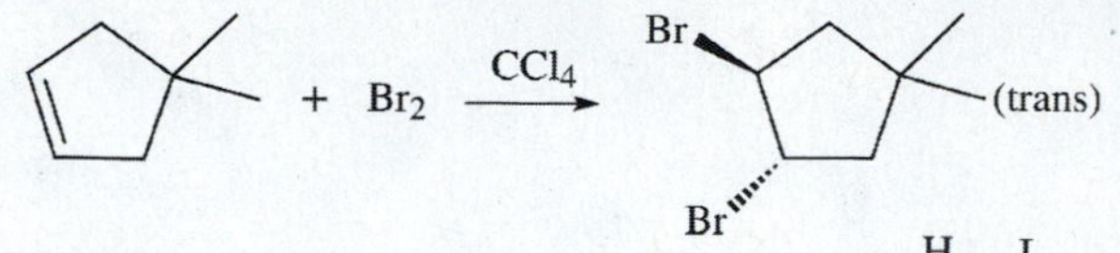
+ Br2
CCl4
Br
Br
(trans)

(b)

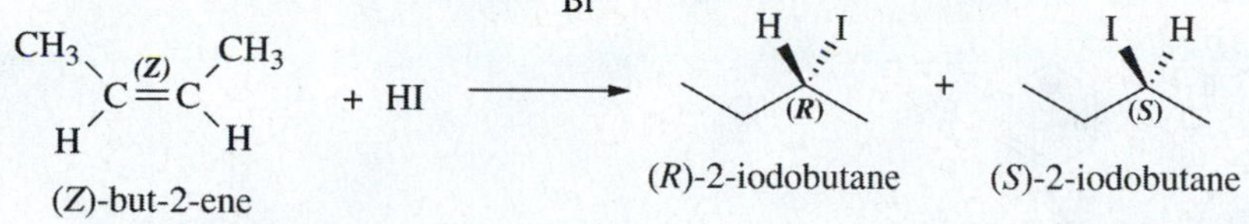
(Z)-but-2-ene
+ HI
(R)-2-iodobutane
(S)-2-iodobutane

(c)

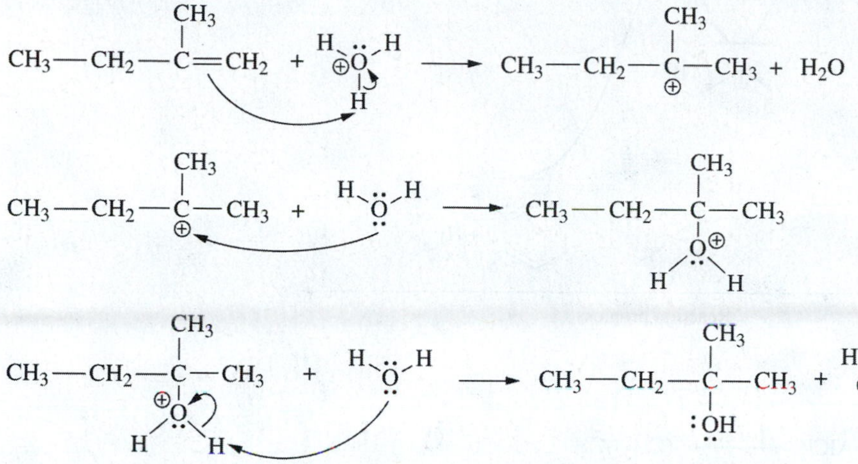

33. **(M)**

(a)

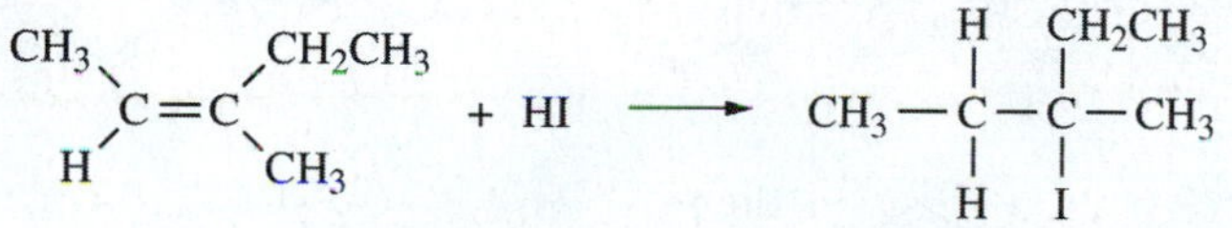
+ HI

(b)

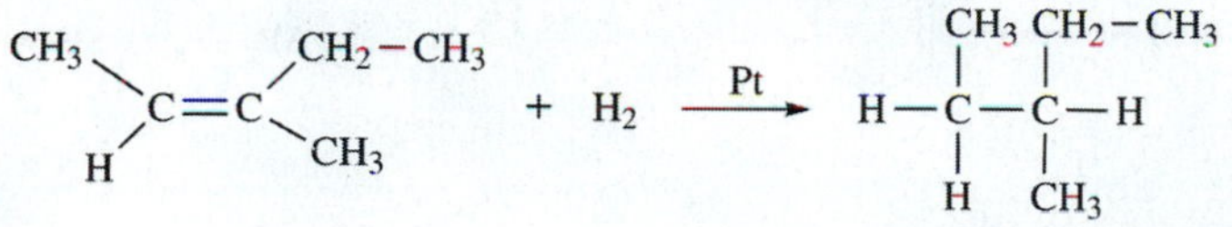
+ H2
Pt

(c)

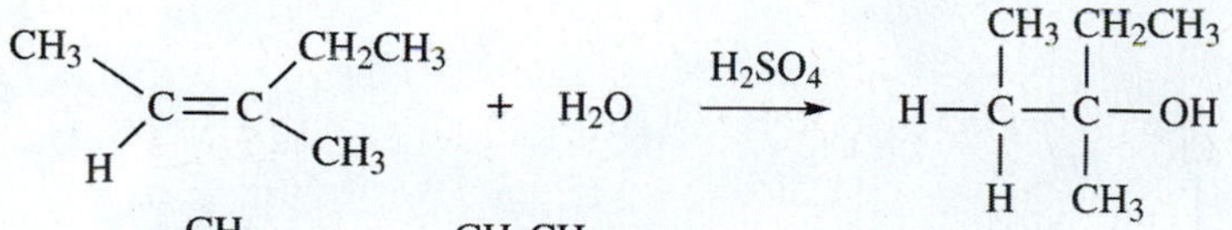
+ H2O
H2SO4

(d)

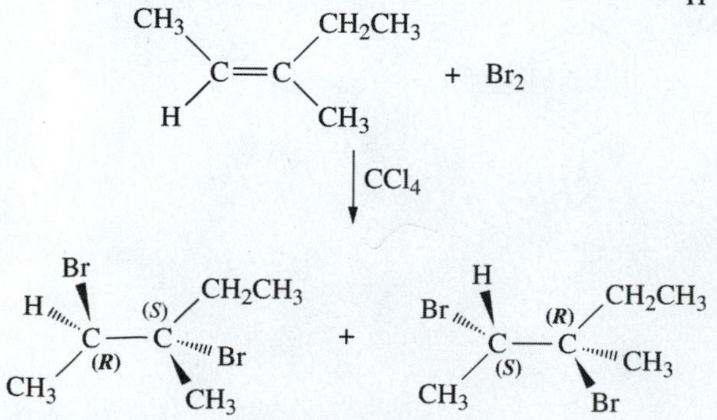
+ Br2
CCl4

Electrophilic Aromatic Substitution

35. **(M)**

37. **(M)**

$$(CH_3)_2CHCl + AlCl_3 \rightleftharpoons (CH_3)_2\overset{\oplus}{C}H + AlCl_4^{\ominus}$$

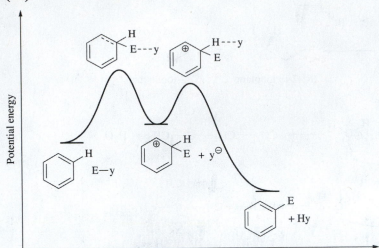

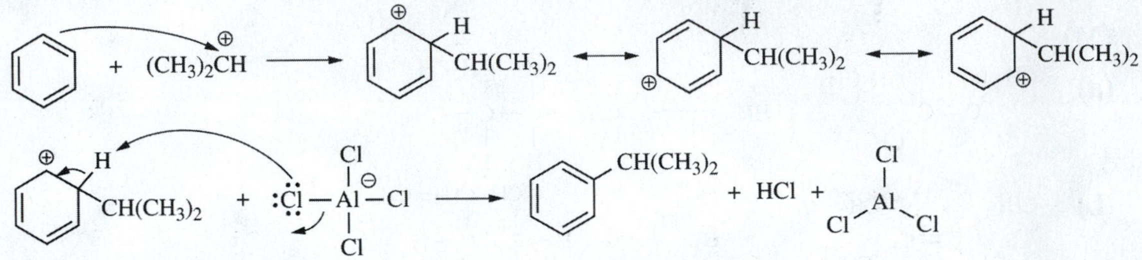

39. **(M)**

(a)

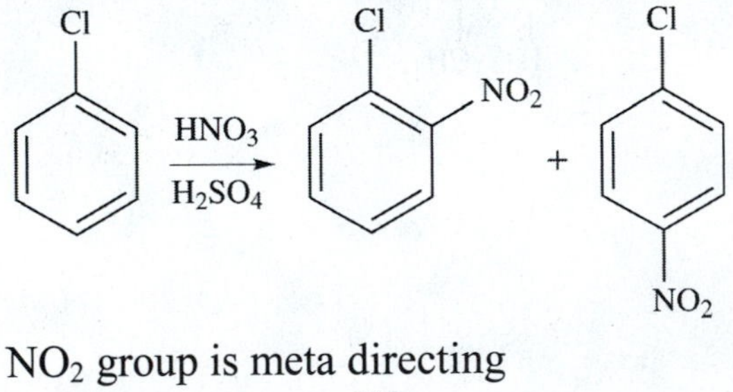

(b) NO_2 group is meta directing

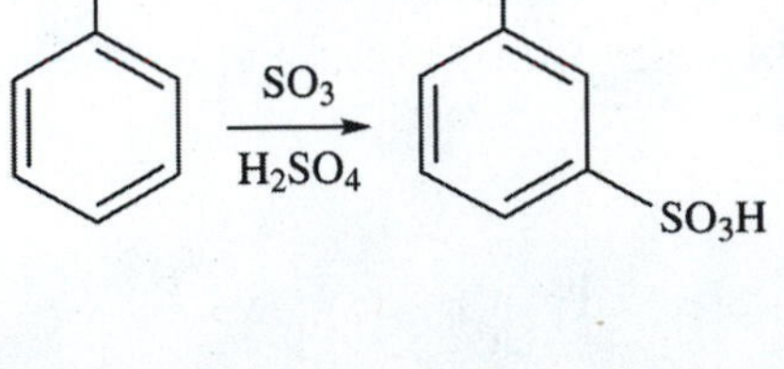

(c) NO_2 group is meta directing and CH_3 group is both ortho and para directing.

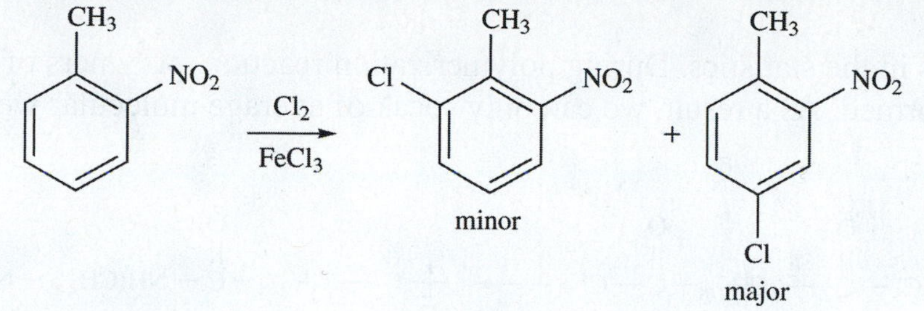

minor

major

Reactions of Alkanes

41. **(M)**

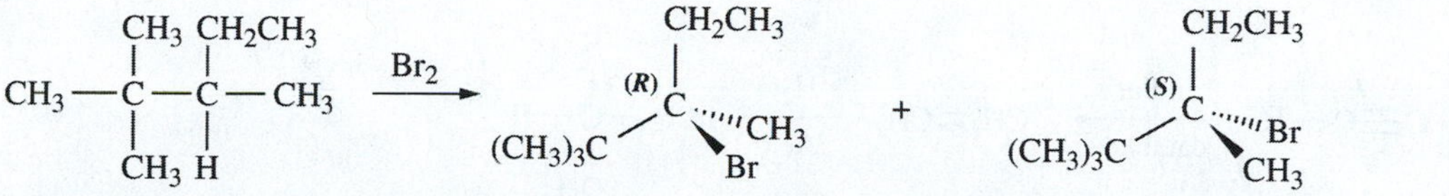

2,2,3-trimethylpentane   (R)-3-bromo-2,2,3-trimethylpentane   (S)-3-bromo-2,2,3-trimethylpentane

43. **(M)**

(a)

Initiation

$$F-F \xrightarrow[\text{light}]{\text{heat or}} F^\bullet + F^\bullet$$

Propagation

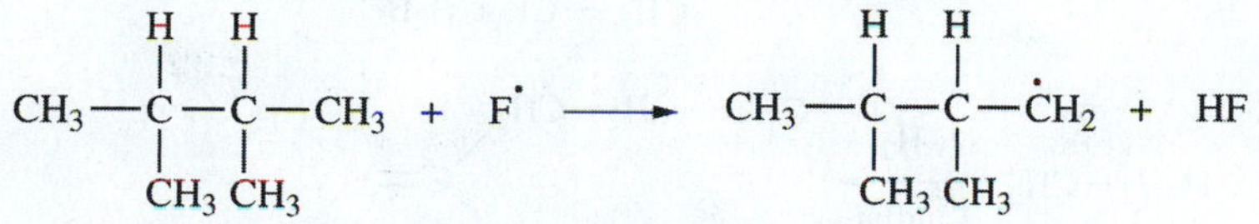

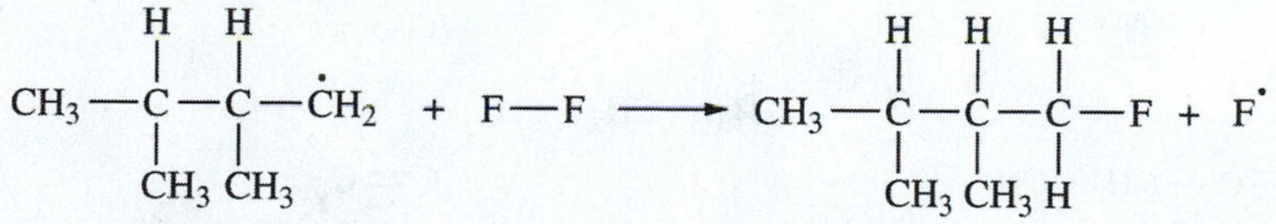

1-fluoro-2,3-dimethylbutane

Termination

$$F^\bullet + F^\bullet \longrightarrow F-F$$

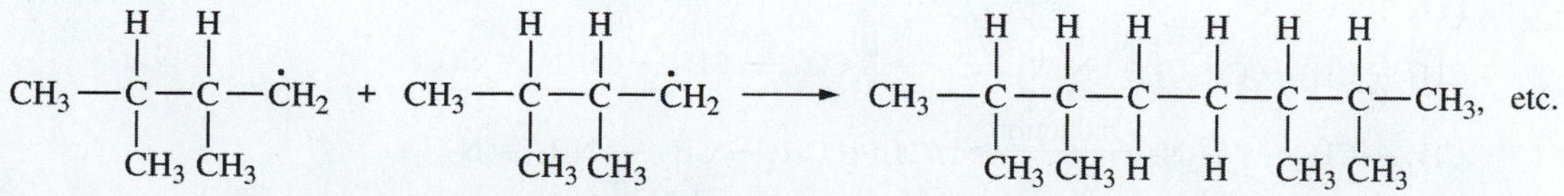

(b) Statistically, there is a higher probability of abstracting an H atom from 12 CH_3 protons to yield 1-fluoro-2,3-dimethybutane than of abstracting an H atom from 2 CH protons to yield 2-fluoro-2,3-dimethylbutane.

Polymerization Reactions

45. **(M)** The reason lies in the statistics. During polymerization reaction, polymers of different chain-lengths are formed. As a result, we can only speak of average molecular weight.

47. **(M)**

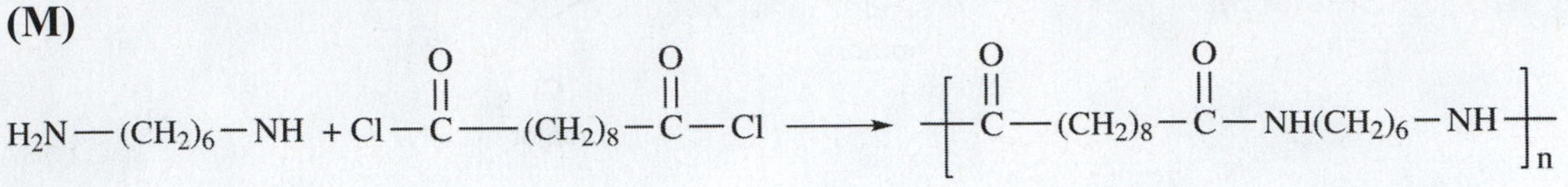

$$H_2N-(CH_2)_6-NH + Cl-\overset{\overset{\displaystyle O}{\|}}{C}-(CH_2)_8-\overset{\overset{\displaystyle O}{\|}}{C}-Cl \longrightarrow \left[\overset{\overset{\displaystyle O}{\|}}{C}-(CH_2)_8-\overset{\overset{\displaystyle O}{\|}}{C}-NH(CH_2)_6-NH\right]_n$$

Synthesis of Organic Compounds

49. **(D)**

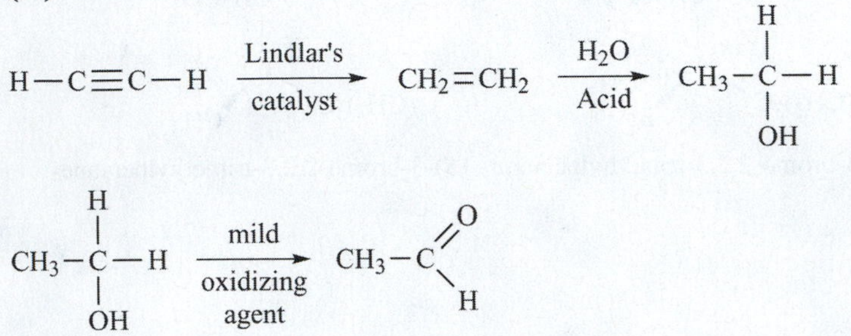

$$H-C\equiv C-H \xrightarrow[\text{catalyst}]{\text{Lindlar's}} CH_2=CH_2 \xrightarrow[\text{Acid}]{H_2O} CH_3-\overset{\overset{\displaystyle H}{|}}{\underset{\underset{\displaystyle OH}{|}}{C}}-H$$

$$CH_3-\overset{\overset{\displaystyle H}{|}}{\underset{\underset{\displaystyle OH}{|}}{C}}-H \xrightarrow[\substack{\text{oxidizing} \\ \text{agent}}]{\text{mild}} CH_3-\overset{\overset{\displaystyle O}{\diagup}}{\underset{\underset{\displaystyle H}{}}{C}}$$

51. **(D)**

$$H-C\equiv C-H \xrightarrow[CH_3CH_2Br]{NaNH_2} H-C\equiv C-CH_2-CH_3 \xrightarrow[CH_3-CH_2CH_2Br]{NaNH_2} \overset{\overset{\displaystyle CH_3}{|}}{CH_2}-CH_2-C\equiv C-\overset{\overset{\displaystyle CH_3}{|}}{CH_2}$$

$$CH_3-CH_2-CH_2-C\equiv C-CH_2-CH_3 \xrightarrow[\substack{\text{Lindlar's} \\ \text{catalyst}}]{H_2} \underset{\text{\textit{cis} isomer}}{\underset{H}{\overset{CH_3-CH_2-CH_2}{C}}=\underset{H}{\overset{CH_2-CH_3}{C}}}$$

$$CH_3-CH_2-CH_2-C\equiv C-CH_2-CH_3 \xrightarrow[NH_3]{Na} \underset{\text{\textit{trans} isomer}}{\underset{H}{\overset{CH_3-CH_2-CH_2}{C}}=\underset{CH_2-CH_3}{\overset{H}{C}}}$$

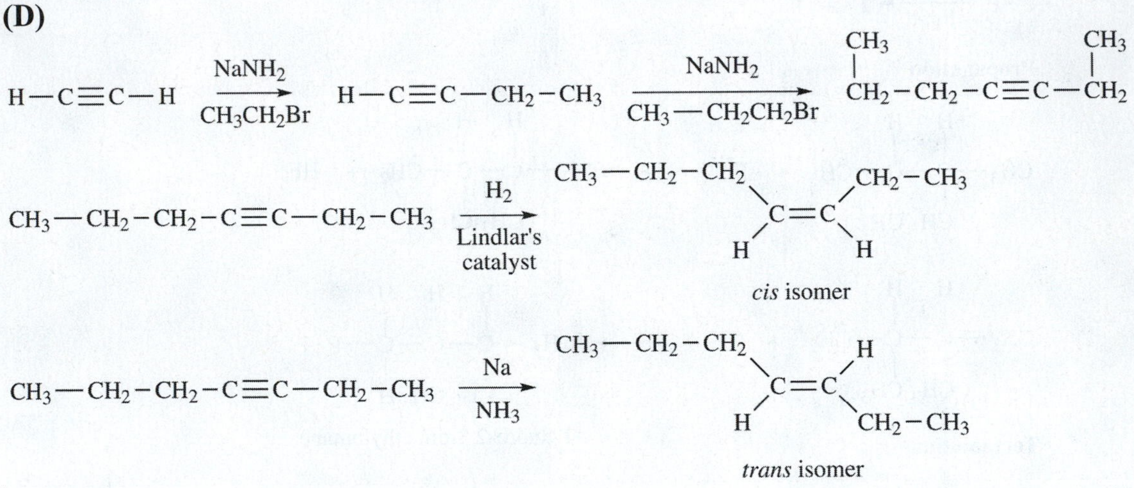

53. **(D)**

$$CH_3-CH_2-CH_2Cl + NaN_3 \longrightarrow CH_3-CH_2-CH_2N_3 + NaCl$$

$$CH_3-CH_2-CH_2N_3 \xrightarrow{\text{reduction}} CH_3-CH_2-CH_2-NH_2 + N_2$$

INTEGRATIVE AND ADVANCED EXERCISES

55. **(M)**

(a) 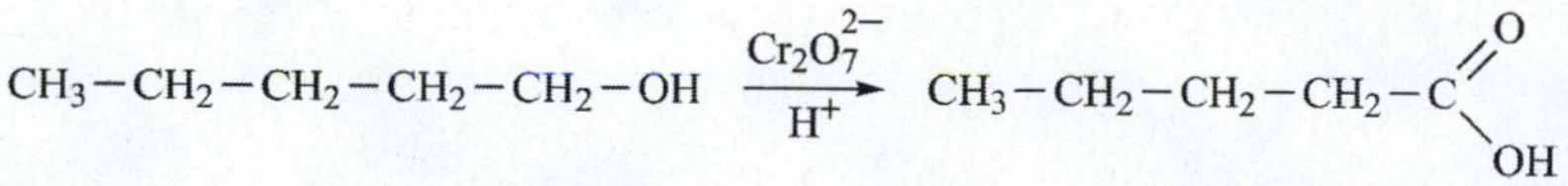 $CH_3-CH_2-CH_2-CH_2-CH_2-OH \xrightarrow[H^+]{Cr_2O_7^{2-}} CH_3-CH_2-CH_2-CH_2-C\!\!\!\begin{array}{c} {}^{\diagup O} \\ {}_{\diagdown OH} \end{array}$

(b) $CH_3-CH_2-CH_2-C\!\!\!\begin{array}{c} {}^{\diagup O} \\ {}_{\diagdown OH} \end{array} + CH_3-CH_2-OH \xrightarrow{H^+} CH_3-CH_2-CH_2-C\!\!\!\begin{array}{c} {}^{\diagup O} \\ {}_{\diagdown O-CH_2-CH_3} \end{array} + H_2O$

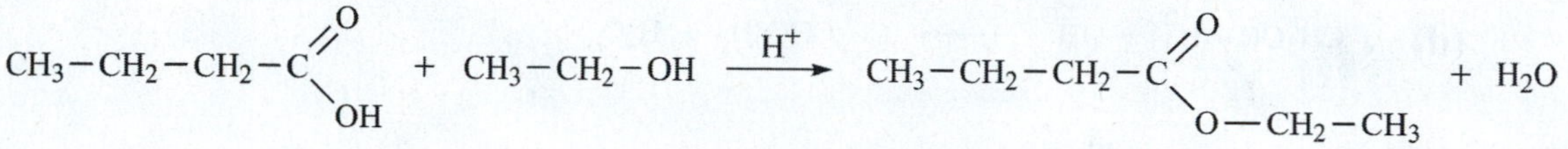

(c)

$$CH_2\!\!=\!\!\overset{\overset{\displaystyle CH_3}{|}}{C}-CH_2-CH_3 + HBr \longrightarrow CH_3-\overset{\overset{\displaystyle CH_3}{|}}{\underset{\underset{\displaystyle Br}{|}}{C}}-CH_2-CH_3 + H_2O$$

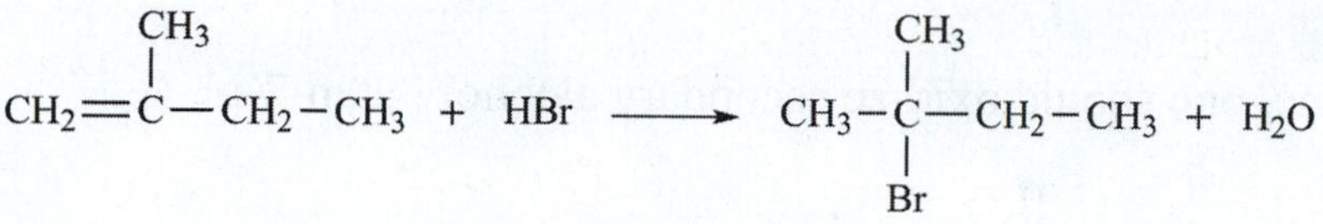

57. **(M)**

$CH_3-CH\!\!=\!\!CH_2$

$\xrightarrow[Pt]{H_2} CH_3-CH_2-CH_3$

$\xrightarrow{Cl_2} CH_3-\overset{\overset{\displaystyle Cl}{|}}{CH}-\underset{\underset{\displaystyle Cl}{|}}{CH_2}$

$\xrightarrow{HCl} CH_3-\overset{\overset{\displaystyle Cl}{|}}{CH}-CH_3$

$\xrightarrow[H^{\oplus}]{H_2O} CH_3-\overset{\overset{\displaystyle OH}{|}}{CH}-CH_3$

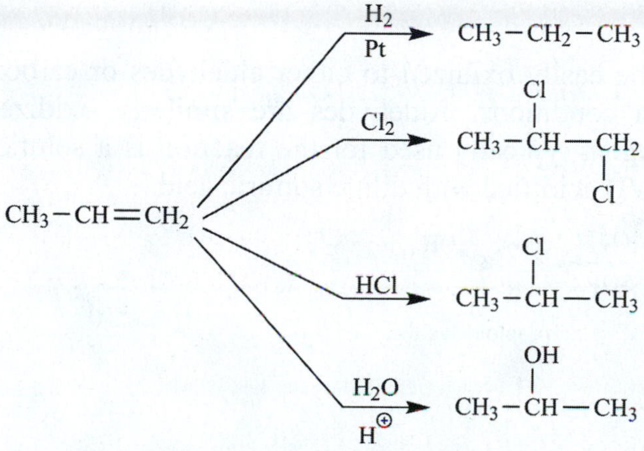

59. **(M)**

(a) Esters are prone to hydrolysis under basic conditions

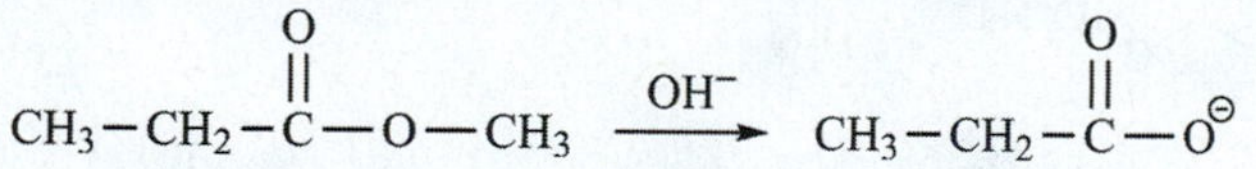

$$CH_3-CH_2-\overset{\overset{\displaystyle O}{||}}{C}-O-CH_3 \xrightarrow{OH^-} CH_3-CH_2-\overset{\overset{\displaystyle O}{||}}{C}-O^{\ominus}$$

(b) Carboxylic acids react with dilute NaOH

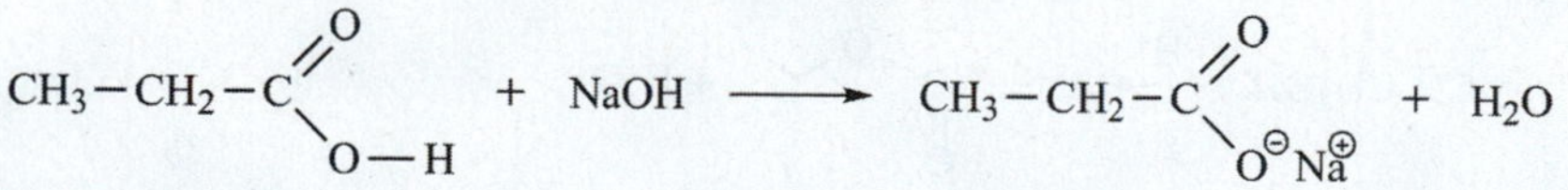

$$CH_3-CH_2-C\!\!\!\begin{array}{c} {}^{\diagup O} \\ {}_{\diagdown O-H} \end{array} + NaOH \longrightarrow CH_3-CH_2-C\!\!\!\begin{array}{c} {}^{\diagup O} \\ {}_{\diagdown O^{\ominus} Na^{\oplus}} \end{array} + H_2O$$

(c) Anion is basic and it will react with dilute HCl

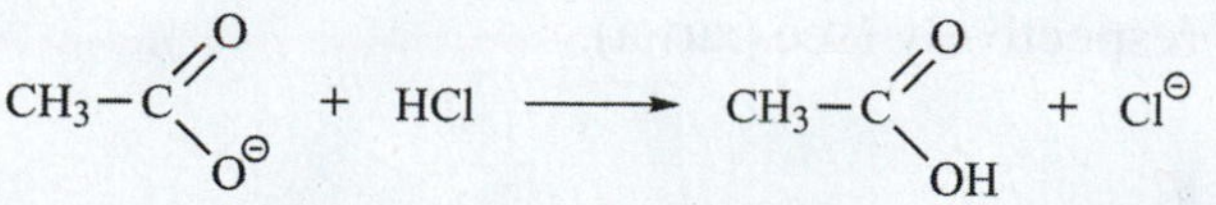

$$CH_3-C\!\!\!\begin{array}{c} {}^{\diagup O} \\ {}_{\diagdown O^{\ominus}} \end{array} + HCl \longrightarrow CH_3-C\!\!\!\begin{array}{c} {}^{\diagup O} \\ {}_{\diagdown OH} \end{array} + Cl^{\ominus}$$

61. **(M)**

(a) $CH_3CH_2NH_2 + HCl \longrightarrow CH_3CH_2NH_3^{\oplus}Cl^{\ominus}$

(b) $(CH_3)_3\overset{\oplus}{N}H\ \overset{\ominus}{Br}$

(c) no reaction

(d) $CH_3CH_2NH_3^{\oplus} + OH^{\ominus} \longrightarrow CH_3CH_2NH_2 + H_2O$

63. **(M)**

To prepare butan-2-one, one should oxidize secondary alcohol butan-2-ol.

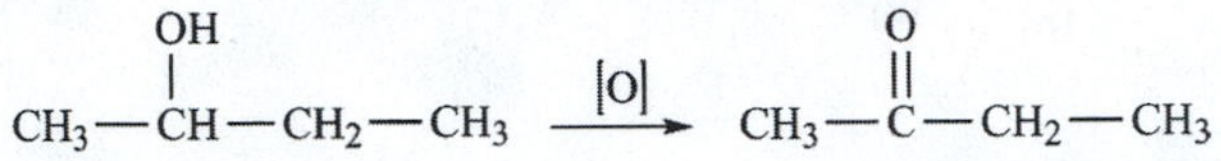

butan-2-one

65. **(M)**

(a) Generally, primary alcohols can be easily oxidized to either aldehydes or carboxylic acids depending on the reaction conditions. Aldehydes are similarly oxidized to carboxylic acids. The oxidizing agent typically used for the reaction is a solution of sodium or potassium dichromate(VI) acidified with dilute sulfuric acid.

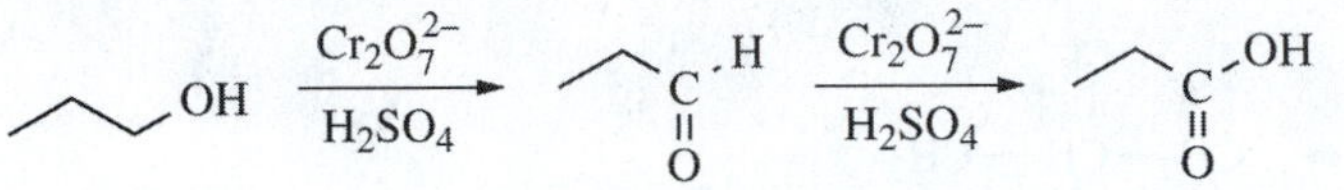

propan-1-ol propionaldehyde propionic acid

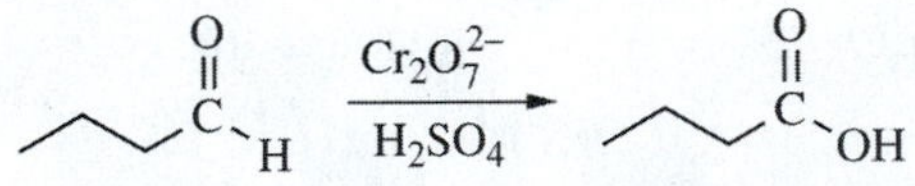

butyraldehyde butyric acid

(b) Propionic acid is a weak monoprotic acid that will react with NaOH. The products of the neutralization reaction are the corresponding sodium salt and water.

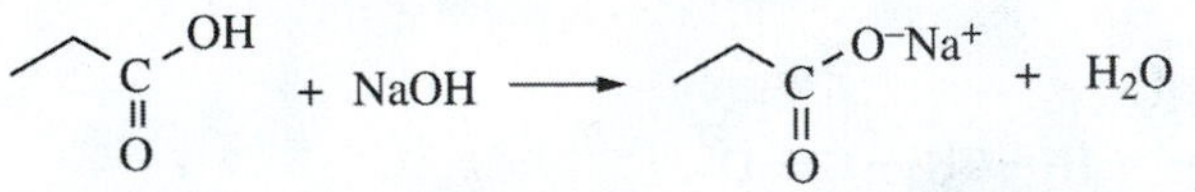

(c) Propionic acid reacts with ethanol to form the ester ethyl propionate. This reaction is called esterification. The additional product is H_2O.

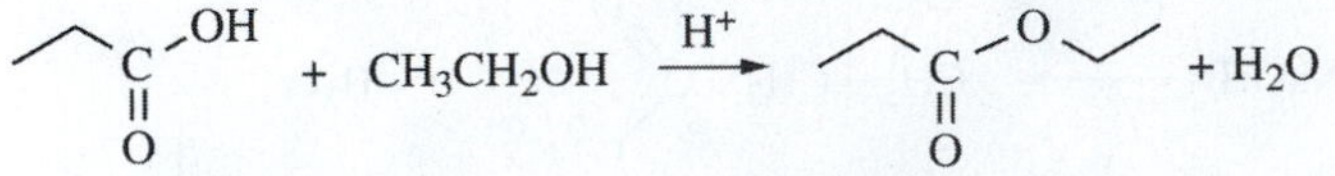

ethyl propionate

(d) Alcohols and aldehydes can be oxidized to carboxylic acids. The products are propionic and butyric acids, respectively (see part a).

(e) Carboxylic acids and aldehydes can be reduced to primary alcohols. Typically, strong reducing agents, such as lithium tetrahydridoaluminate $LiAlH_4$ (carboxylic acids and aldehydes) or sodium tetrahydridoborate $NaBH_4$ (aldehydes), are used.

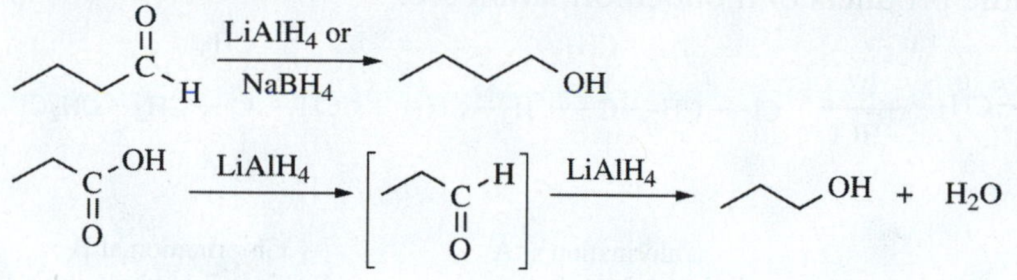

67. **(M)**

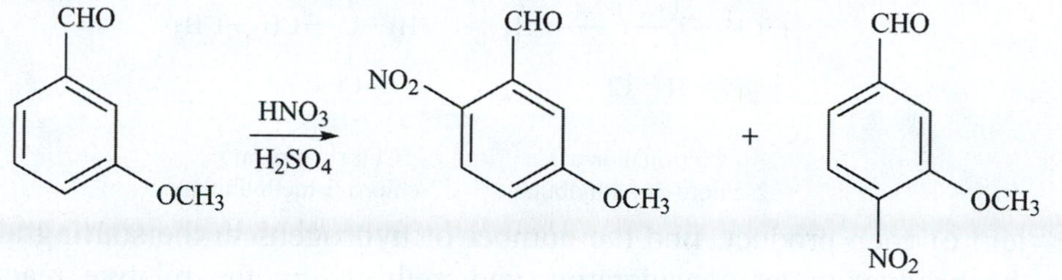

69. **(D)** Halogen group is ortho and para directing

(I)

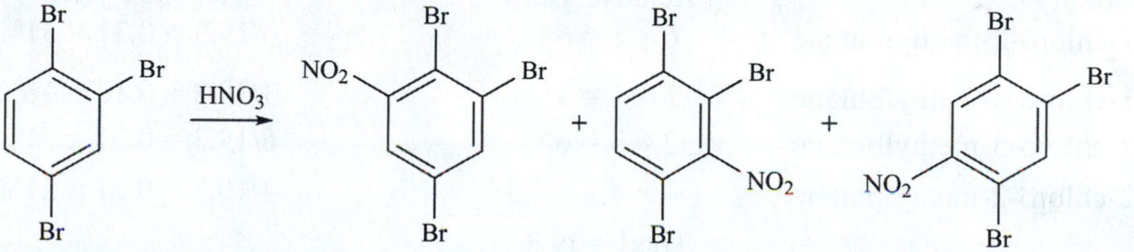

(II)

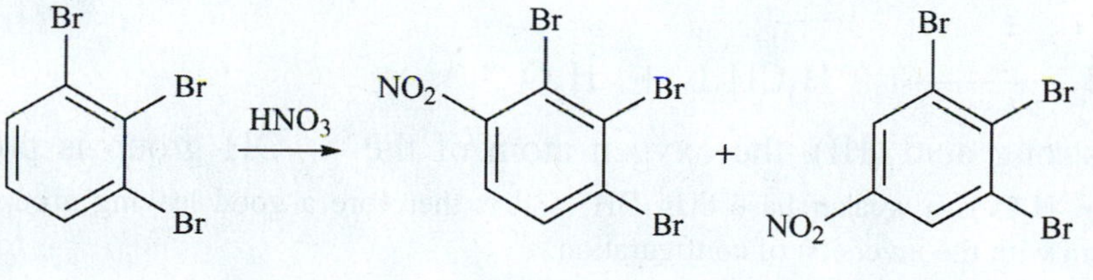

(III)

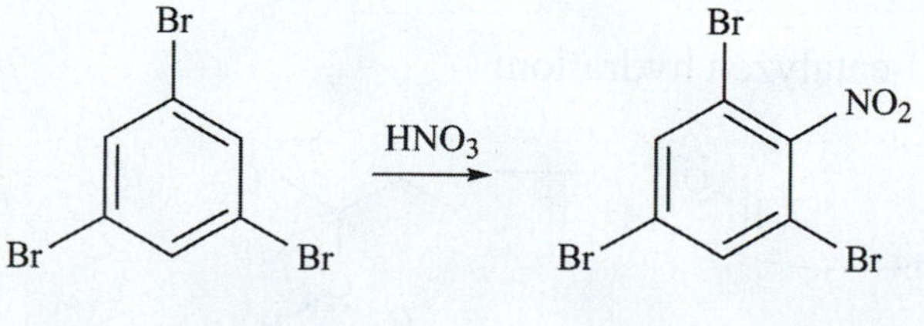

71. **(D)**

2-methylbutane contains nine primary hydrogens, two secondary hydrogens and one tertiary hydrogen. The possible products of monochlorination are:

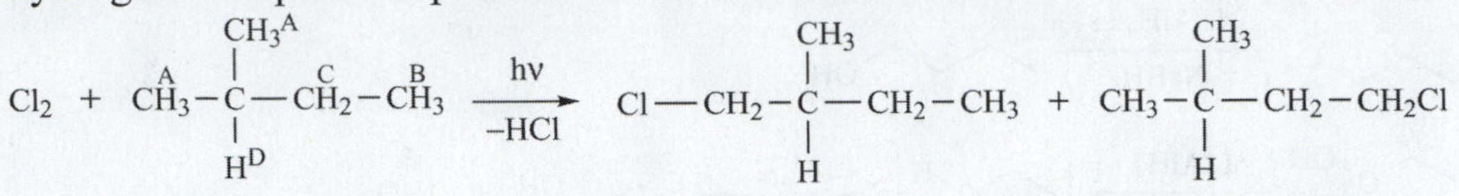

Chlorination at A
1-chloro-2-methylbutane

Chlorination at B
1-chloro-3-methylbutane

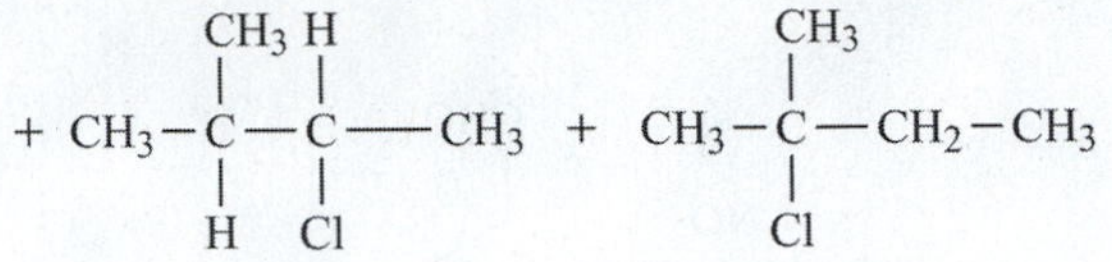

Chlorination at C
2-chloro-3-methylbutane

Chlorination at D
2-chloro-2-methylbutane

To calculate the yield of each product, find the number of hydrogens in the starting alkane that give rise to the product under consideration and multiply by the relative reactivity corresponding to the type of hydrogen

Product	Relative yield	Absolute yield
1-chloro-2-methylbutane	$6 \times 1 = 6$	$6/19.3 = 0.31 = 31\%$
1-chloro-3-methylbutane	$3 \times 1 = 3$	$3/19.3 = 0.16 = 16\%$
2-chloro-3-methylbutane	$2 \times 3 = 6$	$6/19.3 = 0.31 = 31\%$
2-chloro-2-methylbutane	$1 \times 4.3 = 4.3$	$4/19.3 = 0.21 = 21\%$
	Total $= 19.3$	

73. **(M)**

$$CH_3CH_2OH + HI \xrightarrow{\Delta} CH_3CH_2I + H_2O$$

In the presence of strong acid (HI), the oxygen atom of the — OH group is protonated, forming $CH_3CH_2 \overset{\oplus}{-}OH_2$. H_2O is a weaker base than OH^- and is therefore a good leaving group. Reaction occurs via S_N2 mechanism with the inversion of configuration.

75. **(D)** The reaction of interest is the acid-catalyzed hydration:

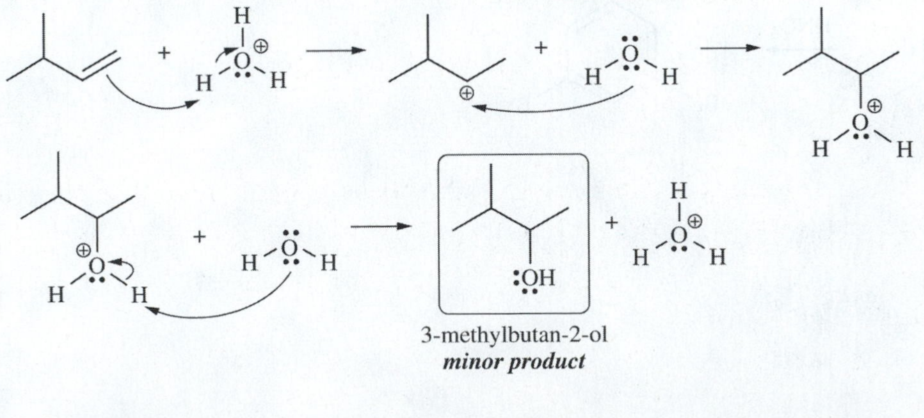

3-methylbutan-2-ol
minor product

740

Carbocation rearrangement occurs by 1,2-hydride shift to give a more stable tertiary carbocation which undergoes attack by water, followed by deprotonation, to give the major product.

2-methylbutan-2-ol
major product

77. **(D)** The desired product can be synthesized using the following sequences of chemical reactions:

$$CH_4 + Cl_2 \xrightarrow{h\nu} CH_3Cl + Cl_2 \xrightarrow{h\nu} CH_2Cl_2$$

$$CH_4 + Cl_2 \xrightarrow{h\nu} CH_3Cl \xrightarrow{OH^-} CH_3OH$$

SELF-ASSESSMENT EXERCISES

79. **(E)**

(a) Nucleophilic substitution corresponds to a substitution (either S_N1 or S_N2) for aliphatic compounds.

Electrophilic aromatic substitution is typical for aromatic compounds (an atom is replaced by an electrophile)

(b) An addition reaction is the opposite of an elimination reaction. In an addition reaction, two or more atoms (molecules) combine to form a larger one.

(c) S_N1 reaction involves the formation of carbocation. S_N2 reaction, on the other hand, is a one-step process in which bond breaking and bond making occur simultaneously at a carbon atom with a suitable leaving group.

(d) E1 reactions are unimolecular elimination reactions that proceed via carbocation intermediates. E2 reactions are bimolecular, one-step reactions that require an antiperiplanar conformation at the time of π-bond formation and β-bond breaking and do not involve carbocations.

81. **(M)**

(a) CN^- is a better nucleophile than Cl^- and it will therefore react faster

(b) Although both substrates are primary alkyl halides, the uncrowded CH_3I reacts faster than the more crowded 1-iodo-2-methyl-butane in an S_N2 reaction.

83. **(M)** Ether formation proceeds via an S_N2 mechanism. The combination of $(CH_3)_2CHONa$ and CH_3I will be better for S_N2 (backside attack) because there is less crowding of the α carbon atom in CH_3I than in $(CH_3)_2CHI$ and no chance of forming an alkene by elimination. The first reaction will favor E-2.

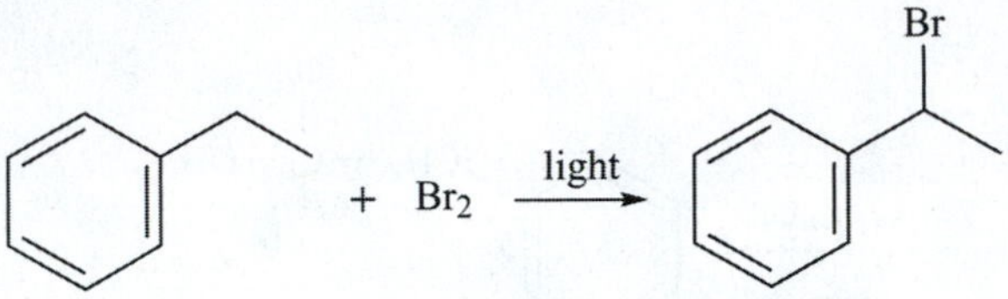

85. **(M)**

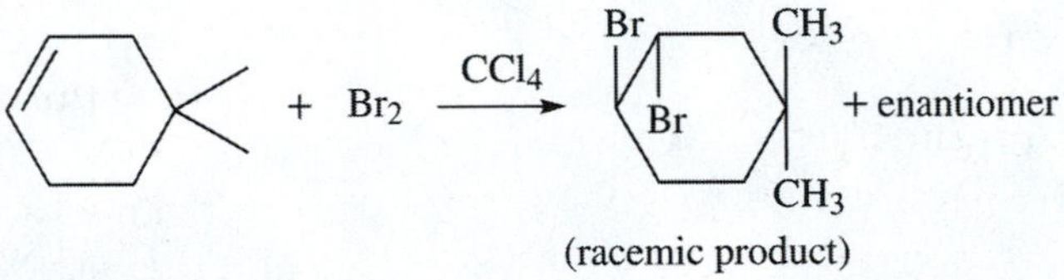

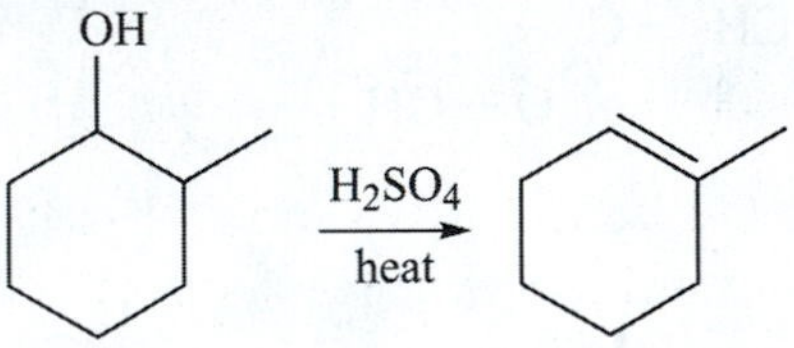

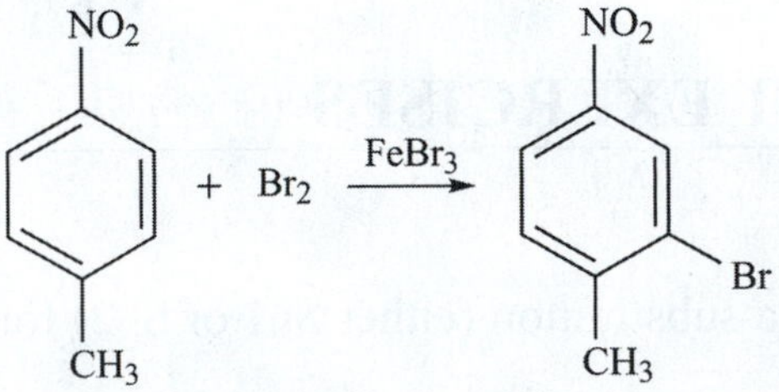

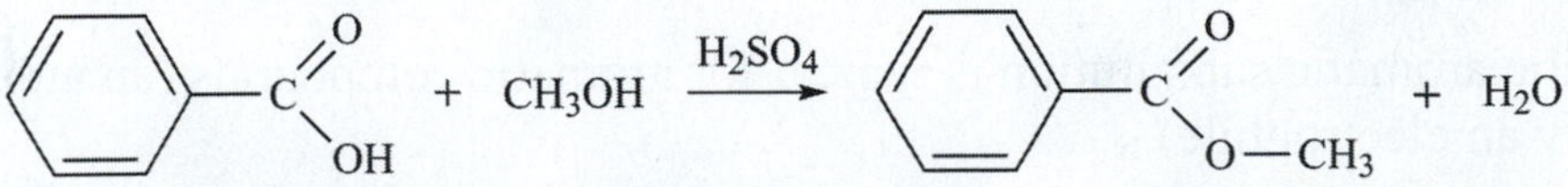

CHAPTER 28
CHEMISTRY OF THE LIVING STATE
PRACTICE EXAMPLES

1A **(E)**

$$H_2N-\overset{\displaystyle O}{\underset{\displaystyle \underset{CH_3}{|}\underset{|}{CHOH}}{\overset{||}{CH-C}}}-NH-\overset{\displaystyle O}{\underset{\displaystyle \underset{CH_3}{|}\underset{|}{CHOH}}{\overset{||}{CH-C}}}-NH-\overset{\displaystyle O}{\underset{\displaystyle \underset{S-CH_3}{|}\underset{|}{(CH_2)_2}}{\overset{||}{CH-C}}}-OH$$

The amino acids are threonine, threonine, and methionine. This tripeptide is dithreonylmethionine.

1B **(E)** The amino acids are serine, glycine, and valine. The N terminus is first.

$$H_2N-\overset{\displaystyle O}{\underset{\displaystyle \underset{OH}{|}\underset{|}{CH_2}}{\overset{||}{CH-C}}}-NH-\overset{\displaystyle O}{\underset{\displaystyle H}{\overset{||}{CH-C}}}-NH-\overset{\displaystyle O}{\underset{\displaystyle \underset{CH_3}{|}\underset{|}{H_3C-CH}}{\overset{||}{CH-C}}}-OH$$

2A **(M)** Because it is a pentapeptide and five amino acids have been identified, no amino acid is repeated. The sequences fall into place, as follows.

	Gly	Cys	Val	Phe	Tyr	
	Gly	Cys				second fragment
		Cys	Val	Phe		third fragment
			Val	Phe		first fragment
				Phe	Tyr	fourth fragment
pentapeptide sequence	Gly	Cys	Val	Phe	Tyr	

2B **(M)** Because it is a hexapeptide and there are five distinct amino acids, one amino acid must appear twice. The fragmentation pattern indicates that the doubled amino acid is glycine. The sequences fall into place if we begin with the N-terminal end.

	Ser	Gly	Gly	Ala	Val	Trp	
	Ser	Gly	Gly				third fragment
		Gly	Gly	Ala			second fragment
				Ala	Val	Trp	fourth fragment
					Val	Trp	first fragment
hexapeptide sequence	Ser	Gly	Gly	Ala	Val	Trp	

INTEGRATIVE EXAMPLE

__A__ **(M)**

$$\Pi = MRT$$

$$\text{mol MyG} = 0.500 \text{ g MyG} \times \frac{1 \text{ mol MyG}}{16900 \text{ g MyG}} = 2.985 \times 10^{-5} \text{ mol MyG}$$

$$M = \frac{2.985 \times 10^{-5} \text{ mol MyG}}{0.025 \text{ L}} = 0.0011834 \text{ M}$$

$$\Pi = (0.0011834 \text{ M})(0.08206 \text{ L} \cdot \text{atm} \cdot \text{K}^{-1})(298 \text{ K}) = 0.02894 \text{ atm}$$

$$\Pi(\text{Pa}) = 0.02894 \text{ atm} \times \frac{101325 \text{ Pa}}{1 \text{ atm}} = 2932 \text{ Pa}$$

To determine the height of the water column, several things should be kept in mind:

1) The pressure unit, 1 Pascal (Pa) is defined as 1 N/m^2 (1 Newton of force per m^2)

2) Force (F) in Newtons is given by $F = mg$. Since m is in units of kg and g in units of m/s^2, Newtons can be expressed in units of $\text{kg} \cdot \text{m/s}^2$.

With the above points in mind,

$$P = \frac{F}{A} = \frac{m \cdot g}{A}$$

Since mass (m) = density (D) × volume (V), we can make the following substitutions:

$$P = \frac{m \cdot g}{A} = \frac{D \times V \times g}{A} = \frac{D \times (w \times l \times h) \times g}{w \times l} = D \times h \times g$$

$$2932 \frac{\text{N}}{\text{m}^2} = 1000 \frac{\text{kg}}{\text{m}^3} \times h \times 9.8 \frac{\text{m}}{\text{s}^2}$$

$$h = 0.300 \text{ m}$$

B **(D)**
(a)

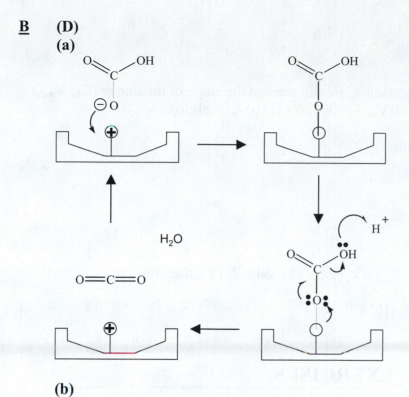

H$_2$O

(b)

Given Data	
Rate (M/s), V	[CO2] (M), S
9.02E − 05	1.25E − 03
1.67E − 04	2.50E − 03
2.90E − 04	5.00E − 03
6.47E − 04	2.00E − 02

→

Lineweaver-Burk Eq.	
1/V	1/S
1.11E + 04	8.00E + 02
5.99E + 03	4.00E + 02
3.45E + 03	2.00E + 02
1.55E + 03	5.00E + 01

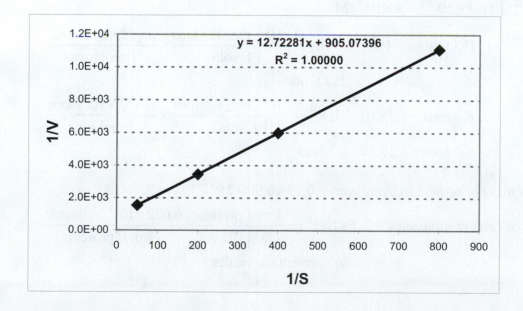

y = 12.72281x + 905.07396
R^2 = 1.00000

The Lineweaver-Burk equation is

$$\frac{1}{V} = \frac{1}{V_{max}} + \frac{K_M}{V_{max}} \cdot \frac{1}{[S]}$$

From the graph and the linear regression, we can see that the slope of the line m (K_M/V_{max}) = 12.7228 s, and the y-intercept b ($1/V_{max}$) = 905.074 (M/s)$^{-1}$. Therefore,

$$V_{max} = 1/905.074 = 0.001105 \text{ M/s}$$

$$\frac{K_M}{V_{max}} = \frac{K_M}{0.001105 \text{ M/s}} = 12.723 \text{ s}$$

$$K_M = 0.01406 \text{ M} = 14.1 \text{ mM}$$

From the problem, we note that $V_{max} = k_2 [E]_0$. The value for k_2, therefore, is:

$$k_2 = \frac{V_{max}}{[E]_0} = \frac{0.001105 \text{ M/s}}{2.3 \times 10^{-9} \text{ M}} = 4.8 \times 10^5 \text{ s}^{-1}$$

EXERCISES

Structure and Composition of the Cell

1. **(M)** The volume of a cylinder is given by $V = \pi r^2 h = \pi d^2 h / 4$.

$$V = \left[3.14159 (1 \times 10^{-6} \text{ m})^2 (2 \times 10^{-6} \text{ m}) \div 4 \right] \times \frac{1000 \text{ L}}{1 \text{ m}^3} = 1.6 \times 10^{-15} \text{ L}$$

The volume of the solution in the cell is $V_{soln} = 0.80 \times 1.6 \times 10^{-15} \text{ L} = 1.3 \times 10^{-15} \text{ L}$.

(a) $[H^+] = 10^{-6.4} = 4 \times 10^{-7}$ M

$$\text{no. H}_3\text{O}^+ \text{ ions} = 1.3 \times 10^{-15} \text{ L} \times \frac{4 \times 10^{-7} \text{ mol H}^+ \text{ ions}}{1 \text{ L soln}} \times \frac{6.022 \times 10^{23} \text{ ions}}{1 \text{ mol ions}}$$

$$= 3 \times 10^2 \text{ H}_3\text{O}^+ \text{ ions}$$

(b) $\text{no. K}^+ \text{ ions} = 1.3 \times 10^{-15} \text{ L} \times \frac{1.5 \times 10^{-4} \text{ mol K}^+ \text{ ions}}{1 \text{ L soln}} \times \frac{6.022 \times 10^{23} \text{ ions}}{1 \text{ mol ions}}$

$$\text{no. K}^+ \text{ ions} = 1.2 \times 10^5 \text{ K}^+ \text{ ions}$$

3. **(E)** mass of protein in cytoplasm = $0.15 \times 0.90 \times 2 \times 10^{-12}$ g = 2.7×10^{-13} g.

$$\text{no. of protein molecules} = 2.7 \times 10^{-13} \text{ g} \times \frac{1 \text{ mol protein}}{3 \times 10^4 \text{ g}} \times \frac{6.022 \times 10^{23} \text{ molecules}}{1 \text{ mol protein}}$$

$$= 5 \times 10^6 \text{ protein molecules}$$

Lipids

5. **(M)**

(a) $C_{15}H_{31}COOH$ is palmitic acid. $C_{17}H_{29}COOH$ is linolenic acid or eleosteric acid. $C_{11}H_{23}COOH$ is lauric acid. Thus, the given compound is glyceryl palmitolinolenolaurate or glyceryl palmitoeleosterolaurate.

(b) $C_{17}H_{33}COOH$ is oleic acid. Thus, the compound is glyceryl trioleate or triolein.

(c) $C_{13}H_{27}COOH$ is myristic acid. Thus, the compound is sodium myristate.

7. **(M)**

(a)

Trilaurin	**Trilinolein**

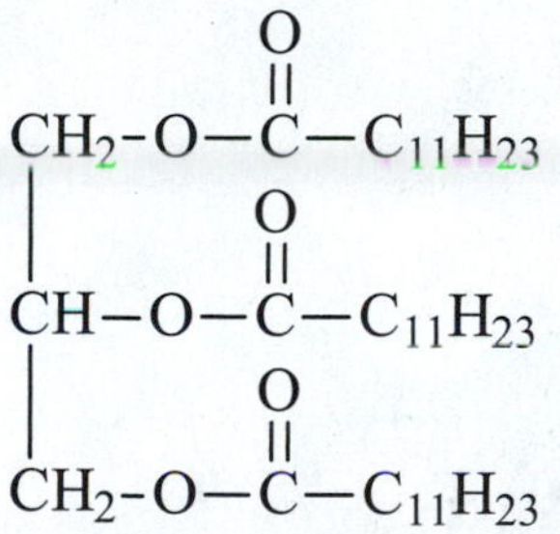

$$CH_2-O-\overset{\displaystyle O}{\overset{\|}{C}}-C_{11}H_{23}$$
$$CH-O-\overset{\displaystyle O}{\overset{\|}{C}}-C_{11}H_{23}$$
$$CH_2-O-\overset{\displaystyle O}{\overset{\|}{C}}-C_{11}H_{23}$$

$$CH_2-O-\overset{\displaystyle O}{\overset{\|}{C}}-C_{17}H_{31}$$
$$CH-O-\overset{\displaystyle O}{\overset{\|}{C}}-C_{17}H_{31}$$
$$CH_2-O-\overset{\displaystyle O}{\overset{\|}{C}}-C_{17}H_{31}$$

A triglyceride or glycerol ester	A triglyceride or glycerol ester
Saturated triglyceride -made using saturated acid	Unsaturated triglyceride -made using unsaturated acid
A fat (usually solid at room temperature)	An oil (usually liquid at room temperature)

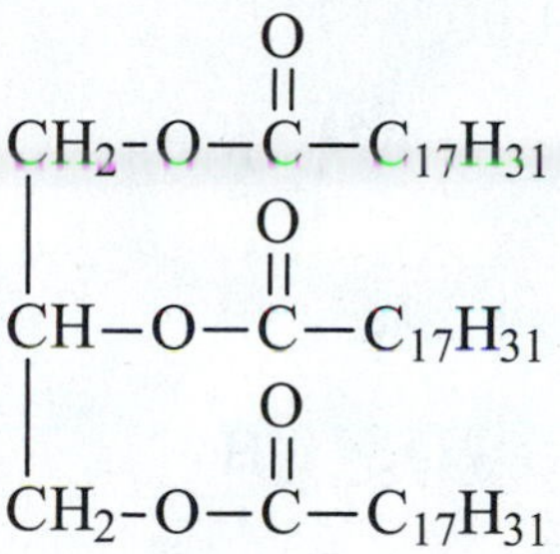

(b) Soaps: salts of fatty acids (from saponification of triglycerides)

Phospholipids: derived from glycerols, fatty acids, phosphoric acid, and a nitrogen containing base (both soaps and phospholipids have hydrophilic heads and hydrophobic tails.)

9. **(E)** Polyunsaturated fatty acids are characterized by a large number of $C=C$ double bonds in their hydrocarbon chain. Stearic acid has no $C=C$ double bonds and therefore is not unsaturated, let alone polyunsaturated. But eleostearic acid has three $C=C$ double bonds and thus is polyunsaturated. Polyunsaturated fatty acids are recommended in dietary programs since saturated fats are linked to a high incidence of heart disease. Of the lipids listed in Table 28-2, safflower oil has the highest percentage of unsaturated fatty acids, predominately linoleic acid, which is an unsaturated fatty acid with two C=C bonds.

11. **(E)**

tripalmitin

saponification products of tripalmitin: sodium palmitate and glycerol

$CH_2OOC(CH_2)_{14}CH_3$
$CHOOC(CH_2)_{14}CH_3$
$CH_2OOC(CH_2)_{14}CH_3$

$CH_2OHCHOHCH_2OH$
$NaOOC(CH_2)_{14}CH_3$

Carbohydrates

13. **(M)**

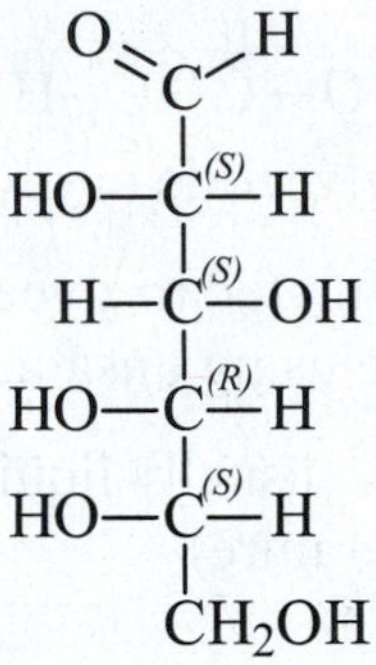

L-Glucose
Aldohexose

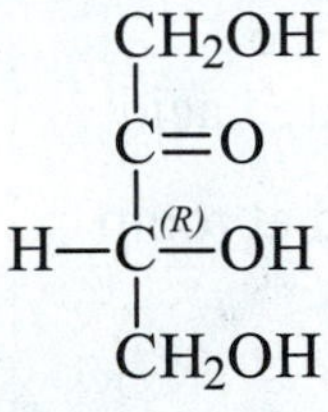

D-Erythrulose
Ketotetrose

Chapter 28: Chemistry of the Living State

15. (M)

(a) D-(−)-arabinose is the optical isomer of L-(+)-arabinose. Its structure is shown below.

(b) A diastereomer of L-(+)-arabinose is a molecule that is its optical isomer, but not its mirror image. There are several such diastereomers, some of which are shown below.

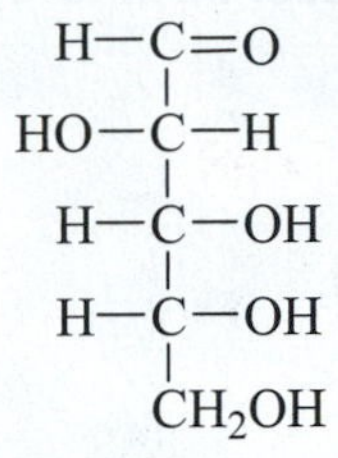

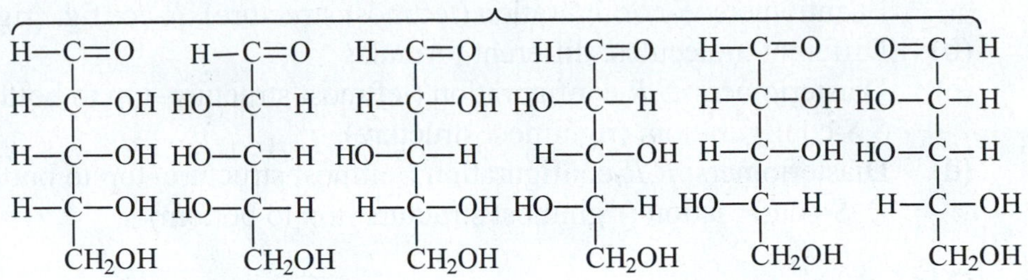

17. (M)

(a) A dextrorotatory compound rotates the plane of polarized light to the right, namely clockwise.

(b) A levorotatory compound is one that rotates the plane of polarized light to the left, namely counterclockwise.

(c) A racemic mixture has equal amounts of an optically active compound and its enantiomer. Since these two compounds rotate polarized light by the same amount but in opposite directions, such a mixture does not exhibit a net rotation of the plane of polarized light.

(d) (R) In organic nomenclature, this designation is given to a chiral carbon atom. First, we must assign priorities to the four substituents on the chiral carbon atom. With the lowest priority group pointing directly away from the viewer, we say that the stereogenic center has an R-configuration if a curved arrow from the group of highest priority through to the one of lowest priority is drawn in a clockwise direction.

19. (E)

A reducing sugar has a sufficient amount of the straight-chain form present in equilibrium with its cyclic form such that the sugar will reduce $Cu^{2+}(aq)$ to insoluble, red $Cu_2O(s)$. Only free aldehyde groups are able to reduce the copper(II) ion down to copper (I).

Next, we need to calculate the mass of Cu_2O expected when 0.500 g of glucose is oxidized in the reducing sugar test:

$$\text{mass } Cu_2O \text{ (g)} = \frac{1 \text{ mol glucose}}{180.2 \text{ g glucose}} \times \frac{2 \text{ mol } Cu^{2+}}{1 \text{ mol glucose}} \times \frac{1 \text{ mol } Cu_2O}{2 \text{ mol } Cu^{2+}} \times \frac{143.1 \text{ g } Cu_2O}{1 \text{ mol } Cu_2O} = 0.397 \text{g } Cu_2O$$

21. **(E)** Enantiomers are alike in all respects, including in the degree to which they rotate polarized light. They differ only in the direction in which this rotation occurs. Since α-glucose and β-glucose rotate the plane of polarized light by different degrees, and in the same direction, they are not enantiomers, but rather diastereomers.

Fischer Projections and *R, S* Nomenclature

23. **(E)**
    (a) Enantiomers: *S*-configuration (leftmost structure), *R*-config. (rightmost structure)
    (b) Different molecules: different formulas
    (c) Diasteriomers: *S,R* configuration (leftmost structure–top to bottom), *S,S*-configuration. (rightmost structure)
    (d) Diasteriomers: *R,R*-configuration (leftmost structure–top to bottom), *R, S*-configuation (rightmost structure–top to bottom)

25. **(M)**

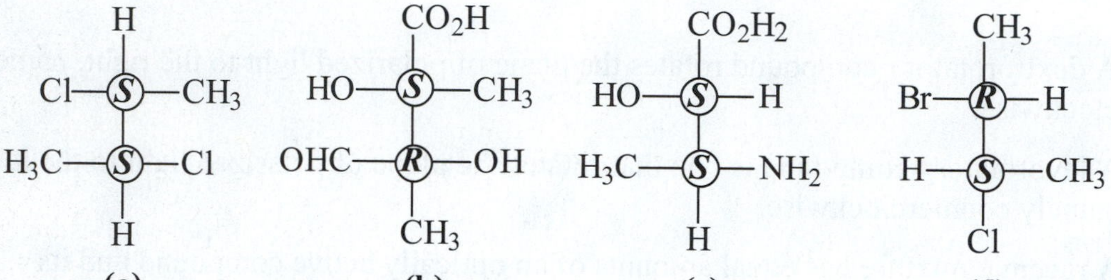

      (a)   **(b)**   **(c)**   **(d)**

Amino Acids, Polypeptides, and Proteins

27. **(E)**
    (a) An α-amino acid has an amine group ($-NH_2$) bonded to the same carbon as the carboxyl group ($-COOH$). For example: glycine $\left(H_2NCH_2COOH\right)$ is the simplest α-amino acid.

    (b) A zwitterion is a form of an amino acid in which the amine group is protonated $\left(-NH_3^+\right)$ and the carboxyl group is deprotonated $\left(-COO^-\right)$. For instance, the zwitterion form of glycine is $^+H_3NCH_2COO^-$.

    (c) The pH at which the zwitterion form of an amino acid predominates in solution is known as the isoelectric point. The isoelectric point of glycine is p$I = 6.03$.

    (d) The peptide bond is the bond that forms between the carbonyl group of one amino acid and the amine group of another, with the elimination of a water molecule between them. The peptide bond between two glycine molecules is shown as a bold dash (▬▬) in the structure below.

(e) Tertiary structure describes how a coiled protein chain further interacts with itself to wrap into a cluster through a combination of salt linkages, hydrogen bonding, and disulfide linkages, to name a few.

29. **(M)** The pI of phenylalanine is 5.74. Thus, phenylalanine is in the form of a cation in 1.0 M HCl (pH = 0.0), an anion in 1.0 M NaOH (pH = 14.0), and a zwitterion at pH = 5.7. These three structures follow.

(a)

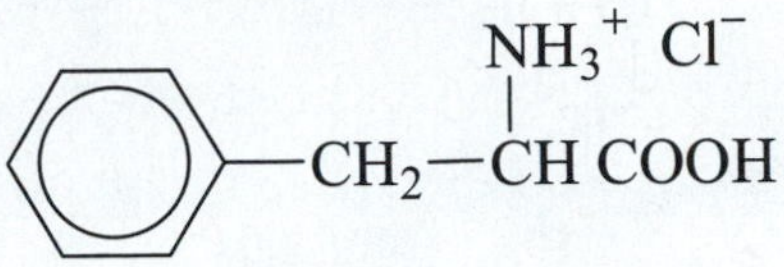

$$NH_3^+ \ Cl^-$$
$$C_6H_5-CH_2-\underset{|}{CH} \ COOH$$

(b)

$$NH_2$$
$$C_6H_5-CH_2-\underset{|}{CH} \ COO^- Na^+$$

(c)

$$NH_3^+$$
$$C_6H_5-CH_2-\underset{|}{CH} \ COO^-$$

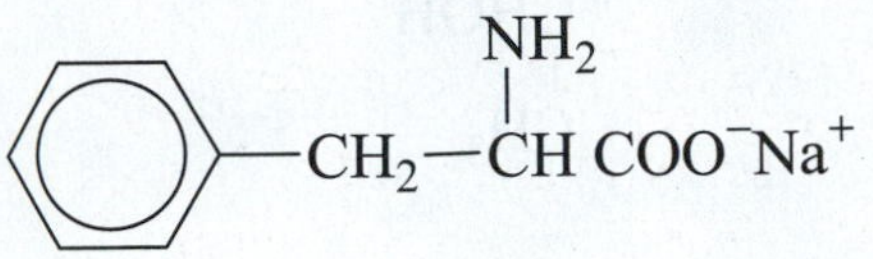

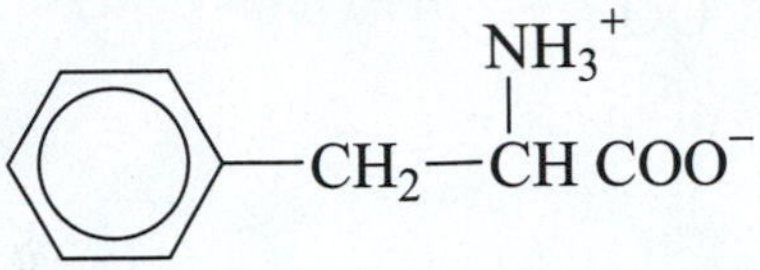

31. **(E)**

(a) alanylcysteine

$$H_2N-\underset{\underset{CH_3}{|}}{CH}-\overset{\overset{O}{||}}{C}-NH-\underset{\underset{\underset{SH}{|}}{CH_2}}{CH}-\overset{\overset{O}{||}}{C}-OH$$

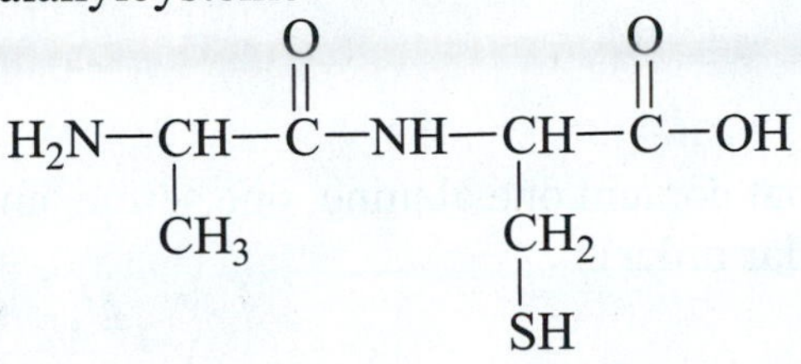

(b) threonylvalylglycine

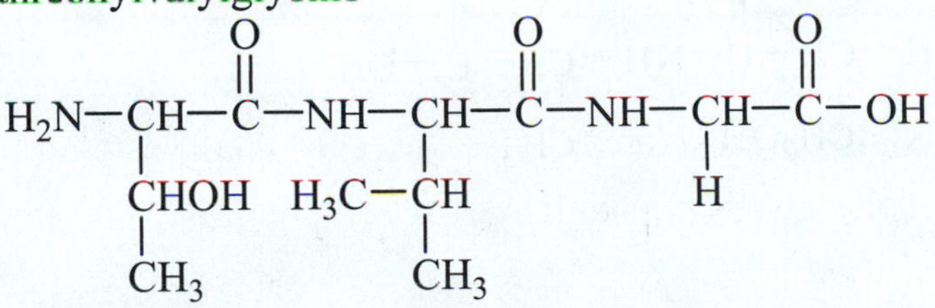

$$H_2N-\underset{\underset{\underset{CH_3}{|}}{CHOH}}{CH}-\overset{\overset{O}{||}}{C}-NH-\underset{\underset{\underset{CH_3}{|}}{H_3C-CH}}{CH}-\overset{\overset{O}{||}}{C}-NH-\underset{\underset{H}{|}}{CH}-\overset{\overset{O}{||}}{C}-OH$$

33. **(M)** pH = 6.3 is near the isoelectric point of proline (6.21). Thus proline will not migrate very effectively under these conditions. But pH = 6.3 is considerably more acidic than the isoelectric point of lysine (pI = 9.74). Thus, lysine is positively charged in this solution and consequently will migrate toward the negatively charged cathode.
Furthermore, pH = 6.3 is much less acidic than the isoelectric point of aspartic acid (pI = 2.96). Aspartic acid, therefore, is negatively charged in this solution and consequently will migrate toward the positively charged anode.

35. **(M)**

(a) in strongly acidic solution

$^+H_3NCH(CHOHCH_3)COOH$

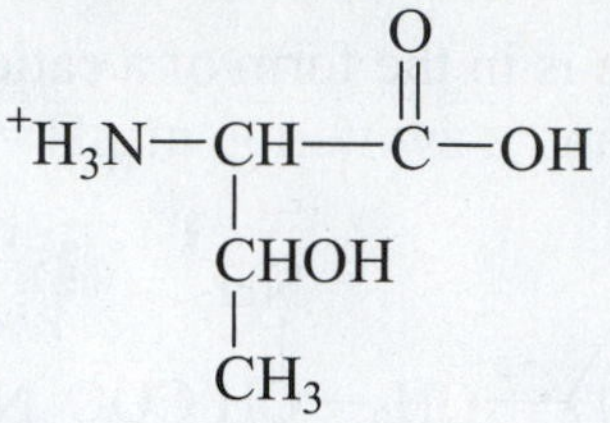

$^+H_3N-\overset{\underset{|}{CHOH}}{\underset{|}{\underset{CH_3}{}}}CH-\overset{O}{\overset{||}{C}}-OH$

(b) at the isoelectric point

$^+H_3NCH(CHOHCH_3)COO^-$

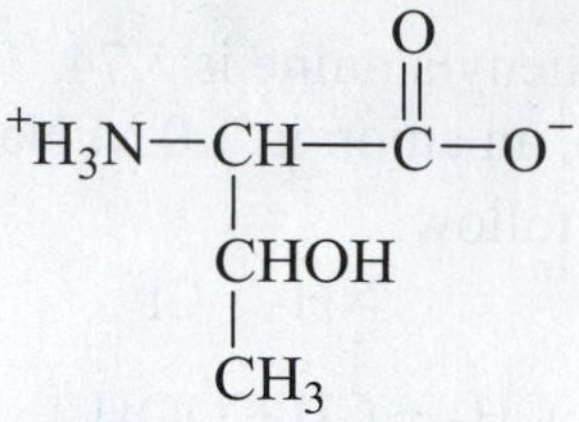

$^+H_3N-\overset{\underset{|}{CHOH}}{\underset{|}{\underset{CH_3}{}}}CH-\overset{O}{\overset{||}{C}}-O^-$

(c) in strongly basic solution $H_2NCH(CHOHCH_3)COO^-$

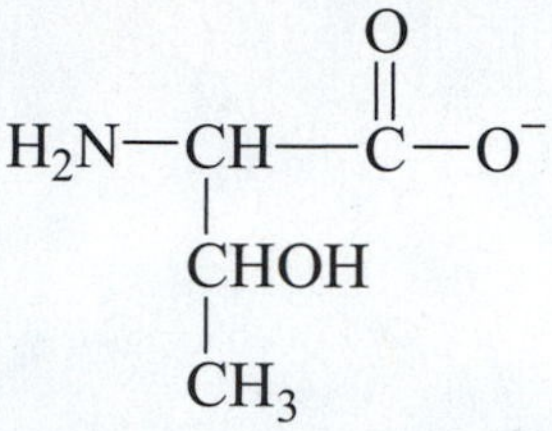

$H_2N-\overset{\underset{|}{CHOH}}{\underset{|}{\underset{CH_3}{}}}CH-\overset{O}{\overset{||}{C}}-O^-$

37. **(D)**

(a) The structures of the six tripeptides that contain one alanine, one serine, and one lysine are drawn below (in no particular order).

Lys-Ser-Ala (1 of 6)

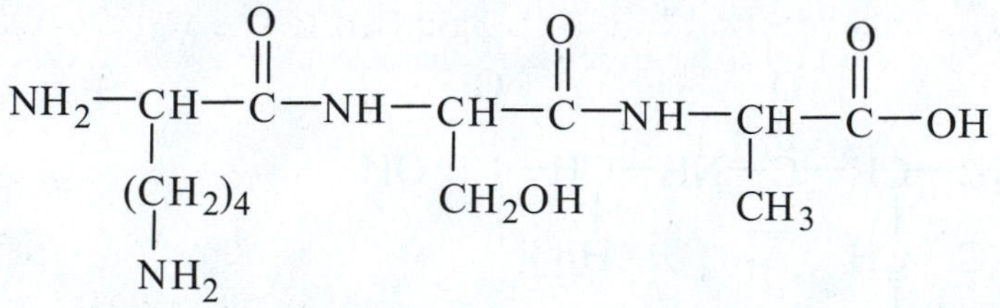

$NH_2-\underset{\underset{NH_2}{\underset{|}{(CH_2)_4}}}{\underset{|}{CH}}-\overset{O}{\overset{||}{C}}-NH-\underset{\underset{CH_2OH}{|}}{CH}-\overset{O}{\overset{||}{C}}-NH-\underset{\underset{CH_3}{|}}{CH}-\overset{O}{\overset{||}{C}}-OH$

Lys-Ala-Ser (2 of 6)

$NH_2-\underset{\underset{NH_2}{\underset{|}{(CH_2)_4}}}{\underset{|}{CH}}-\overset{O}{\overset{||}{C}}-NH-\underset{\underset{CH_3}{|}}{CH}-\overset{O}{\overset{||}{C}}-NH-\underset{\underset{CH_2OH}{|}}{CH}-\overset{O}{\overset{||}{C}}-OH$

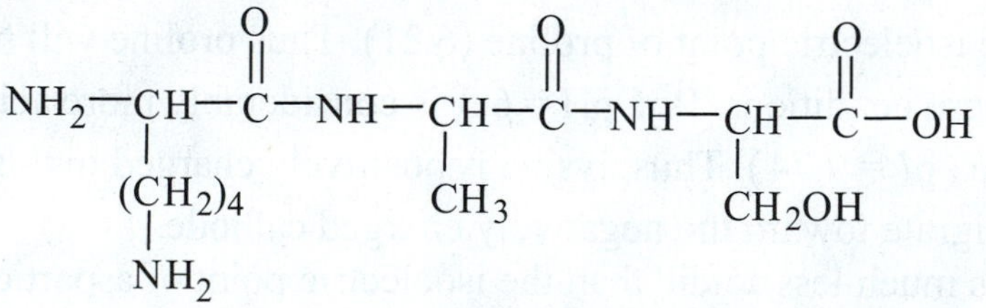

Ser-Lys-Ala (3 of 6)

$NH_2-\underset{\underset{CH_2OH}{|}}{CH}-\overset{O}{\overset{||}{C}}-NH-\underset{\underset{NH_2}{\underset{|}{(CH_2)_4}}}{\underset{|}{CH}}-\overset{O}{\overset{||}{C}}-NH-\underset{\underset{CH_3}{|}}{CH}-\overset{O}{\overset{||}{C}}-OH$

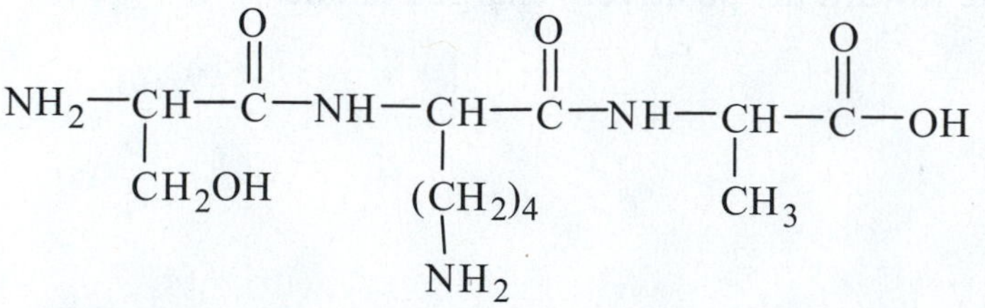

Ser-Ala-Lys (4 of 6)

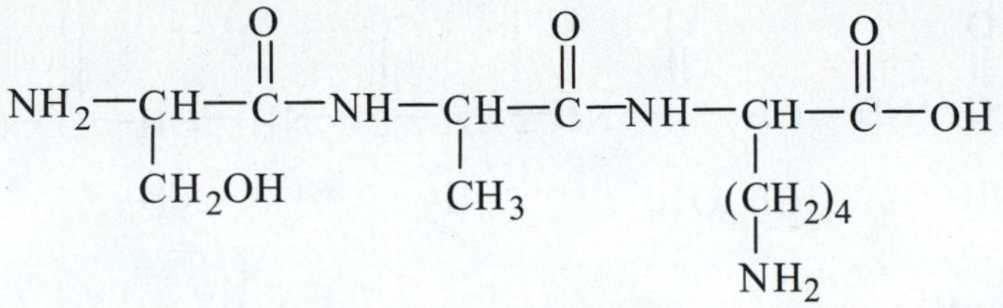

$$NH_2-\underset{\underset{CH_2OH}{|}}{CH}-\overset{\overset{O}{\|}}{C}-NH-\underset{\underset{CH_3}{|}}{CH}-\overset{\overset{O}{\|}}{C}-NH-\underset{\underset{\underset{NH_2}{|}}{(CH_2)_4}}{CH}-\overset{\overset{O}{\|}}{C}-OH$$

Ala-Ser-Lys (5 of 6)

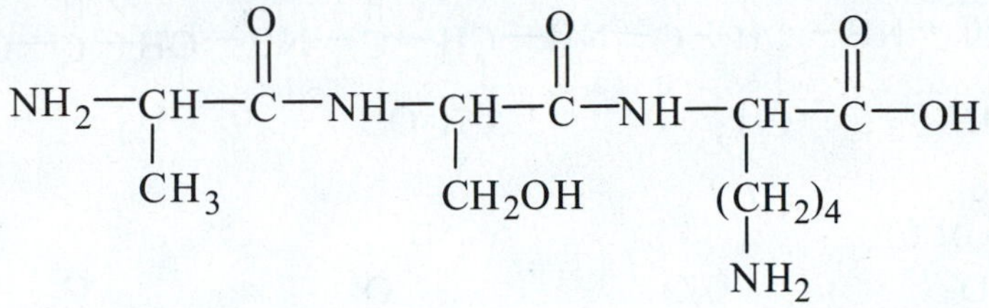

$$NH_2-\underset{\underset{CH_3}{|}}{CH}-\overset{\overset{O}{\|}}{C}-NH-\underset{\underset{CH_2OH}{|}}{CH}-\overset{\overset{O}{\|}}{C}-NH-\underset{\underset{\underset{NH_2}{|}}{(CH_2)_4}}{CH}-\overset{\overset{O}{\|}}{C}-OH$$

Ala-Lys-Ser (6 of 6)

$$NH_2-\underset{\underset{CH_3}{|}}{CH}-\overset{\overset{O}{\|}}{C}-NH-\underset{\underset{\underset{NH_2}{|}}{(CH_2)_4}}{CH}-\overset{\overset{O}{\|}}{C}-NH-\underset{\underset{CH_2OH}{|}}{CH}-\overset{\overset{O}{\|}}{C}-OH$$

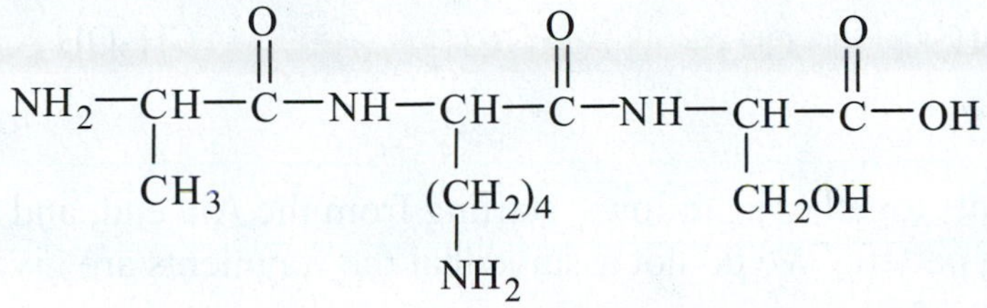

(b) The structures of the six tetrapeptides that contain two serine and two alanine amino acids each follow (in no particular order).

Ala-Ser-Ala-Ser (1 of 6)

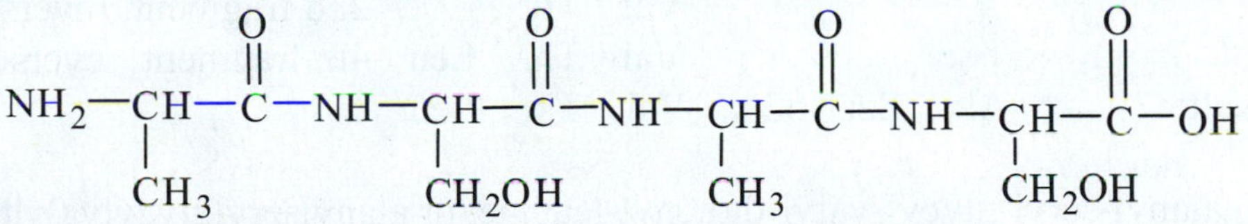

$$NH_2-\underset{\underset{CH_3}{|}}{CH}-\overset{\overset{O}{\|}}{C}-NH-\underset{\underset{CH_2OH}{|}}{CH}-\overset{\overset{O}{\|}}{C}-NH-\underset{\underset{CH_3}{|}}{CH}-\overset{\overset{O}{\|}}{C}-NH-\underset{\underset{CH_2OH}{|}}{CH}-\overset{\overset{O}{\|}}{C}-OH$$

Ala-Ala-Ser-Ser (2 of 6)

$$NH_2-\underset{\underset{CH_3}{|}}{CH}-\overset{\overset{O}{\|}}{C}-NH-\underset{\underset{CH_3}{|}}{CH}-\overset{\overset{O}{\|}}{C}-NH-\underset{\underset{CH_2OH}{|}}{CH}-\overset{\overset{O}{\|}}{C}-NH-\underset{\underset{CH_2OH}{|}}{CH}-\overset{\overset{O}{\|}}{C}-OH$$

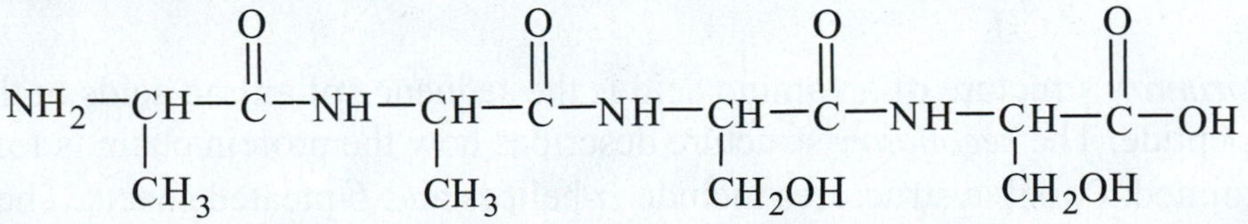

Ala-Ser-Ser-Ala (3 of 6)

$$NH_2-\underset{\underset{CH_3}{|}}{CH}-\overset{\overset{O}{\|}}{C}-NH-\underset{\underset{CH_2OH}{|}}{CH}-\overset{\overset{O}{\|}}{C}-NH-\underset{\underset{CH_2OH}{|}}{CH}-\overset{\overset{O}{\|}}{C}-NH-\underset{\underset{CH_3}{|}}{CH}-\overset{\overset{O}{\|}}{C}-OH$$

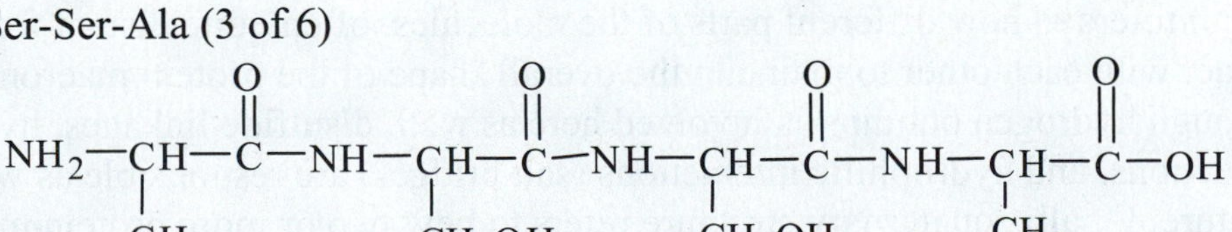

Ser-Ser-Ala-Ala (4 of 6)

$$NH_2-CH-\overset{\overset{\displaystyle O}{\|}}{C}-NH-CH-\overset{\overset{\displaystyle O}{\|}}{C}-NH-CH-\overset{\overset{\displaystyle O}{\|}}{C}-NH-CH-\overset{\overset{\displaystyle O}{\|}}{C}-OH$$

CH₂OH   CH₂OH   CH₃   CH₃

Ser-Ala-Ser-Ala (5 of 6)

$$NH_2-CH-\overset{\overset{\displaystyle O}{\|}}{C}-NH-CH-\overset{\overset{\displaystyle O}{\|}}{C}-NH-CH-\overset{\overset{\displaystyle O}{\|}}{C}-NH-CH-\overset{\overset{\displaystyle O}{\|}}{C}-OH$$

CH₂OH   CH₃   CH₂OH   CH₃

Ser-Ala-Ala-Ser (6 of 6)

$$NH_2-CH-\overset{\overset{\displaystyle O}{\|}}{C}-NH-CH-\overset{\overset{\displaystyle O}{\|}}{C}-NH-CH-\overset{\overset{\displaystyle O}{\|}}{C}-NH-CH-\overset{\overset{\displaystyle O}{\|}}{C}-OH$$

CH₂OH   CH₃   CH₃   CH₂OH

39. **(M)**

(a) We put the fragments together as follows, starting from the Ala end, and then placing them in a matching pattern. We do not assume that the fragments are given with the N-terminal end first.

```
Fragments:   Ala  Ser                      3rd fragment
                  Ser  Gly  Val            1st fragment
                       Gly  Val  Thr       5th fragment
                            Val  Thr       2nd fragment, reversed
                            Val  Thr  Leu  4th fragment, reversed
Result:      Ala  Ser  Gly  Val  Thr  Leu
```

(b) alanyl-seryl-glycyl-valyl-threonyl-leucine, or alanylserylglycylvalylthreonylleucine

41. **(E)**

The *primary* structure of an amino acid is the sequence of amino acids in the chain of the polypeptide. The *secondary* structure describes how the protein chain is folded, coiled, or convoluted. Possible structures include α-helices and β-pleated sheets. These secondary structures are held together principally by hydrogen bonds. The *tertiary* structure of a protein refers to how different parts of the molecules, often quite distant from each other, interact with each other to maintain the overall shape of the protein macromolecule. Although hydrogen bonding is involved here as well, disulfide linkages, hydrophobic interactions, and hydrophilic interactions (salt bridges) are responsible as well for tertiary structure. Finally, quaternary structure refers to how two or more protein molecules pack together into a larger protein complex. Not all proteins have a *quaternary* structure since many proteins have only one polypeptide chain.

43. **(M)**

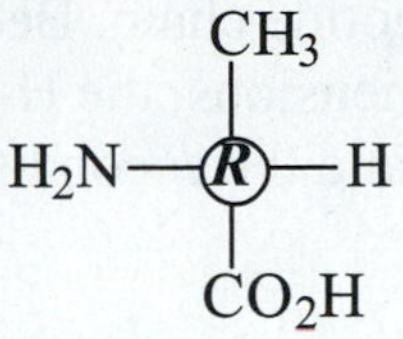

$$H_2N \overset{CH_3}{\underset{CO_2H}{-\overset{|}{R}-}} H \qquad R\text{-alanine}$$

45. **(D)**

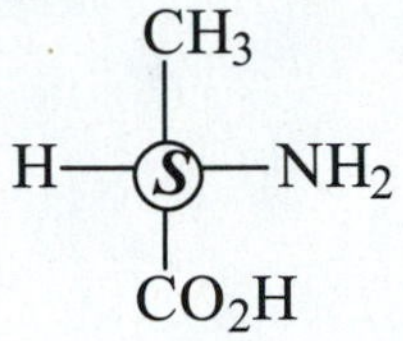

S-alanine 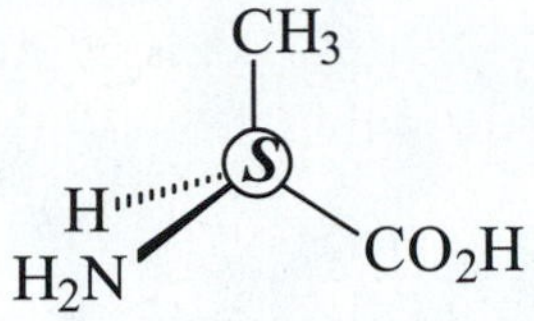   S-alanine 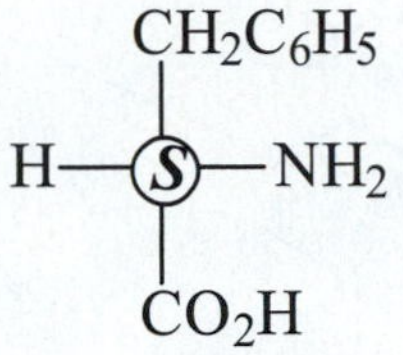   S-phenylalanine 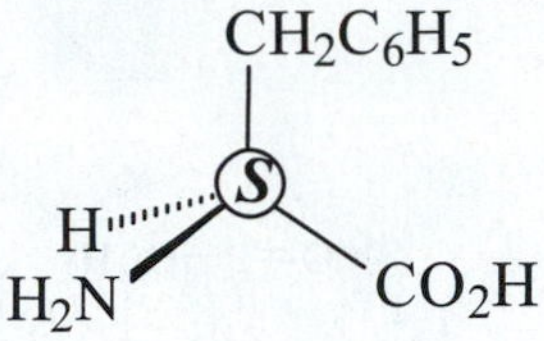   S-phenylalanine

Nucleic Acids

47. **(E)** Two major types of nucleic acids are DNA (deoxyribonucleic acid), and RNA (ribonucleic acid). Both of them contain phosphate groups. These phosphate groups alternate with sugars to form the backbone of the molecule. The sugars are deoxyribose in the case of DNA and ribose in the case of RNA. Attached to each sugar is a purine or a pyrimidine base. The purine bases are adenine and guanine. One pyrimidine base is cytosine. In the case of RNA the other pyrimidine base is uracil, while for DNA the other pyrimidine base is thymine.

49. **(D)** The complementary sequence to AGC is TCG. One polynucleotide chain is completely shown, as is the hydrogen bonding to the bases in the other polynucleotide chain. Because of the distortions that result from depicting a 3-D structure in two dimensions, the H- bonds themselves are distorted (they are all of approximately equal length) and the second sugar phosphate chain has been omitted.

INTEGRATIVE AND ADVANCED EXERCISES

52. **(E)** If we assume that there is only one active site per enzyme and that a silver ion is necessary to deactivate each active site, we obtain the molar mass of the protein as follows.

$$\text{molar mass} = \frac{1.00 \text{ mg}}{0.346 \text{ µmol Ag}^+} \times \frac{1 \text{ µmol}}{10^{-6} \text{ mol}} \times \frac{1 \text{ mol Ag}^+}{1 \text{ mol protein}} \times \frac{1 \text{ g}}{1000 \text{ mg}} = 2.9 \times 10^3 \text{ g/mol}$$

This is a minimum value because we have assumed that 1 Ag^+ ion is all that is necessary to denature each protein molecule. If more than one silver ion were required, the next to last factor in the calculation above would be larger than 1 mol Ag^+/1 mol protein, and the molar mass would increase correspondingly.

56. **(M)** The pI value is the pH at the second equivalence point, much like the pH of HPO_4^{2-} is $pI = \frac{1}{2}(pK_{a_2} + pK_{a_3})$.

$pK_I = \frac{1}{2}(pK_{a_2} + pK_{a_3}) = \frac{1}{2}(8.65 + 10.76) = 9.71$

highly acidic form $^+H_3NCH_2CH_2CH(NH_3^+)COOH$

acidic form $\quad\quad ^+H_3NCH_2CH_2CH(NH_3^+)COO^-$

zwitterion $\quad\quad H_2NCH_2CH_2CH(NH_3^+)COO^-$

$pK_{a_1} = 1.94 \quad\quad pK_{a_2} = 8.65 \quad\quad pK_{a_3} = 10.76$

58. **(D)** Assume all the fragments are given with the N-terminal end listed first. Since there is Arg at the N-terminal end (and only two Arg amino acids), we begin with the fragment that starts with an Arg residue and build the chain from there, making sure to end with an Arg fragment.

Fragments:	Arg	Pro	Pro						
	Pro	Pro	Gly						
		Pro	Gly	Phe					
			Gly	Phe	Ser	Pro			
				Ser	Pro	Phe			
					Pro	Phe	Arg		
						Phe	Arg		
Nonapeptide:	Arg	Pro	Pro	Gly	Phe	Ser	Pro	Phe	Arg

59. (D)

Amino Acid	Messenger RNA	DNA
serine	UCU	AGA
	UCC	AGG
	UCA	AGT
	UCG	AGC
glycine	GGU	CCA
	GGC	CCG
	GGA	CCT
	GGG	CCC
valine	GUU	CAA
	GUC	CAG
	GUA	CAT
	GUG	CAC
alanine	GCU	CGA
	GCC	CGG
	GCA	CGT
	GCG	CGC

One example for the required sequence would be AGA CCA CAA CGA. The redundancy enables the organism to produce a given amino acid in several ways in case of transcription error. For example, CG and any third DNA nucleotide will yield alanine in the polypeptide chain.

62. (M) Reaction of the pentapeptide with DFNB indicates that Met is the N-terminal end. The remaining information provides the following possible sequences of fragments.

(1) Met   Met Gly *or* Met Gly Met

(2) Met   Met

(3) Ser   Met

(4) Met   Met Ser *or* Met Ser Met

The fact that no fragment contains both Ser and Gly is strong evidence that these two amino acids are not adjacent in the chain. In fact, there must be a Met between them. Experiment (2) indicates that two Met residues are adjacent. Experiments (1) and (4), in combination with the fact that Met is the N-terminal end, mean that a Gly-Met-Ser or a Ser-Met-Gly sequence cannot end the chain. Hence,
Met   Gly Met Met Ser   or   Met Ser Met Met Gly

Experiment (3) confirms that the right-hand order above must be correct, for there is no way that a dipeptide with Ser as the N-terminal end can be obtained from the left-hand order above.

65. (D)

(a) $\Delta_r G^{o\prime} = -RT\ln K$

$$\ln K = \frac{-\Delta_r G^{o\prime}}{RT} = \frac{-23000 \, J}{(8.31 J/K \cdot mol)(298.15 \, K)} = -9.3 \qquad K = e^{-9.3} = 9.3 \times 10^{-5}$$

(b) $\ln K = \frac{-\Delta_r G^{o\prime}}{RT} = \frac{-(-30000 \, J)}{(8.31 \, J/K \cdot mol)(298.15 \, K)} = 12.1 \qquad K = e^{12.1} = 1.8 \times 10^5$

(c) $A + ADP + Pi \longrightarrow B + ATP \qquad$ so…

$$\frac{[B]}{[A]} = \frac{[ADP][K][Pi]}{[ATP]} = \frac{1}{400}(1.8 \times 10^5)(0.005) = 2.25 \qquad \frac{coupled}{uncoupled} = \frac{2.25}{9.3 \times 10^{-5}}$$

$$= 2.4 \times 10^4$$

FEATURE PROBLEMS

66. (D)

Glyceryl Tristearate   **Glyceryl Trioleate**

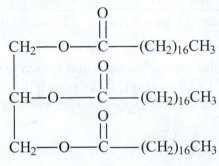

Molar Mass = 891.5 g mol^{-1} Molar Mass = 885.45 g mol^{-1}

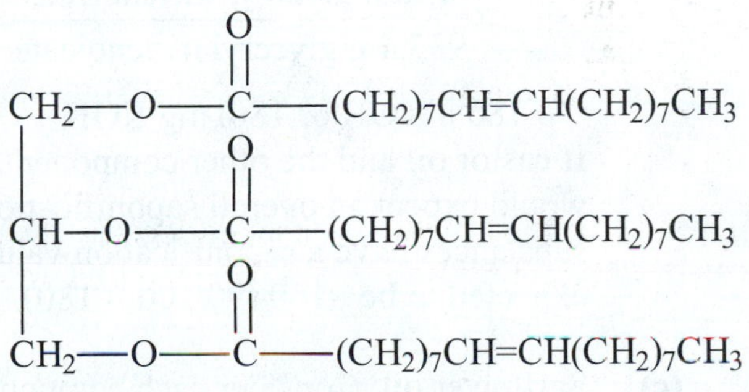

(a) Saponification value of glyceryl tristearate: mass of KOH

$$= 1.00 \text{ g glyceryl tristearate} \times \frac{1 \text{ mol glyceryl tristearate}}{891.5 \text{ g glyceryl tristearate}} \times \frac{3 \text{ mol KOH}}{1 \text{ mol glyceryl tristearate}}$$

$$\times \frac{56.1056 \text{ g KOH}}{1 \text{ mol KOH}} = 0.189 \text{ g KOH or 189 mg KOH} \quad \text{Saponification value} = 189$$

Iodine number for glyceryl trioleate: mass of I_2

$$= 100 \text{ g glyceryl trioleate} \times \frac{1 \text{ mol glyceryl trioleate}}{885.45 \text{ g glyceryl trioleate}} \times \frac{3 \text{ mol } I_2}{1 \text{ mol glyceryl trioleate}}$$

$$\times \frac{253.81 \text{ g } I_2}{1 \text{ mol } I_2} = 86.0 \text{ g } I_2 \quad \text{Iodine number} = 86.0$$

(b) As we do not know the types of substances present in castor oil and their percentages, we will assume that castor oil is the only glyceryl triricnoleate.

$$
\begin{array}{l}
CH_2-O-\overset{\displaystyle O}{\overset{\displaystyle \|}{C}}-(CH_2)_7CH=CHCH_2CHOH(CH_2)_5CH_3 \\[4pt]
CH-O-\overset{\displaystyle O}{\overset{\displaystyle \|}{C}}-(CH_2)_7CH=CHCH_2CHOH(CH_2)_5CH_3 \\[4pt]
CH_2-O-\overset{\displaystyle O}{\overset{\displaystyle \|}{C}}-(CH_2)_7CH=CHCH_2CHOH(CH_2)_5CH_3
\end{array}
$$

Glyceryl Triricnoleate
Molar Mass = 933.4 g mol^{-1}

Iodine number for glyceryl triricnoleate: mass of I_2

= 100 g glyceryl triricnoleate

$$\times \frac{1 \text{ mol glyceryl triricnoleate}}{933.4 \text{ g glyceryl triricnoleate}} \times \frac{3 \text{ mol } I_2}{1 \text{ mol glyceryl triricnoleate}} \times \frac{253.81 \text{g } I_2}{1 \text{ mol } I_2}$$

= 81.6 g I_2 Iodine number = 81.6

Saponification value of glyceryl triricnoleate: mass of KOH = 1.00 g glyceryl triricnoleate

$$\times \frac{1 \text{ mol glyceryl triricnoleate}}{933.4 \text{ g glyceryl triricnoleate}} \times \frac{3 \text{ mol KOH}}{1 \text{ mol glyceryl triricnoleate}} \times \frac{56.1056 \text{ g KOH}}{1 \text{ mol KOH}}$$

= 0.180 g KOH or 180. mg KOH   Saponification value = 180.

If castor oil and the other components have similar saponification values, then we would expect an overall saponification value of 180. However, if the remaining substances have a saponification value of zero, the overall saponification value is expected to be ~160 (90/100 × 180).

(c) Safflower oil: Consider each component

Palmitic acid: Molar mass: $C_3H_5(C_{15}H_{31}CO_2)_3 = 807.34$ g mol^{-1}

Iodine number = 0 (saturated)

Saponification value of palmitic acid: mass of KOH

= 1.00 g palmitic acid

$$\times \frac{1 \text{ mol palmitic acid}}{807.33 \text{ g palmitic acid}} \times \frac{3 \text{ mol KOH}}{1 \text{ mol palmitic acid}} \times \frac{56.1056 \text{ g KOH}}{1 \text{ mol KOH}}$$

= 0.208 g KOH or 208 mg KOH   Saponification value = 208

Stearic acid: Molar mass: $C_3H_5(C_{17}H_{35}CO_2)_3 = 891.49$ g mol^{-1}

Iodine number = 0 (saturated)

Saponification value of stearic acid: mass of KOH

$$= 1.00 \text{ g stearic acid} \times \frac{1 \text{ mol stearic acid}}{891.49 \text{ g stearic acid}} \times \frac{3 \text{ mol KOH}}{1 \text{ mol stearic acid}} \times \frac{56.1056 \text{ g KOH}}{1 \text{ mol KOH}}$$

= 0.189 g KOH or 189 mg KOH   Saponification value = 189

Oleic acid: Molar mass: $C_3H_5(C_{17}H_{33}CO_2)_3 = 885.45$ g mol^{-1}

Iodine number for oleic acid: mass of I_2

$= 100$ g oleic acid $\times \dfrac{1 \text{ mol oleic acid}}{885.45 \text{ g oleic acid}} \times \dfrac{3 \text{ mol } I_2}{1 \text{ mol oleic acid}} \times \dfrac{253.81 \text{ g } I_2}{1 \text{ mol } I_2} = 86.0$ g I_2

Iodine number $= 86.0$

Saponification value of oleic acid: mass of KOH

$= 1.00$ g oleic acid $\times \dfrac{1 \text{ mol oleic acid}}{885.45 \text{ g oleic acid}} \times \dfrac{3 \text{ mol KOH}}{1 \text{ mol oleic acid}} \times \dfrac{56.1056 \text{ g KOH}}{1 \text{ mol KOH}}$

$= 0.190$ g KOH or 190. mg KOH   Saponification value $= 190.$

Linoleic acid: Molar mass: $C_3H_5(C_{17}H_{31}CO_2)_3 = 879.402$ g mol^{-1}

Iodine number for linoleic acid: mass of I_2

$= 100$ g linoleic acid $\times \dfrac{1 \text{ mol linoleic acid}}{879.402 \text{ g linoleic acid}} \times \dfrac{6 \text{ mol } I_2}{1 \text{ mol linoleic acid}} \times \dfrac{253.81 \text{ g } I_2}{1 \text{ mol } I_2}$

$= 173$ g I_2 Iodine number $= 173$

Saponification value of linoleic acid: mass of KOH

$= 1.00$ g linoleic acid $\times \dfrac{1 \text{ mol linoleic acid}}{879.402 \text{ g linoleic acid}} \times \dfrac{3 \text{ mol KOH}}{1 \text{ mol linoleic acid}} \times \dfrac{56.1056 \text{ g KOH}}{1 \text{ mol KOH}}$

$= 0.191$ g KOH or 191 mg KOH   Saponification value $= 191$

Linolenic acid: Molar mass: $C_3H_5(C_{17}H_{29}CO_2)_3 = 873.348$ g mol^{-1}

Iodine number for linolenic acid: mass of I_2

$= 100$ g linolenic acid $\times \dfrac{1 \text{ mol linolenic acid}}{873.354 \text{ g linolenic acid}} \times \dfrac{9 \text{ mol } I_2}{1 \text{ mol linolenic acid}} \times \dfrac{253.81 \text{ g } I_2}{1 \text{ mol } I_2}$

$= 261.\underline{6}$ g I_2 Iodine number $= 261.\underline{6} \approx 262$

Saponification value of linolenic acid: mass of KOH

$= 1.00$ g linolenic acid $\times \dfrac{1 \text{ mol linolenic acid}}{873.354 \text{ g linolenic acid}} \times \dfrac{3 \text{ mol KOH}}{1 \text{ mol linolenic acid}} \times \dfrac{56.1056 \text{ g KOH}}{1 \text{ mol KOH}}$

$= 0.193$ g KOH or 193 mg KOH   Saponification value $= 193$

Summary: Acid	I_2#	Sap#	%	High I_2 # %	Low I_2 # %	High Sap# %	Low Sap# %
Palmitic	0	208	6–7	6	7	7	6
Stearic	0	189	2–3	2	3	2	3
Oleic	86	190	12–14	12	14	12	14
Linoleic	173	191	75–80	78.5	75.5	87.5	76.5
Linolenic	262	193	0.5–1.5	1.5	0.5	1.5	0.5

Hence: The iodine number for safflower oil may range between 150 to 144. The saponification value for safflower oil may range between 211 and 179. (Note:the high iodine number contribution for each acid in the mixture is calculated by multiplying its percentage by its iodine number. The sum of all of the high iodine number contributions for all of the components in the mixture equals the high iodine number for safflower oil. Similar calculations were used to obtain the low iodine number, along with the high/low saponification numbers.

SELF-ASSESSMENT EXERCISES

72. **(E)** The answer is (b), oil. Glyceryl trilinoleate is a triglyceride containing three linolic acid substituents. Linolic acid is unsaturated, making it likely that it is a liquid at room temperature and therefore an oil.

73. **(E)** The answer is (d), neither to the left nor the right. DL-erythrose describes a racemic mixture: one which rotates a plane of polarized light in a dextrorotatory, and the other in a levorotatotory fashion. The two substances cancel each other out in their effect in rotating a plane of polarized light.

74. **(E)** The answer is (a), β-galactose.

75. **(E)** The answer is (d), denaturation of protein. Heating the egg white causes the protein tertiary and secondary structure to change and makes the protein molecules coagulate.

76. **(E)** The answer is (e), ATP.

77. **(E)** The answer is (c). Glycerol is part of the triglycerides.

78. **(E)** The answer is (b), double helix.

79. **(M)** Below is the structure of glyceryltripalmitate:

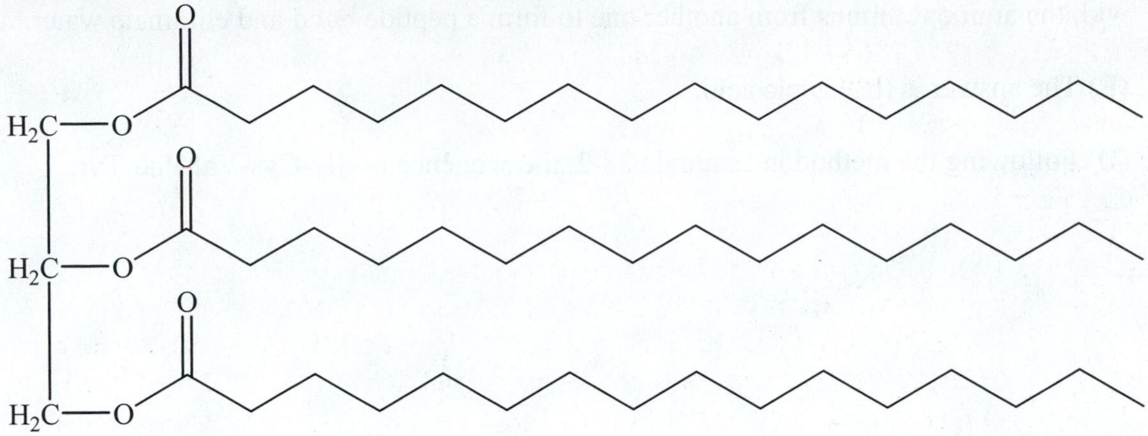

Based on the molecular structure, its MW = 224 (MW of palmitate) × 3 + 42.06 (MW of C_3H_6, glyceride) = 714.06 g/mol.

$$125 \text{ g Gly-tripalmitate(GTP)} \times \frac{\text{mol GTP}}{714.06 \text{ g GTP}} \times \frac{3 \text{ mol palmitate}}{1 \text{ mol GTP}} \times \frac{1 \text{ mol Na(palm.)}}{1 \text{ mol palm.}}$$

$$\times \frac{247 \text{ g Na(palm)}}{1 \text{ mol Na(palm.)}} = 129 \text{ g sodium palmitate soap}$$

80. **(M)** There are two major things to look for to determine if the chain is DNA or RNA: the ribose/deoxyribose sugar and presence of uracil or thymine. Looking at the structure, it is seen that the bases attached to the sugars down the right side of the structure are adenine (purine), uracil (pyrimidine), guanine (purine), and cytosine (pyrimidine). Furthermore, the sugar is ribose, making this an RNA chain.

81. **(M)** The answer is (e), none of these. Consulting Table 28.3, it is seen that the residue does not match that of any amino acid given.

82. **(M)** The answer is (b), 1,4-dichlorobutane. See structure below:

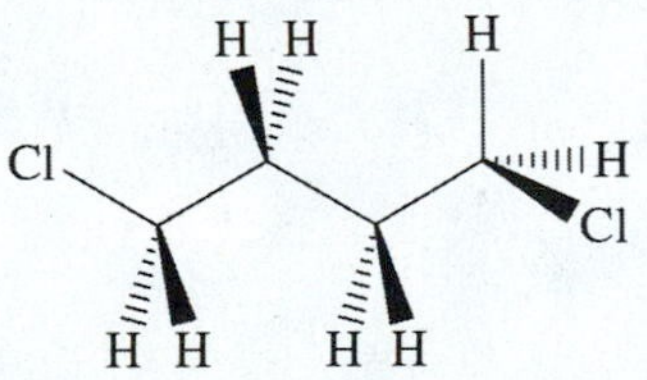

83. **(E)** The answer is (a), glycine. It is the only non-chiral amino acid, because the –R group is just –H.

84. **(E)** The answer is (d), they always rotate plane-polarized light.

85. **(E)** The answer is (c), -CH(CH$_3$)$_2$.

86. **(E)** The answer is (d), condensation. The carboxylic terminus from one amino acid reacts with the amino terminus from another one to form a peptide bond and eliminate water.

87. **(E)** The answer is (b.), Oleic acid.

88. **(D)** Following the method in Example 28-2, the sequence is Gly-Cys-Val-Phe-Tyr.